Handbook of
Behavioral Neurobiology

Volume 7
Reproduction

HANDBOOK OF BEHAVIORAL NEUROBIOLOGY

General Editor:
Frederick A. King
Yerkes Regional Primate Research Center, Emory University, Atlanta, Georgia

Volume 1   Sensory Integration
Edited by R. Bruce Masterton

Volume 2   Neuropsychology
Edited by Michael S. Gazzaniga

Volume 3   Social Behavior and Communication
Edited by Peter Marler and J. G. Vandenbergh

Volume 4   Biological Rhythms
Edited by Jürgen Aschoff

Volume 5   Motor Coordination
Edited by Arnold L. Towe and Erich S. Luschei

Volume 6   Motivation
Edited by Evelyn Satinoff and Philip Teitelbaum

Volume 7   Reproduction
Edited by Norman Adler, Donald Pfaff, and Robert W. Goy

Volume 8   Developmental Psychobiology and Developmental Neurobiology
Edited by Elliott M. Blass

A Continuation Order Plan is available for this series. A continuation order will bring delivery of each new volume immediately upon publication. Volumes are billed only upon actual shipment. For further information please contact the publisher.

Handbook of
Behavioral Neurobiology

Volume 7
Reproduction

Edited by

Norman Adler
University of Pennsylvania
Philadelphia, Pennsylvania

Donald Pfaff
The Rockefeller University
New York, New York

and

Robert W. Goy
Wisconsin Regional Primate Research Center
Madison, Wisconsin

PLENUM PRESS • NEW YORK AND LONDON

Library of Congress Cataloging in Publication Data

Main entry under title:

Reproduction.

  (Handbook of behavioral neurobiology; v. 7)
  Includes bibliographies and index.
  1. Reproduction. 2. Sex (Biology) 3. Reproduction (Psychology) 4. Sexual behavior in
animals. I. Adler, Norman T. II. Pfaff, Donald W., 1939–   . III. Goy, Robert W. IV.
Series. [DNLM: 1. Mammals—physiology. 2. Reproduction. 3. Sex Behavior. W1
HA51I v.7/WQ 205 R426]
QP251.R4442   1985   596′.016   85-9330
ISBN 0-306-41768-5

QP
251
.R442
1985

Contributors

ELIZABETH ADKINS-REGAN, *Department of Psychology and Section of Neurobiology and Behavior, Cornell University, Ithaca, New York*

N. T. ADLER, *Department of Psychology, University of Pennsylvania, Philadelphia, Pennsylvania*

T. O. ALLEN, *Department of Psychology, University of Pennsylvania, Philadelphia, Pennsylvania*

LYNWOOD G. CLEMENS, *Department of Zoology and Neuroscience Program, Michigan State University, East Lansing, Michigan*

DAVID CREWS, *Department of Biology and Psychology and Social Relations and The Museum of Comparative Zoology, Harvard University, Cambridge, Massachusetts*

BARRY J. EVERITT, *Anatomy Department, University of Cambridge, Cambridge, England*

H. H. FEDER, *Institute of Animal Behavior, Rutgers University, Newark, New Jersey*

DAVID A. GOLDFOOT, *Wisconsin Regional Primate Research Center, 1223 Capitol Court, Madison, Wisconsin*

BENJAMIN L. HART, *Department of Physiological Sciences, School of Veterinary Medicine, University of California, Davis, California*

ALFRED JOST, *Laboratory of Developmental Physiology, Collège de France, Place Marcelin Berthelot, Paris, France*

MITZI G. LEEDY, *Department of Physiological Sciences, School of Veterinary Medicine, University of California, Davis, California*

VICTORIA N. LUINE, *The Rockefeller University, New York, New York*

WILLIAM G. LUTTGE, *Department of Neuroscience and Center for Neurobiological Sciences, University of Florida, College of Medicine, Gainesville, Florida*

CARL OLOF MALMNÄS, *Late of the Department of Medical Pharmacology, University of Uppsala, Uppsala, Sweden*

ANNE D. MAYER, *Institute of Animal Behavior, Rutgers—The State University, Newark, New Jersey*

BRUCE S. MCEWEN, *The Rockefeller University, New York, New York*

BENGT J. MEYERSON, *Department of Medical Pharmacology, University of Uppsala, Uppsala, Sweden*

DOAN MODIANOS, *Laboratory of Neurobiology and Behavior, The Rockefeller University, New York, New York*

DEBORAH A. NEFF, *Wisconsin Regional Primate Research Center, 1223 Capitol Court, Madison, Wisconsin*

MICHAEL NUMAN, *Department of Psychology, Boston College, Chestnut Hill, Massachusetts*

DONALD PFAFF, *Laboratory of Neurobiology and Behavior, The Rockefeller University, New York, New York*

JOHN A. RESKO, *Department of Physiology, Oregon Health Sciences University, Portland, Oregon*

JAY S. ROSENBLATT, *Institute of Animal Behavior, Rutgers—The State University, Newark, New Jersey*

HAROLD I. SIEGEL, *Institute of Animal Behavior, Rutgers—The State University, Newark, New Jersey*

RAE SILVER, *Department of Psychology, Barnard College of Columbia University, New York, New York*

INGEBORG L. WARD, *Department of Psychology, Villanova University, Villanova, Pennsylvania*

O. BYRON WARD, *Department of Psychology, Villanova University, Villanova, Pennsylvania*

DAVID R. WEAVER, *Department of Zoology and Neuroscience Program, Michigan State University, East Lansing, Michigan*

RICHARD E. WHALEN, *Department of Psychology, University of California, Riverside, California*

PAULINE YAHR, *Department of Psychobiology, University of California, Irvine, California*

Preface

The subject of this book is reproduction—specifically, the interplay between reproductive physiology (especially neural and endocrine events) and behavior. In presenting this topic, there are two expository goals. The first is to study reproduction at all of the major *levels of biological organization*—from the molecular (e.g., hormone receptors in the brain), through the cellular (e.g., ovarian morphogenesis), systemic (e.g., operation of the hypothalamo-pituitary-ovarian axis), and the organismic levels of organization. Analogously, behavior is treated from the most molecular, elementary, and fundamental components (e.g., copulatory reflexes), through behavior in the reproductive dyad (e.g., analysis of female sexual behavior), to complex social behavior (e.g., the interaction of social context and behavioral sex differences).

To the extent that these levels of biological and behavioral organization represent a "vertical axis" in behavioral neurobiology, a second goal is to treat the "horizontal axis" of biological organization, viz., *time*. There are, therefore, treatments of evolutionary origins (e.g., a phylogenetic survey of psychosexual differentiation), genetic origins in the individual (e.g., sexual organogenesis), ontogenetic development (e.g., behavioral sexual differentiation), and the immediate physiological precursors of behavior (e.g., hormonal and nonhormonal initiation of maternal behavior). In addition to tracing the origins of reproduction and reproductive behavior, one extends the time-line from the behavior to its physiological consequences (e.g., neuroendocrine consequences of sexual behavior).

These two axes of biological explanation *(levels of organization* and *temporal span)* are the principles by which the book is ordered. Part I treats the *development* of sexuality. Jost describes the development of sexual morphology. Resko's chapter explores the developmental changes in steroid hormones. Adkins-Regan presents an evolutionary overview of psychosexual differentiation, and the Wards present a case study of sexual differentiation in the rat—the biological preparation that has produced perhaps the greatest body of information on this topic.

Part II of the book deals with proximal hormonal influences on sexual behavior in adult organisms. Crews and Silver provide extensive coverage of the reproductive physiology and behavior of nonmammalian vertebrates, while Clemens and Weaver concentrate on the precise interactions of hormones and sexual behavior in female mammals. Maternal behavior is treated in the chapter by Rosenblatt, Mayer, and Siegel. The dynamics of steroid hormones released by the gonads are described by Feder.

While maintaining a focus on hormonal effects, Part III of the book deals with the neurological bases of mammalian reproductive behavior. Hart and Leedy present a comparative approach to the neurological control of sexual behavior in males, while Pfaff and Modianos cover the neural mechanisms of female sexual behavior. Meyerson, Malmas, and Everitt survey neuropharmacological and neurochemical studies of sexual behavior. This part concludes with a treatment of the neural mechanisms controlling parental behavior.

Part IV of the book is devoted to the biochemical models of hormone action. Whalen, Yahr, and Luttge describe hormone metabolism and its relevance to sexual behavior. Luine and McEwen analyze the critical role of steroid hormone receptors in the brain and the pituitary.

Part V treats the interaction of social-sexual behavior and reproduction. Allen and Adler trace the consequences of stimuli derived from reproductive behavior through the neuroendocrine system of female mammals. Goldfoot and Neff highlight the complex influences of the social setting on the occurrence and display of sexual behavior.

By surveying reproduction from this variety of neurobiological perspectives, this volume intends to set forth a coherent and comprehensive picture of reproductive behavior, including its causes and consequences. This has been a large undertaking; the work has taken some time to produce. Several of the authors submitted their chapters early on (Chapters 1, 2, 3, 5, 8, 10, 11, 12, 13, 14, and 15), and the editors thank them for their patience. Others in the original list of authors could not meet the deadlines; their chapters were regretfully omitted. There are, however, no gaps in the scope of the volume because a number of colleagues were asked, late in the project, to supply the needed material as quickly as possible. They graciously agreed to the editors' request, sent in the needed chapters quickly, and produced remarkably thorough, insightful scholarship (Chapters 4, 6, 7, 9, and 16). All of the authors have the thanks and gratitude of the editors.

Contents

CHAPTER 3

Elizabeth Adkins-Regan

CHAPTER 4

Sexual Behavior Differentiation: Effects of Prenatal Manipulations in Rats   77

Ingeborg L. Ward and O. Byron Ward

PART II HORMONAL ACTIVATION OF BEHAVIOR

CHAPTER 5

Reproductive Physiology and Behavior Interactions in Nonmammalian
Vertebrates

David Crews and Rae Silver

CHAPTER 6

The Role of Gonadal Hormones in the Activation of Feminine Sexual
Behavior

Lynwood G. Clemens and David R. Weaver

CHAPTER 7

Jay S. Rosenblatt, Anne D. Mayer, and Harold I. Siegel

CHAPTER 8

H. H. Feder

PART III NEUROLOGICAL BASES OF MAMMALIAN REPRODUCTIVE BEHAVIOR

CHAPTER 9

Neurological Bases of Male Sexual Behavior: A Comparative Analysis 373

Benjamin L. Hart and Mitzi G. Leedy

CHAPTER 10

Donald Pfaff and Doan Modianos

CHAPTER 11

Neuropharmacology, Neurotransmitters, and Sexual Behavior in Mammals . 495

Bengt J. Meyerson, Carol Olof Malmnäs, and Barry J. Everitt

CHAPTER 12

Brain Mechanisms and Parental Behavior 537

Michael Numan

PART IV BIOCHEMICAL MODELS OF HORMONAL ACTION

CHAPTER 13

The Role of Metabolism in Hormonal Control of Sexual Behavior 609

Richard E. Whalen, Pauline Yahr, and William G. Luttge

CHAPTER 14

Steroid Hormone Receptors in Brain and Pituitary: Topography and

Victoria N. Luine and Bruce S. McEwen

PART I
Differentiation and Organization of Reproductive Processes

Sexual Organogenesis

Alfred Jost

Introduction

The foundations that underlie our knowledge of sexual differentiation were built long ago. Some remain strong and secure, especially those concerning the sex ducts, the external genitalia, and the relations between the mesonephros and sex structures. Other long-accepted views have failed to be properly verified and dim our understanding of the real events; this assertion applies especially to the problem of gonadal differentiation.

Before entering into particulars, it probably is appropriate to recall some of the anatomical and embryological bases of the structures involved (Figure 1). Some sex structures are homologous in the male and female in that the same primordium differentiates into structures characteristic of either sex; such is the case for the derivatives of the urogenital sinus and of the genital tubercle. Other parts are not strictly homologous in both sexes as they derive from different primordia. Thus, the fetal Wolffian ducts (the excretory ducts of the transitory mesonephric kidney) produce the vasa deferentia, the epididymides, and the seminal vesicles in males; the fetal Müllerian ducts become the tubes and the uterus in females. It should be understood that the terms *Wolffian* and *Müllerian ducts* refer essentially to the inner epithelial lining of these ducts and that the mesenchyme that forms the connective chorion and the muscularis is common to both sexes.

It has long been widely accepted that the testis and the ovary are not strictly homologous structures. It was assumed that the testicular seminiferous cords arose

This chapter reutilizes several sections of a previous paper (Jost, 1981) with alterations and additions to bring it up to date.

ALFRED JOST   Laboratory of Developmental Physiology, Collège de France, Place Marcelin Berthelot, Paris (5), France.

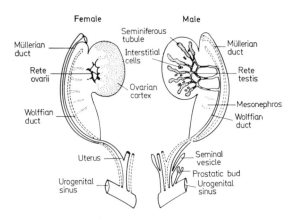

Fig. 1. Composite scheme showing some homologies in the development of male and female internal organs from the undifferentiated condition characterized by the presence on the mesonephros of a double set of ducts (Wolffian and Müllerian) and of a gonadal primordium. In females, the Wolffian ducts and in males the Müllerian ducts disappear, as indicated by the interrupted lines. In the presumptive ovaries, no ovarian structure has appeared at this stage of development. (From "General Outline about Reproductive Physiology and Its Developmental Background" by A. Jost. In H. Gibian and E. J. Plotz [Eds.], *Mammalian Reproduction.* Berlin: Springer-Verlag, 1970. Copyright 1970 by Springer-Verlag. Reprinted by permission.)

from primary sex cords and the ovarian follicles from secondary sex cords, and that both were proliferated by the superficial epithelium. This concept should now be forgotten (see the section on "Gonadal Sex").

In 1946, I began studying the role of the sex glands in the differentiation of the genital apparatus. It appeared that the fetal testes are the major active organs in differentiating the two sexes and that the gonadal sex precedes and controls the sex differentiation of the other parts of the body. Therefore, I proposed (Jost, 1958) to define sex as being the result of a chain of events controlling each other, as seen in Figure 2. Much has been learned since that time, but the scheme remains useful with qualifications. The role of the Y chromosome as the sex-determining

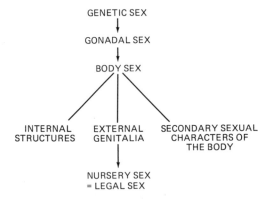

Fig. 2. Chain of events in sexual differentiation. (From "Embyonic Sexual Differentiation [Morphology, Physiology, Abnormalities]" by A. Jost. In H. W. Jones and W. W. Scott [Eds.], *Hermaphroditism, Genital Anomalies and Related Endocrine Disorders.* Baltimore: Williams & Wilkins, 1958. Copyright 1958 by Williams & Wilkins Co. Reprinted by permission.)

chromosome was recognized in 1959, both in mice (Welshons and Russel, 1959) and in man (Ford *et al.*, 1959). Genetic control was also demonstrated for several processes involved in sex differentiation (Table 1); we now know that the realization of a normal "genetic sex" depends on many genes. It is generally assumed that the gene(s) responsible for the morphological differentiation of the gonad into a testis is (are) the major determinant(s) of sex differentiation and that the capacity of the testis to produce its hormones and of the tissues to respond to these hormones depend on separate genes (see Ohno, 1979).

Body Sex

Role of the Fetal Testis

In my experiments reported in 1947 and 1953, I surgically removed the gonads from intrauterine rabbit fetuses at various developmental stages. Figure 3 illustrates some of the results obtained when the fetuses were castrated after the early stages of gonadal differentiation but before the initiation of the sexual differentiation of the other parts of the genital tract. Whatever their genetic sex, all the gonadless fetuses acquired a feminine genital apparatus, including development of the genital ducts and urogenital sinus and the external genitalia (not shown in Figure 3). Figure 3 also clearly shows that the fetal testis has definite effects on the sex structures: it induces the disappearance of the Müllerian ducts, which persist in the absence of testes; it incorporates Wolffian derivatives into the male apparatus; and it masculinizes the other parts. As a whole, the testis prevents the expression of the inherent program for femaleness of the body and produces maleness as shown in Figure 4.

Table 1. Some Genetic Factors Involved in Male Differentiation

Testicular morphology
  Y chromosome
  Autosomal sex-reversal gene in XX,Sxr mice
  Gene(s) responsible for gonadal dysgenesis?
Testicular endocrine function
  Synthesis of testosterone
    21,22 Desmolase
    3β-Hydroxysteroid dehydrogenase-$\Delta 5$ isomerase
    17α Hydroxylase
    17β Hydroxysteroid dehydrogenase
    17,20 Desmolase
  Production of Müllerian inhibiting factor ("anti-Müllerian hormone")?
Target tissues
  Receptors for anti-Müllerian hormone?
  Cytosolic "receptor molecule" for steroid hormones
  5α Reductase (urogenital sinus, external genitalia, etc.)
  Aromatase (brain)

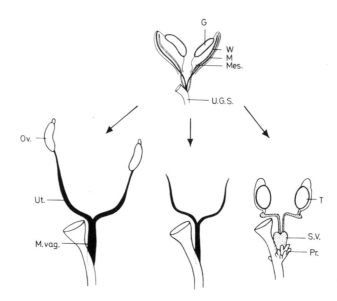

Fig. 3. Schematic presentation of sexual differentiation of the sex ducts in the rabbit embryo. From the undifferentiated condition (top) may arise either the female structure (bottom left); the male structure (bottom right); or the gonadless feminine structure in castrated embryos of either sex (bottom middle). G, gonad; W, Wolffian duct (stippled); M, Müllerian duct; Mes., mesonephros; U.G.S., urogenital sinus; Ov., ovary; Ut., uterine horn; M.vag., Müllerian vagina; T, testis; S.V., seminal vesicle; Pr., prostate. (From "Hormonal Influences in the Sex Development of the Bird and Mammalian Embryo" by A. Jost. *Memoirs of the Society for Endocrinology*, 1960, 7, 49–62. Copyright 1960 by the *Journal of Endocrinology*. Reprinted by permission.)

Experiments done first on guinea pigs (Phoenix *et al.*, 1959) and extensive inquiries done on other animals, especially the rat, have given evidence that, in these animal species, some parts of the brain are influenced by testicular hormone, during a critical phase, in a way similar to that of other sex characters of the body (Figure 5). The hormonal influence exerted during fetal life or neonatally has long-lasting effects that become apparent in adulthood. These effects, as studied in the rat, can be schematically surveyed in the following way (see MacLusky and Naftolin, 1981): (1) permanent suppression of the cyclical release of LH by the adult pituitary, cyclicity being characteristic of females and of neonatally castrated rats and (2) suppression or extensive decrease of the possibility that the animal will display lordosis in adulthood after appropriate hormonal treatment. In addition to these "defeminization" effects, testosterone has "masculinizing" effects, not only on morphology by also on behavior, for example, facilitated copulatory activity. There

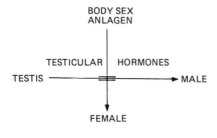

Fig. 4. Interpretive scheme of body sex differentiation. In males, the inherent trend toward femaleness is counteracted and prevented by the testicular hormones, which impose maleness. (From "General Outline about Reproductive Physiology and Its Developmental Background" by A. Jost. In H. Gibian and E. J. Plotz [Eds.], *Mammalian Reproduction*. Berlin: Springer-Verlag, 1970. Copyright 1970 by Springer-Verlag. Adapted by permission.)

are many species differences, which are analyzed elsewhere in this book, but when present, the testicular effect on the nervous structures obeys the scheme in Figure 5.

Although this chapter is devoted to mammalian sex differentiation, it is worth mentioning that in birds the mechanisms of genetic sex determination and of phenotypic differentiation are quite different. In birds, the females are heterozygous (XY or WZ chromosome constitution), whereas the males are homozygous (XX or ZZ chromosome constitution). After early embryonic castration, the sex characters become essentially masculine (Wolff and Wolff, 1951); in females, femininity must be imposed by the embryonic ovaries on a basic male program. The same has been shown to apply to the nervous system in quails: genetic female embryos are "demasculinized" by the ovarian hormone, which reduces the capacity for copulatory behavior in adulthood (see Jost, 1982). The comparison of the mammalian and bird mechanisms of control show how important the genetic and fetal endocrine factors are in determining the adult condition of animals like the rat or the quail.

If we come back to the mammalian control system as schematized in Figure 4, this simple scheme raises several questions: Which testicular chemical signals are involved? How do the target organs respond to chemical signals? Finally, how and why do testes differentiate in genetically male fetuses?

DUAL NATURE OF THE TESTICULAR CHEMICAL SIGNALS

From the very beginning of these experiments, it was observed that testosterone could masculinize the body of the castrated male rabbit fetus, but that it did not induce the regression (often called *inhibition*) of the Müllerian ducts. Castrated males given testosterone had "persisting Müllerian ducts" when sacrificed (Jost, 1947). Similarly, when a fetal testis was grafted on the mesosalpinx of a female rabbit fetus, it could induce persistence of the Wolffian ducts and inhibit the Müllerian duct, at least on one side of the body (Jost, 1947). In contrast, a crystal of testosterone placed in the same location in a female rabbit fetus masculinizes the Wolffian derivatives but it does not suppress the Müllerian ducts (Jost, 1955). In other experiments, it was also noticed that Müllerian ducts from young male rat fetuses, when grafted into the testis of an adult rat, were not inhibited in that location (Jost, 1968). Nor could testosterone inhibit the rat Müllerian ducts cultivated *in vitro* (Jost and Bergerard, 1949; Josso, 1971).

Fig. 5. Scheme similar to that in Figure 4 summarizing the effect of testosterone on the neural structures involved in reproduction and sexual behavior in the rat. (From "Genetic and Hormonal Factors in Sex Differentiation of the Brain" by A. Jost. *Psychoneuroendocrinology*, 1983, *8*, 183–193. Copyright 1983 by Pergamon Press, Ltd. Reprinted by permission.)

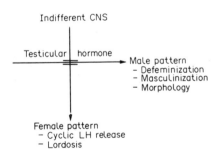

These observations and others led to the assumption that the testis could produce two kinds of signals, one responsible for the regression of the Müllerian ducts, the other one masculinizing the young (Figure 6). The defect of production or the absence of the appropriate effect either of the Müllerian inhibitor (Brook *et al.*, 1973) or of the masculinizing hormone (Wilson *et al.*, 1981) or of both (e.g., after castration) results in characteristic morphological defects of the genital tract.

TESTICULAR ANDROGENS

The major androgen secreted by the fetal testis and responsible for masculinizing the genital tract is testosterone. Very interesting data have been gathered concerning both testosterone production by the fetal testis and testosterone action at the level of target tissues.

TESTOSTERONE PRODUCTION. It has been established that the fetal testis contains testosterone and that, from a certain stage on, testosterone is a circulating hormone in the male fetus (Figure 7). However, the mode of distribution of testosterone in the earliest phases of sex differentiation has to be studied further because, during early phases of sex differentiation, it seems to be distributed locally in the genital tract rather than through the blood stream (Jost, 1947, 1953).

The synthesis of testosterone from precursors by the fetal testis incubated *in vitro* has been studied in several laboratories. The testis of the rabbit fetus is capable of synthesizing testosterone from pregnenolone on Day 19 but not on Day 17 when one enzymatic function is missing, the 3β hydroxy-steroid-dehydrogenase $\Delta 5$ isomerase system, which is limiting for testosterone synthesis (Figure 8). If that function fails to develop, the testis remains permanently incapable of testosterone synthesis. Such failure is one example of the five enzymatic defects preventing testosterone synthesis that have been recognized in humans (Table 1). Total or partial

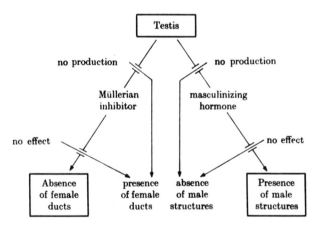

Fig. 6. Scheme summarizing the dual testicular control of the genital tract in males, with indications of the consequences when the fetal testicular hormones are absent or are prevented from acting. (From "General Outline about Reproductive Physiology and Its Developmental Background" by A. Jost. In H. Gibian and E. J. Plotz [Eds.], *Mammalian Reproduction.* Berlin: Springer-Verlag, 1970. Copyright 1970 by Springer-Verlag. Adapted by permission.)

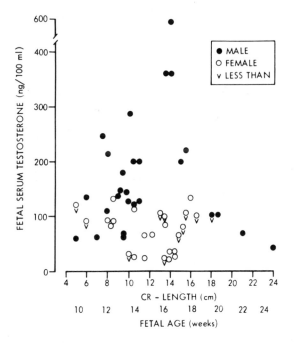

Fig. 7. Fetal serum testosterone concentrations measured by radioimmunoassay in human fetuses according to age or crown-rump length. It is noteworthy that masculinization of the body begins when the fetus is approximately 3 cm in length and is to a large extent completed when the fetal crown length is 8 cm. (From "Studies on Human Sexual Development: II. Fetal and Maternal Serum Gonadotropin and Sex Steroid Concentrations" by F. L. Reyes, R. S. Boroditzky, J. S. D. Winter, and C. Faiman. *Journal of Clinical Endocrinology and Metabolism*, 1974, *38*, 612–617. Copyright 1974 by J. B. Lippincott Company. Reprinted by permission.)

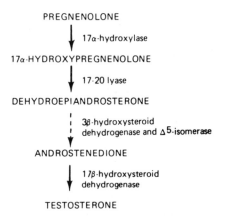

Fig. 8. Testicular androgen synthesis. The rate-limiting reaction in the synthetic pathway of pregnenolone to testosterone occurs during the conversion of dehydroepiandrosterone and androstenedione in fetal rabbit gonads on the 17th day of gestation. (From "Developmental Pattern of Testosterone Synthesis in the Fetal Gonad of the Rabbit" by J. D. Wilson and P. K. Siiteri. *Endocrinology*, 1973, *92*, 1182–1191. Copyright 1973 by J. B. Lippincott Company. Reprinted by permission.)

lack of testosterone synthesis results in totally or partially feminine conformation of the genitalia.

TESTOSTERONE ACTION ON TARGET ORGANS. The first steps of the action of testosterone on target cells have been elucidated. Testosterone enters the cells of androgen-sensitive tissues. In the cells, it may act in its original form, or more frequently, it is first converted by intracellular enzymatic systems into either dihydrotestosterone (DHT) (5α-androstane, 17OH, 3 one) by a 5α reductase system (e.g., in the prostate) or into estradiol, by an aromatase system (e.g., in nervous tissue; see Figure 9). Moreover, the active steroid must bind to a specific receptor protein in order to be effective. Schematically summarized, the "hormone + receptor" complex induces DNA transcription in the nucleus. The newly formed mRNA directs the synthesis of specific new proteins that ultimately lead to the cellular response in an unknown way.

In a series of studies in rabbit (Wilson, 1973) or human fetuses (Siiteri and Wilson, 1974), it was established that the embryonic sex duct system, including the Wolffian ducts, does not convert testosterone into DHT before the internal male ducts are sexually differentiated; testosterone itself is the hormone stabilizing the Wolffian ducts. On the contrary, the undifferentiated urogenital sinus and genital tubercle perform the conversion of testosterone to DHT early in development and probably require DHT for masculinization (Figure 10; see Wilson *et al.*, 1981). The necessity of the conversion of testosterone into DHT explains why patients with defective cellular 5α-reductase activity have external genitalia that are feminine or incompletely masculinized and masculine internal ducts.

In human subjects or mice where the receptor protein is defective, neither testosterone nor DHT exerts any masculinizing activity: The whole body acquires feminine features despite the testicular production of testosterone (Gehring *et al.*, 1971; Bardin *et al.*, 1973; Keenan *et al.*, 1974). However, the Müllerian ducts are absent because the testicular Müllerian inhibiting factor is present and normally active. This condition is well known under the name of *testicular feminization.*

Defective or absent masculine response of the indifferent sex primordia to androgens probably results from still other causes, if further postreceptor steps involved in the cellular response to androgens are defective.

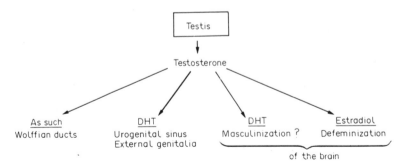

Fig. 9. Scheme showing how testosterone acts on target tissues during sexual differentiation either in its original form or after conversion to DHT or estradiol. (From "Genetic and Hormonal Factors in Sex Differentiation of the Brain" by A. Jost. *Psychoneuroendocrinology*, 1983, *8*, 183–193. Copyright 1983 by Pergamon Press, Ltd. Reprinted by permission.)

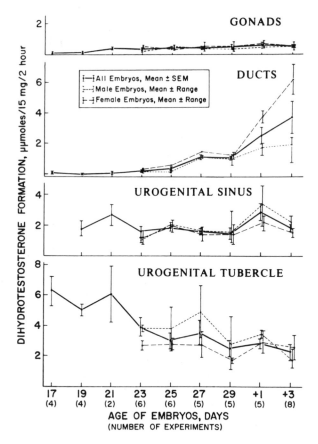

Fig. 10. Dihydrotestosterone formation from testosterone by different parts of the urogenital tract of rabbit fetuses or newborns of various ages incubated *in vitro*. (From "Dihydrotestosterone Formation in Fetal Tissues of the Rabbit and Cat" by J. D. Wilson and I. Lasnitzki. *Endocrinology*, 1971, *89*, 659–668. Copyright 1971 by J. B. Lippincott Company. Adapted by permission.)

A compelling series of experiments made in rats and in a few other species clearly indicate that the "defeminizing" effect of testosterone on the brain is actually produced by an estrogen, probably estradiol, resulting from a complex aromatization of testosterone (see MacLusky and Naftolin, 1981). The effect of testosterone can be prevented by estrogen antagonists, or it can be mimicked by estrogens. Estrogen receptors are present in the developing brain. These data suggest that a so-far-undescribed endocrine defect—absence of aromatase or of estrogen receptors in the brain—might result in behavioral abnormalities in those species in which estrogens act on the differentiation of the nervous structures mediating sex behavior (Jost, 1982).

MÜLLERIAN INHIBITOR (OR ANTI-MÜLLERIAN HORMONE)

Knowledge of the Müllerian inhibitor also called *anti-Müllerian hormone* (Josso *et al.*, 1977) or *Müllerian inhibiting substance* (Donahoe *et al.*, 1982) has been progressing rapidly since 1977.

It should first be noticed that for most of the experiments one still has to rely on the biological and mainly qualitative (rather than truly quantitative) test developed in my laboratory by Régine Picon (1969): inhibition (or induction to regress) of the Müllerian epithelium in *in vitro* cultures of the sex ducts of 14.5-day-old female rat fetuses after removal of the ovarian primordium. A radioimmunoassay for the fetal calf hormone has now become available, but new developments are still necessary (see below).

It should first be noted that the Müllerian inhibitor is produced by the fetal testis before the Müllerian ducts are developed and that it does not prevent them from being laid down concomitantly with early testicular organogenesis, as was seen in rabbit fetuses (Koike and Jost, 1982). The ducts seem to become "sensitive" to that factor only from a certain stage on, perhaps after they have attained a certain stage of differentiation. During normal development *in vivo,* the ducts begin retrogressing in their more anterior part, the first to be established, while the posterior part is still growing down toward the urogenital sinus. The early regression of the anterior part of the Müllerian ducts might be due to an earlier receptivity to/or exposure to higher concentrations of inhibitor, or to both.

On the other side, the Müllerian inhibitor is produced long after the Müllerian ducts have retrogressed, as has been verified in rat (Picon, 1969), in human, and in bovine or porcine fetuses. It seems to be mainly a fetal protein, although it was detected in the rete fluid of the adult boar (Josso *et al.,* 1979) and in the follicular fluid from adult bovine ovary (Vigier *et al.,* 1984).

The prolonged period of production of the so-called anti-Müllerian hormone led me to wonder whether it has no other role (see also Donahoe *et al.,* 1982), for instance, in gonadal differentiation. The answer to that question will probably have to await the time when larger amounts of it become available.

The Müllerian inhibitor is produced by the primordial Sertoli cells of the fetal testis. This fact was convincingly demonstrated when fetal calf Sertoli cells were cultivated *in vitro* (Blanchard and Josso, 1974); accordingly, it is produced by the early rabbit fetal testis at a stage preceding the appearance of the Leydig cells (Koike and Jost, 1982).

Another point of interest is the relative absence of zoological specificity of the Müllerian inhibitor: rat, (Picon, 1969), human (Josso, 1972), rabbit (Picon, 1971), and calf (Josso, 1972) fetal testis inhibited rat Müllerian ducts; fetal rat testis inhibited rabbit (Picon, 1971) or calf (Jost *et al.,* 1972) Müllerian ducts; but rat or mouse fetal testis failed to inhibit the ducts of chick embryos (Weniger, 1965; Josso *et al.,* 1977).

The isolation and identification of the anti-Müllerian hormone or Müllerian inhibiting substance was pursued with perseverance by Nathalie Josso's groupe in Paris and by Patricia Donahoe's group in Boston. The discovery that it is a glycoprotein and can incorporate ^{3}H-fucose *in vitro* was a great help in the isolation from the incubation medium of fetal calf testis. Partial purification on wheat germ lectin (WGA) permitted the obtaining of a monoclonal antibody (Vigier *et al.,* 1982; Donahoe *et al.,* 1982). Final purification was obtained with an immunoaffinity column (Josso *et al.,* 1981): the calf anti-Müllerian hormone (AMH) is a dimer; its molecular weight is 124,000 Daltons. Unfortunately, the antibody used so far reacts with calf AMH, but not with the rat factor; the radioimmunoassay developed recently (Vigier, Legeai, Picard, and Josso, 1982) can be used only in bovines, but new progress is expected quite soon. The perspectives for the future are promising.

ESTROGEN PRODUCTION BY THE FETAL OVARY

Several recent publications opened a new chapter in fetal endocrinology, in showing that the presumptive ovary produces estrogens very early (Mauléon et al., 1977; George et al., 1978; Sholl and Goy, 1978). This progress occurred because of the development of elegant techniques. The physiological role of these estrogens still has to be determined. As already mentioned, the fetal ovary is not necessary for the differentiation of a female genital tract, but in castrated male or female rabbit fetuses, the diameter of the Müllerian duct is smaller than in controls (Figure 3). This reduction might indicate that, normally, the fetal ovaries somewhat stimulate the development of the Müllerian ducts. Current concepts of estrogen production in the adult ovary state that theca cells synthesize androgens that are "aromatized" in the granulosa cells. If the same is valid for the fetal presumptive ovary, a similar cellular functional duality should exist even before the definite ovarian structure, characterized by a definite follicle, is realized.

It should be noticed that the timing and the amount of estrogen formation are such that that estrogen does not "defeminize" the central nervous system of normal females.

GONADAL SEX

It results from the previous data that the differentiation of the gonads, particularly of the testes, is the key step in mammalian sex differentiation. The morphological and biochemical mechanisms resulting in the differentiation of a testis and their genetic control remain to be elucidated.

The gonads develop on the inner aspect of the mesonephros (Figure 1) near to glomeruli or nephric tubules, a region that becomes populated by the primordial germ cells originating from the primitive gut. An indifferentiated gonadal primordium is first laid down, made of somatic cells, including mesenchymal cells with nearby collagen fibers, and interspersed germ cells. Gonadal sex first becomes histologically recognizable when seminiferous cords differentiate in the testicular primordium. The Leydig cells appear during a subsequent developmental phase. In the past, every possible theory concerning the participation and role of the different cell types in gonadal organogenesis was proposed: the advocated candidates were either the superficial epithelium, or the inner mesonephric nephrons, or the mesenchyme between them. For reasons that might be interesting to determine, one scheme has been widely accepted and repeated from textbook to textbook. It assumes that primary and secondary sex cords growing successively inward from the superficial gonadal layer constitute, respectively, the primordium of the testis and of the ovary. Recently, the theory of the nephric origin has been revived (Zamboni *et al.*, 1979).

TESTICULAR ORGANOGENESIS

In 1972, I reported a study of the early development of the rat gonad (Jost, 1972) that met three absolute requirements: (1) gonads of varying and exactly known ages were studied in serial sections (this requirement necessarily precluded electron microscopy as a method in the initial study); (2) it was necessary to see the cytoplasm of the cells and not only nuclei of shrunken cells, as had often been the case in past studies; and (3) the study, as far as possible, revealed the differentiation of individual cells, rather than of more-or-less well-defined populations of cells, such as the "sex cords," the "medulla," and so on. The 1972 study was made with the patient help of Simone Perlman, who sectioned the gonadal region of 198 rat fetuses, 12–14 days old, fixed in several concentrations of several fixative fluids. The genetic sex of the young was ascertained by studying the sex chromatin bodies in the amnion. Since then, the observations made with the light microscope have been confirmed with the electron microscope (Jost *et al.*, 1974; Magre and Jost, 1980) and in cultures *in vitro* (Jost *et al.*, 1981, 1982). The first indication of testicular organogenesis, seen on Day 13, was the differentiation in the depth of the primordium, near the mesonephric structures, of large cells, showing a clear swollen cytoplasm and encompassing the germ cells. These cells were the primordial Sertoli cells. During the next hours, more primordial Sertoli cells differentiated and aggregated. Thus, the seminiferous cords, the precursors of the adult seminiferous tubules, became progressively delineated. A definite basal membrane formed around these structures only from Day 14 on. This process of testicular differentiation could be confirmed experimentally *in vitro*. Indifferentiated primordia from 12.5-day-old male rat fetuses cultured *in vitro* in synthetic media or in media containing chick embryo extract differentiated, in three days, structures similar to seminiferous cords with their large Sertoli cells. In those cultures in which fetal calf serum (or definite fractions of it) was added to the medium, the morphogenesis of seminiferous cords was prevented, but the altered gonad contained scattered clear cells similar to Sertoli cells (Magre *et al.*, 1980, 1981; Jost *et al.*, 1981). It was then verified that these abnormal gonads inhibit *in vitro* the Müllerian ducts from 14.5-day-old rat fetuses. Because the Müllerian inhibiting factor is a marker of the Sertoli cells, the result suggests that physiological differentiation took place at the cellular level, although morphogenesis was prevented (Jost *et al.*, 1981). Testosterone-secreting Leydig cells also differentiate under these conditions (Jost *et al.*, 1984).

Now it should be verified whether the mode of early testicular differentiation reported for rats applies to other animal species. Investigations on rabbit fetuses suggest that, although the first Sertoli cells seem to appear around Day 14 more superficially than in the rat, the essence of the process is the same (Jost and Magre, 1984).

As already mentioned, it is noteworthy that the interstitial cells or Leydig cells differentiate during a subsequent developmental phase, on Day 15 in the rat and on Day 19 in rabbits. This is a general rule in mammals; that is, the Leydig cells appear later than the seminiferous cords.

Ovarian Morphogenesis

Ovarian morphological differentiation occurs still later in the same mass of cells that becomes a testis in males; it is characterized by two successive developmental phases. First, the incipient meiosis of the germ cells occurs and is then followed by the development of follicles around a few germ cells at the diplotene phase of the meiotic prophase. The majority of germ cells degenerate. Experiments suggesting that the rete ovarii triggers meiosis perhaps by producing a meiosis-inducing substance have been reported (Byskov, 1978), whereas the testis could produce a meiosis-preventing substance. However, these influences of somatic cells on the germ cells still need further elucidation.

In the developing ovary, the meiotic germ cells themselves probably play an important role in folliculogenesis, as no follicles develop in those conditions where the germ cells of the early presumptive ovaries degenerate precociously, for example, in XO human fetuses.

Control of Gonadal Differentiation

In mammals, testes develop in those individuals whose chromosomal formula contains a Y chromosome (Welshons *et al.*, 1959), or in some instances of the translocation of a part of the Y chromosome to an X chromosome. However, the exact mechanism of control of gonadal sex by one (or more?) gene located on the Y chromosome still remains unknown.

In the past, it was often assumed that the initial sexual orientation of the gonads resulted from a competition between masculinizing and feminizing inductors or hormones. It was thought that these agents were produced by discrete parts of the gonads and expressed the balance between masculinizing and feminizing genes, believed to be located, respectively, on the Y and on the X chromosomes. However, in mammals, the initial testicular differentiation depends on the presence of a Y chromosome, whatever the number of X chromosomes.

In reevaluating the available evidence, it seemed to me that the simplest way to interpret the data was to accept a parallel between the mechanisms controlling body sex and the sex of the gonad (Jost, 1965, 1970a,b; Jost *et al.*, 1973). In both cases, the primordium contains a feminine program that has to be counteracted in order to impose masculine differentiation (Figure 11). The so-called male organizer in Figure 11 has to be defined.

In 1975, Wachtel *et al.*, (1975a,b) proposed the very interesting theory that gonadal sex was imposed on the undifferentiated gonad by the histocompatibility antigen linked to the presence of a Y chromosome. In mammals, the male has a Y chromosome, and the HY antigen was assumed to impose testicular differentiation

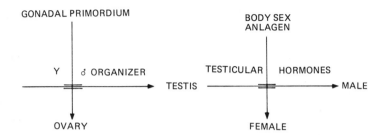

Fig. 11. Interpretive scheme for sexual differentiation in mammals. The gonadal primordium contains its own feminine program, which in males is prevented by a triggering mechanism that depends on the Y chromosome and that results in testicular differentiation. (From "General Outline about Reproductive Physiology and Its Developmental Background" by A. Jost. In H. Gibian and E. J. Plotz [Eds.], *Mammalian Reproduction.* Berlin: Springer-Verlag, 1970. Copyright 1970 by Springer-Verlag. Reprinted by permission.)

on the indifferent gonad and to play the part of the "male organizer" in Figure 11. In birds and in many amphibians, the female has a Y chromosome (often called W), and in these animals, the HY antigen was assumed to induce ovarian organogenesis in females. This theory raised a great deal of interest and stimulated much research. In many cases, a parallel could be demonstrated between the presence of testes and of HY antigen in the cells of adult subjects. In other cases, the hypothesis was contradicted, for example, by HY antigen found in fertile females (Wachtel, 1979). Similarly, the generalization of the concept to mammalian freemartins or to lower vertebrates (Jost, 1979) meets with difficulties. Sex reversal could be produced in male birds or in female heterozygous amphibians (e.g., Xenopus laevis) by estrogens. The HY antigen could be detected in the reversed homozygous males (they have no Y or W chromosome) after sex reversal (Müller *et al.,* 1979; Wachtel *et al.,* 1980) as if it were the result of the presence of ovaries and not its cause.

In mammals, there is no compelling evidence that the HY antigen directs testicular organogenesis. Moreover, future studies will probably have to better define which antigen is concerned: evidence has been given that the histogenic (skin-graft testing) and serological tests for HY antigen may detect two different antigens (Melvold *et al.,* 1977).

In any case, even if a single protein could be responsible for the differentiation of a new cell type, at the origin of a testis, it is unlikely that it also governs the subsequent phases of testicular organogenesis. A cascade of successive influences is probably involved after the initial trigger.

REFERENCES

Bardin, C. W., Bullock, L. P., Sherins, R. J., Mowszovicz, I., and Blackburn, W. R. Androgen metabolism and mechanism of action in male pseudohermaphroditism: A study of testicular feminization. *Recent Progress in Hormone Research,* 1973, *29,* 65–105.

Blanchard, M. G., and Josso, N. Source of the anti-Müllerian hormone synthesized by the fetal testis: Müllerian-inhibiting activity of fetal bovine Sertoli cells in tissue culture. *Pediatric Research,* 1974, *8,* 968–971.

Brook, C. G. D., Wagner, H., Zachmann, M., Prader, A., Armendares, S., Frenk, S., Aleman, P., Najjar, S. S., Slim, M. S, Genton, N., and Bozic, C. Familial occurrence of persistent Müllerian structures in otherwise normal males. *British Medical Journal*, 1973, *1*, 771–773.

Byskov, A. G. Regulation of initiation of meiosis in fetal gonads. 5th Annual Workshop on the Testis, Geilo, Norway. *International Journal of Andrology*, 1978, Suppl. 2, 29–38.

Donahoe, P. K., Budzik, G. P., Trelstad, R., Mudgett-Hunter, M., Fuller, A., Jr., Hutson, J. M., Ikawa, H., Hayashi, A., and MacLaughlin, D. Müllerian-inhibiting substance: An update. *Recent Progress in Hormone Research*, 1982, *38*, 279–330.

Ford, C. E., Polani, P. E., Briggs, J. H., and Bishop, P. M. F. A presumptive human XXY/XX mosaic. *Nature*, 1959, *183*, 1030–1032.

Gehring, U., Tomkins, G. M., and Ohno, S. Effect of the androgen-insensitivity mutation on a cytoplasmic receptor for dihydrotestosterone. *Nature New Biology*, 1971, *232*, 106–107.

George, F. W., Milewich, L., and Wilson, J. D. Oestrogen content of the embryonic rabbit ovary. *Nature*, 1978, *274*, 172–173.

Josso, N. Effect of testosterone and of some of its 17-hydroxylated metabolites on the Müllerian duct of the foetal rat, in organ culture. *Revue Européenne d'Études Cliniques et Biologiques*, 1971, *16*, 694–697.

Josso, N. Evolution of the Müllerian-inhibiting activity of the human testis: Effect of fetal, perinatal and postnatal human testicular tissue on the Müllerian duct of the fetal rat in organ culture. *Biology of the Neonate*, 1972, *20*, 368–379.

Josso, N., Picard, J. Y., and Tran, D. The antimüllerian hormone. *Recent Progress in Hormone Research*, 1977, *33*, 117–167.

Josso, N., Picard, J. Y., Dacheux, J. L., and Courot, M. Detection of anti-Müllerian activity in boar rete testis fluid. *Journal of Reproduction and Fertility*, 1979, *57*, 397–400.

Josso, N., Picard, J. Y, and Vigier, B. Purification de l'hormone antimüllérienne bovine à l'aide d'un anticorps monoclonal. *Comptes Rendus des Séances de l'Académie des Sciences*, Paris, 1981, *293*, 447–450.

Jost, A. Recherches sur la différenciation sexuelle de l'embryon de lapin: III. Rôle des gonades foetales dans la différenciation sexuelle somatique. *Archives d'Anatomie Microscopique et de Morphologie Expérimentale*, 1947, *36*, 271–316.

Jost, A. Problems of fetal endocrinology: The gonadal and hypophyseal hormones. *Recent Progress in Hormone Research*, 1953, *8*, 379–418.

Jost, A. *Biologie des Androgènes chez l'Embryon*. IIIème Réunion des Endocrinologistes de Langue Française. Paris: Masson et Cie, 1955.

Jost, A. Embryonic sexual differentiation (morphology, physiology, abnormalities). In H. W. Jones and W. W. Scott (Eds.), *Hermaphroditism, Genital Anomalies and Related Endocrine Disorders* (2nd ed. in 1971). Baltimore: Williams & Wilkins, 1958.

Jost, A. Hormonal influences in the sex development of the bird and mammalian embryo. *Memoir of the Society for Endocrinology*, 1960, *7*, 49–62.

Jost, A. Gonadal hormones in the sex differentiation of the mammalian fetus. In R. L. De Haan and H. Ursprung (Eds.), *Organogenesis*. New York: Holt, Rinehart, & Winston, 1965.

Jost, A. Modalities in the action of androgens on the foetus. In C. Cassano, M. Finkelstein, A. Klopper, and C. Conti (Eds.), *Research on Steroids: Proceedings of the Third Meeting of the International Study Group for Steroid Hormones*, Vol. 3. Amsterdam: North-Holland Publishing, 1968.

Jost, A. General outline about reproductive physiology and its developmental background. In H. Gibian and E. J. Plotz (Eds.), *Mammalian Reproduction* (21 Colloquium der Gesellschaft für Biologische Chemie, Mosbach). Berlin: Springer-Verlag, 1970a.

Jost, A. Hormonal factors in the sex differentiation of the mammalian foetus. *Philosophical Transactions of the Royal Society of London, B*, 1970b, *259*, 119–130.

Jost, A. Données préliminaires sur les stades initiaux de la différenciation du testicule chez le rat. *Archives d'Anatomie Microscopique et de Morphologie Expérimentale*, 1972, *61*, 415–438.

Jost, A. Basic sexual trends in the development of vertebrates. In *Sex, Hormones, and Behaviour*, Ciba Foundation Symposium 62. New York: Excerpta Medica, Elsevier, 1979.

Jost, A. Fetal sexual differentiation. In N. Kretchmer and J. A. Brasel (Eds.), *Biomedical and Social Bases of Pediatrics*. New York, Paris: Masson, 1981.

Jost, A. Genetic and hormonal factors in sex differentiation of the brain. *Psychoneuroendocrinology*, 1983, *8*, 183–193.

Jost, A., and Bergerard, Y. Culture *in vitro* d'ébauches du tractus génital du foetus de rat. *Comptes Rendus de la Société de Biologie,* Paris, 1949, *143,* 608–609.

Jost, A. and Magre, S. Testicular development phases and dual hormonal control of sexual organogenesis. In M. Serio, M. Motta, M. Zanisi, and R.L. Martini (Eds.), *Sexual Differentiation: Basic and Clinial Aspects.* Serono Symposium 11. New York: Raven Press, 1984.

Jost, A., Vigier, B., and Prepin, J. Freemartins in cattle: The first steps of sexual organogenesis. *Journal of Reproduction and Fertility,* 1972, *29,* 349–379.

Jost, A., Vigier, B., Prepin, J., and Perchellet, J. P.Studies on sex differentiation in mammals. *Recent Progress in Hormone Research,* 1973, *29,* 1–41.

Jost, A., Magre, S., and Cressent, M. Sertoli cells and early testicular differentiation. In R. E. Mancini and L. Martini (Eds.), *Male Fertility and Sterility.* New York: Academic Press, 1974.

Jost, A., Magre, S., and Agelopoulou, R. Early stages of testicular differentiation in the rat. *Human Genetics,* 1981, *58,* 59–63.

Jost, A., Magre, S., Agelopoulou, R., and Chartrain, I. Aspects of gonadal differentiation in mammals. In P. G. Crosignani, B. L. Rubin, and M. Fraccaro (Eds.), *Genetic Control of Gamete Production and Function.* Proceedings of the Serono Clinical Colloquia on Reproduction 3. London: Academic Press; New York: Grune & Stratton, 1982.

Jost, A., Patsavoudi, E., Magre, S., Castanier, M., and Scholler, R. Relations entre organogenèse testiculaire et sécrétion de testostérone par le testicule foetal in vitro. *Pathologie Biologie,* 1984, *32,* 860–862.

Keenan, B. S., Meyer, W. J., III, Hadjian, A. J., Jones, H. W., Jr., and Migeon, C. J. Syndrome of androgen insensitivity in man: Absence of 5α-dihydrotestosterone binding protein in skin fibroblasts. *Journal of Clinical Endocrinology and Metabolism,* 1974, *38,* 1143–1146.

Koike, S., and Jost, A. Production précoce de l'inhibiteur Müllerien par le testicule foetal de lapin. *Comptes Rendus de l'Académie des Sciences,* Paris, 1982, *295,* 701–705.

MacLusky, N. J., and Naftolin, F. Sexual differentiation of the central nervous system. *Science,* 1981, *211,* 1294–1303.

Magre, S., and Jost, A. The initial phases of testicular organogenesis in the rat: An electron microscopy study. *Archives d'Anatomie Microscopique et de Morphologie Expérimentale,* 1980, *69,* 297–318.

Magre, S., Agelopoulou, R., and Jost, A. Cellules de Sertoli et organogenèse du testicule foetal. *Annales d'Endocrinologie,* Paris, 1980, *41,* 531–537.

Magre, S., Agelopoulou, R., and Jost, A. Action du sérum de foetus de veau sur la différenciation *in vitro* au maintien des cordons séminifères du testicule du foetus de rat. *Comptes Rendus de l'Académie des Sciences,* Paris, 1981, *292,* 85–89.

Mauléon, P., Bézard, J., and Terqui, M. Very early and transient 17β-estradiol secretion by fetal sheep ovary: *In vitro* study. *Annales de Biologie Animale, Biologie, Biophysique,* 1977, *17,* 399–401.

Melvold, R. W., Kohn, H. I., Yerganian, G., and Fawcett, D. W. Evidence suggesting the existence of two H-Y antigens in the mouse. *Immunogenetics,* 1977, *5,* 33–41.

Müller, U., Zenzes, M. T., Wolf, U., Engel, W., and Weniger, J.-P. Appearance of H-W(H-Y) antigen in the gonads of oestradiol sex-reversed male chicken embryos. *Nature,* 1979, *280,* 142–144.

Ohno, S. *Major Sex Determining Genes.* Monographs on Endocrinology, No. 11. Berlin: Springer-Verlag, 1979.

Phoenix, C. H., Goy, R. W., Gerall, A. A., and Young, W. C. Organizing action of prenatally administered testosterone propionate on the tissues mediating mating behaviour in the female guinea-pig. *Endocrinology,* 1959, *65,* 369–382.

Picon, R. Action du testicule foetal sur le développement *in vitro* des canaux de Müller chez le rat. *Archives d'Anatomie Microscopique et de Morphologie Expérimentale,* 1969, *58,* 1–19.

Picon, R. Etude comparée de l'action inhibitrice des testicules de lapin et de rat sur les canaux de Müller de ces deux espèces. *Comptes Rendus de l'Académie des Sciences,* Paris, 1971, *272,* 98–101.

Reyes, F. L., Boroditzky, R. S., Winter, J. S. D., and Faiman, C. Studies on human sexual development: II. Fetal and maternal serum gonadotropin and sex steroid concentrations. *Journal of Clinical Endocrinology and Metabolism,* 1974, *38,* 612–617.

Sholl, S. A., and Goy, R. W. Androgen and estrogen synthesis in the fetal guinea pig gonad. *Biology of Reproduction,* 1978, *18,* 160–169.

Siiteri, P. K., and Wilson, J. D. Testosterone formation and metabolism during male sexual differentiation in the human embryo. *Journal of Clinical Endocrinology and Metabolism,* 1974, *38,* 113–125.

Vigier, B., Picard, J. Y., and Josso, N. A monoclonal antibody against bovine anti-Müllerian hormone. *Endocrinology,* 1982, *110,* 131–137.

Vigier, B., Legeai, L., Picard, J. Y, and Josso, N. A sensitive radioimmunoassay for bovine anti-Müllerian hormone allowing its detection in male and freemartin fetal serum. *Endocrinology*, 1982, *111*, 1409, 1411.

Vigier, B., Picard, J. Y., Tran, D., Legeai, L., and Josso, N. Production of anti-Müllerian hormone: Another homology between Sertoli and granulosa cells. *Endocrinology*, 1984, *114*, 1315, 1320.

Wachtel, S. S. H-Y antigen in the mammalian female. *Annales de Biologie Animale, Biochimie, Biophysique*, 1979, *19*, 1231–1237.

Wachtel, S. S., Koo, G. C., and Boyse, E. A. Evolutionary conservation of H-Y ("male") antigen. *Nature*, 1975a, *254*, 270–272.

Wachtel, S. S., Ohno, S., Koo, G. C., and Boyse, E. A. Possible role for H-Y antigen in the primary determination of sex. *Nature*, 1975b, *257*, 235–236.

Wachtel, S. S., Bresler, P. A., and Koide, S. S. Does H-Y antigen induce the heterogametic ovary? *Cell*, 1980, *20*, 859–864.

Welshons, W. J., and Russell, L. B. The Y-chromosome as the bearer of male determining factors in the mouse. *Proceedings of the National Academy of Sciences*, N.Y., 1959, *45*, 560–566.

Weniger, J. P. Étude comparée des actions hormonales des testicules embryonnaires de poulet et de souris en culture *in vitro*. *Archives d'Anatomie Microscopique et de Morphologie Expérimentale*, 1965, *54*, 909–919.

Wilson, J. D. Testosterone uptake by the urogenital tract of the rabbit embryo. *Endocrinology*, 1973, *92*, 1192–1199.

Wilson, J. D., and Lasnitzki, I. Dihydrotestosterone formation in fetal tissues of the rabbit and rat. *Endocrinology*, 1971, *89*, 659–668.

Wilson, J. D., and Siiteri, P. K. Developmental pattern of testosterone synthesis in the fetal gonad of the rabbit. *Endocrinology*, 1973, *92*, 1182–1191.

Wilson, J. D., Griffin, J. E., George, F. W., and Leshin, M. The role of gonadal steroids in sexual differentiation. *Recent Progress in Hormone Research*, 1981, *37*, 1–39.

Wolff, Et., and Wolff, Em. The effects of castration on bird embryos. *Journal of Experimental Zoology*, 1951, *116*, 59–98.

Zamboni, L., Bézard, J., and Mauléon, P. The role of the mesonephros in the development of the sheep fetal ovary. *Annales de Biologie Animale, Biochimie, Biophysique*, 1979, *19*, 1153–1178.

2

Gonadal Hormones during Sexual Differentiation in Vertebrates

JOHN A. RESKO

INTRODUCTION

It is generally agreed that, at least in mammals, sex chromosomes determine the sex of the individual. For "normal" sexual development, however, more is required than the mere presence of these chromosomes. It was recognized early in this century—and has been verified many times since—that chemical substances (hormones) of gonadal origin influence the development of the central nervous system and anlagen of the reproductive tract so that they are either masculine or feminine in character. The influence of gonadal hormones on target tissues is covered elsewhere in this volume. Here, we discuss and bring together what is known about the hormones that are secreted by the gonads during sexual differentiation in vertebrates. The so-called period of sexual differentiation in vertebrates varies from species to species. In long-gestation mammals, such as human beings, rhesus monkeys, and guinea pigs, sexual differentiation takes place prenatally. In species with short gestations, such as rats and mice, sexual development, especially of the central nervous system, continues into the neonatal period. In some species (e.g., some primates), the so-called critical period for hormone action on sexual differentiation is not known. In fact, the term *critical period* may not be a good one because it is

JOHN A. RESKO   Department of Physiology, Oregon Health Sciences University, Portland, Oregon 97201.

too restrictive for hormone actions that may be longer in duration or for actions that vary in time span for different sexual functions.

Evidence for steroid biosynthesis by the fetal gonads has been found in a variety of vertebrate species. In this chapter, we present some of this evidence. Special emphasis is placed on studies of fetal blood plasma or serum in which hormones were measured by sensitive and specific assay procedures. This chapter is intended not as a complete and extensive review of all published reports, but more as a guide into an area of fetal endocrinology that may be useful to students and others.

Steroid Hormone Production in Developing Birds and Amphibia

Morphology of the Embryonic Chick Testes

The chick embryo, because of availability and ease of experimental manipulation, has been used frequently in embryological studies. The incubation period of the chick embryo is 19.5 days. Morphological differentiation of the gonads from undifferentiated structures into identifiable testes or ovaries probably occurs on Day 6.5 of incubation, but times ranging from Day 5 to Day 8 have been noted (Romanoff, 1960). A few interstitial cells appear in the embryonic testes by Day 13 of incubation (Swift, 1916). By Days 15–17, the interstitial cells are found in large amounts. These cells rapidly decrease in number after the chicks hatch.

Biosynthetic Capacity of Chick Gonads for Steroid Production

In this species, it appears that steroid biosynthesis occurs in fetal testes before the interstitial cells are identified as distinct entities. On Day 2, Δ^5-3β-hydroxysteroid dehydrogenase, a key enzyme for steroid biosynthesis, can be identified histochemically in cells of the genital ridge (Woods and Weeks, 1969). Many studies have shown the capacity of chick gonads, from Days 6.0–8.5, to synthesize estrogen and androgen (Weniger and Zeis, 1971; Galli and Wasserman, 1973; Haffen, 1970; Guichard et al., 1973). Testes that are feminized by estrogen treatment seem to lose their capacity for androgen biosynthesis and make estrogens instead (Akram and Weniger, 1967, 1969).

Estimation of androgen content in developing gonads by immunofluorescence suggests that androgens are produced by the gonads of both sexes before differentiation of the gonads occurs (Woods and Podczaski, 1974). At the time of gonadal differentiation, however, higher androgen concentrations are found in the testes and the right ovary than in the left ovary. In the male embryo, the Wolffian ducts persist and become the vasa deferentia of the adult. The Müllerian ducts, on the other hand, cease growing by Day 8, after which they regress. In females, growth of the right Müllerian ducts ceases by Day 8, thereafter regressing. The left Müllerian ducts persist and become the adult oviducts. These data are consistent with early theories of sexual differentiation in this species (Willier, 1942, 1952) that propose that sex steroids are produced by undifferentiated gonads and are responsible for their differentiation. In this case, the testes develop in the male and the right ovary regresses because of the greater capacity for androgen formation and the inhibitory action of this hormone on ovarian development. Testosterone (T) concentrations in plasma from embryonic blood vessels on the chick are shown in Figure 1. Plasma from males contains significantly more T than plasma from females beginning on Day 7.5 of incubation. Testosterone is found in nearby equal quantities in plasma from males and females on Day 5.5, which is 1 day before differentiation of the gonads and before the appearance of interstitial cells in the fetal testes. Patterns of steroid hormones in plasma and the embryonic gonads suggest a picture that is similar to that in mammals in that a sex difference in the secretion of T early in embryonic life is observable. The presence of T in both gonads before differentiation and the relationship between the concentrations of androgen in the right and the left gonads provide stronger evidence than in other species that androgens participate in the differentiation of fetal gonads. Other observations (Woods and Erton, 1978) suggest that estrogens are also biosynthesized by the chick gonad at approximately 3.5 days of incubation. As gestation proceeds, however, the left ovary has greater capacity for estrogen biosynthesis than fetal testes.

EFFECTS OF SEX STEROIDS ON AMPHIBIAN DEVELOPMENT

In many amphibians, sexual differentiation can be affected by experimentation, but the results may vary considerably among species. Parabiotic union of male and female embryos results in masculinization of the ovaries (Dodd, 1960), and injection of sex steroids affects sexual development. Among the Ranidae, total masculinization can be accomplished by low levels of androgen. *Rana agiles* (Vannini, 1941) and *R. temporaria* (Gallien, 1944) can be androgenized by treatment after metamorphosis, and *R. clamitans* (Mintz *et al.*, 1945) can be androgenized after normal sexual differentiation has taken place. The actions of estrogens in high Anura are highly variable. Dodd (1960) has discussed some of the effects of estrogen on development.

The urodeles and low Anura respond to sex steroids in a manner different from that of the high Anura. In these species, estrogens feminize, and androgen effects can be paradoxical (Dodd, 1960). Larva from *R. sylvatica* and *R. pipiens* are androgenized by treatment with estradiol (Dale, 1962).

Fig. 1. Testosterone concentrations in plasma from the chick embryo. Data are presented as means. The numbers in parentheses are the numbers of animals in each group. The standard deviation for each mean value ranged from 1.1 to 5.3 pg/ml of testosterone. (From "Plasma Testosterone Levels in the Chick Embryo" by J. E. Woods, R. M. Simpson, and P. L. Moore. *General and Comparative Endocrinology*, 1975, *27*, 543–547. Copyright 1975 by Academic Press. Reprinted by permission.)

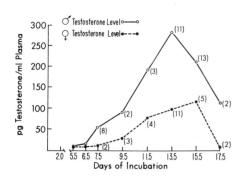

JOHN A. RESKO

Little information is available on steroid production by endocrine tissues of developing amphibia. In *Xenopus laevis,* estrogens have been found in ovaries shortly after metamorphosis (Gallien and Chalumeau-Le Foulgoc, 1960), and 17-ketosteroids have been quantified in excretions during metamorphosis (Rapola, 1963). Breuer *et al.* (1966) showed sex differences in steroid metabolism in larvae. In larvae of *R. sylvatica* and *R. pipiens,* Dale (1962) identified androsterone and estradiol-17β by Stage 26. In the urodele *Pleurodeles waltlii,* Ozon (1963) found active metabolism of estradiol-17β and the presence of hydroxysteroid dehydrogenase in Stage 4 larvae. The Δ^5-3β-hydroxysteroid-dehydrogenases, essential enzymes for androgen biosynthesis, were found in larval testes before metamorphosis (Collenot, 1964) and in Stage 45–54 larvae (Ozon, 1969). At Stage 55, testes are capable of converting pregnenolone to progesterone (P) (Ozon, 1969). Larvae of several amphibian species are capable of metabolizing T and estradiol-17β (Chieffi, 1958). These phenomena are evidence for steroid biosynthesis and metabolism in amphibian larval stages. The relationship of gonadal secretions to sexual differentiation in developing amphibians is not clear.

STEROID HORMONES IN DEVELOPING RODENTS

MORPHOLOGICAL CONSIDERATIONS

Because many of the relationships between steroid hormones and development of the nervous system had been worked out in laboratory rodents, when more sensitive and specific methods for determining steroid hormones were developed, endocrinologists tested the capacity of the fetal gonads to produce steroids, especially T, *in vitro,* and later, they measured these hormones in fetal plasma or serum. Before this point, however, clues to the secretory activity of fetal gonads had been obtained from analysis of their structure. In the laboratory rat, the gonads differentiate on about Day 13.5 (Torrey, 1945). Morphological evidence, such as the presence of Leydig cells, suggests that fetal testes are secretory (Torrey, 1945; Roosen-Runge and Anderson, 1959).The number of Leydig cells in the fetal testes of the rat reaches maximum intensity on Day 18 or 19 of gestation (Lording and De Kretser, 1972; Roosen-Runge and Anderson, 1959), then declines about 2 weeks postpartum. The Leydig cell is thought to be the type of cell in which androgen synthesis occurs. This conjecture has been confirmed in the fetal rat (Niemi and Ikonen, 1961) and mouse (Baillie and Griffiths, 1964) by histochemical demonstration of the 3β-ol-dehydrogenase, an enzyme essential for androgen biosynthesis. The presence of this enzyme within fetal testes can be demonstrated on Day 15 of gestation (Niemi and Ikonen, 1962), and it appears before regression of the gonadal cortex in the rat (Schlegel *et al.,* 1967). Histochemical evidence for the appearance of 3β-ol-dehydrogenase is dependent on the substrate used for this reaction. For example, in fetal mouse testes, 3β-ol-dehydrogenase can be demonstrated on Day 11 of gestation if pregnenolone is used as a substrate; it can be demonstrated on Day 15 if dehydroepiandrosterone is used as a substrate (Baillie and Griffiths, 1964).

STEROID HORMONE PRODUCTION BY FETAL AND NEONATAL GONADS OF SHORT-GESTATION RODENTS

The capacity of fetal and neonatal rat testes to biosynthesize T is shown in Table 1. According to Noumura *et al.* (1966), between Day 14.5 and Day 15.5 of gestation, fetal rat testes rapidly change in their capacity to biosynthesize T from P. This biosynthetic capability continues into the neonatal period but drops off by postnatal Day 40. Warren *et al.* (1975) measured T in fetal rat testes and found significant quantities on Day 15.5 of gestation. The testes contained significant amounts of T in the first few hours of birth; although T was nondetectable in the neonatal period, it was detectable around Day 60. These latter data agree well with

TABLE 1. CAPACITY OF FETAL AND NEONATAL RAT TESTES TO BIOSYNTHESIZE TESTOSTERONE

Age (days)	$P \rightarrow T^a$ (%)	$P \rightarrow A^a$ (%)	T^b (ng/mg)	T^c (ng/mg)	A^c
Prenatal					
13.5	0.2	0.4	—		
14.5	1.8	0.2	—		
15.5	37.6	4.3	0.16		
16.5	58.4	4.5	0.71		
17.5	65.3	5.3	0.82		
18.5	70.7	6.3	2.49		
19.5	69.0	6.0	1.64		
20.5	72.2	6.9	1.66		
21.5	72.3	5.6	—		
Postnatal					
0			0.92		
⅙			0.27		
1	66.8	6.1	ND^d	0.33	0.11
5				0.10	0.03
10				0.07	0.01
15				Trace	ND^d
20			ND^d		
30				Trace	0.01
40	3.9	3.1	ND^d	Trace	0.01
60			0.025	0.03	ND^d
90				0.08	0.01

[a]P = 7-^{3}H-progesterone (4.5 mμM) converted to T (testosterone) or A (androstenedione) (from Noumura *et al.*, 1966).
[b]T = testosterone (ng/mg) of testicular tissue measured by gas–liquid chromatography (from Warren *et al.*, 1973).
[c]T = testosterone (ng/mg) of testicular tissue measured by gas–liquid chromatography (from Resko *et al.*, 1968).
[d]ND = nondetectable.

earlier work by Resko *et al.* (1968), who measured T in testes from Day 1 to Day 90 of postnatal life.

Warren *et al.* (1972) demonstrated that fetal rat testes possess the enzyme necessary to convert acetate to cholesterol and several androgens on Day 19.5 of gestation. Picon (1976) determined the amount of T produced by testes in culture as a function of gestational age. Testes from 13.5-day-old fetuses were unable to produce T on the first day of culture, but they acquired this capacity on the second and third days. Dibutyrl cyclic AMP stimulated T production by the testes beginning on Day 14.5 of fetal life. In late gestation, the biosynthetic capacity of the fetal testes does not appear to be under the control of the anterior pituitary, as decapitation on Day 18.5 of gestation has no effect on the testes' capacity to convert P to T. Decapitation on Day 16.5, however, decreases this capacity (Noumura *et al.*, 1966).

Sexual differentiation of the central nervous system in short-gestation mammals such as the rat continues into the neonatal period. The hormones, especially androgens, to which the neonatal brain is exposed have been the subject of a few investigations. In 1968, we (Resko, Feder, and Goy) recognized the need to obtain information on the concentrations of T and androstenedione (A) in plasma of the neonatal rat. In those days, we quantified T and A by gas–liquid chromatography with electron capture detection. The sensitivity of the assay was in the range of 2 ng of T. Our measurements of T and A in testicular tissue are shown in Table 1. These data suggest differences in the quantities of T produced within the neonatal testes with time after birth. More T was found in testes from postnatal Day 1 rats than in those from Day 5 rats. Analysis of T in blood obtained by cardiac puncture required a pool from 230 rats on Day 1 (36.4 ml) for a single determination. Quantification of T in this pooled plasma indicated that the newborn male rat has about 270 pg/ml of T in its plasma and that by postnatal Day 10 this level drops to about 9 pg/ml. These results were the first indication that T is available in the circulatory system of the newborn male rat for physiological action at a time when exogenous hormone is effective. Comparisons between males and females were not made, however. Döhler and Wuttke (1975) measured T, estradiol-17β, P, and androgen (by a less specific antibody) in newborn rats from Day 1 through adulthood. Here we will discuss data from the first 20 days. (The data have been redrawn in Figure 2.) Estradiol and P concentrations did not differ between sexes. Serum T was higher in males than in females for the first 21 days of life. The sex differences were more pronounced when androgen was measured by a less specific antiserum. These results were interpreted to mean that androgens other than T are found in the circulatory system of the neonatal male rat and that these may be important in sexual differentiation.

More recent studies report plasma levels of testosterone in male and female rat fetuses from Day 17 of gestation and in neonates on Days 1, 3, and 5 postpartum (Weisz and Ward, 1980). The levels of testosterone in male fetuses showed the greatest difference from female fetuses on Days 18 and 19 of gestation. Only on Day 18 of gestation, however, was there no overlap between testosterone concentrations in male and female fetuses. Testosterone concentrations decreased in both

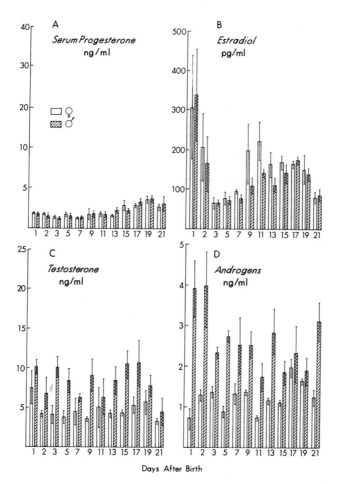

Fig. 2. Sex steroids in serum of the neonatal rat. Measurements were made by radioimmunoassay. Androgens in Panel D were quantified by means of a less specific antiserum than in Panel C. Serum androgen levels were significantly different between males and females at all time periods studied. (From "Changes with Age in Levels of Serum Gonadotropins, Prolactin, and Gonadal Steroids in Prepubertal Male and Female Rats" by K. D. Döhler and W. Wuttke. *Endocrinology,* 1975, *97,* 898–907. Copyright 1975 by J. B. Lippincott Company. Reprinted by permission.)

sexes after birth, but the sex difference remained. Corbier *et al.* (1984) quantified T in rat sera collected at 2-hr intervals after birth. These studies demonstrated not only a significant sex difference in serum T concentrations during the first six hours after birth but also showed a significant surge of T in infant sera 2 hr after birth.

STEROID HORMONE PRODUCTION BY FETAL GONADS OF LONG-GESTATION RODENTS

In a long-gestation rodent, the guinea pig, the ovaries are distinguishable from the testes between Day 22 and Day 26 of gestation. Although interstitial cells can be observed earlier, from Day 26 to Day 30 the interstitial tissue is accompanied by the appearance of distinct Leydig cells (Black and Christensen, 1969). Early in gestation (Days 22–24), smooth endoplasmic reticula (cytoplasmic inclusions that are associated with steroid hormone production) appear in interstitial cells (Black and Christensen, 1969). The appearance of these structures coincides with androgen secretion determined by indirect methods of detection. Using these methods, Ortiz *et al.* (1966) detected androgenic stimulation of rat ventral prostate *in vitro* by 22-day-old fetal guinea pig testes. Older developmental stages (Price *et al.*, 1964a,b) were also studied. Direct measurements of androgen in the fetal guinea pig's circulatory system by techniques such as radioimmunoassay (RIA) are found in few published reports. In our laboratory, we (Buhl, Pasztor, and Resko, 1979) have measured T in mixed umbilical-vessel plasma from 35-day-old fetuses. These preliminary data indicate that fetal males have more T in their circulation than females that are littermates.

TROPHIC CONTROL OF THE FETAL GONADS IN RODENTS

As presented above, there is considerable evidence demonstrating the capacity of fetal rat testes to secrete androgen during the period of sexual differentiation. The nature of trophic influences on neonatal rat testes is not well understood. However, some data do suggest that the neonatal rat pituitary exerts trophic influence over the gonads and that negative feedback between the testes and the hypothalamus-hypophyseal system has been established by this early age. Unilateral orchidectomy of a newborn rat results in hypertrophy of the remaining gonad within 3 days (Yagimura *et al.*, 1969). Removal of the testes results in elevation of gonadotropin levels that were low before castration (Goldman and Gorski, 1971). In these experiments, gonadotropin levels do not rise if exogenous T or estradiol-17β is administered. Similarly, the neonatal testes are capable of responding to exogenous luteinizing hormone (LH) or human chorionic gonadotropin (hCG) with hypertrophy and hyperplasia of the interstitial cells, a fact that suggests participation of the pituitary gland in testicular activity at this stage of development (Arai and Serisawa, 1973).

FUNCTIONAL CONSIDERATIONS OF FETAL RABBIT GONADS

In the rabbit, the fetal testes can be distinguished histologically from ovaries by Day 14 of gestation. Lipsett and Tullner (1965) found that fetal testes but not ovaries produce T from radiolabeled pregnenolone by Day 18 of gestation before interstitial cells differentiate. By Day 17 of gestation, the rabbit placenta contains significant quantities of C^{21} steroids that at least potentially could be used for T biosynthesis by the fetal testes (Wilson and Siiteri, 1973). Correlations of gonadotropin receptors and T content in fetal testes show that hCG binding and T content are higher in fetal testes on Day 18.5 of gestation compared with Day 18.0 (Catt *et al.*, 1975). Fetal ovaries, on the other hand, contain little receptor or T on Day 18.5. These relationships (Figure 3) argue in favor of gonadotropin as the initial stimulus of Leydig cell differentiation and T biosynthesis. Plasma levels of T have

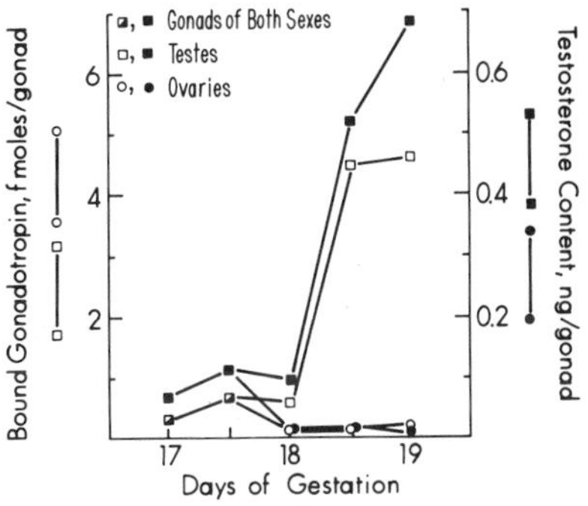

Fig. 3. The relationship between human chorionic gonadotropin binding and testosterone content of fetal rabbit gonads on Days 17–19 of gestation. Each point represents a single determination. (From "LH-hCG Receptors and Testosterone Content during Differentiation of the Testes in the Rabbit Embryo" by K. J. Catt, M. L. Dufau, W. B. Neaves, P. C. Walsh, and J. D. Wilson, *Endocrinology*, 1975, *97*, 1157–1165. Copyright 1975 by J. B. Lippincott Company. Reprinted by permission.)

been quantified in male and female fetuses from Days 20 to 31 of gestation (Veyssiere *et al.*, 1976). In these studies, concentrations of T in males ranged from 132 to 361 pg/ml, whereas females had significantly lower levels (21-116 pg/ml). An interesting observation has been made recently on the fetal ovaries of the rabbit (Milewich *et al.*, 1977; George *et al.*, 1979). Although the fetal ovaries have been shown to differentiate later than the testes in this species, the ovaries acquire the capacity to convert C^{19} steroids to estradiol-17β and to form a variety of other steroid hormones by the side chain cleavage of cholesterol at approximately the same time that the testes begin to secrete T (see Figure 4). These biochemical events (viz., E biosynthetic capacity in fetal ovaries) occur soon after the appearance of secretory granules in Rathke's pouch, the anlagen of the anterior pituitary gland (Campbell, 1966; Schecter, 1970).

In the laboratory rat, Ketelslegers *et al.* (1978) studied the relationship between the postnatal development of gonadotropin receptors in the testes and serum concentrations of T and gonadotropins. Testicular LH receptors increased continuously from postnatal Day 15 to 38. FSH receptors, on the other hand, reached their maximum by Day 15 and declined thereafter. The data suggest that FSH controls testicular sensitivity to LH during postnatal development and appear to be more pertinent to mechanisms of puberty rather than sexual differentiation.

FETAL HORMONES OF DOMESTIC ANIMALS

Table 2 presents the time in gestation when the testes can be distinguished from the ovaries in 10 mammalian species. In the domestic pig, sexual differentiation of the fetal testes takes place on approximately Day 36 of gestation. At this time in gestation, the fetus is about 3.5–4.5 cm in crown-rump (CR) length. Histochemical reactions for enzymes that are necessary for the biosynthesis of androgens ($\Delta^5 3\beta$-hydroxysteroid dehydrogenase) are found at very early stages of fetal development, that is, at a CR length of 1.5 cm (Moon and Raeside, 1972). Similarly, gonads from pig fetuses (2.0–5.2 cm in CR length; Days 25–40 of gestation) were analyzed for their capacity to produce T in culture. The gonad that produced T

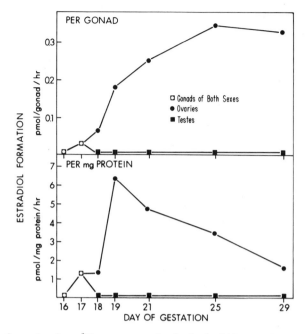

Fig. 4. Estrogen formation from ³H-testosterone by the fetal rabbit ovary in vitro. Incubations were carried out in 0.5 ml of Eagle's medium, at 37°C for 1 hr. Each point represents a mean of two or three determinations. (From "Estrogen Formation by the Ovary of the Rabbit Embryo" by L. Milewich, F. W. George, and J. D. Wilson. *Endocrinology*, 1977, *100*, 187–196. Copyright 1977 by J. B. Lippincott Company. Reprinted by permission.)

earliest was from fetuses 2.2–2.3 cm in CR length (Stewart and Raeside, 1976). Larger quantities of this hormone were produced at later gestational ages (when fetuses were 3.2–3.9 cm in CR length). In fetuses larger than 4.0 cm in CR length, the amounts of T produced *in vitro* declined. The histochemical data for $\Delta^5 3\beta$-hydroxysteroid dehydrogenase activity ascribed this activity in 2.5- to 3.0-cm fetuses to the developing sex cords of the fetal testes and not to the Leydig cells

TABLE 2. GESTATIONAL AGES OF SELECTED SPECIES OF MAMMALS WITH APPROXIMATE TIMES WHEN FETAL TESTES DIFFERENTIATE

		Approximate day of gestation for	
Genus and species	Common name	Differentiation of fetal testes	Delivery[a]
Rattus norvegicus	Rat	13.5	21
Oryctolagus cuniculus	Rabbit	14–15	31
Canis familiaris	Dog	33	63
Cavia porcellus	Guinea pig	22–26	68
Sus scrofa	Pig	36	114
Ovis aries	Sheep	35	151
Macaca mulatta	Rhesus monkey	38–40	168
Homo sapiens	Human	49–52	278
Bos taurus	Cattle	41	284
Equus caballus	Horse	30	334

[a]May change with different breeds.

or the mesenchymal cells of the interstitium (Moon and Raeside, 1972). Measurements of T in fetal plasma by RIA from Day 45 of gestation to delivery indicate that fetal males have more T in their plasma than fetal females and that peak plasma concentrations of T occur between Day 50 and Day 60 of gestation. Female fetuses have small amounts of this hormone in the circulatory system (Meusy-Desolle, 1974.)

In cattle, fetal testes differentiate on approximately Day 41 of gestation. Gonads from preimplantation embryos (30–38.5 days of gestation) cultured *in vitro* produced P but not T or estradiol-17β. Gonads from postimplantation embryos (42 days of gestation, at which time the ovaries and testes were already differentiated) produced T (testes) or estradiol-17β (ovaries). Postimplantation ovaries and testes produced progesterone *in vitro* that increased with age (Shemesh *et al.*, 1978). Kim *et al.* (1972) found marked sex differences in the quantities of T in the fetal circulation between the third and eighth months of gestation. During the second and third months, whole body extracts taken from males contained significantly more T (72 ± 11 pg/ml) than extracts taken from females (36 ± 9.0 pg/ml). In another study, Challis *et al.* (1974) measured T, A, estrone, estradiol-17β, P, and LH in fetal plasma from the third to the ninth month of gestation. Progesterone and A concentrations did not differ between the sexes. During some of the time periods sampled, LH and the estrogens were significantly higher in female than in male fetuses. During the third and fourth months, but not at later times in gestation, LH was higher in female than in male fetuses. The estrogens, however, differed between the sexes in the samples taken during the seventh and eighth months of gestation. Testosterone was significantly higher in male plasma than in female plasma from the third to the seventh month of gestation. The sex differences in quantities of estrogen demonstrated in the above studies seem to indicate that fetal ovaries are capable of producing estrogen. These differences substantiate earlier work by Roberts and Warren (1964), who showed that bovine fetal ovaries (at least 1 month preterm) can biosynthesize estrogen from androgen.

In fetal sheep, T and A have been quantified in the testes from the time of differentiation of the testes (Day 35) until Day 135 of gestation (Attal, 1969). Significant quantities of both androgens were found in fetal testicular tissue. Testosterone and A were found in pools of gonads (seven fetuses, sex unknown) on Day 30 of gestation. These levels were probably, but not certainly, due to the contributions of the undifferentiated testes. Pomerantz and Nalbandov (1975) quantified T in gonads and in plasma of fetal sheep from Day 70 to Day 150 of gestation. On Day 70, high plasma T concentrations (658 ± 145 pg/ml) were found in males. Plasma from females, on the other hand, contained approximately 100 pg/ml. On Day 70 of gestation, sexual differentiation is probably complete in this species. Many other papers report gonad and plasma levels of steroids in fetal sheep. These studies were performed in late gestation and therefore cannot provide us with data that would be useful to a consideration of that early period in gestation when sexual differentiation seems to occur.

Development of the gonads in the fetal horse follows what appears to be a different pattern. Fetal horse testes differentiate on about Day 30 of gestation.

From the 40th to the 120th day of pregnancy, large quantities of gonadotropin are secreted in the pregnant mare by the endometrial cups. The highest serum levels of this hormone are found around the 60th day of pregnancy. Immediately after, but not during, this phase of gonadotropin secretion, large quantities of estrogen are secreted. Estrogen secretion reaches a maximum between the seventh and eighth months of pregnancy. During the time of estrogen secretion, the fetal gonads (both ovaries and testes) show remarkable growth, which is reflected in the large increase in the number of interstitial cells in both gonads. The fetal male and female seem to differ, however, in the distribution of these cells within the gonads. In the female, they are confined primarily to the medulla, whereas in the male they are found throughout the gonad. This difference may arise because cortical and medullary regions are distinguishable only in the fetal ovaries and not in the testes of the horse by light microscopy (Gonzalez-Angulo et al., 1971). At approximately the eighth month, the number of interstitial cells found within the gonads beings to wane. The relationship of these remarkable endocrine events in the pregnant horse to sexual development is not well understood. Most of these events probably occur after differentiation has already taken place, as the fetal testes of the horse differentiate on approximately Day 30 of gestation. Little is known about the capacity of this structure to biosynthesize steroid hormones in this early period. Two studies have tested the capacity of fetal gonads to synthesize hormone (MacArthur et al., 1967; Raeside, 1976). Both studies were performed in late gestation. In the former study, a 9-month-old fetal testis biosynthesized several steroids, including T, A, and estrone, from ^{14}C-acetate. In the latter, both testes and ovaries between the fifth and eighth months of gestation contained dehydroepiandrosterone. Little is known about the biosynthetic capacity or serum levels of steroid hormones during the period between Day 30, when the fetal testes differentiate, and midgestation.

Production of Steroid Hormones in Primate Fetuses

Differentiation of the Fetal Testes

In humans, the fetal testes differentiate between Day 49 and Day 52 of gestation. On Day 52, glandular interstitial cells, the presumed source of steroid hormones, appear. In rhesus monkeys, on the other hand, the testes differentiate on about Day 38 of gestation. In this species, interstitial cells can be recognized on about Day 50 of gestation; they reach a maximum on about Day 85 to Day 90 and thereafter fall off sharply (Van Wagenen and Simpson, 1965). The capacity of primate fetal testes to synthesize steroid hormones is well established, but when this capacity begins is not well known. Genital ridge mesenchyme of human fetuses contains 3β-, 16β-, and 17β-hydroxysteroid dehydrogenases, enzymes that are necessary for steroid biosynthesis (Baillie et al., 1966). At 8 weeks of conceptual age, enzymatic activity usually associated with steroidogenesis is found in human fetal testes (Baillie et al., 1965; Niemi et al., 1967), and substrates such as NADH,

NADPH, and glucose-6-phosphate can be used by the testes (Niemi *et al.*, 1967). Similar results have been obtained with older fetuses (Mancini *et al.*, 1963). In human fetal testes, when the density of interstitial cells is greatest so are the 3β-hydroxysteroid dehydrogenases (Baillie *et al.*, 1965). Although these studies indicate that critical enzymatic steps for the biosynthesis of androgen can take place in fetal testes, they say nothing about the types and quantities of steroids that are produced. Solomon and coworkers (1967) were unable to detect androgenic hormones in perfused fetoplacental units of 18-week-old human fetuses. Coutts *et al.* (1969) found that T and A are biosynthesized from 17α-hydroxyprogesterone by Week 22 of gestation.

CAPACITY OF THE FETAL TESTES FOR STEROID BIOSYNTHESIS

Table 3 presents data pertaining to the capacity of the fetal testes to produce steroids from radioactive precursors (or measurement of endogenous amounts in testicular tissue). In human fetuses, T is not biosynthesized by fetal testes until the eighth week of gestation (Huhtaniemi *et al.*, 1970; Siiteri and Wilson, 1974). At approximately the same time, the 3β-hydroxysteroid dehydrogenases appear in the testes (Baillie *et al.*, 1965; Niemi *et al.*, 1967). Testosterone formation by human fetal testes *in vitro*, with time of gestation, is shown in Figure 5. Earlier work (Block, 1964) showed that T and A are formed from P by human fetal testes during the ninth week of gestation. In 6.5-week-old rhesus monkey fetuses, testes biosynthesize T and A from [^{14}C]-pregnenolone (Resko, 1970). In the same study, performed T and A were found in testes of 14-week-old fetuses. These studies indicate that fetal testes develop the capacity to biosynthesize androgen early in development.

ANDROGENS IN THE SYSTEMIC CIRCULATION OF PRIMATE FETUSES

Quantification of steroid hormones in the fetal circulation of primate fetuses was simplified with the development of specific RIAs for these hormones. In our early studies, in which we measured T and A by gas–liquid chromatography (Resko, 1970), we had to pool plasma from several fetuses of the same gestational age in order to obtain a single hormone determination. In spite of the limitations of this approach, we demonstrated that fetal males possess more T in their systemic circulation than do fetal females during the last half of gestation. Androstenedione levels do not differ between the sexes. Using RIA techniques, we later confirmed these results (Resko *et al.*, 1973) and in addition showed that dihydrotestosterone in plasma from the umbilical artery does not differ between male and female fetuses (Resko, 1977). The mean concentrations of these hormones measured in umbilical artery plasma from Days 59–163 of gestation are shown in Table 4. Testosterone concentrations in plasma from individual 59- to 163-day-old monkey fetuses are shown in Figure 6. Most of the samples from male fetuses contained significantly more T than those from female fetuses. This observation has been confirmed by others (Huhtaniemi *et al.*, 1977), who have shown that fetal monkey testes produce T in reponse to hCG in late gestation and that fetal testes, but not

TABLE 3. STEROID BIOSYNTHESIS BY PRIMATE FETAL TESTES[a]

Precursor	Approximate fetal age (weeks)[b]	Species[c]	Products formed[d]							References[e]
			P^5	P^4	17αOH-P	20α-OL	DHA	A	T	
Pregnenolone	6.5	M	−	−	−	−	−	+	+	(7)
Pregnenolone	6–8	H	−	−	−	−	−	−	−	(10)
Progesterone	8.1–10	H	−	−	+	−	+	−	+	(10)
No precursor (testis content)	8	H	−	−	+	−	−	−	+	(9)
Progesterone	9	H	−	−	+	−	−	−	+	(4)
No precursor (testis content)	10	H	−	−	−	−	−	−	+	(8)
Acetate	11	H	+	−	−	−	−	−	+	(11)
No precursor (testis content)	12	H	−	−	−	−	−	+	+	(6)
No precursor (testis content)	14	M	−	−	−	−	−	+	+	(7)
Progesterone	15	H	−	−	+	−	−	+	+	(2)
Progesterone	16	H	−	−	+	+	−	+	−	(3)
Pregnenolone	21	H	−	−	−	−	+	+	+	(1)
Acetate	26	H	+	+	+	−	−	−	+	(5)

[a] + indicates an identified product; − indicates the product was not found. From "Fetal Hormones and Development of the Central Nervous System in Primates," by J. A. Resko. In J. A. Thomas and R. L. Singhal (Eds.), *Advances in Sex Hormone Research: Vol. 3. Regulatory Mechanisms Affecting Gonadal Hormone Action.* Baltimore: University Park Press, 1977, 139–168. Copyright 1977 by University Park Press. Reprinted by permission.

[b] Earliest fetal age studied.

[c] H = human; M = monkey.

[d] P^5 = pregnenolone; P^4 = progesterone; 17αOH-P = 17αOH-progesterone; 20αOL = 20α-hydroxypregn-4-ene-3-one; DHA = dehydroepiandrosterone; A = androstenedione; T = testosterone.

[e] (1) Acevedo et al., 1961; (2) Block et al., 1962; (3) Acevedo et al., 1963; (4) Block, 1964; (5) Rice et al., 1966; (6) Huhtaniemi et al., 1970; (7) Resko, 1970; (8) Reyes et al., 1973; (9) Diez D'Aux and Murphy, 1974; (10) Siiteri and Wilson, 1974; and (11) Serra et al., 1970.

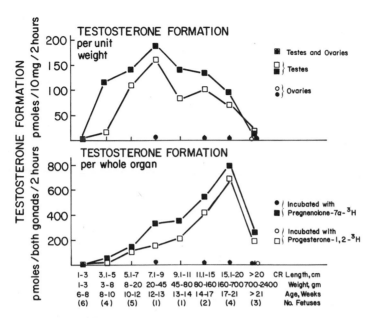

Fig. 5. Testosterone formation from radiolabeled precursors by human fetal gonads *in vitro*. Incubations were carried out in 1 ml of Krebs-Ringer phosphate buffer, pH 7.4, for 2 hr. Each point represents either a single determination or the mean of two to six embryos. (From "Testosterone Formation and Metabolism during Male Sexual Differentiation in the Human Embryo" by P. K. Siiteri and J. D. Wilson. *Journal of Clinical Endocrinology and Metabolism*, 1974, *38*, 113–125. Copyright 1974 by the Endocrine Society. Reprinted by permission.)

ovaries, have binding sites for gonadotropins. There is a trend, however, for T concentrations to be lower in the male as gestation progresses. We now have data that demonstrate sex difference in the concentrations of T in the umbilical artery from Days 40–50 of gestation (Resko *et al.*, 1980). The differences between male and female fetuses in quantity of T observed in the fetal circulation, especially during the early period (Days 40–60 of gestation), probably represent an important component of sexual differentiation. In human beings, the earliest quantification of T in fetal blood was at 11 weeks of age (Reyes *et al.*, 1974). Male fetuses have

TABLE 4. ANDROGENS IN FETAL PLASMA[a] OF THE RHESUS MONKEY

Androgen measured	pg/ml ± SE	
	Male	Female
Testosterone	1202 ± 158[b]	253 ± 18[b]
	(39)	(38)
Androstenedione	2287 ± 199	2061 ± 125
	(32)	(37)
Dihydrotestosterone	767 ± 57	696 ± 37
	(34)	(38)

[a]Umbilical artery plasma from Day 59–163 of gestation.
[b]Testosterone concentrations differed significantly between the sexes ($p <$ .01).

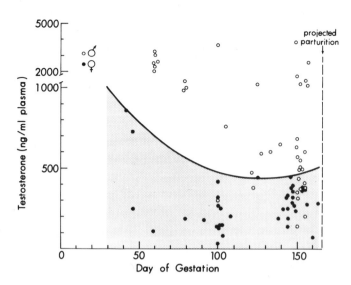

Fig. 6. Testosterone concentrations in umbilical artery plasma of fetal rhesus monkeys. Each point represents data from an individual monkey fetus; (●) from females and (O) from males, at the gestation periods indicated. The shaded area depicts the range of concentrations for females.

more T in their systemic circulation than do females (Abramovich and Rowe, 1973; Diez D'Aux and Murphy, 1974; Reyes *et al.*, 1974). This difference disappears in late gestation (Mizuno *et al.*, 1968; Forest *et al.*, 1971; Reyes *et al.*, 1974) or is greatly reduced (Forest *et al.*, 1973). Declining levels of T in the human fetal male correlate with declining numbers of Leydig cells in the fetal testes in late gestation.

OTHER HORMONES AND SEXUAL DIFFERENTIATION

It is becoming increasingly clear, at least in some species of vertebrates, that sexual differentiation of the reproductive tract and the central nervous system of the male is related to the secretory capacity of the testes. The fetal ovaries appear to produce no steroids during the period of sexual differentiation. Not much is known about the chemical factors associated with differentiation of the gonads. Our focus has been directed toward the secretion of T by the fetal testes, but another testicular secretion, nonsteroidal in nature, appears to play a significant role in the development of typical males. This subtance, Müllerian duct inhibiting factor, prevents the female duct system from developing in the male (Josso, 1971). It has a molecular weight greater than 15,000, and it cannot be a steroid hormone (Josso, 1972). In addition to these more traditional ways of viewing the role of hormones in development, there is another way: perhaps other steroid hormones participate with T in the androgenization process. One of these hormones, P, interferes with the action of T in some mammalian systems. Papers that review the antiandrogenic properties of P have been reviewed (Resko, 1977). In rhesus monkeys, female fetuses, on the average, possess more P in the circulatory system than do males. Table 5 contains data on P concentrations in the circulation of fetal rhesus monkeys. On the average, the concentration of P in fetal female plasma is three

TABLE 5. PROGESTERONE CONCENTRATIONS IN UMBILICAL ARTERY PLASMA OF FETAL RHESUS
MONKEYS (COMPARISON OF MALES AND FEMALES)

Fetal sex	No. of fetuses studied	Mean concentration of progesterone (ng/ml)	Fetuses in each category (%)		
			< 5 ng/ml	5–10 ng/ml	> 10 ng/ml
Females	43	14.8 ± 3.3^{a}	30.2 (13)[b]	37.2 (16)	32.6 (14)
Males	36	5.0 ± 0.8^{a}	61.1 (22)	25.0 (9)	13.9 (5)

[a]Differed significantly ($p < .02$).
[b]Number in parentheses is the number of animals in each category (progesterone concentration).

times greater than the concentration found in plasma from fetal males. In another part of the table, I arbitrarily divided the animals of each sex into three categories: the percentage of animals with less than 5 ng, the percentage with 5–10 ng, and the percentage with more than 10 ng of P/ml of plasma. Approximately 69% of the females possessed concentrations of P greater than 5 ng/ml, whereas only 39% of male values were above 5 ng/ml. Little information is available to suggest that this sex difference in the circulating levels of P plays a role in sexual differentiation, and there is some overlap between males and females. At this time, we can only speculate that P is antiandrogenic, as it is in rodents, and that differentiation is effected by a synergism of androgen with an antiandrogen. Data for progesterone concentrations in plasma from neonatal rats has not been consistent. Döhler and Wuttke (1974, 1975) and Weisz and Ward (1980) did not find sex differences in the quantities of this hormone, whereas Shapiro *et al.* (1976) found higher progesterone levels in female plasma than in male plasma. The concept of a steroid hormone such as P participating in androgen action is not too different from the recently reported phenomenon of regulatory interactions of P with estradiol-17β in target tissues such as the oviduct and the uterus (Brenner *et al.*, 1974; Resko *et al.*, 1976; Tseng and Gurpide, 1975a,b). Progesterone appears to modulate tissue levels of estrogen receptor (Brenner *et al.*, 1974; Tseng and Gurpide, 1975a) as well as tissue concentrations of estrogen (Resko *et al.*, 1976). It also affects enzymes that metabolize estradiol-17β to less active compounds such as estrone (Tseng and Gurpide, 1975b). For differentiation, P may interact with T produced by the fetal testes at the level of the target tissue to produce degrees of androgenization in the male, or it may prevent the circulating androgen of the female from producing its effect.

SUMMARY

Evidence drawn from *in vitro* measurements of the capacity of gonads to biosynthesize steroid hormones or measurements of these hormones by specific RIAs indicates that fetal testicular tissue acquires the capacity to secrete T and other steroids early in development, and that these hormones act on undifferentiated anlagen to effect androgenization. The cellular components of androgenization are not well understood. In other words, we do not know what irreversible biochemical process androgens initiate in development, especially of the central nervous system. As adults, males differ from females in sexual behavior. In some species, females release gonadotropins in a cyclic fashion, whereas males do not. What differences exist between the sexes in the organizational plan of the central nervous system at the cellular level? Apparently, such differences depend on the presence or absence of fetal testes. These unresolved problems are presented as a challenge for future investigation.

References

Abramovich, D. R., and Rowe, P. Fetal plasma testosterone levels at mid-pregnancy and at term. *Journal of Endocrinology*, 1973, *56*, 621–622.

Acevedo, H. F., Axelrod, L. R., Ishikawa, E., and Takaki, F. Steroidogenesis in the human fetal testis: The conversion of pregnenolone-7α-H^3 to dehydroepiandrosterone, testosterone, and 4-androstene-3, 17-dione. *Journal of Clinical Endocrinology and Metabolism*, 1961, *21*, 1611–1613.

Acevedo, H. F., Axelrod, L. R., Ishikawa, E., and Takaki, F. Studies in fetal metabolism: II. Metabolism of progesterone-4-C^{14} and pregnenolone-7α-H^3 in human fetal testes. *Journal of Clinical Endocrinology and Metabolism*, 1963, *23*, 885–890.

Akram, H., and Weniger, J. P. Sécrétion d'estrone et d'oestradiol par le testicule féminisé de l'embryon de poulet. *Comptes Rendus Hebdomadaires des Séances de l'Académie des Sciences; D: Sciences Naturelle (Paris)*, 1967, *264*, 1806–1807.

Akram, H., and Weniger, J. P. Sécrétion d'estrone et d'oestradiol par les gonads embryonnaires d'oiseaux. *General and Comparative Endocrinology*, 1969, *12*, 568–573.

Arai, Y., and Serisawa, K. Effect of gonadotropins on neonatal testicular activity and sexual differentiation of the brain in the rat. *Proceedings of the Society for Experimental Biology and Medicine*, 1973, *143*, 656–660.

Attal, J. Levels of testosterone, androstenedione, estrone and estradiol-17β in the testes of fetal sheep. *Endocrinology*, 1969, *85*, 280–289.

Baillie, A. H., and Griffiths, K. 3β-Hydroxysteroid dehydrogenase in the fetal mouse Leydig cell. *Journal of Endocrinology*, 1964, *31*, 63–66.

Baillie, A. H., Niemi, M., and Ikonen, M. 3β-Hydroxysteroid dehydrogenase activity in the human foetal testes. *Acta Endocrinologica (Kobenhavn)*, 1965, *48*, 429–438.

Baillie, A. H., Ferguson, M. M., and Hart, D. McK. Histochemical evidence of steroid metabolism in the human genital ridge. *Journal of Clinical Endocrinology and Metabolism*, 1966, *26*, 738–741.

Black, V. H., and Christensen, A. K. Differentiation of interstitial and Sertoli cells in fetal guinea pig testes. *American Journal of Anatomy*, 1969, *124*, 211–238.

Block, E. Metabolism of 4-^{14}C-progesterone by human fetal testes. *Endocrinology*, 1964, *74*, 833–845.

Block, E., Tissenbaum, B., and Benirschke, K. The conversion of [4-^{14}C] progesterone to 17α-hydroxyprogesterone, testosterone and Δ^4-androstene-3,17-dione by human fetal testes in vitro. *Biochimica et Biophysica Acta*, 1962, *60*, 182–184.

Brenner, R. M., Resko, J. A., and West, N. B. Cyclic changes in oviductal morphology and residual cytoplasmic estradiol binding capacity induced by sequential estradiol-progesterone treatment of spayed rhesus monkeys, *Endocrinology*, 1974, *95*, 1094–1104.

Breuer, H., Dahm, K., Mikamo, B., and Witschi, E. Differences in steroid metabolism of male and female larvae of *Xenopus laevis*. *Proceedings of the Second International Congress on Hormonal Steroids. Excerpta Medica International Congress*, 1966, Ser. III, p. 215.

Buhl, A. E., Pasztor, L. M., and Resko, J. A. Sex steroids in guinea pig fetuses after sexual differentiation of the gonads. *Biology of Reproduction*, 1979, *21*, 905–908.

Campbell, H. J. The development of the primary plexus of the median eminence of the rabbit. *Journal of Anatomy*, 1966, *199*, 381–387.

Catt, K. J., Dufau, M. L., Neaves, W. B., Walsh, P. C., and Wilson, J. D. LH-hCG receptors and testosterone content during differentiation of the testes in the rabbit embryo. *Endocrinology*, 1975, *97*, 1157–1165.

Challis, F. R. G., Kim, C. K., Naftolin, F., Judd, H. L., Yen, J. S. C., and Benirschke, K. The concentrations of androgens, oestrogens, progesterone, and luteinizing hormone in the serum of foetal calves throughout the course of gestation. *Journal of Endocrinology*, 1974, *60*, 107–115.

Chieffi, G. Experimental sex reversal of amphibian larvae and sex hormones metabolism. *Proceedings of the 15th International Congress of Zoology, London*, 1958, 600–601.

Collenot, A. Mise en évidence histochimique d'une Δ^5-3β-hydroxystéroïde déshydrogénase dans les gonades non différenciées et en cours de différenciation des mâles génétiques de l'Urodèle *Pleurodeles waltlii* Michah. *Comptes Rendus Hebdomadaires des Séances de l'Académie des Sciences; D: Sciences Naturelle (Paris)*, 1964, *259*, 2535–2537.

Corbier, P., Roffi, J., Roda, J., and Kerdelhue, B. Increased activity of the hypothalamic-pituitary-testicular axis in the rat at birth: Implications in the sexual differentiation of the brain? In M. Serio *et al.* (Eds.), *Sexual Differentiation: Basic and Clinical Aspects*. New York: Raven Press, 1984.

Coutts, J. R. T., Macnaughton, M. C., Ling, W., and Solomon, S. The metabolism of 17α-hydroxy-[4-^{14}C] progesterone in the human foetoplacental unit at mid-pregnancy. *Biochemistry Journal*, 1969, *112*, 31P.

Dale, E. Steroid excretion by larval frogs. *General and Comparative Endocrinology*, 1962, *2*, 171–176.

Diez D'Aux, R. C., and Murphy, B. E. P. Androgens in the human fetus. *Journal of Steroid Biochemistry*, 1974, *5*, 207–210.

Dodd, J. M. Gonadal and gonadotrophic hormones in lower vertebrates. In A. S. Parkes (Ed.), *Marshall's physiology of reproduction*, Vol. 1. London: Longman, 1960.

Döhler, K. D., and Wuttke, W. Serum LH, FSH, prolactin and progesterone from birth to puberty in female and male rats. *Endocrinology*, 1974, *94*, 1003–1008.

Döhler, K. D., and Wuttke, W. Changes with age in levels of serum gonadotropins, prolactin, and gonadal steroids in prepubertal male and female rats. *Endocrinology*, 1975, *97*, 898–907.

Forest, M. G., Ances, I. G., Tapper, A. J., and Migeon, C. J. Percentage binding of testosterone, androstenedione, and dehydroisoandrosterone in plasma at the time of delivery. *Journal of Clinical Endocrinology and Metabolism*, 1971, *32*, 417–425.

Forest, M. G., Cathiard, A. M., and Bertrand, J. A. Evidence of testicular activity in early infancy. *Journal of Clinical Endocrinology and Metabolism*, 1973, *37*, 148–151.

Galli, F., and Wasserman, G. F. Steroid biosynthesis by gonads of 7- and 10-day-old chick embryos. *General and Comparative Endocrinology*, 1973, *21*, 77–83.

Gallien, L. Recherches expérimentales sur l'organogénase sexuelle chez les Batraciens amoures. *Bulletin Biologique*, 1944, *78*, 257. Cited in *Marshall's physiology of reproduction*, Vol. 1, Pt. 2. London: Longmans, 1960.

Gallien, L., and Chalumeau-Le Foulgoc, M. T. Mise en évidence de stéroïdes oestrogènes dans l'ovaire juvénile de *Xenopus laevis* Daudin et cycle des oestrogènes au cours de la ponte. *Comptes Rendus Hebdomadaires des Séances de l'Académie des Sciences; D: Sciences Naturelle (Paris)*, 1960, *251*, 460–461.

George, F. W., Simpson, E. R., Milewich, L., and Wilson, J. D. Studies on the regulation of the onset of steroid hormone biosynthesis in fetal rabbit gonads. *Endocrinology*, 1979, *105*, 1100–1106.

Goldman, B. D., and Gorski, R. A. Effects of gonadal steroids on the secretion of LH and FSH in neonatal rats. *Endocrinology*, 1971, *89*, 112–115.

Gonzalez-Angulo, A., Hernandex-Jauregui, P., and Marguez-Monter, H. Fine structure of gonads of the fetal horse *(Equus caballus)*. *American Journal of Veterinary Research*, 1971, *32*, 1665–1766.

Guichard, A., Cedard, L., and Haffen, K. Aspect comparatif de la synthèse de stéroïdes sexuels par les gonades embronnaires de poulet à différents stades du développement (étude en culture organotypique à partir de précurseurs radioactifs). *General and Comparative Endocrinology*, 1973, *20*, 16–28.

Haffen, K. Biosynthesis of steroid hormones by the embryonic gonads of vertebrates. *Advances in Morphology*, 1970, *8*, 285–306.

Huhtaniemi, I., Ikonen, M., and Vikko, R. Presence of testosterone and other neutral steroids in human fetal testes. *Biochemical and Biophysical Research Communications*, 1970, *38*, 715–720

Huhtaniemi, I. P., Korenbrot, C. C., Seron-Ferre, M., Foster, D. B., Parer, J. T., and Jaffe, R. B. Stimulation of testosterone production in vivo and in vitro in the male rhesus monkey fetus in late gestation. *Endocrinology*, 1977, *100*, 839–844.

Josso, N. Interspecific character of the Müllerian-inhibiting substance: Action of the human fetal testis, ovary, and adrenal of the fetal rat müllerian duct in organ culture. *Journal of Clinical Endocrinology and Metabolism,* 1971, *32,* 404–409.

Josso, N. Permeability of membranes to the Müllerian-inhibiting substance synthesized by the human fetal testes in vitro: A clue to its biochemical nature. *Journal of Clinical Endocrinology and Metabolism,* 1972, *34,* 265–270.

Ketelslegers, J.-M., Hetzel, W. D., Sherins, R. J., and Catt, K. J. Developmental changes in testicular gonadotropin receptors: Plasma gonadotropins and plasma testosterone in the rat. *Endocrinology,* 1978, *103,* 212–222.

Kim, C. K., Yen, S. S. C., and Benirschke, K. Serum testosterone in fetal cattle. *General and Comparative Endocrinology,* 1972, *18,* 404–407.

Lipsett, M. B., and Tullner, W. W. Testosterone synthesis by the fetal rabbit gonad. *Endocrinology,* 1965, *77,* 273–277.

Lording, D. W., and De Kretser, D. M. Comparative ultrastructural and histochemical studies of the interstitial cells of the rat testes during fetal and postnatal development. *Journal of Reproduction and Fertility,* 1972, *29,* 261–269.

MacArthur, E., Short, R. V., and O'Donnell, V. J. Formation of steroids by the equine foetal testis. *Journal of Endocrinology,* 1967, *38,* 331–336.

Mancini, R. E., Vilar, O., Lavieri, J. C., Andrada, J. A., and Heinrich, J. J. Development of Leydig cells in the normal human testes: A cytological, cytochemical and quantitative study. *American Journal of Anatomy,* 1963, *112,* 203–214.

Meusy-Dessolle, N. Évolution du taux de testosterone plasmatique au cours de la foetale chez le porc domestique *(Sus scrofa* L.). *Comptes Rendus Hebdomadaires des Séances de l'Académie des Sciences; D: Sciences Naturelle (Paris),* 1974, *278,* 1257–1260.

Milewich, L., George, F. W., and Wilson, J. D. Estrogen formation by the ovary of the rabbit embryo. *Endocrinology,* 1977, *100,* 187–196.

Mintz, B., Foote, C. L., and Witschi, E. Quantitative studies on response of sex characters of differentiated *Rana clamitans* larvae to injected androgens and oestrogens. *Endocrinology,* 1945, *37,* 286–296.

Mizuno, M., Labotsky, J., Lloyd, C. W., Kobayashi, T., and Murasawa, Y. Plasma androstenedione and testosterone during pregnancy and in the newborn. *Journal of Clinical Endocrinology and Metabolism,* 1968, *28,* 1133–1142.

Moon, Y. S., and Raeside, J. I. Histochemical studies on hydroxysteroid dehydrogenase activity of fetal pig gonads. *Biology of Reproduction,* 1972, *7,* 278–287.

Niemi, M., and Ikonen, M. Steroid 3β-ol-dehydrogenase activity in foetal Leydig's cells. *Nature* (London), 1961, *189,* 592–593.

Niemi, M., and Ikonen, M. Cytochemistry of oxidative enzyme systems in the Leydig cells of the rat testis and their functional significance. *Endocrinology,* 1962, *70,* 167–174.

Niemi, M., Ikonen, M., and Hervonen, A. Histochemistry and fine structure of the interstitial tissue in the human foetal testes. *Ciba Foundation Colloquy on Endocrinology,* 1967, *16:* Endocrinology of the Testes, 31–55.

Noumura, T., Weisz, J., and Lloyd, C. W. In vitro conversion of 7-^{3}H-progesterone to androgens by the rat testis during the second half of fetal life. *Endocrinology,* 1966, *78,* 245–253.

Ortiz, E., Price, D., and Zaaijer, J. Organ culture studies of hormone secretion in endocrine glands of fetal guinea pigs: II. Secretion of androgenic hormones in adrenals and testes during early stages of development. *Verhandelingen der Koninklijke Nederlandse Akademii van Wetenschappen.* Proceedings Series C, 1966, *69,* 400–408.

Ozon, R. Analyse in vivo, du métabolisme des oestrogènes au cours de la différenciation sexuelle chez le Triton *Pleurodeles walthii* Michah. *Comptes Rendus Hebdomadaires des Séances de l'Académie des Sciences; D: Sciences Naturelle (Paris),* 1963, *257,* 2332–2335.

Ozon, R. Steroid biosynthesis in larval and embryonic gonads of lower vertebrates. *General and Comparative Endocrinology,* 1969, Suppl. *2,* 135–140.

Picon, R. Testosterone secretion by foetal rat testis in vitro. *Journal of Endocrinology,* 1976, *71,* 231–238.

Pomerantz, K. K., and Nalbandov, A. V. Androgen level in the sheep fetus during gestation. *Proceedings of the Society for Experimental Biology and Medicine,* 1975, *149,* 413–416.

Price, D., Ortiz, E., and Zaaijer, J. J. P. Detection of androgenic secretion by an in vitro technique in the undifferentiated gonad and adrenal cortical tissue of fetal guinea pigs. *American Zoologist,* 1964a, *4,* 416.

Price, D., Ortiz, E., and Zaaijer, J. J. P. Secretion of androgenic hormone by testes and adrenal glands of fetal guinea pig. *American Zoologist,* 1964b, *3,* 553–554.

Raeside, J. I. Dehydroepiandrosterone in the fetal gonads of the horse. *Journal of Reproduction and Fertility,* 1976, *46,* 423–425.

Rapola, J. The adrenal cortex and metamorphosis of *Xenopus laevis* Daudin. *General and Comparative Endocrinology,* 1963, *3,* 412–421.

Resko, J. A. Androgen secretion by the fetal and neonatal rhesus monkey. *Endocrinology,* 1970, *87,* 680–687.

Resko, J. A. Fetal hormones and development of the central nervous system in primates. In J. A. Thomas and R. L. Singhal (Eds.), *Advances in sex hormone research, Vol. 3: Regulatory Mechanisms Affecting Gonadal Hormone Action.* Baltimore: University Park Press, 1977.

Resko, J., Feder, H. H., and Goy, R. W. Androgen concentrations in plasma and testis of developing rats. *Journal of Endocrinology,* 1968, *40,* 485–491.

Resko, J. A., Malley, A., Begley, D., and Hess, D. L. Radioimmunoassay of testosterone during fetal development of the rhesus monkey. *Endocrinology,* 1973, *93,* 156–161.

Resko, J. A., Boling, J. L., Brenner, R. M., and Blandau, R. J. Sex steroids in reproductive tract tissues: Regulation of estradiol concentrations by progesterone. *Biology of Reproduction,* 1976, *15,* 153–157.

Resko, J. A., Ellinwood, W. E., Pasztor, L. M., and Buhl, A. E. Sex steroids in the umbilical circulation of fetal rhesus monkeys from the time of gonadal differentiation. *Journal of Clinical Endocrinology and Metabolism,* 1980, *50,* 900–905.

Reyes, F. I., Winter, J. S. D., and Faiman, C. Studies on human sexual development: I. Fetal gonadal and adrenal sex steroids. *Journal of Clinical Endocrinology and Metabolism,* 1973, *37,* 74–78.

Reyes, F. I., Boroditsky, R. S., Winter, J. S. D., and Faiman, C. Studies on human sexual development: II. Fetal and maternal serum gonadotropin and sex steroid concentrations. *Journal of Clinical Endocrinology and Metabolism,* 1974, *38,* 612–617.

Rice, B. F., Johanson, C. A., and Sternberg, W. H. Formation of steroid hormones from acetate-1-^{14}C by a human fetal testis preparation grown in organ culture. *Steroids,* 1966, *7,* 79–90.

Roberts, J. D., and Warren, J. C. Steroid biosynthesis in the fetal ovary. *Endocrinology,* 1964, *74,* 846–862.

Romanoff, A. L. *The Avian Embryo.* New York: Macmillan, 1960.

Roosen-Runge, E. C., and Anderson, D. The development of the interstitial cells in the testis of the albino rats. *Acta Anatomica,* 1959, *37,* 125–137.

Schecter, J. A light and electron microscopic study of Rathke's pouch in fetal rabbits. *General and Comparative Endocrinology,* 1970, *14,* 53–67.

Schlegel, R. J., Farias, E., Russo, N., More, J., and Gardner, L. Structural changes in the fetal gonads and gonaducts during maturation of an enzyme, steroid 3β-ol-dehydrogenase, in the gonads, adrenal cortex and placenta of fetal rats. *Endocrinology,* 1967, *81,* 565–572.

Serra, G., Perez-Palacios, G., and Jaffe, R. B. De novo testosterone biosynthesis in the human fetal testis. *Journal of Clinical Endocrinology and Metabolism,* 1970, *30,* 128–130.

Shapiro, B. H., Goldman, A. S., Bongiovanni, A. M., and Marino, J. M. Neonatal progesterone and feminine sexual development. *Nature,* 1976, *264,* 795–796.

Shemesh, M., Ailenberg, M., Milaguir, F., Ayalon, N., and Hansel, W. Hormone secretion by cultured bovine pre- and post-implantation gonads. *Biology of Reproduction,* 1978, *19,* 761–767.

Siiteri, P. K., and Wilson, J. D. Testosterone formation and metabolism during male sexual differentiation in the human embryo. *Journal of Clinical Endocrinology and Metabolism,* 1974, *38,* 113–125.

Solomon, S., Bird, C. E., Ling, W., Iwamiya, M., and Young, P. C. M. Formation and metabolism of steroids in the fetus and placenta. *Recent Progress in Hormone Research,* 1967, *23,* 197–347.

Stewart, D. W., and Raeside, J. I. Testosterone secretion by the early fetal pig testes in organ culture. *Biology of Reproduction,* 1976, *15,* 25–28.

Swift, C. H. Origin of the sex cords and definitive spermatogonia in the male chick. *American Journal of Anatomy,* 1916, *20,* 375–410.

Torrey, T. W. The development of the urinogenital system of the albino rat. *American Journal of Anatomy,* 1945, *76,* 375–397.

Tseng, L., and Gurpide, E. Effects of progestins on estradiol receptor levels in human endometrium. *Journal of Clinical Endocrinology and Metabolism*, 1975a, *41*, 402–404.

Tseng, L., and Gurpide, E. Induction of human endometrial estradiol dehydrogenase by progestins. *Endocrinology*, 1975b, *97*, 825–833.

Vannini, E. Rapida azione mascolinizzante del testosterone sulla gonadi di girini di Rana agilis "in metamorfosi." *Reale Accademia d'Italia*, 1941, Fasc. 8, seŕ vii, I. Cited in A. S. Parkes (Ed.), *Marshall's Physiology of Reproduction*, Vol. 1, Pt. 2. London: Longmans, 1960.

Van Wagenen, G., and Simpson, M. Embryology of the ovary and testes. *Homo sapiens* and *Macaca mulatta*. New Haven, Conn.: Yale University Press, 1965.

Veyssiere, G., Berger, M., Jean-Faucher, C., De Turckheim, M., and Jean, C. Levels of testosterone in the plasma, gonads, and adrenals during fetal development of the rabbit. *Endocrinology*, 1976, *99*, 1263–1268.

Warren, D. W., Haltmeyer, G. C., and Eik-Nes, K. B. Synthesis and metabolism of testosterone in the fetal rat testis. *Biology of Reproduction*, 1972, *7*, 94–99.

Warren, D. W., Haltmeyer, G. C., and Eik-Nes, K. B. Testosterone in the fetal rat testis. *Biology of Reproduction*, 1973, *8*, 560–565.

Warren, D. W., Haltmeyer, G. C., and Eik-Nes, K. B. The effect of gonadotropin on the fetal and neonatal rat testis. *Endocrinology*, 1975, *96*, 1226–1229.

Weisz, J., and Ward, I. L. Plasma testosterone and progesterone titers of pregnant rats, their male and female fetuses, and neonatal offspring. *Endocrinology*, 1980, *106*, 306–316.

Weniger, J. P., and Zeis, A. Biosynthèse d'oestrogènes par les ébauches gonadiques de poulet. *General and Comparative Endocrinology*, 1971, *16*, 391–395.

Willier, B. H. Hormonal control of embryonic differentiation in birds. *Cold Spring Harbor Symposia*, 1942, *10*, 135–144.

Willier, B. H. Development of sex hormone activity of the avian gonad. *Annals of the New York Academy of Sciences*, 1952, *55*, 159–171.

Wilson, J. D., and Siiteri, P. K. Developmental pattern of testosterone synthesis in the fetal gonad of the rabbit. *Endocrinology*, 1973, *92*, 1182–1191.

Woods, J. E., and Erton, L. E. The synthesis of estrogens in the gonads of the chick embryo. *General and Comparative Endocrinology*, 1978, *36*, 360–370.

Woods, J. E., and Podczaski, E. S. Androgen synthesis in the gonads of the chick embryo. *General and Comparative Endocrinology*, 1974 *24* 413–423.

Woods, J. E., and Weeks, R. L. Ontogenesis of the pituitary-gonadal axis in the chick embryo. *General and Comparative Endocrinology*, 1969, *13*, 242–254.

Woods, J. E., Simpson, R. M., and Moore, P. L. Plasma testosterone levels in the chick embryo. *General and Comparative Endocrinology*, 1975, *27*, 543–547.

Yagimura, T., Matsuda, A., Murasawa, Y., Kobayashi, T., and Kobayashi, T. Presence of hypothalamo-pituitary testicular axis in the early postnatal period. *Endocrinologia Japonica*, 1969, *16*, 5–10.

3

Nonmammalian Psychosexual Differentiation

Elizabeth Adkins-Regan

Introduction

Nonmammalian vertebrates have traditionally played a significant role in the study of morphological sexual differentiation, but they have been relatively neglected in the study of behavioral sexual differentiation (psychosexual differentiation). Several recent developments, however, have sparked new interest in other vertebrate classes. These include the discovery of gross anatomical sexual dimorphism in songbird brains (Nottebohm and Arnold, 1976), the possibility of major species or class differences in the pattern of psychosexual differentiation (Adkins-Regan, 1981), and a generally greater tendency to view reproductive physiology and behavior as part of overall reproductive strategies to be explained within an evolutionary/ecological context. It is also becoming apparent that certain nonmammalian species are potentially very valuable as models for studying the role of hormones in brain and behavioral development, in part because they lack some problems inherent in mammals. For example, most birds, fish, and amphibians have no external copulatory organs, and so the effects of hormones administered early in life on adult copulatory behavior are not confounded by alterations in copulatory organs. In egg-laying species, there is no maternal gestation to complicate treatments given during embryonic life. Many nonmammalian species are sufficiently independent

Elizabeth Adkins-Regan   Department of Psychology and Section of Neurobiology and Behavior, Cornell University, Ithaca, New York 14853.

of the parents at birth or hatching so that they can be easily reared without parents; thus, a confound is avoided wherein the parents behave differently toward hormone-treated and control offspring (as has been shown to happen in rats, C. L. Moore, 1982).

In this chapter, I attempt to summarize what is currently known about the differentiation of behavior and neural mechanisms for behavior in teleost fish, amphibians, reptiles, and birds. (At this time, nothing is known about psychosexual differentiation in nonteleost fish such as sharks.)

The material has been organized in two ways: first, by species or class, and second, around questions and theoretical issues.

PSYCHOSEXUAL DIFFERENTIATION IN THE DOMESTIC CHICKEN (GALLUS GALLUS)

SEX DIFFERENCES AND THE REPRODUCTIVE BEHAVIOR OF CHICKENS

Domestic chickens of most breeds are externally dimorphic in that males have larger combs and longer necks and tail feathers. The sexes also differ with respect to several distinctive behavior patterns (see Guhl and Fischer, 1969, for a summary of chicken behavior, from which the following is taken). Ordinarily, only males crow. A sexually active male introduced to an unfamiliar female courts the female by waltzing around her or tidbitting (food calling). If the female is receptive, she crouches in reponse to the courting, and the male then mounts the female by standing on her outstretched wings, treads (alternately moves his feet), presses his everted cloaca to hers briefly, and steps off. A male introduced to another male may also waltz and, if a fight escalates, attacks feet or beak first with hackles raised.

These behavior patterns are hormone-dependent. Castration of the male results in a gradual reduction or cessation of copulation, waltzing, and crowing (Goodale, 1913). Injection of androgen into capons (castrated males) reinstates all three (Davis and Domm, 1943). Implantation of testosterone propionate directly into the brains of capons activates sexual behavior, and such experiments indicate that the preoptic area is importantly involved in copulation (Barfield, 1969; Crawford and Glick, 1975). Injection of estrogen into capons also activates copulation but does not activate crowing or waltzing (Domm and Davis, 1942; Goodale, 1918; Guhl, 1949). The effects of estrogen on the receptive behavior of males have not been systematically studied.

Ovariectomy of hens results in elimination of crouching (Allee and Collias, 1940; Domm and Davis, 1942). Androgens do not activate receptivity in either poulards (ovariectomized females) or capons (Domm and Davis, 1942). Injection of androgen into poulards activates some crowing and waltzing but very seldom produces the complete copulatory pattern with cloacal contact (Domm and Davis, 1942).

Thus, adult chickens exhibit sex differences in crowing, waltzing, and mating behavior. Dimorphism in mounting, treading, and achieving cloacal contract (patterns normally seen only in males) and, to a lesser extent, in crowing and waltzing, persists even when gonad dimorphism is eliminated by equating hormone levels during testing.

Effects of Hormonal Manipulations before Hatching on the Behavior of Chickens

As early as 1939, it was shown that male sexual behavior is reduced or suppressed by early treatment with sex steroids (Domm, 1939). In this experiment, brown leghorn chicken eggs were injected with from 0.5 to 1 mg of each of several estrogens between the third and fifth days of incubation. (The incubation period of domestic fowl is about 21 days.) As adults, most of the males eventually crowed, mounted, and treaded but exhibited "low libido." Because the testes of these birds were feminized, it is possible that the reduction in male behavior was due solely to insufficient testicular androgen in adulthood. In contrast to the treated males, treated females behaved in a normal manner.

Later experiments obtained similar suppressive effects on male behavior. Kaufman (1956) injected 1.5 mg of rutoestrol into the airspace of chicken eggs after 24 or 48 hours of incubation. No crowing or copulation was seen in males observed at 4 to 7 months of age. Again, however, the testes were also feminized. Glick (1961) dipped eggs incubated for 3 days in solutions of testosterone propionate in ethyl alochol (the Seltzer technique; Pincus and Hopkins, 1958). Both males that were examined failed to produce offspring. In one male, the semen was normal and fertile when used for artificial insemination, a result suggesting that the testes were reasonably normal. Nonetheless, this bird never mated. Of 10 additional males treated similarly in a later study, 9 failed to mate, but semen samples from 3 out of 4 were fertile (Glick, 1965).

None of the experiments described above can be said to prove that early sex hormones have true organizational effects on avian behavior, because the males were tested as intact animals. A more recent study is considerably less confounded by deleterious effects of the treatments on the testes. Wilson and Glick (1970) dipped chicken eggs incubated for 3 days in solutions of testosterone propionate (TP) or estradiol benzoate (EB) in ethyl alcohol or injected them at various times with oil solutions of these hormones. The birds were given TP at 25 days of age, so that their behavior would be activated precociously, and they were tested at 41 days of age for 8 days. The frequencies of waltzing, attempted mating, and completed mating were recorded. Treatment with either hormone prior to the thirteenth day of incubation greatly suppressed mating attempts. This suppression of male-typical mating patterns was also seen in females. The reductions in waltzing and completed matings were similar but not as striking. When given exogenous estrogen (at an unspecified age), the males treated before hatching with either hormone became receptive. Early-treated females were also receptive. The authors concluded that sex steroids induce behavioral sexual differentiation in chickens, and that the effect of early sex-hormone exposure is chiefly inhibition of the potential to display male-typical behavior in response to exogenous testosterone. This pattern of results suggests that, in contrast to mammals, the male may be the behavioral neutral (anhormonal) sex, and that embryonic ovarian estrogen is responsible for female differentiation.

Neurobiological Mechanisms of Psychosexual Differentiation in Domestic Fowl

Glick and his colleagues have explored several mechanisms by which exposure to sex steroids during embryonic life could suppress the potential for male mating behavior. In one experiment, the cholesterol content of the cerebral hemispheres was found to be greatest in control chicks or chicks from eggs dipped in testosterone on Day 18, and lowest in chicks from eggs dipped on days 3, 6, or 12 (Wilson and Glick, 1970). The authors interpreted this finding to mean that testosterone, when administered during a critical period, prevents a high degree of the myelination that is needed for male behavior. Because treatment at this age does not affect female sexual receptivity, presumably such myelination is not needed for receptivity.

The possibility that the suppression of male behavior by embryonic exposure to testosterone is mediated by alterations in brain enzymes was investigated by Kilgore and Glick (1970). Eggs were dipped on the third day of incubation into ethanol containing either 2.56 gm TP/100 ml (a behaviorally effective dosage) or no TP. Chicks or embryos of various ages were decapitated. For embryos, the right cerebral hemisphere, the right optic lobe, and the right half of the brainstem were analyzed for alkaline phosphatase activity (AP). The left portions were analyzed for acetylcholinesterase activity (AChE). For chicks, the left cerebral hemisphere and the brainstem were analyzed for AChE. For all brain parts in both in both control and TP-treated birds, AP activity decreased from Day 9 to Day 12 of incubation, then increased from Day 12 to Day 21. AP activity was uniformly lower in TP-treated birds than in controls, but the difference was significant only for cerebral hemispheres on Days 9 and 18 and for brainstems on Days 9 and 12. AChE activity increased throughout embryonic development for all brain parts in both control and TP-treated birds. TP did not consistently affect AChE activity except to lower it in brainstems on Day 18. No effects of TP on AChE could be detected in chicks examined on the day of hatching or at 5 to 10 weeks posthatching.

Thus, AP, but not AChE, activity is lowest at around the time that sexual differentiation of the brain probably takes place, and it is lowered by dipping eggs in a TP solution. It is tempting to conclude that AP activity mediates the effects of sex hormones on brain differentiation, but such a conclusion would be premature. Undoubtedly, other important brain functions besides sexual behavior are also maturing at about the same time. Furthermore, other exogenous substances besides TP that are not involved in sex differentiation may well also be capable of lowering AP activity; and conversely, exogenous TP may alter the level of many enzymes and other neurochemicals, only some of which have any causal relationship to sexual differentiation of the brain.

Two studies have attempted (but only with partial success) to determine the relevant anatomical loci for the behavioral changes caused by administering hormones before hatching (Crawford and Glick, 1975; Haynes and Glick, 1974). Based on the fact that lesions in the mammillary region increase sexual behavior in rats (Lisk, 1966), Haynes and Glick (1974) hypothesized that embryonic exposure to testosterone causes hyperactivity of the mammillary nuclei. They lesioned these nuclei in control males and males hatched from eggs dipped in a 1.28% TP solution on the third day of incubation. The lesions failed to increase mating behavior in the males from TP-treated eggs, but as they also failed to increase mating in control males, little can be concluded from this particular experiment.

Crawford and Glick (1975) found that lesions of the red nuclei (midbrain structures) increased the mating behavior of normal males but not of males from TP-treated eggs. They also found that implants of testosterone placed in the preopticus paraventricularis magnocellularis (PPM) nuclei of immature males stimulated more mating attempts in normal males than in males exposed to testosterone during incubation, but that mating attempts in the latter males were nonetheless increased by the implants. These results suggest that the PPM is not completely inactivated by embryonic exposure to testosterone, and that the depressant effect of early TP on mating is not due solely to hyperactivity of the inhibitory red nuclei. Unfortunately, interpretation of the effects of the implants is complicated by the fact that, judging from the combs on the birds, there was leakage from the brain implants into the general circulation.

Sex steroids are believed to act on the brain in the same manner that they act on other target organs—by being selectively bound to receptor molecules in the target tissue and thereby passing into the cell nuclei (McEwen and Pfaff, 1973). Nuclear uptake, then, is one indication that neuronal tissue is capable of functionally responding to steroids. A recent experiment shows that such nuclear uptake of estrogen occurs in the brains of chicks at least as early as Day 10 of incubation. Martinez-Vargas *et al.* (1975) prepared autoradiograms from brains of embryos injected with [³H] estradiol at different stages of incubation. Nuclear concentration of radioactivity was found in 10- and 17-day-old embryos, and in newly hatched chicks, but not in 6-day-old embryos. Nuclear uptake was prevented by prior injection of unlabeled estradiol, as is typical of cells containing limited-capacity estradiol receptors. Nuclear uptake in 10-day-old embryos was greatest in the medial preoptic nucleus, the lateral portion of the nucleus hypothalamicus posterior medialis, and in the ventrolateral portion of the nucleus inferior. The medial preoptic nucleus in particular has been implicated in the control of sexual behavior in adult chickens (Barfield, 1969; Crawford and Glick, 1975). Evidently, the mechanism by which ovarian estrogen or exogenous steroids could permanently alter the sexual behavior areas of the brains of domestic fowl is present by Day 10 of incubation.

PSYCHOSEXUAL DIFFERENTIATION IN JAPANESE QUAIL (*COTURNIX COTURNIX JAPONICA*)

SEX DIFFERENCES AND THE REPRODUCTIVE BEHAVIOR OF QUAIL

Behavioral dimorphism is pronounced in reproductively active domestic Japanese quail (which henceforth will be referred to simply as *quail*). Reproductive behavior strongly resembles that seen in the domestic chicken (both are members of the family Phasianidae). Male quail crow loudly, especially when isolated from females. In the presence of a female, the male struts stiffly; this strutting resembles the waltzing of chickens. Copulation in quail is notable for its short latency and almost comical intensity. In the author's laboratory, the *median* latency to attempt copulation is about 2 sec for sexually experienced males (Adkins and Adler, 1972) and about 6 min for males copulating for the first time and tested in a large, unfamiliar cage (Adkins, unpublished data, 1972). The median postcopulatory refractory period is only 9 min (Adkins, unpublished manuscript, 1972). Males frequently attempt to copulate even with unreceptive females or other males.

Male copulation consists of three components: head grabbing, in which the male grasps the feathers of the head or neck of the female; mounting; and cloacal contact movement, in which the mounted male leans backward with outstretched wings and twists his tail under the female's tail so that cloacal contact can be achieved. Cloacal contact, if achieved, lasts about 1 sec (Adkins, 1974). "Since the CCM [cloacal contact movement] represents our best criterion of a completed copulation, it stands as the most rigorous test of successful sexual performance" (Beach and Inman, 1965, p. 1429).

Sexual receptivity consists of squatting and holding still while the male mounts. No definite posture analogous to lordosis is generally seen. Some females actively solicit copulation from males by running in front of them or gently pecking them.

These sexual behavior patterns are very hormone-dependent. Castration of male quail is followed rapidly (within 8 days) by complete loss of crowing, strutting, and copulation in all individuals (Beach and Inman, 1965). Exposure to a short photoperiod, such as 8 hr of light per day, also eliminates sexual behavior, an effect that is mediated by pronounced testicular regression (Sachs, 1969). Injection of TP or implantation of testosterone pellets into surgically castrated males rapidly reinstates sexual behavior (Beach and Inman, 1965; Wilson and Bermant, 1972). Treatment of "photically castrated" males with TP also reinstates sexual behavior (Adkins and Nock, 1976). "Photically castrated" males injected with EB copulate but do not crow or strut (Adkins and Pniewski, 1978; Wada, 1982). When tested with male partners, they are receptive to the same extent as females injected with the same dosage of estradiol (Adkins and Adler, 1972). Testosterone does not activate receptivity in quail (Adkins and Adler, 1972).

Ovariectomy of female quail eliminates receptivity in all birds; and again, exposure to short days has the same effect, mediated by ovarian regression (Adkins,

1973). Injection of EB into surgically or photically castrated females restores receptivity in a dose-dependent manner (Adkins and Adler, 1972; Adkins and Nock, 1976). Estradiol does not activate any male-typical sexual behavior in females. Injection of TP into surgically or photically castrated females activates some crowing and strutting, but very little copulation (Adkins and Adler, 1972). An occasional female will head-grab or mount, but rarely does a female perform a cloacal contact movement. These low levels of male-typical behavior in female quail persist even when the birds are given testosterone metabolites such as estradiol or dihydrotestosterone (Balthazart and Schumacher, 1983).

Thus, adult quail exhibit behavioral dimorphism in their capacity for male-typical copulatory patterns and, to a lesser extent, for crowing and strutting, that goes beyond gonadal dimorphism. Receptivity, in contrast, does not appear to be dimorphic once hormone levels during testing have been equated in the two sexes.

EFFECTS OF HORMONE MANIPULATIONS BEFORE HATCHING ON REPRODUCTIVE BEHAVIOR

In the earliest relevant study, Wentworth *et al.* (1968) sprayed quail eggs with mestranol (a synthetic estrogen) dissolved in fuel oil on Day 0, 6, or 12 of incubation (the incubation period is 16–18 days). Males hatched from the treated eggs chased females, strutted, and copulated significantly less often than control males. Androgen secretion at the time of testing was depressed, however, so that an organizational effect of hormones on behavioral development was not necessarily demonstrated.

Evidence for an organizational action of early hormones in quail was obtained by Adkins (1975). Eggs were injected on Day 10 of incubation with either 2.5 mg TP, 1.2 mg TP, 0.05 mg EB, the oil vehicle, or nothing. At age 7 weeks (sexual maturity is reached in about 6 weeks), birds were transferred to short days (8 hr light: 16 hr dark). Three weeks after transfer to short days, all birds were injected with either 2.5 mg TP/day, to activate male-typical characteristics, or 0.05 mg EB/day, to activate female-typical characteristics. After 9–11 days of hormone treatment, the birds were tested with both male and female partners. Crowing, strutting, head grabbing, mounting, cloacal contact movements, proctodeal gland surface area, and receptivity (scored on an 8-point scale) were recorded. At the end of the experiment, all birds were killed and examined to verify gonadal sex. (Treatment on Day 10 of incubation occurs too late to permanently change the sex of the gonads; Burns, 1961; Wentworth *et al.*, 1968.)

The results of this experiment were as follows. Embryonic exposure to either TP or EB suppressed male-typical behavior; such males seldom crowed, strutted, or copulated. The higher dosage of TP was more effective, and EB was even more effective—no EB-treated males ever copulated. Failure to mate was seen following adult treatment with EB as well as following adult TP treatment. Females were not masculinized by early exposure to sex hormones and showed the same low levels of male-typical sexual behavior as the control females. Receptivity in females was not affected by injecting the eggs with hormones, nor was receptivity in males. This last result was not surprising, given that there is little sex difference in receptivity (Adkins and Adler, 1972).

To summarize this experiment, the males treated with EB or TP before hatching were no longer significantly different from females with respect to either male-typical or female-typical sexual behavior. Thus, they were feminized (or, more accurately, demasculinized), whereas the females were unaffected. A similar demasculinizing effect of embryonic estradiol or testosterone treatment on male quail behavior was seen in experiments by Whitsett *et al.* (1977) and by Warner *et al.* (1977).

These studies of quail suggest that the male is the behaviorally neutral sex, and that female differentiation results from embryonic exposure to endogenous steroids, presumably ovarian estrogens. Thus, females deprived of their ovarian estrogen during development should differentiate as males and, unlike most normal females, should be capable of responding to TP in adulthood by copulating in the male manner.

Support for this model of quail differentiation comes from an experiment by the author (Adkins, 1976). Because surgical gonadectomy of bird embryos is technically impossible if continued development and hatching are desired, females were instead treated with the chemical antiestrogen nitromifene citrate (CI-628). Eggs were injected on Day 9 of incubation with either 0.1 mg CI-628 or the distilled-water vehicle. When sexually mature, the birds were "photically castrated," injected with 1 mg TP/day, and tested with female partners. CI-628 masculinized the behavior of the females so that treated females displayed more head grabbing and mounting than controls. Three of the CI-628-treated females displayed the complete male copulatory sequence, including cloacal contact. Although the increase in cloacal contact movements was not statistically significant, it was nonetheless striking because in two previous experiments (Adkins, 1975; Adkins and Adler, 1972) in which females were given TP as adults, only 1 of 34 ever performed a cloacal contact. Males, as expected, were unaffected by the CI-628.

Psychosexual Differentiation in Zebra Finches (*Poephila guttata*)

Sex Differences and the Reproductive Behavior of Zebra Finches

A valuable feature of this Australian species is that its reproductive biology, including behavior, is known in both laboratory and field. The birds are highly opportunistic breeders adapted to an unpredictable semiarid environment (Davies, 1978; Sossinka, 1980).

The adults, although monogamous, are sexually dimorphic both in plumage (males have chestnut cheek patches; females do not) and in bill color (males have red bills; females have orange bills). Male-typical behavior patterns include undirected song, directed song, courtship dancing, beak wiping during courtship, mounting, and cloacal contact. Female-typical behavior includes tail quivering, a copulatory solicitation posture. Parental activities and aggressive behavior are performed by both sexes, but there are quantitative sex differences for most activities.

Male courtship, copulation, and aggression have been shown to be reduced by adult castration and restored to preoperative levels by TP replacement (Arnold, 1975a). The androgen dependence of singing depends on whether undirected or directed (courtship) song is measured. Castrated males still sing undirected songs, although at a lower rate and tempo, but directed song is eliminated or greatly reduced by castration and is restored by TP (Arnold, 1975a; Pröve, 1974). Both estrogenic and androgenic metabolites of testosterone are required in order to restore courtship and copulation in castrated males (Harding, 1983). The effect of ovariectomy and hormone replacement on the behavior of females does not appear to have been studied.

The most striking fact about behavioral dimorphism in this species is that only males ever sing; females never do, not even when given large amounts of TP in adulthood (Arnold, 1974). Studies of the neural mechanisms of song in this species have demonstrated a related, striking, and quite unprecedented size dimorphism in areas of the brain that have been shown to be part of the zebra-finch song system (Nottebohm and Arnold, 1976). HVc, RA, and the hypoglossal nucleus of the medulla are much larger in males than in females (male:female ratios 5.01, 5.53, and 1.50, respectively), and area X is apparently absent altogether (or at least undefined) in females. HVc, in addition to being larger in males, also contains more hormone-concentrating cells in males than in females following injection of tritiated testosterone (Arnold and Saltiel, 1979).

EFFECT OF EARLY HORMONE MANIPULATIONS ON ZEBRA FINCHES

This species turns out to be excellent for studies of psychosexual differentiation, not only because of the striking neural dimorphism. Unlike most altricial birds (those that hatch in a relatively undeveloped state), they breed well in the laboratory and prefer to breed colonially. In addition, their development is unusually rapid, about 14 weeks from egg to sexual maturity (Sossinka, 1980).

It has now been shown that sex hormones administered shortly after hatching can influence the development of both singing behavior and the neural song system in zebra finches, a result suggesting that the sex differences described above normally result from organizational hormone actions. In the first such study (Gurney and Konishi, 1980; Gurney, 1982), zebra finch chicks were hatched in an incubator, implanted subcutaneously on the day of hatching with Silastic pellets containing 50 μg of either estradiol (E) or dihydrotestosterone (DHT), and cross-fostered onto Bengalese finch parents. The birds were studied as intact adults. Undirected singing and song development in males were unaffected by the early implants, as were the volumes of the song areas HVc, RA, and DM. In females, in contrast, the volume of both HVc and RA was increased by early E and DHT, and the volume of DM was increased by early DHT. E implants given to adult intact females did not cause these areas to increase in volume, but adult implants of DHT did cause DM to enlarge to male size. Some females given early E or DHT were given DHT or T implants as intact adults and were tested for undirected song. Females that had been given early E sang, and examination of their brains revealed that the adult T and DHT had caused further increases in HVc and RA above those produced by early hormones alone. Females that had been given early DHT did not sing.

These results are extremely important because they show in a direct way how hormones act on developing neural tissue to organize future behavior and determine an adult behavioral sex difference.

A subsequent examination at the cellular level of the brains of finches treated with hormones early in life clarified the nature of the neural organizational effects (Gurney, 1981; Konishi and Gurney, 1982). Within RA there were marked sex differences in both the number of neurons present (males had more neurons) and in the proportion of neurons with large, rather than small, soma diameters (males had larger diameters). Testosterone treatment of females had a dose-related effect on the proportion of neurons in RA that had large, rather than small, soma diameters; thus T masculinized neuron size in RA. E and DHT also maculinized soma diameter in females, but DHT had only a small effect, whereas E had an effect nearly equal to T. All three steroids also masculinized cell number in the RA of females, but whereas T and DHT were roughly equal, E produced a much smaller effect. Thus, the two masculinizing actions of sex hormones on RA neurons differ with respect to steroid specificity, suggesting different biochemical mechanisms.

Behavioral aspects of zebra finch sexual differentiation have been studied in two additional laboratories. One has examined the effect of both prehatching and posthatching administration of low dosages of estradiol or testosterone on male and female finches (Pohl-Apel and Sossinka, 1982a,b; Sossinka and Pohl-Apel, 1982; Sossinka et al., 1981). In all experiments the birds were tested as intact adults. Estradiol-treated males showed normal behavior patterns, but at a lower frequency than controls. Estradiol-treated females, when given testosterone as adults, were masculinized and exhibited all the various male-typical behavior patterns. Posthatching treatment was more effective than prehatching treatment. Prehatching treatment with testosterone had only slight effects on either males or females; posthatching treatment had little effect on males, and masculinized singing in females, but less effectively than early estradiol did.

The other laboratory studied the behavior of male and female finches that were given daily injections of estradiol benzoate for the first two weeks after hatching (Adkins-Regan, 1984; Morris, 1980). After reaching adulthood, birds were gonadectomized and implanted with testosterone propionate. As expected, the nestling treatment masculinized singing in the females, but at the same time, copulation was demasculinized in males. Furthermore, neither male-typical nor female-typical copulatory behavior was masculinized in the females.

That the results from these two laboratories were not identical is presumably due to the fact that in one case the birds were tested in the intact state, thus confounding organizational and activational effects, whereas in the latter case birds were gonadectomized and given hormone replacement. With activational hormone exposure controlled, the results suggest that sexual behavior *per se* differentiates according to the same pattern in zebra finches as in chickens and quail, one in which the male is the neutral sex and early sex hormone exposure primarily demasculinizes males. In contrast, the effects of early hormones on singing are clearly quite different, and follow a mammallike pattern in which early sex hormones primarily masculinize females. Thus, within the same species, the same early hormone treatment can have quite opposite effects on different male-typical behavior patterns, an unusual psychosexual differentiation phenomenon without a clear precedent even in the mammalian literature.

PSYCHOSEXUAL DIFFERENTIATION IN OTHER AVIAN SPECIES

Pigeons and doves have figured importantly in hormone-behavior research generally, but thus far, there has been only one published study of psychosexual differentiation in a columbiform bird. Orcutt (1971) implanted subcutaneous pellets of diethylstilbestrol (DES), a synthetic estrogen, in squabs of different ages. The pellets were removed at various times after implantation or were left in place. When adult, the birds were left as intact animals and paired with males or females for observation. The behavior patterns recorded included bow cooing (a male display); bowing (normally performed more frequently by females); displacement preening (performed more frequently by males); nest calling (performed more frequently by males); nest-material gathering (again more frequent in males); nest building (usually done by the female); begging (a female solicitation display); billing (a female pattern); and squatting (normally seen only in females). The few significant effects of the pellets that were seen all suggested feminization of the males and included reductions in nest cooing and bow cooing.

PSYCHOSEXUAL DIFFERENTIATION IN REPTILES

Thus far, reptiles have been almost completely absent from studies of the behavioral effects of early hormone manipulations. In part, this absence can be explained in terms of practical difficulties: many species are difficult to rear from hatching in captivity, and the time to reach sexual maturity is usually much longer than in mammals of similar sizes. But certain snakes and lizards do have potential, particularly smaller species with large clutch sizes or litters (some reptiles are viviparous).

Thus far, the only study of behavioral differentiation is one by Crews (1983) using garter snakes *(Thamnophis)*. In this species, the males exhibit postnuptial gametogenesis; that is, mating occurs after emerging from the winter hibernaculum, but before testosterone levels and sperm production increase (Crews, 1983). Injection of androgen into adult males does not activate mating behavior, but exposure to cold does. Adult females, even when injected with testosterone or exposed to cold, rarely court other females. In Crews's experiment, male snakes were castrated or sham-operated 8–10 weeks after birth, while still prejuveniles. Half of the castrated males were implanted with testosterone at the time of surgery.

The females were left intact and implanted with testosterone. When tested 4 weeks later (while still prejuveniles), the castrated males implanted with testosterone all courted; the castrated or sham-operated males did not. Furthermore, the testosterone-implanted females also courted. Because normal adult males, but not females, court, and testosterone does not activate courtship in adult males, these results suggest that the adult sex difference in capacity to court might indeed arise from an early sex difference in androgen secretion and even more interestingly, that male-typical behavior might be organized by steroid hormones but activated by temperature (perhaps via a nonsteroid hormone). Clearly, more work with this species will be valuable.

Green anoles *(Anolis carolinensis)* offer some possibilities for differentiation experiments as well. The hormonal activation of reproductive behavior in both males and females is reasonably well understood (see Crews, 1979, for a review). Furthermore, studies of adult sex differences suggest that there is less dimorphism in male-typical behavior than in female-typical behavior (Adkins and Schlesinger, 1979; Mason and Adkins, 1976). A reasonable hypothesis would be that early androgen reduces the capacity for female-typical behavior in males.

In some reptiles, sex appears to be environmentally determined, chiefly by incubation temperature (Bull and Vogt, 1981; Pieau, 1975; Yntema, 1979). But because temperature has this effect during embryonic development, it does not necessarily follow that the process of behavioral differentiation would have to be any different than it is in other reptiles. Thus, the situation is conceptually distinct from environmental determination in those hermaphroditic fish that change sex after reaching maturity (to be discussed below).

PSYCHOSEXUAL DIFFERENTIATION IN AMPHIBIANS

Amphibians have not yet participated in experiments focusing on behavioral differentiation, but it is possible to make some inferences about this process from some of the studies of gonadal differentiation in which individuals were completely and functionally sex-reversed by early steroid treatment. Where such individuals mated with others of their own genetic sex to produce offspring, it is safe to assume that sexual behavior was also at least partially reversed.

The reproductive behavior of anuran amphibians (frogs and toads) consists of courtship vocalizations and mating, in which the male clasps the female reflexively around the "waist" (amplexus). A single clasp may last for over a day. The reproductive behavior of urodeles (salamanders and newts) is not as well known and involves the secretion of pheromones, postural displays, and mating. In many species, fertilization is internal; the male produces a spermatophore that is passed to the female or picked up by her during mating.

Earlier attempts to activate sexual behavior in amphibians with steroid hormones were largely unsuccessful (e.g., Palka and Gorbman, 1973). More recent efforts have been successful in three species. In the frog *Xenopus laevis,* Kelley and Pfaff (1976) restored clasping in castrated males by implanting pellets of testosterone or dihydrotestosterone, and they induced receptivity in ovariectomized females with a combination of estradiol, progesterone, and LHRH. Diakow and Nemiroff (1981) obtained evidence that both vasotocin and prostaglandin stimulate female mating behavior in the frog *Rana pipiens.* In the newt *Taricha granulosa,* Moore (1978) activated sexual behavior with implants of either testosterone or dihydrotestosterone in castrated males that had been sexually responsive prior to castration and also activated sexual behavior in castrated androgen-implanted males with arginine-8 vasopressin (Moore and Zoeller, 1979).

A recent study (Schmidt, 1982) has demonstrated a sex difference in one of the neural vocalization structures (the magnocellular portion of the pretrigeminal nucleus) in the toad *Bufo americanus.* This structure contains larger cells in males than in females, and in females, it can be masculinized by giving TP in adulthood. The fact that this masculinization is complete suggests that embryonic or larval hormones play little role in the development of this sex difference.

Sex differentiation occurs slowly and rather late in amphibians, around the third week after hatching in *X. laevis* (Witschi, 1971), and thus is particularly amenable to experimental intervention. Hormones can be administered easily by putting them in the water the larvae live in. .

Female larvae of four anurans (*Pseudacris* sp., *Rana temporia, R. japonica,* and *R. sylvatica*) are completely and permanently masculinized by developing in water containing androgens (Burns, 1961; Dodd, 1960; Gallien, 1965, 1967). These sex-reversed genetic females look like males, produce sperm, mate with normal genetic females, and produce offspring, all of which are females. This reversal is accomplished by as little as 1 ppm TP in *R. sylvatica* (Gallien, 1955).

Certain urodeles and primitive anurans such as *Xenopus* can also be sex-reversed, but in contrast to higher anurans, it is the male that is most easily reversed. Genetic male larvae of five species (*Ambystoma* sp., *X. laevis, Hynobius* sp., *Triturus helveticus,* and *Pleurodeles waltlii*) are completely and permanently reversed by being reared in water containing estrogens, and they mate with normal males to produce offspring that are all male (Foote, 1964; Gallien, 1955, 1965, 1967). As little as 66 μg/liter of EB reverses all males of *P. waltlii.* Androgens either cause slight temporary masculinization or cause feminization (Dodd, 1960; Gallien, 1955).

PSYCHOSEXUAL DIFFERENTIATION IN TELEOST FISH

As is the case with amphibians, our understanding of early hormonal effects on later behavior in fish is derived from incidental observations made in the course of morphological studies and from the successful production of offspring by sex-reversed individuals.

Fish are highly variable in their reproduction. External sexual dimorphism is marked in some species, and size differences alone can be considerable. In internally fertilizing species that copulate, the male sometimes has a modified anal fin to aid sperm transfer. Reproductive behavior may include courtship displays, copulation or spawning, nest building, or parental care of eggs or young (Hoar, 1962). Courtship and copulation are dependent on gonadal hormones in some species, but in others, gonadectomy seems to have little or no effect on sexual behavior (Aronson, 1957; Baggerman, 1968; Liley, 1969; Reinboth, 1972).

Where gonadal sex differentiation occurs prior to birth or hatching, treatment of fry with sex steroids produces only partial sex reversal; but where differentiation occurs after birth or hatching, complete and functional gonadal reversal can be induced with sex steroids. An excellent example of this phenomenon is found in Yamamoto's work (1969) with the oviparous medaka *Oryzias latipes* (see also Fineman *et al.*, 1974). This species never shows spontaneous sex reversal or intersexuality. If the fry are fed diets containing sex steroids for 10 weeks after hatching, complete sex reversal is achieved. Genetic females treated with androgens mature as males, produce sperm, and successfully fertilize normal females. Genetic males are reversed completely with estrogens. Treatment after primary sex differentiation has occurred never results in complete reversal.

Complete reversal can also be obtained in the guppy *(Poecilia reticulata)*, even if hormone exposure is entirely prenatal (guppies are viviparous). Methyltestosterone given to gravid females for 24 hours results in all-male broods (Dzwillo, 1962). Clemens *et al.* (1966) obtained 90% males by feeding newborns food containing 25 µg/g methyl testosterone for 60 days. The sex-reversed females produced sperm but did not sire young, and tests indicated that this deficiency was behavioral. Only 25% of the sex-reversed females displayed masculine courtship behavior, only 20% displayed gonopodial thrusting, and 75% showed no interest in females. These observations suggest that complete morphological reversal is not necessarily an indicator of total behavioral reversal.

Complete morphological reversal has been obtained in other species as well, but behavioral data are lacking in all cases. Clemens and Inslee (1968) changed genetic female *Tilapia mossambica* (a mouth-brooding cichlid species) into sperm-producing males by administering methyltestosterone to the fry for the first 69 days after hatching. Yamamoto and Kajishima (1968) treated newly hatched goldfish *(Carassius auratus)* with sex hormones for two months and obtained complete sex reversals in both directions. Hackman and Reinboth (1974) put estradiol in the water of young *Hemihaplochromis multicolor* (another mouth-brooding cichlid) and found that estradiol feminized all the males. These feminized genetic males could be bred to normal males and therefore probably displayed female behavior.

As is the case with amphibians, little is known about the behavioral responses of adult fish to heterologous sex hormones; thus, these apparent behavioral sex reversals are difficult to interpret. Male behavior has been activated with androgens in females of several species, including Siamese fighting fish *(Betta splendens),* gouramis *(Trichogaster trichopterus),* swordtails *(Xiphophorus helleri),* guppies, sticklebacks *(Gasterosteus aculeatus), H. multicolor,* and medakas, but the responses of males given estrogens are not yet known (Kramer, 1972; Liley, 1972; Reinboth, 1972; Wai and Hoar, 1963).

The ease with which behavior can be reversed at different ages has not been systematically explored, but evidently, reversal does not necessarily end with sexual maturity. In two species, the Siamese fighting fish and the paradise fish *(Macropodus opercularis)*, adult ovariectomized females have been observed to undergo complete morphological sex reversal (Becker *et al.*, 1975; Lowe and Larkin, 1975; Noble and Kumpf, 1937). These sex-reversed females display male-typical threat, courtship, and mating behavior and produce offspring in matings with normal genetic females. Oddly enough, the ovaries of juvenile *Betta* were not masculinized by testosterone in one study (Reinboth, 1970).

Thus, several fish can readily be experimentally sex-reversed, occasionally even in adulthood. Nonetheless, all of these species, with the possible exception of the swordtail, are normally quite stable sexually, and spontaneous sex reversal occurs only rarely (Atz, 1964). In contrast, some other species of fish change sex as a normal part of their life cycle. This phenomenon is not seen in any birds, reptiles, or amphibians (but is common in invertebrates) and has such intriguing implications for the whole concept of sexual differentiation that a more extensive discussion is appropriate.

HERMAPHRODITISM IN TELEOST FISH

A few definitions will prove useful (see Ghiselin, 1974, for a detailed discussion of terminology). A *hermaphroditic* species (as opposed to a bisexual or *gonochoristic* species) is one in which at least some normal (i.e., nonpathological) individuals possess both male and female germ cells, either *simultaneously* (synchronously) or *successively* (sequentially). In a successive hermaphroditic species, individuals either change from male to female (are *protandrous*) or from female to male (are *protogynous*). Such sex change is behavioral as well as morphological.

Hermaphroditism is found in 15 teleost fish families, 3 freshwater and 12 marine. Of the 79 known hermaphroditic species (the list is sure to grow with more research), 24 are simultaneous hermaphrodites, at least 1 of which is self-fertilizing; 14 are protandrous; and 41 are protogynous (Harrington, 1971; Reinboth, 1970).

In several species of successive hermaphrodites, including some coral reef dwellers, the occurrence of sex reversal in an individual is determined by stimuli from other fish; that is, there is social control of sex reversal (Fishelson, 1975; Fricke and Fricke, 1977; Robertson, 1972; Shapiro, 1980; Warner *et al.*, 1975). In other species, reversal is more directly controlled by nonsocial environmental factors or by genetic timing mechanisms (Atz, 1964; Harrington, 1971; Liem, 1963).

The sexuality of these hermaphroditic fish, however bizarre it may seem by avian or mammalian standards, is such a well-controlled part of their life cycles that it must have significant adaptive value. What functions does it serve, and how did it evolve? Several types of explanations have been put forth in the literature (Ghiselin, 1974; Maynard Smith, 1978; Warner, 1978).

One older view is that hermaphroditism is the phylogenetically primitive condition in vertebrates. The taxonomic distribution of the phenomenon within the teleost fish clearly indicates, however, that the opposite is true: hermaphroditism has arisen from gonochorism several times through convergent evolution (Atz, 1964; Chan, 1970; Ghiselin, 1974).

Another traditional idea is that hermaphroditism is advantageous because it allows self-fertilization, thereby doing away with the need to expend energy finding a mate, engaging in sexual activity, and so on. The fact that few simultaneous hermaphrodites actually engage in self-fertilization (possibly to avoid inbreeding) weakens this explanation.

Low-density models are somewhat more useful. They propose that at low densities, such as in the deep sea or in small temporary pools, finding a mate of the right sex can be a problem, particularly for relatively immobile animals. A simultaneous hermaphrodite is at an advantage because it can mate with *any* other individual, not just half of those encountered. A successive hermaphrodite can change sex if the partner is not the right fit. In general, a patchy environment results in a low effective mating-density (Smith, 1975) and should favor the evolution of hermaphroditism. Coral reefs, although climatically stable, are considered patchy environments with regard to critical resources (Choat and Robertson, 1975), and a number of coral reef species are hermaphrodites. *Amphiprion* illustrates the patchy-environment problem well: because these fish are confined to anemone clumps, the group size is quite limited, but it would be dangerous to leave the safety of the clump to search for a mate (Fricke and Fricke, 1977). But other coral-reef hermaphrodites occur at high densities (Fischer, 1980; Pressley, 1981).

A conceptually related way to understand the maintenance (but not the origin; Fischer, 1980) of simultaneous hermaphroditism in some coral reef serranid fish focuses not on density *per se,* but on the likelihood that an individual living as a nonhermaphrodite male would be able to leave more offspring than the same individual living as a female (Fischer, 1980; Pressley, 1981). If males, for whatever reason (not just low density, but also other time demands such as territorial defense), are unlikely to be able to mate with multiple females, then there is no advantage to being a male. And as a pair of simultaneous hermaphrodites can produce twice as many offspring as a pair of relatively monogamous gonochorists, it is better to be the former than the latter. As Fischer (1980) and Maynard Smith (1978) have pointed out, this argument (and also the low-density models) for simultaneous hermaphroditism is reminiscent of the presumed advantage of parthenogenetic (all-female) reproduction in those species that exhibit it.

The gene-disperal model argues that, in relatively immobile organisms, successive hermaphroditism where offspring change sex at the same age ensures that siblings will not mate with each other. Williams (1975) rejected this explanation on the grounds that it is based on group, rather than individual, selection.

The size-advantage model is meant to apply mainly to successive hermaphrodites that are dimorphic in size. It recognizes that many fish continue to grow after reaching sexual maturity, and that the advantage of being large or small may differ for the two sexes. If large size is an advantage for females (as it would be if fecundity is proportional to size), or if small size is advantageous to males (as it would be if smaller males are more mobile), protandry would be selected for.

Warner (1975) has expanded this size-advantage model, making it more rigorous and more explicitly based on individual selection for reproductive output. Basically, animals should change sex if one sex gains in fertility with age more than the other sex (Warner *et al.*, 1975). If one knows the survival and fecundity as a function of age for each sex, and if one knows the type of mating system, the model allows one to predict the type of hermaphroditism and the age at which sex change will occur. It predicts that random mating populations where female fecundity increases with age should be protandorous, whereas more selective maters where female fecundity decreases with age, or the males compete for mates, would be protogynous. The initial predictions of the model were reasonably well met (Warner, 1975).

The most general model of this kind is that of Charnov and Bull (1977), in which sex should be environmentally determined whenever the fitness of the two sexes is differentially and strongly environmentally determined and there is little control over the environment (i.e, when the habitat is unpredictable).

It is likely that no one explanation or set of ecological factors can account for all, or even most, instances of fish hermaphroditism (Liem, 1968; Williams, 1975), and that each species needs to be studied on its own if we are to understand the function and evolution of its hermaphroditism.

HORMONAL MECHANISMS OF SPONTANEOUS SEX REVERSAL

What are the physiological bases and causes of sex reversal in fish? This question has not yet been satisfactorily answered for any species, but some preliminary attempts have been made.

The gonads of hermaphrodites are histologically similar to those of gonochorists (Chan *et al.*, 1975). In species with both hermaphroditic and gonochoristic males, such as *Thalassoma,* the gonads of the two types of males can be distinguished (Choat and Robertson, 1975). The gonadal hormones of successive hermaphrodites are the same steroids that are found in gonochorists, but it is possible that the relative amounts are different. In a study of hormones in four hermaphroditic species and two gonochoristic species, Idler *et al.* (1976) found that the hermaphrodites produced considerable 11β-OH-testosterone and little 11-ketotestosterone, whereas the gonochorists produced considerable 11-ketotestosterone and little 11β-OH-testosterone. Depending on the species, both types of gonadal tissue may or may not be present prior to reversal (Reinboth, 1970). In either case, reversal is accompanied by the degeneration of one type of tissue and the development of the other. The changeover in gonadal tissue is nicely reflected in the gonadal hormone output in *Monopterus albus,* a protogynous species. As the animal changes from female to male, the plasma level of estradiol decreases and the level of testosterone increases (Chan *et al.*, 1975).

Several physiological factors have been proposed as the proximate trigger for sex change, such as the exhaustion of one type of germ cell, seasonal and age changes in the level of two hypothetical antagonistic gonadal-induction hormones, and changes in the levels of pituitary hormones that differentially favor ovaries over testes (summarized in Atz, 1964; Harrington, 1971). A recurring theme is that seasonal breeding might facilitate sex reversal by temporarily freeing one type of gonadal tissue from inhibition by the other (Fishelson, 1975).

Several attempts have been made to experimentally reverse the sex of hermaphroditic species, generally with sex steroids. Most attempts have met with little success, for example, in *M. albus* (Chan *et al.*, 1975) and *Sparus auratus* (Reinboth, 1970). Occasionally, however, precocious reversal has occurred. A single androgen injection into protogynous *Maena maena* initiated male development and spermatogenesis (Reinboth, 1970). Large female *Anthia squamipinnis* treated with the largest of several doses of testosterone became males (Fishelson, 1975). Estrogen treatment of females undergoing spontaneous reversal stopped the process temporarily, but when the treatment ceased, reversal continued.

Thus, in some species of hermaphrodites and gonochorists (as discussed previously), steroid hormone treatment can cause sex reversal. But whether steroids are a cause of spontaneous reversal or merely a result is unknown, and experimental sex reversal in both gonochoristic and hermaphroditic fish may or may not reflect mechanisms of spontaneous reversal in the successive hermaphrodites (Reinboth, 1970).

QUESTIONS AND THEORETICAL ISSUES

ARE PARADOXICAL EFFECTS OF EARLY HORMONE TREATMENTS DUE TO CONVERSION OF ANDROGEN TO ESTROGEN?

Androgens and estrogens seem to have similar effects on the sexual development of nonmammalian vertebrates, if they have any effect at all. Either both masculinize, or both feminize, depending on the species. Similar paradoxical effects are seen in the development of mammalian sexual behavior (see Chapter 4). It has been suggested that paradoxical feminization of mammals by certain androgens is due to biochemical conversion (aromatization) of these androgens to estrogenic metabolites (Clemens, 1974; McDonald *et al.*, 1970; Naftolin *et al.*, 1972). Only those androgens that are aromatizable should mimic estrogen.

This theory is particularly applicable to some of the avian data. In zebra finches, testosterone, which is aromatizable, masculinizes singing when given to nestlings in a manner similar to, but weaker than, estradiol (Sossinka and Pohl-Apel, 1982), whereas dihydrotestosterone, which is not aromatizable, does not appear to masculinize singing (Gurney, 1982).

A series of experiments has examined aromatization in relation to the psychosexual differentiation of quail (Adkins, 1979; Adkins-Regan *et al.*, 1982). In this species, the minimum dosages that demasculinize males when injected into eggs on Day 10 of incubation are 500 μg TP and 1 μg EB (Adkins, 1979). The following experimental findings all indicate that TP demasculinizes copulatory behavior because it is aromatized (Adkins-Regan *et al.*, 1982): (1) dihydrotestosterone and androsterone, which are nonaromatizable androgens, fail to demasculinize copulatory behavior; (2) TP injected together with the antiestrogen tamoxifen (ICI 46,474) fails to demasculinize copulatory behavior; and (3) TP injected together with an aromatase inhibitor ATD (1,4,6-androstatrien-3,17-dione) fails to demasculinize copulatory behavior. But demasculinization of crowing and strutting by TP does not appear to be due to aromatization, as indicated by these experimental findings: (1) dihydrotestosterone does demasculininze crowing and strutting and (2) tamoxifen does not prevent TP from demasculinizing crowing and strutting.

ARE EARLY HORMONES RESPONSIBLE FOR THE DIFFERENTIATION OF OVULATION CONTROL MECHANISMS?

In rodents, treatment of neonatal females with steroids not only masculinizes or defeminizes behavior but also masculinizes the cyclic LH release mechanism in the hypothalamus and thereby causes an anovulatory syndrome. It is certainly possible that some nonmammals might also be dimorphic for ovulation control as a result of hormonal dimorphism early in life, but so far there are almost no relevant experiments. Female quail hatched from eggs injected with EB or TP do frequently fail to lay eggs, or they lay fewer eggs, but this syndrome is caused by oviduct abnormalities, rather than by a failure to ovulate (Rissman, Ascenzi, Johnson, and Adkins-Regan, 1984). Urbanski and Follett (1982) have found dimorphism in the basal LH levels of gonadectomized quail; plasma levels of males are always higher, regardless of photoperiod. LH levels should therefore be an interesting endpoint in a differentiation study.

DOES ANY BEHAVIOR NOT DIRECTLY RELATED TO REPRODUCTION ALSO DIFFERENTIATE UNDER THE INFLUENCE OF EARLY HORMONES IN NONMAMMALIAN SPECIES?

Only one study related to this issue appears to have been done. Mauldin *et al.* (1975) dipped eggs incubated for 3 days in solutions of testosterone propionate. When the chicks were 3 weeks old, the time required for each chick to move from a 5-cm circle through a door in a wire screen to be near a group of chicks (the flock), a highly desirable goal for a young chick, was recorded on 5 consecutive days. The chicks treated with the higher of two TP dosages took longer than the control chicks to reach the flock on all but the first day of testing; that is, they did not show as much improvement from day to day as did the controls. Unfortunately, the relevance of this finding to psychosexual differentiation is not entirely clear, as the sex of the chicks is not given in the report, and as no information is provided about whether the behavioral end-point, a learning task, is sexually dimorphic.

HOW USEFUL IS THE ORGANIZATION/ACTIVATION MODEL FOR UNDERSTANDING NONMAMMALIAN PSYCHOSEXUAL DIFFERENTIATION?

The concept of organizing effects of early hormones, together with the distinction between organizing and activating actions, has been quite valuable as a theoretical framework for the mammalian experiments and has successfully guided such research for 20 years (since the publication of Young *et al.*, 1964). But how does it fit the comparative data? Here, I will examine each of its features or implications separately.

ARE EARLY HORMONE EFFECTS PERMANENT? In the experiments that have focused on behavioral consequences (i.e., the avian studies), the time period between treatment and testing has been long enough so that the behavior does seem to have been permanently sex-reversed. But no experiments have systematically asked whether there might be some eventual "recovery," for example, after several seasonal cycles, repeated sexual experience, neural aging, or repeated hormone treatments. Nonetheless, the early hormone effects are clearly vastly different in duration from the behavioral changes seen when adults are treated. The experiments on sex reversal in amphibians and nonhermaphroditic fish also seem to involve permanent behavioral (and morphological) changes.

DO THE BEHAVIORAL EFFECTS OF EARLY HORMONES SEEM TO REFLECT ORGANIZATION OR JUST THRESHOLD CHANGES? The results of the experiments described previously with quail and chickens can be adequately explained by a model in which embryonic hormones merely change the sensitivity of neural circuits to activational hormones, rather than organizing anything *de novo* (see Beach, 1971). But Gurney's work with zebra finches (Gurney and Konishi, 1980; Gurney, 1982) provides compelling evidence that true neural organization may be occurring, and it constitutes the best evidence to date for such organization in any species, mammal or nonmammal.

DO THE BEHAVIORAL EFFECTS OF EARLY HORMONES OCCUR ONLY DURING LIMITED CRITICAL PERIODS? Psychosexual differentiation in rodents is one of the best examples of a critical-period phenomenon in behavioral development. The literature suggests that psychosexual differentiation in birds also can be altered only during a limited period. Chickens hatched from eggs injected with TP before Day 13 of incubation, but not after, are demasculinized (Wilson and Glick, 1970; Benoff, 1979). Treatment very early in incubation (Day 3) is also effective, a result suggesting either that the critical period is at least 10 days long, or (more likely) that hormones administered before the onset of the critical period remain in the embryo long enough to overlap with it. Quail hatched from eggs injected with EB before Day 12 of incubation are demasculinized, but injection after Day 12 has no effect on copulatory behavior, and injection on Day 12 demasculinizes some but not all birds (Adkins, 1979). Thus, the critical period in both of these highly precocial species seems to end before hatching.

This conclusion has been questioned by a study that compared the adult behavior of quail gonadectomized 6–12 hr after hatching with that of sham-operated controls (Hutchison, 1978). Unfortunately, at the time of testing, the controls were apparently intact and exposed only to endogenous hormones, whereas the experimental groups were gonadectomized and receiving exogenous hormones; thus, the result confounded the activational and organizational effects of hormones. More recently, Schumacher and Balthazart (1984) gonadectomized quail 1 day or 1, 2, 4, or 6 weeks after hatching. At age 2 months, the birds were implanted with testosterone and tested. Males were unaffected by early castration, but females ovariectomized before age 4 weeks showed somewhat more male-typical behavior than would be expected. These results suggest that demasculinization of females by their ovarian estrogen might continue after hatching. If this is true, it is difficult to explain why the injection of males with EB after the eleventh day of incubation fails to demasculinize them (Adkins, 1979).

In zebra finches, singing by females is more extensively masculinized by posthatching than by prehatching estrogen injection (Pohl-Apel and Sossinka, 1982a). Males castrated after 9 days of age do not seem to be feminized (Arnold, 1975b). Thus, there does appear to be a critical period for masculinization that occurs largely after hatching, which is not unexpected in this highly altricial species. A reasonable prediction would be that most or all altricial birds have behavioral critical periods that extend into the posthatching period or are primarily posthatching.

In amphibians and fish, the kind of information is seldom available that would be needed to support or to refute critical periods. Usually there is no systematic variation in age at the time of treatment, and the extent to which the behavior of adults can be hormonally reversed (i.e., whether there is any true behavioral dimorphism) is not known. If adults can be behaviorally reversed as readily as larvae or fry can, then it would be unnecessary to envision a separate organizational period early in life.

There are two groups of fish for which it is already clear that the usual organization/activation model is incorrect or unnecessary. One consists of those species (*Betta splendens* and *Macropodis opercularis*) in which the ovariectomy of adult females is followed by apparently complete behavioral and morphological sex reversal (Lowe and Larkin, 1975; Noble and Kumpf, 1937; Becker *et al*, 1975). The other consists of the successive hermaphrodites that were discussed previously. Sexual differentiation in these fish is clearly unlike differentiation in other fish and in other vertebrates in important ways. Not only does it occur more than once in the life of the animal, but the second period of differentiation occurs in adulthood and is somewhat labile temporally. As Harrington (1971) points out, in gonochorists, the sex genes become dormant after initial differentiation, but in successive hermaphrodites, they do not. Whether sex differentiation is confined to these two periods or could be artificially induced at any time is not yet clear because the physiological trigger is generally unknown. Are there two organizational periods (in the mammalian sense), or is there none?

ORGANIZATIONAL VERSUS ACTIVATIONAL HORMONES. In mammals, hormones can even organize behavior that itself does not require hormones for activation, for example, play behavior in rhesus monkeys (Goy and McEwen, 1980). Thus far, there are no clear analogous examples in nonmammals; but there are cases that make it clear that the *same* hormone(s) need not be both the organizer and the activator. For example, in quail, estrogen treatment of male embryos feminizes crowing, but in adult males, estrogen never activates crowing (Adkins and Pniewski, 1978).

CONCLUSION. Thus far, the organization/activation model can adequately account for avian psychosexual differentiation but will probably not be correct for certain other vertebrate classes. As a methodological guide, however, it is always valuable. That is, the effects of early hormones on *differentiation* can never be conclusively demonstrated without controlling for activational hormone levels at the time of testing. When adult hormone levels of early-treated animals are actually measured, they are usually lower than normal (e.g., Mashaly and Glick, 1979), and so all animals must as adults be gonadectomized and given replacement therapy.

Is the Empirically Determined Pattern of Hormone Secretion of Embryos or Young Consistent with Conclusions Derived from the Behavioral Data?

The ideal data to answer this question would come from simultaneous measurements of androgens and estrogens in the plasma of the same individuals at different times during development. For practical reasons, such experiments have seldom been done. What is generally available are measurements of either androgens or estrogens in different individuals killed or bled at different ages.

Thus far, most experiments have examined either chickens or quail, in which estrogens seem to be the major organizing hormones. In chickens, the gonads differentiate morphologically beginning at about 6.5 days of incubation; in quail, they begin differentiation at about 5.5 days. Ozon (1965), using spectofluorimetry, determined total estrogens in the blood of 10- and 21-day female chicken embryos and found that the levels were high, higher than those of adult laying hens. More recently, Woods and Brazzill (1981), using RIA, determined the plasma estradiol levels of male and female chickens embryos from incubation age 7.5 days to 17.5 days. Female levels were higher than male levels (e.g., 975 vs. 683 pg/ml plasma on Day 11.5), and they increased during this period. Measurements of androgens using RIA indicate significant levels in both chicken and quail embryos (Gasc and Thibier, 1979; Ottinger and Bakst, 1981; Tanabe *et al.*, 1979). In chickens, males produce more testosterone than females do (Woods *et al.*, 1975), but in both sexes, peak levels are much lower than peak levels of estradiol (Woods and Brazzill, 1981). Thus, the existing data on the hormonal output of chicken embryos are consistent with a model in which the male is the neutral sex and ovarian estrogen causes feminine development; plasma levels of estradiol in quail are not yet known.

Endogenous androgen levels are surprisingly high in both chicken and quail embryos. It is reasonable to ask whether this androgen contributes to psychosexual differentiation. One experiment examined this possibility (Adkins-Regan, in press). Quail eggs were injected on Day 9 of incubation with oil or ATD (an aromatization inhibitor). As adults, males were exposed to short days, injected with testosterone propionate, and tested for male-typical behavior. ATD increased the level of male-typical copulatory behavior. Thus, normal male quail are actually slightly demasculinized by their own androgen during embryonic development, and this process is mediated by aromatization.

Gonadal development in altricial birds has received little attention. So far there do not appear to be any published *in vivo* measurements of hormone levels in embryos or very young nestlings of any altricial bird. But hormonal data are available at slightly later ages in one species, the zebra finch. Males were repeatedly sampled every five days from 8 to 75 days posthatching (Pröve, 1983). Four steroids were measured: testosterone, estradiol, dihydrotestosterone, and progesterone. There were peaks of testosterone at Days 18–21 and 34–37 and after Day 70, whereas the dihydrotestosterone levels were uniformly low. Estradiol was highest from Day 38 to Day 49, and progesterone was highest from Day 14 to Day 25 and after Day 70. Because the critical period for behavioral differentiation occurs earlier than most of these measurements (Pohl-Apel and Sossinka, 1982a), it is not yet possible to tell whether hormone production is consistent with the results of the behavioral experiments.

DO SEX CHROMOSOMES OR GENES HAVE A DIRECT (NONHORMONALLY MEDIATED) ROLE IN PSYCHOSEXUAL DIFFERENTIATION?

So far, the answer to this question is a clear "no." Male quail treated with estradiol as embryos show sexual behavior that is quantitatively and qualitatively indistinguishable from that of females tested under equivalent adult hormonal conditions (Adkins, 1975). Wilson and Glick's (1970) male chickens exposed to estradiol as embryos attempted to mate even less often than untreated females. Female zebra finches exposed to estradiol shortly after hatching grow up to be capable of singing in a manner quite similar to that of males (Gurney and Konishi, 1980; Pohl-Apel and Sossinka, 1982b). The spectrum of dimorphic behavior that has been examined is, of course, quite incomplete; but so far, it seems that males of two avian species can be completely sex-reversed by sex hormones with respect to courtship and copulation, and females of a third species can be sex-reversed with respect to singing. Thus, a direct role for the sex chromosomes in the development of these behaviors need not be postulated.

In certain species of fish and amphibians, the genetic sex is no deterrent at all to manipulation of phenotypic sex with steroids; and complete reversal of either sex of the medaka is possible by appropriate hormone administration (Yamamoto, 1969). This finding suggests that phenotypic sex may be totally determined by the internal environment, and it argues against any role of the sex genes or chromosomes in the differentiation of sex, outside their role in directing gonadal differentiation.

ARE PATTERNS OF PSYCHOSEXUAL DIFFERENTIATION CORRELATED WITH MORPHOLOGICAL DIFFERENTIATION OR WITH SEX CHROMOSOME TYPE IN THE DIFFERENT VERTEBRATE CLASSES?

Comparing behavioral differentiation with two other dimensions of vertebrate sexuality—development of sex structures and sex-determining mechanisms—turns out to be quite illuminating. In this section, I briefly summarize these relationships and try to indicate their significance. For a more extensive discussion from this perspective, see Adkins-Regan (1981). For detailed reviews of morphological differentiation, see Armstrong and Marshall (1964), Burns (1961), Jost (1979), Price *et al.* (1975), and Van Tienhoven (1968). As the gonaducts differentiate according to different rules from other sex structures (see Adkins-Regan, 1981), they are not included in this discussion. For reviews of chromosomal sex-determining mechanisms, see Atz (1964), Beatty (1964, 1970), Dodd (1960), Gorman (1973), Morescalchi (1973), Ohno (1967), and Yamamoto (1969).

BIRDS. If pairs of chicken or duck gonads are cultured together, the ovary dominates the testis; that is, the testis is feminized, even if both gonads are undifferentiated when first cultured (Jost, 1960; Wolff, 1959). Exogenous estrogens feminize the embryonic testes of quail, chickens, turkeys *(Meleagris gallopavo),* and gulls *(Larus argentatus)* (Taber, 1964), but such feminization is seldom permanent. Androgens, in contrast, have little or no effect on ovarian development (Taber, 1964).

Duck embryos castrated prior to sexual differentiation develop a male-typical genital tubercle and syrinx, regardless of genetic sex (Wolff, 1959), and birds spontaneously lacking gonads have masculine secondary sex characters (Taber, 1964). The differentiation of explanted secondary sex characters *in vitro* also follows a masculine course, unless estrogens are added to the culture medium, in which case the differentiation is feminine (Van Tienhoven, 1968). Injection of estrogens into chicken and duck eggs feminizes the secondary sex characters, whereas androgens have little or no effect or feminize them (Burns, 1961; Romanoff, 1960; Taber, 1964). Thus, the male is the neutral form for morphological differentiation in these birds.

In some animals, the female has two homologous sex chromosomes (XX) whereas the male has a heterologous pair (XY). In others, the female has the heterologous pair. The sex possessing the homologous pair is the homogametic sex. In all birds studied so far (at least 86 species), the male is the homogametic sex and the female is the heterogametic sex (Rayt-Chaudhuri, 1973).

REPTILES. The most pertinent data come from experiments with lizards. Early castration of *Lacerta vivipara* embryos results in retention of the masculine genital tubercles by females (Dufaure, 1966), and thus, the male is the neutral form for this secondary sex character in this species. Exogenous androgens weakly masculinize the secondary sex characters of female embryos, but estrogens extensively feminize the cloaca and the outer genital region of male embryos (Dantschakoff, 1938; Dufaure, 1966; Forbes, 1964). In *Lacerta,* as in many other lizards, the male is the homogametic sex. Other species of lizards, including *Anolis carolinensis,* have female homogamety, but the effect of sex steroids on morphological differentiation in these species is not known.

Embryos of *Cnemidophorus uniparens,* an all-female parthenogenetic lizard species, can be extensively masculinized by treatment with testosterone (Crews *et al.,* 1982); the sex-determining mechanism of this species is unknown.

The few studies of sex differentiation in reptiles other than lizards indicate that early sex-steroid treatment feminizes more readily than it masculinizes (Forbes, 1964). Of these species, where the sex determination type has been determined, the male has been homogametic (Gorman, 1973; Engel *et al.,* 1981).

AMPHIBIANS. In experiments employing gonad grafts or parabiosed pairs, the testis dominates the ovary in both anuran (frog–toad) and urodele (salamander–

newt) amphibians (Foote, 1964; Gallien, 1967). Exogenous androgens strongly masculinize the ovaries of larval females of the higher anurans (Ranidae and Hylidae). In urodeles and primitive anurans, the opposite kind of effect is seen: estrogens feminize the larval testes (Foote, 1964; Gallien, 1955).

Gonadectomy prior to differentiation in *Xenopus* or *Triturus* results in undifferentiated (femalelike) cloacal glands and absence of the nuptial pads, a male secondary sex character (Burns, 1961; Witschi, 1971). Androgen masculinizes the cloacal glands and pads of both sexes.

As a class, amphibians, like reptiles, contain both types of sex-determining mechanisms. Urodeles and *Xenopus,* a primitive frog, have male homogamety, whereas the higher anurans have female homogamety.

TELEOST FISH. Gonads of a number of species are readily sex-reversed by exogenous steroids (Atz, 1964) and, in one species, *Oryzias latipes,* can be reversed in either direction (Yamamoto, 1969). Some fish do not seem to have a homogametic-heterogametic system of sex determination at all, but in those that do, the female is homogametic (Yamamoto, 1969).

OVERVIEW

Table 1 provides a somewhat oversimplified summary of the relationships among psychosexual differentiation, morphological differentiation, and sex-determining mechanisms in vertebrates. Some interesting generalizations emerge from this table.

One conclusion that could be drawn is that behavioral and morphological differentiation occur in a parallel manner: they seem to follow the same rules. It is as if in embryonic or fetal sex differentiation the brain is just another secondary-sex target organ. Thus, in mammals, the female is the neutral sex, and steroids masculinize females, morphologically and behaviorally, whereas in those birds for which both kinds of differentiation have been studied, the male is the neutral sex, and steroids feminize males, morphologically and behaviorally.

Another conclusion that could be drawn is that relationships exist among the three aspects of sexuality shown in Table 1. The correlation that emerges is that the homogametic sex is commonly also the neutral sex for morphological and behavioral differentiation, suggesting possible causality. The relationship between sex differentiation and sex determination has been pointed out previously with respect to sexual morphology (Burns, 1961; Foote, 1964; Gallien, 1965; Jost, 1960; Mittwoch, 1975; Van Tienhoven, 1968; Witschi, 1959). Table 1 suggests that it may apply to sexual behavior as well.

The most striking exception to this second conclusion is singing in the zebra finch, because the behavioral data so far imply that the female is the neutral sex. The pattern of morphological differentiation in this species is unknown. It is unlikely that zebra finches are an exception to the rule of avian male homogamety, because all of the other 23 members of the zebra finch order (Passeriformes) that have been examined have male homogamety (Ray-Chaudhuri, 1973). Furthermore, while song differentiates as if the female were the neutral sex, male-typical copu-

TABLE 1. VERTEBRATE SEXUALITY

Class	Genus	Homogametic sex	Morphological[a] neutral sex	Behavioral[a] neutral sex
Osteichthyes	*Carassius*	F	?[b]	?[b]
	Oryzias	F	?[b]	?[b]
	Poecilia	F	F	F
	Hemihaplochromis	F	M	M
	Tilapia	F	F	F
Amphibia	*Hynobius*	F?	M	M
	Ambystoma	M	M	M
	Pleurodeles	M	M	M
	Triturus	M?	F?[c]	M
	Xenopus	M	F?[c]	M
	Pseudacris	F	F	F
	Rana	F	F	F
Reptilia	*Emys*	M	M	?
	Testudo	M	M?	?
	Anolis	F	M?	?
	Sceloporus	F	M?	?
	Lacerta	M	M	?
	Thamnophis	M	?	F?
Aves	*Gallus*	M	M	M
	Coturnix	M	M?	M
	Anas	M	M	?
	Columba	M	?	M?
	Poephila	M?	?	M[d]
Mammalia[e]	*Didelphis*	F	F?	?
	Several	F	F	F

[a]Determined either directly, by examining development in the absence of gonads, or indirectly, by examining hormone-induced reversal. A question mark by itself means that no
information is available; a question mark after F or M means that sex can easily be
reversed, but it is not known whether the other sex could also easily be reversed.
[b]Both sexes are easily reversed.
[c]The different sources do not agree.
[d]For copulatory behavior (F for singing).
[e]See Burns (1961). *Didelphis* is the Virginia opossum.

latory behavior differentiates as if the male were the neutral sex (Adkins-Regan,
1984).

CONCLUSIONS

In all nonmammalian species for which behavior has been the target of investigation, sex hormones administered at some time in early life have had permanent
effects on sexual behavior that suggest that these hormones direct normally occurring psychosexual differentiation. These permanent effects appear to reflect neural
alterations or even organization, rather than peripheral changes. In other words,
early hormones do appear to produce sex differences in the brain. All indications
are that the brain differentiates according to the same rules as the secondary sex
characters, even though the rules (e.g., which sex is the neutral sex) vary among

species. Psychosexual differentiation is correlated with sex determination type so that the homogametic sex tends to be the neutral sex.

It is worth asking if the vertebrate data suggest any evolutionary "trends" of the sort that used to be fashionable in comparative studies of behavior. Clearly, different taxa have found it advantageous to settle on different modes of sex differentiation; and the so-called higher classes are more uniform within themselves with respect to sex differentiation and determination than are fish, amphibians, or reptiles. It is tempting to see a trend away from the great sexual liability in lower vertebrates toward the greater stability in birds and mammals. As Atz (1964) has pointed out, however, nonhermaphroditic species of fish are not so much unstable within species as variable across species. Although several freshwater species can easily be sex-reversed by experimental treatments, there is nothing unstable about the sex of these fish in a normal environment, and spontaneous intersexuality is rare in these species. Ease of experimental sex reversal may actually reflect the likelihood that steroids, rather than the degree of sexual stability, are the gonadal inducers. A more reasonable possibility is that, in lower vertebrates, the potential for permanent changes in sexual behavior is not confined to a critical period early in life, and that evolution has resulted in a progressive decrease in the length of the period for reversal.

Given that so many vertebrates have never been studied, and that progress in nonmammalian psychosexual differentiation is likely to be slow, it is important to consider what issues or species should have the highest priority. The following are some biased, but hopefully helpful, suggestions:

1. Hormonal organization of the songbird brain, in order to determine the details of the neuronal changes, the mechanisms by which steroids cause these changes, and how these changes translate into behavior.
2. Actual measurement of steroid levels during differentiation, particularly in altricial birds and in ectothermic vertebrates.
3. Psychosexual differentiation in birds with different mating systems, particularly polyandrous species with reversed sexual dimorphism (Jenni, 1974).
4. Psychosexual differentiation in reptiles with different types of sex determination (e.g., *Anolis carolinensis* with female homogamety vs. *Lacerta vivipara* with male homogamety).
5. Effects of heterologous hormones in adults of those species of amphibians and fish whose young have already been successfully sex-reversed.
6. Psychosexual differentiation in the platyfish *Xiphophorus maculatus,* which includes both female homogametic populations and male homogametic populations (Dodd, 1960).
7. Effects of early hormone treatments in spontaneously successive hermaphrodites.
8. Investigation of the possibility that nonsteroidal hormones might contribute to psychosexual differentiation.
9. Further theoretical work on the adaptiveness, if any, of different patterns of sex differentiation and determination. For example, why is the male the

homogametic and neutral sex in some species, whereas the female is homogametic and neutral in others?

In short, it is hoped that future research will apply the comparative approach
intelligently, taking advantage of its power to uncover the principles of and the
meanings behind vertebrate sexuality.

REFERENCES

Adkins, E. K. Functional castration of the female Japanese quail. *Physiology and Behavior*, 1973, *10*, 619.

Adkins, E. K. Electrical recording of copulation in quail. *Physiology and Behavior*, 1974, *13*, 475.

Adkins, E. K. Hormonal basis of sexual differentiation in the Japanese quail. *Journal of Comparative and Physiological Psychology*, 1975, *89*, 61.

Adkins, E. K. Embryonic exposure to an antiestrogen masculinizes behavior of female quail. *Physiology and Behavior*, 1976, *17*, 357.

Adkins, E. K. Effect of embryonic treatment with estradiol or testosterone on sexual differentiation of the quail brain: Critical period and dose-response relationships. *Neuroendocrinology*, 1979, *29*, 178–185.

Adkins, E. K., and Adler, N. T. Hormonal control of behavior in the Japanese quail. *Journal of Comparative and Physiological Psychology*, 1972, *81*, 27.

Adkins, E. K., and Nock, B. Behavioral responses to sex steroids of gonadectomized and sexually regressed quail. *Journal of Endocrinology*, 1976, *68*, 49.

Adkins, E. K., and Pniewski, E. E. Control of reproductive behavior by sex steroids in male quail. *Journal of Comparative and Physiological Psychology*, 1978, *92*, 1169–1178.

Adkins, E. K., and Schlesinger, L. Androgens and the social behavior of male and female lizards, *Anolis carolinensis*. *Hormones and Behavior*, 1979, *13*, 139–152.

Adkins-Regan, E. Early organizational effects of hormones: An evolutionary perspective. In N. T. Adler (Ed.), *Neuorendocrinology of Reproduction*. New York: Plenum Press, 1981.

Adkins-Regan, E. Sexual differentiation of behavior in the zebra finch. Paper presented at the Conference on Reproductive Behavior, Pittsburgh, June 1984.

Adkins-Regan, E. Exposure of embryos to an aromatization inhibitor increases copulatory behaviour of male quail. *Behavioural Processes*, in press.

Adkins-Regan, E., Pickett, P., and Koutnik, D. Sexual differentiation in quail: Conversion of androgen to estrogen mediates testosterone-induced demasculinization of copulation but not other male characteristics. *Hormones and Behavior*, 1982, *16*, 259–278.

Allee, W. C., and Collias, N. The influence of estradiol on the social organization of flocks of hens. *Endocrinology*, 1940, *27*, 87.

Armstrong, C. N., and Marshall, A. J. (Eds.). *Intersexuality in Vertebrates Including Man*. London: Academic Press, 1964.

Arnold, A. P. *Behavioral Effects of Androgen in Zebra Finches (Poephila guttata) and a Search for Its Sites of Action*. Ph.D. dissertation, Rockefeller University, 1974.

Arnold, A. P. The effects of castration and androgen replacement on song, courtship, and aggression in zebra finches *(Poephila guttata)*. *Journal of Experimental Zoology*, 1975a, *191*, 309–325.

Arnold, A. P. The effects of castration on song development in zebra finches *(Poephila guttata)*. *Journal of Experimental Zoology*, *191*, 1975b, 261–277.

Arnold, A. P., and Saltiel, A. Sexual difference in pattern of hormone accumulation in the brain of a songbird. *Science*, 1979, *205*, 702–704.

Aronson, L. R. Reproductive and parental behavior. In M. E. Brown (Ed.), *Physiology of Fishes*, Vol. 2. New York: Academic Press, 1957.

Atz, J. W. Intersexuality in fishes. In C. N. Armstrong and A. J. Marshall (Eds.), *Intersexuality in Vertebrates Including Man*. London: Academic Press, 1964.

Baggerman, B. Hormonal control of reproductive and parental behavior in fishes. In E. J. W. Barrington and C. B. Jorgensen (Eds.), *Perspectives in Endocrinology: Hormones in the Lives of Lower Vertebrates*. London: Academic Press, 1968.

Balthazart, J., and Schumacher, M. Testosterone metabolism and sexual differentiation in quail. In J. Balthazart, E. Pröve, and R. Gilles (Eds.), *Hormones and Behaviour in Higher Vertebrates.* Berlin: Springer-Verlag, 1983.

Barfield, R. J. Activation of copulatory behavior by androgen implanted into the preoptic area of the male fowl. *Hormones and Behavior,* 1969, *1,* 37.

Beach, F. A. Hormonal factors controlling the differentiation, development, and display of copulatory behavior in the ramstergig and related species. In E. Tobach, L. R. Aronson, and E. Shaw (Eds.), *The Biopsychology of Development.* New York: Academic Press, 1971.

Beach, F. A., and Inman, N. G. Effects of castration and androgen replacement on mating in male quail. *Proceedings of the National Academy of Sciences,* 1965, *54,* 1426.

Beatty, R. A. Chromosome deviations and sex in vertebrates. In C. N. Armstrong and A. J. Marshall (Eds.), *Intersexuality in Vertebrates Including Man.* London: Academic Press, 1964.

Beatty, R. A. Genetic basis for the determination of sex. *Philosophical Transactions of the Royal Society* (London) B, 1970, *259,* 3.

Becker, P., Roland, H., and Reinboth, R. An unusual approach to experimental sex inversion in the teleost fish, *Betta* and *Macropodus.* In R. Reinboth (Ed.), *Intersexuality in the Animal Kingdom.* New York: Springer-Verlag, 1975.

Benoff, F. H. Testosterone-induced precocious sexual behavior in chickens differing in adult mating frequency. *Behavioral Processes,* 1979, *4,* 35.

Bull, J. J., and Vogt, R. C. Temperature-sensitive periods of sex determination in emydid turtles. *Journal of Experimental Zoology,* 1981, *218,* 435.

Burns, R. K. Role of hormones in the differentiation of sex. In W. C. Young (Ed.), *Sex and Internal Secretions.* Baltimore: Williams & Wilkins, 1961.

Chan, S. T. H. Natural sex reversal in vertebrates. *Philosophical Transactions of the Royal Society* (London) B, 1970, *259,* 59.

Chan, S. T. H., O, W.-S., and Hui, S. W. B. The gonadal and adenohypophysial functions of natural sex reversal. In R. Reinboth (Ed.), *Intersexuality in the Animal Kingdom.* New York: Springer-Verlag, 1975.

Charnov, E. L., and Bull, J. When is sex environmentally determined? *Nature,* 1977, *266,* 828.

Choat, J. H., and Robertson, D. R. Protogynous hermaphroditism in fishes of the family Scaridae. In R. Reinboth (Ed.), *Intersexuality in the Animal Kingdom.* New York: Springer-Verlag, 1975.

Clemens, H. P., and Inslee, T. The production of unisexual broods by *Tilapia mossambica* sex-reversed with methyl testosterone. *Transactions of the American Fish Society,* 1968, *97,* 18.

Clemens, H. P., McDermitt, C., and Inslee, T. The effect of feeding methyl testosterone to guppies for sixty days after birth. *Copeia,* 1966, 280.

Clemens, L. G. Neurohormonal control of male sexual behavior. In W. Montagna and W. A. Sadler (Eds.), *Reproductive Behavior.* New York: Plenum Press, 1974.

Crawford, W. C., and Glick, B. The function of the preoptic, mammilaris lateralis and ruber nuclei in normal and sexually inactive male chickens. *Physiology and Behavior,* 1975, *15,* 171.

Crews, D. Endocrine control of reptilian reproductive behavior. In C. Beyer (Ed.), *Endocrine Control of Sexual Behavior.* New York: Raven, 1979.

Crews, D. Control of male sexual behaviour in the Canadian red-sided garter snake. In J. Balthazart, E. Pröve, and R. Gilles (Eds.), *Hormones and Behaviour in Higher Vertebrates.* Berlin: Springer-Verlag, 1983.

Crews, D., Gustafson, J. E., and Tokarz, R. Psychobiology of parthenogenesis in reptiles. In R. Huey, E. Pianka, and T. Schoener (Eds.), *Lizard Ecology.* Cambridge: Harvard University Press, 1982.

Dantschakoff, V. Über chemische Werkzeuge bei der Realisation normal bestimmter embryonaler geschlechtlicher Histogenese bei Reptilien. *Archiv für Entwicklungsmechanik der Organismen,* 1938, *138,* 465.

Davies, S. J. J. F. The timing of breeding by the zebra finch at Mileura, Western Australia. *Ibis,* 1978, *119,* 369–372.

Davis, D. E., and Domm, L. V. The influence of hormones on the sexual behavbior of domestic fowl. In *Essays in Biology.* Berkeley: University of California Press, 1943.

Diakow, C., and Nemiroff, A. Vasotocin, prostaglandin, and female reproductive behavior in the frog, *Rana pipiens. Hormones and Behavior,* 1981, *15,* 86.

Dodd, J. M. Genetic and environmental aspects of sex determination in cold-blooded vertebrates. *Memoirs of the Society for Endocrinology,* 1960, 7, 17.

Domm, L. V. Intersexuality in adult Brown Leghorn males as a result of estrogenic treatment during early embryonic life. *Proceedings of the Society for Experimental Biology and Medicine*, 1939, *42*, 310.

Domm, L. V., and Davis, D. E. The effect of sex hormones on sexual behavior of domestic fowl. *Anatomical Record*, 1942, *82*, 493.

Dufaure, J.-P. Recherches descriptives et expérimentales sur les modalités et facteurs du développement de l'appareil génital chez le lézard vivipare (*Lacerta vivipara* Jacquin). *Archives d'Anatomie Microscopique et de Morphologie Experimentale*, 1966, *55*, 437.

Dzwillo, M. Über künstliche Erzeugung funktioneller männlicher und weiblicher Genotyps bei *Lebistes reticulatus*. *Biologisches Zentralblatt*, 1962, *81*, 575.

Engel, W., Klemme, B., and Schmid, M. H-Y antigen and sex determination in turtles. *Differentiation*, 1981, *20*, 152.

Fineman, R., Hamilton, J., Chase, G., and Bolling, D. Length, weight and secondary sex character development in male and female phenotypes in three sex chromosomal genotypes (XX, XY, YY) in the killifish, *Oryzias latipes*, *Journal of Experimental Zoology*, 1974, *189*, 227.

Fischer, E. A. The relationship between mating system and simultaneous hermaphroditism in the coral reef fish, *Hypoplectrus nigricans* (Serranidae), *Animal Behaviour*, 1980, *28*, 620.

Fishelson, L. Ecology and physiology of sex reversal in *Anthias squamipinnis* (Peters), (Teleostei: Anthiidae). In R. Reinboth (Ed.), *Intersexuality in the animal kingdom*. New York: Springer-Verlag, 1975.

Foote, C. L. Intersexuality in amphibians. In C. N. Armstrong and A. J. Marshall (Eds.), *Intersexuality in Vertebrates Including Man*. London: Academic Press, 1964.

Forbes, T. R. Intersexuality in reptiles. In C. N. Armstrong and A. J. Marshall (Eds.), *Intersexuality in Vertebrates Including Man*. London: Academic Press, 1964.

Fricke, H., and Fricke, S. Monogamy and sex change by aggressive dominance in a coral reef fish. *Nature*, 1977, *266*, 830.

Gallien, L. G. The action of sex hormones on the development of sex in amphibia. *Memoirs of the Society for Endocrinology*, 1955, *4*, 188.

Gallien, L. G. Genetic control of sexual differentiation in vertebrates. In R. L. DeHaan and H. Ursprung (Eds.), *Organogenesis*. New York: Holt, Rinehart and Winston, 1965.

Gallien, L. G. Developments in sexual organogenesis. In M. Abercrombie and J. Brachet (Eds.), *Advances in Morphogenesis*, Vol. 6. New York: Academic Press, 1967.

Gasc, J.-M., and Thibier, M. Plasma testosterone concentration in control and testosterone-treated chick embryos. *Experientia*, 1979, *35*, 1411–1412.

Ghiselin, M. T. *The Economy of Nature and the Evolution of Sex*. Berkeley: University of California Press, 1974.

Glick, B. The reproductive performance of birds hatched from eggs dipped in male hormone solutions. *Poultry Science*, 1961, *40*, 1408.

Glick, B. Embryonic exposure to testosterone propionate will adversely influence future mating behavior in male chickens, *Federation Proceedings*, 1965, *24*, 700.

Goodale, H. D. Castration in relation to the secondary sexual characters in Brown Leghorns. *American Naturalist*, 1913, *34*, 127.

Goodale, H. D. Feminized male birds. *Genetics*, 1918, *3*, 276.

Gorman, G. C. The chromosomes of the reptilia: A cytotaxonomic interpretation. In A. B. Chiarelli and E. Capanna (Eds.), *Cytotaxonomy and Vertebrate Evolution*. London: Academic Press, 1973.

Goy, R. W., and McEwen, B. S. *Sexual Differentiation of the Brain*. Cambridge, Mass.: M.I.T. Press, 1980.

Guhl, A. M. Heterosexual dominance and mating behavior in chickens. *Behaviour*, 1949, *2*, 106.

Guhl, A. M., and Fischer, C. L. The behavior of chickens. In E. S. E. Hafez (Ed.), *The Behavior of Domestic Animals*. Baltimore: Williams & Wilkins, 1969.

Gurney, M. E. Hormonal control of cell form and number in the zebra finch song system. *Journal of Neuroscience*, 1981, *1*, 658–673.

Gurney, M. E. Behavioral correlates of sexual differentiation in the zebra finch song system. *Brain Research*, 1982, *231*, 153.

Gurney, M. E., and Konishi, M. Hormone-induced sexual differentiation of brain and behavior in zebra finches. *Science*, 1980, *208*, 1380–1383.

Hackman, E., and Reinboth, R. Delimitation of the critical stage of hormone-influenced sex differentiation in *Hemihaplochromis multicolor* (Hilgendorf) (Cichlidae). *General and Comparative Endocrinology*, 1974, *22*, 42.

Harding, C. F. Hormonal specificity and activation of social behaviour in the male zebra finch. In J. Balthazart, E. Pröve, and R. Gilles (Eds.), *Hormones and Behaviour in Higher Vertebrates*. Berlin: Springer-Verlag, 1983.

Harrington, R. W. How ecological and genetic factors interact to determine when self-fertilizing hermaphrodites of *Rivulus marmoratus* change into functional secondary males, with a reappraisal of the modes of intersexuality among fishes. *Copeia*, 1971, 389.

Haynes, R. L., and Glick, B. Hypothalamic control of sexual behavior in the chicken. *Poultry Science*, 1974, *53*, 27.

Hoar, W. S. Reproductive behavior of fish. *General and Comparative Endocrinology*, 1962, Suppl. 1, 206.

Hutchison, R. E. Hormonal differentiation of sexual behavior in Japanese quail. *Hormones and Behavior*, 1978, *11*, 363–387.

Idler, D. R., Reinboth, R., Walsh, J. M., and Truscott, B. A comparison of 11-hydroxytestosterone and 11-ketotestosterone in blood of ambisexual and gonochoristic teleosts. *General and Comparative Endocrinology*, 1976, *30*, 517.

Jenni, D. A. Evolution of polyandry in birds. *American Zoology*, 1974, *14*, 129.

Jost, A. Hormonal influences in the sex development of bird and mammalian embryos. *Memoirs of the Society for Endocrinology*, 1960, *7*, 49.

Jost, A. Basic sexual trends in the development of vertebrates. In *Sex, Hormones and Behavior*. (Ciba Foundation Symposium No. 62) Amsterdam: Excerpta Medica, 1979.

Kaufman, L. Experiments on sex modification in cocks during their embryonal development. *World's Poultry Science Journal*, 1956, *12*, 41.

Kelley, D. B., and Pfaff, D. W. Hormone effects on male sex behavior in adult South African clawed frogs, *Xenopus laevis*. *Hormones and Behavior*, 1976, *7*, 159.

Kilgore, L., and Glick, B. Testosterone's influence on brain enzymes in the developing chick. *Poultry Science*, 1970, *49*, 16.

Konishi, M., and Gurney, M. E. Sexual differentiation of brain and behaviour. *Trends in Neurosciences*, 1982, *5*, 20–23.

Kramer, D. L. The role of androgens in the parental behavior of the blue gourami, *Trichogaster trichopterus* (Pisces, Belontiidae). *Animal Behaviour*, 1972, *20*, 798.

Liem, K. F. Sex reversal as a natural process in the synbranchiform fish *Monopterus albus*. *Copeia*, 1963, 303.

Liem, K. F. Geographical and taxonomic variation in the pattern of natural sex reversal in the teleost fish order Synbranchiformes. *Journal of Zoology* London, 1968, *156*, 225.

Liley, N. R. Hormones and reproductive behavior in fishes. In W. S. Hoar and D. J. Randall (Eds.), *Fish Physiology*. New York: Academic Press, 1969.

Liley, N. R. Effects of estrogens and other steroids on the sexual behavior of the female guppy, *Poecilia reticulata*. *General and Comparative Endocrinology*, 1972, Suppl. 3, 542.

Lisk, R. D. Inhibitory centers in sexual behavior in the male rat. *Science*, 1966, *152*, 669.

Lowe, T. P., and Larkin, J. R. Sex reversal in *Betta splendens* Regan with emphasis on the problem of sex determination. *Journal of Experimental Zoology*, 1975, *191*, 25.

Martinez-Vargas, M. C., Gibson, D. B., Sar, M., and Stumpf, W. E. Estrogen target sites in the brain of the chick embryo. *Science*, 1975, *190*, 1307.

Mashaly, M. M., and Glick, B. Comparison of androgen levels in normal males *(Gallus domesticus)* and in males made sexually inactive by embryonic exposure to testosterone propionate. *General and Comparative Endocrinology*, 1979, *38*, 105–110.

Mason, P., and Adkins, E. K. Hormones and social behavior in the lizard, *Anolis carolinensis*. *Hormones and Behavior*, 1976, *7*, 75.

Mauldin, J. M., Wolfe, J. L., and Glick, B. The behavior of chickens following embryonic treatment with testosterone propionate. *Poultry Science*, 1975, *54*, 2133.

Maynard Smith, J. *The Evolution of Sex*. Cambridge: Cambridge University Press, 1978.

McDonald, P., Beyer, C., Newton, F., Brien, B., Baker, R., Tan, H. S., Sampsom, C., Kitching, P., Greenhill, R., and Pritchard, D. Failure of 5α-dihydrotestosterone to initiate sexual behavior in the castrated male rat. *Nature*, 1970, *227*, 964.

McEwen, B. S., and Pfaff, D. W. Chemical and physiological approaches to neuroendocrine mechanisms: Attempts at integration. In W. F. Ganong and L. Martini (Eds.), *Frontiers in Neuroendocrinology*. New York: Oxford University Press, 1973.

Mittwoch, U. Chromosomes and sex differentiation. In R. Reinboth (Ed.), *Intersexuality in the Animal Kingdom*. New York: Springer-Verlag, 1975.

Moore, C. L. Maternal behavior of rats is affected by hormonal condition of pups. *Journal of Comparative and Physiological Psychology*, 1982, *96*, 123.

Moore, F. L. Differential effects of testosterone plus dihydrotestosterone on male courtship of castrated newts, *Taricha granulosa. Hormones and Behavior*, 1978, *11*, 202.

Moore, F. L., and Zoeller, R. T. Endocrine control of amphibian sexual behavior: Evidence for a neurohormone-androgen interaction. *Hormones and Behavior*, 1979, *13*, 207.

Morescalchi, A. Amphibia. In A. B. Chiarelli and E. Capanna (Eds.), *Cytotaxonomy and Vertebrate Evolution*. London: Academic Press, 1973.

Morris, J. B. *The Effect of Early Steroid Hormone Administration on the Adult Sexual Behavior of the Zebra Finch, Poephila guttata*. Master's thesis, Cornell University, 1980.

Naftolin, F., Ryan, K. J., and Petro, Z. Aromatization of androstenedione by the anterior hypothalamus of adult male and female rats. *Endocrinology*, 1972, *90*, 295.

Noble, G. K., and Kumpf, K. F. R. Sex reversal in the fighting fish, *Betta splendens. Anatomical Record*, 1937, *70*, 97.

Nottebohm, F., and Arnold, A. P. Sexual dimorphism in vocal control areas of the songbird brain. *Science*, 1976, *194*, 211.

Ohno, S. *Sex Chromosomes and Sex-Linked Genes*. Berlin: Springer-Verlag, 1967.

Orcutt, F. S. Effects of oestrogen on the differentiation of some reproductive behaviours in male pigeons *(Columba livia). Animal Behaviour*, 1971, *19*, 277.

Ottinger, M. A., and Bakst, M. R. Peripheral androgen concentrations and testicular morphology in embryonic and young male Japanese quail. *General and Comparative Endocrinology*, 1981, *43*, 170–177.

Ozon, R. Mise en évidence d'hormones stéröides oestrogènes dans le sang de la poule adulte et chez l'embryon de poulet. *Comptes Rendus de l'Académie des Sciences* (Paris), 1965, *261*, 5664.

Palka, Y., and Gorbman, A. Pituitary and testicular influenced sexual behavior in male frogs, *Rana pipiens. General and Comparative Endocrinology*, 1973, *21*, 148.

Pieau, C. Temperature and sex differentiation in embryos of two chelonians, *Emys orbicularis* L. and *Testudo graeca* L. In R. Reinboth (Ed.), *Intersexuality in the Animal Kingdom*. New York: Springer-Verlag, 1975.

Pincus, G., and Hopkins, T. F. The effects of various estrogens and steroid substances on sex differentiation in the fowl. *Endocrinology*, 1958, *62*, 112.

Pohl-Apel, G., and Sossinka, R. Der Einfluss früher Hormongaben auf den Entwicklungsverlauf und die Ausbildung primärer und sekundärer Geschlechtsmerkmale einschliesslich Sexualverhalten beim Zebrafinken, *Taeniopygia guttata castanotis* (Estrildidae): III. Wirkung von prä- und postnatalen Östrogengaben. *Verhandlungen der Deutschen Zoologischen Gesellschaft*, 1982a, *75*, 326 (abstract).

Pohl-Apel, G., and Sossinka, R. Männchen-typischer Gesang bei weiblichen Zebrafinken *(Taeniopygia guttata castanotis). Journal für Ornithologie*, 1982b, *123*, 211.

Pressley, P. H. Pair formation and joint territoriality in a simultaneous hermaphrodite: The coral reef fish *Serranus tigrinus. Zeitschrift für Tierpsychologie*, 1981, *56*, 33.

Price, D., Zaaijer, J. J. P., Ortiz, E., and Brinkmann, A. C. Current views on embryonic sex differentiation in reptiles, birds and mammals. In E. J. W. Barrington (Ed.), *Trends in Comparative Endocrinology*. American Zoological Suppl., 1975.

Pröve, E. Der Einfluss von Kastration und Testosteronsubstitution auf das Sexualverhalten männlicher Zebrafinken *(Taeniopygia guttata castanotis* Gould). *Journal of Ornithology*, 1974, *115*, 338–347.

Pröve, E. Hormonal correlates of behavioural development in male zebra finches. In J. Balthazart, E. Pröve and R. Gilles (Eds.), *Hormones and Behaviour in Higher Vertebrates*. Berlin: Springer-Verlag, 1983.

Ray-Chaudhuri, R. Cytotaxonomy and chromosome evolution in birds. In A. B. Chiarelli and E. Capanna (Eds.), *Cytotaxonomy and Vertebrate Evolution*. London: Academic Press, 1973.

Reinboth, R. Intersexuality in fishes. *Memoirs of the Society of Endocrinology*, 1970, *18*, 515.

Reinboth, R. Some remarks on secondary sex characters, sex and sexual behavior in teleosts. *General and Comparative Endocrinology*, 1972, Suppl. 3, 565.

Rissman, E. F., Ascenzi, M., Johnson, P., and Adkins-Regan, E. Effect of embryonic treatment with oestradiol benzoate on reproductive morphology, ovulation and oviposition and plasma LH concentrations in female quail *(Coturnix coturnix japonica). Journal of Reproduction and Fertility,* 1984, *71,* 211–217.

Robertson, D.R. Social control of sex reversal in a coral-reef fish. *Science,* 1972, *177,* 1007.

Romanoff, A. L. *The Avian Embryo.* New York: Macmillan, 1960.

Sachs, B. D. Photoperiodic control of reproductive behavior and physiology of the male Japanese quail *(Coturnix coturnix japonica). Hormones and Behavior,* 1969, *1,* 7.

Schmidt, R. S. Masculinization of toad pretrigeminal nucleus by androgens. *Brain Research,* 1982, *244,* 190.

Schumacher, M., and Balthazart, J. The postnatal demasculinization of sexual behavior in the Japanese quail *(Coturnix coturnix japonica). Hormones and Behavior,* 1984, *18,* 298.

Shapiro, D. Y. Serial female sex changes after simultaneous removal of males from social groups of a coral reef fish. *Science,* 1980, *209,* 1136.

Smith, C. L. The evolution of hermaphroditism in fishes. In R. Reinboth (Ed.), *Intersexuality in the Animal Kingdom.* New York: Springer-Verlag, 1975.

Sossinka, R. Ovarian development in an opportunistic breeder, the zebra finch *Poephila guttata castanotis. Journal of Experimental Zoology,* 1980, *211,* 225–230.

Sossinka, R., and Pohl-Apel, G. Der Einfluss früher Hormongaben auf den Entwicklungsverlauf und die Ausbildung primärer and sekundärer Geschlechtsmerkmale einschliesslich Sexualverhalten beim Zebrafinken, *Taeniopygia guttata castanotis* (Estrildidae): II. Wirkung von postnatalen Testosteron- und Antiandrogengaben. *Verhandlungen der Deutschen Zoologischen Gesellschaft,* 1982, *75,* 333 (abstract).

Sossinka, R., Pohl-Apel, G., and Hall, M. R. Der Einfluss früher Hormongaben auf den Entwicklungsverlauf und die Ausbildung primärer and sekundärer Geschlechtsmerkmale einschliesslich Sexualverhalten beim Zebrafinken, *Taeniopygia guttata castanotis* (Estrildidae): I. Wirkung von pränatalen Testosterongaben. *Verhandlungen der Deutschen Zoologischen Gesellschaft,* 1981, *202,* 202.

Taber, E. Intersexuality in birds. In C. N. Armstrong and A. J. Marshall (Eds.), *Intersexuality in Vertebrates Including Man.* London: Academic Press, 1964.

Tanabe, Y., Nakamura, T., Fujioka, K., and Doi, O. Production and secretion of sex steroid hormones by the testes, the ovary, and the adrenal glands of embryonic and young chickens *(Gallus domesticus). General and Comparative Endocrinology,* 1979, *39,* 26–33.

Urbanski, H. F., and Follett, B. K. Sexual differentiation of the photoperiodic response in Japanese quail. *Journal of Endocrinology,* 1982, *92,* 279.

Van Tienhoven, A. *Reproductive Physiology of Vertebrates.* Philadelphia: Saunders, 1968.

Wada, M. Effects of sex steroids on calling, locomotor activity, and sexual behavior in castrated male Japanese quail. *Hormones and Behavior,* 1982, *16,* 147.

Wai, E. H., and Hoar, W. S. The secondary sex characters and reproductive behavior of gonadectomized sticklebacks treated with methyl testosterone. *Canadian Journal of Zoology,* 1963, *41,* 611–628.

Warner, R. L., Cain, J. R., Moreng, G. R., and Maniscalco, V. J. Reproductive organs and behavior of Japanese quail following treatment of embryos with steroids and antisteroids. *Anatomical Record,* 1977, *187,* 742.

Warner, R.R. The adaptive significance of sequential hermaphroditism in animals. *American Naturalist,* 1975, *109,* 61.

Warner, R. R. The evolution of hermaphroditism and unisexuality in aquatic and terrestrial vertebrates. In E. S. Reese and F. J. Lighter (Eds.), *Contrasts in Behavior.* New York: Wiley, 1978.

Warner, R. R., Robertson, D. R., and Leigh, E. G. Sex change and sexual selection. *Science,* 1975, *190,* 633.

Wentworth, B. C., Hendricks, B. G., and Sturtevant, J. Sterility induced in Japanese quail by spray treatment of eggs with mestranol. *Journal of Wildlife Management,* 1968, *32,* 879.

Whitsett, J. M., Irvin, E. W., Edens, F. W., and Thaxton, J. P. Demasculinization of male Japanese quail by prenatal estrogen treatment. *Hormones and Behavior,* 1977, *8,* 254–260.

Williams, G. C. *Sex and Evolution.* Princeton, N.J.: Princeton University Press, 1975.

Wilson, J. A., and Glick, B. Ontogeny of mating behavior in the chicken. *American Journal of Physiology,* 1970, *218,* 951.

Wilson, M. I., and Bermant, G. An analysis of social interactions in Japanese quail, *Coturnix coturnix japonica. Animal Behaviour,* 1972, *20,* 252.

Witschi, E. Age of sex-determining mechanisms in vertebrates. *Science,* 1959, *130,* 372.

ELIZABETH ADKINS-
REGAN

Witschi, E. Mechanisms of sexual differentiation. In M. Hamburgh and E. J. W. Barrington (Eds.), *Hormones in Development*. New York: Appleton-Century-Crofts, 1971.

Wolff, E. Endocrine function of the gonad in developing vertebrates. In A. Gorman (Ed.), *Comparative Endocrinology*. New York: Wiley, 1959.

Woods, J. E., and Brazzill, D. M. Plasma 17β-estradiol levels in the chick embryo. *General and Comparative Endocrinology*, 1981, *44*, 37–43.

Woods, J. E., Simpson, R. M., and Moore, P. L. Plasma testosterone levels in the chick embryo. *General and Comparative Endocrinology*, 1975, *27*, 543–547.

Yamamoto, T.-O. Sex differentiation. In W. S. Hoar and D. J. Randall (Eds.), *Fish Physiology*. New York: Academic Press, 1969.

Yamamoto, T.-O., and Kajishima, T. Sex hormone induction of sex reversal in the goldfish and evidence for male heterogamety. *Journal of Experimental Zoology*, 1968, *146*, 163.

Yntema, C. L. Temperature levels and periods of sex determination during incubation of eggs of *Chelydra serpentina*. *Journal of Morphology*, 1979, *159*, 17–28.

Young, W. C., Goy, R. W., and Phoenix, C. H. Hormones and sex behavior. *Science*, 1964, *143*, 212–218.

Sexual Behavior Differentiation
Effects of Prenatal Manipulations in Rats

Ingeborg L. Ward and O. Byron Ward

Introduction

The first concrete demonstration that fetal testicular hormones have permanent influences on reproductive behavior potentials of mammals was provided by a report published in 1959 by Phoenix, Goy, Gerall, and Young. Inspired by several isolated studies appearing in the late 1930s (Dantchakoff, 1938a,b,c), Phoenix *et al.* showed that the administration of testosterone to pregnant guinea pigs resulted in female offspring that displayed high levels of malelike mounting behaviors when tested in adulthood. Moreover, these females were severely impaired in their ability to perform the typical female sexual pattern. This seminal paper founded an entirely new research area within the field of behavioral endocrinology. Before that time, considerable information was available on the mechanism by which various components of sexual morphology emerged, but virtually nothing was known about the differentiation of behavior into male and female patterns.

Since the initial demonstration by Phoenix *et al.* (1959), literally hundreds of studies replicating and extending these observations have been published. These studies have addressed an infinite variety of questions, including what behaviors are affected by the exposure of developing organisms to testicular steroids; which are the critical perinatal stages in different animal species during which steroids

Ingeborg L. Ward and O. Byron Ward   Department of Psychology, Villanova University, Villanova, Pennsylvania 19085. Supported by Grant HD-04688 from the National Institute of Child Health and Human Development and by Research Scientist Development Award II 1-K2-MH00049 from the National Institute of Mental Health.

exert their organizing action; what metabolites of testosterone are utilized by different target tissues destined to mediate behavior; and precisely what peripheral, central, cellular, and subcellular fractions of the nervous system are involved in the process of behavior differentiation.

Considerable progress has been made in the past 23 years, and an all-inclusive coverage of more than a small part of this literature would require volumes. The laboratory rat has been the species of choice for the majority of these studies, and this review focuses primarily on factors that influence the hormonal milieu during prenatal ontogeny in this species. Fortunately, several excellent reviews have recently been published that include coverage of topics not emphasized by the present chapter. The reader is referred to Baum (1979) for a more comparative emphasis; to Beatty (1979) for coverage of nonreproductive sexually dimorphic behaviors; to Gorski (1979, 1982) for a discussion of perinatal hormonal influences on sex differences in central nervous system structures; to McEwen (1982), who included material on cellular and intracellular mechanisms; to Goy and Goldfoot (1975) for coverage of material on nonhuman primate sexual differentiation; and to Dörner (1980) and Goy and McEwen (1980) for more general reviews.

SEXUAL DIFFERENTIATION THEORY

The present formulation of the sexual differentiation theory is only slightly different from that first presented by Phoenix *et al.* (1959). It appears that complete masculinization of a variety of sexually dimorphic behaviors shown by infant, juvenile and adult organisms requires exposure to species-specific patterns of testicular steroids at very precise stages of perinatal development. If androgens are not present in sufficient amounts at the appropriate points in time, the organism is behaviorally feminized, regardless of genetic or gonadal sex. This organizing action of androgens during perinatal life is distinct from the activating function exerted by gonadal hormones from the time of puberty onward. Although some sexually dimorphic behaviors do not require further hormonal stimulation—for example, urination postures of dogs (Beach, 1975) and play behavior patterns in juvenile primates and rodents (Goy, 1970, 1978; Olioff and Stewart, 1978; Meaney and Stewart, 1981; Beatty *et al.*, 1981)—most are exhibited only after the gonads at puberty introduce sufficient hormone into the circulation to activate the potentials established during perinatal life. As appears to be the case for reproductive morphology, it is also true that establishment or suppression of sexual behavior potentials cannot be altered once the critical organizational period has passed.

AROMATIZATION HYPOTHESIS

One major unresolved question is whether testosterone is the actual steroid that exerts the organizing action on neural target tissue. In the rat, estrogen, a aromatized metabolite of testosterone, has been implicated in one or more of the steps involved in the normal process of sexual behavior differentiation. The so-called aromatization hypothesis was first suggested by studies that found that injection of estradiol into neonatally castrated males or into newborn female rats masculinizes and defeminizes their sexual behavior potentials (Wilson, 1943; Koster, 1943; Whalen and Nadler, 1963; Levine and Mullins, 1964; Feder and Whalen, 1965; Whalen and Edwards, 1967; Gerall, 1967; Mullins and Levine, 1968; Hendricks, 1969; Edwards and Thompson, 1970; Gerall *et al.*, 1972; Hendricks and Weltin, 1976). Furthermore, the masculinization and defeminization of behavior that neonatal androgen treatment typically produces is attenuated if estrogen receptor blockers or drugs that prevent the aromatization process are given in conjunction with the testosterone (McDonald and Doughty, 1972, 1973; McEwen *et al.*, 1977; Vreeburg *et al.*, 1977; Södersten, 1978; Booth, 1978; Clemens and Gladue, 1978; Davis *et al.*, 1979; McEwen *et al.*, 1979; Gladue and Clemens, 1980; Fadem and Barfield, 1981). As the fetal and neonatal rat brain is known to contain aromatizing enzymes (Reddy *et al.*, 1974; Weisz *et al.*, 1982; George and Ojeda, 1982), it seems reasonable to postulate that some target cells take up circulating testosterone and convert it to estradiol intracellularly. It is the interaction of estradiol with critical intracellular events that proponents of the aromatization hypothesis feel underlies the process of sexual behavior differentiation. The extent to which this scenario is true or is generally applicable to more than a few isolated rodent species remains to be established.

CRITICAL PERIOD: POSTNATAL OR PRE-PLUS POSTNATAL

The aromatization hypothesis is only one of several controversies that have arisen over the years regarding the extent to which reproductive physiology and behavior are altered by the presence of androgenic steroids during perinatal life. One obstacle in the way of the resolution of these controversies is the widely accepted notion that the critical process occurs primarily or exclusively during neonatal life in the most widely studied laboratory species, the rat. As the literature reviewed below demonstrates, this notion is a misconception that may have retarded progress in teasing apart various interacting factors underlying sexual behavior differentiation.

The working hypothesis has been that the gonads of normal genetic males secrete androgenic steroids into the circulation at particular points during perinatal life. These hormones are taken up by appropriate target tissue, setting into motion as-yet-unknown processes, the end result of which is the masculinization and defeminization of adult behavior potentials. The gonads of fetal and neonatal females appear to be largely nonsecretory (Noumura *et al.*, 1966; Schlegel *et al.*, 1967; Quattropani and Weisz, 1973; Warren *et al.*, 1973, 1975). Thus, the same target tissues in normal genetic females develop in a very different hormonal milieu from that in males, a milieu characterized by relatively low levels of androgen compared to the levels in males. Two basic experimental strategies have been utilized to test this theory. In the first, females are exposed to androgen during what are believed to be the critical developmental stages. In the second, normal endogenous androgen levels of males are altered through chemical, surgical, or environmental interventions.

POSTNATAL EFFECTS

In research utilizing the rat, it quickly became apparent that very dramatic and profound alterations were obtainable if hormonal manipulations were made neonatally. Single injections of the synthetic androgen testosterone propionate (TP) into female rats any time within the first 5 days after birth permanently blocked the cyclic release of gonadotrophic hormones by the hypothalamic pituitary axis. Furthermore, such females did not display the characteristic estrous behavior pattern (Figure 1); instead, they showed higher levels of the male intromission pattern than did untreated control females (Barraclough, 1959, 1961; Barraclough and Gorski, 1961, 1962; Goy *et al.*, 1962; Gorski and Barraclough, 1961, 1963; Harris and Levine, 1965; Gerall, 1967; Gorski, 1968; Whalen and Edwards, 1967; Ward and Renz, 1972). Thus, postnatally androgenized females are infertile, unreceptive, and capable of high levels of male copulatory behavior.

The data from neonatally castrated male rats complement those from androgenized females. If ovaries are implanted into adult or pubertal males in whom the testes have been removed within 5 days of birth, corpora lutea are formed (Pfeiffer, 1936; Harris, 1963; Gorski and Wagner, 1965; Gorski, 1966, 1967; Arai and Gorski, 1974; Marić *et al.*, 1974). If castration is delayed until after Day 5, only follicles are found in later implanted ovaries. This finding demonstrates that the cyclic female pattern of gonadotrophin release is retained in genetic males if androgen is removed in time. Further, neonatally castrated males are capable of high levels of female solicitation and lordosis behavior if primed with estrogen and progesterone prior to testing (Grady *et al.*, 1965; Gerall *et al.*, 1967). If given androgen in adulthood, neonatal castrates show high levels of mounting, but the ejaculatory potential is eliminated and the ability to show intromissions is severely impaired

Fig. 1. Lordosis pattern of an estrous female being mounted by a vigorous male.

(Larsson, 1966; Hart, 1968; Nadler, 1969; Beach *et al.*, 1969; Gray *et al.*, 1976;
Grady *et al.*, 1965; Gerall *et al.*, 1967). This impressive series of studies clearly
implicates the neonatal period as being important in the process of sexual differentiation in the rat.

81

SEXUAL BEHAVIOR
DIFFERENTIATION

STRATEGIES FOR MANIPULATING ANDROGENIC HORMONE LEVELS IN FETUSES

Demonstrations that the neonatal period is important for sexual differentiation in the rat do not preclude the possibility that the process may already have
been ongoing during fetal development. This possibility has to be investigated by
making manipulations in fetal males and females comparable to those performed
in neonates. Technically, such investigation raises a number of difficulties that have
to be surmounted before the empirical question can be adequately addressed.
Obviously, it is much more difficult to manipulate fetuses protected within the
uterine environment than it is to manipulate newly born pups.

The most widely used strategy for mimicking in females the high androgen
levels to which fetal males are normally exposed is to inject the mother with TP
during specific days of the pregnancy. Although this procedures is effective, as evi-
denced by severe masculinization of the females' secondary reproductive structures, only a fraction of the hormone being injected actually reaches the fetuses.
This point was made particularly effectively by Vreeburg *et al.* (1981). These investigators infused radioactively tagged testosterone into pregnant guinea pigs and
measured the amount of label recovered in the fetuses. The percentage binding of
(^{3}H) testosterone by male or female fetal plasma was found to be considerably less
than in the mothers. Thus, large dosages are required if the female fetuses are to
be affected. However, androgens themselves influence the course of the pregnancy. If TP is given too early, the fetuses are reabsorbed. If the treatment is continued too long, the processes of parturition and lactation are blocked, and the
animals must be delivered by cesarean section and fostered to a normal mother. If
the hormone dosage is too large, the mother aborts or the fetuses die (Swanson
and van der Werff ten Bosch, 1965; Gerall and Ward, 1966; Fels and Bosch, 1971).
Thus, the effects on behavior of only those treatments compatible with the maintenance of pregnancy can be assessed. The extent to which these duplicate in
female fetuses the process normally ongoing in males is difficult to gauge.

It is no easier to castrate fetal males. *In utero* surgical castration of male rats
surviving to be tested in adulthood has not been reported. The more common
approach has been to attempt chemical castration by giving antiandrogenic drugs
to mothers in the hope that effective dosages will cross the placental barrier and
enter into the circulation of the male fetuses. The most commonly used drugs are
cyproterone acetate (CA) and flutamide. Of these, flutamide is the better tool
because it does not have the progestational characteristics of CA that make it necessary to deliver CA-treated litters surgically.

Nevertheless, the picture that has emerged from the use of these two experimental strategies, coupled with two less invasive approaches, has led to the growing
conviction that the process of sexual behavior differentiation in the rat spans the

first few days before as well as after birth. Thus, a review of the existing literature is relevant.

PRENATAL DIFFERENTIATION OF FEMALE BEHAVIOR AND REPRODUCTIVE FUNCTIONING IN FEMALES. Evidence exists both for and against the possibility that differentiation of the cyclic pattern of gonadal hormone release, as well as of estrous behavior, begins during fetal development in the rat. Normal patterns of cyclic motor activity, as well as ovaries containing a normal complement of corpora lutea, have been reported in prenatally androgenized females (Revesz *et al.*, 1963; Kennedy, 1964; Gummow, 1975; Nadler, 1969; Whalen *et al.*, 1966). This finding is contrasted with reports that such females either have fewer corpora lutea or completely lack them when the ovaries are examined in adulthood (Swanson and van der Werff ten Bosch, 1965; Fels and Bosch, 1971; Ward and Renz, 1972; Popolow and Ward, 1978). Similarly, normal (Revesz *et al.*, 1963; Gummow, 1975; Whalen *et al.*, 1966), as well as severely impaired, estrous behavior (Kennedy, 1964; Gerall and Ward, 1966; Nadler, 1969; Ward and Renz, 1972; Dunlap *et al.*, 1978a; Popolow and Ward, 1978; Huffman and Hendricks, 1981) has been obtained in prenatally androgenized females by different investigators.

PRENATAL DIFFERENTIATION OF MALE BEHAVIOR IN FEMALES. The effects of the testosterone exposure of gonadal females on their potential for male copulatory behaviors illustrate an important point. Various sexually dimorphic behaviors and even different components of some complex patterns may have critical periods that are slightly different in duration or timing. The male copulatory pattern of the rat consists of three conspicuous features: mounting of the lure, intromissions, and ejaculation. Each component appears to evolve in different time frames during perinatal development and thus can be differentially influenced by the behavior-organizing actions of gonadal steroids. For example, there are a number of studies reporting that prenatal exposure of females to androgens increases both the number of mount and the number of intromission responses shown in adulthood (Gerall and Ward, 1966; Ward and Renz, 1972; Nadler, 1969; Popolow and Ward, 1978). Others have failed to find any potentiation of either of these male behavior patterns in prenatally androgenized females (Whalen *et al.*, 1969; Gummow, 1975; Huffman and Hendricks, 1981). Neonatal exposure achieved by injecting the females directly for the first few days after birth enhances intromission, but not mounting responses (Whalen and Edwards, 1967; Ward and Renz, 1972; Harris and Levine, 1965). However, a limited number of postnatally androgenized females show the male ejaculatory pattern (Ward, 1969; Thomas *et al.*, 1980, 1982), a behavior not seen in females exposed to TP only during prenatal development. Continuous treatment during both the prenatal and the postnatal periods yields a female rat that, when given testosterone in adulthood, shows copulatory behaviors indistinguishable from those of normal males (Dörner, 1968; Whalen and Robertson, 1968; Ward, 1969; Dörner and Seidler, 1973; Sachs *et al.*, 1973; Pollak and Sachs, 1975; Gray *et al.*, 1976). Different critical periods for the mounting, intromission, and ejaculatory responses in the rat can also be demonstrated by depriving males of endogenous androgen during perinatal ontogeny.

gent results in patterns of cyclicity, estrous behavior, and male copulatory potentials of prenatally androgenized females could possibly be explained by methodological differences. There is little consistency among studies in a number of
potentially critical procedural details, such as dosage and type of drug or hormone,
stage of gestation at which the prenatal treatment was initiated or terminated, and,
perhaps most important, the route of administration of the test compound. Only
a few studies are available in which direct systematic comparisons were made of
any of these variables. Swanson and van der Werff ten Bosch (1965) found that
subcutaneous injection of a single large dose of TP (10–25 mg) into mothers on
Day 19, 20, 21, or 22 of pregnancy resulted in anovulatory ovaries in the female
offspring, if the ovaries were autopsied at 21 weeks of age. Smaller dosages were
not effective, nor was an effect obtained if the hormone was injected intraperito-
neally, or into the uterus itself (Swanson and van der Werff ten Bosch, 1964; Fels
and Bosch, 1971). However, if the fetus was injected directly or TP was inserted
into the amniotic sac in large (e.g., 0.5–1 mg) but not small (e.g., 20 µg) amounts,
the ovaries were anovulatory in adulthood. Large dosages of androgen injected
intramuscularly into the mother have also yielded pronounced effects on the integrity of the ovary of the female offspring (Ward and Renz, 1972; Popolow and
Ward, 1978). The data on ovarian cyclicity indicate that the route by which androgen is introduced into the circulation of the fetuses is a major factor determining
whether lasting effects are obtained. Less systematic research exists comparing the
efficacy of various procedures on the differentiation of behavior potentials.

One other factor whose potential impact on the outcome of prenatal manipulation studies has been minimally investigated should be considered. This is the
age of the offspring at which the effects of the prenatal treatment are assessed. It
is well known that the injection of a single small dose of androgen (e.g., 10 µg TP)
into neonatal females induces the so-called delayed anovulatory syndrome (DAS).
DAS animals appear to have normal ovarian functioning and behavior if evaluated
at an early age. However, if autopsy or behavioral testing is delayed until the
females are 90–120 days old, profound defects in both cyclicity and estrous behavior are obtained (Swanson and van der Werff ten Bosch, 1964; Gorski, 1968; Harlan and Gorski, 1977; Napoli and Gerall, 1970). Similarly, neonatally castrated
males show more lordosis behavior when testing is initiated at 50–60 days than at
90–120 days of age (Dunlap *et al.*, 1972, 1973).

That a similar phenomenon may occur in prenatally androgenized females
exposed to less than the minimal amount of hormone needed for full masculinization and defeminization is suggested by two studies. In the first (Swanson and
van der Werff ten Bosch, 1965), the ovaries of prenatally androgenized females
were evaluated at 8 weeks and at 21 weeks of age. At the earlier age, the gonad
was of normal weight and histological appearance, but by 21 weeks, most animals
had small anovulatory ovaries. In the second study, Dunlap, Gerall, and Carlton
(1978a) injected prenatally androgenized females and neonatally gonadectomized
males with estrogen and progesterone beginning at 40, 80, or 120 days of age. A
steady decline in receptivity scores occurred as a function of age. It is possible that

some of the discrepant results reported by various investigators reflect differences in the age of the animals when reproductive functioning and behavior were evaluated.

PRENATAL DIFFERENTIATION OF FEMALE BEHAVIOR IN MALES. Manipulation of endogenous androgen levels in fetal males yields results that are a little more consistent than were those for prenatally androgenized females, but areas of discrepancy exist in this literature also. Males chemically castrated during only prenatal development by injecting the mothers with CA or flutamide show high levels of lordotic behavior when the males were treated with estradiol in adulthood (Ward, 1972a; Gladue and Clemens, 1978, 1982). Similar effects on lordosis behavior are obtained with the aromatase inhibitor 1,4,6-androstatriene-3,17-dione (ATD) (Clemens and Gladue, 1978; Gladue and Clemens, 1980; Whalen and Olsen, 1981). All three drugs increase the males' sensitivity to estrogen in adulthood. However, activation of the full estrous behavior pattern in normal female rats requires the synergistic action of the two ovarian hormones, estrogen and progesterone. Both ATD and flutamide treatment during fetal development increase adult sensitivity to progesterone in males. Apparently, CA does not (Ward, 1972a). There is disagreement on whether soliciting behavior can be affected with such prenatal manipulations. CA-treated males (Ward, 1972a) and some (Whalen and Olsen, 1981) but not all (Clemens and Gladue, 1978) ATD-exposed males show increased levels of the hopping, darting, and ear-wiggling component of the female's estrous behavior pattern. In all cases, the quality of the estrous pattern observed in prenatally manipulated males is not as good as in neonatally castrated males. By the same token, the female pattern exhibited by neonatally castrated males is not equivalent to that of normal females (Grady et al., 1965; Whalen and Edwards, 1966; Gerall et al., 1967; Beach et al., 1969; Hendricks, 1972; Gerall and Dunlap, 1973; Hendricks and Duffy, 1974).

PRENATAL DIFFERENTIATION OF MALE BEHAVIOR IN MALES. Few studies have assessed the effects of prenatal drug manipulations on male copulatory patterns in males. The available data indicate that CA or flutamide treatment of the mother reduces the number of intromissions shown by the male offspring in adulthood (Nadler, 1969; Clemens et al., 1978). Whether (Gladue and Clemens, 1980) or not (Whalen and Olsen, 1981) prenatal ATD treatment reduces male copulatory behavior remains to be resolved.

ADDITIVE EFFECTS OF PRE-PLUS POSTNATAL MANIPULATIONS

It is quite clear that, in both sexes, the effects on adult sexual behavior are more profound if the hormone or drug manipulation is made during both prenatal and postnatal development than if it is restricted to either period alone (e.g., Neuman and Elger, 1966; Ward, 1969; Nadler, 1969; Gladue and Clemens, 1982; Whalen and Olsen, 1981; Gray et al., 1976; Popolow and Ward, 1978).

Studies utilizing designs in which androgen levels are potentiated during one stage and reduced during the other yield a subtractive effect. For example, the receptivity scores of females or neonatally castrated males were reduced if the animals had been exposed to exogenous TP prenatally (Dunlap *et al.*, 1978a). Conversely, neonatally androgenized females showed much less impairment of lordosis if they had been given flutamide prenatally (Gladue and Clemens, 1982).

EXCHANGE OF ANDROGEN AMONG FETUSES

There are additional animal models that indicate that the critical period for sexual differentiation in the rat begins during prenatal ontogeny. In 1971, Clemens and Coniglio published an abstract that demonstrated a positive relationship between malelike mounting behavior in normal female rats and the number of male littermates with whom they had shared the uterus. The mechanism mediating this masculinization of behavior potentials was postulated to be the passage of testicular secretions from the fetal males to their female siblings.

There are two theories regarding the mode of transfer of hormone among fetuses occupying the same uterine horn. The first is that diffusion occurs across amniotic membranes in close proximity to one another. This theory would explain the tendency of females to show increasing degrees of behavioral as well as morphological masculinization as a function of their proximity to males *in utero* (Clemens, 1974; Clemens *et al.*, 1978). However, the finding that equivalent masculinization occurs in females contiguous to one male and those separated from the nearest male by another female sibling prompted the suggestion that, perhaps, the exchange is effected through the vasculature (Meisel and Ward, 1981). This theory seems likely, as indicated by data gathered on females grouped according to the number of males upstream from them with regard to the direction of blood flow in the uterine vein and artery. Those with one or more males upstream had larger anogenital distances and higher levels of mounting than females that had developed with no male upstream. This demonstration is particularly compelling in the group flanked by one male and one female. If the male was on the upstream side, the females were more masculinized than if the male was directly contiguous but on the downstream side with regard to uterine blood flow. That female rats can be partially masculinized *in utero* by male littermates is also suggested by a recent study (Tobet *et al.*, 1982) in which females were injected neonatally with low dosages (3.5 μg) of TP. The delayed anovulatory syndrome occurred at an earlier age in those females that had been flanked *in utero* by a male sibling on either side than in females flanked by female siblings.

Altogether, there is strong evidence that the course of sexual differentiation of some female rats normally is influenced *in utero* by androgens produced by the male siblings. This conclusion is further supported by studies in which fetal females were exposed to the antiandrogens CA and flutamide or the aromatase inhibitor ATD. These drugs markedly reduced the females' ability to show male copulatory patterns and increased sensitivity to the lordosis-activating properties of estrogen and progesterone in adulthood (Ward and Renz, 1972; Gladue and Clemens, 1978, 1982; Clemens *et al.*, 1978; Clemens and Gladue, 1978).

BEHAVIORAL SYNDROME

The prenatal organization of sexual behavior potentials is also indicated by the prenatal stress syndrome. This behavioral syndrome is characteristic of the majority of adult male rats whose mothers were exposed to intense environmental stressors during the last week of gestation. Specifically, such males show enhanced lordotic behavior potentials but either are unable to ejaculate or do so only after extensive therapy (Ward, 1972b, 1977). The types of stressors found to be effective include exposing the mothers to restraint coupled with high-intensity light (Ward, 1972b, 1977; Herrenkohl and Whitney, 1976; Dahlöf et al., 1977; Whitney and Herrenkohl, 1977; Dunlap et al., 1978b; Meisel et al., 1979; Götz and Dörner, 1980; Rhees and Fleming, 1981); to social crowding (Dahlöf et al., 1977); to malnutrition (Rhees and Fleming, 1981); or to a conditioned emotional response (Masterpasqua et al., 1976).

Although the process of sexual differentiation is known to continue through the first few days postpartum in the rat, neonatal stress appears to have no effect on developing males. Rats exposed to stressful handling procedures or vibration for the first 10 days after birth show normal adult sexual patterns. Combining prenatal with postnatal stress produces an effect no different from that of the prenatal treatment alone (Ward, 1972b, 1977). If prenatally stressed males are cross-fostered to normal mothers at birth, the syndrome persists, a result indicating that the behavioral alterations are caused by the prenatal treatment rather than by maternal factors operating neonatally (Herrenkohl and Whitney, 1976; Whitney and Herrenkohl, 1977; Meisel, 1980).

ALTERATIONS IN PLASMA TESTOSTERONE LEVELS AND TESTICULAR ENZYME ACTIVITY

The alterations in behavior seen in prenatally stressed males appear to have a hormonal basis. As shown in Figure 2, plasma testosterone titers in normal male and female rat fetuses have a characteristic pattern during the last week of gestation (Weisz and Ward, 1980). There are no significant sex differences on Day 17 of gestation. Between Days 17 and 18, fetal males show a marked rise. This elevation persists through Day 19 of fetal life, after which male testosterone titers drop to levels only slightly but significantly higher than those of female littermates. Testosterone levels are considerably lower in neonatal than in fetal rats. Now consider the case of prenatally stressed fetuses, shown in Figure 3 (Ward and Weisz, 1980). On Day 17 of gestation, when normal males do not differ from female littermates, stressed males have abnormally high levels of testosterone. On Days 18 and 19, stressed males not only fail to show an elevation in plasma testosterone but actually experience a decrease. Thereafter, there are no significant differences between control and stressed fetuses. A similar pattern is seen if testicular enzyme activity is considered (Orth et al., 1983). Quantification in fetal Leydig cells of Δ^5-3β-hydroxysteroid dehydrogenase (3β-HSD), a critical steroid in the synthesis of tes-

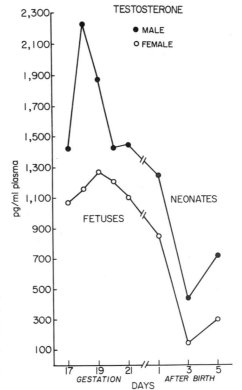

Fig. 2. Plasma testosterone levels of male and female rats from Days 17–21 of gestation and Days 1–5 postpartum. (From "Plasma Testosterone and Progesterone Titers of Pregnant Rats, Their Male and Female Fetuses, and Neonatal Offspring" by J. Weisz and I. L. Ward. *Endocrinology*, 1980, *106*, 306–316. Copyright 1980 by The Endocrine Society. Adapted by permission.)

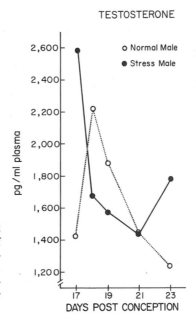

Fig. 3. Plasma testosterone levels of prenatally stressed and control males from Days 17–21 of gestation and the day of birth (Day 23). (From "Maternal Stress Alters Plasma Testosterone in Fetal Males" by I. L. Ward and J. Weisz. *Science*, 1980, *207*, 328–329. Copyright 1980 by the American Association for the Advancement of Science. Adapted by permission.)

tosterone, reveals abnormally high levels of activity in the testes of stressed as compared to control males on Days 16 and 17 of gestation. These high levels are followed by lower than normal activity on Days 18 and 19. The abnormal ontogenetic pattern of plasma testosterone and Leydig-cell enzyme activity led us to suggest that in prenatally stressed male fetuses there is a desynchronization in the rate of the synthesis and the release of testicular secretions and the development of target tissue in the central nervous system mediating sexual behaviors (Ward and Weisz, 1980, 1984). It is entirely possible that the abnormal behavioral syndrome seen in prenatally stressed males is a direct consequence of the weak androgenic milieu in which CNS structures developing on Days 18 and 19 of gestation evolve. Although considerable sexual differentiation is still ongoing at later time periods when prenatally stressed males have normal plasma testosterone levels, early deficiencies may increase the threshold to the masculinizing properties of lower testosterone titers typically present in normal neonatal males.

It is of interest that prenatally stressed fetuses have consistently lower levels of brain aromatase activity than untreated fetuses (Weisz et al., 1982). Whether these lower levels play a role in the abnormal sexual behavior differentiation that the males experience remains to be clarified. The aromatization hypothesis would predict that insufficient levels of this enzyme during perinatal development should result in the incomplete masculinization of adult behavior potentials. However, no sex differences in aromatase activity were obtained either in normal or in stressed fetal or neonatal rats (George and Ojeda, 1982; Weisz et al., 1982).

The reduction in testosterone titers of prenatally stressed males is limited to a few days of gestation. This might explain why the alterations in behavior and morphology are less extreme than those typically found in males treated prenatally with antiandrogenic drugs (see previous section). Athough males with this syndrome display high levels of lordotic behavior if injected in adulthood with either androgens (e.g., TP) or estrogen and progesterone, the quality, as well as the amount, of the resulting estrous pattern is not on the order of those of a receptive female. There is little evidence available to indicate whether other androgen-dependent sexually dimorphic behaviors are affected. The existing data suggest that home-cage emergence behavior is feminized (Masterpasqua et al., 1976). Open-field behavior may (Masterpasqua et al., 1976) or may not be altered (Meisel et al., 1979), and active avoidance acquisition is like that of normal males (Meisel et al., 1979).

No Persisting Modification of Morphology

Similarly, the effects on various androgen-dependent morphological structures is minimal. During fetal development and neonatal life, prenatally stressed males have lighter body weights and reduced anogenital distances (Ward, 1976; Dahlöf et al., 1978), but this effect is transient. When evaluated in adulthood, these males are anatomically indistinguishable from untreated controls (Ward, 1977).

A few prenatally stressed males eventually show the ejaculatory pattern if exposed to various types of "therapy." The most effective of these are injections of TP over a prolonged period in adulthood (Ward, 1977) or cohabitation with a female (Dunlap *et al.*, 1978b). If such males are tested repeatedly under optimal hormonal and stimulus conditions, four discrete subpopulations emerge. A few prenatally stressed males are normal; that is, they ejaculate and show little lordotic behavior. A few show no sexual behavior whatever. A third group display high levels of lordotic behavior but do not mount females. A fourth group, about 40%, are bisexual; that is, the sexual behavior emitted is stimulus-bound. They mount and ejaculate when placed with an estrous female but, within minutes, display the lordosis pattern if the lure female is replaced with a vigorous stud male (Ward, 1977).

The reasons for the heterogeneity in the sexual behavior potentials of prenatally stressed males is not clear at this time, but at least one factor has been identified that potentiates or attenuates the ejaculatory behavior potential of these males. Beginning at the time of weaning (22 days of age), Dunlap *et al.* (1978b) reared control and prenatally stressed males either in total social isolation or in all-male groups. When tested with estrous females in adulthood, very few of either of the isolated groups copulated. The highest intromission rate was shown by the socially reared control group, and the socially reared stressed males showed intermediary levels. The males then were rehoused in individual home cages containing two prepuberal females. Retesting two weeks later with estrous adult females revealed that this "therapy" resulted in markedly increased intromission rates in all groups except prenatally stressed males reared in social isolation. The failure of this latter group to show any improvement in male copulatory behavior suggests an interaction between prenatal hormonal factors and prepuberal social stimuli in the organization of adult sexual behavior potentials. Apparently, males with less than optimal exposure to masculinizing hormones during fetal development benefit from being reared in a favorable social milieu and are more responsive to therapies that promote copulation in control noncopulators. On the other hand, if such males are deprived of adequate social stimulation at an early age, the deficit in adult copulatory potentials is potentiated, and the ability to respond to therapies is blocked. Not investigated was the question of whether socialization alters the heightened female behavior potential that characterizes prenatally stressed males. This important possibility needs to be assessed, as does the more general question of the extent to which social stimuli interact with perinatal hormonal events to potentiate or attenuate the behavioral effects that either treatment alone would be expected to induce.

The prenatal stress syndrome offers an animal model in which fetal males experience a limited alteration in the androgenic milieu in which the CNS develops. Apparently, intense environmental stimuli alter testosterone levels not only in adult males, as has been shown repeatedly (Bardin and Peterson, 1967; Gray *et al.*, 1978; Collu *et al.*, 1979; Taché *et al.*, 1980), but also in fetuses. In adults, the consequences of fluctuating testosterone levels are temporary. In fetuses, the effects are on neural organization and thus are lasting. However, because various behavioral components evolve at different times in the perinatal period, the behavioral alterations experienced by prenatally stressed males are probably limited by the intensity and the timing of the stressor applied to the mother and by the rate at which she and the fetuses adapt to it (Ward and Weisz, 1984). Very little information is available on any of these latter parameters.

PRENATAL TESTOSTERONE SYNTHESIS INHIBITION BY OPIATES

After exposure to an androgen receptor blocker (CA or flutamide) during fetal development, adult male rats show incomplete masculinization of both external genitalia and sexual behaviors. Thus, a dissociation in sexual differentiation occurs between genetic and gonadal sex, which are male, and genital and behavioral sex, which are feminized. Following exposure to prenatal stress, the failure of masculinization is apparent in behavioral potentials, but not in external morphology. The dissociation is between behavior, which is feminized and demasculinized and external anatomy, which is homologous with gonadal and genetic expectations. Prenatal stress apparently works by partially attenuating the synthesis of testosterone, as indicated by the absence of both the testosterone and the 3β-HSD surge that normal male rats experience on Days 18 and 19 of gestation. This finding has led us to investigate other treatments that inhibit testosterone synthesis to ascertain whether fetal exposure would produce a behavioral syndrome similar to that characterizing prenatally stressed males.

Opiates, in relatively small doses, suppress testosterone levels in adult male rats (Cicero *et al.*, 1976). This change in testosterone is thought to be mediated by an opiate action on the LHRH-LH axis (Cicero, 1980), resulting in insufficient stimulation of testosterone synthesis. Opiates such as morphine pass the placental barrier in appreciable amounts, and because of an immature blood–brain barrier, they concentrate in the fetal brain (Johannesson *et al.*, 1972). If the fetal hypothalamic-pituitary-gonadal axis responds to opiates in a fashion similar to the response in adults, *in utero* exposure to morphine might modify the sexual differentiation of males.

PRENATAL OPIOID SYNDROME

Small doses of morphine, given to pregnant rats every 8 hr beginning on the fifteenth day of gestation, significantly reduced the percentage of their male offspring able to show the ejaculatory response in adulthood and increased their potential to exhibit lordosis (Ward *et al.*, 1983). Thus, the consequences of fetal morphine exposure in reproductive behavior closely resemble those of prenatal stress.

No gross modifications in anatomy or physiology were discernible in adult male rats exposed to low doses of morphine *in utero* (Ward *et al.*, 1983). Body weight, anogenital distance, penile length, testes weight, and plasma levels of LH and testosterone were not significantly different from those of control males.

EFFECTS ON FETAL PLASMA TESTOSTERONE AND TESTICULAR ENZYME ACTIVITY

Morphine, administered on the same schedule and in the same amount as in the behavioral study (Ward *et al.*, 1983) suppressed the rise in 3β-HSD activity normally seen on the 18th and 19th days of gestation (Badway *et al.*, 1981). This finding suggests that opiate exposure partially inhibits testosterone synthesis in the fetus, as it does in the adult male rat (Cicero, 1980). This hypothesis gains further support from a report in which the synthetic opioid methadone was administered to pregnant rats from Day 14 to Day 19 of gestation. Plasma testosterone levels were depressed in the male fetuses of these mothers on Day 20 of gestation (Singh *et al.*, 1980).

POSSIBLE ROLE OF ENDOGENOUS OPIATE RELEASE IN MEDIATING THE PRENATAL STRESS SYNDROME

Apparently both prenatal stress and prenatal morphine administration change the androgenic environment of the male fetus sufficiently to compromise the development of normal sexual behavior potentials, but insufficiently to alter adult anatomy or physiology. The similar consequences of prenatal opiate exposure and prenatal stress in conjunction with the finding that, in adults, stress triggers a release of β-endorphin (Guillemin *et al.*, 1977) suggest that the effects of gestational stress might be mediated by endogenous opiates. If fetal males were exposed to increased endogenous opiates following stress to their mother, the opiate exposure might explain the loss of the surge in plasma testosterone and testicular enzymes seen in prenatally stressed male fetuses. Limited data exist in support of this possibility.

Because changes in the activity of 3β-HSD closely parallel the normal developmental pattern of plasma testosterone (Orth and Weisz, 1980; Weisz and Ward, 1980), 3β-HSD can be used as an index of the secretory potential of the developing Leydig cells of fetal rats. Fetal rats from mothers exposed to stress (i.e., restraint and light) starting on the fourteenth day of gestation show a marked reduction in 3β-HSD activity on the nineteenth day of gestation. However, as seen in Figure 4, if pregnant females are injected with the opiate receptor blocker naltrexone before each stress session, 3β-HSD activity is not reduced (Ward *et al.*, 1983). This finding supports the conclusion that environmental stress alters sexual differentiation in fetal males by stimulating the release of endogenous opiates.

CONCLUSIONS

During the perinatal period of sexual differentiation in the rat, maximum sex differences in plasma testosterone titers occur during the prenatal period, approximately 4 to 5 days before birth. Blocking androgen receptors, the targets of this

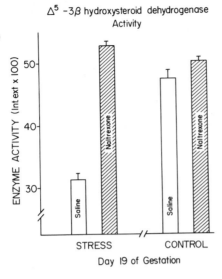

Δ^5 –3β hydroxysteroid dehydrogenase
Activity

Fig. 4. Enzyme activity ($\Delta^5 - 3\beta$ hydroxysteroid dehydrogenase) in Leydig cells of 19-day-old stressed and control fetal male rats whose mothers were injected with saline or naltrexone. (From "A Possible Role of Opiates in Modifying Sexual Differentiation" by O. B. Ward, J. M. Orth, and J. Weisz. In M. Schlumpf and W. Lichtensteiger [Eds.], *Drugs and Hormones in Brain Development*. Basel: Karger, 1983. Copyright 1983 by S. Karger AG, Basel. Adapted by permission.)

prenatal testosterone surge, by the administration of CA or flutamide results in males with impaired male behavior potentials and enhanced lordosis behavior. Conversely, the administration of testosterone to fetuses that lack the endogenous source for an androgen surge (i.e., gonadal females) disrupts cyclicity in the hypothalamic-pituitary-ovarian axis, impairs the potential for lordotic behavior, and potentiates male copulatory patterns. For the same reason, females developing in uterine positions that expose them to androgens produced by male littermates are somewhat masculinized.

Alterations in the synthesis of testosterone in male fetuses induced by prenatal stress or exposure to opiates results in a failure of masculinization of behavior without residual modifications in external anatomical structures. The importance of this finding is that there has been a tendency to assume that, unless individuals with a diversity of sexual orientations also have discernible modifications in adult reproductive morphology or physiology, the etiology of their behavioral predispositions must be entirely social. The finding that behavioral sex can be dissociated from genital sex suggests that abnormal fluctuations of gonadal hormones during fetal life may contribute to the etiology of atypical behavioral orientations, even when adult anatomical and physiological parameters are homologous with genetic and gonadal sex.

REFERENCES

Arai, Y., and Gorski, R. A. Possible participation of pituitary testicular feedback regulation in the sexual differentiation of the brain in the male rat. In M. Kawakami (Ed.), *Rhythms in Neuroendocrine Activity*. Tokyo: Igaku Shoin, 1974.

Badway, D., Orth, J., and Weisz, J. Effect of morphine on Δ^5-3β-ol-steroid dehydrogenase in Leydig cells of fetal rats: A quantitative cytochemical study. *Anatomical Record*, 1981, *199*, 15a.

Bardin, C. W., and Peterson, R. E. Studies of androgen production by the rat: Testosterone and androstenedione content of blood. *Endocrinology*, 1967, *80*, 38–44.

Barraclough, C. A. Induction of ovulation in the sterile rat by hypothalamic stimulation. *Anatomical Record*, 1959, *133*, 248.

Barraclough, C. A. Production of anovulatory, sterile rats by single injections of testosterone propionate. *Endocrinology*, 1961, *68*, 62–67.

Barraclough, C. A., and Gorski, R. A. Evidence that the hypothalamus is responsible for androgen-induced sterility in the female rat. *Endocrinology*, 1961, *68*, 68–79.

Barraclough, C. A., and Gorski, R. A. Studies on mating behaviour in the androgen-sterilized female rat in relation to the hypothalamic regulation of sexual behaviour. *Journal of Endocrinology*, 1962, *25*, 175–182.

Baum, M. J. Differentiation of coital behavior in mammals: A comparative analysis. *Neuroscience and Biobehavioral Reviews*, 1979, *3*, 265–284.

Beach, F. A. Hormonal modification of sexually dimorphic behavior. *Psychoneuroendocrinology*, 1975, *1*, 3–23.

Beach, F. A., Noble, R. G., and Orndoff, R. K. Effects of perinatal androgen treatment on responses of male rats to gonadal hormones in adulthood. *Journal of Comparative and Physiological Psychology*, 1969, *68*, 490–497.

Beatty, W. W. Gonadal hormones and sex differences in nonreproductive behaviors in rodents: Organizational and activational influences. *Hormones and Behavior*, 1979, *12*, 112–163.

Beatty, W. W., Dodge, A. M., and Traylor, K. L. Temporal boundary of the sensitive period for hormonal organization of social play in juvenile rats. *Physiology and Behavior*, 1981, *26*, 241–243.

Booth, J. E. Effects of the aromatization inhibitor androst-4-ene-3,6,17-trione on sexual differentiation induced by testosterone in the neonatally castrated rat. *Journal of Endocrinology*, 1978, *79*, 69–76.

Cicero, T. J. Effects of exogenous and endogenous opiates on the hypothalamic-pituitary-gonadal axis in the male. *Federation Proceedings*, 1980, *39*, 2551–2554.

Cicero, T. J., Wilcox, C. E., Bell, R. D., and Meyer, E. R. Acute reductions in serum testosterone levels by narcotics in the male rat: Stereospecificity, blockage by naloxone and tolerance. *Journal of Pharmacology and Experimental Therapeutics*, 1976, *198*, 340–346.

Clemens, L. G. Neurohormonal control of male sexual behavior. In W. Montagna and W. A. Sadler (Eds.), *Reproductive Behavior*. New York: Plenum Press, 1974.

Clemens, L. G., and Coniglio, L. Influence of prenatal litter composition on mounting behavior of female rats. *American Zoologist*, 1971, *11*, 617–618.

Clemens, L. G., and Gladue, B. A. Feminine sexual behavior in rats enhanced by prenatal inhibition of androgen aromatization. *Hormones and Behavior*, 1978, *11*, 190–201.

Clemens, L. G., Gladue, B. A., and Coniglio, L. P. Prenatal endogenous androgenic influences on masculine sexual behavior and genital morphology in male and female rats. *Hormones and Behavior*, 1978, *10*, 40–53.

Collu, R., Taché, Y., and Ducharme, J. R. Hormonal modifications induced by chronic stress in rats. *Journal of Steroid Biochemistry*, 1979, *11*, 989–1000.

Dahlöf, L.-G., Hård, E., and Larsson, K. Influence of maternal stress on offspring sexual behaviour. *Animal Behaviour*, 1977, *25*, 958–963.

Dahlöf, L.-G., Hård, E., and Larsson, K. Influence of maternal stress on the development of the fetal genital system. *Physiology and Behavior*, 1978, *20*, 193–195.

Dantchakoff, V. Rôle des hormones dans la manifestation des instincts sexuel. *Comptes Rendus Hebdomadaires des Séances de l'Académie des Sciences*, 1938a, *206*, 945–947.

Dantchakoff, V. Sur les effets de l'hormone mâle dans une jeune cobaye fémelle traité depuis un stade embryonnaire (inversions sexuelles). *Comptes Rendus des Séances de la Société de Biologie et de ses Filiales* (Paris), 1938b, *127*, 1255–1258.

Dantchakoff, V. Sur les effets de l'hormone mâle dans un jeune cobaye male traité depuis un stade embryonnaire (production d'hypermales). *Comptes Rendus des Séances de la Société de Biologie et de ses Filiales* (Paris), 1938c, *127*, 1259–1262.

Davis, P. G., Chaptal, C. V., and McEwen, B. S. Independence of the differentiation of masculine and feminine sexual behavior in rats. *Hormones and Behavior*, 1979, *12*, 12–19.

Dörner, G. Hormonal induction and prevention of female homosexuality. *Journal of Endocrinology*, 1968, *42*, 163–164.

Dörner, G. Sexual differentiation of the brain. *Vitamins and Hormones*, 1980, *38*, 325–381.

Dörner, G., and Seidler, C. Influence of high-dosage perinatal androgen on the sexual behavior, gonadotropin secretion, and sexual organs of the rat. *Archives of Sexual Behavior*, 1973, *2*, 267–272.

Dunlap, J. L., Gerall, A. A., and Hendricks, S. E. Female receptivity in neonatally castrated males as a function of age and experience. *Physiology and Behavior*, 1972, *8*, 21–23.

Dunlap, J. L., Gerall, A. A., and McLean, L. Enhancement of female receptivity in neonatally castrated males by prepuberal ovarian transplants. *Physiology and Behavior*, 1973, *10*, 701–705.

Dunlap, J. L., Gerall, A. A., and Carlton, S. F. Evaluation of prenatal androgen and ovarian secretions on receptivity in female and male rats. *Journal of Comparative and Physiological Psychology*, 1978a, *92*, 280–288.

Dunlap, J. L., Zadina, J. E., and Gougis, G. Prenatal stress interacts with prepuberal social isolation to reduce male copulatory behavior. *Physiology and Behavior*, 1978b, *21*, 873–875.

Edwards, D. A., and Thompson, M. L. Neonatal androgenization and estrogenization and the hormonal induction of sexual receptivity in rats. *Physiology and Behavior*, 1970, *5*, 1115–1119.

Fadem, B. H., and Barfield, R. J. Neonatal hormonal influences on the development of proceptive and receptive feminine sexual behavior in rats. *Hormones and Behavior*, 1981, *15*, 282–288.

Feder, H. H., and Whalen, R. E. Feminine behavior in neonatally castrated and estrogen-treated male rats. *Science*, 1965, *147*, 306–307.

Fels, E., and Bosch, L. R. Effect of prenatal administration of testosterone on ovarian function in rats. *American Journal of Obstetrics and Gynecology*, 1971, *111*, 964–969.

George, F. W., and Ojeda, S. R. Changes in aromatase activity in the rat brain during embryonic, neonatal, and infantile development. *Endocrinology*, 1982, *111*, 522–529.

Gerall, A. A. Effects of early postnatal androgen and estrogen injections on the estrous activity cycles and mating behavior of rats. *Anatomical Record*, 1967, *157*, 97–104.

Gerall, A. A., and Dunlap, J. L. The effect of experience and hormones on the initial receptivity in female and male rats. *Physiology and Behavior*, 1973, *10*, 851–854.

Gerall, A. A., and Ward, I. L. Effects of prenatal exogenous androgen on the sexual behavior of the female albino rat. *Journal of Comparative and Physiological Psychology*, 1966, *62*, 370–375.

Gerall, A. A., Hendricks, S. E., Johnson, L. L., and Bounds, T. W. Effects of early castration in male rats on adult sexual behavior. *Journal of Comparative and Physiological Psychology*, 1967, *64*, 206–212.

Gerall, A. A., Dunlap, J. L., and Hendricks, S. E. Effect of ovarian secretions on female behavioral potentiality in the rat. *Journal of Comparative and Physiological Psychology*, 1972, *82*, 449–465.

Gladue, B. A., and Clemens, L. G. Androgenic influences on feminine sexual behavior in male and female rats: Defeminization blocked by prenatal antiandrogen treatment. *Endocrinology*, 1978, *103*, 1702–1709.

Gladue, B. A., and Clemens, L. G. Masculinization diminished by disruption of prenatal estrogen biosynthesis in male rats. *Physiology and Behavior*, 1980, *25*, 589–593.

Gladue, B. A., and Clemens, L. G. Development of feminine sexual behavior in the rat: Androgenic and temporal influences. *Physiology and Behavior*, 1982, *29*, 263–267.

Gorski, R. A. Localization and sexual differentiation of the nervous structures which regulate ovulation. *Journal of Reproduction and Fertility*, Suppl. 1, 1966, 67–88.

Gorski, R. A. Localization of the neural control of luteinization in the feminine male rat (FALE). *Anatomical Record*, 1967, *157*, 63–69.

Gorski, R. A. Influence of age on the response to paranatal administration of a low dose of androgen. *Endocrinology*, 1968, *82*, 1001–1004.

Gorski, R. A. Long-term hormonal modulation of neuronal structure and function. In F. O. Schmitt and F. G. Worden (Eds.), *The Neurosciences: Fourth Study Program*. Cambridge, Mass.: M.I.T. Press, 1979.

Gorski, R. A. Comparative aspects of sexual differentiation of the brain. In M. M. Grumbach, P. C. Sizonenko, and M. L. Aubert (Eds.), *The Control of the Onset of Puberty*, Vol. 2. New York: Academic Press, 1982.

Gorski, R. A., and Barraclough, C. A. In differential effectiveness of small dosages of testosterone propionate in the induction of sterility in the female rat. *Anatomical Record*, 1961, *139*, 304.

Gorski, R. A., and Barraclough, C. A. Effects of low dosages of androgen on the differentiation of hypothalamic regulatory control of ovulation in the rat. *Endocrinology*, 1963, *73*, 210–216.

Gorski, R. A., and Wagner, J. W. Gonadal activity and sexual differentiation of the hypothalamus. *Endocrinology*, 1965, *76*, 226–239.

Götz, F., and Dörner, G. Homosexual behaviour in prenatally stressed male rats after castration and oestrogen treatment in adulthood. *Endokrinologie*, 1980, *76*, 115–117.

Goy, R. W. Early hormonal influences on the development of sexual and sex-related behavior. In F. O. Schmitt (Ed.), *The Neurosciences: Second Study Program*. New York: Rockefeller University Press, 1970.

Goy, R. W. Development of play and mounting behavior in female rhesus monkeys virilized prenatally with esters of testosterone and dihydrotestosterone. In D. J. Chivers and J. Herbert (Eds.), *Recent Advances in Primatology*, Vol. 1. New York: Academic Press, 1978.

Goy, R. W., and Goldfoot, D. A. Neuroendocrinology: Animal models and problems of human sexuality. *Archives of Sexual Behavior*, 1975, *4*, 405–420.

Goy, R. W., and McEwen, B. S. *Sexual differentiation of the brain*. Cambridge, Mass.: M.I.T. Press, 1980.

Goy, R. W., Phoenix, C. H., and Young, W. C. A critical period for the suppression of behavioral receptivity in adult female rats by early treatment with androgen. *Anatomical Record*, 1962, *142*, 307.

Grady, K. L., Phoenix, C. H., and Young, W. C. Role of the developing rat testis in differentiation of the neural tissues mediating mating behavior. *Journal of Comparative and Physiological Psychology*, 1965, *59*, 176–182.

Gray, G. D., Davis, H. N., and Dewsbury, D. A. Masculine sexual behavior in male and female rats following perinatal manipulation of androgen: Effects of genital anesthetization and sexual experience. *Hormones and Behavior*, 1976, *7*, 317–329.

Gray, G. D., Smith, E. R., Damassa, D. A., Ehrenkranz, J. R. L., and Davidson, J. M. Neuroendocrine mechanisms mediating the suppression of circulating testosterone levels associated with chronic stress in male rats. *Neuroendocrinology*, 1978, *25*, 247–256.

Guillemin, R., Vargo, T., Rossier, J., Minick, S., Ling, N., Rivier, C., Vale, W., and Bloom, F. β-Endorphin and adrenocorticotropin are secreted concomitantly by the pituitary gland. *Science*, 1977, *197*, 1367–1369.

Gummow, L. J. Postnatal androgenization influences social behavior of adult rats tested in standard male and female sexual paradigms. *Behavioral Biology*, 1975, *13*, 385–399.

Harlan, R. E., and Gorski, R. A. Steroid regulation of luteinizing hormone secretion in normal and androgenized rats at different ages. *Endocrinology*, 1977, *101*, 741–749.

Harris, G. W. Castration of the new-born male rat and lack of sexual differentiation of the brain. *Journal of Physiology*, 1963, *169*, 117–118.

Harris, G. W., and Levine, S. Sexual differentiation of the brain and its experimental control. *Journal of Physiology*, 1965, *181*, 379–400.

Hart, B. L. Neonatal castration: Influence on neural organization of sexual reflexes in male rats. *Science*, 1968, *160*, 1135–1136.

Hendricks, S. E. Influence of neonatally administered hormones and early gonadectomy on rats' sexual behavior. *Journal of Comparative and Physiological Psychology*, 1969, *69*, 408–413.

Hendricks, S. E. Androgen modification of behavioral responsiveness to estrogen in the male rat. *Hormones and Behavior*, 1972, *3*, 47–54.

Hendricks, S. E., and Duffy, J. A. Ovarian influences on the development of sexual behavior in neonatally androgenized rats. *Developmental Psychobiology*, 1974, *7*, 297–303.

Hendricks, S. E., and Weltin, M. Effect of estrogen given during various periods of prepuberal life on the sexual behavior of rats. *Physiological Psychology*, 1976, *4*, 105–110.

Herrenkohl, L. R., and Whitney, J. B. Effects of prepartal stress on postpartal nursing behavior, litter development and adult sexual behavior. *Physiology and Behavior*, 1976, *17*, 1019–1021.

Huffman, L., and Hendricks, S. E. Prenatally injected testosterone propionate and sexual behavior of female rats. *Physiology and Behavior*, 1981, *26*, 773–778.

Johannesson, T., Steele, W. J., and Becker, B. A. Infusion of morphine in maternal rats at near term: Maternal and foetal distribution and effects on analgesia, brain DNA, RNA and protein. *Acta Pharmacologica et Toxicologica* (Copenhagen), 1972, *31*, 353–368.

Kennedy, G. C. Mating behaviour and spontaneous activity in androgen-sterilized female rats. *Journal of Physiology*, 1964, *172*, 393–399.

Koster, R. Hormone factors in male behavior of the female rat. *Endocrinology*, 1943, *33*, 337–348.

Larsson, K. Effects of neonatal castration upon the development of the mating behavior of the male rat. *Zeitschrift für Tierpsychologie*, 1966, *23*, 867–873.

Levine, S., and Mullins, R. Estrogen administered neonatally affects adult sexual behavior in male and female rats. *Science*, 1964, *144*, 185–187.

Marić, D., Tadić, R., and Miline, R. The influence of the gonads on the functional development of the hypothalamo-hypophysial system of the male rat. *Neuroendocrinology*, 1974, *15*, 92–98.

Masterpasqua, F., Chapman, R. H., and Lore, R. K. The effects of prenatal psychological stress on the sexual behavior and reactivity of male rats. *Developmental Psychobiology*, 1976, *9*, 403–411.

McDonald, P. G., and Doughty, C. Inhibition of androgen-sterilization in the female rat by administration of an anti-oestrogen. *Journal of Endocrinology*, 1972, *55*, 455–456.

McDonald, P. G., and Doughty, C. Androgen sterilization in the neonatal female rat and its inhibition by an estrogen antagonist. *Neuroendocrinology*, 1973, *13*, 182–188.

McEwen, B. S. Gonadal steroid influences on brain development and sexual differentiation. In R. Greep (Ed.), *Reproductive Physiology*, Vol. 4. Baltimore: University Park Press, 1982.

McEwen, B. S., Lieberburg, I., Chaptal, C., and Krey, L. C. Aromatization: Important for sexual differentiation of the neonatal rat brain. *Hormones and Behavior*, 1977, *9*, 249–263.

McEwen, B. S., Lieberburg, I., Chaptal, C., Davis, P. G., Krey, L. C., MacLusky, N. J., and Roy, E. J. Attenuating the defeminization of the neonatal rat brain: Mechanisms of action of cyproterone acetate, 1,4,6-androstatriene-3,17-dione and a synthetic progestin, R5020. *Hormones and Behavior*, 1979, *13*, 269–281.

Meaney, M. J., and Stewart, J. Neonatal androgens influence the social play of prepubescent rats. *Hormones and Behavior*, 1981, *15*, 197–213.

Meisel, R. L. *Effects of Prenatal Stress and Uterine Position on the Sexual Behavior and Morphological Differentiation of Male and Female Rats* M.S. thesis, Villanova University, 1980.

Meisel, R. L., and Ward, I. L. Fetal rats are masculinized by male littermates located caudally in the uterus. *Science*, 1981, *213*, 239–242.

Meisel, R. L., Dohanich, G. P., and Ward, I. L. Effects of prenatal stress on avoidance acquisition, open-field performance and lordotic behavior in male rats. *Physiology and Behavior*, 1979, *22*, 527–530.

Mullins, R. F., and Levine, S. Hormonal determinants during infancy of adult sexual behavior in the female rat. *Physiology and Behavior*, 1968, *3*, 333–338.

Nadler, R. D. Differentiation of the capacity for male sexual behavior in the rat. *Hormones and Behavior*, 1969, *1*, 53–63.

Napoli, A. M., and Gerall, A. A. Effect of estrogen and anti-estrogen on reproductive function in neonatally androgenized female rats. *Endocrinology*, 1970, *87*, 1330–1337.

Neuman, F., and Elger, W. Permanent changes in gonadal function and sexual behavior as a result of early feminization of male rats by treatment with an antiandrogenic steroid. *Endokrinologie*, 1966, *50*, 209–225.

Noumura, T., Weisz, J., and Lloyd, C. W. *In vitro* conversion of 7-³H-progesterone to androgen by the rat testis during the second half of fetal life. *Endocrinology*, 1966, *78*, 245–253.

Olioff, M., and Stewart, J. Sex differences in the play behavior of prepubescent rats. *Physiology and Behavior*, 1978, *20*, 113–115.

Orth, J., and Weisz, J. Development of Δ^5-3 β hydroxysteroid dehydrogenase and glucose-6-phosphate dehydrogenase activity in Leydig cells of the fetal rat testis: A quantitative cytochemical study. *Biology of Reproduction*, 1980, *22*, 1201–1209.

Orth, J. M., Weisz, J., Ward, O. B., and Ward, I. L. Environmental stress alters the developmental pattern of Δ^5-3β-hydroxysteroid dehydrogenase activity in Leydig cells of fetal rats: A quantitative cytochemical study. *Biology of Reproduction*, 1983, *28*, 625–631.

Pfeiffer, C. A. Sexual differences of the hypophyses and their determination by the gonads. *American Journal of Anatomy*, 1936, *58*, 195–225.

Phoenix, C. H., Goy, R. W., Gerall, A. A., and Young, W. C. Organizing action of prenatally administered testosterone propionate on the tissues mediating mating behavior in the female guinea pig. *Endocrinology*, 1959, *65*, 369–382.

Pollak, E. I., and Sachs, B. D. Masculine sexual behavior and morphology: Paradoxical effects of perinatal androgen treatment in male and female rats. *Behavioral Biology*, 1975, *13*, 401–411.

Popolow, H. B., and Ward, I. L. Effects of perinatal androstenedione on sexual differentiation in female rats. *Journal of Comparative and Physiological Psychology*, 1978, *92*, 13–21.

Quattropani, S. L., and Weisz, J. Conversion of progesterone to estrone and estradiol *in vitro* by the ovary of the infantile rat in relation to the development of its interstitial tissue, *Endocrinology*, 1973, *93*, 1269–1276.

Reddy, V. V. R., Naftolin, F., and Ryan, K. J. Conversion of androstenedione to estrone by neural tissues from fetal and neonatal rats. *Endocrinology*, 1974, *94*, 117–121.

Revesz, C., Kernaghan, D., and Bindra, D. Sexual drive of female rats 'masculinized' by testosterone during gestation. *Journal of Endocrinology*, 1963, *25*, 549–550.

Rhees, R. W., and Fleming, D. E. Effects of malnutrition, maternal stress, or ACTH injections during pregnancy on sexual behavior of male offspring. *Physiology and Behavior*, 1981, *27*, 879–882.

Sachs, B. D., Pollak, E. I., Krieger, M. S., and Barfield, R. J. Sexual behavior: Normal male patterning in androgenized female rats. *Science*, 1973, *181*, 770–772.

Schlegel, J. J., Farias, E., Russo, N. C., Moore, J. R., and Gardner, L. I. Structural changes in the fetal gonads and gonaducts during maturation of an enzyme, steroid 3β-ol-dehydrogenase, in the gonads, adrenal cortex and placenta of fetal rats. *Endocrinology*, 1967, *81*, 565–572.

Singh, H. H., Purohit, V., and Ahluwalia, B. S. Effect of methadone treatment during pregnancy on the fetal testes and hypothalamus in rats. *Biology of Reproduction*, 1980, *22*, 480–485.

Södersten, P. Effects of anti-oestrogen treatment of neonatal male rats on lordosis behaviour and mounting behaviour in the adult. *Journal of Endocrinology*, 1978, *76*, 241–249.

Swanson, H. E., and van der Werff ten Bosch, J. J. The "early androgen" syndrome: Differences in response to prenatal and postnatal administration of various doses of testosterone propionate in female and male rats. *Acta Endocrinologica*, 1964, *47*, 37–50.

Swanson, H. E., and van der Werff ten Bosch, J. J. The "early-androgen" syndrome: Effects of pre-natal testosterone propionate. *Acta Endocrinologica*, 1965, *50*, 379–390.

Taché, Y., Ducharme, J. R., Charpenet, G., Haour, F., Saez, J., and Collu, R. Effect of chronic inter-mittent immobilization stress on hypophyso-gonadal function of rats. *Acta Endocrinologica*, 1980, *93*, 168–174.

Thomas, D. A., McIntosh, T. K., and Barfield, R. J. Influence of androgen in the neonatal period on ejaculatory and postejaculatory behavior of the rat. *Hormones and Behavior*, 1980, *14*, 153–162.

Thomas, D. A., Barfield, R. J., and Etgen, A. M. Influence of androgen on the development of sexual behavior in rats: I. Time of administration and masculine copulatory responses, penile reflexes, and androgen receptors in females. *Hormones and Behavior*, 1982, *16*, 443–454.

Tobet, S. A., Dunlap, J. L., and Gerall, A. A. Influence of fetal position on neonatal androgen induced sterility and sexual behavior in female rats. *Hormones and Behavior*, 1982, *16*, 251–258.

Vreeburg, J. T. M., van der Vaart, P. D. M., and van der Schoot, P. Prevention of central defeminization but not masculinization in male rats by inhibition neonatally of oestrogen biosynthesis. *Journal of Endocrinology*, 1977, *74*, 375–382.

Vreeburg, J. T. M., Woutersen, P. J. A., Ooms, M. P., and van der Werff ten Bosch, J. J. Androgens in the fetal guinea-pig after maternal infusion of radioactive testosterone. *Journal of Endocrinology*, 1981, *88*, 9–16.

Ward, I. L. Differential effect of pre- and postnatal androgen on the sexual behavior of intact and spayed female rats. *Hormones and Behavior*, 1969, *1*, 25–36.

Ward, I. L. Female sexual behavior in male rats treated prenatally with an anti-androgen. *Physiology and Behavior*, 1972a, *8*, 53–56.

Ward, I. L. Prenatal stress feminizes and demasculinizes the behavior of males. *Science*, 1972b, *175*, 82–84.

Ward, I. L. Exogenous androgen activates female behavior in noncopulating, prenatally stressed male rats. *Journal of Comparative and Physiological Psychology*, 1977, *91*, 465–471.

Ward, I. L., and Renz, F. J. Consequences of perinatal hormone manipulation on the adult sexual behavior of female rats. *Journal of Comparative and Physiological Psychology*, 172, *78*, 349–355.

Ward, I. L., and Weisz, J. Maternal stress alters plasma testosterone in fetal males. *Science*, 1980, *207*, 328–329.

Ward, I. L., and Weisz, J. Differential effects of maternal stress on circulating levels of corticosterone, progesterone and testosterone in male and female rat fetuses and their mothers. *Endocrinology*, 1984, *114*, 1635–1644.

Ward, O. B. Alterations of androgen-sensitive morphology in fetal males by prenatal stress. *Eastern Conference on Reproductive Behavior*. Saratoga Springs, New York, 1976 (abstract).

Ward, O. B., Orth, J. M., and Weisz, J. A possible role of opiates in modifying sexual differentiation. In M. Schlumpf and W. Lichtensteiger (Eds.), *Drugs and Hormones in Brain Development*. Basel: Karger, 1983.

Warren, D. W., Haltmeyer, G. C., and Eik-Nes, K. B. Testosterone in the fetal rat testis. *Biology of Reproduction*, 1973, *8*, 560–565.

Warren, D. W., Haltmeyer, G. C., and Eik-Nes, K. B. The effect of gonadotrophins on the fetal and neonatal testis. *Endocrinology*, 1975, *96*, 1226–1229.

Weisz, J., and Ward, I. L. Plasma testosterone and progesterone titers of pregnant rats, their male and female fetuses, and neonatal offspring. *Endocrinology*, 1980, *106*, 306–316.

Weisz, J., Brown, B. L., and Ward, I. L. Maternal stress decreases steroid aromatase activity in brains of male and female rat fetuses. *Neuroendocrinology*, 1982, *35*, 374–379.

Whalen, R. E., and Edwards, D. A. Sexual reversibility in neonatally castrated male rats. *Journal of Comparative and Physiological Psychology*, 1966, *62*, 307–310.

Whalen, R. E., and Edwards, D. A. Hormonal determinants of the development of masculine and feminine behavior in male and female rats. *Anatomical Record*, 1967, *157*, 173–180.

Whalen, R. E., and Nadler, R. D. Suppression of the development of female mating behavior by estrogen administered in infancy. *Science*, 1963, *141*, 273–274.

Whalen, R. E., and Olsen, K. L. Role of aromatization in sexual differentiation: Effects of prenatal ATD treatment and neonatal castration. *Hormones and Behavior*, 1981, *15*, 107–122.

Whalen, R. E., and Robertson, R. T. Sexual exhaustion and recovery of masculine copulatory behavior in virilized female rats. *Psychonomic Science*, 1968, *11*, 319–320.

Whalen, R. E., Peck, C. K., and LoPiccolo, J. Virilization of female rats by prenatally administered progestin. *Endocrinology*, 1966, *78*, 965–970.

Whalen, R. E., Edwards, D. A., Luttge, W. G., and Robertson, R. T. Early androgen treatment and male sexual behavior in female rats. *Physiology and Behavior*, 1969, *4*, 33–39.

Whitney, J. B., and Herrenkohl, L. R. Effects of anterior hypothalamic lesions on the sexual behavior of prenatally-stressed male rats. *Physiology and Behavior*, 1977, *19*, 167–169.

Wilson, J. G. Reproductive capacity of adult female rats treated prepuberally with estrogenic hormone. *Anatomical Record*, 1943, *86*, 341–363.

PART II
Hormonal Activation of Behavior

Reproductive Physiology and Behavior Interactions in Nonmammalian Vertebrates

DAVID CREWS AND RAE SILVER

INTRODUCTION

Reproduction is an intricate and complex process, the investigation of which can involve the study of physiology, behavior, and individual and evolutionary history. Vertebrates exhibit a diversity of reproductive strategies in each of these aspects. Examples include internal and external fertilization; viviparous, ovoviviparous, and oviparous modes of reproduction; cystic and noncystic patterns of gametogenesis; and bisexual and unisexual reproduction, including hermaphroditism and parthenogenesis. Indeed, at least nine genera of nonmammalian vertebrates exhibit unisexual reproduction in which diploid or triploid ova either develop spontaneously or are stimulated to develop by sperm from bisexual species (gynogenetic reproduction) or by gametic fusion with the sperm of bisexual species (hybridogenesis).

Associated with each of these strategies are corresponding morphological, physiological, and behavioral adaptations. Thus, in some species, there are no

This chapter is dedicated to Frank A. Beach to honor his long active interest in and support of research in comparative behavioral endocrinology.

DAVID CREWS   Departments of Biology and Psychology and Social Relations and The Museum of Comparative Zoology, Harvard University, Cambridge, Massachusetts 02138.   RAE SILVER   Department of Psychology, Barnard College of Columbia University, New York, New York 10027.   Supported in part by NIMH 58572, NSF 75-13796, and NIMH Research Scientist Development Award 00135 to David Crews and by NIMH 02384 to Rae Silver. The review of literature is current to January 1978.

external morphological differences between the sexes, whereas in others extreme sexual dimorphisms occur. Whether or not the sexes differ externally, in most vertebrates they differ internally. However, two exceptions, simultaneous and sequential functional hermaphroditism, have been reported in fish and amphibians. In the functionally hermaphroditic species, individuals possess both female and male primary and secondary sex characters and are capable of self-fertilization (although cross-fertilization between individuals is more common). In the sequentially hermaphroditic species, individuals are either protogynous (first female and then male) or protoandrous (first male and then female).

Behavioral adaptations are equally varied. For example, the sexes may come together in response to an external cue and may simply release the gametes into the water. Alternatively, some species require prolonged interactions between the male and the female to synchronize reproductive processes and to ensure fertilization. These behavioral interactions serve not only to facilitate gonadal growth but also to establish pair bonds that enhance the reproductive success of the individuals.

Behavioral interactions integrating reproductive processes are not necessarily restricted to the single breeding pair or even to a single species. Social interactions can synchronize breeding in an entire population. In the case of parasitic species, observations of the reproductive behavior of other species may stimulate readiness to breed.

This diversity of reproductive adaptations is undoubtedly reflected in equally diverse hormonal and neural controlling mechanisms. Although only a very few species have been studied, the other chapters in the present book demonstrate that a good deal is known about mammalian reproductive processes. However, if one is to understand the evolutionary history of physiological influences on reproductive behavior in vertebrates, nonmammalian species must be investigated. Unfortunately, few investigators have taken advantage of the rich array of breeding adaptations exhibited by the "lower" vertebrates, and hence, little is known about their underlying mechanisms.

Although much of our knowledge of vertebrate reproductive physiology has been gathered from studies of mammals, there are sufficient structural and functional similarities between mammals and nonmammals to indicate that many of the mechanisms regulating reproduction are probably basic to vertebrates in general. For example, the interaction and interdependence of the hypothalamus, pituitary, and gonads have long been recognized as a fundamental feature of vertebrate reproductive endocrinology. Similarly, there is now abundant evidence that, in all vertebrate classes, neuroendocrine mechanisms controlling physiological processes also modulate behavioral events. The reader should keep in mind, however, that each of the surviving vertebrate groups has undergone millions of years of independent evolution in a variety of environments, each demanding specialized adaptations, and that a diversity of controlling mechanisms would be expected. Thus, as will be apparent in this review, principles developed from mammalian research do not apply *a priori* to the "lower" vertebrates.

In this chapter, we will review briefly the literature on the regulation of mating behavior in nonmammalian vertebrates, discuss in some depth those species about which substantial experimental analysis is available, and point out promising areas for further study. It should be remembered throughout that the reproductive process involves a sequence of interdependent stages, each one a consequence of preceding events and an essential prerequisite for the events to follow.

TELEOST FISHES

Several studies indicate that hormones of the hypothalamo-pituitary-gonadal axis regulate reproductive behavior in fish. Although gonadal hormones play a major role in controlling reproductive behavior, there is some evidence that anterior pituitary and neurohypophyseal hormones also directly regulate reproductive responses in fish. In an excellent review of the literature, Liley (1969) noted that relatively little is known about the hormonal control of fish reproductive behavior. Although there has been some progress in this area (cf. Liley, 1980) our knowledge of the hormone–behavior relationships in fish has not advanced significantly in the last 10 years.

THE HYPOTHALAMO-PITUITARY-GONADAL AXIS

Teleost fish possess a hypothalamo-pituitary-gonadal (H-P-G) axis similar in structure and function to that of mammals and other tetrapod vertebrates (Hoar, 1969; Ball and Baker, 1969; Dodd et al., 1971; Chester-Jones et al., 1972; Donaldson, 1973; Peter, 1973, 1977; Schreibman et al., 1973; DeVlaming, 1974; Schreibman, 1978). There is good evidence of hypothalamic control of pituitary function in teleost fish. In the carp *(Crypritnus carpio)*, systemic injection of hypothalamic extract causes an increase in pituitary gonadotropin (GTH) release both *in vivo* (Breton and Weil, 1973) and *in vitro* (Breton et al., 1972a, 1975) conditions. A gonadotropin-releasing factor (Gn RH) has also been found in hypothalamic extracts of goldfish (*Carassius auratus;* Crimm et al., 1976) and rainbow trout (*Salmo gairdeneri;* Breton et al., 1972b). Recent experiments have demonstrated that, in several fish species, administration of synthetic luteinizing hormone-releasing hormone (LH-RH) causes granule depletion of pituitary gonadotropin cells (Kaul and Vollrath, 1974; Hirose and Ishida, 1974; Lam et al., 1976; Chan, 1977). In the brown trout *(S. trutta),* LH-RH injections result in an increase in plasma GTH levels within 15 minutes (Crim and Cluett, 1974). Injection of LH-RH into sexually inactive Japanese medaka *(Oryzias latipes)* stimulates ovarian development (Chan, 1977).

Gonadotropin-releasing hormone-producing cells are believed to be localized in the nucleus preopticus (NPO) region of the hypothalamus (Peter, 1973, 1977;

Figure 1). Gonadotropin-releasing hormone is released into the pituitary portal system and transported to the adenohypophysis (Zambrano, 1972). In addition, there is evidence of direct innervation of the pituitary gonadotropin-producing cells (Schreibman, 1978); for example, Kandel (1964) and, more recently, Hayward (1974) have demonstrated that axons originating from cells located in the NPO of goldfish extend into the pituitary. Destruction of the nucleus lateralis tuberis (NLT) results in a decrease in gonadosomatic index (gonadal weight/body weight × 100) in the goldfish (Peter, 1970; see also Zambrano, 1972). The NLT is believed to be the site of control of GTH secretion. Implantation of antiestrogens (clomiphene citrate or ICI 46474) in the NLT or the pituitary of gravid female goldfish significantly increases serum GTH levels, whereas implantation of steroid agonists in the NPO are without effect, a result suggesting that the pituitary and the NLT region are sites for negative feedback effects (Billard and Peter, 1977; see also Sage and Bromage, 1970).

A close correlation between changes in activity in gonadotropin-producing cells of the anterior pituitary and in the gonadal state (gametogenesis and steroidogenesis) of fish has long been noted (Schreibman *et al.*, 1973; Schreibman, 1978). Metuzals *et al.* (1968) and Polder (1971) were able to correlate histological and cytological changes in the gonadotropes located in the proximal pairs distalis of the adenohypophysis of the cichlid fish *Aequidens portalegrensis* with behavioral changes during the reproductive cycle. In salmonid fishes, plasma GTH levels increase during gonadal recrudescence and, in females, are highest at the time of ovulation (Crimm *et al.*, 1975; see also DeVlaming, 1974). Castration elevates circulating immunoreactive GTH levels in rainbow trout, indicating further that the hypothalamo-pituitary axis in fish is under negative feedback control by gonadal steroids (Billard *et al.*, 1976, 1977).

Several studies have been aimed at identifying the active gonadotropic factor in fish pituitaries (reviewed by Donaldson, 1973). For example, salmon pituitary

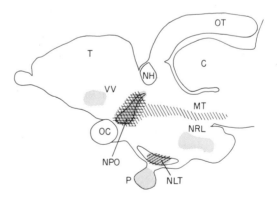

Fig. 1. A parasagittal view of a generalized teleost fish brain showing regions known to be involved in the control of gonadotropin secretion (///), spawning behavior (≡), and spermiation (\\\). See text for details. Stippled areas indicate sites of steroid uptake. Abbreviations: C, cerebellum; MT, midbrain tegmentum; NH, nucleus habenularis; NLT, nucleus lateralis tuberis; NPO, nucleus preopticus; OT, optic tectum; OC, optic chiasm; P, pituitary; T, telencephalon; NRL, nucleus recessus lateralis; VV, area ventralis telencephali pars ventralis.

gonadotropin maintains gonadal activity in hypophysectomized catfish (Nayyar, *et al.*, 1976; see also Sundararaj *et al.*, 1976) and the guppy, *Poecilia reticulata* (Liley and Donaldson, 1969), and induces testicular growth in immature trout (Drance *et al.*, 1976). Bioassay and biochemical data indicate that the teleost pituitary, unlike that of mammals, elaborates a single LH-like gonadotropin (Donaldson, 1973; Burzawa-Gerard *et al.*, 1976), although this may not be true of all fish species (Ball and Baker, 1969; Idler *et al.*, 1976; Haider and Blüm, 1976). A number of experiments have demonstrated that mammalian LH, but not FSH, stimulates gonadal development in intact and hypophysectomized fish (Hoar, 1969; Wiebe, 1969; Pickford *et al.*, 1972; Nayyar *et al.*, 1976). Finally, injection of methallibure, a chemical that suppresses pituitary gonadotropin secretion via its influence on catecholamine pathways mediating the release of hypothalamic Gn RH (Kitay and Westfall, 1977), has been shown to suppress spermatogenesis and serum GTH levels in a number of fish species (Pandey, 1970; Wiebe, 1968; Hoar *et al.*, 1967; Singh *et al.*, 1977); the effect of methallibure can be reversed by simultaneous treatment with mammalian LH, FSH, human chorionic gonadotropin (hCG), or pituitary extract (Wiebe, 1969; Billard *et al.*, 1971; Hyder, 1972; Donaldson, 1973; Singh *et al.*, 1977), a finding suggesting that it operates at the pituitary rather than at the gonadal level.

Gonadal production of steroids by fishes has been reviewed by Gottfried and van Mullem (1967), Ozon (1972a,b), and Lambert and van Oordt (1974). Estrogens (estradiol-17β, estrone, and estriol), progesterone, and androgens (testosterone and 11-ketotestosterone) appear to be specific for teleost fish. Few studies, however, have been directed toward the identification and quantification of circulating hormones in fish. Estradiol-17β has been monitored in the peripheral plasma and has been found to be correlated with ovarian maturity in *Tilapia aurea* (Yaron *et al.*, 1977); ovariectomy results in a precipitous and permanent decline in estradiol levels and parameters related to plasma vitelloprotein, whereas exogenous estradiol treatment of ovariectomized fish stimulates synthesis and release of vitelloproteins. Idler and co-workers (1976) monitored plasma steroid levels in ambisexual (functional hermaphrodites) and gonochoristic (bisexual) teleosts. They reported large quantities (687–1884 ng/100 ml) of 11β-hydroxytestosterone but an absence of 11-ketotestosterone in ambisexual species; in the two gonochoristic species studied, the typical teleost pattern was observed, that is, high (221–1137 ng/100 ml) concentrations of 11-ketotestosterone and low 11–12 ng/100 ml) levels of 11β-hydroxytestosterone. In the winter flounder *(Pseudopleuronectes americanus)*, testosterone levels in males change only slightly during the annual cycle, but 11-ketotestosterone concentrations rise dramatically (to 17.9 μg/100 ml) near the time of spawning. In the female, testosterone levels are highest prior to spawning (10–15 μg/100 ml), whereas 11-ketotestosterone remains low throughout the year (0.01–0.2 μg/100 ml; Campbell *et al.*, 1976). In female plaice *(Pleuronectes platessa)*, plasma concentrations of estradiol-17β and testosterone are at their highest level prior to spawning (16.2 and 6.2 μg/100 ml, respectively); males show similar but less distinct seasonal changes in steroid concentrations (Wingfield and Crimm, 1977).

Localized uptake and retention of steroid hormones by the central nervous system (CNS) have been demonstrated in two species of fish (Figure 1). In the green sunfish *(Lepomis cyanellus)* and the paradise fish *(Macropodus opercularis)*, testosterone and estradiol are concentrated in the NPO, the NLT, the ventralis ventralis nucleus, the nucleus recessus lateralis, and the pars distalis of the pituitary (Morell *et al.,* 1975b; Davis *et al.,* 1977). The NLT region has been implicated in feedback regulation of pituitary hormone secretion (Peter, 1970, 1973, 1977).

In summary, the organization and regulation of the H-P-G axis of fish is similar to that of "higher" vertebrates.

HORMONE–BEHAVIOR INTERACTIONS IN TELEOST FISH

Traditionally, reproductive behavior has been considered to be under the control of gonadal steroid hormones. Theoretically, however, the possible influence of hypothalamic and hypophyseal hormones has long been appreciated. Indeed, work with mammals in the 1970s (Moss and McCann, 1973; Pfaff, 1973; Crowley *et al.,* 1976) implicated hypothalamic releasing hormones in the regulation of the female rat's sexual behavior.

Although few species have been investigated in any detail, available evidence indicates that, in fishes, hormones elaborated by the brain and the pituitary as well as by the gonads directly influence reproductive behavior. As the role of hormones in the control of reproductive behavior of fishes has been reviewed by Liley (1969, 1980; see also Fiedler, 1974), we will concentrate on work in the 1970s.

ROLE OF PITUITARY AND GONADAL HORMONES IN THE CONTROL OF MALE SEXUAL BEHAVIOR. More is known about the courtship behavior of the three-spined stickleback *(Gasterosteus aculeatus)* than of any other species of fish (reviewed by Wootton, 1977). Much of our knowledge of the physiological correlates of the striking behavioral changes seen during the reproductive cycle of this fish has been provided by Baggerman and Hoar and their coworkers.

The stickleback is a small, seasonally breeding fish, reproducing in fresh water in the early spring. Gonadal recrudescence is initiated by the seasonal increase in day length, which is believed to stimulate GTH secretion; fish maintained under long photoperiods (18L:6D) mature rapidly and exhibit almost continual reproductive activity, whereas fish exposed to short photoperiods (6L:18D) never become sexually active (Baggerman, 1957, 1966, 1968, 1969, 1972; Hoar, 1965). As the spawning season approaches, males establish breeding territories and acquire the well-known red nuptial coloration. The male builds a nest out of pieces of grass and algae, which he glues together with a mucous secretion produced by a modified portion of the kidney; both male sexual coloration and kidney mucous secretion are androgen-dependent secondary sexual characters (Ikeda, 1933). After nest building, the territorial males attract and court females, leading them to the nest site while performing a highly stereotyped sequence of displays known as the *zig zag dance.* Spawning is followed by a parental care phase in which the male fans (aerates) the eggs. Thus, there is a predictable sequence of events during the reproductive cycle, beginning with migration and ending with parental behavior (Figure 2).

Castration of males during the presexual phase results in increased aggressive behavior and decreased sexual behavior (Hoar, 1962a,b, 1965; Wai and Hoar, 1963; see also Baggerman, 1968). If maintained under long photoperiods, castrated as well as intact males (and females) exhibit higher levels of aggressive behavior than when maintained under short photoperiods (Hoar, 1962a,b). Administration of mammalian LH stimulates aggressive behavior in mature and immature intact fish (Hoar, 1962a,b; see also Blüm and Fiedler, 1968). Methallibure treatment leads to a decline in aggressive behavior in sexually active males (Carew, 1968), further suggesting that pituitary gonadotropin is causally involved in controlling the level of aggression in male sticklebacks. Finally, androgen replacement therapy (methyltestosterone) leads to a decline in territoriality (via negative feedback on GTH secretion?) and an increase in sexual behavior (Wai and Hoar, 1963).

These results have led Hoar and Baggerman to hypothesize that pituitary gonadotropin(s) controls the presexual phase (territorial behavior), whereas gonadal steroids regulate male courtship and spawning behavior. The transition from pituitary control of aggressive behavior to gonadal steroid control of courtship behavior is believed to occur during nest building, the transition phase between territory establishment and courtship of the female. Evidence that sup-

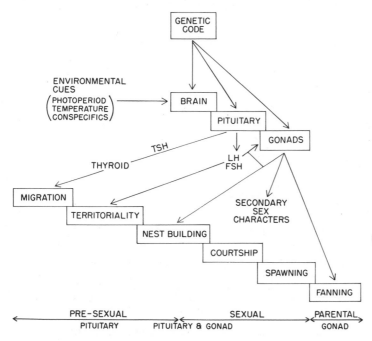

Fig. 2. Sequence of behavioral events in the reproductive cycle of the male three-spined stickleback *(Gasterosteus aculeatus)* and the presumed role of pituitary and gonadal hormones in their regulation. (From "The Endocrine System as a Chemical Link between the Organism and Its Environment" by W. S. Hoar. *Transactions of the Royal Society of Canada,* 1965, *3,* 175–200. Copyright 1965 by the Royal Society of Canada. Adapted by permission.)

ports this hypothesis is limited, however. Castration during nest building leads to a rapid decline in aggressive behavior and cessation of male reproductive behavior; androgen replacement therapy reinstates nest building more rapidly if castrated males are maintained under long photoperiods than if they are housed under short photoperiods (Baggerman, 1966, 1968; Hoar, 1962a,b). This finding has led to the suggestion of a synergism between photoperiodically induced gonadotropin secretion and testicular hormones in the control of nest-building behavior.

As pointed out by Wootton (1977), this interpretation must be viewed with caution because the evidence that long day lengths stimulate gonadotropin secretion is indirect, and other neural and hormonal changes besides GTH secretion could be produced by this lighting regimen. Further, the finding that exogenous GTH elicits aggressive behavior in the intact immature fish does not necessarily mean that behavior is influenced directly, as androgen secretion may have occurred. Indeed, examination of Hoar's data indicates that GTH treatment of sexually immature males stimulated kidney mucous secretion, an androgen-dependent response, and that a small number of animals went on to build nests. Recent experiments have demonstrated that treatment with the antiandrogen cyproterone acetate during the presexual phase reduces aggressive behavior and nuptial coloration as well as inhibiting kidney modification and the later stages of spermatogenesis (Rouse et al., 1977).

Several studies of other teleost fish species indicate that aggressive behavior is maintained and sexual behavior diminished after castration. In male sunfish (Lepomis spp.), reproductive behavior is under androgen control; gonadectomized male longears (Lepomis megalotis peltastes) and pumpkinseeds (L. gibbosus) are highly aggressive but fail to build nests, whereas treatment with methyltestosterone restores nest building (Smith, 1969). In the blue gourami (Trichogaster trichopterus), castrated males fight as vigorously as intact males until a stable dominance hierarchy is established. However, castrates fail to build nests or to perform spawning and prespawning movements unless they are also treated with methyltestosterone (Johns and Liley, 1970). In male platyfish (Xiphophorus maculatus), castration results in a decline in aggressive behavior and a reduction in sexual behavior (Chizinsky, 1968). In contrast, Tavolga (1955) reported that castration in the gobiid fish (Bathygobius soporator) results in a marked decline in aggressive behavior toward other males, with no decrement in courtship behavior; hypophysectomy results in cessation of both aggressive and sexual behavior. Similar effects of gonadectomy were reported by Noble and Kumpf (1936) for the jewel fish (Hemichromis bimaculatus) and the Siamese fighting fish (Betta splendens; see also Aronson, 1959); it should be kept in mind, though, that the completeness of castration is critical, as microscopic remnants of testicular tissue are capable of maintaining reproductive behavior (Johns and Liley, 1970). Davis et al. (1976b) reported that, in the paradise fish, both surgically castrated males and males receiving methallibure treatment continue to exhibit spawning behavior. As pointed out by Davis, removal of methallibure from the diet during the week of behavioral testing could have resulted in recovery of gonadotropin secretion and hence could have stimulated gonadal recrudescence and sexual behavior (see also Hyder et al., 1974).

HORMONES AND RESPONSIVENESS TO EXTERNAL STIMULI. There is some evidence that sensitivity to external stimuli varies with hormonal state in fishes. Working with gobiid fish, Tavolga (1955) found that intact males courted only gravid females, whereas castrated males courted males, gravid females, and nongravid females indiscriminately; testicular regeneration resulted in the reappearance of a discrimination response toward gravid and nongravid females. Wootton (1970) found that female sticklebacks showed a decrease in the threshold for detecting red (the ventral color of sexually active males) during the breeding season.

Tavolga (1955) demonstrated further that males, nongravid females, and gravid females presented in a glass container are courted by intact males, a finding suggesting that recognition of the gravid females is based on a chemical cue. The pheromone is presumably produced by the ovary at the time of ovulation. Sexual pheromones have been demonstrated in a number of species, including the channel catfish *Ictalurus punctatus* (Timms and Kleerekoper, 1972), *Astyanax mexicanus* (Wilkens, 1972), and goldfish (*Carassius auratus;* Partridge *et al.,* 1976). Interestingly, Partridge *et al.* (1976) reported that only males that have spermiated respond to postovulatory females.

CONTROL OF SPAWNING BEHAVIOR. In considering the endocrine control of reproductive behavior in oviparous "lower" vertebrates, it is important to distinguish the growth and release of gametes from the gonad (ovulation and spermiation) from the release of gametes from the body (oviposition and sperm release). Although these events are usually closely associated temporally, this is not always the case, and research with fish indicates that ovulation and spermiation may have different neural and hormonal controlling mechanisms from those of oviposition and sperm release (spawning). For example, in the Japanese medaka, ovulation takes place between 1 and 5 A.M, and if a sexually active male is present, the eggs are oviposited between 4 and 7 A.M. (Egami and Nambu, 1961). If the preovulatory female is separated from the male, ovulation still occurs at the usual time, but the ovulated eggs are retained in the ovarian lumen until they are reabsorbed or until the female is exposed to a courting male. Similarly, in the goldfish, ovulation is spontaneous, whereas oviposition and spawning behavior require the presence of a male and green plants (Hoar, 1965).

It is also important that the term *spawning* be operationally defined for each species investigated, as the causal bases of the behavior may differ between species (see Table 1). For example, the "spawning reflex" exhibited by cyprinodont species (killifishes) is not preceded by pair formation and can be induced hormonally in isolated gonadectomized, hypophysectomized males and females. Administration of fish and mammalian neurohypophyseal preparations (crude extracts or purified vasopressin, arginine vasopressin, arginine vasotocin, and oxytocin), as well as of synthetic oxytocin, elicits the spawning reflex in male and female *Fundulus heteroclitus* (Wilhelmi *et al.,* 1955; Ramaswami and Lakshman, 1958; Macey *et al.,* 1974; Pickford and Strecker, 1977); *Gambusia* spp. (Ishii, 1961); *Oryzias* spp. (Egami, 1959); *Rhodeus* (Egami and Ishii, 1962); and *Jordanella floridae* (Crawford, 1975, cited in Liley, 1980). In an attempt to localize the region of the brain involved in the spawning reflex in *F. heteroclitus,* Macey *et al.* (1974) lesioned different areas of

TABLE 1. INFLUENCE OF HYPOPHYSEAL EXTRACTS AND PURIFIED HORMONES ON SPAWNING
BEHAVIOR IN TELEOST FISHES

Species	Substance administered	Response	Authority
Cryprinodontiformes			
Oryzias latipes	Neurohypophyseal extract	Spawning	Egami, 1959
Fundulus heteroclitus	Oxytocin, vasopressin, neurohypophyseal extract	Spawning	Wilhelmi *et al.*, 1955; Macey *et al.*, 1974
Gambusia sp.	Neurohypophyseal extract	Spawning	Ishii, 1961
Ostariophysi (Cypriniformes)			
Misgurnus anguillicaudatus	Neurohypophyseal extract	No response	Egami and Ishii, 1962
Cyprinus spp.	Anterior pituitary extract, mammalian	Spawning	Sinha, 1971
Puntius gonionotus	gonadotropin, neurohypophyseal hormones	No response	Sinha, 1971
Heteropneustes	Mammalian FSH and LH	Spawning	Sundararaj, 1957
Carassius auratus	LH-RH	Spawning	Lam *et al.*, 1975
	Prostaglandin (PGF$_2\beta$)	Spawning	Stacey, 1976

the forebrain prior to injection of porcine pituitary suspension or synthetic arginine vasotocin. Only lesions that destroyed the NPO adversely affected the spawning response (see Figure 1); incomplete lesions of the NPO or lesions in other parts of the diencephalon or the telencephalon did not block the occurrence of the spawning reflex when neurohypophyseal hormones were injected. The NPO, which has afferents similar to those known for the supraoptic and paraventricular nuclei of mammals (Peter, 1973), has also been implicated through direct brain stimulation studies in the control of sperm release in the sunfish (Demski *et al.*, 1975) and the bluegill (Demski and Knigge, 1971).

In contrast to the Cyprinodontidae, spawning behavior in other teleost species may involve a coordinated sequence of behavioral displays between the male and the female that are necessary for successful oviposition and fertilization. For example, in the goldfish, ovulation is spontaneous and is triggered by a rise in water temperature; gravid (vitellogenic) females do not ovulate unless the water temperature is raised to 20°C (Yamamoto *et al.*, 1966; Yamamoto and Yamazaki, 1967; Pandey and Hoar, 1972). Ovulation is under pituitary control and can be induced by synthetic LH-RH (Lam *et al.*, 1975). Both clomiphene citrate (Pandey and Hoar, 1972; Pandey *et al.*, 1973) and prostaglandins (Stacey and Pandey, 1975; see also Peter and Billard, 1976) are also effective in inducing ovulation in intact but not in hypophysectomized females. If, however, a sexually active male and a suitable spawning substrate are not present, female goldfish do not exhibit spawning behavior following normal or hormonally induced ovulation.

In an elegant series of experiments, Stacey and Liley (1974) demonstrated that, if oviposition is prevented by a glass plug in the ovipore of an ovulated female, spawning behavior is prolonged. Further experiments indicated that the placement of recently ovulated eggs in the ovarian lumen via the ovipore induces spawning in females with active ovaries. In females with regressed ovaries, neither estradiol alone nor saline plus eggs was effective in inducing spawning. However, injection of estradiol in combination with eggs induced spawning in sexually inactive females.

The mechanism whereby injected ovulated eggs induce spawning behavior has also been explored. Ovulation results in stretching the ovarian lumen and the oviduct. In mammals, uterine distention and cervical stimulation are associated with the release of prostaglandins (PG), which induce lordosis behavior via their action on LH-RH and LH (Rodriguez-Sierra and Komisaruk, 1977). Stacey (1976) investigated the role of PG in egg-induced spawning behavior in the goldfish. Spawning behavior is inhibited if indomethacin (IM), an inhibitor of PG biosynthesis, is administered prior to or coincident with the injection of ovulated eggs. Intraperitoneal injection of the prostaglandin $PGF_{2\alpha}$ is effective in overcoming IM blockade of spawning behavior in induced females; $PGF_{2\alpha}$ alone is capable of inducing spawning behavior in untreated intact females within 5–120 min of injection. Hypophysectomized females fail to respond to injections of PG, but two weeks' pretreatment with salmon pituitary gonadotropin restores the spawning response to $PGF_2\alpha$.

In the herring *(Clupea harengus)*, spawning occurs en masse in shallow water and appears to be stimulated by milt released by other spawning males. Stacey and Hourston (1982) reported that both male and female herring increase swimming speed and protrude the genital papilla on contact with suspended milt. If a smooth substrate free of detritus is contacted, the spawning fish orients the belly toward the substrate and deposits a ribbon of eggs (or milt). Herring pretreated with IM (10 μg/g) fail to respond to suspended milt, a finding suggesting that, in the herring, as in the goldfish, PGs are involved in the regulation of spawning behavior.

Although these experiments suggest that distention of the oviduct following ovulation results in the release of PGs that consequently influence behavior, the role of endogenous PGs is not yet clear. Further, it is not known whether PGs influence gonadotropin secretion by acting at both the hypothalamic (Chobsieng et al., 1975) and the pituitary (Sato et al., 1975) levels. Injection of PGs (PGE_2 and $PGF_{2\alpha}$) into the third ventricle of female goldfish significantly suppresses serum GTH concentrations (Peter and Billard, 1976). A direct behavioral effect of PGs has also been demonstrated recently in mammals. Exogenous PGE_2 facilitates female receptivity (lordosis) in estrogen-primed ovariectomized rats (Dudley and Moss, 1976; Hall et al., 1975; Rodriguez-Sierra and Komisaruk, 1977), and Rodriguez-Sierra and Komisaruk (1978) suggested that PGE_2 synergizes with LH-RH to stimulate female sexual receptivity. Comparable experiments have not been conducted with fish, although Lam et al. (1975), working with preovulatory goldfish, noted eggs in tanks following LH-RH treatment. Urotensin from the caudal neurosecretory system has also been implicated in the regulation of fish spawning behavior (reviewed by Berlind, 1973; Lederis, 1973).

In a series of studies, Liley has investigated the endocrine control of sexual receptivity in the female guppy. Female guppies undergo cycles of sexual receptivity that are correlated with ovarian activity. Females are maximally receptive shortly after parturition (Liley, 1966; Liley and Wishlow, 1974; see also Carlson, 1969). Early experiments, however, indicated that ovariectomized virgin females were as responsive to male courtship as intact virgins (Liley, 1968). This persistence of sexual receptivity in ovariectomized females led Liley to conclude that the ovary was not essential for the performance of sexual behavior in this species. Subsequent experiments indicated that the high level of sexual behavior observed following ovariectomy is restricted to the first test period and that sexual responsiveness declines with repeated testing (Liley and Wishlow, 1974). Ovariectomized females are highly responsive initially to male courtship on first encounter whether tested 2, 8, 10, or 24 days after surgery (Liley, 1972; Liley and Wishlow, 1974). Hormones play a critical role in the maintenance of receptivity after initial testing (Liley, 1972). Estrogen replacement therapy reinstates receptivity in nonresponsive ovariectomized females; estradiol-17β and diethylstilbesterol (DES) are slightly more potent behaviorally than the other estrogens, estriol and estrone.

Unlike the case with male aggressive behavior (see next section), there is little evidence that pituitary gonadotropins may be involved in the regulation of female sexual behavior. Hypophysectomy abolishes female sexual responsiveness in the guppy (Liley, 1968), but pretreatment of hypophysectomized females with salmon pituitary gonadotropin reinstates sexual behavior (Liley and Donaldson, 1969). This action is most likely due to gonadotropin stimulation of ovarian hormone secretion.

Other workers have studied the effect of ovariectomy on female fish. Ovariectomized *Xiphophorus helleri* (Noble and Kumpf, 1936) and *Poecilia* sp. (Ball, 1960) continued to be sexually attractive, but it is not reported whether these females were also receptive to the male's courtship.

NEURAL CONTROL OF AGGRESSIVE AND COURTSHIP BEHAVIOR. Relatively few experiments have been done on the neural control of aggressive and courtship behavior in fish (for reviews, see Segaar, 1965; Aronson, 1970; Aronson and Kaplan, 1968; Bernstein, 1969; Bruin, 1977).

Kamrin and Aronson (1954), working with the male platyfish, reported that forebrain ablation markedly reduced male courtship behavior. Similar results have been reported for the green sunfish (*L. cyanellus;* Hale, 1956); guppy (Segaar, 1965); Siamese fighting fish (*Betta splendens;* Bruin, 1977); and the cichlid fishes *Tilapia macrocephala* (Aronson, 1948), *T. mossambica* (Overmier and Gross, 1974), and *Hemihaplochromis philander* (Ribbink, 1972). Interestingly, in all of these experiments, it was noted that some males continued to mate successfully.

In the stickleback, large frontal or lateral telencephalic lesions result in decreased aggression and reduced the duration of sexual activity. Lesions in the mediocaudal aspect of the forebrain, on the other hand, increase aggression, and sexual behavior becomes persistant (Segaar and Nieuwenhuys, 1963). The reverse seems to be true for the Siamese fighting fish. In this species, radiofrequency lesions of the mediocaudal area of the dorsal telencephalon result in a decrease in both frontal and lateral aggressive displays, and withdrawal movements increase in duration (Bruin, 1977). Lesions of the frontal area of the dorsal telencephalon result in opposite changes: both frontal and lateral displays increase, and withdrawal is abolished. These findings led Bruin to conclude that the frontal portion of the dorsomedial telencephalon inhibits aggressive behavior in male *B. splendens,* whereas the most caudal portion of the dorsomedial telencephalon has a facilitatory function.

Davis and co-workers have conducted an extensive investigation of the role of the forebrain in the control of reproductive behavior in the paradise fish *(Macropodus opercularis)*. Removal of the telencephalon anterior and dorsal to the NPO disrupts courtship and nest-building behavior in sexually active males (Davis *et al.,* 1976a; Schwagmeyer *et al.,* 1977). Bilateral telencephalic ablation was also found to decrease the gonadosomatic index of males (Kassel *et al.,* 1976), a result indicating that the lesions destroyed hypothalamic connections that regulate hypophyseal secretions (see also Peter, 1970, 1973).

Finally, electrical stimulation of the NPO and other brain areas known to concentrate androgens (see above) elicits nest-building and courtship behavior but inhibits aggression in the bluegill *L. macrochirus* (Demski and Knigge, 1971); stimulation of the medioventral torus semicircularis elicits vocalizations in the toadfish *Opsanus beta* (Demski and Gerald, 1972). Electrical stimulation of the NPO area elicits vertical banding, a component of male sexual display, in the green sunfish (Bauer and Demski, 1974).

GENETIC CONTROL OF THE HYPOTHALAMO-PITUITARY-GONADAL AXIS. It is obvious that both inter- and intraspecific variations in reproductive behavioral physiology are in part due to individual differences in genotype (cf. Shire, 1976), yet there have been relatively few investigations of the underlying genetic factors or of how their expression may be influenced by the organism's internal and external environment. Recent research has revealed that the platyfish *(Xiphophorus maculatus)* represents a unique system for investigating the genetic control of the development of the H-P-G axis.

The onset of sexual maturity in the platyfish is controlled by a single gene locus *(P)* that determines the age at which the gonadotropic zone of the anterior pituitary differentiates and becomes physiologically active (Kallman and Schreibman, 1973; Kallman *et al.,* 1973; Schreibman and Kallman, 1977). Further, the alleles for differentiation are linked to a variety of pigment genes that serve as genetic markers of the *P* locus. Thus, there exists an early-maturing strain p^e, identified by a red iris color that matures at approximately 12.5 weeks, and a late-maturing strain p^l, identified by a red body color, which typically matures at 26.0 weeks; heterozygotes mature at intermediate times (Kallman and Schreibman, 1973).

Sexual maturation in male platyfish is marked by a decrease in the growth rate and by the transformation of the anal fin into a rod-shaped intromittent organ, the gonapodium, in several clearly defined stages (Kallman and Schreibman, 1973). Transformation of the anal fin is androgen-dependent in a quantitative manner and reflects testicular growth and steroidogenesis (Pickford and Atz, 1957). Further, Kallman and Schreibman (1973; Kallman *et al.,* 1973; Schreibman and Kallman, 1977) have demonstrated that gametogenesis is correlated with the appearance of neurosecretory material in the hypothalamus and the development and differentiation of the gonadotropic zone of the pituitary. These findings led Kallman and Schreibman to suggest that pituitary differentiation (or hypothalamic controlling mechanisms) become functional at a genetically determined time.

There is also evidence that the precise timing of sexual maturation is influenced by social factors. Using the same strains of the platyfish as Kallman and Schreibman, Sohn (1977) reported that, in pairs of early-maturing fish taken from the same brood, the fish that are initially larger, and hence dominant, significantly retard sexual maturation of the subordinate fish, as indicated by differentiation of the gonapodium, until the subordinate fish has *surpassed* the dominant fish in body size. Because body growth stops at sexual maturity in males, this "disinhibition" of the subordinate fish can occur at a variable time, depending on the size of the larger, behaviorally dominant fish. In fact, preliminary experiments indicate that by constantly increasing the size of the larger, dominant fish, it is possible to produce a sexually mature but much larger than normal, p^r male (Sohn and Crews, 1977).

Given this information, questions regarding the mechanisms of action on both the behavioral and the physiological levels can now be asked. For example, at what level of the H-P-G axis is the behavioral inhibition of the maturation of the subordinate fish occurring? In this regard, it is interesting that LH-RH levels appear to increase dramatically at sexual maturation (Schreibman, personal communication, 1977) and thus may stimulate pituitary organization. Is attainment of sexual maturity genetically determined or dependent on the individual's achieving a certain body weight before the maturational process can be initiated? Experiments indicate that sexual maturation is initiated only after a certain critical body weight is reached (Sohn and Crews, 1977). Or, finally, is the social inhibition of sexual maturation reported by Sohn (1977) due to visual, olfactory, or tactile stimuli? It has been observed that, when reared in pairs, the dominant fish repeatedly chases the subordinate fish, a finding suggesting that inhibition of sexual maturation in the subordinate fish is due to these behavioral interactions.

Amphibians

The literature on physiological aspects of amphibian reproductive biology and their controlling mechanisms has been reviewed, and the reader is directed to these sources (brain–behavior interactions: Jørgensen, 1968, 1970, 1974; Jørgensen and Larsen, 1966; Lofts, 1974; van Oordt, 1974; reproductive physiology: Lofts, 1968, 1974; Redshaw, 1972; Follett and Redshaw, 1974). Oksche and Ueck (1976) reviewed the evidence for other neuroendocrine systems (e.g., aminergic neurons in the circumventricular organs) that appear to be involved in pituitary regulation in amphibians. A great deal of work has also been done on the reproductive behavior of amphibians (Noble, 1931; Salthe and Mecham, 1974; Taylor and Guttman, 1977). The purpose of this section is to integrate the physiological and behavioral data on control of reproductive events.

There is clear evidence that the hypothalamo-pituitary-gonadal axis (H-P-G) in amphibians functions according to principles established for other tetrapod vertebrates. Dierickx (1965, 1966, 1967b) has been able to localize the gonadotropin-controlling center in the frog *Rana temporaria* by sectioning the base of the brain at different levels and using ovarian growth and ovulation as indices of pituitary activity. Cuts made just anterior to the pituitary inhibited gonadotropin release (Figure 3, Level A), whereas cuts just posterior to the optic chiasm (Figure 3, Level B) did not interfere with gonadal growth. On the basis of these findings, Dierickx (1965) concluded that neurons located in the ventral infundibular nucleus (VIN; also known as the parvicellular area of the pars ventralis of the tuber cinereum) secrete Gn RH (see also Peute and Meij, 1973). Recent immunofluorescence studies, however, indicate that Gn RH-producing cells are located in the preoptic area and that Gn RH is transported via fiber systems to the VIN (Goos *et al.*, 1976; Doerr-Schott and Dubois, 1976; Alpert *et al.*, 1976).

The magnocellular preoptic nucleus (MPOA) of amphibians is homologous to the supraoptic and paraventricular nuclei of mammals (Oksche and Ueck, 1976) and appears to be involved in the control of ovulation in amphibians. Hypothalamic cuts posterior to the optic chiasm inhibit ovulation, whereas cuts that leave intact the hypothalamo-hypophyseal tract between the MPOA and the VIN allow normal ovulation (Dierickx, 1967b). Lesions of the MPOA do not influence normal seasonal gonadal development or steroid secretion (as indicated by the development of secondary sex characters) but interfere with ovulation (Dierickx, 1967d).

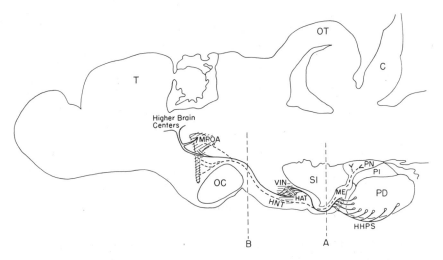

Fig. 3. A parasagittal view of the neuroendocrine-regulating regions in the amphibian brain. Experiments indicate that the magnocellular preoptic nucleus (MPOA) is involved in ovulation (///) and the production of pituitary gonadotropin-releasing hormone, which is transported to the ventral infundibular nucleus (VIN) (≡). See text for details. Other abbreviations: C, cerebellum; HAT, hypothalamo-adenophypophyseal tract; HHPS, hypothalamo-hypophyseal portal system; HNT, hypothalamo-neurohypophyseal tract; ME, median eminence; OC, optic chiasm; OT, optic tectum; PD, pars distalis; PI, pars intermedia; PN, pars nervosa; SI, saccus infundibuli; T, telencephalon.

Jørgensen (1968, 1970) has reported similar results in the toad *Bufo bufo* following transection of the ventral hypothalamus between the pituitary and the VIN. Anterior deafferentation of the ventral hypothalamus at this level prevents normal ovarian recrudescence following spawning. Transection of hypothalamo-hypophyseal connections by ectopic pituitary autotransplantation to sites other than the median eminence (ME) region results in testicular regression and atrophy of the nuptial pads in the crested newt (*Triturus cristatus carnifex;* Mazzi, 1970) and the green frog (*Rana esculenta;* Rastogi and Chieffi, 1972); animals with the pars distalis autotransplanted to the region of the ME exhibit normal spermatogenic activity and development of secondary sex characters.

Direct evidence that the amphibian pituitary is under the stimulatory control of the hypothalamus is available. Incubation of frog *(R. pipiens)* pituitaries with hypothalamic homogenates stimulates the release of bioassayable LH-like gonadotropin (Thornton and Geschwind, 1974). Deery (1974) and Alpert *et al.,* (1976) have extracted a substance immunologically identical to synthetic LH-RH from the hypothalamus of the South African clawed frog *(Xenopus laevis)* and *Rana* species. Electrical stimulation of the VIN induces spermiation in *R. esculenta* (Stutinsky and Befort, 1964).

There have been relatively few studies of the effects of purified or synthetic Gn RH on pituitary function in amphibians. A single injection (1 µg) of synthetic Gn RH stimulates ovulation in PMS-primed *X. laevis;* gonadal recrudescence (vitellogenesis) is initiated by a series of injections (three times daily for 4 days; Thornton and Geschwind, 1974). Similarly, Mazzi and co-workers (1974) found that chronic administration of synthetic LH-RH (0.5 µg/day) to animals with ectopic pituitary transplants maintains gonadotropin secretion, measured by gametogenesis in the crested newt; perfusion of the pituitary *in situ* with 6.0 µg LH-RH stimulates ovulation in females (Vellano *et al.,* 1974). Licht (1974b) demonstrated that a single injection of synthetic FSH/LH-RH induces spermiation in intact tree frogs *(Hyla regilla;)* the median effective dose was between 35 and 75 ng. Hypophysectomy abolishes this response, whereas pituitary replacement restores it.

Histochemical and physiological studies have identified the gonadotropin-producing cells in the amphibian pituitary (reviewed by Hoar, 1966; van Oordt, 1968, 1974); the activity of these cell types is correlated with gametogenesis, and surgical or pharmacological gonadectomy leads to their hypertrophy and degranulation. Hormone replacement therapy prevents hypertrophy of the gonadotropes following castration. Steroid hormone administration leads to cellular regression in intact animals, a result indicating a gonadal steroid feedback controlling mechanism similar to that observed in mammals.

Early workers proposed that two different cell types in the amphibian anterior pituitary elaborated two gonadotropins: an FSH-like molecule, produced by the B_2 cells and responsible for the initial stages of gametogenesis, and an LH-like molecule, produced by B_3 cells and responsible for the final maturation of the gametes, ovulation and spermiation, and gonadal steroid secretion (see van Oordt, 1974, for review). Although recent biochemical identification of an FSH and an LH molecule in the pituitaries of the bullfrog *Rana catesbeiana* (Licht and Papkoff, 1974), the tiger salamander *Ambystoma tigrinum* (Licht *et al.,* 1975), and the leopard frog *R. pipiens* (Farmer *et al.,* 1977) would tend to support this conclusion, recent histochemical and immunocytochemical investigations have not confirmed the two gonadotrope–two gonadotropin hypothesis. It is now believed that both FSH and LH-like molecules are produced by B_2 cells (cf. Evennett and Thornton, 1971; Doerr-Schott, 1976; see van Oordt, 1974, for review).

There exists an extensive literature showing that gonadal activity in amphibians is regulated by pituitary secretions. Hypophysectomy of sexually active anuran and urodele amphibians results in gonadal regression, cessation of gametogenesis, and a decline in plasma levels of gonadal steroid hormones. Implantation of whole pituitaries or injection of pituitary homogenates or extracts prevents gonadal atrophy following hypophysectomy, but this response varies seasonally: winter-dormant frogs are less sensitive to gonadotropin stimulation than frogs at the start of the spermatogenic cycle (reviewed by van Oordt, 1968, 1974; Lofts, 1968, 1974; see also Lehman, 1977).

In hypophysectomized *Rana temporaria,* injection of mammalian FSH stimulates the initiation of spermatogenesis but has no apparent effect on spermiation or interstitial cell activity (steroidogenesis); administration of mammalian LH typically induces spermiation and "androgen production" (not measured directly) but does not stimulate spermatogenesis (Lofts, 1961). Mammalian gonadotropins do not exhibit the same degree of specificity of function, however, in other amphibian species. For example, in *Xenopus laevis,* both mammalian FSH and LH increase activity of the steroid biosynthetic enzyme, 3β-hydroxysteroid dehydrogenase (3β-HSD). In other experiments, Muller (1977a,b) has used a completely homologous system to investigate GTH regulation of testicular function in the bullfrog *R. catesbeiana.* Injection of bullfrog LH into intact and hypophysectomized bullfrogs increases plasma concentrations of DTH and T, whereas bullfrog FSH has little or no effect on steroid levels (Muller, 1977a; see also Wada *et al.,* 1976). Similarly, long-term administration of bullfrog LH stimulates development of the Wolffian ducts in hypophysectomized bullfrogs, whereas bullfrog FSH is without effect (Muller, 1976). There is also some evidence that testicular androgens may synergize with pituitary gonadotropins in the control of spermatogenesis in amphibians (e.g., Moore, 1975; Rastogi *et al.,* 1976).

Jørgensen and co-workers have studied the pituitary control of the ovarian cycle of the toad *Bufo bufo.* Daily injections of mammalian FSH or hCG stimulate ovarian growth in hypophysectomized as well as intact toads, whereas mammalian LH is much less potent in this regard (Kjaer and Jørgensen, 1971; Jørgensen, 1973, 1974, 1975). Injection of estradiol 17β into postspawning toads retards growth of the next generation of oocytes but does not influence recruitment of oocytes for the subsequent follicular cycle, a finding suggesting that the initial stages of oogenesis are independent of gonadotropins (Jørgensen, 1974).

In some of their actions, gonadotropins do not stimulate the gonad directly but act via gonadotropin-induced steroidogenesis (cf. Licht and Crews, 1976). For example, gonadotropin-induced vitellogenesis is mediated by estrogen stimulation of yolk protein synthesis in the liver (Redshaw, 1972; Follett and Redshaw, 1974).

Similarly, bioassay experiments reveal that the ovulation response in female frogs, long known to be induced by mammalian LH, can also be induced by steroid hormones (Schuetz, 1967). The mechanism of gonadotropin stimulation of oocyte maturation and ovulation is mediated by ovarian progestational steroids, in particular by progesterone (Wright, 1971; Merrian, 1972; Thornton, 1972; Belle *et al.,* 1975; Wasserman and Smith, 1978).

The amphibian gonad is known to produce estrogens, progestins, and androgens via the same biosynthetic pathways as in mammals (Redshaw and Nicholls, 1971; Thornton, 1972; Ozon, 1972a,b: Ozon and Stocker, 1974). The site of steroid production in the amphibian testis has been the subject of a great deal of controversy. Leydig, or interstitial, cells are prominent in the anuran testis and undergo seasonal fluctuations in volume, precursor material content (cholesterol and lipids), and histochemistry (steroid dehydrogenase activity), which are correlated closely with seasonal variations in androgen-dependent secondary sexual structures (reviewed by Lofts, 1968, 1974). The Sertoli cells also undergo a seasonal cycle in activity and are believed to have the capacity to synthesize steroids (cf. van Oordt and Brands, 1970), but the steroidogenic function of these cells *in vivo* has yet to be established. In urodele amphibians, the "glandular tissue" of the testis consists of transformed Sertoli cells and hypertrophied connective tissue of the lobule walls and is believed to be the site of androgen production (Ozon, 1967; Picheral, 1970; Muller, 1976).

Estradiol 17β, T, and DHT have been identified in the blood of amphibians (Rivarola *et al.,* 1968; Polzonetti-Magni *et al.,* 1970; D'Istria *et al.,* 1974; Vellano and Bona, 1975; Gavaud, 1975; Muller, 1976). In *R. esculenta,* plasma androgens are at their lowest concentration immediately following the breeding season (April–June) and remain low throughout the summer (1–2 ng/ml). Plasma androgens increase during the winter, reaching their highest levels (12–19 ng/ml) just before breeding (D'Istria *et al.,* 1974). In the bullfrog, Muller (1977a) identified two peaks in plasma androgen concentrations. The first peak occurs prior to the breeding season in February–March and is characterized by more DHT than T present in the plasma; during the breeding season itself (June–July), the ratio of DHT to T is reversed and there is more T in the plasma relative to DHT.

Secondary sex characters in amphibians are under steroid hormone control. For example, castration results in loss of sexual pigmentation, decline in tail fin height, and atrophy of seminal vesicles and of nuptial pads in anurans and urodeles, whereas androgen replacement therapy (T or DHT) restores or maintains activity of these structures (cf. Greenberg, 1942; D'Istria *et al.,* 1972; N'Diaye *et al.,* 1974; Singhas and Dent, 1975). In sexually active male plethodontid salamanders, the hedonic glands located on the chin, behind the eye, and at the base of the tail produce a substance that is used during the courtship sequence. The activity of the glands varies seasonally, and Pool and Dent (1977) demonstrated that the hedonic glands in the red-spotted newt *(Notopthalmus viridescens)* are under hormonal control; interestingly, testosterone alone has no effect, whereas treatment with prolactin and testosterone in combination transforms the regressed gland of castrated males to the breeding state. The antiandrogen cyproterone acetate causes regression of the nuptial pads, degranulation of gonadotropic basophils, and lower androgen titer in intact male frogs (Rastogi and Chieffi, 1975) and salamanders (Della Corte *et al.*, 1973; Moore and Muller, 1977).

Uptake of sex steroids by cells in the CNS of *Xenopus laevis* has been reported recently (see Figure 4). Although both the male and female brain bind both estradiol and testosterone, there is a difference in the pattern and distribution of steroid-concentrating sites. Estradiol is concentrated in a number of brain regions, including the striatum, ventral-lateral septum, amygdala, anterior preoptic area (APOA), and the VIN (Morrell *et al.*, 1975a,b; see also Kelley and Pfaff, 1977). The major areas of testosterone uptake are the APOA and the VIN, and occasional testosterone-labeled cells occur in the tectum, the torus semicircularis, and the ventral thalamus (Kelley *et al.*, 1975). The striatum, the septum, and the amygdala appear to concentrate estradiol exclusively, whereas two regions, one ventral to the cerebellum in the dorsal tegmental area of the medulla and the other in the region of the IX–X motor nucleus, concentrate testosterone exclusively. The APOA and VIN are believed to be the sites of Gn RH-producing cells (see p. 115).

PHYSIOLOGICAL FACTORS CONTROLLING AMPHIBIAN REPRODUCTIVE BEHAVIOR

The reproductive behavior of amphibians has long interested behavioral biologists, and the reader should consult Noble (1931), Bogert (1960), Salthe and Mecham (1974), and Wells (1977, 1978) for excellent reviews of this literature.

Vocalizations play a prominent role in reproduction in the anuran amphibians (frogs and toads). There are apparently three basic call types in the vocal repertoire of anurans: mating calls, territorial calls, and the release call. The mating call is the

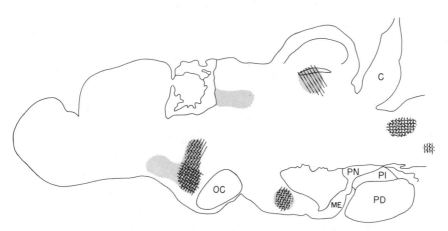

Fig. 4. A parasagittal view of the steroid-concentrating regions in the amphibian brain and their relationship to neural areas known to be involved in the control of reproductive behavior. Experiments indicate that the trigeminoisthmic tegmentum, the medial preoptic area, and the "arcuate" nucleus are involved in calling behavior: release call (≡), mating call (///); that the ventral preoptic area and the torus semicircularis are involved in orientation behavior (\\\); and that the clasping reflex is situated in deep medullary areas (|||). See text for details. Shaded areas indicate regions that have been reported to concentrate testosterone (dark stipple) and estradiol (light stipple). See Figure 3 for identification of structures.

most commonly heard call and is produced by breeding males. It is characteristically species-typical and serves as a premating reproductive-isolating mechanism among sympatric congeners. The mating call orients males and females to the breeding site and is used by females to discriminate and select specific males with which to mate. It has also been suggested that the breeding chorus produced by the aggregated males may facilitate female reproductive development in addition to synchronizing the vocalizations of the males, leading to a more intense chorus. The territorial call is distinct from the mating call and appears to maintain spacing of individual males at the breeding site. The release call or croak is produced by a sexually unreceptive female or a male when clasped by a sexually active male; gravid females do not emit the release call until after they have oviposited. Although species differ in the time spent in amplexus, mating is completed within 24 hours in most species.

Unlike anurans, vocalizations in urodele amphibians (salamanders) are rare. Instead, the urodeles rely on olfactory, visual, and tactile stimuli for species and mate recognition. For example, in *Triturus* spp., the male uses his tail to waft toward the female pheromones produced by secondary sexual glands located in the cloaca. Gravid females orient and approach the male, but this response is abolished if the nares are blocked or if the olfactory nerves are cut. In plethodontid salamanders, the males possess hedonic mental glands, located in and around the chin, which are rubbed over the female's snout during courtship. In some species, the female excretes an attractant pheromone whose production is correlated with her ovarian state. There is evidence that sexual pheromones in urodeles, like vocalizations in anurans, are species-typical and function as reproductive-isolating mechanisms.

NEURAL MECHANISMS CONTROLLING CALLING BEHAVIOR: THE RELEASE CALL. Aronson and Noble (1945) were the first to examine specifically the neural mechanisms involved in anuran vocalizations. They reported that complete removal of the forebrain, the diencephalon, the optic tectum, the cerebellum, and the anterior tegmentum did not prevent the elicitation of the release call in male *R. pipiens*, whereas localized lesion of the inferior colliculus (= torus semicircularis) abolished this behavior. Schmidt conducted an extensive series of experiments with representatives of several genera (including *Hyla, Rana,* and *Bufo*) on the central mechanisms of frog vocalizations, using both lesion and stimulation techniques. Unlike Aronson and Noble, Schmidt (1966a) found that frogs eventually recover (10 days to 6 weeks) and emit normal release calls when grasped following complete removal of the inferior colliculus, but that they fail to do so if the trigeminoisthmic tegmentum (the floor of the medulla under the cerebellum) is destroyed (see Figure 4); electrical stimulation of this area, in particular the main sensory nucleus V, evokes normal release calls. Colombo *et al.* (1972) noted an increase in spindle alphalike EEG activity in unreceptive female toads when clasped by males. Interestingly, these changes are similar to those reported by Segura and co-workers as being hormone-induced (see p. 124).

Diakow (1977) examined the stimuli initiating and inhibiting the release croak in *R. pipiens*. The release croak can be elicited by manually grasping the frog just behind the pectoral regions, whereas denervation of the skin of the trunk abolishes the release call in response to manual stimulation. These findings led Diakow to suggest that the trunk stimulation during clasping is responsible for the initiation of the release call in nongravid females.

Noble and Aronson's observation (1942) that distension of the abdomen by intraperitoneal injection of physiological saline causes recently oviposited—and hence no longer receptive—females to allow males to mount and clasp led to the hypothesis that intra-abdominal pressure from the ovulated egg mass inhibits the release croak in gravid females. Diakow (1977) showed that artificial distension of both males and females reduced the frequency of release croaks in response to manual stimulation or clasping by sexually active males. Distended females were also observed to remain in amplexus for longer periods than control females. In view of recent experiments with mammals (Rodriguez-Sierra and Komisurak, 1977, 1978) and fish (Stacey, 1976), it would be of interest to determine whether prostaglandins are responsible for this inhibition of the release call and maintenance of amplexus.

NEURAL MECHANISMS CONTROLLING CALLING BEHAVIOR: THE MATING CALL. Whereas the trigeminoisthmic tegmentum appears to be the main locus of anuran vocalization, patterned calling such as the species-typical mating calls emitted by breeding male frogs requires additional input from the magnocellular preoptic area and the "arcuate" nucleus via the ventral thalamus (Schmidt, 1966a, 1968, 1974b). Bilateral lesions of the MPOA and the arcuate nucleus abolish mating calling, and electrical stimulation of these areas reliably elicits vocalizations in response to tapes of conspecific calls that are similar acoustically to the spontaneous mating calls of breeding males. Schmidt (1968, 1969) suggested that the MPOA acts as an "androgen-sensitive activator" of more posterior mating-calling centers, including the main sensory nuclei V.

On the basis of these experiments, Schmidt (1971, 1973, 1974b) proposed a model of the neural mechanisms involved in mating calling anurans. This model postulates the existence of a motor coordination center (an "efferent vocal center") in the region of the hypoglossal and vagus nuclei that is responsible for the generation of patterned calling. A sensory correlation center (the "afferent vocal center") near the main sensory nuclei V receives and integrates sensory input and transmits it to the more posterior motor-coordination site. It is this latter center that the androgen-concentrating neurons in the POA activate to trigger mating calling during the breeding season.

Wada and Gorbman (1977a) electrically evoked mating calls from freely moving *R. pipiens*. The most effective site for stimulation of sexual behavior was found to be the rostral part of the POA, an area known to concentrate labeled T in *X. laevis* (Kelly *et al.*, 1975) and to be responsive to local implantation of T (Wada and Gorbman, 1977b). The discrepancy between Schmidt (1968, 1973, 1974a) and Wada and Gorbman (1977a) regarding the postulated anatomical locus of mating calling in *R. pipiens* may be due to differences in stimulation parameters. In Schmidt's experiments, electrodes were placed in the ventral magnocellular region of the preoptic nucleus, and the stimulation currents were relatively higher than those used in Wada and Gorbman's study (100 Hz, 0.3–0.5 msec duration, 300–900 μA intensity versus 100 Hz, 0.5 msec, 20–200 μA). Thus, as Wada and Gorbman argued, the possible spreading of electrical stimulation from the electrode site in Schmidt's experiments may have made it difficult to localize precisely the sex vocalization center within the preoptic area.

NEURAL MECHANISMS INVOLVED IN ORIENTATION AND CLASPING BEHAVIOR. When a gravid female anuran hears the mating call of a conspecific male, she orients and moves toward the source of the sound. Schmidt (1968, 1969, 1971) reported that the ventral MPOA is essential for orientation behavior whereas the telencephalon and the dorsal part of the preoptic area are not (see Figure 4).

Interestingly, Schmidt (1968) noted that only those females that had ovulated exhibited the orientation response and suggested that this behavior is mediated by neuroendocrine events accompanying oviducal distension. Subsequent experiments, however, revealed that females that had had the oviduct ligated, a process that prevents the eggs from entering the oviducts, continued to orient, a result indicating that oviducal distension is not essential for this behavior (Schmidt, 1971). The possibility remains that deep visceral receptors are involved in the orienting responses, as in the *Rana* release call (Diakow, 1977).

Beginning with Spallanzini's observation in 1786 that decapitated male frogs continue to clasp and fertilize females, there have been several reports regarding the neural control of anuran clasping behavior; the reader is referred to Aronson and Noble (1945) and Beach (1967) for excellent reviews of this early literature. In a classic study, Aronson and Noble (1945) systematically investigated the neural mechanisms controlling male clasping behavior in *R. pipiens*. Removal of the entire forebrain, diencephalon, optic tectum, cerebellum, and anterior tegmentum failed to interfere with clasping behavior, although forebrainless males failed to release the female after oviposition; recent experiments have also implicated the ventral posterior torus semicircularis in releasing behavior (Schmidt, 1974a). Lesions of the tegmentum at the level of motor nucleus IV effectively abolish clasping behavior, a result indicating that clasping is primarily a spinal reflex that is modified by input from higher brain centers. Following brain transection behind the cerebellum or caudal to the medulla oblongata, males exhibit abnormal clasping movements and fail to assume the normal amplectic posture.

Hutchison (1964; Hutchison and Poynton, 1964) investigated the neural mechanisms controlling clasping behavior in *Xenopus laevis*. Transection of the brain at different levels reveals that in *Xenopus*, as in *Rana*, clasping behavior is a spinal reflex. Brain transection at the level of the anterior medulla does not interfere with clasping behavior, whereas transection between the posterior medulla and the spinal cord abolishes this response. Unlike female *Rana* (Aronson and Noble, 1945), however, female *Xenopus* with transections in the anterior medulla exhibit rigid clasping behavior.

In contrast, Segura *et al.* (1971b) reported that, in the toad *Bufo arenarum*, transection of the neuroaxis at every level from the rostral border of the mesencephalon to the vertex of the IV ventricle results in abnormal, rigid clasping. As both discrete electrolytic lesions and microinjection of 3 M KCl in the raphe nucleus of the reticular formation produce similar responses, it appears that, in the toad, clasping is under inhibitory control from areas more anterior than those reported for frogs.

HORMONAL CONTROL OF SEXUAL BEHAVIOR. Although the neuroendocrine control of pituitary function in amphibians has been established, the hormonal bases of amphibian sexual behavior are as yet poorly understood. It has long been known that clasping and calling, components of male sexual behavior, are hormone-dependent. For example, male sexual behavior is abolished by hypophysectomy or castration (see Dodd, 1960, for review of early research; Schmidt, 1966b; Palka and Gorbman, 1973; Wada and Gorbman, 1977b; Kelley and Pfaff, 1976), and a number of investigators have noted that implantation of whole pituitaries or injection of pituitary homogenates or purified mammalian gonadotropins rapidly reinstate sexual behavior in intact (cf. Schmidt, 1966b; Palka and Gorbman, 1973; Trottier and Armstrong, 1975; Humphrey, 1977) but not in castrated (Palka and Gorbman, 1973; Kelley and Pfaff, 1976) male amphibians. Reimplantation of the testes in pituitary-stimulated castrates restores calling behavior (Schmidt, 1966b; Palka and Gorbman, 1973; see Figure 5), but administration of a wide variety of steroid hormones (including TP, DHT, estradiol, EB, progesterone, androstendione, androsterone, and corticosterone) is totally ineffective in this regard (Wolf, 1939; Blair, 1946; Schmidt, 1966b; Palka and Gorbman, 1973; Wada and Gorbman, 1977b; Moore and Muller, 1977).

Because pituitary injection or administration of purified mammalian or amphibian gonadotropin elevates plasma T and DHT levels (Wada *et al.*, 1976; Muller, 1977b) and there is evidence of seasonal variation in the level of circulating androgens (D'Istria *et al.*, 1974; Wada and Gorbman, 1977b; Muller, 1976), the

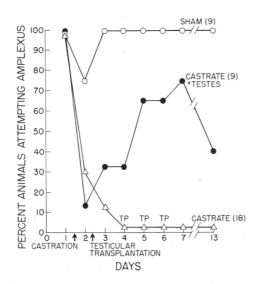

Fig. 5. Frequency of attempted amplexus in male frogs *(Rana pipiens)* following castration and testicular grafts. (From "Pituitary and Testicular Influenced Sexual Behavior in Male Frogs *Rana Pipiens*" by Y. S. Palka and A. Gorbman. *General and Comparative Endocrinology*, 1973, *21*, 148–151. Copyright 1973 by Academic Press. Reprinted by permission.)

failure of systematically administered androgens to stimulate male sexual behavior in ranid frogs is puzzling. Gorbman and his co-workers suggested that a second, at present unidentified, testicular factor is necessary for the expression of sexual behavior. Interestingly, this lack of success in eliciting amphibian sexual behavior with exogenous steroids appears to be restricted to a few species of ranid frogs and urodeles. Kelley and Pfaff (1976) reported that the free T and DHT, but not the estradiol, reliably induce clasping in castrated *Xenopus laevis*. In addition to a possible species difference, Kelley and Pfaff suggested three procedural factors (length of observation, form of hormone administered, and disturbance during testing) that may give rise to the differences in results.

In view of these generally negative results with injections, it is significant that both intracranial implants of T and electrical stimulation in the rostral part of the POA induce mating behavior in *R. pipiens* (Wada and Gorbman, 1977a,b). This region corresponds closely to those areas found by Schmidt (1968, 1973) to be involved in male calling behavior and by Kelley *et al.*, (1975) to concentrate androgens.

Hormonal modulation of central nervous function has also been reported by Segura and co-workers. The female toad *Bufo arenarum* exhibits a circannual rhythm in EEG activity that is correlated with the sexual cycle (Segura and de Juan, 1966). High-amplitude, fusiform, alphalike activity is characteristic of female toads during the breeding season, whereas low-amplitude, fast, desynchronized activity is predominant in reproductively inactive females. After ovariectomy of breeding females, the frequency of spindle activity decreases progressively and is totally abolished within 7 days (Segura *et al.*, 1971a). Unilateral ovariectomy (Segura *et al.*, 1971a) or administration of EB to bilaterally ovariectomized females (Colombo *et al.*, 1972; see also Oshima and Gorbman, 1969a,b) prevents the decline in EEG activity.

Reptiles

The reptiles represent a pivotal group in vertebrate evolution. Because both the birds and the mammals arose from reptilian stock approximately 250 million and 350 million years ago, respectively, studies of reptilian reproduction should shed light on the possible origins and variations in neuroendocrine control mechanisms.

The Hypothalamo-Pituitary-Gonadal Axis

The following briefly summarizes the available information on the reptilian hypothalamo-pituitary-gonadal axis. For more detailed information, the reader should consult reviews by Lofts (1968, 1969, 1972); Callard (Callard *et al.*, 1972a,b,c; Callard and Lance, 1978; Lance and Callard, 1978; Callard and Callard, 1978); Licht (1974a; Licht and Pearson, 1978; Licht *et al.*, 1977); Oksche (1976); Fox (1977); and Crews (1978a).

The neural and vascular connections between the hypothalamus and the pituitary in reptiles are similar to those of other vertebrates. There is also evidence that pituitary gonadotropin secretion is modulated via positive and negative feedback control (cf. Gesell and Callard, 1972). For example, Lisk (1967) reported that TP to EB stereotaxically implanted in the median eminence region of the hypothalamus of males (TP) and females (EB) inhibits normal seasonal gonadal recrudescence in *Dipsosaurus d. dorsalis*. In more systematic investigations, Callard and McConnell (1969) and Callard *et al.*, (1972a,c) found that both EB and estrogen undecyclate implanted in the hypothalamus of *Sceloporus cyanogenys* effectively inhibited ovulation; follicular development was not arrested, however, a result suggesting that ovulation is either inhibited or delayed via a negative neuroendocrine feedback mechanism. Hypothalamic progesterone implants, on the other hand, induce follicular atresia and consequently block oviducal development (Callard *et al.*, 1972a,c; Callard and Doolittle, 1972; see also Klicka and Mahmoud, 1977).

Both hypothalamic extracts and synthetic mammalian LH-RH have been shown to stimulate the release of LH from dispersed turtle pituitary cells *in vitro* and to increase both plasma progesterone and immunoreactive LH levels *in vivo* (Callard and Lance, 1978).

Cyclic gonadal changes have been correlated with histological changes in the reptilian pituitary gland (see, for review, Saint-Girons, 1970; Licht, 1974a; Licht and Pearson, 1978). Acidophilic, basophilic, and chromophobic cells have all been identified in the anterior lobe of the pituitary gland of reptiles (cf. Pearson *et al.*, 1973). Basophils are most abundant both immediately preceding and during the height of the breeding season, but acidophils outnumber the basophils during the refractory and other sexually quiescent periods. This close correlation between the proportion of basophilic cells in the pars distalis (PD) and gonadal cyclicity has led to the belief that the basophils are GTH-producing cells (Pearson *et al.*, 1973, 1976). However, attempts to separate the basophils further into FSH- and LH-producing cells (cf. Saint-Girons, 1970) have thus far proved unsuccessful in reptiles (Licht and Pearson, 1977).

Hypophysectomy of both breeding and sexually quiescent male reptiles results in a complete cessation of spermatogenesis, degeneration and sloughing of the spermatocytes, accumulation of cellular debris in the tubuli luminae, phagocytosis of the remaining germinal cells, collapse of the seminiferous tubules, reduction of interstitial cell size, reduction of testis volume and weight, and atrophy of the renal sex segment (Licht, 1974a; Crews, 1978a). Hypophysectomy also causes a marked reduction in testicular phospholipids and an increase on concentration of neural lipids; these effects parallel changes characteristic of the period preceding and including the quiescence phase in the normal testicular cycle.

In working with ectothermic vertebrates, it is important to keep in mind that their physiological systems are strongly influenced by temperature. For example, removal of the adenohypophysis of adult, sexually active male *Anolis carolinensis* maintained at 31°C causes rapid regression in spermatogenic activity and testis size and weight, whereas hypophysectomized lizards maintained at 20°C show only slight modification in testicular histology and weight at the end of three weeks

(Licht and Pearson, 1969). These testicular changes are not characteristic of intact animals maintained at the same temperature (Pearson *et al.*, 1976).

There have been relatively few investigations into the effects of hypophysectomy on ovarian development and cytology in reptiles, but reports indicate that hypophysectomy of breeding female turtles, snakes, and lizards results in follicular atresia and oviducal atrophy (for review, see Crews, 1978a). Injections of whole pituitaries, pituitary extracts, or exogenous gonadotropins stimulate gonadal recrudescence in hypophysectomized and sexually inactive snakes and lizards (reviewed by Licht, 1974a; Fox, 1977; Crews, 1978a). It is well documented that mammalian LH, hCG, FSH, and PMS all stimulate testicular recrudescence and oogenesis-vitellogenesis in sexually quiescent reptiles, but that FSH and PMS are the more potent gonadotropins in their effects on both male and female gonadal recrudescence (reviewed by Licht, 1974a; Licht *et al.*, 1977). Although distinct FSH and LH molecules exist in the pituitary of chelonian and crocodilian reptiles (but not squamate reptiles), and common binding sites for radioiodinated FSH and LH have been demonstrated recently in reptilian gonadal tissues in both heterologous and homologous radioligand assays (Licht and Midgley, 1976a,b; Adachi and Ishii, 1977), experiments with these preparations have failed to uncover a specific steroidogenic function of the LH molecule (Licht *et al.*, 1977). There is no experimental evidence that an LH–FSH synergism, as has been described in mammals, exists in reptiles (Crews and Licht, 1975b).

3β-HSD has been identified in the reptilian testis, and the site of steroid biosynthesis has been tentatively identified. In snakes, the Leydig cells are believed to be the major sites of testosterone production (Callard, 1967). Tsui (1976) established that the circumtesticular cells of the lizard *Cnemidophorus tigris* are similar to the interstitial Leydig cells and are the sites of androgen biosynthesis in this species. The Sertoli cells are also believed to produce androgens (Lofts, 1972).

The enzymes 3β-HSD and 21-hydroxylase have been localized histochemically in the granulosa cells and the theca interna of the preovulatory follicles and the corpus luteum (cf. Colombo and Yaron, 1976; Colombo *et al.*, 1974; Crews and Licht, 1975a); these cells are known to be the site of progesterone production in the turtle preovulatory follicle (Crews and Licht, 1975a).

Seasonal histological changes in the testes are correlated with changes in steroid biosynthesis. The relationship between androgen production, cyclical testicular recrudescence, and interstitial cell activity has been explored by Lofts and his colleagues (cf. Lofts, 1968, 1969, 1972). Their investigations indicate that the cyclic buildup of cholesterol-rich lipids (presumed precursors to steroids) in the interstitial cells (the "seasonal interstitium lipid cycle") is inversely correlated with seasonal androgen production. Similar relationships between seasonal histochemical changes in the Leydig cells, androgen production, and the spermatogenic cycle have been reported in snakes, turtles, and lizards (Lance and Callard, 1978; Callard and Callard, 1978).

Callard and co-workers (Callard *et al.*, 1976; Callard and Callard, 1978) have monitored seasonal changes in plasma androgen levels in the freshwater painted turtle *Chrysemys picta*. Plasma testosterone levels are highest at the time of emergence from hibernation in March (individual levels range from in excess of 1000 pg/ml to 13,276 pg/ml) and April (440–7400 pg/ml) when mating occurs. Testosterone levels are lowest during the summer months (136–280 pg/ml) and begin to increase again during late spermatogenesis (November–December) to levels comparable to those found in the spring. A similar seasonal cycle in total 17β-hydroxysteroid production (from a minimum concentration of 5000 pg/ml to a maximum of greater than 30,000 pg/ml during the breeding season) has been reported by Bourne and Seamark (1975) for the Australian skink *tiliqua scincoides*. Judd *et al.*, (1976) reported a mean testosterone level of 10, 167 pg/ml ($\pm$ 2916 SE) during the breeding season in *Iguana iguana;* androstenedione levels were very low or nondetectable.

Callard and Lance (1978) have also determined seasonal variation in ovarian hormone levels in the turtle *C. picta*. Plasma estradiol parallels ovarian (follicular) growth with a preovulatory surge in E titer and a second, smaller rise that is correlated with ovarian recrudescence in the fall (see Figure 6). Concentrations of immunoreactive "LH" and progesterone are also maximal at the time of ovulation. All three hormone levels decline following oviposition. A similar pattern in circulating hormone levels has been noted for the green sea turtle *Chelonia mydas* (Callard and Lance, 1978). Interestingly, plasma T also exhibits a seasonal cycle that

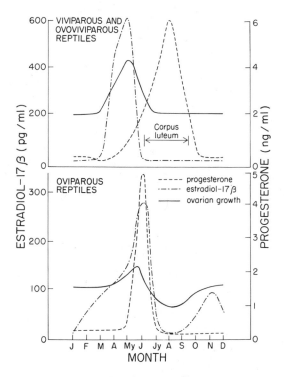

Fig. 6. Relationship between ovarian growth and pattern of hormone production in live-bearing (viviparous and ovoviviparous) reptiles and egg-laying (oviparous) reptiles. (From "Phylogenetic Trends in the Hormonal Control of Gonadal Steroidogenesis" by V. Lance and I. P. Callard. In P. Pang and A. Epple [Eds.], *Evolution of Vertebrate Endocrines*. Lubbock: Texas Tech Press, 1978. Copyright 1978 by Texas Tech Press. Adapted by permission.)

complements the pattern of E production. In *C. picta*, plasma T increases markedly during the fall period of ovarian recrudescence (mean level of 1838 ± 678 pg/ml), and a minor peak occurs during the spring (420 ± 75 pg/ml).

The pattern of progesterone production during the reproductive cycle in oviviparous and viviparous reptiles differs from that observed in oviparous reptiles (see Callard and Lance, 1978, for review; Figure 6). Instead of the preovulatory rise in progesterone followed by a decline correlated with formation and regression of the corpus luteum that is characteristic of egg-laying reptiles, live-bearing lizards and snakes have the highest concentrations of progesterone during midpregnancy.

Gonadectomy has many of the same effects on the morphology and physiology of reptiles as have been reported in other vertebrate classes. Gonadectomy is followed by the disappearance of sexual coloration, regression of femoral pores, and a marked decrease in the secretory activity of the epithelial cells lining the renal segment, epididymal tubules, and oviduct (for review see Licht, 1974a; Crews, 1978a).

Exogenous steroids stimulate the development of secondary sex characters in intact, sexually active reptiles and restore sexual activity in castrates (Licht, 1974a; Crews, 1978a). For example, in the male, the modified portion of the kidney, the renal sex segment, is under androgen control, as are other secondary accessory structures (Fox, 1977). In the female, the oviduct responds to estrogen in a dose-dependent manner (Callard and Klotz, 1973), and there is an estrogen–progesterone synergism in the regulation of oviduct growth (cf. Yaron, 1972). Progesterone also has an antigonadal action. If injected in the midstages of vitellogenesis, progesterone inhibits follicular maturation and oviduct development in *Sceloporus cyanogenys* (Callard *et al.*, 1972a,c).

Autoradiographic studies indicate that estradiol is concentrated in specific telencephalic and diencephalic structures in the lizard brain, including the medial and lateral preoptic areas, the amygdala, the medial and lateral septum, and the torus semicircularis (see Figure 7). Distribution of DHT and T uptake is generally similar to that of estradiol, although in most areas the number and intensity of androgen-labeled cells is clearly less (Morrell *et al.*, 1977).

ENDOCRINE AND NEURAL CONTROL OF REPTILIAN REPRODUCTIVE BEHAVIOR

Reptiles display a rich array of social signals and complex social systems (see reviews by Evans, 1961; Carpenter, 1978; Crews, 1978a; Ferguson, 1978; Jenssen, 1977, 1978; Stamps, 1977). The lizards have been the most extensively studied, and the behavioral repertoire and social structure of a number of species have been described. These investigations reveal that each species exhibits certain highly stereotyped, species-typical displays that are used in both intra- and inter-specific communication. In the iguanid lizards, the most thoroughly studied group, these displays are usually in the form of up-and-down bobbing movements coordinated with the exposure of brilliantly colored patches of skin. Complex vocalizations have also been described by Marcellini (1977) for nocturnal lizards, and there is some

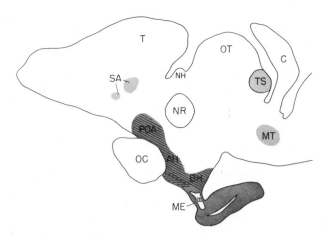

Fig. 7. A parasagittal view of the neuroendocrine-regulating regions in the lizard brain and their relationship to neural areas known to be involved in the control of reproductive behavior. Stippled areas indicate steroid-concentrating sites. Implantation of steroid hormones indicates that the basal hypothalamus–median eminence region (≡) is involved in the feedback control of gonadotropin secretion and that the preoptic area (///) is involved in the regulation of reproductive behavior. See text for details. Abbreviations: AH, anterior hypothalamus; C, cerebellum; ME, median eminence; MT, midbrain tegmentum; NH, nucleus habenularis; NR, nucleus rotundus; OC, optic chiasm; OT, optic tectum; P, pituitary; PH, posterior hypothalamus; SA, strioamygdaloid area (not in plane of section); T, telencephalon; TS, torus semicircularis; III, third ventricle.

evidence that pheromonal communication may also occur (cf. Greenberg, 1943; Evans, 1961; Hunsaker, 1962; Cole, 1966; Madison, 1977; DeFazio *et al.*, 1977).

Snakes and turtles were long believed to have limited behavioral repertoires and to lack a well-defined social structure. Recent investigations, however, reveal that this is definitely not the case (cf. Auffenberg, 1977; Carpenter, 1977; Carpenter *et al.*, 1976; Crews, 1976; Ross and Crews, 1977, 1978). In addition to pheromonal communication, snakes exhibit a number of ritualized visual and tactile displays in agonistic and sexual encounters. Turtles have also been found to have a highly complex signal repertoire in several sense modalities; vocalizations, visual displays, tactile stimulation, and scent marking all appear to play a role in turtle reproduction. The reproductive behavior of crocodilians has also been described recently and is found to include postural displays, audible and subaudible vocalizations, narial geysering of water, and violent splashing (Garrick and Lang, 1977; Garrick *et al.*, 1978).

Endocrine Control of Male Reproductive Behavior. Several investigators have noted that castration results in a loss of sexual behavior in reptiles (see Dodd, 1960, and Crews, 1978a, for review of early literature) and consequently prevents facilitation by male courtship of environmentally induced ovarian recrudescence in winter-dormant females (Crews *et al.*, 1974).

A good deal of work has been done on the lizard *Anolis carolinensis*. In this species, courtship behavior declines rapidly following castration, but aggressive behavior decreases more gradually or may even increase initially; the castration-induced decline in behavior can be accelerated by moving the animals into unfamiliar cages (Crews, 1974). Androgen replacement therapy (either daily injection, crystalline implants, or silastic capsules) restores reproductive behavior in the opposite order; that is, aggressive behavior returns before courtship behavior appears. This order of appearance of behavioral displays is the same as that observed during the seasonal reproductive cycle (Crews, 1975). Further, aggressive behavior is elicited before courtship behavior appears following the implantation of T into the medial POA of castrates (see Figure 7). These findings suggest that agressive behavior and sexual behavior have different hormone thresholds or that androgens have different sites of action in *A. carolinensis.*

It has been suggested that it is the metabolites of testosterone, primarily estrogens, that activate sexual behavior in the male (see reviews by Ryan *et al.,* 1972; Naftolin *et al.,* 1975). According to this view, testosterone serves as a prohormone or precursor and is converted to estrogens by aromatase enzymes in the limbic system (primarily the septum), the preoptic area, and the anterior hypothalamus before binding to intracellular receptors and entering the cell nucleus. Callard and coworkers (Callard *et al.,* 1977; Callard and Ryan, 1977) localized aromatase enzymes in the hypothalamus and the strioamygdaloid complex of the turtle *(C. picta.)* Incubation of these regions with labeled androgen (androstenedione) yields estrone and estradiol (Callard and Ryan, 1977); interestingly, the strioamygdaloid area synthesized more estrogen than the entire preoptic anterior hypothalamic area. These findings have led the authors to propose that androgen aromatization is a phylogenetically ancient property of the vertebrate central nervous system. Support for this hypothesis, however, is as yet inconclusive, as there have been few direct tests of the function of androgens and estrogens *in vivo* in reptiles. Evans (1952) reported that EB stimulates growth of the middle foreclaw (a male secondary sex character) in the juvenile slider turtle *Pseudemys scripta troostii.*

The effect of free T, E, and DHT on the reproductive behavior of castrated male *A. carolinensis* has been examined (Crews *et al.,* 1978). Results indicate that the latency to first aggressive and sexual displays in staged encounters is significantly reduced by T but not by E or DHT; T also restores challenge and courtship frequencies to near precastration levels, whereas both E and DHT are without effect. These results cannot be due to differences in access to administered hormones; autoradiographic studies have demonstrated that T, E, and DHT are taken up and retained by the reptilian central nervous system (see p. 128). Further, measurement of circulating levels of hormones indicates that silastic pellets elevate blood levels above normal.

Adkins (personal communication, 1977) found that injection of DHTP stimulates reproductive behavior in castrated *A. carolinensis.* This difference in results may be due to procedural differences or hormone regimens.

Several workers have shown that administration of DHT and E in combination stimulates copulatory behavior in castrated male rats, whereas either DHT or E alone is ineffective (Baum and Vreeburg, 1973; Feder *et al.,* 1974). Such results have led to the suggestion that the central action of E on male reproductive behavior requires active and developed peripheral tissues. Support for this hypothesis was provided by Davis and Barfield (1977), who demonstrated that intracranial implants of EB in the POA induce sexual behavior in castrated male rats only if DHT is administered systematically.

Dihydrotestosterone is a potent stimulator of secondary sex structures in reptiles (Dufaure and Gignon, 1975; Crews, 1978a), yet experiments in which castrated *A. carolinensis* received DHT and E simultaneously yielded equivocal results. In two separate experiments, half of the castrates responded sexually after DHT + E implantation (Crews *et al.*, 1978). Further, although each responding male differed in the time of onset of reproductive behavior, the males would suddenly exhibit the complete reproductive pattern, including mounting, within a single testing period; this behavior markedly contrasts with that of males receiving T therapy, which exhibit a gradual restoration of successive components of reproductive behavior over a period of days. There is also some indication that the responsiveness of males to DHT + E treatment is correlated with precastration levels of behavior.

ENDOCRINE CONTROL OF FEMALE SEXUAL RECEPTIVITY. Evidence from other vertebrates, primarily mammals, indicates that ovarian hormone secretion is cyclic and is correlated with follicular maturation (cf. Young, 1961). Several studies have shown that female sexual receptivity in iguanid lizards is associated with follicular development and ovulation. The temperate-zone lizard *A. carolinensis* has been the most extensively studied species in this regard and so is discussed here in detail.

Like the well-studied laboratory mouse and rat, anoline lizards exhibit sexual receptivity cyclically. During the breeding season, a female *A. carolinensis* lays a single egg every 10–14 days; this pattern of ovarian activity is generated by the growth (yolk accumulation) of a single follicle; ovulation occurs when the follicle reaches 8.0 mm in diameter (Crews, 1975). As this egg is shelled in the ipsilateral oviduct, the largest follicle in the contralateral ovary matures and is ovulated (see Figure 8). This cyclic alteration between ovaries is not inherent in the gonads but is due to the disproportionate availability of circulating GTH to the largest follicle (Jones, 1975). Thus, the remaining ovary of unilaterally ovariectomized females

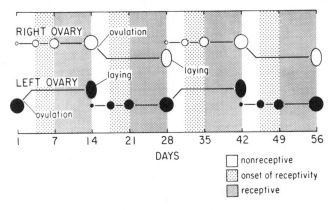

Fig. 8. During the breeding season, female sexual receptivity is cyclic and is correlated with the pattern of ovarian activity and follicular maturation in the lizard *Anolis carolinensis*. See text for details. (From "The Annotated Anole: Studies on the Control of Lizard Reproduction" by D. Crews. *American Scientist*, 1977, *65*, 428–434. Copyright 1977 by the *American Scientist*. Adapted by permission.)

undergoes compensatory hypertrophy and ovulates every 15–17 days instead of once every 23 days, as is typical of intact females in the laboratory (Valenstein, Sheline, and Crews, unpublished, 1977; see also Jones *et al.*, 1977).

Correlational studies have demonstrated that in anoline lizards, female sexual behavior is associated with follicular development and ovulation (Crews, 1973a; Stamps, 1975). Unmated females stand for a courting male and allow him to mate only during the latter half of the follicular cycle. Thus, female sexual receptivity in *Anolis* is rhythmical in nature in a way that corresponds to these cyclic ovarian changes (see Figure 8). In *A. carolinensis,* females can be sexually receptive for up to 7 days every 2 weeks but, because copulation terminates sexual receptivity (see p. 134), females are normally in estrus for only 1–2 days. The transition from non-receptivity to receptivity occurs when the growing follicle is between 3.5 and 6.0 mm in diameter. The exact point of transition, however, varies from female to female, a finding suggesting possible individual differences in neural sensitivity to follicular secretions (estrogen).

There have been very few experimental investigations of the endocrine control of female sexual receptivity in reptiles. In *A. carolinensis,* ovariectomy abolishes sexual receptivity, and the administration of exogenous estrogens to ovariectomized females or reproductively inactive females induces estrous behavior (Noble and Greenberg, 1940; Mason and Adkins, 1976; Crews, 1975, 1978a; Valenstein and Crews, 1977). A single injection (subcutaneous) of 0.8 μg EB stimulates receptive behavior in an adult ovariectomized *A. carolinensis* within 24 hr; this dose appears to be close to the threshold dose for inducing receptivity in this species, as the majority of females receiving lower doses are not receptive to male courtship (Crews, 1978a; Valenstein and Crews, 1977). Estrone benzoate is much less potent than EB in its effects on female behavior, and progesterone alone is totally ineffective (Crews, 1975, 1978a). Noble and Greenberg (1941) reported, and Mason and Adkins (1976) have since confirmed, that TP also induces estrous behavior as well as stimulating oviduct development in ovariectomized *A. carolinensis;* whether the androgen acts via its conversion to estrogen remains to be determined.

In many mammals, progesterone has both facilitatory and inhibitory effects on female sexual receptivity (cf. Feder, 1977; Feder and Marrone, 1977). For example, if progesterone is administered at an appropriate time after estrogen priming in guinea pigs, the progesterone synergizes with the estrogen to facilitate the onset of receptivity. If, on the other hand, progesterone is administered simultaneously with or before estrogen, receptivity is inhibited. Studies of the pattern of endogenous ovarian hormone secretion in mammals reveal that these facilitatory and inhibitory effects are not pharmacological but have a physiological role in the control of female sexual behavior.

Recent experiments have shown that progesterone synergizes with estrogen to facilitate the estrogenic effects on female behavior in the lizard, *A. carolinensis,* as it does in a number of mammalian species (McNicol and Crews, 1979). Ovariectomized females do not normally respond sexually to a single injection of 0.4 μg EB but stand for the male if this injection is followed 24 hr later by a single injection of 60 μg of progesterone.

Sexually receptive female *A. carolinensis* that have been allowed to mate are almost immediately unreceptive to further male courtship (Crews, 1973a; see p. 134). This inhibition of receptivity persists only until the follicle is ovulated and does not prevent receptivity during subsequent follicular cycles. Mating inhibition of female receptivity appears to depend on the duration of intromission and/or ejaculation by the male. The transition from the sexually receptive to the sexually inhibited state is dependent on the presence of the ovary and thus presumably reflects a deviation from the normal secretion pattern of GTH and/or ovarian hormones. Because the onset of heat often results in mating, it is important to understand the physiology underlying mating-induced termination of estrus and how it may differ from the loss of receptivity in unmated females following ovulation.

Females in estrogen-induced estrus show the same behavioral effects of mating as females in natural estrus (Valenstein and Crews, 1977). However, mating inhibition of female receptivity can be overridden if high doses of estrogen are administered. For example, if, immediately following mating, females receive a subcutaneous 10-mm silastic implant of EB (0.025 in O.D. × 0.047 in I.D.), they are not receptive to courting males; the percentage of inhibited females begins to decline at 6 hr postmating, and all females are once again receptive 24 hr after the first mating.

Experiments also indicate that the presence of the ovaries is critical to coition-induced inhibition of female receptivity. Ovariectomized females in EB-induced estrus are significantly more likely to remate 24 hr after copulation than are sham-operated animals receiving identical dosages of estrogen (Valenstein and Crews, 1977).

These studies indicate that copulation-induced abbreviation of estrous behavior following mating in female *A. carolinensis* can be divided into an *initiation* phase, which is induced by neurally mediated stimuli arising from intromission, and a *maintenance* phase, in which the nonreceptive state is maintained by a change in ovarian hormone production.

Hormonal control of female rejection behavior has been implicated in other iguanid lizards. Female *Holbrookia maculata* (Clarke, 1965), *Crotophytus collaris* (Cooper and Ferguson, 1972a,b, 1973; Ferguson, 1976), *C. wislizenii* (Medica *et al.*, 1973), and *Sceloporus virgatus* (Vinegar, 1972) exhibit a rejection behavior following ovulation that effectively discourages males from mating. Coincident with this change in behavior, gravid females rapidly acquire bright orange spots on the sides and flanks, which they display to the courting male. This color change is under hormonal control and can be induced by exogenous progesterone but not by estradiol-17β in ovariectomized *Crotophytus* (Cooper and Ferguson, 1972a,b; Ferguson, 1976; Medica *et al.*, 1973).

PERIPHERAL CONTROL OF REPRODUCTIVE BEHAVIOR. The squamate reptiles (lizards and snakes) are unique among vertebrates in having bilaterally symmetrical intromittent organs called *hemipenes* (Cope, 1900; Dowling and Savage, 1960). In these reptiles, the vasa deferentia do not join into a common urogenital sinus as in other vertebrates but are separate and transport sperm to a seminal groove on the surface of the ipsilateral hemipenis (Crews, 1978b). During mating, the male pushes his tail beneath the female's to oppose the cloacae. Intromission is achieved by eversion of the hemipenis closest to the female; thus, a male mounting on the left side of a female everts his right hemipenis (Crews, 1973b). Males alternate in their use of the hemipenes during successive copulations (Crews, 1978b).

Experiments with the lizard *A. carolinensis* indicate that sensory feedback from the hemipenis is important in the male's initial orientation during mounting as well as in the termination of copulation (Crews, 1973b, 1978b). For example, removal of one hemipenis causes males to mount females only on the side that allows them to use the remaining hemipenis. Bilaterally hemipenectomized males, on the other hand, continue to alternate their copulation posture; although these males assume a normal copulation posture, they often fail to maintain cloacal contact and never initiate separation from the female (Crews, 1973b). There is also evidence that visceral feedback is important in male orientation; unilaterally castrated males tend to mount and copulate with the hemipenis contralateral to the castrated side, a finding suggesting that proprioceptive feedback from the testes monitors the ability of each testis to provide an adequate number of mature sperm for each copulation.

Sensory stimuli also appear to be involved in the regulation of female sexual receptivity. In both snakes and lizards, the act of mating immediately terminates sexual receptivity in estrous females (Ross and Crews, 1977, 1978; Crews, 1973b; Valenstein and Crews, 1977); ovariectomized female *A. carolinensis* given estrogen replacement therapy are also unreceptive following copulation but are sexually receptive once again in 6 hr (Crews, unpublished, 1977).

Several investigators have noted that female lizards receiving shorter than normal copulation continue to be receptive to males (Greenberg and Noble, 1944; Crews, 1973b; Stamps, 1975), a finding suggesting that the inhibition of receptivity depends on the duration of intromission and/or ejaculation by the male. Further evidence that copulation-induced abbreviation of estrous behavior is mediated by neural stimuli arising from male intromission is the observation that females allowed to mate with bilaterally hemipenectomized males continue to be receptive (Crews, 1973b) whereas cloacal probing abolishes receptivity in unmated females. (Crews, unpublished, 1977; see also Ross and Crews, 1977, 1978).

CENTRAL NERVOUS CONTROL OF SOCIAL BEHAVIOR. There has been an increasing number of investigations of the central nervous control of reptilian social behavior. Removal of the anterior two-thirds of the telencephalon has no apparent influence on responsiveness in lizards (Diebschlag, 1938), but complete decerebration (including removal of the amygdaloid complex in the posterior telencephalon) results in a loss of aggressive and flight reactions (Goldby and Gamble, 1957; see Figure 8). Keating *et al.* (1970) reported that bilateral lesions of the amygdaloid complex (including the nucleus sphericus, the nucleus ventromedialis, and the posterior dorsal ventricular ridge) in the South American caiman *(Caiman sklerops)* decrease attack and retreat responses to handling; the decline in response frequency was correlated with the amount of damage to the ventromedial nucleus. Electrical stimulation of these same sights results in flight responses, whereas stimulation of areas in the diencephalon (not identified) yields attack behavior. Similarly, in the lizard *Sceloporus occidentalis,* only bilateral lesions of the amygdaloid nuclei abolish aggressive behavior normally shown by dominant males to intruders (Tarr, 1977); following similar lesions, subordinate animals fail to respond to the challenge displays of socially dominant males. Sugarman (1974; Sugarman and Demski, 1978) reported that electrical stimulation of hypothalamic and midbrain areas evoke gular extension, a component of agonistic behavior, in the collared lizard *Crotaphytus collaris;* the areas exhibiting the lowest threshold were located in the midbrain and the medullary reticular formation, but responses were also evoked from the POA and the basal hypothalamus (see also Distel, 1976). Finally, stimulation of areas immediately external to the central nucleus of the torus semicircularis elicits vocalizations in the Tokay gecko *Gekko gekko* (Kennedy, 1975).

Taking advantage of the almost complete decussation of the optic tracts in *A. carolinensis* (Butler and Northcutt, 1971), Greenberg (1977, 1978) demonstrated dramatic effects on aggressive behavior following unilateral lesions of the dorsal ventricular ridge (DVR). He reported that, when the eye ipsilateral to the lesioned side is covered, males fail to challenge intruding lizards; these same males behave normally (aggressively) when the contralateral eye is covered, however.

It should be noted that all of the brain areas shown to have influences on social behavior concentrate steroid hormones (Morrell *et al.,* 1977; see p. 128). Indeed, implantation of testosterone in the medial POA of castrate male lizards *(A. carolinensis)* stimulates reproductive behavior within 1 week of implantation (Morgentaler and Crews, 1978). Radiofrequency lesions of this area dramatically reduce social behavior in both intact and androgen-treated castrated male *A. carolinensis* (Wheeler and Crews, 1978).

Hormonal Control of Pheromone Production. Pheromones have been found to be a major social signal in turtle and snake species (Evans, 1961; Madison, 1977). For example, during courtship, male tortoises (*Gopherus* spp.) scrape the enlarged medial scale of the forelimb across the subdentary glands and present the leg to another individual (Auffenberg, 1966); Auffenberg (1977) suggested that the secretions produced by this gland communicate sexual and possibly social status as well as species identity. Further, there is some evidence that the production of this pheromone(s) is under hormonal control; size and secretory activity are correlated with the breeding season, and Weaver (1970) has shown the gland to be under androgen control.

Male garter snakes (genus *Thamnophis*) discriminate the reproductive state of the female on the basis of a pheromone released from the dorsal skin; production of this attractant pheromone is estrogen-dependent (Crews, 1976) and is enhanced by shedding (Kubie *et al.,* 1978). In *T. sirtalis,* female attractivity is correlated with dosage with EB in both intact and ovariectomized females (Crews, 1976; see also Ross and Crews, 1977, 1978). Choice-test experiments have demonstrated that males discriminate and court only estrogen-injected females (Crews, 1976).

Recent studies have demonstrated that female *Thamnophis* are not attractive to other males after copulation (Devine, 1977; Ross and Crews, 1977, 1978). This change in female attractivity is a consequence of the deposition of a mating plug by the male. Experiments with castrated males and with intact males that have had the vas deferens ligated between the renal sex segment and the hemipenes indicate that the renal-sex-segment secretions, which form the mating plug, contain a pheromone that renders females unattractive to males (Ross and Crews, 1978). The sexual segment of the kidney undergoes a seasonal cycle of secretory activity that corresponds to the spermatogenic cycle and that is androgen-dependent (see Prasad and Reddy, 1972, for review).

The interested reader should consult the following references for more recent reviews of reptilian reproductive behavior: Crews (1979, 1980, 1983); Crews and Greenberg (1981); Greenberg and Crews (1982); and Ingle and Crews (1985).

BIRDS

Beginning with Rowan's (1925) classical demonstration of the stimulatory effects of light on gonadal recrudescence in the junco *Junco hyemalis,* the study of interrelationships between reproductive physiology and behavior in birds has yielded significant insights into our understanding of vertebrate reproduction. For example, research which began over 45 years ago with ducks *(Anas platyrhynchos)* has resulted in the elucidation of the neuroendocrine basis of the photoperiodic response shown by seasonally breeding vertebrates (reviewed by Benoit and Assenmacher, 1959; Benoit, 1961; Farner and Follett, 1966; Lofts and Murton, 1973). Studies of the white-throated sparrow *(Zonotrichia albicollis)* in the late 1960s led to the discovery of an endogenous "clock" with a circadian periodicity that interacts with biotic and/or physical factors in a synergistic fashion to influence the timing of major behavioral and physiological changes during the year (reviewed by Meier, 1975); the importance of endogenous biological rhythms in the control of behavior has since been demonstrated in representatives of all vertebrate classes. Other researchers working with the canary *(Serinus serinus)* and the ring dove *(Streptopelia risoria)* have demonstrated that, within the breeding season itself, there exists a complex interaction of environmental, experiential, and physiological factors that regulate the sequence and the coordination of male and female species-typical behavior patterns necessary for successful reproduction (see reviews by Hinde, 1965; Lehrman, 1965; Kelley and Pfaff, 1977, Cheng, 1979; Silver, 1978). Finally, the detailed analysis of bird song and its physiological bases has provided breakthroughs in the understanding of the neural and hormonal basis of sex differences as well as contributing to our understanding of the development of laterality of brain function (Nottebohm and Arnold, 1976; Arnold *et al.,* 1976). Thus, studies of the control of avian reproduction have provided a useful experimental model for analyzing the interaction of the endogenous and exogenous factors controlling reproduction.

In the following sections, we review the anatomy, physiology, and behavior of reproduction in birds. Gaps in our understanding are pointed out, and major advances are discussed in detail. We do not cover the literature on the photoperiodic control of reproductive cycles in birds, as this topic has been reviewed extensively by Follett (1978); Assenmacher *et al.* (1973); Follett and Davies (1975); Farner and Lewis (1971); and Menaker (1971).

THE HYPOTHALAMO-PITUITARY-GONADAL AXIS

In order to understand functional aspects of the neuroendocrine system in birds, it is necessary to review features of the neurosecretory system that are characteristic of this class (Figure 9). The following features are recognized: (1) Nerve cells in the avian hypothalamus are distributed diffusely, making it difficult to isolate particular nuclei and neurosecretory cells. (2) The median eminence (ME) is morphologically distinct from the rest of the brain, and an anterior and a posterior neurohemal area can be distinguished. (3) In many birds, the pars distalis (PD) is completely separated from the neurohypophysis and the rest of the brain and is a less compact unit than in mammals. An epithelial stalk connecting the PD and the buccal epithelium may occur. (4) There is no pars intermedia. (5) In contrast to the arrangement in mammals, the primary capillaries of the hypothalamohypophyseal portal system in the ME are superficial, and the blood vessels leading from the anterior and posterior ME to the cephalic and caudal lobes of the PD are also widely separated. This distribution of blood vessels may be significant in view of the distribution of cell types in the different lobes and may provide a mechanism whereby the CNS provides specific stimuli to different portions of the adenohypophysis (for review, see Assenmacher, 1973; Farner and Oksche, 1962; Farner *et al.*, 1967; Kobayashi and Wada, 1973; Schreibman, 1978).

Although there is no doubt that in birds, as in other vertebrates, releasing hormones are synthesized in the hypothalamus (for review, see Kobayashi and

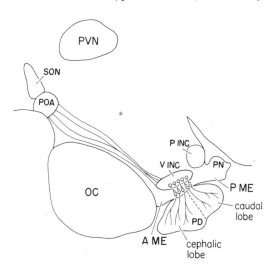

Fig. 9. A parasagittal view of a generalized bird brain showing the major neural and neurosecretory pathways of the hypothalomohypophyseal system. Abbreviations: A ME, anterior median eminence; OC, optic chiasm; PD, pars distalis; P INC, posterior dorsal infundibular nuclear complex; P ME, posterior median eminence; PN, pars nervosa; POA, preoptic area; PVN, paraventricular nucleus; SON, supraoptic nucleus; V INC, ventral infundibular nuclear complex.

Wada, 1973), the diffuse distribution of nerve cells and fiber tracts in this brain region has made it difficult to localize the site of production of specific releasing hormones. Even though the site of production of LH-RH is not yet known, the site of LH-RH release and the neural areas controlling LH-RH release have been clarified with the use of several techniques.

Electrolytic lesions or surgical isolation of the infundibular complex of the basal hypothalamus indicate that this region is involved in the GTH secretion of the Japanese quail *Coturnix coturnix japonica* (Sharp and Follett, 1969; Oliver and Baylé, 1973; Stetson, 1972a,b,c; Davies and Follett, 1975b). If fragments of the adenohypophysis are implanted in the ventral infundibular complex (V INC) of the hypothalamus of male quail, the cells remain active, whereas fragments placed more dorsally regress (Sharp, 1972). Electrical stimulation of V INC results in a significant rise in plasma LH (Davies and Follett, 1975a), whereas complete deafferentation of the tuberal hypothalamus prevents testicular growth and results in low circulating levels of LH (Davies and Follett, 1975b; see Figure 9).

Discrete cuts of the anterior and the anterolateral hypothalamic connections effectively block GTH secretion, whereas posterior cuts are ineffective (Davies and Follett, 1975c). The cut fibers appear to originate in the preoptic region (POA) of the anterior hypothalamus; lesions of the POA or knife cuts rostral to this area have no apparent effect. Bilateral lesions of the ventromedial hypothalamus block ovulation in turkeys (Opel, 1977). There is some evidence that areas in the posterior hypothalamus may also play a role in pituitary regulation; electrical stimulation of the posterior (dorsal) infundibular nuclear complex induces LH release in amounts comparable to that following electrical stimulation of the POA (Davies and Follett, 1975a).

The immunofluorescence technique reveals that in male ducks, LH-RH-containing neurons are located in the periventricular region, especially in the dorsal part of the nucleus preopticus paraventricularis magnocellularis (Bons *et al.*, 1977). Tuberal hypothalamic implants of puromycin block testicular growth in quail, a result suggesting that this is the site of synthesis of gonadotropin-releasing factors (Wada, 1974). Median eminence extracts are known to stimulate ovulation in chickens (cf. Clark and Fraps, 1967). The effective factor in the ME is chromatographically and immunologically identical to synthetic LH-RH, which has been shown to induce LH secretion in quail (Bicknell and Follett, 1975; Davies, 1976), chickens *Gallus domesticus* (Furr *et al.*, 1973; Bonney *et al.*, 1974), and turkey *Meleagris gallopavo* (Burke and Cogger, 1977).

Finally, it should be noted that, in birds, neurohypophyseal hormones have been implicated in the regulation of the PD (for review, see Kobayashi and Wada, 1973) and behavior (Kihlström and Danninge, 1972). It is hypothesized that neurohypophyseal hormones travel via the anterior median eminence to the PD, where they may be biologically active. Much of the supporting evidence is based on correlated seasonal changes in hypothalamic and gonadal secretory systems. For example, Konishi and Kato (1967) found that the neurosecretory system of Japanese quail is inactive (and the materials are stored) in the winter period of sexual inactivity, and that it is depleted in the summer period of sexual activity. This rhythm can be altered by manipulation of the photoperiod (Konishi, 1967). Thus, neurosecretory material is more abundant in birds kept under short days than in birds housed in continuous light. Also, in hens, the posterior lobe hormones are known to facilitate oviposition (Nalbandov, 1966).

Two distinct cell types, beta (presumably FSH) and gamma (presumably LH), have been demonstrated in the pituitary of duck, chicken, pigeon *(Columba livia)*, quail, and white-crowned sparrow *Zonotrichia leucophrys gambelli* (for review, see Assenmacher, 1973). In many avian species, these cell types are located in different regions within the PD, with beta cells lying in the cephalic lobe and the gamma cells lying in the caudal lobe.

Attempts to measure the gonadotropic potency of the cephalic and caudal lobes of the avian PD have yielded inconsistent results. The GTH activity of the two lobes was found to be equal in photoperiodically stimulated quail, as measured by uptake of ^{32}P in the chick testis (Follett and Farner, 1966; Tixier-Vidal *et al.*, 1968). In domestic hens, the GTH activity of the cephalic lobe is three times as potent in nonlaying individuals as in those sacrificed at the end of broodiness (Nakajo and Imai, 1961). When the cephalic, middle, or caudal aspect of the PD from cockerels is implanted in the chorioallantoic membrane of 10-day-old hypophysectomized (by decapitation) embryos, FSH activity is associated with the caudal portion (Brasch and Betz, 1971).

The question of whether FSH, LH, or both gonadotropins control spermatogenesis and steroidogenesis in Aves has been addressed by several investigators. Godden and Scanes (1975) suggest that spermatogenesis is under the control of FSH and steroidogenesis under the control of LH. In both *in vivo* and *in vitro* experiments, avian LH has been found to be highly specific in its action (Brown *et al.*, 1975; Licht *et al.*, 1977; see also review by Lofts and Murton, 1973). Similar specificity of action has been demonstrated (e.g., Ishii and Furuya, 1975) in experiments using mammalian gonadotropins. For example, injection of ovine LH in intact hens caused a premature rise in plasma progesterone but not estrogen (Shahabi *et al.*, 1975). However, there is evidence that spermatogenesis (and steroidogenesis) may not be dependent on a specific gonadotropin. For example, in birds, as in other vertebrate classes, FSH has steroidogenic capability and LH has gametogenic activity (for review, see Licht *et al.*, 1977). Also, a combination of T and ovine LH or FSH is more effective than T alone in maintaining testicular weight (Brown and Follett, 1977), a finding suggesting that both pituitary and gonadal hormones are necessary for full testicular function. Thus, although the presence of distinct pituitary cell types, as well as the recent separation and identification of functionally distinct FSH and LH molecules in Galliformes, suggests the presence of two gonadotropins in the avian pituitary, further research is needed to determine the functional specificity of each.

There is substantial evidence that GTH production is associated with steroidogenic activity of the gonads in birds. For example, Greeley and Meyer (1953) first demonstrated increased GTH activity in the pituitary of sexually active male ring-necked pheasants *(Phasianus colchicus)*. In quail, GTH content of the anterior pituitary increases under stimulatory long photoperiods (Follett and Farner, 1966;

Tanaka *et al.*, 1965). Pituitary gonadotropes in the domestic mallard reared under natural photoperiods exhibit an annual cycle in activity that is correlated with the testicular cycle (Tixier-Vidal *et al.*, 1962; Assenmacher and Tixier-Vidal, 1965). Finally, it is well established that hypophysectomy produces gonadal atrophy in both sexes (see Assenmacher, 1973, for a review of the early literature), whereas treatment of hypophysectomized animals with exogenous GTH maintains gonadal development (e.g., Mitchell, 1970).

The testes of birds, like those of reptiles, consist of a paired ovoid structure surrounded by a thick fibrous outer covering, the tunica albuginea. Unlike those of mammals, each testis is housed in an extra-abdominal sac located in the body cavity near the anterior end of the kidney and is composed of a mass of convoluted seminiferous tubules, which are anastomatic and not separated by septa. The avian testis undergoes marked seasonal variation in size. Unlike in some temperate reptiles in which spermatogenesis begins after the postbreeding refractory period, in birds the testes collapse soon after the breeding season and remain quiescent until the subsequent season.

The interstitial Leydig cells are generally agreed to be the primary site of testosterone production (van Tienhoven, 1968; Temple, 1974). For example, in seasonally breeding birds, there is a well-defined seasonal cycle of interstitial cell activity with rhythmic accumulation and depletion of cholesterol-positive lipid droplets (Lofts, 1968) and of 3β-HSDH (Lifts and Murton, 1973). Thus, the maximal height of epididymal epithelial cells occurs when the interstitial cell depletion of lipids is greatest in California quail *Lophortyx californicus* (Jones, 1970). Garnier and Attal (1970) reported two seasonal peaks in peripheral plasma testosterone levels that are correlated with maximal average size of interstitial cells, one occurring at the beginning of the period of sexual activity and the other at the time of testicular refractoriness. In male starlings *(Sturnus vulgaris)*, plasma testosterone levels are high in the breeding period (4.2 ng/10 ml) and low in the postbreeding period (0.5 ng/10 ml), and these T levels are correlated with the condition of the interstitial cells, testicular development, and spermatogenesis (Temple, 1974).

There is also evidence of steroid synthesis by the Sertoli cells of the seminiferous tubules. Sertoli cells contain steroid dehydrogenases (Woods and Domm, 1966) and show a seasonal cycle in content of cholesterol-positive lipid droplets similar to that of the interstitial tissue. Lofts and Bern (1972) suggested that perhaps FSH acts on the Sertoli cells to stimulate local androgen secretion, which in turn influences spermatogenesis. The distinct function, if any, of steroid secretion by Sertoli cells and by interstitial cells awaits further research.

Factors regulating ovarian development and steroidogenesis have been explored. In all birds except some members of the Accipitrinae, Falconinae, and Cathartidae, only the left ovary and oviduct are functional. If the left ovary is removed, compensatory development of the right gonad occurs. The functional left ovary undergoes marked seasonal changes in size and appearance. In the nonbreeding period, the ovary is reduced to a small, compact structure resembling a miniature cluster of grapes. During the breeding period, maturing follicles become markedly enlarged and vascularized. During oogenesis, the oocyte is surrounded initially by a single layer of granulosa cells, and as it matures, multilayered thecal cells form around the outside. The thecal cells become glandular, respond positively to tests for cholesterol, and shows 3β-hydroxysteroid dehydrogenase (3β-HSDH) activity (Chieffi and Botte, 1965; Boucek and Savard, 1970). In the chicken, the granulosa cells develop an agranular endoplasmic reticulum and small sudanophilic granules indicative of steroidogenic activity (Wyburn *et al.*, 1966; Dahl, 1971), whereas the theca interna is devoid of lipids and cholesterol. After ovulation, the granulosa cells show 3β-HSDH activity (Armstrong *et al.*, 1977) and become inflated with lipids for approximately 72 hr, but luteinization does not occur and no corpus luteum forms (Wyburn *et al.*, 1966). There are, however, species differences in the fate of the ruptured follicle. In chickens, it is rapidly phagocytized, whereas in ducks, it may persist for months; the cells of the ruptured follicle wall become agranular and lipoidal, and it has been suggested that this structure is a source of postovulatory steroids.

The precise site of ovarian steroid biosynthesis in birds is unclear. Four cell types—theca, granulosa, interstitial cells of the preovulatory and degenerating postovulatory follicle, and the corpus atreticum—all possess glandular characteristics suggesting an endocrine function. Estrogen production is correlated with seasonal changes in thecal cells and oviduct growth (Marshall and Coombs, 1957), and *in vitro* incubation of thecal tissue with cholesterol yields estrogen (Botte *et al.*, 1966), a result indicating that the theca interna may be the site of estrogen synthesis. The largest quantity of estradiol in fowl is present in the ovarian vein and numerous small ovarian follicles (Senior and Furr, 1975). Follicular progesterone is reportedly highest in the preovulatory follicle and declines rapidly following ovulation (Furr, 1969, 1973), a finding suggesting that progesterone is secreted by the large follicles, but Ma's data (cited by Day and Nalbandov, 1977) show that the concentration of progesterone in the follicular wall is greater in the postovulatory follicle than in preovulatory follicles of any size. The atretic follicle has also been suggested as a site of progesterone production (Marshall and Coombs, 1957). There is some evidence that in birds, as in mammals, androgens are secreted by ovarian interstitial cells (Benoit, 1950; Taber, 1951; Marshall and Coombs, 1957; Woods and Domm, 1966). *In vitro* studies of quail ovary indicate that formation of 5β-reduced androgens from C19 precursors occurs at all stages of follicular development, whereas 11-oxygenated androgen levels are low in the ovarian stroma with previtellogenic follicles, high during vitellogenesis, and minimal after ovulation (Colombo and Colombo Belvedere, 1977). Studies of steroid concentration in the follicular wall and in plasma of hens during the ovulatory cycle indicate that the steroid-secreting capacity of cells changes during follicular maturation (Shahabi *et al.*, 1975; Imai and Nalbandov, 1971, 1978).

The advent of techniques for measuring circulating levels has made possible studies of changes in hormone levels during breeding. Not unexpectedly, these studies show that androgens are present in the plasma of male chickens (Schrocksnadel and Bator, 1971); ducks (Jallageas and Attal, 1968; Jallageas and Assenmacher, 1970); pigeons (Jallageas and Attal, 1968); and several passeriforms (Wingfield and Farner, 1975). It is also clear that androgenic hormones are substantially higher during the breeding season than in the nonbreeding period in starlings (Temple, 1974; Kerlan and Jaffe, 1974) and ducks (Garnier and Attal, 1970; Garnier, 1971; Balthazart and Hendrick, 1976). Determination of androgen levels within a breeding cycle in male ring doves indicates that plasma testosterone increases four fold during the 5- to 7-day period of courtship and declines to baseline levels during incubation and brooding (Feder *et al.*, 1977; see Figure 10). Plasma progesterone levels do not change during the cycle (Silver *et al.*, 1974; Silver, 1978).

Avian gonadotropic hormones have been measured by means of both heterologous (Bagshawe *et al.*, 1968; Croix *et al.*, 1974, Godden *et al.*, 1975) and homologous (Follett *et al.*, 1972; Wentworth *et al.*, 1976) systems. Plasma LH increases from 3 ng/ml in the nonbreeding period to over 15 ng/ml in the breeding season in white-crowned sparrows, and a similar pattern is seen in house sparrows, *Passer domesticus* (Murton, 1975). In ducks, seasonal changes have been demonstrated in plasma FSH (Balthazart and Hendrick, 1976) and LH (Haase *et al.*, 1975). The half-life of LH in Japanese quail is the same in different physiological states (short days, long days, castration plus long days), a finding indicating that seasonal differences in circulating LH are due solely to a change in the rate of LH secretion (Davies *et al.*, 1976).

In female birds, much research has been aimed at the commercially important chicken (Nalbandov, 1976). The chicken usually lays one egg a day in a sequence or a clutch of four eggs, then skips a day before the egg-laying sequence is repeated. The ovarian follicles are arranged in hierarchical size order, and only the largest ovarian follicle ovulates on any day of the clutch sequence. The sequence of hormonal changes preceding ovulation has been studied. Plasma E, P, T, and LH all peak 4–7 hr prior to ovulation and fall soon after they reach their peak in chickens (Furr *et al.*, 1973; Kappauf and van Tienhoven, 1972; Lagüe *et al.*, 1975; Peterson and Common, 1971, 1972; Shahabi *et al.*, 1975; Peterson *et al.*, 1973; Schrocksnadel and Bator, 1971; Jallageas and Attal, 1968; Haynes *et al.*, 1973; Wilson and Sharp, 1973). The follicular wall content of E and T is inversely related to follicle size, whereas P shows no such relation (Shahabi *et al.*, 1975). A similar pattern of hormone secretion is reported in ring doves, where plasma E (Korenbrot *et al.*, 1974), P (Silver *et al.*, 1974), T (Feder *et al.*, 1977) and LH (Cheng and Follett, 1976) increase during the period of courtship, especially prior to egg laying, and remain at baseline values during the remainder of the breeding cycle (Silver, 1978; see Figure 11).

When plasma from individual hens is sampled, increases in plasma P appear to coincide with increases in LH (Furr *et al.*, 1973), whereas increases in E precede the LH peak by about 2 hr (Senior and Cunningham, 1974). Plasma T also rises before LH (Etches and Cunningham, 1977), and exogenously administered T stimulates LH release at certain phases of follicular development (Wilson and Sharp, 1976), whereas antisera to T are more effective than antisera to E or P in blocking ovulation (Furr and Smith, 1975). These data suggest that, in the chicken, T initiates the pituitary LH release necessary for ovulation. In gonadectomized animals, in contrast to laying fowl, LH release is episodic (Wilson and Sharp, 1975).

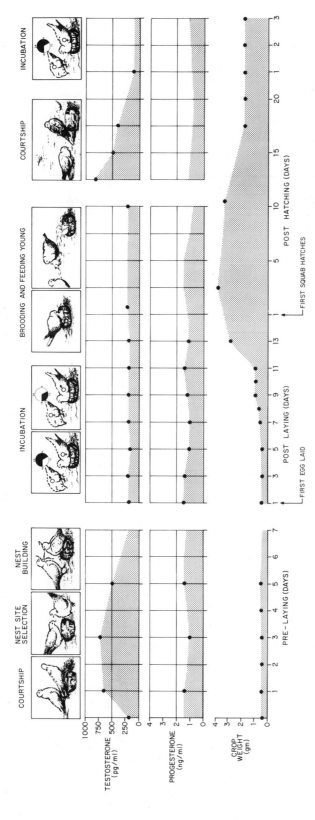

Fig. 10. Changes in the male ring dove's behavior, steroid hormone secretion, and crop growth. Estrogen is not detectable in the male's plasma, and systematic changes in plasma progesterone have not been detected during the cycle. In the male, the secretion of testosterone is related to courtship and nest-oriented behavior. The transition from courtship to incubation is not mediated by changes in testosterone secretion. Crop growth begins during incubation and increases when squab hatch; unlike in the female, some crop development persists at the time that a new cycle of courtship and incubation begins. See the text for further details.

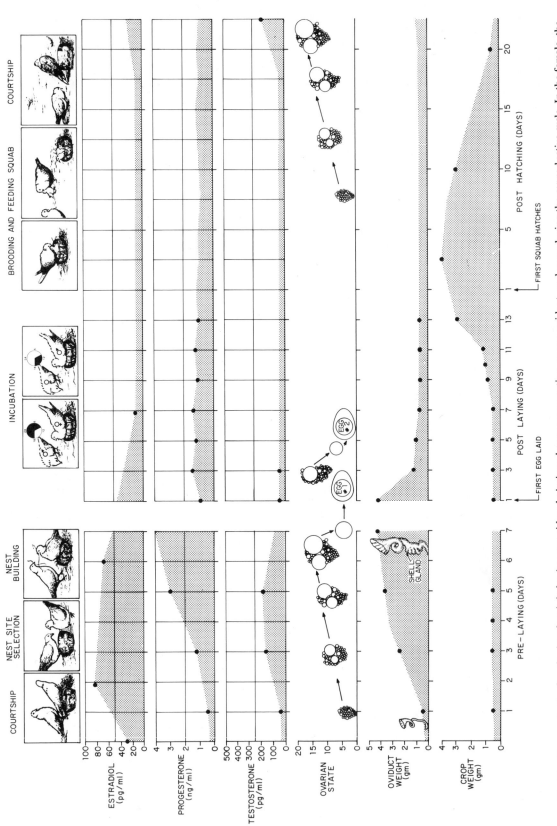

Fig. 11. Changes in the female ring dove's behavior, steroid and pituitary hormone secretion, ovary, oviduct, and crop during the reproductive cycle. In the female, the secretion of estrogen and progesterone produces ovarian development and is causally related to the transition from courtship to nest-building and incubation behavior. Crop growth begins during incubation, increases when the squab hatch, and returns to the undeveloped state when courtship begins anew. See the text for further details.

Testosterone is known to be effective in suppressing gonadotropin secretion in male birds (Wilson, 1970; Cusick and Wilson, 1972; Stetson, 1972b; Follett, 1973; Davies, 1976). In castrated quail intraperitoneal implants of silastic capsules bearing T or TP cause a decline in plasma LH proportional to the size of the implant (Desjardins and Turek, 1977). Massa *et al.* (1977b) examined the effects of a variety of 5α and 5β reduced metabolites of testosterone on plasma LH in castrated male quail and found that several of the 5α reduced metabolites are as effective as T in mediating feedback effects on LH.

Implants of 17β-estradiol in the preoptic or supraoptic regions or in the infundibular area block ovulation in Japanese quail, a result indicating that estradiol prevents LH release (Stetson, 1972a). Systemically administered estradiol increases and progesterone decreases pituitary LH content, whereas T has no effect (Heald *et al.,* 1968), a result contrasting with the effect of T on LH release in hens (Wilson and Sharp, 1976).

The function of the steroid feedback system of immature female mallards has been explored. Intact mallards show a decline in LH between Day 8 and Day 28 whereas birds ovariectomized on Day 22 show higher plasma LH levels than intact controls (Storey and Nicholls, 1977).

Steroid uptake in the brain has been studied in five avian species (see Figure 12): chicken, chaffinch *(Fringilla coelebs),* duck, ring dove, and zebra finch *(Poephila guttata)* (Morrel *et al.,* 1975b; Zigmond, 1975). After injection of ³H-testosterone,

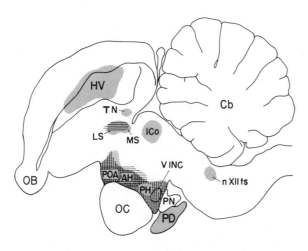

Fig. 12. A parasagittal view of the steroid-concentrating regions in the avian brain and their relationship to neural areas known to be involved in the control of reproductive behavior. Stippled areas indicate steroid-concentrating sites. Implantation of steroids and pituitary fragments, electrical stimulation, and lesions indicate that the basal hypothalamus–preoptic area (|||) is involved in feedback control of gonadotropin secretion. Implants of steroid hormones, lesions, or electrical stimulation indicates that the preoptic area, the basal hypothalamus, and the septum (≡) are involved in the regulation of sexual behavior. (See text for details.) Abbreviations: AH, anterior hypothalamus; Cb, cerebellum; HV, ventral hyperstriatum; ICo, nucleus intercollicularis; LS, lateral septum; MS, medial septum; nXIIts, nucleus of the hypoglossal nerve, tracheosyringeal portion; OB, olfactory bulb; OC, optic chiasm; PD, pars distalis; PH, posterior hypothalamus; POA, preoptic area; TN, nucleus taeniae; V INC, ventral infundibular nucleus.

labeled cells are seen in the preoptic area, the midline hypothalamic nuclei, and the septal region of the limbic system in chicken (Meyer, 1973); Barfield *et al.,* 1979); zebra finch (Arnold *et al.,* 1976; and ring dove (Stern, 1972; Zigmond *et al.,* 1972a, 1973). In the chick, T-labeled cells are found in nucleus intercollicularis and in nucleus isthmo-opticus (Meyer *et al.,* 1976), though precise anatomical and physiological data on the function of these areas are not available. Similarly, in the zebra finch, T is concentrated in the nucleus hyperstriatum ventrale, the nucleus intercollicularis, and the tracheosyringeal component of the hypoglossal motor nucleus (Arnold *et al.,* 1976).

Studying the intact chicken, Wood-Gush *et al.* (1977) found that E, P, and T are taken up in a similar pattern by cells in the ventral hyperstriatum of the telencephalon. Both E- and T-labeled cells were also found in the dorsal hyperstriatum and the nucleus intercalatus of the hyperstriatum.

In studies of ^{3}H-estradiol uptake in male and female ring doves, Martinez-Vargas *et al.* (1975, 1976) identified labeled cells in the preoptic-hypothalamic regions and in the nucleus taenia and the nucleus intercollicularis. Rhees *et al.* (1972) reported that, in 3-week-old ducks, corticosterone and progesterone are localized in the septum, the hippocampus, portions of the hyperstriatum, the septomesencephalic and quintofrontal tracts, and the pituitary. The area of P-labeled cells in the duck is more extensive and more posterior than in the chicken (Wood-Gush *et al.,* 1977), although this discrepancy could be due to prevailing experimental conditions (endogenous secretion of hormones, age of subjects) or species differences in the two studies.

ENDOCRINE AND NEURAL CONTROL OF AVIAN REPRODUCTIVE BEHAVIOR

A substantial body of information indicates that a variety of behaviors, including aggression, courtship, mating, and vocalization, are influenced by reproductive hormones. There is also evidence that external stimuli alter behaviors both directly and, by influencing the secretion of reproductive hormones, indirectly. In the following sections, the older well-established literature is cited only briefly to provide background information, and more recent findings are elaborated on.

ENDOCRINE CONTROL OF AGGRESSIVE BEHAVIOR. Substantial evidence, based on correlational data, the effects of castration and hormone replacement, and neural implants of hormones, indicates that androgens mediate aggressive responses in birds (for review, see Armstrong, 1947; Allee, 1952; Collias, 1950; Guhl, 1961; Silver *et al.,* 1978). For example, the onset of seasonal aggressive behavior is correlated with gonadal development and androgen-dependent secondary sexual structures (e.g., in bank swallow *Riparia riparia;* Petersen, 1955). Using the more direct technique of radioimmunoassay to monitor plasma androgens, Harding and Follett (1979; personal communication, 1979) found that androgen levels are significantly different (higher in some and lower in other individuals) in male red-winged blackbirds *(Agelaius phoeniceus)* captured during aggressive encounters than in control males captured while foraging. Furthermore, males that emitted more alarm vocalizations in response to an intruder had lower levels of T than did males emitting few alarm vocalizations, a finding suggesting that androgen levels in birds are altered as a consequence of the outcome of the aggressive encounter.

Experimental manipulation of hormone levels was first done by Allee and his co-workers(1939), who found that following administration of TP, low-ranking hens showed an elevation of social status. These observations have been confirmed in a variety of species in experiments where exogenous androgens were given to intact or castrated animals (California quail, Emlen and Lorenz, 1942; ring doves, Bennett, 1940; satin bowerbirds, *Ptilonorhynchus violaceus,* Marshall, 1954; zebra finches, Arnold, 1975a; hens, Allee and Foreman, 1955; Japanese quail, Kuo, 1960; starlings, Davis, 1957; Selinger and Bermant, 1967; see Silver *et al.,* 1978, for review).

Androgenic hormones do not act in isolation; they interact with other factors in producing aggressive responses. Thus, Kuo (1960) found that both intact (56%) and castrated (25%) quail could be trained to fight and that some males (17%) retained their ability to fight after gonadectomy. Similarly, testosterone injections alone do not alter aggressive responding in pigeons with previously established dominance relationships unless they are also conditioned to peck (Lumia, 1972).

A number of investigators have suggested that pituitary gonadotropic hormones directly influence aggressive behaviors in birds; however, the validity of the data showing that LH increases aggressiveness in starlings (Davis, 1963; Mathewson, 1961) suffers from poor experimental design and data analysis (van Tienhoven, 1968). Studies showing a direct action of FSH on aggressive responses in male pigeons (Murton and Westwood, 1975) and of LH on aggressive responses in male red-billed weaverbirds, *Quelea quelea* (Butterfield and Crook, 1968; Crook and Butterfield, 1968, 1970), must also be interpreted with caution, as the gonadotropic hormone was administered to animals bearing testes. Also, LH levels are decreased in male red-winged blackbirds caught during an aggressive encounter (Harding and Follett, 1979). In a study using ovariectomized female weaverbirds, Lazarus and Crook (1973) suggested that LH increases aggressive responding in the breeding season, but not at other times, in females that are incompletely and only unilaterally ovariectomized. Finally, Balthazart and Hendrick (1976) argued that gonadotropins stimulate aggressive responses in ducks on the basis that these behaviors are shown at times of the year when plasma levels of pituitary hormones are high and androgen secretions are low. However, it is not known whether this kind of aggression is the same as or different from that seen when androgens are high and gonadotropins are low. In summary, although a direct behavioral effect of gonadotropic hormones has been claimed repeatedly, convincing data have yet to be gathered.

ENDOCRINE CONTROL OF SEXUAL BEHAVIOR. Courtship and mating in male birds are androgen-dependent (for review see Collias, 1950; Guhl, 1961; Guyomarc'h, 1976; Lehrman, 1961; Silver *et al.,* 1978). Sexual behavior declines following castration, and the full spectrum of species-typical responses is restored by androgen replacement therapy (cf. ring doves, Hutchison, 1970; quail, Adkins and Adler, 1972; Schein *et al.* 1972; Beach and Inman, 1965). The antiandrogen cyproterone acetate reduces copulatory behavior and cloacal gland size in quail (Adkins and

Mason, 1974) and reduces courtship behavior in male and female doves (Silver, 1977). The effect of testosterone on the courtship behavior of ring doves and chicks can also be blocked by progesterone administered systemically (Erickson *et al.*, 1967) or intrahypothalamically (Hutchison, 1975; Komisaruk, 1967; Meyer and Angelo, 1975). These studies open the way for a number of questions about the precise mechanism of steroid action and about the role of changing levels of androgenic secretions that are seen during a breeding period.

Fluctuations in androgen levels have been noted within the breeding period in several species. For example, the male ring dove shows an increase in testosterone secretion within 4 hr of pairing with a female in a laboratory breeding cage (Feder *et al.*, 1977; See Figure 10). This elevated level of testosterone is maintained throughout the courtship period and declines to baseline levels (characteristic of isolated males) at the onset of incubation. Androgen levels remain low until the next period of courtship some 4–5 weeks later.

Some of the factors involved in the initiation, maintenance, and decline of courtship-phase androgen secretion in males have been explored (O'Connell *et al.*, 1981a,b). If two males are paired, or if a deaf male is placed with a female, no increase in plasma androgen ensues. If a male is placed with a new female daily, for 9 consecutive days, androgen levels remain elevated for a prolonged period. Finally, if a male is introduced to a novel, incubating female on the third day of courtship with his own mate, his androgen levels decline to baseline values by the next day (Day 4 after introduction to the breeding situation). Taken together, these data show that androgen secretion in the male is initiated and maintained by stimuli or behavior associated with courtship and declines following exposure to stimuli or behavior associated with incubation. The sensory and neural pathways involved in these responses remain to be explored.

Changes in plasma androgens could mediate events during courtship or during incubation (Hutchison and Katongole, 1975; Silver and Feder, 1973; Silver and Buntin, 1973). If a male is castrated and paired with a female about 12 hr later, he shows courtship behavior at a markedly reduced level (Saad and Silver, unpublished, 1978). Furthermore, males treated with the antiandrogen cyproterone acetate show reduced levels of bow cooing and nest cooing (Silver, 1977). Taken together, the data suggest that the levels of androgen secretion during the courtship phase of the breeding cycle in ring doves influence the amount of courtship shown by the male. There appears to be no relationship between fluctuations in androgen levels and the behavioral transition from courtship to incubation. Males paired soon after castration (Saad and Silver, unpublished, 1978) or intact or castrated males given daily injections of testosterone (100 μg/day) all incubate when their mates lay eggs (Silver and Feder, 1973). Hutchison (1969, 1975) has explored the way in which androgen titer's affect brain mechanisms mediating the behavior or male ring doves by using hypothalamic T implants. Because the rate of hormone release is proportional to the implant surface area, this technique can be used to determine whether the various components of courtship display have different thresholds of elicitation. Low-diffusion implants restore nest-oriented courtship, whereas medium- and high-diffusion implants restore more complete courtship.

Interestingly, high-diffusion implants induce the greatest number of aggressive responses toward a stimulus female. This sequence parallels the loss of behaviors after castration, where aggressive bowing disappears first (1–3 days), and nest-oriented components disappear later (15–20 days; Hutchison, 1970), but this sequence is the *reverse* of the sequence seen during normal seasonal development. It may be the case that an understanding of the effects of gonadotropins on aggressive and courtship behavior will clarify this problem (see page 147). The effectiveness of androgen implants is inversely related to the time since castration (Hutchison, 1974a,b).

Eider ducks *(Somateria mollissima)* show seasonal and tide-related cycles of display during the period of pair formation (Gorman, 1974, 1977). Androgen levels (T and DHT) are correlated with seasonal increases in display, but no such correlation between hormone secretion and behavior exists for tidal cycles of display. In addition, no difference in androgen levels is detected before and after coitus. It should be noted that the male eider duck does not show parental care although pairing is prolonged, lasting about 9 months. The courtship display involves up to 30 males displaying around a single female. In mallards, social displays and sexual behavior are exhibited primarily during the early morning, and plasma testosterone levels are highest at this time (Balthazart, 1975, 1976). It remains to be established whether this relationship is causal.

The testes of the pied flycatcher *(Ficedula hypoleuca)* normally regress at about the time the young hatch. In contrast to untreated controls, males given testosterone show prolonged testicular development and territorial behavior and fail to feed young (Silverin, 1977), a finding suggesting that, in this species, testicular regression is important for the development of the male's feeding of young.

Attempts have been made to clarify the role of androgens and estrogens in the regulation of male sexual behavior. In ring doves of both sexes, a dose–response study shows that TP, but not EB, is effective in stimulating male-typical behaviors, such as bow cooing and hop charging. Conversely, EB, but not TP, induces the female-typical response of squatting (sexual receptivity). Responses that are normally shown by both sexes, including wing flipping and nest soliciting, are induced at a lower dose by the homotypical hormone in each sex; that is, EB is more effective in the female, and TP is more effective in the male (Cheng and Lehrman, 1975).

In contrast, Hutchison (1970) suggested that estrogen is more effective than testosterone in stimulating nest soliciting in castrated males, but he did not do a dose–response analysis. In budgerigars *(Melopsittacus undulatus)*, estrogen induces "soliciting for copulation" in both males and females, whereas both E and T stimulate mounting attempts in males (Brockway, 1974). To determine whether the conversion of androgen to estrogen is important in the activation of male sexual behavior, Adkins (1976) examined the effectiveness of antiestrogens in quail. CI-628 reduces sexual receptivity in E-treated female and male quail, and it blocks TP-stimulated copulation in males, a finding suggesting that estrogens may be functionally important in stimulating the behaviors essential for copulation in each sex. The role of aromatization in avian brain tissue requires further analysis. Thus, although estrogen is not detectable in the plasma of male doves (Korenbrot *et al.*, 1974), aromatase activity has been reported in chickens in the preoptic area, the hypothalamus, and the nucleus taeniae, an area corresponding to the amygdala in mammals (Callard *et al.*, 1978a,b).

There has been interest in the role of other androgen metabolites in the regulation of sexual behavior (Adkins and Nock, 1976; Cheng and Lehrman, 1973). Following a systemic injection of ³H-testosterone, both T and DHT are concentrated in the hypothalamus of the ring dove (Stern, 1972), and equal amounts of T and DHT are seen in cell nuclei following injection of radioactive T in the chaffinch (Zigmond *et al.*, 1972b). *In vitro* studies indicate that 5β-DHT is an important metabolite of T in the chick and the European starling (Nakamura and Tanabe, 1974), and in the male European starling, T is converted to both 5α-DHT and 5β-DHT in the anterior pituitary, the hypothalamus, and the hyperstriatum (Massa *et al.*, 1977a). Interestingly, the levels of 5β-DHT are substantially greater than those of 5α-DHT. The behavioral effects of T and 5α-DHT have been studied in doves. Testosterone given systemically or intrahypothalamically is effective in restoring courtship behavior, whereas neither 5α-DHT (doves and pigeons; Hutchison, 1976; Pietras and Wenzel, 1974) nor 5β-DHT (doves; Saad and Silver, unpublished data, 1976) is effective.

Female sexual behavior is dependent on ovarian hormone secretion. Female sexual behavior is abolished by ovariectomy (Cheng, 1973b) or by housing females in short photoperiods, the equivalent of functional castration (Adkins, 1973; Noble, 1973). Estrogen replacement therapy restores all female behavior patterns except nest building in ovariectomized ring doves (Cheng, 1973a), and LH-RH has been shown to synergize with subthreshold doses of estrogen to induce nest-soliciting and squatting behavior in ovariectomized-hypophysectomized ring doves (Cheng, 1977). In intact ring doves (Lehrman, 1958), canaries (Warren and Hinde, 1959), peach-faced lovebirds, *Agapornis roseicollis* (Orcutt, 1965), intact budgerigars (Brockway, 1969a), and ovariectomized budgerigars (Hutchison, 1975), estrogen treatment stimulates nest-oriented behavior, but in ovariectomized ring doves, both estrogen and progesterone are necessary (Cheng and Silver, 1975). It is interesting that in ring doves, the estrogen-induced sequence of recovery of behaviors in gonadectomized females parallels the order of their appearance in the normal breeding cycle (Cheng, 1973a).

Neural Control of Courtship and Mating Behavior. The neural control of avian reproductive behavior has been studied by means of hormone implantation, lesion, and electrical brain stimulation of discrete areas of the CNS.

Implantation of androgen in the preoptic region elicits aggressive behavior in castrated ring doves (Barfield, 1971a; Hutchison, 1974a,b) but not in capons (Barfield, 1969). Capons bearing TP implants in the lateral forebrain do show aggressive behavior (Barfield, 1969). Precocious copulation by male chicks can be induced by T implants in the medial POA (Gardner and Fisher, 1968), and implantation of progesterone in the some area inhibits androgen-induced precocious copulation (Meyer, 1972). Courtship behavior is restored in castrated ring doves and capons by T implants in the AH-POA (Hutchison, 1967, 1970, 1974a,b; Barfield,

1969, 1971b). Progesterone implants in this region block some components of courtship behavior in intact, sexually active ring doves (Komisurak, 1967). Finally, T implants in these areas, as well as in the ventral neostriatum intermediale, induce nest-material collecting in male doves (Erickson and Hutchison, 1977).

Electrolytic lesions in the POA and the mammillary bodies of male fowl suppress copulatory behavior (Barfield, 1965; Perek *et al.*, 1973; Haynes and Glick, 1974). In the male chick, anterior hypothalamic lesions disrupt copulatory behavior (Meyer and Salzen, 1970). Lesions of the lateral septum produce an increase in the nest-oriented behavior of castrated male ring doves given low doses of TP, whereas androgen-dependent bow cooing is not affected (Cooper and Erickson, 1976).

Phillips and Youngren (1971) examined the effects of electrical stimulation of the brain at 1500 loci in the chicken and found that coordinated attack responses are elicited only in medial and basal forebrain regions. Similar results are seen in ducks (Maley, 1969), gulls, *Larus argentatus* and *fuscus* (Delius, 1973), and pigeons (Goodman and Brown, 1966). Sites in the paleostriatum elicit attack in pigeons (Goodman and Brown, 1966), and stimulation of the lateral forebrain induces aggressive behavior in ducks and chickens (Phillips, 1964; Putkonen, 1967).

Neural regulation of reproductive behavior in female birds has received scant attention. We are aware of only three studies. Gibson and Cheng (1979) reported that lesions of the anterior POA produce gonadal atrophy and a decline in female sexual behavior, whereas lesions in the posterior medial hypothalamus (PMH) abolish courtship behavior without accompanying gonadal atrophy. In ovariectomized doves, estrogen replacement does not restore sexual behavior after PMH lesions. Cohen and Cheng (1981) reported that E implants in nucleus intercollicularis (ICo) stimulate nest cooing in ovariectomized doves, whereas lesions in ICo block nest-cooing behavior. Lesioned females do not show normal follicular growth in response to a courting male. Given that ICo receives input from vocal control nuclei (RA; see Figure 13) in other species (Nottebohm *et al.*, 1976; Brown, 1965, 1971), the lesion effects may be due to the female's inability to coo. Finally, Wood-Gush and Gentle (1978) have shown that the characteristic sequence of nest-oriented behavior patterns preceding egg laying in the hen are disrupted following lesions of the ventral hyperstriatum. This behavior is under ovarian hormonal control (Wood-Gush and Gilbert, 1973), and the ventral hyperstriatum is an area of estrogen and progesterone uptake (Wood-Gush *et al.*, 1977).

NEURAL AND ENDOCRINE CONTROL OF SONG PRODUCTION. In a number of avian species, castration eliminates or reduces vocalization and singing behavior, whereas androgen treatment of intact or castrated birds enhances vocalization (for review, see Arnold, 1975b). Song development parallels gonadal development in zebra finches (Sossinka, 1975). Castration of young (9–17 days) male zebra finches does not prevent development of normal song, although castrates develop song more slowly than controls (Arnold, 1975b). Castration of adults reduces singing rate (Arnold 1975a; Pröve, 1974), and testosterone injection reverses these changes (Arnold, 1975a). In young male finches, androgen treatment advances the onset of signing (Sossinka *et al.*, 1975). In juvenile pigeons, plasma LH rises just prior to

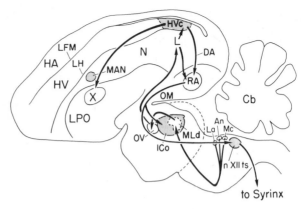

Fig. 13. A parasagittal view of the steroid-concentrating regions in the generalized song-bird brain and their relationship to neural areas known to be involved in the control of song. Abbreviations: An, nucleus angularis of the cochlear nucleus; Cb, cerebellum; DA, tractus archistriatalis dorsalis; HA, accessory hyperstriatum; HV, ventral hyperstriatum; HVc, caudal portion of ventral hyperstriatum; ICo, nucleus intercollicularis; L, field L; La, nucleus laminaris of the cochlear nucleus; LFM, lamina frontalis suprema; LH, lamina of the hyperstriatum; LPO, parolfactory lobe; MAN, nucleus magnocellularis of the anterior neostriatum; Mc, nucleus magnocellularis of the cochlear nucleus; MLd, nucleus mesencephalicus lateralis pars dorsalis; N, neostriatum; nXIIts, nucleus of the hypoglossal nerve, tracheosyringeal portion; OM, occipitomesencephalic tract; OV, nucleus ovoidalis; RA, nucleus robustus archistriatalis; X, area X. (From "Hormone Concentrating Cells in Vocal Control and Other Areas of the Brain in Zebra Finch [*Poephila guttata*]" by A. P. Arnold, F. Nottebohm, and D. W. Pfaff. *Journal of Comparative Neurology*, 1976, *165*, 487–512. Copyright 1976 by the *Journal of Comparative Neurology*. Adapted by permission.)

the time of breaking of voice at about 2 months, long before sexual maturity, which occurs at 4–5 months (Abs *et al.,* 1977).

In an extensive series of studies using both behavioral and anatomical techniques, Nottebohm and his co-workers (1976; see Figure 13) traced the neural pathways regulating song. In the canary, efferent pathways have been traced from the caudal hyperstriatum ventriale (HVc) to area X, and a round, large-celled archistriatal nucleus (RA). The RA projects ipsilaterally to the midbrain nucleus ICo and to the thalamus, as well as sending fibers to the motor nucleus of the syrinx nXIIts. Lesions of HVc or RA produce marked deficits in song production, whereas unilateral lesions of area X or ICo have no effect. Interestingly, autoradiographic studies of testosterone uptake indicate that HVc, ICo, and nXIIts are all sites of steroid concentration (Arnold *et al.,* 1976). Although behavioral studies indicate that lesions of the left hemisphere disrupt song more than do right-hemisphere lesions (canary; Nottebohm *et al.,* 1976), comparable steroid uptake occurs in both hemispheres (Arnold *et al.,* 1976).

Androgen treatment increases singing in both sexes in some species (female and male canaries, Herrick and Harris, 1957; chaffinches, Hooker, 1968; white-crowned sparrows, Konishi, 1965; white-throated sparrows, Falls, 1969), whereas in female zebra finches testosterone is ineffective (Nottebohm and Arnold, 1976). There are dramatic differences in the size of the brain areas controlling song that correlate with sex differences in singing behavior (Nottehohm and Arnold, 1976).

In both zebra finches and canaries, two vocal control areas, HVc and RA, and also area X are larger in males than in females.

Some of these brain areas have been implicated in the control of song in other species (see review by Phillips and Peek, 1975). Electrical stimulation of ICo induces vocalization in pigeons (Delius, 1971b; Popa and Popa, 1933), red-winged blackbirds (Brown, 1965, 1971; Newman, 1970, 1972), Stellar's jay, *Cyancitta stelleri* (Brown, 1973), chickens (Phillips and Youngren, 1971, 1973), and gulls, *Larus argentatus* and *L. fuscus* (Delius, 1971a,b), whereas the preoptic-anterior hypothalamic region is effective in ring doves and pigeons (Akerman, 1966a,b).

INTERACTION OF INTERNAL AND EXTERNAL STIMULI. The hormonal and neural regulation of sexual behavior can best be understood in the context of the role of environmental stimuli in the regulation of physiological state and in the elicitation of sexual behavior.

Research with ring doves provides a good example of the interdependence of behavioral and physiological responses of the sexes. Stimuli provided by the intact courting male induce ovarian development that leads to ovulation (Lehrman, 1959; Lehrman *et al.*, 1961; Erickson, 1970; Cheng, 1974). The contribution of contact and of visual and vocal cues has been explored. If the mates are separated by a glass plate, latency to ovulate is not altered in pigeons (Matthews, 1939) or ring doves (Erickson and Lehrman, 1964; Barfield, 1971b), a finding indicating that contact is not necessary for ovarian development. (Of course, the eggs are infertile.) If a female hears but does not see other courting doves, she rarely lays eggs, but if she can hear the colony and see another female (Collias, 1950) or see herself in a mirror (Lott and Brody, 1966), she often ovulates. Furthermore, colony sounds alone can produce some stimulation of the female's reproductive system (Lott *et al.*, 1967; Lehrman and Friedman, 1969), as can visual stimuli alone (Lambe and Erickson, 1973). The stimulatory effect of hearing the male, however, requires the active participation of the female (Friedman, 1977). A female who sees another female being courted does not undergo ovarian growth. The degree of responsiveness to the male depends on the reproductive state of the female (Cheng, 1974).

In budgerigars, male vocalization as well as the sight of other pairs stimulates ovarian development and oviposition provided that nesting facilities are available (Vaugien, 1951; Ficken *et al.*, 1960; Brockway, 1962, 1964). However, hearing the male is sufficient to stimulate ovarian growth in isolated females (Brockway, 1964, 1969b). Brockway has also identified the particular component of the male's vocalizations that stimulates the female gonadal state. Of the six different vocalizations performed by the male budgerigar, only the "soft warble," which normally is closely associated with precopulatory behavior, stimulates ovarian activity and oviposition in females (Brockway, 1965, 1969b). In canaries, the quality of the male's song, evaluated in terms of the size of the repertoire, produces differential responses in females (Kroodsma, 1976); females who hear a tape-recorded repertoire made up of 35 syllables lay more eggs and build nests faster than do females that hear 5-syllable repertoires.

The type of behavior exhibited by the male can have a pronounced effect on female reproductive physiology. Prolonged aggressive display by a male ring dove delays ovulation (Hutchison and Lovari, 1976). Low-ranking, less aggressive males pair earlier than high-ranking, more aggressive males in pigeons housed in groups (Castoro and Guhl, 1958).

Each of the above experiments indicates that the male influences the female's reproductive physiology. The converse is also true. Male ring doves paired with a female, but not those paired with another male, showed an increase in plasma androgen levels (Feder et al., 1977). Also, males show compensatory hypertrophy after unilateral gonadectomy if they are housed in isolation under long photoperiods (14L:10D), but not if housed under short photoperiods (8L:16D). However, if males on short photoperiods are exposed to a female, then compensatory testicular hypertrophy does occur (Cheng, 1976). In starlings, spermatogenesis is maintained for longer periods in birds housed in 12L:12D if the males are housed with females (Schwab and Lott, 1969) than if they are housed in all-male groups. A male dove secreting sufficient androgen nest-builds, but the time at which he does so depends on the behavior and the ovarian condition of the female. Thus, males paired with E- and P-treated females (which spend a lot of time at the next site) carry nesting material immediately on pairing, whereas males paired with control females (which spend little time at the nest site) carry very little (Martinez-Vargas and Erickson, 1973; Erickson and Martinez-Vargas, 1975).

Finally, doves proceeding through a reproductive cycle with their mates show a transition from courtship to incubation behavior. Although the transition from *not* incubating to incubating appears to depend on endogenous changes within the male (Lehrman, 1965; Silver et al., 1973), the transition from courting to *not* courting seems to be influenced by the stimulus complex presented by the female. Males proceeding through a reproductive cycle with their mates and tested at selected intervals show courtship behavior when placed with a stimulus female irrespective of the stage of the cycle they had reached with their mate (Silver and Barbiere, 1977). Thus, the male's display of courtship during the cycle is influenced by the external stimulus situation presented by the female. Confirming this conclusion, field observations indicate that the time taken for starting a new nest after destruction of the clutch bears no relation to the stage of incubation of the destroyed clutch (see Lehrman, 1959, for review).

External factors such as photoperiod, rainfall, and stimuli from the nest and eggs have long been known to influence reproduction in birds (van Tienhoven and Planck, 1973). For example, exposure to long daylengths (14L:10D) increases pituitary and ovarian hormone secretion and also increases nest building in canaries (Follett et al., 1973). This effect of photoperiod on nest-building behavior can be dissociated from the effect on pituitary and gonadal hormone secretion. Thus, the response still occurs in photorefractory females whose pituitaries do not respond to long days (Hinde et al., 1974; Steel and Hinde, 1972a,b). In addition, if given a standard dose of E, ovariectomized females show more nest building if they are maintained on long photoperiods than if they are maintained on short photoperiods (Steel and Hinde, 1972a; Hinde et al., 1974). By comparing canaries housed in skeleton (e.g., 8L:5D:1L:10D) long days to those housed in short (9L:15D) days (total number of hours of light being equal), it was shown that the daylength-dependent nest-building response to E treatment is mediated by a circadian rhythm of photosensitivity (Steel and Hinde, 1976). Also, T-treated, castrated male ring doves held on long photoperiods (16L:8D) show higher levels of nest building than males held in short photoperiods (8L:16D), a finding indicating that the photoperiod affects building behavior through some mechanism other than changes in endogenous androgen levels (McDonald and Liley, 1978).

Female ring doves exposed to a courting male have greater follicular and oviduct development and greater soliciting and nest-building behavior when housed in long daylengths (16L:8D) than when housed in short (8L:16D) daylengths (Liley, 1976a). Hormone treatment (E, P, or E and P) of females produced greater sexual and nest-oriented responses in females housed in long photoperiods than in females housed in short photoperiods (Liley, 1976b).

Some desert birds breed at irregular intervals following rainfall. Marshall and Disney (1957) demonstrated that the stimulating factor in the African red-billed weaver is the availability of nesting material. For example, groups maintained in outdoor cages provided with food and with "rain" from a sprinkler built nests only if green grass was also provided. In contrast, Friedkalns and Bennett (1977) showed that, in the Australian zebra finch *(Taeniopygia guttata castanotis)*, male gonadal growth is induced primarily by humidity and is affected to a smaller extent by the availability of water and green grass.

Tactile stimuli from the nest are known to influence ovulation in canaries (Hinde, 1958; Hinde *et al.*, 1963). If females are deprived of nesting material and/or nest cups, egg laying is delayed in canaries (Hinde and Warren, 1959; Warren and Hinde, 1961); Bengalese finches (Slater, 1969); and ring doves (Lehrman *et al.*, 1961). Doves whose nests are destroyed daily build more than those given full nest bowls (White, 1975). In Bengalese finches, where the male usually carries nesting material to the female, egg laying is delayed if the pair is separated and only the male has access to nesting material (Slater, 1969).

In an attempt to elucidate the stimulus control of nesting behavior, Wood-Gush noted that when a hen lays a soft-shelled egg, no nesting behavior is shown. Usually, the mating behavior of the hen follows a predictable sequence, consisting of random pacing, calling, nest examination, nest entry, and sitting on the nest until the egg is laid (Wood-Gush and Gilbert, 1970). These behaviors can be induced by treatment with E and P (Wood-Gush and Gilbert, 1969, 1973). Nesting behavior is not dependent on the presence of the egg in the oviduct (Wood-Gush, 1963) nor on neural stimuli from the ovary (Wood-Gush and Gilbert, 1970), but it is influenced by the ovary as shown by ligation, excision, or cocaine injection of the ruptured ovulatory follicle (Wood-Gush and Gilbert, 1964, Gilbert and Wood-Gush, 1965, 1968). The data suggest that the newly ovulated follicle produces a blood-borne substance that is important in stimulating the hen's nesting behavior and oviduct. Day and Nalbandov (1977) reported that, in hens, levels of the prostaglandin F (PGF) are low throughout the 26-hr laying cycle, until 4–6 hr before ovulation, when it shows a four fold increase. The postovulatory follicle shows a

100-fold increase in PGF about 24 hr after its own ovulation and 2 hr prior to the next ovulation and oviposition. Indomethacin administration delays oviposition but does not alter steroidogenesis. These data suggest that substances from the postovulatory follicle may control the time of oviposition; unfortunately, the effects on concomitant nesting behavior were not analyzed in the latter study.

In other contexts, stimuli associated with the egg can influence physiology and behavior. If birds incubate infertile eggs, then the time that the female takes to lay new eggs is reduced in comparison with the time taken by females incubating fertile eggs (Silver and Gibson, 1980; Allen, 1979). The fertility of eggs influences the timing of recycling both before and after oviposition (Cheng, 1977). Thus, females laying infertile eggs as a result of having mated with castrated males recycle sooner than females that have laid fertile eggs but are given infertile eggs for incubation.

The interested reader should consult the following references for recent reviews of avian reproduction: Balthazart (1983), Silver (1983), and Silver and Cooper (1983).

CONCLUDING REMARKS

In this chapter, we have reviewed the structure and function of the hypo-thalamo-pituitary-gonadal axis in nonmammalian vertebrates and have shown how each component of this axis plays an important role in the regulation of species-typical behavior patterns. These studies demonstrate that there is both conservation and variation in psychoneuroendocrine controlling mechanisms. For example, in all vertebrate classes investigated, steroid hormones are concentrated in similar discrete areas of the CNS, which in turn integrate internal and external stimuli mediating gonadotropin hormone secretion and reproductive behavior so that the sequence of physiological and behavioral events necessary for successful reproduction is ensured. Although hormonal secretion at each level of the HPG axis is responsive to changes in the internal and external environments in all vertebrates, the specific stimuli responsible for these changes vary among species. Similarly, although reproductive behavior is under hormone control, the form of behavior elicited varies dramatically among species. The comparative behavioral endocrinological approach not only provides valuable insight into the origins and adaptations of reproductive behavior but also offers clues to the mechanisms of hormone action in vertebrates.

Acknowledgments

We wish to thank John Buntin, Bob Goy, Richard E. Jones, Darcy B. Kelley, Charles H. Muller, and Martin P. Schreibman for reading individual sections of this review, and Deedra McClearn for valuable comments on the entire manuscript.

Abs, M., Ashton, S. A., and Follett, B. K. Plasma luteinizing hormone in juvenile pigeons around the time of breaking of the voice. *Hormones and Behavior,* 1977, *9,* 189–192.

Adachi, T., and Ishii, S. Binding of rat follicle-stimulating hormone to the turtle ovary *in vitro* and inhibition of the binding by gonadotropin preparations of various vertebrates. *General and Comparative Endocrinology,* 1977, *33,* 1–7.

Adkins, E. K. Functional castration of the female Japanese quail. *Physiology and Behavior,* 1973, *10,* 619–621.

Adkins, E. K. The effects of the antiestrogen Cl-628 on sexual behavior activated by androgen or estrogen in quail. *Hormones and Behavior,* 1976, *7,* 417–429.

Adkins, E. K., and Adler, N. T. Hormonal control of behavior in the Japanese quail. *Journal of Comparative Physiology and Psychology,* 1972, *81,* 27–36.

Adkins, E. K., and Mason, P. Effects of Cyproterone acetate in the male Japanese quail. *Hormones and Behavior,* 1974, *5,* 1–6.

Adkins, E. K., and Nock, B. L. The effects of the antiestrogen C1-628 on sexual behavior activated by androgen or estrogen in quail. *Hormones and Behavior,* 1976, *7,* 417–429.

Akerman, B. Behavioural effects of electrical stimulation in the forebrain of the pigeon: I. Reproductive behaviour. *Behaviour,* 1966a, *26,* 323–338.

Akerman, B. Behavioural effects of electrical stimulation in the forebrain of the pigeon: II. Protective behaviour. *Behaviour,* 1966b, *26,* 339–350.

Allee, W. L. Dominance and hierarchy in societies of vertebrates. In P. Grassé (Ed.), *Structure et Physiologie des Sociétés Animals.* Paris: CNRS, 1952.

Allee, W. L., and Foreman, D. Effects of an androgen on dominance and subordinance in six common breeds of *Gallus gallus. Physiological Zoology,* 1955, *28,* 89–115.

Allee, W. L., Collias, N. E., and Lutherman, L. Z. Modification of the social order in flocks of hens by the injection of testosterone propionate. *Physiological Zoology,* 1939, *12,* 412–440.

Allen, T. O. Oviposition of fertile eggs and postponement of subsequent egg-laying in ring doves *(Streptopelia risoria). Journal of Reproduction and Fertility,* 1979, *55,* 61–64.

Alpert, L. C., Brawer, J. R., Jockson, I. M. D., and Reichlen, S. Localization of LHRH in neurons in frog brain *(Rana pipiens* and *Rana catesbiana). Endocrinology,* 1976, *98,* 910–921.

Armstrong, D. G., Davidson, M. F., Gilbert, A. B., and Wells, J. W. Activity of 3-β-hydroxysteroid dehydrogenase in the post-ovulatory follicle of the domestic fowl *(Gallus domesticus). Journal of Reproduction and Fertility,* 1977, *49,* 261–266.

Armstrong, E. A. *Bird Display and Behaviour.* London: Lindsay Drummond, 1947.

Arnold, A. P. The effects of castration and androgen replacement on song, courtship, and aggression in zebra finches *(Poephila guttata). Journal of Experimental Zoology,* 1975a, *191,* 309–326.

Arnold, A. P. The effects of castration on song development in zebra finches *(Poephila guttata). Journal of Experimental Zoology,* 1975b, *191,* 261–278.

Arnold, A. P., Nottebohm, F., and Pfaff, D. W. Hormone concentrating cells in vocal control and other areas of the brain in zebra finch *(Poephila guttata). Journal of Comparative Neurology,* 1976, *165,* 487–512.

Aronson, L. R. Problems in the behavior and physiology of a species of African mouthbreeding fish. *Transactions of the New York Academy of Science,* 1948, *2,* 33–43.

Aronson, L. R. Hormones and reproductive behavior: Some phylogenetic considerations. In A Gorbman (Ed.), *Comparative Endocrinology.* New York: Wiley, 1959.

Aronson, L. R. Functional evolution of the forebrain in lower vertebrates. In L. R. Aronson, E. Tobach, D. S. Lehrman, and J. S. Rosenblatt (Eds.), *Development and Evolution of Behavior.* San Francisco: W. H. Freeman, 1970.

Aronson, L. R., and Kaplan, H. Function of the teleostean forebrain. In D. Ingle (Ed.), *The Central Nervous System and Fish Behavior.* Chicago: University of Chicago Press, 1968.

Aronson, L. R., and Noble, G. K. The sexual behavior of Anura: II. Neural mechanisms controlling mating in the male leopard frog, *Rana pipiens. Bulletin of the American Museum of Natural History,* 1945, *86,* 87–139.

Assenmacher, I. Reproductive endocrinology: The hypothalamo-hypophysial axis. In D. S. Farner (Ed.), *Breeding Biology of Birds.* Washington, D.C.: National Academy of Sciences, 1973.

Assenmacher, I., and Tixier-Vidal, A. Hypothalamic-pituitary relations. *Excerpta Medica International Congress Series*, 1965, *83*, 131–145.

Assenmacher, I., Daniel, J. Y., and Jallageas, M. Annual endocrine rhythms and environment. In V. H. T. James (Ed.), *Endocrinology*, Vol I. Amsterdam: Excerpta Medica, 1973.

Auffenberg, W. Display behavior in tortoises. *American Zoolist, 1977, 17, 241–250.*Auffenberg, W. On the courtship of Gopherus polyphemus. *Herpetologica, 1966, 22, 113–117.*

Baggerman, B. An experimental study on the time of breeding and migration in the three-spined stickleback *(Gasterosteus aculeatus L.). Archives of Neurology and Zoology*, 1957, *12*, 105–318.

Baggerman, B. On the endocrine control of reproductive behaviour in the male three-spined stickleback *(Gasterosteus aculeatus L.). Symposia of the Society for Experimental Biology*, 1966, *20*, 427–456.

Baggerman, B. Hormonal control of reproductive and parental behaviour in fishes. In E. J. W. Barrington and C. B. Jørgensen (Eds.), *Perspectives in Endocrinology*. New York: Academic Press, 1968.

Baggerman, B. Influence of photoperiod and temperature on the timing of the breeding season in the stickleback, *Gasterosteus aculeatus. General Comparative Endocrinology*, 1969, *13*, 491.

Baggerman, B. Photoperiodic responses in the stickleback and their control by a daily rhythm of photosensitivity. *General Comparative Endocrinology*, Suppl. 3, 1972, 466–476.

Bagshawe, K. D., Orr, A. H., and Godden, J. Cross-reaction in radioimmunoassay between human chorionic gonadotropin and plasma from various species. *Journal of Endocrinology*, 1968, *42*, 513–518.

Ball, J. N. Reproduction in female bony fishes. *Symposia of the Zoological Society of London*, 1960, *1*, 105–135.

Ball, J. N., and Baker, B. I. The pituitary gland: Anatomy and histophysiology. In W. S. Hoar and J. D. Randall (Eds.), *Fish Physiology*, Vol. 2. New York: Academic Press, 1969.

Balthazart, J. On the daily distribution of displays and sexual behaviour in semi-wild mallards. IRCS *Medical Science*, 1975, *3*, 404.

Balthazart, J. Daily variations of behavioural activities and of plasma testosterone levels in the domestic duck *Anas platyrhynchos. Journal of Zoology London*, 1976, *180*, 155–173.

Balthazart, J. Hormonal correlates of behavior. In D. S. Farner, J. R. King, and K. C. Parkes (Eds.), *Avian Biology*, Vol. VII. New York: Academic Press, 1983.

Balthazart, J., and Hendrick, J. Annual variation in behavior, testosterone and plasma FSH levels in the Rouen duck, *Anas platyrhynchos. General and Comparative Endocrinology*, 1976, *28*, 171–183.

Barfield, R. J. Effects of preoptic lesions on the sexual behavior of male domestic fowl. *American Zoologist*, 1965, *5*, 686–687 (abstract).

Barfield, R. J. Activation of copulatory behavior by androgen implanted into the preoptic area of the male fowl. *Hormones and Behavior*, 1969, *1*, 37–51.

Barfield, R. J. Activation of sexual and aggressive behavior by androgen implanted into the male ring dove brain. *Endocrinology*, 1971a, *89*, 1470–1476.

Barfield, R. J. Gonadotrophic hormone secretion in the female ring dove in response to visual and auditory stimulation by the male. *Journal of Endocrinology*, 1971b, *49*, 305–310.

Barfield, R. J., Ronay, G., and Pfaff, D. W. Autoradiographic localization of androgen-concerntrating cells in the brain of the male domestic fowl. *Neuroendocrinology*, 1979, *28*, 217–227.

Bauer, D. H., and Demski, L. S. Vertical banding evoked by electrical stimulation of the brain in the green sunfish *(Lepomis cyanellus). American Zoologist*, 1974, *14*, 1282 (abstract).

Baum, M. J., and Vreeburg, J. T. Copulation in castrated male rats following combined treatment of estradiol and dihydrotestosterone. *Science*, 1973, *182*, 283–285.

Beach, F. A. Cerebral and hormonal control of reflexive mechanisms involved in copulatory behavior. *Physiological Review*, 1967, *54*, 297–315.

Beach, F. A. Sexual attractivity, proceptivity and receptivity. *Hormones and Behavior*, 1976, 7, 105–138.

Beach, F. A., and Inman, N. G. Effects of castration and androgen replacement on mating in male quail. *Proceedings of the National Academy of Sciences of the U.S.A.*, 1965, *54*, 1426–1431.

Belle, R., Schorderet-Slatkine, S., Drury, K. C., and Ozon, R. *In vitro* progesterone binding to *Xenopus laevis* oocytes. *General and Comparative Endocrinology*, 1975, *25*, 339–345.

Bennett, M. A. The social hierarchy in ring doves: II. The effects of treatment with testosterone mopionate. *Ecology*, 1940, *21*, 148–165.

Benoit, J. Reproduction caractères sexuel et hormones: Déterminisme du cycle saisonnier. In P.-P. Grassé (Ed.), *Traité de zoologie*, Vol. 15. Paris: Masson, 1950.

Benoit, J. Opto-sexual reflex in duck: Physiological and histochemical aspects. *Yale Journal of Biology and Medicine*, 1961, *34*, 97–116.

Benoit, J., and Assenmacher, I. The control by visible radiations of the gonadotropic activity of the duck hypophysis. *Recent Progress in Hormone Research*, 1959, *15*, 143–164.

Berlind, A. Caudal neurosecretory system: A physiologist's view. *American Zoologist*, 1973, *13*, 759–770.

Bernstein, J. J. Anatomy and physiology of the central nervous system. In W. S. Hoar and J. D. Randall (Eds.), *Fish Physiology*, Vol. 3. New York: Academic Press, 1969.

Bicknell, R. J., and Follett, B. K. A quantitative assay for luteinizing hormone–releasing hormone (LH-RH) using dispersed pituitary cells. *General and Comparative Endocrinology*, 1975, *26*, 141–152.

Billard, R., and Peter, R. E. Gonadotropin release after implantation of anti-estrogens in the pituitary and hypothalamus of goldfish, *Carassius auratus*. *General and Comparative Endocrinology*, 1977, *32*, 213–220.

Billard, R., Breton, B., and Escaffre, A. M. Maintien et restauration de la spermatogénèse par un extrait acétonique hypophysaire de carpe chez le cyprin *(Carassius auratus)* traité au méthallibure. *Annales de Biologie Animal Biochemie Biophysick*, 1971, *11*, 167–174.

Billard, R., Richard, M., and Breton, B. Stimulation de la secretion gonadotrope hypophysaire après castration chez la truite arc-enciel: Variation de la reponse au cours du cycle reproducteur. *Comptes Rendus de l'Académie des Sciences*, 1976, *D283*, 171–174.

Billard, R., Richard, M., and Breton, B., Stimulation of gondotropin secretion after castration in rainbow trout. *General and Comparative Endocrinology*, 1977, *33*, 163–165.

Blair, A. P. The effects of various hormones on primary and secondary sex characters of juvenile *Bufo fowleri*. *Journal of Experimental Zoology*, 1946, *103*, 365–400.

Blüm, V., and Fiedler, K. Hormonal control of reproductive behavior in the some cichlid fish. *General and Comparative Endocrinology*, 1968, *5*, 186–196.

Bogert, C. M. The influence of sound on the behavior of amphibians and reptiles. In *Animal Sounds and Communication*. American Institute of Biological Sciences Publications, 1960, *7*, 137–320.

Bonney, R. C., Cunningham, F. J., and Furr, B. J. A. Effect of synthetic luteinizing hormone releasing hormone on plasma luteinizing hormone in the female domestic fowl *Gallus domesticus*. *Journal of Endocrinology*, 1974, *63*, 539–547.

Bons, N., Kerdelhue, B., and Assenmacher, I. Evidence for LH-RH containing neurons in the anterior hypothalamus of the duck *(Anas platyrhynchos)*: An immunofluorenscence study. *Ninth Conference of European Comparative Endocrinologists*, Giesser, 1977.

Botte, V., Delrio, G., and Lupo di Prisco, P. Conversion of neutral steroid hormones "in vitro" by the theca cells of the growing ova and by the post-ovulatory follicles of the laying hen. *Excerpta Medica Foundation International Congress Series*, 1966, *111* (abstract 738).

Boucek, R. J., and Savard, K. Steroid formation by the avian ovary *in vitro (Gallus domesticus)*. *General and Comparative Endocrinology*, 1970, *15*, 6–11.

Bourne, A. R., and Seamark, R. F. Seasonal changes in 17β hydroxysteroids in the plasma of a male lizard *(Tiliqua rogosa)*. *Comparative Biochemistry and Physiology*, 1975, *50B*, 535–536.

Brasch, M., and Betz, T. W. The hormonal activities associated with the cephalic and caudal regions of cockerel pars distalis. *General and Comparative Endocrinology*, 1971, *16*, 241–256.

Breton, B., and Weil, C. Effets du LH/FSH-RH synthétique et d'extraits hypothalamiques de carpe sur la sécrétion d'hormone gonadotrope *in vivo* chez la carpe *(Cryprinus carpio L.)*. *Comptes Rendus de l'Académie des Sciences, Séries D*, 1973, *277*, 2061–2064.

Breton, B., Jalabert, B., Billard, R., and Weil, C. Stimulation *in vitro* de la libération d'hormone gonadotrope hypophysaire par un facteur hypothalamique chez la carpe *Cyprinus carpio* L. *Sciences, Séries D*, 1972a, *273* 2591–2594.

Breton, B., Weil. C., Jalabert, B., and Billard, R. Activité rèciproque des facteurs hypothalamique de bélier *(ovis aries)* et de poissons téléostéens sur la sécrétion *in vitro* des hormones C-HG et LH respectivement pars les hypophyses de carpe et de bélier. *Comptes Rendus de l'Académie des Sciences, Séries D*, 1972b, *274*, 2530–2533.

Breton, B., Jalabert, B., and Weil, C. Caracterisation partielle d'un facteur hypothalamique de liberation des hormones gonadotropes chex la carpe *(Cyprinus carpio L.)* étude *in vitro*. *General and Comparative Endocrinology*, 1975, *25*, 405–415.

Brockway, B. F. The effects of nest-entrance positions and male vocalizations on reproduction in budgerigars. *Living Bird, First Annual of the Cornell Laboratory of Ornithology*, 1962, 93–101.

Brockway, B. F. Social influences on reproductive physiology and ethology of budgerigars *(Melopsittacus undulatus)*. *Animal Behavior*, 1964, *12*, 493–501.

Brockway, B. F. Stimulation of ovarian development and egg laying by male courtship vocalization in budgerigars *(Melopsittacus undulatus)*. *Animal Behavior*, 1965, *13*, 575–578.

Brockway, B. F. Hormonal and experiential factors influencing the nestbox oriented behaviour of budgerigars *(Melopsittacus undulatus)*. *Behaviour*, 1969a, *35*, 1–26.

Brockway, B. F. Roles of budgerigar vocalization in the integration of breeding behaviour. In R. A. Hinde (Ed.) *Bird Vocalizations*. Cambridge: Cambridge University Press, 1969b.

Brockway, B. F. The influence of some experiential and genetic factors, including hormones, on the visible courtship behaviour of budgerigars *(Melopsittacus)*. *Behaviour*, 1974, *51*, 1–18.

Brown, J. L. Vocalizations evoked from the optic lobe of a songbird. *Science*, 1965, *149*, 1002–1003.

Brown, J. L. An exploratory study of vocalization areas in the brain of the redwinged blackbird *(Agelaius phoeniceus)*. *Behaviour*, 1971, *39*, 91–127.

Brown, J. L. Behavior elicited by electrical stimulation of the brain of Stellar's jay. *Condor*, 1973, *75*, 1–16.

Brown, N. L., and Follett, B. K. Effects of androgens on the testes of intact and hypophysectomized Japanese Quail. *General and Comparative Endocrinology*, 1977, *33*, 267–277.

Brown, N. L., Bayle, J. D., Scanes, C. G., and Follett, B. K. Chicken gonadotrophins: Their effects on the testes of immature and hypophysectomized Japanese quail. *Cell and Tissue Research*, 1975, *156*, 499–520.

Bruin, J. P. C, de. Telencephalic functions in the behaviour of the Siamese fighting fish, *Betta splendens* Regan (Pisces, Anabantidae). Amsterdam, Unpublished thesis, 1977.

Burke, W. H., and Cogger, E. A. The effect of luteinizing hormone releasing hormone (LHRH) on serum L.H. and ovarian growth in turkeys. *Poultry Science*, 1977, *56*, 239–242.

Burzawa-Gerard, E., Goncharov, B., Dumas, A., and Fontaine, Y. A. Further biochemical studies on carp gonadotropin (c-GTH): Biochemical and biological comparison of c-GTH and a gonadotropin from *Acipenser stellatus* Pall. (Chondrostei). *General and Comparative Endocrinology*, 1976, *29*, 498–505.

Butler, A. B., and Northcutt, R. G. Retinal projections in *Iguana iguana* and *Anolis carolinensis*. *Brain Research*, 1971, *26*, 1–13.

Butterfield, P. A., and Crook, J. H. Annual cycle of nest building and agonistic behaviour in captive *Quelea quelea* with reference to endocrine factors. *Animal Behavior*, 1968, *16*, 308–317.

Callard, G. V., and Ryan, K. J. Androgen metabolism in explant cultures of adult turtle brain. *Proceeding of the Society for the Study of Reproduction. Annual Meetings*, 1977 (abstract 135).

Callard, G. V., Petro, Z., and Ryan, K. J. Identification of aromatase in the reptilian brain. *Endocrinology*, 1977, *100*, 1214–1218.

Callard, G. V., Petro, Z., and Ryan, K. J. Conversion of androgen to estrogen and other steroids in the vertebrate brain. *American Zoologist*, 1978a, *18*, 511–523.

Callard, G. V., Petro, Z., and Ryan, K. J. Phylogenetic distribution of aromatase and other androgen-converting enzymes in the central nervous system. *Endocrinology*, 1978b, *103*, 2283–2290.

Callard, I. P. Testicular steroid synthesis in the snake, *Natrix sipedon pictiventris*. *Journal of Endocrinology*, 1967, *37*, 105–106.

Callard, I. P., and Callard, G. V. Testicular regulation in lower vertebrates. *Biology of Reproduction*, 1978, *18*, 16–43.

Callard, I. P., and Doolittle, J. P. The influence of intrahypothalamic implants of progesterone on ovarian growth and function in the ovoviviparous lizard, *Sceloporus cyanogenys*. *Comparative Physiology and Biochemistry*, 1972, *44A*, 625–630.

Callard, I. P., and Klotz, K. L. Sensitivity parameters of estrogen action in the iguanid lizard *Dipsosaurus dorsalis*. *General and Comparative Endocrinology*, 1973, *21*, 314–321.

Callard, I. P., and Lance, V. The control of reptilian follicular cycles. In J. H. Calaby and C. H. Tyndale-Biscoe (Eds.), *Evolution of Reproduction*. Canberra: Australian Academy of Sciences, 1978.

Callard, I. P., and McConnell, W. F. Effects of intrahypothalamic estrogen implants on ovulation in *Sceloporus cyanogenys*. *General and Comparative Endocrinology*, 1969, *13*, 496.

Callard, I. P., Bayne, C. G., and McConnell, W. F. Hormones and reproduction in the female lizard *Sceloporus cyanogenys*. *General and Comparative Endocrinology*, 1972a, *18*, 175–194.

Callard, I. P., Chan, S. W. C., and Potts, M. A. The control of the reptilian gonad. *American Zoologist*, 1972b, *12*, 273–287.

Callard, I. P., Doolittle, J., Banks, W. C., and Chan, S. W. C. Recent studies on the control of the reptilian ovarian cycle. *General and Comparative Endocrinology Supplement*, 1972c, *3*, 65–75.

Callard, I. P., Callard, G. V., Lance, V., and Eccles, S. Seasonal changes in testicular structure and function and the effects of gonadotropins in the freshwater turtle, *Chrysemys picta*. *General and Comparative Endocrinology*, 1976, *30*, 347–356.

Campbell, C. M., Walsh, J. M., and Idler, D. R. Steroids in the plasma of the winter flounder (*Pseudopleuronectes americanus* Walbaum): A seasonal study and investigation of steroid involvement in oocyte maturation. *General and Comparative Endocrinology,* 1976, *29,* 14–20.

Carew, B. A. M. Some effects of methallibure (I.C.I. 33,828) on the stickleback *Gasterosteus aculeatus* L. University of British Columbia, unpublished M.Sc. thesis, 1968.

Carlson, D. R. Female sexual receptivity in *Gambusia affinis* (Baird and Girard). *Texas Journal of Science,* 1969, *21,* 167–173.

Carpenter, C. C. Communication and displays of snakes. *American Zoologist,* 1977, *17,* 217–223.

Carpenter, C. C. Variation and evolution of stereotyped behavior in reptiles: I. A survey of reptile behavior patterns relating primarily to agonistic behavior, courtship and mating which are thought or implied to be sterotyped. In C. Gans and D. W. Tinkle (Eds.), *Biology of the Reptilia.* New York: Academic Press, 1978.

Carpenter, C. C., Gillingham, J. C., and Murphye, J. B. The combat ritual of the rock rattlesnake *(Crotalus lepidus). Copeia,* 1976, 764–780.

Castoro, P. L., and Guhl, A. M. Pairing behavior of pigeons related to aggressiveness and territory. *The Wilson Bulletin,* 1958, *70,* 57–69.

Chan, K. K-S. Effect of synthetic luteinizing-releasing hormone (LH-RH) on ovarian development in Japanese medaka, *Oryzias latipes. Canadian Journal of Zoology,* 1977, *55,* 155–160.

Cheng, M.-F. Effect of estrogen on behavior of ovariectomized ring doves *(Streptopelis risoria). Journal of Comparative Physiology and Psychology,* 1973a, *83,* 234–239.

Cheng, M.-F. Effect of ovariectomy on the reproductive behavior of female ring doves *(Streptopelia risoria). Journal of Comparative Physiology and Psychology,* 1973b, *83,* 221–233.

Cheng, M.-F. Ovarian development in the female ring dove in response to stimulation by intact and castrated male ring doves. *Journal of Endocrinology,* 1974, *63,* 43–53.

Cheng, M.-F. Interaction of lighting and other environmental variables on activity of hypothalamo-hypophyseal-gonadal system. *Nature,* 1976, *263,* 148–149.

Cheng, M.-F. Egg fertility and prolactin as determinants of reproductive recycling in doves. *Hormones and Behavior,* 1977, *9,* 85–98.

Cheng, M.-F. Progress and prospect in ring dove research: A personal view. In J. S. Rosenblatt, R. A. Hinde, E. Shaw, and C. Beer (Eds.), *Advances in the Study of Behavior,* Vol. 9. New York: Academic Press, 1979.

Cheng, M.-F., and Follett, B. K. Plasma luteinizing hormone during the breeding cycle of the female ring dove. *Hormones and Behavior,* 1976, *7,* 199–205.

Cheng, M.-F., and Lehrman, D. S. Relative effectiveness of diethylstilbesterol and estradiol benzoate in inducing female behavior patterns of ovariectomized ring doves *(Streptopelia risoria). Hormones and Behavior,* 1973, *4,* 123–127.

Cheng, M.-F., and Lehrman, D. S. Gonadal hormone specificity in the sexual behavior of ring doves. *Psychoneuroendocrinology,* 1975, *1,* 95–102.

Cheng, M.-F., and Silver, R. Estrogen-progesterone regulation of nest building and incubation behavior of female ring doves *(Streptopelia risoria). Journal of Comparative Physiology and Psychology,* 1975, *88,* 256–263.

Chester-Jones, I., Bellamy, D., Chan, D. K. O., Follett, B. K., Henderson, I. W., Phillips, J. G., and Snart, R. S. Biological actions of steroid hormones in nonmammalian vertebrates. In D. R. Idler (Ed.), *Steroids in Nonmammalian Vertebrates.* New York: Academic Press, 1972.

Chieffi, G., and Botte, V. The distribution of some enzymes involved in the steroidogenesis of hen's ovary. *Experientia,* 1965, *21,* 16–17.

Chizinsky, W. Effects of castration upon sexual behavior in male platyfish, *Xiphophorus maculatus. Physiological Zoology,* 1968, *41,* 466–475.

Chobsieng, P., Naor, Z., Kock, Y., Zor, U., and Linder, H. R. Stimulatory effect of prosta glandin E_2 on LH release in the rat: Evidence for hypothalamic site of action. *Neuroendocrinology,* 1975, *17,* 12–17.

Clark, C. E., and Fraps, R. M. Induction of ovulation in chickens with median eminence extracts. *Poultry Science,* 1967, *46,* 1245–1246.

Clarke, R. F. An ethological study of the iguanid genera *Callisaurus, Cophosaurus,* and *Holbrookia. Emporia State Research Studies,* 1965, *13,* 1–66.

Cohen, J., and Cheng, M. F. The role of the midbrain in courtship behavior of the female ring dove *(Streptopelia risoria):* Evidence from radiofrequency lesion and hormone implant studies. *Brain Research,* 1981, *207,* 279–301.

Cole, C. J. Femoral gland in lizards: A review. *Herpetoligica*, 1966, *22*, 199–206.

Collias, N. E. Hormones and behavior with special reference to birds and the mechanism of hormone action. In E. S. Gordon (Ed.), *A Symposium on Steroid Hormones*. Madison: University of Wisconsin Press, 1950.

Colombo, L., and Colombo Belvedere, P. Biosynthesis of 5β-reduced and 11-oxygenated androgens by the ovary of the Japanese Quail at different stages of the follicular cycle. *Ninth Conference of European Comparative Endocrinologists*, Giessen, 1977 (abstract).

Colombo, J. A., and Segura, E. T. Effects of estradiol on brain electrogram of the female toad *Bufo arenarum* Hensel during the breeding season. *General and Comparative Endocrinology*, 1972, *18*, 268–275.

Colombo, L., and Yaron, Z, Steroid 21-hydroxylase activity in the ovary of the snake *Storeria dekayi* during pregnancy. *General and Comparative Endocrinology*, 1976, *28*, 403–412.

Colombo, J. A. Negroni, J. A., and Segura, E. T. Behavioral and EEG patterns of the female toad *Bufo arenarum* during artificial clasping. *Physiology and Behavior*, 1972, *9*, 21–25.

Colombo, L., Yaron, Z., Daniels, E., and Belvedere, P. Biosynthesis of 11-deoxycorticosterone by the ovary of the yucca night lizard, *Xantusia vigilis*. *General and Comparative Endocrinology*, 1974, *24*, 331–337.

Cooper, R. L., and Erickson, C. J. Effects of septal lesions on the courtship behavior of male ring doves *(Streptopelia risoria)*. *Hormones and Behavior*, 1976, *7*, 441–450.

Cooper, W. E., Jr., and Ferguson, G. W. Relative effectiveness of progesterone and testosterone as inductors of orange spotting in female collared lizards. *Herpetologica*, 1972a, *28*, 64–65.

Cooper, W. E., Jr., and Ferguson, G. W. Steroids and color change during gravidity in the lizard *Crotaphytus collaris*. *General and Comparative Endocrinology*, 1972b, *18*, 69–72.

Cooper, W. E., Jr., and Ferguson, G. W. Estrogenic priming of color change induced by progesterone in the collared lizards, *Crotaphytus collaris*. *Herpetologica*, 1973, *29*, 107–110.

Cope, E. D. The crocodilians, lizards and snakes of North America. *Report of the United States National Museum, 1898*, 153 (1900).

Crews, D. Behavioral correlates to gonadal state in the lizard, *Anolis carolinensis*. *Hormones and Behavior*, 1973a, *4*, 307–313.

Crews, D. Coition-induced inhibition of sexual receptivity in female lizards *(Anolis carolinensis)*. *Physiology and Behavior*, 1973b, *11* 463–468.

Crews, D. Effects of castration and subsequent androgen replacement therapy on male courtship behavior and environmentally-induced ovarian recrudescence in the lizard *Anolis carolinensis*. *Journal of Comparative and Physiological Psychology*, 1974, *87*, 963–969.

Crews, D. Psychobiology of reptilian reproduction. *Science*, 1975, *189*, 1059–1065.

Crews, D. Hormonal control of male courtship behavior and female attractivity in the garter snake *(Thamnophis sirtalis sirtalis)*. *Hormones and Behavior*, 1976, *7*, 451–460.

Crews, D. The annotated anole: Studies on the control of lizard reproduction. *American Scientist*, 1977, *65*, 428–434.

Crews, D. Control of reptilian reproductive cycles. In C. Beyer (Ed.), *Endocrine Control of Sexual Behavior*. New York: Raven Press, 1978a.

Crews, D. Hemipenile preference: Stimulus control of male mounting behavior in the lizard, *Anolis carolinensis*. *Science*, 1978b, *199*, 195–196.

Crews, D. The neuroendocrinology of reproduction in reptiles. *Biology of Reproduction*, 1979, *20*, 51–73.

Crews, D. Interrelationships among ecological, behavioral, and neuroendocrine processes in the reproductive cycle of *Anolis carolinensis* and other reptiles. In J. S. Rosenblatt, R. A. Hinde, C. G. Beer, and M.-C. Busnel (Eds.), *Advances in the Study of Behavior*, Volume 11. New York: Academic Press, 1980.

Crews, D. Regulation of reptilian reproductive behavior. In J.-P. Ewert, R. Capranica, and D. Ingle (Eds.), *Advances in Vertebrate Neuroethology*. New York: Plenum Press, 1983.

Crews, D., and Greenberg, N. Function and causation of social signals in lizards. *American Zoologist*, 1981, *21*, 273–294.

Crews, D., and Licht, P. Site of progesterone production in the reptilian ovarian follicle. *General and Comparative Endocrinology*, 1975a, *27*, 553–556.

Crews, D., and Licht, P. Stimulation of reptilian ovarian steroidogenesis *in vitro* by mammalian, reptilian, and amphibian gonadotropins. *General and Comparative Endocrinology*, 1975b, *27*, 71–83.

Crews, D. P., Rosenblatt, J. S., and Lehrman, D. S. Effects of unseasonal environmental regime, group presence, group composition and males' physiological state on ovarian recrudescence in the lizard, *Anolis carolinensis. Endocrinology,* 1974, *94,* 541–547.

Crews, D. P., Traina, V., Wetzel, F. T., and Muller, C. Hormonal control of male reproductive behavior in the lizard, *Anolis carolinensis:* Role of testosterone, digydrotestosterone and estradiol. *Endocrinology,* 1978, *103,* 1814–1821.

Crim, L. W., and Cluett, D. M. Elevation of plasma gonadotropin concentration in response to mammalian gonadotropin releasing hormone (GRH) treatment of the brown trout as determined by radioimmunoassay. *Endocrine Research Communication,* 1974, *1,* 101–110.

Crim, L. W., Watts, E. G., and Evans, D. M. The plasma gonadotropin profile during sexual maturation in a variety of salmonid fishes. *General and Comparative Endocrinology,* 1975, *27,* 62–70.

Crim, L. W., Peter, R. E., and Billard, R. Stimulation of gonadotropin secretion by intraventricular injection of hypothalamic extracts in the goldfish, *Carassius auratus. General and Comparative Endocrinology,* 1976, *30,* 77–82.

Croix, D., Hendrick, J. C., Balthazart, J., and Franchimont, P. Dosage radioimmunologique de l'hormone folliculo-stimulant (FSH) hypophysaire de canard à l'aide d'un système de mammifères. *Comptes Rendus. Societe de-Biologie,* 1974, *168,* 136–140.

Crook, J. H., and Butterfield, P. A. Effects of testosterone propionate and luteinizing hormone on agonistic and nest building behaviour of *Quelea quelea. Animal Behavior,* 1968, *16,* 370–384.

Crook, J. H., and Butterfield, P. A. Gender role in the social system of *Quelea.* In J. H. Crook (Ed.), *Social Behavior in Birds and Mammals.* New York: Academic Press, 1970.

Crowley, W. R., Rodriguez-Sierra, J. F., and Komisaruk, B. R. Hypophysectomy facilitates sexual behavior in female rats. *Neuroendocrinology,* 1976, *20,* 328–338.

Cusick, E. K., and Wilson, F. E. On the control of spontaneous testicular repression in tree sparrows *(Spizella arborea). General and Comparative Endocrinology,* 1972, *19,* 441–456.

Dahl, E. The fine structure of the granulosa cells in the domestic fowl and the rat. *Zeitschrift für Zellforschung und Mikroskopische Anatomie, West Germany,* 1971, *119,* 58–67.

Davies, D. T. Steroid feedback in the male and female Japanese quail. *Journal of Endocrinology,* 1976, *70,* 513–514.

Davies, D. T., and Follett, B. K. Electrical stimulation of the hypothalamus and luteinizing hormone secretion in Japanese quail. *Journal of Endocrinology,* 1975a, *67,* 431–438.

Davies, D. T., and Follett, B. K. The neuroendocrine control of gonadotrophin release in the Japanese quail: I. The role of the tuberal hypothalamus. *Proceedings of the Royal Society London B,* 1975b, *191,* 285–301.

Davies, D. T., and Follett, B. K. The neuroendocrine control of gonadotrophin release in the Japanese quail: II. The role of the anterior hypothalamus. *Proceedings of the Royal Society London B,* 1975c, *191,* 303–315.

Davies, D. T., Baylé, J. -D., Bicknell, R. J., and Aston, S. A. The half-life of luteinizing hormone in intact and castrated quail during photoperiodic stimulation. *Endocrine Research Communications,* 1976, *3,* 243–256.

Davis, D. E. Aggressive behavior in castrated starlings. *Science,* 1957, *126,* 253.

Davis, D. E. The hormonal control of aggressive behavior. *Proceedings XIII International Ornithological Congress,* 1963, *2,* 994–1003.

Davis, P. G., and Barfield, R. J. Masculine and feminine behavior in the male. *Program Eastern Conference on Reproductive Behavior,* June 5–8, 1977 (abstract).

Davis, R. E., Kassel, J., and Schwagmeyer, P. Telencephalic lesions and behavior in the teleost, *Macropodus opercularis:* Reproduction, startle response, and operant behavior in the male. *Behavioral Biology,* 1976a, *18,* 165–178.

Davis, R. E., Mitchell, M., and Dolson, L. The effects of methallibure on conspecific visual reinforcement, social display frequency, and spawning in the paradise fish, *Macropodis opercularis* (L.) Belontiidae. *Physiology and Behaviors,* 1976b, *17,* 47–52.

Davis, R. E., Morrell, J. I., and Pfaff, D. W. Autoradiographic localization of sex steroid concentrating cells in the brain of the teleost *Macropodus opercularis* (Osteichthyes: Belontiidae). *General and Comparative Endocrinology,* 1977, *33,* 496–505.

Day, S. L., and Nalbandov, A. V. Presence of prostoglandin F (PGF) in hen follicles and its physiological role in ovulation and oviposition. *Biology of Reproduction,* 1977, *16,* 486–494.

Deery, D. J. Determination by radioimmunoassay of the luteinizing hormone-releasing hormone (LHRH) content of the hypothalamus of the rat and some lower vertebrates. *General and Comparative Endocrinology,* 1974, *24,* 280–285.

DeFazio, A., Simon, C. A., Middendorf, G. A., and Romano, D. Substrate licking in an iguanid lizard: A response to novel situations in *Sceloporus jarrovi. Copeia,* 1977, 706–709.

Delius, J. D. Foraging behavior patterns of herring gulls elicited by electrical forebrain stimulation. *Experientia,* 1971a, *27,* 1287–1289.

Delius, J. D. Neural substrates of vocalizations in gulls and pigeons. *Experimental Brain Research,* 1971b, *12,* 64–80.

Delius, J. D. Agnostic behaviour of juvenile gulls, a neuroethological study. *Animal Behavior,* 1973, *21,* 236–246.

Della Corte, F., Angelini, F., Galgano, M., and Marinucci, A. The ultra structive of the pars distalis of the hypophysis in males of *Triturus cristatus carnifex* Laur. treated with the anti androgen cyproterone acetate. *Zeitschrift für Zellforschung,* 1973, *137,* 209–221.

Demski, L. S., and Gerald, J. W. Sound production evoked by electrical stimulation of the brain in toadfish *(Opsanus beta). Animal Behavior,* 1972, *20,* 507–513.

Demski, L. S., and Knigge, K. M. The telencephalon and hypothalamus of the bluegill *(Lepomis macrochirus):* Evoked feeding, aggressive and reproductive behavior with representative frontal sections. *Journal of Comparative Neurology,* 1971, *143,* 1–16.

Demski, L. S., Bauer, D. H., and Gerald, J. W. Sperm release evoked by electrical stimulation of the fish brain: A functional-anatomical study. *Journal of Experimental Zoology,* 1975, *191,* 215–232.

Desjardins, C., and Turek, F. W. Effects of testosterone on spermatogenesis and luteinizing hormone release in Japanese quail. *General and Comparative Endocrinology,* 1977, *33,* 293–303.

Devine, M. C. Copulatory plugs, restricted mating opportunities and reproductive competition among male garter snakes. *Nature,* 1977, *267,* 345–346.

DeVlaming, V. L. Environmental and endocrine control of teleost reproduction. In C. B. Schreck (Ed.), *Control of Sex in Fishes.* Blacksburg: Virginia Polytechnic Institute and State University, Extension Division, 1974.

Diakow, C. Initiation and inhibition of the release croak of *Rana pipiens. Physiological Behavior,* 1977, *19,* 607–610.

Diebschlag, E. Beobacktungen und Versuche an intakten und grosshirnlosen Eidechsen und Ringelnattern. *Zoologischer Anzeiger,* 1938, *124,* 30–49.

Dierickx, K. The origin of the aldehyde-fuchsin-negative nerve fibres of the median eminence of the hypophysis: A gonadotropic centre. *Zeitschrift für Zellforschung,* 1965, *66,* 504–518.

Dierickx, K. Experimental identification of the hypothalamic gonadotropic centre. *Zeitschrift für Zellforschung,* 1966, *74,* 53–79.

Dierickx, K. The function of the hypophysis without preoptic neurosecretory control. *Zeitschrift für Zellforschung,* 1967a, *78,* 114–130.

Dierickx, K. The gonadotropic centre of the tuber cinereum hypothalami and ovulation. *Zeitschrift für Zellforschung,* 1967b, *77,* 188–203.

Distel, H. Behavior and electrical brain stimulation in the green iguana, *Iguana iguana* L.: I. Schematic brain atlas and stimulation device. *Brain, Behavior and Evolution,* 1976, *13,* 421–450.

D'Istria, M., Botte, V., Delrio, G., and Chieffi, G. Implication of testosterone and its metabolites in the hormonal regulation of thumb pads of *Rana esculenta. Steriods Lipids Research,* 1972, *3,* 321–327.

D'Istria, M., Delrio, G., Botte, V., and Chieffi, G. Radioimmunoassay of testosterone, 17β-oestradiol and oestrone in the male and female plasma of *Rana esculenta* during the sexual cycle. *Steriods Lipids Research,* 1974, *5,* 42–48.

Dodd, J. M. Gonadal and gonadotrophic hormones in lower vertebrates. In A. S. Parkes (Ed.), *Marshall's Physiology of Reproduction,* Vol. 1, Part 2. London: Longmans, 1960.

Dodd, J. M., Follett, B. K., and Sharp, P. J. Hypothalamic control of pituitary function in submammalian vertebrates. *Advanced Comparative Physiology and Biochemistry,* 1971, *4,* 114–223.

Doerr-Schott, J. Immunohistochemical detection by light and electron microscopy, of pituitary hormones in cold-blooded vertebrates: I. Fish and amphibians. *General and Comparative Endocrinology,* 1976, *28,* 487–512.

Doerr-Schott, J., and Dubois, M. P. Immunoreactive neurons in the brain of *Xenopus Laevis. General and Comparative Endocrinology,* 1976, *29,* 246–247 (abstract).

Donaldson,.E. M. Reproductive endocrinology of fishes. *American Zoologist,* 1973, *13,* 909–927.

Dowling, H. G., and Savage, J. M. A guide to the snake hemipenis: A survey of basic structure and systematic characteristics. *Zoologica*, 1960, *45*, 17–28.

Drance, M. G., Hollenberg, M. J., Smith, M., and Wylie, V. Histological changes in trout testis produced by injections of salmon pituitary gonadotropin. *Canadian Journal of Zoology*, 1976, *54*, 1285–1293.

Dudley, C. A., and Moss, R. L. Prostaglandin E_2: Facilitatory action on the lordotic response. *Journal of Endocrinology*, 1976, *71*, 457–458.

Dufaure, J.-P., and Gignon, A. Action des hormones androgènes sur l'épididyme d'un reptile lacertilien, *Lacerta vivipare* Jacquin: Effects de la testostérone et de ses principaux métabolites en culture organotypique. *General and Comparative Endocrinology*, 1975, *25*, 112–120.

Egami, N. Preliminary note on the induction of the spawning reflex and oviposition in *Oryzias latipes* by the administration of neurohypophyseal substances. *Annotationes Zoologicae Japon*, 1959, *32*, 13–17.

Egami, N., and Ishii, S. Hypophyseal control of reproductive functions in teleost fishes. *General and Comparative Endocrinology*, 1962, *Suppl. 1*, 248–253.

Egami, N., and Nambu, M. Factors initiating mating behavior and oviposition in the fish, *Orzyias latipes*. *Journal of the Faculty of Science, University of Tokyo*, 1961, *9*, 263–278.

Emlen, J. T., and Lorenz, K. Z. Pairing responses of free-living Valley Quail to sex hormone pellet implants. *Aukland*, 1942, *59*, 369–378.

Erickson, C. J. Induction of ovarian activity in female ring doves by androgen treatment of castrated males. *Journal of Comparative Physiology and Psychology*, 1970, *71*, 210–215.

Erickson, C. J., and Hutchison, J. B. Induction of nest-material collecting in male Barbary doves by intracerebral androgen. *Journal of Reproduction and Fertility*, 1977, *50*, 9–16.

Erickson, C. J., and Lehrman, D. S. Effect of castration of male ring doves upon ovarian activity of females. *Journal of Comparative Physiology and Psychology*, 1964, *58*, 164–166.

Erickson, C. J., and Martinez-Vargas, M. -C. The hormonal basis of cooperative nest-building. In P. Wright, P. G. Caryl, and D. M. Vowles (Eds.), *Neural and Endocrine Aspects of Behavior in Birds*. Amsterdam: Elsevier, 1975.

Erickson, C. J., Bruder, R. H., Homisauruk, B. R., and Lehrman, D. S. Selective inhibition by progesterone of androgen-induced behavior in male ring doves *(Streptopelia risoria)*. *Endocrinology*, 1967, *81*, 39–44.

Etches, R. J., and Cunningham, F. J. The plasma concentrations of testosterone and LH during the ovulation cycle of the hen *(Gallus domesticus)*. *Acta endocrinologica*, 1977, *84*, 357–366.

Evans, L. T. Endocrine relationships in turtles: II. Claw growth in the slider, *Pseudemys scripta troostii*. *Anatomical Record*, 1952, *112*, 251–263.

Evans, L. T. Structure as related to behavior in the organization of populations in reptiles. In W. F. Blair (Ed.), *Vertebrate Speciation*. Austin: University of Texas Press, 1961.

Evennett, P. J., and Thornton, V. F. The distribution and development of gonadotropic activity in the pituitary of *Xenopus laevis*. *General and Comparative Endocrinology*, 1971, *16*, 606–607.

Falls, J. B. Functions of territorial song in the white-crowned sparrow. In R. Hinde (Ed.), *Bird Vocalizations*. New York: Cambridge University Press, 1969.

Farmer, S. W., Licht, P., Papkoff, H., and Daniels, E. L. Purification of gonadotropins in the leopard frog *(Rana pipiens)*. *General and Comparative Endocrinology*, 1977, *32*, 158–162.

Farner, D. S., and Follett, B. K. Light and other environmental factors affecting avian reproduction. *Journal of Animal Science*, 1966, *25*, *Suppl.* 90–118.

Farner, D. S., and Lewis, R. A. Photoperiodism and reproductive cycles in birds. *Photophysiology*, 1971, *6*, 325–370.

Farner, D. S., and Oksche, A. Neurosecretion in birds. *General and Comparative Endocrinology*, 1962, *2*, 113–147.

Farner, D. S., Wilson, F. E., and Oksche, A. Avian Neuroendocrine mechanisms. In L. Martini and W. F. Ganong (Eds.), *Neuroendocrinology*, Vol. 2. New York: Academic Press, 1967.

Feder, H. H. Regulation of sexual behavior by hormones in female nonprimates. In J. Money and H. Musaph (Eds.), *Handbook of Sexology*. Elsevier, Holland: North-Holland Biomedical Press, 1977.

Feder, H. H., and Marrone, B. L. Progesterone: Its role in the central nervous system as a facilitator and inhibitor of sexual behavior and gonadotropin release. *Annals of the New York Academy of Sciences*, 1977, 331–354.

Feder, H. H., Naftolin, F., and Ryan, K. J. Male and female responses in male rats given estradiol benzoate and 5α-androstan-17β-ol-3-one propionate. *Endocrinology*, 1974, *94*, 136–141.

Feder, H. H., Storey, A., Goodwin, D., Reboulleau, L., and Silver, R. Testosterone and "5α-Dihydro-testosterone" levels in peripheral plasma of male and female ring doves *(Streptopelia risoria)* during the reproductive cycle. *Biology and Reproduction,* 1977, *16,* 666–677.

Ferguson, G. W. Color Change and reproductive cycling in female collared lizards *(Crotaphytus collaris).* *Copeia,* 1976, 491–494.

Ferguson, G. W. Variation and evolution of stereotyped behavior of reptiles: II. Social displays of reptiles: Communications value, ultimate causes of variation, taxonomic significance and heritability of population differences. In C. Gans and D. W. Tinkle (Eds.), *Biology of the Reptilia,* Vol. 7. New York: Academic Press, 1978.

Ficken, R. W., van Tienhoven, A., Ficken, M. S., and Sibley F. C. Effect of visual and vocal stimuli on breeding in the budgerigar *(Melopsitacus undulatus). Animal Behavior,* 1960, *8,* 104–106.

Fiedler, K. Hormonale Kontrolle des Verhaltens bei Fischen. *Fortschritte der Zoologie,* 1974, *22,* 268–309.

Follett, B. K. The neuroendocrine regulation of gonadotrophin secretion in avian reproduction. In D. S. Farner (Ed.), *Breeding Biology of Birds.* Washington, D.C.: National Academy of Sciences, 1973.

Follett, B. K. Photoperiodism and seasonal breeding in birds and mammals. In D. G. E. Lamming and D. B. Creighton (Eds.), *Control of Ovulation.* London: Butterworths, 1978.

Follett, B. K., and Davies, D. T. Photoperiodicity and the neuroendocrine control of reproduction in birds. *Symposia of the Zoological Society of London,* 1975, *35,* 199–224.

Follett, B. K., and Farner, D. S. Pituitary gonadotropins in the Japanese quail *(Coturnix coturnix japonica)* during photoperiodically induced gonadal growth. *General and Comparative Endocrinology,* 1966, *7,* 125–131.

Follett, B. K., and Redshaw, M. R. The physiology of vitellogenesis. In B. Lofts (Ed.), *Physiology of the Amphibia,* Vol. 2. New York: Academic Press, 1974.

Follett, B. K., Scares, C. G., and Cunningham, F. J. A radioimmunoassay for avian luteinizing hormone. *Journal of Endocrinology,* 1972, *52,* 359–378.

Follett, B. K., Hinde, R. A., Steel, E., and Nicholls, T. J. The influence of photoperiod on nest-building, ovarian development and LH secretion in canaries *(Serinus canarius). Journal of Endocrinology,* 1973, *59,* 151–162.

Fox, H. The urogenital system of reptiles. In C. Gans and T. S. Parsons (Eds.) *Biology of the Reptilia,* Vol. 6. London and New York: Academic Press, 1977.

Friedkalns, J., and Bennett, R. K. Environmental factors regulating gonadal growth in the zebra finch *(Taeniopygia guttata castanotis). Ninth Conference of European Comparative Endocrinologists,* Giessen, 1977 (abstract).

Friedman, M. B. Interactions between visual and vocal courtship stimuli in the neuroendocrine response of female doves. *Journal of Comparative Physiology and Psychology,* 1977, *91,* 1408–1416.

Furr, B. J. A. Identification of steroids in the ovaries and plasma of laying hens and the site of production of progesterone in the ovary. *General and Comparative Endocrinology,* 1969, *13* (abstract 56).

Furr, B. J. A. Radioimmunoassay of progesterone in peripheral plasma of the domestic fowl in various physiological states and in follicular venous plasma. *Acta Endocrinologica Copenhagen,* 1973, *72,* 89–100.

Furr, B. J. A., and Smith, G. K. Effects of antisera against gonadal steriods on ovulation in the hen *Gallus domesticus. Journal of Endocrinology,* 1975, *66,* 303–304.

Furr, B. J. A., Bonney, R. C., England, R. J., and Cunningham, F. J. Luteinizing hormone and progesterone in peripheral blood during the ovulatory cycle of the hen *Gallus domesticus. Journal of Endocrinology,* 1973, *75,* 159–169.

Gardner, J. E., and Fisher, A. E. Induction of mating in male chicks following preoptic implantation of androgen. *Physiology and Behavior,* 1968, *3,* 709–712.

Garnier, D. H. Variations de la testostérone du plasma périphérique chez le Carnard Pékin au cours du cycle annuel. *Comptes Rendus de l'Académie des Sciences Paris,* 1971, *272* Série D, 1665–1668.

Garnier, H. D., and Attal, J. Variation de la testostérone du plasma testiculaire et des cellules interstitielles chez le Canard Pékin au cours du cycle annuel. *Competes Rendus de l'Académie de Sciences Paris,* 1970, *270,* Série D, 2472–2475.

Garrick, L. D., and Lang, J. W. Social signals and behaviors of adult alligators and crocodiles. *American Zoologist,* 1977, *17,* 225–239.

Garrick, L. D., Lang, J. W., and Herzog, H. A. Jr. Social signals of adult American alligators. *Bulletin of the American Museum of Natural History,* 1978, *160,* 153–192.

Gavaud, J. An experimental study of the influence of external factors on spermatogenesis and steroid-ogenesis of *Nectophrynoides occidentalis* Angel. *Journal de Physiologie* (Paris), 1975, *70*, 549–559.

Gesell, M. S., and Callard, I. P. The hypothalamic-hypophysial neurosecretory system in the iguanid lizard, *Dipsosaurus dorsalis:* A qualitative and quantitative study. *General and Comparative Endocrinology*, 1972, *19*, 397–404.

Gibson, M. J., and Cheng, M. F. Neural mediation of estrogen-dependent courtship behavior in female ring doves. *Journal of Comparative and Physiological Psychology*, 1979, 93(5), 855–867.

Gilbert, A. B., and Wood-Gush, D. G. M. Control of the nesting behaviour of the domestic hen: III. The effect of cocaine in the postovulatory follicle. *Animal Behavior*, 1965, *13*, 284–285.

Gilbert, A. B., and Wood-Gush, D. G. M. Control of the nesting behaviour of the domestic hen: IV. Studies on the pre-ovulatory follicle. *Animal Behavior*, 1968, *16*, 168–170.

Godden, P. M. M., and Scanes, C. G. Studies on the purification and properties of avian gonadotrophins. *General and Comparative Endocrinology*, 1975, *27*, 538–542.

Godden, P., Scanes, C. G., and Sharp, P. J. Variation in the level of follicle stimulating hormone in the circulation of birds, as determined by homologous radioimmunoassay. *Journal of Endocrinology*, 1975, *67*, 20P (abstract).

Goldby, F., and Gamble, H. F. The reptilian cerebral hemispheres. *Biological Review*, 1957, *32*, 383–420.

Goodman, I. J., and Brown, J. L. Stimulation of positively and negatively reinforcing sites in the avian brain. *Life Sciences*, 1966, *5*, 693–704.

Goos, H. J. Th., Ligtenberg, P., and van Oordt, P. G. W. J. Immunohistochemical study on GTH-RH in the frog, *Rana esculenta. General and Comparative Endocrinology*, 1976, *29*, 246 (abstract).

Gorman, M. L. The endocrine basis of pair formation behaviour in the male Eider *(Somateria mollissima). Ibis*, 1974, *116*. 451–465.

Gorman, M. L. Sexual behaviour and plasma androgen concentrations in the male eider duck *(Somateria mollissima), Journal of Reproduction and Fertility*, 1977, *49*, 225–230.

Gottfried, H., and van Mullem, P. J. On the histology of the interstitium and the occurrence of steroids in the stickleback (*Gasterosteus aculeatus* L.) testis. *Acta Endocrinologicae*, 1967, *56*, 1–15.

Greeley, F., and Meyer, R. K. Seasonal variations in testis-stimulating activity of male pheasant pituitary glands. *Aukland*, 1953, *70*, 350–358.

Greenberg, B. Some effects of testosterone on the sexual pigmentation and other sex characters of the cricket frog *(Acris Gryllus). Journal of Experimental Zoology*, 1942, *91*, 435–451.

Greenberg, B. Social behavior of the westerbanded gecko, *Coleonyx varietagus* Baird. *Physiological Zoology*, 1943, *16*, 110–122.

Greenberg, B., and Noble, G. K. Social behavior of the American chameleon (*Anolis carolinensis* Voigt). *Physiological Zoology*, 1944, *17*, 392–439.

Greenberg, N. A neuroethological study of display behavior in the lizard *Anolis carolinensis* (Reptilia, Lacertilia, Iguanidae). *American Zoologist*, 1977, *17*, 191–201.

Greenberg, N. Ethological consideration in the experimental study of lizard behavior. In N. Greenberg and P. D. MacLean (Eds.), *The Behavior and Neurology of Lizards.* Rockville, Md.: NIMH, 1978.

Greenberg, N., and Crews, D. Physiological ethology of aggrssion in amphibians and reptiles. In B. Svare (Ed.), *Hormones and Aggression.* New York: Plenum Press, 1982.

Guhl, A. M. Gonadal hormones and social behavior in infrahuman vertebrates. In W. C. Young (Ed.), *Sex and Internal Secretions.* Baltimore: Williams and Wilkins, 1961.

Guyomarc'h, J. -C. Le comportement des gallinacés: Un bilan et une orientation méthodologique. *Annales de Médecine Vétérinaire*, 1976, *120*, 377–393.

Haase, E., Sharp, P. J., and Paulke, E. Annual cycle of plasma luteinizing hormone concentrations in wild mallard drakes. *Journal of Experimental Zoology*, 1975, *194*, 553–558.

Haider, S. G., and Blüm, V. Investigations on the evolution of the pituitary gonadotrophin system in lower vertebrates: Presentation of an evolutionary theory. *General and Comparative Endocrinology*, 1976, *29*, 251.

Hale, E. B. Effects of forebrain lesions on the aggressive behavior of green sunfish *Lepomis cyanellus. Physiological Zoology*, 1956, *29*, 107–127.

Hall, N. R., Luttge, W. G., and Berry, R. B. Intracerebral prostaglandin E_2: Effect upon sexual behavior, open field activity and body temperature in ovariectomized female rats. *Prostaglandins*, 1975, *10*, 877–888.

Harding, C., and Follett, B. K. Hormone changes triggered by aggression in a natural population of blackbirds. *Science*, 1979, *203*, 918–920.

Haynes, N. B., Cooper, K. J., and Kay, M. J. Plasma progesterone concentration in the hen in relation to the ovulatory cycle. *British Poultry Science*, 1973, *14*, 349–357.

Haynes, R. L., and Glick, B. Hypothalamic control of sexual behavior in the chicken. *Poultry Science*, 1974, *53*, 27–38.

Hayward, J. N. Physiological and morphological identification of hypothalamic magnocellular neuroendocrine cells in goldfish preoptic nucleus. *Journal of Physiology*, 1974, *239*, 103–124.

Heald, P. J., Rookledge, K. A., Furnival, B. E., and Watts, G. D. The effects of gonadal hormones on the levels of pituitary luteinizing hormone in the domestic fowl. *Journal of Endocrinology*, 1968, *41*, 313–318.

Herrick, E. H., and Harris, J. O. Singing in female canaries. *Science*, 1957, *125*, 1299–1300.

Hinde, R. A. The nest-building behaviour of domesticated canaries. *Proceedings of the Zoological Society of London*, 1958, *131*, 1–48.

Hinde, R. A. Interaction of internal and external factors in integration of canary reproduction. In F. A. Beach (Ed.), *Sex and Behavior*. New York: Wiley, 1965.

Hinde, R. A., and Warren, R. P. The effect of nest building on later reproductive behaviour in domesticated canaries. *Animal Behavior*, 1959, 7, 35–41.

Hinde, R. A., Bell, R. Q., and Steel E. Changes in sensitivity of the canary brood patch during the natural breeding season. *Animal Behavior*, 1963, *11*, 553–560.

Hinde, R. A., Steel, E., and Follett, B. K. Effect of photoperiod on oestrogen-induced nest-building in ovariectomized or refractory female canaries *(Serinus canarius)*. *Journal of Reproduction and Fertility*, 1974, *40*, 383–399.

Hirose, K., and Ishida, R. Induction of ovulation in the ayu, *Plecoglossus altivelis*, with LH-releasing hormone (LH-RH). *Bulletin of the Japanese Society of Scientific Fisheries*, 1974, *40*, 1235–1240.

Hoar, W. S. Hormones and the reproductive behaviour of the male three-spined stickleback *(Gasterosteus aculeatus)*. *Animal Behavior*, 1969a, *10*, 247–266.

Hoar, W. S. Reproductive behavior of fish. *General and Comparative Endocrinology*, 1962b, *Suppl. 1*, 206–216.

Hoar, W. S. The endocrine system as a chemical link between the organism and its environment. *Transactions of the Royal Society of Canada*, 1965, *3*, 175–200.

Hoar, W. S. Hormonal activities of the pars distalis in cyclostomes, fish and amphibia. In G. W. Harris and B. T. Donovan (Eds.), *The Pituitary Gland*, Vol. 1. Washington, D.C.: Butterworth, 1966.

Hoar, W. S. Reproduction. In W. S. Hoar and D. J. Randall (Eds.), *Fish Physiology*, Vol. 3. New York: Academic Press, 1969.

Hoar, W. S., Wiebe, J., and Wai, E. H. Inhibition of the pituitary gonadotropic activity of fishes by a dithiocarbamoylhydrazine derivative (ICI 33, 828). *General and Comparative Endocrinology*, 1967, *8*, 101–109.

Hooker, B. Birds. In T. Sebeok (Ed.), *Animal Communication*. Bloomington: Indiana University Press, 1968.

Humphrey, R. R. Factors influencing ovulation in the Mexican axolotl as revealed by induced spawnings. *Journal of Experimental Zoology*, 1977, *199*, 209–214.

Hunsaker, D. Ethological isolating mechanism in the *Sceloporus torquatus* group of lizards. *Evolution*, 1962, *16*, 62–74.

Hutchison, J. B. Investigations on the neural control of clasping and feeding in *Xenopus laevis* (Daudin). *Behaviour*, 1964, *24*, 47–66.

Hutchison, J. B. Initiation of courtship by hypothalamic implants of testosterone propionate in castrated doves *(Streptopelia risoria)*. *Nature* (London), 1967, *222*, 176–177.

Hutchison, J. B. Changes in hypothalamic responsiveness to testosterone in male barbary doves *(Streptopelia risoria)*. *Nature*, 1969, *222* 176–177.

Hutchison, J. B. Differential effects of testosterone and oestradiol on male courtship in Barbary doves *(Streptopelia risoria)*. *Animal Behavior*, 1970, *18*, 41–51.

Hutchison, J. B. Differential hypothalamic sensitivity to androgen in activation of reproductive behavior. In F. O. Schmitt and F. G. Worden (Eds.), *The Neurosciences: Third Study Program*. Cambridge Mass.: M.I.T. Press, 1974a.

Hutchison, J. B. Post castration decline in behavioral responsiveness to intrahypothalamic androgen in doves. *Brain Research*, 1974b, *80*, 1–10.

Hutchison, J. B. Target cells for gonadal steroids in the brain: Studies on steroid-sensitive mechanisms of behaviour. In P. Wright, P. G. Caryl, and D. M. Vowles (Eds.), *Neural and Endocrine Aspects of Behaviour in Birds*. Amsterdam: Elsevier Publishing Company, 1975.

Hutchison, J. B. Hormones and brain mechanisms of sexual behaviour: A possible relationship between cellular and behavioural events in doves. In P. S. Davies (Ed.) *Perspectives in Experimental Biology, Vol. 1: Zoology.* New York: Pergamon Press, 1976.

Hutchison, J. B., and Katongole, C. B. Plasma testosterone in courting and incubating male Barbary doves *(Streptopelia risoria). Journal of Endocrinology,* 1975, *65,* 275–276.

Hutchison, J. B., and Lovari, S. Effects of male aggressiveness on behavioural transitions in the reproductive cycle of the Barbary dove. *Behaviour,* 1976, *59* (3–4), 296–318.

Hutchison, J. B., and Poynton, J. C. A neurological study of the clasp reflex in *Xenopus laevis* (Deaudin). *Behaviour,* 1964, *22,* 41–63.

Hutchison, R. E. Influence of oestrogen on the initiation of nesting behaviour in female budgerigars. *Journal of Endocrinology,* 1975, *64,* 417–428.

Hyder, M. Endocrine regulation of reproduction in *Tilapia. General and Comparative Endocrinology,* 1972, *Suppl. 3,* 729–740.

Hyder, M., Aruna, V. S., Campbell, C. M., and Dadzie, S. Methallibure studies on *Tilapia:* II. Effect of *Tilapia* pituitary homoginate (TPH), human chorionic gonadotropin (HCG) and testosterone propionate (TP) on the testes of methallibure-treated *Tilapia nigra. General and Comparative Endocrinology,* 1974, *23,* 245–255.

Idler, D. R., Reinboth, R., Walsh, J. M., and Truscott, B. A comparison of 11-hydroxytestosterone and 11-ketotestosterone in blood of ambisexual and gonochoristic teleosts. *General and Comparative Endocrinology,* 1976, *30,* 517–521.

Ikeda, K. Effect of castration on the secondary sexual characteristics of andromous three-spined stickleback, *Gasterosteus aculeatus aculeatus* (L.) *Japanese Journal of Zoology,* 1933, *5,* 135–137.

Imai, K., and Nalbandov, A. V. Changes in FSH activity of anterior pituitary glands and of blood plasma during the laying cycle of the hen. *Endocrinology,* 1971, *88,* 1465–1470.

Imai, K., and Nalbandov, A. V. Plasma and follicular steroid levels of laying hens after administration of gonadotrophins. *Biology and Reproduction,* 1978, *19,* 779–784.

Impekoven, M. Prenatal parent-young interactions in birds and their long-term effects. In J. S. Rosenblatt, R. A. Hinde, E. Shaw, and C. Beer (Eds.), *Advances in the Study of Behavior,* Vol. 7. New York: Academic Press, 1976.

Ingle, D., and Crews, D. Vertebrate neuroethology: Definitions and paradigms. *Annual Review of Neuroscience,* 1985, *8,* 457–494.

Ishii, S. Artificial induction of parturition in the top minnow, *Gambusia* sp. (abstract, in Japanese). *Zoological Magazine* (Tokyo), 1961, *70,* 3–4.

Ishii, S., and Furuya, T. Effects of purified chicken gonadotropins on the chick testis. *General and Comparative Endocrinology,* 1975, *25,* 1–8.

Jallageas, M., and Assenmacher, I. Testostérone de canard photostimulé ou soumis à des injections répétées de testostérone. *Compte Rendus. Société de Biologie,* 1970, *164,* 2338–2341.

Jallageas, M., and Attal, J. Dosage par chromatographe en phase gazeuze de la testostérone plasmatique non-conjuguée chez le canard, le caille, le pigeon. *Comptes Reudus de l'Académie des Sciences,* 1968, *267,* 341–343.

Jenssen, T. A. Evolution of anoline lizard display behavior. *American Zoologist,* 1977, *17,* 203–215.

Jenssen, T. A. Display diversity of anoline lizards and problems of interpretation. In N. Greenberg and P. D. MacLean (Eds.), *Behavior and Neurology of Lizards: An Interdisciplinary Conference.* Rockville, Md.: NIMH, 1978.

Johns, L. S., and Liley, N. R. The effects of gonadectomy and testosterone treatment on the reproductive behavior of the male blue gourami *Trichogaster trichopterus. Canadian Journal of Zoology,* 1970, *48,* 977–987.

Jones, R. E. Effects of season and gonadotropin on testicular interstitial cells of California quail. *Aukland,* 1970, *87,* 729–737.

Jones, R. E. Endocrine control of clutch size in reptiles: IV. Estrogen-induced hyperemia and growth in immature ovaries of *Anolis carolinensis. General and Comparative Endocrinology,* 1975, *25,* 211–222.

Jones, R. E., Fitzgerald, K. T., and Tokarz, R. R. Endocrine control of clutch size in reptiles: VII. Compensatory ovarian hypertrophy following unilateral ovariectomy in *Sceloporus occidentalis. General and Comparative Endocrinology,* 1977, *31,* 157–159.

Jorgensen, C. B. Central nervous control of adenohypophysial functions. In E. J. W. Barrington and C. B. Jørgensen (Eds.), *Perspectives in Endocrinology.* New York: Academic Press, 1968.

Jørgensen, C. B. Hypothalamic control of hypophyseal functions in anurans. In L. Martini, M. Mota, and F. Fraschini (Eds.), *The Hypothalamus*. New York: Academic Press, 1970.

Jørgensen, C. B. Pattern of recruitment of oocytes to second growth phase in normal toads, and in hypophysectomized toads, *Bufo bufo bufo* (L.), treated with gonadotropin (HCG). *General and Comparative Endocrinology*, 1973, *21*, 152–159.

Jørgensen, C. B. Integrative functions of the brain. In B. Lofts (Ed.), *Physiology of the Amphibia*, Vol. 2. New York: Academic Press, 1974.

Jørgensen, C. B. Factors controlling the annual ovarian cycle in the toad *Bufo bufo bufo* (L.). *General and Comparative Endocrinology*, 1975, *25*, 264–273.

Jørgensen, C. B., and Larsen, L. O. Neuroendocrine mechanisms in lower vertebrates. In L. Martini and W. F. Ganong (Eds.), *Neuroendocrinology*, Vol. 2. New York: Academic Press, 1966.

Judd, H. L., Laughlin, G. A., Bacon, J. P., and Benirschke, K. Circulating androgen and estrogen concentrations in lizards *(Iguana iguana)*. *General and Comparative Endocrinology*, 1976, *30*, 391–395.

Kallman, K. D., and Schreibman, M. P. A sex-linked gene controlling gonadotrop differentiation and its significance in determining the age of sexual maturation and size of the platyfish, *Xiphophorus maculatus*. *General and Comparative Endocrinology*, 1973, *21*, 287–304.

Kallman, K. D., Schreibman, M. P., and Borkowski, V. Genetic control of gonadotrop differentiation in the platyfish, *Xiphophorus maculatus* (Poeciliidae). *Science*, 1973, *181*, 678–680.

Kamrin, R. P., and Aronson, L. R. The effects of forebrain lesions on mating behavior in the male platyfish, *Xiphophorus maculatus*. *Zoologica*, 1954, *39*, 133–140.

Kandel, E. R. Electrical properties of hypothalamic neuroendocrine cells. *Journal of General Physiology*, 1964, *47*, 691–717.

Kappauf, B., and van Tienhoven, A. Progesterone concentrations in peripheral plasma of laying hens in relation to the time of ovulation. *Endocrinology*, 1972, *90*, 1350–1355.

Kassel, J., Davis, R. E., and Schwagmeyer, P. Telencephalic lesions and behavior in the teleost, *Macropodus opercularis*: Further analysis of reproductive and operant behavior in the male. *Behavioral Biology*, 1976, *18*, 179–188.

Kaul, S., and Vollrath, L. The goldfish pituitary: I. Cytology. *Cell Tissue Research*, 1974, *154*, 211–230.

Keating, E. G., Kormann, L. A., and Horel, J. A. The behavioral effects of stimulating and ablating the reptilian amygdala *(Caiman sklerops)*. *Physiological Behavior*, 1970, *5*, 55–59.

Kelley, D. B., and Pfaff, D. W. Hormone effects on male sex behavior in adult South African clawed frogs, *Xenopus laevis*. *Hormones and Behavior*, 1976, *7*, 159–182.

Kelley, D., and Pfaff, D. W. Generalizations from comparative studies on neuroanatomical and endocrine mechanisms for sex behavior. In J. B. Hutchison (Ed.), *Biological Determinants of Sexual Behavior*. Chichester; England: Wiley, 1977.

Kelley, D. B., Morrell, J. I., and Pfaff, D. W. Autoradiographic localization of hormone-concentrating cells in the brain of an amphibian, *Xenopus laevis*: I. Testosterone. *Journal of Comparative Neurology*, 1975, *164*, 47–62.

Kennedy, M. C. Vocalization elicited in a lizard by electrical stimulation of the midbrain. *Brain Research*, 1975, *91*, 321–325.

Kerlan, J. T., and Jaffe, R. M. Plasma testosterone during the testicular cycle of the redwinged blackbird *(Agelaius phoeniceus)*. *General and Comparative Endocrinology*, 1974, *22*, 428–432.

Kihlström, J. E., and Danninge, I. Neurohypophysial hormones and sexual behavior in males of the domestic fowl *(Gallus domesticus* L.) and the pigeon *(Columba livia* Gmel.). *General and Comparative Endocrinology*, 1972, *18*, 115–120.

Kitay, D. S., and Westfall, T. C. Dose related effected of methallibure (ICI 33, 828), *in vitro*, upon hypothalamic catecholamine uptake in the teleost, *Poecilia latipinna*. *General and Comparative Endocrinology*, 1977, *31*, 402–406.

Kjaer, K., and Jørgensen, C. B. Effects of mammalian gonadotropins on ovary in the toad *Bufo bufo bufo* (L.). *General and Comparative Endocrinology*, 1971, *17*, 424–431.

Klicka, J., and Mahmoud, I. Y. The effects of hormones on the reproductive physiology of the painted turtle, *Chrysemys picta*. *General and Comparative Endocrinology*, 1977, *31*, 407–413.

Kobayashi, H., and Wada, M. Neuroendocrinology in birds. In D. S. Farner and J. R. King (Eds.), *Avian Biology*, Vol. 3. New York: Academic Press, 1973.

Komisaruk, B. R. Effects of local brain implants of progesterone on reproductive behavior in ring doves. *Journal of Comparative Physiology and Psychology*, 1967, *64*, 219–224.

Konishi, M. The role of auditory feedback in the control of vocalization in the white-crowned sparrow. *Zeitschrift für Tierpsychologie*, 1965, *22*, 770–783.

Konishi, T. Neurosecretory activities in the anterior median eminence in relation to photoperiodic testicular response in young Japanese quail *(Coturnix coturnix japonica)*. *Endocrinologia Japonica*, 1967, *14*, 60–68.

Konishi, T., and Kato, M. Light-induced rhythmic changes in the hypothalamic neurosecretory activity in Japanese quail, *Coturnix coturnix japonica*. *Endocrinologia Japonica*, 1967, *14*, 239–245.

Korenbrot, C. C., Schomberg, S. W., and Erickson, C. J. Radioimmunoassay of plasma estradiol during the breeding cycle of ring doves *(Streptopelia risoria)*. *Endocrinology*, 1974, *94*, 1126–1132.

Kroodsma, D. E. Reproductive development in a female songbird: Differential stimulation by quality of male song. *Science*, 1976, *192*, 574–575.

Kubie, J. L., Cohen, J., and Halpern, M. Shedding of estradiol treated garter snakes enhances their sexual attractiveness and that of untreated penmates. *Animal Behavior*, 1978, 562–570.

Kuo, Z. Y. Studies on the basic factors in animal fighting: III. Hormonal factors affecting fighting in quails. *Journal of Genetic Psychology*, 1960, *96*, 217–223.

Laguë, P. C., van Tienhoven, A., and Cunningham, F. J. Concentrations of estrogens, progesterone and LH during the ovulatory cycle of the laying chicken *(Gallus domesticus)*. *Biology and Reproduction*, 1975, *12*, 590–598.

Lam, T. J., Pandey, S., and Hoar, W. S. Induction of ovulation in goldfish by synthetic luteinizing hormone-releasing hormone (LH-RH). *Canadian Journal of Zoology*, 1975, *53*, 1189–1192.

Lam, T. J., Pandey, S., Nagahama, Y. and Hoar, W. S. Effect of synthetic luteinizing-releasing hormone (LH-RH) on ovulation and pituitary cytology of the goldfish *Carassius auratus*. *Canadian Journal of Zoology*, 1976, *54*, 816–824.

Lambe, D. R., and Erickson, C. J. Ovarian activity of female ring doves *(Streptopelia risoria)* exposed to marginal stimuli from males. *Journal of Comparative Physiology and Psychology*, 1973, *1*, 281–283.

Lambert, J. G. D., and van Oordt, P. G. W. J. Ovarium hormones in teleosts. *Fortschritte der Zoologie*, 1974, *22*, 340–349.

Lance, V., and Callard, I. P. Phylogenetic trends in the hormonal control of gonadal steroidogenesis. In P. Pang and A. Epple (Eds.), *Evolution of Vertebrate Endocrines*. Lubbock: Texas Technical University Press, 1978.

Lazarus, J., and Crook, J. H. The effects of luteinizing hormone, oestrogen and ovariectomy on the agonistic behaviour of female *Quelea quelea*. *Animal Behavior*, 1973, *21*, 49–60.

Lederis, K. Current studies on urotensins. *American Zoologist*, 1973, *13*, 771–773.

Lehman, G. C. Environmental influence on ovulation and embryonic development in *Rana pipiens*. *Journal of Experimental Zoology*, 1977, *199*, 51–56.

Lehrman, D. S. Effect of female sex hormones on incubation behavior in the ring dove *(Streptopelia risoria)*. *Journal of Comparative Physiology and Psychology*, 1958, *51*, 142–145.

Lehrman, D. S. Hormonal responses to external stimuli in birds. *Ibis*, 1959, *101*, 478–496.

Lehrman, D. S. Hormonal regulation of parental behavior in birds and infrahuman mammals. In W. C. Young (Ed.), *Sex and Internal Secretions*. Baltimore: Williams and Wilkins, 1961.

Lehrman, D. S. Interaction between internal and external environments in the regulation of the reproductive cycle of the ring dove. In F. A. Beach (Ed.), *Sex and Behavior*. New York: Wiley, 1965.

Lehrman, D. S., and Friedman, M. Auditory stimulation of ovarian activity in the ring dove. *Animal Behavior*, 1969, *17*, 494–497.

Lehrman, D. S., Brody, P. N., and Wortis, R. P. The presence of the mate and of nesting material as stimuli for the development of incubation behavior and for gonadotropin secretion in the ring dove *(Streptopelia risoria)*. *Endocrinology*, 1961, *68*, 507–516.

Licht, P. Endocrinology of the reptilia: The pituitary system. *Chemical Zoology*, 1974a, *9*, 399–448.

Licht, P. Induction of sperm release in frogs by mammalian gonadotropin-releasing hormone. *General and Comparative Endocrinology*, 1974b, *23*, 352–354.

Licht, P., and Crews, D. Gonadotropin stimulation of *in vitro* progesterone production in reptilian and amphibian ovaries. *General and Comparative Endocrinology*, 1976, *29*, 141–151.

Licht, P., and Midgley, A. R., Jr. Competition for the *in vitro* binding of radioiodinated human follicle-stimulating hormone in reptilian, avian, and mammalian gonads by nonmammalian gonadotropins. *General and Comparative Endocrinology*, 1976a, *30*, 364–371.

Licht, P., and Midgley, A. R., Jr. *In vitro* binding of radioiodinated human follicle-stimulating hormone to reptilian and avian gonads: Radioligand studies with mammalian hormones. *Biology and Reproduction*, 1976b, *15*, 195–205.

Licht, P., and Papkoff, H. Separation of two distinct gonadotropins from the pituitary gland of the bullfrog *(Rana catesbiana)*. *Endocrinology*, 1974, *94*, 1587–1594.

Licht, P., and Pearson, A. K. Effects of adenohypophysectomy on testicular function in the lizard *Anolis carolinensis*. *Biology and Reproduction*, 1969, *1*, 107–119.

Licht, P. and Pearson, A. K. Structure and function of the reptilian pituitary gland. *International Review of Cytology*, 1978, *7*, 239–286.

Licht, P., Farmer, S. W., and Papkoff, H. The nature of the pituitary gonadotropins and their role in ovulation in a urodele amphibian *(Ambystoma trigrimum)*. *Life Sciences*, 1975, *17*, 1049–1054.

Licht, P., Papkoff, H., Farmer, S., Muller, C., Tsui, H. W., and Crews, D. Evolution of gonadotropin structure and function. *Recent Progress in Hormone Research*, 1977, *33*, 169–248.

Liley, N. R. Ethological isolating mechanisms in four sympatric species of poeciliid fishes. *Behaviour Supplement*, 1966, *13*, 1–107.

Liley, N. R. The endocrine control of reproductive behaviour in the female guppy, *Poecilia reticulata* Peters. *Animal Behavior*, 1968, *16*, 318–331.

Liley, N. R. Hormones and reproductive behaviour in fishes. In W. S. Hoar and D. J. Randall (Eds.), *Fish Physiology*, Vol. 3. New York: Academic Press, 1969.

Liley, N.R. The effects of estrogens and other steroids on the sexual behavior of the female guppy, *Poecilia reticulata*. *General Comparative Endocrinology*, 1972, *Suppl. 3.*, 542–552.

Liley, N. R. Physiological maturation and reproductive behaviour of female doves *(Streptopelia risoria)* held under long and short photoperiods. *Canadian Journal of Zoology*, 1976a, *54*, 343–354.

Liley, N. R. The role of estrogen and progesterone in the regulation of reproductive behaviour in female ring doves *(Streptopelia risoria)* under long vs. short photoperiods. *Canadian Journal of Zoology*, 1976b, *54*, 1409–1422.

Liley, N. R. Patterns of hormonal control in the reproductive behaviour of fish, and their relevance to fish management and culture programs. In J. E. Bardach, J. J. Magnuson, R. C. May, and J. M. Reinhart (Eds.), *Fish Behaviour and Its Use in the Capture and Culture of Fish*. Manila, Phillipines: International Center for Living Aquatic Resource Management, 1980.

Liley, N. R., and Donaldson, E. M. The effects of salmon pituitary gonadotropin on the ovary and the sexual behavior of the female guppy, *Poecilia reticulata*. *Canadian Journal of Zoology*, 1969, *47*, 569–573.

Liley, N. R., and Wishlow, W. The interaction of endocrine and experiential factors in the regulation of sexual behaviour in the female guppy *Poecilia reticulata*. *Behaviour*, 1974, *48*, 185–214.

Lisk, R. D. Neural control of gonad size by hormone feedback in the desert iguana *Dipsosaurus dorsalis dorsalis*. *General and Comparative Endocrinology*, 1967, *8*, 258–266.

Lofts, B. The effects of follicle-stimulating hormone and luteinizing hormone on the testis of hypophysectomized frogs *(Rana temporaria)*. *General and Comparative Endocrinology*, 1961, *1*, 179–189.

Lofts, B. Patterns of testicular activity. In E. J. W. Barrington and C. B. Jørgensen (Eds.), *Perspectives in Endocrinology: Hormones in the Lives of Lower Vertebrates*. New York: Academic Press, 1968.

Lofts, B. Seasonal cycles in reptilian testes. *General and Comparative Endocrinology*, 1969, *Suppl. 2*, 147–155.

Lofts, B. The sertoli cell. *General and Comparative Endocrinology*, 1972, *Suppl. 3.*, 636–648.

Lofts, B. Reproduction. In B. Lofts (Ed.), *Physiology of the Amphibia*, Vol. 2. New York: Academic Press, 1974.

Lofts, B., and Bern, H. A. The functional morphology of steroidogenic tissues. In D. R. Idler (Ed.), *Steroids in Non-mammalian Vertebrates*. New York: Academic Press, 1972.

Lofts, B., and Murton, R. K. Reproduction in birds. In D. S. Farner and J. R. King (Eds.), *Avian Biology*, Vol. 3. New York: Academic Press, 1973.

Lott, D. F., and Brody, P. N. Support of ovulation in the ring dove by auditory and visual stimuli. *Journal of Comparative Physiology and Psychology*, 1966, *62*, 311–313.

Lott, D. F., Scholz, S. D., and Lehrman, D. S. Extroceptive stimulation of the reproductive system of the female ring dove *(Streptopelia risoria)* by the mate and by the colony milieu. *Animal Behavior*, 1967, *15*, 433–347.

Lumia, A. R. The relationships among testosterone, conditioned aggression and dominance in male pigeons. *Hormones and Behavior*, 1972, *3*, 277–286.

Macey, M. J., Pickford, G. E., and Peter, R. E. Forebrain localization of the spawning reflex response to exogenous neurohypophysial hormones in the killifish, *Fundulus heteroclitus*. *Journal of Experimental Zoology*, 1974, *190*, 269–280.

Madison, D. M. Chemical communication in amphibians and reptiles. in D. Müller-Schwarze and M. M. Mozell (Eds.), *Chemical Signals in Vertebrates*. New York: Plenum Press, 1977.

Maley, M. J. Electrical stimulation of agonistic behaviour in the mallard. *Behaviour*, 1969, *34*, 138–160.

Marcellini, D. Acoustic and visual display behavior of gekkonid lizards. *American Zoologist,* 1977, *17,* 251–260.

Marshall, A. J. *Bowerbirds: Their Displays and Breeding Cycles.* London: Oxford University Press, 1954.

Marshall, A. J., and Coombs, C. J. F. The interaction of environmental, internal and behavioural factors in the rook *(Corvus f. frugilegus). Linnaeus Proceedings of the Zoolological Society of London,* 1957, *128,* 545–589.

Marshall, A. J., and Disney, H. J. de S. Experimental induction of the breeding season in a xerophilous bird. *Nature,* 1957, 647–649.

Martinez-Vargas, M. C., and Erickson, C. J. Social and hormonal determinants of nest building in the ring dove *(Streptopelia risoria). Behaviour,* 1973, *45,* 12–37.

Martinez-Vargas, M. C., Stumpf, W. E., and Sar, M. Estrogen localization in the dove brain: Phylogenetic considerations and implications for nomenclature. In W. E. Stumpf and L. D. Grant (Eds.), *Anatomical Neuroendocrinology.* Basel: Karger, 1975.

Martinez-Vargas, M. C., Stumpf, W. E., and Sar, M. Anatomical distribution of estrogen target cells in the avian CNS: A comparison with the mammalian CNS. *Journal of Comparative Neurology,* 1976, *167,* 83–104.

Mason, P., and Adkins, E. L. Hormones and sexual behavior in the lizard *Anolis carolinensis. Hormones and Behavior,* 1976, *7,* 75–86.

Massa, R., Cresti, L., and Martini, L. Metabolism of testosterone in the anterior pituitary gland and the central nervous system of the European starling *(Sturnus vulgaris). Journal of Endocrinology,* 1977a, *75* 347–354.

Massa, R., Davies, D. T., and James, R. The feedback effects of 5α- and 5β-reduced metabolites of testosterone on plasma LH in the Japanese quail. *Ninth Conference of European Comparative Endocrinologists,* 1977b, Giessen (abstract).

Mathewson, S. Gonadotropic hormones affect aggressive behaviour in starlings. *Science,* 1961, *134,* 1522–1523.

Matthews, L. H. Visual stimulation and ovulation in pigeons. *Proceedings of the Royal Society of London, Ser, B,* 1939, *126,* 557–560.

Mazzi, V. The hypothalamus as a thermodependent neuroendocrine center in urodeles. In L. Martini, M. Motta, and F. Fraschini (Eds.), *The Hypothalamus.* New York: Academic Press, 1970.

Mazzi, V., Vellano, C., Colucci, D., and Merlo, A. Gonadotropin stimulation by chronic administration of synthetic luteinizing hormone-releasing hormone in hypophysectomized pituitary grafted male newts. *General and Comparative Endocrinology,* 1974, *24,* 1–9.

McDonald, P. A., and Liley, N. R. The effects of photoperiod on androgen-induced reproductive behavior in male ring doves *(Streptopelia risoria). Hormones and Behavior,* 1978, *10,* 85–96.

McNicol, D., Jr., and Crews, D. Estrogen/progesterone synergy in the control of femal sexual receptivity in the lizard, *Anolis carolinensis.* General and Comparative Endocrinology, 1979, *38,* 68–74.

Medica, P. A., Turner, F. B., and Smith, D. D. Hormonal induction of color change in female leopard lizards, *Crotaphytus wislizenii. Copeia,* 1973, *1973,* 658–661.

Meier, A. H. Chronoendocrinology of vertebrates. In B. E. Eleftheriou and R. L. Sprout (Eds.), *Hormonal Correlates of Behavior,* Vol. 2. New York: Plenum Press, 1975.

Menaker, M. Rhythms, reproduction and photoreception. *Biology of Reproduction,* 1971, *4,* 295–308.

Merriam, R. W. On the mechanism of action of gonadotropic stimulation of oocyte maturation in *Xenopus laevis. Journal of Experimental Zoology,* 1972, *180,* 421–426.

Metuzals, J., Ballintijn-deVries, G., and Baerends, G. P. The correlation of histological changes in the adenohypophysis of the cichlid fish *Aequidens portalegrensis* (Hensel) with behaviour changes during the reproductive cycle. *Koninkl. Nederl. Akad. Wetensch. Amsterdam, Proc. Serc. C.,* 1968, *71,* 391–410.

Meyer, C. C. Inhibition of precocial copulation in the domestic chick by progesterone brain implant. *Journal of Comparative Physiology and Psychology,* 1972, *79*(1), 8–12.

Meyer, C. C. Testosterone concentration in the male chick brain: An autoradiographic survey. *Science,* 1973, *180,* 1381–1383.

Meyer, C. C., and Angelo, R. L. Suppression of precocial copulation by progesterone implants in the male chick forebrain. *Journal of Comparative Physiology and Psychology,* 1975, *88,* 687–692.

Meyer, C. C., and Salzen, E. A. Hypothalamic lesions and sexual behavior in the domestic chick. *Journal of Comparative Physiology and Psychology,* 1970, *73,* 365–376.

Meyer, C. C., Parker, D. M., and Salzen, E. A. Androgen-sensitive midbrain sites and visual attention in chicks. *Nature,* 1976, *259,* 689–690.

Mitchell, M. E. Treatment of hypophysectomized hens with partially purified avian FSH. *Journal of Reproduction and Fertility*, 1970, *22*, 223–241.

Moore, F. L. Spermatogenesis in larval *Ambystoma tigrinum:* Positive and negative interactions of FSH and testosterone. *General and Comparative Endocrinology*, 1975, *26*, 525–533.

Moore, F. L., and Muller, C. H. Androgens and male mating behavior in rough-skinned newts, *Taricha granulosa*. *Hormones and Behavior*, 1977, *9*, 309–320.

Morgantaler, A., and Crews, D. The role of the anterior hypothalamus-preoptic area in the regulation of reproductive behavior in the lizard, *Anolis carolinensis:* Implantation studies. *Hormones and Behavior*, 1978, *11*, 61–73.

Morrell, J. I., Kelley, D. B., and Pfaff, D. W. Autoradiographic localization of hormone-concentrating cells in the brain of an amphibian, *Zenopus laevis:* II. Estradiol. *Journal of Comparative Neurology*, 1975a, *164*, 63–78.

Morrell, J. I., Kelley, D. B., and Pfaff, D. W. Sex steroid binding in the brains of vertebrates: Studies with light microscopic autoradiography. in K. M. Knigge, D. E. Scott, M. Kobayashi, and S. Ishii (Eds.), *The Ventricular System in Neuroendocrine Mechanisms*. Proceedings of the Second Brain-Endocrine Interaction Symposium. Basel: Karger, 1975b.

Morrell, J. I., Crews, D., Ballin, A., and Pfaff, D. W. Autoradiographic localization of ^{3}H-estradiol, ^{3}H-testosterone, and ^{3}H-dihydrotestosterone in the brain of the lizard, *Anolis carolinensis*. *Society for Neuroscience Annual Meetings*, 1977 (abstract).

Moss, R. L., and McCann, S. M. Induction of mating behavior in rats by luteinizing hormone-releasing factor. *Science*, 1973, *181*, 177–179.

Muller, C. H. *Steroidogenesis and spermatogenesis in the male bullfrog, Rana catesbeinana: Regulation by purified bullfrog gonadotropins*. Ph.D. thesis, University of California, Berkeley, 1976.

Muller, C. H. *In vitro* stimulation of 5α-dihydrotestosterone and testosterone secretion from bullfrog testis by nonmammalian and mammalian gonadotropins. *General and Comparative Endocrinology*, 1977a, *33*, 109–121.

Muller, C. H. Plasma 5α-dihydrotestosterone and testosterone in the bullfrog, *Rana catesbeinana:* Stimulation by bullfrog LH. *General and Comparative Endocrinology*, 1977b, *33*, 122–132.

Murton, R. K. Ecological adaptation in avian reproductive physiology. *Symposia of the Zoological Society of London*, 1975, *35*, 149–175.

Murton, R. K., and Westwood, N. J. Integration of gonadotrophin and steroid secretion, spermatogenesis and behaviour in the reproductive cycle of male pigeon species. In P. Wright, P. G. Caryl, and D. M. Vowler (Eds.), *Neural and Endocrine Aspects of Behaviour in Birds*. Amsterdam: Elsevier, 1975.

Naftolin, F., Ryan, K. J. Davis, I. J., Reddy, V. V., Flores, F., Petro, Z., and Kuhn, M. The formation of estrogens by central neuroendocrine tissues. *Recent Progress in Hormone Research*, 1975, *31*, 295–319.

Nakajo, S., and Imai, K. Gonadotropin content in the cephalic and the caudal lobe of the anterior pituitary in laying, nonlaying and broody hens. *Poultry Science*, 1961, *40*, 739–744.

Nakamura, T., and Tanabe, Y. *In vitro* metabolism of steroid hormones by chicken brain. *Acta Endocrinologica*, 1974, *75*, 410–416.

Nalbandov, A. V. Hormonal activities of the pars distalis in reptiles and birds. In G. W. Harris and B. T. Donovan (Eds.), *The Pituitary Gland*, Vol. 1. London: Butterworths, 1966.

Nalbandov, A. V. *Reproductive Physiology of Mammals and Birds* (3rd ed.). San Francisco: W. H. Freeman, 1976.

Nayyar, S. K., Keshavanath, P., and Sundararaj, B. I. Maintenance of spermatogenesis and seminal vesicles in the hypophysectomized catfish *Heteropneustes fossilis* (Bloch): Effects of ovine on salmon gonadotropin, and testosterone. *Canadian Journal of Zoology*, 1976, *54*, 285–292.

N'Diaye, A., Sandoz, D., Boisvieux-Ulrich, E., and Ozon, R. Action des androgènes chez l'amphibien anoure *Discoglossus pictus* (Ottb): III. Effets de la castration et action des hormones androgènes sur les ultrastructures de la vesicule seminale. *General Comparative Endocrinology*, 1974, *24*, 286–304.

Newman, J. D. Midbrain regions relevant to auditory communication in songbirds. *Brain Research*, 1970, *22*, 259–261.

Newman, J. D. Midbrain control of vocalizations in redwinged blackbirds *(Agelaius phoeniceus)*. *Brain Research*, 1972, *48*, 227–242.

Noble, G. K. *The Biology of the Amphibia*. New York: McGraw-Hill, 1931.

Noble, G. K., and Aronson, L. R. The sexual behavior of anura: I. The normal mating pattern of *Rana pipiens*. *Bulletin of the American Museum of Natural History*, 1942, *80*, 127–142.

Noble, G. K., and Greenberg, B. Testosterone propionate, a bisexual hormone in the American cha-meleon. *Proceedings of the Society for Experimental Biology and Medicine*, 1940, *44*, 460–462.

Noble, G. K., and Greenberg, B. Effects of seasons, castration and crystalline sex hormones upon the urogenital system and sexual behavior of the lizard *(Anolis carolinensis):* I. The adult female. *Journal of Experimental Zoology*, 1941, *88*, 451–479.

Noble, G. K., and Kumpf, K. F. The sexual behavior and secondary sexual characters of gonadectomized fish. *Anatomical Record*, 1936, *67*, 113.

Noble, R. Hormonal control of receptivity in female quail *(Coturnix coturnix japonica)*. *Hormones and Behavior*, 1973, *4*, 61–72.

Nottebohm, F., and Arnold, A. P. Sexual dimorphism in vocal control areas of the songbird brain. *Science*, 1976, *194*, 211–213.

Nottebohm, F., Stokes, T. M., and Leonard, C. M. Central control of song in the canary *(Serinus canarius)*. *Journal of Comparative Neurology*, 1976, *165*, 457–486.

O'Connell, M. E., Reboulleau, C., Feder, H. H., and Silver, R. Social interactions and androgen levels in birds: I. Female characteristics associated with increased plasma androgen levels in the male ring dove *(Streptopelia risoria)*. *General and Comparative Endocrinology*, 1981a, *44*, 454–463.

O'Connell, M. E., Silver, R., Feder, H. H., and Reboulleau, C. Social interactions and androgen levels in birds: II. Social factors associated with a decline in plasma androgen levels in male ring doves *(Streptopelia risoria)*. *General and Comparative Endocrinology*, 1981b, *44*, 464–469.

Oksche, A. The neuroanatomical basis of comparative neuroendocrinology. *General and Comparative Endocrinology*, 1976, *29*, 225–239.

Oksche, A., and Ueck, M. The nervous system. In B. Lofts (Ed.), *The Physiology of the Amphibia*, Vol. 3. New York: Academic Press, 1976.

Oliver, J., and Baylé, J. D. Photoperiodically evoked potentials in the gonadotropic areas of the quail hypothalamus. *Brain Research*, 1973, *64*, 103–121.

Opel, H. On hypothalamic areas regulating ovulation in the turkey. *Proceedings of the 10th Annual Meeting of the Society for the Study of Reproduction*, 1977, 79.

Orcutt, F. S. Estrogen stimulation of nest material cutting in the immature peach-faced lovebird *(Agapornis roseicollis)*. *American Zoologist*, 1965, *5*, 197 (abstract #28).

Oshima, K., and Gorbman, A. Effect of estradiol on NaCl-evoked olfactory bulbar potentials in goldfish: Dose-response relationships. *General and Comparative Endocrinology*, 1969a, *13*, 92–97.

Oshima, K., and Gorbman, A. Effects of sex hormones on photically evoked potentials in frog brain. *General and Comparative Endocrinology*, 1969b, *12*, 397–404.

Overmier, J. B., and Gross, D. Effects of telencephalic ablation upon nest-building and avoidance behaviors in East African mouthbreeding fish, *Tilapia mossambica*. *Behavioral Biology*, 1974, *12*, 211–222.

Ozon, R. Synthese *in vitro* des hormones steroids dans le testicule et l'ovaire de l'amphibien urodele *Pleurodeles waltlii* Michah. *General and comparative endocrinology*, 1967, *8*, 214–227.

Ozon, R. Androgens in fishes, amphibians, reptiles and birds. In D. R. Idler (Ed.), *Steroids in Nonmammalian Vertebrates*. New York: Academic Press, 1972a.

Ozon, R. Estrogens in fishes, amphibians, reptiles, and birds. In D. R. Idler (Ed.), *Steroids in Nonmammalian Vertebrates*. New York: Academic Press, 1972b.

Ozon, R., and Stocker, C. Formation *in vitro* de 5α-dihydrotestosterone par le testicule de *Discoglossus pictus*. *General and Comparative Endocrinology*, 1974, *23*, 224–236.

Palka, Y. S., and Gorbman, A. Pituitary and testicular influenced sexual behavior in male frogs *Rana pipiens*. *General and Comparative Endocrinology*, 1973, *21*, 148–151.

Pandey, S. Effects of methallibure on the testes and secondary sex characters of the adult and juvenile guppy *Poecilia reticulata* Peters. *Biology and Reproduction*, 1970, *2*, 239–244.

Pandey, S., and Hoar, W. S. Induction of ovulation in goldfish by clomiphene citrate. *Canadian Journal of Zoology*, 1972, *50*, 1679–1680.

Pandey, S., Stacey, N., and Hoar, W. S. Mode of action of clomiphene citrate in inducing ovulation of goldfish. *Canadian journal of Zoology*, 1973, *51*, 1315–1316.

Partridge, B. L., Liley, N. R., and Stacey, N. E. The role of pheromones in the sexual behaviour of the goldfish. *Animal Behavior*, 1976, *24*, 291–299.

Pearson, A. K., Licht, P., and Zambrano, D. Ultrastructure of the pars distalis of the lizard *Anolis carolinensis* with special reference to the identification of the gonadotropic cell. *Zeitschrift Zellforschung*, 1973, *137*, 293–312.

Pearson, A. K., Tsui, H. W., and Licht, P. Effect of temperature on spermatogenesis, on the production and action of androgens and on the ultrastructure of gonadotropic cells in the lizard *Anolis carolinensis*. *Journal of Experimental Zoology*, 1976, *195*, 291–304.

Perek, M., Ravona, H., Snapir, N., and Luxemburg, D. Sexual behavior of white leghorn cocks bearing lesions in various locations. *Physiology and Behavior*, 1973, *10*, 479–484.

Peter, R. E. Hypothalamic control of thyroid and gonadal activity in the goldfish, *Carassius auratus*. *General and Comparative Endocrinology*, 1970, *14*, 334–356.

Peter, R. E. neuroendocrinology of teleosts. *American Zoologist*, 1973, *13*, 743–755.

Peter, R. E. The preoptic nucleus in fishes: A comparative discussion of function-activity relationships. *American Zoologist*, 1977, *17*, 775–785.

Peter, R. E., and Billard, R. Effects of third ventricle injection of prostaglandins on gonadotropin secretion in the goldfish, *Carassius auratus*. *General and Comparative Endocrinology*, 1976, *30*, 451–456.

Peterson, A. J. The breeding cycle of the bank swallow. *The Wilson Bulletin*, 1955, *67*, 235–286.

Peterson, A. J., and Common, R. H. Progesterone concentration in peripheral plasma of laying hens as determined by competitive protein-binding assay. *Canadian Journal of Zoology*, 1971, *49*, 599–604.

Peterson, A. J., and Common, R. H. Estrone and estradiol concentrations in peripheral plasma of laying hens as determined by radioimmunoassay. *Canadian Journal of Zoology*, 1972, *50*, 395–404.

Peterson, A. J., Henneberry, G. O., and Common, R. H. Androgen concentration in the peripheral plasma of laying hens. *Canadian Journal of Zoology*, 1973, *51*, 753–758.

Peute, J., and Meij, J. C. A. Ultrastructural and functional aspects of the nucleus infundibularis ventralis in the green frog, *Rana esculenta*. *Zeitschrift für Zellforschung und Mikroskopische Anatomie*, 1973, *144*, 191–217.

Pfaff, D. W. Luteinizing hormone-releasing factor potentiates lordosis behavior in hypophysectomized ovariectomized female rats. *Science*, 1973, *182*, 1148–1149.

Phillips, R. E. "Wildness" in the mallard duck: Effects of brain lesions and stimulation on "escape behavior" and reproduction. *Journal of Comparative Neurology*, 1964, *122*, 139–155.

Phillips, R. E., and Peek, F. W. Brain organization and neuromuscular control of vocalization in birds. In P. Wright, P. Caryl, and D. M. Vowles (Eds.), *Neural and Endocrine Aspects of Behaviour in Birds*. Amsterdam: Elsevier, 1975.

Phillips, R. E., and Youngren, O. M. Brain stimulation and species-typical behaviour: Activities evoked by electrical stimulation of the brains of chickens *(Gallus gallus)*. *Animal Behavior*, 1971, *19*, 757–799.

Phillips, R. E., and Youngren, O. M. Electrical stimulation of the brain as a tool for the study of animal communication. *Brain, Behavior, and Evolution*, 1973, *8*, 253–286.

Picheral, B. Tissues secreting steroid hormones in urodele amphibians: IV. Electron and light microscopic studies of the glandular tissues of the testis and the interrenal gland in hypophysectomized *Pleurodeles waltii* Michah. *Zeitschrift Zellforsch Mikroskop Anatomie*, 1970, *107*(1), 68–86.

Pickford, G. E., and Atz, J. W. *The Physiology of the Pituitary Gland of Fishes*. New York: New York Zoological Society, 1957.

Pickford, G. E., and Strecker, E. L. The spawning reflex response of the killifish, *Fundulus heteroclitus*: Isotocin is relatively inactive in comparison with arginine vasotocin. *General and Comparative Endocrinology*, 1977, *32*, 132–137.

Pickford, G. E., Lofts, B., Bara, G. and Atz, J. W. Testis stimulation of hypophysectomized male killifish, *Fundulus heteroclitus*, treated with mammalian growth hormone and/or luteinizing hormone. *Biology and Reproduction*, 1972, *7*, 370–386.

Pietras, R. J., and Wenzel, B. M. Effects of androgens on body weight, feeding and courtship behavior in the pigeon. *Hormones and Behavior*, 1974, *5*, 289–302.

Polder, J. J. W. On gonads and reproduction behavior in the cichlid fish *Aquidens portalegrensis* (Hensel). *Netherlands Journal of Zoology*, 1971, *21*, 265–271.

Polzonetti-Magni, A., Lupo di Prisco, C., Rastogi, R. K., Bellini-Cardellini, L., and Chieffi, G. Estrogens in the plasma of *Rana esculenta* during the annual cycle and following ovariectomy. *General and Comparative Endocrinology*, 1970, *14*, 212–213.

Pool, T. B., and Dent, J. N. The ultrastructure and the hormonal control of product synthesis in the hedonic glands of the red-spotted newt *Notopthalmus viridescens*. *Journal of Experimental Zoology*, 1977, *201*, 177–202.

Popa, G. T., and Popa, F. G. Certain functions of the midbrain in pigeons. *Proceedings of the Royal Society of London, Series B*, 1933, *113*, 191–195.

Prasad, M. R. N., and Reddy, P. R. K. Physiology of the sexual segment of the kidney in reptiles. *General and Comparative Endocrinology*, 1972, *Suppl. 3*, 649–662.

Pröve, Von E. Der Einfluss von Kastraion und Testosteronsubstitution auf das Sexualverhalten männlicher Zebrafinken *(Taeniopygia guttata castanotis)*. *Journal für Ornithologie*, 1974, *115*, 338–347.

Putkonen, P. T. S. Electrical stimulation of the avian brain. *Annales Academiae Scientiarum Fennicae, Series A, V Medica*, 1967, *130*, 95.

Ramaswami, L. S., and Lakshman, A. B. Spawning catfish with mammalian hormones. *Nature*, 1958, *182*, 122.

Rastogi, R. K., and Chieffi, G. Hypothalamic control of the hypophyseal gonadotropic function in the adult male green frog, *Rana esculenta L. Journal of Experimental Zoology*, 1972, *181*, 263–270.

Rastogi, R. K., and Chieffi, G. The effects of antiandrogens and antiestrogens in nonmammalian vertebrates. *General and Comparative Endocrinology*, 1975, *26*, 79–91.

Rastogi, R. K., Iela, L., Saxena, P. K., and Chieffi, G. The control of spermatogenesis in the green frog, *Rana esculenta. Journal of Experimental Zoology*, 1976, *196*, 151–166.

Redshaw, M. R. The hormonal control of the amphibian ovary. *American Zoologist*, 1972, *12*, 289–306.

Redshaw, M. R., and Nicholls, T. J. Oestrogen biosynthesis by ovarian tissue of the South African clawed toad, *Xenopus laevis* Daudin. *General Comparative Endocrinology*, 1971, *16*, 85–96.

Rhees, R. W., Abel, J. H., Jr., and Haack, D. W. Uptake of tritiated steroids in the brain of the duck *(Anas platyrhynchos)*. An autoradiographic study. *General Comparative Endocrinology*, 1972, *18*, 292–300.

Ribbink, A. J. The behaviour and brain function of the cichlid fish *Hemihaplochromis philander*. *Zoologica Africana*, 1972, *7*, 21–41.

Rivarola, M. A., Snipes, C. A., and Migeon, C. J. Concentration of androgens in systemic plasma of rats, guinea pigs, salamanders and pigeons. *Endocrinology*, 1968, *82*, 115–121.

Rodriguez-Sierra, J. F., and Komisaruk, B. Effects of prostaglandin E^2 and indomethacin on sexual behavior in the female rat. *Hormones and Behavior*, 1977, *9*, 281–289.

Rodriguez-Sierra, J. F., and Komisaruk, B. R. Lordosis induction in the rat by prostaglandin E$_2$ systemically and intracranially in the absence of the ovaries. *Prostaglandin*, 1978, *15*, 513–523.

Ross, P., Jr., and Crews, D. Influence of the seminal plug on mating behaviour in the garter snake. *Nature*, 1977, *267*, 344–345.

Ross, P., Jr., and Crews, D. Stimuli influencing mating behavior in the garter snake, *Thamnophis radix*. *Behavioral Ecolology and Sociobiology*, 1978, *4*, 133–142.

Rouse, E. F., Coppenger, C. J., and Barnes, P. R. The effect of an androgen inhibitor on behavior and testicular morphology in the stickleback *Gasterosteus aculeatus*. *Hormones and Behavior*, 1977, *9*, 8–18.

Rowan, W. Relation of light to bird migrations and developmental changes. *Nature* (London), 1925, *115*, 494–495.

Ryan, K. J., Naftolin, F., Reddy, V., Flores, F., and Petro, Z. Estrogen Formation in the brain. *American Journal of Obstetrics and Gynecology*, 1972, *114*, 454–460.

Sage, M., and Bromage, N. R. The activity of the pituitary cells of teleost *Peoecilia* during the gestation cycle and the control of the gonadotropic cells. *General and Comparative Endocrinology*, 1970, *14*, 127–136.

Saint-Girons, H. The pituitary gland. In C. Gans and T. S. Parsons (Eds.), *Biology of the Reptilia*, Vol. 3. New York: Academic Press, 1970.

Salthe, S. N., and Mecham, J. S. Reproductive and courtship patterns. In B. Lofts (Ed.), *Physiology of the Amphibia*, Vol. 2. New York: Academic Press, 1974.

Sato, T., Hirono, M., Jyiyo, T., Iesaka, T., Taya, K., and Igarashi, M. Direct action of prostaglandins on the rat pituitary. *Endocrinology*, 1975, *96*, 45–49.

Schein, M. W., Diamond, M., and Carter, C. S. Sexual performance levels of male Japanese quail *(Coturnix coturnix japonica)*. *Animal Behavior*, 1972, *20*, 61–67.

Schmidt, R. S. Central mechanisms of frog calling. Behaviour, 1966a, *26*, 251–285.

Schmidt, R. S. Hormonal mechanisms of frog mating calling. *Copeia*, 1966b, 637–644.

Schmidt, R. S. Preoptic activation of frog mating behaviour. *Behaviour*, 1968, *30*, 239–257.

Schmidt, R. S. Preoptic activation of mating call orientation in female anurans. *Behaviour*, 1969, *35*, 114–127.

Schmidt, R. S. A model of the central mechanisms of male anuran acoustic behaviour. *Behaviour*. 1971, *39*, 288–317.

Schmidt, R. S. Central mechanisms of frog calling. *American Zoologist*, 1973, *13*, 1169–1177.

Schmidt, R. S. Neural correlates of frog calling: Independence from peripheral feedback. *Journal of Comparative Physiology*, 1974a, *88*, 321–333.

Schmidt, R. S. Neural correlates of frog calling: Trigeminal tegmentum. *Journal of Comparative Physiology*, 1974b, *92*, 229–254.

Schneider, H. Acoustic behavior and physiology of vocalization in the European tree frog. In D. H. Taylor and S. I. Guttman (Eds.), *Reproductive Biology of Amphibians*. New York: Plenum Press, 1977.

Schreibman, M. P. The adenohypophysis: Structure and function interrelationships during evolution. In P. Pang and A. Epple (Eds.), *Evolution of Vertebrate Endocrines*. Lubbock: Texas Technical University Press, 1978.

Schreibman, M. P., and Kallman, K. D. The genetic control of the pituitary gonadal axis in the platyfish, *Xiphophorus maculatus*. *Journal of Experimental Zoology*, 1977, *200*, 277–294.

Schreibman M. P. Leatherland, J. F., and McKeown, B. A. Functional morphology of the teleost pituitary gland. *American Zoologist*, 1973, *13*, 719–742.

Schrocksnadel, H., and Bator, A. Plasma testosterone levels in cocks and hens. *Steroids*, 1971, 18, 359–365.

Schuetz, A. W. Action of hormones on germinal vesicle breakdown in frog *(Rana pipiens)* oocytes. *Journal of Experimental Zoology*, 1967, *166*, 347–354.

Schwab, R. G., and Lott, D. F. Testis growth and regression in starlings *(Sturnus vulgaris)* as a function of the presence of females. *Journal of Experimental Zoology*, 1969, *171*, 39–42.

Schwagmeyer, P., Davis, R. E., and Kassel, J. Telencephalic lesions and behaviour in the teleost *Macropodus opercularis* (L.): Effects of telencephalon and olfactory bulb ablation on spawning and foam-nest building. *Behavioral Biology*, 1977, *20*, 463–470.

Segaar, J. Behavioral aspects of degeneration and regeneration in fish brain: A comparison with higher vertebrates. *Progress in Brain Research*, 1965, *14*, 143–231.

Segaar, J. and Nieuwenhuys, R. New etho-physiological experiments with male *Gasterosteus aculeatus*, with anatomical comment. *Animal Behavior*, 1963, *11*, 331–344.

Segura, E. T., and de Juan, A. R. Electroencephalographic studies in toads. *Electroencephalography and Clinical Neurophysiology*, 1966, *21*, 373–380.

Segura, E. T. Colombo, J. A., and deHardy, D. J. Effects of castration on brain electrogram of the female toad *Bufo arenarum* Hensel. *General and Comparative Endocrinology*, 1971a, *16*, 298–303.

Segura, E. T., de Juan, A. O. R., Colombo, J. A., and Kacelnik, A. The sexual clasp as a reticularly controlled behavior in the toad, *Bufo arenarum* Hensel. *Physiology and Behavior*, 1971b, *7*, 157–160.

Selinger, H., and Bermant, G. Hormonal control of aggressive behaviour in Japanese quail *(Coturnix coturnix japonica)*. *Behaviour*, 1967, *28*, 255–268.

Senior, B. E., and Cunningham, F. J. Oestradiol and luteinizing hormone during the ovulatory cycle of the hen. *Journal of Endocrinology*, 1974, *60*, 201–202.

Senior, B. E., and Furr, B. J. A. A preliminary assessment of the source of estrogen within the ovary of the domestic fowl, *Gallus domesticus*. *Journal of Reproduction and Fertility*, 1975, *43*, 241–247.

Shahabi, N. A., Norton, H. W., and Nalbandov, A. V. Steroid levels in follicles and the plasma of hens during the ovulatory cycle. *Endocrinology*, 1975, *96*, 962–968.

Sharp, P. J. Pituitary implants in the hypothalamus of coturnix quail. *Journal of Endocrinology*, 1972, *53*, 329–330.

Sharp, P. J., and Follett, B. K. The effect of hypothalamic lesions on gonadotrophin release in Japanese quail *(Coturnix coturnix japonica)*. *Neuroendocrinology*, 1969, *5*, 205–218.

Shire, J. G. M. The forms, uses and significance of genetic variation in endocrine systems. *Biological Review*, 1976, *51*, 105–141.

Silver, R. Effects of the anti-androgen cyproterone acetate on reproduction in male and female ring doves. *Hormones and Behavior*, 1977, *9*, 371–379.

Silver, R. The parental behavior of ring doves: The intricately coordinated behavior of the male and female is based on distinct physiological mechanisms in the sexes. *American Scientist*, 1978, 66, 209–215.

Silver, R. Parental care: Hormonal and non-hormonal control mechanisms. In R. Rosenblum and M. Moltz (Eds.), *Symbiosis in Parent–Offspring Interactions*. New York: Plenum Press. 1983.

Silver, R., and Barbiere, C. Display of courtship and incubation behavior during the reproductive cycle of the male ring dove *(Streptopelia risoria)*. *Hormones and Behavior*, 1977, *8*, 8–21.

Silver, R., and Buntin, J. The reproductive cycle of the male ring dove: I. Influence of adrenal hormones, *Journal of Comparative Physiology and Psychology*, 1973, *84*, 453–463.

Silver, R., and Cooper, M. Avian behavioral endocrinology, *Bioscience*, 1983, *33*, 567–572.

Silver, R., and Feder, H. H. Reproductive cycle of the male ring dove: II. Role of gonadal hormones in incubation behavior. *Journal of Comparative Physiology and Psychology*, 1973, *84*, 464–471.

Silver, R., and Givson, M. J. Termination of incubation in doves: Influence of egg fertility and absence of mate. *Hormones and Behavior*, 1980, *14*, 93–106.

Silver, R. Feder, H. H., and Lehrman, D. S. Situational and hormonal determinants of courtship, aggressive, and incubation behavior in male ring doves *(Streptopelia risoria)*. *Hormones and Behavior*, 1973, *4*, 163–172.

Silver, R., Reboulleau, C., Lehrman, D. S., and Feder, H. H. Radioimmunoassay of plasma progesterone during the reproductive cycle of male and female ring doves *(Streptopelia risoria)*. *Endocrinology*, 1974, *94*, 1547–1554.

Silver, R., O'Connell, M., and Saad, R. Androgens in birds. In C. Beyer (Ed.), *Endocrine Control of Sexual Behavior*. New York: Raven Press, 1978.

Silverin, B. Prolongation of the sexual activity in the male pied flycatcher *(Fidecula hypoleuca)* and its effects in a breeding population. *Ninth Conference of European Comparative Endocrinologists*, Giessen, 1977 (abstract).

Singh, T. P., Raizada, R. B., and Singh, A. J. Effect of methallibure on gonadotropic content, ovarian ^{32}P uptake and gonadosomatic index in the freshwater catfish, *Heteropneustes fossilis* (Bloch). *Journal of Endocrinology*, 1977, *72*, 321–237.

Singhas, C. A., and Dent, J. N. Hormonal control of the tail fin and of the nuptial pads in the male red-spotted newt. *General and Comparative Endocrinology*, 1975, *26*, 382–393.

Sinha, V. R. P. Induced spawning in carp with fractionated fish pituitary extract. *Journal of Fish Biology*, 1971, *3*, 263–272.

Slater, P. J. B. The stimulus to egg-laying in the Bengalese finch. *Journal of Zoology, London*, 1969, *158*, 427–440.

Smith, R. J. F. Control of prespawning behavior of sunfish *(Lepomis gibbosus* and *L. megalotis)*: I. Gonadal androgen. *Animal Behavior*, 1969, *17*, 279–285.

Sohn, J. J. Socially induced inhibition of genetically determined maturation in the platyfish, *Xiphophorus maculatus*. *Science*, 1977, *195*, 199–201.

Sohn, J. J., and Crews, D. Size mediated onset of genetically determined maturation in the platyfish, *Xiphophorus maculatus*. *Proceedings of the National Academy of Science*, 1977, *74*, 4547–4548.

Sossinka, R. Quantitative Untersuchungen zur sexuellen Reifung des Zebrafinken, *Taeniopygia castano-tis*. *Verhandlungen der Deutschen Zoologischen Gesellschaft*, 1974, 344–347, Stuttgart, 1975.

Sossinka, R., Pröve, E., and Kalberlah, H. H. Der Enfluss von Testosteron auf den Gesangssinn beim Zebrafinken *(Taeniopygia guttata castanotis)*. *Zeitschrift für Tierpsychologie*, 1975, *39*, 259–264.

Stacey, N. E. Effects of indomethacin and prostaglandins on the spawning behaviour of female goldfish. *Prostaglandins*, 1976, *12*, 113–120.

Stacey, N. E., and Hourston, A. S. Observations on spawning and feeding behavior of captive Pacific herring. *Canadian Journal of Fisheries and Aquatic Sciences*, 1982, *39*, 412–416.

Stacey, N. E., and Liley, N. R. Regulation of spawning behaviour in the female goldfish. *Nature*, 1974, *247*, 71–72.

Stacey, N. E., and Pandey, S. Effects of indomethacin and prostaglandins on ovulation of goldfish. *Prostaglandins*, 1975, *9*, 597–607.

Stamps, J. A. Courtship patterns, estrus periods, and reproductive condition in a lizard, *Anolis aeneus*. *Physiology and Behavior*, 1975, *14*, 531–535.

Stamps, J. A. Spacing patterns in lizards. In C. Gans (Ed.), *Biology of the Reptilia: Behavior and Ecology*. New York: Academic Press, 1977.

Steel, E., and Hinde, R. A. Influence of Photoperiod on oestrogenic induction of nest-building in canaries. *Journal of Endocrinology*, 1972a, *55*, 265–278.

Steel, E., and Hinde, R. A. Influence of photoperiod on PMSG-induced nest-building in canaries. *Journal of Reproduction and Fertility*, 1972b, *31*, 425–431.

Steel, E., and Hinde, R. A. Effect of a skeleton photoperiod on the daylength-dependent response to oestrogen in canaries *(Serinus canarius)*. *Journal of Endocrinology*, 1976, *70*, 247–254.

Stern, J. M. Androgen accumulation in hypothalamus and anterior pituitary of male ring doves: influence of steroid hormones. *General and Comparative Endocrinology*, 1972, *18*, 439–449.

Stetson, M. H. Feedback regulation by oestradiol of ovarian function in Japanese quail. *Journal of Reproduction and Fertility*, 1972a, *31*, 205–213.

Stetson, M. H. Hypothalamic regulation of gonadotropin release in female Japanese quail. *Zeitschrift für Zellforschung und Mikroskopische Anatomie*, 1972b, *130*, 411–428.

Stetson, M. H. Hypothalamic regulation of testicular function in Japanese quail. *Zeitschrift für Zellforschung und Mikroskopische Anatomie*, 1972c, *130*, 389–410.

Storey, C. R., and Nicholls, T. J. Plasma luteinizing hormone levels following ovariectomy of juvenile female mallards *(Anas platyrhyncos)*. *General and Comparative Endocrinology*, 1977, *33*, 8–12.

Stutinsky, F., and Befort, J. J. Effets des stimulations electriques du diencephale de *Rana esculenta* male. *General and Comparative Endocrinology*, 1964, *4*, 370–379.

Sugarman, R. A. Gular extension evoked by electrical stimulation in collared lizrds *(Crotaphytus collaris)*. *American Zoologist*, 1974, *14*, 1282 (abstract).

Sugarman, R. A., and Demski, L. S. Agnostic behavior elecited by electrical stimulation of the brain in western collared lizards, *Crotaphytus* collaris. *Brain, Behavior and Evolution*, 1978, *15*, 446–449.

Sundararaj, B. I. *Studies on the structural correlation of fish pituitary and gonads and experiments on inducing spawning in fishes.* Unpublished Ph.D. thesis, 1957, University of Mysore (quoted in Ramaswami and Lakshman, 1958).

Sundararaj, B. I., Nayyar, S. K., Burzawa-Gerard, E., and Fontaine, Y. A. Effects of carp gonadotropin on ovarian maintenance, maturation, and ovulation in hypophysectomized catfish, *Heteropneustes fossilis* (Bloch). *General and Comparative Endocrinology*, 1976, *30*, 472–476.

Taber, E. Androgen secretion in the fowl. *Endocrinology*, 1951, *48*, 6–16.

Tanaka, K. M., Mather, F. B., Wilson, W. D., and McFarland, L. Z. Effects of photoperiods on early growth of gonads and on potency of gonadotropins of the anterior pituitary in *Coturnix*. *Poultry Science*, 1965, *44*, 662–665.

Tarr, R. S. The role of the amygdala in the intraspecies aggressive behavior of the iguanid lizard, *Sceloporus occidentalis*. *Physiology and Behavior*, 1977, *18*, 1153–1158.

Tavolga, W. N. Effects of gonadectomy and hypophysectomy on prespawning behavior in males of the gobiid fish, *Bathygobius soporator*. *Physiological Zoology*, 1955, *28*, 218–233.

Taylor, D. H., and Guttman, S. I. *The Reproductive Biology of Amphibians.* New York: Plenum Press, 1977.

Temple, S. A. Plasma testosterone titers during the annual reproductive cycle of starlings *(Sturnus vulgaris)*. *General and Comparative Endocrinology*, 1974, *22*, 470–479.

Thornton, V. F. A progesterone-like factor detected by bioassay in the blood of the toad *(Bufo bufo)* shortly before induced ovulation. *General and Comparative Endocrinology*, 1972, *18*, 133–139.

Thornton, V. F., and Geschwind, I. I. Hypothalamic control of gonadotropin release in amphibia: Evidence from studies of gonadotropin release *in vitro* and *in vivo*. *General and Comparative Endocrinology*, 1974, *23*, 294–301.

Timms, A. M., and Kleerekoper, H. The locomotor response of male *Ictalurus punctatus*, the channel catfish, to a pheromone released by the ripe female of the species. *Transactions of the American Fisheries Society*, 1972, *102*, 302–310.

Texier-Vidal, A., Herlant, M., and Benoit, J. La préhypophyse du canard Pekin au cours du cycle annuel. *Archives of Biology* (Liège), 1962, *73*, 317–368.

Tixier-Vidal, A., Follett, B. K., and Farner, D. S. The anterior pituitary of the Japanese quail, *Coturnix coturnix japonica. Zeitschrift für Zellforschung Mikroskopische Anatomie*, 1968, *92*, 610–635.

Trottier, T. M., and Armstrong, J. B. Hormonal stimulation as an aid to artificial insemination in *Ambystoma mexicanum. Canadian Journal of Zoology*, 1975, *53*, 171–173.

Tsui, H. W. Stimulation of androgen production by the lizard testis: Site of action of ovine FSH and LH. *General and Comparative Endocrinology*, 1976, *28*, 386–394.

Valenstein, P., and Crews, D. Hormonal and nonhormonal factors controlling mating-induced termination of behavioral estrus in the female lizard, *Anolis carolinensis. Hormones and Behavior*, 1977, *9*, 362–370.

van Oordt, P. G. W. J. The analysis and identification of the hormone-producing cells of the adenohypophysis. In E. J. W. Barrington and C. B. Jørgensen (Eds.), *Perspectives in Endocrinology*. New York: Academic Press, 1968.

van Oordt, P. G. W. J. Cytology of the adenohypophysis. In B. Lofts (Ed.), *Physiology of the Amphibia*, Vol. 2. New York: Acdemic Press, 1974.

van Oordt, P. G. W. J., and Brands, F. The sertoli cells in the testis of the common frog, *Rana temporaria. Journal of Endocrinology*, 1970, *48*, 1 (abstract).

van Tienhoven, A. *Reproductive Physiology of Vertebrates.* Philadelphia: Saunders, 1968.

van Tienhoven, A., and Planck, R. J. The effect of light on avian reproductive activity. In R. O. Greep (Ed.), *Handbook of Physiology*, Vol. 2, Part 1: *Female Reproductive System*. Washington, D.C.: American Physiological Society, 1973.

Vaugien, L. Ponte induite chez la perruche ondulée maintenue à l'obscurité et dans l'ambience des volières. *Comptes Rendus de l'Académie Sciences* (Paris), 1951, *232*, 1706–1708.

Vellano, C., and Bona, A. Some biochemical data on seasonal variations in steroidogenesis in the testes and fat bodies of *Triturus cristatus carnifex* Laur. *della Accademia Atti delle Scienze di Torino*, 1975.

Vellano, C., Bona, A., Mazzi, V., and Colucci, D. The effect of synthetic luteinizing hormone releasing hormone on ovulation in the crested newt. *General and Comparative Endocrinology*, 1974, *24*, 338–340.

Vinegar, M. B. The function of breeding coloration in the lizard, *Sceloporus virgatus*. *Copeia*, 1972, 660–664.

Wada, M. Effect of hypothalamic implantation of puromycin on photostimulated testicular growth in the Japanese quail *(Coturnix coturnix japonica)*. *General and Comparative Endocrinology*, 1974, *22*, 54–61.

Wata, M., and Gorbman, A. Mate calling induced by electrical stimulation in freely moving leopard frogs, *Rana pipiens*. *Hormones and Behavior*, 1977a, *9*, 141–149.

Wada, M., and Gorbman, A. Relation of mode of administration of testosterone to evocation of male sex behavior in frogs. *Hormones and Behavior*, 1977b, *8*, 310–319.

Wada, M., Wingfield, J. C., and Gorbman, A. Correlation between blood level of androgens and sexual behavior in male leopard frogs, *Rana pipiens*. *General and Comparative Endocrinology*, 1976, *29*, 72–77.

Wai, E. H., and Hoar, W. S. The secondary sex characters and reproductive behavior of gonadectomized sticklebacks treated with methyl testosterone. *Canadian Journal of Zoology*, 1963, *41*, 611–628.

Warren, R. P., and Hinde, R. A. The effect of oestrogen and progesterone on the nest-building of domesticated canaries. *Animal Behavior*, 1959, *7*, 209–213.

Warren, R. P., and Hinde, R. A. Does the male stimulate oestrogen secretion in female canaries? *Science*, 1961, *133*, 1354–1355.

Wasserman, W. J., and Smith, L. D. Oocyte maturation in nonmammalian vertebrates. In R. E. Jones (Ed.), *The Vertebrate Ovary*. New York: Plenum Press, 1978.

Weaver, W. G. Courtship and combat behavior in *Gopherus berlandieri*. *Bulletin of the Florida State Museum, Biological Sciences*, 1970, *15*, 1–43.

Wells, K. D. The social behaviour of anuran amphibians. *Animal Behavior*, 1977, *25*, 666–693.

Wells, K. D. The courtship of frogs. In D. H. Taylor and S. I. Guttman (Eds.), *The Reproductive Biology of Amphibians*. New York: Plenum Press, 1978.

Wentworth, B. C., Burke, W. H., and Birrenkott, G. P. A radioimmunoassay for turkey luteinizing hormone. *General and Comparative Endocrinology*, 1976, *29*, 119–127.

Wheeler, J. M., and Crews, D. The role of the anterior hypothalamus-preoptic area in the regulation of reproductive behavior in the lizard, *Anolis carolinensis:* Lesion studies. *Hormones and Behavior*, 1978, *11*, 42–60.

White, S. J. Effects of stimuli emanating from the nest on the reproductive cycle in the ring dove: II. Building during the pre-laying period. *Animal Behavior*, 1975, *23*, 869–882.

Wiebe, J. P. Inhibition of pituitary gonadotropin activity in the viviparous sea perch *Cymatogaster aggregata* Gibbons by a dithiocarbamoylhydrazine derivative (ICI 33, 828). *Canadian Journal of Zoology*, 1968, *46*, 751–758.

Wiebe, J. P. Endocrine control of spermatogenesis and oogenesis in the viviparous sea perch *Cymatogaster aggregata* Gibbons. *General and Comparative Endocrinology*, 1969, *12*, 267–275.

Wilhelmi, A. E., Pickford, G. E., and Sawyer, W. H. Initiation of the spawning reflex response in *Fundulus* by the administration of fish and mammalian neurohypophyseal preparations and synthetic oxytocin. *Endocrinology*, 1955, *57*, 243–252.

Wilkens, H. Über Präadaptionen für das Höhlenleben, untersucht am Laichverhalten ober- und unterirdischer Populationen des *Astyanax mexicanus* (Pisces). *Zoologischer Anzeiger*, 1972, *188*, 1–11.

Wilson, F. E. The tubero-infundibular region of the hypothalamus: A focus of testosterone sensitivity in the male tree sparrow *(Spizella arborea)*. In W. Bargmann and B. Scharrer (Eds.), *Aspects of Neuroendocrinology*. Berlin: Springer-Verlag, 1970.

Wilson, S. C., and Sharp, P. J. Variations in plasma LH levels during the ovulatory cycle of the hen, *Gallus domesticus*. *Journal of Reproduction and Fertility*, 1973, *35*, 561–564.

Wilson S. C., and Sharp, P. J. Episodic release of luteinizing hormone in the domestic fowl. *Journal of Endocrinology*, 1975, *64*, 77–86.

Wilson, S. C., and Sharp, P. J. Effects of androgens, oestrogens and deoxycorticosterone acetate on plasma concentrations of luteinizing hormone in laying hens. *Journal of Endocrinology*, 1976, *69*, 93–102.

Wingfield J. C., and Crimm, A. S. Seasonal changes in plasma cortisol, testosterone and estradiol-7β in the plaice, *Pleuronectes platessa L. General and Comparative Endocrinology*, 1977, *31*, 1–11.

Wingfield, J. C., and Farner, D. S. The determination of five steroids in avian plasma radioimmunoassay and competitive protein-binding. *Steroids*, 1975, *26*, 311–326.

Wolf, O. M. Effect of testosterone propionate injections into castrate male frogs, *Rana pipiens. Anatomical Record*, 1939, *75*, Suppl. 55.

Wood-Gush, D. G. M. The control of the nesting behaviours of the domestic hen: I. The role of the oviduct. *Animal Behavior*, 1963, *11*, 293–299.

Wood-Gush, D. G. M., and Gentle, M. J. The hyperstriatum and nesting behaviour in the domestic hen. *Animal Behavior*, 1978, *26*(4), 1157–1164.

Wood-Gush, D. G. M., and Gilbert, A. B. The control of the nesting behaviour of the domestic hen: II. The role of the ovary. *Animal Behavior*, 1964, *12*, 451–453.

Wood-Gush, D. G. M., and Gilbert, A. B. Oestrogen and the pre-laying behaviour of the domestic hen. *Animal Behavior*, 1969, *17*, 586–589.

Wood-Gush, D. G. M., and Gilbert, A. B. The nesting behaviour of hens with ovarian transplants. *Animal Behavior*, 1970, *18*, 52–54.

Wood-Gush, D. G. M., and Gilbert, A. B. Some hormones involved in the nesting behaviour of hens. *Animal Behaviour*, 1973, *21*, 98–103.

Wood-Gush, D. G. M., Langley, G. A. S., Leitch, A. F., Gentle, M. J., and Gilbert, A. B. An autoradiographic study of sex steroids in the chicken telencephalon. *General and Comparative Endocrinology*, 1977, *31*, 161–168.

Woods, J. W., and Domm, L. V. A histochemical identification of the androgen producing cells in the gonads of the domestic fowl and albino rat. *General and Comparative Endocrinology*, 1966, *7*, 559–570.

Wootton, R. J. Aggression in the early phases of the reproductive cycle of the male three-spined stickleback *(Gasterosteus aculeatus). Animal Behavior*, 1970, *18*, 740–746.

Wootton, R. J. *The Biology of the Sticklebacks.* New York: Academic Press, 1977.

Wright, P. A. 3-keto-Δ^4 steroid: Requirement for ovulation in *Rana pipiens. General and Comparative Endocrinology*, 1971, *16*, 511–519.

Wyburn, G. M., Johnston, H. S., and Aitken, R. N. C. Fate of the granulosa cells in the hen's follicle. *Zeitschrift für Zellforschung Mikroskopische Anatomie*, 1966, *72*, 53–65.

Yamamoto, K., and Yamazaki, F. Hormonal control of ovulation and spermiation in goldfish. *Gunma Symposia on Endocrinology*, 1967, *4*, 131–145.

Yamamoto, K., Nagahama, Y., and Yamazaki, F. A method to induce artificial spawning of goldfish all through the year. *Bulletin of the Japanese Society of Scientific Fisheries*, 1966, *32*, 977–983.

Yaron, Z. Endocrine aspects of gestation in viviparous reptiles. *General and Comparative Endocrinology*, 1972, *Suppl. 3*, 663–674.

Yaron, Z., Terkatin-Shimony, A., Shaham, Y., and Salzer, H. Occurrence and biological activity of estradiol-17 in the intact and ovariectomized *Tilapia aurea* (Cichlidae, Teleostei). *General and Comparative Endocrinology*, 1977, *33*, 45–52.

Young, W. C. The hormones and mating behavior. In W. C. Young (Ed.), *Sex and Internal Secretions*, Vol. 2. Baltimore: Williams and Wilkins, 1961.

Zambrano, D. Innervation of the teleost pituitary. *General and Comparative Endocrinology*, 1972, *Suppl. 3*, 22–31.

Zigmond, R. E. Target cells for gonadal steroids in the brain: Studies on hormone binding and metabolism. In P. G. Caryl and D. M. Vowles (Eds.), *Neural and Endocrine Aspects of Behaviour in Birds.* New York: Elsevier, 1975.

Zigmond, R. E., Nottebohm, F., and Pfaff, D. W. Distribution of androgen-concentrating cells in the brain of the chaffinch. Proceedings of the IVth International Congress of Endocrinology. *Excerpta Medica Foundation International Congress Series*, 1972a, *256* (abstract 340).

Zigmond, R. E., Stern, J. M., and McEwen, B. S. Retention of radioactivity in cell nuclei in the hypothalamus of the ring dove after injection of ³H-testosterone. *General and Comparative Endocrinology*, 1972b, *18*, 450–453.

Zigmond, R. E., Nottebohm, F., and Pfaff, D. W. Androgen-concentrating cells in the midbrain of a song bird. *Science*, 1973, *179*, 1005–1007.

The Role of Gonadal Hormones in the Activation of Feminine Sexual Behavior

Lynwood G. Clemens and David R. Weaver

Historical Concepts of Feminine Sexual Behavior

For early scholars, the first questions about sexual behavior were largely ones of localization: Where does the sexual impulse come from? The poet Virgil associated the copulatory activity of cattle with the presence of the gadfly and thought that the gadfly noise or bite might cause sexual activity. In 1901, W. Heape introduced the term *estrus* for the recurrent periods of sexual excitement in animals, which are also called *heat*. (The word *oïstros* is the Greek name for the gadfly.)

Early scholars thought that the sexual impulse built up from sensations or activity resulting from the distension of some bodily organ, such as the gonads, the seminal vesicles, or the uterus. In support of these concepts were experiments like those of Tarchanoff (1887), who reported that draining the seminal vesicles of the male frog *Rana temporaria* while he was clasped to the female resulted in the termination of sexual clasping (amplexus). If the vesicles were then refilled with a foreign substance (e.g., milk), amplexus was resumed. Tarchanoff concluded that

Lynwood G. Clemens and David R. Weaver   Department of Zoology and Neuroscience Program, Michigan State University, East Lansing, Michigan 48824.   This work was supported by NIH Grant HD 06760 to Lynwood G. Clemens and an NSF Predoctoral Fellowship to David R. Weaver.

the seminal vesicles were the starting point of centripetal impulses that set into motion sexual activity.

Later work by Steinach (1894) showed that the seminal vesicles of *R. temporaria* were empty before coitus, only becoming gradually filled during coitus, and thus the sexual impulse could not arise there. Steinach further reported that removal of the seminal vesicles did not abolish amplexus but might shorten it somewhat. He concluded that it was in the swollen testes that the sexual impulse originated.

The term *hormone* was introduced into the scientific literature in 1902 by Bayliss and Starling to refer to substances that are secreted into the blood and have their action at a distance from the gland where they originate. This concept gave scientists interested in behavior something new to consider, and in 1910, Steinach rejected the idea that neural impulses produced from the distension of some peripheral organ were the "initiator" of sexual activity. Instead, on the basis of his experiments, he proposed that peripherally generated hormones act centrally to initiate sexual activity. Steinach injected sexually inactive male frogs with one of three preparations: crushed testes from another inactive male, crushed testes from a sexually active male, or central nervous tissue from a sexually active male. Injection of either of the latter two preparations led to the induction of sexual clasping, sometimes within a period of hours. He concluded that the injected material acted on the brain to remove an inhibition that was in operation during sexually quiescent periods. Steinach suggested that there was a process of "erotization" *(Erotisieurung)* of the neural center by material that originated in the testes.

In spite of this early indication of the influence of blood-borne factors on the brain, the idea that a peripheral organ, and not the brain, was directly responsible for producing sexual drive continued for some time:

> Some organ, other than the gonads, but in structural and functional dependence on the testes, effects a tumescence of tension in itself or in another tissue or organ; this tension or tumescence initiates the afferent impulses which stimulate activity. (Nissen, 1929, p. 526)

It was not until the mid-1930s that the concept of a peripheral origin of impulses for female sexual activity was seriously weakened. A psychologist, Josephine Ball (1934), removed the uterus and vagina of immature female rats and found that this removal did not prevent the animals from showing estrous behavior when they reached adulthood. The same result was obtained with cats in the following year (Bard, 1935).

In the late 1930s, studies were published showing that specific brain areas are necessary for sexual responses to occur (e.g., Bard, 1939; Dempsey and Rioch, 1939). Ablation of some regions abolished female sexual responses even though adequate ovarian hormones were present. These studies helped focus attention on the brain as a site for mediating the effects of ovarian hormones on female sexual behavior (see Beach, 1981, for review).

Much of the work in the 1930s and the 1940s concentrated on the relation of estrogen and progesterone to behavior and to the changes in ovarian physiology throughout the estrous cycle. In 1936, Dempsey *et al.* showed that induction of sexual receptivity in female guinea pigs required the action of these two ovarian hormones. They found that nonreceptive (ovariectomized) females could be brought into a state of sexual receptivity similar to that seen in normally receptive females by progesterone injection after a priming period of estrogen treatment. Although estrogen alone induced some sexual activity, the administration of progesterone 48 hr after treatment with a small dose of estrogen resulted in the female's becoming highly sexually active within 4–5 hr. Similar demonstrations were subsequently made for the laboratory rat (Boling and Blandau, 1939), the mouse (Ring, 1944), and the hamster (Frank and Fraps, 1945).

There followed from these studies volumes of work on the nature of estrogen and progesterone action. In this chapter, we try to summarize some of the current concepts that relate these ovarian hormones to feminine sexual behavior. In order to do so, we first summarize some of the basic principles of steroid hormone structure. This summary is followed by a summary of the relation of female sexual behavior to the cyclic nature of ovarian secretion. In subsequent sections, we summarize experimental studies that relate gonadal hormone action to female sexual behavior. To do this, it is necessary to divide sexual behavior theoretically into three distinct categories (attractivity, proceptivity, and receptivity) in the manner of Beach (1976).

STRUCTURE AND BIOSYNTHESIS OF GONADAL STEROIDS

In order to more fully understand the influence of various gonadal hormones on sexual behavior, it is necessary to appreciate the interrelationships of their synthesis and structure.

STRUCTURE OF GONADAL STEROIDS

Cholesterol is the precursor for all steroid hormones. (Feder, 1981a). The "gonadal" steroids (progestins, androgens, and estrogens) are produced from cholesterol by enzymatic cleavage of carbon atoms from its 28-carbon structure. Progestins consist of 21 carbon atoms, and androgens and estrogens consist of 19 and 18 carbons, respectively. All of these compounds have the steroid nucleus (three 6-carbon rings and an attached 5-carbon ring), as do the other derivatives of cholesterol, such as the corticosteroids. Because of the common enzymatic-biosynthetic pathway by which these steroids arise, a tissue that produces one of them is capable of producing the others. One exception to this general rule is the adrenal cortex, which is unique in having 11-β-hydroxylase, the enzyme necessary for the synthesis of glucocorticoids and mineralocorticoids. Although testes are recognized for their ability to produce androgens, especially testosterone, their ability to secrete estrogens (Baird *et al.*, 1973; DeJong *et al.*, 1973) is frequently overlooked, as is the ability of the adrenal cortex to produce "gonadal" steroids. The relative amounts of each steroid produced by each tissue reflect the relative activities of the biosynthetic enzymes in that tissue. Each steroid-producing tissue produces several different steroids, but the relative amounts of these hormones vary greatly by tissue, so that the testes can indeed be characterized as the main source of androgen in the males of most species.

In addition to their carbon backbone, steroid hormones have a variety of substituent groups. Their location and stereochemical orientation produces variation in the structure that can produce large changes in the biological activity of the steroid molecule.

Although space does not permit a complete treatment here of the International Union of Pure and Applied Chemistry (IUPAC) nomenclature system, a quick introduction to the terms used in naming the positions within the structure of the steroids will help in the discussion of biosynthetic pathways that follows. The rings of the steroid nucleus are designated by letter from *A* to *D*, with the numbering of the individual carbon atoms as shown in Figure 1. The carbon atoms attached to the C-10 and C-13 positions in some steroids are designated C-19 and C-18, respectively. Additional carbon atoms are present in the form of an aliphatic chain attached at C-17 in the progestins and cholesterol; the numbering of these atoms starts at C-20. The location of substituent groups on the carbon backbone is denoted by giving the number of the carbon to which it is attached. The stereochemical orientation is given relative to the C-10 position; that is, a substituent is designated as α if it is transrelative to C-10 and β if it is cis to C-10. Double bond locations are given by referring to the lower number of the carbon atoms involved in the bond and by adding the suffix *-ene* instead of *-ane* (e.g., pregn-4-ene instead of pregnane). Finally, substituent groups are identified by either a prefix or a suffix added to the root, as in pregn-4-ene-3,20-dione, which identifies the 2 keto groups (C = 0) in positions 3 and 20 with the suffix *di-one*. The *4* in this example tells us that there is a "substituent" involving the 4-carbon, and the *-ene* tells us that this substituent is a double bond. Alternatively, the *4* could be placed before the word (e.g., 4-pregnene-3,20-dione). The common or trivial name for this steroid, by the way, is progesterone.

There are usually several correct names for any one steroid molecule because of the possible differences in the location of substituent group numbers or names within the root word.

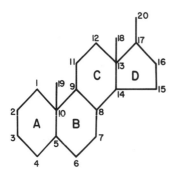

Fig. 1. Alphabetical and numerical designations for the carbon rings and atoms of a generalized steroid molecule.

Biosynthesis of Gonadal Steroids

187

GONADAL
HORMONES IN
FEMININE SEXUAL
BEHAVIOR

Many workers (Dorfman and Ungar, 1965; Feder, 1981a; Luttge, 1979; Steinberger, 1970) have addressed the problem of determining the origins of the gonadal steroids in the intact animal. The generally accepted scheme of steroid hormone biosynthesis follows.

Following its production, a series of enzymes act on cholesterol to cleave 6 carbons from the side chain attached to the C-17 position. The resulting 21-carbon structure is a progestin. Specifically, it has an acetyl group attached at C-17, a hydroxyl group at C-3 (β position), and a double bond from C-5 to C-6. This compound is pregnenolone. Progesterone can readily be made from pregnenolone by oxidation of the 3-hydroxyl to a keto group and shifting the double bond from the 5-6 position to the 4-5 position. The rest of the family of progestins are variations on one of these two basic structures made by altering the side groups at several locations. Glucocorticosteroids, which have hydroxyl groups on C-11 and C-21, are produced from the progestins (see Figure 2). Androgens are produced from either progesterone or pregnenolone following removal of the two carbons attached to the C-17 position. Progestins hydroxylated at the 17α position are intermediates in these pathways. A large number of other androgenic compounds, including testosterone, can be produced directly from the androgen androstenedione. In addition, removal of C-19 from androstenedione and aromatization of the A-ring (e.g., removal of hydrogen atoms and formation of double bonds to produce an aromatic or benzenelike A ring) produces estrone, an estrogen. Similarly, aromatization and C-19 cleavage from testosterone results in estradiol-17β (which we will refer to simply as *estradiol*). Other estrogenic compounds are derived from one of these two (estradiol and estrone), with the exception of estriol, which can also arise via an alternate pathway from the androgen dehydroepiandrosterone. In addition to occurring in steroid hormone-producing organs, aromatization of androgens to estrogens also occurs in nervous tissue (Naftolin *et al.*, 1975).

A more schematic representation of the biosynthetic pathways involved in steroidogenesis is presented in Figure 2. Symbols used to designate the various hormones are decoded in Table 1, which provides the common and systematic names of the major gonadal hormones of interest to reproductive biologists.

STEROID STRUCTURE AND ESTRUS INDUCTION

Although estrogenic substances vary greatly in their potency, they are all capable of stimulating female sexual behavior in a variety of mammalian species when given in doses sufficient to compensate for their potency differences (see Morali and Beyer, 1979). Esterified forms of estrogens (e.g., with benzoic or propionic acid attached to the estrogen molecule by an ester linkage) are more potent than the free forms (Feder and Silver, 1974; Powers, 1975; Robinson and Reardon, 1961), presumably because of an increase in their biological half-lives. Among the free (nonesterified) estrogens, estradiol is the most potent in stimulating receptiv-

LYNWOOD G.
CLEMENS AND
DAVID R. WEAVER

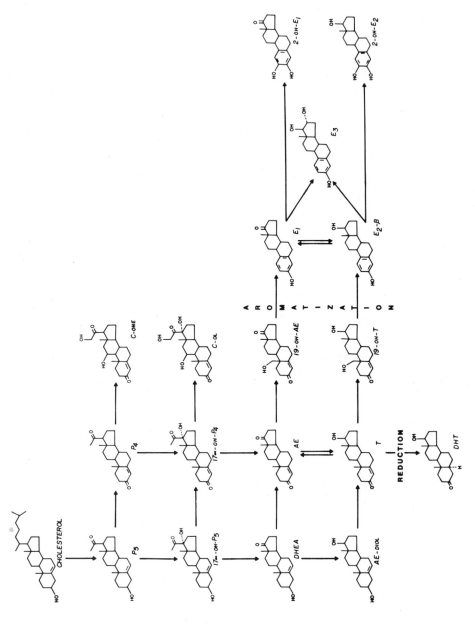

Fig. 2. Biosynthetic pathways involved in the production of corticosteroids, progestins, androgens and estrogens from their common precursor, cholesterol.

TABLE 1. STEROID HORMO

189

GONADAL
HORMONES IN
FEMININE SEXUAL
BEHAVIOR

Common name	Abbreviation	Systematic name
Pregnenolone	P5	Pregn-5-en-3β-ol-20-one
Progesterone	P4	Pregn-4-en-3,20-dione
17 Alpha hydroxypregnenolone	17-α-OH-P5	Pregn-5-en-3β,17α-diol-20-one
17 Alpha hydroxyprogesterone	17-α-OH-P4	Pregn-4-en-17α-ol-3,20 dione
Corticosterone	C-one	Pregn-4-en-11β,21-diol-3,20-dione
Cortisol	C-ol	Pregn-4-en-11β,17α,21-triol-3,10-dione
Dehydroepiandrosterone	DHEA	Androst-5-en-3β-ol-17-one
Androstenedione	AE	Androst-4-en-3,20-dione
Androstenediol	AEdiol	Androst-5-en-3β,17β-diol
Testosterone	T	Androst-4-en-17β-ol-3-one
Dihydrotestosterone	DHT	5-Androstan-17β-ol-e-one
19 Hydroxy androstenedione	19-OH-AE	Androst-4-en-19-ol-3,17-dione
19 Hydroxy testosterone	19-OH-T	Androst-4-en-17β,19-diol-3-one
Estrone	E1	Estra-1,3,5(10)-trien-3-ol-17-one
2 Hydroxy estrone (catechol-E1)	2-OH-E1	Estra-1,3,5(10)-trien-2,3-diol-17-one
Estradiol-17α	E2α	Estra-1,3,5(10)-trien-3,17α-diol
Estradiol-17β	E2β	Estra-1,3,5(10)-trien-3,17β-diol
2 Hydroxy estradiol (catechol-E2)	2-OH-E2	Estra-1,3,5(10)-trien-2,3,17β-triol
Estriol	E3	Estra-1,3,5(10)-trien-3,16α,17β-triol

ity, followed by estrone and then estriol (Beyer *et al.*, 1971; Feder and Silver, 1974; Robinson and Reardon, 1961). Synthetic estrogens also vary in their ability to activate receptivity (e.g., see Robinson and Reardon, 1961).

Hydroxylated metabolites of estradiol and estrone (catechol estrogens) vary in their ability to activate female sexual behavior in rodents when administered prior to the injection of progesterone. This variability is in part a function of the route of administration; it also seems to depend on the presence of other estrogens. Marrone *et al.* (1977) found 2-hydroxyestradiol alone to be ineffective in stimulating receptivity in ovariectomized female rats or guinea pigs when implanted in the hypothalamus. When 2-hydroxyestradiol and estradiol were implanted simultaneously, however, the level of receptivity was greater than that seen with estradiol implants alone. Similarly, Luttge and Jasper (1977) found intracranial implants of 2-hydroxyestradiol to be ineffective, but intravenous 2-hydroxyestradiol or estradiol in combination with intracranial 2-hydroxyestradiol was more effective than intracranial 2-hydroxyestradiol or intravenous estradiol alone. In a more recent study (Ball *et al.*, 1980), continuous microinfusion of 2-hydroxyestradiol over a 7-day period was found to be ineffective in stimulating receptivity in female rats. These studies indicate that 2-hydroxyestradiol, although incapable of stimulating receptivity when given alone, may possess enough estrogenic activity to act in combination with more potent estrogens (e.g., estradiol) to stimulate receptivity following progesterone treatment. In contrast to the variable results for 2-hydroxyestradiol., Ball *et al.* (1980) found 4-hydroxyestradiol to be as effective as estradiol in stimulating receptivity when administered by osmotic minipumps for 7 days and followed by progesterone injection on the eighth day. Furthermore, when injected into the jugular vein, 4-hydroxyestradiol was more effective than estradiol by the same route.

A wide variety of androgens have also been tested for their ability to facilitate female sexual behavior in several species (Morali and Beyer, 1979; Young, 1961). Those androgens in which aromatization of the A-ring is possible are more potent in inducing estrus in ovariectomized females than nonaromatizable androgens, a finding indicating that the active form may be an estrogenic metabolite (Beyer and Komisaruk, 1971). Aromatizable androgens are also capable of synergism with progesterone, in much the same way as are estrogens (see Morali and Beyer, 1979). Androgenic activity (e.g., ability to stimulate accessory glands and penile spines in the male) is not related to the estrus-inducing activity, a finding indicating that the receptors mediating the two effects vary in their specificity. In addition, antiestrogens are able to block the estrus-inducing activity of some aromatizable androgens used in combination with progesterone, a finding further implicating estrogenic metabolites of the androgens as the effective substances in the activation of female sexual behavior by androgens (Beyer and Vidal, 1971; Whalen et al., 1972).

ESTROUS CYCLICITY AND HORMONE LEVELS IN INTACT FEMALES

In this section, we briefly consider the normal cyclic nature of gonadal steroids and sexual behavior in intact females of a variety of species. This discussion will help to lay the groundwork for considering experimental evidence concerning the influence of gonadal steroids on various aspects of feminine sexual activity and their mechanism of action (Feder, 1981b; Fox and Laird, 1970; Morali and Beyer, 1979; van Tienhoven, 1968).

SPONTANEOUS OVULATION

Female mammals that ovulate spontaneously have alternating periods of follicular and luteal development. These alternating states of the ovary produce cyclic alterations in the levels of gonadal steroid hormones, specifically estrogen and progesterone. These fluctuations in steroid levels, in turn, are often, but not always, associated with cyclic changes in the willingness of the female to copulate. As the ovarian follicles approach maturity under the influence of the gonadotropic hormones follicle-stimulating hormone (FSH) and luteinizing hormone (LH), estrogen production rises very rapidly. This estrogen acts on the hypothalamus via positive feedback to stimulate the LH peak, which triggers ovulation. Progesterone secretion from the corpus luteum (CL) ceases prior to the rise in estradiol levels; this cessation has a permissive function that allows the LH peak, as progesterone inhibits LH secretion (Hauger et al., 1977). In addition to stimulation of the LH peak, estrogen stimulates sexual receptivity via action within several brain areas, among them the hypothalamus. The result is that copulation occurs during the periovulatory period when the probability of a successful mating (e.g., one that results in pregnancy) is greatest. In many species, plasma estrogen levels decline rapidly during the period of sexual receptivity, and the animal remains in estrus even though estrogen has returned to low levels (see Figure 3). Experiments that further investigate this observation are discussed in the sections on latency and duration of hormone action.

In most species, plasma progesterone levels are low throughout the periovulatory period. Rodents are the exception in that a large, brief increase in progesterone occurs shortly after the preovulatory gonadotropin surge (see Figure 4; Butcher *et al.*, 1974; Croix and Franchimont, 1975; Saidapur and Greenwald, 1978). In these species (rat, hamster, mouse, and guinea pig), progesterone markedly facilitates sexual behavior when combined with the previous priming effect of estrogen (Boling and Blandau, 1939; Collins *et al.*, 1938; Frank and Fraps, 1945; Ring, 1944). Some primates show a minor increase in progesterone at this time (Karsch and Sutton, 1976; Knobil, 1973), but it is not believed to be of behavioral significance. Menstrual cycles for rhesus and human females are illustrated in Figure 5.

Following ovulation, corpora lutea (CLs) form and produce progesterone. This progesterone prepares the uterine epithelium for implantation of the blastocyst in the event that fertilization has occurred. The postovulatory (luteal) rise in progesterone seems to terminate estrus in some species. Luteal progesterone also generally inhibits gonadotropin levels (Baird and Scaramuzzi, 1976; Hauger *et al.*, 1977).

The active (secretory) life of the CL depends on whether pregnancy is established: In the absence of pregnancy, the CLs regress after a length of time that varies by species. Copulatory stimulation during estrus, however, may modify the life of the CL. For example, nonmated female rats do not form functional CLs and return to estrus in 4 or 5 days (depending on strain). Sterile-mated females, on the other hand, form functional CLs and pseudopregnancy results, lasting approximately 14 days.

Destruction of the CLs, or luteolysis, leads to a decline in plasma progesterone levels. The removal of the inhibitory effect of progesterone on estrogen-stimulated gonadotropin release allows gonadotropin and estrogen levels to increase, and another cycle begins. In sheep and cattle, luteal progesterone serves a "priming" function important to the display of sexual behavior in response to the increased estrogen levels that occur before ovulation (Young, 1961).

INDUCED OVULATION (REFLEX OVULATION)

Females of some mammalian species do not ovulate spontaneously. Instead, they require copulatory stimulation during behavioral estrus to induce ovulation. The cat and the rabbit are the most extensively studied reflex ovulators.

The female cat (queen) is seasonally polyestrous, with periods of receptivity occurring every 12–21 days in the absence of mating during the mating season. As in spontaneous ovulators, estrogen from waves of developing follicles produces receptivity, but estrogen by itself is incapable of stimulating an LH surge leading to ovulation in the cat (Figure 6). Copulatory stimulation, however, produces an

LYNWOOD G.
CLEMENS AND
DAVID R. WEAVER

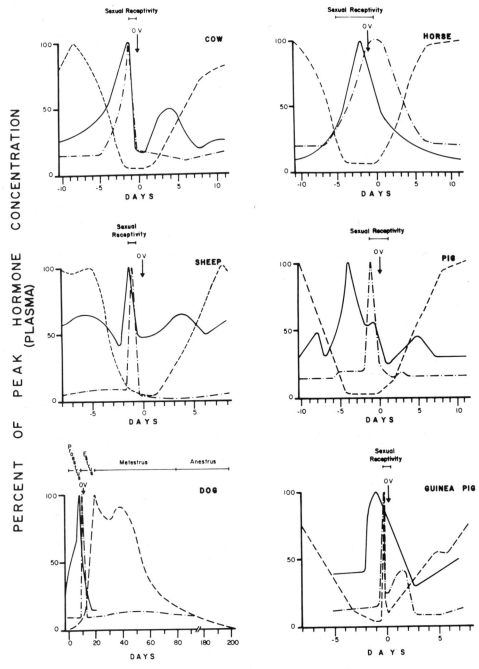

Fig. 3. Peripheral plasma levels of estradiol (—), progesterone (- - -), and luteinizing hormone (-··-) during the estrous cycle in the cow, horse, sheep, pig, dog, and guinea pig. Values are expressed relative to the peak values attained during the cycle. Graphs are based on the following references: cow (Glencross *et al.*, 1973; Hansel *et al.*, 1973; Schams *et al.*, 1977; Shemesh *et al.*, 1972); mare (Evans and Irvine,, 1975; Noden *et al.*, 1975; Pattison *et al.*, 1974; Stabenfeldt and Hughes, 1977); ewe (Cumming *et al.*, 1973; Pant *et al.*, 1977); sow (Anderson 1980; Hansel *et al.*, 1973); bitch (Concannon *et al.*, 1975; Phemister *et al.*, 1973; Stabenfeldt and Schille, 1977); guinea pig (Blatchley *et al.*, 1976; Challis *et al.*, 1971; Croix and Franchimont, 1975).

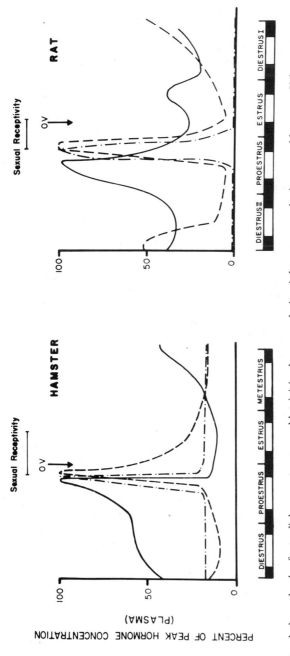

Fig. 4. Peripheral plasma levels of estradiol, progesterone, and luteinizing hormone during 4-day estrous cycles in rats and hamsters. Values are expressed relative to the peak value attained during the cycle. Legend as in Figure 3. References: rat (Butcher *et al.*, 1974; Smith *et al.*, 1975; Sodersten and Eneroth, 1981); hamster (Bast and Greenwald, 1974; Kent, 1968; Saidapur and Greenwald, 1978).

LYNWOOD G.
CLEMENS AND
DAVID R. WEAVER

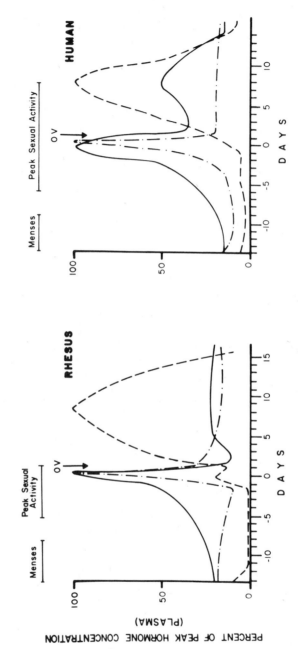

Fig. 5. Peripheral plasma levels of estradiol, progesterone, and luteinizing hormone during menstrual cycles in rhesus monkeys and humans. Values are expressed relative to the peak values attained during the cycle. Legend as in Figure 3. References: rhesus monkey (Karsch and Sutton, 1976; Knobil, 1973; Weick et al., 1973; Wilson et al., 1982); human (Adams et al., 1978; Guerrero et al., 1976; Jha et al., 1978; Udry and Morris, 1968).

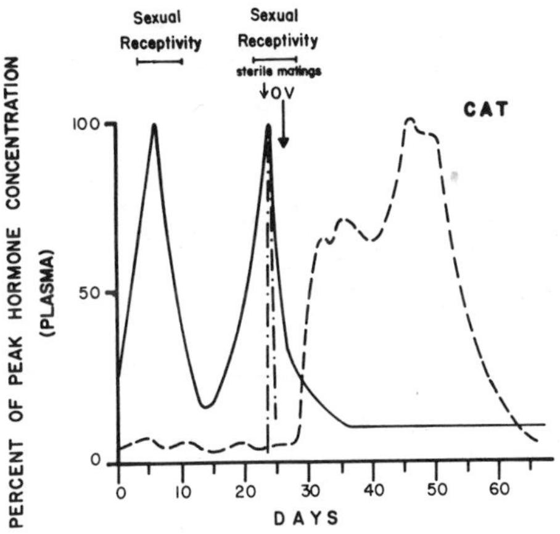

Fig. 6. Peripheral plasma levels of estradiol, progesterone, and luteinizing hormone in an induced ovulator, the cat. Values are expressed relative to the peak value attained during the cycle. Two heat cycles are represented. In the first, the queen was not mated and ovulation did not occur. In the second cycle of sexual receptivity, the queen was mated several times; as a result of mating, an LH peak occurred that resulted in ovulation. Corpora lutea were formed and produced progesterone through pseudopregnancy, which lasted over 30 days. Legend as in Figure 3. (References: Paape *et al.*, 1975; Schille *et al.*, 1979; Verhage *et al.*, 1976; Wildt *et al.*, 1980.

LH peak within 12 hr, and ovulation follows 26–58 hr later (Wildt *et al.*, 1980). Estradiol levels drop rapidly after copulation; progesterone increases 3–4 days later (Schille *et al.*, 1979; Verhage *et al.*, 1976). The persistence of receptivity until the rise in progesterone allows repeated coital stimulation, which increases the likelihood of ovulation (Wildt *et al.*, 1980).

Overlapping waves of follicles develop in female rabbits (Hill and White, 1933) and produce a relatively constant plasma level of estradiol. Does accept the male (buck) at any time, then, because the constant estradiol stimulation results in more-or-less persistent estrus. As in the cat, repeated coital stimulation is most effective in inducing ovulation. Repeated mounting with pelvic thrusting induces ovulation, but penile insertion is not necessary (Staples, 1967). Gonadal hormones and gonadotropin levels increase rapidly after an ovulatory coital stimulation, and ovulation occurs within 10–12 hr (Hilliard, 1973; Hilliard and Eaton, 1971; Mills *et al.*, 1981). The "estrous cycle" of the doe, then, consists only of alternating periods of sexual receptivity and pregnancy or pseudopregnancy.

The panels of Figures 3, 4, 5, and 6 allow comparison of the relationship between peripheral plasma hormone levels and the occurrence of sexual receptivity and ovulation in a variety of species. Only the major ovarian hormones (estradiol and progesterone) and the gonadotropic hormone LH are presented here, as these are considered the most important for female sexual behavior and ovulation across these species. Other hormones may be important in some species, but limitations of space do not permit their inclusion.

It should be noted that, in many species, estrous cycles occur only during a limited portion of the year. Seasonal changes in day length are used by many species to determine when the season of estrous cyclicity will occur (see review by

Elliott and Goldman, 1981). In this way, conception occurs only during the appropriate period of the year, and young are, correspondingly, born at a time when their chances for survival are maximal (e.g., in the spring).

ANALYSIS OF FEMININE SEXUAL BEHAVIOR

It is convenient to divide feminine sexual behavior into three components (Beach, 1976): attractivity, proceptivity, and receptivity. As referred to by Beach, these three categories represent theoretical constructs that are inferred from behavioral observations. This type of conceptualization is a continuation of the trend toward refining the quantitative analysis of sexual behavior that began in the 1930s and 1940s; it has stimulated a great deal of new research and has "rearranged" some of our views concerning the dynamics of feminine sexual behavior.

ATTRACTIVITY

In functional terms, *attractivity* refers to those qualities (behavioral or otherwise) that assist the male in identifying the "female's sex and reproductive status, synchronizing and orienting the male's coital responses, and finally promoting the emission of sperm while the penis is in the vagina" (Beach, 1976, p. 107). Responses of the male are generally used to measure a female's "attractiveness" as a sexual stimulus. Thus, one may time or count the visits that a male makes to a female. In some cases the female may be restrained by a cage or by being tethered to one spot, or she may be free to move about. The frequency of copulation by males with a specific female, given a choice of several females, provides another measure of her "attractiveness."

The specific stimulus features of the female that serve to attract the male are varied. In some cases, the male may be attracted by odor cues; in others, by visual cues or by specific behaviors of the female (proceptivity). Ejaculation depends heavily on the contact stimuli provided by the vagina. Hence, a female's stimulus value "covers the full range of stimulation which she provides, extending from her ability to attract a distant male to those characteristics of her vagina which contribute to the occurrence of the ejaculatory reflex" (Beach, 1976, p. 107).

INFLUENCE OF OVARIAN HORMONES ON ATTRACTIVITY. In the discussion that follows, we have separated primate species from nonprimate species because the principles that hold for one group do not always generalize easily to the other. To a great extent, the female primate is emancipated from rigorous hormonal control of her sexual behavior. Consequently, we feel it helpful to maintain a separate discussion of these species.

Primate Species. An increase in the sexual activity of rhesus monkeys is associated with the follicular phase of the menstrual cycle, with a peak in ejaculation frequency (a measure of attractivity) occurring around midcycle. A progressive decline in ejaculation frequency occurs through the subsequent luteal phase (Goy and Resko, 1972; Michael and Zumpe, 1970; Michael *et al.*, 1972, 1982). This rhythmic variation in sexual activity is abolished by bilateral ovariectomy (Michael *et al.,* 1967, 1982). Treatment of ovariectomized females with estrogen increases male contact, mounting, intromission, and ejaculation (Michael *et al.,* 1982; Wallen and Goy, 1977). The amount of time the male spends near the female is also increased (Johnston and Phoenix, 1976). Withdrawal of estrogen produces a decline in the frequency of both mounting and ejaculation (Baum *et al.,* 1977b). Thus, it would appear that during periods of estrogen dominance (e.g., during the follicular phase of the menstrual cycle), ejaculation is more likely.

Administration of progesterone, on the other hand, decreases attractivity and thus brings about a decrease in ejaculation by the male (Baum *et al.,* 1977b; Michael *et al.,* 1968). Finally, administration of exogenous estrogen followed by progesterone (in quantities that mimic the high estrogen levels seen during the follicular phase, followed by the progesterone dominance during the luteal phase of the menstrual cycle) produces rhythmic variation in the sexual activity of ovariectomized females similar to the variation seen in intact rhesus females (Michael *et al.,* 1982). The relationship of hormones to attractivity and ejaculation frequency is discussed further in the section dealing with the stimulus basis for attractivity.

In free-ranging baboons (*Papio anubis, P. hamadryas, and P. ursinus*), sexual activity increases during the follicular phase of the menstrual cycle and declines during the luteal phase. Associated with the increased sexual interaction is an increase in the size and color of the sexual skin. Some researchers have suggested that this coloration may add to the attractivity of the female (see Beach, 1976, for review). Detumescence of the sexual swelling occurs during the luteal phase.

Although variation in attractivity throughout the menstrual cycle seems well established for some primate species, this is not true for all primates. In the stumptail macaque *(Macaca arctoides)* sexual activity (mounting, intromission, or ejaculation) did not vary with the menstrual cycle in animals that were paired on a daily basis (Slob *et al.,* 1978a,b). Although ovariectomy resulted in an overall decrease in ejaculation frequency, some pairs continued to show multiple ejaculation up to 6 months after ovariectomy.

An increase in sexual activity occurs around the time of ovulation in the great apes (the gorilla and the orangutan). In the gorilla, the frequency of male solicitations (attractivity) increases around the time of ovulation (Nadler *et al.,* 1983). This increase in the attractivity of the female gorilla was associated with an increase in circulating estradiol levels. In the orangutan, the male initiated copulation on each day of testing throughout the cycle when pairs were mated on a daily basis in captivity. This forceful initiation of sexual activity by the male does not indicate any change in the female's attractivity.

When free-ranging, ovariectomized chacma baboons *(P. ursinus)* were given estrogen, there was an increase in sexual swelling in the female and an increase in the mounting of females by the males in the troop. The addition of a progestational hormone reduced the sexual swelling as well as sexual activity. However, there was no quantitative relation between the degree of swelling and the amount of mounting behavior (Saayman, 1972).

Rhythmic variation in the sexual activity of humans has also been reported during the menstrual cycle (Adams *et al.*, 1978; Udry and Morris, 1968). The peak in sexual activity, including the highest incidence of orgasm in the female, occurs around midcycle, when estradiol levels are high.

Nonprimate Species. Two examples, one from experimentation with laboratory rats and one from dogs, illustrate the concept of attractivity in nonprimate species. Additional discussion of the concept can be found in Beach (1976).

When male rats were given the choice of approaching either an estrous or a diestrous female in a restraining cage, the males spent more time near the estrous females (Carr *et al.*, 1966; Stern, 1970). Male rats also preferred ovariectomized females that had been given estrogen and progesterone over ovariectomized females given no gonadal hormone (Carr *et al.*, 1965).

Male beagles spent more time near caged females when the females were in estrus than when they were in anestrus (LeBoeuf, 1967). Attraction seemed greatest when the females were near the middle of their estrous cycle; at this time, "males whined, barked and pawed the cages of an estrous female or attempted to make contact by licking, sniffing or pawing her body" (LeBoeuf, 1967, p. 314).

Both estrogen and progesterone influence the attractivity of the female beagle. When spayed female beagles were treated with estrogen, their attractivity increased, as measured by the frequency of mount with thrust behavior by the male (Beach and Merari, 1968). Several days after estrogen treatment was discontinued, mount frequency began to decline; treatment with progesterone at this time led to an abrupt increase in mounting within 4 hr after progesterone treatment.

STIMULUS FACTORS INVOLVED IN ATTRACTIVITY. As mentioned earlier, the concept of attractivity as summarized by Beach (1976) includes both behavioral and nonbehavioral cues. In this section, we enumerate some of these stimulus factors.

Nonbehavioral Factors. Perhaps the most ubiquitous stimulus associated with attractivity is odor. These "pheromonal" chemical stimuli play a role in the sexual behavior of most mammals that have been studied, among them ungulates, rodents, carnivores, and primates.

Rams preferentially approach ewes that have been treated with progesterone and estrogen compared to ewes not in heat; they are capable of making this distinction between estrous and nonestrous ewes on the basis of olfactory stimuli (Lindsay, 1965). Although anosmic rams mated normally with unrestrained females, the rams' loss of ability to detect odors reduced their copulatory frequency if the female was tethered (Lindsay and Fletcher, 1972). This reduction in mating presumably resulted from the inability of the estrous ewe to seek the male when she was tethered (proceptive behavior) and the inability of the ram to use olfactory cues to detect her estrous state from a distance.

When male rats were given a choice between investigating the odor of an estrous female and investigating the odor of a nonestrous female, they chose the odor of the estrous female (Carr *et al.*, 1965; Stern, 1970). Male rats were attracted to and preferred urine from estrous females over that from nonestrous females (LeMagnen, 1952; Pfaff and Pfaffman, 1969).

Although male hamsters are attracted to vaginal secretions, they do not distinguish between estrous and nonestrous secretions (Landauer, *et al.*, 1978). Although this behavior might seem paradoxical, the observation that female hamsters show the greatest amount of vaginal marking during diestrus has led to the suggestion that odor is being used at this time as a broadcast or advertisement of impending receptivity and fertility (Johnston, 1977). Thus, given the context of hamster social behavior, attraction to female odor at any time during her nonpregnant status would be advantageous for the male.

Although the bull apparently receives both olfactory and gustatory stimulation by licking the vulva of the cow when she is in heat, it is not clear whether olfaction is used by the bull to locate an estrous cow (Craig, 1981).

Vaginal secretions from estrogen-treated female beagles are highly attractive to the male. When cotton balls were applied to the vaginae of hormone-treated females or nonestrous females, the cotton balls from estrogen-treated females elicited more sniffing from male beagles than the cotton balls from anestrous females (Beach and Merari, 1970). Injection of the estrogen-treated bitch with progesterone increased the attractiveness of the cotton pledgets within 4 hr after the female was injected. Cotton balls applied to estrogen-progesterone-treated females elicited intense responses from the males: instead of just sniffing the cotton stimulus, the males licked and chewed the cotton balls.

> This reaction is particularly significant, in view of the reaction of males to females. When a female is in anestrus or especially when she is in proestrus, the male may sniff the vulva repeatedly but vigorous and prolonged licking does not occur. In contrast, males exposed to an estrous female frequently lick the vulva so vigorously that the bitch's hindquarters are practically lifted off the ground. This appears to provide an important source of stimulation for the female as well as the male. (Beach and Merari, 1970, p. 11)

The increased attractiveness of the female following both estrogen and progesterone is also shown by an increase in male mounting following the progesterone injections (Beach and Merari, 1968).

In some nonhuman primates, olfactory stimuli are an important part of the female's attractivity. In rhesus monkeys, estrogen administered intravaginally resulted in an increase in male mounting that was dose-related (Michael and Saayman, 1968), but changes in female presentations (proceptive behavior) were absent. Subcutaneous (s.c.) administration of estrogen led to a greater enhancement of sexual activity: "Ejaculation times, sexual performance indices, and male success ratios were all more consistently improved with the s.c. route than when corresponding doses were administered intravaginally" (Michael and Saayman, 1968, p. 245). Thus, vaginally administered estrogen increased attractivity, whereas proceptivity and receptivity were enhanced more by the s.c. route, which would allow greater exposure of the brain to the estrogen. Vaginal secretions also appear to be important in the chimpanzee; the male normally sniffs and tastes the external labia of the female when she is in estrus (McGinnis, 1973, cited by Beach, 1976).

There is some controversy about the precise nature of the active factor in vaginal secretions and the conditions under which they are attractive to the rhesus. It has been reported that male rhesus bar-press to gain access to and copulate with estrogen-treated females (Michael and Keverne, 1968, 1970) and that extracts of vaginal secretions from estrogen-treated females facilitate copulation by male monkeys when the material is applied to nonestrogenized, ovariectomized females (Keverne and Michael, 1971; Michael and Keverne, 1970; Michael *et al.*, 1971). Other investigators, however, have not always obtained similar results (Goldfoot *et al.*, 1976c). When the immediate copulatory history of the donor and the recipient females was controlled so that ejaculate from prior copulations was not present at the time the vaginal secretion sample was taken, application of vaginal lavages from estrogen-treated females did not facilitate male sexual behavior (Goldfoot *et al.*, 1976c). However, two of three males showed increased sexual behavior with lavage recipients when the vaginal lavage was contaminated with ejaculate that had been in the vagina for 24 hr. The reader is urged to study this series of articles along with the following references before drawing conclusions regarding vaginal odors and attractivity in rhesus monkeys (Goldfoot *et al.*, 1966a; Keverne, 1976a; Michael *et al.*, 1971, 1976).

Progesterone decreases attractivity in the rhesus (Zumpe and Michael, 1968); this reduction may be due to a reduction (or induction) of a vaginal factor (Baum *et al.*, 1977b). *It is not clear whether this factor is an olfactory cue that the male detects or whether some other aspect of the vagina is changed.* The idea that the influence of progesterone is limited to vaginal influence stems from the finding that intravaginal application of progesterone in quantities that do not alter plasma levels of progesterone results in decreased mounting and ejaculation by the male (Baum *et al.*, 1977a,b). However, when the same amount (250 μg) was given systemically, there was no effect on male behavior.

Other cues, besides olfaction, play a role in attractivity. In primates, the coloration of the female sexual skin attracts a good deal of attention by males (Saayman, 1972). Grooming the genitalia after intromission may provide gustatory stimuli for male rodents that convey information on the state of the female's vagina and serve an attractive function.

Behavior Cues in Attractivity. The female mammal engages in numerous behaviors that serve to bring her reproductive status to the attention of the male; these behaviors are termed *proceptive behaviors,* and they are discussed in detail in a subsequent section. However, because these behaviors may also serve as stimuli that attract the male, it is important that they be given attention here under the concept of attractivity as well. Only a few examples are needed to illustrate the point.

When cows come into heat, they begin mounting other cows. This behavior may bring them to the attention of the bull (Craig, 1981). In many primates, the female approaches the male and remains near him as she approaches midcycle (the time of ovulation). The mounting of males by females during the periovulatory period of the cycle has been reported in rhesus monkeys (Carpenter, 1942; Goy and Resko, 1972), in baboons (DeVore, 1965), and in pig-tailed macaques (Bernstein, 1967).

ATTRACTIVITY AND INDIVIDUAL DIFFERENCES. Although two females may both be sexually receptive, their attractivity to a particular male may differ. When male beagles were given the opportunity to visit estrous females tethered in an open field, males showed consistent preferences for particular females (LeBoeuf, 1967). These preferences were reflected in individual differences in the amount of time a male spent near a particular tethered estrous female when given the choice of being near or at a distance from her (see Figure 7). Eddie, for example, chose to spend the majority of his test time with Peg, close to her. He chose to stay away from Blanche, however, a behavior indicating that her attractivity was lower to him. Ken showed the opposite pattern (interacting with Blanche and avoiding Peg in their separate tests), a finding that demonstrates the individual nature of these preferences.

There are numerous reports in primates showing that males have "favorite females" (Phoenix, 1973). When a male rhesus is caged with several females, he often copulates exclusively with only one of the females (Michael *et al.*, 1972), even when all of them are in the same estrous condition. Similar preferences are reported for stumptail macaques as well (Slob *et al.*, 1978a). These preferences often override the hormonal status of the female (Herbert, 1970; Slob *et al.*, 1978a).

PROCEPTIVITY

Proceptive behavior refers to the

> appetitive activities shown by females in response to stimuli received from males. The arbitrary distinction . . . between a female's appetitive and consummatory responses is useful, but its limitations must be noted. In the actual mating sequence, appetitive and consummatory reaction often alternate; and further-more the same response, e.g., assumption of the coital posture, can be appetitive in one circumstance and part of the consummatory complex in another. (Beach, 1976, pp. 114–115)

The female in estrus is most attractive to the male, and at the same time, she is most attracted to the male. In functional terms, the state of estrus serves to direct both male and female partners toward each other. Proceptive behaviors by the female place her in proximity to the male at a time when her attractivity and fertility are highest.

Although some proceptive behaviors appear to enhance the female's attractivity to the male, others do not fit easily into that category. Beach (1976) chose to include as proceptive behavior all appetitive responses that seem to be evoked by stimuli received from the male and that "increase the probability of masculine sexual behavior directed toward the female" (p. 116).

Beach noted five specific types of behavior that can be classified as proceptive:

1. *Affiliative behavior.* The estrous female approaches and remains in proximity to the male.
2. *Solicitational behavior.* This behavior can take the form of coital posturing or specialized gestures.

LYNWOOD G.
CLEMENS AND
DAVID R. WEAVER

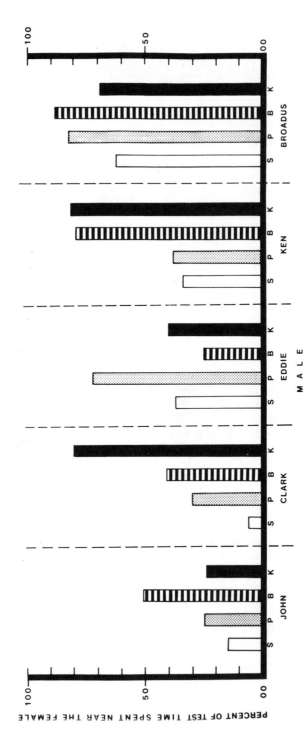

Fig. 7. The percentage of test time spent by each of five male beagles with each of four tethered, estrous females. The four females, Spot(S), Peg (P), Blanche (B), and Kate (K), were tested individually with each male. (After LeBoeuf, 1967.)

3. *Alternating approach and withdrawal.* The female approaches the male and then retreats for some distance. If the male does not follow, she may then return and retreat again. This behavior is plainly stimulating to many males.
4. *Physical contact responses.* In many species the female contacts the male or pushes herself into the male.
5. *Mounting by the female.* During estrus, the incidence of mounting by the female increases in many species (Beach, 1968). This behavior often increases the level of excitement in both the male and the female.

INFLUENCE OF OVARIAN HORMONES ON PROCEPTIVE BEHAVIOR IN PRIMATE SPECIES. As a thorough review of this topic is provided elsewhere (Beach, 1976), only a few of the more salient and recent studies need be discussed here.

In rhesus monkeys, there is an increase in female affiliative behavior at midcycle near the time of ovulation, when estrogen levels are high. When male rhesus were restrained in a small cage, it was found that the time that the female spent close to the restrained male increased as she approached midcycle (Czaja and Bielert, 1975). Other behaviors that are normally associated with sexual arousal ("hand slaps," when the female slaps the floor; "head ducks"; and "head bobs") also increased at midcycle (Czaja and Bielert, 1975).

A similar tendency toward increased affiliative behavior and increased solicitational behavior was seen when male rhesus were tethered and the behavior of the female was measured (Pomerantz and Goy, 1983). During the follicular stage of the cycle, females approached and solicited the males at a higher rate and spent more time sitting near the males than when the females were in the luteal phase of their menstrual cycle.

In a paradigm in which the female was trained to release the male by pressing a lever, it was found that the mean time for release of the male was shortest at midcycle for female rhesus and longer during the late luteal phase of the cycle (Bonsall *et al.*, 1978; Keverne, 1976b; Michael and Bonsall, 1977).

The hormonal basis for this cyclicity in release latency is further supported by the finding that ovariectomy resulted in cessation of lever pressing to gain access to a male in four of five rhesus females (Keverne, 1976b; Michael *et al.*, 1972). Subsequent administration of estrogen restored lever pressing to levels that approached those seen in intact females when release latency is averaged over the whole cycle (Keverne, 1976b; see Figure 8).

These studies, in which restraint is imposed on the male, allow for better investigation of the female's role in initiating sexual interactions because she must either engage in an operant task or in some way take the initiative. In each case, the proceptive behavior of the female increases at a time when estrogen levels have increased (see Eaton, 1973). When the male is not restrained, it is not always possible to see changes in the female's proceptive behavior as a function of her menstrual cycle. As a female's attractivity increases toward midcycle, the male becomes more active sexually, and it is difficult to determine the female's proceptive "poten-

LYNWOOD G.
CLEMENS AND
DAVID R. WEAVER

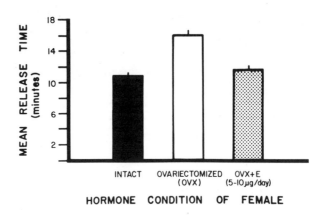

Fig. 8. In female rhesus monkeys trained to release a caged male by pressing a lever, latency to release of the male depended on the female's hormonal condition. (After Keverne, 1976b.)

tial" when the male partner is active (see Baum *et al.*, 1977a). The assessment of receptivity is also complicated by this midcycle increase in attractivity (see below).

In the gorilla and the orangutan, female solicitations are restricted to the periovulatory period (Nadler, 1976, 1977; Nadler *et al.*, 1983). At the present time, there is a need for experiments with the great apes in which the female is better able to govern the mating situation. Under such conditions, one might see more clearly the effect and the frequency of proceptive behaviors. The male orangutan is highly active sexually when tested on a daily basis, and this high activity limits the extent to which one can measure the female's behavior.

INFLUENCE OF OVARIAN HORMONES ON PROCEPTIVE BEHAVIOR IN NONPRIMATE SPECIES. Female rodents often approach the male when in estrus. When hamsters were given the opportunity to approach a member of the opposite sex, both males and females showed approach behavior. When the number of times the male approaches the female plus the number of times the female approaches the male is regarded as 100%, it was found that a female in diestrus approached the male about 50% of the time. When the female reached estrus, however, female approach frequency "rises significantly to over 70%" (Steel, 1980, p. 260). In addition to approaching the male, the estrous female hamster showed a strong inclination to remain near the male. In a 300-sec test, estrous females spent over 225 sec within 6 cm of a caged male when given the choice of being alone, near a diestrous female, or near a male (Beach *et al.*, 1976). During diestrus, females spent less than 75 sec near either the male or the female (see Figure 9).

In the hamster, female attraction to the male appears to be heavily dependent on auditory stimuli from the male. Females that were partially deafened by destruction of the tympanum and the ossicular chain failed to discriminate between caged males and diestrous females (Beach *et al.*, 1976). Male hamsters emit high-frequency calls (ultrasonic vocalizations) that are most likely responsible for the females' ability to distinguish the males. Under different test conditions, it has been shown that both estrous and diestrous females are attracted more to male body odors than to female body odors and that the attraction to male body odors may

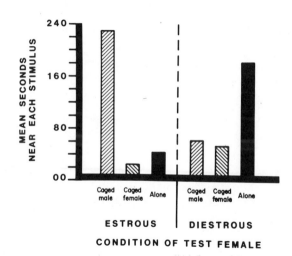

Fig. 9. Variation in proceptivity over the estrous cycle of female hamsters. When given the opportunity to spend time near a caged male or a caged diestrous female or away from both stimulus animals, estrous female hamsters remained near the caged male. Diestrous females, in contrast, chose to spend the majority of test time away from both caged stimulus animals. (Data from Beach *et al.*, 1976.)

be greater when the female is in estrus than when she is in diestrus (Johnston, 1979). Consequently, the female hamster can respond to both auditory and olfactory stimuli from the male.

When female rats were given a choice between approaching a caged male or a caged female in a runway choice situation, the test females in proestrus or estrus chose to approach the male more often than when they were in diestrus (Eliasson and Meyerson, 1975). After priming with estradiol benzoate, similarly, ovariectomized females preferentially spent more time in the vicinity of a sexually active male than with a castrated male rat. This preference was further enhanced after treatment with progesterone (Edwards and Pfeifle, 1983). Similarly, approach frequency toward a tethered, sexually active male was also increased further in females by estradiol plus progesterone treatment than by estradiol alone.

During estrus, the laboratory rat shows a variety of proceptive behaviors, including "hopping" and "darting" movements. These behaviors occur in response to the presence of the male. The darting movements occur as short runs away from the male that terminate abruptly in a crouching stance. In the hopping behavior, the female lands on all four feet and then crouches. Both behaviors appear to have attractive effects on the male (Madlafousek and Hlinak, 1971; Madlafousek *et al.*, 1976).

Another behavior included in the proceptive category is "ear wiggling." Female rats in estrus often vibrate the head, causing the ears to "wiggle." Although this behavior may have some attractive function for the male, it also may produce important stimulation for the female. Rapid oscillation of the head excites the anterior and posterior semicircular canals of the labyrinth. This excitation results in increased stimulation of the lateral vestibulospinal tract, which is considered by some to be important for lordosis (Pfaff, 1980, p. 192).

LYNWOOD G.
CLEMENS AND
DAVID R. WEAVER

Although the female rat copulates with the male following estrogen treatment alone, the display of solicitation (hopping, darting, and ear wiggling) is heavily dependent on estrogen plus progesterone (Edwards and Pfeifle, 1983; Fadem *et al.*, 1979; Hardy and DeBold, 1971b; Hlinak and Madlafousek, 1981; Tennent *et al.*, 1980; Whalen, 1974). When ovariectomized females were given a constant amount of estrogen and graded doses of progesterone, the incidence of solicitations increased with the dose of progesterone (Fadem *et al.*, 1979; Tennent *et al.*, 1980; see Figure 10).

When females were tested with males that were restrained at one end of the cage, the latency for the estrogen-primed female to return to the male following intromission or ejaculation decreased with increasing dosages of progesterone (Fadem *et al.*, 1979; see Figure 11).

The hormonal regulation of proceptivity, although most extensively studied in rodents, has also been demonstrated in other species. The ewe shows soliciting and ram-seeking behavior when in heat; this behavior consists of rubbing up against the ram. Ram-seeking behavior by the ewe is enhanced by treatment with progesterone and estrogen (Lindsay and Fletcher, 1972). The sow, when in heat, mounts the boar and spends more time near the boar than when she is in diestrus (Signoret, 1970; Hafez and Signoret, 1969).

Approach behavior by estrous females may be specific to particular males. Two examples illustrate this point.

When montane voles *(Microtus montanous)* were tested in an elongated area with a male tethered at each end and a neutral area in between, females copulated preferentially with one of the males. When one of the males was castrated, the female mated preferentially with the intact male (Webster *et al.*, 1982). The female's choice was associated with the amount of copulatory stimulation. It was found that preferred males were more likely to achieve intromission when they mounted, and they achieved more thrusts per intromission than nonpreferred males (Webster *et al.*, 1982). Because the montane vole is an induced ovulator,

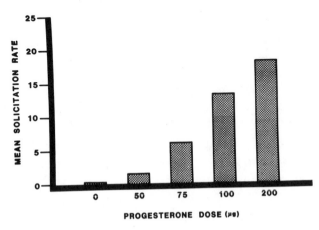

Fig. 10. Solicitation behavior increased with increasing progesterone dose in estrogen-primed female rats. (After Fadem *et al.*, 1979.)

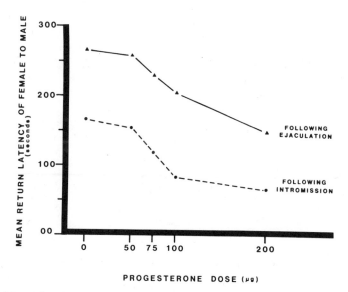

PROGESTERONE DOSE (µg)

Fig. 11. The latency for estrogen-progesterone-primed female rats to return to a tethered male follow-ing intromission or ejaculation by the male decreased with increasing progesterone dose. (After Fadem *et al.*, 1979.)

achieving appreciable amounts of copulatory stimulation is a requisite for fertiliza-tion and implantation. (In other induced ovulators such as cats and rabbits, the amount of copulatory stimulation is proportional to the probability of ovulation; Staples, 1967; Wildt *et al.*, 1980).

Female beagles show an increase in the time spent near the male when the female is in estrus (LeBoeuf, 1967). Estrous females given the choice of approach-ing a male or a female approached the male more and spent more time near the male than the female. In addition, female beagles showed preferences for specific males. Given the choice of approaching any of several tethered males, the female often spent more time with a specific male rather than visiting all three equally (Beach, 1970; Beach and LeBoeuf, 1967; see Figure 12). The particular stimulus characteristics that influenced the female's choice are unknown.

In addition to approaching the male, estrous bitches also "present" to the male by turning their hindquarters toward him (Beach and Merari, 1970). A female beagle in heat also mounts the male and shows pelvic thrusting. These behaviors of approach, present, and mount do not occur in ovariectomized females but can be induced with estrogen treatments.

RECEPTIVITY

INFLUENCE OF OVARIAN HORMONES ON SEXUAL RECEPTIVITY. In nonprimate species, ovarian hormones play an important role in the induction of sexual recep-tivity. For some species (particularly those referred to as *induced ovulators*), sexual behavior is largely dependent on estrogen, whereas in some spontaneously ovulat-ing species both estrogen and progesterone are important. In many primates, how-

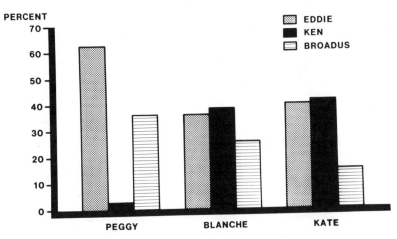

Fig. 12. Individual differences in the percentage of total visiting time spent by each estrous beagle bitch (Peggy, Blanche, and Kate) with each of three different males. These differences reflect discrimination in the exhibition of proceptivity by the females. (Figure redrawn from Beach, 1970, with permission.)

ever, sexual behavior may continue throughout the menstrual cycle and may not be diminished by ovariectomy.

Receptivity in Primate Species. Although it is often reported that sexual activity increases at midcycle for many species of primates (e.g., Carpenter, 1942), this may not reflect a change in the female's sexual receptivity. Laboratory studies of several macaque species have shown that receptivity (as measured by the female's acceptance ratio) does not vary greatly with changes in ovarian steroids.

Measurement of the acceptance ratios for Japanese monkeys (*Macaca fuscata;* Eaton, 1973) and stumptail macaques (*M. arctoides;* Slob *et al.,* 1978a) failed to show significant changes in receptivity during the menstrual cycle. Ovariectomized stumptail females continued to copulate, and no change in acceptance ratios was noted following ovariectomy (Slob *et al.,* 1978b). Furthermore, the treatment of ovariectomized stumptail females with exogenous estrogen failed to alter acceptance ratios. Similar treatment of rhesus females *(M. mulatta)* also failed to alter acceptance ratios (Baum *et al.,* 1977a; Trimble and Herbert, 1968; Wallen and Goy, 1977). These studies, in demonstrating the independence of the female's acceptance ratio and her hormonal state, indicate that the variations in sexual activity that occur during the menstrual cycle may be a function more of the female's attractivity than of her receptivity.

There is some evidence to suggest that androgens may influence feminine sexual activity in rhesus females. In one study, removal of the major source of androgens in the female, the adrenal glands, increased female refusals of the male and decreased acceptance ratios (Everitt *et al.,* 1972). Subsequent treatment with androstenedione increased acceptance ratios in one of three adrenalectomized females. Testosterone implanted directly into the anterior hypothalamus–preoptic area also reversed the effects of adrenalectomy (Everitt and Herbert, 1975). In some studies, however, testosterone treatment failed to alter acceptance ratios (Wallen and Goy, 1977), and high doses even reduced acceptance ratios in the rhesus (Trimble and Herbert, 1968). Although some have noted that androgens may be important in the regulation of feminine sexual behavior, it is important to remember that both androstenedione and testosterone can be aromatized to estradiol. This fact, along with the finding that the nonaromatizable androgen dihydrotestosterone reduced acceptance ratios (Wallen and Goy, 1977), suggests that caution should be used in the interpretation of androgen influences on sexual receptivity. For example, it has been suggested that receptivity may be enhanced by the appropriate balance between 5-α-reduced androgens (such as dihydrotestosterone) and the aromatized products of testosterone. Hence, high doses of testosterone might lead to diminished receptivity because of the subsequent production of higher amounts of reduced androgens (Wallen and Goy, 1977).

Thus, in attempting to account for the variations in sexual activity during the menstrual cycle, it is difficult to see these changes as a change in the female's receptivity *per se*. It seems that most of the hormone-induced alteration in sexual behavior during the primate menstrual cycle is dependent on changes in the female's attractivity (Baum *et al.*, 1977a).

The same conclusion also pertains to species other than the macaques. In talapoin monkeys *(Miopithecus talapoin)*, sexual activity has also been reported to be greatest at midcycle (Scruton and Herbert, 1970), but this activity was largely generated by the male's interest in the female (i.e., female attractivity). Following ovariectomy, there was very little sexual interaction (Dixson and Herbert, 1977). Administration of estrogen resulted in a increase in sexual activity, with the female showing increased proceptivity. In the prosimian species *Galago crassicaudatus*, sexual behavior is restricted entirely to the period of vaginal estrus (Eaton *et al.*, 1973).

Receptivity of Nonprimate Species. Experimental analysis of the relation of estrogen and progesterone to sexual behavior in normal, cycling female rats indicates that estrogen acts as a priming agent for the subsequent influence of progesterone. The temporal relationship of these hormones to sexual receptivity is shown in Figure 13. During the 48-hr period preceding the onset of sexual receptivity, estrogen levels gradually rise and reach a peak late in the afternoon of proestrus. At about this time, progesterone levels begin to rise rapidly, reaching a peak during the early evening of proestrus. The onset of sexual receptivity is associated with this rise in progesterone levels.

If the ovaries are removed 3 hr before lights-out on the day of proestrus, thereby eliminating the surge in progesterone that precedes the time of sexual receptivity, the female does not become receptive. If ovariectomy is followed by exposure to exogenous progesterone, there is little or no decrease in receptivity as measured by the proportion of mounts to which the female responds with lordosis, the postural response characteristic of receptive female rodents (Powers, 1970). A similar study has demonstrated the necessity of the proestrous ovarian-progesterone surge for the induction of receptivity in hamsters (Ciaccio and Lisk, 1971).

The priming effects of estrogen on sexual receptivity in normal female rats can also be detected by means of the paradigm of ovarian removal at specific times in the cycle as described above. When the ovaries were removed 17 hr before the lights went off on the day of proestrus, females achieved very low lordosis scores

LYNWOOD G.
CLEMENS AND
DAVID R. WEAVER

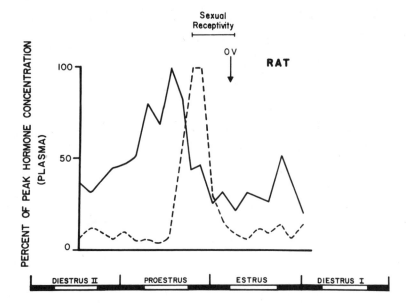

Fig. 13. Temporal relationship of estradiol (————) and progesterone (- - - -) levels in peripheral plasma to sexual receptivity in female rats. Hormone levels are expressed relative to the peak values attained during the cycle. OV, ovulation. (After Butcher *et al.*, 1974; Sodersten and Eneroth, 1981.)

even though progesterone was administered at the time of the normal rise in progesterone. Removal of the ovary 11 hr before lights-out (after the release of estradiol from the ovary) had little effect on lordosis as long as progesterone was administered at the appropriate time (Moreines and Powers, 1977).

Although sexual behavior can be induced in ovariectomized rats with estrogen alone (Davidson *et al.*, 1968; Edwards *et al.*, 1968; Pfaff, 1970), subsequent injections of progesterone result in higher levels of receptivity than would be seen with estrogen alone (Beach, 1942; Boling and Blandau, 1939). The level of sexual activity induced is related to both the dose of estrogen and the dose of progesterone (Whalen, 1974). Thus, for a given level of estrogen, the probability of lordosis is increased with increased amounts of progesterone. Similarly, with a constant amount of progesterone, the probability of lordosis is increased with increasing doses of estradiol (Whalen, 1974). The progesterone facilitation of estrogen effects has also been shown in the guinea pig (Dempsey *et al.*, 1936); the mouse (Ring, 1944); the hamster (Frank and Fraps, 1945); and the dog (Beach and Merari, 1968).

Hormonal Influences on Sexual Receptivity in Induced Ovulators. Females of some species do not ovulate unless or until they copulate with a male: Cats, rabbits, brown lemmings, prairie voles, and ferrets are some of the reflex ovulators that have been studied for their behavior. In some spontaneously ovulating species such as rats and guinea pigs, progesterone appears just prior to the time of ovulation and plays a major role in bringing the female into heat. In other spontaneously ovulating species, estrogen alone is capable of inducing receptivity without a progesterone surge (e.g., the horse and the dog; see Figure 3). The estrogen responsible for receptivity is produced by mature ovarian follicles that soon rupture as ovulation occurs. In this way, sexual receptivity coincides with the time of maximal fertility. In the reflexly ovulating species, the stimuli derived from mating synchronize ovulation with the arrival of sperm. It is not surprising, then, to learn that progesterone does not play a major role in the facilitation of sexual behavior in induced ovulators. Cats given daily injections of estrogen become receptive in about 3 days (Michael and Scott, 1964). In both the brown and the collared lemming (*Lemmus trimucronatus* and *Decrostonyx groenlandicus*), sexual receptivity was not enhanced by progesterone (Hasler and Banks, 1976; Huck *et al.*, 1979). Exposure to progesterone for more than 24 hr inhibited receptivity in the brown lemming (Huck *et al.*, 1979). In the prairie vole *(Microtus ochrogaster),* estrogen induced receptivity within 48 hr; treatment with progesterone inhibited receptivity (Carter, 1982). Ferrets *(Muxtella furol)* became receptive following treatment with estrogen alone (Baum and Schretlen, 1978), as did rabbits (McDonald *et al.*, 1970).

The comparison of females of reflexly ovulating species with those of spontaneously ovulating species reveals that, in spontaneous ovulators, the occurrence of sexual receptivity is dependent on either the burst of progesterone or the rapid increase in estrogen that precedes ovulation, or both, whereas in induced ovulators, synchronization of fertility and sexual activity is achieved by making ovulation dependent on coital stimulation.

HORMONAL MECHANISMS INVOLVED IN SEXUAL RECEPTIVITY

Latency to Estrogen Facilitation of Receptivity. Following the injection of estrogen, there is a latent period of 16–24 hr in the rat until progesterone treatment is capable of inducing lordosis. In one study, ovariectomized rats were given a single intravenous (i.v.) injection of estradiol-17β (100 μg) and tested for lordosis at varying times afterward (Green *et al.*, 1970). When progesterone was present prior to the estrogen treatment, females began showing receptivity 16 hr after estrogen infusion. From Hour 16 on, there was a linear increase in the probability of lordosis until a maximal response was reached at 24 hr. If progesterone was not present prior to the i.v. estradiol treatment, an increase in lordosis was not apparent until 22 hr after the estradiol injection. The latency from estrogen treatment to lordosis does not appear to be related to estrogen dose in the rat (Quadagano *et al.*, 1972); the cat (Peretz, 1968); the rabbit (McDonald *et al.*, 1970); the guinea pig (Collins *et al.*, 1938); or the pig (Signoret, 1967).

Duration of Receptivity. The length of time a female rat remains receptive to the male can be related to the amount of estrogen exposure. When female rats were ovariectomized 6 hr before the luteinizing hormone peak and were given progesterone, the duration of the behavioral period of receptivity was shortened, even though ovariectomy at the time that the female would normally be in maximal estrus has no effect on the level of lordosis response (Moreines and Powers, 1977). Other studies have shown that the amount of estrogen administered is positively correlated with the duration of receptivity in the rat (Quadagno *et al.*, 1972); the cat (Peretz, 1968); the rabbit (McDonald *et al.*, 1970); the guinea pig (Joslyn and Feder, 1971); and the pig (Signoret, 1967).

The duration of receptivity is also affected by copulatory activity. In numerous species, copulation may shorten the period of time that the female continues to copulate; this is true for chinchillas, albino rats, golden hamsters, rice rats, albino mice, guinea pigs, ewes, and cats (Bignami and Beach, 1968; Blandau *et al.*, 1941; Carter and Schien, 1971; Dewsbury, 1970; Edwards, 1970; Goldfoot and Goy, 1970; Hardy and DeBold, 1971a, 1972; Parson and Hunter, 1967; Whalen, 1963). In the golden hamster, the unmated female may remain in heat for up to 16 hr (Kent, 1968). However, under conditions of continuous mating, the golden hamster becomes unreceptive within about 3 hr (Carter and Schien, 1971). This reduction in female receptivity appears to be related to the stimulation that the female receives both from the mounting by the male and from the vaginal and cervical stimulation resulting from intromission (Carter, 1973). When males were allowed to mount the female without achieving intromission, the receptivity of the female was decreased slightly. However, if mechanical stimulation was also applied to the vaginae of these females, a precipitous drop in receptivity was noted (Carter, 1973). Thus, duration of receptivity appears to be related both to hormonal state and to the stimulation that the female receives during copulation. Copulatory activity results in stimulation that terminates sexual receptivity.

The duration of the lordosis response itself is directly related to the amount of estrogen administered (Hardy and DeBold, 1971b). With larger amounts of estrogen, the female holds the lordosis response longer. However, the duration of the lordosis in hormone-treated females never did reach that shown by females in natural estrus (Hardy and DeBold, 1971b).

Duration of Estrogen Exposure Required to Induce Receptivity. There is disagreement regarding the duration of estrogen exposure necessary to induce sexual behavior in animals primed with estrogen followed by progesterone. At one extreme is the view that estrogen must be present throughout the entire period from the first exposure to estrogen until the injection of progesterone. Alternatively, estrogen may be seen as having a "triggering" effect: The initial estrogen exposure sets off a chain of events that brings about receptivity. In addition to the "maintenance" and "triggering" mechanisms of estrogen action, an intermediate mechanism has been proposed: Following initial exposure, estrogen may need to be present only during critical periods prior to progesterone exposure in order to fully facilitate sexual receptivity. The attempts to resolve this conflict suffer from methodological problems (see below), however, so that their interpretation is difficult.

Early studies supported a triggering action of estrogen in the induction of receptivity. Brief (30-min) exposure to estrogen via temporary subcutaneous implantation of an estradiol-filled silastic capsule induced receptivity in female guinea pigs when followed by progesterone injection (Bullock, 1970).

It has also been reported that exposure of the ovariectomized female rat to a 10-mm silastic capsule bearing estradiol-17β for 30 min induced receptivity when progesterone was administered 72 hr later (Johnston and Davidson, 1979). When

5-mm silastic capsules (half the size of those used by Johnston and Davidson) were used, however, it was found that exposure of the female to the capsule for 20 hr or more was required for maximum receptivity when progesterone was given 48 hr after the implants and the females were tested for lordosis 6 hr later (Sodersten *et al.,* 1981).

Near-maximal levels of lordosis could also be achieved by dividing the estrogen exposure into two pulses of estradiol. Implantation and removal of the silastic implants during the first and last 4 hr of the 20-hr period reproduced the effect of 20 hr of continuous exposure. "However, if the first estradiol pulse was advanced 4 hours . . . or if the second estradiol pulse was delayed 4 hours, . . . the behavioral effect was significantly reduced." (Sodersten *et al.,* 1981, p. 58). Somewhat similar effects have also been reported by others (see Parsons *et al.,* 1982). Based on these studies using pulse administration, it has been suggested that estradiol must be present at critical times between the onset of estrogen stimulation and its functional effect (Clark and Roy, 1983; Harris and Gorski, 1978; Parsons *et al.,* 1982; Sodersten *et al.,* 1981).

Although the particular temporal relations differ somewhat from those in the rat, parallel findings have been reported for the guinea pig (see Feder *et al.,* 1979, for review).

The conclusion that estrogen need not be present during the entire time from initial treatment to the induction of receptivity is based on the notion that estrogen does not remain in the target tissues for more than a few hours beyond the time that the silastic implant is removed. Although most of the estrogen dissipates rapidly from the brain following a single pulse, detectable amounts of estradiol remain in the brain for up to 24 hr. When a portion of the remaining estradiol was displaced by treatment with an antiestrogen at 12 or 18 hr after estrogen treatment, there was a significant decrease in lordosis (Blaustein *et al.,* 1979). It has been suggested that the very small amounts of estradiol that remain in the brain between the estradiol pulses may indeed have a behavioral effect. Simply removing the silastic implant after several hours does not ensure that all estrogen has been removed. At this time, the concept that estradiol achieves its effect only at specific "critical" times must be viewed with particular caution. The problem is further complicated by dose considerations (see Clark and Roy, 1983).

Latency for Progesterone Facilitation of Lordosis. Following estrogen priming, treatment with progesterone subcutaneously results in a facilitation of lordosis within the next 6 hr (Gorzalka and Whalen, 1977). When progesterone was administered intravenously, the onset of lordosis occurred within 1 hr (Kubli-Garfias and Whalen, 1977; Lisk, 1960; Meyerson, 1972). Some authors have reported increases in lordosis within the first hour after crystalline progesterone is implanted directly in the brain (Luttge and Hughes, 1976; Ross *et al.,* 1971; Yanase and Gorski, 1976), whereas others have not seen a significant increase until 4–5 hr after intracerebral implantation (Powers, 1972; Rubin and Barfield, 1983). At this time, it is not clear why there is a difference between these reports in the latency for intracerebral progesterone effects, but it has been suggested that, with lower amounts of estrogen priming, the latency for progesterone facilitation may be longer (Rubin and Barfield, 1983).

In the guinea pig, the latency to heat following implantation of crystalline progesterone in the ventromedial hypothalamus was 1.4 hr, compared to 3.6 hr following systemically administered progesterone (Morin and Feder, 1974).

Inhibitory Influences of Progesterone. Progesterone appears to inhibit response to a subsequent dose of progesterone in rodents, a phenomenon called *sequential inhibition* (see Morin, 1977, for review). Following a period of estrogen priming, the administration of progesterone facilitates lordosis, but the administration of progesterone again 24 hr later does not facilitate lordosis in rats, hamsters, or guinea pigs (Blaustein and Wade, 1977b; DeBold *et al.*, 1976, 1978b; Zucker, 1966). However, if the first of the two progesterone treatments is omitted, the injection of progesterone 24 hr later facilitates lordosis. Thus, the loss of response to a second dose of progesterone is not due to dissipation of the estrogen priming (see Table 2). There is considerable evidence to indicate that the loss of response to the second dose of progesterone is associated with a reduction in cytoplasmic progestion receptors in specific areas of the brain. Treatment with estrogen produces an increase in cytoplasmic progestin receptors in the hypothalamus and the preoptic area of the brain (Blaustein and Feder, 1979b; MacLusky and McEwen, 1978; Moguilewsky and Raynaud, 1979a,b). Subsequent treatment with progesterone causes a decrease in these cytoplasmic receptors and a transient increase in nuclear receptors (Blaustein, 1982a,b; Blaustein and Feder, 1980) as the hormone–receptor complexes move into the nucleus. It is thought that translocation of these complexes into the nucleus is a necessary step for progesterone to influence the cells via interaction with the genome (O'Malley and Means, 1974). After 24 hr, cytoplasmic receptor levels remain depressed, and treatment with proges-

TABLE 2. INHIBITION OF RECEPTIVITY BY PROGESTERONE[a]

	Treatment at 0 hr	Treatment at 40 hr	Level of receptivity at 44 hr	Treatment at 64 hr	Level of receptivity in test at 64 hr or 68 hr
1	EB (2 μg)	Oil	Low	—	Low
2	EB (2 μg)	P (0.5 mg)	High	—	Low
3	EB (2 μg)	Oil	Low	P (0.5 mg)	High
4	EB (2 μg)	P (0.5 mg)	High	P (0.5 mg)	Low
5	EB (2 μg)	P (0.5 mg) + EB (10 μg)	High	P (0.5 mg)	High
6	EB (10 μg)	Oil	Low	P (0.5 mg)	High
7	EB (10 μg)	P (0.05 mg)	—	P (0.5 mg)	Low
8	EB (10 μg)	P (0.05 mg)	—	P (20 mg)	High

[a]Priming with estradiol benzoate (EB, 2 μg) alone is insufficient to induce receptivity in ovariectomized female guinea pigs (Line 1), but the injection of progesterone (P, 0.5 mg) at either 40 or 64 hr after EB brings about high levels of receptivity within 4 hr (Lines 2 and 3). If progesterone is injected at both 40 and 64 hr, however, the animals are behaviorally unresponsive to the second progesterone injection, demonstrating sequential inhibition (Line 4). Sequential inhibition can be overcome by administration of additional EB at the time of the first progesterone injection (Line 5). Sequential inhibition also occurs with a higher initial EB dose (10 μg) and a lower first P dose (0.05 mg) (Lines 6 and 7). A high (20 mg) progesterone dose is able to overcome this inhibition (Line 8). (Data from Blaustein, 1982a,b.)

terone does not facilitate lordosis, nor does the progesterone cause an increase in nuclear receptors (Blaustein and Feder, 1979a, 1980). Two treatments appear capable of overcoming the loss of responsiveness to a second progesterone treatment: (1) simultaneous treatment with estrogen (see Blaustein, 1982a, for references) or (2) treatment with a very high dose of progesterone (Blaustein, 1982b; Hansen and Sodersten, 1979). The estrogen treatment appears to increase cytoplasmic progestin receptors and thereby to increase sensitivity to the progesterone (Blaustein, 1982a), and the high dose of progesterone appears to produce a significant translocation of progestin receptors to the nucleus (Blaustein, 1982b), even though cytoplasmic receptors are low. (See also "Note Added in Proof" on p. 227.)

In addition to inhibiting sexual receptivity when administered sequentially after estrogen treatment as just discussed, progesterone also inhibits receptivity when administered at the same time as the estrogen-priming injections. This concurrent inhibition of estrogen priming by progesterone has been observed in rats (Blaustein and Wade, 1977a, 1978; Edwards *et al.,* 1968); in guinea pigs (Wallen *et al.,* 1975; Zucker, 1966); in hamsters (DeBold *et al.,* 1976, 1978a,b; Zucker, 1966); and in sows and ewes (Signoret, 1971). When ovariectomized female rats were injected with estradiol benzoate plus progesterone at 0 hr and then with progesterone alone at 42 hr, for example, they were considerably less receptive when tested at approximately 48 hr than females that received estradiol benzoate alone at 0 hr and progesterone at 42 hr (Blaustein and Wade, 1977a).

As with sequential inhibition, concurrent inhibition by progesterone can be overcome in the rat by increasing the estrogen-priming dose or decreasing the inhibiting progesterone dose (Blaustein and Wade, 1977a). Progesterone administered concurrently with estrogen apparently blocks the priming effect of estrogen, but it does not do so by interfering with the uptake and retention of estrogen in brain cell nuclei (Blaustein and Wade, 1977a; Marrone and Feder, 1977) as do antiestrogens (Roy and Wade, 1977). Estrogen is presumably able to interact with the genome and to induce the formation of cytoplasmic progesterone receptors, then, but the concurrently administered progesterone may translocate these receptors as they are formed. As a result, the cytoplasmic progestin receptor pool may be limited (as it is after the first progesterone injection in a sequential inhibition regimen), and the final progesterone dose cannot result in the accumulation of the critical level of nuclear progestin receptors required to activate lordosis. It should be apparent from this explanation (the progestin receptor hypothesis) that concurrent and sequential inhibition are thought to be the same mechanism observed in two different experimental approaches (Blaustein and Wade, 1977b).

A note of caution should be added here. The mere arrival of progestin receptors in the nucleus of specific brain cells does not thereby ensure that sexual receptivity will result in the estrogen-primed animal. For example, although increased nuclear progestin receptors in the hypothalamic preoptic region can be demonstrated in estrogen-primed male guinea pigs, high levels of sexual receptivity are not seen (Blaustein *et al.,* 1980). Thus, although the association of progestins with nuclear chromatin material may be necessary for some actions of progesterone, such an association is not sufficient. The association of a steroid with the chromatic material is only one in a sequence of many steps that eventually result in behavioral change (see next section). Consequently, it should come as no surprise that direct correlations between receptors and behavior do not occur under all circumstances.

Protein Synthesis and the Induction of Receptivity by Ovarian Hormones. Steroid hormones are believed to bind to cytoplasmic receptors in target cells, and then the hormone–receptor complex translocates into the cell nucleus, where it regulates the expression of genes. Because the DNA in a gene is transcribed into an RNA message and this messenger RNA then serves as the blueprint in the construction of a protein, it becomes apparent that steroid hormones act to control which proteins the cell produces. The formation of progestin receptors in certain brain regions after estradiol treatment (Blaustein and Feder, 1979b; MacLusky and McEwen, 1978; Moguilewsky and Raynaud, 1979a) is one example of a steroid hormone inducing the production of a specific protein. Other proteins whose levels in brain tissue are influenced by estrogen treatment include the enzymes and receptors for several neurotransmitters, among them acetylcholine and the monoamines (Dohanich *et al.*, 1982; Luine and Rhodes, 1983; Luine *et al.*, 1975; Rainbow *et al.*, 1980). Thus, the influence of gonadal steroids on nerve cell function that leads to receptivity may require protein synthesis. This hypothesis is supported by the finding that drugs that interfere with DNA-dependent RNA synthesis (actinomycin-D) or protein synthesis (cycloheximide, anisomycin) impair the expression of sexual behavior induced by estrogen or estrogen–progesterone priming (Meinkoth *et al.*, 1979; Meisel and Pfaff, 1983; Quadagno *et al.*, 1971; Rainbow *et al.*, 1982). In most of these studies, direct application of the drug to the ventromedial hypothalamus (VMH) was used. The VMH has been identified as a primary site of action for gonadal steroid action in the induction of receptivity (see Rubin and Barfield, 1983, for references).

CONCLUSIONS

Feminine sexual behavior in the mammal is comprised of a complex set of processes that serve to enhance the probability of copulation at or around the time of ovulation. At the physiological level, one can see several modes by which the synchronization of ovulation and sexual behavior occurs. In some species, the act of copulation itself produces stimulation that causes ovulation, thereby ensuring the presence of the ovum at the time that the sperm arrive in the female's reproductive tract. In other species, the synchronization of ovum and sperm is achieved by the ovarian hormones produced during the development of the ovum and ovulation.

Although in most species an increase in copulatory activity occurs around the time of ovulation, the specific mechanisms that produce this increase are not identical from one species to the next, and for some species, the mechanisms remain unclear.

The conceptual division of female sexual behavior into three components (attractivity, proceptivity, and receptivity) has been useful in analyzing the means by which gonadal hormones influence sexual behavior. Within this context, the increase in copulatory activity seen around the time of ovulation in most species can be attributed, in varying degrees, to hormonal influences on attractivity and proceptivity as well as on receptivity. Experimental approaches designed to separate these conceptual processes have provided new insights into female sexual behavior. The tripartite analysis has also been a useful approach in species in which it has been difficult to demonstrate reliable changes in female behavior as a result of fluctuations in ovarian hormones, as in several primate species. This problem may reflect the possibility that, in some primate species, the female is not as rigidly influenced by ovarian hormones as females in other species. In addition, the hormones themselves have nonbehavioral influences on attractivity that affect the male and thereby increase his sexual activity. As the male becomes sexually active, it becomes difficult to determine whether the behavior shown by the female is a response to hormonal influences within the female or whether she is responding to the activity of the male.

The mechanisms by which hormones achieve their influence on copulatory behavior in the female involve actions of the steroid hormones, both on the central nervous system and on peripheral structures. For a hormone to have its influence, it is often transported to the nucleus of the target cell, where it then acts on the genetic material. Recent investigations are beginning to provide a clearer understanding of how steroid hormones interact with the cell nuclear material. These studies often, but not always, reveal significant correlations between alterations in behavior and the presence of a particular steroid hormone in cell nuclei.

Studies that relate sexual behavior to the mechanisms of steroid hormone action have demonstrated the importance of the synthesis of specific proteins to the induction of sexual receptivity. As the techniques of molecular biology, neurotransmitter psychopharmacology, and neuroanatomy are refined and applied to the study of the relationship of hormones and sexual behavior, our understanding of the molecular mechanisms involved in the induction of female sexual behavior will continue to grow.

Acknowledgments

We would like to thank Drs. Gail Richmond and Tony Nunez for helpful comments on the manuscript. We are also indebted to Mr. David Brigham for preparing the figures and to Ms. Ruth Ann Reynolds for typing the manuscript.

REFERENCES

Adams, D. B., Gold A. R., and Burt, A. D. Rise in female-initiated sexual activity at ovulation and its suppression by oral contraceptives. *New England Journal of Medicine,* 1978, *299,* 1145–1150.

Anderson, L. L. Pigs. In E. S. E. Hafez (Ed.), *Reproduction in Farm Animals*. Philadelphia: Lea and Febiger, 1980.

Baird, D. T., and Scaramuzzi, R. J. Changes in the secretion of ovarian steroids and pituitary luteinizing hormone in the periovulatory period in the ewe: The effect of progesterone. *Journal of Endocrinology*, 1976, *70*, 237–245.

Baird, D. T., Galraith, A., Fraser, I. S., and Neusman, J. E. The concentration of oestrone and oestradiol-17β in spermatic venous blood in man. *Journal of Endocrinology*, 1973, *57*, 285–288.

Ball, J. Sex behavior of the rat after removal of the uterus and vagina. *Journal of Comparative Psychology*, 1934, *18*, 419–422.

Ball, P., Naish, S. J., and Naftolin, F. Catecholoestrogens and the induction of sexual receptivity in the ovariectomized rat. *Acta Endocrinologica*, 1980, *Suppl. 234*, 103–104.

Bard, P. Effects of denervation of the genitalia on the oestral behavior of cats. *American Journal of Physiology*, 1935, 113, 5–6.

Bard, P. Central nervous mechanisms for emotional behaviour patterns in animals. *Research Publications of the Association for Research in Nervous and Mental Diseases*, 1939, *19*, 190–218.

Bast, J. D., and Greenwald, G. S. Serum profiles of follicle-stimulating hormone, luteinizing hormone and prolactin during the estrous cycle of the hamster. *Endocrinology*, 1974, *94*, 1295–1299.

Baum, M. J., and Schretlen, P. J. Oestrogenic induction of sexual behavior in ovariectomized ferrets housed under short or long photoperiods. *Journal of Endocrinology*, 1978, *78*, 295–296.

Baum, M. J., Everitt, B. J., Herbert, J., and Keverne, E. B. Hormonal basis of proceptivity and receptivity in female primates. *Archives of Sexual Behavior*, 1977a, *6*, 173–192.

Baum, M. J., Keverne, E. B., Everitt, B. J., Herbert, J. and De Greef, W. J. Effects of progesterone and estradiol on sexual attractivity of female rhesus monkeys. *Physiology and Behavior*, 1977b, *18*, 659–670.

Bayliss, W. M., and Starling, E. H. The mechanism of pancreatic secretion. *Journal of Physiology*, 1902, *28*, 325–353.

Beach, F. A. Importance of progesterone to induction of sexual receptivity in spayed female rats. *Proceeding of the Society for Experimental Biology and Medicine*, 1942, *51*, 369–371.

Beach, F. A. Factors involved in the control of mounting behavior by female mammals. In M. Diamond (Ed.), *Perspectives in Reproduction and Sexual Behavior*. Bloomington: Indiana University Press, 1968.

Beach, F. A. Hormonal effects on socio-sexual behavior in dogs. In *Colloquium der Gesellschaft für Biologische Chemie*. Berlin: Springer-Verlag, 1970.

Beach, F. A. Sexual attractivity, proceptivity, and receptivity in female mammals. *Hormones and Behavior*, 1976, *7*, 105–138.

Beach, F. A. Historical origins of modern research on hormones and behavior. *Hormones and Behavior*, 1981, *15*, 325–376.

Beach, F. A., and LeBoeuf, B. J. Coital behaviour in dogs: I. Preferential mating in the bitch. *Animal Behavior*, 1967, *15*, 546–558.

Beach, F. A., and Merari, A. Coital behavior in dogs: IV. Effects of progesterone on mating and other forms of social behavior in the bitch. *Proceedings of the National Academy of Sciences (Washington)*, 1968, *61*, 442–446.

Beach, F. A., and Merari, A. Coital behavior in dogs: V. Effects of estrogen and progesterone on mating and other forms of social behavior in the bitch. *Journal of Comparative and Physiological Psychology*, 1970, *70*, 1–22.

Beach, F. A., Stern, B., Carmichael, M., and Ransom, E. Comparisons of sexual proceptivity and receptivity in female hamsters. *Behavioral Biology*, 1976, *18*, 473–487.

Bernstein, I. S. A field study of the pigtail monkey *(Macaca nemistrina)*. *Primates*, 1967, *8*, 217–228.

Beyer, C., and Komisaruk, B. Effects of diverse androgens on estrous behavior, lordosis reflex and genital tract morphology in the rat. *Hormones and Behavior*, 1971, *2*, 217–225.

Beyer, C., and Vidal, N. Inhibitory action of MER-25 on androgen-induced oestrus behavior in the ovariectomized rabbit. *Journal of Endocrinology*, 1971, *51*, 401–402.

Beyer, C., Morali, G., and Vargas, R. Effect of diverse estrogens on estrous behavior and genital tract development in ovariectomized rats. *Hormones and Behavior*, 1971, *2*, 273–277.

Bignami, G., and Beach, F. A. Mating behaviour in the chinchilla. *Animal Behavior*, 1968, *16*, 45–53.

Blandau, R. J., Boling, J. L., and Young, W. C. The length of heat in the albino rat as determined by the copulatory response. *Anatomical Record*, 1941, *79*, 453–463.

Blatchley, F. R., Donovan, B. T., and Ter Haar, M. B. Plasma progesterone and gonadotrophin levels during the estrous cycle of the guinea pig. *Biology of Reproduction*, 1976, *15*, 29–38.

Blaustein, J. D. Alteration of sensitivity to progesterone facilitation of lordosis in guinea pigs by modulation of hypothalamic progestin receptors. *Brain Research*, 1982a, *243*, 287–300.

Blaustein, J. D. Progesterone in high doses may overcome progesterone's desensitization effect on lordosis by translocation of hypothalamic progestin receptors. *Hormones and Behavior*, 1982b, *16*, 175–190.

Blaustein, J. D., and Feder, H. H. Cytoplasmic progestin binding in guinea pig brain: Characteristics and relationship to the induction of sexual behavior. *Brain Reserach*, 1979a, *169*, 481–497.

Blaustein, J. D., and Feder, H. H. Cytoplasmic progestin receptors in female guinea pig brain and their relationship to refractoriness in expression of female sexual behavior. *Brain Research*, 1979b, *177*, 489–498.

Blaustein, J. D., and Feder, H. H. Nuclear progestin receptors in guinea pig brain measured by an *in vitro* exchange assay after hormonal treatments that affect lordosis. *Endocrinology*, 1980, *106*, 1061–1069.

Blaustein, J. D., and Wade, G. N. Concurrent inhibition of sexual behavior, but not brain [^{3}H]estradiol uptake, by progesterone in female rats. *Journal of Comparative and Physiological Psychology*, 1977a, *91*(4), 742–751.

Blaustein, J. D., and Wade, G. N. Sequential inhibition of sexual behavior by progesterone in female rats: Comparison with a synthetic antiestrogen. *Journal of Comparative and Physiological Psychology*, 1977b, *91*, 752–760.

Blaustein, J. D., and Wade, G. N. Progestin binding by brain and pituitary cell nuclei and female rat sexual behavior. *Brain Research*, 1978, *140*, 360–367.

Blaustein, J. D., Dudley, S. D., Gray, J. M., Roy, E. J., and Wade, G. N. Long term retention of estradiol by brain cell nuclei and female rat sexual behavior. *Brain Research*, 1979, *173*, 355–359.

Blaustein, J. D., Ryer, H. I., and Feder, H. H. A sex difference in the progestin receptor system of guinea pig brain. *Neuroendocrinology*, 1980, *31*, 403–409.

Boling, J. L., and Blandau, R. J. The estrogen-progesterone induction of mating responses in the spayed female rat. *Endocrinology*, 1939, *25*, 359–364.

Bonsall, R. W., Zumpe, D., and Michael, R. P. Menstrual cycle influences on operant behavior of female rhesus monkeys. *Journal of Comparative and Physiological Psychology*, 1978, *92*, 846–855.

Bullock, D. Induction of heat in ovariectomized guinea pigs by brief exposure to estrogen and progesterone. *Hormones and Behavior*, 1970, *1*, 137–143.

Butcher, R. L., Collins, W. E., and Fugo, N. W. Plasma concentration of LH, FSH, prolactin, progesterone and estradiol-17β throughout the 4-day estrous cycle of the rat. *Endocrinology*, 1974, *94*, 1704–1708.

Carpenter, C. R. Sexual behavior of free ranging rhesus monkeys *(Macaca mulatta):* I. Specimens, procedures and behavioral characteristics of estrus. II. Periodicity of estrus, homosexual, autoerotic and nonconformist behavior. *Journal of Comparative Psychology*, 1942, *33*, 113–162.

Carr, W. J., Loeb, L. S., and Dissinger, M. L. Responses of rats of sex odors. *Journal of Comparative and Physiological Psychology*, 1965, *59*, 370–377.

Carr, W. J., Loeb, L. S., and Wylie, N. R. Responses to feminine odors in normal and castrated male rats. *Journal of Comparative and Physiological Psychology*, 1966, *62*, 336–338.

Carter, C. S. Stimuli contributing to the decrement in sexual receptivity of female golden hamsters *(Mesocricetus auratus)*. *Animal Behaviour*, 1973, *21*, 827–834.

Carter, C. S. *Estrus induction and pair bonding in the prairie vole.* Paper presented at the Conference on Reproductive Behavior, East Lansing, Michigan, 1982.

Carter, C. S., and Schein, M. W. Sexual receptivity and exhaustion in the female golden hamster. *Hormones and Behavior*, 1971, *2*, 191–200.

Challis, J. R. G., Heap, R. B., and Illingworth, D. V. Concentrations of oestrogen and progesterone in the plasma of non-pregnant, pregnant and lactating guinea pigs. *Journal of Endocrinology*, 1971, *51*, 333–345.

Ciaccio, L. A., and Lisk, R. D. Hormonal control of cyclic estrus in the female hamster. *American Journal of Physiology*, 1971, *221*, 936–942.

Clark, A. S., and Roy, E. J. Behavioral and cellular responses to pulses of low doses of estradiol-17β. *Physiology and Behavior*, 1983, *30*, 561–565.

Collins, V. J., Boling, J. L., Dempsey, E. W., and Young, W. C. Quantitative studies of experimentally induced sexual receptivity in the spayed guinea pig. *Endocrinology*, 1938, *23*, 188–196.

Concannon, P. W., Hansel, W., and Visek, W. J. The ovarian cycle of the bitch: Plasma estrogen, LH and progesterone. *Biology of Reproduction*, 1975, *13*, 112–121.

Craig, J. V. *Domestic Animal Behavior: Causes and Implications for Animal Care and Management*. Englewood Cliffs, N.J.: Prentice-Hall, 1981.

Croix, D., and Franchimont, P. Changes in the serum levels of the gonadotrophins, progesterone and estradiol during the estrous cycle of the guinea pig. *Neuroendocrinology*, 1975, *19*, 1–11.

Cumming, I. A., Buckmaster, J. M., Blockey, M. A. D. Goding, J. R., Winfield, C. G., and Baxter, R. W. Constancy of interval between luteinizing hormone release and ovulation in the ewe. *Biology of Reproduction*, 1973, *9*, 24–29.

Czaja, J. A., and Bielert, C. Female rhesus sexual behavior and distance to a male partner: Relation to stage of the menstrual cycle. *Archives of Sexual Behavior*, 1975, *4*, 583–597.

Davidson, J. M., Smith, E. R., Rodgers, C. H., and Bloch, G. J. Relative thresholds of behavioral and somatic responses to estrogen. *Physiology and Behavior*, 1968, *3*, 277–289.

DeBold, J. F., Martin, J. V., and Whalen, R. E. The excitation and inhibition of sexual receptivity in female hamsters by progesterone: Time and dose relationships, neural localization and mechanisms of action. *Endocrinology*, 1976, *99*, 1519–1527.

DeBold, J. F., Morris, J. L., and Clemens, L. G. The inhibitory actions of progesterone: Effects on male and female sexual behavior of the hamster. *Hormones and Behavior*, 1978a, *11*, 18–41.

DeBold, J. F., Ruppert, P. H., and Clemens, L. G. Inhibition of estrogen-induced sexual receptivity of female hamsters: Comparative effects of progesterone, dihydrotestosterone and an estrogen antagonist. *Pharmacology, Biochemistry, and Behavior*, 1978b, *9*, 81–86.

DeJong, F. H., Hey, A. H., and van der Molen, H. J. Effect of gonadotrophins on the secretion of oestradiol-17β and testosterone by the rat testis. *Journal of Endocrinology*, 1973, *57*, 277–284.

Dempsey, E. W., and Rioch, D. M. The localization in the brain stem of the oestrous responses of the female guinea pig. *Journal of Neurophysiology*, 1939, *2*, 9–18.

Dempsey, E. W., Hertz, R., and Young, W. C. The experimental induction of oestrus (sexual receptivity) in the normal and ovariectomized guinea pig. *American Journal of Physiology*, 1936, *116*, 201–209.

DeVore, I. *Primate Behaviour*. New York: Holt, 1965.

Dewsbury, D. A. Copulatory behaviour of rice rats *(Oryzomys palustris)*. *Animal Behavior*, 1970, *18*, 266–275.

Dixson, A. F., and Herbert, J. Gonadal hormones and sexual behavior in groups of monkeys *(Miopithecus talapoin)*. *Hormones and Behavior*, 1977, *8*, 141–154.

Dohanich, G. P., Witcher, J. A., Weaver, D. R., and Clemens, L. G. Alteration of muscarinic binding in specific brain areas following estrogen treatment. *Brain Research*, 1982, *241*, 347–350.

Dorfman, R. I., and Ungar, F. *Metabolism of Steroid Hormones*. New York: Academic Press, 1965.

Eaton, G. G. Social and endocrine determinants of sexual behavior in simian and prosimian females. In C. H. Phoenix (Ed.), *Symposium of the IVth International Congress of Primatology*, Vol. 2, *Primate Reproductive Behavior*. Basel: Karger, 1973.

Eaton, G. G., Slob, A. K., and Resko, H. A. Cycles of mating behaviour, oestrogen and progesterone in the thick-tailed bushbaby *(Galago crassicaudatus crassicaudatus)* under laboratory conditions. *Animal Behavior*, 1973, *21*, 309–315.

Edwards, D. A. Induction of estrus in female mice: Estrogen-progesterone interactions. *Hormones and Behavior*, 1970, *1*, 229–304.

Edwards, D. A., and Pfeifle, J. K. Hormonal control of receptivity, proceptivity and sexual motivation. *Physiology and Behavior*, 1983, *30*, 437–443.

Edwards, D. A., Whalen, R. E., and Nadler, R. D. Induction of estrus: Estrogen-progesterone interactions. *Physiology and Behavior*, 1968, *3*, 29–33.

Eliasson, M., and Meyerson, B. J. Sexual preference in female rats during the estrous cycle, pregnancy and lactation. *Physiology and Behavior*, 1975, *14*, 705–710.

Elliott, J. A., and Goldman, B. D. Seasonal reproduction: Photoperiodism and biological clocks. In N. T. Adler (Ed.), *Neuroendocrinology of Reproduction: Physiology and Behavior*. New York: Plenum Press, 1981.

Evans, M. J., and Irvine, C. H. G. Serum concentrations of FSH, LH and progesterone during the oestrous cycle and early pregnancy in the mare. *Journal of Reproduction and Fertility Supplement*, 1975, *23*, 193–200.

Everitt, B. J., and Herbert, J. The effects of implanting testosterone propionate into the central nervous system on the sexual behavior of adrenalectomised female rhesus monkeys. *Brain Research*, 1975, *86*. 109–120.

Everitt, B. J., Herbert, J., and Hamer, J. D. Sexual receptivity of bilaterally adrenalectomised female rhesus monkeys. *Physiology and Behavior*, 1972, *8*, 409–415.

Fadem, B. H., Barfield, R. J., and Whalen, R. E. Dose-response and time-response relationships between progesterone and the display of patterns of receptive and proceptive behavior in the female rat. *Hormones and Behavior*, 1979, *13*, 40–48.

Feder, H. H. Essentials of steroid structure, nomenclature, reactions, biosynthesis, and measurements. In N. T. Adler (Ed.), *Neuroendocrinology of Reproduction: Physiology and Behavior*. New York: Plenum Press, 1981a.

Feder, H. H. Estrous cyclicity in mammals. In N. T. Adler (Ed.), *Neuroendocrinology of Reproduction: Physiology and Behavior*. New York: Plenum Press, 1981b.

Feder, H. H., and Silver, R. Activation of lordosis in ovariectomized guinea pigs by free and esterified forms of estrone, estradiol-17β and estriol. *Physiology and Behavior*, 1974, *13*, 251–255.

Feder, H. H., Landau, I. T., and Walker, W. A. Anatomical and biochemical substrates of the actions of estrogens and anti-estrogens on brain tissues that regulate female sex behavior of rodents. In C. Beyer (Ed.), *Endocrine Control of Sexual Behavior*. New York: Raven Press, 1979.

Fox, R. R., and Laird, C. W. Sexual cycles. In E. S. E. Hafez (Ed.), *Reproduction and Breeding Techniques for Laboratory Animals*. Philadelphia: Lea and Febiger, 1970.

Frank, A. H., and Fraps, R. M. Induction of estrus in the ovariectomized golden hamster. *Endocrinology*, 1945, *37*, 357–361.

Glencross, R. G., Munro, I. B., Senior, B. E., and Pope, G. S. Concentrations of oestradiol-17β, oestrone and progesterone in jugular venous plasma of cows during the oestrous cycle and in early pregnancy. *Acta Endocrinologica*, 1973, *73*, 374–384.

Goldfoot, D. A., and Goy R. W. Abbreviation of behavioral estrus in guinea pigs by coital and vagino-cervical stimulation. *Journal of Comparative Physiological Psychology*, 1970, 72, 426–434.

Goldfoot, D. A., Goy, R. W., Kravetz, M. A., and Freeman, S. K. Reply to Keverne. *Hormones and Behavior*, 1976a, *7*, 376–378.

Goldfoot, D. A., Goy, R. W., Kravetz, M. A., and Freeman, S. K. Reply to Michael, Bonsall and Zumpe. *Hormones and Behavior*, 1976b, *7*, 373–375.

Goldfoot, D. A., Kravetz, M. A., Goy, R. W., and Freeman, S. K. Lack of effect of vaginal lavages and aliphatic acids on ejaculatory responses in rhesus monkeys: Behavioral and chemical analyses. *Hormones and Behavior*, 1976c, 7, 1–27.

Gorzalka, B. B., and Whalen, R. E. The effects of progestins, mineralocorticoids, glucorticoids, and steroid solubility on the induction of sexual receptivity in rats. *Hormones and Behavior*, 1977, *8*, 94–99.

Goy, R. W., and Resko, J. A. Gonadal hormones and behavior of normal and pseudohermaphroditic nonhuman female primates. *Recent Progress in Hormone Research*, 1972, *28*, 707–732.

Green, R., Luttge, W. G., and Whalen, R. E. Induction of receptivity in ovariectomized female rats by a single intravenous injection of estradiol-17β. *Physiology and Behavior*, 1970, *5*, 137–141.

Guerrero, R., Aso, R., Brenner P. G., Cekan, Z., Landgren, B. M., Hagenfeldt, K., and Diczfaluzy, E. Studies on the pattern of circulating steroids in the normal menstrual cycle. *Acta Endocrinologica* (Kbh), 1976, *81*, 133–149.

Hafez, E. S. E., and Signoret, J. P. The behaviour of swine. In E. S. E. Hafez (Ed.), *The Behaviour of Domestic Animals* (2nd ed.). Baltimore: Williams & Wilkins, 1969.

Hansel, W., Concannon, P. W., and Lukaszewska, J. H. Corpora lutea of the large domestic animals. *Biology of Reproduction*, 1973, *8*, 222–245.

Hansen, S., and Sodersten, P. Reversal of progesterone inhibition of sexual behavior in ovariectomized rats by high doses of progesterone. *Journal of Endocrinology*, 1979, *80*, 381–388.

Hardy, D. F., and DeBold, J. F. Effects of mounts without intromission upon the behavior of female rats during the onset of estrogen-induced heat. *Physiology and Behavior*, 1971a, *7*, 643–645.

Hardy, D. F., and DeBold, J. F. The relationship between levels of exogenous hormones and the display of lordosis by the female rat. *Hormones and Behavior*, 1971b, *2*, 287–297.

Hardy, D. F., and DeBold, J. F. Effects of coital stimulation upon behavior of the female rat. *Journal of Comparative and Physiological Psychology*, 1972, *78*, 400–408.

Harris, J., and Gorski, J. Evidence for a discontinuous requirement for estrogen in stimulation of deoxyribonucleic acid synthesis in the immature rat uterus. *Endocrinology*, 1978, *103*, 240–245.

Hasler, J. F., and Banks, E. M. The behavioral and somatic effects of ovariectomy and replacement therapy in female collared lemmings (*Dicrostonyx groenlandicus*). *Hormones and Behavior*, 1976, *7*, 59–74.

Hauger, R. L., Karsch, F. J., and Foster, D. L. A new concept for control of the estrous cycle of the ewe based on the temporal relationships between luteinizing hormone, estradiol and progesterone in peripheral serum and evidence that progesterone inhibits tonic LH secretion. *Endocrinology,* 1977, *101,* 807–817.

Heape, W. The "sexual season" of mammals and the relation of the "pro-oestrum" to menstruation. *Quarterly Journal of Microbiological Science,* 1901, *44,* 1–40.

Herbert, J. Hormones and reproductive behavior in rhesus and talapoin monkeys. *Journal of Reproduction and Fertility Supplement,* 1970, *11,* 119–140.

Hill, M., and White, W. E. The growth and regression of follicles in the oestrous rabbit. *Journal of Physiology* (London), 1933, *80,* 174–178.

Hilliard, J. Corpus luteum function in guinea pigs, hamsters, rats, mice and rabbits. *Biology of Reproduction,* 1973, *8,* 203–221.

Hilliard, J., and Eaton, L. W., Jr. Estradiol-17β, progesterone and 20α-hydroxypregn-4-en-3-one in rabbit ovarian venous plasma: II. From mating through implantation. *Endocrinology,* 1971, *89,* 522–526.

Hlinak, Z., and Madlafousek, J. Estradiol treatment and precopulatory behavior in ovariectomized female rats. *Physiology and Behavior,* 1981, *26,* 171–176.

Huck, U. W., Carter, C. S., and Banks, E. M. Estrogen and progesterone interactions influencing sexual and social behavior in the brown lemming, *Lemmus trimucronatus. Hormones and Behavior,* 1979, *12,* 40–49.

Jha, P., Jain, A. K., Rahman, S. A., Dugwekar, Y. G., and Laumas, K. R. Serum LH, FSH, progesterone and estradiol-17β levels of normally menstruating Indian women. *Indian Journal of Medical Research,* 1978, *67,* 57–65.

Johnson, D. F., and Phoenix, C. H. Hormonal control of female sexual attractiveness, proceptivity, and receptivity in rhesus monkeys. *Journal of Comparative and Physiological Psychology,* 1976, *90,* 473–483.

Johnston, P. G., and Davidson, J. M. Priming action of estrogen: Minimum duration of exposure for feedback and behavioral effects. *Neuroendocrinology,* 1979, *28,* 155–159.

Johnston, R. E. The causation of two scent-marking behaviour patterns in female hamsters *(Mesocricetus auratus). Animal Behavior,* 1977, *25,* 317–327.

Johnston, R. E. Olfactory perferences, scent marking, and "proceptivity" in female hamsters. *Hormones and Behavior,* 1979, *13,* 21–39.

Joslyn, W. D., and Feder, H. H. Facilitatory and inhibitory effects of supplementary estradiol benzoate given to ovariectomized, estrogen primed guinea pigs. *Hormones and Behavior,* 1971, *2,* 307–314.

Karsch, F. J., and Sutton, G. P. An intra-ovarian site for the luteolytic action of estrogen in the rhesus monkey. *Endocrinology,* 1976, *98,* 553–561.

Kent, G. C., Jr. Physiology of reproduction. In R. A. Hoffman, P. F. Robinson, and H. Magalhaes (Eds.), *The Golden Hamster: Its Biology and Use in Medical Research.* Ames: Iowa State University Press, 1968.

Keverne, E. B. Reply to Goldfoot *et al. Hormones and Behavior,* 1976a, *7,* 369–372.

Keverne, E. B. Sexual receptivity and attractiveness in the female rhesus monkey. In J. S. Rosenblatt, R. A. Hinde, E. Shaw, and C. Beer (Eds.), *Advances in the Study of Behavior,* Vol. 7. New York: Academic Press, 1976b.

Keverne, E. G., and Michael, R. P. Sex attractant properties of ether extracts of vaginal secretions from rhesus monkeys. *Journal of Endocrinology,* 1971, *51,* 313–322.

Knobil, E. On the regulation of the primate corpus luteum. *Biology of Reproduction,* 1973, *8,* 246–258.

Kubli-Garfias, C., and Whalen, R. E. Induction of lordosis behavior in female rats by intravenous administration of progestins. *Hormones and Behavior,* 1977, *9,* 380–386.

Landauer, M. R., Banks, E. M., and Carter, C. S. Sexual and olfactory preferences of naive and experienced male hamsters. *Animal Behavior,* 1978, *28,* 611–621.

LeBoeuf, B. J. Interindividual associations in dogs. *Behaviour,* 1967, *29,* 268–295.

LeMagnen, J. Les phénomènes olfacto-sexuals chez le rat blanc. *Archives des Sciences Physiologiques,* 1952, *6,* 295–332.

Lindsay, D. R. The importance of olfactory stimuli in the mating behaviour of the ram. *Animal Behavior,* 1965, *13,* 75–78.

Lindsay, D. R., and Fletcher, I. C. Ram-seeking activity associated with oestrous behaviour in ewes. *Animal Behavior,* 1972, *20,* 452–456.

Lisk, R. D. A comparison of the effectiveness of intravenous, as opposed to subcutaneous injections of progesterone for the induction of estrous behavior in the rat. *Canadian Journal of Biochemistry and Physiology*, 1960, *38*, 1381–1383.

Luine, V. N., and Rhodes, J. C. Gonadal hormone regulation of MAO and other enzymes in hypothalamic areas. *Neuroendocrinology*, 1983, *36*, 235–241.

Luine, V. N., Khylchevskaya, R. I., and McEwen, B. S. Effect of gonadal steroids on activities of monoamine oxidase and choline acetylase in rat brain. *Brain Research*, 1975, *86*, 293–306.

Luttge, W. G. Endocrine control of mammalian male sexual behavior: An analysis of the potential role of testosterone metabolites. In C. Beyer (Ed.), *Endocrine Control of Sexual Behavior*. New York: Raven Press, 1979.

Luttge, W. G., and Hughes, J. R. Intracerebral implantation of progesterone: re-examination of the brain sites responsible for facilitation of sexual receptivity in estrogen-primed ovariectomized rats. *Physiology and Behavior*, 1976, *17*, 771–775.

Luttge, W. G., and Jasper, T. W. Studies on the possible role of 2-OH-estradiol in the control of sexual behavior in female rats. *Life Sciences*, 1977, *20*, 419–426.

MacLusky, N. J., and McEwen, B. S. Oestrogen modulates progestin receptor concentrations in some rat brain regions but not in others. *Nature*, 1978, *274*, 276–278.

Madlafousek, J., and Hlinak, Z. The first copulations of adult males change their dependence on the female precopulatory behavior (in rats). *Ceskoslovenska Psychologie*, 1971, *15*, 1–11.

Madlafousek, J., Hlinak, Z., and Beran, J. Decline of sexual behavior in castrated male rats: Effects of female precopulatory behavior. *Hormones and Behavior*, 1976, *7*, 245–252.

Marrone, B. L., and Feder, H. H. Characteristics of [³H]-progestin uptake and effects of progesterone on [³H]-estrogen uptake in brain, anterior pituitary and peripheral tissues of male and female guinea pigs. *Biology of Reproduction*, 1977, *17*, 42–57.

Marrone, V. B. L., Rodriguez-Sierra, J. F., and Feder, H. H. Role of catechol estrogens in activation of lordosis in female rats and guinea pigs. *Pharmacology, Biochemistry and Behavior*, 1977, *7*, 13–17.

McDonald, P., Vidal, N., and Beyer, C. Sexual behavior in the ovariectomized rabbit after treatment with different amounts of gonadal hormones. *Hormones and Behavior*, 1970, *1*, 161–172.

Meinkoth, J., Quadagno, D. M., and Bast, J. P. Depression of steroid induced sex behavior in the ovariectomized rat by intracranial injection of cycloheximide: Preoptic area compared to the ventromedial hypothalamus. *Hormones and Behavior*, 1979, 12: 199–204.

Meisel, R. L., and Pfaff, D. W. *Effects of intracranial anisomycin on sexual behavior in female rats.* Paper presented at the Conference on Reproductive Behavior, Medford, Massachusetts, 1983.

Meyerson, B. Latency between intravenous injection of progestins and the appearance of estrous behavior in estrogen-treated ovariectomized rats. *Hormones and Behavior*, 1972, *3*, 1–9.

Michael, R. P., and Bonsall, R. W. Peri-ovulatory synchronisation of behaviour in male and female rhesus monkeys. *Nature*, 1977, *265*, 463–465.

Michael, R. P., and Keverne, E. B. Pheromones and the communication of sexual status in primates. *Nature* (London), 1968, *218*, 746–749.

Michael, R. P., and Keverne, E. B. Primate sex pheromones of vaginal origin. *Nature* (London), 1970, *225*, 84–85.

Michael, R. P., and Saayman, G. S. Differential effects on behaviour of the subcutaneous and intravaginal administration of oestrogen in the rhesus monkey *(Macaca mulatta)*. *Journal of Endocrinology*, 1968, *41*, 231–246.

Michael, R. P., and Scott, P. P. The activation of sexual behaviour in cats by the subcutaneous administration of oestrogen. *Journal of Physiology* (London), 1964, *171*, 254–264.

Michael, R. P., and Zumpe D. Sexual initiating behaviour by female rhesus monkeys *(Macaca mulatta)* under laboratory conditions. *Behaviour*, 1970, *36*, 168–185.

Michael, R. P., Herbert, J., and Welegalla. J. Ovarian hormones and the sexual behavior of the male rhesus monkey *(Macaca mulatta)* under laboratory conditions. *Journal of Endocrinology*, 1967, *39*, 81–98.

Michael, R. P., Saayman, G. S., and Zumpe, D. The suppression of mounting behaviour and ejaculation in male rhesus monkeys *(Macaca mulatta)* by administration of progesterone to their female partners. *Journal of Endocrinology*, 1968, *41*, 421–431.

Michael, R. P., Keverne, E. B., and Bonsall, R. W. Pheromones: Isolation of male sex attractants from a female primate. *Science*, 1971, *172*, 964–966.

Michael, R. P., Zumpe, D., Keverne, E. B., and Bonsall, R. W. Neuroendocrine factors in the control of primate behavior. *Recent Progress in Hormone Research*, 1972, *28*, 665–706.

Michael, R. P., Bonsall, R. W., and Zumpe, D. "Lack of effects of vaginal fatty acids, etc.": A reply to Goldfoot *et al. Hormones and Behavior,* 1976, *7,* 365–368.

Michael, R. P., Zumpe, D., and Bonsall, R. W. Behavior of rhesus monkeys during artificial menstrual cycles. *Journal of Comparative and Physiological Psychology,* 1982, *96,* 875–885.

Mills, T., Copland, A., and Osteen, K. Factors affecting the postovulatory surge of FSH in the rabbit. *Biology of Reproduction,* 1981, *25,* 530–535.

Moguilewsky, M., and Raynaud, J. Estrogen-sensitive progestin-binding sites in the female rat brain and pituitary. *Brain Research,* 1979a, *164,* 165–175.

Moguilewsky, M., and Raynaud, J. The relevance of hypothalamic and hypophyseal progestin receptor regulation in the induction and inhibition of sexual behavior in the female rat. *Endocrinology,* 1979b, *105,* 516–522.

Morali, G., and Beyer, C. Neuroendocrine control of mammalian estrous behavior. In C. Beyer (Ed.), *Endocrine Control of Sexual Behavior.* New York: Raven Press, 1979.

Moreines, J., and Powers, J. B. Effects of acute ovariectomy on the lordosis response of female rats. *Physiology and Behavior,* 1977, *19,* 277–283.

Morin, L. P. Theoretical review: Progesterone: Inhibition of rodent sexual behavior. *Physiology and Behavior,* 1977, *18,* 701–715.

Morin, L. P., and Feder, H. H. Hypothalamic progesterone implants and facilitation of lordosis behavior in estrogen-primed ovariectomized guinea pigs. *Brain Research,* 1974, *70,* 81–93.

Nadler, R. D. Sexual behavior of captive lowland gorillas. *Archives of Sexual Behavior,* 1976, *5,* 487–502.

Nadler, R. D. Sexual behavior of captive orangutans. *Archives of Sexual Behavior,* 1977, *6,* 457–475.

Nadler, R. D., Collins, D. C., Miller, L. C., and Graham, C. E. Menstrual cycle patterns of hormones and sexual behavior in gorillas. *Hormones and Behavior,* 1983, *17,* 1–17.

Naftolin, F., Ryan, K. J., Davies, I. J., Reddy, V. V., Flores, F., Petro, Z., Kuhn, M., White, R. J., Takaoka, Y., and Wolin, L. The formation of estrogens by central neuroendocrine tissues. *Recent Progress in Hormone Research,* 1975, *31,* 295–319.

Nissen, H. W. The effects of gonadectomy, vasotomy, and injection of placental and archic extracts on the sex behavior of the white rat. *Genetic Psychology Monographs,* 1929, *5,* 451–547.

Noden, P. A., Oxender, W. D., and Hafs, H. D. The cycle of oestrous, ovulation and plasma levels of hormones in the mare. *Journal of Reproduction and Fertility Supplement,* 1975, *23,* 181–192.

O'Malley, B. W., and Means, A. R. Female steroid hormones and target cell nuclei. *Science,* 1974, *183,* 610–620.

Paape, S. R., Shille, V. M., Seto, H., and Stabenfeldt, G. H. Luteal activity in the pseudo-pregnant cat. *Biology of Reproduction,* 1975, *13,* 470–474.

Pant, H. C., Hopkinson, C. R. N., and Fitzpatrick, R. J. Concentration of oestradiol, progesterone, luteinizing hormone and follicle-stimulating hormone in the jugular venous plasma of ewes during the oestrous cycle. *Journal of Endocrinology,* 1977, *73,* 147–255.

Parson, S. D., and Hunter, G. L. Effect of the ram on duration of oestrus in the ewe. *Journal of Reproduction and Fertility,* 1967, *14,* 61–70.

Parsons, B., McEwen, B. S., and Pfaff, D. W. A discontinuous schedule of estradiol treatment is sufficient to activate progesterone-facilitated feminine sexual behavior and to increase cytosol receptors for progestins in the hypothalamus of the rat. *Endocrinology,* 1982, *110,* 613–619.

Pattison, M. L., Chen, C. L., Kelley, S. T., and Brandt, G. W. Luteinizing hormone and estradiol in peripheral blood of mares during the estrous cycle. *Biology of Reproduction,* 1974, *11,* 245–250.

Peretz, E. Estrogen dose and the duration of the mating period in cats. *Physiology and Behavior,* 1968, *3,* 41–43.

Pfaff, D. W. Nature of sex hormone effects on rat sex behavior: Specificity of effects and individual patterns of response. *Journal of Comparative and Physiological Psychology,* 1970, *73,* 349–358.

Pfaff, D. W. *Estrogens and Brain Function.* New York: Springer-Verlag, 1980.

Pfaff, D. W., and Pfaffman, C. Olfactory and hormonal influences on the basal forebrain of the male rat. *Brain Research,* 1969, *15,* 137–156.

Phemister, R. D., Holst, P. A., Spano, J. S., and Hopwood, M. L. Time of ovulation in the beagle bitch. *Biology of Reproduction,* 1973, *8,* 74–82.

Phoenix, C. H. Ejaculation by male rhesus as a function of the female partner. *Hormones and Behavior,* 1973, *4,* 365–370.

Pomerantz, S., and Goy, R. W. Proceptive behavior of female rhesus monkeys during tests with tethered males. *Hormones and Behavior,* 1983, *17,* 237–248.

Powers, J. B. Hormonal control of sexual receptivity during the estrous cycle of the rat. *Physiology and Behavior,* 1970, *5,* 831–835.

Powers, J. B. Facilitation of lordosis in ovariectomized rats by intracerebral progesterone implants. *Brain Research,* 1972, *48,* 311–325.

Powers, J. B. Anti-estrogenic suppression of the lordosis response in female rats. *Hormones and Behavior,* 1975, *6,* 379–392.

Quadagno, D. M., Shryne, J., and Gorski, R. W. The inhibition of steroid-induced sexual behavior by intrahypothalamic actinomycin-D. *Hormones and Behavior,* 1971, *2,* 1–10.

Quadagno, D. M., McCullough, J., and Langan, R. The effect of varying amounts of exogenous estradiol benzoate on estrous behavior in the rat. *Hormones and Behavior,* 1972, *3,* 175–179.

Rainbow, T. C., Davis, P. G., and McEwen, B. S. Anisomycin inhibits the activation of sexual behavior by estrogen and progesterone. *Brain Research,* 1980, *194,* 548–555.

Rainbow, T. C., McGinnis, M. Y., Davis, P. G., and McEwen, B. S. Application of anisomycin to the lateral ventromedial nucleus of the hypothalamus inhibits the activation of sexual behavior by estradiol and progesterone. *Brain Research,* 1982, *233,* 417–423.

Ring, J. R. The estrogen-progesterone induction of sexual receptivity in the spayed female mouse. *Endocrinology,* 1944, *34,* 269–275.

Robinson, T. J., and Reardon, T. F. The activity of a number of oestrogens as tested in the spayed ewe. *Journal of Endocrinology,* 1961, *23,* 97–107.

Ross, F., Claybaugh, C., Clemens, L. G., and Gorski, R. A. Short latency induction of estrous behavior with intracerebral gonadal hormones in ovariectomized rats. *Endocrinology,* 1971, *89,* 32–38.

Roy, E. J., and Wade, G. N. Binding of [^{3}H]estradiol by brain cell nuclei and female rat sexual behavior: Inhibition by antiestrogens. *Brain Research,* 1977, *126,* 73–87.

Rubin, B. S., and Barfield, R. J. Progesterone in the ventromedial hypothalamus facilitates estrous behavior in ovariectomized, estrogen-primed rats. *Endocrinology,* 1983, *113,* 797–804.

Saayman, G. S. Effects of ovarian hormones upon the sexual skin and mounting behavior in the free-ranging chacma baboon *(Papio ursinus). Folia Primatologica,* 1972, *17,* 197–303.

Saidapur, S. K., and Greenwald, G. S. Peripheral blood and ovarian levels of sex steroids in the cyclic hamster. *Biology of Reproduction,* 1978, *18,* 401–408.

Schams, D., Schallenberger, E., Hoffman, B., and Karg, H. The oestrous cycle of the cow: Hormonal patterns and time relationships concerning oestrous, ovulation, and electrical resistance of the vaginal mucus. *Acta Endocrinologica,* 1977, *86,* 180–192.

Schille, V. M., Lundstrom, K. E., and Stabenfeldt, G. H. Follicular function in the domestic cat as determined by estradiol-17β concentrations in plasma: Relation to estrous behavior and cornification of exfoliated vaginal epithelium. *Biology of Reproduction,* 1979, *21,* 953–963.

Scruton, D., and Herbert, J. The menstrual cycle and its effects on behaviour in the Talapoin monkey *(Miopithecus talapoin). Journal of Zoology,* 1970, *162,* 419–436.

Shemesh, M., Ayalon, N., and Lindner, H. R. Oestradiol levels in the peripheral blood of cows during the oestrous cycle. *Journal of Endocrinology,* 1972, *55,* 73–78.

Signoret, J. P. Durée du cycle oestrien et de l'oestrus chez la Truie. Action de benzoate d'oestradiol chez la fémelle ovaliectomisée. *Annales de Biologie Animale Biochimie Biophysique,* 1967, *7,* 407–421.

Signoret, J. P. Reproductive behaviour in pigs. *Journal of Reproduction and Fertility Supplement,* 1970, *11,* 105–117.

Signoret, J. P. Étude de l'action inhibitrice de la progestérone sur l'apparition du comportement sexuel induit par injection d'oestrogènes chez la Truie et la Brebis ovariectomisées. *Annales de Biologie Animale Biochimie Biophysique,* 1971, *11,* 489–494.

Slob, A. K., Baum, M. J., and Schenck, P. E. Effects of the menstrual cycle, social grouping, and exogenous progesterone on heterosexual interaction in laboratory housed stumptail macaques *(M. arctoides). Physiology and Behavior,* 1978a, *21,* 915–921.

Slob, A. K., Wiegand, S. J., Goy, R. W., and Robinson, J. A. Heterosexual interactions in laboratory-housed stumptail macaques *(Macaca arctoides):* Observations during the menstrual cycle and after ovariectomy. *Hormones and Behavior,* 1978b, *10,* 193–211.

Smith, M. S., Freeman, M. E., and Neill, J. D. The control of progesterone secretion during the estrous cycle and early pregnancy in the rat: Prolactin, gonadotropin and steroid levels associated with rescue of the corpus luteum. *Endocrinology*, 1975, *96*, 219–226.

Sodersten, P., and Eneroth, P. Serum levels of oestradiol-17β and progesterone in relation to sexual receptivity in intact and ovariectomized rats. *Journal of Endocrinology*, 1981, *89*, 45–54.

Sodersten, P., Eneroth, P., and Hansen, S. Induction of sexual receptivity in ovariectomized rats by pulse administration of oestradiol-17β. *Journal of Endocrinology*, 1981, *89*, 55–62.

Stabenfeldt, G. H., and Hughes, J. P. Reproduction in horses. In H. H. Cole and P. T. Cupps (Eds.), *Reproduction in Domestic Animals* (3rd ed.). New York: Academic Press, 1977.

Stabelfeldt, G. H., and Schille, V. M. Reproduction in the dog and cat. In H. H. Cole and P. T. Cupps (Eds.), *Reproduction in Domestic Animals* (3rd ed.). New York: Academic Press, 1977.

Staples, R. E. Behavioural induction of ovulation in the oestrous rabbit. *Journal of Reproduction and Fertility*, 1967, *13*, 429–435.

Steel, E. Changes in female attractivity and proceptivity throughout the oestrous cycle of the syrian hamster *(Mesocricetus auratus)*. *Animal Behavior*, 1980, *28*, 256–265.

Steinach, E. Untersuchungen zur vergleichenden Physiologie der mannlichen Geschlechtsorgane: I. Über den Geschlechtstrieb und den Geschlechtstakt bei Froschen. *Pflüger's Archiv für die Gesammte Physiologie des Menschen und der Tiere*, 1894, *56*, 304–330.

Steinach, E. Geschlechtstreib und echt sekundare Geschlechtsmerkmale als Folge der innersekretorischen Funktion der Keimdrusen: II. Über die Entstehung des Umklammersreflexes bei Froschen. *Zentralblatt für Physiologie*, 1910, *24*, 551–570.

Steinberger, E. Biosynthesis of androgens. In H. Gibian and E. J. Plotz (Eds.), *Mammalian Reproduction*. New York: Springer-Verlag, 1970.

Stern, J. J. Responses of male rats to sex odors. *Physiology and Behavior*, 1970, *5*, 519–525.

Tarchanoff, J. R. Zur Physiologie des Geschlechtsapparate des Frosches. *Pflüger's Archiv für die Gesammte Physiologie des Menschen und der Tiere*, 1887, *40*, 330–351.

Tennent, B. J., Smith, E. R., and Davidson, J. M. The effects of estrogen and progesterone on female rat proceptive behavior. *Hormones and Behavior*, 1980, *14*, 65–75.

Trimble, M. R., and Herbert, J. The effect of testosterone or oestradiol upon the sexual and associated behaviour of the adult female rhesus monkey. *Journal of Endocrinology*, 1968, *42*, 171–185.

Udry, J. R., and Morris, N. M. Distribution of coitus in the menstrual cycle. *Nature* (London), 1968, *220*, 593–596.

van Tienhoven, A. *Reproductive Physiology of Vertebrates*. Philadelphia: W. B. Saunders, 1968.

Verhage, H. G., Beamer, N. B., and Brenner, R. M. Plasma levels of estradiol and progesterone in the cat during polyestrus, pregnancy and pseudopregnancy. *Biology of Reproduction*, 1976, *14*, 579–585.

Wallen, K., and Goy, R. W. Effects of estradiol benzoate, estrone, and propionates of testosterone or dihydrotestosterone on sexual and related behaviors of ovariectomized rhesus monkeys. *Hormones and Behavior*, 1977, *9*, 228–248.

Wallen, K., Goy, R. W., and Phoenix, C. H. Inhibitory actions of progesterone on hormonal induction of estrus in female guinea pigs. *Hormones and Behavior*, 1975, *6*, 127–138.

Webster, D. G., Williams, M. H., and Dewsbury, D. A. Female regulation and choice in the copulatory behavior of montane voles *(Microtus montanus)*. *Journal of Comparative and Physiological Psychology*, 1982, *96*, 661–667.

Weick, R. F., Dierschke, D. J., Karsch, F. J., Butler, W. R., Hotchkiss, J., and Knobil, E. Periovulatory time courses of circulating gonadotrophic and ovarian hormones in the rhesus monkey. *Endocrinology*, 1973, *93*, 1140–1147.

Whalen, R. E. Sexual behaviour of cats. *Behaviour*, 1963, *20*, 321–342.

Whalen, R. E. Estrogen-progesterone induction of mating in female rats. *Hormones and Behavior*, 1974, *5*, 157–162.

Whalen, R. E., Battie, C., and Luttge, W. G. Anti-estrogen inhibition of androgen induced sexual receptivity in rats. *Behavioral Biology*, 1972, *7*, 311–320.

Wildt, D. E., Seager, D. W. J., and Chakraborty, P. K. Effect of copulatory stimuli on incidence of ovulation and on serum luteinizing hormone in the cat. *Endocrinology*, 1980, *107*, 1212–1217.

Wilson, M. R., Gordon, T. P., and Collins, D. C. Age differences in copulatory behavior and serum 17β-estradiol in female rhesus monkeys. *Physiology and Behavior*, 1982, *28*, 733–737.

Yanase, M., and Gorski, R. A. Sites of estrogen and progesterone facilitation of lordosis behavior in the spayed rat. *Biology of Reproduction*, 1976, *15*, 536–543.

Young, W. C. The hormones and mating behavior. In W. C. Young (Ed.), *Sex and Internal Secretions* (3rd ed.). Baltimore: Williams & Wilkins, 1961.

Zucker, I. Facilitatory and inhibitory actions of progesterone on sexual responses of spayed guinea pigs. *Journal of Comparative and Physiological Psychology,* 1966, *62,* 376–381.

Zumpe, D., and Michael, R. P. The clutching reaction and orgasm in the rhesus monkey *(Macaca mulatta). Journal of Endocrinology,* 1968, *40,* 117–123.

NOTE ADDED IN PROOF

Studies published since this chapter was written indicate a revision of the terminology for steroid hormone receptors is necessary. What we have referred to as "cytoplasmic receptors" in the text are isolated in the cytosolic fraction of cell homogenates and are properly called *cytosolic receptors.* Recent studies indicate that cytosolic estrogen receptors are in fact located in the nucleus of cells that have not been disrupted by homogenization (W. J. King and G. L. Greene, *Nature* 307 [1984] 745–747; W. V. Welshons, M. E. Lieberman, and J. Gorksi, *Nature* 307 [1984] 747–749); the same may be true for progesterone receptors. Thus, the phrases *cytoplasmic receptors* and *translocation into the nucleus* are misleading. It is more appropriate to say that "cytosolic receptors" are "activated" by steroid molecules and undergo conformational changes to become "nuclear" receptors. The difference between *cytosolic* and *nuclear* receptors, then, is a difference in activation state and affinity for chromatin; it is not necessarily a change in location as was indicated by the terminology used in the text.

7

Maternal Behavior among the Nonprimate Mammals

JAY S. ROSENBLATT, ANNE D. MAYER, AND HAROLD I. SIEGEL

INTRODUCTION

The aims of this chapter are to present an overview of the research on maternal behavior among the nonprimate mammals, the theoretical issues that exist in this field of study, and the problems that are currently being investigated. Our emphasis is on the causal mechanisms underlying the various phases of maternal behavior rather than on the adaptive radiations in this principal line of evolutionary specialization (Eisenberg, 1981). Although the problems that are being investigated in this area are drawn from whatever knowledge we have of the natural history of species, in order to gain adequate control over the many factors that contribute to the regulation of maternal behavior, it has been found necessary to study laboratory species and domesticated animals. This necessity is determined by our need for precise descriptions of mother–young interactions, on the basis of which analysis of these interactions can proceed, and for accurate measurements of circulating levels of hormones during pregnancy and lactation, which enable us to remove and introduce hormones in order to study their effects on maternal behavior. Fortunately, whatever studies exist comparing laboratory and domesticated species

JAY S. ROSENBLATT, ANNE D. MAYER, AND HAROLD I. SIEGEL   Institute of Animal Behavior, Rutgers—The State University, Newark, New Jersey 07102.   The research from this laboratory reported in this chapter and the writing were supported by NIMH grant MH-08604 and a Biomedical Research Support Grant to Jay S. Rosenblatt.

with wild representatives of the same species suggest that maternal behavior has not been substantially altered by these more constrained living conditions. This suggestion holds out hope that our studies will be relevant to natural populations not only of the species that are being studied but also to populations that have not yet come under laboratory study.

Maternal behavior may be defined as any behavior that the female (or the male, in which case we refer to it as *Parental behavior*) performs in relation to the young. This functional definition is adequate for most purposes, and it allows for the possibility that we shall continually add to this pattern as we discover additional kinds of behavior that pertain to the young.

It is also useful to distinguish maternal behavior from other kinds of behavior that the female performs if we are interested in describing her behavior, not in terms of its function in relation to the young, but in relation to some underlying grouping of causal factors that are not well defined as yet, but that act when she is performing maternal behavior and are absent when she is not. We may say that all ways in which the female's behavior changes during the time when she is providing maternal care constitute her maternal behavior. This definition allows for the possibility that behavior not immediately involved in the care of the young, but in some way associated in the female with the maternal condition underlying the care of the young, can be included within the study of maternal behavior.

This is an expanding definition; for example, it has become apparent that important changes in the female's behavior related to the forthcoming delivery occur during pregnancy, when, in fact, she is not actually taking care of young. These, too, must be studied if we are to understand how maternal behavior arises during the reproductive cycle. Moreover, maternal behavior appears early in the life of females of many species, long before they can adequately take care of young or even bear young. The study of this phenomenon will give us important insights into the ontogenetic origins of maternal behavior.

Our studies have led us to view maternal behavior as a developmental study and as one in which several levels of analysis, behavioral and physiological, are necessary if we are to fully understand this complex behavior. These two approaches have been combined in this review: At the onset of maternal behavior, during pregnancy, and at parturition, endocrine processes predominate, but as maternal behavior is established and the relationship of the mother to her young is formed, behavioral processes come to the fore. Of course, during each phase of maternal behavior, there is a combination of physiological and behavioral processes that governs the performance of maternal care.

This chapter starts with a brief survey of the evolution of reproduction among the mammals with some comments on the patterns of maternal care. This beginning is followed by four principal sections that deal, first, with the onset of maternal behavior during pregnancy and near term; second, with the postpartum regulation of maternal behavior, particularly behavioral synchrony between mother and young; third, with lactation and maternal behavior; and fourth with weaning. There has been no attempt to present a comprehensive review of the literature in any of these areas.

Evolution of Modes of Reproduction among the Mammals

It has been proposed (Eisenberg, 1981) that the evolutionary history of parental care and reproduction in the mammals can be traced to a therapsid reptilian ancester that hatched altricial newborn from small, numerous eggs having little yolk and requiring, in addition to incubation by the homeothermic female, nourishing by secretions from modified sebaceous glands. Lactation was thus established, but the small yolk limited incubation time, and the altricial condition of the young became stabilized among the monotremes with the additional development of the pouch to permit the female to remain mobile during incubation as, for example, in the spiny anteater.

The subsequent advance made by the marsupials involved retention of the egg within the uterus and the establishment of ovoviviparity. Uterine life is, however, limited both by dependence on yolk, in the main, instead of maternal nutrition as among the eutheria, and by lack of defense against the immunorejection system of the mother (Lilligraven, 1975). Embryonic life is confined within the length of the estrous cycle among the marsupials. Within this constraint, several different patterns have been developed (Sharman *et al.,* 1966). In one pattern, gestation terminates *before* the end of the estrous cycle, and the subsequent estrus is inhibited by the suckling of the newborn in the pouch, or attached to the nipples outside in pouchless species. Not until suckling is reduced or the young are lost does the female resume cycling (provided weaning occurs outside the period of seasonal anestrous of the species) and become pregnant once again to repeat the cycle. In a second principal mode of reproduction among the marsupials, gestation terminates a short time *after* the occurrence of estrus, mating, and the initiation of a second pregnancy. The fertilized egg enters a period of diapause and ceases embryonic development while the young are in the pouch suckling. Embryonic development is resumed when suckling is reduced, and within a short time, there is a second young, which occupies the pouch after the first has left the pouch but continues to suckle. Each young sucks from a different nipple, and the milk each receives is different and suited to the needs of its age (Lincoln and Renfree, 1981; Sharman, 1961).

Eutherians made the important advance of developing placental tissue derived from the embryonic trophoblast that allows for an efficient transfer of nutrients from the mother and protects the embryo from reacting to the maternal immunorejection response. This advance enables them to retain the young in the uterus for a longer period than a single estrous cycle, as in the marsupials, and results in the birth of more advanced altricial young and, ultimately precocial young. With young more advanced at birth, the lactation period can be shortened.

The eutherian mode of reproduction requires modifications of the reproductive cycle to establish mechanisms for prolonging the life of the corpus luteum to provide progesterone in order to maintain pregnancy. Many different modifications have evolved to serve this purpose, and they are described in an excellent review (Amoroso *et al*, 1979).

PATTERNS OF MATERNAL CARE

The mammals have become specialized in phylogeny in the care of their young. Maternal care reaches its most advanced development in this order and particularly in the eutherian mammals, which comprise the largest number of orders with the greatest degree of radiation of new species, genera, and families. Among the mammals, the monotremes represent transitional forms in maternal care largely because of the retention of reptilian egg-laying (ovoparity) among the two families and three species that are extant. There are two basic patterns of maternal care: In one pattern, found in the spiny anteater, the single egg is carried in a special pouch that develops at the time of egg delivery, and the young, when it hatches, lives in the pouch and obtains milk from a milk-secreting gland that lacks a nipple. The second pattern, found in the duck-billed platypus, consists of egg laying by the female in a grass-lined den that she has constructed. The female remains with the eggs and incubates them until they hatch. The young are then cared for in the sealed burrow containing the nest and are provided with milk from hair surrounding the mother's mammary glands. After hatching, in both of the above species, and after the young has left the pouch in the spiny anteater, the mother nurses her young, at variable intervals of up to 1½–2days, and, in the latter species, she shows solicitous attentiveness to her young, which is characteristic of mammalian mothers (Ewer, 1968).

Among the marsupial mammals in which ovoviviparity is the mode of reproduction, there are two basic patterns of maternal care. The majority of marsupials (e.g., the red-necked wallaby and the red kangeroo), those having either permanent or temporary pouches, carry their embryonic young in the pouch for varying periods, usually for several months to the greater part of a year. When the young reach a given age, they leave the pouch but return to the mother to nurse; remaining in her vicinity, they often climb back into the pouch when danger arises. In other species (e.g., the koala and the bush-tailed phalanger), at this time the young cling to the mother's fur and are carried by her in this manner. A number of marsupials (e.g., the woolly opossum) do not have pouches, and the young, also embryonic at birth, cling to the mother's nipples and are transported by her in this manner. After they release the nipples at a later age, they cling to her back or sides and are transported from place to place (Morris, 1965).

Patterns of maternal care among the marsupials are adapted to the mother's need to remain mobile in order both to obtain enough food in the usually barren environments in which they live and to escape predators. The young, particularly, are adapted to cling to the mother, either in the pouch or outside it, rooted to a nipple; should they become dislodged, the mother does not make any effort to reattach them, and they invariably perish.

Among the eutherian mammals the widest variety of species-typical patterns of maternal care have evolved. These can, however, be classified into three basic types, which are clearly related to the developmental status of the newborn and the sedentary or mobile life of the mother, living either underground, on the surface, or at some height in trees or caves.

A sample distribution among the various mammalian orders of newborn varying in status from extremely altricial to extremely precocial, with three additional intermediate grades along this continuum, is shown in Figure 1 (Eisenberg, 1981). It indicates that there is no relationship between newborn status and the level of intellectual development attained in adulthood by members of the order. In each of the orders, the developmental status of the newborn appears to be quite uniform among different species, with some exceptions. Obviously, it is to the mode of life that one must turn to find relationships between the developmental status of the newborn and the patterns of maternal care.

Giving birth to altricial young imposes a pattern of maternal care that involves depositing the young in a nest and providing them with nutrition, warmth, and protection by frequent visits to the next or by occasional visits and sufficient pro-

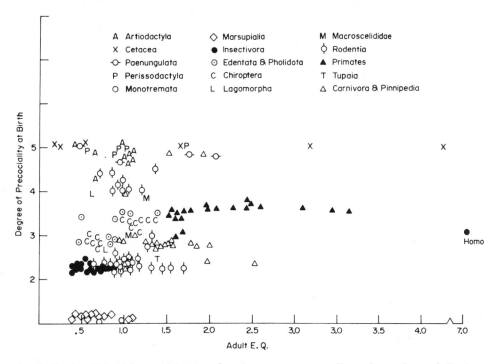

Fig. 1. Grades of altricial-precocial status of newborn among mammalian orders and encephalization quotient (E.Q.). (1) Eyes closed; embyonic. (2) Eyes closed; naked; roll; sometimes can cling (2.5, Fur shows). (3) Eyes barely closed or just open; furred; can crawl well (3.5, Eyes open, can cling). (4) Eyes open; furred; can stand. (5) Eyes open; furred; can walk and follow or swim. (From *The Mammalian Radiations* by J. F. Eisenberg. Chicago: University of Chicago Press, 1981. Copyright 1981 by the University of Chicago Press. Reprinted by permission.)

visioning to allow the young to be protected by the choice of a well-hidden nest site. The mother usually constructs the nest at a site chosen before the young are born, and in addition to nursing the young, she licks them to stimulate elimination. In many species, but not all, she retrieves them to the nest if they should wander from it. In many of these species, the mother exhibits a form of nest defense characterized by aggressive attacks directed at conspecifics that approach her nest and young (maternal aggression).

The above pattern of maternal behavior is characteristic of most rodent species, lagomorphs, carnivores, the guinea pig, and insectivores, with suitable modification based on a large number of special factors in each species.

Precocial young are capable of following the mother shortly after parturition, though they may be nested at birth and for a few days afterward. Among the ungulates, there are two types of patterns as depicted in Figure 2 (Lent, 1974). There are the "hider-type" species (e.g., the roe deer, Thomson's gazelle, and the Uganda kob), in which the young remain in hiding at the nest site and are visited at long intervals by the mother, who spends most of the time away from the nest foraging for food. The mother, however, returns to nurse the young. Hider-type young later become the "follower type," the pattern that is practiced from the beginning by many species (e.g., sheep, caribou, wildebeest, and chamois). In these species, the mother nurses her young and maintains a constant vigil, leading it away from danger and protecting it from predators. Maternal care usually consists of licking the young, leading them, and exhibiting maternal aggression. This care is often characterized by an exclusive bond between the mother and her young that is formed soon after birth in the seclusion of the delivery site away from the social group. This pattern of maternal behavior is found among sheep, goats, elephants, giraffes, many cetacea, and pinnipeds, with important variations depending on the ecological niche of the species and its lifestyle.

A third pattern of maternal behavior is based on the clinging of the newborn to the mother or the carrying of the newborn by the mother. This pattern is found in the semiprecocial primate species chiefly, but also in other orders (e.g., the North African elephant shrew). In some species, mothers may alternate between carrying the young and depositing them in a nest (e.g., the crib among humans).

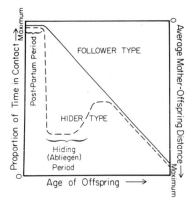

Fig. 2. Two types of early patterns of behavior among mothers and young in the ungulates. Note left ordinate increases from bottom to top and right ordinate increases from top to bottom. (From "Mother-Infant Relationships by Ungulates" by P. C. Lent. In V. Geist and F. Walther [Eds.], *The Behavior of Ungulates and Its Relation to Management*, Vol. 1. IUCN Publications, New Series No. 24, 1974. Copyright 1974 by the International Union for Conservation of Nature and Natural Resources. Reprinted by permission.)

In still others, the mother may deposit her young in a tree nest (e.g., the lesser mouse, the lemur, and the aye-aye).

Onset of Maternal Behavior during Pregnancy and Near Term

Behavioral Changes during Pregnancy and Near Term

Although pregnancy has long been considered a period during which few changes occur in the female's behavior related to her forthcoming delivery and care of young, recent studies in a number of species have indicated that this may not be the case. In this section, we review the recent research on this subject in a number of rodents, in the rabbit and the cat, and among several ungulates for which evidence exists.

Among the rodents, mice and hamsters initiate nest building shortly after mating, during the first 4 or 5 days of pregnancy, long before the females give birth on the 19th or 16th days of pregnancy, respectively (Koller, 1952, 1955; Lisk, 1971; Lisk *et al.*, 1969; Richards, 1969). Other species, such as the rat and the gerbil (and among the lagamorphs, the rabbit), initiate nest building much closer to actual parturition, a day or two before delivery (Elwood, 1975, 1977; Rosenblatt and Siegel, 1975; Wallace, 1973; Zarrow *et al.*, 1962). In the rat, a rise in nest building has been observed starting 36–48 hr before parturition; often, a nest results, but quite frequently, the female deposits nest material at one site and then carries it from there to another site, with the result that no nest is actually constructed until the young appear during parturition. In the rabbit, loosening of the ventral hair precedes the initiation of nest building by a short time, and the hair is then used to line the female's nest in preparation for the young (Zarrow *et al.*, 1962).

Scent marking is common in a number of small mammals, including the rabbit, gerbil, and hamster. In the rabbit and hamster, chin or flank scent-marking declines during pregnancy; in the rabbit, it reaches a low point just before parturition and then rises again during postpartum estrus (Leonard, 1972; Soares and Diamond, 1982). Because scent marking functions to attract males in these species, its decline may discourage males from approaching the female during pregnancy. However, scent marking increases during pregnancy in the gerbil; in this species, it is used to mark the young postpartum (Wallace, 1973).

When young are presented to females during late pregnancy, items of maternal behavior may appear that normally are not seen until the female gives birth. Although it does not represent the natural situation, of course, this procedure is useful for tracing the development of maternal responsiveness during pregnancy and for correlating this development with changes in the pattern of endocrine secretions that regulates pregnancy and initiates parturition.

In certain strains of rats (e.g., Charles River Sprague-Dawley), all components of maternal behavior observed after parturition have been observed during late pregnancy and especially near term; in other strains (e.g., Long Evans and the Wistar; Kristal, 1980; Slotnick *et al.,* 1973), fewer females may be responsive before delivery. In a recent study in which females of a responsive strain were briefly exposed to newborn pups during late pregnancy, it was found that the first rise in the proportion of females retrieving and tending the pups, from 6% to 46%, occurred between Day 20 and Day 21, that is, between 24 and 48 hr before parturition (Mayer and Rosenblatt, 1984). The search for hormonal antecedents of this onset of maternal behavior, however, should focus on the period between Day 19 and Day 20, allowing for a latency of hormone action of between 12 and 24 hr (see Terkel and Rosenblatt, 1972). Before delivery, nearly 100% of females of this strain retrieved, gathered, and crouched over even 4- to 9-day-old pups, although these older pups are less effective stimuli for eliciting maternal behavior than newborn (Mayer and Rosenblatt, 1980; Stern and MacKinnon, 1978). In one study (Karli, 1955), a grown mouse was the object of maternal retrieving and attempts to crouch over it by late pregnant females.

How females respond to pups during the prepartum period depends, in part, on their previous experience in having given birth and having reared litters. It has been shown (Moltz and Wiener, 1966) that experienced (i.e., multiparous) females were able to withstand the effects of ovariectomy on Day 20 of pregnancy, which removed the most important source of the steroid hormones estradiol and progesterone; pups were delivered surgically to prevent difficulties of parturition from occurring after removal of the ovaries, and the females exhibited normal maternal behavior. These findings were taken as evidence that maternal experience enables a female to respond to hormone levels lower than those that normally are present and that are needed by inexperienced females to stimulate maternal behavior.

In a study of ours (Mayer and Rosenblatt, 1984), the effects of experience were investigated in intact females during late pregnancy. Multiparous females killed pups much less frequently than females in their first pregnancies (53% vs 6%), although among the latter all of the killing occurred more than 3½ hr prior to parturition. Multiparous females also became maternal toward pups earlier in pregnancy than inexperienced females. Whereas the latter were mostly unresponsive to older pups until around 3½ hr before parturition, multiparous females became highly responsive to older pups between Day 21 and Day 22, and 84% retrieved them when tested 3½–6 hr before delivering their own. However, as hormonal factors began to play a more important role with the approach of parturition, the differences in responsiveness between inexperienced and multiparous females disappeared, and nearly all showed maternal behavior almost immediately when offered pups.

Another important aspect of maternal behavior, aggression against intruders, was also observed in near-term females. A previous study (Erskine *et al.,* 1978) had not found that females attacked intruders until after parturition. In the present study (Mayer and Rosenblatt, 1984), females were tested during the day on which they were to deliver, starting, therefore, about 6–8 hr before parturition. There was an increase in aggression against young male intruders starting around 5 hr before parturition, but the greatest increase occurred after about 3½ hr before parturition. Only 17% of the females exhibited aggression before 3½ hr prepartum, whereas 83% did so after that time; experienced and inexperienced females did not differ in this respect. In the earlier study, therefore, females may have been tested too early to allow detection of the change in their prepartum aggressive behavior.

The situation during pregnancy among mice is complicated by the fact that strains differ in the degree of maternal responsiveness to pups that they display as virgins. In some strains (Beniest-Noirot, 1958; Noirot, 1972), females are immediately responsive to pups even as virgins, and in others, either they are less responsive to pups but do not cannibalize them to any great extent (Saito and Takahashi, 1979, 1980) or they cannibalize them unless they have recently given birth (Jakubowski and Terkel, 1982).

In a highly responsive strain in which nonpregnant females exhibit all components of maternal behavior within a few minutes of their first exposure to pups, there is an initial *decline* in responsiveness to pups shortly after mating, and the nonpregnant level of responsiveness is not regained until the end of pregnancy (Noirot and Goyens, 1971). In strains in which females are less responsive as virgins, late-pregnant females retrieve more rapidly and in a larger percentage of cases than virgins, and they gather more nest materials (Fraser and Barnett, 1975). The increase in maternal responsiveness (Saito and Takahashi, 1980) develops gradually during pregnancy: Initially, virgins readily perform two items of maternal behavior, licking and retrieving pups, and then the number of maternal behavior components performed increases between the third and seventeenth days. Nest building and crouching over the young begin to appear in a majority of sequences. Cannibalism was rare at the start of pregnancy, then rose to 10% of the observations by midpregnancy, and finally it practically disappeared by the seventeenth day of pregnancy.

In an early study, it was found that female hamsters show the onset of maternal behavior toward young pups around 2½ hr before parturition (Siegel and Greenwald, 1975). Normally, however, when hamster females that have not just given birth are presented with young pups, they cannibalize them, although on rare occasions they exhibit maternal behavior. In tracing the development of maternal responsiveness during pregnancy, therefore, one observes first a decrease in pup killing and then females' beginning to retrieve and attend to pups. At times, there is a precarious balance between the two opposite tendencies: A female retrieves a pup only to cannibalize it, or she cannibalizes one or two pups before retrieving the next ones. Two studies have shown that maternal responsiveness increases from the fourteenth day on: Given optimal testing conditions of three very young pups placed in the nest on Day 14 of pregnancy, 10% of the females were immediately maternal toward them; 100% of the females tested on the fifteenth day were maternal (Buntin, Jaffe, and Lisk, 1984; Siegel *et al.*, 1983).

Cannibalism of pups also declines in gerbil females near term (Elwood, 1977). The decline begins early in pregnancy and continues until parturition, but maternal responses are almost entirely absent until after parturition.

In the rat, in addition to maternal responses shown directly to young placed nearby during late gestation, there are subtle changes indicating an increased responsiveness to stimuli associated with young or to aspects of the pregnancy state. During pregnancy (24–48 hr before delivery), female rats begin to orient to and move toward the cries of pups located outside their cages, and female cats show a selective neurophysiological response to recorded kitten cries as against monotonous tones (Koranyi *et al.*, 1976). Also, the late-pregnant female rat (1–2 days before parturition) begins to discriminate between her own nest odors and those of a nonpregnant female, spending more time with her own nest material than with that of the nonpregnant female. However, she does not distinguish her own pregnancy nest odors from those of another pregnant female, clearly preferring these odors to the nest odors of a nonpregnant female (Bauer, 1983). Because her own pups will eventually acquire the nest odors either from the mother herself or from the nest material, this developing preference for these nest odors may make it easier for the female to lose her initial aversion to pup odors, and thereby to reduce her odor-based avoidance of them (Fleming and Rosenblatt, 1974b,c: Mayer and Rosenblatt, 1975). Finally, late-pregnant females increase their consumption of protein and fat but maintain carbohydrate intake at premating levels (Richter and Barelare, 1938; Leshner *et al.*, 1972). Those that do not, provide inadequate nursing (Scott *et al.*, 1948).

As a consequence of changes due to pregnancy, late-pregnant females show an increase in ultrasonic calling from Day 21 that reaches a peak just before parturition (Lewis and Schriefer, 1985). Whether this entire period represents a protracted beginning of labor or is based on other sources of stress is not clear (Fuchs, 1978).

There is also a gradual increase in self-licking of the mammary regions along the nipple lines and the genital region in mid- to late-pregnant female rats (Roth and Rosenblatt, 1967). By collaring females with a wide rubber collar and thus preventing them from licking themselves, it was found that the mammary glands depend on this licking for the development of glandular secretory tissue (lobular-alveolar development) and for the initiation of milk synthesis (Herrenkohl and Campbell, 1976; McMurtry and Anderson, 1971; Roth and Rosenblatt, 1968).

Still further, there is the development of placentophagia (i.e., eating of placentas) during pregnancy in the rat (Kristal, 1980). This behavior, which normally appears only at parturition, has been found in a small proportion of nonpregnant virgins (4%–20%). This proportion does not change appreciably until the last moments of pregnancy, during delivery, when all females eat placentas; stressing pregnant females, however, has been reported to cause an increase to 50% of the proportion of females that eat placentas after midpregnancy (Kristal *et al.*, 1981).

Of a somewhat different nature is the alteration of thermal homeostasis in rats during pregnancy (Wilson and Stricker, 1979). During pregnancy, there is imposed on the female an increased level of heat production because of hyperphagia and consequent increased metabolism, the added thermal burden of fetal metabolism, the elevated progesterone levels of pregnancy, which are usually associated with heat production, and a reduction in heat loss resulting from the female's increased girth (Figure 3). This hyperthermia must be reduced if the developing fetuses are not to suffer harmful consequences. The pregnant female accomplishes this reduc-

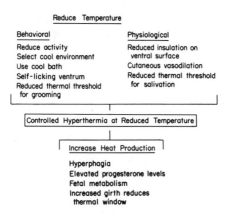

Fig. 3. Factors involved in thermal homeostasis in pregnant rats. Those factors that reduce temperature are above and those that increase it are below. (Based on data from Wilson and Stricker, 1979.)

tion in several ways: She seeks cool environments, she reduces her behavioral activity, and she licks her ventrum, spreading saliva that cools her as it evaporates. This
latter response, which occurs normally at a given temperature, occurs at lower temperatures (i.e., there is a reduced threshold for licking and for the release of saliva
for self-grooming) in pregnant females. There are also physiological mechanisms
that lower the pregnant female's body temperature, including reduced insulation
of her ventral surface and increased cutaneous vasodilation, both of which enable
her to dissipate heat more rapidly.

After females give birth, they continue to be hyperthermic, partly because of
the hormones secreted during lactation. The measures they take to control the
hyperthermia during the postpartum period differ from those taken during pregnancy. Their interaction with their young in the nest causes an increase in body
temperature which forces them to leave the young until their body temperature
falls and they again can enter the nest and be in contact with the young (Leon *et
al.,* 1978; Woodside *et al.,* 1980, 1981; Woodside and Leon, 1980). We shall discuss this behavior more fully in a later section.

Among the ungulates, studies of behavioral changes during pregnancy and
near term are scarce, and those that do exist have focused almost entirely on the
period immediately prepartum. One review (Lent, 1974) indicates that, among the
herd-living animals of this group, there is a tendency for the female to isolate herself from the group before she delivers; this tendency may increase with subsequent births. In the red deer, just before calving time, the female leaves the area
of high population density and occupies a secluded and isolated delivery site, which
later serves as the nursing site and affords the young protection, by concealment
and seclusion, from predators and conspecifics; at the same time, it allows the
mother–young relationship to be established without interference (Clutton-Brock
and Guinness, 1975). In captivity, this species shows restlessness and loss of appetite prior to parturition (Arman, 1974), as does the reticulated giraffe observed
before, during, and after parturition in the Buffalo Zoological Gardens (Kristal and
Noonan, 1979). In addition to these changes, red deer, free-ranging on a farm,

were observed to become very aggressive and to bellow in late pregnancy as parturition became imminent (Arman *et al.*, 1978). Little is known about behavioral changes during pregnancy in seals, but they are described as less aggressive than postparturient females with young (McCann, 1982).

Nest building of an elaborate sort has been described during late pregnancy in the domestic pig and the European wild board (Frädrich, 1974; Naaktgeboren, 1979). A depression the length of the female's body is dug with the snout. It is filled with layers of dry grass, leaves, and small sticks and then shaped to the female's body and covered again with layers of sticks, leaves, grass, and hay.

Although it has been reported anecdotally that late-pregnant goats and sheep attempt to steal young from their mothers (Hersher *et al.*, 1963), an observation implying that they display a high level of maternal responsiveness before delivery, this behavior has not been studied experimentally. It has been reported that a merino ewe, at the point of labor, was interested in the placental fluids and newly born lambs of other ewes (Alexander, 1960). This report is in accord with a very thorough study in ewes of the Préalpes du Sud race of the attractiveness of amniotic fluid to the ewe at various stages of the reproductive cycle (Levy, 1980–1981). Food mixed with either water or amniotic fluid was offered simultaneously to ewes, and their willingness to eat the food mixed with amniotic fluid, relative to that mixed with water, was used as a measure of their preference for the fluid. At all stages of the estrous cycle, amniotic fluid was aversive to the ewes, and it was only slightly less aversive during late pregnancy, when dexamethasone was given to bring on delivery. Only when parturition began, 39–49 hr after the dexamethasone, did the ewes show a strong perference for food mixed with amniotic fluid; this preference was maintained for a relatively brief period and began to fade by 4 hr after parturition.

A somewhat similar finding has been reported in the beagle (Dunbar *et al.*, 1981). On the day prior to their own delivery, bitches were presented with their own amniotic fluid from a previous birth daubed on a foster puppy, and their retrieval and licking behavior directed toward this puppy, as compared to that directed toward a puppy daubed with plain water, was noted. During four tests totaling 20 min, the females only briefly investigated the puppies, equally, but did not lick or retrieve either of them. When tested with the same puppies after parturition, the one with amniotic fluid was licked 40 times more frequently than the one daubed with water. The mothers were not tested during delivery, as the ewes were, so that it could be determined how precipitously the attraction to amniotic fluid developed.

Among ewes, particularly experienced ewes, there is an increase in maternal responsiveness toward newborn lambs, starting around 10 days before parturition, that continues to rise until parturition (Poindron and LeNeindre, 1979).

Among the *marsupials*, there are no discernible behavioral differences between pregnant and nonpregnant females, both undergoing an estrous cycle until parturition intervenes in some species. Preparturition pouch-cleaning has been reported in the gray kangaroo, the tammar wallaby, and the opossum (Berger, 1970; Hamilton, 1963; Poole and Pilton, 1965; Sharman and Calaby, 1964).

HORMONAL BASIS OF THE ONSET OF MATERNAL BEHAVIOR

There is a question whether this section should precede or follow the section on parturition, and the answer lies in whether the hormonal stimulation of maternal behavior requires the endocrine changes of parturition or is already completed before these changes occur. Stated differently, do animals require parturition in order to become maternal, or are they maternal before parturition in order to enable them to respond appropriately to their young as they appear during delivery? We believe that the latter formulation is more likely to be true, although parturition also plays an important role in maternal behavior, a subject that we discuss in a later section.

ENDOCRINE CONTROL OF PREGNANCY AND PARTURITION

The endocrine control of pregnancy and parturition in the species for which data are available is a highly complex topic that is beyond the scope of this chapter. Of particular importance in all species are the hormones estradiol and progesterone, which are secreted by the ovaries, the placenta, or both in different species. A general, comparative review and theory of the role of these hormones in pregnancy and parturition was proposed in a volume (Heap *et al.*, 1977) that also includes detailed reviews of many topics pertinent to our subject (Ciba Foundation Symposium 47, "The Fetus and Birth"). In addition, two volumes that appeared in the 1970s contain chapters of interest to any one who wishes to learn more about the hormonal control of pregnancy and parturition: Ellendorff *et al.* (1979) and Klopper and Gardner (1973).

In an earlier chapter on this subject (Rosenblatt and Siegel, 1981), we published the endocrine profiles during pregnancy and at parturition of the principal domestic and laboratory species that have been used in the study of maternal behavior. The common pattern among all of these species is the rise in circulating levels of progesterone early in pregnancy that is maintained throughout most of pregnancy until the decline, which occurs shortly before parturition. During most of pregnancy, circulating estradiol is maintained at low levels, but shortly before parturition, the level increases sharply and the ratio of circulating levels of estradiol to the levels of progesterone is reversed from that which was present during most of pregnancy. There is an exception in the hamster, among the subprimate mammals, in that there is a fall in circulating levels of both progesterone and estradiol at the end of pregnancy.

Another hormone of interest is prolactin, a hormone released by the anterior pituitary gland. It is maintained at low circulating levels throughout most of pregnancy until the levels rise sharply, shortly before parturition, in anticipation of the initiation of lactation. Endocrine profiles of circulating levels of estradiol, progesterone, and prolactin during pregnancy are shown for the rat, mouse, hamster, rabbit, and sheep in Figures 4 to 8.

ENDOCRINE STIMULATION OF MATERNAL BEHAVIOR

An endocrine basis for maternal behavior among eutherian mammals is strongly suggested by a number of factors: The onset of maternal behavior is asso-

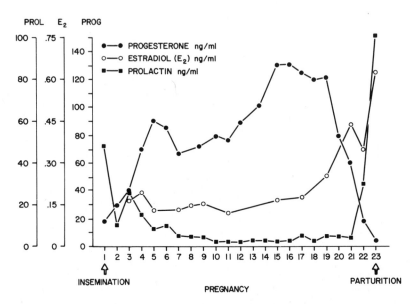

Fig. 4. Circulating levels of estradiol (E_2), progesterone (Prog), and prolactin (Prol) during pregnancy in the rat. Based on Pepe and Rothchild (1974) progesterone; Shaikh (1971) estradiol; Morishige *et al.* (1973) prolactin. (From "Progress in the Study of Maternal Behavior in the Rat: Hormonal, Nonhormonal, Sensory, and Developmental Aspects" by J. S. Rosenblatt, H. I. Siegel, and A. D. Mayer. In J. S. Rosenblatt, R. A. Hinde, C. G. Beer, and M.-C. Busnel [Eds.], *Advances in the Study of Behavior,* Vol. 10. New York: Academic Press, 1979. Copyright 1979 by Academic Press, Inc. Reprinted by permission.)

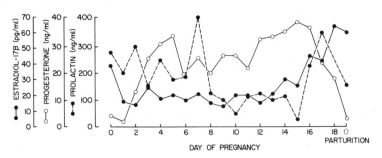

Fig. 5. Circulating levels of estradiol-17β, progesterone, and prolactin during pregnancy in the mouse. Based on McCormack and Greenwald (1974) estradiol-17β and progesterone; Murr *et al.* (1974) prolactin. (From "Factors Governing the Onset of and Maintenance of Maternal Behavior among Nonprimate Mammals: The Role of Hormonal and Nonhormonal Factors" by J. S. Rosenblatt and H. I. Siegel. In D. J. Gubernick and P. H. Klopfer [Eds.], *Parental Care in Mammals.* New York: Plenum Press, 1981. Copyright 1981 by Plenum Press. Reprinted by permission.)

ciated with a combination of hormonal changes that are somewhat unique in the reproductive cycle of the female. Moreover, there is little else to account for the dramatic changes that occur in the female's behavior between the time she begins parturition and her initial responses to her newly emerging young; the change in the female appears to be internally activated rather than to be stimulated solely by the events of parturition. In confirmation are many observations that late-pregnant female rats delivered by cesarean section, and therefore unable to be stimulated by the process of delivery itself (however, subjected to the same hormonal stimulation as normally delivering mothers), nevertheless respond almost immediately to pups when they are fully recovered from the anesthetic used for the delivery (Mayer and Rosenblatt, 1980). It is remarkable to observe a female goat awakening

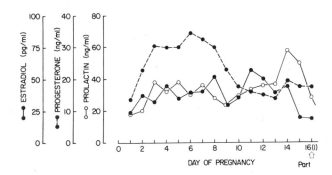

Fig. 6. Circulating levels of estradiol, progesterone, and prolactin during pregnancy in the hamster. Based on Baranczuk and Greenwald (1974) estradiol and progesterone; Bast and Greenwald (1974) prolactin. (From "Factors Governing the Onset of and Maintenance of Maternal Behavior among Nonprimate Mammals: The Role of Hormonal and Nonhormonal Factors" by J. S. Rosenblatt and H. I. Siegel. In D. J. Gubernick and P. H. Klopfer [Eds.], *Parental Care in Mammals.* New York: Plenum Press, 1981. Copyright 1981 by Plenum Press. Reprinted by permission.)

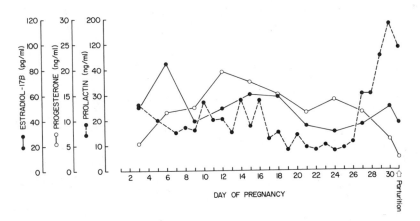

Fig. 7. Circulating levels of estradiol-17β, progesterone, and prolactin during pregnancy in the rabbit. Based on Challis *et al.,* (1973) estradiol-17β and progesterone; McNeilly and Friesen (1978) prolactin. (From "Factors Governing the Onset of and Maintenance of Maternal Behavior among Nonprimate Mammals: The Role of Hormonal and Nonhormonal Factors" by J. S. Rosenblatt and H. I. Siegel. In D. J. Gubernick and P. H. Klopfer [Eds.], *Parental Care in Mammals.* New York: Plenum Press, 1981. Copyright 1981 by Plenum Press. Reprinted by permission.)

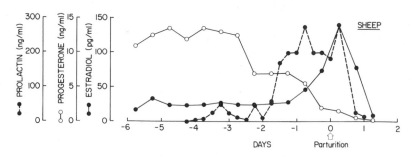

Fig. 8. Circulating levels of estradiol, progesterone, and prolactin during the last 6 days of pregnancy in sheep. Based on Chamley *et al.* (1973) estradiol, progesterone, and prolactin. (From "Factors Governing the Onset of and Maintenance of Maternal Behavior among Nonprimate Mammals: The Role of Hormonal and Nonhormonal Factors" by J. S. Rosenblatt and H. I. Siegel. In D. J. Gubernick and P. H. Klopfer [Eds.], *Parental Care in Mammals.* New York: Plenum Press, 1981. Copyright 1981 by Plenum Press. Reprinted by permission.)

from a cesarean-section delivery, stretching her neck toward her newborn kid, lying nearby, with which she has had no prior contact, and immediately beginning to lick it, just as she would have done had she given birth to it normally (Rosenblatt, unpublished observation, 1979).

In the rat, the observational evidence is, in addition, supported by experiments other than cesarean-section delivery. An early effort attempted to mix the blood of one female that gave birth and exhibited maternal behavior with that of another undergoing estrous cycling (i.e., by forming a parabiotic union) to see whether the latter female would begin to display maternal behavior (Stone, 1925). This experiment failed, but a later experiment in which blood was taken from females that had just given birth and were displaying maternal behavior and was given to females at various times during the estrous cycle was more successful in stimulating a somewhat rapid (i.e., latency of about 2 days) onset of maternal behavior (Terkel and Rosenblatt, 1968; Koranyi *et al.,* 1976). An improvement in this procedure was even more successful and resulted in a more rapid stimulation of maternal behavior in about 14 hr (Terkel and Rosenblatt, 1972). Females were implanted with chronic heart catheters with a long tube emerging from the back of the neck that enabled us to join together the circulatory systems of two females and to pump blood from one to another in a procedure that was labeled *cross-transfusion.* After 6 hr of cross-transfusion of blood between a mother that had just given birth and a nonpregnant intact female, at which time both had shared equal amounts of the other's blood for 3 hr, the cross-transfusion was terminated, and females were observed for their behavior toward pups that had been given to them at the start of the transfusion. Nearly 90% of the recipient females became responsive in less than 24 hr when blood was cross-transfused soon after parturition, but blood cross-transfused at a later time (24 hr postpartum) or earlier (27 hr prepartum) was ineffective in stimulating maternal behavior.

No comparable studies have been performed on other mammalian species; nevertheless, it is taken as axiomatic that the onset of maternal behavior among the eutherian mammals is based on hormonal stimulation.

The situation among metatherian mammals, in which the duration of pregnancy does not extend beyond the length of a single estrous cycle and may be considerably shorter, is not as clear as among the eutherian mammals, which have been studied much more extensively. First, it is not simple to decide how much maternal responsiveness is present at parturition in these species. In pouch-bearing species, the newly born young climb into the pouch with minimal aid from the mother. Similarly, in those species in which at birth the young grasp nipples and cling to the mother, the initiative exerted by the mother is not readily apparent. In both pouch-bearing and clinging species, if the young fail to reach and grasp a nipple they are not given assistance by the mother and are left to die.

The extent of maternal care by the parturient tammar wallaby mother is cleaning of the pouch (Russell and Giles, 1973), but this can be stimulated in cycling females by placing a newborn young in the female's pouch.

It is consistent with these observations that, in the Virginia opossum, circulating levels of estrogen and progesterone in pregnant females throughout the gestation period of 12.5 days do not differ from those of nonpregnant females for the equivalent period undergoing an estrous cycle (Harder and Fleming, 1981). Hormones that were not measured in this study might, of course, differ between pregnant and nonpregnant cycling females. However, even the mammary gland development that occurs during the estrous cycle is similar in pregnant and nonpregnant female bush opossums during the 17 days that precede parturition and begins to differ only when the newborn begin to suckle (Sharman, 1961).

In all of these species, the mother is not faced with attending to young on her own initiative or at their insistence until the young are far more grown than they are at parturition. In some species, this may not be until the young are around 6 months of age and can leave the pouch or can release the nipple and cling to the mother in a more voluntary fashion (see Russell, 1973). It may be that it is this period of maternal care that is comparable to the immediately postpartum period seen in eutherian mammals, and it is at this time that the possibility of a hormonal basis of maternal behavior should be sought.

ANALYTICAL STUDIES OF THE HORMONAL BASIS OF MATERNAL BEHAVIOR

Although there is a more or less uniform pattern of circulating levels of estradiol, progesterone, and prolactin during pregnancy in a wide variety of mammalian species, this is no assurance that hormonal stimulation of maternal behavior is also uniform in these species. It is necessary to analyze the specific hormones of the combination of hormones that are responsible for the onset of maternal behavior in each species. To date, this analysis has been begun in only a handful of studies; in this section, we review, in summary form, the present status of findings in these species.

The largest number of studies, and therefore the greatest amount of progress, has been made in the laboratory rat on which studies of the hormonal basis of maternal behavior date back to the 1920s. A view proposed by early investigators (Riddle *et al.,* 1942) was that prolactin was the "maternal hormone" in the rat. This view has not stood up under later attempts to duplicate the original findings (see Rosenblatt and Siegel, 1981); instead, more recent studies have shown that a combination of hormones administered in a definite pattern (see Figure 9) can successfully stimulate the rapid onset of maternal behavior in the ovariectomized rat (Moltz *et al.*, 1970; Zarrow *et al.*, 1971). In the schedule of hormone treatment shown in the upper section of this figure, the latencies for maternal behavior were 34–40 hr and in the schedule shown in the lower section of the figure, the latencies were even shorter.

These studies established that the combination of estradiol, progesterone, and prolactin administered in a pattern that was intended to duplicate the circulating levels of these hormones found during pregnancy (Figure 4) stimulated maternal behavior.

Because pituitary, ovarian, and placental hormonal interrelationships are quite complicated at the end of pregnancy when maternal behavior arises in the rat, it has thus far proved difficult to directly intervene at this time to determine which hormones are actually stimulating maternal behavior. One preparation that has been used, therefore, is that described above: the ovariectomized female treated with hormones and tested for maternal behavior toward the end of the treatment. We have developed a second preparation for studying the hormonal basis for the onset of maternal behavior (Roseblatt and Siegel, 1975; Siegel and Rosenblatt, 1975b). It is based on an earlier finding (Weisner and Sheard, 1933) that terminating pregnancy prematurely late in gestation in the rat initiates maternal behavior shortly afterward. Underlying this phenomenon is the fact that, dur-

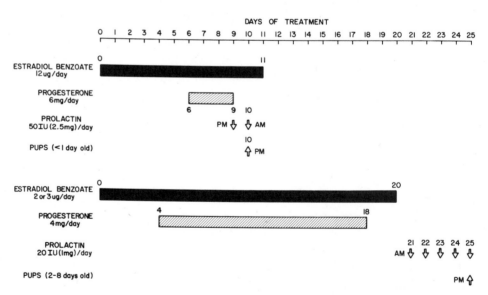

Fig. 9. Hormone treatments and test conditions used by (upper figure) Moltz *et al.* (1970) and (lower figure) Zarrow *et al.* (1971) to stimulate maternal behavior in ovariectomized rats. (From "Progress in the Study of Maternal Behavior in the Rat: Hormonal, Nonhormonal, Sensory, and Developmental Aspects" by J. S. Rosenblatt, H. I. Siegel, and A. D. Mayer. In J. S. Rosenblatt, R. A. Hinde, C. G. Beer, and M.-C. Busnel [Eds.], *Advances in the Study of Behavior*, Vol. 10. New York: Academic Press, 1979. Copyright 1979 by Academic Press, Inc. Reprinted by permission.)

ing pregnancy, estrous cycling is suspended, but following pregnancy termination, either naturally at term or experimentally by using hysterectomy (i.e., removing uteri and fetuses plus placentas) or cesarean-section delivery (i.e., removing fetuses and placentas only), females undergo an estrous cycle (it is called *postpartum estrus* when it occurs immediately after parturition); circulating levels of estradiol-17B rise followed by a rise in progesterone and by the appearance of estrous behavior. Our studies using hysterectomy to terminate pregnancy showed that, from mid-pregnancy onward, hysterectomy was followed by the appearance of maternal behavior with latencies that ranged from 3 or 4 days to 24–48 hr close to term (Figure 10; Rosenblatt and Siegel, 1975). On the other hand, females that were not hysterectomized and that remained pregnant did not exhibit maternal behavior with short latencies. The hysterectomy, therefore, was the starting point for hormonal changes that stimulated maternal behavior. Now, instead of having to depend on gradual and uncontrolled changes in endocrine secretions, we could initiate these changes with the hysterectomy, and they occurred more rapidly. We made an important assumption in these early studies that has in large part been supported by later studies: We assumed that the hormonal changes following hysterectomy were the same as those that occur at the end of pregnancy and, therefore, that the onset of maternal behavior in the two situations was based on the same hormonal mechanism.

When the ovaries were removed at the time of hysterectomy in each of the groups of pregnant animals, the onset of maternal behavior, compared to that in groups that were only hysterectomized, was delayed. This finding indicated that, after hysterectomy, the ovarian secretions were instrumental in stimulating the short-latency maternal behavior. This conclusion led us to our next experiments, in which we hysterectomized and ovariectomized (HO) pregnant females on the sixteenth day of pregnancy and injected them with estradiol benzoate (EB) and, in some cases, with EB and progesterone (P) to stimulate sexual behavior that we had observed in our hysterectomized females about 48 hr after surgery. EB restored the short-latency onset of maternal behavior, and P did not have any discernible effect (Figure 11). Similar studies were then performed on nonpregnant females (Siegel and Rosenblatt, 1975b): Estrous-cycling females were ovariectomized and hysterectomized and then given EB, and maternal behavior appeared with relatively short latencies compared to those in HO females (Figure 12).

Fig. 10. Cumulative percentage of groups retrieving on days following hysterectomy. Pregnant females hysterectomized on the 10th (10 H), 13th (13 H), 16th (16 H), and 19th (19 H) days of pregnancy. Pups were presented at 48 hr postsurgery; latencies measured from the beginning of pup exposure. (From "Hysterectomy-Induced Maternal Behavior during Pregnancy in the Rat" by J. S. Rosenblatt and H. I. Siegel, *Journal of Comparative and Physiological Psychology*, 1975, *89*, 685–700. Copyright 1975 by the American Psychological Association. Reprinted by permission.)

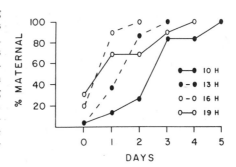

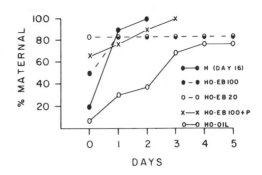

Fig. 11. Cumulative percentages of 16-day pregnant rats showing retrieving after hysterectomy (H) or hysterectomy-ovariectomy (HO) plus estradiol benzoate (EB) treatment of 100 or 20 µg/kg and progesterone (P) at 0.5 mg. Retrieving tests started 48 hr after surgery and EB injection. P was injected 4 hr prior to first test. (From "Progress in the Study of Maternal Behavior in the Rat: Hormonal, Nonhormonal, Sensory, and Developmental Aspects" by J. S. Rosenblatt, H. I. Siegel, and A. D. Mayer. In J. S. Rosenblatt, R. A. Hinde, C. G. Beer, and M.-C. Busnel [Eds.], *Advances in the Study of Behavior*, Vol. 10. New York: Academic Press, 1979. Copyright 1979 by Academic Press, Inc. Reprinted by permission.)

Because hysterectomy is not a "natural" stimulus for the termination of pregnancy, we have also used prostaglandin $F_2\alpha$ ($PGF_2\alpha$), normally secreted by the uterus at term, which is believed to play an important role in terminating pregnancy in the rat (Strauss *et al.*, 1975). At 16 and 19 days of pregnancy, the use of $PGF_2\alpha$ terminated pregnancy in about 30 hr, and there was a rapid onset of maternal behavior shortly afterward (Rodriguez-Sierra and Rosenblatt, 1982).

Once we had gained this foothold on the problem of the hormonal basis of maternal behavior, further analysis was required to clarify the possible role of prolactin and the influence of progesterone. Prolactin release can be significantly reduced by injecting EB-treated females with either ergocornine (ergocornine hydrogen maleate), ergocryptine (2-bromo-α-ergocryptine), or apomorphine (apomorphine hydrochloride). EB stimulated maternal behavior as readily in females treated with these prolactin-release-blocking substances as without them (Numan *et al.*, 1977; Rodriguez-Sierra and Rosenblatt, 1977).

The situation with progesterone is more complex. During pregnancy, there is a prolonged period, prior to the rise in estrogen secretion, when there are high circulating levels of progesterone, which then decline as estrogen levels rise. Two

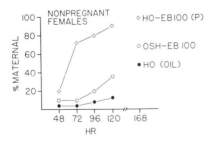

Fig. 12. Cumulative percentage of nonpregnant rats showing retrieving after hysterectomy-ovariectomy (HO) or ovariectomy sham hysterectomy (OSH). Groups were treated with 100 µg/kg estradiol benzoate (EB) or oil at surgery. Progesterone (P) was given 44 hr after EB. Retrieving tests started 48 hr postoperatively. (From "Progress in the Study of Maternal Behavior in the Rat: Hormonal, Nonhormonal, Sensory, and Developmental Aspects" by J. S. Rosenblatt, H. I. Siegel, and A. D. Mayer. In J. S. Rosenblatt, R. A. Hinde, C. G. Beer, and M.-C. Busnel [Eds.], *Advances in the Study of Behavior*, Vol. 10. New York: Academic Press, 1979. Copyright 1979 by Academic Press, Inc. Reprinted by permission.)

questions arise: First, is the decline of progesterone a necessary condition for the action of estrogen, and second, does the prolonged period of exposure to high levels of progesterone play any role in the estrogen stimulation of maternal behavior? Summarizing a number of studies reported from different laboratories including our own, with respect to the first question, we can say that overlap of high levels of progesterone with estrogen during the first 24 hr of estrogen activity significantly blocks estrogen stimulation of maternal behavior (Bridges and Feder, 1978; Numan, 1978; Siegel and Rosenblatt, 1975c, 1978). The decline in progesterone at the end of pregnancy, just at the time when estrogen is rising, therefore, is a necessary condition for estrogen action. If high levels of progesterone are maintained by injecting progesterone on Day 21 of pregnancy, half of the females fail to exhibit maternal behavior following cesarean-section delivery, which is necessary under these conditions (Moltz *et al.*, 1969). Anytime high levels of circulating progesterone are maintained while estrogen levels are rising, maternal behavior is inhibited in a large proportion of females.

Yet the prolonged period of high circulating levels of progesterone during pregnancy aids estrogen in the arousal of maternal behavior. Progesterone implants, which are given to nonpregnant ovariectomized females for 13–16 days and which are then removed before estrogen is given, facilitate the action of the estrogen, and either lower doses of estrogen are able to stimulate maternal behavior than when progesterone has not been given to the females or latencies are shorter to the same estrogen does (Bridges and Russell, 1981; Doerr *et al.*, 1981).

Recalling the earlier effective hormone treatments (Figure 9), we can now say that the lengthy period of progesterone administration, its termination, and the subsequent administration of estrogen were probably the crucial aspects of the hormonal stimulation; the prolactin treatment probably did not play any role in these treatments.

Several studies have suggested that oxytocin, a hormone released during gestation and suckling, which acts on the uterus to facilitate vigorous uterine contractions during delivery and on the smooth muscles to express milk from the mammary alveoli may be involved in the onset on maternal behavior (Pedersen and Prange, 1979; Pedersen *et al.*, 1982). In these studies, ovariectomized females are treated with EB and, 48 hr later, are given a single injection of oxytocin directly into the intracerebral ventricle of the brain. A significant proportion exhibit maternal behavior when exposed to pups immediately after the oxytocin injection, more than when a control substance is injected. The effectiveness of this treatment has been confirmed in one strain of rat (Zivic Miller), by Fahrbach, Morrell, and Pfaff (1984) but two other laboratories using different strains have not been able to replicate these findings (Rubin, Menniti, and Bridges, 1983; Swanson and Bolwerk, 1983). The issue, therefore, remains unsettled until further research on this problem is done.

Our findings that maternal aggression in the rat arises prepartum rather than postpartum, as reported earlier (Erskine *et al.*, 1978), raise the possibility that the hormonal basis of this behavior may be similar to that of maternal behavior in general, with which it is temporally related at its onset (Mayer and Ronseblatt, 1984).

Two fairly successful attempts to stimulate maternal aggression hormonally have been reported. In one, late-pregnant HO females, given estradiol benzoate, exhibited levels of maternal aggression equal to those of lactating mothers, and in a second study, the hormone regime shown in the upper section of Figure 9 was used with ovariectomized females, with the result that nearly 40% of the females exhibited maternal aggression, a percentage equal to that of lactating females in this study (Krehbiel and LeRoy, 1979; LeRoy, 1977).

The hormonal basis of maternal nest-building in the rabbit has received considerable study, but other aspects of maternal behavior, such as maternal aggression and nursing (rabbits do not retrieve their young), have not been studied. Several procedures have been used, including pregnancy termination, by various means, at different times during pregnancy; hormone treatment of ovariectomized females; and, to study the role of prolactin, hypophysectomy and the use of the prolactin-release-blocking agent ergocornine (see detailed review in Rosenblatt and Siegel, 1981). These studies have shown that prolactin, secreted in large amounts shortly before parturition (Figure 7), is crucial in stimulating the near-parturition onset of maternal nest-building (Zarrow *et al.*, 1971). Ovariectomized female rabbits that are treated with estrogen and progesterone for a prolonged period (more than 18 days) exhibit maternal nest-building, but if they are hypophysectomized, they fail to do so. Also, hypophysectomy performed on females during the last third of pregnancy results in failure to build nests after the abortion has terminated the pregnancy (Anderson *et al.*, 1971). Ergocornine also prevented the onset of maternal nest-building when parturition took place, whereas replacement of prolactin restored this behavior, a finding further implicating prolactin in the onset of nest building. The roles of estrogen and progesterone in the onset of maternal nest-building and the relationship between the endocrine control of nest building and other maternal behavior in the rabbit remain to be studied.

In the mouse, the prepartum onset of maternal nest-building has been attributed to the joint action of estrogen and progesterone (Lisk, 1971). As Figure 5 shows, circulating progesterone increases between Day 1 and Day 5 of pregnancy, this increase being preceded by a slight rise in estradiol on Day 3 and a decline on Day 4 and thereafter. Treatment of ovariectomized females with a low level of estrogen followed by progesterone results in the onset of nest-building at a level above that which normally occurs (Lisk, 1971). Other components of maternal behavior (in the responsive strains that have been studied) appear to be largely nonhormonal in nature, as they appear in virgins without hormonal stimulation. Prolactin and progesterone have been described as enhancing maternal behavior but as not being crucial to its appearance (Voci and Carlson, 1973).

Among certain strains of mice, the onset of maternal aggression arises early prepartum, increases throughout pregnancy, is maintained after parturition, and declines during the third week (Noirot *et al.*, 1975). In other strains, it appears that maternal aggression arises during the first 48 hr postpartum and reaches a peak at 3 days (Svare and Gandelman, 1973). Because it is these strains that have been studied for the hormonal onset, we shall review these findings. A good deal of research, summarized in several reviews (Rosenblatt and Siegel, 1981; Svare, 1981;

Gandelman, 1980), has shown that maternal aggression does not arise spontaneously at or after parturition. The female requires a 48-hr period immediately postpartum, during which she is suckled by her young, before she exhibits maternal aggression. Although one might expect that suckling, which causes the release of prolactin, might act in this manner, it has been shown by several methods that prolactin secretion plays no role in the onset of this behavior. No association between the circulating levels of prolactin and maternal aggression has been found (Broida et al., 1981), and when ergocornine and ergocryptine were given to females to block prolactin release starting at delivery, there was no effect on maternal aggression. At later times, maternal aggression was not affected by hypophysectomy performed before the ninth day postpartum (Mann et al., 1980; Svare et al., 1982).

Where a prepartum onset of nest building in the hamster has been reported (Richards, 1969; Daly, 1972), it has been attributed to circulating levels of estrogen and progesterone during pregnancy (Figure 6). Implants of estrogen and progesterone acting in ovariectomized females for 18 days, 2 days longer than the 16-day gestation period of the hamster, were effective in stimulating maternal nest-building. However, the other components of maternal behavior in the hamster (retrieving, crouching over young, etc.) were not stimulated in ovariectomized females by daily injections of estradiol benzoate and progesterone (cited in Siegel and Rosenblatt, 1980). Moreover, preventing the near-term decline in either progesterone or estradiol or both of these hormones by injecting them into near-term females had no effect on the onset of maternal behavior. None of the pregnancy-termination procedures that elicited maternal behavior in rats has been effective in stimulating the onset of maternal behavior in hamsters (Siegel and Rosenblatt, 1984).

Maternal aggression in the hamster has its beginning early in pregnancy and increases to near-maximum levels at midpregnancy, continuing throughout lactation. It has been related to prolactin secretion on the basis of studies done during the postpartum period (Wise and Pryor, 1977). Females that are ovariectomized to remove estrogen and progesterone from circulation, without effect on maternal aggression or maternal care, and then are given ergocornine to block the release of prolactin, show almost a total absence of maternal aggression. Other aspects of maternal behavior are also affected: Pup killing is increased, there are poorer nests, and less time is spent with the pups. Prolactin does not fully restore maternal aggression nor inhibit pup killing in a large proportion of the females.

Little is known of the hormonal regulation of gestation in the gerbil, and therefore, it is difficult to assign the behavioral changes during this period to a hormonal background. The increase in scent marking during pregnancy may be associated with an increase in estrogen at the end of pregnancy, as ovariectomy 2 days before parturition prevents this increase, and in nonpregnant females, the ovariectomy-induced decline in scent marking is restored by estrogen (Owen and Thiessen, 1973, 1974). The more important change from cannibalism to maternal care begun during pregnancy, has not yet been studied hormonally. Gerbils that are hysterectomized and hysterectomized-ovariectomized two days before parturition both show marked deficiencies in maternal behavior, suggesting either an effect of surgery *per se* or some common hormonal process affected equally by both surgical procedures (Wallace, 1973).

Several other hormonal procedures have been employed in efforts to unravel the hormonal basis of maternal behavior in the gerbil, but none has veen very successful. Progesterone has been given starting about 6 days before delivery, to prevent the presumed terminal decline in this hormone, without effect on maternal behavior following surgical delivery. A modification of the procedure shown in the upper part of Figure 9 was used with effects that included pup killing, *reduced* sniffing and licking of pups, and no important display of maternal care (Wallace, 1973).

Among ungulates, little is known about the hormonal basis for the onset of maternal behavior in the pig, cow, and goat, even though the endocrine picture in these species during pregnancy and at its termination (see Rosenblatt and Siegel, 1981) resembles that of the rat and other rodents, for which at least some information about the hormonal basis of maternal behavior is available. Only the ewe has been studied extensively, and this species, too, shows a hormone picture during pregnancy that resembles that seen in the rat (Figure 8). Initial explorations of the hormonal basis of maternal behavior measured maternal reponsiveness at times during the reproductive cycle when circulating levels of estradiol were higher than at other times (reviewed by Poindron and Le Neindre, 1980). It should be stated at the outset that in all of the studies cited, unless otherwise specified, the ewes were multiparous; many of the hormonal effects on maternal behavior are not seen in nulliparous ewes. The periods when circulating levels of estradiol are higher than at other times include estrus, at about 140 days of gestation (10 days before parturition), and at parturition itself; at all of these times, maternal behavior was increased toward newborn lambs. The greatest increase occurred at parturition, and here, estradiol levels are at peak values (Poindron and Le Neindre, 1980). Equally significantly, when tests were conducted while circulating levels of progesterone were high (Days 45, 75, and 105 of gestation), females were unresponsive to lambs.

On the basis of this study, direct treatment of ovariectomized ewes was initiated: Long- and short-term treatments (30 days and 7 days, respectively) with EB or a combination of progesterone and EB were both successful in stimulating 50%–60% of the females to accept newborn lambs (Poindron and Le Neindre, 1979). A 1-day treatment with either EB or progesterone was even more successful in stimulating maternal behavior: At the end of 24 hr, more than 80% of females receiving EB were maternal, as well as about 45% of these receiving progesterone. In ewes, progesterone can be converted to estradiol, and this conversion may account for the effectiveness of this hormone.

As in the rat, estrogen might be effective in the ewe through the secretion of prolactin. This possibility was examined in a study in which high levels of circulating prolactin were maintained in two groups of females, and low levels were produced by ergocryptine in a third. In addition, estradiol levels were maintained at high levels in two groups, one with a high level of prolactin and the other with very low levels. Ewes with the following combinations of estradiol and prolactin were therefore tested for maternal behavior at 24 hr, the first time they were exposed to their own 24-hr-old lambs: (1) high estradiol, low prolactin; (2) high prolactin, high estradiol; and (3) high prolactin, low estradiol. Groups 1 and 2 exhibited maternal behavior in 77% and 61% of the cases, whereas ewes with high prolactin levels but low estradiol levels (Group 3) were maternal in only 14% of the cases.

One further study showed that estrogen may substitute, to some extent, for lamb stimulation in maintaining maternal behavior after parturition. Ewes with reduced contact with lambs during the first 24 hr after parturition were delivered with either dexamethasone or estradiol benzoate treatment, with the consequence that, in the latter group, estrogen levels were high for this period. At the end of 12 hr, a larger percentage of high-estrogen females were responsive to the lambs; the differences were greater than 30% in each group (Figure 13).

Among the carnivores, only the domestic cat, the puma (Bonney *et al.,* 1981), and the beagle have been studied with respect to the endocrine control of pregnancy, and all show the same general pattern as rodents and the ungulates studied with respect to circulating levels of estrogen and progesterone. The dog also shows the terminal rise in prolactin (see review in Rosenblatt and Siegel, 1981). Bitches exhibit maternal behavior following mating-induced pseudopregnancy lasting the full length of actual pregnancy. Toward the end of pseudopregnancy, as in normal pregnancy, the female searches for a secluded parturition site, and there she

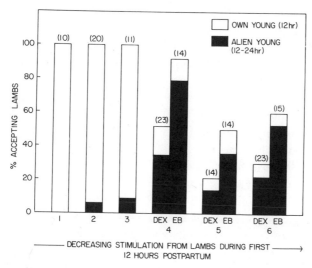

Fig. 13. Early postpartum maintenance of maternal behavior in the ewe. During the first 12 hr postpartum, the ewes had the following kinds of contact or no contact with their own lambs: (1) full contact, including suckling; (2) full contact without suckling; (3) lambs nearby, but no licking or suckling; (4) lamb at 1 m distance; (5) lamb enclosed in transparent box allowing sight and hearing; and (6) completely separated from lamb. Dex, dexamethasone-induced parturition, normal levels of plasma estradiol; EB, estradiol-benzoate-induced parturition, elevated levels of plasma estradiol. Tests with own and alien lambs were conducted at 12 hr postpartum. Only ewes that spontaneously accepted their own lambs were tested with alien lambs afterward. Numbers in parentheses are sample sizes. Data from Poindron and Le Neindre (1980). (From "Factors Governing the Onset of and Maintenance of Maternal Behavior among Nonprimate Mammals: The Role of Hormonal and Nonhormonal Factors" by J. S. Rosenblatt and H. I. Siegel. In D. J. Gubernick and P. H. Klopfer [Eds.], *Parental Care in Mammals.* New York: Plenum Press, 1981. Copyright 1981 by Plenum Press. Reprinted by permission.)

"delivers" her young and "cares" for them for several weeks (Smith and McDonald, 1974). In the case of pseudopregnancy, the female may adopt young or "mother" small soft objects. In the cat, a similar pattern of preparturitional establishment of a secluded nest site has been observed (Rosenblatt, 1963). Little else is known about the hormonal basis of maternal behavior in these species.

POSTPARTUM REGULATION OF MATERNAL BEHAVIOR: BEHAVIORAL SYNCHRONY BETWEEN MOTHER AND YOUNG

Parturition marks the beginning of the transition between the hormonal onset of maternal behavior and the maintenance of maternal care by the stimulation that the mother receives from her young. In the early part of the postpartum period, the transition is completed: Maternal behavior is firmly established on this basis, and its further development is based on the behavioral synchrony that emerges in the relationship between the mother and her young (Rosenblatt, 1965; Rosenblatt and Siegel, 1981). Behavioral synchrony has two aspects: First, it refers to the fact that the mother's behavior is synchronized with the needs and behavioral capacities of the young. When the young are most in need of maternal care and are least able to obtain it on their own initiative, shortly after birth, the mother takes the initiative in providing this care. As the young become more capable, behaviorally and physiologically (e.g., by developing thermoregulation), the mother relaxes her care, and the initiative shifts to the young (Rosenblatt *et al.*, 1962; Rosenblatt, 1965). Second, the detailed interactions between mother and young involve reciprocal stimulation that is synchronized in a complementary fashion (Hinde, 1979), the behavior of each "fitting into" or complementing the behavior of the other. The separate behavioral interactions between mother and young have discrete underlying bases in each member, making the relationship between the two a composite of interactions and at the same time possessing a more general aspect that is common to the many kinds of interactions.

Starting with the earliest mammals, the young have played an important role in regulating the reproductive physiology of the mother. This is particularly evident in two areas: in ovarian functioning during lacatation and in lactation itself. Both of these endocrine functions are regulated by suckling and related stimuli from the young. In those species in which there is a postpartum estrus immediately following parturition, the single estrous cycle is followed by a prolonged period of diestrus, often referred to as the *suckling diestrus.* And in those species in which there is no postpartum estrus, there is, nevertheless, a similar suspension of estrous cycling during lactation. Lacation is maintained by the action of sucking in causing the release of prolactin and other hormones (the "lactogenic complex"). Sucking acts on the mammary glands to stimulate the synthesis of milk and on the posterior pituitary to cause the release of oxytocin, the hormone responsible for "milk letdown," which makes milk available to the young (see section on lactation).

The regulation of the behavioral and physiological functioning of the mother by the young is profound, extending to all levels of her functioning. There is an equally profound regulation of the functioning of the young by the mother as described in some excellent reviews of the mother–young relationship in the rat (Hofer, 1981, 1982). All aspects of the physiological and behavioral functioning of the young are tied to stimulation from the mother, and therefore, the young suffer profoundly when they are separated from her. The extent of dependence of the young on the mother has prompted several investigators to characterize the relationship as parasitic (Galef, 1981) or symbiotic (Alberts and Gubernick, 1983). Both emphasize, at least in part, that the behavior of the young is keyed to obtain the resources that they need from the mother (e.g., food, warmth, and protection), and the mother is responsive to those stimuli from the young, which signal their needs. On her part, she also obtains needed resources from the young. There is coalescence in the young of behavioral (and physiological) mechanisms relating them to the mother and those that are related to their needs, and the same may be said for the mother: Those behavioral processes by means of which she relates to the young are one side of the coin, the other side of which represents those mechanisms by means of which she fulfills her own needs.

In this section we deal with a number of topics: (1) the mother's dependence on the young to maintain her maternal responsiveness postpartum; (2) the possible nonhormonal basis of postpartum maternal responsiveness; (3) some mechanisms of interaction between mother and young; and (4) parturition as a transition phase.

There are special problems that pertain to given species but not others, as, for example, the formation of a specific relationship between mother and young among many species of ungulates, sea lions, and seals (Lent, 1974; Trillmich, 1981), which is discussed under the above topics. It should be noted that information on all of these topics for most species is largely nonexistent. We shall have occasion frequently to refer to studies on the rat, for which information is available on most of the above topics.

MOTHER'S DEPENDENCE ON THE YOUNG TO MAINTAIN HER MATERNAL RESPONSIVENESS

Among many species, the mother depends on stimulation from the young for the maintenance of her maternal responsiveness once the behavior has been initiated under the influence of hormonal stimulation. In the rat, removing the young during parturition for periods of 2–4 days, then returning the young to test the maternal responsiveness of the female, revealed the mother's dependence. On the morning of the third day, about 30% of the females were still responsive to pups, but on the morning of the fifth day, none exhibited maternal behavior immediately (Rosenblatt and Lehrman, 1963; Rosenblatt, 1965). This dependence was greater immediately after parturition than several days later, when this new basis of maternal responsiveness was more firmly established. If pups were removed at the end of the third, ninth, or fourteenth days for a period of 4 days, on the return of 5- to 10-day old pups to each group of females, 40%–60% of the females exhibited maternal behavior, and if pups were left with the females, they were reared to weaning. In contrast, pups left with mothers whose separation occurred at parturition died shortly thereafter.

The stimulation that rat mothers receive from their pups that enables them to maintain their responsiveness has been studied (Fleming and Cummings, 1982) with the use of females that were delivered by cesarean section to prevent contact during parturition. Of the various kinds of contact between mothers and pups (e.g., smells and sounds of pups only), only direct contact with the pups was effective in maintaining maternal responsiveness 7 days later.

Placentophagia, a behavior that normally occurs at parturition in the rat and nearly all other mammals, is very likely stimulated in the rat by hormones at the end of pregnancy and perhaps by other features of pregnancy termination, including distention of the reproductive tract (Kristal, 1980). Once it has been performed, it is maintained if females receive pup stimulation during the rearing of their litters, and it can be elicited in more than 75% of them 3 weeks later when the females are undergoing estrous cycling (Kristal and Graber, 1976).

Among mice, the experience of giving birth and eating placentas led to an increase in the number of females that ate placentas as late as 10 days postpartum (Kristal, 1980). In one strain (BALB/cBY), only brief contact with the pups immediately postpartum was sufficient to increase the percentage from 25% to 75%, but in a second strain (C57BL/6BY), a more extended period of contact and nursing pups was required to enable 50% of the females to maintain this behavior.

The greater dependence of recently parturient mothers on pup stimulation for maintaining maternal responsiveness has also been shown in the hamster (Siegel and Greenwald, 1978). Mothers were allowed 1-, 24-, or 48-hr contact with their pups immediately postpartum before the pups were removed. Each of these groups was then subdivided into independent subgroups that were tested at daily intervals (each subgroup being tested only once) with young pups. The mothers allowed 1-hr contact continued to be responsive to pups for three additional days; they began to cannibalize them on the fourth day (i.e., the fifth day postpartum). Those allowed 24-hr contact with their pups maintained responsiveness for 1 additional day before they began to cannibalize pups. Forty-eight hours of contact enabled the females to respond maternally up to 14 days, until day 16 of lactation. In the hamster, therefore, the mother requires 24- to 48-hr exposure to her young before her behavior is firmly enough established to sustain an extended separation from them.

This conclusion is supported by a study (Buntin et al., 1981) in which hamsters were given 36- to 48 hr contact with their litters, immediately postpartum, and then were tested with newborn pups 3–6 weeks later for their retention. In both groups, 30% of the females were immediately maternal, and the remainder became maternal again after 48–72 hr of periodic testing.

Among the ungulates, dependence on the young for the maintenance of maternal responsiveness has been studied chiefly in the ewe, but there are related studies among goats. In these species, maternal reponsiveness is rapidly narrowed to the mother's own young; other young are rejected when they attempt to nurse, and only the mother's own young is permitted to nurse (see Rosenblatt and Siegel,

1981). To study the course of maternal responsiveness itself, mothers are prevented from developing this specificity by removal of their young at parturition. In the goat, it has been reported that a mother whose kid was removed at birth was no longer responsive to it when it was returned 2 hr later (Klopfer *et al.,* 1964). Later studies have challenged this finding from two points of view: The first is whether the phenomenon can be replicated and the second is whether some other factor may intervene to cause the female to reject her kid when it is returned to her after 1 or 2 hr. In two studies, females were found to accept their kids after 1 or 2 hr (Gubernick, 1980, 1981; Gubernick *et al.,* 1979; Hemmes, 1969), and in a series of studies (Gubernick, 1981), it has been shown that kids, not marked by other females in the interval when they are separated from their mothers, do not elicit rejection by their own mothers. Evidently, female goats are able to maintain their responsiveness beyond the 2-hr period previously found, but how long after parturition they remain responsive has not yet been determined (see a later section for a fuller discussion of the problem).

Among ewes, responsiveness to lambs in females whose young are removed at parturition declines over the first 12 hr of 25% of postparturient mothers, and it remains at this low level for an additional 12 hr. During the first 12 hr, ewes whose lambs had been removed at birth were given one of six different degrees of lamb contact to determine which features of the lamb maintain maternal responsiveness (Poindron and Le Neindre, 1980). The ewes given no contact were responsive in only 25% of the cases, and this percentage was not improved by allowing them only to see a lamb in a transparent walled box (Figure 13). Fifty percent were maternal when lambs were present in an open box at a 1-m distance, and all were responsive when lambs were nearby, even though they were enclosed in a perforated box that allowed the ewe only to receive lamb odors. When they were in an open box nearby or were free to approach the ewes, lambs maintained maternal behavior in all of the ewes.

Nonhormonal Basis of Postpartum Maternal Behavior

In the rat, the shift from the hormonal onset of maternal behavior prepartum to dependence on pup stimulation for its maintenance postpartum is accomplished by a shift to a nonhormonal basis of maternal behavior (Rosenblatt *et al.,* 1979). Although a similar shift to dependence on stimulation from the young postpartum can be shown in other species (see Rosenblatt and Siegel, 1981), it does not necessarily follow that there is also a shift to a nonhormonal basis for maternal behavior in these species. Additional criteria must be met to establish what is essentially a null hypothesis: the absence of control by hormonal factors. In the rat, these criteria, which have been reviewed elsewhere in detail (Rosenblatt *et al.,* 1979), are only summarized here.

Following parturition in the rat, there is a complete change in the pattern of circulating hormones from that which gave rise to maternal behavior. It is unlikely that estrogen plays a role in postpartum maternal behavior, as it does in the onset: Estrogen secretion is maintained at a low level postpartum, and moreover, removal of the ovaries shortly after parturition and of the pituitary gland during the first few days does not affect maternal behavior in any discernible way, although lactation is eliminated by the latter operation (Bintarningsih *et al.*, 1958; Erskine *et al.*, 1980; Obias, 1957). Prolactin-release-blocking agents also do not reduce maternal behavior postpartum (Numan *et al.*, 1972; Stern, 1977). Also, progesterone administered postpartum has no effect on maternal behavior, although it blocks the onset in half the females if given prepartum (Moltz *et al.*, 1969).

Perhaps one of the strongest arguments that maternal behavior postpartum may be nonhormonal is the fact that maternal behavior can be elicited from nonpregnant female rats under circumstances in which it is quite certain that it is nonhormonal in nature. By exposing nonpregnant females to young pups continuously for several days (i.e., sensitization), maternal behavior is elicited that is not prevented by hypophysectomizing or ovariectomizing the females in advance (Cosnier and Couturier, 1966; Rosenblatt, 1967; Wiesner and Sheard, 1933). Sensitization is not affected by increasing the circulating levels of prolactin (Baum, 1978), nor can sensitized females stimulate maternal behavior in recipient animals that are cross-transfused blood from them (Terkel and Rosenblatt, 1971, 1972). Finally, the maternal behavior of sensitized females is not disrupted by becoming pregnant: If it were dependent on hormones, the change in the pattern of circulating hormones that occurs during pregnancy would, very likely, disrupt their maternal behavior (Bridges, 1978; Cohen and Bridges, 1981).

The maternal behavior that appears under these circumstances very closely resembles that of lactating females that have been stimulated by hormones prepartum (Rosenblatt, 1967; Fleming and Rosenblatt, 1974a; Reisbick *et al.*, 1975). There are differences, however, particularly in the difficulty of eliciting nest defense (Erskine, 1978; LeRoy and Krehbiel, 1978) and in the failure to elicit placentophagia (i.e., eating of placentas; L. C. Peters, personal communication, 1980) by sensitizing females. Differences in T-maze performance between sensitized and lactating females have been shown to be based largely on the testing procedures (Bridges *et al.*, 1972; Cohen and Bridges, 1978; Mayer and Rosenblatt, 1979). These several differences may not be important, however, as the nonhormonal phase of maternal behavior is always preceded by the hormonal phase, during which placentophagia and nest defense are initiated, and they need only be maintained by pup stimulation postpartum; there is evidence that this is the case, at least with respect to nest defense (Erskine *et al.*, 1980).

The fact that nonpregnant females can be stimulated directly, without the intervention of hormones, to exhibit maternal behavior that resembles that shown by lactating females during both the period of maintenance and the period of decline (Fleming and Rosenblatt, 1974a; Reisbick *et al.*, 1975) adds strong support to the idea that the postpartum maternal behavior of rats is nonhormonally mediated (Figure 14). If this is the case, then the regulation of maternal behavior undergoes an important shift from a hormonal to a nonhormonal basis shortly after parturition. This shift is depicted in Figure 15, which also shows a proposed transition phase intervening between these two principal bases. This transition is one in which hormonal stimulation, having reached a peak during prepartum and

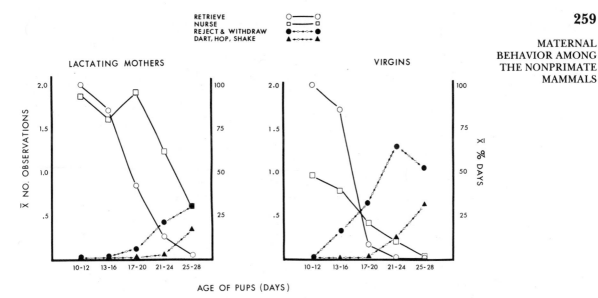

Fig. 14. Maternal behavior of sensitized virgins and lactating mothers from Day 10 to Day 38 of care of young. Lactating mothers and sensitized females were tested with foster young increasing in age by 1 day each day. Mean frequencies or percentages of each group exhibiting the behavioral items for each 4-day period are shown. (From "Decline of Maternal Behavior in the Virgin and Lactating Rat" by S. Reisbick, J. S. Rosenblatt, and A. D. Mayer, *Journal of Comparative and Physiological Psychology*, 1975, *89*, 722–732. Copyright 1975 by the American Psychological Association. Reprinted by permission.)

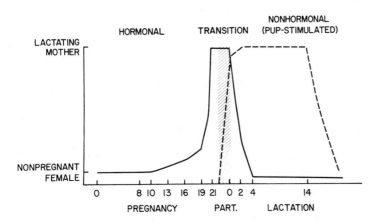

Fig. 15. Scheme of the regulation of maternal behavior during the maternal behavior cycle of the female rat. The ordinate shows levels of maternal behavior represented by lactating mothers (high) and nonpregnant female (low), and the abscissa shows three phases: pregnancy, parturition, and lactation. (From "Maternal Behavior in the Laboratory Rat" by J. S. Rosenblatt and H. I. Siegel. In R. W. Bell and W. P. Smotherman (Eds.), *Maternal Influences and Early Behavior*, Chapter 7, Figure 1, p. 156. Jamaica, N.Y.: Spectrum Publications, 1981. Copyright 1981 by Spectrum Publications, Inc. Reprinted by permission of the publisher.)

parturition, declines and nonhormonal stimulation from the pups increasingly becomes the basis for the maternal behavior exhibited by the female. The transition phase is discussed more fully in the next section on parturition.

Whether sensitization of nonpregnant females is a general phenomenon among the mammals, and whether, therefore, there exists at least this evidence to support the concept of a nonhormonal phase of maternal behavior in other mammalian species, it is too early to say. Among those strains of mice that are responsive to pups almost immediately on their first presentation, it has been strongly suggested, on the basis of studies employing postpartum hypophysectomy, that postpartum maternal behavior is largely nonhormonal (Leblond, 1938, 1940; Leblond and Nelson, 1937). In the less responsive strains of mice, nonpregnant females are cannibalistic even when experienced (Jakubowski and Terkel, 1982), but that finding does not rule out the possibility that, once they have become maternal on the basis of hormonal stimulation, they do not remain so on the basis of nonhormonal stimulation, at least for the duration of lactation until weaning has occurred. In the mouse, postpartum maternal aggression survives hypophysectomy and use of agents that block prolactin release, and all indications are that it is strongly tied to pup stimulation for its maintenance (Mann *et al.*, 1980; Svare *et al.*, 1982).

Among hamsters, several investigators have been successful in inducing maternal behavior in nonpregnant females by pup stimulation, and latencies, surprisingly, have been shorter than those found in rats (Buntin *et al.*, 1979; Siegel and Rosenblatt, 1978; Swanson and Campbell, 1979). They have ranged between 18 and 33 hr in the different studies. There is little doubt that pups regulate maternal behavior postpartum, as the study (Siegel and Greenwald, 1978) cited earlier indicated, but with respect to maternal aggression, a recent study shows that there is no decline even up to weaning, as in other species (e.g., rat and mouse), and that removing pups from the mother for as long as 6 hr does not reduce her maternal aggression as it does in these other species (Siegel *et al.*, 1983). Moreover, as discussed earlier, blocking the release of prolactin toward the end of the first week did eliminate maternal aggression and interfered seriously with maternal behavior (Wise and Pryor, 1977). All these facts indicate that, as late as the end of the first week, maternal behavior in the hamster may still be under hormonal control. Whether this control changes after that time remains to be discovered in future studies.

Experienced ewes that are exposed to newborn lambs several times during 24 hr begin to show maternal behavior in 30% of the cases. Whether this behavior is hormonally based or is nonhormonal has not been determined, nor has it been determined whether the postpartum maintenance of maternal responsiveness, which is dependent on stimulation from young, is also nonhormonally regulated. Nonhormonal control is likely, as there is a dramatic change in the pattern of circulating hormones after parturition, a change making it improbable that the hormone that stimulates maternal behavior continues to do so in the same manner. Definitive evidence, however, is lacking.

The onset of maternal behavior and the early transition to young-regulated maintenance, though very important for the subsequent course of maternal care, are, nevertheless, only the beginning of the interaction between the mother and her young. They set the stage for the further development of this interaction based on developmental changes in the young and changes in the maternal responsiveness and the behavior of the mother. This, of course, is a vast topic in itself that we can deal with only sketchily in this chapter.

Among a wide variety of species, the general course of the interaction of the mother and her young can be characterized in terms of the relative contributions of the mother and the young at different times from parturition to weaning. This "measure" of the interaction includes all of the perceptual, motor, and motivational capacities of the developing young, the maternal condition of the female, and the current state of the relationship between the two.

In the rat, nursings during the first 2 weeks are largely initiated by the mother, which approaches the nest containing the litter and settles on top of the pups, licking them and stimulating them to begin to search for nipples and to attach to and suckle from them. This and other changes in the mother's behavior toward her young as related to developmental changes in the young are shown in Figure 16 (Rosenblatt, 1965; see also Lee and Williams, 1977). Starting on the fifteenth day, the young begin to approach the mother outside the nest, and they initiate feeding by attaching to nipples and pinning her to the floor. During a third phase, the weaning phase, the mother increasingly avoids and rejects the approaches of the young by darting around the cage, by pressing her body against the floor or walls, and by actively preventing them from approaching her ventrum (Figure 14; Reisbick *et al.*, 1975). When the cage permits her to leave the litter behind, she does so for increasing periods.

The mother–young relationship in the cat exhibits a similar pattern of changes in the initiation of feeding during the 2-month suckling period, as shown in Figure 17 (Rosenblatt *et al.*, 1962). These observations, more detailed than in the rat, show that the transition from mother-initiated to young-initiated feedings occurs gradually: The patterns numbered II to IV represent mutually facilitated feedings in which the young approach the mother and the mother, responding to this approach, facilitates nursing by lying, sitting, or remaining in place, enabling the young to attach to nipples. In the final phase, the young initiate feedings almost entirely, and the mother avoids them until they "corner" her and force her to feed them.

The relationship of these developments in the feeding relationship between the mother and the young to other aspects of the behavioral interaction is shown in Figure 18, which provides a profile of the shelf-going behavior of the mother and the suckling and play behavior of the kittens for a single litter of three kittens (Rosenblatt, 1963). At a time when feedings are initiated by the young, the mother increases her shelf-going behavior, which puts her out or reach of the kittens.

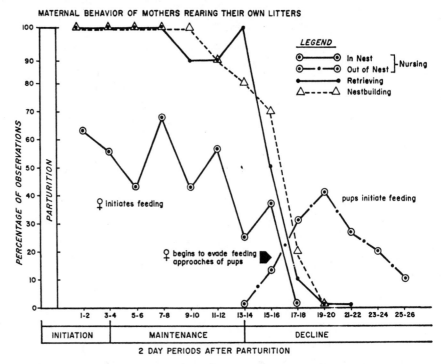

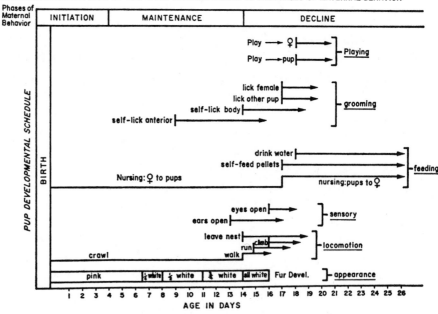

Fig. 16. Maternal behavior of the female rat (upper graph) correlated with measures of pup development (lower graph). Maternal nest-building, retrieving, and nursing behavior are shown in relation to changes in physical and behavioral characteristics of the pups.

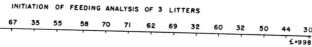

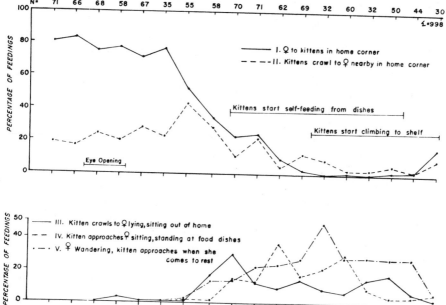

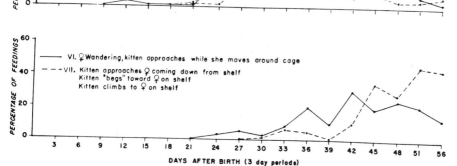

Fig. 17. The initiation of nursing-suckling in the cat during the 8-week nursing period, based on daily observations, grouped into 3-day periods, of 3 litters, of 3 kittens each. (From "Early Socialization in the Domestic Cat as Based on Feeding and Other Relationships between Female and Young" by J. S. Rosenblatt, G. Turkewitz, and T. C. Schneirla. In B. M. Foss [Ed.], *Determinants of Infant Behavior*. London: Methuen, 1961. Copyright 1961 by Methuen & Co. Reprinted by permission.)

They, in turn, begin to take food from dishes, starting a development that eventuates in weaning of the young. When the mother does become available to the young, the young have already fed from dishes and are not hungry. She therefore reduces her milk production, so that they become even more dependent on dish food rather than her milk; they approach her even less frequently and less persistently to suckle from her.

The mother–young relationship in the dog, taken from several sources, is presented in somewhat grosser detail in Figure 19, but it does show the main features of the interactions that occur at different stages (Rosenblatt, 1965). Care behavior (i.e., contact with the puppies, nursing and licking them) occurs with high frequency early, when the young, highly altricial in their capacities, mainly suckle from

JAY S. ROSENBLATT
ET AL.

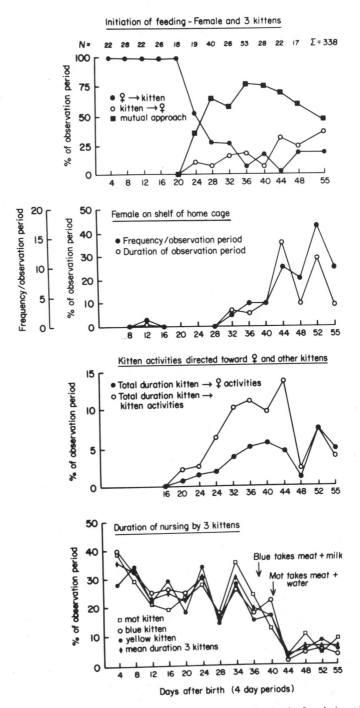

Fig. 18. Relationship among initiation of feeding (top graph), shelf-going by female (next lower graph), kitten play activity (next lower graph), and nursing duration (bottom graph) for a single litter of three kittens. Based on daily 3-hr observations grouped into 4-day periods. (From Rosenblatt, unpublished, 1961.)

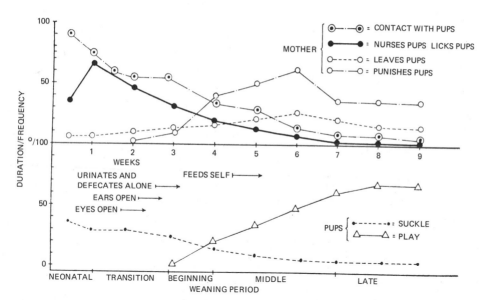

Fig. 19. Selected items of maternal behavior (upper graph) and pup development (lower graph) during the 2-month litter period among dogs. Changes in maternal behavior are shown in relation to changes in the physical and behavioral characteristics of the offspring. Figure based upon data reported by Rheingold (1963), Scott (1958), and Fuller and DuBuis (1962).

the mother. It is likely that, sometime around the end of the third week or the beginning of the fourth, the puppies begin to initiate suckling from the mother as their play activity increases. The decline in suckling is associated with an increase in play and an increase in the frequency with which the mother leaves the puppies and even punishes them, usually when they approach her for feeding.

Among ungulates, the precocial status of the young at birth, which enables them to stand and walk and to respond to the mother on the basis of visual and auditory stimuli, in addition to proximal stimuli, results in an abbreviated early period of mother-initiated approaches for feeding and a prolongation of the period of young-initiated approaches. In the roe deer, for example (Espmark, 1969), it is only during the first week and a half that the mother initiates feeding by approaching the young. During the first two days, this consists of the mother approaching the lying fawn, sniffing it, and licking it, in response to which the fawn stands up, searches, and grasps a nipple and suckles (Figure 20). From the third day onward, during the first week and a half, the mother comes within a few steps of the fawn, then stops and stands with her neck outstretched toward the lying fawn. After a few seconds, the fawn gets up and walks to the mother, making nose-to-nose contact with her, followed by searching for the nipple, locating it, grasping it, and sucking. In gradual steps, the fawn takes the initiative in the feeding relationship: First, the fawn stands at the resting site, moves about in circles, and makes rhythmical high-pitched bleats, to which the mother responds by quickly going to the fawn and nursing. Then, in the next weeks, the fawn stands up and walks, sniffing and grazing, now and then, showing a searching behavior that eventually brings it into visual contact with the mother. The fawn then walks up to her and imme-

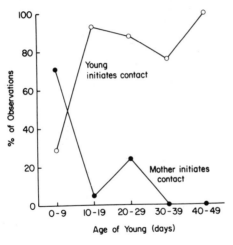

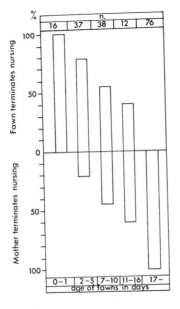

Fig. 20. Contacts between mother and young in the roe deer from birth to 7 weeks of age. Contacts initiated by the mother consist of the mother approaching the fawn and either sniffing and licking it or standing nearby, reaching toward it with her head. Contacts initiated by the fawn consist of standing, making high-pitched calls, searching for the mother, or walking to the mother nearby. (Based on data from Espmark, 1969.)

Fig. 21. Termination of nursing-suckling by mother and fawn during the first 4 months after parturition. Frequencies are expressed as percentages of the total number of terminations observed, shown on top. (From "Mother-Young Relations and Development of Behavior in Roe Deer [*Capreolus capreolus* L.]" by Y. Espmark, *Viltrey*, 1969, *6*, 461–540. Copyright 1969 by Yngve Espmark. Reprinted by permission.)

diately receives care from her eventuating in nursing. If the mother remains out of sight, the fawn makes the high-pitched vocalizations and the mother comes to it. Another, later mode of feeding occurs almost entirely at the fawn's initiative: The mother is nearby grazing or resting, and the fawn, not having nursed for between 7 and 12 hr, approaches the mother and suckles.

Several other measures of the interaction between the mother and the fawn also portray the changing role of each of the participants, the mother declining in her initiative toward the fawn as her maternal responsiveness wanes, and the young increasingly taking the initiative. Initially, during the first day (Figure 21), the fawn terminates nursings almost exclusively, the mother being willing to nurse as long as the fawn continues, but over the next two weeks, the mother increasingly terminates the nursing (which the young has initiated; see Figure 20), until by the middle of the third week she terminates all nursings.

Initially the mother also licks her fawn continuously while nursing it (Figure 22), but over the next 4–5 weeks, this behavior declines and is replaced by sporadic licking and then no licking at all.

A similar course of mother–young interaction has been observed in the right whale off the coast of Argentina during the breeding season and the following year (Taber and Thomas, 1982). During the entire first year, the mother and the single young remain in close proximity, but the initiative for maintaining this proximity shifts from the mother during most of the year to the young in the preweaning and separation period. After they have been at sea for 6 months, following the 4-month early breeding season, they return to the breeding area, and at this time, the initiative for maintaining close proximity shifts to the young during the remaining few weeks in which the young continues to suckle, prior to the permanent separation of the mother from her young.

The relationship between the mother and her young in the species we have described, in reality, consists of a mosaic of behavioral interactions, each having its own proximal causal mechanism that may extend over time and undergo changes. Several of these mechanisms have been analyzed in some detail in the mother–young relationship in the rat, and they give substance to the concept of behavioral synchrony.

One of the earliest observations of maternal retrieving or transport of young (Sturman-Hulbe and Stone, 1929), when a fan was played over the nest, has recently been reanalyzed (Brewster and Leon, 1980) as shown in Figure 23. Nests were partially destroyed or flooded so that the females were stimulated to pick up

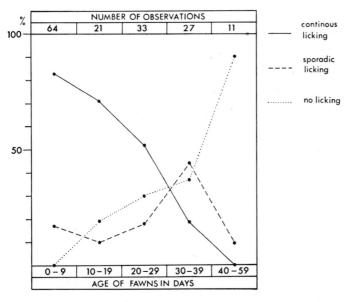

Fig. 22. Licking of fawns by mother while nursing. The frequency is expressed as a percentage of the total number of observations. (From "Mother-Young Relations and Development of Behavior in Roe Deer [*Capreolus capreolus* L.]" by Y. Espmark, *Viltrey*, 1969, *6*, 461–540. Copyright 1969 by Yngve Espmark. Reprinted by permission.)

JAY S. ROSENBLATT
ET AL.

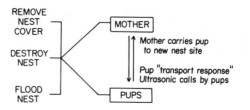

Fig. 23. Behavioral interactions between mother and young in the rat, mediating transport of young to a new nest site. On left are stimuli that elicit transport and on the right the interaction they initiate. (Based on data from Brewster and Leon, 1980.)

their pups in their mouths and transport them to the new nest sites. The survival value for the pups of being transported by the mother to a new nest site is obvious and this behavior occurred in nearly all instances of nest disturbance. Because these disturbances to the nest had direct effects on the mother as well as on the pups, we can ask in what way the transport was initiated and how it was carried out.

The pups played little role in the transport to the new nest sites apart from exhibiting the "transport response"—a response of becoming completely relaxed and hanging loosely from the mother's mouth when she picked them up, preferentially by the neck. Anesthetized pups were picked up as readily as awake pups, finding indicating that the individual pups did not signal their distress by ultrasonic calls, thereby eliciting the mother's response. She was capable, nevertheless, of responding to their calls and would pick up a litter in which there were calling pups in preference to a litter that was not calling. We must conclude that, in this instance, the synchrony between mother and young was based largely on the mother's reaction to the environmental disturbance, but when she reacted to this disturbance by transporting her young to a new site, their needs were met, with only a minimal contribution by them.

The recycling of water between mother and young is a somewhat more complex form of synchrony (Friedman *et al.*, 1981). It involves behavioral and physiological interactions, as shown in Figure 24. The mother constitutes the only source of water for the young: 70% of a daily nursing (at 10 days of age) consists of water in the amount of 30 ml. This water loss must be replenished by the mother, and two sources are available to her: outside sources and recycling of water from the

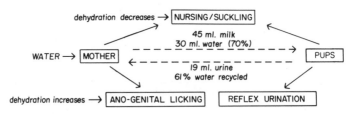

Fig. 24. Behavioral interactions between mother and young in the rat, mediating water recycling. On top is the nursing-suckling interaction, during which the mother loses water to the pups, and on bottom is the licking interaction, during which the mother regains water from the pups. (Based on data from Friedman *et al.*, 1981.)

pups through genital licking and the elicitation of reflex urination. The mother in this study retrieved 19 ml of urine in this manner, which is about 61% of the water loss incurred during a single day's nursing.

In the mother, there is a relationship between water balance and both nursing and genital licking of the pups. Dehydration reduces nursing and increases maternal licking of the pups. Water loss is therefore limited, and at the same time, water intake is increased. Because the mother initiates nursing during the first two weeks (Rosenblatt and Lehrman, 1963) and enters the nest to lick the young, it is she who plays the major role in behaviorally regulating water intake and loss by the pups as well as her own water economy. The pups contribute by grasping nipples and sucking whenever they are given the opportunity by the mother; they facilitate water recycling by urinating only when licked by the mother. They are unable to urinate spontaneously but readily expose their genital region when the mother begins to lick them.

The behavioral and physiological synchrony between the mother and her young during nest contacts and nursing represents an even more complex form of synchrony than that of water recycling. It has been proposed (Leon *et al.,* 1978; Woodside and Leon, 1980; Woodside *et al.,* 1981) that the principal causes of this synchrony are parallel changes in body temperature between mother and young during periods of contact. At these times, body temperatures rise because of the transmission of heat to the young and the eventual buildup of heat on the mother's ventrum (and ultimately her preoptic area), which terminates the nesting-nursing session. This cycle can be seen in Figure 25, where at 22°C (middle figure) there is an increase in maternal ventrum and core temperatures while the mother is in contact with the pups, a parallel increase in litter temperatures (i.e., nest temperature), a decrease in maternal temperature when she gets off the pups, and a parallel decrease in litter temperature. At higher (bottom figure) and lower (top figure) ambient temperatures, nesting sessions are respectively shorter, because of the more rapid buildup of the mother's temperature, or longer, because the mother's body temperature rises more slowly. During intervals between nesting-nursing sessions, body temperature declines more slowly in the warm ambient temperature than in the cool ambient temperature; thus, the overall duration of a nesting-nonnesting session remains the same under all ambient temperature conditions, but the relative proportion of the time spent in one or the other activity varies inversely.

The balance of factors contributing to heat increase, on the one hand, and heat reduction, on the other, is shown in Figure 26. Increased heating is based, first of all, on the physiological demands of lactation, a high metabolic rate that results from hyperphagia and reduced insulation of the ventrum by the enlargement of an insulating mammary-gland layer. Hyperthermia is aggravated when the female makes contact with pups that are at or above ambient temperature: Prolonged contact with pups reduces heat dissipation through the ventrum even further.

Heat reduction occurs through dissipation via the tail and vascularized, relatively furless body surfaces and by alterations in hormonal secretions that result in

JAY S. ROSENBLATT
ET AL.

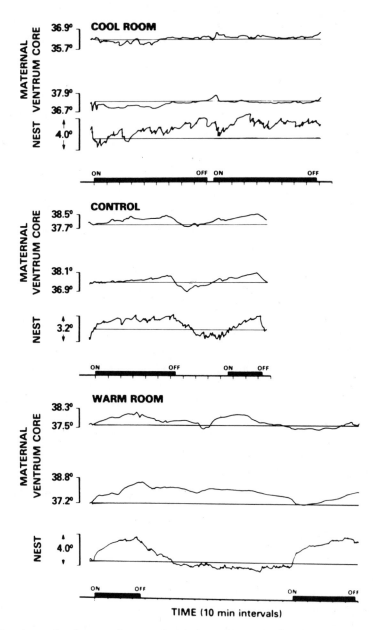

Fig. 25. Sample epochs of maternal core and ventral temperatures and nest temperatures during and between nesting bouts of mothers in cool (18°C), warm (26°C) or control (22°C) rooms. Bout duration is indicated by a heavy black line. A thin line is drawn from the temperature at the onset of the first bout to facilitate a visual comparison of the ongoing temperature changes. (From "Thermal Control of Mother-Young Contact in Rats" by M. Leon, P. G. Croskerry, and G. K. Smith, *Physiology and Behavior* 1978, *21*, 793–811. Copyright 1978 by Pergamon Press, Ltd. Reprinted by permission.)

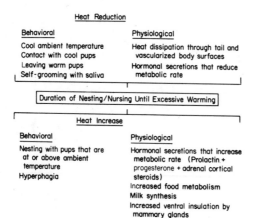

Fig. 26. Thermal factors regulating the duration of nesting-nursing contact in the female rat during lactation. The upper part of the figure shows the behavioral and physiological factors that reduce heat, and the bottom part shows those that increase heat. (Based on data from Leon *et al.*, 1978, and Woodside *et al.*, 1981.)

a reduction in metabolic rate. The ovarian hormones in the presence of the adrenal glands stimulated by prolactin appear to maintain the lactating mother's high metabolic rate, and changes in these hormonal secretions are likely to underly the onset and the termination of thermal regulation of nesting-nursing (Leon *et al.*, 1978). Heat reduction through behavioral means occurs when the mother seeks a cool environment and also when she makes contact with cool pups, as well as when she leaves warm pups with which she has been in contact. The mother also reduces her body temperature by prolonged bouts of self-grooming, spreading saliva, which evaporates, over the exposed surfaces of her body.

The lack of thermoregulation by the pups is crucial to the mother's regulation of her nesting-nursing contacts with them. When she leaves them, their body temperatures decline. Although the decline is retarded somewhat by their tendency to huddle, which reduces heat loss by the litter (Alberts and Brunjes, 1978; Cosnier, 1965), eventually the decline ranges over 4°C in warm and cool ambient temperatures (see Figure 25). It is this decline that enables the female once again to nest and nurse with the pups when her own body temperature has fallen. As the pups become older and begin to regulate their own body temperatures, nesting-nursing bouts become shorter because pup temperature increase is faster.

Among sheep, following the establishment of the bond between the ewe and lamb, there develops a nearly exclusive nursing relationship. How this develops and what it consists of has been studied (Poindron, 1974, 1976; Poindron and Le Neindre, 1980). Lambs tend to approach their mothers for suckling and to stand sucking in a limited number of patterns (Figure 27). The ewe's own lamb performs most of its feedings by passing in front of her; she sniffs its anal region to identify it as her own, and then the lamb stands parallel to the ewe's body with its hind end facing her, enabling her further to sniff it. Variations of this basic pattern involve the lamb's either not passing in front of the ewe but allowing her to sniff it when it is suckling or passing in front of her but suckling with its hind end out of her

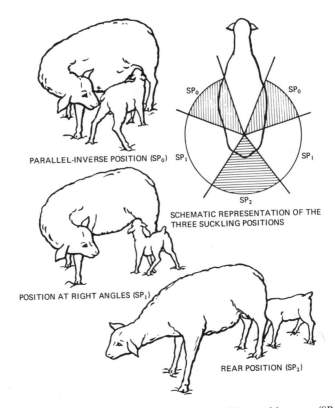

PARALLEL-INVERSE POSITION (SP₀)

SCHEMATIC REPRESENTATION OF THE
THREE SUCKLING POSITIONS

POSITION AT RIGHT ANGLES (SP₁)

REAR POSITION (SP₂)

Fig. 27. Illustrations of three suckling positions by lambs. Suckling position zero (SP_0), SP_1, and SP_2 are also shown schematically in the upper right diagram. SP_0 and SP_1 are used by the mother's own lamb and SP_2 by alien lambs. (From "Mother-Young Relationships in Intact or Anosmic Ewes at the Time of Sucking" by P. Poindron, *Biology of Behavior*, 1976, *2*, 161–177. Copyright 1976 by Masson S. A. Paris. Adapted by permission.)

sniffing range. One pattern employed exclusively by alien young and very rarely by her own young involves approaching her from the rear and suckling between her rear legs without the ewe's sniffing the lamb. These suckling patterns and the predominance of the first pattern develop over the first 2–3 weeks after birth. Their development shows that the exclusive bond between mother and young is not based only on the ewe's behavior but is a factor that governs the lamb's behavior increasingly during this period: The lamb acts differently when attempting to suckle from its own mother and from an alien mother.

These suckling patterns obviously serve to enable the ewe to identify its own lamb for suckling. They are therefore the product of ewe–lamb interaction. This fact was shown clearly when ewes were made anosmic before parturition by olfactory bulb removal (Poindron, 1976). Under these circumstances, the predominant pattern of suckling described above was not the sole principal pattern, nor was it used only by the ewe's own lamb. Abbreviated patterns of approach to the ewe were much more frequent (Figure 28), and both the principal pattern and the secondary patterns were used in a large proportion of sucklings by alien lambs. The fact that the female could no longer identify her own young and did not use olfac-

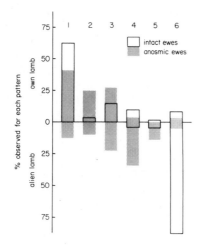

Fig. 28. Proportion of various suckling patterns observed in intact and anosmic ewes. Line bars represent suckling patterns used by the mother's own single lamb and the shaded bars show the patterns used by an alien single lamb. Patterns 1–4 include suckling in parallel inverse position (SP$_0$—see Figure 27), Pattern 5 includes suckling in a perpendicular position (SP$_1$), and Pattern 6 includes suckling in a rear position (SP$_2$). Pattern 1 includes, also, the lamb's passing in front of the ewe and the ewe's sniffing of the lamb. Pattern 3 also includes sniffing by the ewe, and Pattern 5 may not include sniffing by the ewe. (From "Mother-Young Relationships in Intact or Anosmic Ewes at the Time of Sucking" by P. Poindron, *Biology of Behavior,* 1976, *2,* 161–177. Copyright 1976 by Masson S. A. Paris. Adapted by permission.)

tion in responding to the lambs allowed the young to develop different patterns of suckling based, one would presume, on the ewe's visual recognition of the young, which was not as exclusive as her olfactory recognition of them.

In the gray seal, a similar pattern of exclusive nursing of the mother's own pup exists, but how the two come together for feeding is somewhat different (Fogden, 1971). The mother remains offshore during intervals between nursings (Figure 29), and the pup sleeps where it was last fed, which constitutes the pair's territory. When it awakens, the pup raises its head and calls with increasing vigor, and the mother climbs out of the water near the pup. Before permitting any contact, she first smells the pup, and then they both touch muzzles, exchanging olfactory contact. Suckling occurs at the initiative of the mother, which rolls to one side, presenting her erect nipples; the pup then searches, locates, and grasps the nipple and sucks.

Parturition: The Transition Phase

Parturition is a pivotal event in the reproductive cycle of the female and, in particular, in the pre- and postpartum transition in the regulation of maternal behavior. As we have seen, during parturition the regulation of maternal behavior undergoes a shift from its hormonal basis in the endocrine events of late pregnancy to reliance on stimulation from the young. In the rat, and perhaps in other species (e.g., sheep and goat), this transition corresponds to a shift from hormonal to non-hormonal regulation of maternal behavior. It is important, therefore, to examine

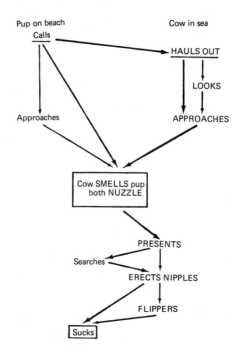

Fig. 29. Nursing–suckling interaction between mother and pup in the Grey seal. See text for explanation. (From "Mother-Young Behavior at Grey Seal Breeding Beaches" by S. C. L. Fogden, *Journal of Zoology*, London, 1971, *164*, 61–92. Copyright 1971 by the Zoological Society of London. Reprinted by permission.)

the events of parturition itself to see how this transition is mediated by the behavior of the mother during delivery.

The control of parturition depends largely on the circulating levels of ovarian hormones, chiefly estrogen and progesterone, but the fine tuning and the temporal organization of delivery depend as well on prostaglandin, oxytocin, relaxin, and, in some species, adrenal cortical hormones (Heap *et al.*, 1977; Liggins, 1973). The relative contribution of fetal and maternal sources of these secretions varies in different species; the reader is referred to several volumes that review the literature on the hormonal basis of parturition in laboratory and domestic animal species (Ellendorff *et al.*, 1979; Klopfer and Gardner, 1973). Our aim here is to deal specifically with parturition behavior, but a knowledge of the underlying organic events of parturition is indispensable, of course.

Descriptions of parturition behavior are available for a large number of species, ranging from small rodents, lagomorphs, and carnivores to many species of ungulates, pinnipeds, and dolphins. Explanations for many of the species differences in the mother's behavior during parturition are based on such factors as the position of the fetus; whether its legs are tucked under its body or outstretched; whether presentations are head first or are evenly divided between head- and tail-end births; the ease or difficulty with which the fetus passes through the pelvic girdle or the absence of such an impediment to delivery; whether the site of delivery is on land, in a tree, or under water; and whether the mother can reach the fetus to lick it during delivery or must await its delivery before she can turn around to lick it (Naaktgeboren, 1979).

The behavioral events of parturition initially are based on the internal organic events of delivery, but as parturition progresses, especially among species with more than one or two fetuses, the presence of newly born young and the expulsion of the afterbirth increasingly play a role in the female's behavior. Parturition as a whole may be divided into three principal periods: the dilation period, during which the cervical and vaginal opening dilates; the expulsion period, during which fetuses are delivered through the forceful contractions of the uterus, aided by voluntary abdominal contractions and the assumption of a posture that facilitates passage of the fetus through the birth canal; and the final, afterbirth period, during which the afterbirth is delivered. This last period may be interspersed with the expulsion period in species (e.g., rat, mouse, and cat) in which afterbirths are delivered between deliveries of fetuses. Most descriptions of parturition are of the expulsion and afterbirth periods.

Parturition in the rat and cat has been described in terms of four phases that occur in the delivery of a single fetus during the expulsion period (Rosenblatt and Lehrman, 1963; Schneirla *et al.*, 1963). In the first phase, the *contraction phase*, there are waves of contraction and the female lies with her body pressed against the floor, or she assumes a near "lordotic" posture, with hind legs extended backward and raised; at times, during this phase, females remain haunched in the nest or home site. This phase may last a long time, but it is brought to an end when the *delivery phase* begins, during which the fetus begins to appear in the birth canal and the female may begin to have contact with it, licking the fluids and membranes covering it, and licking the surrounding parts of her own genital region. The *post-delivery phase* follows, during which the female may assist in the delivery of the fetus by pulling it forward from the vaginal opening between her hind legs, using her mouth and forepaws, meanwhile licking and rotating it. Then, the female turns to the placenta, which has trailed after the fetus, and begins to lick and eat it. The female licks the birth fluids from her ventrum, the fetus and the placenta, and the surrounding floor area. The next phase in the *interval between births*, which may be brief if the contraction phase of the next delivery begins immediately.

As deliveries proceed in multiparous species such as the rat and cat, the female's attention and her behavior are directed increasingly from her body to the newly born and emerging fetuses. In a large study in which more than 120 primiparous and multiparous female rats were observed during parturition (Dollinger *et al.*, 1980), it was found that females alternated in their attention between the newly born pups and the ongoing delivery. They licked the pups and then interrupted this activity to attend to the uterine contractions, the vaginal stimulation from an emerging fetus, and the fluids spread on their bodies. During the latter half of parturition, there was an increase in birth-oriented activities (grooming the body, pulling the fetus from the birth canal, and eating the placenta) as the females kept pace with the increasing rate of pup and placental deliveries. At the end of the deliveries, the mothers turned their attention, once again, to the pups and lay down among them in what has been labeled the *postpartum resting interval* (Schneirla *et al.*, 1963).

Among the ungulates, the roe deer gives birth lying on her side as the young is expelled. The mother begins to lick the fawn, still surrounded by embryonic membranes, and for the next 30 min, she licks it continuously. The fawn, meanwhile, makes futile attempts to stand, finally crawls toward the udder region, gets hold of a teat, and begins to suck for the first time at 32 min (Espmark, 1969).

In the red deer, observed in the wild and in semicaptivity, females delivered in a lying position in two-thirds to three-fourths of the instances (Arman, 1974; Arman *et al.*, 1978). Contractions lasted nearly 2 hr on the average before calving took place, and nearly an hour intervened between the time the calf first appeared at the vaginal opening and final delivery. Nearly all births were trouble-free; the front feet appeared first, followed by the head, and the amniotic sac broke within a few minutes of delivery, freeing the calf, which began to search for the udder within 20 min of delivery and was able to attach and suck about 25 min after delivery, usually while the mother was on her side. The amniotic fluid, membranes, and placenta were immediately licked up or eaten when they appeared. The calves began to move and vocalize soon after birth, and the mothers responded by approaching and licking them.

The reticulated giraffe was observed during parturition and has been described in great detail (Kristal and Noonan, 1979). During acute dilation, the mother paced continuously, and as the amniotic sac made its appearance, the mother turned and licked it. After 15 min, fluid ran down the female's legs, and the calf's forelegs emerged. The mother bore down at 25 min and licked the sac during this contraction phase as the calf's snout became visible. At 45 min, the calf's head emerged, its mouth open, and fluid ran out of its mouth. The sac ruptured at 50 min, and the mother turned and smelled it over the next 5 min; as the mother bore down frequently, the calf's shoulders emerged, and at 55 min, the calf fell the short distance to the floor and the mother turned and began to lick it as the calf tried to raise its head. During the next 15 min, the mother alternated between licking the calf and pulling the amniotic membranes from the calf and eating them. About 1 hr and 25 min after birth, during which time it had made several attempts to stand and was licked continually, the calf finally attached to a nipple and sucked. The afterbirth was extruded over the next 3 hr and was only partially consumed.

During the birth of a stellar sea lion (Vania, 1965), the delivery occurred without assistance through licking from the mother, which spent most of the time "bearing down" and turned to the pup only when it had been delivered. It has also been reported among pigs that the mother lies on her side as she delivers her litter and does little to aid the young in either cleaning themselves or locating and attaching to the nipples after birth (Naaktegeboren, 1979). The same has been reported among uniparous camels (Lent, 1974).

With few exceptions, what is common in the behavior of females of many species during parturition, which is of significance to our understanding of the role of parturition in the organization of maternal behavior, is the female's gradual introduction to the newborn as it emerges from the birth canal. This introduction is mediated by the mother's extensive licking of the birth fluids that have spread to her own body, the fetus, and the entire surrounding site of parturition. The description of parturition in the hamster (Rowell, 1961), in which the mother is seen as treating the pup emerging from her vaginal opening as an extension of her own body, licking it and her own body alternately, is characteristic of a large number of species. During parturition, the mother experiences a heightened level of excitement, and the newborn, having emerged into the air and breathing for the first time, is also highly excited and moving about, often vocalizing. It responds to licking by the mother by pushing toward her, and any contact with her also stimulates an approach response (Kristal and Noonan, 1979). Both are therefore in an optimal condition for responding to one another immediately following parturition, before the female enters a period of rest and before the young also become less active, after their first feeding.

Among rats, it has been proposed that during parturition the female begins to establish a relationship with her newborn that provides the sensory stimuli for the maintenance of her maternal behavior in the face of declining hormonal influence (Rosenblatt *et al.*, 1979). Support for this concept comes from a series of studies (Bridges, 1975, 1977) in which mothers were tested for maternal behavior 25 days after their last contact with their own pups, when they were currently undergoing estrous cycling. The maternal behavior that they exhibited in the tests at that time was nonhormonally based. Groups of mothers were given different amounts of postpartum experience with their own litters ranging from 4–6 hr to 21 days (Figure 30), and later, they all exhibited maternal behavior in less than 2 days of exposure to young pups (top panel of Figure 30). Without experience (bottom panel of Figure 30), they required more than 4 days of exposure to pups. Most important for our present purposes were the findings regarding females that were exposed to pups only during parturition, either for the entire duration or for only a portion of parturition (second panel from the top of Figure 30). These females had latencies for responding to pups that were about the same as those of females with much longer contact with their litters. If, however, the females were not permitted to make contact with their pups during parturition, or if they were delivered by Cesarean section and were not given pups immediately afterward, their latencies at the end of 25 days were equal to those of females that had had *no* experience with pups. The hormonal stimulation of late pregnancy was insufficient to establish the nonhormonal basis of maternal behavior without actual contact with pups, which normally occurs for the first time during parturition.

In the rat, however, unlike some other species that we shall discuss, females remain responsive to pups, on the basis of hormonal stimulation, for at least 24 hr after parturition and even longer. If pups are presented during this period, even if the female has not previously had contact with them postpartum, she generally adopts them and mothers them. Although parturition is normally the optimal time for this process to occur, it can occur at a later time. The same appears to be true in the cat and dog, both of which adopt young for some time after parturition. It may be true of many nest-building species that bear altricial newborn, in which the

JAY S. ROSENBLATT
ET AL.

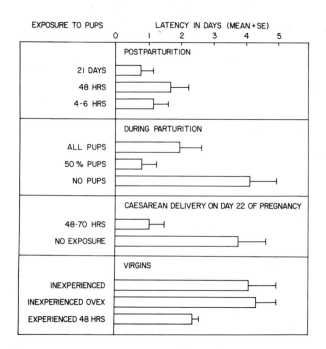

Fig. 30. Effects of parturition and postparturition contact with pups on latencies for induction of maternal behavior 25 days later. Period of exposure to pups shown on the ordinate and latencies on the upper abscissa. Cesarean-section-delivered females and virgins shown for comparison. Based on data reported by Bridges (1975, 1977). (From "Progress in the Study of Maternal Behavior in the Rat: Hormonal, Nonhormonal, Sensory, and Developmental Aspects" by J. S. Rosenblatt, H. I. Siegel, and A. D. Mayer. In J. S. Rosenblatt, R. A. Hinde, C. G. Beer, and M.-C. Busnel (Eds.), *Advances in the Study of Behavior*, Vol. 10. New York: Academic Press, 1979. Copyright 1979 by Academic Press, Inc. Reprinted by permission.)

process of forming a behavioral bonding, on which the postpartum maintenance depends, is a slow and gradual one extending over several days (Rosenblatt, unpublished observations, 1963).

Among sheep, goats and cows, and very likely among many ungulates and species with precocial young, in which an exclusive bond is formed between the mother and her offspring, parturition plays a more important role in the establishment of the bond, and therefore in generating the conditions for the maintenance of maternal behavior postpartum (Lent, 1974). Early reports on the goat indicated that the mother acquired the basis for responding to kids during early licking of the newborn, and the basis for responding specifically to her own kid during later licking, all during the first few minutes after its birth (Klopfer *et al.*, 1964; Klopfer and Gamble, 1966; Klopfer, 1971). Mothers whose kids were removed from them at parturition, before they had licked them, failed to respond to them when they were returned as early as 1 hr later, but if the mothers had licked their kids during the first 5 min, before the kids were removed, they accepted the kids on their return and they rejected alien kids.

In these studies, what we have referred to as the transition period is called the *critical period* for the attachment of the goat mother to her own young, and as we have seen, it is believed to be very brief in this species. The first indication that this critical period may not coincide with the transition period came from a study in which mothers were made anosmic before parturition (Klopfer and Gamble, 1966). These mothers were also given 5 min of contact with their kids during parturition, and then the young were removed for 3 hr. At the end of this period, these anosmic mothers accepted their own kids, but they also accepted alien kids because they were unable to form an exclusive bond with their own kids. Their level of responsiveness to alien young was as high as that to their own young in a previous study in which the bond was an exclusive one (Klopfer *et al.*, 1964). This finding indicated that there are two separate processes involved, one in which maternal responsiveness is maintained by contact with the young immediately postpartum, which may be the "true" transition period, and another in which an exclusive bond is formed with the mother's own kid during the "critical period" referred to earlier. One study did not find any decline in maternal responsiveness, even though the kids had been removed at parturition, before they were licked (Hemmes, 1969).

More recent studies in the goat have suggested that the early studies were unknowingly influenced in their results by the fact that kids that were removed from their mothers, while awaiting return to them, were often given to other mothers, which, it is now known, marked these young in ways that are not yet understood completely, so that, on the kids' return to their own mothers, they were rejected because of the alien odors that they carried (Gubernick, 1980, 1981; Gubernick *et al.*, 1979).

In one study, mothers were given 5-min contact with their own kids; then they were separated from them for 1 hr, at the end of which they were tested for acceptance of their own kid and alien kids. They accepted their own kid *and* alien kids that had had no contact with their own mothers but rejected alien kids that had had 24-hr contact with their mothers (Gubernick *et al.*, 1979). A second study attempted to determine the source of maternal labeling, that is, whether it is a result of maternal licking, of nursing, or of a combination of both sources. Mothers were given 5-min contact with their own kids at parturition and then separated from them for 1 hr. At the end of this period, the mothers were tested for their responses to their own kids and alien kids that had had the following exposures to their own mothers: licking only, no contact with the mother but being bottle-fed on mother's milk, licking and mother's milk by nursing, unlabeled, and being fed skim milk. The mother's own kid was always accepted, and unlabeled kids were accepted nearly 70% of the time, but all other groups were rejected in more than 70% of the instances (Figure 31). These studies indicate that the critical period for the acceptance of alien young may approach more closely the period of postpartum maternal responsiveness if the young are not labeled by the alien mother, and they sharpen the distinction between the critical period for the formation of an exclusive bond between the mother and her kid and the duration of maternal responsivness in the postpartum transition period.

In the ewe, the critical period for the formation of an exclusive bond between the mother and her lamb lasts about 2 hr: Mothers allowed contact with their own lambs accept alien newborn for about 2 hr (which explains, in part, why they usu-

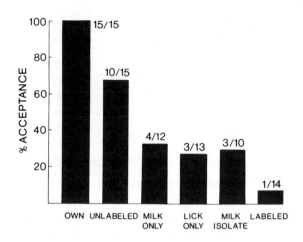

Fig. 31. Percentage of goat mothers accepting labeled and unlabeled kids after 5-min contact with their own kid at parturition then separation for 1 hr. Unlabeled kids were kept apart from their mothers from birth and were not fed their mother's milk; labeled kids were fed and licked by their mothers; licked kids were only licked; and milk isolates were fed milk from their mothers but were not licked. (From "Parent and Infant Attachment in Mammals" by D. J. Gubernick. In D. J. Gubernick and P. H. Klopfer [Eds.], *Parental Care in Mammals.* New York: Plenum Press, 1981. Copyright 1981 by Plenum Press. Reprinted by permission.)

ally accept the second and third lambs of multiple births, which are usually born at half-hour intervals); then they reject them. Older lambs are rejected after 30 min (Poindron and Le Neindre, 1980). This period, however, does not represent the duration of responsiveness to young based on the hormonally stimulated onset of maternal behavior, but the critical period for accepting alien newborn is already the product of the interaction between this factor and postpartum experience with the newborn. A more valid measure of the duration of maternal responsiveness is obtained by removing lambs at parturition and tracing the maintenance of responsiveness. When this test was done, responsiveness was found to decline over the first 12 hr to 25% of responsive mothers, but if newborn were used for testing instead of the mothers' own young, then 50% of the females accepted the young. As maternal responsiveness declines, the exclusive bond is more difficult to form and ewes accept both their own and older young (Figure 13). The transition period in the ewe, therefore, lasts for at least 24 hr, at a lower level, of course, than that which was present at parturition. Although under normal circumstances olfactory stimuli from the young play an important role in the maintenance of maternal responsiveness and are the basis for the formation of an exclusive bond with the young, they are not essential, as prepartum anosmic mothers retain their responsiveness for as long as intact mothers (Poindron and Le Neindre, 1980; Rosenblatt and Siegel, 1981).

It has long been suspected that the act of delivery itself, apart from the hormonal stimulation and the interaction with the newly born offspring, might play some role in the onset or maintenance of maternal behavior. In the rat, uterine distention with hypertonic saline, simulating an aspect of delivery, reduced latencies for sensitizing pseudopregnant females (Graber and Kristal, 1977). More definitive evidence has been reported in the ewe (Keverne *et al.*, 1983). Under several circumstances, the application of vaginal-uterine stimulation either by means of a vaginal vibrator or insertion of a rubber balloon into the uterus and its inflation, both for 5 min, had striking effects on maternal behavior. In nonpregnant multiparous ewes, vaginal stimulation following priming with subthreshold doses of progesterone and estrogen resulted in the appearance of all aspects of maternal behavior (licking, acceptance at udder, low-pitched bleats, and absence of aggressive behavior toward young as well as high-pitched bleats on removal of the lamb) in tests with newborn lambs immediately after the stimulation. Control females not given vaginal stimulation initially were totally unresponsive to the young, but they too became highly responsive when subsequently given vaginal stimulation (Figure 32).

A second experiment explored the effect of uterine stimulation on the ewe's critical period for accepting an alien newly born lamb when rearing its own lamb. After 2 hr of exposure to its own lamb immediately following parturition, the lamb

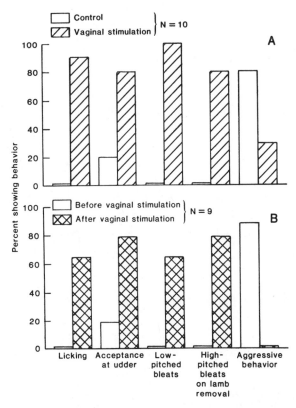

Fig. 32. Effect of vaginal stimulation on nonpregnant ewes given subthreshold priming doses of progesterone-estrogen. (A) Comparison of stimulated and nonstimulated ewes tested with newborn lambs immediately after vaginal stimulation. All differences between the groups are significant. (B) Control group after vaginal stimulation compared to its own performance before vaginal stimulation. All differences are significant. (From "Vaginal Stimulation: An Important Determinant of Maternal Bonding in Sheep" by E. B. Keverne, F. Levy, P. Poindron, and D. R. Lindsay, *Science*, 1983, *219*, 81–83. Copyright 1983 by the American Association for the Advancement of Science. Reprinted by permission.)

was removed and an alien newborn was introduced. The ewe was aggressive and disturbed by the alien lamb and did not permit it to nurse from her because the ewe had already formed an exclusive bond with its own offspring. At this point, uterine stimulation was applied for 5 min, and the ewe was immediately tested again with the newborn; this time 9 of 11 ewes showed intense licking accompanied by the emission of low-pitched bleats indicative of acceptance of the lamb. This finding suggests that, in addition to its facilitating effect on basic maternal responsiveness, vaginal-uterine stimulation may affect the duration of the critical period during which the ewe is willing to accept a newborn while rearing a previous newborn. Both of these effects may depend on vaginal-uterine influences on responses to olfactory stimulation and olfactory learning in the ewe, as the authors suggested, but they may also be based on the release of some additional vaginal-uterine-stimulated secretion that potentiates maternal responsiveness in the ewe.

LACTATION AND MATERNAL BEHAVIOR

It is the distinguishing characteristic of mammals that until they are weaned, the young are fed by means of the secretion of milk from milk-synthesizing mammary glands. Species differ, however, in the frequency of nursing, the pattern of nursing, and the rate of growth during the suckling period. In the rat and the cat, suckling may occur two to three times per 2-hr period, and this is the case for many nest-living altricial young among the small mammals. There are exceptions; for example, the rabbit and the hare nurse their offspring once per day, and the tree shrew nurses its young once every 48 hr. Among the ungulates, nursing periods vary according to whether the young are left at a nest site (hiders) and depend on the mother's returning to them at intervals, or whether they follow the mother (followers) and can nurse on their own initiative (Lent, 1974). In the roe deer, a hider species (Espmark, 1969), there may be three or four nursings per day during the initial days; this frequency increases to four to six, with an even greater increase in the duration of nursing from 20 to 40 min per nursing and then to 70 min at a later age. A survey of nursing frequencies and duration among the ungulates (Lent, 1974), which does not distinguish hider from follower types, shows a range of more than 10 nursings per day to only 1 or 2 per day. Among seals, the mother remains with the newborn for about a week, nursing it on a regular basis; then she leaves it for intervals of 1–3 days, returning to nurse and remaining with it for 0.5–3 days (Trilmich, 1981). A similar pattern of nursing occurs among sea lions: The mother remains with her newborn for 4–7 days after birth, then returns nightly to nurse it (Trillmich, 1981).

A profile of the composition of the milk of various orders has been compiled (Eisenberg, 1981) showing the percentage of concentration of water, fat, protein, and carbohydrate (Figure 33). There is a tendency for species in which the mother nurses infrequently, many of which have young that grow rapidly and attain a large body size and weight at an early age, to produce milk rich in fat and protein (e.g., cetacea and pinnipeds); exceptions are rabbits and many rodent species with rich

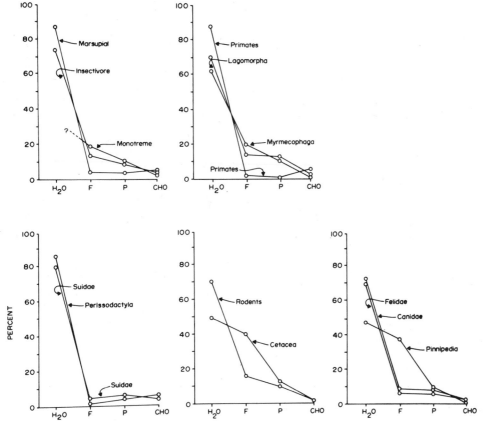

Fig. 33. Composition of various mammalian milks. F, fat; P, protein; CHO, carbohydrate. (From *The Mammalian Radiations* by J. F. Eisenberg Chicago: University of Chicago Press, 1981. Copyright 1981 by the University of Chicago Press. Reprinted by permission.)

milk and relatively small body size and weight but rapid rates of growth during infancy. Some theorists have proposed that one can predict the rate of nursing on the basis of these profiles, those species with milk low in fat and protein content requiring more frequent nursing than those with milk high in fat and protein content.

The nursing-suckling relationship between the mother and the young is at the center of their behavioral interaction, although other characteristics of the mother (thermal, tactile, olfactory, visual, and auditory) play a role in the attraction that she holds for the young, and other activities and stimuli from the young are important in the mother's response to them. Nevertheless, it is through their suckling, initially, that the young exert their primary influence on the endocrine secretions of the mother during lactation and thereby regulate her reproductive functioning. The nature of this influence is twofold: On the one hand, suckling stimulates the release of hormones—chiefly prolactin, but other anterior pituitary hormones as well—that promote the synthesis of milk, and oxytocin, which causes milk letdown, making the milk available to the young for withdrawal. On the other hand, it inhibits the release of gonadotropic hormones and alters the pituitary response to gona-dotropic releasing factors, thereby suspending estrous cycling for several weeks or months beyond the initial postpartum estrus in some species. Both of these actions of suckling are already present in the marsupials and show a parallel evolution in them and the eutherian mammals.

It was proposed in the early part of this century that maternal behavior was a direct outcome of mammary gland stimulation arising from the internal pressure generated by accumulating milk in the gland. This drive-stimulus interpretation is still believed by many people, but the evidence that exists, largely in the rat (and, surprisingly, among human mothers who are not nursing their babies), suggests that pressure within the mammary gland is not the basis of maternal responsiveness in the rat. Prospective rat mothers whose mammary glands and nipples are removed prior to parturition nevertheless exhibit maternal behavior that is indis-tinguishable from that of lactating mathers (Moltz *et al.*, 1967; Wiesner and Sheard, 1933). Moreoever, treatments that prevent lactation (i.e. hypophysectomy and the use of prolactin-release-blocking agents) appear also to have no effect on maternal behavior in the rat. From another point of view, the fact that nonpreg-nant females that are sensitized exhibit maternal behavior in the absence of mam-mary gland development and function further indicates the independence of maternal behavior from lactation.

Current thinking proposes just the opposite of what had been proposed ear-lier: It is maternal behavior that is the basis for the maintenance of lactation once it has been established during pregnancy, rather than the reverse. As a result of her maternal behavior, the female allows pups to suckle, and this suckling pro-motes further development of the mammary glands and lactation. This may be especially true among the marsupials. It was found that mammary gland develop-ment in the bush opossum did not differ during the 17-day period of pregnancy in pregnant and nonpregnant females, both undergoing estrous cycles, but following parturition and the initiation of suckling, the mammary glands of the mothers underwent further development and milk production in response to suckling stim-ulation (Sharman, 1961). Moreover, in several species of kangaroo, each of the mammary glands responds to the suckling stimulus of the young attached to it: When two young of different ages are suckling, the older joey receives milk of a different composition than the younger one, and its gland is much more developed and responds with milk ejection at higher concentrations of oxytocin (Lincoln and Renfreee, 1981). Finally, females may be induced to lactate by the placing in their pouches of young that attach to nipples and suckle (Sharman, 1961).

In this section, we discuss this latter formulation of the situation rather than the former one, which has little basis in evidence. Our discussion focuses on the rat because research on our topic is most developed with respect to this animal; however, similar findings, in part, have been reported in cows (Goodman *et al.*, 1979). We will not discuss the entire range of neuroendocrine-regulating mecha-nisms that underly the secretion of prolactin and oxytocin in response to suckling in the rat; the reader is referred to an excellent comprehensive review by Gros-venor and Mena (1974).

The release of prolactin and related hormones of the lactogenic complex is dependent on suckling, as is the release of oxytocin (Mena *et al.*, 1980). It has been shown that there is an immediate rise in circulating levels of prolactin (Terkel *et al.*, 1972) on the initiation of sucking, and after a latency of 15 min, oxytocin is released periodically at the same time (Deis, 1968). Although the release of these hormones is under the direct influence of suckling stimulation during the initial phase of maternal care, as time progresses the release of prolactin comes under the influence of exteroceptive stimuli that are associated with suckling, chiefly the odors and sounds of pups (Grosvenor, 1965; Grosvenor *et al.*, 1970; Mena and Grosvenor, 1971; Grosvenor and Mena, 1974). Oxytocin may also be released by the exteroceptive stimuli that are generated during suckling and that are heard by another lactating female (Deis, 1968).

Exteroceptive release of prolactin starting at the end of the second week means that the mother is stimulated to release prolactin and thus to synthesize milk for later feedings simply by being in the presence of her pups prior to suckling. At first, pups located beneath the female are effective stimuli, but later, they need only be alongside her and out of contact (Mena and Grosvenor, 1972). Ultimately, however, lactation is inhibited by exteroceptive stimuli from the pups through a mechanism that does not prevent prolactin release but inhibits its action on milk synthesis at the mammary gland (Grosvenor and Mena, 1973).

Weaning

Weaning represents the dissolution of the bond between the mother and the young and either the dispersal of the young or the continued association of mother and young in a new relationship based on the termination of maternal care or its persistence in a much weakened form on the part of the mother and the attainment of independent functioning on the part of the young. Weaning occurs by stages, and different aspects of the mother–young relationship are dissolved at different times. In many species, especially those in which the young are subject to predation pressure until they are somewhat mature and in which foraging is a highly complex process and requires a long period of practice and experience, the young remain with the mother or in the family group for many years before they are fully weaned and capable of living apart from the mother.

Although there are many descriptive studies of weaning, there are few studies that have analyzed the component processes contributing to weaning either in the mother or in the young (see Galef, 1981). In this section, reference is made to weaning in the rat and in the cat, where there exists some experimental evidence, particularly with respect to the termination of nursing and suckling.

In the rat, the young begin to be weaned about the end of the third week, and weaning is completed by the end of the fourth week (Galef, 1981). The young are capable of eating solid food beginning around the 15th or 16th day, but they continue to suckle as their main food source until they are 25 days old and may suckle until they are older. On the mother's part, nursing declines at the end of the third week (see Figure 16), and avoidance and rejection of the young begin between the 17th and 21st days (Reisbick *et al.,* 1975).

The process of decline in maternal behavior can be viewed either as a passive one, in which the stimuli from the young that normally elicit maternal behavior and maintain maternal responsiveness gradually disappear as the young develop, or as an active one, in which the behavioral interaction between the mother and the young plays a role. By removing litters from mothers on the 9th or 14th days, these alternative possibilities were examined: If the decline were a passive process, then maternal behavior would decline more rapidly when the litter was absent because pup stimuli would have been abruptly removed, but if it were an active process, as defined above, then removing the litter prematurely would slow the decline in maternal behavior. The decline was slowed when the litter was removed at either age, and it did not reach as low a level as when the litter was present (Rosenblatt and Lehrman, 1963). This finding suggests that weaning is an active process in which the young play a role.

Are pups ready to be weaned as soon as they are able to take solid food on the 15th day, or is there an optimal period for weaning, during which they are actually weaned, between 20 and 25 days of age? Weaning seems to be forced by a reduction in suckling caused by the mother's avoidance behavior and her frequent rejection of suckling approaches by the young. In one study, therefore, pups were separated from their mothers to eliminate suckling from the 15th to the 20th day (they were fed milk-wet mash during this period), from the 20th to the 25th day, or from the 15th to the 25th day (Williams *et al.,* 1980). The pups were tested, either when deprived overnight of feeding or nondeprived, by being placed with an anesthetized mother for suckling. Those separated from the mother for 5 days starting at 15 days readily suckled on Day 20 and for 17–23 min of the 30-min test. Those tested on Day 25 with a prior 5-day separation suckled only 2–8 min, depending on whether they were nondeprived or deprived, and those tested at the same age with 10 days of prior separation from the mother suckled for less than 1 min under both deprivation conditions. Control animals that remained with the mother sucked nearly 15 min if they were deprived at 25 days of age. There is, therefore, an optimal period during which experiences in the litter, presumably, can reduce the pup's suckling, and this period occurs between 20 and 25 days of age.

Another important factor—namely, reduced milk availability—has been found to play a role in the weaning of young from suckling in the rat (Henning, 1982). When 6-day-old pups were transferred to 13-day lactating mothers, they weaned themselves to solid food when they reached the age of 19–21 days (Figure 34). Their foster mothers were in their 26th to 28th days of lactation and were providing less milk than the pups would normally be receiving at that age. Their hunger was not satisfied, so they turned to eating solid food. On the other hand, when 13-day-old pups were transferred to 6-day lactating mothers, they did not increase their intake of solid food to as great an extent as the former pups because they were suckling from mothers in their 13th day of lactation who provided adequate amounts of milk. Their intake was about equal to that of nontransferred pups.

Fig. 34. Proportion of solid food eaten by rat pups between 15 and 23 days old under three experimental conditions: 6d pup → 13d dam = pups 6 days of age transferred to mothers 13 days postpartum; 13d pups → 6d dam = pups 13 days of age transferred to mothers 6 days postpartum; control = pups transferred to mothers matched for days postpartum. The 6d pup to 13d dam had a significantly lower value on Day 17 and significantly higher values on Day 21 and Day 23. All other comparisons were not significant. (From "Role of Milk-Borne Factors in Weaning and Intestinal Development" by S. J. Henning, *Biology of Neonate,* 1982, *4,* 265–272. Copyright 1982 by S. Karger AG, Basel. Reprinted by permission.)

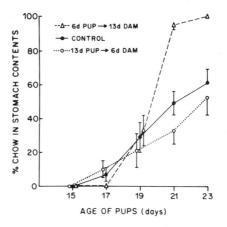

The relationship between maternal milk supply and pup intake of solid food was suggested earlier (Galef, 1981) on the basis of studies in the rat that indicated that the absolute food requirements of the pup increase steadily while the milk supply diminishes after the 15th day (Babicky *et al.,* 1970, 1973), requiring, therefore, that the pup turn to solid food to supplement its food intake.

For both the mother and the young, therefore, weaning is an active process in the rat. Suckling is actively discouraged by the mother, but what discourages the mother in her responsiveness to the young is not yet clear. In special circumstances, in which rat mothers have become pregnant during postpartum estrus and are carrying another litter while taking care of their current young, the mother may prevent the young from nursing a few days before she gives birth to the new litter, and she may actively attack young that have already been weaned when they approach the nest once the new litter has been born (Rowland, 1981; Gilbert *et al.,* 1983). This behavior may depend on hormonal stimulation during the pregnancy, but this possibility does not account for the behavior of mothers that are not pregnant.

The weaning process in the relationship between a mother and her litter of three kittens is depicted, descriptively, in Figure 18. As we discussed earlier, the mother gradually reduces her initiation of nursing as the kittens increasingly approach her to initiate suckling. An important change takes place in her behavior around the 28th to the 32nd day, when she begins to use the cage shelf and escapes from the kittens below, which as yet cannot climb to the shelf. This development in her is preceded, however, by an earlier rise in play activity among the kittens, which begins around Days 20–24 and accelerates between Day 24 and Day 32. Play is directed to other kittens chiefly, but it is also directed at the female's face, tail, body, and so on, and she appears to be annoyed by it. Her excursions to the cage shelf occur almost invariably when play is in progress, and she returns to the floor when the kittens have ceased playing and have fallen asleep. As the mother spends more time on the shelf, she is less available for nursing, and at around 40 days in the litter shown, but at earlier ages in other litters, there is a decline in suckling that is immediately preceded by the beginning of feeding and drinking from dishes.

The pattern of interrelationships between feeding initiation, the female's going to the shelf, the initiation and rise of social play among the kittens, and the decline in suckling, preceded by the initiation of taking food and water from the other sources, was observed in six multikitten litters. This pattern suggests, but, of course, does not prove, that the process of weaning begins with the mother's response to the play activity of the young, a measure of their social maturity and sensorimotor capacities. She leaves the kittens more frequently and for longer intervals, forcing them to find food by themselves and thus reducing their readiness to suckle from her when she is available. It has been shown that during the first 21 days, suckling among kittens is not regulated by food intake, but after that time, kittens that are either not hungry or are hungry but have not received milk during previous suckling periods reduce their subsequent suckling behavior (Koepke and Pribram, 1971).

Summary

It is difficult to summarize the rather large body of research on maternal behavior among the subprimate mammals that we have reviewed. What emerges from this review, however, are glimmerings of generalizations that may become clearer when additional data are available on a greater variety of species and in greater detail in those species that have already been studied. One of these generalizations concerns the division of maternal behavior into two principal phases. These phases correspond to the periods of pregnancy and postparturition. In the eutherian mammals, at least, maternal responsiveness arises as an outgrowth of the physiological, chiefly endocrine, processes of pregnancy. Its onset is timed to enable the mother to respond maternally to her newborn when they appear during parturition. During this phase, maternal behavior is hormonally based, and the onset is synchronized with the onset of parturition and the beginning of lactation, both of which are also based on the hormonal changes of late pregnancy.

The second phase of maternal responsiveness is based on stimulation that the female receives from the young. In several species, but not in all that have been studied, this phase is nonhormonally based (except for lactation, of course). Because the female's behavior is oriented toward the young, it is appropriate that it should be based on stimuli from them that are expressive of their behavioral capacities and biological needs. A nonhormonal basis for postpartum maternal behavior provides an adaptive advantage because it enables females of species that exhibit estrus immediately postpartum or shortly thereafter to become pregnant and still to engage in maternal care of their young. This ability increases the rate of reproduction considerably: In the rat, it enables a female that has just given birth to become pregnant and to produce a second litter in about 38 days, compared to the 48 days that would otherwise be required if the female's endocrine system were engaged in maternal behavior. This advantage represents a saving of more than 20% of the duration of the maternal behavior cycle.

One consequence of the mother's dependence on stimulation from her young for the maintenance of her maternal responsiveness is her need to respond to the changes in this stimulation as the young develop. We have analyzed this interaction between mother and young in terms of the concept of behavioral synchrony. This concept simultaneously characterizes the outcome of the interaction and the nature of the interaction. The outcome is that the mother's behavior complements that of the young in provision for their needs at all stages of their development. When they are helpless, the mother takes the initiative in providing care; as they become more capable of taking care of themselves, they take the initiative. They approach the mother for feeding and later seek other sources of food. The mother, in turn, reduces her care of them.

The nature of the interaction consists of the causal mechanisms that underlie it. At one level of analysis, these are the sensory stimuli that are exchanged between mother and young. At a deeper level, however, they are the individual physiological and behavioral processes that are involved in the interaction. Several of these processes in the rat were analyzed: the process of thermoregulation in the mother and the young and the recycling of water. This close interdependence calls to mind the analogical terms *parisitism* and *symbiosis,* which characterize the relationship. Such terms imply correctly that the mother–young relationship is deeply rooted in individual processes that have been mutually selected for between mother and young during the evolution of maternal care in each species and among the subprimate mammals in general. The broader dual nature of the concept of behavioral synchrony is perhaps preferable to the terms *parisitism* and *symbiosis,* which are specialized in their evolutionary meanings and refer only to certain characteristics of the relationship.

The existence of two principal phases in the organization of maternal behavior poses the problem of the transition from one phase to the next during the reproductive cycle. This transition phase is centered on parturition, but it may extend beyond it for limited periods in different species. During the transition phase, the hormonally based level of maternal responsiveness established prepartum recedes, and the postpartum responsiveness, which is based on stimulation from the young and which is nonhormonal in some species, increases. In species with altricial newborn, there does not intervene a further phase, which is present among many ungulate species with precocial newborn. This phase is the critical period for the formation of an individually specific relationship between the mother and her young. In sheep and goats, in which this period has been studied most extensively, this critical period lasts for about 2 hr, but the period of maternal responsiveness lasts for at least half a day. The development of an individually specific relationship is already the product of the interaction between the mother's level of maternal responsiveness and her experience with her newborn. If either of these factors is altered, the critical period is affected and may be lengthened or shortened accordingly. In the rat and other species with altricial newborn, maternal responsivesness may persist for several days following parturition.

Our view of parturition has undergone a change since we began our studies of maternal behavior in the rat. At one time, parturition was viewed as the time when maternal behavior was initiated and, therefore, when maternal responsiveness first reached a high level. Contact between the mother and her young during parturition was seen as contributing to the establishment of maternal behavior. With the discovery that, in the rat and other species, a high level of maternal responsiveness is already present before parturition, this early view has had to be revised. Now we believe that the prepartum establishment of maternal responsiveness is necessary for the female to respond maternally toward her newborn when they appear during parturition. Rat strains differ as to when this high level of responsiveness is established, and this difference is likely to occur among individuals of the same strain as well. The physiological changes that precede parturition and those that accompany it have a sufficiently strong impact on maternal behavior so that individual differences may be largely overcome; the high level of responsiveness found in females following cesarean-section delivery just before term, when contact with newborn during parturition has been precluded, is evidence of the strength of these physiological changes.

The prepartum onset of maternal behavior does not rule out a role for the behavioral interaction that occurs between mother and young during parturition. This role, however, can be better viewed in terms of the transitional phase discussed above. During parturition, the next phase of the maternal behavior cycle is initiated. The interaction between the mother and her offspring during parturition is the basis for this phase. Although parturition plays a special role in this transition process, it is not a unique role because postparturition contact can often substitute for the contact that normally occurs during parturition.

Acknowledgments

We wish to thank Cindy Banas for the illustrations and Winona Cunningham for secretarial assistance. Publication number 376 of the Institute of Animal Behavior.

References

Alberts, J. R., and Brunjes, P. C. Ontogeny of thermal and olfactory determinants of huddling in the rat. *Journal of Comparative and Physiological Psychology,* 1978, *92,* 897–906.

Alberts, J. R., and Gubernick, D. J. Reciprocity and resource exchange: A symbiotic model of parent-offspring relations. In H. Moltz and L. Rosenblum (Eds.), *Symbiosis in parent-young interactions.* New York: Plenum, 1983.

Alexander, G. Maternal behaviour in the Merino ewe. *Proceedings of the Australian Society of Animal Production,* 1960, *3,* 105–114.

Amoroso, E. C., Heap, R. B., and Renfree, M. B. Hormones and the evolution of viviparity. In E. J. W. Barrington (Ed.), *Hormones and Evolution,* Vol. 2. New York: Academic Press, 1979.

Anderson, C. O., Zarrow, M. X., Fuller, G. B., and Denenberg, V. H. Pituitary involvement in maternal nest-building in the rabbit. *Hormones and Behavior,* 1971, *2,* 183–189.

Arman, P. A note on parturition and maternal behavior in captive red deer (*Cervus elaphus* L.). *Journal of Reproduction and Fertility,* 1974, *37,* 87–90.

Arman, P., Hamilton, W. J., and Sharman, G. A. M. Observations on the calving of free-ranging tame red deer (*Cervus elaphus*). *Journal of Reproduction and Fertility,* 1978, *54,* 279–283.

Babicky, A., Ostadolova, I., Parizek, J., Kolar, J., and Bibr, B. Use of radioisotope techniques for determining the weaning period in experimental animals. *Physiologia Bohemoslovaca*, 1970, *19*, 457–467.

Babicky, A., Parizek, J., Ostadolova, I., and Kolar, J. Initial solid food intake and growth of young rats in nests of different sizes. *Physiologia Bohemoslovaca*, 1973, *22*, 557–566.

Baranczuk, R., and Greenwald, G. S. Plasma levels of oestrogen and progesterone in pregnant and lactating hamsters. *Journal of Endocrinology*, 1974, *63*, 125–135.

Bast, J. D., and Greenwald, G. S. Daily concentrations of gonadotrophins and prolactin in the serum of pregnant or lactating hamsters. *Journal of Endocrinology*, 1974, *63*, 527–532.

Bauer, J. H. Effects of maternal state on the responsiveness to nest odors of hooded rats. *Physiology and Behavior*, 1983, *30*, 229–232.

Baum, M. J. Failure of pituitary transplants to facilitate the onset of maternal behavior in ovariectomized virgin rats. *Physiology and Behavior*, 1978, *20*, 87–89.

Beniest-Noirot, E. Analyse du comportement dit "maternel" chez la souris. *Monographies Françaises Psychologie*, 1958, *1*, Centre National de Recherche Scientifique, Paris.

Berger, P. J. *The Reproductive Biology of the Tammar Wallaby, Macropu eugenii Desmarest (Marsupiala)*. Doctoral dissertation, 1970, Tulane University.

Bintarningsih, Lyons, W. R., Johnson, R. E., and Li, C. H. Hormonally-induced lactation in hypophysectomized rats. *Endocrinology*, 1958, *63*, 540–547.

Bonney, R. C., Moore, H. D. M., and Jones, D. M. Plasma concentrations of oestradiol-17 beta and progesterone, and laparoscopic observations of the ovary in the puma *(Felis concolor)* during oestrus, pseudopregnancy and pregnancy. *Journal of Reproduction and Fertility*, 1981, *63*, 523–531.

Brewster, J., and Leon, M. Facilitation of maternal transport by Norway rat pups. *Journal of Comparative and Physiological Psychology*, 1980, *94*, 80–88.

Bridges, R. S. Long-term effects of pregnancy and parturition upon maternal responsiveness in the rat. *Physiology and Behavior*, 1975, *14*, 245–249.

Bridges, R. S. Parturition: Its role in the long term retention of maternal behavior in the rat. *Physiology and Behavior*, 1977, *18*, 487–490.

Bridges, R. S. Retention of rapid onset of maternal behavior during pregnancy in primiparous rats. *Behavioral Biology*, 1978, *24*, 1113–117.

Bridges, R. S., and Feder, H. H. Effects of various progestins and deoxycorticosterone on the hormonal inhibition of maternal behavior in the ovariectomized-hysterectomized primigravid rat. *Hormones and Behavior*, 1978, *10*, 30–39.

Bridges, R. S., and Russell, D. W. Steroidal interactions in the regulation of maternal behaviour in virgin female rats: Effects of testosterone, dihydrotestosterone, oestradiol, progesterone and aromatase inhibitor 1,4,6-androstatriene-3,17-dione. *Journal of Endocrinology*, 1981, *90*, 31–40.

Bridges, R., Zarrow, M. X., Gandelman, R., and Denenberg, V. H. Differences in maternal responsiveness between lactating and sensitized rats. *Developmental Psychobiology*, 1972, *5*, 127–137.

Broida, J., Michael, S. D., and Svare, B. Plasma prolactin levels are not related to the initiation, maintenance, and decline of postpartum aggression in mice. *Behavioral and Neural Biology*, 1981, *32*, 121–125.

Buntin, J. D., Jaffe, S., and Lisk, R. D. Changes in responsiveness to newborn in pregnant, nulliparous golden hamsters. *Physiology and Behavior*, 1984, *32*, 437–439.

Buntin, J. D., Catanzaro, C., and Lisk, R. D. Facilitatory effects of pituitary transplants on intraspecific aggression in female hamsters. *Hormones and Behavior*, 1981, *15*, 214–225.

Challis, J. R. G., Davies, I. J., and Ryan, K. J. The concentrations of progesterone, estrone and estradiol-17 beta in the plasma of pregnant rabbits. *Endocrinology*, 1973, *93*, 971–976.

Chamley, W. A., Buckmaster, J. M., Cerini, M. E., Cumming, I. A., Goding, J. R., Obst, J. M., Williams, A., and Winfield, C. Changes in the levels of progesterone, corticosteroids, estrone, estradiol-17 beta and prolactin in the peripheral plasma of the ewe during late pregnancy and at parturition. *Biology of Reproduction*, 1973, *9*, 30–35.

Clutton-Brock, T. H., and Guinness, F. E., Behaviour of red deer (*Cervus elaphus* L.) at calving time. *Behaviour*, 1975, *55*, 287–300.

Cohen, J., and Bridges, R. S. Retention of maternal behavior in nulliparous and primiparous rats: Effects of duration of previous maternal experience. *Journal of Comparative and Physiological Psychology*, 1981, *95*, 450–459.

Cosnier, J. *Le Comportement Grégaire du Rat d'Élevage (Étude Éthologique)*. Unpublished doctoral dissertation, 1965, University of Lyon.

Cosnier, J., and Couturier, C. Comportement maternel provoqué chez les rattes adultes castrées. *Compte Rendu des Séances de la Société de Biologie*, 1966, *160*, 789–791.

Daly, M. The maternal behaviour cycle in golden hamsters. *Zeitschrift für Tierpsychologie*, 1972, *31*, 289–299.

Deis, R. P. The effect of an exteroceptive stimulus on milk ejection in lactating rats. *Journal of Physiology*, 1968, *197*, 37–46.

Doerr, H. K., Siegel, H. I., and Rosenblatt, J. S. Effects of progesterone withdrawal and estrogen on maternal behavior in nulliparous rats. *Behavioral and Neural Biology*, 1981, *32*, 35–44.

Dollinger, M. J., Holloway, Jr., W. R., and Denenberg, V. H. Parturition in the rat *(Rattus norvegicus):* Normative aspects and the temporal patterning of behaviours. *Behavioral Processes*, 1980, *5*, 21–37.

Dunbar, I., Ranson, E., and Buehler, M. Pup retrieval and maternal attraction to canine amniotic fluids. *Behavioral Processes*, 1981, *6*, 249–260.

Eisenberg, J. F. *The Mammalian Radiations: An Analysis of Trends in Evolution, Adaptation, and Behavior.* Chicago: University of Chicago Press, 1981.

Ellendorff, F., Taverne, M., and Smidt, D. *Physiology and Control of Parturition in Domestic Animals.* Amsterdam: Elsevier, 1979.

Elwood, R. W. Paternal and maternal behaviour in the Mongolian gerbil. *Animal Behavior*, 1975, *23*, 766–772.

Elwood, R. W. Changes in the responses of male and female gerbils *(Meriones unguiculatus)* towards test pups during pregnancy in the female. *Animal Behavior*, 1977, *25*, 46–51.

Erskine, M. S. *Hormonal and Experiential Factors Associated with the Expression of Aggression during Lactation in the Rat.* Ph.D. dissertation, 1978, University of Connecticut.

Erskine, M. S., Barfield, R. J., and Goldman, B. D. Intraspecific fighting during late pregnancy and lactation in rats and effects of litter removal. *Behavioral Biology*, 1978, *23*, 206–218.

Erskine, M. S., Barfield, R. J., and Goldman, B. D. Postpartum aggression in rats: I. Effects of hypophysectomy. *Journal of Comparative and Physiological Psychology*, 1980, *94*, 484–494.

Espmark, Y. Mother–young relations and development of behaviour in roe deer *(Capreolus capreolus* L.). *Viltrey*, 1969, *6*, 461–540.

Ewer, R. F. *Ethology of Mammals.* New York: Plenum Press, 1968.

Fahrbach, S. E., Morrell, J. I., and Pfaff, D. W. Oxytocin induction of short-latency maternal behavior in nulliparous, estrogen-primed female rats. *Hormones and Behavior*, 1984, *18*, 267–286.

Fleming, A., and Rosenblatt, J. S. Maternal behavior in the virgin and lactating rat. *Journal of Comparative and Physiological Psychology*, 1974a, *86*, 957–972.

Fleming, A. S., and Rosenblatt, J. S. Olfactory regulation of maternal behavior in rats: I. Effects of olfactory bulb removal in experienced and inexperienced lactating and cycling females. *Journal of Comparative and Physiological Psychology*, 1974b, *86*, 221–232.

Fleming, A. S., and Rosenblatt, J. S. Olfactory regulation of maternal behavior in rats: II. Effects of peripherally induced anosmia and lesions of the lateral olfactory tract in pup-induced virgins. *Journal of Comparative and Physiological Psychology*, 1974c, *86*, 233–246.

Fogden, S. C. L. Mother-young behaviour at Grey seal breeding beaches. *Journal of Zoology*, London, 1971, *164*, 61–92.

Frädrich, H. A comparison of behaviour in the Suidae. In V. Geist and F. Walther (Eds.), *The Behavior of Ungulates and Its Relation to Management*, Vol. 1. IUCN Publications, New Series No. 24, 1974.

Fraser, D. G., and Barnett, S. A. Effects of pregnancy on parental and other activities of laboratory mice. *Hormones and Behavior*, 1975, *6*, 181–188.

Friedman, M. I., Bruno, J. P., and Alberts, J. R. Physiological and behavioral consequences in rats of water recycling during lactation. *Journal of Comparative and Physiological Psychology*, 1981, *95*, 26–35.

Fuchs, A.-R. Hormonal control of myometrial function during pregnancy and parturition. *Acta Endocrinologica*, Supplement, 1978, *221*, 1–69.

Fuller, J. L., and DuBuis, E. M. The behaviour of dogs. In E. S. E. Hafez (Ed.), *The Behaviour of Domestic Animals.* London: Baillière, Tindall and Cox, 1962.

Galef, B. G., Jr. The ecology of weaning: Parasitism and the achievement of independence by altricial mammals. In D. J. Gubernick and P. H. Klopfer (Eds.), *Parental Care in Mammals.* New York: Plenum Press, 1981.

Gandelman, R. Determinants of maternal aggression in mice. In R. W. Bell and W. P. Smotherman (Eds.), *Maternal Influences and Early Behavior.* New York: Spectrum, 1980.

Gilbert, A. N., Burgoon, D. A., Sullivan, K. A., and Adler, N. T. Mother-weanling interactions in Norway rats in the presence of a successive litter produced by postpartum mating. *Physiology and Behavior,* 1983, *30,* 267–271.

Goodman, G. T., Tucker, H. A., and Convey, E. M. Presence of the calf affects secretion of prolactin in cows. *Proceedings of the Society for Experimental Biology and Medicine,* 1979, *161,* 421–424.

Graber, G. C., and Kristal, M. B. Uterine distention facilitates the onset of maternal behavior in pseudopregnant but not in cycling rats. *Physiology and Behavior,* 1977, *19,* 133–137.

Grosvenor, C. E. Evidence that exteroceptive stimuli can release prolactin from the pituitary gland of the lactating rat. *Endocrinology,* 1965, *76,* 340–342.

Grosvenor, C. E., and Mena, F. Evidence that suckling pups, through an exteroceptive mechanism, inhibit the milk stimulatory effect of prolactin in the rat during late lactation. *Hormones and Behavior,* 1973, *4,* 209–222.

Grosvenor, C. E., and Mena, F. Neural and hormonal control of milk secretion and milk ejection. *Lactation,* 1974, *1,* 227–276.

Grosvenor, C. E., Maiweg, H., and Mena, F. A study of actors involved in the development of the exteroceptive release of prolactin in the lactating rat. *Hormones and Behavior,* 1970, *1,* 111–120.

Gubernick, D. J. Maternal "imprinting" or maternal "labelling" in goats? *Animal Behavior,* 1980, *28,* 124–129.

Gubernick, D. J. Parent and infant attachment in mammals. In D. J. Gubernick and P. H. Klopfer (Eds.), *Parental Care in Mammals.* New York: Plenum Press, 1981.

Gubernick, D. J., Jones, K. C., and Klopfer, P. H. Maternal "imprinting" in goats? *Animal Behavior,* 1979, *27,* 314–315.

Hamilton, W. J. The success story of the opossum. *Natural History,* 1963, *72,* 16–25.

Harder, J. D., and Fleming, M. A. Estradiol and progesterone profiles indicate a lack of endocrine recognition of pregnancy in the opossum. *Science,* 1981, *212,* 1400–1402.

Heap, R. B., Galil, A. A. A., Harrison, F. A., Jenkin, G., and Perry, J. S. Progesterone and oestrogen in pregnancy and parturition: Comparative aspects and hierarchical control. *Ciba Foundation Symposium 47* (new series), 1977, 127–157.

Hemmes, R. B. *The Ontogeny of the Maternal-Filial Bond in the Domestic Goat.* Doctoral dissertation, 1969, Duke University.

Henning, S. J. Role of milk-borne factors in weaning and intestinal development. *Biology of Neonate,* 1982, *4,* 265–272.

Herrenkohl, L. R., and Campbell, C. Mechanical stimulation of mammary gland development in virgin and pregnant rats. *Hormones and Behavior,* 1976, *7,* 183–197.

Hersher, L., Richmond, J. B., and Moore, A. U. Maternal behavior in sheep and goats. In H. L. Rheingold (Ed.), *Maternal Behavior in Mammals.* New York: Wiley, 1963.

Hinde, R. A. Towards understanding relationships. *European Monographs in Social Psychology,* Vol. 18. New York: Academic Press, 1979.

Hofer, M. A. Parental contributions to the development of their offspring. In D. J. Gubernick and P. H. Klopfer (Eds.), *Parental Care in Mammals.* New York: Plenum Press, 1981.

Hofer, M. A. Seeing is believing: A personal perspective on research strategy in developmental psychobiology. *Developmental Psychobiology,* 1982, *15,* 399–408.

Jakubowski, M., and Terkel, J. Infanticide and caretaking in non-lactating *Mus musculus:* Influences of genotype, family group and sex. *Animal Behavior,* 1982, *30,* 1029–1035.

Karli, P. C. Social reactions of pregnant and lactating rats. *Gestation: Transactions of the Second Conference,* March 8, 9, and 10, 1955, Princeton, N. J., edited by C. A. Villee.

Keverne, E. B., Levy, F., Poindron, P., and Lindsay, D. R. Vaginal stimulation: An important determinant of maternal bonding in sheep. *Science,* 1983, *219,* 81–83.

Klopfer, P. H. Mother love: What turns it on? *American Scientist,* 1971, *59,* 404–407.

Klopfer, P. H., and Gamble, J. Maternal "imprinting" in goats: The role of chemical senses. *Zeitschrift für Tierpsychologies,* 1966, *23,* 588–592.

Klopfer, P. H., Adams, D.K., and Klopfer, M. S. Maternal imprinting in goats. *Proceedings of the National Academy of Science, U.S.A.,* 1964, *52,* 911–914.

Klopper, A., and Gardner, J. *Endocrine Factors in Labour.* London: Cambridge University Press, 1973.

Koepke, J. E., and Pribram, K. H. Effect of milk on the maintenance of sucking in kittens from birth to six months. *Journal of Comparative and Physiological Psychology,* 1971, *75,* 363–377.

Koller, G. Der Nestbau der weissen Maus und seine hormonale Auslösung. *Verhandlungen der Deutschen Zoologischen Gesellschaft,* Freiburg, 1952, 160–168.

Koller, G. Hormonale und psychische Steuerung beim Nestbau weiser Mäuse. *Verhandlungen der Deutschen Zoologischen Gesellschaft,* Freiburg, 1955, 123–132.

Koranyi, L., Lissak, K., Tamasy, V., and Kamaras, L. Behavioral and electrophysiological attempts to elucidate central nervous system mechanisms responsible for maternal behavior. *Archives of Sex Behavior,* 1976, *5,* 503–510.

Krehbiel, D. A., and LeRoy, L. M. The quality of hormonally stimulated maternal behavior in ovariectomized rats. *Hormones and Behavior,* 1979, *12,* 243–252.

Kristal, M. Placentophagia: A biobehavioral enigma (or De gustibus non disputandum est). *Neural Biobehavioral Review,* 1980, *4,* 141–150.

Kristal, M. B., and Graber, G. C. Placentophagia in nonpregnant rats: Influence of estrous cycle state and birthplace. *Physiology and Behavior,* 1976, *17,* 599–606.

Kristal, M. B., and Noonan, M. Perinatal maternal and neonatal behaviour in the captive reticulated giraffe. *South African Journal of Zoology,* 1979, *14,* 103–107.

Kristal, M. B., Peters, L. C., Franz, J. R., Whitney, J. F., Nishita, J. K. and Steuer, M. A. The effect of pregnancy and placentophagia in Long-Evans rats. *Physiology and Behavior,* 1981, 27, 591–595.

Leblond, C. P. Extra-hormonal factors in maternal behavior. *Proceedings of the Society for Experimental Biology,* New York, 1938, *38,* 66–70.

Leblond, C. P. Nervous and hormonal factors in the maternal behavior of the mouse. *Journal of Genetics and Psychology,* 1940, *57,* 327–344.

Leblond, C. P., and Nelson, W. O. Maternal behavior in hypophysectomized male and female mice. *American Journal of Physiology,* 1937, *120,* 167–172.

Lee, M. H. S., and Williams, D. I. A longitudinal study of mother-young interaction in the rat: The effects of infantile stimulation, diurnal rhythms, and pup maturation. *Behaviour,* 1977, *63,* 241–261.

Lent, P. C. Mother-infant relationships in ungulates. In V. Geist and F. Walther (Eds.), *The Behaviour of Ungulates and Its Relation to Management,* Vol. 1. IUCN Publications, New Series No. 24, 1974.

Leon, M., Croskerry, P. G., and Smith, G. K. Thermal control of mother-young contact in rats. *Physiology and Behavior,* 1978, *21,* 793–811.

Leonard, C. M. Effects of neonatal (day 10) olfactory bulb lesion on social behavior of female golden hamsters *(Mesocricetus auratus). Journal of Comparative Physiology and Psychology,* 1972, *80,* 208–215.

LeRoy, L. M. *Induction of Maternal Behavior in the Rat.* Eastern Conference on Reproductive Behavior, University of Connecticut, Storrs, June 5–8, 1977.

LeRoy, L. M., and Krehbiel, D. A. Variations in maternal behavior in the rat as a function of sex and gonadal state. *Hormones and Behavior,* 1978, *11,* 232–247.

Leshner, A. I., Siegel, H. I., and Collier, G. Dietary self-selection by pregnant and lactating rats. *Physiology and Behavior,* 1972, *8,* 151–154.

Levy, F. *Existence et Control de l'Attraction par le Liquide Amniotique chez la Brebis (Ovis aries).* Unpublished thesis, 1980–1981, University of Paris.

Lewis, P. R., and Schriefer, J. A. Ultrasound production by pregnant rats. *Behavioral and Neural Biology,* 1985.

Liggins, G. C. Hormonal interactions in the mechanisms of parturition. In A. Klopper and J. Gardner (Eds.), *Endocrine Factors in Labour.* London: Cambridge University Press, 1973.

Lillegraven, J. A. Biological considerations of the marsupial-placental dichotomy. *Evolution,* 1975, *29,* 707–722.

Lincoln, D. W., and Renfree, M. B. Mammary gland growth and milk ejection in the agile wallaby, *Macropus agilis,* displaying concurrent asynchronous lactation. *Journal of Reproduction and Fertility,* 1981, *63,* 193–203.

Lisk, R. D. Oestrogen and progesterone synergism and elicitation of maternal nestbuilding in the mouse *(Mus musculus.). Animal Behavior,* 1971, *19,* 606–610.

Lisk, R. D., Prelow, R. A., and Friedman, S. A. Hormonal stimulation necessary for eliciting of maternal nest building in the mouse. *Animal Behavior,* 1969, *17,* 730–738.

Mann, M., Michael, S. D., and Svare, B. Ergot drugs suppress plasma prolactin and lactation but not aggression in parturient mice. *Hormones and Behavior,* 1980, *14,* 319–328.

Mayer, A. D., and Rosenblatt, J. S. Olfactory basis for the delayed onset of maternal behavior in virgin female rats: Experiential effects. *Journal of Comparative and Physiological Psychology,* 1975, *89,* 701–710.

Mayer, A. D., and Rosenblatt, J. S. Hormonal influences during the ontogeny of maternal behavior in female rats. *Journal of Comparative and Physiological Psychology,* 1979, *93,* 879–898.

Mayer, A. D., and Rosenblatt, J. S. Hormonal interaction with stimulus and situational factors in the initiation of maternal behavior in nonpregnant rats. *Journal of Comparative and Physiological Psychology*, 1980, *94*, 1049–1059.

Mayer, A. D., and Rosenblatt, J. S. Prepartum changes in maternal responsiveness and nest defense in the rat. *Journal of Comparative Psychology*, 1984, *98*, 177–188.

McCann, T. S. Aggressive and maternal activities of female southern elephant seals *(Mirounga leonina)*. *Animal Behavior*, 1982, *30*, 268–276.

McCormack, J. T., and Greenwald, G. S. Progesterone and oestradiol-17β concentrations in the peripheral plasma during pregnancy in the mouse. *Journal of Endocrinology*, 1974, *62*, 101–107.

McMurtry, J. P., and Anderson, R. R. Prevention of self-licking on mammary gland development in pregnant rats. *Proceedings of the Society for Experimental Biology and Medicine*, 1971, *137*, 354–356.

McNeilly, A. S., and Friesen, H. G. Prolactin during pregnancy and lactation in the rabbit. *Endocrinology*, 1978, *102*, 1548–1554.

Mena, F., and Grosvenor, C. E. Release of prolactin in rats by exteroceptive stimulation: Sensory stimuli involved. *Hormones and Behavior*, 1971, *2*, 107–116.

Mena, F., and Grosvenor, C. E. Effect of suckling and of exteroceptive stimulation upon prolactin release in the rat during late lactation. *Journal of Endocrinology*, 1972, *52*, 11–22.

Mena, F., Pacheco, P., Whitworth, N. S., and Grosvenor, C. E. Recent data concerning the secretion and function of oxytocin and prolactin during lactation in the rat and rabbit. In C. Valverde-Rodriquez and H. Arachiza (Eds.), *Frontiers in Hormone Research*, Vol. 6. Basel: Karger, 1980.

Moltz, H., and Wiener, E. Ovariectomy: Effects on the maternal behavior of the primiparous and multiparous rat. *Journal of Comparative and Physiological Psychology*, 1966, *62*, 383–387.

Moltz, H., Geller, D., and Levin, R. Maternal behavior in the totally mammectomized rat. *Journal of Comparative and Physiological Psychology*, 1967, *64*, 225–229.

Moltz, H., Levin, R., and Leon, M. Differential effects of progesterone on the maternal behavior of primiparous and multiparous rat. *Journal of Comparative and Physiological Psychology*, 1969, *67*, 36–50.

Moltz, H., Lubin, M., Leon, M., and Numan, M. Hormonal induction of maternal behavior in the ovariectomized nulliparous rat. *Physiology and Behavior*, 1970, *5*, 1373–1377.

Morishige, W. K., Pepe, G. J., and Rothchild, I. Serum luteinizing hormone (LH), prolactin and progesterone levels during pregnancy in the rat. *Endocrinology*, 1973, *92*, 1527–1530.

Morris, D. *The Mammals: A Guide to Living Species*. New York: Harper and Row, 1965.

Murr, S. M., Bradford, G. E., and Geschwind, I. I. Plasma luteinizing hormone, follicle-stimulating hormone and prolactin during pregnancy in the mouse. *Endocrinology*, 1974, *94*, 112–116.

Naaktgeboren, C. Behavioural aspects of parturition. *Animal Reproductive Science*, 1979, *2*, 155–166.

Noirot, E. The onset of maternal behavior in rats, hamsters, and mice. In D. S. Lehrman, R. A. Hinde, and E. Shaw (Eds.), *Advances in the Study of Behavior*, Vol. 4. New York: Academic Press, 1972.

Noirot, E., and Goyens, J. Changes in maternal behavior during gestation in the mouse. *Hormones and Behavior*, 1971, *2*, 207–215.

Noirot, E., Goyens, J., and Buhot, M.-C. Aggressive behavior of pregnant mice towards males. *Hormones and Behavior*, 1975, *6*, 9–17.

Numan, M. Progesterone inhibition of maternal behavior in the rat. *Hormones and Behavior*, 1978, *11*, 209–231.

Numan, M., Leon, M., and Moltz, H. Interference with prolactin release and the maternal behavior of female rats. *Hormones and Behavior*, 1972, *3*, 29–38.

Numan, M., Rosenblatt, J. S., and Komisaruk, B. R. The medial preoptic area and the onset of maternal behavior in the rat. *Journal of Comparative and Physiological Psychology*, 1977, *91*, 146–164.

Obias, M. D. Maternal behavior of hypophysectomized gravid albino rats and development and performance of their progeny. *Journal of Comparative and Physiological Psychology*, 1957, *50*, 120–124.

Owen, J., and Thiessen, D. D. Regulation of scent marking in the female Mongolian gerbil *(Meriones unguiculatus)*. *Physiology and Behavior*, 1973, *11*, 441–445.

Owen, K., and Thiessen, D. Estrogen and progesterone interaction in the regulation of scent marking in the female Mongolian gerbil *(Meriones unguiculatus)*. *Physiology and Behavior*, 1974, *12*, 351–356.

Pedersen, C. A., and Prange, A. J., Jr. Induction of maternal behavior in virgin rats after intracerebroventricular administration of oxytocin. *Proceedings of the National Academy of Science*, 1979, *76*, 6661–6665.

Pedersen, C. A., Ascher, J. A., Monroe, Y. L., and Prange, A. J., Jr. Oxytocin induces maternal behavior in virgin female rats. *Science*, 1982, *216*, 648–650.

Pepe, G. J., and Rothchild, I. A comparative study of serum progesterone levels in pregnancy and in various types of pseudopregnancy in the rat. *Endocrinology,* 1974, *95,* 275–279.

Poindron, P. Étude de la relation mére-jeune chez des brebis (*Ovis aries),* lors de l'allaitement. *Compte Rendu Hebdomadaire Séances Académie Science* Série D, 1974, *278,* 2691–2694.

Poindron, P. Mother-young relationships in intact or anosmic ewes at the time of sucking. *Biology of Behavior,* 1976, *2,* 161–177.

Poindron, P., and Le Neindre, P. Hormonal and behavioural basis for establishing maternal behaviour in sheep. In L. Zichella and P. Pancheri (Eds.), *Psychoneuroendocrinology in Reproduction.* Amsterdam: Elsevier, 1979.

Poindron, P., and Le Neindre, P. Endocrine and sensory regulation of maternal behaviour in the ewe. In J. S. Rosenblatt, R. A. Hinde, C. Beer, and M.-C. Busnel (Eds.), *Advances in the Study of Behavior.* New York: Academic Press, 1980.

Poole, W. E., and Pilton, P. E. Reproduction in the grey kangaroo Macropus canguru. *S.S.I.R.O. Wildlife Research,* 1965, *9,* 218–234.

Reisbick, S., Rosenblatt, J. S., and Mayer, A. D. Decline of maternal behavior in the virgin and lactating rat. *Journal of Comparative and Physiological Psychology,* 1975, *89,* 722–732.

Rheingold, H. L. Maternal behavior in the dog. In H. L. Rheingold (Ed.), *Maternal Behavior in Mammals.* New York: Wiley, 1963.

Richards, M. P. M. Effects of oestrogen and progesterone on nest building in the golden hamster. *Animal Behaviour,* 1969, *17,* 356–361.

Richter, C. P., and Barelare, B. Nutritional requirements of pregnant and lactating rats studied by the self-selection method. *Endocrinology,* 1938, *23,* 15–24.

Riddle, O., Lahr, E. L., and Bates, R. W. The rôle of hormones in the initiation of maternal behavior in rats. *American Journal of Physiology,* 1942, *137,* 299–317.

Rodriguez-Sierra, J., and Rosenblatt, J. S. Does prolactin play a role in estrogen-induced maternal behavior in rats: Apomorphine reduction of prolactin release. *Hormones and Behavior,* 1977, *9,* 1–7.

Rodriguez-Sierra, J. F., and Rosenblatt, J. S. Pregnancy termination by prostaglandin $F_2\alpha$ stimulates maternal behavior in the rat. *Hormones and Behavior,* 1982, *16,* 343–351.

Rosenblatt, J. S. The basis of synchrony in the behavioral interaction between the mother and her offspring in the laboratory rat. In B. M. Foss (Ed.), *Determinants of Infant Behavior,* Vol. 3. London: Methuen, 1965.

Rosenblatt, J. S. Nonhormonal basis of maternal behavior in the rat. *Science,* 1967, *156,* 1512–1514.

Rosenblatt, J. S., and Lehrman, D. S. Maternal behavior of the laboratory rat. In H. L. Rheingold (Ed.), *Maternal Behavior in Mammals.* New York: Wiley, 1963.

Rosenblatt, J. S., and Siegel, H. I. Hysterectomy-induced maternal behavior during pregnancy in the rat. *Journal of Comparative and Physiological Psychology,* 1975, *89,* 685–700.

Rosenblatt, J. S., and Siegel, H. I. Factors governing the onset and maintenance of maternal behavior among nonprimate mammals: The role of hormonal and nonhormonal factors. In D. J. Gubernick and P. H. Klopfer (Eds.), *Parental Care in Mammals.* New York: Plenum Press, 1981.

Rosenblatt, J. S., Turkewitz, G., and Schneirla, T. C. Development of suckling and related behavior in neonate kittens. In E. L. Bliss (Ed.), *Roots of Behavior,* New York: Hoeber, 1962.

Rosenblatt, J. S., Siegel, H. I., and Mayer, A. D. Progress in the study of maternal behavior in the rat: Hormonal, nonhormonal, sensory, and developmental aspects. In J. S. Rosenblatt, R. A. Hinde, C. G. Beer, and M.-C. Busnel (Eds.), *Advances in the Study of Behavior,* Vol. 10. New York: Academic Press, 1979.

Roth, L. L., and Rosenblatt, J. S. Changes in self-licking during pregnancy in the rat. *Journal of Comparative and Physiological Psychology,* 1967, *63,* 397–400.

Roth, L. L., and Rosenblatt, J. S. Self-licking and mammary development during pregnancy in the rat. *Journal of Comparative and Physiological Psychology,* 1968, *42,* 363–378.

Rowell, T. E. Maternal behaviour in non-maternal golden hamsters (*Mesocricetus auratus*). *Animal Behavior,* 1961, *9,* 11–15.

Rowland, D. L. Effects of pregnancy on the maintenance of maternal behavior in the rat. *Behavioral and Neural Biology,* 1981, *31,* 225–235.

Russell, E. M. Mother-young relations and early behavioural development in the marsupials Macropus eugenii and Megaleia rufa. *Zeitschrift für Tierpsychologie,* 1973, *33,* 163–203.

Rubin, B. S., Menniti, F. S. and Bridges, R. S. Intracerebroventricular administration of oxytocin and maternal behavior in rats after prolonged and acute steroid treatment. 1983, *17,* 45–53.

Russell, E. M., and Giles, D. C. The effects of young in the pouch on pouch cleaning in the tammar wallaby, *Macropus eugenii* desmarest (Marsupialia). *Behaviour*, 1973, *51*, 19–37.

Saito, T. R., and Takahashi, K. W. Studies on the maternal behavior in the mouse: II. Difference in the maternal behavior pattern among different strains. *Japanese Journal of Animal Reproduction*, 1979, *25*, 117–119.

Saito, T. R., and Takahashi, K. W. Studies on the maternal behavior in the mouse: III. The maternal behavior of pregnant mice. *Japanese Journal of Animal Reproduction*, 1980, *26*, 43–45.

Schneirla, T. C., Rosenblatt, J. S., and Tobach, E. Maternal behavior in the cat. In H. L. Rheingold (Ed.), *Maternal Behavior in Mammals*. New York: Wiley, 1963.

Scott, E. M., Smith, S. J., and Verner, E. L. Self-selection of diet: VII. The effect of age and pregnancy on selection. *Journal of Nutrition*, 1948, *35*, 281–286.

Shaikh, A. A. Estrone and estradiol levels in the ovarian venous blood from rats during the estrous cycle and pregnancy. *Biology of Reproduction*, 1971, *5*, 297–307.

Sharman, G. B. The initiation and maintenance of lactation in the marsupial, *Trichosurus vulpecula*. *Journal of Endocrinology*, 1961, *25*, 375–385.

Sharman, G. B., and Calaby, J. H. Reproductive behavior in the red kangaroo *(Megaleia rufa)* in captivity. *C.S.I.R.O. Wildlife Research*, 1964, *9*, 58–85.

Sharman, G. B., Calaby, J. H., and Poole, W. E. Patterns of reproduction in female diprotodont marsupials. In I. W. Rowlands (Ed.), *Comparative Biology of Reproduction in Mammals*. New York: Academic Press, 1966.

Siegel, H. I., and Greenwald, G. S. Prepartum onset of maternal behavior in hamsters and the effects of estrogen and progesterone. *Hormones and Behavior*, 1975, *6*, 237–245.

Siegel, H. I., and Greenwald, G. S. Effects of mother-litter separation on later maternal responsiveness in the hamster. *Physiology and Behavior*, 1978, *21*, 147–149.

Siegel, H. I., and Rosenblatt, J. S. Estrogen-induced maternal behavior in hysterectomized-ovariectomized virgin rats. *Physiology and Behavior*, 1975a, *14*, 465–471.

Siegel, H. I., and Rosenblatt, J. S. Hormonal basis of hysterectomy-induced maternal behavior during pregnancy in the rat. *Hormones and Behavior*, 1975b, *6*, 211–222.

Siegel, H. I., and Rosenblatt, J. S. Progesterone inhibition of estrogen-induced maternal behavior in hysterectomized-ovariectomized virgin rats. *Hormones and Behavior*, 1975c, *6*, 223–230.

Siegel, H. I., and Rosenblatt, J. S. Duration of estrogen stimulation and progesterone inhibition of maternal behavior in pregnancy-terminated rats. *Hormones and Behavior*, 1978, *11*, 12–19.

Siegel, H. I., and Rosenblatt, J. S. Hormonal and behavioral aspects of maternal care in the hamster: A review. *Neuroscience and Biobehavioral Review*, 1980, *4*, 17–26.

Siegel, H. I., Clarke, M. C., and Rosenblatt, J. S. Maternal responsiveness during pregnancy in the hamster *(Mesocricetus auratus)*. *Animal Behavior*, 1983, *31*, 497–502.

Slotnick, B. M., Carpenter, M. L., and Fusco, R. Initiation of maternal behavior in pregnant nulliparous rats. *Hormones and Behavior*, 1973, *4*, 53–59.

Smith, M. S., and McDonald, L. E. Serum levels of luteinizing hormone, and progesterone during the estrous cycle, pseudopregnancy and pregnancy in the dog. *Endocrinology*, 1974, *94*, 404–412.

Soares, M. J., and Diamond, M. Pregnancy and chin marking the rabbit, *Oryctolagus cuniculas*. *Animal Behavior*, 1982, *30*, 941–943.

Stern, J. M. Effects of ergocryptine on postpartum maternal behavior, ovarian cyclicity and food intake in rats. *Behavioral Biology*, 1977, *21*, 134–140.

Stern, J. M., and MacKinnon, D. A. Sensory regulation of maternal behavior in rats: Effects of pup age. *Developmental Psychobiology*, 1978, *11*, 579–586.

Stone, C. P. Preliminary note on maternal behavior of rats living in parabioses. *Endocrinology*, 1925, *9*, 505–512.

Strauss, J. F., Sokoloski, J., Caploe, P., Duffy, P., Mintz, G., and Stambaugh, R. L. On the role of prostaglandins in parturition in the rat. *Endocrinology*, 1975, *96*, 1040–1043.

Sturman-Hulbe, M., and Stone, C. P. Maternal behavior in the albino rat. *Journal of Comparative Psychology*, 1929, *9*, 203–237.

Svare, B. R. Maternal aggression in mammals. In D. J. Gubernick and P. H. Klopfer (Eds.), *Parental Care in Mammals*. New York: Plenum Press, 1981.

Svare, B., and Gandelman, R. Postpartum aggression in mice: Experiential and environmental factors. *Hormones and Behavior*, 1973, *4*, 323–334.

Svare, B., Mann, M. A., Broida, J., and Michael, S. D. Maternal aggression exhibited by hypophysectomized parturient mice. *Hormones and Behavior*, 1982, *16*, 445–461.

Swanson, H. H. and Bolwerk, E. The role of oxytocin in the induction of maternal behaviour in virgin female rats. Paper presented at International Ethological Conference, August 29–September 6, 1983, Brisbane, Australia.

Swanson, L. J., and Campbell, C. S. Maternal behavior in the primiparous and multiparous golden hamster. *Zeitschrift für Tierpsychologie,* 1979, *50,* 96–104.

Taber, S., and Thomas, P. Calf development and mother-calf spatial relationships in southern right whales. *Animal Behavior,* 1982, *30,* 1073–1083.

Terkel, J., and Rosenblatt, J. S. Maternal behavior induced by maternal blood plasma injected into virgin rats. *Journal of Comparative and Physiological Psychology,* 1968, *65,* 479–482.

Terkel, J., and Rosenblatt, J. S. Aspects of nonhormonal maternal behavior in the rat. *Hormones and Behavior,* 1971, *2,* 161–171.

Terkel, J., and Rosenblatt, J. S. Humoral factors underlying maternal behavior at parturition: Cross transfusion between freely moving rats. *Journal of Comparative and Physiological Psychology,* 1972, *80,* 365–371.

Terkel, J., Blake, C. A., and Sawyer, C. H. Prolactin levels in lactating rats after suckling or exposure to ether. *Endocrinology,* 1972, *91,* 49–53.

Trillmich, F. Mutual mother-pup recognition in Galapagos fur seals and sea lions: Cues used and functional significance. *Behaviour,* 1981, *78,* 21–42.

Vania, J. Birth of a stellar sea lion pup. *Bioscience,* 1965, *15,* 794–795.

Voci, V. E., and Carlson, N. R. Enhancement of maternal behavior and nest building following systemic and diencephalic administration of prolactin and progesterone in the mouse. *Journal of Comparative and Physiological Psychology,* 1973, *83,* 388–393.

Wallace, P. *Hormonal Influences on Maternal Behavior in the Female Mongolian Gerbil (Meriones unguiculatus).* Ph.D. dissertation, 1973, University of Texas at Austin.

Wiesner, B. P., and Sheard, N. M. *Maternal Behavior in the Rat.* London: Oliver and Boyd, 1933.

Williams, C. L., Hall, W. G., and Rosenblatt, J. S. Changing oral cues in suckling of weaning-age rats: Possible contributions to weaning. *Journal of Comparative and Physiological Psychology,* 1980, *94,* 472–483.

Wilson, N. E., and Stricker, E. M. Thermal homeostasis in pregnant rats during heat stress. *Journal of Comparative and Physiological Psychology,* 1979, *93,* 585–594.

Wise, D. A., and Pryor, T. L. Effects of ergocornine and prolactin on aggression in the postpartum golden hamster. *Hormones and Behavior,* 1977, *8,* 30–39.

Woodside, B., and Leon, M. Thermoendocrine influences on maternal nesting behavior in rats. *Journal of Comparative and Physiological Psychology,* 1980, *94,* 41–60.

Woodside, B., Pelchat, R., and Leon, M. Acute elevation of the heat load of mother rats curtails maternal bouts. *Journal of Comparative and Physiological Psychology,* 1980, *94,* 61–68.

Woodside, B., Leon, M., Attard, M., Feder, H. H., Siegel, H. I., and Fischette, C. Prolactin-steroid influences on the thermal basis for mother-young contact in Norway rats. *Journal of Comparative and Physiological Psychology,* 1981, *95,* 771–780.

Zarrow, M. X., Gandelman, R., and Denenberg, V. H. Prolactin: Is it an essential hormone for maternal behavior in mammals? *Hormones and Behavior,* 1971, *2,* 343–354.

Zarrow, M. X., Sawin, P. B., Ross, S., and Denenberg, V. H. Maternal behavior and its endocrine basis in the rabbit. In E. L. Bliss (Ed.), *Roots of Behavior.* New York: Harper and Row, 1962.

8

Peripheral Plasma Levels of Gonadal Steroids in Adult Male and Adult, Nonpregnant Female Mammals

H. H. FEDER

The introduction, about 20 years ago, of sensitive and specific assay methods has permitted the measurement of plasma androgens, estrogens, and progestins in the systemic circulation in representatives of all vertebrate classes. The present survey deals with such measurements in adult mammals. The survey begins with a review of the systemic levels of plasma androgens in males of various mammalian orders. The second section considers estrogen and progestin levels in adult, nonpregnant females. The third section deals with the value of plasma hormone measurements in neurobiological research.

ANDROGENS IN THE SYSTEMIC CIRCULATION OF ADULT MALE MAMMALS

Methods such as gas–liquid chromatography (GLC), double-isotope derivative determinations (DIDD), competitive protein binding (CPB), and radioimmunoassay (RIA) have been used to measure androgens in the systemic plasma (or serum)

H. H. FEDER Institute of Animal Behavior, Rutgers University, Newark, New Jersey 07102.   Preparation on this chapter was supported by NIH 04467 and NIMH-K02-29006.

of adult male mammals. Gustafson and Shemesh (1976) provided a tabulation of the data available through 1975.

There are few data on androgens in prototheria and metatheria (but see Lincoln, 1978, for testosterone levels in five species of marsupials) as well as in certain orders of eutheria. Among eutherians, circulating androgens do not appear to have been measured in dermopterans ("flying lemurs"), edentates (armadillos and sloths), pholidotes (pangolins and scaly anteaters), tubulidentates (aardvarks), sirenians (manatees and dugongs), and pinnipeds (seals and sealions).

INSECTIVORA AND CHIROPTERA

The musk shrew *(Suncus murinus)* is an insectivore with nonscrotal testes. The testes are permanently abdominal and reside in the coelomic cavity. This position results in relatively elevated testicular temperatures, yet resistance to these temperature effects ensures normal spermatogenesis. The animals reach anatomical sexual maturity at 33 days of age and can mate successfully by 51 days of age. Hasler *et al.,* (1975) studied, with RIA, testosterone (T) levels in the systemic plasma of musk shrews through development and into adulthood. Plasma T levels were very low until about 20 days of age, rose to 1.5 ng/ml between Day 25 and Day 30, dipped to 1.0 ng/ml at 33 days, and rose again to 2.2 ng/ml at Day 45. Older adults ($>$ 100 days of age) had T levels of about 1.5 ng/ml. The primary source of T is the testis, as shown by the fact that castration reduced plasma T levels to $<$ 0.1 ng/ml.

Another insectivore, the mole *Talpa europaea,* has been assayed for plasma T by RIA (Racey, 1978). Moles were caught in Lancashire, England, where the onset of *T. europaea* spermatogenesis and mating behavior occurs in March–April, continuing for only about two months. Levels of T were 6.6 ng/ml in February, 10.3 ng/ml in March, and 2.6 to 5.0 ng/ml in April. In May, T levels rose to 30.8 ng/ml, then declined to 8.1 ng/ml in July. T levels remained between 1.3 and 3.9 ng/ml from August through December. Thus, elevated T levels are clearly associated with spermatogenesis and mating in this species. The proximate cause of the testicular cycle in moles is not known, but Racey hypothesized that photoperiodic cues are involved. It is noteworthy that viable spermatozoa can remain in the epididymis for 3 months after spermatogenesis ceases. Racey suggested that this may permit males to impregnate females that come into first estrus late in the year or that come into a second estrus. This notion implies that males retain sexual interest and sexual behavioral capacity after T has declined to quite low levels. For both musk shrews and moles, it is unknown whether there is a diurnal rhythm of androgen secretion, and whether there are episodic bursts of androgen over the course of a day.

Whereas work with the insectivore *T. europeaea* indicates the presence of a breeding season in which spermatogenesis and mating occur concurrently, the situation with the species of chiropterans discussed next is quite different. In these chiropterans, seasonal spermatogenesis and mating occur sequentially rather than concurrently. That is, spermatogenesis proceeds prior to mating. Mating onset occurs after the period of active spermatogenesis, with the stored, still viable, sperm released during copulation.

Studies of circulating androgen in three species of hibernating chiropterans have been reported. Racey and Tam (1974) reported data on androgen levels for the bat *Pipistrellus pipistrellus*. Whole blood T was measured by GLC, and the values were about 80 ng/ml in August (period of spermatogenesis) and 80 to 120 ng/ml in February and November (period of hibernation). Note that these values are for whole blood; plasma values might be about double those reported for whole blood. Androstenedione (AD) levels in blood ranged from nondetectable to 9.8 ng/ml. Thus, in *Pipistrellus*, the systemic T levels are extremely high (e.g., compare with the values reported for nonchiropterans) and remain elevated even during the hibernation period.

In captive *Nyctalus noctula*, the T levels (as measured by RIA) rose from 4.7 ng/ml to 134 ng/ml from May to August (as spermatogenesis progressed). The levels remained relatively high (22–65 ng/ml) during autumn (when mating took place) and then decreased (to about 15 ng/ml) during the winter months (the period of hibernation and sperm storage). These data of Racey (1974) are intriguing in several respects. First, as androgen levels rise during the course of spermatogenesis, these bats become increasingly aggressive. The increased aggressiveness is, in part, signified by prolonged high-pitched vocalizations. Work with nonmammalian and nonchiropteran mammalian species indicates a role for androgen in vocalization (e.g., Floody *et al.*, 1979), and similar studies of the role of androgens in bat vocalization may be of interest. A second feature of *N. noctula* is the enormously high (134 ng/ml) level of T seen in systemic plasma in August. This level should be compared to reported values for nonchiropteran mammals. A third feature is that, during the hibernation period, sexual accessory organs are maintained (in terms of weight and fructose content), spermatozoa survive in the cauda epididymis, and copulatory behavior can be observed if the bats are aroused from torpor. Apparently, the maintenance of the accessory organs, spermatozoa, and neural tissues that mediate sexual behavior requires a lower level of androgen than spermatogenesis and aggressive behavior. However, comparisons of androgen thresholds for copulatory and aggressive behaviors do not appear to have been made in noctule bats.

Gustafson and Shemesh (1976) measured systemic plasma T by RIA in a wild population of brown bats *(Myotis lucifugus)*. Baseline T levels (2.5–11.7 ng/ml) occurred during early spermatogenesis (May–mid-July), during the breeding period (September–May), and throughout hibernation (late September–April). However, during the latter part of spermatogenesis (mid-July–late August), T levels rose as high as 59.1 ng/ml. The authors suggested that this T surge is required to stimulate the accessory glands fully. The authors agreed with Racey's contention, made on the basis of work with noctule bats (1974), that lower levels of T are required for the maintenance of sexual accessories, spermatozoa, and copulatory behavior than for initiating growth of the sexual accessory tissues. However, both Racey (1974) and Gustafson and Shemesh (1976) pointed out that baseline levels of T are insufficient to maintain the sexual accessory tissues when bats emerge from hibernation in spring. These studies imply that the low metabolic rate characteristic of torpor inhibits the involution of sexual accessory tissue (Courrier, 1927; Racey and Tam, 1974), perhaps by preserving androgenic stimulation of these tissues.

There are several ways in which the data on bats could be expanded. Studies on the frequency and amplitude of episodic bursts of T (if these occur during the course of a day) could be made at various times of the year and could be related to gonadotropin regulation. Data on circadian rhythmicity in plasma T levels throughout an annual cycle would also be of interest, as would a systematic study of the relation of plasma T levels to specific components of aggressive and copulatory behavior. Such studies might indicate that the massive levels of T that occur during late summer not only promote spermatogenesis and initiate growth of sexual accessory tissue but also prime the CNS so that copulation can be expressed in the presence of lower levels of T during the breeding season.

Nonhuman Primates

Androgen levels in two prosimian species were determined by RIA by Van Horn et al. (1976). In the diurnal prosimian *Lemur catta* (ring-tailed lemur) T levels were highest (13.5 ng/ml) during darkness and lowest (2.8 ng/ml) during the light phase. In the nocturnal prosimian *Galago crassicaudatus* (bushbaby), there was no significant difference in T level during dark (2.6 ng/ml) and light (3.9 ng/ml) phases.

A survey of several species of Old and New World monkeys was carried out by Snipes *et al.* (1969) using a DIDD technique. Scattered studies of individual species can also be found. For example, African green monkeys (vervets; Cercopithecus Sp.) showed diurnal variations in serum assayed for T and dihydrotestosterone (DHT) by RIA (Beattie and Bullock, 1978) during the summer months. Values ranged from 6.8 to 12.4 ng/ml, with peak concentrations occurring at night. Serum estradiol levels were on the order of 0.01 ng/ml.

The squirrel monkey *(Saimiri sciureus)* showed seasonal variation in sexual behavior that coincided with variations in plasma T level. Mendoza *et al.* (1978a) used an RIA method to investigate these hormone variations and found T levels as high as 60.6 ng/ml in January–March and as low as 3.4 ng/ml in June–August. Squirrel monkeys apparently have little DHT in the peripheral circulation.

The rhesus macaque *(Macaca mulatta)* has been the most extensively studied nonhuman primate. GLC, RIA, and DIDD studies indicate that T levels in the adult range from about 5.0 to 17.7 ng/ml and AD levels from 0.7 to 1.4 ng/ml (Resko, 1967, 1971; Snipes *et al.*, 1969; Resko and Phoenix, 1972; Rose *et al.*, 1972; Michael *et al.*, 1974; Goodman *et al.*, 1974; Dierschke *et al.*, 1975). In addition, Snipes *et al.*, (1969) found very high levels of dehydroepiandrosterone (probably of adrenal origin) in peripheral plasma (21.2 ng/ml). DHT levels appear to be relatively low in rhesus macaque peripheral plasma. Perachio *et al.* (1977) found no significant difference between samples chromatographed to separate DHT from T and nonchromatographed samples subjected to an RIA system in which DHT is a significant cross-reactant. However, Robinson *et al.* (1975), using a direct RIA measurement of DHT, found levels of 1 to 3 ng/ml compared to T levels of 5 to 10 ng/ml. Incidentally, these authors found that old (> 20 year) rhesus monkeys had circulating DHT and T levels that were not significantly lower than those of younger adults.

It is likely that a great deal of the variability in reported T levels is attributable to factors such as diurnal rhythms, seasonal rhythms, and social interactions. An early study of diurnal rhythms in peripheral T levels used a CPB method and found the highest T levels at 22.00 h (15.5 ng/ml) (Plant and Michael, 1971). A more complete study, using RIA and CPB procedures, confirmed this result. T levels were highest at 22.00 hr (17.8 ng/ml) and lowest at 16.00 hr (7.8 ng/ml) in adult males kept in a room illuminated from 06.15 to 20.15 hr daily (Michael *et al.*, 1974). Furthermore, the authors discussed evidence that male rhesus monkeys show five to seven episodic T peaks within a 24-hr period. Perachio *et al.* (1977) also found the highest T levels throughout most of the lights-off period and noted that castration abolished this nightly elevation of T level. However, these authors also presented evidence of a diurnal pattern of adrenal T secretion, with an early-evening nadir characterizing this pattern. This study also used methodological refinements not included in previous studies of diurnal T variations (Goodman *et al.*, 1974; Michael *et al.*, 1974). The refinements consisted of using chair-adapted, catheterized monkeys placed in a chamber that isolated them from the environmental stimuli associated with blood sampling. In this way, it is possible to obviate the problem of arousing the animals from sleep in order to collect blood samples.

The effects of season on T levels in *M. mulatta* were first studied by Resko (1967), who found no significant relationship in monkeys kept in laboratory conditions under "no special lighting regimen." Plant *et al* (1974) used an RIA method to show seasonal variations in T in animals kept in 14-hr light per day. T levels were 10.1 ng/ml in December and declined progressively to 5.9 ng/ml in April of the following year. T levels remained depressed during spring and early summer and then increased progressively to reach a peak of 11.7 ng/ml in September. Of the eight animals studied, six showed evident seasonal effects; such effects were either weak or absent in the remaining two subjects. The data are difficult to interpret because they may have been confounded by allowing the males to have access to females at unspecified times, because the standard errors of the mean were very large, and because samples were taken over the course of only 1 year. However, the same laboratory (Michael and Bonsall, 1977) subsequently carried out a 3-year study and obtained the same seasonal pattern. Gordon *et al.* (1976) did a 3-year RIA study of plasma T and also concluded that the levels were highest in September–October–November (12 ng/ml), whereas the March–April levels were as low as 2 ng/ml. All seven subjects used in this study showed these seasonal patterns. T concentrations rose prior to an observed seasonal increase in mounting and reached peak levels coincidentally with the period of maximum mounting behavior. The annual decline in T levels occurred in parallel with a decrease in the number of sexually receptive females, a finding suggesting that stimuli from receptive females may contribute to the regulation of seasonal T variation in males. A long-term study of males isolated from females would be required to determine whether seasonal factors directly influence T levels in males. A study by Beck and Wuttke (1979) used animals kept in individual cages under "controlled conditions." Presumably, these males were not permitted to copulate with females (but there is no indication of whether males were housed in olfactory, auditory, or visual isolation from females). Males in this study exhibited a seasonal cycle of T, with the lowest serum levels in March–June and higher levels in July–November. Aside from the possible interaction between social factors and seasonal factors in the regulation of T levels, there also appears to be an interaction between diurnal rhythms and seasonal rhythms. Rose *et al.* (1978) showed that, during the autumn breeding season, there were higher concentrations of T at 10.00 hr and 22.00 hr compared with those observed during the summer.

Although copulation does not appear to induce a rise in peripheral plasma T levels in rhesus macaques (Dixson, *et al.*, 1976), other social and environmental factors may affect this parameter. For example, Rose *et al.* (1975) claimed that defeat in agonistic encounters results in decreased plasma T, and that a positive correlation exists between an animal's plasma T and its dominance rank among monkeys caged in all-male groups (Rose *et al.*, 1971). However, *M. mulatta* and *M. fuscata* kept in heterosexual groups were found not to exhibit a significant positive correlation between T levels and dominance rank (Gordon *et al.*, 1976; Eaton and Resko, 1974).

Studies of macaque species other than *M. mulatta* are few. The work of Eaton and Resko (1974) suggests a plasma T level of about 11.7 ng/ml in *M. fuscata* (Japanese macaque), and work by Slob *et al.* (1979) suggests a level of about 6.7 to 14.8 ng/ml to *M. arctoides* (stumptail macaque), with a seasonal effect evident. Serum T levels were not influenced by multiple ejaculations in stumptails (Goldfoot *et al.*, 1975).

T levels in three baboons were 1.8, 1.2, and 0.2 ng/ml (Snipes *et al.*, 1969), and data on adult chimpanzees suggest a plasma T lvel of 4.9 ng/ml (McCormack, 1971).

Work with nonhuman primates suggests several problems worthy of further investigation. First, studies of *M. mulatta* indicate that the aromatization of androgen is not an absolute requirement for male sexual behavior (see Phoenix, 1978). Is this true for nonhuman primates in general? Second, work with *M. mulatta* indicates that although male sex behavior is highly androgen-dependent (Wilson *et al.*, 1972), sexual activity can persist in some males long after castration and in the presence of very low quantities (0.2 to 0.6 ng/ml) of plasma T (Resko and Phoenix, 1972). Furthermore, T concentrations in intact males are not correlated with the rank-order of sex behavior scores among individual males (Gordon *et al.*, 1976). What are the factors that regulate such differences among individuals in their behavioral responses to T or the lack of T? Do such factors operate differently in different species of nonhuman primates? Third, what neuroendocrine factors differentiate macaques, in which no postcopulatory rise in T seems to occur, from the many nonprimate species in which copulation induces a transient rise in peripheral plasma T?

HUMANS

305

PLASMA LEVELS OF
GONADAL
STEROIDS IN
MAMMALS

A very large number of studies of peripheral plasma androgens has been published. The results of several of the RIA studies were tabulated by Overpeck *et al.* (1978). Levels of T in adult males are about 4.6 to 6.4 ng/ml, AD levels are about 1.1 ng/ml, and DHT levels are about 0.3 to 0.6 ng/ml. Another major androgen in the peripheral circulation is dehydroepiandrosterone (presumably of adrenal origin), with mean levels ranging from 1.8 to 5.5 ng/ml. Etiocholanolone is also present at levels of 0.4 to 2.5 ng/ml (the original papers for all the data above are Graef, 1971; Hasan *et al.*, 1973; Abraham *et al.*, 1973; Demisch *et al.*, 1973; Barberia and Thorneycroft, 1974; Youssefnejadian *et al.*, 1974; Purvis, *et al.*, 1975; Lee and Migeon, 1975; Apter *et al.*, 1976; Vermeulen, 1976; Fiorelli *et al.*, 1976; Rosenfield *et al.*, 1977; many other papers have been published on the same topic).

It is now evident that there are rapid fluctuations in T level, probably as a result of the pulsatile release of androgen by the testes and the adrenals. For example, Murray and Corker (1973), Naftolin *et al* (1973), Alford *et al.* (1973), and Smith *et al.* (1974) have shown rapid oscillations (4- to 20-min frequencies) in plasma T levels in men.

Superimposed on these episodic oscillations is a circadian androgen pattern characterized by an increasing T level during sleep and a maximum level during the early morning hours (Dray *et al.*, 1965; Lipsett *et al.*, 1966; Resko and Eik-Nes, 1966; Southern *et al.*, 1967; Nieschlag and Ismail, 1970; Okamoto *et al.*, 1971; Evans *et al.*, 1971; Faiman and Winter, 1971; Rowe *et al.*, 1974; Leymarie *et al.*, 1974; Schiavi *et al.*, 1974; Rubin *et al.*, 1975; some authors have failed to find such a pattern: Kirschner *et al.*, 1965; Gordon *et al.*, 1968; Boon *et al.*, 1973). The circadian increases in T were not blocked by dexamethasone, whereas circadian increases in AD and dehydroepiandrosterone were blocked when this agent was given just before the subject retired. This finding suggests that circadian T increases have a testicular source, whereas the circadian patterns of the other two androgens have an adrenal source (Judd *et al.*, 1973). An interesting and important aspect of the sleep–wake pattern of T is its potential role in the onset of puberty (Parker *et al.*, 1975; Judd *et al.*, 1977). Apparently, as the pubertal process unfolds, there is enhanced luteinizing-hormone (LH) release during sleep (Boyar *et al.*, 1972). Although there is not a 1:1 correspondence between LH episodic oscillations and T episodic oscillations, it is proposed that the increased frequency of LH oscillations seen during sleep as puberty progresses causes increased T secretion, thereby promoting sexual maturation. A systematic comparative endocrinological study of the relation of sleep patterns to the onset of puberty in males of various species has not been carried out and might be of considerable interest.

Even longer cycles may be superimposed on the episodic and circadian oscillations in T levels. For example, Doering *et al.* (1975) found that 12 of 20 adult males had cycles of plasma T with periods ranging between 8 and 30 days (clustering around 20–22 days). The mean amplitudes of these cycles were 9% to 28% of the subjects' mean T levels. Circannual changes in plasma T have been investigated in France and Holland, and there appears to be a peak in summer and early autumn and a nadir in winter and early spring (Smals *et al.*, 1976; Reinberg *et al.*, 1978). The timing of the circadian T peak also is said to vary during the course of the year (Reinberg *et al.*, 1975). The issue of whether circadian and circannual T rhythms are due to endogenous or to environmental factors is not resolved.

However, there is ample evidence that environmental factors can influence plasma T levels. For example, heterosexual intercourse results in a transient increase in plasma T (Fox *et al.*, 1972), whereas masturbation results in little or no increase in plasma T levels (Purvis *et al.*, 1976). The viewing of sexually arousing movies is also claimed to cause transient increases in plasma LH and T (La Ferla *et al.*, 1978; Pirke *et al.*, 1974). On the other hand, psychological stress may cause decreases in plasma T level (Kreuz *et al.*, 1972). However, these and related studies must be accepted with caution because episodic, circadian, and circannual rhythms are often not taken into consideration.

A variety of other factors may also influence androgen levels in men. T (gonadal source?) levels dropped and AD (adrenal source?) levels increased in most participants after a "noncompetitive" but presumably stressful marathon run (Dessypris *et al.*, 1976). Under conditions in which men go from rest to activity (Lipsett *et al.*, 1966; Briggs *et al.*, 1973) or engage in less stressful exercise than marathon running (Sutton *et al.*, 1973, 1974), there is an increase in plasma androgens or no significant change in plasma androgens (Carstensen *et al.*, 1973). Dessypris *et al.* (1976) made the point that the physical condition of the subject may influence the direction or the magnitude of a change in plasma androgen level after exercise.

Obese men have significantly lower plasma T levels than nonobese control subjects (Glass *et al.*, 1977). However, the decreased T levels are not associated with the expected symptoms of hypogonadism. Apparently, plasma proteins that bind T are also decreased in obese subjects, and the result is a concentration of plasma *unbound* in T in obese subjects not very different from the concentration in nonobese subjects (Kley *et al.*, 1979). It is of interest that obese men have elevated levels of estradiol and estrone in peripheral plasma (Schneider *et al.*, 1979).

A single study failed to find effects of high altitude on plasma T levels in Sherpas of Nepal (Bangham and Hackett, 1978).

The influence of aging on plasma T levels has received the attention of investigators. In general, the findings are that T and DHT decrease with age (the reported onset of the decrease varies from 45 to 70 years of age) and that the plasma protein binders of T increase with age; the result is a significantly lowered concentration of unbound T in aged males (Vermeulen *et al.*, 1972; Nieschlag *et al.*, 1973; Pirke and Doerr, 1973; Stearns *et al.*, 1974; Lewis *et al.*, 1976b). Four of six men aged 72–92 years showed a clear circadian rhythm of plasma T and DHT with a peak in the early morning hours (Koliyannis *et al.*, 1974). Stearns *et al.* (1974) made the point "that provided general health is good, libido and potency may be well maintained in elderly men" even though plasma levels of unbound T are significantly reduced. Although it is reasonable to suspect that total and/or unbound T levels decline with age in men, the degree to which age *per se* contributes to any decrease in androgen cannot be said to be known with certainty. Although researchers generally state that they make their measurements on healthy subjects not using medications, it is likely that elderly subjects, more than a younger group of men, will have had some serious ailments, or that they may suffer from some ailment or imbalance not subtly enough evaluated.

LAGOMORPHA

Plasma androgens in male rabbits have been determined in a number of laboratories. The average adult levels of plasma T were about 5 ng/ml in New Zealand rabbits (Saginor and Horton, 1968, DIDD; also see Smith and Hafs, 1973, CPB and RIA). Saginor and Horton (1968) and Haltmeyer and Eik-Nes (1969) gave convincing evidence that copulation led to a sixfold increase in peripheral plasma T within 30 min; even the presence of a female led to a greater than fourfold increase in T over resting levels. The function of this female-induced rise in T is still in question. Suggestions include the idea that anticipation of copulation increases T levels, and that increased T levels facilitate male sexual behavior (Saginor and Horton, 1968), and the idea that female-induced increases in T level cause an increase in sperm count (Haltmeyer and Eik-Nes, 1969). There is evidence of episodic release of T in male rabbits, with four to six bursts per 24-hr period. The range between valley and peak during these episodes was 0.5 to 10 ng/ml (Rowe *et al.*, 1973; Moor and Younglai, 1975). Although there was no indication of a diurnal rhythm of T in rabbits, there was an indication in these RIA studies of a seasonal variation in T. For example, in the Moor and Younglai (1975) study, T levels were somewhat lower in December–June than in July–November. These data fit with data indicating that lower testicular weights in domestic rabbits occur in the summer. Longitudinal studies of T levels were carried out by Berger *et al.* (1976, 1979). Both studies indicate that T began to increase at Day 50 (1.5 ng/ml), reached a peak at Day 90 (4–5 ng/ml), and then declined to a plateau on Days 150–240 (1.7–2.1 ng/ml). These data, collected on New Zealand rabbits with the aid of GLC and RIA methods, are interesting in another respect. Earlier workers (Falvo and Nalbandov, 1974) found that the chromatographic separation of samples prior to RIA led to lower levels of androgen measured as T than in samples not subjected to chromatography prior to RIA. This finding suggested a significant contribution to plasma androgen concentration by a steroid that cross-reacts in the RIA systems commonly used for T. Although DIDD had shown AD concentrations to be almost 50% those of T in male rabbits (Saginor and Horton, 1968), AD is not a likely candidate because it does not cross-react significantly in the RIA systems commonly used for T. A more likely candidate is 5α-dihydrotestosterone (DHT), which does cross-react in these systems. Indeed, Schanbacher and Ewing (1975) and Berger *et al.* (1979) demonstrated peripheral plasma DHT levels in rabbits that were almost 50% of those of T during adulthood. Additionally, Schanbacher and Ewing (1975) measured the 3α-ol and 3β-ol derivatives of DHT. The concentrations of these steroids were about 0.5 ng/ml each (in the same study, T was 1.2 ng/ml and DHT was 0.5 ng/ml).

Androgens have been studied in several types of rodents. We will consider first the data from species not commonly used in laboratory research and then proceed to the more frequently studied male rodents.

The adult male *Praomys (Mastomys) natalensis* has peripheral serum T levels of 4.2 ng/ml and DHT levels of 0.4 ng/ml, as measured by RIA (Lewis *et al.*, 1976a). The species is of interest because the female possesses a well-developed prostate despite having low serum T levels in adulthood (0.4 ng/ml).

Collared lemmings *(Dicrostonyx groenlandicus)* were raised in the laboratory and caged in all-male groups as adults. Dominant males had significantly higher (1.6 ng/ml) levels of T than subordinates (0.5 or 0.9 ng/ml, depending on whether the subordinate was active or passive) or controls maintained in isolation (0.7 ng/ml) (Buhl *et al.*, 1978). The effects of the presence of a female on T levels in dominant and submissive males were not assessed. A study of the interaction between male social status and the effects of the presence of an estrous female on T might be of interest.

Male Syrian hamsters *(Mesocricetus auratus)* kept under a 14:10 L:D lighting schedule had peripheral plasma T levels of about 5 ng/ml; when the hamsters were moved into a constant dark environment, the T levels gradually dropped to < 1 ng/ml. Reexposure of the hamsters to a 14:10 L:D schedule eventually resulted in levels of T as high as 10 ng/ml (Berndtson and Desjardins, 1974). Bartke *et al.* (1975) also used RIA and found plasma T levels of 1.6 ng/ml in hamsters maintained in 14:10 L:D and 0.7 ng/ml in hamsters kept in 5:19 L:D. Although both studies demonstrated the dependence of the Syrian hamster testicular-pituitary-hypothalamic axis on photoperiodic cues (also see Turek and Campbell, 1979), no explanation was offered for the discrepancy in T levels between the two studies under the 14:10 L:D condition. One possible reason for the discrepancy between the Berndtson and Desjardins (1974) and Bartke *et al.* (1975) study may be the existence of a circadian rhythm of T, with the two sets of experimenters not collecting blood at equivalent times of day. This thought is prompted by the finding of Hoffmann and Nieschlag (1977) of a pronounced circadian rhythm of T in the serum of the Siberian hamster *(Phodopus sungorus)* kept under natural lighting conditions. T levels were highest just before sunset (3.7 ng/ml) and significantly lower (0.2–1.5 ng/ml) at other times of day. Another possible complicating factor in T measurements in male Syrian hamsters is that a reflex increase in T may be induced by pheromonal factors from the female (Macrides *et al.*, 1974).

Laboratory-raised strains of house mice *(Mus musculus)* have been rather extensively studied. A longitudinal RIA study of Swiss strain CD-1 mice by Jean-Faucher *et al.* (1978) revealed that males raised in cohabitation with females had plasma T levels of 0.2 ng/ml (Day 20), 5.8 ng/ml (Days 30–40), and 0.4 ng/ml (Day 90). The first fertile matings in this strain occurred at Day 40. Other longitudinal studies of T levels (McKinney and Desjardins, 1973; Barkley and Goldman, 1977; Selmanoff *et al.*, 1977) also showed a pubertal rise in T, but the evidence for a postpubertal decline in T was not consistent. Discrepancies in absolute T values among the studies might be attributable to differences in RIA techniques (e.g., whether or not DHT and T were separated chromatographically prior to RIA), to housing conditions (e.g., whether males cohabited with females or not), or to strain differences. The considerable influence of genetic factors on T levels in mice is suggested by results from three studies: adult DBA/2J males versus C57BL/10J males (T = 4.6 vs. 1.3 ng/ml at about 70 days of age) (Bartke, 1974); CD-1 males versus YS males (T = 18.7 vs. 1.0 ng/ml in adulthood) (Bartke *et al.,* 1973); and C57BL/6J genetically obese males versus C57/BL/6J lean males (T = 2.0 vs. > 8.0 ng/ml at 78 days of age) (Swerdloff *et al.,* 1976). However, even among animals with similar genetic constitutions, there may be dramatic interindividual variation. Bartke *et al.* (1973) observed variations as great as 0.3 to 44.4 ng/ml T in CD-1 mice maintained under identical conditions. Similar variability was noted in Rockland-Swiss male mice (Barkley and Goldman, 1977). Such data render suspect the value of single determinations of T in mice (e.g., Ivanyi *et al.* 1972; Falvo and Nalbandov, 1974; Bartke *et al.,* 1977) and suggest the existence of episodic, pulsatile surges of T in mice. This likelihood is strengthened by the existence of great intraindividual variability in T levels in adult mice (Bartke and Dalterio, 1975). Social factors have significant effects on T levels in mice. Short-term exposure of males to females or female urinary odors induced increased T levels, but males kept in continuous cohabitation with females did not show elevated T titers (Macrides *et al.,* 1975; Maruniak and Bronson, 1976; Maruniak *et al.,* 1978). Coquelin and Bronson (1979) made the interesting point that the male hypothalamic-pituitary-testicular sites may become habituated to stimuli associated with a particular female. These authors also used a cannulation procedure that permitted repeated short-interval blood sampling, a difficult but valuable procedure in hormone measurements in mice. Another social factor that may influence T levels in mice is intermale aggression. McKinney and Desjardins (1973) found slightly reduced T levels in subordinate, as compared to dominant or singly caged, mice. Selmanoff *et al.,* (1977) found higher serum T levels in isolated and grouped nonfighting mice than in paired fighting mice. However, these authors found no correlation between dominance status and serum T levels. Finally, plasma T in mice may be influenced by disease and/or aging. Bronson and Desjardins (1977) found lower T levels in aged (18–30 months old) DBF$_1$ male mice. This decrease was attributed to two factors: (1) debilitative, pathological changes in some aged mice and (2) loss of episodic LH release in a subpopulation of the remaining healthy, aged mice studied. Nelson *et al.* (1975) and Eleftheriou and Lucas (1974) found no decreases in plasma T level in aged mice provided that data from unhealthy, aged individuals were excluded.

Predictably, the rat *(Rattus norvegicus)* has been the most extensively studied rodent. A recent longitudinal RIA study found T levels of ~ 1 ng/ml (20–40 days of age), 4 to 6 ng/ml (55–60 days of age), and 3 ng/ml (120 days of age) (Smith *et al.,* 1977). A plethora of DIDD, GLC, CPB, and RIA studies indicate that plasma levels of T in laboratory rats range between about 1.7 and 7.7 ng/ml in adults (Bardin and Peterson, 1967; Rivarola *et al.,* 1968; Resko *et al.,* 1968; Knorr *et al.,* 1970; Kniewald *et al.,* 1971; Grewal *et al.,* 1971; Grota, 1971; Amatayakul *et al.,*

1971; Hafiez *et al.*, 1972; Bartke *et al.*, 1973, 1977; Free *et al.*, 1973; Robel *et al.*, 1973; Badr and Bartke, 1974; Cooper and Waites, 1974; Falvo and Nalbandov, 1974; Harris and Bartke, 1974; Frankel *et al.*, 1975), and from 1.8 to 6.2 ng/ml at sexual maturity (Lee *et al.*, 1975; Saksena *et al.*, 1975). Some of the forementioned studies also showed that plasma DHT levels were low in rats, a result confirmed by Pujol *et al.* (1976) (T = 3.2 ng/ml; DHT = 0.2 ng/ml T; for other studies of DHT see Robel *et al.*, 1973; Coyotupa *et al.*, 1973; Gupta *et al.*, 1975; for plasma assay of androsterone and 5α-androstane-3α,17β-diol, see Moger, 1977). Additionally, evidence of episodic fluctuations of T level has been presented (e.g., Bartke *et al.*, 1973). There seems to be general agreement that there is a daily rhythmicity in T level (Kinson and Liu, 1973a,b; Wilson *et al.*, 1976; Hostetter and Piacsek, 1977; Mock *et al.*, 1978; Keating and Tcholakian, 1979). However, there is considerable disagreement over the precise pattern of this rhythmicity, probably because the various studies differ in the strain of rat used, age of animals, housing conditions, and method of blood collection. Keating and Tcholakian (1979) indicated T elevations during the dark period, as well as lowest T concentrations during the late dark and early light hours in freely moving, conscious rats. Their data and the data of Kinson and Liu (1973a,b) suggest that the rhythmic pattern of T is not endogenous, but Mock *et al.* (1978) found daily rhythmic changes in T in rats kept in constant darkness, a finding suggesting an endogenous rhythm. Daily T rhythms were abolished by hypothalamic deafferentation (Kalra and Kalra, 1977a). Androgens other than T (DHT, androsterone, and 5α-androstane-3α,17β-diol) appeared to undergo the same daily pattern of change as T (Kalra and Kalra, 1977b; Moger and Murphy, 1977). A shifting circannual rhythm of T has also been noted, even under controlled conditions of housing (Kinson and Liu, 1973a; Mock *et al.*, 1975, Mock and Frankel, 1978). The physiological significance of a circannual T rhythm in laboratory rats is obscure, but the considerable (as much as fivefold) differences in T levels at various times of year pose a potential problem in the interpretation of plasma T data in rats. Social factors may also influence T levels in male rats. Purvis and Haynes (1972) and Kamel *et al.* (1975) showed that visual and/or olfactory cues associated with females caused the release of T in sexually experienced male rats. T levels rose more noticeably when the male was exposed to an estrous, rather than an anestrous, female. An even more dramatic increase in T occurred a short time after copulation (experienced males > inexperienced males), but there was no evidence that sexual activity led to chronically elevated T levels (Kamel *et al.*, 1975). Further studies established that (1) T levels were as elevated after attempts to mate with unreceptive females as after mating with estrous females; (2) male rats that showed no interest in females did not exhibit elevations in T after exposure to females; and (3) the peak in female-induced T elevation occurred 30–60 min after the initiation of mating, and this T peak was preceded by a rise in serum LH (Kamel *et al.*, 1977; Kamel and Frankel, 1978). Another factor influencing T levels in rats that will be considered here is aging. Several groups of workers, using various strains, have found that plasma T levels decline significantly in rats > 13–24 months of age (Lloyd, 1972; Ghanadian *et al.*, 1975; Chan *et al.*, 1977; Gray, 1978). Although T assays have been performed on a number of strains of rats (Fischer, Long-Evans, Wistar, and so on), there appears to be no systematic study of strain differences in T in rats.

In guinea pigs *(Cavia porcellus)*, T and AD have been measured in peripheral plasma. Resko (1970) found that adult males of the Topeka type (> 90 days of age) had T levels of 1.2 to 6.7 ng/ml and AD levels of 0.3 to 0.9 ng/ml as measured by GLC. The mean T level was approximately 2.0 ng/ml. This T is of testicular origin, as demonstrated by the fact that, within 6 hr after castration, T declined to undetectable levels. Another extensive longitudinal study, also using GLC, indicated that T (and AD) levels were 6.0 (4.0) ng/ml at Day 50, 5.5 (3.5) ng/ml at Day 70, 4.5 (1.5) ng/ml at Day 90, 3.5 (1.2) ng/ml at 6–22 months, and 1.8 (1.8) ng/ml at 32 months in Dunkin-Hartley strain guinea pigs (Rigaudière *et al.,* 1976). These data suggest a T peak at puberty (Days 55–82 in guinea pigs), lower but stable levels during adulthood, and a further drop in T during senescence. DIDD measurements of T in Hartley-strain adult guinea pigs of unspecified age indicated a T level of 3.1 ng/ml, in agreement with the range suggested by GLC studies (Rivarola *et al.,* 1968). In the same study, AD levels were 1.7 ng/ml and dehydroepiandrosterone levels were 3.0 ng/ml. With an RIA procedure, Harding and Feder (1976) measured levels of 2.1 ng/ml in Hartley-strain guinea pigs and indicated that the circulating levels of DHT were low. Harding and Feder also demonstrated that the presence of an estrous female caused a slight elevation in T in proven copulators, as well as a slight decline in T in noncopulators. However, none of the studies with adult guinea pigs examined the possibility that episodic bursts of T occur during the course of a day. That such episodic bursts occur in adults is probable, because work with immature guinea pigs revealed the presence of these surges (Brain and Peddie, 1979).

In general, the work with rodents, although abundant in quantity, is limited in scope. Highly inbred strains of animals kept under controlled laboratory conditions have been analyzed and reanalyzed, but studies of hormones in relation to naturally occurring environmental changes and in the context of normal social and sexual interactions are fewer in number.

Cetacea

There appears to be only one report of circulating androgen levels in cetaceans. Harrison and Ridgway (1971) measured plasma T in a captive fertile bottlenose dolphin *(Tursiops truncatus)* by CPB. The highest values, obtained in April–May and September–October, were 16 to 24 ng/ml; the baseline levels in December were about 2 ng/ml. However, no statements can be made about seasonal variation on the basis of monthly samples taken from one animal. Additional samples will be difficult to obtain because sexual maturity is not attained in this species until at least 10 years of age.

Carnivora

The stoat *(Mustela erminea)* has a well-defined breeding season extending from May through July. Gulamhusein and Tam (1974) used GLC to show that T levels in a small number of stoats were about 11 ng/ml in December, 25 ng/ml in March and April, 4 ng/ml in May, 11 to 15 ng/ml in June–July, and 5 ng/ml in late August. The authors suggested that the high March–April levels are associated with the late stages of spermatogenesis and initiation of the growth of accessory tissues. The smaller T peak in June–July is thought to be associated with the onset of male sexual behavior and the maintenance of the sexual accessories. This seasonal cycle of T is probably dependent on photoperiodic cues.

The male ferret (Mustela putorius) also shows a seasonal cycle of testicular activity. The breeding season extends from March to August, with males probably capable of breeding by mid-February (several weeks prior to the onset of estrus in the female). Three studies examined plasma T levels in this species. Neal, et al. (1977) used RIA to study annual cycles of plasma T levels and found that, from August to December, 86% of the samples contained less than 5 ng/ml T. A rise in plasma T began in January, and peak levels of 47.5 ng/ml were recorded in April, May, and June. In this study, T and DHT were not separated chromatographically prior to RIA. Neal and Murphy (1977) demonstrated that male ferrets can respond to exogenous LH with increased T levels outside the breeding season. Rieger and Murphy (1977) studied mature ferrets in July (when T levels were undergoing a seasonal decline) and found that there were three peaks of T during the course of a day. There were also episodic peaks of DHT, but the bursts did not coincide with those of T. Interestingly, the ratio of DHT to T varied from as low as 1 to 4 to as high as 1 to 1. Thus, failure to separate T and DHT chromatographically can lead, at least in this species, to considerable overestimation of T levels.

A third mustelid, the mink (Mustela vison), was studied by Nieschlag and Bieniek (1975). These authors assayed T and AD by an RIA of the peripheral plasma of this seasonal breeder during the first year of life. Prior to puberty in July–October, the levels of T were 1 to 2 ng/ml, with AD levels of only about 0.5 ng/ml. At puberty, in November, T and AD were each ~ 4 ng/ml. Values of T and A each rose in December and January to 6 ng/ml. A peak of about 12 ng/ml occurred in February for both T and AD. By April, T and AD levels were about 0.5 ng/ml. The peak in androgen levels were associated with the breeding season. The mink seems to have an unusually high ratio of AD to T in the peripheral circulation, but the functional significance of this high ratio is unknown.

The black bear (Ursus americanus) is a seasonally breeding carnivore. McMillin et al. (1976) used an RIA method to determine plasma T in 21 wild bears over the course of a year. These authors found that T was highest from March through mid-July (1.5 ng/ml) and significantly lower from mid-July through November (0.5 ng/ml). The breeding season occurs in June or early July and is therefore associated with elevated T levels. Of further interest is the finding that bears that have spent the winter in isolation in their dens, and that have lost almost 20% of their body weight during this winter period, exhibit elevations of T level (to 1.6 ng/ml) before emerging from their dens in the spring. Thus, an environmental factor other than exposure to females or nutritional state probably accounts for the seasonal cycle of T levels in black bears. Systematic study of the possible roles of photoperiod and temperature has not yet been carried out in black bears.

The red fox *(Vulpes vulpes)* has been studied by Joffre and Joffre (1975) and Joffre (1977), using an RIA technique. The authors found a significant seasonal variation, with the highest T values (0.9–1.1 ng/ml) in December–February, lower levels in March–May (0.2–0.5 ng/ml), and the lowest levels in June–November (< 0.1–0.4 ng/ml). These data nicely correspond with the fact that the mating season extends from December to February, and that testicular involution is evident from May to September. The authors also demonstrated that peak levels of T were somewhat lower in young males experiencing their first breeding season than in adult males during the breeding season.

Analysis of peripheral plasma (or serum) androgens has also been carried out in domestic dogs. The adult peripheral plasma levels of T in mongrels were reported to be 1.9 ng/ml by CPB (Kelch *et al.*, 1972). The data of Comhaire *et al.* (1974) are quite similar in that T levels of 0.8 ng/ml in whole blood samples were found (by means of RIA) in "unselected" adult males. Other RIA studies of plasma T in adult beagles indicate levels of about 1.5–3.0 ng/ml (Jones *et al.*, 1976) and 3.5 to 4.2 ng/ml (Hart and Ladewig, 1979). However, these values may be somewhat inflated because dogs have been shown to contain measurable quantities of DHT in the peripheral circulation. A DIDD study by Tremblay *et al.* (1972) indicated peripheral plasma concentrations of 1.2 ng/ml T and 0.3 ng/ml DHT in mongrel dogs. Another study, using GLC, indicated T levels of 2.5 ng/ml and DHT levels of 0.3 ng/ml (Folman *et al.*, 1972). De Palatis *et al.* (1978) remarked on the existence of episodic bursts of androgen in mongrel dogs, and their RIA study failed to provide evidence for a diurnal rhythm in peripheral T levels. The episodic T peaks came approximately 50 min after bursts of LH and ranged as high as 10 to 12 ng/ml. It is somewhat surprising to find that apparently no attempts have been made to compare androgen levels among various inbred strains of dogs and to correlate such findings with varying degrees of pugnacity and other behavioral and physiological characteristics.

PROBOSCIDEA AND HYRACOIDEA

Plasma T has been measured in a proboscidean by Jainudeen *et al.* (1972). The authors used a CPB assay of plasma obtained from 11 captive Asiatic elephants *(Elephas maximus*; age range: 22–45 years). These elephants typically display a group of characteristics subsumed under the term *musth*. The characteristics include aggressiveness, enlargement of the temporal glands, and a continuous dribbling of urine from the penis. When in full musth, the elephants had very high T levels (29.6–65.4 ng/ml), a finding suggesting that musth is androgen-dependent. Levels of T were much lower when the elephants showed no sign of musth (< 0.2–1.4 ng/ml), were just coming into musth (4.3–13.7 ng/ml), or were in the period just after musth (< 0.2–1.2 ng/ml). It is noteworthy that the males "copulated readily with oestrous females" while in the nonmusth phase and continued showing copulatory behavior in the premusth condition. During premusth, the males also showed frequent erections and "masturbated by striking the erect penis against the belly." It was not possible to observe copulatory behavior during musth because the animals become too dangerous to handle. The authors suggested that the copulatory behavior shown before the musth period induces increased androgen secretion. The increased androgen levels then cause the onset of musth. If this idea could be tested and verified, it would provide an unusual example of copulation inducing a long-term increase in androgen levels (musth lasts about 2–3 months). Most other cases of copulation-induced androgen elevations in males are of much shorter duration (e.g., see data for rabbits, bulls, and guinea pigs). However, the role of seasonal factors in musth is not clear.

Levels of T in the peripheral plasma do not appear to have been measured in the African elephant, a species in which the existence of a musth period is not established.

One study of plasma T levels, as measured by CPB, in a hyracoid has been reported by Neaves (1973). The rock hyrax *(Procavia habessinica)* is an African seasonal breeder. T levels were 27.0 ng/ml during the 3-month breeding season and 5.3 ng/ml during the nonbreeding season.

PERISSODACTYLA

Sexual activity and sperm output can occur throughout the entire year in stallions. However, copulatory behavior, sperm output, and seminal volume are maximal in the spring through the midsummer (reviewed in Stabenfeldt and Hughes, 1977). Berndtson *et al.* (1974) measured peripheral plasma T from five mature quarter-horse stallions. Each stallion was bled three times a month for 13 months, and the T was determined by RIA. Highest plasma T levels were found in May (3.2 ng/ml), and the lowest values were found in October, December, and January (< 1.5 ng/ml). Overall, T levels were higher during the spring–midsummer period than during other times of the year (but see Thompson *et al.*, 1977, for a study in which seasonal T variation did not reach significance). A study of thoroughbred stallions indicated that there is a diurnal, as well as a seasonal, cycle in plasma T levels. Sharma (1976) showed that plasma T, as measured by RIA, was higher (3.5 ng/ml) at 0800 hr than at 1600–2400 hr (1.2–1.7 ng/ml) during the month of March. Diurnal variations in T in domestic breeds are also suggested by the work of Ganjam and Kenney (1975).

The foregoing data from domestic breeds are in close agreement with data from free-roaming wild horses of Montana–Wyoming *(Equus caballus)*. In two papers in which CPB was utilized (Kirkpatrick *et al.,* 1976, 1977), it was shown that wild stallions have T levels of about 1.6 to 3.0 ng/ml, that a diurnal variation exists (nadir between 1900 and 2400 hr and peak between 0600 and 1200 hr), and that seasonal variation also exists (nadir in September, peak in May). The possibilities that stallions exhibit episodic bursts of T during the day and that sexual activity may cause increases in T levels in stallions were not examined. Intermale aggression among wild stallions was said to be highest during the breeding season, when T levels were elevated (Kirkpatrick *et al.,* 1977).

The data on androgen levels in stallions are remarkably consistent across studies. They indicate a rather low level of peripheral plasma T, and even the elevations of T level in spring–midsummer are quite undramatic. This finding may be related to the fact that reproductive function is not totally curtailed at any time of the year. Thus, stallions do not require a massive surge of T (as seen in some true seasonal breeders) to "start up the machinery" for spermatogenesis, accessory gland growth, and the priming of CNS tissues that regulate gonadotropin release and sexual behavior. It would be interesting to know whether these "startup" processes require larger quantities of T at puberty in stallions.

Additional data on Equidae include the finding that mean T levels in donkeys in the tropics rose from 0.4 ng/ml in November to 1.8 ng/ml in late January (Gombe and Katongole, 1977).

ARTIODACTYLA

Peripheral plasma androgens have been measured in a variety of nondomesticated species of Artiodactyla. For example, reindeer (*Rangifer tarandus*) show levels of T that vary from 1 ng/ml in August to 30 to 60 ng/ml in mid-Setpember to barely detectable in late October. These data were collected by GLC and CPB on two adult (5-year-old) reindeer (Whitehead and McEwan, 1973). Because the main breeding season of reindeer in northern latitudes is in September, the androgen data fit well with physiological and behavioral events associated with rut. Leader-Williams (1979a,b) studied a larger group of reindeer on a subantarctic island. Their breeding season occurred in late March–early April. During this period of the year, T levels, as measured by RIA, were 8 ng/ml in 2 to 3-year olds and 19.3 ng/ml in reindeer over 3 years of age. T levels at other times of year were low or undetectable. In caribou in northern latitudes, T levels were 30 to 35 ng/ml during rut (which occurs in October) and low or undetectable the remainder of the year (Whitehead and McEwan, 1973). The roe deer (*Capreolus capreolus*) shows an increase in T during rut in mid-July. However, these RIA data indicate T levels of only about 4.2 to 6.4 ng/ml during rut in roe deer. During the remainder of the year, the T levels were even lower (Giménez *et al.*, 1975). Mirarchi *et al.* (1977) and McMillin *et al.* (1974) studied T levels in white-tailed deer (*Odocoileus virginianus*). In the McMillin *et al.* study, RIA determinations indicated T levels of 2.1 ng/ml in August–November nad 0.3 ng/ml in December–July. These deer breed in November. However, from August to September, the males engage in "shadow-fighting" with saplings. The males show interest in females and may pursue them as early as 2 weeks before the onset of heat. These prerut behaviors probably depend on rising titers of T prior to November. During rut itself, the T levels are elevated. Interestingly, during the rut period, there is a severe decrease in food intake. This indicates that the rise in T during rut is not dependent on nutritional factors. One wonders whether the rise in T may actually have a causative role in decreasing food intake. T can be aromatized to estrogen, and estrogen is known to cause reductions in food intake in other mammalian species (Wade, 1976). This question may be relevant to deer, because aromatization has already been suggested as playing a role in the induction of male sexual behavior in red deer (Fletcher, 1978).

Although the circannual aspect of plasma T in reindeer, caribou, roe deer, and white-tailed deer seems established by the foregoing data, there are no studies of episodic and diurnal variations, nor are there studies of the effect on T of stimuli emitted by conspecific females. These aspects of T levels have been studied in domesticated Artiodactyla, particularly in bulls, rams, boars, and goats.

The work with bulls indicates some degree of variability in the results among and within studies. This may be due to the episodic release of T and/or to strain differences. Smith and Hafs (1973) reported levels of about 6 ng/ml T by CPB and RIA. These levels are in the same range reported for French Friesian bulls by Thibier (1975), using RIA (4.3 ng/ml); for Angus $\times$ Hereford bulls by Rawlings *et al.* (1978) by RIA (6.4 ng/ml); and for Herefords by Schanbacher (1979). These authors, as well as Katongole *et al.* (1971), Rawlings *et al.* (1972), Thibier (1976), and Schams *et al.* (1978), attributed variability in plasma T levels to episodic bursts of T secretion. The range of T in bulls as a result of these bursts is between < 1 and 20 ng/ml. There appears to be some disagreement over the time of onset of episodic bursts of T. Rawlings *et al.* (1978) first noted episodic variations at 70 weeks of age, but LaCroix and Pelletier (1979) suggested that testicular responses to episodic LH surges can be detected at about 17 weeks of age. There is controversy over the existence of a circadian rhythm of T (Sanwal *et al.*, 1974; Thibier, 1976). There is some evidence that stimuli associated with the female cause a reflex increase in T levels. Katongole *et al.* (1971) claimed that T increased in Ayrshire and Guernsey bulls after the bulls saw or copulated with cows, but these increases were evident only if T levels were low at the time of exposure to the female. Smith *et al.* (1973) also noted a postejaculatory increase in T in 3 to 6-year old bulls, but this response was not as evident in younger adults (1.5–2.5 years of age). Rawlings *et al.* (1972) measured AD as well as T in 5 to 11-month old Holstein bulls. AD levels varied from < 0.1 to 7.1 ng/ml and T levels from 2.4 to 4.8 ng/ml. At 6 and 12 months of age, AD was 1.8 ng/ml and T 0.5 ng/ml. However, it is difficult to arrive at a meaningful ratio of T to AD because of the variability in both steroids. Previous work by Lindner (1959) and Lindner and Mann (1960) indicated that the testicular ratio of T to AD is much larger in adult than in prepubertal bulls, but whether this change in ratio is a causative factor in the onset of puberty in bulls is questionable.

Data on rams also indicate variability in T levels suggestive of episodic secretion. Katongole (1971) measured T levels of 6 to 20 ng/ml in Suffolk rams by CPB (bleedings carried out in October–January). Cooper and Waites (1974) found levels of about 2.0 ng/ml T in Clun Forest rams bled in March and assayed by RIA. In more detailed studies of the episodic nature of T levels, Purvis *et al.* (1974) found four to eight peaks of T per 24-hr period, and Falvo *et al.* (1975) found that T surges occurred after episodic bursts of LH secretion. More recent studies have confirmed the episodic nature of T levels in a variety of strains (Lincoln *et al.*, 1977; Foster *et al.*, 1978). Lincoln *et al.* (1977) also found a circadian rhythm of T that was related to increased gonadotropin release during the dark phase of the 24-hr cycle. The picture is made more complicated by the seasonal variation in episodic T secretion. Katongole *et al.* (1974) used CPB in Suffolk rams to demonstrate that the magnitude and frequency of episodic T secretion vary with the season. Lincoln (1976) and Lincoln and Peet (1977) used artificial photoperiods to reproduce the annual cycle and to study episodic T secretion in Soay rams by RIA. The seasonal differences in episodic T release were reflected in overall circannual differences in T levels. Thus, the overall T levels in January–September were 0.5 to 10 ng/ml, and the overall T levels in October–December were 3 to 28 ng/ml in Suffolk rams (Katongole *et al.*, 1974). A similar T pattern was reported in Suffolk and Finnish Landrace strains (Schanbacher and Lunstra, 1976; T values 1 to 4 ng/ml in January–July, 3 to 9 ng/ml in September–December, by RIA) and in Finnish Landrace and Managra synthetic (Line-M) rams (the September–December T peak for sexually active Finnish Landrace was 24 to 30 ng/ml; for the Managra synthetic strain, 6 to 20 ng/ml, by RIA, as measured by Sanford *et al.*, 1977). Sanford *et al.* (1977) also provided evidence that seasonally increased mating behavior was correlated with seasonally increased T levels (see Lincoln and Davidson, 1977, for correlations between T levels and sexual and aggressive behaviors). The apparent primary cue for seasonally increased T levels is a decrease in day length (Lincoln and Davidson, 1977; Barrell and Lapwood, 1979; Schanbacher and Ford, 1979), but temperature changes may also play some role (Gomes and Joyce, 1975). Because ewes utilize the same primary cue in their timing of reproductive activity, male and female sheep show similar seasonal cycles of gonadal secretion. However, some spermatogenesis occurs throughout the year in males. A recent report indicates that mature rams kept near estrous ewes had higher T levels than rams isolated from ewes and kept in all-male groups. Plasma T was also found to increase after copulation in some of these Clun × Hampshire rams (Illius *et al.*, 1976). Incidentally, no effect of the ewes on T levels was seen in immature rams. (For additional data on rams, see Wetteman and Desjardins, 1973; Davies *et al.*, 1977).

Androgens in the peripheral plasma of boars have been examined. In adult miniature pigs (Gottingen strain), Elsaesser *et al.* (1973) used CPB to find T levels of 4 to 7 ng/ml and DHT levels of 0.6 to 0.8 ng/ml in subjects older than 5 weeks. Brock and Wetteman (1979) found T levels of 0.6 to 11.5 ng/ml by RIA in Yorkshire boars. Colenbrander *et al.* (1978) found T levels of 1.9 ng/ml in adult Dutch Landrace × Yorkshire boars. Brock and Wetteman also noted that there were about three episodic peaks of T per 24 hr, but no diurnal variation in T levels. The latter result was disputed by two groups (Ellendorff *et al.*, 1975; Claus and Giménez, 1977); both groups noted diurnal variations in T. An interesting and unusual feature of androgens in pigs is the presence of significant quantities of 16-unsaturated compounds. These have little or no androgenic activity, but they possess a strong odor and may serve as pheromones or facilitators of estrus detection (Reed *et al.*, 1974). In adult Norwegian Landrace boars, one of these 16-unsaturated androgens, 5α-androst-16-en-3-one, has a 2.5 times higher concentration in peripheral plasma than T (Andresen, 1976). In the same study, it was claimed that use of the boars for mating caused increases in 5α-androst-16-en-3-one and T (Andresen, 1976), but Ellendorff *et al.* (1975) failed to detect copulation-induced increases in T in male miniature pigs.

The goat has received somewhat less attention, in terms of T measurements, than bulls, rams, or boars. Saumande and Rouger (1972) measured T in alpine goats by GLC and found a marked seasonal variation (also see Racey *et al.*, 1975). According to Saumande and Rouger, the levels of T were < 1–5 ng/ml at all times of year except the late summer and autumn, when the T levels ranged from 10 to 20 ng/ml. The androstenedione and epitestosterone levels did not exceed 0.5 ng/ml at any time of year. Seasonal patterns were also seen in pygmy goats, and circadian rhythms and episodic bursts of T were said to exist in this RIA study (Muduuli *et al.*, 1979).

ESTROGENS AND PROGESTINS IN ADULT, NONPREGNANT FEMALE MAMMALS

The measurement techniques (GLC, DIDD, CPB, and RIA) used for determinations of androgens in males have also been used for assays of estrogens and progestins in the peripheral plasma of females. Because fluctuations in estrogens and progestins are generally quite dramatic during the course of estrous or menstrual cycles, there has been less emphasis on the episodic and circadian patterns of these steroids in females than is the case for androgens in males. Overpeck *et al.* (9178) provided a tabulation of RIA data in females (and males) that covers much of the literature for rhesus monkeys, chimpanzees, humans, hamsters, mice, and rats through 1977.

There is a lack of data on estrogens and progestins in the peripheral plasma of prototheria and metatheria, with the following exceptions. There is an interesting CPB study of the tammar wallaby *(Macropus eugenii)*. The tammar wallaby was found to have very low levels of total progestins and of progesterone (< 1 ng/ml) in the peripheral plasma at all stages of the reproductive cycle (Lemon, 1972). The quokka, a small wallaby *(Setonix brachyurus)* had levels of plasma progesterone of 0.6 ng/ml in the early stages of the estrous cycle and of 2.5 ng/ml at the peak of the luteal phase, by RIA (Cake *et al.*, 1980). Among eutherians, data on dermopterans, pholidotes, tubulidentates, sirenians, and pinnipeds are lacking, but there is an RIA report of plasma progesterone levels (∼ 10 ng/ml) in the edentate *Dasypus novemcinctus* (nine-banded armadillo) during spontaneous ovulation (Peppler and Stone, 1976).

Throughout the remainder of this section, estradiol is denoted by E2 (distinctions between 17β and 17α forms will be made only occasionally), estrone by E1, estriol by E3, and progesterone by P. Where data on androgens in females are available, they are mentioned. Data on steroids during pregnancy and pseudopregnancy are not discussed because their inclusion would make this chapter almost endless. Issues that make the interpretation of steroid data difficult, and that have already been discussed in the section on males, are not reiterated.

INSECTIVORA AND CHIROPTERA

319

PLASMA LEVELS OF
GONADAL
STEROIDS IN
MAMMALS

The Asian musk shrew *(Suncus murinus)*, a representative of the most primitive eutherians, the insectivores, has no discernible vaginal estrous cycle (the vaginal smear consists predominantly of nucleated epithelial cells) and no period of seasonal reproductive quiescence. Musk shrews are reflex ovulators. Estradiol levels in musk shrews are below the limits of detectability by RIA; P levels are 20% higher in sexually receptive, as compared to nonreceptive, shrews (Hasler, 1972, cited by Dryden and Anderson, 1977). Dryden and Anderson (1977) showed that neither ovariectomy nor estrogen treatment altered uterine and vaginal weights and histologies. Furthermore, ovariectomized shrews can exhibit mating behavior. These surprising findings suggest that (1) reproductive status in females of this species is not controlled by estrogen or that (2) musk shrews are extraordinarily sensitive to the actions of low levels of estrogen produced by the adrenal gland. It is worth noting that males of this species respond to castration and to T treatment with alterations in prostatic and epididymal weight. It might be of interest to determine whether females respond to exogenous T and whether males respond to exogenous estrogen.

Few investigators have measured circulating steroids in bats. Oxberry (1979) measured "estrogen" (E2 + E1 + E3?) and P by RIA in *Antrozous pallidus*, a species that hibernates and exhibits delayed ovulation. In these bats, estrus and copulations occur in late summer and early autumn, and the bats enter hibernation soon thereafter. Intermittent arousal and copulation may occur throughout hibernation. Spermatozoa are stored in the female reproductive tract until after permanent arousal in the spring, when (delayed) ovulation, fertilization, and gestation occur. In late summer, estrogen levels were $\sim$ 40 to 50 pg/ml; these values declined to baseline levels ($\sim$ 20 pg/ml) during hibernation. On spring arousal, the estrogen levels increased from baseline to $\sim$ 90 pg/ml and then fell immediately after ovulation. Baseline levels of P ($\sim$ 3 ng/ml) obtained from the onset of estrus in late summer throughout hibernation. After spring arousal, ovulation, and corpus luteum formation, P levels rose to $\sim$ 20 ng/ml.

Jerrett (1979) measured plasma P by RIA in the nonhibernating bat *Tadarida brasiliensis mexicana* (Mexican free-tailed bat). In this species, copulation, ovulation, and fertilization occur in late February–mid-March. At this time, P levels were 10 to 15 ng/ml. A dramatic increase in P levels occurred, beginning at the onset of pregnancy in March and culminating with parturition in July. After cessation of lactation in mid-August, P levels remained at baseline throughout diestrus (10 ng/ml or less).

Other workers have measured steroids in the plasma of *Macrotus californicus* (California leaf-nosed bat) (Burns and Easley, 1977; Krutzsch, 1978). It would appear that the Chiroptera offer rich possibilities for futher endocrinological work because this taxa is one of the most diverse among mammals. Numerous specializations in reproductive function occur in various species of bats, including delayed ovulation, delayed implantation, delayed embryonic development, ovarian-uterine asymmetry, menstruation, hibernation, and seasonality (Jerrett, 1979, and references therein).

Nonhuman Primates

Among the prosimians, E2 and P have been measured by RIA in ring-tailed lemurs *(Lemur catta)* kept under controlled photoperiods (Van Horn and Resko, 1977). Mean peak plasma concentrations of E2 (120 pg/ml) were observed on the first day of fully cornified vaginal smears. This estrogen peak apparently facilitated female sexual receptivity about 26 hr later, and this receptivity persisted for about 10–22 hr (also see Evans and Goy, 1968). Basal E2 levels during the remainder of the cycle were about 10 to 20 pg/ml. On the day of the E2 peak, P levels were low (3.3 ng/ml), but they rose about 3–4 days later to approximately 10 ng/ml and continued to rise to a luteal peak of 92.5 ng/ml at 25 days after the E2 peak. Progesterone values declined during the remaining 15 days of the estrous cycle. This primate species is interesting in several respects: (1) there is a period of sexual receptivity that is limited to the time around ovulation; (2) E2, but not P, facilitates the expression of this receptivity; (3) the ovarian cycle is activated by short day lengths and inhibited by long day lengths; and (4) partner preference is evident.

Another prosimian, the bushbaby *Galago crassicaudatus,* has an estrous cycle of 44 days, with female sexual receptivity restricted to about 6 days of the cycle around the time of ovulation (Eaton *et al.,* 1973). Estradiol evidently facilitates receptivity in bushbabies.

In the suborder Simiae, the marmoset is unusual (but not unique) in not having a detectable menstrual cycle, However, marmosets also differ from representatives of the suborder Prosimii because they do not have an obvious behavioral estrous cycle. Preslock *et al.* (1973) measured serum progestins (by CPB) and estrogens (by RIA) in the marmosets *Saguinis oedipus oedipus* (n = 5) and *Saguinis fuscicollis illigeri* (n = 1). In the *S. oedipus* females, the basal immunoreactive estrogen levels were 2300 pg/ml, with a peak of 6200 pg/ml. These estrogen levels are many times higher than those reported for other mammalian species. It is unclear whether these high levels reflect a true species effect, or whether more refined measurements of individual estrogens, rather than the sum of immunoreactive estrogens, would bring estrogen levels in marmosets into line with those reported for other mammals. Basal levels of progestins were 140 ng/ml, with peak levels of 272 ng/ml. The interpeak interval for the progestins was 15.5 days. Again, these levels are higher than those reported for other nonpregnant mammals. These high levels of estrogen and progestin in marmosets may be a true species effect, a reflection of nonspecific measurement technique, or a reflection of the extensive and continuous luteinization of the ovary in this New World species.

Another New World primate, the capuchin monkey *(Cebus appela)* has been studied with RIA procedures (Nagle *et al.,* 1979). Capuchin monkeys have a menstrual-cycle length of about 21 days. Estradiol levels were 70 to 100 pg/ml during the early follicular phase, rose to a midcycle peak of 503.3 pg/ml, and declined thereafter to baseline values ($\sim$ 50 pg/ml). Plasma P levels were 3 to 10 ng/ml during the follicular phase; there was no discernible preovulatory P surge; and after ovulation and corpus luteum formation, P levels rose steadily to a peak of 69.9 ng/ml (attained 6 days after the E2 peak). About 4 days after attainment of peak values of P, there was a discernible drop in plasma P level, and this decline continued to baseline ($\sim$ 3 ng/ml) as the corpus luteum degenerated. It is of interest to note that E2 and P levels in the capuchin monkey are higher than those found in Old World primates, but not higher than the values reported for another New World primate, the marmoset. It is difficult to ascribe the relatively high values in the capuchin monkey to faulty measurement, because the cross-reaction in the estrogen RIA system was only 1.7% for E1 and 0.2% for E3. In the progestin RIA system, no difference in P values was seen between samples that were chromatographed (to separate P from 5α-dihydroprogesterone, 20α-dihydroprogesterone, and 17α-dihydroprogesterone) and samples that were not chromatographed. This finding suggests that New World primates may have higher levels of ovarian hormone in the circulation than Old World primates, but the physiological significance of this possible difference is not known.

The squirrel monkey *(Saimiri sciureus)* is another New World species. Squirrel monkeys do not menstruate, and they have a short estrous cycle (7–12.5 days) that is strongly influenced by seasonal factors (Mendoza *et al.*, 1978b). Monthly mean E2 levels in group-housed laboratory squirrel monkeys ranged from 36.4 to 139.6 pg/ml from April to December, and from 165.6 to 423.7 pg/ml in January to March. Variations among individual females were appreciable, with each female's highest levels ranging from 123 pg/ml to 1207 pg/ml, in a study using a reliable and specific RIA for E2. Variations in E2 and P during the course of squirrel monkey estrous cycles were measured by Wolf *et al.* (1977).

The most thoroughly studied nonhuman primate is the rhesus monkey *(Macaca mulatta)*. Rhesus monkeys have menstrual cycles of about 28 days' duration. RIA measurements indicate a midcycle peak of E2 of 69 to 360 pg/ml (E1 = 30–280 pg/ml). During the luteal phase of the cycle, the E2 levels were 18–93 pg/ml (E1 = 15–80 pg/ml), and during the follicular phase prior to the midcycle peak, the E2 levels were 42 to 80 pg/ml (E1 = 17–40 pg/ml) (Hotchkiss *et al.*, 1971; Yamaji *et al.*, 1971; Bosu *et al.*, 1972, 1973; Knobil, 1972; Weick *et al.*, 1973; Spies *et al.*, 1974; Resko *et al.*, 1975; Czaja *et al.*, 1977; Johnson and Phoenix, 1978). Progesterone levels, measured by CPB and RIA, were highest during the luteal phase, with values ranging from 4 to 11 ng/ml at the peak of luteal activity. During the nonluteal phases of the cycle, P levels were much lower (0.1–0.5 ng/ml) (Neill *et al.*, 1967b; Johansson *et al.*, 1968; Kirton *et al.*, 1970; Karsch *et al.*, 1973; Niswender and Spics, 1973; Resko *et al.*, 1974, 1975; Spies *et al.*, 1974; Foster, 1977; Czaja *et al.*, 1977; Clark *et al.*, 1978; Johnson and Phoenix, 1978). Protein-deficient rhesus monkeys have subnormal levels of P in the circulation (Gupta and Anand, 1971), and postmenopausal and ovariectomized rhesus have low levels of P (0.1–0.5 ng/ml) (Hess and Resko, 1973; Spies *et al.*, 1974). Although there is an increase in total progestins just prior to ovulation in rhesus monkeys, this may be attributable to a preovulatory rise in 17α-hydroxyprogesterone and/or P (Johansson *et al.*, 1968; Bosu *et al.*, 1972). The circulating levels of 17α-hydroxyprogesterone are quantitatively significant (1.0–5.0 ng/ml at midcycle and 0.5–1.0 ng/ml during the luteal and early follicular phases) (Bosu *et al.*, 1972, 1973). The progestin 20α-dihydroprogesterone is quantitatively less important, with levels ranging from 0 to about 1.0 ng/ml (Resko *et al.*, 1974). Plasma T levels in rhesus have been measured by RIA (Hess and Resko, 1973; Johnson and Phoenix, 1978). In both studies, the peak T values (0.36 and 0.84 ng/ml, respectively) occurred coincidentally with peak E2 levels at midcycle just prior to ovulation, whereas lower values (0.2 and 0.7 ng/ml, respectively) obtained at other times in the menstrual cycle. There appear to be no studies of AD levels during the menstrual cycle of rhesus monkeys. There is some evidence of a diurnal rhythm in ovarian steroid in rhesus monkeys (Spies *et al.*, 1974).

Other species of macaques have also been studied. The Japanese macaque *(Macaca fuscata)* is unusual among macaques in having a well-defined breeding season under both natural and laboratory conditions. In Japanese macaques maintained under controlled laboratory conditions in Kyoto, the menstrual cycles began in December, recurred four to five times at 28-day intervals, and ceased in April. RIA evaluations of E2 during the breeding season indicated a peak in the late follicular phase (150–250 pg/ml) of each cycle and a luteal phase peak of P (2.0–5.3 ng/ml) (Aso *et al.*, 1977). The causes of the onset of the amenorrhea of the nonbreeding season in Japanese macaques are not clear, especially when it is noted that the seasonal activation and deactivation of the reproductive system occur under laboratory conditions of controlled lighting, temperature, and humidity and a constant food supply. It is possible that study of the seasonal amenorrhea of Japanese macaques could lead to better understanding of certain causes of amenorrhea in women. Progesterone levels in the pigtail macaque *(Macaca nemestrina)* have been measured by CPB (Bullock *et al.*, 1972). This species has a menstrual cycle of about 33 days' duration and a luteal phase P peak of 15 to 20 ng/ml. However, baseline P values were fairly substantial ($\sim$ 2 ng/ml), and the authors acknowledged that their estimates of P may be inflated. The crab-eating or cynomolgus monkey *(Macaca fascicularis)* has a 29- to 31-day menstrual cycle. Early-follicular-phase E2 levels, as measured by RIA, gradually increase from 50 pg/ml to 150 pg/ml and reach a late follicular phase peak of 225 to 300 pg/ml. Progesterone levels, measured by RIA and CPB, reach a maximum of 7 to 10 ng/ml during the luteal phase and drop to about 1 ng/ml or less at nonluteal phases of the cycle, with no evidence of a preovulatory surge of P (Stabenfeldt and Hendrickx, 1973; Goodman *et al.*, 1977; Shaikh *et al.*, 1978). The bonnet monkey *(Macaca radiata)*, with a menstrual cycle length of 26–28 days, shows a peak of E2, as measured by RIA, of 250 pg/ml during the late follicular phase, 24–48 hr prior to ovulation (Lasley *et al.*, 1974; also see Parkin and Hendrickx, 1975, for estimates of total immunoreactive estrogens). Progesterone, as measured by CPB, reached a luteal phase peak of 4.3 ng/ml and nonluteal phase P levels of < 1 ng/ml (Stabenfeldt and Hendrickx, 1972). Although this study indicated no preovulatory rise in P, another study, in which total progestins were measured, suggested a preovulatory increase. This increase may be attributable to 17α-hydroxyprogesterone (Parkin and Hendrickx, 1975). Estradiol and P levels were also measured in stump-tailed macaques *(Macaca arctoides)* by RIA. Peak E2 levels ($\sim$ 500 pg/ml) were attained 1 day prior to the midcyle LH surge. Highest levels of P ($\sim$ 6 ng/ml) occurred during the luteal phase. It is noteworthy that serum P increased (to about 0.4 ng/ml) significantly above baseline 0–12 hr prior to the attainment of peak midcycle LH values, but shortly after initiation of the LH surge (Wilks, 1977; Wilks *et al.*, 1980).

The baboon *Papio hamadryas* has a menstrual cycle of about 32 days' mean duration. Estradiol levels, as measured by RIA, reached a peak of 250 pg/ml (a coincident peak of E1 was 100 pg/ml). This E2 peak precedes (by 1 day) the preovulatory surge of LH and is probably crucial to the LH surge. During the day of the LH surge, E2 levels decline to 150 pg/ml and then decline further (to $\sim$ 50 pg/ml) during the remainder of the cycle, except for a gradual rise beginning at 4 days prior to the next LH peak. Baseline E1 levels are on the order of 50 pg/ml. Baseline levels of P and 20α-dihydroprogesterone occurred during the follicular phase ($\sim$ 0.3 ng/ml), and neither progestin exhibited a preovulatory increase. Levels of both progestins increased after ovulation, with peak values (4.0 ng/ml and 2.3 ng/ml, respectively) attained 6 days after the LH surge. Interestingly, a peak of 17α-hydroxyprogesterone occurred during the preovulatory day of the LH surge (1.3 ng/ml), and baseline values ($\sim$ 0.5 ng/ml) were evident during the remainder of the cycle. The mean pregnenolone values hovered around 1.0 to 1.4 ng/ml throughout the cycle. Mean T levels ranged from 0.15 to 0.23 ng/ml, and the mean AD levels from 1.2 to 2.0 ng/ml over the entire cycle; the highest levels of both androgens were generally associated with the periovulatory period. No systematic variations in DHT and dehydroepiandrosterone levels were evident during the cycle (Goncharov *et al.*, 1976). Essentially, the same overall results for estrogens and progestins were reported by Kling and Westfahl (1978) for *P. cynocephalus*. However, *P. cynocephalus* exhibited a clearer increase in periovulatory T, and lower levels of AD, than *P. hamadryas*. A report of steroid levels in a third species of baboon, *P. anubis,* has also been cited (Wildt, Doyle, Stone and Harrison, 1977, cited in Kling and Westfahl, 1978). The effects of immobilization stress on hormone levels in *P. hamadryas* have been studied by Goncharov *et al.* (1979). Decreases in E2 and E1 during the follicular phase, and in P and 20α-dihydroprogesterone during the luteal phase, were evident. In male baboons, T and DHT levels were decreased after immobilization stress. (For additional data on baboons, see Stevens *et al.*, 1970; Koyama *et al.*, 1977).

Among the great apes, the chimpanzee *(Pan troglodytes)* has a menstrual cycle with a median duration of 34 days. Serum E2 levels peaked (352 pg/ml) just before the midcycle LH surge, with a secondary rise (181 pg/ml) during the luteal phase. Baseline E2 levels at menses were about 10 to 54 pg/ml (Reyes *et al.*, 1975). This study and another (Graham *et al.*, 1972) suggest peak P levels of 5 to 15 ng/ml during the luteal phase of the cycle, with very low values at other stages of the cycle, including the periovulatory stage.

Lowland gorillas *(Gorilla gorilla)* have menstrual cycles of about 31–32 days' duration. Peak E2 levels, as measured by RIA, were 200–500 pg/ml at 1–4 days prior to the midcycle LH surge. A decline in E2 levels for 2–3 days then occurred, and this was followed by a second E2 rise (34%–57% of the late-follicular-phase peak) associated with corpus luteum formation. Estradiol levels were low at menses and remained less than 100 pg/ml through the early follicular phase. During most of the follicular phase, the P levels were < 0.5 ng/ml. Significant elevations in P occurred during the luteal phase, with peak levels (10.8–12.2 ng/ml) attained 6–7 days after the LH surge; the P levels declined to near baseline by 10 days after the LH surge, as the corpus luteum degenerated. There was a significant midcycle elevation of T (0.35–0.50 ng/ml) that was about double the level found at other stages of the menstrual cycle (Nadler *et al.*, 1979). There appear to be no studies of peripheral plasma steroids in the third member of the apes, the orangutans.

HUMANS

There are many DIDD, CPB, GLC, and RIA studies of peripheral plasma levels of sex steroids in women. During the early follicular phase, E2 levels are about 30–50 pg/ml (E1 = 29 pg/ml; E3 = 8 pg/ml). As follicular development progresses, there is a midcycle peak of E2 to about 200–400 pg/ml (E1 = 80–200 pg/ml; E3 = 150 pg/ml), and the increased E2 provokes the surge of gonadotropin that results in ovulation. Estradiol levels usually decline by the time the gonadotropin surge occurs, but after ovulation and corpus luteum formation, E2 levels vary from 50 to 250 pg/ml, with most authors indicating the existence of a luteal phase elevation of E2 (E1 = 80–140 pg/ml; E2 = 11–120 pg/ml) (Svendsen and Sørensen, 1964; Baird and Guevara, 1969; Korenman *et al.*, 1970; Corker *et al.*, 1970; Mikhail *et al.*, 1970; Robertson *et al.*, 1971; Abraham *et al.*, 1972; Holmdahl and Johansson, 1972; Dupon *et al.*, 1973; Powell and Stevens, 1973; Hawkins and Oakey, 1974; Kim et al., 1974; Skinner *et al.*, 1974; Ribeiro *et al.*, 1974; Baird and Fraser, 1974; Rotti *et al.*, 1975; Legros, *et al.*, 1975; Guerrero *et al.*, 1976; Frölich *et al.*, 1976; and many others). There is apparently no difference between E2 levels during the day and during the night in the follicular phase (Horsky *et al.*, 1973).

Progesterone levels in women are about 0.2 to 1.8 ng/ml during the follicular phase. There may be an increase of P prior to the midcycle LH surge (Hoff and Quigley, 1980), but the highest concentrations of P during the menstrual cycle occur in the luteal phase (3.8 ng/ml to peaks of about 14.0–20.0 ng/ml) (Riondel *et al.*, 1965; van der Molen *et al.*, 1965; Neill *et al.*, 1967a; Yoshimi and Lipsett, 1968; Johansson, 1969; Johansson and Gemzell, 1969; Johansson and Wide, 1969; Cargille *et al.*, 1969; Johansson and Gemzell, 1971; Johansson *et al.*, 1971; West *et al.*, 1973; Dighe and Hunter, 1974; Jänne *et al.*, 1974; Apter *et al.*, 1976; Pauerstein *et al.*, 1978; and many others). No diurnal variations in plasma P occur during the luteal phase of the mesntrual cycle, although diurnal variations in plasma P appear prior to the midcycle LH surge and in the third trimester of pregnancy (Hoff and Quigley, 1980; Runnebaum *et al.*, 1972). Another progestin present in significant quantities in the peripheral plasma of women is 17α-hydroxyprogesterone. The concentration of this steroid during the follicular phase is only about 0.3–0.7 ng/ml, but the titer increases to about 2.0 ng/ml at the time of the preovulatory LH peak and, after a transient postovulatory decline, remains in the range of about 2 ng/ml during the life of the corpus luteum (Abraham *et al.*, 1972; Holmdahl and Johansson, 1972; West *et al.*, 1973; Thorneycroft *et al.*, 1974; Jänne *et al.*, 1974; Hami *et al.*, 1975; Wicklings, 1975; Aedo *et al.*, 1976). Pregnenolone and 17α-hydroxypregnenolone have also been measured in women, with only the first-mentioned steroid exhibiting cyclic variations (somewhat higher in the postovulatory than in the preovulatory phases) (Strott *et al.*, 1970; Abraham *et al.*, 1973; McKenna et al., 1974; Apter *et al.*, 1976; Guerrero *et al.*, 1976; Aedo *et al.*, 1976). These steroids can attain appreciable levels (1.2–3.8 ng/ml). 20α-Dihydroprogesterone and 5α-dihydroprogesterone vary cyclically in the same pattern as P. Follicular-phase levels of these reduced metabolites of P are 0.2 to 0.4 and 0.2 ng/ml, respectively, whereas the luteal-phase levels are 3.4 to 3.8 ng/ml and 1.4–2.8 ng/ml, respectively (Wu *et al.*, 1974; Aedo *et al.*, 1976; Milewich *et al.*, 1977). The physiological significance of these two progestins is a matter of speculation.

Several androgens have been measured in the peripheral plasma of women. Values of T during the follicular phase, midcycle, and luteal phase all fall in the range of 0.3 to 0.6 ng/ml, and AD values range from 0.9 to 3.0 ng/ml during these phases. Although peaks of T and AD during the menstrual cycle are not dramatic, there seems to be agreement that both steroids are statistically significantly elevated at midcycle. Levels of DHT are much lower than those of T or AD (0.1–0.3 ng/ml), but dehydroepiandrosterone and etiocholanolone (both apparently of adrenal origin) titers are 2.5 to 7.8 ng/ml and 2.5 ng/ml, respectively, during the follicular phase. Most workers do not detect cyclic variations in DHT, dehydroepiandrosterone, or etiocholanolone, but there are some reports of cyclic variation in DHT and in etiocholanolone (Coyotupa *et al.*, 1972; Guerrero *et al.*, 1976). Androsterone levels in women are about 0.5 ng/ml (Graef, 1971; Youssefnejadian *et al.*, 1973; West *et al.*, 1973; Judd and Yen, 1973; Abraham *et al.*, 1973; Kim *et al.*, 1974; Goebelsmann *et al.*, 1974; Barberia and Thorneycroft, 1974; Abraham, 1974; Ribeiro *et al.*, 1974; Apter *et al.*, 1976; Guerrero *et al.*, 1976). There is a report that, in female patients with anorexia nervosa, the T levels are markedly supernormal (Baranowska and Zgliczyński, 1979), but not all investigators have found the same correlation (Nillius *et al.*, 1975; Hurd *et al.*, 1977). There is also a claim that lesbians have higher levels of T than heterosexual women during the menstrual cycle (Gartrell *et al.*, 1977).

After menstrual cycles cease in women, the plasma concentration of many of the steroids declines. Postmenopausal values have been recorded for E2 (10–34 pg/ml), E1 (30–42 pg/ml), E3 (6–7 pg/ml), P (0.09 ng/ml), and 17α-hydroxyprogesterone (1.8 ng/ml) (estrogens: Powell and Stevens, 1973; Judd *et al.*, 1974, 1976; Yen *et al.*, 1975; Rotti *et al.*, 1975; progestins: Abraham *et al.*, 1973; Dighe and Hunter, 1974; McKenna *et al.*, 1974; Wicklings, 1975; androgens: Abraham *et al.*, 1973; Barberia and Thorneycroft, 1974; Kim *et al.*, 1974). The endocrinology of the menopausal transition has also been studied (e.g., Sherman *et al.*, 1976).

There is a substantial literature on steroids and puberty in human females (e.g., Angsusingha *et al.*, 1974; Apter and Vihko, 1977; Penny *et al.*, 1977; Apter *et al.*, 1978), and it is interesting to note that there is a circadian rhythm of plasma E2 (highest values during the daytime hours of 1400–1600 and lowest values during sleep) in premenarchal girls (Boyar *et al.*, 1976).

H. H. FEDER

The domestic rabbit is a reflex ovulator that breeds at any time of year and experiences irregular bouts of persistent behavioral estrus. There is no estrous cycle in the strict sense (Asdell, 1964). The ovarian follicles develop in overlapping waves, so that, as some follicles begin to degenerate, others have begun to mature. This arrangement favors a rather constant level of estrogen production and sexual receptivity during the breeding period. Peripheral plasma E2 levels are on the order of 60 pg/ml (E1 concentrations are even higher, $\sim$ 75 pg/ml). After coitus, but prior to ovulation, there is a transient increase in the concentration of E2 in ovarian vein plasma (Eaton and Hilliard, 1971) that is not apparent in the peripheral circulation (Wu *et al.*, 1977). During estrus and before coitus, the P and 20α-dihydroprogesterone levels in peripheral plasma are low ($<$ 1 ng/ml and $\sim$ 2 ng/ml), but soon after coitus (or HCG injection), both progestins increase significantly in concentration (to preovulatory peaks of 5 ng/ml and 12 ng/ml, respectively). Testosterone and AD levels are about .05 ng/ml, and coitus during estrus (or HCG injection) induces a preovulatory rise in peripheral plasma T (to 0.10 ng/ml), but not AD. Ovulation occurs about 10–12 hr after coitus, at which time the steroids return to basal levels.

RODENTIA

The Syrian or golden hamster *(Mesocricetus auratus),* a rodent frequently studied in the laboratory, has a 4-day estrous cycle (1 day of vaginal estrus, 2 days of diestrus, and 1 day of proestrus). Plasma E1, as measured by RIA, varies from 13.8 to 25.1 pg/ml during the day of vaginal estrus (just after ovulation has occurred), from 15.4 to 45.7 pg/ml on the first day of diestrus, from 46.8 to 100.8 pg/ml on the second day of diestrus (with higher values occurring in the afternoon and evening than during the morning of this day), and from 24.2 to 186.6 pg/ml on the day of proestrus (lordosis behavior and the surge of LH occur on the afternoon–evening of this day). On the day of proestrus, E2 levels were lowest in the morning hours (38.3 pg/ml) and highest before the LH surge in the afternoon (186.6 pg/ml), and they returned to baseline (24.2 pg/ml) by late evening. In contrast to the very distinct changes in E2 levels during the estrous cycle, changes in E1 over the cycle were not obviously correlated with behavioral and pituitary secretory events. Estrone levels varied between 36.9 and 70.1 pg/ml during vaginal estrus, between 20.9 and 114.6 pg/ml during the first day of diestrus, between 46.5 and 72.7 pg/ml during second day of diestrus, and between 47.7 and 99.4 pg/ml during the day of proestrus. The data suggest that hamsters have very high circulating levels of estrogens (compare with rats and guinea pigs, for example). However, it is of interest to note that, even after ovariectomy, E2 levels are 23.2 pg/ml and E1 levels are 42.2 pg/ml. This finding indicates either that adrenal estrogen secretion in hamsters is substantial, or that substances other than E2 and E1 were being measured with the RIA system used in this study (Baranczuk and Greenwald, 1973; also see Saidapur and Greenwald, 1978). The increased E2 levels during diestrus and proestrus cause the release of a surge of LH on the afternoon of proestrus (Labhsetwar, 1972). After the LH surge is initiated, there is a dramatic increase in preovulatory P (Lukaszewska and Greenwald, 1970; Leavitt and Blaha, 1970; Norman and Greenwald, 1971; Saksena and Shaikh, 1974; Ridley and Greenwald 1975; Saidapur and Greenwald, 1978). Peak levels of P during the afternoon–early evening preovulatory surge are about 11.0 to 18.2 pg/ml, with lower levels (1.6–6.6 pg/ml) of P at earlier times of day during proestrus. The preovulatory P is secreted primarily by the ovarian interstitium, with perhaps some contribution by the follicles (Norman and Greenwald, 1971; Leavitt et al., 1971). Preovulatory P synergizes with the previously secreted E2 to facilitate sexual receptivity (Frank and Fraps, 1945). Progesterone levels during the day of vaginal estrus have been reported to be about 9.8 to 18.9 ng/ml, during the first day of diestrus, about 3.6 to 14.3 ng/ml, and during the second day of diestrus between 10.8 and 19.9 ng/ml (Leavitt and Blaha, 1970; Lukaszewska and Greenwald, 1970). However, in a more recent RIA report, the figures reported were 5.1 ng/ml, 5.2 ng/ml, and 1.5 ng/ml at 0900 hr of each of the three days, respectively (Saidapur and Greenwald, 1978). A more detailed study of the corpora lutea of hamsters revealed degeneration after the evening of the first day of diestrus (Terranova and Greenwald, 1978). The hamster adrenal cortex releases some P, but less than the amount released by the rat adrenal cortex (Brown et al., 1976). Androgens have also been measured in cyclic hamsters. Serum T levels increase during the afternoon of proestrus (0.14–0.21 ng/ml) in parallel with the surge of preovulatory P. Serum T levels during the remainder of the cycle range from 0.06 to 0.09 ng/ml. Serum AD levels change very little over the entire cycle, with values of 1.0 to 1.9 ng/ml representing the entire range of change (Saidapur and Greenwald, 1978). The effects of aging on steroid hormone levels during the hamster estrous cycle have received little attention. Blaha and Leavitt (1974) showed a wider variation in the P levels of old hamsters than in young, mature hamsters at the time of fertilization. Parkening et al. (1978) failed to detect differences in the levels of P, E2, E1, and 20α-dihydroprogesterone 3–5 hr after the expected time of ovulation in young (3- to 5-month-old) and old (14- to 17-month-old) hamsters.

There are relatively few studies of steroid hormones in nonpregnant mice. Plasma P has been measured in prairie deermice (Peromyscus maniculatus) under laboratory conditions (Albertson et al., 1975). CPB revealed P levels of 7.2 ng/ml at proestrus, 14.6 ng/ml at vaginal estrus, and 26.7–40.5 ng/ml in the days of diestrus that followed (the number of diestrous days ranged from 1–5). Changes in P levels during the preovulatory period of proestrus were not monitored in this spontaneously ovulating species. Somewhat lower values of P were reported for Mus musculus. For example, Virgo and Bellward (1974), McCormack and Greenwald (1974), and Campbell et al. (1976) reported P levels of 2.0 to 3.0 ng/ml during proestrus and 4.9 to 6.0 ng/ml during early diestrus. Levels of E2 were higher during proestrus (23.0–39.0 pg/ml) than during early diestrus (4.0–5.0 pg/ml).

Rats (Rattus norvegicus) have been the most extensively studied rodents, and GLC, CPB, and RIA methods have been used for measurement of their steroid levels. Laboratory rats usually have estrous cycles of 4–5 days' duration (1 day of vaginal estrus, 1 day of proestrus, and 2–3 days of diestrus). There is agreement that the E2 levels are highest at proestrus (35.0–52.0 pg/ml) and are crucial for the LH surge and the onset of lordosis behavior that precede ovulation. After ovulation, on the day of vaginal estrus, the E2 levels have declined (1.6–10.0 pg/ml). There is a slight increase on the first day of diestrus, and there is a significant increase in E2 by the afternoon of the second day of diestrus (9.0–18.0 pg/ml). Estrone levels are 40.0 to 80.0 pg/ml throughout the estrous cycle (figures for E2 and E1 taken from Dupon and Kim, 1973; Kalra and Kalra, 1974; Hawkins et al., 1975; Shaikh and Shaikh, 1975; additional studies of E2 that indicate the same pattern but somewhat different absolute values are Brown-Grant et al., 1970; Naftolin et al., 1972; Butcher et al., 1974; Nequin et al., 1975a; Kalra and Kalra, 1977b). The ovulatory surge of LH that occurs on the afternoon of proestrus immediately induces a surge of preovulatory P (36.7–40.0 ng/ml). By the time ovulation has occurred and the vaginal smear is estrus, P levels are 9.3 to 15.0 ng/ml. There is a small, but significant, rise in P levels on the first day of diestrus (13.2–30.0 ng/ml), but by the second and third days of diestrus, P values have declined to 13.0 to 15.0 ng/ml (data taken from Feder et al., 1968b; Horikoshi and Suzuki, 1974; essentially the same pattern of data has been described by many other workers, including Barraclough et al., 1971; Piacsek et al., 1971; Kalra and Kalra, 1974; Butcher et al., 1974; Smith et al., 1975; Nequin et al., 1975b; Shaikh and Shaikh, 1975; Kalra and Kalra, 1977b). It is now established that the adrenal gland is capable of contributing significantly to the level of P found in the peripheral circulation (Feder et al., 1968a; Barraclough et al., 1971). Progestins other than P have also been measured in rat plasma. The concentration of 20α-dihydroprogesterone far exceeds that of P during the estrous cycle. For example, Nequin et al., (1975b) reported peak levels of 20α-dihydroprogesterone of about 230 ng/ml during proestrus. The patterning of 20α-dihydroprogesterone was similar to that of P (highest at proestrus, lower during the first day of diestrus, and lowest during the second day of diestrus). Horikoshi and Suzuki (1974) found that 20α-dihydroprogesterone increased daily (except on the day of proestrus) between 1600 and 2200 hr. Another progestin, 17α-dihydroprogesterone, varied only between 1.7 and 2.7 ng/ml (Shaikh and Shaikh, 1975), whereas the pregnenolone values were even lower ($\sim$ 0.6 ng/ml) (Bergon et al., 1975). Androgen concentrations in peripheral plasma also appear to vary with the stage of the estrous cycle. Testosterone and AD levels are highest on the day of proestrus (0.16–0.18 ng/ml and 0.14–0.21 ng/ml, respectively) and lower on other days of the cycle (0.08–0.13 ng/ml and 0.09–0.14 ng/ml respectively) (Dupon and Kim, 1973; Falvo et al., 1974). Gay and Tomacari (1974) suggested that T secreted during proestrus is responsible for elevated FSH levels in blood during the morning of vaginal estrus.

Studies of aging rats indicate that aging rats in constant estrus (incidence highest at 19 months of age) have very low, constant P values ($\sim$ 5 ng/ml) and constant E2 levels of about half those of young females in proestrus. Old females in persistent diestrus (highest incidence about 27 months of age) have constant P levels of about half those of young females in proestrus and E2 levels about three-eighths those of young females in proestrus (Huang et al., 1978). Lu et al. (1979) extended these observations and found moderately lowered levels of T and AD and drastically lowered levels of 20α-dihydroprogesterone in aged, constant-estrus rats compared with young rats in proestrus. In aged, persistent-diestrus rats, 20α-dihydroprogesterone levels equaled those of young rats in proestrus. A study by Miller and Riegle (1980) shows that aging cyclic rats have significantly lower serum P values than young cyclic rats.

Many external factors, such as photoperiod, season, olfactory cues, stress and crowding, and tactile cues associated with intromission by the male, are known to influence pituitary function and thereby to alter steroid hormones in the plasma of commonly studied myomorphs such as rats, mice, and hamsters. The literature on this subject is too vast to review here, but the reader is referred to the following papers and reviews: photoperiod: Turek and Campbell (1979); season: Ramaley and Bunn (1972); olfactory cues: Aron (1979); stress and crowding: Christian (1971); tactile cues associated with mating: Adler (1978).

In contrast to the myomorphs discussed above, the montane vole *(Microtus montanus)* is a reflex, rather than spontaneous, ovulator. Gray *et al.* (1976) measured P by RIA and found levels of 9.0 ng/ml during diestrus, 14.0 ng/ml during estrus in unmated voles, and 22.0 ng/ml during estrus and 1 hr after mating.

The guinea pig *(Cavia porcellus)* has a much longer estrous cycle (16–18 days' duration) than rodents such as mice, hamsters, and rats. One factor that underlies this extended cycle length is the corpus luteum of the guinea pig. This species is unique among laboratory rodents in that it forms functional corpora lutea in the absence of mating. In other respects, the guinea pig cycle is similar to the estrous cycles of other rodents. In guinea pigs, E2 levels begin to increase on about Day 10 postovulation and reach a peak on Day 15 or 16 ($\sim$ 70 pg/ml), the period of proestrus. At other times of the cycle, E2 levels remain at about 20 to 30 pg/ml (Croix and Franchimont, 1975). However, changes in the systemic levels of estrogen have been difficult to detect even by sensitive methods (Challis *et al.*, 1971; Sasaki and Hanson, 1974), and there is evidence in these papers that E1 levels are higher than E2 levels in guinea pigs. The preovulatory LH surge induced by estrogen is immediately followed by, and probably causes, a transitory, but substantial, release of preovulatory P (Feder *et al.*, 1968a; Croix and Franchimont, 1975; Blatchley *et al.*, 1976) from nonluteal (i.e., interstitial or follicular) ovarian sources (Feder and Marrone, 1977). This preovulatory P has a peak value of about 3.6 ng/ml and facilitates the onset of sexual receptivity (Feder *et al.*, 1968a). The period of sexual receptivity terminates about 8–10 hr later, almost simultaneously with ovulation, and at this time, P levels are low ($\sim$ 0.1 ng/ml). In all of these respects, there is no substantial difference in endocrine patterns among guinea pigs, hamsters, and rats. By about 2 days after ovulation, P levels increase and continue to rise, reaching a plateau at about 5 to 12 ng/ml postestrus. The peak values of P during this luteal phase of the cycle are about 2.5 to 3.7 ng/ml, as measured by GLC and RIA methods (Feder *et al.*, 1968a; Challis *et al.*, 1971; Croix and Franchimont, 1975; Blatchley *et al.*, 1976; Hossain *et al.*, 1979). By Day 13 postestrus, P levels drop precipitously (to about 0.3 ng/ml) and remain depressed until the onset of the next estrus. At no time in the estrous cycle does the guinea pig have significant quantities of 20α-dihydroprogesterone in the circulation. Ovariectomy results in a very low level of P in the circulation, a finding suggesting that the guinea pig adrenal secretes very little of this steroid (Feder *et al.*, 1968a). For additional data, see Garris and Mitchell (1979).

Although the timing of ovulation is somewhat influenced by environmental lighting, ovulation can occur at any time of day in guinea pigs (Donovan and Lockhart, 1972). No studies of the effects of lighting or other environmental factors on hormone levels in guinea pigs appear to have been carried out.

Studies of plasma steroids in hystricomorphs other than the guinea pig are sparse. Protgesterone levels are reported for the nonpregant green acouchi *Myoprocta pratti* (Rowlands *et al.*, 1970) and for the coypu *Myocastor coypus* (Rowlands and Heap, 1966).

CETACEA

I was unable to find reports on the circulating levels of gonadal steroids in cetaceans. However, Fisher and Harrison (1970) and Harrison and Ridgway (1971) have described reproductive cycles of two species, the common porpoise *(Phocoena phocoena)* and the bottlenose dolphin *(Tursiops truncatus)*.

CARNIVORA

Plasma P levels have been measured in nonpregnant stoats *(Mustela erminea)* with a GLC method (Gulamhusein and Thawley, 1974). Levels were about 2.5 ng/ml throughout the year. Females of this species come into heat in early summer, at which time they are impregnated. However, implantation is delayed until March, and parturition occurs 4 weeks later. It is interesting to note that the plasma P values do not rise significantly above nonpregnant levels in pregnant stoats until the time of implantation nears.

The progesterone levels in ferrets *(Mustela putorius)* in estrus or anestrus were too low to detect by CPB assay (< 0.8 ng/ml) (Blatchley and Donovan, 1972).

Minks *(Mustela vison)* are usually said to be reflex ovulators, but Møller (1975) found that some minks ovulated "spontaneously" and formed corpora lutea. These corpora lutea were associated with peripheral plasma P levels of about 27 to 33 ng/ml. In females that had not ovulated, the P levels were about 3 to 9 ng/ml, as measured by CPB.

Studies of plasma P have been made in other mustelids during delayed implantation (e.g., spotted skunk: Mead and Eik-Nes, 1969; European badger: Bonnin *et al.*, 1978).

There do not appear to be measurements of plasma steroids in female bears, but there are some interesting descriptions of the reproductive biology of female grizzly bears (Craighead *et al.*, 1969).

Steroid levels have been determined in various spontaneously ovulating canid species. Wolves *(Canis lupus)* are seasonal breeders, exhibiting a single estrus between late January and early April. During the anestrous period (June–December), the E2 levels were 5 to 20 pg/ml and the P levels were < 1 to 2 ng/ml. During the breeding season, the mean duration of proestrus was 15.7 days, estrus was 9.0 days, and the luteal phase was 63 days. E2 levels during the breeding season were 10 to 20 pg/ml (proestrus), 30 to 70 pg/ml (late proestrus), and 10 to 30 pg/ml (luteal phase). P levels increased to a peak of 22 to 40 ng/ml at 10–13 days after ovulation (ovulation probably occurs 1–9 days after the onset of estrus) (Seal *et al.*, 1979). Although steroid levels have also been measured in the blue fox *(Alopex lagopus)* (Møller, 1973) and the raccoon dog *(Nyctereutes procyonoides)* (Valtonen *et al.*, 1978), the bulk of the work with canids has been carried out on various breeds of domestic dogs *(Canis familiaris)*, most often beagles.

The breeding period in dogs lasts an average of 7–8 months and can occur in any season of the year (Stabenfeldt and Shille, 1977). The proestrous phase lasts about 5–12 days, the estrous phase about 5–12 days, and the diestrous period of luteal function about 30–90 days. There is general agreement that immunoassayable estrogen begins to rise during proestrus (~ 25 pg/ml), and the peak levels of 46 to 113 pg/ml are seen approximately 48–96 hr before ovulation, during the late proestrous or early estrous period (Bell *et al.*, 1971; Jones *et al.*, 1973; Hadley, 1975; Nett *et al.*, 1975; Edqvist *et al.*, 1975; Concannon *et al.*, 1975). It is probable that the increased estrogen triggers the ovulatory surge of LH that occurs about the first day of estrus (± 1 day). At the time of the LH surge, the estrogen levels decline. Some (Phemister *et al.*, 1973; Austad *et al.*, 1976) but not all (Smith and McDonald, 1974) studies indicate that the P levels rise concomitantly with or just after the onset of the LH surge but prior to ovulation. This preovulatory progesterone may play a role in facilitating estrous behavior (Concannon *et al.*, 1977). P levels increase dramatically after ovulation, attaining peak values (18–35 ng/ml) 10–25 days postovulation (Smith and McDonald, 1974; Concannon *et al.*, 1975, 1977; Graf, 1978). In dogs, the formation of functional corpora lutea does not require mating, and it is remarkable that corpora lutea in unmated dogs have a life span that equals or sometimes even exceeds that of corpora lutea in pregnant dogs. Another unusual neuroendocrine feature in dogs is that estrous behavior continues for several days after ovulation even though progesterone levels are rising and estrogen levels have declined (Concannon *et al.*, 1977). This finding suggests that dogs are relatively insensitive to the inhibitory effects of P on sexual behavior, but systematic studies of this notion have not been carried out.

Studies of episodic and circadian patterns of plasma estrogen and progestin in dogs do not appear to have been carried out. The environmental factors involved in seasonality have not been delineated, and the effects of stimuli from the male on the female's endocrine secretions have not been explored.

Steroid levels have been measured by RIA during the estrous cycle of the captive lion *(Panthera leo)*. Three lionesses isolated from males showed E2 surges ranging from 19 to 108 pg/ml and P surges ranging from 17 to 282 ng/ml. When females were placed with males, female sexual behavior was often correlated with increased E2 levels. The surges of E2 were often followed by surges of P about 1 week later, with maintenance of progesterone-producing corpora lutea lasting 2–6 weeks. The data suggest that although copulation can cause reflex ovulation in lionesses, ovulation can also occur in the absence of copulatory stimuli from the male (Schmidt *et al.,* 1979).

Domestic cats *(Felis cattus)* are usually considered seasonal breeders (but see Jemmett and Evans, 1977). In temperate zones, they undergo anestrus in fall and early winter. During the periods of sexual activity, there is a proestrous stage (1–3 days), followed by an estrous stage ($\sim$ 6–10 days). In unmated females, there is usually an interval of 2–3 weeks between the multiple estrous periods of the breeding season (Stabenfeldt and Shille, 1977; Wildt *et al.,* 1978). Verhage *et al.* (1976) measured E2 and P by RIA in four cats. They found peaks of E2 ($\sim$ 60 pg/ml) at about 16-day intervals in unmated cats kept under a controlled light:dark schedule. During the troughs between these peaks, E2 levels of about 8 pg/ml were found. These observations were supported by Shille *et al.* (1979), who also noted that, as ovarian follicular development progressed and E2 levels increased, an increasing proportion of cats displayed estrous behavior. P was undetectable during the polyestrous period (Verhage *et al.,* 1976). Because the cat is a reflex ovulator, no corpora lutea are formed unless there is coital stimulation (other factors such as olfactory stimuli do not appear to have been tested). Cats, like dogs, show a sharp decrease in estrogen levels and a gradual increase in P immediately after ovulation and may continue to display sexual behavior during the postovulatory period (Paape *et al.,* 1975).

PROBOSCIDEA

Some time ago, Hanks and Short (1972) reported that so little P was secreted by elephant corpora lutea, that the hormone was difficult to detect in peripheral plasma even during pregnancy. Later, Plotka *et al.* (1975) measured P by RIA in nonpregnant African *(Loxodonta africana)* and Asian *(Elaphus maximus)* elephants and found levels of only 0.02–0.2 ng/ml. The E2 (+ cross-reacting E1 and E3) levels were 9–37 pg/ml. Levels of P were only somewhat higher even during pregnancy (0.4–0.5 ng/ml). This finding suggests that the elephant is extraordinarily sensitive to P, but studies that explore the biochemical basis of this sensitivity have not been carried out.

PERISSODACTYLA

Most mares are seasonally polyestrous, although some show ovarian cycles throughout the year. Reproductive inactivity is usually associated with the winter months, and decreases in photoperiod are apparently a major factor influencing gonadal activity (Oxender *et al.,* 1977; Stabenfeldt and Hughes, 1977). The mean length of a cycle between spontaneous ovulations is about 20–22 days (behavioral estrus begins approximately 5–6 days before ovulation and ends about 2 days after ovulation), and corpus luteum function begins about 1 day after ovulation, with peak activity 6 days postovulation and luteal regression about Day 15 postovulation (Stabenfeldt and Hughes, 1977; Asa *et al.,* 1979).

Estrogens and progestins have been measured in the peripheral circulation of cyclic mares. Serum E2 levels, as measured by RIA, increased from 20.0- pg/ml 5 days prior to ovulation (late diestrus) to a peak of 141 pg/ml 2 days before ovulation, then gradually decreased to diestrous levels within 3 days after ovulation. However, these may be overstimates of E2 because the antibody used cross-reacted with E1 and E3 (Pattison *et al.*, 1974). For additional studies of plasma estrogens, see Nett *et al.* (1973) and Noden *et al.* (1975). Progestin levels in mares have been measured by several workers. There is general agreement that P begins to rise within 24 hr after ovulation and attains maximal levels about 6–10 days postovulation (8–22 ng/ml). As the corpus luteum wanes and estrous behaviors begin to be exhibited, P levels fall below 1 ng/ml (Short, 1959; Smith *et al.*, 1970; Plotka *et al.*, 1971; Stabenfeldt *et al.*, 1972; Sharp and Black, 1973; Squires *et al.*, 1974). Estrous behavior, characterized by the display of increased approaching and following of stallions by mares, urination, presenting, clitoral winking, and tail raising, as well as the occurrence of copulation (Asa *et al.*, 1979), can apparently occur after ovariectomy or during seasonal anovulatory periods, albeit with somewhat lower frequency than among intact mares in the breeding season (Asa *et al.*, 1980b). It is hypothesized that sexual activity outside the breeding season contributes to the year-round maintenance of social bonds in harem groups. It is further suggested that adrenal steroids (androgens?) play a role in maintaining the estrous behavior of ovariectomized mares (Asa *et al.*, 1980a).

ARTIODACTYLA

The roe deer *(Capreolus capreolus)* is a seasonal breeder, with estrus occurring in July–August. Nonpregnant roe deer have mean total estrogen (E1 + E2 + E3) levels of 57–107 pg/ml, with the highest values (75–107 pg/ml) occurring from January to June. P levels in nonpregnant roe deer were 0.1 ng/ml in July, 1.7 ng/ml in August, 1.9 to 3.0 ng/ml in August–March, 1.1 ng/ml in April, and 0.1 ng/ml in May. In nonpregnant does kept separate from the buck, ovulation occurred and the corpus luteum, once formed, remained active until the second half of September (Hoffmann *et al.*, 1978). It should also be mentioned that the roe deer is the only artiodactyl known to exhibit delayed implantation (this phenomenon is also found in certain marsupials, bats, carnivores, insectivores, and endentates). In contrast to the case of most other species of delayed implanters, the corpus luteum of the roe deer secretes appreciable quantities of P during the diapause (Sempéré, 1977).

Seasonal changes in estrogen and P have been studied by RIA in white-tailed deer *(Odocoileus virginianus)*. In this species, breeding usually occurs in November, in Minnesota, with cycles of 24 to 26-day duration repeated two or three times during the breeding season if mating does not occur. Estrogen (E1 + E2 + E3) levels did not appear to change significantly over the course of a year in nonpregant females (~ 30 pg/ml). P levels were about 3.9 ng/ml during December–February, 1.94 ng/ml during March–May, and 0.5 ng/ml or less from June to November in nonpregnant, nonlactating does (Plotka *et al.*, 1977).

Many investigators have measured steroid hormones in cattle. Cattle have an estrous cycle of 18–25 days (mean ~ 21 days for cows and mean ~ 20 days for heifers) and ovulate spontaneously. Some work on estrogen (Henricks et al., 1971b; Christensen et al., 1971; Mason et al., 1972; Shemesh et al., 1972) suggested that the circulating levels were quite high, but later work with specific RIA and protein-binding techniques suggests that E2 levels reach a peak of 6 to 14 pg/ml about 0.5 days prior to the onset of estrous behavior (Echternkamp and Hansel, 1971; Wetteman et al., 1972; Glencross et al., 1973; Dobson and Dean, 1974; Dobson, 1978). These peak levels of E2 are thought to trigger LH release and ovulation after the onset of estrous behavior (estrous behavior duration is about 12–24 hr). The E2 levels fall during estrus, averaging about 4 to 5 pg/ml 2–11 days postestrus, then begin increasing about 4 days prior to the next estrus. A transient elevation of E2 to about 9 pg/ml is sometimes reported during the luteal phase. E1 levels showed little cyclic variation, with an average concentration of 4 pg/ml. Interestingly, the α-isomer of E2 was present at fairly high levels in peripheral plasma (20–30 pg/ml), exceeding the peak levels of the biologically active β-isomer of E2 (14 pg/ml) (Dobson and Dean, 1974). There are many studies of P levels in a variety of breeds of dairy and beef cattle. Although the studies do not always agree on the absolute levels of P during the cycle, there is general agreement that P levels are minimal from 2 days before estrus to 2 days after estrus; the P levels increase steeply between Day 3 and Day 7 postestrus, increase gradually for another 8–10 days, and abruptly fall 3 or 4 days prior to the next estrus. Important features of the progestin pattern in cattle are the lack of a preovulatory rise in P or in 17α-hydroxyprogesterone (Kazama and Hansel, 1970; Henricks et al., 1971a; Thibier et al., 1973; Thibier and Saumande, 1975; a report that suggests a rise in preovulatory progestins in cattle is open to methodological question: Ayalon and Shemesh, 1974) and the correspondence between the P level and the life of the corpus luteum (Schomberg et al., 1967; Plotka et al., 1967; Pope et al., 1969; Stabenfeldt et al., 1969b; Henricks et al., 1970; Hansel and Snook, 1970; Garverick et al., 1971; Shemesh et al., 1971; Sprague et al., 1971; Wetteman et al., 1972; Dobson et al., 1975; Kindahl et al., 1976; Herriman et al., 1979). The most reliable measurements by RIA suggest that the peak levels of P during the luteal phase are in the range of 4–10 ng/ml. Testosterone has also been measured in cow peripheral plasma (Shemesh and Hansel, 1974; Kanchev et al., 1976; Herriman et al., 1979). Peak values of T occur from the 5th to the 15th days postestrus and range from about 12 to 18 pg/ml plasma. There appears to be a dearth of information about episodic and circadian patterns (see review by Robinson, 1977, for evidence of a circadian pattern of ovulation, and Miller and Alliston, 1974, for lack of effect of circadian temperature changes on plasma P levels) of ovarian steroids in cattle. The effects of various environmental factors (photoperiod, rainfall, pasture growth, and the presence of a bull) on ovarian function have been reviewed by Robinson (1977), but additional studies of environmental factors on peripheral hormone levels may be of value. Lamond et al. (1971) provided evidence for breed differences (Angus vs. Hereford) in plasma P in cattle, indicating a genetic source of variation in the length of the proestrous trough of plasma P. For additional data, see Schams *et al.* (1977).

The sheep is an extensively studied artiodactyl. Robertson (1977) pointed out that the annual season of sexual activity of ewes varies from a monoestrous condition in some wild types, through a seasonal polyestrous state in most domesticated breeds, to the ability of certain tropical breeds to reproduce at almost all times of year. In northern latitudes, the estrous season usually extends from September to February, with individual estrous cycles having a duration of 16.5 to 17.5 days. The onset of estrous behavior (usually defined as the onset of the ewe's acceptance of mounting by a ram) has a duration of approximately 30 hr and precedes spontaneous ovulation by 24–27 hr (Robertson, 1977). There appear to be at least three waves of follicular growth during a single estrous cycle, with only the last wave culminating in ovulation. The waves of follicular growth in the ewe are accompanied by increases in estrogen. Hauger *et al.* (1977) suggested that the E2 levels are maximal (25 pg/ml) about 5–7 days after estrus (first wave of follicular growth), and that a smaller peak (9 pg/ml) occurs about 15 days after estrus and about 2 days prior to the next estrus. During the rest of the cycle, the E2 levels do not exceed 3 pg/ml. Other investigators have also assayed estrogen in the systemic circulation (Scaramuzzi, *et al.*, 1970; Yuthasastrakosol *et al.*, 1975; Pant *et al.*, 1977). The report by Pant *et al.* suggests that E2 (and cross-reacting E1 and E3) levels peak at about 14–21 pg/ml approximately 2 days before estrus. A major role of the increased levels of E2 late in the cycle is undoubtedly the stimulation of LH release. Baird *et al.* have noted that AD secretion in ewes follows the same temporal pattern as E2 (Baird *et al.*, 1976a,b). About three days after the induction of ovulation by LH and the beginning of corpus luteum formation, appreciable quantities of P appear in the systemic circulation (Short and Moore, 1959; Plotka and Erb, 1967; Thorburn *et al.*, 1969; Stabensfeldt *et al.*, 1969c; Obst and Seamark, 1970; Allison and McNatty, 1972; Cunningham *et al.*, 1975; Pant *et al.*, 1977; Hauger *et al.*, 1977). A midluteal phase peak of approximately 2.4–3.7 ng/ml plasma P has been recorded. Progesterone levels have declined to about 0.6 ng/ml by 16 days after ovulation. This decline in plasma P is correlated with the demise of the corpora lutea. Progesterone levels remain very low (~ 0.1–0.5 ng/ml) as the next estrous period approaches and for about 2 days after estrus (Thorburn et al., 1969; Stabenfeldt *et al.*, 1969c; Pant *et al.*, 1977), is presumably because the ovaries of sheep (and cows and pigs) do not have large proportions of interstitial tissue, a potential source of preovulatory P (Hansel *et al.*, 1973). Recall that a preovulatory surge in P is an important feature of the estrous cycles of rats, hamsters, and guinea pigs because it helps to facilitate female sexual receptivity. However, in sheep, it is the rise and fall of luteal P, rather than a rise in preovulatory P, that facilitates (with estrogen) sexually receptive behavior. Interestingly, sheep without recently active corpora lutea (e.g., pubertal sheep at first ovulation or sheep undergoing the first ovulation of a breeding season) often ovulate without manifesting estrous behavior. This "silent estrus" is presumably caused by a lack of progesterone-priming of the estrogen-sensitive neural tissues that mediate sexual behavior (Robertson, 1977).

There is one report that claims to demonstrate a diurnal rhythm of plasma P during the luteal phase in sheep. Higher concentrations of plasma P were said to occur during daylight than during darkness (McNatty *et al.,* 1973), but another report failed to show such a relationship (Thorburn *et al.,* 1969). Both reports indicated considerable within-sheep and within-day variations in plasma P concentrations, an indication suggesting episodic secretion.

There is evidence of episodic patterns of E2 in sheep. Baird *et al.* (1976b) found that ewes produced pulses of LH about every 2 hr on Days 12 and 14 of the estrous cycle. Within 5 min of each pulse of LH, E2 secretion increased, whereas fluctuations in P levels were not related to the LH surges. Baird (1978) examined the question of how estrogen levels rise in the 2 days prior to the onset of estrous behavior. He found that on Days 15 and 16 of the cycle, LH pulses increase in frequency (to more than once per hour) but have a smaller amplitude than during earlier stages of the cycle. Despite the smaller amplitude of LH pulses, more E2 is secreted in response to each pulse than at earlier stages of the cycle. It is likely that the severe depression of P levels at this time contributes to the increased stimulatory effect of LH on E2 secretion. The data also suggest that the apparently smooth rise in plasma estrogen levels that begins about a day before estrous behavior is really composed of summated bursts of E2 in response to repeated episodic pulses of LH that occur with increasing frequency. At the onset of estrus, the LH pulses no longer cause increases in E2.

During the anestrous period, the ovaries are relatively, but not completely, quiescent. They secrete measurable quantities of E2 and AD in response to endogenous pulses of LH that occur about once every 5 hr (Scaramuzzi and Baird, 1977). Ovarian estrogens apparently contribute to the inhibition of LH release during anestrus (Legan *et al.,* 1977).

The transition between anestrous and breeding phases is regulated, at least in part, by a fall in the ratio of light to dark during a 24-hr period (Robertson, 1977). Legan *et al.* (1977) and Legan and Karsch (1979) made the important observation that there is seasonal variation in the response of the hypothalamo-hypophyseal axis to the inhibitory effects of E2 on LH release. Ovariectomy results in high serum LH at all times of year. When ovariectomized ewes are given silastic implants of E2, anestrus serum LH levels are depressed, but the same implants fail to suppress serum LH levels during the breeding season. Thus, anestrous ewes appear to be more sensitive to the inhibitory effects of E2 on LH secretion than ewes during the breeding season.

Domesticated pigs of various breeds have been the subject of a few endocrinological studies. Pigs are spontaneous ovulators with estrous cycles of 21 days (usual range, 18–24 days) occurring throughout the year without apparent seasonality. Just before the female becomes receptive to the boar (estrus), she becomes restless, seeks the attention of the herdsman or boar, may mount other females, and may emit malelike growling sounds (Dziuk, 1977). Mean total plasma estrogens were measured by RIA in Duroc gilts (Henricks *et al.,* 1972; Guthrie *et al.,* 1972). Estrogen levels were low (< 20 pg/ml) during the cycle until about 16–17 days after estrus; then, they increased to a maximum (40–50 pg/ml) 1–2 days prior to the onset of the next estrus. Eiler and Nalbandov (1977) used an RIA method and reported estrogen levels of 20–30 pg/ml during the course of the estrous cycle, but preestrous maxima were found in only about half the subjects, presumably because the sampling was too infrequent. Shearer *et al.* (1972) used a CPB method to measure estrogens in Landrace × large white gilts and reported E2 levels of 5 to 37 pg/ml throughout the cycle except for an increase to 85 pg/ml 1 day before estrus. These values are almost undoubtedly overestimates. There is excellent agreement among laboratories and methodologies with regard to the circulating P levels in a variety of breeds, including Yorkshire, Landrace, Duroc, and Göttingen (miniature pigs). These studies show an initial rise in P about 2–3 days after estrus, followed by a steep increase up to Days 6–7 postestrus, with mean maximal levels (∼ 30–35 pg/ml) attained 9–14 days postestrus. After this time, there is a precipitous decline, and the P levels remain quite low (0.5–3.0 ng/ml) for the week preceding the next estrus (Tillson and Erb, 1967; Short, 1961; Stabenfeldt *et al.*, 1969a; Edqvist and Lamm, 1971; Shearer *et al.*, 1972; Henricks *et al.*, 1972; Guthrie *et al.*, 1972; Parvizi *et al.*, 1976; Gleeson and Thorburn, 1973; Eiler and Nalbandov, 1977). The basic form of the pig estrous cycle is similar to that of cattle and sheep: Estrogen levels rise prior to estrus and induce estrous behavior as well as a surge of LH that results in ovulation. There is no preovulatory surge of P, but P is produced after ovulation by the newly formed copora lutea. The demise of the corpora lutea about 14–15 days postestrus is associated with a decline in P levels (Guthrie *et al.*, 1972; Hansel *et al.*, 1973). The effects of various environmental stimuli on the levels of ovarian hormones in plasma have received scant attention in this species.

PLASMA STEROID DATA: USES AND LIMITATIONS

In writing the foregoing sections, I took as my task the provision of an annotated bibliography that would present a reasonably complete picture of what is known about the concentrations of gonadal steroids in the peripheral plasma of adult male and adult, nonpregnant female mammals. These data are of obvious value for understanding the reproductive physiology of mammalian species. For example, if one is to perform experiments in which steroid replacement therapy is used for gonadectomized subjects, one would like to design the therapy to mimic the concentration and patterning of the missing endogenous hormone. This design is important if artifacts due to under- or overdosing are to be avoided. Another way in which the study of peripheral levels of endogenous gonadal steroids can be of value is in the assessment of the effects of internal and external factors on gonadal activity. For example, changes in peripheral plasma gonadal steroid concentrations may be used as evidence that an environmental factor or an administered substance has altered gonadal function. The potency of such factors then can be compared between sexes and among strains and species. Despite these potential uses of plasma steroid data, there are severe limitations on the ways in which these data can be interpreted. In the following paragraphs, I highlight some of the factors that complicate the interpretation of plasma steroid data and, in the interests of brevity, illustrate these remarks with a very selective group of references to work on mammals. Complications having to do with measurement techniques, with the circadian rhythms of steroid secretion, and with the intracellular metabolism of prohormones have been alluded to in the text and are not emphasized further.

BIOCHEMICAL FACTORS THAT COMPLICATE INTERPRETATION OF PLASMA STEROID DATA

BINDING OF STEROIDS TO SERUM PROTEINS. Steroids are bound by a variety of serum proteins, the spectrum of which varies among species. For example, in all mammals, albumin is a low-affinity, high-capaicty binder of progesterone and other steroids. Orosomucoid is another binder of progesterone. In humans, and most other mammals, corticosteroid binding globulin (CBG) also binds progesterone, but with higher affinity and specificity than albumin. The guinea pig possesses an additional serum protein, not present in other species, that binds progesterone with even higher affinity and greater specificity than CBG (Milgrom *et al.*, 1970; Tan and Murphy, 1974; Westphal *et al.*, 1977). This progesterone-binding protein (PBP) occurs in the serum of pregnant guinea pigs, and its presence reduces progesterone clearance during pregnancy by 10-fold (Illingworth *et al.*, 1970). In humans, there is a sex-hormone-binding globulin (SBG) that binds 5α-DHT, testosterone, and estradiol with higher affinity and greater specificity than does albumin (Corvol *et al.*, 1971). Rabbit plasma also contains a sex-hormone-binding globulin (Mahoudeau *et al.*, 1973; Danzo and Eller, 1975). Distribution of steroids between free and unbound factions varies according to physiological state (Burke and Anderson, 1972). Rats apparently lack this binding protein (DeMoor *et al.*, 1969), but during perinatal life, rats possess an estradiol-binding plasma protein (EBP) that is absent in guinea pigs (Raynaud *et al*, 1971; Savu *et al.*, 1974; Plapinger *et al.*, 1977). Clearly, the types of serum protein binders for steroids may differ not only among species but within a species according to sex, age, and physiological state. The functions of these binders include the transport of steroids to target tissues and the protection of steroids from chemical or enzymatic attack, resulting in a decreased metabolic clearance rate of the steroids. Binding of steroids to these serum proteins is thought to inactivate the steroids; activation occurs when the steroid–protein complex is dissociated and the unbound steroid is made available to target cells. Because typical measurements of peripheral plasma steroid levels do not differentiate between unbound steroid (a small proportion of the total steroid in circulation) and protein-bound steroid and do not differentiate among various types of protein binding in serum, it is sometimes difficult to predict the physiological effect of a given concentration of total steroid in the circulation. For example, estrogen levels in perinatal female rats may be elevated (Weisz and Gunsalus, 1973), but the presence of EBP markedly diminishes the potency of endogenous estrogens and protects the organism from their effects (Plapinger and McEwen, 1978). The net effect of the presence of steroid-binding serum proteins is difficult to predict. On the one hand, the high-affinity proteins protect the steroids and delay their metabolic clearance, and on the other hand, the binding of steroid to these serum proteins inactivates the steroid. It is particularly important to bear in mind that exogenous chemical treatment may affect the steroid-binding proteins without affecting the total plasma steroid level, a result giving one a false sense of security that the treatment is without effect on steroid function (e.g., Horth *et al.*, 1979). The possibility that environmental factors influence the concentrations of various serum steroid-binding proteins, thereby altering steroid potency, has not been examined.

BINDING OF STEROIDS TO INTRACELLULAR PROTEINS. Once a gonadal steroid reaches a target tissue, it is said to passively diffuse through the cell membrane and to bind to specific, high-affinity receptor proteins in the cytoplasm. The steroid–receptor complex is then translocated to the nucleus, where it becomes associated with receptor sites on the chromatin and induces RNA and protein synthesis (Gorski and Gannon, 1976). It should be obvious that this mechanism of steroid action presents certain problems for the interpretation of plasma steroid levels. For example, the same concentration of free plasma steroid may have quantitatively different effects because the binding characteristics, concentration, and compartmentalization of intracellular receptor may vary among species (Plapinger *et al.*, 1977), individuals, tissues (Feder *et al.*, 1979), regions of tissues (Pfaff and Keiner, 1973; Sar and Stumpf, 1977), and sexes (Whalen and Massicci, 1975). Changes in steroid binding may occur with age (Peng and Peng, 1973), season (Boyd and Spelsberg, 1979), previous exposure or lack of exposure to steroid hormones (Martel and Psychoyos, 1976); the ingestion of phytoestrogens (Tang and Adams, 1978), and many other factors. These are examples of the well-known principle that responses to a hormonal stimulus depend not only on the concentration and characteristics of the hormone, but also on the characteristics of the substrate, or soma, on which the hormone acts (Young, 1969). It should also be apparent that target cell nuclei may retain a hormone and continue to react with it many hours after titers of the hormone have presumably declined to low levels in the peripheral circulation (Walker and Feder, 1977). Thus, even a simplified model of the nuclear mechanism of steroid action conveys the idea that the biological activity of a steroid in a target tissue cannot be predicted from a knowledge of plasma hormone concentrations. When it is realized that a comprehensive model of steroid action is much more complicated than is indicated above, this idea is reinforced. For example, there is evidence of selectivity of steroid movement into the cell membrane (Pietras and Szego, 1977); evidence of multiple cytoplasmic receptors for a single hormone in a given tissue (Eriksson *et al.*, 1978; Smith *et al.*, 1979); and evidence of nonnuclear actions of steroids (Kelly *et al.*, 1977).

If it is obviously erroneous to attempt to predict biological activity from plasma steroid levels, it is also a mistake to assume that concentration of intracellular steroid–receptor complexes is always predictive of biological effects of steroids. Although there are many examples of positive correlations between the steroid–receptor complex concentrations and biological responsiveness (e.g., Moguilewsky and Raynaud, 1979; Blaustein and Feder, 1979), there are other examples of failure to find such correlations. For example, sex differences in levels of estrogen-induced hypothalamic-progestin receptor in guinea pigs do not account for sex differences in their behavioral responsiveness to progesterone in any obvious way (Blaustein *et al.*, 1980). Female-musk-shrew uterus has estrogen receptors but fails to respond to estrogen (Dryden and Anderson, 1977). Factors such as the availability of acceptor sites and the blockade of events after protein synthesis are potentially important considerations in this context.

INTERACTIONS OF STEROIDS WITH OTHER STEROIDS AND OTHER HORMONES. In this review, we have dealt with unconjugated steroids. However, steroids also occur in conjugated forms as sulfates or glucuronides, and these conjugated forms might possess biological activity. This circumstance is another complicating factor in the interpretation of plasma steroid concentrations because conjugated and unconjugated forms of a steroid might enhance or suppress one another's biological activity.

Circulating steroids of one type may enhance or suppress the actions of other types of gonadal steroids. These activities may occur at multiple levels (e.g., actions on SBG, intracellular receptors, or steroid-metabolizing enzymes) and may render a given plasma concentration of one steroid more or less biologically potent than might be expected in the absence of another steroid. In addition to the well-documented interactions between estrogens and progestins in reproductive tissues (Feder and Marrone, 1977), there are also numerous examples of interactions between androgens and estrogens (e.g., DeBold *et al.*, 1978). Several authors have argued that, in certain cases, it is the relative proportion of two steroids, rather than the absolute quantity of either, that is crucial to a biological effect. This point is nicely brought home in an article by Forbes (1968).

The problem of hormone interactions becomes even deeper because steroids from nongonadal sources also influence the biological activity of circulating gonadal steroids (Szego and Roberts, 1953), and because nonsteroidal hormones enhance or dampen the action of steroids. For example, the pituitary protein hormone ACTH synergizes with testosterone to promote male copulatory behavior in rabbits and to promote yawning response (Bertolini *et al.*, 1975). The peptide hormone GnRH synergizes with estradiol to promote the female lordosis responses in rats (Moss and McCann, 1973; Pfaff, 1973). Oxytocin accelerates the onset of estradiol-induced maternal behavior in certain strains of rats (Pedersen and Prange, 1979). Other factors, such as prostaglandins, facilitate steroid-induced lordosis responses in rats (Rodriguez-Sierra and Komisaruk, 1977) and abbreviate such responses in guinea pigs (Marrone *et al.*, 1979).

All of these considerations about hormone interactions must temper any interpretation of plasma steroid data. In fact, when we say that X ng/ml of steroid in plasma is associated with some biological response, what we mean is that X ng/ml steroid *against a background of activity by other hormones* is associated with the response. Failure to keep in mind this proviso about the background against which hormones act may lead to errors not only in the interpretation of data on endogenous steroid levels, but also to overconfidence in the results of experiments in which endocrine glands are ablated and replacement therapy is given.

Physiological Factors That Complicate Interpretation of Plasma Steroid Data

341

PLASMA LEVELS OF
GONADAL
STEROIDS IN
MAMMALS

STRAIN DIFFERENCES. In the previous sections, it was noted that there may be significant differences in plasma steroid levels among strains of a species (e.g., see data on male mice). There is also evidence of strain differences in steroid metabolism (Ford *et al.,* 1979) and of strain differences in responsiveness to the actions of steroid hormones (Goy and Young, 1957; Luttge and Hall, 1973). Given these important differences, it is erroneous to describe steroid levels in a particular strain as though the strain represented the species as a whole (for an excellent discussion, see Shire, 1976). In fact, the study of hormone action would greatly benefit from a closer scrutiny of the relation between genotype and hormones. For example, the fascinating finding that, in some strains of mice male, copulatory behavior does not decline after castration but does decline after androgen replacement therapy (McGill, 1978) may provide an insight into the mechanism of androgen action that goes beyond the consideration of strain differences. This finding may, perhaps, help us to understand why copulatory behavior in certain species of bats increases as androgen levels decline (see the first section of this chapter; I thank Bruce Goldman for drawing my attention to this possibility). In a more general way, such information may lead to an understanding of how the presence of a hormone can promote responses in certain cases, whereas the withdrawal of the hormone may promote the same types of responses in other instances. Another example of the value of work with particular strains is seen in the work of Bartke (1974). He found that male mice of one strain (DBA/2J) had higher plasma testosterone levels than another strain (C57BL/10J). However, in castrates, the sensitivity of the seminal vesicles to exogenous androgenic stimulation was considerably greater in the C57 than in the DBA strain (seminal vesicles weigh about the same in untreated intact subjects of both strains). Bartke argued that increased sensitivity of the seminal vesicles in the C57 animals represents a mechanism of physiological compensation for the abnormally low level of endogenous testosterone in this strain. This same notion of compensatory mechanisms was advanced by Feder *et al.* (1974) in a cross-species analysis. That is, hormonal sensitivity of neural tissues regulating female sexual behavior was inversely related to the plasma concentration of the ovarian hormone in rats, hamsters, and guinea pigs.

SUPRATHRESHOLD SECRETION OF HORMONES. In 1977, Demassa *et al.* reported that subcutaneous implants of testosterone in castrated male rats supported normal ejaculatory behavior even when the implants produced plasma testosterone levels that were $< 10\%$ of those found in untreated, intact subjects. Thus, it is apparent that rats (and presumably other species) *normally* produce more testosterone than is absolutely required for maintenance of ejaculatory behavior. The authors pointed out that correlations are absent between levels of plasma testosterone (once these are adequate) and levels of ejaculatory behavior in rats and other species because the amount of androgen normally produced far exceeds the minimum requirements for expression of this behavior. This "safety factor" is an important item to consider whenever one attempts to correlate a biological response with a plasma hormone level. A related problem is the difference among tissues in their threshold of responsiveness to a hormone. Thus, testosterone levels may be 10 times beyond the level required to activate the tissues mediating ejaculatory behavior, but 30 times beyond that required for another response, and 2 times that required for a third type of response. A table in which are listed plasma the testosterone levels in males of various species may give us an impression that we know something about the relative requirements for testosterone in these species. However, this impression would be a mistaken one because we would be lacking crucial information about the "safety factors" for various responses in each species.

STATUS OF RESPONSIVE TISSUES. Factors such as time of day, season, age, and time since last exposure to hormone influence the responsiveness of tissue to hormones, thereby precluding a simple formulation of relationships between plasma hormone levels and tissue responses. For example, Hansen *et al.* (1979) found that ovariectomized rats exposed to constant plasma levels of estradiol showed a daily rhythm of lordosis behavior, the highest levels of lordosis occurring during the dark portion of the daily light:dark cycle. Photoperiodic regulation of the sensitivity of neural tissues to steroids has been demonstrated in a number of species (Turek and Campbell, 1979), including rams (Lincoln and Davidson, 1977) and Syrian hamsters exposed to short photoperiods (Ellis *et al.*, 1979). In addition, the amount of testosterone required to prevent hibernation in Turkish hamsters varies according to physiological state (Hall and Goldman, 1979). Aging decreases responsiveness to estrogen's negative-feedback effects on LH in female rats (Gray *et al.*, 1980) but either has no effect (Peng *et al.*, 1977) or increases responsivenss to estrogen's actions on lordosis behavior (Cooper, 1977). Beach and Orndoff (1974) presented data indicating that responsiveness to estrogen in female rats decreases after prolonged estrogen deprivation. There is even a claim that responsiveness to the effects of estrogen and progesterone on lordosis behavior is enhanced in rats when the uterus is removed (Siegel *et al.*, 1978). Changes in the responsiveness of target tissues to steroids may be due to a variety of causes, including alternations in intracellular receptors. Whatever the causes, the fact remains that variations in tissue responsiveness complicate attempts to relate plasma hormone levels to biological responses.

ENVIRONMENTAL AND PSYCHOLOGICAL FACTORS THAT COMPLICATE INTERPRETATION OF PLASMA STEROID DATA

SOCIAL OR ENVIRONMENTAL CONTEXT. Individual subjects with similar plasma hormone levels do not necessarily exhibit similar patterns or frequencies of hormone-dependent behaviors. Indeed, the same individual may show different hormone-dependent behavioral patterns and frequencies in the absence of any change in hormone levels. One factor associated with this variability is the situation or context within which the behavior occurs. This point has been demonstrated in several ways. Phoenix (1973) noted that androgen-dependent ejaculatory behavior in individual male rhesus monkeys varied according to the female with which the male was tested, a finding suggesting that "pair compatability may be as important as hormonal stimulation in regulating male sexual performances" (275–277). Zumpe and Michael (1979) made the interesting suggestion that aggression in male rhesus monkeys (presumably at least partially androgen-dependent) may be directed toward a female partner when she is not sexually attractive but may be redirected away from the partner onto the environment when she is sexually attractive. It is doubtful that a change in androgen level in the male would be predictably associated with the direction of the male's aggressive behaviors. Johnson and Phoenix (1978) discussed the possibility that the patterning and degree of proceptive behavior in rhesus monkeys and other primates may depend on the testing situation or on species-specific sexual interactions; for examples, menstrual cycle variations are more likely to occur if the completion of copulation depends on a behavioral decision of the female (e.g., lever pressing to attain access to a male). Brown-Grant (1975) found that steroid-dependent lordosis behavior in perinatally "defeminized" adult rats was strongly influenced by the amount of time the rats were given to acclimate to the test arena. All of these illustrations demonstrate that hormone-dependent processes are greatly influenced by the social or environmental setting in which the process occurs, a finding again precluding easy correlations between plasma hormone levels and hormone-dependent responses. There are even indications, yet to be verified, that social setting may influence target tissue metabolism of plasma hormones (Dessi-Fulgheri *et al.*, 1975).

EXPERIENCE. Thompson and Edwards (1971) suggested that spayed female mice given the experience of sex behavior tests without later hormone replacement therapy showed facilitated responses to the induction of female sex behavior by hormone treatment. Barkley and Goldman (1977) showed that male mice that won or lost an aggressive encounter showed significantly greater sex-organ development than males in encounters in which no fighting occurred. In an older study (Rosenblatt and Aronson, 1958), the rapidity with which postcastration declines in the male sexual behavior of cats occurred depended on the degree of prior sexual experience of the subjects. In some of these instances, it is possible that experiential factors acted via the release of hormones, but in other instances, the experiential alteration of steroid-dependent behaviors may take place through nonhormonal mechanisms. It is conceivable that experiential and hormonal inputs summate in such a way that an animal with appropriate experiential background can "get by" with a smaller amount of hormonal stimulation than an animal with less experience. A summation of this type is seen in the lordosis behavior of rats after estrogen treatment and tactile stimulation (Komisaruk and Diakow, 1973). Because experiential factors may substitute for, or otherwise influence the strength of, hormonal factors, it is again apparent that simple correlations between plasma hormone levels and biological (particularly behavioral) responses are unlikely to be seen.

Mood. The idea that mood state modifies responsiveness to circulating hormones in the absence of a change in the level of these hormones has not been addressed in a systematic way, but it is an idea that deserves study. Nock *et al.* (1981) found that the blockade of noradrenergic transmission in female guinea pigs causes a decrease in hypothalamic responsiveness to estrogen–progesterone treatment, presumably by blocking the induction of intracellular progestin receptors by estrogen. This type of relationship suggests that mood alterations that result in variations in neurotransmission may affect an organism's responsiveness to circulating hormones. This effect would present yet another obstacle to establishing correlations between plasma hormone concentrations and biological responses.

CONCLUDING REMARKS

When one considers the limitations of interpretation that must be placed on plasma-hormone-concentration data, it is almost a shock to encounter the voluminous literature on plasma hormone levels. However, most authors interpret their data with appropriate caution. It is in the area of comparative endocrinology that we appear to risk the greatest likelihood of error and to have the least systematic data. This may seem a strange statement to make after so many references have been made to so many mammalian species. But the fact is that, although comparisons are often made among these species (I have not always avoided this temptation in the present survey), there are no systematic comparative studies because rigorous formulation of the criteria to be used in such comparative studies have not been established. The direction in which we must head, if a truly comparative picture of hormone levels is to be achieved, is outlined in an editorial by Yates (1979) entitled "Comparative Physiology: Compared to What?"

Acknowledgments

The literature search for this manuscript was completed in 1979, and the article was submitted in 1980. I am very grateful to Nancy Roberts for her intelligent assistance with the bibliography. I am also delighted to be able to thank Steve Feder for his excellent assistance with the references. I thank Winona Cunningham and Nancy Jachim for typing. Contribution number 381 of the IAB.

REFERENCES

Abraham, G. E. Ovarian and adrenal contribution to peripheral androgens during the menstrual cycle. *Journal of Clinical Endocrinology and Metabolism*, 1974, *39*, 340–346.

Abraham, G. E., Odell, W. D., Swerdloff, R. S., and Hopper, K. Simultaneous radioimmunoassay of plasma FSH, LH, progesterone, 17-hydroxyprogesterone, and estradiol-17β during the menstrual cycle. *Journal of Clinical Endocrinology*, 1972, *34*, 312–318.

Abraham, G. E., Buster, J. E., Kyle, F. W., Corrales, P. C., and Teller, R. C. Radioimmunoassay of plasma pregnenolone, 17-hydroxypregnenolone and dehydroepiandrosterone under various physical conditions. *Journal of Clinical Endocrinology and Metabolism*, 1973, *37*, 140–144.

Adler, N. T. On the mechanisms of sexual behaviour and their evolutionary constraints. In J. B. Hutchison (Ed.), *Biological Determinants of Sexual Behaviour*. New York: Wiley, 1978.

Aedo, A.-R., Landgren, B.-M., Cekan, Z., and Diczfalusy, E. Studies on the pattern of circulating steroids in the normal menstrual cycle. *Acta Endocrinologica*, 1976, *82*, 600–616.

Albertson, B. D., Bradley, E. L., and Terman, C. R. Plasma progesterone concentrations in prairie deermice *(Peromyscus maniculatus bairdii)* from experimental laboratory populations. *Journal of Reproduction and Fertility*, 1975, *42*, 407–414.

Alford, F. P., Baker, H. W. G., Burger, H. G., de Kretser, D. M., Hudson, B., Johns, M. W., Masterton, J. P., Patel, Y. C., and Rennie, G. C. Temporal patterns of integrated plasma hormone levels during sleep and wakefulness: II. Follicle-stimulating hormone, luteinizing hormone, testosterone and estradiol. *Journal of Clinical Endocrinology and Metabolism*, 1973, *37*, 848–854.

Allison, A. J., and McNatty, K. P. Levels of progesterone in peripheral blood plasma of Romney and Merino ewes during the oestrous cycle. *New Zealand Journal of Agricultural Research*, 1972, *15*, 825–830.

Amatayakul, K., Ryan, R., Uozumi, T., and Albert, A. A reinvestigation of testicular-anterior pituitary relationships in the rat: I. Effects of castration and cryptorchidism. *Endocrinology*, 1971, *88*, 872–880.

Andresen, Ø. Concentrations of fat and plasma 5α-androstenone and plasma testosterone in boars selected for rate of body weight gain and thickness of back fat during growth, sexual maturation and after mating. *Journal of Reproduction and Fertility*, 1976, *48*, 51–59.

Angsusingha, K., Kenny, F. M., Nankin, H. R., and Taylor, F. H. Unconjugated estrone, estradiol and FSH and LH in prepubertal and pubertal males and females. *Journal of Clinical Endocrinology and Metabolism*, 1974, *39*, 63–68.

Apter, D., and Vihko, R. Serum pregnenolone, progesterone, 17-hydroxyprogesterone, testosterone and 5α-dihydrotestosterone during female puberty. *Journal of Clinical Endocrinology and Metabolism*, 1977, *45*, 1039–1048.

Apter, D., Jänne, O., Karvonen, P., and Vihko, R. Simultaneous determination of five sex hormones in human serum by radioimmunoassay after chromatography of Lipidex-5000. *Clinical Chemistry*, 1976, *22*, 32–38.

Apter, D., Viinikka, K., and Vihko, R. Hormonal pattern of adolescent menstrual cycles. *Journal of Clinical Endocrinology and Metabolism*, 1978, *47*, 944–954.

Aron, C. Mechanisms of control of the reproductive function by olfactory stimuli in female mammals. *Physiological Reviews*, 1979, *59*, 129–284.

Asa, C. S., Goldfoot, D. A., and Ginther, O. J. Sociosexual behavior and the ovulatory cycle of ponies *(Equus caballus)* observed in harem groups. *Hormones and Behavior*, 1979, *13*, 49–65.

Asa, C. S., Goldfoot, D. A., Garcia, M. C., and Ginther, O. J. Dexamethasone suppression of sexual behavior in the ovariectomized mare. *Hormones and Behavior*, 1980a, *14*, 55–64.

Asa, C. S., Goldfoot, D. A., Garcia, M. C., and Ginther, O. J. Sexual behavior in ovariectomized and seasonally anovulatory pony mares *(Equus caballus)*. *Hormones and Behavior*, 1980b, *14*, 46–54.

Asdell, S. A. *Patterns of Mammalian Reproduction*. Ithaca, N.Y.: Cornell University Press, 1964.

Aso, T., Tominaga, T., Oshima, K., and Matsubayashi, K. Seasonal changes of plasma estradiol and progesterone in the Japanese monkey *(Macaca fuscata fuscata)*. *Endocrinology*, 1977, *100*, 745–750.

Austad, R., Lunde, A., and Sjaastad, Ø. V. Peripheral plasma levels of oestradiol-17β and progesterone in the bitch during the oestrous cycle, in normal pregnancy and after dexamethasone treatment. *Journal of Reproduction and Fertility*, 1976, *46*, 129–136.

Ayalon, N., and Shemesh, M. Pro-oestrous surge in plasma progesterone in the cow. *Journal of Reproduction and Fertility*, 1974, *36*, 239–243.

Badr, R. M., and Bartke, A. Effect of ethyl alcohol on plasma testosterone level in mice. *Steroids*, 1974, *23*, 921–928.

Baird, D. T. Pulsatile secretion of LH and ovarian estradiol during the follicular phase of the sheep estrous cycle. *Biology of Reproduction*, 1978, *18*, 359–364.

Baird, D. T., and Fraser, I. S. Blood production and ovarian secretion rates of estradiol-17β and estrone in women throughout the menstrual cycle. *Journal of Clinical Endocrinology and Metabolism*, 1974, *38*, 1009–1017.

Baird, D. T., and Guevara, A. Concentration of unconjugated estrone and estradiol in peripheral plasma in nonpregnant women throughout the menstrual cycle, castrate and postmenopausal women and in men. *Journal of Clinical Endocrinology and Metabolism*, 1969, *29*, 149–156.

Baird, D. T., Land, R. B., Scaramuzzi, R. J., and Wheeler, A. G. Endocrine changes associated with luteal regression in the ewe: The secretion of ovarian oestradiol, progesterone and androstenedione and uterine prostaglandin $F_{2\alpha}$ throughout the oestrous cycle. *Journal of Endocrinology*, 1976a, *69*, 275–286.

Baird, D. T., Swanston, I., and Scaramuzzi, R. J. Pulsatile release of LH and secretion of ovarian steroids in sheep during the luteal phase of the estrous cycle. *Endocrinology*, 1976b, *98*, 1490–1496.

Bangham, C. R. M., and Hackett, P. H. Effects of high altitude on endocrine function in the Sherpas of Nepal. *Journal of Endocrinology*, 1978, *79*, 147–148.

Baranczuk, R., and Greenwald, G. S. Peripheral levels of estrogen in the cyclic hamster. *Endocrinology*, 1973, *92*, 805–812.

Baranowska, B., and Zgliczyński, S. Enhanced testosterone in female patients with anorexia nervosa: Its normalization after weight gain. *Acta Endocrinologica*, 1979, *90*, 328–335.

Barberia, J. M., and Thorneycroft, I. H. Simultaneous radioimmunoassay of testosterone and dihydro-testosterone. *Steroids*, 1974, *23*, 757–766.

Bardin, C. W., and Peterson, R. E. Studies of androgen production by the rat: Testosterone and androstenedione content of the blood. *Endocrinology*, 1967, *80*, 38–44.

Barkley, M. S., and Goldman, B. D. A quantitative study of serum testosterone, sex accessory organ growth, and the development of intermale aggression in the mouse. *Hormones and Behavior*, 1977, *8*, 208–218.

Barraclough, C. A., Collu, R., Massa, R., and Martini, L. Temporal interrelationships between plasma LH, ovarian secretion rates and peripheral plasma progestin concentrations in the rat: Effects of nembutal and exogenous gonadotropins. *Endocrinology*, 1971, *88*, 1437–1447.

Barrell, G. K., and Lapwood, K. R. Effects of pinealectomy on the secretion of luteinizing hormone, testosterone and prolactin in rams exposed to various lighting régimes. *Journal of Endocrinology*, 1979, *80*, 397–405.

Bartke, A. Increased sensitivity of seminal vesicles to testosterone in a mouse strain with low plasma testosterone levels. *Journal of Endocrinology*, 1974, *60*, 145–148.

Bartke, A., and Dalterio, S. Evidence for episodic secretion of testosterone in laboratory mice. *Steroids*, 1975, *26*, 749–756.

Bartke, A., Steele, R. E., Musto, N., and Caldwell, B. V. Fluctuations in plasma testosterone levels in adult male rats and mice. *Endocrinology*, 1973, *92*, 1223–1228.

Bartke, A., Croft, B. T., and Dalterio, S. Prolactin restores plasma testosterone levels and stimulates testicular growth in hamsters exposed to short day-length. *Endocrinology*, 1975, *97*, 1601–1604.

Bartke, A., Smith, M. S., Michael, S. D., Peron, F. G., and Dalterio, S. Effects of experimentally-induced chronic hyperprolactinemia on testosterone and gonadotropin levels in male rats and mice. *Endocrinology*, 1977, *100*, 182–186.

Beach, F. A., and Orndoff, R. K. Variation in the responsiveness of female rats to ovarian hormones as a function of preceding hormonal deprivation. *Hormones and Behavior*, 1974, *5*, 201–205.

Beattie, C. W., and Bullock, B. C. Diurnal variation of serum androgen and estradiol-17β in the adult male green monkey (*Ceropithecus* sp.). *Biology of Reproduction*, 1978, *19*, 36–39.

Beck, W., and Wuttke, W. Annual rhythms of luteinizing hormone, follicle-stimulating hormone, prolactin and testosterone in the serum of male rhesus monkeys. *Journal of Endocrinology*, 1979, *83*, 131–139.

Bell, E. T., Christie, D. W., and Younglai, E. V. Plasma oestrogen levels during the canine oestrous cycle. *Journal of Endocrinology*, 1971, *51*, 225–226.

Berger, M., Chazaud, J., Jean-Faucher, Ch., De Turckheim, M., Veyssiere, G., and Jean, Cl. Developmental patterns of plasma and testicular testosterone in rabbits from birth to 90 days of age. *Biology of Reproduction*, 1976, *15*, 561–564.

Berger, M., Carre, M., Jean-Faucher, Ch., De Turckhrim, M., Veyssiere, G., and Jean, Cl. Changes in the testosterone to dihydrotestosterone ratio in plasma and testes of maturing rabbits. *Endocrinology*, 1979, *104*, 1450–1454.

Bergon, L., Gallant, S., and Brownie, A. C. Serum 11-deoxycorticosterone levels in adrenal-regeneration hypertension under conditions of quiescence and stress. *Steroids*, 1975, *25*, 323–342.

Berndtson, W. E., and Desjardins, C. Circulating LH and FSH levels and testicular function in hamsters during light deprivation and subsequent photoperiodic stimulation. *Endocrinology*, 1974, *95*, 195–205.

Berndtson, W. E., Pickett, B. W., and Nett, T. M. Reproductive physiology of the stallion: IV. Seasonal changes in the testosterone concentration of peripheral plasma. *Journal of Reproduction and Fertility,* 1974, *39,* 115–118.

Bertolini, A., Gessa, G. L., and Ferrari, W. Penile erection and ejaculation: A central effect of ACTH-like peptides in mammals. In M. Sandler and G. L. Gessa (Eds.), *Sexual Behavior: Pharmacology and Biochemistry.* New York: Raven Press, 1975.

Blaha, G. C., and Leavitt, W. W. Ovarian steroid dehydrogenase histochemistry and circulating progesterone in aged golden hamsters during the estrous cycle and pregnancy. *Biology of Reproduction,* 1974, *11,* 153–161.

Blatchley, F. R., and Donovan, B. T. Peripheral plasma progestin levels during anoestrus, oestrus and pseudopregnancy and following hypophysectomy in ferrets. *Journal of Reproduction and Fertility,* 1972, *31,* 331–333.

Blatchley, F. R., Donovan, B. T., and Ter Haar, M. B. Plasma progesterone and gonadotrophin levels during the estrous cycle of the guinea pig. *Biology of Reproduction,* 1976, *15,* 29–38.

Blaustein, J. D., and Feder, H. H. Cytoplasmic progestin-receptors in guinea pig brain: Characteristics and relationship to the induction of sexual behavior. *Brain Research,* 1979, *169,* 481–497.

Blaustein, J. D., Ryer, H. I., and Feder, H. H. A sex difference in the progestin receptor system of guinea pig brain. *Neuroendocrinology,* 1980, *31,* 403–409.

Bonnin, M., Canivenc, R., and Ribes, Cl. Plasma progesterone levels during delayed implantation in European badger *(Meles meles). Journal of Reproduction and Fertility,* 1978, *52,* 55–58.

Boon, D. A., Keenan, R. E., and Slaunwhite, W. R., Jr. Plasma testosterone in men: Variation but not circadian rhythm. *Steroids,* 1973, *20,* 269–278.

Bosu, W. T. K., Holmdahl, T. H., Johansson, E. D. B., and Gemzell, C. Peripheral plasma levels of oestrogens, progesterone and 17α-hydroxyprogesterone during the menstrual cycle of the rhesus monkey. *Acta Endocrinologica,* 1972, *71,* 755–764.

Bosu, W. T. K., Johansson, E. D. B., and Gemzell, C. Peripheral plasma levels of oestrone, oestradiol-17β and progesterone during ovulatory menstrual cycles in the rhesus monkey with special reference to the onset of menstruation. *Acta Endocrinologica,* 1973, *74,* 732–742.

Boyar, R. M., Finkelstein, J. W., Roffwarg, H. P., Kapen, S., Weitzman, E. D., and Hellman, L. Synchronization of augmented luteinizing hormone secretion with sleep during puberty. *New England Journal of Medicine,* 1972, *287,* 582–586.

Boyar, R. M., Wu, R. H. K., Roffwarg, H., Kapen, S., Weitzman, E. D., Hellman, L., and Finkelstein, J. W. Human puberty: 24-hour estradiol patterns in pubertal girls. *Journal of Clinical Endocrinology and Metabolism,* 1976, *43,* 1418–1421.

Boyd, P. A., and Spelsberg, T. C. Seasonal changes in the molecular species and nuclear binding of the chick oviduct progesterone receptor. *Biochemistry,* 1979, *18,* 3685–3690.

Brain, P. C., and Peddie, M. J. Plasma testosterone levels in serial samples from immature male guinea-pigs. *Journal of Endocrinology,* 1979, *80,* 35p.

Briggs, M. H., Garcia-Webb, P., and Cheung, T. Androgen and exercise. *British Medical Journal,* 1973, *3,* 49–50.

Brock, L. W., and Wettemann, R. P. Variations in serum testosterone in boars. *Journal of Animal Science,* 1979, *42,* 244–245.

Bronson, F. H., and Desjardins, C. Reproductive failure in aged CBF male mice: Interrelationships between pituitary gonadotropic hormones, testicular function, and mating success. *Endocrinology,* 1977, *101,* 939–945.

Brown, G. P., Courtney, G. A., and Marotta, S. F. Progesterone secretion of adrenal glands of hamsters and comparison of ACTH influence in rats and hamsters. *Steroids,* 1976, *28,* 275–282.

Brown-Grant, K. A re-examination of the lordosis response in female rats given high doses of testosterone propionate or estradiol benzoate in the neonatal period. *Hormones and Behavior,* 1975, *6,* 351–378.

Brown-Grant, K., Exley, D., and Naftolin, F. Peripheral plasma oestradiol and luteinizing hormone concentrations during the oestrous cycle of the rat. *Journal of Endocrinology,* 1970, *48,* 295–296.

Buhl, A. E., Hasler, J. F., Tyler, M. C., Goldberg, N., and Banks, E. M. The effects of social rank on reproductive indices in groups of male collared lemmings *(Dicrostonyx groenlandicus). Biology of Reproduction,* 1978, *18,* 317–324.

Bullock, D. W., Paris, C. A., and Goy, R. W. Sexual behaviour, swelling of the sex skin and plasma progesterone in the pigtail macaque. *Journal of Reproduction and Fertility,* 1972, *31,* 225–236.

Burke, C. W., and Anderson, D. C. Sex-hormone-binding gloculin is an oestrogen amplifier. *Nature*, 1972, *240*, 38.

Burns, J. M., and Easley, R. G. Hormonal control of delayed development in the California leaf-nosed bat, *Macrotus californicus*. *General and Comparative Endocrinology*, 1977, *32*, 163–166.

Butcher, R. L., Collins, W. E., and Fugo, N. W. Plasma concentration of LH, FSH, prolactin, progesterone and estradiol-17β throughout the 4-day estrous cycle of the rat. *Endocrinology*, 1974, *94*, 1704–1708.

Cake, M. H., Owen, F. J., and Bradshaw, S. D. Difference in concentration of progesterone in plasma between pregnant and non-pregnant quokkas *(Setonix brachyurus)*. *Journal of Endocrinology*, 1980, *84*, 153–159.

Campbell, C. S., Ryan, K. D., and Schwartz, N. B. Estrous cycles in the mouse: Relative influence of continuous light and the presence of a male. *Biology of Reproduction*, 1976, *14*, 292–299.

Cargille, C. M., Ross, G. T., and Yoshimi, T. Daily variations in plasma follicle stimulating hormone, luteinizing hormone and progesterone in the normal menstrual cycle. *Journal of Clinical Endocrinology*, 1969, *29*, 12–19.

Carstensen, H., Amér, B., Amér, I., and Wide, L. The postoperative decrease of plasma testosterone in man, after major surgery, in relation to plasma FSH and LH. *Journal of Steroid Biochemistry*, 1973, *4*, 45–55.

Challis, J. R., Heap, R. B., and Illingworth, D. V. Concentration of oestrogen and progesterone in the plasma of non-pregnant, pregnant and lactating guinea pigs. *Journal of Endocrinology*, 1971, *51*, 333–345.

Chan, S. W. C., Leathem, J. H., and Esashi, T. Testicular metabolism and serum testosterone in aging male rats. *Endocrinology*, 1977, *101*, 128–133.

Christensen, D. S., Wiltbank, J. N., and Hopwood, M. L. Blood hormone levels during the bovine estrous cycle. *Journal of Animal Science*, 1971, *33*, 251.

Christian, J. J. Population density and reproductive efficiency. *Biology of Reproduction*, 1971, *4*, 248–294.

Clark, J. R., Dierschke, D. J., and Wolf, R. C. Hormonal regulation of ovarian folliculogenesis in rhesus monkeys: I. Concentrations of serum luteinizing hormone and progesterone during laparoscopy and patterns of follicular development during successive menstrual cycles. *Biology of Reproduction*, 1978, *17*, 779–783.

Claus, R., and Giménez, T. Diurnal rhythm of 5α-androst-16-en-3-one and testosterone in peripheral plasma of boars. *Acta Endocrinologica*, 1977, *84*, 200–206.

Colenbrander, B., de Jong, F. H., and Wensing, C. J. G. Changes in serum testosterone concentrations in the male pig during development. *Journal of Reproduction and Fertility*, 1978, *53*, 377–380.

Comhaire, F., Mattheeuws, D., and Vermeulen, A. Testosterone and oestradiol in dogs with testicular tumours. *Acta Endocrinologica*, 1974, *77*, 408–416.

Concannon, P. W., Hansel, W., and Visek, W. J. The ovarian cycle of the bitch: Plasma estrogen, LH and progesterone. *Biology of Reproduction*, 1975, *13*, 112–121.

Concannon, P., Hansel, W., and McEntee, K. Changes in LH, progesterone and sexual behavior associated with preovulatory luteinization in the bitch. *Biology of Reproduction*, 1977, *17*, 604–613.

Cooper, R. L. Sexual receptivity in aged female rats. Behavioral evidence for increased sensitivity to estrogen. *Hormones and Behavior*, 1977, *9*, 321–333.

Cooper, T. G., and Waites, G. M. H. Testosterone in rete testis fluid and blood of rams and rats. *Journal of Endocrinology*, 1974, *62*, 619–629.

Coquelin, A., and Bronson, F. H. Release of luteinizing hormone in male mice during exposure to females: Habituation of the response. *Science*, 1979, *206*, 1099–1101.

Corker, C. S., Exley, D., and Naftolin, F. Assay of 17 beta-oestradiol by competitive protein binding methods. *Acta Endocrinologica*, 1970, *Supplement 147*, 305.

Corvol, P. L., Chrambach, A., Rodbard, D., and Bardin, C. W. Physical properties and binding capacity of testosterone-estradiol-finding globulin in human plasma, determined by polyacrylamide gel electrophoresis. *Journal of Biological Chemistry*, 1971, *246*, 3435–3443.

Courrier, R. Etude sur le déterminisme des caractères sexuels secondaires chez quelques mammifères à activité testiculaire périodique. *Archives of Biology*, 1927, *37*, 173–334.

Coyotupa, J., Parlow, A. F., and Abraham, G. E. Simultaneous radioimmunoassay of plasma testosterone and dihydrotestosterone. *Analytical Letters*, 1972, *5*, 329.

Coyotupa, J., Parlow, A. F., and Kovacic, N. Serum testosterone and dihydrotestosterone levels following orchiectomy in the adult rat. *Endocrinology*, 1973, *92*, 1579–1581.

Craighead, J. J., Hornocker, M. G., and Craighead, F. C., Jr. Reproductive biology of young female grizzly bears. *Journal of Reproduction and Fertility*, 1969, *Supplement 6*, 447–475.

Croix, D., and Franchimont, P. Changes in the serum levels of the gonadotrophins, progesterone and estradiol during the estrous cycle of the guinea pig. *Neuroendocrinology*, 1975, *19*, 1–11.

Cunningham, N. F., Symons, A. M., and Saba, N. Levels of progesterone LH and FSH in the plasma of sheep during the oestrous cycle. *Journal of Reproduction and Fertility*, 1975, *45*, 177–180.

Czaja, J. A., Robinson, J. A., Eisele, S. G., Scheffler, G., and Goy, R. W. Relationship between sexual skin colour of female rhesus monkeys and midcycle plasma levels of oestradiol and progesterone. *Journal of Reproduction and Fertility*, 1977, *49*, 147–150.

Damassa, D. A., Smith, E. R., Tennent, B., and Davidson, J. M. The relationship between circulating testosterone levels and male sexual behavior in rats. *Hormones and Behavior*, 1977, *8*, 275–286.

Danzo, B. J., and Eller, B. C. Steroid-binding proteins in rabbit plasma: Separation of testosterone-binding globulin (TeBG) from corticosteroid-binding globulin (CBG), preliminary characterization of TeBG, anc changes in TeBG concentration during sexual maturation. *Molecular and Cellular Endocrinology*, 1975, *2*, 351–368.

Davies, R. V., Main, S. J., and Setchell, B. P. Seasonal changes in plasma follicle-stimulating hormone, luteinizing hormone and testosterone in rams. *Journal of Endocrinology*, 1977, *72*, 12P.

DeBold, J. F., Ruppert, P. H., and Clemens, L. G. Inhibition of estrogen-induced sexual receptivity of female hormone: Comparative effects of progesterone, dihydrotestosterone and an estrogen antagonist. *Pharmacology, Biochemistry and Behavior*, 1978, *9*, 81–86.

Demisch, K., Magnet, W., Neubauer, M., Ehlers, E., and Schoffling, K. Unconjugated 5-androstene-3alpha, 17beta-diol in human plasma. *Acta Endocrinologica (Suppl.) (Kbh.)*, 1973, *173*, 113.

DeMoor, P., Steeno, O., Heyns, W., and Van Baelen, H. Steroid binding beta-globulin in plasma: Pathophysiological data. *Annales d'Endocrinologie*, 1969, *30*, 233.

DePalatis, L., Moore, J., and Falvo, R. E. Plasma concentrations of testosterone and LH in the male dog. *Journal of Reproduction and Fertility*, 1978, *52*, 201–207.

Dessi-Fulgheri, F., Lupo di Prisco, C., and Verdarelli, P. Influence of long-term isolation on the production and metabolism of gonadal sex steroids in male and female rats. *Physiology and Behavior*, 1975, *14*, 495–499.

Dessypris, A., Kuoppasalmi, K., and Adlercreutz, H. Plasma cortisol, testosterone, androstenedione and luteinizing hormone (LH) in a non-competitive marathon run. *Journal of Steroid Biochemistry*, 1976, *7*, 33–37.

Dierschke, D. J., Walsh, S. W., Mapletoft, R. J., Robinson, J. A., and Ginther, O. J. Functional anatomy of the testicular vascular pedicle in the rhesus monkey: Evidence for the local testosterone concentrating mechanism. *Proceedings for the Society of Experimental Biology and Medicine*, 1975, *148*, 236–242.

Dighe, K. K., and Hunter, W. M. A solid-phase radioimmunoassay for plasma progesterone. *Biochemical Journal*, 1974, *143*, 219–231.

Dixson, A. F., Phoenix, C. H., and Resko, J. A. Effects of sexual-behavior and electroejaculation on levels of testosterone, luteinizing-hormone and cortisol in circulation of male rhesus monkeys. *Journal of Endocrinology*, 1976, *71*, 100P (abst.).

Dobson, H. Plasma gonadotrophins and oestradiol during oestrus in the cow. *Journal of Reproduction and Fertility*, 1978, *52*, 51–53.

Dobson, H., and Dean, P. D. G. Radioimmunoassay of oestrone, oestradiol-17α and -17β in bovine plasma during the oestrous cycle and the last stages of pregnancy. *Journal of Endocrinology*, 1974, *61*, 479–486.

Dobson, H., Cooper, M. J., and Furr, B. J. A. Synchronization of oestrus with I.C.I. 79,939, an analogue of $PGE_{2\alpha}$, and associated changes in plasma progesterone, oestradiol-17β and LH in heifers. *Journal of Reproduction and Fertility*, 1975, *42*, 141–144.

Doering, C. H., Kraemer, H. C., Brodie, H. K. H., and Hamburg, D. A. A cycle of plasma testosterone in the human male. *Journal of Clinical Endocrinology and Metabolism*, 1975, *40*, 492–500.

Donovan, B. T., and Lockhart, A. N. Light and the timing of ovulation in the guinea-pig. *Journal of Reproduction and Fertility*, 1972, *30*, 207–211.

Dray, F., Reinberg, A., and Sebaoun, J. Rythme biologique de la testostérone libre du plasma chez l'homme adulte sain: Existence d'une variation circadienne. *Comptes Rendus·Hebdomadaires des Séances de l'Académie des Sciences Paris* (Série D), 1965, *261*, 573–576.

Dryden, G. L., and Anderson, J. N. Ovarian hormone: Lack of effect on reproductive structures of female Asian musk shrews. *Science*, 1977, *197*, 782–783.

Dupon, C., and Kim, M. H. Peripheral plasma levels of testosterone, androstenedione and oestradiol during the rat oestrous cycle. *Journal of Endocrinology*, 1973, *59*, 653–654.

Dupon, C., Hosseinian, A., and Kim, M.H. Simultaneous determination of plasma estrogens, androgens, and progesterone during the human menstrual cycle. *Steroids*, 1973, *22*, 47–61.

Dziuk, P. J. Reproduction in pigs. In H. H. Cole and P. T. Cupps (Eds.), *Reproduction in Domestic Animals* (3rd ed.), New York: Academic Press, 1977.

Eaton, G. G., and Resko, J. A. Plasma testosterone and male dominance in a Japanese macque *(Macaca fuscata)* troop compared with repeated measures of testosterone in laboratory males. *Hormones and Behavior*, 1974, *5*, 251–259.

Eaton, G. G., Slob, A., and Resko, J. A. Cycles of mating behaviour, oestrogen and progesterone in the thick-tailed bushbaby *(Galago crassicaudatus crassicaudatus)* under laboratory conditiions. *Animal Behavior*, 1973, *21*, 309–315.

Eaton, L. W., Jr., and Hilliard, J. Estradiol-17β, progesterone and 20α-hydroxypregn-4-en-3-one in rabbit ovarian venous plasma: I. Steroid secretion from paired ovaries with and without corpora lutea; effect of LH. *Endocrinology*, 1971, *89*, 105–111.

Echternkamp, S. E., and Hansel, W. Plasma estrogens, luteinizing hormone, and corticoid in postpartum cows. *Journal of Dairy Science*, 1971, *54*, 800.

Edqvist, L-.E., and Lamm, A. M. Progesterone levels in plasma during the oestrous cycle of the sow measured by a rapid competitive protein binding technique. *Journal of Reproduction and Fertility*, 1971, *25*, 447–449.

Edqvist, L-.E., Johansson, E. D. B., Kasström, H., Olsson, S.-E., and Richkind, M. Blood plasma levels of progesterone and oestradiol in the dog during the oestrous cycle and pregnancy. *Acta Endocrinologia (Kbh.)*, 1975, *78*, 554–564.

Eiler, H., and Nalbandov, A. V. Sex steroids in follicular fluid and blood plasma during the estrous cycle of pigs. *Endocrinology*, 1977, *100*, 331–338.

Eleftheriou, B. E., and Lucas, L. A. Age-related changes in testes, seminal vesicles, and plasma testosterone levels in male mice. *Gerontologia*, 1974, *20*, 231–238.

Ellendorff, F., Parvizi, N., Pomerantz, D. K., Hartjen, A., König, A., Smidt, D., and Elsaesser, F. Plasma luteinizing hormone and testosterone in the adult male pig: 24 hour fluctuations and the effect of copulation. *Journal of Endocrinology*, 1975, *67*, 403–410.

Ellis, G. B., Losee, S. H., and Turek, F. W. Prolonged exposure of castrated male hamsters to a nonstimulatory photoperiod: Spontaneous change in sensitivity of the hypothalamic-pituitary axis to testosterone feedback. *Endocrinology*, 1979, *104*, 631–635.

Elsaesser, F., Pomerantz, D. K., Ellendorff, F., Kreikenbaum, K., and König, A. Plasma LH, testosterone and DHT in the pig from birth to sexual maturity. *Acta Endocrinologica (Kbh.)*, 1973, *Suppl. 173*, 148 (abstr.).

Eriksson, H., Upchurch, S. Hardin, J. W., Peck, E. J., Jr., and Clark, J. H. Heterogeneity of estrogen receptors in the cytosol and nuclear fractions of the rat uterus. *Biochemical and Biophysical Research Communications*, 1978, *81*, 1–7.

Evans, C. S., and Goy, R. W. Social behaviour and reproductive cycles in captive ring-tailed lemurs *(Lemur catta.) Journal of Zoology*, 1968, *156*, 181–197.

Evans, J. I., MacLean, A. W., Ismail, A. A. A., and Love, D. Concentrations of plasma testosterone in normal men during sleep. *Nature*, 1971, *229*, 261–262.

Faiman, C., and Winter, J. S. D. Diurnal cycles in plasma FSH, testosterone and cortisol in men. *Journal of Clinical Endocrinology and Metabolism*, 1971, *33*, 186–192.

Falvo, R. E., and Nalbandov, A. V. Radioimmunoassay of peripheral plasma testosterone in males from eight species using a specific antibody without chromatography. *Endocrinology*, 1974, *95*, 1466–1468.

Falvo, R. E., Buhl, A., and Nalbandov, A. V. Testosterone concentrations in the peripheral plasma of androgenized female rats and in the estrous cycle of normal female rats. *Endocrinology*, 1974, *95*, 26–29.

Falvo, R. E., Buhl, A. E., Reimers, T. J., Foxcroft, G. R., Dunn, M. H., and Dziuk, P. J. Diurnal fluctuations of testosterone and LH in the ram: Effect of HCG and gonadotrophin-releasing hormone. *Journal of Reproduction and Fertility*, 1975, *42*, 503–510.

Feder, H. H., and Marrone, B. L. Progesterone: Its role in the central nervous systemic as a facilitator and inhibitor of sexual behavior and gonadotropin release. *Annals of the New York Academy of Science*, 1977, *286*, 331–354.

Feder, H. H., Resko, J. A., and Goy, R. W. Progesterone concentrations in the arterial plasma of guinea-pigs during the oestrous cycle. *Journal of Endocrinology*, 1968a, *40*, 505–513.

Feder, H. H., Resko, J. A., and Goy, R. W. Progesterone levels in the arterial plasma of preovulatory and ovariectomized rats. *Journal of Endocrinology*, 1968b, *41*, 563–569.

Feder, H. H., Siegel, H. I., and Wade, G. N. Uptake of 6,7-^{3}H estradiol-17β in ovariectomized rats, guinea pigs and hamsters: Correlation with species differences in behavioral responsiveness to estradiol. *Brain Research*, 1974, *71*, 93–103.

Feder, H. H., Landau, I. T., and Walker, W. A. Anatomical and biochemical substrates of the actions of estrogens and antiestrogens on brain tissues that regulate female sex behavior of rodents. In C. Beyer (Ed.), *Endocrine Control of Sexual Behavior*. New York: Raven Press, 1979.

Fiorelli, G., Borrelli, D., Forti, G., Gonnelli, P., Razzagli, M., and Serio, M. Simultaneous determination of androstenedione testosterone and 5α-dihydrotestosterone in human spermatic and peripheral venous plasma. *Journal of Steroid Biochemistry*, 1976, *7*, 113–116.

Fisher, H. D., and Harrison, R. J. Reproduction in the common porpoise *(Phocoena phocoena)* of the North Atlantic. *Journal of Zoology*, London, 1970, *161*, 471–486.

Fletcher, T. J. The induction of male sexual behavior in red deer *(Cervus elaphus)* by the administration of testosterone to hinds and estradiol-17β to stags. *Hormones and Behavior*, 1978, *11*, 74–88.

Floody, O. R., Walsh, C., and Flanagan, M. T. Testosterone stimulates ultrasound production by male hamsters. *Hormones and Behavior*, 1979, *12*, 164–171.

Folman, Y., Haltmeyer, G. C., and Eik-Nes, K. B. Production and secretion of 5α-dihydrotestosterone by the dog testis. *American Journal of Physiology*, 1972, *222*, 653–656.

Forbes, T. R. Synergisms and antagonisms of testosterone-progesterone and estradiol-17β-estrone-progesterone mixtures in a uterine bioassay in the mouse. *Endocrinology*, 1968, *83*, 411–413.

Ford, H. C., Lee, E., and Engel, L. L. Circannual variation and genetic regulation of hepatic testosterone hydroxylase activities in inbred strains of mice. *Endocrinology*, 1979, *104*, 857–861.

Foster, D. L. Luteinizing hormone and progesterone secretion during sexual maturation of the rhesus monkey: Short luteal phases during the initial menstrual cycles. *Biology of Reproduction*, 1977, *17*, 584–590.

Foster, D. L., Mickelson, I. H., Ryan, K. D., Coon, G. A., Drongowski, R. A., and Holt, J. A. Ontogeny of pulsatile luteinizing hormone and testosterone secretion in male lambs. *Endocrinology*, 1978, *102*, 1137–1146.

Fox, C. A., Ismail, A. A. A., Love, D. N., Kirkham, K. E., and Loraine, J. A. Studies on the relationship between plasma testosterone levels and human sexual activity. *Journal of Endocrinology*, 1972, *52*, 51–58.

Frank, A. H., and Fraps, R. M. Induction of estrus in the ovariectomized golden hamster. *Endocrinology*, 1945, *37*, 357–361.

Frankel, A. I., Mock, E. J., Wright, W. W., and Kamel, F. Characterization and physiological validation of a radioimmunoassay. *Steroids*, 1975, *25*, 73–98.

Free, M. J., Jaffe, R. A., Jain, S. K., and Gomes, W. R. Testosterone concentrating mechanism in the reproductive organs of the male rat. *Nature*, 1973, *244*, 24–26.

Frölich, M., Brand, E. C., and van Hall, E. V. Serum levels of unconjugated aetiocholanolone, androstenedione, testosterone, dehydroepiandrosterone, aldosterone, progesterone and oestrogens during the normal menstrual cycle. *Acta Endocrinologica*, 1976, *81*, 548–562.

Ganjam, V. K., and Kenney, R. M. Androgens and estrogens in normal and cryptorchid stallions. *Journal of Reproduction and Fertility*, 1975, *Supplement 23*, 67–73.

Garris, D. R., and Mitchell, J. A. Intrauterine oxygen tension during the estrous cycle in the guinea pig: Its relation to uterine blood volume and plasma estrogen and progesterone levels. *Biology of Reproduction*, 1979, *21*, 149–159.

Gartrell, N. K., Loriaux, D. L., and Chase, T. N. Plasma testosterone in homosexual and heterosexual women. *American Journal of Psychiatry*, 1977, *134*, 1117–1119.

Garverick, H. A., Erb, R. E., Niswender, G. D., and Callahan, C. J. Reproductive steroids in the bovine: III. Changes during the estrous cycle. *Journal of Animal Science*, 1971, *32*, 946–956.

Gay, V. L., and Tomacari, R. L. Follicle-stimulating hormone secretion in the female rat: Cyclic release is dependent on circulating androgen. *Science*, 1974, *184*, 75–77.

Ghanadian, R., Lewis, J. G., and Chisholm, G. D. Serum testosterone and dihydrotestosterone changes with age in rats. *Steroids*, 1975, *25*, 753–762.

Gimémez, T., Barth, D., Hoffman, B., and Karg, H. Blood levels of testosterone in the roe deer *(Capreolus capreolus)* in relationship to the season. *Acta Endocrinologica (Kbh.)*, 1975, *Suppl. 193*, 59 (abstr.).

Glass, A. R., Swerdloff, R. S., Bray, G. A., Dahms, W. T., and Atkinson, R. L. Low serum testosterone and sex-hormone-binding-globulin in massively obese men. *Journal of Clinical Endocrinology and Metabolism,* 1977, *45,* 1211–1219.

Gleeson, A. R., and Thorburn, G. D. Plasma progesterone and prostaglandin-F concentrations in cyclic sow. *Journal of Reproduction and Fertility,* 1973, *32,* 343.

Glencross, R. G., Munro, I. B., Senior, B. E., and Pope, G. S. Concentrations of oestradiol-17β, oestrone and progesterone in jugular venous plasma of cows during the oestrous cycle and in early pregnancy. *Acta Endocrinologica,* 1973, *73,* 374–384.

Goebelsmann, U., Arce, J. J., Thorneycroft, I. H., and Mishell, D. R., Jr. Serum testosterone concentrations in women throughout the menstrual cycle and following HCG administration. *American Journal of Obstetrics and Gynecology,* 1974, *119,* 445–452.

Goldfoot, D. A., Slob, A. K., Scheffler, G., Robinson, J. A., Wiegand, S. J., and Cords, J. Multiple ejaculations during prolonged sexual tests and lack of resultant serum testosterone increases in male stumptail macaques *(M. arctoides). Archives of Sexual Behavior,* 1975, *4,* 547–560.

Gombe, S., and Katngole, C. B. Plasma testosterone levels in donkeys in the tropics. *Journal of Endocrinology,* 1977, *74,* 151–152.

Gomes, W. R., and Joyce, M. C. Seasonal changes in serum testosterone in adult rams. *Journal of Animal Science,* 1975, *41,* 1373–1375.

Goncharov, N., Aso, T., Cekan, Z., Pachalia, N., and Diczfalusy, E. Hormonal changes during the menstrual cycle of the baboon *(Papio hamadryas). Acta Endocrinologica,* 1976, *82,* 396–412.

Goncharov, N. P., Taranov, A. G., Antonichev, A. V., Gorlushkin, V. M., Aso, T., Cekan, S. Z., and Diczfalusy, E. Effect of stress on the profile of plasma steroids in baboons *(Papio hamadryas). Acta Endocrinologica,* 1979, *90,* 372–384.

Goodman, A. L., Descalzi, C. D., Johnson, D. K., and Hodgen, G. D. Composite pattern of circulating LH, FSH, estradiol, and progesterone during the menstrual cycle in cynomolgus monkeys. *Proceedings of the Society for Experimental Biology and Medicine,* 1977, *155,* 479–481.

Goodman, R. L., Hotchkiss, J., Karsch, F. J., and Knobil, E. Diurnal variations in serum testosterone concentrations in the adult male rhesus monkey. *Biology of Reproduction,* 1974, *11,* 624–630.

Gordon, R. D., Spinks, J., Dulmanis, A., Hudson, B., Halberg, F., and Bartter, F. C. Amplitude and phase relations of several circadiar rhythms in human plasma and urine: Demonstration of rhythm for tetrahydrocortisol and tetrahydrocorticosterone. *Clinical Science,* 1968, *35,* 307–324.

Gordon, T. P., Rose, R. M., and Bernstein, I. S. Seasonal rhythm in plasma testosterone levels in the rhesus monkey *(Macaca mulatta):* A three year study. *Hormones and Behavior,* 1976, *7,* 229–243.

Gorski, J., and Gannon, F. Current models of steroid-hormone action—Critique. *Annual Review of Physiology,* 1976, *38,* 425–450.

Goy, R. W., and Young, W. C. Strain differences in the behavioral responses of female guinea pigs to alpha-estradiol benzoate and progesterone. *Behaviour,* 1957, *10,* 340–354.

Graef, V. A method for the determination of free etiocholanolone in plasma. *Zeitschrift für Klinische Chemie und Klinische Biochemie,* 1971, *9,* 238–241.

Gräf, K.-J. Serum oestrogen, progesterone and prolactin concentrations in cyclic, pregnant and lactating beagle dogs. *Journal of Reproduction and Fertility,* 1978, *52,* 9–14.

Graham, C. E., Collins, D. C., Robinson, H., and Preedy, J. R. K. Urinary levels of estrogens and pregnanediol and plasma levels of progesterone during the menstrual cycle of the chimpanzee: Relationship to the sexual swelling. *Endocrinology,* 1972, *91,* 13–24.

Gray, G. D. Changes in the levels of luteinizing hormone and testosterone in the circulation of ageing male rats. *Journal of Endocrinology,* 1978, *76,* 551–552.

Gray, G. D., Davis, H. N., Kenney, A. McM., and Dewsbury, D. A. Effect of mating on plasma levels of LH and progesterone in montane voles *(Microtus montanus). Journal of Reproduction and Fertility,* 1976, *47,* 89–91.

Gray, G. D., Tennent, B., Smith, E. R., and Davidson, J. M. Luteinizing hormone regulation and sexual behavior in middle-aged female rats. *Endocrinology,* 1980, *107,* 187–194.

Grewel, T., Mickelson, O., and Hafs, H. D. Androgen secretion and spermatogenesis in rats following semistarvation. *Proceedings of the Society for Experimental Biology and Medicine,* 1971, *138,* 723–727.

Grota, L. J. Effects of age and experience on plasma testosterone. *Neuroendocrinology,* 1971, *8,* 136–143.

Guerrero, R., Aso, T., Brenner, P. F., Cekan, Z., Landgren, B.-M., Hagenfeldt, K., and Diczfalusy, E. Studies on the pattern of circulating steroids in the normal menstrual cycle: I. Simultaneous assays of progesterone, pregenolone, dehydroeipandrosterone, testosterone, dihydrotestosterone, andro-stenedione, oestradiol and oestrone. *Acta Endocrinologica*, 1976, *81*, 133–149.

Gulamhusein, A. P., and Tam, W. H. Reproduction in the male stoat, *Mustela erminea. Journal of Reproduction and Fertility*, 1974, *41*, 303–312.

Gulamhusein, A. P., and Thawley, A. R. Plasma progesterone levels in the stoat. *Journal of Reproduction and Fertility*, 1974, *36*, 405–408.

Gupta, S. R., and Anand, B. K. Effect of protein deficiency on plasma progesterone levels during the menstrual cycle of adult rhesus monkeys. *Endocrinology*, 1971, *89*, 652–658.

Gupta, D., Zarzycki, J., and Rager, K. Plasma testosterone and dihydrotestosterone in male rats during sexual maturation and following orchidectomy and experimental bilateral cryptorchidism. *Steroids*, 1975, *25*, 33–42.

Gustafson, A. W., and Shemesh, M. Changes in plasma testosterone levels during the annual reproductive cycle of the hibernating bat, *Myotis lucifugus lucifugus* with a survey of plasma testosterone levels in adult male vertebrates. *Biology of Reproduction*, 1976, *15*, 9–24.

Guthrie, H. D., Henricks, D. M., and Handlin, D. L. Plasma estrogen, progesterone and luteinizing hormone prior to estrus and during early pregnancy in pigs. *Endocrinology*, 1972, *91*, 675–679.

Hadley, J. C. Total unconjugated oestrogen and progesterone concentrations in peripheral blood during the oestrous cycle of the dog. *Journal of Reproduction and Fertility*, 1975, *44*, 445–451.

Hafiez, A. A., Lloyd, C. W., and Bartke, A. The role of prolactin in the regulation of testis function: The effects of prolactin and luteinizing hormone on the plasma levels of testosterone and androstenedione in hypophysectomized rats. *Journal of Endocrinology*, 1972, *52*, 327–332.

Hall, V., and Goldman, B. Effects of gonadal steroid hormones on hibernation in the Turkish hamster *(Mesocricetos brandti) Journal of Comparative Physiology*, 1980, *135*, 107–114.

Haltmeyer, G. C., and Eik-Nes, K. B. Plasma levels of testosterone in male rabbits following copulation. *Journal of Reproduction and Fertility*, 1969, *19*, 273–277.

Hami, M., Rosler, A., and Rabinowitz, D. A nonchromatographic radioimmunoassay for 17α-hydroxy-progesterone. *Journal of Clinical Endocrinology and Metabolism*, 1975, *40*, 863–867.

Hanks, J., and Short, R. V. The formation and function of the corpus luteum in the African elephant, *Loxodonta africana. Journal of Reproduction and Fertility*, 1972, *29*, 79–89.

Hansel, W., and Snook, R. B. Pituitary ovarian relationships in the cow. *Journal of Dairy Science*, 1970, *53*, 945–961.

Hansel, W., Concannon, P. W., and Lukaszewska, J. H. Corpora lutea of the large domestic animals. *Biology of Reproduction*, 1973, *8*, 222–245.

Hansen, S. Södersten, P., Eneroth, P., Srebro, B., and Hole, K. A sexually dimorphic rhythm in oestradiol-activated lordosis behaviour in the rat. *Journal of Endocrinology*, 1979, *83*, 267–274.

Harding, C. F., and Feder, H. H. Relation between individual differences in sexual behavior and plasma testosterone levels in the guinea pig. *Endocrinology*, 1976, *98*, 1198–1205.

Harris, M. E., and Bartke, A. Concentration of testosterone in testis fluid of the rat. *Endocrinology*, 1974, *95*, 701–706.

Harrison, R. J., and Ridgway, S. H. Gonadal activity in some bottlenose dolphins *(Tursiops truncatus). Journal of Zoology, London*, 1971, *165*, 355–366.

Hart, B. L., and Ladewig, J. Effects of medial preoptic-anterior hypothalamic lesions on development of sociosexual behavior in dogs. *Journal of Comparative and Physiological Psychology*, 1979, *93*, 556–573.

Hasan, S. H., Weber, B., Neumann, F., and Friedrich, E. Plasma testosterone levels in childhood and puberty, measured by radioimmunoassay. *Acta Endocrinologica (Suppl.) (Kbh).*, 1973, *173*, 168.

Hasler, M. J., Falvo, R. E., and Nalbandov, A. V. Testicular development and testosterone concentrations in the testis and plasma of young male shrews *(Suncus murinus). General and Comparative Endocrinology*, 1975, *25*, 36–41.

Hauger, R. L., Karsch, F. J., and Foster, D. L. A new concept for control of the estrous cycle of the ewe based on the temporal relationships between luteinizing hormone, estradiol and progesterone in peripheral serum and evidence that progesterone inhibits tonic LH secretion. *Endocrinology*, 1977, *101*, 807–817.

Hawkins, R. A., and Oakey, R. E. Estimation of oestrone sulphate, oestradiol-17β and oestrone in peripheral plasma: Concentrations during the menstrual cycle and in men. *Journal of Endocrinology*, 1974, *60*, 3–17.

Hawkins, R. A., Freedman, B., Marshall, A., and Killen, E. Oestradiol-17β and prolactin levels in rat peripheral plasma. *British Journal of Cancer*, 1975, *32*, 179–185.

Henricks, D. M., Dickey, J. F., and Niswender, G. D. Serum luteinizing hormone and plasma progesterone levels during the estrous cycle and early pregnancy in cows. *Biology of Reproduction,* 1970, *2,* 346–351.

Henricks, D. M., Dickey, J. F., and Hill, J. R. Plasma estrogen and progesterone levels in cows prior to and during estrus. *Endocrinology,* 1971a, *89,* 1350–1355.

Henricks, D. M., Lamond, D. R., Hill, J. R., and Dickey, J. F. Plasma total estrogens and progesterone concentrations during proestrus and after mating in beef heifers. *Society for the Study of Reproduction,* 1971b, 4th annual meeting (abstr.), 13.

Henricks, D. M., Guthrie, H. D., and Handlin, D. L. Plasma estrogen, progesterone and luteinizing hormone levels during the estrous cycle in pigs. *Biology of Reproduction,* 1972, *6,* 210–218.

Herriman, I. D., Harwood, D. J., Mallinson, C. B., and Heitzman, R. J. Plasma concentrations of ovarian hormones during the oestrous cycle of the sheep and cow. *Journal of Endocrinology,* 1979, *81,* 61–64.

Hess, D. L., and Resko, J. A. The effects of progesterone on the patterns of testosterone and estradiol concentrations in the systemic plasma of the female rhesus monkey during the intermenstrual period. *Endocrinology,* 1973, *92,* 446–453.

Hoff, J. D., and Quigley, M. E. Hormonal changes associated with the midcycle LH surge: A multiphasic rise in progesterone (abstr.). *Endocrine Society 62nd Annual Meeting,* 1980, 230.

Hoffman, K., and Nieschlag, E. Circadian rhythm of plasma testosterone in the male djungarian hamster *(Phodopus sungorus). Acta Endocrinologica,* 1977, *86,* 193–199.

Hoffman, B., Barth, D., and Karg, H. Progesterone and estrogen levels in peripheral plasma of the pregnant and nonpregnant roe deer *(Capreolus capreolus). Biology of Reproduction,* 1978, *19,* 931–935.

Holmdahl, T. H., and Johansson, E. D. B. Peripheral plasma levels of 17α-hydroxyprogesterone, progesterone and oestradiol during normal menstrual cycles in women. *Acta Endocrinologica,* 1972, *71,* 743–754.

Horikoshi, H., and Suzuki, Y. On circulating sex steroids during the estrous cycle and the early pseudopregnancy in the rat with special reference to its luteal activation. *Endocrinologia Japonica,* 1974, *21,* 69–79.

Horsky, J., Stroufava, A., Jouja, V., and Presl. J. Nyctohemeral rhythm of urinary LH and plasma estradiol-17-beta levels in normal cycle of women. *Physiologica Bohemoslovaca,* 1973, *22,* 69.

Horth, C. E., Lobo, P. J., Shelton, J. R., Asbury, M. J., Clarke, J. M., and Venning, G. R. Effects of spironolactone on the plasma binding and unbound levels of testosterone and oestradiol in healthy men. *Journal of Endocronology,* 1979, *78,* 67P (abstr.).

Hossain, M. I., Lee, C. S., Clarke, I. J., and O'Shea, J. D. Ovarian and luteal blood flow, and peripheral plasma progesterone levels, in cyclic guinea-pigs. *Journal of Reproduction and Fertility,* 1979, *57,* 167–174.

Hostetter, M. W., and Piacsek, B. E. Patterns of pituitary and gonadal hormone secretion during a 24 hour period in the male rat. *Biology of Reproduction,* 1977, *16,* 495–498.

Hotchkiss, J., Atkinson, L. E., and Knobil, E. Time course of serum estrogen and luteinizing hormone (LH) concentrations during the menstrual cycle of the rhesus monkey. *Endocrinology,* 1971, *89,* 177–183.

Huang, H. H., Steger, R. W., Bruni, J. F., and Meites, J. Patterns of sex steroid and gonadotropin secretion in aging female rats. *Endocrinology,* 1978, *103,* 1855–1859.

Hurd, H. P., Palumbo, P. J., and Ghabib, H. Hypothalamic-endocrine dysfunction in anorexia nervosa. *Proceedings of the Mayo Clinic,* 1977, *52,* 711–716.

Illingworth, D. V., Heap, R. B., and Perry, J. S. Changes in the metabolic clearance rate of progesterone in the pregnant guinea-pig. *Journal of Endocrinology,* 1970, *48,* 409–417.

Illius, A. W., Haynes, N. B., and Lamming, G. E. Effects of ewe proximity on peripheral testosterone levels and behavior in the ram. *Journal of Reproduction and Fertility,* 1976, *48,* 25–32.

Ivanyi, P., Hampl, R., Starka, L., and Mickova, M. Genetic association between H-2 gene and testosterone metabolism in mice. *Nature,* 1972, *238,* 280–281.

Jainudeen, M. R., Katongole, C. B., and Short, R. V. Plasma testosterone levels in relation to musth and sexual activity in the male Asiatic elephant, *Elephas maximus. Journal of Reproduction and Fertility,* 1972, *29,* 99–103.

Jänne, O., Apter, D., and Vihko, R. Assay of testosterone, progesterone, and 17α-hydroxyprogesterone in human plasma by radioimmunoassay after separation on hydroxyalkoxypropyl Sephadex. *Journal of Steroid Biochemistry*, 1974, *5*, 155–162.

Jean-Faucher, Ch., Berger, M., de Turckheim, M., Veyssiere, G., and Jean, Cl. Developmental patterns of plasma and testicular testosterone in mice from birth to adulthood. *Acta Endocrinologica*, 1978, *89*, 780–788.

Jemmett, J. E., and Evans, J. M. A survey of sexual behavior and reproduction of female cats. *Journal of Small Animal Practice*, 1977, *18*, 31–37.

Jerrett, D. P. Female reproductive patterns in nonhibernating bats. *Journal of Reproduction and Fertility*, 1979, *56*, 369–378.

Joffre, M. Relationship between testicular blood flow, testosterone secretion and spermatogenic activity in young and adult wild red foxes *(Vulpes vulpes). Journal of Reproduction and Fertility*, 1977, *51*, 35–40.

Joffre, M., and Joffre, J. Variations de las testostérone mis au cours de la période prépubèr du Renardeau et au cours du cycle génital saisonnier du Renard mâle adulte *(Vulpes vulpes)* en captivité. *Compte Rendus de l'Académie des Sciences* (Série D), 1975, *281*, 819–821.

Johansson, E. D. B. Progesterone levels in peripheral plasma during the luteal phase of the normal human menstrual cycle measured by a rapid competitive protein binding technique. *Acta Endocrinologica*, 1969, *61*, 592–606.

Johansson, E. D. B., and Gemzell, C. The relation between plasma progesterone and total urinary oestrogens following induction of ovulation in women. *Acta Endocrinologica*, 1969, *62*, 89–97.

Johansson, E. D. B., and Gemzell, C. Plasma levels of progesterone during the luteal phase in normal women treated with synthetic oestrogens (RS 2874, F 6103 and ethinyloestradiol). *Acta Endocrinologica*, 1971, *68*, 551–560.

Johansson, E. D. B., and Wide, L. Periovulatory levels of plasma progesterone and luteinizing hormone in women. *Acta Endocrinologica*, 1969, *62*, 82–88.

Johansson, E. D. B., Neill, J. D., and Knobil, E. Periovulatory progesterone concentration in the peripheral plasma of the rhesus monkey with a methodologic note on the detection of ovulation. *Endocrinology*, 1968, *82*, 143–148.

Johansson, E. D. B., Wide, L., and Gemzell, C. Luteinizing hormone (LH) and progesterone in plasma and LH and oestrogens in urine during 42 normal menstrual cycles. *Acta Endocrinologica*, 1971, *68*, 502–512.

Johnson, D. F., and Phoenix, C. H. Sexual behavior and hormone levels during the menstrual cycles of rhesus monkeys. *Hormones and Behavior*, 1978, *11*, 160–174.

Jones, G. E., Boyns, A. R., Cameron, E. H. D., Bell, E. T., Christie, D. W., and Parkes, M. F. Plasma oestradiol, luteinizing hormone and progesterone during the oestrous cycle in the beagle bitch. *Journal of Endocrinology*, 1973, *57*, 331–332.

Jones, G. E., Baker, K., Fahmy, D. R., and Boyns, A. R. Effect of luteinizing hormone releasing hormone on plasma levels of luteinizing hormone, oestradiol and testosterone in the male dog. *Journal of Endocrinology*, 1976, *68*, 469–474.

Judd, H. L., and Yen, S. S. Serum androstenedione and testosterone levels during the menstrual cycle. *Journal of Clinical Endocrinology and Metabolism*, 1973, *36*, 475–481.

Judd, H. L., Parker, D. C., Rakoff, J. S., Hopper, B. R., and Yen, S. S. Elucidation of mechanism(s) of the nocturnal rise of testosterone in men. *Journal of Clinical Endocrinology and Metabolism*, 1973, *38*, 134–141.

Judd, H. L., Judd, G. E., Lucas, W. E., and Yen, S. S. Endocrine function of the postmenopausal ovary: Concentration of androgens and estrogens in ovarian and peripheral vein blood. *Journal of Clinical Endocrinology and Metabolism*, 1974, *39*, 1020–1024.

Judd, H. L., Lucas, W. E., and Yen, S. S. Serum 17β-estradiol and estrone levels in postmenopausal women with and without endometrial cancer. *Journal of Clinical Endocrinology and Metabolism*, 1976, *43*, 272–278.

Judd, H. L., Parker, D. C., and Yen, S. S. Sleep-wake patterns of LH and testosterone release in prepubertal boys. *Journal of Clinical Endocrinology and Metabolism*, 1977, *44*, 865–869.

Kalra, S. P., and Kalra, P. S. Temporal interrelationships among circulating levels of estradiol, progesterone, and LH during the rat estrous cycle: Effects of exogenous progesterone. *Endocrinology*, 1974, *95*, 1711–1718.

Kalra, P. S., and Kalra, S. P. Circadian periodicities of serum androgens, progesterone, gonadotropins and luteinizing hormone-releasing hormone in male rats: The effects of hypothalamic deafferentation, castration and adrenalectomy. *Endocrinology*, 1977a, *101*, 1821–1827.

Kalra, P. S., and Kalra, S. P. Temporal changes in the hypothalamic and serum luteinizing hormone-releasing hormone (LH-RH) levels and the circulating ovarian steroids during the rat oestrous cycle. *Acta Endocrinologica*, 1977b, *85*, 449–455.

Kamel, F., and Frankel, A. I. Hormone release during mating in the male rat: Time course, relation to sexual behavior, and interaction with handling procedures. *Endocrinology*, 1978, *103*, 2172–2179.

Kamel, F., Mock, E. J., Wright, W. W., and Frankel, A. I. Alterations in plasma concentrations of testosterone, LH, and prolactin associated with mating in the male rat. *Hormones and Behavior*, 1975, *6*, 277–288.

Kamel, F., Wright, W. W., Mock, E. J., and Frankel, A. I. The influence of mating and related stimuli on plasma levels of luteinizing hormone, follicle stimulating hormone, prolactin, and testosterone in the male rat. *Endocrinology*, 1977, *101*, 421–429.

Kanchev, L. N., Dobson, H., Ward, W. R., and Fitzpatrick, R. J. Concentrations of steroids in bovine peripheral plasma during the oestrous cycle and the effect of betamethasone treatment. *Journal of Reproduction and Fertility*, 1976, *48*, 341–345.

Karsch, F. J., Krey, L. C., Weick, R. F., Dierschke, D. J., and Knobil, E. Functional luteolysis in the rhesus monkey: The role of estrogen. *Endocrinology*, 1973, *92*, 1148–1152.

Katongole, C. B. A competitive protein-binding assay for testosterone in the plasma of the bull and the ram. *Journal of Endocrinology*, 1971, *51*, 303–312.

Katongole, C. B., Naftolin, F., and Short, R. V. Relationship between blood levels of luteinizing hormone and testosterone in bulls, and the effects of sexual stimulation. *Journal of Endocrinology*, 1971, *50*, 457–466.

Katongole, C. B., Naftolin, F., and Short, R. V. Seasonal variations in blood luteinizing hormone and testosterone levels in rams. *Journal of Endocrinology*, 1974, *60*, 101–106.

Kazama, N., and Hansel, W. Preovulatory changes in the progesterone level of bovine peripheral blood plasma. *Endocrinology*, 1970, *86*, 1252–1256.

Keating, R. J., and Tcholakian, R. K. In vivo patterns of circulating steroids in adult male rats: I. Variations in testosterone during 24- and 48-hour standard and reverse light/dark cycles. *Endocrinology*, 1979, *104*, 184–188.

Kelch, R. P., Jenner, M. R., Weinstein, R., Kaplan, S. L., and Grumbach, M. M. Estradiol and testosterone secretion by human, simian, and canine testes in males with hypogonadism and in male pseudohermaphrodites with the feminizing testes syndrome. *Journal of Clinical Investigation*, 1972, *51*, 824–830.

Kelly, M. J., Moss, R. L., and Dudley, C. A. The effects of microelectrophoretically applied estrogen, cortisol and acetylcholine on medial preoptic-septal unit activity throughout the estrous cycle of the female rat. *Experimental Brain Research*, 1977, *30*, 53–64.

Kim, M. H., Hosseinian, A. H., and Dupon, C. Plasma levels of estrogens, androgens and progesterone during normal and dexamethasone-treated cycles. *Journal of Clinical Endocrinology and Metabolism*, 1974, *39*, 706–712.

Kindahl, H., Edqvist, L.-E., Bane, A., and Granström, E. Blood levels of progesterone and 15-keto-13,14-dihydro-prostaglandin $F_{2\alpha}$ during the normal oestrous cycle and early pregnancy in heifers. *Acta Endocrinologica*, 1976, *82*, 134–149.

Kinson, G. A., and Liu, C.-C. Diurnal variation in plasma testosterone of the male laboratory rat. *Hormones and Metabolism Research*, 1973a, *5*, 233–234.

Kinson, G. A., and Liu, C.-C. Effects of blinding and pinealectomy on diurnal variations in plasma testosterone. *Experientia*, 1973b, *29*, 1415–1416.

Kirkpatrick, J. F., Vail, R., Devous, S., Schwend, S., Baker, C. B., and Wiesner, L. Diurnal variation of plasma testosterone in wild stallions. *Biology of Reproduction*, 1976, *15*, 98–101.

Kirkpatrick, J. F., Wiesner, L., Kenney, R. M., Ganjam, V. K., and Turner, J. W., Jr. Seasonal variation in plasma androgens and testosterone in the North American wild horse. *Journal of Endocrinology*, 1977, *72*, 237–238.

Kirschner, M. A., Lipsett, M. B., and Collins, D. R. Plasma ketosteroids and testosterone in man: A study of the pituitary-testicular axis. *Journal of Clinical Investigation*, 1965, *44*, 657–665.

Kirton, K. T., Niswender, G. D., Midgley, A. R., Jr., Jaffee, R. B., and Forbes, A. D. Serum luteinizing hormone (LH) and progesterone concentrations during the menstrual cycle of the rhesus monkey. *Journal of Clinical Endocrinology and Metabolism*, 1970, *30*, 105–110.

Kley, H. K., Solbach, H. G., McKinnan, J. C., and Krüskemper, H. L. Testosterone decrease and oestrogen increase in male patients with obesity. *Acta Endocrinologica,* 1979, *91,* 553–563.

Kling, O. R., and Westfahl, P. K. Steroid changes during the menstrual cycle of the baboon *(Papio cynocephalus)* and human. *Biology of Reproduction,* 1978, *18,* 392–400.

Kniewald, Z., Zanisi, M., and Martini, L. Studies on the biosynthesis of testosterone in the rat. *Acta Endocrinologica,* 1971, *68,* 614–624.

Knobil, E. Hormonal control of the menstrual cycle and ovulation in the rhesus monkey. *Acta Endocrinologica (Suppl.) (Kbh.),* 1972, *166,* 137–144.

Knorr, D. W., Vanha-Perttula, T., and Lipsett, M. B. Structure and function of rat testis through pubescence. *Endocrinology,* 1970, *86,* 1298–1304.

Koliyannis, A., Hag, N. U., Judge, T., and Grant, J. K. Plasma 17-hydroxyandrogens in elderly human subjects. *Journal of Endocrinology,* 1974, *61,* R31 (abstr).

Komisaruk, B. R., and Diakow, C. Reflex intensity in rats in relation to the estrous cycle, ovariectomy, estrogen administration and mating behavior. *Endocrinology,* 1973, *93,* 548–557.

Korenman, S. G., Tulchinsky, D., and Eaton, L. W., Jr. Radio-ligand procedures for estrogen assay in normal and pregnancy plasma. *Acta Endocrinologica,* 1970, *Suppl. 147,* 291.

Koyama, T., de la Pena, A., and Hagino, N. Plasma estrogen, progestin, and luteinizing hormone during the normal menstrual cycle in the baboon: Role of luteinizing hormone. *American Journal of Obstetrics and Gynecology,* 1977, *127,* 67–72.

Kreuz, L. E., Rose, R. M., and Jennings, J. R. Suppression of plasma testosterone levels and psychological stress. *Archives of General Psychiatry,* 1972, *26,* 479–482.

Krutzsch, P. H. The structure and function of the ovary of the female California leaf-nosed bat, *Macrotus californicus. Anatomical Record,* 1978, *187,* 631.

Labhsetwar, A. P. Role of estrogens in spontaneous ovulation: Evidence for positive feedback in hamsters. *Endocrinology,* 1972, *90,* 941–946.

LaCroix, A., and Pelletier, J. Short-term variations in plasma LH and testosterone in bull calves from birth to 1 year of age. *Journal of Reproduction and Fertility,* 1979, *55,* 81–85.

LaFerla, J. J., Anderson, D. L., and Schalch, D. S. Psychoendocrine response to sexual arousal in human males. *Psychosomatic Medicine,* 1978, *40,* 166–172.

Lamond, D. R., Henricks, D. M., Hill, J. R., Jr., and Dickey, J. F. Breed differences in plasma progesterone concentrations in the bovine during proestrus. *Biology of Reproduction,* 1971, *5,* 258–261.

Lasley, B., Hendricks, A. G., and Stabenfeldt, G. H. Estradiol levels near the time of ovulation in the bonnet monkey *(Macaca radiata). Biology of Reproduction,* 1974, *11,* 237–244.

Leader-Williams, N. Abnormal testes in reindeer, *Rangifer tarandus. Journal of Reproduction and Fertility,* 1979a, *57,* 127–130.

Leader-Williams, N. Age-related changes in the testicular and antler cycles of reindeer, *Rangifer tarandus. Journal of Reproduction and Fertility,* 1979b, *57,* 117–126.

Leavitt, W. W., and Blaha, G. C. Circulating progesterone levels in the golden hamster during the estrous cycle, pregnancy, and lactation. *Biology of Reproduction,* 1970, *3,* 353–361.

Leavitt, W. W., Bosley, C. G., and Blaha, G. C. Source of ovarian preovulatory progesterone. *Nature (New Biology),* 1971, *234,* 283.

Lee, P. A., and Migeon, C. J. Puberty in boys: Correlation of plasma levels of gonadotropins (LH, FSH), androgens (testosterone, androstenedione, dehydroepiandrosterone, and its sulfate), estrogens (estrone and estradiol), and progestins (progesterone and 17-hydroxy-progesterone). *Journal of Clinical Endocrinology and Metabolism,* 1975, *41,* 556–562.

Lee, V. W. K., DeKretser, D. M., Hudson, B., and Wang, C. Variations in serum FSH, LH, and testosterone levels in male rats from birth to sexual maturity. *Journal of Reproduction and Fertility,* 1975, *42,* 121–126.

Legan, S. J., and Karsch, F. J. Neuroendocrine regulation of the estrous cycle and seasonal breeding in the ewe. *Biology of Reproduction,* 1979, *20,* 74–85.

Legan, S. J., Karsch, F. J., and Foster, D. L. The endocrine control of seasonal reproductive function in the ewe: A marked change in response to the negative feedback action of estradiol on luteinizing hormone secretion. *Endocrinology,* 1977, *101,* 818–824.

Legros, J. J., Franchimont, P., and Burger, H. Variations of neurohypophysial function in normally cycling women. *Journal of Clinical Endocrinology and Metabolism,* 1975, *41,* 54–59.

Lemon, M. Peripheral plasma progesterone during pregnancy and the oestrous cycle in the tammar wallaby, *Macropus eugenii. Journal of Endocrinology,* 1972, *55,* 63–71.

Lewis, J. G., Ghanadian, R., and Chisholm, G. D. Serum androgens in male and female *Praomys (Mastomys) natalensis. Journal of Endocrinology,* 1976a, *68,* 27 P (abstr.).

Lewis, J. G., Ghanadian, R., and Chisholm, G. D. Serum 5α-dihydrotestosterone and testosterone changes with age in man. *Acta Endocrinologica,* 1976b, *82,* 444–448.

Leymarie, P., Roger, M., Castanier, M., and Scholler, R. Circadian variations of plasma testosterone and estrogens in normal men: A study by frequent sampling. *Journal of Steroid Biochemistry,* 1974, *5,* 167–171.

Lincoln, G. A. Seasonal variation in the episodic secretion of luteinizing hormone and testosterone in the ram. *Journal of Endocrinology,* 1976, *69,* 213–226.

Lincoln, G. A. Plasma testosterone profiles in male macropodid marsupials. *Journal of Endocrinology,* 1978, *77,* 347–351.

Lincoln, G. A., and Davidson, W. The relationship between sexual and aggressive behaviour, and pituitary and testicular activity during the seasonal sexual cycle of rams, and the influence of photoperiod. *Reproduction and Fertility,* 1977, *49,* 267–276.

Lincoln, G. A., and Peet, M. J. Photoperiodic control of gonadotrophin secretion in the ram: A detailed study of the temporal changes in plasma levels of follicle-stimulating hormone, luteinizing hormone and testosterone following an abrupt switch from long to short days. *Journal of Endocrinology,* 1977, *74,* 355–367.

Lincoln, G. A., Peet, M. J., and Cunningham, R. A. Seasonal and circadian changes in the episodic release of follicle-stimulating hormone, luteinizing hormone and testosterone in rams exposed to artificial photoperiods. *Journal of Endocrinology,* 1977, *72,* 337–349.

Lindner, H. R. Androgens in the bovine testis and spermatic vein blood. *Nature,* 1959, *183,* 1605–1606.

Lindner, H. R., and Mann, T. Relationship between the content of androgenic steroids in the testes and the secretory activity of the seminal fesicles in the bull. *Journal of Endocrinology,* 1960, *21,* 341–360.

Lipsett, M. D., Wilson, H., Kirschner, M. A., Korenman, S. G., Fishman, L. M., Sarfaty, G. A., and Bardin, C. W. Studies on Leydig cell physiology and pathology: Secretion and metabolism of testosterone. *Recent Progress in Hormone Research,* 1966, *22,* 245–281.

Lloyd, B. J. Plasma testosterone and accessory sex glands in normal and cryptorchid rats. *Journal of Endocrinology,* 1972, *54,* 285–396.

Lu, K. H., Hopper, B. R., Vargo, T. M., and Yen, S. S. C. Chronological changes in sex steroid, gonadotropin and prolactin secretion in aging female rats displaying different reproductive states. *Biology of Reproduction,* 1979, *21,* 193–203.

Lukaszewska, J. H., and Greenwald, G. S. Progesterone levels in the cyclic and pregnant hamster. *Endocrinology,* 1970, *86,* 1–9.

Luttge, W. G., and Hall, N. R. Differential effectiveness of testosterone and its metabolites in the induction of male sexual behavior in two strains of albino mice. *Hormones and Behavior,* 1973, *4,* 31–43.

Macrides, F., Bartke, A., Fernandez, F., and D'Angelo, W. Effects of exposure to vaginal odor and receptive females on plasma testosterone in the male hamster. *Neuroendocrinology,* 1974, *15,* 355–364.

Macrides, F., Bartke, A., and Dalteriod, S. Strange females increase plasma testosterone levels in male mice. *Science,* 1975, *189,* 1104–1106.

Mahoudeau, J. A., Corvol, P., and Bricaire, H. Rabbit testosterone-binding globulin: II. Effect on androgen metabolism *in vivo. Endocrinology,* 1973, *92,* 1120–1125.

Marrone, B. L., Rodriguez-Sierra, J. F., and Feder, H. H. Differential effects of prostaglandins on lordosis behavior in female guinea pigs and rats. *Biology of Reproduction,* 1979, *20,* 853–861.

Martel, D., and Psychoyos, A. Endometrial content of nuclear estrogen receptor and receptivity for ovoimplantation in the rat. *Endocrinology,* 1976, *99,* 470–475.

Maruniak, J. A., and Bronson, F. H. Gonadotropic responses of male mice to female urine. *Endocrinology,* 1976, *99,* 963–969.

Maruniak, J. A., Goquelin, A., and Bronson, F. H. Release of LH in response to female urinary odors: Characteristics of response in young males. *Biology of Reproduction,* 1978, *18,* 251–255.

Mason, B. D., Krishamurti, C. R., and Kitts, W. D. Oestrone and oestradiol in jugular vein plasma during the oestrous cycle of the cow. *Journal of Endocrinology,* 1972, *55,* 141–146.

McCormack, J. T., and Greenwald, G. S. Progesterone and oestradiol-17β concentrations in the peripheral plasma during pregnancy in the mouse. *Journal of Endocrinology,* 1974, *62,* 101–107.

McCormack, S. A. Plasma testosterone concentration and binding in the chimpanzee: Effect of age. *Endocrinology,* 1971, *89,* 1171–1177.

McGill, T. E. Genetic factors influencing the action of hormones on sexual behaviour. In J. B. Hutchison (Ed.), *Biological Determinants of Sexual Behaviour*. New York: Wiley, 1978.

McKenna, T. J., DiPietro, D. L., Brown, R. D., Strott, C. A., and Liddle, G. W. Plasma 17-hydroxy-pregnenolone in normal subjects. *Journal of Clinical Endocrinology and Metabolism*, 1974, *39*, 833–841.

McKinney, T. D., and Desjardins, C. Intermale stimuli and testicular function in adult and immature house mice. *Biology of Reproduction*, 1973, *9*, 370–378.

McMillin, J. M., Seal, U. S., Kennlyne, K. D., Erickson, A. W., and Jones, J. E. Annual testosterone rhythm in the adult white-tailed deer *(Odocoileus virginianus borealis)*. *Endocrinology*, 1974, *94*, 1034–1040.

McMillin, J. M., Seal, U. S., Rogers, L., and Erickson, A. W. Annual testosterone rhythm in the black bear *(Ursus americanus)*. *Biology of Reproduction*, 1976, *15*, 163–167.

McNatty, K. P., Revfeim, K. J. A., and Young, A. Peripheral plasma progesterone concentrations in sheep during the oestrous cycle. *Journal of Endocrinology*, 1973, *58*, 219–225.

Mead, R. A., and Eik-Nes, K. B. Seasonal variation in plasma levels of progesterone in western forms of the spotted skunk. *Journal of Reproduction and Fertility*, 1969, *Suppl. 6*, 397–403.

Mendoza, S. P., Lowe, E. L., Davidson, J. M., and Levine, S. Annual cyclicity in the squirrel monkey *(Samiri sciureus):* The relationship between testosterone, fatting, and sexual behavior. *Hormones and Behavior*, 1978a, *11*, 295–303.

Mendoza, S. P., Lowe, E. L., Resko, J. A., and Levine, S. Seasonal variations in gonadal hormones and social behavior in squirrel monkeys. *Physiology and Behavior*, 1978b, *20*, 515–522.

Michael, R. P., and Bonsall, R. W. A 3-year study of an annual rhythm in plasma androgen levels in male rhesus monkeys *(Macaca mulatta)* in the constant laboratory environment. *Journal of Reproduction and Fertility*, 1977, *49*, 129–131.

Michael, R. P., Setchell, K. D. R., and Plant, T. M. Diurnal changes in plasma testosterone and studies on plasma corticosteroids in non-anaesthetized male rhesus monkeys *(Macaca mulatta)*. *Journal of Endocrinology*, 1974, *63*, 325–335.

Mikhail, G., Wu, C. H., Ferin, M., and Vandewiele, R. L. Radioimmunoassay of plasma estrone and estradiol. *Steroids*, 1970, *15*, 333–352.

Milewich, L., Gomez-Sanchez, C., Crowley, G., Porter, J. C., Madden, J. D., and MacDonald, P. C. Progesterone and 5α-pregnane-3,20-dione in peripheral blood of normal young women: Daily measurements throughout the menstrual cycle. *Journal of Clinical Endocrinology and Metabolism*, 1977, *45*, 617–622.

Milgrom, E., Atger, M., and Baulieu, E.-E. Progesterone binding plasma protein (PBP). *Nature*, 1970, *228*, 1205–1206.

Miller, A. E., and Riegle, G. D. Temporal changes in serum progesterone in aging female rats. *Endocrinology*, 1980, *106*, 1579.

Miller, H. L., and Alliston, C. W. Bovine plasma progesterone levels at programmed circadian temperatures. *Life Sciences*, 1974, *14*, 705–710.

Mirarchi, R. E., Scanlon, P. F., Kirkpatrick, R. L., and Schreck, C. B. Androgen levels and antler development in captive and wild white-tailed deer. *Journal of Wildlife Management*, 1977, *41*, 178–183.

Mock, E. J., and Frankel, A. I. A shifting circannual rhythm in serum testosterone concentration in male laboratory rats. *Biology of Reproduction*, 1978, *19*, 927–930.

Mock, E. J., Kamel, F., Wright, W. W., and Frankel, A. I. Seasonal rhythm in plasma testosterone and luteinizing hormone of the male laboratory rat. *Nature*, 1975, *256*, 61–63.

Mock, E. J., Norton, H. W., and Frankel, A. I. Daily rhythmicity of serum testosterone concentration in the male laboratory rat. *Endocrinology*, 1978, *103*, 1111–1121.

Moger, W. H. Serum 5α-androstane-3α,17β-diol, androsterone, and testosterone concentrations in the male rat: Influence of age and gonadotropin secretion. *Endocrinology*, 1977, *100*, 1027–1032.

Moger, W. H., and Murphy, P. R. Serum 5α-androstane-3α,17β-diol, androsterone and testosterone concentrations in the immature male rat: Influence of time of day. *Journal of Endocrinology*, 1977, *75*, 177–178.

Moguilewsky, M., and Raynaud, J. P. The relevance of hypothalamic and hypophyseal progestin receptor regulation in the induction and inhibition of sexual behavior in the female rat. *Endocrinology*, 1979, *105*, 516–522.

Møller, O. M. Progesterone concentrations in the peripheral plasma of the blue fox *(Alopex lagopus)* during pregnancy and the oestrous cycle. *Journal of Endocrinology*, 1973, *59*, 429–538.

Møller, O. M. Plasma progesterone before and after ovariectomy in unmated pregnant mink, *Mustela vison. Journal of Reproduction and Fertility,* 1975, *37,* 367–372.

Moor, B. C., and Younglai, E. V. Variations in peripheral levels of LH and testosterone in adult male rabbits. *Journal of Reproduction and Fertility,* 1975, *42,* 259–266.

Moss, R. L., and McCann, S. M. Induction of mating behavior in rats by luteinizing hormone-releasing factor, *Science,* 1973, *181,* 177–179.

Muduuli, D. S., Sanford, L. M., Palmer, W. M., and Howland, B. E. Secretory patterns and circadian and seasonal changes in luteinizing hormone, follicle stimulating hormone, prolactin and testosterone in the male pygmy goat. *Journal of Animal Science,* 1979, *49,* 543–553.

Murray, M. A. F., and Corker, C. S. Levels of testosterone and luteinizing hormone in plasma samples taken at 10-minute intervals in normal men. *Journal of Endocrinology,* 1973, *56,* 157–158.

Nadler, R. D., Graham, C. E., Collins, D. C., and Gould, K. G. Plasma gonadotropins, prolactin, gonadal steroids, and genital swelling during the menstrual cycle of lowlands gorillas. *Endocrinology,* 1979, *105,* 290–296.

Naftolin, F., Brown-Grant, K., and Corker, C. S. Plasma and pituitary luteinizing hormone and peripheral plasma oestradiol concentrations in the normal oestrous cycle of the rat and after experimental manipulation of the cycle. *Journal of Endocrinology,* 1972, *53,* 17–30.

Naftolin, J., Judd, H. L., and Yen, S. S. Pulsatile patterns of gonadotropins and testosterone in man: The effects of clomiphene with and without testosterone. *Journal of Clinical Endocrinology and Metabolism,* 1973, *36,* 285–288.

Nagle, C. A., Denari, J. H., Quiroga, S., Riarte, A., Merlo, A., Germino, N. I., Gómez-Argaña, F., and Rosner, J. M. The plasma pattern of ovarian steroids during the menstrual cycle in capuchin monkeys *(Cebus apella). Biology of Reproduction,* 1979, *21,* 979–983.

Neal, J., and Murphy, B. D. Response of immature, mature nonbreeding and mature breeding ferret testis to exogenous LH stimulation. *Biology of Reproduction,* 1977, *16,* 244–248.

Neal, J., Murphy, B. D., Moger, W. H., and Oliphant, L. W. Reproduction in the male ferret: Gonadal activity during the annual cycle, recrudescence and maturation. *Biology of Reproduction,* 1977, *17,* 380–385.

Neaves, W. B. Changes in testicular Leydig cells and in plasma testosterone levels among seasonally breeding rock hyrax. *Biology of Reproduction,* 1973, *8,* 451–466.

Neill, J. D., Johansson, E. D. B., Datta, J. K., and Knobil, E. Relationship between the plasma levels of luteinizing hormone and progesterone during the normal menstrual cycle. *Journal of Clinical Endocrinology,* 1967a, *27,* 1167–1173.

Neill, J. D., Johansson, E. D. B., and Knobil, E. Levels of progesterone in peripheral plasma during the menstrual cycle of the rhesus monkey. *Endocrinology,* 1967b, *81,* 1161–1164.

Nelson, J. F., Latham, K. R., and Finch, C. E. Plasma testosterone levels in C57BL/6J male mice: Effects of age and disease. *Acta Endocrinologica,* 1975, *80,* 744–752.

Nequin, L. G., Alvarez, J. A., and Campbell, C. S. Alternations in steroid and gonadotropin release resulting from surgical stress during the morning of proestrus in 5-day cyclic rats. *Endocrinology,* 1975a, *97,* 718–724.

Nequin, L. G., Alvarez, J., and Schwartz, N. B. Steroid control of gonadotropin release. *Journal of Steroid Biochemistry,* 1975b, *6,* 1007–1012.

Nett, T. M., Holtan, D. W., and Estergreen, V. L. Plasma estrogens in pregnant and postpartum mares. *Journal of Animal Science,* 1973, *37,* 962–970.

Nett, T. M., Akbar, A. M., Phemister, R. D., Holst, P. A., Reichert, L. E., Jr., and Niswender, G. D. Levels of luteinizing hormone, estradiol and progesterone in serum during the estrous cycle and pregnancy in the beagle bitch. *Proceedings of the Society for Experimental Biology and Medicine,* 1975, *148,* 134–139.

Nieschlag, E., and Bieniek, H. Endocrine testicular function in mink during the first year of life. *Acta Endocrinologica (Kbhvn.),* 1975, *79,* 375–379.

Nieschlag, E., and Ismail, A. A. A. Bestimmung der Tagesschwankungen des Plasma-Testosterons normaler Männer durch kompetitive Proteinbindung. *Klinische Wochenschrift,* 1970, *48,* 53–54.

Nieschlag, E., Kley, K. H., and Wigelmann, W. Age dependence of the endocrine testicular function in adult men. *Acta Endocrinologica,* 1973, *S177,* 122 (abstr.).

Nillius, S. J., Fries, H., and Wide, L. Successful induction of follicular maturation and ovulation by prolonged treatment with LH-releasing hormone in women with anorexia nervosa. *American Journal of Obstetrics and Gynecology,* 1975, *122,* 921–928.

Niswender, G. D., and Spies, H. G. Serum levels of luteinizing hormone, follicle-stimulating hormone and progesterone throughout the menstrual cycle of rhesus monkeys. *Journal of Clinical Endocrinology and Metabolism*, 1973, *37*, 326–328.

Nock, B., Blaustein, J. D., and Feder, H. H. Changes in noradrenergic transmission after the concentration of cytoplasmic progestin receptors in hypothalamus. *Brain Research*, 1981, *207*, 371–396.

Noden, P. A., Oxender, W. D., and Hafs, H. D. The cycle of oestrus, ovulation, and plasma levels of hormones in the mare. *Journal of Reproduction and Fertility*, 1975, *Supplement 23*, 189–192.

Norman, R. L., and Greenwald, G. S. Effects of phenobarbital, hypophysectomy, and X-irradiation on preovulatory progesterone levels in the cyclic hamster. *Endocrinology*, 1971, *89*, 598–605.

Obst, J. M., and Seamark, R. F. Plasma progesterone concentrations during the reproductive cycle of ewes grazing Yarloop clover. *Journal of Reproduction and Fertility*, 1970, *22*, 545–547.

Okamoto, M., Setaishi, C., Nakagawa, K., Horiuchi, Y., Moriya, K., and Itoh, S. Diurnal variations in the levels of plasma and urinary androgens. *Journal of Clinical Endocrinology and Metabolism*, 1971, *32*, 846–851.

Overpeck, J. G., Colson, S. H., Hohmann, H. R., Applestine, M. S., and Reilly, J. F. Concentrations of circulating steroids in normal prepubertal and adult male and female humans, chimpanzees, rhesus monkeys, rats, mice, and hamsters: A literature survey. *Journal of Toxicology and Environmental Health*, 1978, *4*, 785–803.

Oxberry, B. A. Female reproductive patterns in hibernating bats. *Journal of Reproduction and Fertility*, 1979, *56*, 359–367.

Oxender, W. D., Noden, P. A., and Hafs, H. D. Estrus, ovulation, and serum progesterone, estradiol, and LH concentrations in mares after an increased photoperiod during winter. *American Journal of Veterinary Research*, 1977, *38*, 203–207.

Paape, S. R., Shille, V. M., Seto, H., and Stabenfeldt, G. H. Luteal activity in pseudopregnant cat. *Biology of Reproduction*, 1975, *13*, 470–474.

Pant, H. C., Hopkinson, C. R. N., and Fitzpatrick, R. J. Concentration of oestradiol, progesterone, luteinizing hormone and follicle-stimulating hormone in the jugular venous plasma of ewes during the oestrous cycle. *Journal of Endocrinology*, 1977, *73*, 247–255.

Parkening, T. A., Saksena, S. K., and Lau, I. F. Postovulatory levels of progesterones, oestrogens, luteinizing hormone and follicle-stimulating hormone in the plasma of aged golden hamsters exhibiting a delay in fertilization. *Journal of Endocrinology*, 1978, *78*, 147–148.

Parker, D. C., Judd, H. L., Rossman, L. G., and Yen, S. S. C. Pubertal sleep-wake patterns of episodic LH, FSH and testosterone release in twin boys. *Journal of Clinical Endocrinology and Metabolism*, 1975, *40*, 1099–1109.

Parkin, R. F., and Hendrickx, A. G. The temporal relationship between the preovulatory estrogen peak and the optimal mating period in rhesus and bonnet monkeys. *Biology of Reproduction*, 1975, *13*, 610–616.

Parvizi, N., Elsaesser, F., Smidt, D., and Ellendorff, F. Plasma luteinizing hormone and progesterone in the adult female pig during the oestrous cycle, late pregnancy and lactation, and after ovariectomy and pentobarbitone treatment. *Journal of Endocrinology*, 1976, *69*, 193–203.

Pattison, M. L., Chen, C. L., Kelley, S. T., and Brandt, G. W. Luteinizing hormone and estradiol in peripheral blood of mares during estrous cycle. *Biology of Reproduction*, 1974, *11*, 245–250.

Pauerstein, C. J., Eddy, C. A., Croxatto, H. D., Hess, R., Siler-Khodr, T. M., and Croxatto, H. B. Temporal relationships of estrogen, progesterone, and luteinizing hormone levels to ovulation in women and infrahuman primates. *American Journal of Obstetrics and Gynecology*, 1978, *130*, 878–886.

Pedersen, C. A., and Prange, A. J., Jr. Induction of maternal behavior in virgin rats after intracerebroventricular administration of oxytocin. *Proceedings of the National Academy of Sciences, USA*, 1979, *76*, 6661–6665.

Peng, M. T., and Peng, Y. M. Changes in the uptake of tritiated estradiol in the hypothalamus and hypophysis of old female rats. *Fertility and Sterility*, 1973, *24*, 534–539.

Peng, M. T., Chuong, C. F., and Peng, Y. M. Lordosis response in senile female rats. *Neuroendocrinology*, 1977, *24*, 317–324.

Penny, R., Parlow, A. F., Olanbiwonnu, N. O., and Frasier, S. D. Evolution of the menstrual patterns of gonadotrophin and sex steroid concentrations in serum. *Acta Endocrinologica*, 1977, *84*, 729–737.

Peppler, R. D., and Stone, S. G. Plasma progesterone levels in the female armadillo during delayed implantation and gestation: Preliminary report. *Journal of Animal Science*, 1976, *26*, 501–504.

Perachio, A. A., Alexander, M., Marr, L. D., and Collins, D. C. Diurnal variations of serum testosterone levels in intact and gonadectomized male and female rhesus monkeys. *Steroids*, 1977, *29*, 21–33.

Pfaff, D. W. Luteinizing hormone-releasing factor potentiates lordosis behavior in hypophysectomized ovariectomized female rats. *Science*, 1973, *182*, 1148–1149.

Pfaff, D., and Keiner, M. Atlas of estradiol-concentrating cells in the central nervous system of the female rat. *Journal of Comparative Neurology*, 1973, *151*, 121–158.

Phemister, R. D., Holst, P. A., Spano, J. S., and Hopwood, M. L. Time of ovulation in the beagle bitch. *Biology of Reproduction*, 1973, *8*, 74–82.

Phoenix, C. H. Ejaculation by male rhesus as a function of the female partner. *Hormones and Behavior*, 1973, *4*, 365–370.

Phoenix, C. H. Steroids and sexual behavior in castrated male rhesus monkeys. *Hormones and Behavior*, 1978, *10*, 1–9.

Piacsek, B. E., Schneider, T. C., and Gay, V. L. Sequential study of luteinizing hormone (LH) and "progestin" secretion on the afternoon of proestrus in the rat. *Endocrinology*, 1971, *89*, 39–45.

Pietras, R. J., and Szego, C. M. Specific binding sites for oestrogen at the outer surfaces of isolated endometrial cells. *Nature*, 1977, *265*, 69–72.

Pirke, K. M., and Doerr, P. Age related changes and interrelationships between plasma testosterone, oestradiol and testosterone-binding globulin in normal adult males. *Acta Endocrinologica*, 1973, *74*, 792–800.

Pirke, K. M., Kockott, G., and Dittmar, F. Psychosexual stimulation and plasma testosterone in man. *Archives of Sexual Behavior*, 1974, *3*, 577–584.

Plant, T. M., and Michael, R. P. Diurnal variations in plasma testosterone levels of adult male rhesus monkeys. *Acta Endocrinologica*, 1971, *S155*, 69 (abstr.).

Plant, T. M., Zumpe, D., Sauls, M., and Michael, R. P. An annual rhythm in the plasma testosterone of adult male rhesus monkeys maintained in the laboratory. *Journal of Endocrinology*, 1974, *62*, 403–404.

Plapinger, L., and McEwen, B. S. Gonadal steroid-brain interactions in sexual differentiation. In J. B. Hutchison (Ed.), *Biological Determinants of Sexual Behaviour*. New York: Wiley, 1978.

Plapinger, L., McEwen, B. S., Landau, I. T., and Feder, H. H. Characteristics of estradiol binding macromolecules in fetal and adult guinea pig brain cytosols. *Biology of Reproduction*, 1977, *16*, 586–599.

Plotka, E. D., and Erb, R. E. Levels of progesterone in peripheral blood plasma during the oestrous cycle of the ewe. *Journal of Animal Science*, 1967, *26*, 1363–1365.

Plotka, E. D., Erb, R. E., Callahan, C. J., and Gomes, W. R. Levels of progesterone in peripheral blood plasma during the estrous cycle of the bovine. *Journal of Dairy Sciences*, 1967, *50*, 1158–1160.

Plotka, E. D., Witherspoon, D. M., and Goetsch, D. D. Peripheral plasma progesterone levels during the estrous cycle of the mare. *Federation Proceedings*, 1971, *30*, 419.

Plotka, E. D., Seal, U. S., Schobert, E. E., and Schmoller, G. C. Serum progesterone and estrogens in elephants. *Endocrinology*, 1975, *97*, 485–487.

Plotka, E. D., Seal, U. S., Schmoller, G. C., Karns, P. D., and Keenlyne, K. D. Reproductive steroids in the white-tailed deer *(Odocoileus virginianus borealis):* I. Seasonal changes in the female. *Biology of Reproduction*, 1977, *16*, 340–343.

Pope, G. S., Gupta, S. K., and Munro, I. B. Progesterone levels in the systemic plasma of pregnant, cycling and ovariectomized cows. *Journals of Reproduction and Fertility*, 1969, *20*, 369–381.

Powell, J. E., and Stevens, V. C. Simple radioimmunoassay of five unconjugated ovarian steroids in a single sample of serum or phasma. *Clinical Chemistry*, 1973, *19*, 210–215.

Preslock, J. P., Hampton, S. H., and Hampton, J. K., Jr. Cyclic variations of serum progestins and immunoreactive estrogens in marmosets. *Endocrinology*, 1973, *92*, 1096–1101.

Pujol, A., Bayard, F., Louvet, J.-P., and Boulard, C. Testosterone and dihydrotestosterone concentrations in plasma, epididymal tissues, and seminal fluid of adult rats. *Endocrinology*, 1976, *98*, 111–113.

Purvis, K., and Haynes, N. B. Environmental factors and testicular hormone production in the male rat. *Journal of Reproduction and Fertility*, 1972, *31*, 490–491.

Purvis, K., Illius, A. W., and Haynes, N. B. Plasma testosterone concentrations in the ram. *Journal of Endocrinology*, 1974, *61*, 241–253.

Purvis, K., Brenner, P. F., Landgren, B. M., Cekan, Z., and Diczfalusy, E. Indices of gonadal functions in the human male: I. Plasma levels of unconjugated steroids and gonadotrophins under normal and pathological conditions. *Clinical Endocrinology (Oxford)*, 1975, *4*, 237–246.

Purvis, K., Landgren, B.-M., Cekan, Z., and Diczfalusy, E. Endocrine effects of masturbation in men. *Journal of Endocrinology*, 1976, *70*, 439–444.

Racey, P. A. The reproductive cycle in male noctule bats, *Nyctalus noctula. Journal of Reproduction and Fertility*, 1974, *41*, 169–182.

Racey, P. A. Seasonal changes in testosterone levels and androgen-dependent organs in male moles (*Talpa europaea*). *Journal of Reproduction and Fertility*, 1978, *52*, 195–200.

Racey, P. A., and Tam, W. H. Reproduction in male *Pipistrellus pipistrellus (Mammalia: Chiroptera). Journal of Zoology, London*, 1974, *172*, 101–122.

Racey, P. A., Rowe, P. H., and Chesworth, J. M. Changes in the luteinizing hormone-testosterone system of the male goat during the breeding season. *Journal of Endocrinology*, 1975, *65*, 8p (abstr.).

Ramaley, J. A., and Bunn, E. L. Seasonal variations in the onset of puberty in rats. *Endocrinology*, 1972, *91*, 611–613.

Rawlings, N. C., Hafs, H. D., and Swanson, L. V. Testicular and blood plasma androgens in Holstein bulls from birth through puberty. *Journal of Animal Science*, 1972, *34*, 435–440.

Rawlings, N. C., Fletcher, P. W., Henricks, D. M., and Hill, J. R. Plasma luteinizing hormone (LH) and testosterone levels during sexual maturation in beef bull calves. *Biology of Reproduction*, 1978, *19*, 1108–1112.

Raynaud, J.-P., Mercier-Fodard, C., and Baulieu, E.-E. Rat estradiol binding plasma protein (EBP). *Steroids*, 1971, *18*, 767–788.

Reed, H. C. B., Melrose, D. R., and Patterson, R. L. S. Androgen steroids as an aid to the detection of oestrus in pig artificial insemination. *British Veterinary Journal*, 1974, *130*, 61–67.

Reinberg, A., Lagoguey, M., Chauffournier, J.-M., and Cesselin, F. Circannual and circadian rhythms in plasma testosterone in five healthy young Parisian males. *Acta Endocrinologica*, 1975, *80*, 732–743.

Reinberg, A., Lagoguey, M., Cesselin, F., Touitou, Y., Legrand, J.-C., Delassalle, A., Antreassian, J., and Lagoguey, A. Circadian and circannual rhythms in plasma hormones and other variables of five healthy young human males. *Acta Endocrinologica*, 1978, *88*, 417–427.

Resko, J. A. Plasma androgen levels of the rhesus monkey: Effects of age and season. *Endocrinology*, 1967, *81*, 1203–1212.

Resko, J. A. Androgens in systemic plasma of male guinea pigs during development and after castration in adulthood. *Endocrinology*, 1970, *86*, 1444–1447.

Resko, J. A. Sex steroids in adrenal effluent plasma of the ovariectomized rhesus monkey. *Journal of Clinical Endocrinology and Metabolism*, 1971, *33*, 940–948.

Resko, J. A., and Eik-Nes, K. B. Diurnal testosterone levels in peripheral plasma of human male subjects. *Journal of Clinical Endocrinology*, 1966, *26*, 573–576.

Resko, J. A., and Phoenix, C. H. Sexual behavior and testosterone concentrations in the plasma of the rhesus monkey before and after castration. *Endocrinology*, 1972, *91*, 499–503.

Resko, J. A., Feder, H. H., and Goy, R. W. Androgen concentrations in plasma and testis of developing rats. *Journal of Endocrinology*, 1968, *40*, 485–491.

Resko, J. A., Norman, R. L., Niswender, G. D., and Spies, H. G. The relationship between progestins and gonadotropins during the late luteal phase of the menstrual cycle in rhesus monkeys. *Endocrinology*, 1974, *94*, 128–135.

Resko, J. A., Koering, M. J., Goy, R. W., and Phoenix, C. H. Preovulatory progestins: Observations on their source in rhesus monkeys. *Journal of Clinical Endocrinology and Metabolism*, 1974, *41*, 120–125.

Reyes, F. I., Winter, J. S. D., Faiman, C., and Hobson, W. C. Serial serum levels of gonadotropins, prolactin and sex steroids in the nonpregnant and pregnant chimpanzee. *Endocrinology*, 1975, *96*, 1447–1455.

Ribeiro, W. O., Mishell, D. R., Jr., and Thorneycroft, I. H. Comparison of the patterns of androstenedione, progesterone, and estradiol during the human menstrual cycle. *American Journal of Obstetrics and Gynecology*, 1974, *119*, 1026–1032.

Ridley, K., and Greenwald, G. S. Progesterone levels measured every two hours in the cyclic hamster. *Proceedings of the Society for Experimental Biology and Medicine*, 1975, *149*, 10–12.

Rieger, D., and Murphy, B. D. Episodic fluctuation in plasma testosterone and dihydrotestosterone in male ferrets during the breeding season. *Journal of Reproduction and Fertility*, 1977, *51*, 511–514.

Rigaudière, N., Pelardy, G., Roberg, A., and Delost, P. Changes in the concentrations of testosterone and androstenedione in the plasma and testis of the guinea-pig from birth to death. *Journal of Reproduction and Fertility*, 1976, *48*, 291–300.

Riondel, A., Tait, J. F., Tait, S. A. S., Gut, M., and Little, B. Estimation of progesterone in human peripheral blood using ³⁵S-Thiosemicarbazide. *Journal of Clinical Endocrinology*, 1965, *25*, 229–242.

Rivarola, M. A., Snipes, C. A., and Migeon, C. J. Concentrations of androgens in systemic plasma of rats, guinea pigs, salamanders and pigeons. *Endocrinology*, 1968, *82*, 115–122.

Robel, P., Corpéchot, C., and Baulieu, E. E. Testosterone and androstanolone in rat plasma and tissues. *FEBS Letters*, 1973, *33*, 218–220.

Robertson, D. M., Mešter, J., and Kellie, A. E. The measurement of oestradiol and progesterone in plasma from normal, infertile and comiphene treated women. *Acta Endocrinologica*, 1971, *68*, 523–533.

Robertson, H. A. Reproduction in the ewe and the goat. In H. H. Cole and P. T. Cupps (Eds.), *Reproduction in Domestic Animals (3rd ed.)*. New York: Academic Press, 1977.

Robinson, J. A., Scheffler, G., Eisele, S. G., and Goy, R. W. Effects of age and season on sexual behavior and plasma testosterone and dihydrotestosterone concentrations of laboratory-housed male rhesus monkeys *(Macaca mulatta)*. *Biology of Reproduction*, 1975, *13*, 203–210.

Robinson, T. J. Reproduction in cattle. In H. H. Cole and P. T. Cupps (Eds.), *Reproduction in Domestic Animals (3rd ed.)*. New York: Academic Press, 1977.

Rodriguez-Sierra, J. F., and Komisaruk, B. R. Effects of prostaglandin E2 and indomethacin on sexual behavior in the female rat. *Hormones and Behavior*, 1977, *9*, 281–289.

Rose, R. M., Holaday, J. W., and Bernstein, I. S. Plasma testosterone, dominance rank and aggressive behaviour in male rhesus monkeys. *Nature*, 1971, *231*, 366–368.

Rose, R. M., Gordon, T. P., and Bernstein, I. S. Plasma testosterone levels in the male rhesus: Influences of sexual and social stimuli. *Science*, 1972, 643–646.

Rose, R. M., Bernstein, I. S., and Gordon, T. P. Consequences of social conflict of plasma testosterone levels in rhesus monkeys. *Psychosomatic Medicine*, 1975, *37*, 50–61.

Rose, R. M., Gordon, T. P., and Bernstein, I. S. Diurnal variation in plasma testosterone and cortisol in rhesus monkeys living in social groups. *Journal of Endocrinology*, 1978, *76*, 67–74.

Rosenblatt, J. S., and Aronson, L. R. The decline of sexual behavior in male cats after castration with special reference to the role of prior sexual experience. *Behaviour*, 1958, *12*, 285–338.

Rosenfield, R. L., Jones, T., and Fang, V. S. The relationship between plasma testosterone and mean LH levels in men. *Journal of Clinical Endocrinology and Metabolism*, 1977, *45*, 30–34.

Rotti, K., Stevens, J., Watson, D., and Longcope, C. Estriol concentrations in plasma of normal non-pregnant women. *Steroids*, 1975, *25*, 807–816.

Rowe, P. H., Shenton, J. C., and Glover, T. D. Testosterone levels in peripheral blood plasma of the rabbit under normal experimental conditions. *Acta Endocrinologica (Kbhvn.)*, 1973, *Suppl. 177*, 125 (abstr.).

Rowe, P. H., Lincoln, G. A., Racey, P. A., Lehane, J., Stephenson, M. J., Shenton, J. C., and Glover, T. D. Temporal variations of testosterone levels in the peripheral blood plasma of men. *Journal of Endocrinology*, 1974, *61*, 63–73.

Rowlands, I. W., and Heap, R. B. Histological observations on the ovary and progesterone levels in the coypu, *Myocastor coypus*. *Symposia of the Zoological Society London*, 1966, *15*, 355–352.

Rowlands, I. W., Tam, W. H., and Kleiman, D. G. Histological and biochemical studies on the ovary and of progesterone levels in the systemic blood of the green acouchi *(Myoprocta pratti)*. *Journal of Reproduction and Fertility*, 1970, *22*, 533–545.

Rubin, R. T., Gowin, P. R., Lubin, A., Poland, R. E., and Pirke, K. M. Nocturnal increase of plasma testosterone in men: Relation to gonadotropins and prolactin. *Journal of Clinical Endocrinology and Metabolism*, 1975, *40*, 1027–1033.

Runnebaum, B., Rieben, W., Munstermann, A.-M. B.-V., and Zander, J. Circadian variations in plasma progesterone in the luteal phase of the menstrual cycle and during pregnancy. *Acta Endocrinologica*, 1972, *69*, 731–738.

Saginor, M., and Horton, R. Reflex release of gonadotropin and increased plasma testosterone concentration in male rabbits during copulation. *Endocrinology*, 1968, *82*, 627–630.

Saidapur, S. K., and Greenwald, G. S. Peripheral blood and ovarian levels of sex steroids in the cyclic hamster. *Biology of Reproduction*, 1978, *18*, 401–408.

Saksena, S. K., and Shaikh, A. A. Effect of intrauterine devices on preovulatory LH and progesterone levels in the cyclic hamster. *Journal of Reproduction and Fertility*, 1974, *38*, 205–210.

Saksena, S. K., Lau, I. F., Bartke, A., and Chang, M. C. Effect of indomethacin on blood plasma levels of LH and testosterone in male rats. *Journal of Reproduction and Fertility*, 1975, *42*, 311–317.

Sanford, L. M., Palmer, W. M., and Howland, B. E. Changes in the profiles of serum LH, FSH and testosterone and in mating performance and ejaculate volume in the ram during the ovine breeding season. *Journal of Animal Science*, 1977, *45*, 1382–1391.

Sanwal, P. C., Sundby, A., and Edqvist, L. E. Diurnal variation of peripheral plasma levels of testosterone in bulls measured by a rapid radioimmunoassay procedure. *Acta Veterinaria Scandinavica*, 1974, *15*, 90–99.

Sar, M., and Stumpf, W. E. Distribution of androgen target cells in rat forebrain and pituitary after ^{3}H-dihydrotestosterone administration. *Journal of Steroid Biochemistry*, 1977, *8*, 1131–1135.

Sasaki, Y., and Hanson, G. C. Correlation between the activities of enzymes involved in glucose oxidation in corpus luteum and the concentration of sex steroids in systemic plasma during the reproductive cycle of the guinea pig. *Endocrinology*, 1974, *95*, 1213–1218.

Saumande, J., and Rouger, Y. Variations saisonnières des taux d'androgènes dans le plasma de sang périphérique chez le Bouc. *Comptes Rendus Hebdomadaires des Séances de l'Académie des Sciences Paris* (Série D), 1972, *274*, 89–92.

Savu, L., Vallette, G., Nunez, E., Azria, M., and Jayle, M. F. Étude comparative de la liaison entre proteins sériques et les oestrogènes libres au cours du développement de diverses espèces animales. In R. Masseyeff (Ed.), *L'Alphafoetoproteine*. Paris: INSERM, 1974.

Scaramuzzi, R. J., and Baird, D. T. Pulsatile release of luteinizing hormone and the secretion of ovarian steroids in sheep during anestrus. *Endocrinology*, 1977, *101*, 1801–1806.

Scaramuzzi, R. J., Caldwell, B. V., and Moor, R. M. Radioimmunoassay of LH and estrogen during the estrous cycle of the ewe. *Biology of Reproduction*, 1970, *3*, 110–119.

Schams, D., Schallenberger, E., Hoffmann, B., and Karg, H. The oestrous cycle of the cow: Hormonal parameters and time relationships concerning oestrus, ovulation, and electrical resistance of the vaginal mucus. *Acta Endocrinologica*, 1977, *86*, 180–192.

Schams, D., Gombe, S., Schallenberger, E., Reinhardt, V., and Claus, R. Relationships between short-term variations of LH, FSH, prolactin and testosterone in peripheral plasma of prepubertal bulls. *Journal of Reproduction and Fertility*, 1978, *54*, 145–148.

Schanbacher, B. D. Testosterone secretion in cryptorchid and intact bulls injected with gonadotropin-releasing hormone and luteinizing hormone. *Endocrinology*, 1979, *104*, 360–364.

Schanbacher, B. D., and Ewing, L. L. Simultaneous determination of testosterone, 5α-androstan-17β-ol-3-one, 5α-androstane-3α, 17β-diol and 5α-androstane-3β, 17β-diol in plasma of adult male rabbits by radioimmunoassay. *Endocrinology*, 1975, *97*, 787–792.

Schanbacher, B. D., and Ford, J. J. Photoperiodic regulation of ovine spermatogenesis: Relationship to serum hormones. *Biology of Reproduction*, 1979, *20*, 719–726.

Schanbacher, B. D., and Lunstra, D. D. Seasonal changes in sexual activity and serum levels of LH and testosterone in Finnish Landrace and Suffolk rams. *Journal of Animal Science*, 1976, *43*, 644–650.

Schiavi, R. C., Davis, D. M., White, D., Edwards, A., Igel, G., and Fisher, C. Plasma testosterone during nocturnal sleep in normal men. *Steroids*, 1974, *24*, 191–201.

Schmidt, A. M., Nadal, L. A., Schmidt, M. J., and Beamer, N. B. Serum concentrations of oestradiol and progesterone during the normal oestrous cycle and early pregnancy in the lion *(Panthera leo)*. *Journal of Reproduction and Fertility*, 1979, *57*, 267–272.

Schneider, G., Kirschner, M. A., Berkowitz, R., and Ertel, N. H. Increased estrogen production in obese men. *Journal of Clinical Endocrinology and Metabolism*, 1979, *48*, 633–638.

Schomberg, D. W., Coudert, S. P., and Short, R. V. Effects of bovine luteinizing hormone and human chorionic gonadotrophin on the bovine corpus luteum in vivo. *Journal of Reproduction and Fertility*, 1967, *14*, 227–285.

Seal, U. S., Plotka, E. D., Packard, J. M., and Mech, L. D. Endocrine correlates of reproduction in the wolf: I. Serum progesterone, estradiol and LH during the estrous cycle. *Biology of Reproduction*, 1979, *21*, 1057–1066.

Selmanoff, M. K., Goldman, B. D., and Ginsburg, B. E. Serum testosterone, agonistic behavior, and dominance in inbred strains of mice. *Hormones and Behavior*, 1977, *8*, 107–119.

Sempéré, A. Plasma progesterone levels in the roe deer, *Capreolus capreolus*. *Journal of Reproduction and Fertility*, 1977, *50*, 365–366.

Shaikh, A. A., and Shaikh, S. A. Adrenal and ovarian steroid secretion in the rat estrous cycle temporally related to gonadotropins and steroid levels found in peripheral plasma. *Endocrinology*, 1975, *96*, 37–44.

Shaikh, A. A., Naqvi, R. H., and Shaikh, S. A. Concentrations of oestradiol-17β and progesterone in the peripheral plasma of the cynomolgus monkey *(Macaca fascicularis)* in relation to the length of the menstrual cycle and its component phases. *Journal of Endocrinology,* 1978, *79,* 1–7.

Sharma, O. P. Diurnal variations of plasma testosterone in stallions, *Biology of Reproduction,* 1976, *15,* 158–162.

Sharp, D. C., and Black, D. L. Changes in peripheral plasma progesterone throughout the oestrous cycle of the pony mare. *Journal of Reproduction and Fertility,* 1973, *33,* 535–538.

Shearer, I. J., Purvis, K., Jenkin, G., and Haynes, N. B. Peripheral plasma progesterone and oestradiol-17β levels before and after puberty in girls. *Journal of Reproduction and Fertility,* 1972, *30,* 347–360.

Shemesh, M., and Hansel, W. Measurement of bovine plasma testosterone by radioimmunoassay (RIA) and by a rapid competitive protein binding (CPB) assay. *Journal of Animal Science,* 1974, *39,* 720–724.

Shemesh, M., Lindauer, H. R., and Ayalon, N. Competitive protein-binding assay of progesterone in bovine jugular venous plasma during the oestrous cycle. *Journal of Reproduction and Fertility,* 1971, *26,* 167–174.

Shemesh, M., Ayalon, N., and Lindner, H. R. Oestradiol levels in the peripheral blood of cows during the oestrous cycle. *Journal of Endocrinology,* 1972, *55,* 73–78.

Sherman, B. M., West, J. H., and Korenman, S. G. The menopausal transition: Analysis of LH, FSH, estradiol and progesterone concentrations during menstrual cycles of older women. *Journal of Clinical Endocrinology and Metabolism,* 1976, *42,* 629–636.

Shille, V. M., Lundström, K. E., and Stabenfeldt, G. H. Follicular function in the domestic cat as determined by estradiol-17β concentrations in plasma: Relation to estrous behavior and cornification of exfoliated vaginal epithelium. *Biology of Reproduction,* 1979, *21,* 953–963.

Shire, J. G. M. The forms, uses and significance of genetic variation in endocrine systems. *Biological Reviews,* 1976, *51,* 105–141.

Short, R. V. Progesterone in blood: IV. Progesterone in the blood of mares. *Journal of Endocrinology,* 1959, *19,* 207–210.

Short, R. V. Progesterone. In C. H. Gray and A. L. Bacharach (Eds.), *Hormones in Blood,* Vol. 1. New York: Academic Press, 1961.

Short, R. V., and Moore, N. W. Progesterone in blood: V. Progesterone and 20α-hydroypregn-4-en-3-one in the placenta and blood of ewes. *Journal of Endocrinology,* 1959, *19,* 288–293.

Siegel, H. I., Ahdieh, H. B., and Rosenblatt, J. S. Hysterectomy-induced facilitation of lordosis behavior in the rat. *Hormones and Behavior,* 1978, *11,* 273–278.

Skinner, L. G., England, P. C., Cottrell, K. M., and Selwood, R. A. Proceedings: Serum oestradiol-17-beta in normal premenopausal women and in patients with benign and malignant breast disease. *British Journal of Cancer,* 1974, *30,* 176–177.

Slob, A. K., Ooms, M. P., and Vreeburg, J. T. M. Annual changes in serum testosterone in laboratory housed male stumptail macaques *(M. arctoides). Biology of Reproduction,* 1979, *20,* 981–984.

Smals, A. G. H., Kloppenborg, P. W. C., and Benraad, T. J. Circannual cycle in plasma testosterone levels in man. *Journal of Clinical Endocrinology,* 1976, *42,* 979–982.

Smith, E. R., Damassa, D. A., and Davidson, J. M. Feedback regulation and male puberty: Testosterone-luteinizing hormone relationships in the developing rat. *Endocrinology,* 1977, *101,* 173–180.

Smith, I. D., Bassett, J. M., and Williams, T. Progesterone concentrations in the peripheral plasma of the mare during the oestrous cycle. *Journal of Endocrinology,* 1970, *47,* 523–524.

Smith, K. D., Tcholakian, R. K., Chowdhury, M., and Steinberger, E. Rapid oscillations in plasma levels of testosterone, luteinizing hormone, and follicle-stimulating hormone in men. *Journal of Fertility and Sterility,* 1974, *25,* 965–975.

Smith, M. S., and McDonald, L. E. Serum levels of luteinizing hormone and progesterone during the estrous cycle, pseudopregnancy and pregnancy in the dog. *Endocrinology,* 1974, *94,* 404–412.

Smith, M. S., Freeman, M. E., and Neill, J. D. The control of progesterone secretion during the estrous cycle and early pseudopregnancy in the rat: Prolactin, gonadotropin and steroid levels associated with rescue of the corpus luteum of pseudopregnancy. *Endocrinology,* 1975, *96,* 219–226.

Smith, O. W., and Hafs, H. D. Competitive protein binding and radioimmunoassay for testosterone in bulls and rabbits: Blood serum testosterone after injection of LH or prolactin in rabbits. *Proceedings of the Society for Experimental Biology and Medicine,* 1973, *142,* 804–810.

Smith, O. W., Mongkonpunya, K., Hafs, H. D., Convey, E. M., and Oxender, W. D. Blood serum testosterone after sexual preparation or ejaculation, or after injections of LH or prolactin in bulls. *Journal of Animal Science*, 1973, *37*, 979–984.

Smith, R. G., Clarke, S. G., Zalta, E., and Taylor, R. N. Two estrogen receptors in reproductive tissue. *Journal of Steroid Biochemistry*, 1979, *10*, 31–35.

Snipes, C. A., Forest, M. G., and Migeon, C. J. Plasma androgen concentrations in several species of Old and New World monkeys. *Endocrinology*, 1969, *85*, 941–945.

Southern, A. L., Gordon, G. G., Tochimoto, S., Pinzon, G., Lane, D. R., and Stypulkowski, W. Mean plasma concentration, metabolic clearance and basal plasma production rates of testosterone in normal young men and women using a constant infusion procedure: Effect of time of day and plasma concentration on the metabolic clearance rate of testosterone. *Journal of Clinical Endocrinology*, 1967, *27*, 686–694.

Spies, H. G., Mahoney, C. J., Norman, R. L., Clifton, D. K., and Resko, J. A. Evidence for a diurnal rhythm in ovarian steroid secretion in the rhesus monkey. *Journal of Clinical Endocrinology and Metabolism*, 1974, *39*, 341–351.

Sprague, E. A., Hopwood, M. L., Niswender, G. D., and Wiltbank, J. N. Progesterone and luteinizing hormone levels in peripheral blood of cycling beef cows. *Journal of Animal Science*, 1971, *33*, 99–103.

Squires, E. L., Wentworth, B. C., and Ginther, O. J. Progesterone concentration in blood of mares during the estrous cycle, pregnancy and after hysterectomy. *Journal of Animal Science*, 1974, *39*, 759–767.

Stabenfeldt, G. H., and Hendrickx, A. G. Progesterone levels in the bonnet monkey *(Macaca radiata)* during the menstrual cycle and pregnancy. *Endocrinology*, 1972, *91*, 614–619.

Stabenfeldt, G. H., and Hendrickx, A. G. Progesterone studies in the *Macaca fascicularis*. *Endocrinology*, 1973, *92*, 1296–1300.

Stabenfeldt, G. H., and Hughes, J. P. Reproduction in horses. In H. H. Cole and P. T. Cupps (Eds.), *Reproduction in Domestic Animals* (3rd ed.). New York: Academic Press, 1977.

Stabenfeldt, G. H., and Shille, V. M. Reproduction in the dog and cat. In H. H. Cole and P. T. Cupps (Eds.), *Reproduction in Domestic Animals* (3rd ed.). New York: Academic Press, 1977.

Stabenfeldt, G. H., Akins, E. L., Ewing, L. L., and Morrissette, M. C. Peripheral plasma progesterone levels in pigs during the oestrous cycle. *Journal of Reproduction and Fertility*, 1969a, *20*, 443–449.

Stabenfeldt, G. H., Ewing, L. L., and McDonald, L. E. Peripheral plasma progesterone levels during the bovine oestrous cycle. *Journal of Reproduction and Fertility*, 1969b, *19*, 433–442.

Stabenfeldt, G. H., Holt, J. A., and Ewing, L. I. Peripheral plasma progesterone levels during the ovine estrous cycle. *Endocrinology*, 1969c, *85*, 11–15.

Stabenfeldt, G. H., Hughes, J. P., and Evans, J. W. Ovarian activity during the estrous cycle of the mare. *Endocrinology*, 1972, *90*, 1379–1384.

Stearns, E. L., MacDonnell, J. A., Kaufman, B. J., Padua, R., Lucman, T. S., Winter, J. D. S., and Faiman, C. Declining testicular function with age. *The American Journal of Medicine*, 1974, *57*, 761–766.

Stevens, V. C., Sparks, S. J., and Powell, J. E., Levels of estrogens, progestogens and luteinizing hormone during the menstrual cycle of the baboon. *Endocrinology*, 1970, *87*, 658–666.

Strott, C. A., Bermudez, J. A., and Lipsett, M. B. Blood levels and production rates of 17-hydroxypregnenolone in man. *Journal of Clinical Investigation*, 1970, *49*, 1999–2007.

Sutton, J. R., Coleman, M. J., Casey, J., and Lazarus, L. Androgen responses during physical exercise. *British Medical Journal*, 1973, *1*, 520–522.

Sutton, J. R., Coleman, M. J., and Casey, J. H. The adrenal cortical contribution to serum androgens in physical exercise. *Medicine and Science in Sports*, 1974, *6*, 72.

Svendsen, R., and Sørensen, B. The plasma concentration of unconjugated oestrone and 17β-oestradiol during the normal menstrual cycle. *Acta Endocrinologica*, 1964, *47*, 245–254.

Swerdloff, R. S., Batt, R. A., and Bray, G. A. Reproductive hormonal function in the genetically obese (ob/ob) mouse. *Endocrinology*, 1976, *98*, 1359–1364.

Szego, C. M., and Roberts, S. Steroid action and interaction in uterine metabolism. *Recent Progress in Hormone Research*, 1953, *8*, 419–469.

Tan, S. Y., and Murphy, B. E. P. Specificity of the progesterone-binding globulin of the guinea pig. *Endocrinology*, 1974, *94*, 122–127.

Tang, B. Y., and Adams, N. R. Changes in oestradiol-17β binding in the hypothalami and pituitary glands of persistently infertile ewes previously exposed to oestrogenic subterranean clover: Evidence of alterations to oestrogen receptors. *Journal of Endocrinology*, 1978, *78*, 171–177.

Terranova, P. F., and Greenwald, G. S. Steroid and gonadotropin levels during the luteal-follicular shift of the cyclic hamster. *Biology of Reproduction*, 1978, *18*, 170–175.

Thibier, M. Peripheral plasma testosterone concentrations in bulls around puberty. *Journal of Reproduction and Fertility*, 1975, *42*, 567–569.

Thibier, M. Diurnal testosterone and 17α-hydroxyprogesterone in peripheral plasma of young postpubertal bulls. *Acta Endocrinologica*, 1976, *81*, 623–634.

Thibier, M., and Saumande, J. Oestradiol-17β, progesterone and 17α-hydroxyprogesterone concentrations in jugular venous plasma in cows prior to and during oestrous. *Journal of Steroid Biochemistry*, 1975, *6*, 1433–1437.

Thibier, M., Castanier, M., Tea, N. T., and Scholler, R. Concentration plasmatiques de la 17α-hydroxyprogestérone au cours du cycle de la vache. *Comptes Rendus Hebdomadaires des Séances de l'Académie des Sciences Paris* (Serie D), 1973, *276*, 3049–3052.

Thompson, D. L., Jr., Pickett, B. W., Berndtson, W. E., Voss, J. L., and Nett, T. M. Reproductive physiology of the stallion: VIII. Artificial photoperiod, collection interval and seminal characteristics, sexual behavior and concentrations of LH and testosterone in serum. *Journal of Animal Science*, 1977, *44*, 656–664.

Thompson, M. L., and Edwards, D. A. Experiential and strain determinants of the estrogen-progesterone induction of sexual receptivity in spayed female mice. *Hormones and Behavior*, 1971, *2*, 299–305.

Thorburn, G. D., Bassett, J. M., and Smith, I. D. Progesterone concentration in the peripheral plasma of sheep during the oestrous cycle. *Journal of Endocrinology*, 1969, *45*, 459–469.

Thorneycroft, I. H., Sribyatta, B., Tom, W. K., Nakamura, R. M., and Mishell, D. R., Jr. Measurement of serum LH, FSH, progesterone 17-hydroxyprogesterone and estradiol-17β levels at 4-hour intervals during the periovulatory phase of the menstrual cycle. *Journal of Clinical Endocrinology and Metabolism*, 1974, *39*, 754–758.

Tillson, S. A., and Erb, R. E. Progesterone concentration in peripheral blood plasma of the domestic sow prior to and during early pregnancy. *Journal of Animal Science*, 1967, *26*, 1366–1368.

Tremblay, R. R., Forest, M. G., Shalf, J., Martel, J. G., Kowarski, A., and Migeon, C. J. Studies on the dynamics of plasma androgens and on the origin of dihydrotestosterone in dogs. *Endocrinology*, 1972, *91*, 556–561.

Turek, F. W., and Campbell, C. S. Photoperiodic regulation of neuroendocrine-gonadal activity. *Biology of Reproduction*, 1979, *20*, 32–50.

Valtonen, M. H., Rajakoski, E. J., and Lähteenmäki, P. Levels of oestrogen and progesterone in the plasma of the racoon dog *(Nyctereutes procyonoides)* during oestrus and pregnancy. *Journal of Endocrinology*, 1978, *76*, 549–550.

van der Molen, J. H., Runnebaum, B., Nishizawa, E. E., Kristensen, E., Kirschbaum, T., Wiest, W. G., and Eik-Nes, K. B. On the presence of progesterone in blood plasma from normal women. *Journal of Clinical Endocrinology and Metabolism*, 1965, *25*, 170–176.

Van Horn, R. N., and Resko, J. A. The reproductive cycle of the ring-tailed lemur *(Lemur catta)*: Sex steroid levels and sexual receptivity under controlled photoperiods. *Endocrinology*, 1977, *101*, 1579–1586.

Van Horn, R. N., Beamer, N. B., and Dixson, A. F. Diurnal variations of plasma testosterone in two prosimian primates *(Galago crassicaudatus crassicaudatus* and *Lemur catta)*. *Biology of Reproduction*, 1976, *15*, 523–528.

Verhage, H. G., Beamer, N. B., and Brenner, R. M. Plasma levels of estradiol and progesterone in the cat during polyestrus, pregnancy and pseudopregnancy. *Biology of Reproduction*, 1976, *14*, 579–585.

Vermeulen, A. Testosterone and 5α-androsten-17β-ol-3-one (DHT) levels in man. *Acta Endocrinologica*, 1976, *83*, 651–664.

Vermeulen, A., Rubens, R., and Verdonck, L. Testosterone secretion and metabolism in male senescence. *Journal of Clinical Endocrinology*, 1972, *34*, 730–735.

Virgo, B. B., and Bellward, G. D. Serum progesterone levels in the pregnant and postpartum laboratory mouse. *Endocrinology*, 1974, *95*, 1486–1490.

Wade, G. N. Sex hormones, regulatory behaviors and body weight. In J. S. Rosenblatt, R. A. Hinde, E. Shaw, and C. G. Beer (Eds.), *Advances in the Study of Behavior*, Vol. 6. New York: Academic Press, 1976.

Walker, W. A., and Feder, H. H. Anti-estrogen effects on estrogen accumulation in brain cell nuclei: Neurochemical correlates of estrogen action on female sexual behavior in guinea pigs. *Brain Research*, 1977, *134*, 467–478.

Weick, R. F., Dierschke, D. J., Karsch, F. J., Butler, W. R., Hotchkiss, J., and Knobil, E. Periovulatory time courses of circulating gonadotropic and ovarian hormones in the rhesus monkey. *Endocrinology*, 1973, *93*, 1140–1147.

Weisz, J., and Gunsalus, P. Estrogen levels in immature female rats: True or spurious—ovarian or adrenal? *Endocrinology*, 1973, *93*, 1057–1065.

West, C. D., Mahajan, D. K., Chavré, V. J., Nabors, C. J., and Tyler, F. H. Simultaneous measurement of multiple plasma steroids by radioimmunoassay demonstrating episodic secretion. *Journal of Clinical Endocrinology and Metabolism*, 1973, *36*, 1230–1236.

Westphal, U., Stroupe, S. D., and Cheng, S.-L. Progesgerone binding to serum proteins. *Annals of the New York Academy of Sciences*, 1977, *286*, 10–28.

Wettemann, R. P., and Desjardins, C. Relationship of LH to testosterone in rams. *Journal of Animal Science*, 1973, *36*, 212.

Wettemann, R. P., Hafs, H. D., Edgerton, L. A., and Swanson, L. V. Estradiol and progesterone in blood serum during the bovine estrous cycle. *Journal of Animal Science*, 1972, *34*, 1020–1024.

Whalen, R. E., and Massicci, J. Subcellular analysis of the accumulation of estrogen by the brain of male and female rats. *Brain Research*, 1975, *89*, 255–264.

Whitehead, P. E., and McEwan, E. H. Seasonal variation in the plasma testosterone concentration of reindeer and caribou. *Canadian Journal of Zoology*, 1973, *51*, 651–658.

Wicklings, E. J. Radioimmunoassay of 17-hydroxyprogesterone in serum with or without thin-layer chromatographic purification. *Hormone Research*, 1975, *6*, 78–84.

Wildt, D. E., Guthrie, S. C., and Seager, S. W. J. Ovarian and behavioral cyclicity of the laboratory maintained cat. *Hormones and Behavior*, 1978, *10*, 251–257.

Wilks, J. W. Endocrine characterization of the menstrual cycle of the stumptailed monkey *(Macaca arctoides)*. *Biology of Reproduction*, 1977, *16*, 474–478.

Wilks, J. W., Marciniak, R. D., Hildebrand, D. L., and Hodgen, G. D. Perovulatory endocrine events in the stumptailed monkey *(Macaca arctoides)*. *Endocrinology*, 1980, *107*, 237–244.

Wilson, M., Plant, T. M., and Michael, R. P. Androgens and the sexual behaviour of male rhesus monkeys. *Journal of Endocrinology*, 1972, *52*, R2 (abstr.).

Wilson, M. J., McMillin, J. M., Seal, U. S., and Ahmed, K. Circadian variation of serum testosterone in the adult male rat with a late morning acrophase. *Experientia*, 1976, *32*, 944–945.

Wolf, R. C., O'Connor, R. F., and Robinson, J. A. Cyclic changes in plasma progestins and estrogens in squirrel monkeys. *Biology of Reproduction*, 1977, *17*, 228–231.

Wu, C. H., Prazak, L., Flickinger, G. L., and Mikhail, G. Plasma 20α-hydroxypregn-4-en-3-one in the normal menstrual cycle. *Journal of Clinical Endocrinology and Metabolism*, 1974, *39*, 536–539.

Wu, C. H., Blasco, L., Flickinger, G. L., and Mikhail, G. Ovarian function in the preovulatory rabbit. *Biology of Reproduction*, 1977, *17*, 304–308.

Yamaji, T., Dierschke, D. J., Hotchkiss, J., Bhattacharya, A. N., Surve, A. H., and Knobil, E. Estrogen induction of LH release in the rhesus monkey. *Endocrinology*, 1971, *89*, 1034–1041.

Yates, F. E. Comparative physiology: Compared to what? *American Journal of Physiology: Regulatory, Integrative, and Comparative Physiology*, 1979, *6*, R1–R2.

Yen, S. S. C., Martin, P. L., Burnier, A. M., Czekala, N. M., Greaney, M. O., Jr., and Callantine, M. R. Circulating estradiol, estrone, and gonadotropin levels following administration of orally active 17β-estradiol in postmenopausal women. *Journal of Clinical Endocrinology and Metabolism*, 1975, *40*, 518–521.

Yoshimi, T., and Lipsett, M. B. The measurement of plasma progesterone. *Steroids*, 1968, *11*, 527–540.

Young, W. C. Psychobiology of sexual behavior in the guinea pig. In D. S. Lehrman, R. A. Hinde, and E. Shaw (Eds.), *Advances in the Study of Behavior*, Vol. 2. New York: Academic Press, 1969.

Youssefnejadian, E., Collins, W. P., and Somerville, I. F. Radioimmunoassay of plasma androsterone. *Steroids*, 1973, *22*, 63–72.

Youssefnejadian, E., Virdee, S. S., and Somerville, I. F. Radioimmunoassay of plasma aetiocholanolone. *Journal of Steroid Biochemistry*, 1974, *5*, 529–532.

Yuthasastrakosol, P., Palmer, W. M., and Howland, B. E. Luteinizing hormone, oestrogen and progesterone levels in peripheral serum of anoestrous and cyclic ewes as determined by radioimmunoassay. *Journal of Reproduction and Fertility,* 1975, *43,* 57–65.

Zumpe, D., and Michael, R. P. Relation between the hormonal status of the female and direct and
redirected aggression by male rhesus monkeys *(Macaca mulatta). Hormones and Behavior,* 1979, *12,*
269–279.

PART III

Neurological Bases of Mammalian Reproductive Behavior

9

Neurological Bases of Male Sexual Behavior

A Comparative Analysis

BENJAMIN L. HART AND MITZI G. LEEDY

INTRODUCTION

The sexual behavior of most of the commonly studied mammalian species is highly stereotyped. To the uninitiated, this stereotyping would suggest the possibility of a rather close correspondence between the functioning of particular neuronal systems and the different aspects of sexual behavior, as well as a degree of uniformity across various species in the neural areas that mediate the behaviors. Some work, indeed, supports this viewpoint. However, many studies show that various aspects of male sexual behavior are controlled by diverse neural areas and that there are noteworthy differences among species.

Besides being stereotyped, sexual behavior is also sexually dimorphic, a fact suggesting the possibility of a corresponding dimorphism in the neural mechanisms underlying the behavior. Indeed, there are a number of intriguing reports of microscopic, sexually dimorphic differences in parts of the basal forebrain and the spinal cord that mediate sexual responses. However, this structure–function relationship has to be examined with the recognition that, on occasion, males show behavioral responses characteristic of females, and normal females display mascu-

BENJAMIN L. HART AND MITZI G. LEEDY   Department of Physiological Sciences, School of Veterinary Medicine, University of California, Davis, California 95616.   Research in the studies by the authors and preparation of this chapter have been supported by Grant No. MH 12003 from the National Institute of Mental Health, U.S. Public Health Service.

line behavior, including even the behavioral pattern of ejaculation. There must, therefore, be a dual representation of the neural circuitry for male and female sexual behavior in both sexes. In a theoretical sense, it is possible that the neural circuitry associated with female behavior in males is vestigial or involuted but capable of functioning under certain conditions. Alternatively, the neural circuitry for sexual behavior might be similar in magnitude in males and females. But one could imagine different inhibitory or facilitatory influences that would keep the neural circuitry underlying behavior typical of the opposite sex from being readily displayed. The sexual dimorphism would be in the facilitatory or inhibitory areas, not in the neural circuitry itself.

Laboratory rodents are still the mainstay of neuropsychological investigations, but work on the cat, the dog, the monkey, and the ungulates allows for some discussion of comparative differences and similarities. Also, there are some reports from the human clinical literature that add to the perspective. In terms of using animals as models for understanding neural mechanisms in the sexual behavior of human males, different species serve different purposes, and no particular species appears to be always the most appropriate. Future work will involve many species, as some of the more intricate, or at least more interesting, questions require the comparative approach.

The first part of this review deals with peripheral sensory functions, particularly the olfactory system. The effector mechanisms involved in erection and ejaculation are also discussed. These areas are covered because the activation of central neural structures—and the behavioral expression of central neural processing—ultimately involves these peripheral aspects.

The discussion of central neural processes begins with a review of work on spinal mechanisms because this can be considered the first level of central neural processing. Findings of laboratory investigations on brain mechanisms are presented on the basis of anatomical location. Most of the recent research seems to be related to either mapping neuronal systems or exploring the types of influences that certain brain areas have on behavior.

Because gonadal hormones have such a pronounced influence on sexual behavior, an understanding of the role of any neural system in male sexual behavior requires some information regarding the possible activation or suppression of neural structures by hormones. The hormonal aspect of the review will be limited, but we will convey the notion that sexual behavior is modulated by the activity of hormones at several neuronal sites.

SENSORY MECHANISMS

In terms of the involvement of the special senses in sexual behavior, the olfactory system has received almost all the attention. The reason is obvious; males of practically all species spend considerable time smelling the genitalia of females or urine eliminated by females. In all nonprimates studied, males are able to discriminate by olfaction between receptive and nonreceptive females. The visual system is brought into play in detecting species-typical receptive stances and presentation movements of females. In some species, such as felids, mating calls attract the two sexes to one another. In rats, ultrasonic vocalizations play a role in coordinating mating (Barfield and Geyer, 1975; Barfield *et al.*, 1979; Geyer and Barfield, 1980). It has been known for a long time, however, that, at least for rats and rabbits, none of the special senses are crucial for male copulatory behavior (Beach, 1942; Brooks, 1937; Stone, 1925).

The somatosensory system, as it involves the male genitalia, is also quite important. Tactile stimualtion resulting from a female licking the genital region appears to sexually excite male dogs, for example, much as one might expect from olfactory or visual stimuli. In animals, such as rats, cats, and dogs, that achieve intromission by trial-and-error thrusting of the penis against the perineum of the female, penile cutaneous receptors are important in detecting the vaginal orifice. The ultimate stimulation of the penis occurs during intromission. Such stimulation evokes deeper penetration and fuller erection, activates the ejaculatory responses, and triggers or generates the experience of orgasm. It is logical, therefore, that genital sensory mechanisms have received considerable attention.

Finally, the interoceptive system has received some thought, mainly in the way of negative evidence for the concept that the storage of fluids in accessory sex organs and the stretching of the organ walls enhance a sexual drive. Removal of seminal vesicles, ventral prostate lobes, and coagulating glands does not affect sexual activity in male rats, for example (Beach and Wilson, 1963; Larsson and Swedin, 1971).

The following discussion of sensory systems is restricted to neurological investigations of the olfactory system and genital sensory mechanisms.

Main and Accessory Olfactory System

In terms of the involvement of olfaction in male sexual behavior, probably the most important thing to emphasize is that in most mammals there are five olfactory systems (Graziadei, 1977), of which the main olfactory and the accessory (vomeronasal) olfactory systems appear to be most important (Wysocki, 1979). These two systems are distinctly different in terms of peripheral receptor location, receptor sensitivity, and central projections.

The major olfactory system receives input from receptor cells located in the olfactory mucosa. Olfactory nerves from these receptors terminate in the main olfactory bulbs, and from there, secondary projections terminate in the anterior olfactory nuclei, the olfactory tubercles, the prepyriform cortex, the entorhinal area, and the anterior part of the corticomedial amygdala (Broadwell, 1975; Devor, 1976; Heimer, 1968; Price, 1973; Scalia and Winans, 1975).

The concept of the vomeronasal projection representing a separate system was initially outlined by Herrick (1921) and was later emphasized by Raisman (1972) and Scalia and Winans (1975). The vomeronasal (Jacobson's) organ is a blind pouch that is dorsal to the hard palate and that communicates, depending on the species, with the nasal and/or oral cavities by means of the nasopalatine duct. The mucosa of this organ is superficially similar to that of the main olfactory system. In the mouse, at least, there is a difference in the histochemical distribution of enzymes found between the olfactory and the vomeronasal mucosae (Cuschieri, 1974).

Vomeronasal nerves that arise from receptor cells in the mucosa terminate in the accessory olfactory bulb, which, in turn, projects specifically to the medial amygdaloid nucleus and the posteromedial part of the cortical amygdaloid nucleus. Accessory-olfactory-tract axons also enter the stria terminalis and terminate in its bed nucleus (Scalia and Winans, 1975). It might be well to mention that the accessory olfactory system, including the vomeronasal organ, is absent in humans, as well as in Old World monkeys and apes (Wysocki, 1979).

Although anatomical, neurological, and behavioral studies have yielded little concrete evidence about the specific or unique functions of the vomeronasal organ, evidence is accumulating that the vomeronasal organ is particularly involved in the chemoreception of nonvolatile sex attractants that reach the organ when the animal contacts the vaginal secretions or urine of females. Documentation of this concept is now quite abundant in rodents. Typical of such studies is one in male hamsters showing that a deafferentation of the main olfactory, but not the vomeronasal, system rendered them unable to detect airborne vaginal secretions from estrous female hamsters (Powers *et al.*, 1979). If the hamsters were induced to contact the secretions by placing secretions on anesthetized castrated male hamsters, however, the subjects were able to detect the sex attractants. In guinea pigs, males show a much stronger preference for urine from females than from males. Discrimination between male and female urine is diminished, however, when the subjects cannot contact the urine (Beauchamp *et al.*, 1980). In the same laboratory, it was found that a fluorescein dye, when placed in urine, reached the vomeronasal organ of males that were allowed to contact the urine. The dye apparently reached the organs through the external nares, probably when the animals licked their muzzles (Wysocki *et al.*, 1980).

In ungulates, such as goats, cattle, and sheep, it has been argued for several years that the lip-curl or flehmen response of males, which frequently occurs when they contact females or their recently voided urine, is involved in the transfer of substances from the oral cavity to the vomeronasal organ through the nasopalatine duct (Estes, 1972; Knappe, 1964). Actual documentation of this hypothesis was missing until it was found that a fluorescein dye placed in the oral cavity was introduced into the middle and posterior parts of the vomeronasal organ during flehmen (Ladewig and Hart, 1980). Flehmen, as well as the involvement of the vomeronasal organ, is apparently not required for olfactory discrimination of estrous female goats, and in fact, male goats flehmen more frequently to diestrous urine samples than to estrous urine samples (Ladewig *et al.*, 1980). It is proposed that in goats, and possibly other ungulates that perform flehmen, the vomeronasal organ is used to confirm the diagnosis of the presence or absence of a sex attractant made by the main olfactory organ, or to detect low levels of a sex attractant, as might be involved in discriminating diestrous urine or vaginal secretions from proestrous material (Hart, 1983a). Interestingly, flehmen behavior is under the same kind of gonadal androgen control as mating behavior (Hart, 1983a). For example, flehmen was found to occur more frequently in deer during the rutting season than at other times (Muller-Schwarze, 1979). In tropical male goats, castration reduced the occurrence of flehmen during mating tests (Hart and Jones, 1975).

Most of the studies designed to explore the effects of anosmia on male sexual behavior have involved the production of anosmia by peripheral means, which may or may not eliminate accessory as well as main olfactory function, or the production of anosmia by olfactory bulbectomy, which eliminates the function of both chemoreceptive systems. In general, these procedures have little effect on sexual behavior in the laboratory situation. Peripheral anosmia does not impair the copulatory performance of sexually experienced dogs (Hart and Haugen, 1972) or rhesus monkeys (Goldfoot *et al.*, 1978). In cats (Aronson and Cooper, 1974) and sheep (Fletcher and Lindsay, 1968), olfactory bulbectomy does not impair sexual activity in sexually experienced animals. In male rats, there is some disruption of sexual behavior following bulbectomy, especially if the animals are sexually naive before the operation is performed (Bermant and Taylor, 1969; Larsson, 1969, 1971). Olfactory bulbectomy results in longer intromission and ejaculation latencies, and in a lowered probability of ejaculation in each test. In contrast, copulatory behavior is virtually eliminated in male hamsters (Murphy and Schneider, 1970; Powers and Winans, 1973) and mice (Rowe and Edwards, 1972) following bulbectomy. In these animals, impairment of sexual behavior is evidently due partially to loss of vomeronasal organ input as well as to loss of the main olfactory system.

A demonstration of the importance of the vomeronasal organ in male sexual behavior comes from work showing that sectioning the vomeronasal nerves impaired sexual behavior in approximately one half of the male hamsters tested (Powers and Winans, 1975; Winans and Powers, 1977). When those animals that continued to mate were made anosmic by peripheral treatment with zinc sulphate, they, too, stopped copulating. Depending on the individual animal, either one of the olfactory systems might have maintained sexual activity, but the elimination of both systems resulted in loss of copulatory behavior.

Numerous studies have revealed the importance of the olfactory bulbs in several nonolfactory behavioral activities, including habituation, reactivity, aggression, active and passive avoidance, temperature regulation, water balance, and endocrine function (Edwards, 1974; Wenzel, 1974). The bulbs are properly considered part of the limbic system, and damage to them affects limbic system function in ways unrelated, or only partially related, to the loss of olfactory information. Cain (1974) pointed out that complete bilateral olfactory bulbectomy destroys about 4% of the central nervous tissue rostral to the spinal cord in the rat. Nonetheless, the work reviewed above suggests that the olfactory bulbs are not essential to the mediation of male sexual behavior, apart from relaying the main olfactory or vomeronasal information to the basal forebrain.

The fact that in most species totally anosmic (bulbectomized) animals mate under laboratory situations does not, of course, mean that olfaction or olfactory-bulb neural activity is unimportant in the expression or the development of sexual behavior. Perhaps a much clearer picture of the importance of olfaction in normal male reproductive behavior could be gained by studying animals in natural or seminatural settings or in situations in which social experience or developmental factors are taken into account. For example, combining bulbectomy (Wilhelmsson and Larsson, 1973), or even peripheral anosmia (Thor and Flannelly, 1977), with social isolation markedly suppresses sexual responding in male rats.

GENITAL SENSORY MECHANISMS

Sensory information conveyed to the central nervous system through olfaction and other special senses obviously plays a role in arousing sexual interest in the male and in coordinating movement and orientation of the male toward the female prior to copulation. Stimulation of the penis, acquired by copulatory movements, is usually associated with penile erection and ejaculation. However, erection and even ejaculation can occur in the absence of copulatory or masturbatory stimulation of the penis. The most common example is the phenomenon of nocturnal erection and emission in human males (Ramsey, 1943; Kinsey *et al.*, 1948). Expulsion of seminal fluid along with erection and pelvic thrusting has been observed during sleep in cats (Aronson, 1949). Seminal plugs have been observed in drop pans beneath the wire mesh floors of cages housing rats when the animals are prevented from engaging in genital grooming or stimulation (Beach *et al.*, 1966; Orbach, 1961; Orbach *et al.*, 1967). Interestingly, spontaneous seminal emissions that occur in the absence of genital stimulation are suppressed following coital ejaculations in men (Kinsey *et al.*, 1948) and in rats (Beach, 1975; Van Dis and Larsson, 1970).

Sensory receptors located on the penis can theoretically serve a number of functions. For one thing, there are receptors conveying sensory information about pain from infection, traumatic injury, and possibly excessive sexual stimulation. Mechanoreceptors of both the rapidly and the slowly adapting types (Kitchell *et al.*, 1982) have been found and obviously play a role in the location of the vulvar orifice and in the coordination of copulatory movements. Temperature receptors also exist and in some species may play a critical role in detection of the vagina and in copulatory attempts. It is well known, for example, that the temperature of the artificial vagina is extremely important in inducing the copulatory thrusting of bulls during the collection of semen. With a cold artificial vagina, the male will not thrust.

Finally, there are presumably the receptors that, when stimulated, trigger or cause the release of the ejaculatory response. Such receptors have not been described histologically or examined electrophysiologically for any species. It might be that receptors that convey "ordinary" tactile or pressure information to the central nervous system become converted to special copulatory receptors capable of facilitating or triggering ejaculation once tumescence occurs. This concept is suggested by the observation by Kitchell *et al.* (1955) of encapsulated endings in the clitoris of female sheep, in which long, fine, wavy fibers emerge from one end to terminate as free nerve endings in the overlying stratified squamous epithelium. Similar sensory endings have been seen in the vulva of the cat and in the palate of humans (Garven and Gairns, 1952). A functional interpretation of this structural arrangement is that tumescence in the genitalia alters sensitivity in these genital corpuscles from that of "ordinary" cutaneous sensation to one of sexual significance. The encapsulated portion, when stimulated, could possibly modulate impulses traveling through the fiber that it surrounds.

The most distal end of the penis, the glans, is covered by an epithelium that in dogs, rats, and humans is highly distensible. In fact, the epithelium could be considered very similar to the transitional epithelium lining the inside of the urinary bladder and the urethra (Hart, 1970a). In species such as these, the glans penis contains many venous sinuses, and during erection, the sinuses become greatly distended, causing stretching of the epithelium. In the rat (Phoenix *et al.*, 1976) and the cat (Aronson and Cooper, 1967), papillae are located on the epithelium of the glans, and the stretching presumably results in a protuberance of these papillae. Encapsulated nerve endings have been anatomically associated with the papillae of the rat penis (Beach and Levinson, 1950).

In species that do not have a vascular type of engorgement, the distal end of the penis becomes rigid. Because the erectile spaces are limited by a strong, fibrous connective-tissue capsule, the surface epithelium in such animals as goats, cattle, and sheep does not undergo stretching.

It is known that, in rats (Beach and Levinson, 1950; Phoenix *et al.*, 1976) and cats (Aronson and Cooper, 1967), epithelial papillae over the surface of the glans undergo degeneration following castration. In species such as rabbits, dogs, and humans, in which there are no specialized epithelial derivatives covering the epithelium, it is not known if castration affects the histology of the epithelium. In the rat, and presumably in other species, castration also results in a decrease in penile size. Thus, whether or not papillae are present, it is conceivable that castration could affect the type of sensory information relayed to the central nervous system from the receptors associated with the epithelium.

The only experiment relevant to this consideration is a study by Cooper and Aronson (1974), in which it was found that the sensitivity threshold and the pattern of neuronal discharge, which were evoked by stimulating the epithelium of the glans penis with von Frey hairs, did not differ between intact and castrated male cats. It would also be useful to know whether this finding is true when the penis is erect.

Sectioning the dorsal nerve of the penis, which supplies much of the sensory area of the penis, appears to prevent erection and ejaculation in cats (Aronson and Cooper, 1968), rhesus monkeys (Herbert, 1973), and rats (Lodder and Zeilmaker, 1976). Some investigators have reported that sectioning the dorsal nerve or the pudendal nerve (from which the dorsal nerve originates) in the rat impairs, but does not eliminate, intromissions and ejaculations (Dahlof and Larsson, 1976; Larsson and Sodersten, 1973), but these results may be due to a surgical procedure that leaves some nervous innervation intact.

Assuming that sectioning the dorsal nerve does prevent the elicitation of ejaculation, it seems most likely that the receptors involved are located in the body of the penis, proximal to the glans but distal to the paired crura of the root. In tests with both spinal male rats and male dogs, only stimulation to the body of the penis, and not stimulation of the glans, evokes penile movement or erection and ejaculation (Hart, 1967a, 1968c). Also, penile anesthetization in spinal male rats reduces but does not eliminate penile reflexes (Hart, 1972b). Male rats with the glans penis removed are still capable of showing all copulatory patterns, including the ejaculatory response (Spaulding and Peck, 1974).

Because animals are highly motivated to seek copulatory stimulation, receptor mechanisms associated with the penis are undoubtedly involved in the reinforcement process of sexual activity. Thus, impairment of genital sensation might be expected to attenuate sexual interest or motivation apart from blocking erection and ejaculation. Aronson and Cooper (1968, 1977) noted that their cats with penile deafferentation became very disoriented in their mounting. Monkeys with sectioned dorsal nerves developed disoriented or abnormal thrusting patterns (Herbert, 1973). Eventually, over a period of months and years, there was a decline in mounting in both cats and monkeys. Further evidence of the reinforcing value of penile stimulation on sexual behavior is presented by Lodder (1976) in a study showing that penile deafferentation prior to any heterosexual contact greatly impaired the acquisition of mounting behavior by male rats.

GENITAL EFFECTOR MECHANISMS

The precopulatory or appetitive aspects of male sexual activity, including orientation toward the female, mounting, and thrusting, are basically manifestations of skeletal muscle activity, presumably mediated or controlled by forebrain structures. These responses are all influenced by concurrent sensory stimulation, early sexual experience, inherited predisposition, sexual satiation, and, of course, gonadal hormones. The coordination of all of these influences is highly complex and will probably defy direct and detailed neurological analysis for decades. Nonetheless, this has been an attractive area for the development of neurobehavioral models. Beach (1956) utilized the concept of a sexual arousal mechanism to deal with precopulatory behavior and a copulatory mechanism to deal with responses related to intromission and ejaculation. More recently, Sachs and Barfield, (1976) have reviewed the efforts of others and presented their own ideas in attempting to integrate new findings into an analysis of sexual arousal and performance, especially in the male rat. The peripheral effector mechanisms of erection and ejaculation have been of some concern in the various models of sexual behavior and are the topic of this section.

PENILE ERECTION

To at least some degree in all species, penile erection is due to engorgement of the sinusoids in the erectile bodies, which are the corpus cavernosum penis and the corpus spongiosum penis. The paired corpora cavernosa penis helps form the root of the penis. It is covered by a heavy fibrous capsule or tunic; thus, when it becomes completely engorged, it becomes rigid. This erectile body extends from the ischial tuberosity of the pelvic bone to a penile bone (os bone) in those species that have a bone, and its engorgement results in the penis's becoming rigid from the tip of the penile bone to the pelvis (Hart and Melese-d'Hospital, 1983). In species without a penile bone, the corpus cavernosum penis extends thorughout most of the penis, as it does in humans. The softer, more vascular and expansive corpus spongiosum surrounds the tip of the penile bone or the corpus cavernosum penis. Expansion of this highly vascular erectile tissue, as in the formation of the flared glands (cup) in the rat (Hart, 1968c; Sachs and Garinello 1978) or the bulbus glandis in the dog (Hart and Kitchell, 1966), is the aspect of erection most easily observed in some animals.

It is an established concept that erection is initiated by dilation of helicine arteries in the erectile bodies, allowing blood to flow into the erectile sinusoids. Valvelike mechanisms impede the drainage of blood from the erectile bodies as they begin to engorge. Erection is not necessarily complete, however, when pressure in the erectile bodies approaches that of systemic blood pressure. It is now known that complete engorgement, at least in some species, is achieved by contraction of the striated bulbospongiosus and ischiocavernosus muscles (against a closed venous return system), which increases the pressure in the erectile bodies to several-fold that of the systemic blood pressure. This effect of striated muscle contraction has been shown in the goat, the stallion, and the bull (Beckett *et al.,* 1972, 1973, 1974, 1975).

In the male rat, vascular filling of the corpus spongiosum causes some degree of erection of the glans penis. However, contraction of the bulbospongiosus muscle, which surrounds a bulblike expansion of the corpus spongiosum at the base of the penis, forces blood into the distal glans, causing the flared-cup erection (Hart and Melese-d'Hospital, 1983; Sachs, 1982).

Another way in which striated muscles can participate in the erectile process is by occluding venous drainage. In the dog, for example, the ischiourethral muscle, which originates on the pelvic ischial tuberosity and inserts on a fibrous ring surrounding the dorsal vein of the penis, accelerates the rate of engorgement of the bulbus glandis. Sectioning this muscle prevents complete engorgement of the bulbus in the dog so that there is no genital lock (Hart, 1972a).

Contractions of the striated muscles can also cause movements of all of the penis, or just part of it. This has been recently noted in the male rat, where penile flips are generated by contraction of the ischiocavernosus muscles; these muscles originate on the pelvic bone and insert near the attachment of the penile bone to the corpus cavernosum penis. The contraction of these muscles appears to extend the flexure of the penis where the penile bone attaches (Hart and Melese-d'Hospital, 1983).

ANDROGEN INFLUENCES ON PENILE MUSCLES. Because striated muscle is involved in the attainment of complete erection in some species, it is significant that the penile muscles seem to be particularly sensitive to androgen influences. This has been best documented in the levator ani muscle of the rat, which is readily influenced by castration and androgen replacement (Hayes, 1965). Specific binding of androgens by this muscle, but not skeletal muscle in general, have been reported (Jung and Baulieu, 1972). There are a number of different metabolic influences of androgen in the levator ani that do not occur in skeletal muscle in general (Bergamini, 1975).

Of more significance is the finding that motor neurons of the striated penile muscles in the rat concentrate testosterone (Breedlove and Arnold, 1980). One would therefore expect testosterone to modulate the contraction activity of these striated muscles. This is evidently the case. It has been known for a long time that the systemic administration and withdrawal of testosterone alters the number of penile erections and flips that can be recorded on penile reflex tests in rats (Hart, 1967b). This modulation of penile reflexes can occur within a few hours (Hart *et al.,* 1983).

AUTONOMIC NERVOUS SYSTEM. The fact that the process of erection is influenced not only by autonomic nervous system control of smooth muscle in the helicine arteries but also by somatic efferent control of striated penile muscles may have confused the interpretation of experiments testing the effects of autonomic denervation on erection and mating behavior.

In animal studies, a fairly uniform finding is that sectioning of the hypogastric nerves that presumably carry sympathetic innervation to the genitalia, or sympathectomy itself, has little noticeable effect on erection, intromission, or ejaculatory behavior. This is true of rats (Larsson and Swedin, 1971); guinea pigs (Bacq, 1930); dogs (Beach, 1969); cats (Root and Bard, 1974); and rabbits (Hodson, 1964). Some human patients (ranging from 10% to about 28%) undergoing different types of sympathectomy suffer disturbance of erection (Bues *et al.,* 1957; Whitelaw and Smithwick, 1951), but this disturbance seems explainable on the basis of secondary factors due to surgical intervention (Larsson and Swedin, 1971). Because the sympathetic nervous system does play a role in emptying the accessory sex organs and in contraction of the vas deferens during ejaculation, it is not surprising that animals sustaining sympathectomy become infertile (Swedin, 1971).

The pelvic nerve is believed to carry parasympathetic innervation to the sexual organs. The conventional viewpoint is that relaxation of smooth muscle in the walls of the helicine arteries is caused by parasympathetic activation. In a study on the rat, transection of the pelvic nerve led to no noticeable effect on erection or mating behavior, whereas pudendal nerve transection prevented erection and hence intromission and ejaculation (Lodder and Zeilmaker, 1976). The investigators concluded that erection in the rat must be processed thorugh parasympathetic fibers traveling in the pudendal nerve. The study of Semans and Langworthy (1939) of several decades ago showed that, in the cat, the retention of either the sympathetic or parasympathetic system was sufficient to allow for erection.

Evidence from work on dogs, ungulates, and rats now demonstrates that contraction of the striated penile muscles potentiates or causes erection; hence, the somatic efferent nervous system and also the autonomic nervous system are involved in erection.

Ejaculation involves a series of neural reactions causing expulsion of seminal fluid as well as a stereotyped behavioral response. It is the one aspect of sexual behavior that is most easily conceptualized as a reflex. Yet expulsion of seminal fluid with some postural adjustments and body movements can be routinely elicited reflexively by penile stimulation in spinal males in just one species, the canine. The reflexive mechanisms involved in ejaculation are dealt with in the next section.

Orgasm may be experienced in human males independent of ejaculation. Bors and Comarr (1960) noted that orgasm has been reported by patients with sympathectomy as long as innervation of striated muscles of the pelvic floor and skin of the penis remain intact. Thus, orgasm is perceived in the absence of seminal emission and with denervation of the smooth muscle of the genital tract. On the other hand, orgasm can also be experienced, in spite of deafferentation of the striated pelvic floor muscles and anesthesia of the penis, as long as sympathetically mediated functions such as seminal emission remain intact.

Using normal human male volunteers, Kollberg *et al.* (1962) and Peterson and Stener (1970) utilized electromyography to study the activity of striated muscles on the pelvic floor during ejaculation and orgasm. Figure 1 shows electromyographic activity in some of these muscles during orgasm and subsequent ejaculation. Interestingly, the sensation of orgasm reported by the subject in Figure 1 occurred when a muscle, which the authors indicated was *probably* the deep transverse perineal, was contracting tonically, and ejaculation occurred during the rhythmic contraction of the urethral sphincter. Because these muscles are undoubtedly supplied with afferent endings, it is possible that the normal sensory effects of orgasm and ejaculation are associated with contraction of these muscles. Activation of afferent endings in the striated muscles would represent the peripheral manifestation of orgasm, as compared with a more central process that might be mediated in the brain. The notion that the sensation of orgasm is associated with contractions of

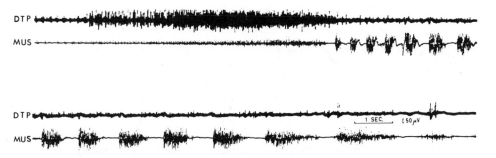

Fig. 1. Electromyogram of the striated urethral sphincter (MUS) and one other muscle (DTP), which was probably the deep transverse perineal, in a human male. The sensation of orgasm reported by the subject occurred when the latter muscle was contracting tonically. (From "Preliminary Results of Electromyographic Study of Ejaculation" by S. Kollberg, I. Petersén, and I. Stener, *Acta Chirologica Scandinavia*, 1962; *123*, 478–483. Copyright 1962 by *Acta Chirologica Scandinavia*. Reprinted by permission.)

the striated genital muscles is at least as plausible as the more conventional notion of associating these sensations with contractions of smooth muscle along the genital tract (Kuntz, 1953).

One other interesting aspect of these electromyographic recordings is the possibility of separating the physiological counterparts of orgasm and ejaculation. It is evident from studies of human patients (Beach *et al.,* 1966; Bors and Comarr, 1960; Money, 1961) that orgasm is not always accompanied by seminal emission.

Just after the occurrence of ejaculation, there is an interval in which males have no interest in copulating and seem to be incapable of achieving another ejaculation (this is called the postejaculatory refractory period). There is reason to believe that some of this refractoriness is spinal in nature and is related to the elicitation of the ejaculatory reflex. This possibility is dealt with in the next section. The refractory period is also a function of supraspinal mechanisms and can be attenuated by such factors as the introduction of a new female and by arousing stimuli such as peripheral electrical shock (Sachs and Barfield, 1976). As is also discussed later, lesions of certain rostral midbrain areas can dramatically shorten the refractory period, a finding that suggests the involvement of certain brain-stem neural systems in the refractory process. Electroejaculation, at least as studied in rats (Arvidsson and Larsson, 1967) and sheep (Pepelko and Clegg, 1965), does not affect the male's ability to copulate and ejaculate immediately afterward. This finding indicates that certain specific neural events must occur, and not just the expulsion of seminal fluid, to produce the refractory phenomenon.

SPINAL REFLEXES

As the first step in examining the central-processing integration link between sensory and effector functions, it is logical to look at reflexive aspects of sexual behavior. Sensory information reaching the spinal cord from the genitalia is obviously conveyed to the brain. By direct or collateral input, genital sensory neurons also reflexively activate peripheral responses, particularly erection and, in some instances, also ejaculation. The occurrence of erection, in turn, leads to a different configuration of genital afferent input into the brain because the receptors are obviously stretched or otherwise affected by erection.

Systematic observations on human patients revealing that erection could be evoked in paraplegic and quadriplegic individuals date back to the postwar period of the mid-1940s (Munro *et al.,* 1948; Talbot, 1949, 1955; Zeitlin *et al.,* 1957). A further interest in understanding and improving the sexual function of casualties from recent military conflicts has led to a more detailed evaluation of sexual reflexes in men with spinal cord injury (Bors and Comarr, 1960; Comarr, 1970; Comarr and Gunderson, 1975; Comarr and Vigue, 1978a,b; Tsuji *et al.,* 1961).

One conclusion from this more recent work on human patients, and from other studies on laboratory animals, is that erection is easily evoked in spinal individuals, but ejaculation (or seminal expulsion) is uncommon. Only in the dog have ejaculatory contractions and seminal expulsion been regularly elicited by genital stimulation of the chronically maintained spinal subject. Ejaculation is rarely reported in human paraplegics with a complete transection.

In chronically maintained spinal male rats, ejaculation rarely occurs in tests for genital reflexes. However, mild electrical shock, applied bilaterally to the skin of the flank, occasionally evokes an ejaculatory complex consisting of seminal expulsion followed by penile erections and flips (Hart and Odell, 1980, 1981). The ejaculatory reaction appears to be inhibited by penile stimulation because when the penile sheath is retracted—an action that delivers some tactile stimulation to the body of the penis—electrical shocks are much less likely to evoke ejaculation. This observation would explain why the reflex tests employed in numerous previous studies in rats seldom produced an ejaculatory reaction. Seminal expulsion (usually spontaneous) has been observed in spinal male guinea pigs (Bacq, 1930) and male cats (Dusser de Barenne and Koskoff, 1932), but only within a few hours after transection. Such seminal expulsion would seem to be due to stimulation of the distal stump of the cord resulting from the surgical procedure and not to a reflexive response.

GENERAL COMMENTS

Before going into individual species, a few more introductory comments are appropriate. All of the work on spinal male rats and dogs has involved transection of the spinal cord in the midthoracic region (Hart, 1978a,b). The reason is that this level is well above those segments involved in penile erection and movement but caudal enough to allow the animal to move about with its front legs and easily eat, drink, and groom. However, transection at the midthoracic level leaves the motor neurons innervating the front legs and neck muscles separated from reflexive elements below the transection level. Thus, the reflexive movements of the front legs and the head associated with copulatory responses are not evident.

The mechanism by which the brain seems to modulate sexual reflexes is also of relevance. Undoubtedly, both facilitation and inhibition are involved. Most evidence suggests that inhibition and disinhibition, rather than facilitation, are involved. Manual stimulation of normal spinally intact animals rarely evokes the same intensity of erection and limb movements as that seen during copulation, apparently because sexual reflexes are normally held under tonic inhibition. During copulation, there is a reduction or disinhibition of this supraspinal inhibition (Beach, 1967; Hart, 1967a, 1968a). A midthoracic transection eliminates the tonic inhibition, and the intensity of sexual reflexes more closely matches the responses seen during copulation. This is evident in human paraplegics, in whom there is the frequent occurrence of spontaneous erections, reflecting a reduction of descending tonic inhibition once the spinal cord is severed.

The possible effects of gonadal hormones in altering the genital sensory and effector mechanisms involved in sexual function have been alluded to. These areas are associated with the afferent and the efferent arms of sexual reflexes, and one would expect reflex activities to be correspondingly affected by androgen. However, there is also evidence that neural elements within the spinal cord are modulated by androgen as well. This consideration receives some attention below.

In both spinally transected and spinally intact male rats, the occurrence of penile reflexes is evidently governed by a spinal pacemaker that appears to operate independently of a major supraspinal modulation (Hart, 1968c; Sachs and Garinello, 1979). Whether the pacemaker, which results in the occurrence of clusters of responses every 2–3 min, has any major role in copulatory behavior remains to be seen. The timing characteristic of the pacemaker can be governed to some degree by the application of mild electrical shock to the flank (Hart and Odell, 1980).

SEXUAL REFLEXES OF MALE DOGS

The first suggestion that movements of the trunk and limbs during copulation might be manifestations of spinal reflexive activity should be attributed to Sherrington. In a contribution to a textbook (1900) he refers to unpublished observations made on spinal male dogs in his laboratory. Touching the preputial skin over the penis evoked extension of the legs, and stimulating the portion of the penis proximal to the glans elicited ventral curvature of the spinal column, causing the penile bone to be thrust forward. Muller (1902, 1906) noticed that erection and sperm production were maintained in dogs with cord transection and that reflexogenic erections were present as long as there was a functioning sacral and coccygeal cord.

The more recent work on spinal male dogs represented an attempt to replicate and extend Sherrington's and Muller's observations by utilizing electromyography to monitor ejaculatory contractions of penile muscles (Hart, 1967a; Hart and Kitchell, 1966) and by comparing sexual reflexes of spinal dogs with ejaculatory responses in intact dogs during copulation (Hart, 1972a). In Figure 2, sporadic contractions of the ischiourethral and the ischiocavernosus and rhythmic contractions of the bulbospongiosus muscles are shown during copulation in an intact male dog and during elicitation of an ejaculatory response in a spinal dog.

Based on (1) movements of the posterior trunk and hind legs, (2) penile erection, and (3) seminal expulsion, four different sexual reflexes are evident in chronic spinal male dogs (Hart, 1967a). One is a response, involving shallow thrusting along with partial erection, that is elicited by rubbing the preputial sheath over the glans of the penis. This reflex does not include any seminal expulsion. A reflex of a different type is elicited by touching the coronal area of the glans penis and is characterized by lordosis of the back, together with a strong extension of the rear legs and penile detumescence, if the penis is erect at the time of stimulation.

These two reflexes appear to play different roles in copulation. The reflex characterized by pelvic thrusting was evidently seen by Sherrington (1900) in his earlier work and probably contributes to the male dog's ability to achive complete intromission once partial intromission is obtained. The reflex, characterized by lordosis and rapid detumescence, may be important in inducing detumescence if the penile erection has progressed too far during the male's mounting and thrusting

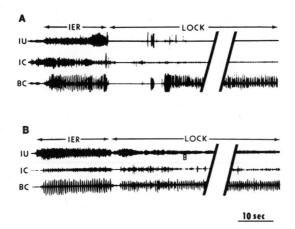

Fig. 2. Electromyographic recordings from the ischiourethral (IU), ischiocavernosus (IC), and bulbospongiosus (BC) muscles of male dogs. A. Recordings during copulation in an intact dog during an intense ejaculatory reaction (IER) and initial part of the copulatory lock (from Hart, 1972a). B. Recordings from a spinal male dog during elicitation of an IER and simulated lock. The break in the record indicates that a long section has been removed. (From "Sexual Reflexes and Mating Behavior in the Male Dog" by B. L. Hart, *Journal of Comparative and Physiological Psychology* 1967, *64*, 388–399. Copyright 1967 by the American Psychological Association. Reprinted by permission.)

for him to achieve intromission, as complete engorgement of the glans penis (and bulb) would prevent intromission.

Two other reflexes in the spinal male dog involve erection and ejaculation. Applying pressure and rubbing the body of the penis just proximal to the glans evokes an intense ejaculatory reaction that has a duration of 15–30 sec. It is characterized by pelvic movement, alternate stepping of the back legs, and rapid penile engorgement and expulsion of seminal fluid, which occurs along with rhythmic contractions of the bulbocavernosus muscle. This reaction has a similar duration and pattern in both intact and spinal subjects (Figure 2). If the penis is stimulated again, the response does not occur before the passage of several minutes. This transient lack of responsiveness is the best evidence that a postejaculatory refractory interval occurs within a reflexive system. The refractory interval ranges from 5 to 30 min, which is less than the customary postejaculatory interval of intact dogs (Hart, 1967a).

The fourth sexual reflex of spinal male dogs is evident after the subsidence of the intense ejaculatory reaction if stimulation of the penis proximal to the glans is continued and, in addition, pressure is applied to the penile tip. This reflex is referred to as a simulated lock because it appears to play a role in the genital lock. No leg movement occurs during this reflex, but full erection of the penis can be maintained for several minutes. Seminal expulsion and rhythmic contractions of the bulbocavernosus muscle continue on an intermittent basis. This response seemed to "run down" after 10–30 min (Figure 2).

Testosterone has a pronounced effect on the latter reflex. For example, when testosterone was withdrawn, the simulated-lock duration was reduced from 11.6 to 4.2 min in 60 days (Hart, 1968b). This change in simulated lock parallels the reduction in genital-lock duration as seen in spinally intact dogs following castration (Hart, 1968b).

SEXUAL REFLEXES OF MALE RATS

Unlike in dogs, erection and ejaculation in the rat cannot be evoked by simply rubbing part of the penis (Hart, 1968c). However, when the preputial sheath is held behind the glans, a series of clusters of genital responses occurs every 2–3 min. The responses cannot be elicited any more frequently or regularly by any other type of genital stimulation. Retracting the sheath puts some pressure on the body of the penis but allows for virtually no stimulation of the glans itself. The clusters of genital responses evoked by this type of stimulation begin with three or four erections, which are either partial or full. During some of the erections, the penis becomes so engorged that it resembles a flared cup. The erectile process occurring during each erection takes about 1 sec, and some erections are followed immediately by a quick dorsal flip of the glans. These "quick flips" have a duration of less than 1 sec. The quick flips are accompanied by no body or pelvic movement. After one to three quick flips, there usually occurs a third response consisting of erection of the glans followed immediately by a long, extended flip of the glans lasting 1–2 sec. These responses are referred to as *long flips*. A long flip is accompanied by ventral flexion of the pelvis. When additional pressure is applied to the sides of the penis during a long flip, the duration and intensity of the long flip may be increased. Erections, quick flips, and very occasionally long flips can be evoked in intact male rats held in the supine position (Hart, 1968c; Kurtz and Santos, 1979; Sachs and Garinello, 1978).

Compared with sexual reflexes in the male dog, the reflexes of the male rat appear to be fragmentary. This is not what one would expect because, if anything, the central nervous system of the male rat should be less encephalized than that of the male dog, from which fact it would follow that the reflexes would form a more complete component of copulatory behavior. One method of producing a more complete reflex system for behavioral analysis is to perform the spinal transection on neonates. This procedure allows various somatic reflex systems, such as standing, walking, and foot placing, to develop more completely than those seen in rats transected in adulthood (Stelzner *et al.*, 1975). However, neonatal transection did not lead to any enhancement or alternation of penile reflexes (Hart, 1980a).

One could argue that penile reflexes displayed by the spinal male rat are not fragmentary or incomplete in comparison to those of the dog, but that we lack an understanding of their role in copulatory behavior. A key point is that reflexes are undoubtedly tied to the species-typical copulatory patterns. This is evident from a series of recent studies. Mild electric shock applied to the flank region of spinal male rats occasionally produces ejaculation (apparently through induction of some excitatory state within the lower spinal cord). These ejaculatory responses are followed by several penile reactions and flips (Hart and Odell, 1981). Because erections and flips follow, rather than precede, ejaculation, they would appear to be important in packing and sealing seminal plugs against the vaginocervical junction while seminal material is coagulating. This temporal arrangement of erections and ejaculation in the rat contrasts with the dog, in which erection precedes or is concomitant with ejaculation.

Another observation bearing on species-typical behavioral differences is that, in spinal male rats, when the penile sheath is retracted (producing penile stimulation), ejaculation is unlikely to be evoked by electrical shock (Hart and Odell, 1981). Penile stimulation has an inhibitory effect on the occurrence of the ejaculatory reflex. The functional importance of this spinal inhibition in the copulatory pattern of rats may be that, during intromissions, retraction of the sheath and penile stimulation are inhibitory to ejaculation. This inhibition produces a series of obligatory intromissions that precede ejaculation. Of course, ejaculation must eventually occur during one of the intromissions, so undoubtedly there must accumulate in the brain an excitatory potential from several intromissions that eventually becomes sufficient to suppress the spinal inhibitory effect. Ejaculation would then occur during the subsequent intromission.

One might ask why male rats have these obligatory intromissions that result from the spinal inhibitory process. The importance of multiple intromissions in the species-specific mating pattern of the male rat has been documented in that intromissions must precede ejaculations to produce a progestational state and to facilitate transcervical sperm transport in the female (Adler, 1978). The necessity of intromissions to the male is that a male can find himself in company with another male rat in mating with a particular female. If one male has just ejaculated and deposited a seminal plug, the second male must remove his competitor's seminal plug before he can deposit his own sperm near the cervical opening. Intromissions without ejaculation serve beautifully to dislodge the first male's plug. And if the dislodging occurs soon enough, the fertilizing potential of the first male is reduced (Adler, 1978; Dewsbury, 1981; Matthews and Adler, 1978). Penile flips and erections undoubtedly occur in conjunction with intromission. If the penis is held rigid, as it would be during a long flip response, a seminal plug would be more easily dislodged. The production of a cuplike erection within the vagina would then help to pull the plug posteriorly by an action not unlike that of a miniature toilet plunger (Hart, 1983b).

To complete the analysis, it has been learned through electromyographic analysis and muscle excision experiments that the bulbocavernosus muscle is essential for the production of penile cups and that the ischiocavernosus muscle potentiates the display of penile flips. Removal of the bulbocavernosus and ischiocavernosus muscles causes the disappearance of cups and the reduction of penile flips, respectively (Hart and Melese-d'Hospital, 1983; Sachs, 1982). The removal of either of these muscles in male rats that are allowed to copulate with females interferes with the animal's ability to dislodge the seminal plug left by another male (Wallach and Hart, 1983). The bulbocavernosus muscle seems to be particularly important in this regard, a finding suggesting that the toilet plunger effect of erectile cups is indeed important. Removal of the bulbocavernosus muscle also disturbs the male rat's ability to deposit a tight and well-formed seminal plug (Sachs, 1982; Wallach and Hart, 1983). Removal of the ischiocavernosus muscle, with disruption of penile flips or penile extension, has only a minor effect on the tightness of the seminal plugs (Wallach and Hart, 1983).

These observations on the role of spinally mediated penile responses in copulatory behavior and sperm competition in the male rat reveal a completely different picture from the one we would expect in attempting to extrapolate from work on spinal dogs. Much of the specificity of the copulatory behavior pattern is coded at the spinal level.

What is intriguing about the study of sexual reflexes is that there is a pronounced effect of systemic withdrawal and administration of testosterone on the occurrence of these reflexes in both dogs (Hart, 1968a) and rats (Davidson et al., 1978; Hart, 1967b, 1979). The research is the most definitive in rats, in which spinal motor neurons that innervate the striated penile muscles have been shown to concentrate testosterone and its metabolite, dihydrotestosterone (Breedlove and Arnold, 1980). Thus, the link between the alteration of systemic levels of testosterone and its effects on penile flips and erections appears to be partially through the influence of the androgen on the motor neurons. Because it is known that sexually relevant cues can produce a transient increase in testosterone secretion (Graham and Desjardins, 1980), and that testosterone elevation potentiates penile reflexes, one could make the case that sexual excitement leads to potentiated reflexes that enhance a male rat's ability to dislodge seminal plugs left in the vagina by other males and to deposit his own plugs as tightly as possible so as to resist dislodgment (Hart, 1983b).

Of course, neurons in the brain also concentrate testosterone; thus, the picture emerges of two neuronal steroid target sites that mediate male copulatory behavior (Hart et al., 1983). The dual system applies to the anatomical separation and also to the fact that brain neurons are responsive to the estrogenic metabolite of testosterone, whereas spinal motor neurons are not. In keeping with this concept of dual target sites, one might expect to find differences in activation thresholds or latencies. These differences do, indeed, exist: The spinal neuronal system underlying penile responses is much more sensitive to changes in testosterone than is the brain, which mediates the motivational and appetitive aspects of sexual behavior (Hart et al., 1983).

Brain Mechanisms

Approaches and Problems

A great diversity of brain structures is involved in male sexual behavior. This diversity is evident at one level from the multitude of sensory systems that influence the behavior. We have dealt with the olfactory and the vomeronasal systems in some detail and have touched on the role of other special senses in the arousal of sexual interest. It is also evident that mildly painful stimulation increases sexual activity (Sachs and Barfield, 1976), as it does a number of other "consummatory" responses in rats (Antelman et al., 1975). Visceral afferent input from the pelvic genital organs plays a role in the sensations of orgasm and ejaculation, if not in adding, as well, to sexual motivation. Tactile stimulation of the penis seems to increase sexual interest, and it activates the responses of erection and ejaculation under normal circumstances.

The complexity of brain structures involved is also indicated by the types of responses that are evident during sexual activity. In addition to autonomic activation, there is somatic efferent activation of striated muscles throughout the body. Further, there is now evidence of increased testosterone and LH production in some species following copulation (Kamel *et al.*, 1977) and even in anticipation of copulation (Graham and Desjardins, 1980).

The approaches used to study brain mechanisms in behavior have involved lesions produced by aspiration, electrocoagulation (direct current or radiofrequency current), and stereotaxically directed knife cuts for examining pathways. Brain stimulation has been carried out in acute laboratory experiments on anesthetized animals and through the use of chronically implanted electrodes for stimulation in restrained or unrestrained conscious animals. A few investigators have employed electrical recording techniques to study localized activities in brain areas related to sexual function.

The neuroendocrine aspects of brain control of sexual behavior have involved examining the effects of the implantation of solid steroids in different areas, as well as determining which brain structures take up hormones or hormone metabolites from the circulation.

In the paragraphs below, we comment briefly on some of the advantages, technical difficulties, and conceptual problems involved in the lesion, stimulation, and neuroendocrine approaches.

BRAIN STIMULATION. Investigators have been interested in a number of behavioral parameters with regard to stimulation experiments. In unanesthetized animals, these include observations on stimulus-bound elicitation of normal and exaggerated sexual behavior. In anesthetized subjects, stimulation has been used to map areas from which penile erection and ejaculation can be evoked.

Roberts (1969) called attention to the difficulties in interpreting results of the stimulation approach. Electrical current at the tip of the electrode acts indiscriminately on neurons and axons projecting to a number of other brain areas. In the immediate zone of electrical stimulation, neural function is undoubtedly highly distorted. Only a small proportion of the neural system on one side of the brain is activated by focal electrical stimulation. In general, stimulation reveals the output effects from the stimulated area rather than the neural processing that goes on within the area. Along a different line, Wyss and Goldstein (1976) pointed out that lesion artifacts may occur in brain stimulation experiments.

Even with the above difficulties, many of the stimulation experiments fit very well conceptually with the lesion studies. For example, in all species studied, lesions of the medial preoptic-anterior hypothalamic area eliminate or reduce male copulatory behavior. Correspondingly, electrical stimulation of the medial preoptic area facilitates or activates mating behavior in rats (Malsbury, 1971; Merari and

Ginton, 1975; Van Dis and Larsson, 1971; Vaughan and Fisher, 1962). However, stimulation of the posterior hypothalamus (Caggiula, 1970; Stephan *et al.,* 1971), the lateral preoptic area (Madlafousek *et al.,* 1970), and the medial forebrain bundle, as it extends into the midbrain (Eibergen and Caggiula, 1973), also activates sexual behavior. Only stimulation of the medial preoptic region has been reported to result in a dramatic increase in sexual activity.

Some of the most interesting work on brain stimulation has been conducted by Perachio and his colleagues, using freely moving, socially interacting male rhesus monkeys (Perachio and Alexander, 1975; Perachio *et al.,* 1969, 1973, 1979). Mounting was induced by electrical stimulation of the preoptic area, the lateral hypothalamus, and the dorsomedial nucleus of the hypothalamus. With stimulation of the dorsomedial hypothalamus or the lateral hypothalamus, it was possible to evoke copulation similar to the multiple-mounting pattern of the rhesus ejaculatory sequence. With the onset of stimulation, the stimulated male usually approached the female directly and grasped her flanks, while simultaneously gripping the lower portion of her legs in the ankle clasp of the normal sexual mount. Penile erection developed rapidly, and insertion was usually achieved in one or two thrusts. As the train of stimulation continued, pelvic thrusting occurred.

During test sessions of 1 hr, a well-adapted male exhibited several ejaculations. A stimulus train delivered immediately after an ejaculation was usually ineffective in producing any sexual behavioral response. From 1 to 5 min following an ejaculation, stimulation elicited mounting, but without pelvic thrusting or erection. After a longer interval, the stimulus trains evoked mounting with penile erection, intromission, and pelvic thrusting. Repeated ejaculatory sequences could be elicited by this pattern of hypothalamic stimulation, so that an increased number of ejaculations, compared to the spontaneous rate, could be produced.

Not all stimulation sites have proved to be equally effective in the rhesus monkey experiments. For example, stimulation of the medial preoptic area was quite effective in eliciting mounts. However, despite repeated stimulation, ejaculation did not occur. Thus, stimulation of the medial preoptic area may be more involved in controlling the initiation of copulatory behavior. Stimulation of the putamen, in the rostral area, also evoked mounting, intromission, and thrusting, but only with an estrous female. On the other hand, preoptic stimulation sometimes produced a set of copulatory mounts even if the female was not in estrus.

Stimulation of the anterior, ventromedial nucleus and posterior portions of the hypothalamus resulted in penile erection in socially isolated or restrained males, although stimulation of these same sites was ineffective in eliciting sexual mounting with receptive females.

In anesthetized squirrel and rhesus monkeys, the preoptic locus is one of the most effective sites for evoking penile erection by electrical stimulation (Robinson and Mishkin, 1966). Erection, of at least some degree, can also be evoked from stimulation of several regions, including the caudate nucleus, the hypothalamus, the amygdala, and the pons (MacLean, *et al.,* 1963; MacLean and Ploog, 1962; Robinson and Mishkin, 1968).

NEUROENDOCRINE ASPECTS. It is not our intention to deal in detail with the action of hormones on tissue or the effects of behavioral events on hormone regulation. However, our conception of neural mediation of sexual behavior is influenced by the notion that specific areas of the brain appear to be acted on by gonadal androgen for the maintenance and activation of sexual behavior (as well as of some other masculine behavioral patterns). It is known that particular areas of the hypothalamus, the preoptic region, and the nuclei related to the amygdalostrial system concentrate testosterone (Pfaff, 1968; McEwen et al., 1970; Sar and Stumpf, 1973).

The activation of copulatory responses in castrated male rats by the implantation of small amounts of testosterone propionate into the basal forebrain led investigators (Christensen and Clemens, 1974; Davidson, 1966; Johnston and Davidson, 1973; Lisk, 1967) initially to postulate that the preoptic area is the most androgen-responsive area for the activation of male sexual behavior. Other studies (Kierniesky and Gerall, 1973; Smith et al., 1977) indicate that there are several such androgen-responsive areas in the hypothalamus of the rat, and that the medial preoptic area is only slightly more effective than other areas.

As in other approaches to understanding brain involvement in sexual behavior, there are inherent problems in interpreting the results of cerebral hormone-implantation experiments. Smith et al. (1977) called attention to the possibility that the implanted steroid may be absorbed to such an extent that systemic circulation can deliver material, sufficient to activate behavior, to all parts of the brain. Thus, one would not be observing localized effects. There appears invariably to be some absorption from the site of implantation, so the crucial question is the threshold of the neural tissue for small amounts of steroid in the systemic circulation. Vascular supply to brain areas differs enough to make implants in some areas inadequate controls for absorption effects.

Another problem is diffusion of the steroid from the implanted site to adjacent brain areas; this diffusion may travel along nerve tracts, where resistance may be less. Of course, a lump of steroid implanted in the brain probably does not reach the individual "target" neurons as well as in the natural perfusion of neurons by a steroid or its metabolite through cerebral circulation or cerebrospinal fluid. Just because a steroid activates some degree of behavior in a particular brain locus does not mean that other such target loci are not also important.

We have already noted that various aspects of the genital sensory system, the effector mechanisms, and the central spinal reflex elements are influenced by androgen. Thus, one should not expect completely normal male sexual behavior in castrates with just one cerebral implant of testosterone, even if the androgen is in the most effective target tissue. Incomplete restoration of sexual behavior does indeed seem to be the case with testosterone implants in the brain of male rats (Davidson, 1977).

LESION APPROACHES. It is perhaps worth emphasizing what is obvious to workers in the field. Because a behavioral function is altered by removing or cutting part of the brain does not mean that the area is the center for the function. The analogy made by Gregory (1962) to removing a spark plug from an automobile engine is entertaining if not instructive. With the spark plug out, there may be vibration, spitting, and coughing of the engine as well as loss of power. Attributing the functions of inhibiting vibration, spitting, and coughing to the spark plug, or even considering it a source of engine power, is obviously unreasonable, as is labeling a part of the brain as a male sex center because its removal eliminates copulatory activity.

Several authors have outlined the problems in interpreting lesion studies. The fact that lesions made with stainless steel electrodes can leave irritative (stimulating) iron deposits is well known. Schoenfield and Hamilton (1977) have called attention to the fact that secondary changes in the brain induced by a lesion may be involved in an alteration of behavior. In addition to the direct changes of necrosis and anterograde and retrograde degeneration, there are indirect changes, including transneuronal degeneration, regeneration, axonal sprouting, denervation supersensitivity, and alteration of neurochemical pools. Merely implanting an electrode (cannula) in the brain may lead to hemorrhagic vascular damage that may alter brain function (Boast *et al.*, 1976). In interpreting the human clinical literature, it is important to realize that traumatic lesions may cause a destruction of brain tissue analogous to that caused in the extirpation experiments on animals. On the other hand, subsequent formation of scar tissue may result in an irritative focus acting more like brain stimulation.

This discussion of problems in various research approaches is not meant to cast doubt on the consistent findings that are contributing to the overall concept of the brain mediation of sexual behavior. Rather, we would like to underline the obvious, that is, that several technical approaches are necessary to build up the conceptual picture. Furthermore, the problems alluded to indicate that, in developing our overall view of brain function, there are going to be conflicting and inconsistent results that possibly do not match with what seems to be an otherwise useful picture. Being cognizant of the problems in the interpretation of various experimental approaches allows us a way of dealing with some of the inconsistencies.

Cerebral Cortex

The neocortex plays an obvious role in male sexual behavior apart from general sensory and motor functions necessary to any species-typical behavior. Various types of early experience and heterosexual contacts have been shown to have profound influences on male sexual behavior. These are too numerous to mention here, but some examples might be useful. Male dogs raised in semi-isolation, having little contact with other dogs, show sexual excitement as adults in the presence of estrous females, but the behavior is characterized by a high frequency of incorrectly oriented mounts (Beach, 1968a). Investigators working with cats are aware of the rather long adaptation time that male cats need, when first introduced to the laboratory, before they mate regularly (Whalen, 1963). Sexually experienced male rats usually suffer less impairment of mating performance than sexually inexperienced rats after penile deafferentation (Dahlof and Larsson, 1976; Lodder,

1976) or olfactory bulbectomy (Larsson, 1975; Bermant and Taylor, 1969). It is logical to assume that the cerebral cortex plays a role in these and other experiential aspects of sexual activity. The tendency has been to ascribe to the neocortex functions related to arousing or maintaining sexual excitement or interest. In addition, it has been argued that, in mammals with more highly evolved brains, the neocortex is more important in the overall mediation of sexual behavior than in species with less developed brains (Beach, 1958).

No particular cortical area seems crucial to the mediation of sexual behavior, but with removal of sufficient cortex, males do stop copulating. Generally, the larger the area removed, the greater the decrement in sexual interest and the greater the disorientation in mounting. Early work of Beach (1940) showed that a decline in male rat sexual behavior occurred in proportion to the amount of neocortex removed. This finding seemed to correspond to the equipotentiality, mass-action concept of Lashley (1929, 1943). Subsequent work by Larsson (1962, 1964), however, indicated that the cortex was not entirely equipotential and that smaller lesions than those performed by Beach, especially in the frontal cortex, would eliminate sexual behavior.

In a series of experiments on cats (Beach *et al.*, 1955, 1956; Zitrin *et al.*, 1956), it was apparent that the frontal cortex was particularly important. It appeared that the removal of a proportion of frontal neocortex smaller than in Beach's earlier experiments on rats would eliminate copulatory behavior in cats.

Aronson and Cooper (1977) later reported on some further experiments in cats. They found that ablation of the primary somatic sensory area produced some minor episodes of disorientation, but that mating behavior was still essentially unchanged. After ablating both somatic sensory areas and the four paired association areas, it was found that some cats almost completely lost sexual interest, whereas others continued to mate, albeit with episodes of major disorientation. The authors interpreted the effects of cortical lesions in producing disorientation as reflecting the degree to which the genitalia are desensitized. Langworthy (1943)—who also noticed the disorientation effects, following extirpation of the frontal area of the cortex, on male cat mounting behavior—attributed the effects to motor impairment, as did Beach *et al.* (1955).

It should be noted that, in cats, there is apparently a projection area for the penis and associated anogenital region in the primary and secondary somatic sensory areas (Aronson and Cooper, 1977; Kullanda, 1960). The cortical extirpation work does indicate that this area of the brain is not crucial in the performance of copulatory behavior. From the studies now completed on the rat and the cat, it appears that the cortex is no more important in one species than in another, despite differences in evolutionary brain development.

LIMBIC SYSTEM

Considerable attention has been directed toward examining the role that structures associated with the limbic system might have in sexual activity. This approach was greatly stimulated by the observation of Kluver and Bucy (1939) in rhesus monkeys of a tendency toward "hypersexuality" that followed bilateral temporal-lobe ablation. This behavioral change seemed to be part of a syndrome including visual agnosia; oral compulsive behavior; profound changes in emotional behavior, with loss of fear and aggressiveness; and changes in feeding behavior, with acceptance of meat as food. About the same time, Papez (1937) noted that there are some grossly visible pathways in the brain that interconnected temporal lobe structures with parts of the hypothalamus, the thalamus, and the cingulate cortex. Papez felt that this system was important in mediating emotional behavior, and MacLean (1949, 1958), in encouraging the use of the term *limbic system* rather than *rhinencephalon*, emphasized that the system might serve an important role in sexual behavior.

Overall concepts of the limbic system treat it as a visceral brain subserving species-specific behavioral patterns such as sexual behavior, or as an interface between an overlying cerebral-cortex integrator and an underlying brain stem implementor (Powell and Hines, 1974). From the lesion and stimulation studies performed, there appears to be no evidence that major parts of the system, such as the amygdala, the hippocampus, the septum, the mammillary bodies, the anterior thalamic nucleus, and the cingulate cortex, function in a unitary fashion. For the most part, lesions in most of the structures associated with the limbic system do not markedly alter male sexual behavior. This is true of the septal region (Michal, 1973) the habenular nuclei (Modianos *et al.*, 1974), the mammillary region, and the pyriform cortex (Gianatonio *et al.*, 1970), to mention results from a few studies. In this section, we only deal with the hippocampal-fornix region and the amygdalostrial system.

HIPPOCAMPAL-FORNIX SYSTEM. A prominent structure of the limbic system is the hippocampus, which projects to the basal forebrain primarily through the fornix (Anderson *et al.*, 1973). In male rats, on which most of the work dealing with hippocampal lesions has been performed, hippocampal damage in either the dorsal or the ventral area has had little effect on behavior. A moderate increase in sexual activity has been suggested in some studies (Bermant *et al.*, 1968; Dewsbury *et al.*, 1968; Kim, 1960), although a moderate decrease has also been reported (Michal, 1973). In guinea pigs, Sainsbury and Jason (1976) noticed a reduction in sexual behavior after fimbria-fornix lesions. The lesions reportedly interrupted the CA_3 area (subfield) output of the hippocampus, leaving outputs of other areas intact. The defect was characterized by an apparent interruption in the sequencing of behavior. The lesioned animals had an enhanced tendency to engage in eating and digging following a sexual mount rather than to continue with another sexual response, as was more characteristic of the controls. The authors suggested that the lesioned animals perhaps had an inability to persist in motor behaviors related to a particular motivation.

AMYGDALOSTRIAL SYSTEM. The effects of temporal lobe lesions on sexual behavior, seen first in monkeys and later in cats, have been largely attributed to the destruction of the basolateral nucleus of the amygdala or the pyriform cortex (Green *et al.*, 1957; Kling, 1974; Schreiner and Kling, 1953, 1956). Gerall (1971) mentioned, however, that the removal of other temporal lobe structures in conjunction with these areas could either facilitate or inhibit the changes in sexuality for which these critical areas might be responsible.

The increases in the sexual behavior of monkeys following temporal lobe lesions seem to be limited to the male (Spies et al., 1976). To some extent, they are androgen-dependent, at least in cats, in that the lesion-provoked, unusual sexual behavior declines after castration and is restored by androgen replacement (Schreiner and Kling, 1954). In monkeys, the changes in sexual behavior observed are an increase in masturbation and the mounting of other males and an increase in heterosexual behavior. There have been reports of attempts at copulation with animals of other species.

There has been some attempt to compare these changes in monkeys with those in human males following bilateral temporal-lobe excision for the treatment of severe epilepsy. A report by Terzian and Dall Ore (1955) of a 19-year-old boy whose frequency of masturbation increased following temporal lobe removal attracted some attention in this regard. Blummer (1970) noted that only a few surgically treated epileptic patients exhibit unusual sexual behavior postoperatively. However, when such episodes occur, they do appear to share some of the characteristics of the behavior of lesioned monkeys. There are some reports of marked increases in sexual activity in human female patients that are reported to be manifestations of the Kluver–Bucy syndrome (Mohan et al., 1975; Shraberg and Weisberg, 1978), but these clinical reports are hard to reconcile with work on monkeys showing no Kluver–Bucy phenomenon in females (Spies et al., 1976).

Some of the more striking changes in sexual behavior following amygdalectomy have been reported in male cats. Lesioned animals have an increased tendency to mount other males, even of other species, as well as stuffed toys or the arm of the experimenter (Green et al., 1957). One lesioned male cat that was given a lethal injection of a barbiturate because of recurring epileptic seizures following surgery continued to attempt copulation with a teddy bear until the anesthetic took effect. When the experimenter tried to remove the teddy bear, the cat seemed to revive and continued his mounting and thrusting for a few seconds, at which point he died.

There has been some question about whether amygdaloid lesions actually increase sexual activity and also about whether the inappropriate sexual activity is possibly a reflection of laboratory conditions, as it is known that male cats mount diverse objects in the laboratory environment (Hagamen et al., 1963; Michael, 1961). One attempt to clarify this issue was pursued by Aronson and Cooper (1979), who found that lesions of the basolateral amygdaloid nuclei and the surrounding cortex resulted in no changes in the frequency or duration of intromission or mounting. Thus, there were no indications of hypersexuality. However, all of the subjects mounted inappropriate objects more frequently. These objects were a stuffed toy, a tranquilized rabbit, and a sedated male cat. There were no postoperative changes in the total mounting time of all the objects combined.

The change in sexual activity following amygdaloid or temporal lobe lesions is not a consistent finding in all species. For example, dogs have not shown the changes characteristic of cats (Fuller et al., 1957).

A number of studies have focused on the effects of various types of amygdaloid lesions on the sexual behavior of male rats. Because of the previous tendency to refer to the effects of amygdalectomy in cats and monkeys as producing hypersexuality, the murine studies have dealt more with the question of lesion effects in facilitating or impairing behavior, rather than with the question of inappropriate responses. Bermant *et al.* (1968) did look for an indication that lesioned male rats had an increased tendency to mount other males and found none. Studies employing large amygdaloid lesions have reported a moderate transient impairment of sexual activity (Bermant *et al.*, 1968; Michal, 1973; Rasmussen *et al.*, 1960).

There have been attempts to analyze lesions that involve smaller, more discrete, regions of the amygdala in rats. Lesions in the basolateral region of the amygdala—the area in which lesions were placed by Aronson and Cooper (1979) in cats—had little or no effect on mating performance with receptive females (Giantonio *et al.*, 1970; Harris and Sachs, 1975).

The lack of any noticeable tendency of basolateral lesions to induce aberrant sexual behavior in rats may be a reflection of the fact that, under laboratory conditions, male rats are not normally as prone as cats to mount inappropriate objects, and it may be possible to produce the syndrome of indiscriminate mounting only in species in which this behavior has some degree of baseline occurrence. The change in sexual behavior in monkeys and cats following amygdaloid lesions is possibly part of an overall motivational syndrome. It has been proposed, for example, that amygdalectomized animals show indiscriminate behavior in general (Weiskrantz, 1956).

In rats, lesions in the corticomedial amygdaloid area result in a marked impairment of copulatory behavior, characterized by a very prolonged latency from the initiation of copulation to ejaculation (Giantonio, *et al.*, 1970; Harris and Sachs, 1975). Also, the ejaculatory performance of rats with corticomedial amygdaloid lesions can be influenced by the hormonal state of the stimulus female (Perkins *et al.*, 1980). When male rats were paired with females that had received estrogen and progesterone, the ejaculation latencies were comparable to those of control animals. However, when the female was brought into estrus with estrogen alone, prolonged ejaculation latencies were seen in the lesioned males. Presumably, the estrogen-plus-progesterone females showed higher levels of soliciting behavior, which could compensate for the effects of the lesions.

Lesions of the corticomedial nucleus of the amygdala severely reduce male sexual behavior in the hamster, whereas lesions of the basolateral amygdala have no effect on mating behavior (Lehman *et al.*, 1980). Hamsters with the corticomedial amygdaloid lesions also showed a reduction in the rate of investigation of the female's anogenital region. These authors concluded that the effect of the lesions is due to disruption of the mediation of olfactory and vomeronasal chemosensory information by the corticomedial amygdala.

The effect of corticomedial amygdaloid lesions in impairing the sexual behavior of male rodents is probably quite significant. The corticomedial area is one of the primary projections of the main and accessory olfactory systems (Scalia and Winans, 1975). Interestingly, lesions of the corticomedial area of rats produce more severe deficits in sexual behavior than does olfactory bulbectomy. The corticomedial amygdaloid area projects, by way of stria terminalis, primarily to the bed nucleus of the stria terminalis (de Olmos, 1972; Swanson, 1976). The bed nucleus, in turn, projects to the medial preoptic area. This is a considerably different projection than that of the basolateral amygdaloid area, which is to the lateral preoptic area, the hypothalamus, and the thalamus through the ventral amygdalofugal pathway (Leonard and Scott, 1971; de Olmos, 1972). Lesions in the latter pathway do not disrupt copulatory behavior (Paxinos, 1974) in rats. After placing lesions in the stria terminalis, Giantonio *et al.,* (1970) found only moderate interruption of copulatory behavior, but in using the knife-cut procedure, Paxinos (1976) reported that copulatory behavior was severely disrupted. Lesions of the bed nucleus of the stria terminalis also impair male rat copulatory behavior, primarily by increasing the number of intromissions necessary for ejaculation (Emery and Sachs, 1976). Several studies have shown that lesions of the medial preoptic-anterior hypothalamic continuum eliminate or severely disrupt copulatory behavior. Thus, for rats at least, we now have evidence that lesions along any point of the system—corticomedial amygdala, stria terminalis, bed nucleus of the stria terminalis, and medial preoptic area—severely impair copulatory behavior. Interestingly, all three of the nuclear areas just mentioned are major neural sites for the concentration of labeled testosterone. It is very likely, as Emery and Sachs (1976) suggested, that each nuclear area has its own modulating effect on sexual behavior.

From the standpoint of generalizing to other species, it is difficult to interpret the rather detailed and analytical work on the limbic system and the sexual behavior of rodents. The aberrant sexual responses of monkeys and cats after large amygdaloid lesions do not seem to occur in rodents, or at least, they have not been evident in the testing situations. On the other hand, the suppressive effects seen after lesions in the corticomedial amygdalostrial system in rats have not been noticed in cats or monkeys with total amygdalectomy. It is possible, of course, that there are major species differences in the effects of these brain manipulations on male sexual behavior.

Work on the medial preoptic-anterior hypothalamic area described below illustrates that, at least with regard to this region, there can be a great deal of uniformity among species in response to lesions. In fact, the comparative approach has yielded information that is complementary across species in terms of understanding the more subtle behavioral aspects of the functions of this area. The fact that the medial preoptic region seems to be a nodal point for inputs from the olfactory, septal, and amygdalostrial systems (through the bed nucleus) and for output through the medial forebrain bundle indicates that possibly greater species similarities exist in the mediation of sexual behavior by the limbic system and related areas than is currently apparent.

MEDIAL PREOPTIC-ANTERIOR HYPOTHALAMIC CONTINUUM

Following the initial observations by Soulairac and Soulairac (1956) that some lesions in the medial preoptic region abolished sexual behavior in male rats and did not result in degenerative changes in the testes, Heimer and Larsson (1967) conducted an extensive study of this phenomenon. They found that large bilateral lesions of the medial preoptic-anterior hypothalamic (MP-AH) continuum immediately and completely eliminated male rat copulatory behavior (intromissions and ejaculations). Smaller lesions resulted in greatly reduced copulatory frequency. The effects of the smaller, but not the larger, lesions could be compensated for by the administration of exogenous testosterone.

Several subsequent investigations (Chen and Bliss, 1974; Lisk, 1968; Giantonio et al., 1970) have confirmed these findings and have documented that the deficits in mating behavior are not due to lesions in the suprachiasmatic nuclei or the interference with gonadotropin control of gonadal androgen secretion (Kamel and Frankel, 1978). The mounting displayed by female rats given exogenous androgen is also eliminated by MP-AH lesions, although the lesions do not disturb female sexual activity (Singer, 1968).

The most long-term study of the effects of MP-AH lesions in male rats has been by Ginton and Merari (1977). They were particularly interested in whether animals lesioned in adulthood, and showing a partial deficit, would eventually recover and copulate normally. They also gathered data dealing with the question of whether the effect of MP-AH lesions was only on the initiation of sexual performance and not on the continuation of sexual activity once mounting occurred. They found no recovery of sexual function in lesioned animals in as many as 8 months of postoperative testing. Furthermore, if there was some recovery (sporadic mating), this was evident within 1 month after surgery. The authors also noted that lesioned animals, which on occasion intromitted, often did not continue mating to ejaculation even with extended testing. Their findings are consistent with those on some of the rats in the Giantonio et al. (1970) study that both the initiation and the continuation of sexual behavior is impaired by MP-AH lesions. It is mentioned below that studies on dogs and monkeys reveal the same effects.

The possibility of a type of compensatory brain plasticity with regard to the mediation of sexual behavior has been explored. It is known that lesioning the brains of some animals in sequential stages allows for a recovery of function in a behavioral system that is not seen in the more usual procedure of performing single-stage bilateral lesions (Finger et al., 1973). One study has shown, however, that sequential, bilateral medial preoptic-anterior hypothalamic lesions made in adult male cats lead to the same marked impairment of copulatory behavior as bilateral lesions made simultaneously (Hart, 1980b).

The question of sparing of behavioral function following damage to the brains of young animals has been examined mostly with regard to the cerebral cortex or cortically derived structures, rather than noncortical areas (Johnson and Almli, 1978; Rosner, 1974). Twiggs et al. (1978) and Leedy et al. (1980) reported that male rats that received large MP-AH lesions prepubertally exhibited normal male sexual behavior if allowed to interact with peers of both sexes during development. However, sparing was not complete, as these animals, which showed normal levels of male behavior under standard testing conditions, had fewer ejaculations to exhaustion and showed a more rapid decline in sexual behavior following castration. Subjects receiving lesions prepubertally but not allowed to interact with peers during development did not show sparing of function. These observations illustrate one of the very few examples of sparing or recovery of function in a noncortical brain area and in a behavioral system that is largely innate. This type of brain plasticity appears to be specific to the rat, however, as such sparing of sexual function was not evident with lesions made in prepubertal dogs (Hart and Ladewig, 1979) and cats (Leedy and Hart, 1985b).

Some important aspects of understanding the role of the MP-AH area in male behavior have not, and probably cannot, be approached in the male rat. One concern deals with the extent to which MP-AH lesions alter the performance of copulatory activity from the standpoint of the supraspinal activation of erection and ejaculation. Another concern is the extent to which sexually dimorphic behavior in general is altered by MP-AH lesions.

Work on cats (Hart *et al.*, 1973), dogs (Hart, 1974c), and goats (Hart, 1985) has shown that extensive electrolytic and radiofrequency lesions in the same MP-AH region as that studied in rats immediately reduce or eliminate copulatory responses. Those animals with impaired copulatory behavior continue to copulate sporadically in the postoperative period. The most effective anterior-posterior locus for the impairment of copulatory behavior is the junction of the medial preoptic and anterior hypothalamic area at the posterior edge of the anterior commissure (Figure 3). On the ventral plane, these lesions are between the optic chiasm (above the suprachiasmatic nucleus) and the anterior commissure. It is interesting that the same interior brain landmarks, as approached using x-ray ventriculography, have been found to be the most effective lesion landmarks for rhesus monkeys as well (Slimp *et al.*, 1978) (Figure 4). In all species studied, only relatively large lesions result in complete elimination of copulatory responses.

In dogs, MP-AH lesions do not block erection and ejaculation, as elicited manually (Hart, 1974c). Also in dogs, lesions that alter male sexual behavior markedly reduce the frequency of leg-lift urine marking behavior. In male cats, MP-AH lesions reduce urine spraying, which is also a sexually dimorphic type of urine marking (Hart and Voith, 1978). In dogs, such lesions do not alter aggressive dominance behavior. Assuming that the aggressive behavior examined was sexually dimorphic, the observations indicate that the MP-AH region is involved in the mediation of some, but not all, masculine behavioral patterns. The same concept was evident in goats, in which MP-AH lesions impair copulatory behavior but not the male-typical flehmen and scent urination behavior (Hart, 1985).

In further work, male beagle puppies were lesioned at 10–11 weeks of age, which is prior to the appearance of the adult patterns of sexual and urine-marking behavior (Hart and Ladewig, 1979). Sociosexual interactions were observed during the juvenile period, and it was found that mounting by lesioned male puppies was greatly reduced in comparison to that of sham-lesioned controls. There was no difference, however, in the development of urine-marking behavior between lesioned and sham-lesioned subjects, nor was there any difference in the number of fights initiated by animals in the two groups. As adults, only 1 of the 10 MP-AH

BENJAMIN L. HART
AND MITZI G.
LEEDY

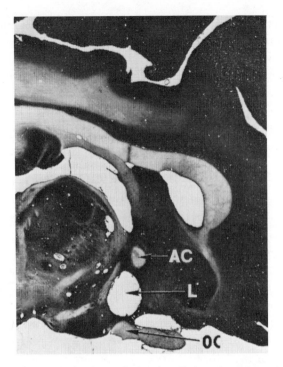

Fig. 3. Sagittal section through a lesion in a male dog that eliminated copulatory behavior and reduced urine marking. Lesions anterior or posterior to this site were less effective. AC, anterior commissure; OC, optic chiasm; L, lesion. (From "The Medial Preoptic-Anterior Hypothalamic Area and Sociosexual Behavior of Male Dogs: A Comparative Neuropsychological Analysis" by B. L. Hart, *Journal of Comparative and Physiological Psychology*, 1974, *86*, 328–349. Copyright 1974 by the American Psychological Association. Reprinted by permission.)

lesioned males copulated with females, a finding revealing a more profound effect of the lesions when placed in juveniles than when placed in adults.

A more recent study of the effects of prepubertal lesions was performed in cats (Leedy and Hart, 1985b). Care was taken to provide optimal conditions for the practice of sociosexual behavior during development, but still, the impairment in copulatory behavior was greater than is reported for similar lesions placed in adult male cats. As in dogs, there was sparing of urine-marking (spraying) behavior. Male dogs with MP-AH lesions placed prepubertally showed an accelerated and enhanced testosterone secretion (Hart and Ladewig, 1980). A similar trend was found in the cat.

The effects of MP-AH lesions in rhesus monkeys have added still another dimension to the understanding of the role of this area of the brain in sexual behavior. As mentioned above, the same types of MP-AH lesions that eliminate copulatory activity in rats, cats, and dogs also reduce or eliminate copulatory activity in male rhesus monkeys (Slimp *et al.*, 1978). However, in monkeys, the sexually dimorphic response of yawning during mating tests by the male subjects was not altered by lesions that eliminated or reduced copulation. In lesioned monkeys with impaired copulatory behavior, masturbation in the home cage was undisturbed. This fact was apparent inasmuch as ejaculatory deposits in the home-cage drop

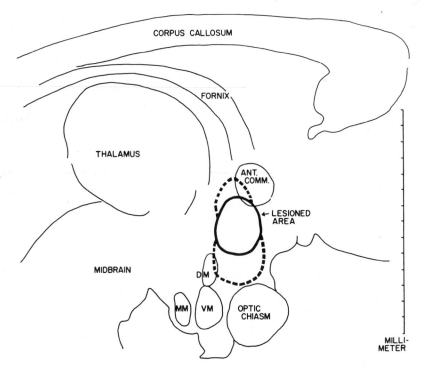

Fig. 4. Sagittal drawing representing the relationship of lesions to brain landmarks approximately 1.5 mm from the midline in rhesus monkeys with impairment of copulatory behavior. The area common to all lesioned subjects with impairment of copulatory behavior is indicated by the dotted line. MM, mammillary COMM., anterior commissure. (From "Heterosexual, Autosexual, and Social Behavior of Adult Male Rhesus Monkeys with Medial Preoptic-Anterior Hypothalamic Lesions" by J. C. Slimp, B. L. Hart, and R. W. Goy, *Brain Research*, 1978, *142*, 105–122. Copyright 1978 by *Brain Research*. Reprinted by permission.)

pans and the frequency of video-taped masturbation were the same for lesioned as for sham-lesioned males. In rhesus monkeys, therefore, it appears that the MP-AH region is specifically involved in the mediation of heterosexual copulation but is not vital to the performance of another common form of sexual activity, masturbation. In lesioned monkeys, it was found that often a copulatory series seemed to be initiated but not carried through. This finding is similar to observations on some MP-AH lesioned dogs (Hart, 1974c) and rats (Giantonio *et al.,* 1979; Ginton and Merari, 1977). In the lesioned monkeys, there were no increases in the frequency of threats displayed or aggression toward females, and grooming activity actually increased. Thus, the interaction of male rhesus monkeys with their partners indicated that the sexual deficits were not secondary to some gross social maladjustments.

The serum levels of testosterone in MP-AH–lesioned male rhesus monkeys were monitored at periodic intervals by radioimmunoassay. The levels of testosterone in lesioned males were found to be at least as high as in sham-lesioned males, a finding showing, again, that sexual impairment was not due to a reduction of gonadal androgen. Work on the monkey also reveals that MP-AH lesions do not cause a type of functional castration by removing androgen-sensitive neurons. The best evidence is observations on the differential effects of lesions and castration on other types of sexually dimorphic behavior. Following castration, yawning during mating tests declines, as does the occurrence of ejaculatory plugs in the home cage (Phoenix, 1974; Phoenix et al., 1973). However, yawning is not affected by MP-AH lesions. One would not expect to see masturbation continue in castrated animals as it does in lesioned subjects, but there are no direct observations to support this expectation. Observations on dogs are consistent with the same concept. For example, castration does not reduce urine-marking behavior in a novel pen (Hart, 1974a), but following MP-AH lesions, urine-marking behavior is markedly impaired (Hart, 1974c).

The MP-AH area does not appear to control the precopulatory behaviors in the male. Giantonio et al. (1970) and Heimer and Larsson (1967) reported that MP-AH lesions in male rats did not eliminate pursuit or anogenital sniffing of the female, even though copulatory behavior was rarely initiated. Also, Bean et al. (1981) found that mice with MP-AH lesions continued to respond to an estrous female with 70-KHz courtship vocalizations. Additionally, the increase in LH and testosterone levels seen when male rats are placed in a familiar testing chamber, or with an estrous female (Graham and Desjardins, 1980; Kamel et al., 1977), are not disrupted by MP-AH lesions, although mating behavior is eliminated (Ryan and Frankel, 1978).

There is little in the human clinical literature that relates directly to the analytical studies of MP-AH lesions in animals. A few reports are consistent with the notion that the MP-AH area is crucial to the performance of sexual behavior by males. In patients convicted of sexual offenses, including "pedophilic homosexuality" and "violent hypersexuality," unilateral lesions were placed in areas extending from the ventromedial nucleus of the hypothalamus rostrally to the medial preoptic area (Dieckmann and Hassler, 1975; Muller et al., 1973; Roeder et al., 1970). The patients receiving the unilateral preoptic lesions were reported to have experienced a reversal in their motivation toward their previous abnormal behavior and a reduction of sexual interest. These clinical observations are somewhat contradicted, however, by the work in rats (Lisk, 1968) and cats (Hart, 1980b) showing that unilateral preoptic lesions have little suppressive effect on sexual behavior.

Reports by Meyers (1961, 1962) on deficits in human sexual behavior following brain lesions are more in line with the animal work described below dealing with possible inputs and outputs of the MP-AH area. Meyers had performed two-stage bilateral lesions of the ansa lenticularis to relieve abnormal motor signs. After the second operation, his patients suffered complete loss of sexual interest and even inability to achieve erection. Although it is still not known exactly what areas of the brain were affected by these lesions (Meyers, personal communication, 1972), the procedure used by Meyers was believed by him to possibly damage the anterior hypothalamus and the medial forebrain bundle, an important output pathway from the MP-AH area (see next setion).

In summary, observations of rodents, carnivores, and primates document that the MP-AH area is the single most critical region for the mediation of male sexual behavior in mammals. Lesions of this area, when placed in adults, uniformly eliminate, or markedly reduce, copulatory behavior. Some, but not all, male-typical behavioral patterns are also affected along with sexual behavior. Species differences appear to come into play with regard to the plasticity of this area, and evidence from comparative studies of MP-AH lesions placed pubertally suggests that the rodent brain is more closely interrelated with other brain structures (and capable of greater plasticity) than are the more highly evolved brains of carnivores.

PATHWAYS ASSOCIATED WITH THE OUTPUT OF MEDIAL PREOPTIC–ANTERIOR HYPOTHALAMIC SYSTEM

The importance of the MP-AH area in male sexual behavior has stimulated interest in learning more about the pathways that connect this area to other brain regions. One of the important fiber systems in the hypothalamus that would seem to be involved in MP-AH output (and undoubtedly some input as well) is the medial forebrain bundle. In rats, lesions in this tract in the anterior, middle, and posterior hypothalamus severely disrupt copulatory behavior (Hitt *et al.*, 1970, 1973; Caggiula *et al.*, 1973). The fact that lesions in this tract in more than one anteroposterior location have the same effect on behavior indicates that the lesion effects are due to the interruption of a pathway rather than to the destruction of neurons.

Medial-forebrain-bundle lesions seem to impair primarily the initiation rather than the execution of sexual activity, in that once intromissions begin, the copulatory sequence proceeds at a normal pace (Szechtman *et al.*, 1978). The deficit in behavior caused by such lesions is not due to a reduced responsiveness to general arousal, as replacing the female, handling, and tail shock, which normally produce behavioral arousal, still fail to induce copulation (Caggiula *et al.*, 1974).

Anatomical studies analyzing projections of the medial preoptic region using tritiated amino acid autoradiography, which allows for the tracing of small-calibered axons, show that this area projects to several limbic forebrain and midbrain areas, partially through the medial forebrain bundle (Conrad and Pfaff, 1975; Swanson, 1976). Because the medial forebrain bundle is located lateral to the MP-AH area, fibers must traverse laterally to reach the bundle. Parasagittal knife cuts that extend from the preoptic region through the ventromedial nucleus, separating mediolateral connections, severely disrupt copulatory behavior, whereas cuts anterior and immediately caudal to the MP-AH area do not disturb mating activity (Paxinos, 1974; Paxinos and Bindra 1972; Szechtman *et al.*, 1978).

Additional evidence that the impairment in sexual behavior seen following medial-forebrain-bundle damage is due to its connections with the MP-AH has been given by Hendricks and Scheetz (1973). They found that unilateral medial forebrain lesions coupled with contralateral MP-AH lesions resulted in an impairment of male sexual behavior comparable to that seen following bilateral MP-AH lesions.

The medial forebrain bundle has also been implicated in the control of urine marking in male rats (Scouten *et al.*, 1980). Urine deposition, although not eliminated, was reduced by about 50%, and this behavior, which is testosterone-dependent, could not be increased by testosterone treatment. Copulatory behavior was also severely reduced in these animals.

Work on other species implicating the medial forebrain bundle in male sexual behavior is limited. Berquist (1970) found that the sexual responses of the male opossum, which were evoked by medial preoptic electrical stimulation, were markedly reduced by medial-forebrain-bundle lesions. In male guinea pigs, lesions of the anterior hypothalamus between the optic chiasm and the median eminence abolished copulatory behavior (Phoenix, 1961). The lesions were large, but Phoenix found no correspondence between any particularly damaged nuclear group and the disruption of sexual behavior. One interpretation of this study is that the lesions damaged either medial-forebrain-bundle fibers or fibers traversing to the medial forebrain bundle from the MP-AH area. The reports by Meyers (1961, 1962), mentioned previously, of the loss of sexual interest and ability in human subjects following knife cuts in the anterior hypothalamic area might also be explained on the same basis.

ROSTRAL MIDBRAIN AREA

If the effects of MP-AH lesions in abolishing male copulatory behavior are impressive, then lesions in another part of the brain that would greatly enhance sexual activity are equally or more impressive. Heimer and Larsson (1964) reported a dramatic reduction of the postejaculatory refractory period of male rats by large lesions in the rostral midbrain. The effects of these lesions seemed to be similar to the effects achieved in a few animals with MP-AH electrical stimulation (Van Dis and Larsson, 1971; Malsbury, 1971; Merari and Ginton, 1975; Vaughan and Fisher, 1962) and suggested that the rostral midbrain lesions removed some inhibitory influences from the system mediating sexual behavior. Initial efforts to replicate the Heimer and Larsson findings produced inconsistent results (Lisk, 1969; Giantonio et al., 1970; Chen and Bliss, 1974). Three subsequent studies documented that the effect on sexual behavior of midbrain lesions is a dramatic one, but the studies have not clarified what is the exact brain locus involved.

Barfield et al. (1975) found that lesions in the ventral part of the rostral midbrain markedly reduced the postejaculatory refractory period. Their lesions were considerably smaller than those employed by Heimer and Larsson. The effect of their lesions was attributed to interruption of biogenic amine pathways, but they did not test for a reduction in any particular amine. In a similar experiment, Clark et al. (1975) found an extreme reduction in the refractory period following the placement of discrete lesions in the middle portion of the rostral midbrain. They attributed the effects of their lesions to an interruption of the dorsal norepinephrine bundle, and this hypothesis was substantiated by a reduction in telencephalic norepinephrine. In a study by Walker et al. (1981), lesions were centered on the dorsal norepinephrine bundle. They also found reduced refractory periods and lower levels of telencephalic norepinephrine. However, they found no positive correlations between the reductions in norepinephrine and copulatory behavior.

Thus, they concluded that some other systems, besides norepinephrine, may mediate the reduced refractory periods.

Unfortunately, there are no reports of rostral-midbrain-lesion effects on the sexual behavior of any other species. Certainly, a comparative approach might help elucidate the role of rostral midbrain areas in sexual behavior as it has for the MP-AH area.

NEURAL INVOLVEMENT IN HETEROTYPICAL BEHAVIOR

It is well known that, in several species, sexual responses typical of the opposite sex are occasionally displayed by both males and females. Beach (1968b) reviewed the comparative literature dealing with mounting behavior by females, pointing out that there are clear species differences in the predisposition of females to mount. Although mounting is common, the occurrence of the ejaculatory response is quite rare in normal females.

The display of feminine receptive postures by males and a willingness to stand while being mounted by other males, although not necessarily common, has been observed in normal rats, cats, ferrets, and monkeys, among other species. Even elements of the sexual reflexes of dogs that are characteristically displayed by one sex can be observed in spinal animals of the opposite sex (Hart 1970b). Thus, spinal male dogs given estrogen show a lateral-curvature "presentation" response on stimulation of the genital region homologous to the vulvar surface of the female; and spinal female dogs exhibit rhythmic contractions in a muscle homologous to the male bulbospongiosus muscle when the clitoris is rubbed.

The predisposition of animals to display mainly masculine or feminine sexual behavior is generally related to early exposure of the central nervous system (and the genitalia) to androgen. That early androgen section affects the neural structures controlling sexual behavior is apparent in cytoarchitectural studies in rodents. These studies focused on the medial preoptic region, but the effects may be more general. At one level, the sexually dimorphic differences in volume of intensely staining cellular components of the medial preoptic area are so great as to be evident on observation of brain sections by the naked eye (Gorski *et al.*, 1978). At the ultrastructural level, Raisman and Field (1973) found a sexual dimorphism in synaptic subtypes among nonamygdaloid inputs in the preoptic area in rats. Greenough *et al.* (1977) observed sex differences in the dendritic branching pattern in the dorsomedial preoptic area of male and female hamsters. The neuronal density, cell size, and volume of the darkly stained area of the preoptic area are also greater in male rats than in females (Hsu *et al.*, 1980). These consistent findings concerning the preoptic area have led to the labeling of the darkly stained area as the "sexually dimorphic nucleus of the preoptic area" (Gorski *et al.*, 1980).

These morphological differences between males and females are dependent on neonatal hormone treatment, in that the castration of males or testosterone administration to females influences the size and density of the nucleus along the lines that one would expect (Gorski *et al.*, 1978; Hsu *et al.*, 1980; Raisman and Field, 1973). In a study of the proliferation of neuronal processes (neurites) from neurons of the mouse hypothalamus and preoptic area *in vitro,* Toran-Allerand (1976) found a much more intense proliferation of neurites in cultures exposed to androgen or estrogen.

Sexual dimorphism has also been found outside the preoptic area. Male rats have more postsynaptic connections in the medial amygdaloid nucleus (Nishizuka and Arai, 1981), and female rats have a subset of faster conducting axons projecting from the ventromedial nucleus of the hypothalamus to the midbrain central gray (Sakuma and Pfaff, 1981).

Again, it must be emphasized that even the demonstration of sex-related neuroanatomical differences in the neural structures involved in mediating sexual behavior does not imply that females do not have the necessary neural circuitry for the display of masculine behavioral patterns, or that males do not have the circuitry for the female responses, such as lordosis. These known sex-related anatomical differences, and additional ones that will undoubtedly be elucidated in the future, are clearly involved in readiness or predisposition, and not in the ability of an animal to engage in masculine or feminine behavior. A couple of experiments illustrate this point. The ejaculatory response is virtually never seen in normal female rats. It is, however, frequently displayed by females (untreated neonatally) receiving long-term estrogen treatment (Emery and Sachs, 1975) or chronic androgen treatment and paced during mating tests with electrical shock (Krieger and Barfield, 1976). Conversely, male ferrets, castrated as adults and given estrogen, show levels of receptivity that are equal to those seen in females (Baum and Gallagher, 1981). One way of looking at the male–female difference in the neural control of behavior is that the interplay of facilitative and inhibitory forces on descending systems is different in males and females, and that these differences may be a function of cytoarchitectural makeup in specific brain areas.

If there is any area of the brain that is as crucial in the display of female sexual behavior in females as the MP-AH area is for male sexual behavior in males, it is the ventromedial nucleus of the hypothalamus and the adjacent areas. This brain area seems to be the critical estradiol-sensitive area for the hormonal activation of female sexual responses in female rats (Barfield and Chen, 1977; Davis *et al.,* 1979) and also in male rats (Davis and Barfield, 1979). Electrical stimulation of this area facilitates the lordosis response in female rats primed with a low dose of estrogen (Pfaff and Sakuma, 1979). Lesions in the ventromedial and/or anterior hypothalamus disrupt female sexual behavior in female rats (Nance *et al.,* 1977a; Singer, 1968); hamsters (Malsbury *et al.,* 1977); guinea pigs (Goy and Phoenix, 1963); cats (Leedy and Hart, 1985a; Sawyer and Robinson, 1956), and sheep (Clegg *et al.,* 1958).

Lesions in the ventromedial region in females have also resulted in a dramatic increase in male sexual responses in female rats (Nance *et al.,* 1977a). Along the same lines, lesions of the MP-AH area enhance several measures of female copulatory behavior in male rats (Christensen *et al.,* 1977). In female rats, lesions in the MP-AH area enhance the ability of females to show lordosis (Nance *et al.,* 1977a; Powers and Valenstein, 1972), and in both male and female guinea pigs, such lesions facilitate lordotic behavior (Rodriguez-Sierra and Terasawa, 1979). From the standpoint of the interaction of the MP-AH and ventromedial areas in male and female sexual behavior of rodents, the picture appears to be reciprocal. This concept does not hold with other species. For example, with male cats it has been shown that MP-AH lesions potentiate femalelike sexual behavior (Hart and Leedy, 1983). However in female cats, ventromedial hypothalamic lesions do not facilitate male behavior (Leedy and Hart, 1985a).

Although increases in heterotypical behavior may be seen following lesions that eliminate homotypical behavior, the areas involved are not necessarily coupled so that a decline in homotypical behavior is followed by a proportionate increase in heterotypical behavior. For example, in the study by Hart and Leedy (1983) on male cats, no correlation was found between those subjects with the greatest increase in the level of femalelike behavior and those with the greatest decrease in male sexual behavior. Similarly, Christensen *et al.* (1977) noted that small lesions in the dorsomedial preoptic area that impaired male sexual activity in male rats did not increase the lordotic responding in these animals.

According to microanatomical findings, the septal area represents one of the most important inputs into the MP-AH area (Swanson, 1976). In light of the above discussion of the contrasting effects of ventromedial hypothalamic and MP-AH lesions on male and female behavior, some work on the effects of septal lesions takes on particular interest. The septal area is one of the brain regions in which the effects produced by lesions are, in general, greater in one sex than in the other. For example, female rats show a greater magnitude of lesion effects related to water- and food-motivated operant behavior than males (Kondo and Lorens, 1971; Lorens and Kondo, 1971).

In female rats with septal lesions, the occurrence of lordosis after estrogen–progesterone treatment is enhanced (Nance *et al.*, 1975a). In male rats, lesions of the septal area also facilitate the display of lordosis, but only if the males are chronically exposed to estrogens shortly after surgery (Gorski *et al.*, 1979; Nance *et al.*, 1975b, 1977b; Stewart and Atkinson, 1977). Gorski *et al.* suggested that there may be hormonal modifications of the recovery process from septal lesions to allow for facilitation of lordotic responding in male rats. Another possibility is that administering estrogen just after surgery allows for the production of some more profound changes in the MP-AH system, or inputs into it, so that the facilitation of lordosis is a reflection of altered MP-AH, rather than septal, function. The fact that lordosis can be facilitated in female rats but not in male rats with septal lesions alone (i.e., no chronic estrogen treatment afterward) may be another consequence of the fact that the MP-AH cytoarchitecture differs substantially in males and females.

Concluding Comment

Recent experimental work on the neurological bases of male sexual behavior has brought us to the threshold of tracing out within the brain, in reasonable detail, a system of nuclei and pathways from sensory input to behavioral output. The eventual picture or map of the system will be complex because different aspects of the behavior—sensory function, effector mechanisms, and integrative processes at the spinal and cephalic level—necessarily involve many structures throughout the central nervous system. At most of the neural loci for the central mediation of male sexual behavior, various types of nonsexual behavior are also mediated or controlled. Hence, there is an interface between sexual behavior and other behavioral systems at all levels. There is also an interface between neural areas involved primarily in "female" sexual behavior and "male" sexual behavior, so that when the system underlying one sex is suppressed, the system for the other sex is often released, although this effect is not invariant and these systems appear not to be reciprocally coupled.

Species differences, as well as similarities, in regard to behavioral changes following lesions are striking. With surprising regularity, lesions of the MP-AH continuum eliminate or markedly reduce copulatory behavior in males of every species studied, including rodents, carnivores, ungulates, and primates. Notable examples of species differences are the marked suppression of male sexual behavior in a couple of rodent species, but not in other animals, following olfactory bulbectomy; the occurrence of Kluver–Bucy-like changes in monkeys and cats, but not in rodents, after amygdaloidectomy; and the plasticity of the MP-AH area to prepubertal lesions in rodents, but not in carnivores.

Species specificity in spinal sexual reflexes reveals that some of the species-typical aspects of mating behavior in males are even programmed at the spinal level. The species differences in the neural mediation of male sexual behavior will continue to be a source both of frustration and of research effort. It would be gratifying for those of us who are most comfortable with unifying hypotheses if these species differences can eventually be related to differences in ecological adaptation and specialized evolution of the brain.

REFERENCES

Adler, N. T. Social and environmental control of reproductive processes in animals. In T. E. McGill, D. A. Dewsbury, and B. N. Sachs (Eds.), *Sex and Behavior*. New York: Plenum Press, 1978.

Anderson, P., Bland, B. H., and Dudard, J. D. Organization of the hippocampal output. *Experimental Brain Research*, 1973, *17*, 152–168.

Antelman, S. M., Szechtman, H., Chin, P., and Fisher, A. E. Tail pinch-induced eating, gnawing and locking behavior in rats: Dependence on the nigrostriatal dopamine system. *Brain Research*, 1975, *99*, 319–337.

Aronson, L. R. Behavior resembling spontaneous emissions in the domestic cat. *Journal of Comparative and Physiological Psychology*, 1949, *49*, 226.

Aronson, L. R., and Cooper, M. L. Penile spines of the domestic cat: Their endocrine-behavior relations. *The Anatomical Record*, 1967, *157*, 71–78.

Aronson, L. R., and Cooper, M. L. Desensitization of the glans penis and sexual behavior in cats. In M. Diamond (Ed.), *Perspectives in Reproduction and Sexual Behavior*. Bloomington: University of Indiana Press, 1968.

Aronson, L. R., and Cooper, M. L. Olfactory deprivation and mating behavior in sexually experienced male cats. *Behavioral Biology*, 1974, *11*, 459–480.

Aronson, L. R., and Cooper, M. L. Central versus peripheral genital densensitization and mating behavior in male cats: Tonic and phasic effects. *Annals of the New York Academy of Science*, 1977, *290*, 299–313.

Aronson, L. R., and Cooper, M. L. Amygdaloid hypersexuality in male cats reexamined. *Physiology and Behavior*, 1979, *22*, 257–265.

Arvidsson, T., and Larsson, K. Seminal discharge and mating behavior in male rat. *Physiology and Behavior*, 1967, *2*, 341–343.

Bacq, Z. M. Impotence of the male rodent after sympathetic denervation of the genital organs. *American Journal of Physiology*, 1930, *96*, 321–330.

Barfield, R. J., and Chen, J. Activation of estrous behavior in ovariectomized rat by intracerebral implants of estradiol benzoate. *Endocrinology*, 1977, *101*, 1716–1725.

Barfield, R. J., and Geyer, L. A. The ultrasonic postejaculatory vocalization and the postejaculatory refractory period of the male rat. *Journal of Comparative and Physiological Psychology*, 1975, *88*, 723–734.

Barfield, R. J., Wilson, C., and McDonald, P. G. Sexual behavior: Extreme reduction of postejaculatory refractory period of midbrain lesions in male rats. *Science*, 1975, *189*, 147–149.

Barfield, R. J. Auerbach, P., Geyer, L. A., and McIntosh, T. K. Ultrasonic vocalization in rat sexual behavior. *American Zoologist*, 1979, *19*, 469–480.

Baum, M. J., and Gallagher, C. A. Increasing dosages of estradiol benzoate activate equivalent degrees of sexual receptivity in gonadectomized male and female ferrets. *Physiology and Behavior*, 1981, *26*, 751–753.

Beach, F. A. Effect of cortical lesions upon the copulatory behavior of male rats. *Journal of Comparative and Physiological Psychology*, 1940, *29*, 193–245.

Beach, F. A. Analysis of stimuli adequate to elicit mating behavior in the sexually inexperienced male rat. *Journal of Comparative Psychology*, 1942, *33*, 163–208.

Beach, F. A. Characteristics of masculine "sex drive." In M. R. Jones (Ed.), *The Nebraska Symposium on Motivation*. Lincoln: University of Nebraska Press, 1956.

Beach, F. A. Evolutionary aspects of psychoendocrinology. In A. Rose and G. G. Simpson (Eds.), *Behavior and Evolution*. New Haven, Conn.: Yale University Press, 1958.

Beach, F. A. Cerebral and hormonal control of reflexive mechanisms involved in copulatory behavior. *Psychological Review*, 1967, *47*, 289–316.

Beach, F. A. Coital behavior of dogs: III. Effects of early isolation on mating in males. *Behavior*, 1968a, *30*, 217–238.

Beach, F. A. Factors involved in the control of mounting behavior by female mammals. In M. Diamond (Ed.), *Reproduction and Sexual Behavior*. Bloomington: Indiana University Press, 1968b.

Beach, F. A. Coital behavior in dogs: VII. Effects of sympathectomy in males. *Brain Research*, 1969, *15*, 243–245.

Beach, F. A. Variables affecting "spontaneous" seminal emision in rats. *Physiology and Behavior*, 1975, *15*, 91–95.

Beach, F. A., and Levinson, G. Effects of androgen on the glans penis and mating behavior of castrated male rats. *Journal of Experimental Zoology*, 1950, *144*, 159–171.

Beach, F. A., and Wilson, J. R. Mating behavior in male rats after removal of seminal vesicles. *Proceedings of the National Academy of Sciences*, 1968, *49*, 624–626.

Beach, F. A., Zitrin, A., and Jaynes, J. Neural mediation of mating in male cats: II. Contributions of the frontal cortex. *Journal of Experimental Zoology*, 1955, *130*, 381–402.

Beach, F. A., Zitrin, A., and Jaynes, J. Neural mediation of mating in male cats: I. Effects of unilateral and bilateral removal of the neocortex. *Journal of Comparative and Physiological Psychology*, 1956, *49*, 321–327.

Beach, F. A., Westbrook, W. H., and Clemens, L. G. Comparisons of the ejaculatory responses in men and animals. *Psychosomatic Medicine*, 1966, *28*, 749–763.

Bean, N. J., Nunez, A. A., and Conner, R. Effects of medial preoptic lesions on male mouse ultrasonic vocalizations and copulatory behavior. *Brain Research Bulletin*, 1981, *6*, 109–112.

Beauchamp, G. L., Wellington, J. L., Wysocki, C. J., Brand, J. G., Kubie, J. L., and Smith, A. B. Chemical communication in the guinea pig: Urinary components of low volatility and their access to the vomeronasal organ. In D. Müller-Schwarze and R. M. Silverstein (Eds.), *Chemical Signals in Vertebrates and Aquatic Animals*. New York: Plenum Press, 1980.

Beckett, S. D., Hudson, R. S., Walker, D. F., Vachon, R. I., and Reynolds, T. M. Corpus cavernosum penis pressure and external penile muscle activity during erection in the goat. *Biology of Reproduction*, 1972, *7*, 359–364.

Beckett, S. D., Hudson, R. S., Walker, D. F., Reynolds, T. M., and Vachon, R. I. Blood pressure and penile muscle activity in the stallion during coitus. *American Journal of Physiology*, 1973, *25*, 1072–1075.

Beckett, S. D., Walker, D. F., Hudson, R. S., Reynolds, T. M., and Vachon, R. I. Corpus cavernosum penis pressure and penile muscle activity in the bull during coitus. *American Journal of Veterinary Research*, 1974, *35*, 761–764.

Beckett, S. D., Purohit, R. C., and Reynolds, T. M. The corups cavernosum penis pressure and external penile muscle activity in the goat during coitus. *Biology of Reproduction*, 1975, *12*, 289–292.

Bergamini, E. Different mechanisms in testosterone action on glycogen metabolism in rat perineal and skeletal muscles. *Endocrinology*, 1975, *96*, 77–84.

Bermant, F., and Taylor, L. Interactive effects of experience and olfactory bulb lesion in male rat copulation. *Physiology and Behavior*, 1969, *4*, 13–17.

Bermant, G., Glickman, S. E., and Davidson, J. M. Effects of limbic lesions on copulatory behavior of male rats. *Journal of Comparative and Physiological Psychology*, 1968, *65*, 118–125.

Berquist, E. H. Output pathways of hypothalamic mechanisms for sexual, aggressive, and other motivated behaviors in the opossum. *Journal of Comparative and Physiological Psychology*, 1970, *70*, 389–398.

Blummer, D. Hypersexual episodes in temporal lobe epilepsy. *American Journal of Psychiatry*, 1970, *126*, 1099–1106.

Boast, C. A., Reid, S. A., Johnson, P., and Zornetzer, S. F. A caution to brain scientists: Unsuspected hemorrhagic vascular damage resulting from mere electrode implantation. *Brain Research*, 1976, *103*, 527–534.

Bors, E., and Comarr, A. E. Neurological disturbances of sexual function with special reference to 529 patients with spinal cord injury. *Urology Survey*, 1960, *10*, 191.

Breedlove, S. M., and Arnold, A. P. Hormone accumulation in a sexually dimorphic motor nucleus in the rat spinal cord. *Science*, 1980, *210*, 564–566.

Broadwell, R. D. Olfactory relationships of the telencephalon and diencephalon in the rabbit: I. An autoradiographic study of the errerent connections of the main and acessory olfactory bulbs. *Journal of Comparative Neurology*, 1975, *163*, 329–345.

Brooks, C. Mc. The role of the cerebral cortex and of various sense organs in the excitation and execution of mating activity in the rabbit. *American Journal of Physiology*, 1937, *120*, 544–553.

Bues, E., Alnor, P., and Peter, D. Sexualfunktionstörungen nach lumbaler Grenzstrangresektion. *Der Chirurg*, 1957, *28*, 103–107.

Caggiula, A. R. Analysis of the copulation reward properties of posterior hypothalamic stimulation in male rats. *Journal of Comparative and Physiological Psychology*, 1970, *7*, 399–415.

Caggiula, A. R., Antelman, S. M., and Zigmond, M. J. Disruption of copulation in male rats after hypothalamic lesions: A behavioral, anatomical and neurochemical analysis. *Brain Research*, 1973, *59*, 273–287.

Caggiula, A. R., Antelman, S. M., and Zigmond, M. J. Ineffectiveness of sexually arousing stimulation after hypothalamic lesions in the rat. *Physiology and Behavior*, 1974, *12*, 313–316.

Cain, D. P. The role of the olfactory bulb in limbic mechanisms. *Psychological Bulletin*, 1974, *81*, 654–671.

Chen, J. J., and Bliss, D. K. Effects of sequential preoptic and mammillary lesions on male rat sexual behavior. *Journal of Comparative and Physiological Psychology*, 1974, *87*, 841–847.

Christensen, L. W., and Clemens, L. G. Intrahypothalamic implants of testosterone or estradiol and resumption of masculine sexual behavior in long-term castrated male rats. *Endocrinology*, 1974, *95*, 984–990.

Christensen, L. W., Nance, D. M., and Gorski, R. A. Effects of hypothalamic and preoptic lesions on reproductive behavior in male rats. *Brain Research Bulletin*, 1977, *2*, 137–141.

Clark, T. K., Caggiula, A. R., McConnell, A., and Antelman, S. M. Sexual inhibition is reduced by rostral midbrain lesions in the male rat. *Science*, 1975, *190*, 169–171.

Clegg, M. T., Santolucito, J. A., Smith, J. D., and Ganong, W. F. The effect of hypothalamic lesions on sexual behavior and estrous cycles in the ewe. *Endocrinology*, 1958, *62*, 790–797.

Comarr, A. E. Sexual function among patients with spinal cord injury. *Urologia Internationalis*, 1970, *25*, 134–168.

Comarr, A. E., and Gundersen, B. B. Sexual function in traumatic paraplegia and quadriplegia. *American Journal of Nursing*, 1975, *75*, 250.

Comarr, A. E., and Vigue, M. Sexual counseling among male and female patients with spinal cord and/ or cauda equina injury. *American Journal of Physical Medicine*, 1978a, *57*, 107–122.

Comarr, A. E., and Vigue, M. Sexual counseling among male and female patients with spinal cord and/ or cauda equina injury. *American Journal of Physical Medicine*, 1978b, *57*, 215–227.

Conrad, L. C. A., and Pfaff, D. W. Axonal projections of medial preoptic anterior hypothalamic neurons. *Science*, 1975, *190*, 1112–1114.

Cooper, K. K., and Aronson, L. R. Effects of castration on neuronal afferent responses from the penis of the domestic cat. *Physiology and Behavior*, 1974, *12*, 93–107.

Cuschieri, A. Enzyme histochemistry of the olfactory mucosa and vomeronasal organ in the mouse. *Journal of Anatomy*, 1974, *118*, 477–489.

Dahlof, L. G., and Larsson, K. Interactional effects of pudendal nerve section and social restriction on male rat sexual behavior. *Physiology and Behavior*, 1976, *16*, 757–762.

Davidson, J. M. Activation of the male rat's sexual behavior by intracerebral implantation of androgen. *Endocrinology*, 1966, *79*, 783–794.

Davidson, J. M. Neurohormonal bases of male sexual behavior. In R. O. Greep (Ed.), *International Review of Physiology: Reproductive Physiology II*, Vol. 13. Baltimore: University Park Press, 1977.

Davidson, J. M. Stefanik, M. L., Sachs, B. D., and Smith, E. R. Role of androgens in sexual reflexes of the male rat. *Physiology and Behavior*, 1978, *21*, 141–146.

Davis, P. G., and Barfield, R. J. Activation of masculine sexual behavior by intracranial estradiol benzoate implants in male cats. *Neuroendocrinology*, 1979, *28*, 217–227.

Davis, P. G., McEwen, B. S., and Pfaff, D. Localized behavioral effects of tritiated estradiol implants in the ventromedial hypothalamus of female rats. *Endocrinology*, 1979, *104*, 898–903.

de Olmos, J. S. The amygaloid projection field in the rat as studied with the cupric-silver method. In B. E. Eleftheriou (Ed.), *Advances in Behavioral Biology:* Vol. 2. *The Neurobiology of the Amygdala*. New York: Plenum Press, 1972.

Devor, M. Fiber trajectories of olfactory bulb efferents in the hamster. *Journal of Comparative Neurology*, 1976, *166*, 31–48.

Dewsbury, D. A. On the function of the multiple-intromission, multiple-ejaculation copulatory patterns of rodents. *Bulletin of the Psychonomic Society*, 1981, *18*, 221–223.

Dewsbury, D., Goodman, E., Salis, P., and Burnell, B. Effects of hippocampal lesions on the copulatory behavior of male rats. *Physiology and Behavior*, 1968, *3*, 651–656.

Dieckmann, G., and Hassler, R. Unilateral hypothalamotomy in sexual delinquents. *Confinia Neurologia*, 1976, *37*, 177–186.

Dusser de Barenne, J. G., and Koskoff, Y. D. Flexor rigidity of the hind legs and priapism in the "secondary" spinal preparation of the male cat. *American Journal of Physiology*, 1932, *102*, 75–86.

Edwards, D. A. Non-sensory involvement of the olfactory bulbs in the mediation of social behaviors. *Behavioral Biology*, 1974, *11*, 287–302.

Eibergen, R. D., and Caggiula, A. R. Ventral midbrain involvement in copulatory behavior of the male rat. *Physiology and Behavior*, 1973, *10*, 435–442.

Emery, D. E., and Sachs, B. D. Ejaculatory pattern in female rats without androgen treatment. *Science*, 1975, *190*, 484–486.

Emery, D. E., and Sachs, B. D. Copulatory behavior in male rats with lesions in the bed nucleus of the stria terminalis. *Physiology and Behavior*, 1976, *17*, 803–806.

Estes, R. D. The role of the vomeronasal organ in mammalian reproduction. *Extrait de Mammalia*. 1972, *36*, 315–341.

Finger, S., Walbran, B., and Stein, D. G. Brain damage and behavioral recovery: Serial lesion phenomena. *Brain Research*, 1973, *63*, 1–18.

Fletcher, I. C., and Lindsay, D. R. Sensory involvement in the mating behavior of domestic sheep. *Animal Behaviour*, 1968, *16*, 410–414.

Fuller, J. L., Rosvold, H. E., and Pribiam, K. H. The effect of an affective and cognitive behavior in the dog of lesions of the pyriform-amgdala-hippocampal complex. *Journal of Comparative and Physiological Psychology*, 1957, *50*, 89.

Garven, H. S. D., and Gairns, F. W. The silver diamine method of staining peripheral nerve elements and the interpretation of the results, with a modification of the Bielschowsky-Gros method for frozen sections. *Quarterly Journal of Experimental Physiology*, 1952, *37*, 131–144.

Gerall, A. A. Role of the nervous system in reproductive behavior. In E. S. E. Hafez, (Ed.), *Comparative Reproduction of Nonhuman Primates*. Springfield, Ill.: Charles C Thomas, 1971.

Geyer, L., and Barfield, R. J. Regulation of social contact by the female rat during the postejaculatory interval. *Animal Learning and Behavior*, 1980, *8*, 679–685.

Giantonio, G. W., Lund, N. L., and Gerall, A. A. Effect of diencephalic and rhinencephalic lesions on the male rat's sexual behavior. *Journal of Comparative and Physiological Psychology*, 1970, *73*, 38–46.

Ginton, A., and Merari, A. Long range effects of MPOA lesions on mating behavior in the male rat. *Brain Research*, 1977, *120*, 158–163.

Goldfoot, D. A., Essock-Vitale, S. M., Asa, C., Thornton, J. E., and Leshner, A. L. Anosmia in male rhesus monkeys does not alter copulatory activity with cycling females. *Science*, 1978, *199*, 1095–1096.

Gorski, R. A., Gordon, J. H., Shryne, J. E., and Southam, A. M. Evidence for a morphological sex difference within the medial preoptic area of the rat brain. *Brain Research*, 1978, *148*, 333–346.

Gorski, R. A., Christensen, L. W., and Nance, D. M. The induction of heterotypical sexual behavior in the rat. *Psychoneuroendocrinology*, 1979, *4*, 311–328.

Gorski, R. A., Harlan, R. E., Jacobson, C. D., Shryne, J. E., and Southam, A. M. Evidence for the existence of a sexually dimorphic nucleus in the preoptic area of the rat. *Journal of Comparative Neurology*, 1980, *193*, 529–539.

Goy, R. W., and Phoenix, C. H. Hypothalamic regulation of female sexual behaviour; establishment of behavioural oestrus in spayed guinea-pigs following hypothalamic lesions. *Journal of Reproduction and Fertility*, 1963, *5*, 23–40.

Graham, J. M., and Desjardins, C. Classical conditioning: Induction of luteinizing hormone and testosterone secretion in anticipation of sexual activity. *Science*, 1980, *210*, 1039–1041.

Graziadei, P. P. C. Functional anatomy of the mammalian chemo-receptor system. In D. Müller-Schwarze and M. M. Mozell (Eds.), *Chemical Signals in Vetebrates*. New York: Plenum Press, 1977.

Green, J. D., Clemente, C. D., and de Groot, J. Rhinencephalic lesions and behavior in cats. *Journal of Comparative Neurology*, 1957, *108*, 505–545.

Greenough, W. R., Carter, C. S., Steerman, C., and deVoogd, T. J. Sex differences in dendritic pattern in hamster preoptic area. *Brain Research*, 1977, *126*, 63–72.

Gregory, R. L. The logic of localization of function in the central nervous system. In E. E. Bernard and M. R. Kare (Eds.), *Biological Prototypes and Synthetic Systems*, Vol. 1. New York: Plenum Press, 1962.

Hagamen, W. D., Zitzmann, D. K., and Reeves, A. G. Sexual mounting of diverse objects in a group of randomly selected unoperated male cats. *Journal of Comparative and Physiological Psychology*, 1963, *56*, 298–302.

Harris, V. S., and Sachs, B. D. Copulatory behavior in male rats following amygdaloid lesions. *Brain Research*, 1975, *86*, 514–518.

Hart, B. L. Sexual reflexes and mating behavior in the male dog. *Journal of Comparative and Physiological Psychology*, 1967a, *64*, 388–399.

Hart, B. L. Testosterone regulation of sexual reflexes in spinal male rats. *Science*, 1967b, *155*, 1282–1284.

Hart, B. L. Alteration of quantitative aspects of sexual reflexes in spinal male dogs by testosterone. *Journal of Comparative and Physiological Psychology*, 1968a, *66*, 726–730.

Hart, B. L. Role of prior experience in the effects of castration on sexual behavior of male dogs. *Journal of Comparative and Physiological Psychology*, 1968b, *66*, 719–725.

Hart, B. L. Sexual reflexes and mating behavior in the male rat. *Journal of Comparative and Physiological Psychology*, 1968c, *65*, 453–460.

Hart, B. L. The male reproductive system. In A. C. Anderson (Ed.), *The Beagle as an Experimental Dog*. Ames: Iowa State University Press, 1970a.

Hart, B. L. Mating behavior in the female dog and the effects of estrogen on sexual reflexes. *Hormones and Behavior*, 1970b, *1*, 93–104.

Hart, B. L. The action of extrinsic penile muscles during copulation in the male dog. *The Anatomical Record*, 1972a, *173*, 1–6.

Hart, B. L. Sexual reflexes in the male rat after anesthetization of the glans penis. *Behavioral Biology*, 1972b, *7*, 127–130.

Hart, B. L. Environmental and hormonal influences on urine marking behavior in the adult male dog. *Behavioral Biology*, 1974a, *11*, 167–176.

Hart, B. L. Gonadal androgen and sociosexual behavior of male mammals: A comparative analysis. *Psychological Bulletin*, 1974b, *81*, 383–400.

Hart, B. L. The medial preoptic-anterior hypothalamic area and sociosexual behavior of male dogs: A comparative neuropsychological analysis. *Journal of Comparative and Physiological Psychology*, 1974c, *86*, 328–349.

Hart, B. L. Hormones, spinal reflexes and sexual behavior. In J. B. Hutchison (Ed.), *Biological Determinants of Sexual Behavior*. New York: Wiley, 1978a.

Hart, B. L. Reflexive aspects of copulatory behavior. In T. E. McGill, D. A. Dewsbury, and B. Sachs (Eds.), *Sex and Behavior: Status and Prospectus*. New York: Plenum Press, 1978b.

Hart, B. L. Activation of sexual reflexes of male rats by dihydrotestosterone but not by estrogen. *Physiology and Behavior*, 1979, *23*, 107–109.

Hart, B. L. Neonatal spinal transection in male rats: Differential effects on penile reflexes and other reflexes. *Brain Research*, 1980a, *197*, 242–246.

Hart, B. L. Sequential medial preoptic-anterior hypothalamic lesions have same effect on copulatory behavior of male cats as simultaneous lesions. *Brain Research*, 1980b, *185*, 423–428.

Hart, B. L. Flehmen behavior and vomeronasal organ function. In D. Müller-Schwarze and R. M. Silverstein (Eds.), *Chemical Signals in Vertebrates*, Vol. 3. New York: Plenum Press, 1983a.

Hart, B. L. Role of testosterone secretion and penile reflexes in sexual behavior and sperm competition in male rats: A theoretical contribution. *Physiology and Behavior*, 1983b, *31*, 823–827.

Hart, B. L. Medial preoptic-anterior hypothalamic lesions and sociosexual behavior of male goats. *Physiology and Behavior*, 1985.

Hart, B. L., and Haugen, C. M. Scent marking and sexual behavior maintained in anosmic male dogs. *Communications in Behavioral Biology*, 1972, *6*, 131–135.

Hart, B. L., and Jones, T. O. A. C. Effects of castration on sexual behavior of tropical male goats. *Hormones and Behavior*, 1975, *6*, 247–258.

Hart, B. L., and Kitchell, R. L. Penile erection and contraction of penile muscles in the spinal and intact dog. *The American Journal of Physiology*, 1966, *210*, 257–261.

Hart, B. L., and Ladewig, J. Effects of medial preoptic-anterior hypothalamic lesions on development of sociosexual behavior in dogs. *Journal of Comparative and Physiological Psychology*, 1979, *93*, 566–573.

Hart, B. L., and Ladewig, J. Accelerated and enhanced testosterone secretion in juvenile male dogs following medial preoptic-anterior hypothalamic lesions. *Neuroendocrinology*, 1980, *30*, 20–24.

Hart, B. L., and Leedy, M. G. Female sexual responses in male cats facilitated by olfactory bulbectomy and medial preoptic-anterior hypothalamic lesions. *Behavioral Neuroscience*, 1983, *97*, 608–614.

Hart, B. L., and Melese-d'Hospital, P. Penile mechanisms and the role of the striated penile muscles in penile reflexes. *Physiology and Behavior*, 1983, *31*, 807–813.

Hart, B. L., and Odell, V. Effects of intermittent electric shock on penile reflexes of male rats: Implications for theories of reflex plasticity and pacing of copulatory behavior. *Behavioral and Neural Biology*, 1980, *30*, 394–398.

Hart, B. L., and Odell, V. Elicitation of ejaculation and penile reflexes of spinal male rats by peripheral electric shock. *Physiology and Behavior*, 1981, *26*, 623–626.

Hart, B. L., and Voith, V. L. Changes in urine spraying, feeding and sleep behavior of cats following medial preoptic-anterior hypothalamic lesions. *Brain Research*, 1978, *145*, 406–409.

Hart, B. L., Haugen, C. M., and Peterson, D. M. Effects of medial preoptic-anterior hypothalamic lesions on mating behavior of male cats. *Brain Research*, 1973, *54*, 177–191.

Hart, B. L., Wallach, S. J. R., and Melese d'Hospital, P. Differences in responsiveness to testosterone of penile reflexes and copulatory behavior of male rats. *Hormones and Behavior*, 1983, *17*, 274–283.

Hayes, K. J. The so-called "levator ani" of the rat. *Acta Endocrinologica*, 1965, *48*, 337–347.

Heimer, L. Synaptic distribution of centripetal and centrifugal nerve fibers in the olfactory system of the rat: An experimental anatomical study. *Journal of Anatomy*, 1968, *103*, 413–432.

Heimer, L., and Larsson, K. Drastic changes in the mating behaviour of male rats following lesions in the junction of diencephalon and mesencephalon. *Experientia*, 1964, *20*, 460–461.

Heimer, L., and Larsson, K. Impairment of mating behavior in male rats following lesions in the preoptic-anterior hypothalamic continuum. *Brain Research*, 1967, *3*, 248–263.

Hendricks, S. E., and Scheetz, H. A. Interaction of hypothalamic structures in the mediation of male sexual behavior. *Physiology and Behavior*, 1973, *10*, 711–716.

Herbert, J. The role of the dorsal nerves of the penis in the sexual behaviour of the male rhesus monkey. *Physiology and Behavior*, 1973, *10*, 293–300.

Herrick, C. J. The connections of the vomeronasal nerve, accessory olfactory bulb and amygdala in Amphibia. *Journal of Comparative Neurology*, 1921, *33*, 213–280.

Hitt, J. C., Hendricks, S. E., Ginsberg, S. I., and Lewis, J. H. Disruption of male, but not female, sexual behavior in rats by medial forebrain bundle lesions. *Journal of Comparative and Physiological Psychology,* 1970, *73,* 377–384.

Hitt, J. C., Byron, D. M., and Modianos, D. T. Effects of rostral medial forebrain bundle and olfactory tubercule lesions upon sexual behavior of male rats. *Journal of Comparative and Physiological Psychology,* 1973, *82,* 24–30.

Hodson, N. Role of hypogastric nerves in seminal emission in the rabbit. *Journal of Reproduction and Fertility,* 1964, *7,* 113–122.

Hsu, H. D., Chen, F. N., and Peng, M. T. Some characteristics of the darkly stained area of the medial preoptic area of rats. *Neuroendocrinology,* 1980, *31,* 327–330.

Johnson, P., and Almli, C. R. Age, brain damage and performance. In S. Finger (Ed.), *Recovery from Brain Damage: Research and Theory.* New York: Plenum Press, 1978.

Johnston, P., and Davidson, J. M. Intracerebral androgens and sexual behavior in the male rat. *Hormones and Behavior,* 1973, *3,* 345–357.

Jung, I., and Baulieu, E. E. Testosterone cytosol "receptor" in the rat levator ani muscle. *Nature-New Biology,* 1972, *237,* 24.

Kamel, F., and Frankel, A. The effect of medial preoptic area lesions on sexually stimulated hormone release in the male rat. *Hormones and Behavior,* 1978, *10,* 10–21.

Kamel, R., Wright, W. W., Mock, E. J., and Frankel, A. I. The influence of mating and related stimuli on plasma levels of luteinizing hormone, follicle stimulating hormone, prolactin and testosterone in the male rat. *Endocrinology,* 1977, *10,* 421–429.

Kierniesky, N., and Gerall, A. A. Effects of testosterone propionate implanted in the brain on the sexual behavior and peripheral tissue of the male rat. *Physiology and Behavior,* 1973, *11,* 633–640.

Kim, C. Sexual activity of male rats following ablation of hippocampus. *Journal of Comparative and Physiological Psychology,* 1960, *53,* 553–557.

Kinsey, A. C., Pomeroy, W. B., and Martin, C. E. *Sexual Behavior in the Human Male.* Philadelphia: Saunders, 1948.

Kitchell, R. L., Campbell, B., Quilliam, T. A., and Larson, L. L. Neurological factors in the sexual behavior of domestic animals. *Proceedings of the 92nd Annual Meeting of the American Veterinary Medical Association,* 1955, 177–189

Kitchell, R. L., Gilanpour, H., and Johnson, R. D. Electrophysiologic studies of penile mechanoreceptors in the rat. *Experimental Neurology,* 1982, *75,* 229–244.

Kling, A. Differential effects of amygdalectomy in male and female nonhuman primates. *Archives of Sexual Behavior,* 1974, *3,* 129–134.

Kluver, H., and Bucy, P. C. Preliminary analysis of functions of the temporal lobes in monkeys. *Archives of Neurological Psychiatry,* 1939, *42,* 979–1000.

Knappe, H. Zur Funktion des Jacobsonschen Organs (organon vomeronsale Jacobsoni). *Der Zoologische Garten,* 1964, *28,* 188–194.

Kollberg, S., Petersén, I., and Stener, I. Preliminary results of electromyographic study of ejaculation. *Acta Chirologica Scandinavia,* 1962, *123,* 478–483.

Kondo, C. Y., and Lorens, S. A. Sex differences in the effects of septal lesions. *Physiology and Behavior,* 1971, *6,* 481–485.

Krieger, M. S., and Barfield, J. R. Masculine sexual behavior: Pacing and ejaculatory patterns in female rats induced by electrical shock. *Physiology and Behavior,* 1976, *16,* 671–675.

Kullanda, K. M. Representation of the internal organs in the cat and dog cerebral and cerebellar cortes: III. Representation of pudendal nerves in the cat cerebal cortex. *Bulletin of Experimental Biology and Medicine,* 1960, *49,* 7–9.

Kuntz, A. *The Autonomic Nervous System* (4th Ed.). Philadelphia: Lea and Febiger, 1953.

Kirtz, R. G., and Santos, R. Supraspinal influences on the penile reflexes of the male rat: A comparison of the effects of copulation, spinal transection, and cortical spreading depression. *Hormones and Behavior,* 1979, *12,* 73–94.

Ladewig, J., and Hart, B. L. Flehmen and vomeronasal organ function in male goats. *Physiology and Behavior,* 1980, *24,* 1067–1071.

Ladewig, J., Price, E. O., and Hart, B. L. Flehmen in male goats: Role in sexual behavior. *Behavioral and Neural Biology,* 1980, *30,* 312–322.

Langworthy, O. R. Behavior disturbances related to decomposition of reflex activity caused by cerebral injury: An experimental study of the cat. *Journal of Neuropathology and Experimental Neurology,* 1943, *3,* 87–99.

Larsson, K. Mating behavior in male rats after cerebral cortical ablation: I. Effects of lesions in the dorsolateral and the median cortex. *Journal of Experimental Zoology*, 1962, *151*, 167–176.

Larsson, K. Mating behavior in male rats after cerebral cortical ablation: II. Effects of lesions in the frontal lobes compared to lesions in the posterior half of the hemispheres. *Journal of Experimental Zoology*, 1964, *155*, 203–214.

Larsson, K. Failure of gonadal and gonadotrophic hormones to compensate for impaired sexual function in anosmic male rats. *Physiology and Behavior,* 1969, *4*, 733–737.

Larsson, K. Impaired mating performances in male rats after anosmia induced peripherally and centrally. *Brain, Behavior and Evolution*, 1971, *4*, 463–471.

Larsson, K. Sexual impairment of inexperienced male rats following pre- and postpubertal olfactory bulbectomy. *Physiology and Behavior*, 1975, *14*, 195–199.

Larsson, K., and Sodersten, P. Mating in male rats after section of the dorsal penile nerve. *Physiology and Behavior*, 1973, *10*, 567–571.

Larsson, K., and Swedin G. The sexual behavior of male rats after bilateral section of the hypogastric nerve and removal of the accessory genital glands. *Physiology and Behavior*, 1971, *6*, 251–253.

Lashley, K. S. *Brain Mechanisms and Intelligence.* Chicago: University of Chicago Press, 1929.

Lashley, K. S. Studies of cerebral function in learning: XII. Loss of the maze habit after occipital lesions in blind rats. *Journal of Comparative Neurology*, 1943, *79*, 431–462.

Leedy, M. G., and Hart, B. L. Female and male sexual responses in female cats with ventromedial hypothalamic lesions. *Behavioral Neuroscience*, 1985a, still in press.

Leedy, M. G., and Hart, B. L. Medial preoptic-anterior hypothalamic lesions in prepubertal male cats: Effects on juvenile and adult sociosexual behavior. *Physiology and Behavior*, 1985b, in press.

Leedy, M. G., Vela E. A., Popolow, H. B., and Gerall, A. A. Effect of prepubertal medial preoptic area lesions on male rat sexual behavior. *Physiology and Behavior*, 1980, *24*, 341–346.

Lehman, M. N., Winans, S. S., and Powers, J. B. Medial nucleus of the amygdala mediates chemosensory control of male hamster sexual behavior. *Science*, 1980, *210*, 557–559.

Leonard, C. M., and Scott, J. W. Origin and distribution of the amygdalofugal pathways in the rat: An experimental neuroanatomical study. *Journal of Comparative Neurology*, 1971, *141*, 313–330.

Lisk, R. D. Neural localization for androgen activation of copulatory behavior in the male rat. *Endocrinology*, 1967, *80*, 754–761.

Lisk, R. D. Copulatory activity of the male rat following placement of preoptic-anterior hypothalamic lesions. *Experimental Brain Research*, 1968, *5*, 306–313.

Lisk, R. D. Reproductive potential of the male rat: Enhancement of copulatory levels following lesions of the mammillary body in sexually non-active and active animals. *Journal of Reproduction and Fertility*, 1969, *19*, 353–356.

Lodder, J. Penile deafferentation and the effect of mating experience on sexual motivation in adult male rats. *Physiology and Behavior*, 1976, *17*, 571–573.

Lodder, J., and Zeilmaker, G. H. Effects of pelvic nerve and pudendal nerve transection on mating behavior in the male rat. *Physiology and Behavior*, 1976, *16*, 745–751.

Lorens, S. A., and Kondo, C. Y. Differences in the consumatory and operant behaviors of male and female septal rats. *Physiology and Behavior*, 1971, *6*, 487–491.

MacLean, P. D. Psychosomatic disease and the "visceral brain": Recent developments bearing on the Papez theory of emotion. *Psychosomatic Medicine*, 1949, *11*, 338–353.

MacLean, P. D. Contrasting functions of limbic and neocortical systems of the brain and their relevance to psychophysiological aspects of medicine. *American Journal of Medicine*, 1958, *25*, 611–626.

MacLean, P. D., and Ploog, D. W. Cerebral representation of penile erection. *Journal of Neurophysiology*, 1962, 29–55.

MacLean, P. D., Denniston, R. H., and Dua, S. Further studies on cerebral representation of penile erection: Caudal thalamus, midbrain, and pons. *Journal of Neurophysiology*, 1963, *26*, 273–293.

Madlafousek, J., Freund, K., and Grofova, I. Variables determining the effect of electrostimulation in the lateral preoptic area on the sexual behavior of male rats. *Journal of Comparative and Physiological Psychology*, 1970, *72*, 28–44.

Malsbury, C. W. Facilitation of male rat copulatory behavior by electrical stimulation of the medial preoptic area. *Physiology and Behavior*, 1971, *7*, 797–805.

Malsbury, C. W., Kow, L. M., and Pfaff, D. W. Effects of medial hypothalamic lesions on the lordosis responses and other behaviors in female golden hamsters. *Physiology and Behavior*, 1977, *19*, 223–237.

Matthews, M. K., and Adler, N. T. Systematic interrelationship of mating, vaginal plug position and sperm transport in the rat. *Physiology and Behavior*, 1978, *20*, 303–309.

McEwen, B., Pfaff, D. W., and Zigmond, R. E. Factors influencing sex hormone uptake by rat brain regions: II. Effects of neonatal treatment of hypophysectomy or testosterone uptake. *Brain Research*, 1970, *21*, 17–28.

Merari, A., and Ginton, A. Characteristics of exaggerated sexual behavior induced by electrical stimulation of the medial preoptic area in male rats. *Brain Research*, 1975, *86*, 97–108.

Meyers, R. Evidence of a locus of the neural mechanisms of libido and penile potency in the septo-fornico-hypothalamic region of the human brain. *Transactions of the American Neurological Association*, 1961, *86*, 81–85.

Meyers, R. Three cases of myoclonus alleviated by bilateral ansotomy, with a note on postoperative albido and impotence. *Journal of Neurosurgery*, 1962, *19*, 71–81.

Michael, R. P. Hypersexuality in non-brain damaged male cats. *Science*, 1961, *134*, 553–554.

Michal, E. K. Effects of limbic lesions on behavior sequences and courtship behavior behavior of male rats *(Rattus norvegicus). Behaviour*, 1973, *44*, 264–285.

Modianos, D. T., Hitt, J. C., and Flexman, J. E. Habenular lesions produced decrements in feminine, but not masculine, sexual behavior in rats. *Behavioral Biology*, 1974, *10*, 75–87.

Mohan, K. J., Salo, M. W., and Nagaswami, S. A case of limbic system dysfunction with hypersexuality and fugue state. *Diseases of the Nervous System*, 1975, *36*, 621–624.

Money, J. Components of eroticism in man: II. The orgasm and genital somesthesia. *Journal of Nervous and Mental Disease*, 1961, *132*, 289–297.

Muller, L. R. Klinische and experimentelle Studien über die innervation der Blase, des Mastdarmes und des Genitalapparates. *Deutsche Zeitschrift für Nervenheilkunde*, 1902, *21*, 86–154.

Muller, L. R. Über die Exstirpation der unteren Hälfte des Rückenmarks und deren Folgeerscheinungen. *Deutsche Zeitschrift für Nervenheilkunde*, 1906, *30*, 413–423.

Muller, D., Roeder, R., and Orthner, H. Further results of stereotaxis in the human hypothalamus in sexual deviations: First use of this operation in addiction to drugs. *Neurochirurgia*, 1973, *16*, 113–126.

Muller-Schwarze, D. Flehmen in the context of mammalian urine communications. In F. J. Ritter (Ed.), *Chemical Ecology: Odor Communications in Animals*. Amsterdam; Elsevier, 1979.

Munro, D. H., Horne, H. W., and Paull, D. P. The effect of injury to the spinal cord and cauda equina on the sexual potency of men. *New England Journal of Medicine*, 1948, *239*, 903–911.

Murphy, M., and Schneider, G. Olfactory bulb removal eliminates mating behavior in the male golden hamster. *Science*, 1970, *167*, 302–304.

Nance, D. M., Shryne, J. E., and Gorski, R. A. Effects of septal lesions on behavioral sensitivity of female rats to gonadal hormones. *Hormones and Behavior*, 1975a, *6*, 59–64.

Nance, D. M., Shryne, J., and Gorski, R. A. Facilitation of female sexual behavior in male rats by septal lesions: An interaction with estrogen. *Hormones and Behavior*, 1975b, *6*, 289–299.

Nance, D. M., Christensen, L. W., Shryne, J. E., and Gorski, R. A. Modifications in gonadotropin control and reproductive behavior in the female rat by hypothalamic and preoptic lesions. *Brain Research Bulletin*, 1977a, *2*, 307–312.

Nance, D. M., Shryne, J. E., Gordon, J. H., and Gorski, R. A. Examination of some factors that control the effects of septal lesions on lordosis behavior. *Pharmacology, Biochemistry and Behavior*, 1977b, *6*, 227–234.

Nishizuka, M., and Arai, Y. Sexual dimorphism in synaptic organization in the amygdala and its dependence on neonatal hormone environment. *Brain Research*, 1981, *212*, 31–38.

Orbach, J. Spontaneous ejaculation in the rat. *Science*, 1961, *134*, 1072.

Orbach, J., Miller, M., Billimoria, A., and Sohlkhan, N. Spontaneous seminal ejaculation and genital grooming in rats. *Brain Research*, 1967, *5*, 520–523.

Papez, J. W. A proposed mechanism of emotion. *Archives of Neurological Psychiatry*, 1937, *38*, 725–743.

Paxinos, G. The hypothalamus: Neural systems involved in feeding, irritability, aggression, and copulation in male rats. *Journal of Comparative and Physiological Psychology*, 1974, *87*, 110–119.

Paxinos, G. Interruption of septal connections: Effects on drinking, irritability, and copulation. *Physiology and Behavior*, 1976, *17*, 81–88.

Paxinos, G., and Bindra, D. Hypothalamic knife cuts: Effects on eating, drinking, irritability, aggression, and copulation in the male rat. *Journal of Comparative and Physiological Psychology*, 1972, *79*, 219–229.

Pepelko, W., and Clegg, M. Studies of mating behavior and some factors influencing the sexual response in the male sheep, *Ovis aries*. *Animal Behaviour*, 1965, *13*, 249–259.

Perachio, A. A., and Alexander, M. The neural basis of agression and sexual behavior in the rhesus monkey. In G. H. Bourne (Ed.), *The Rhesus Monkey*, Vol. 1. New York: Academic Press, 1975.

Perachio, A. A., Alexander, M., and Robinson, B. W. Sexual behavior evoked by telestimulation. *Proceedings of the 2nd International Congress of Primatology*, 1969, *3*, 68–74.

Perachio, A. A., Alexander, M., and Marr, L. D. Hormonal and social factors affecting evoked sexual behavior in rhesus monkeys. *American Journal of Physical Anthropology*, 1973, *38*, 227–232.

Perachio, A. A., Marr, L. D., and Alexander, M. Sexual behavior in male rhesus monkeys elicited by electrical stimulation of preoptic and hypothalamic areas. *Brain Research*, 1979, *177*, 127–144.

Perkins, M. S., Perkins, M. N., and Hitt, J. C. Effects of stimulus female on sexual behavior of male rats given olfactory tubercle and corticomedial amygdaloid lesions. *Physiology and Behavior*, 1980, *25*, 495–500.

Peterson, I., and Stener, I. An electromyographical study of the striated urethral sphincter, the striated anal sphincter and the levatory ani muscle during ejaculation. *Electromyography*, 1970, *10*, 23.

Pfaff, D. Autoradiographic localization of radioactivity in rat brain after injection of tritiated sex hormones. *Science*, 1968, *162*, 1355–1356.

Pfaff, D. W., and Sakuma, Y. Facilitation of the lordosis reflex of female rats from the ventromedial nucleus of the hypothalamus. *Journal of Physiology*, 1979, *288*, 189–202.

Phoenix, C. H. Hypothalamic regulation of sexual behavior in male guinea pigs. *Journal of Comparative and Physiological Psychology*, 1961, *54*, 72–77.

Phoenix, C. H. Effects of dihydrotestosterone on sexual behavior of castrated male rhesus monkeys. *Physiology and Behavior*, 1974, *12*, 1045–1055.

Phoenix, C. H., Slob, A. K., and Goy, R. W. Effects of castration and replacement therapy on sexual behavior of adult male rhesuses. *Journal of Comparative and Physiological Psychology*, 1973, *84*, 472–481.

Phoenix, C. H. Copenhaver, K. H., and Brenner, R. M. Scanning electron microscopy of penile papillae in intact and castrated rats. *Hormones and Behavior*, 1976, *7*, 217–227.

Powell, E. W., and Hines, G. The limbic system: An interface. *Behavioral Biology*, 1974, *12*, 149–164.

Powers, J. B., and Valenstein E. S. Sexual receptivity: Facilitation by medial preoptic lesions in female rats. *Science*, 1972, *175*, 1003–1005.

Powers, J. B., and Winans, S. S. Sexual behavior in peripherally anosmic male hamsters. *Physiology and Behavior*, 1973, *10*, 361–368.

Powers, J. B., and Winans, S. S. Vomeronasal organ: Critical role in mediating sexual behavior of the male hamster. *Science*, 1975, *187*, 961–963.

Powers, J. B., Fields, R. B., and Winans, S. S. Olfactory and vomeronasal system participation in male hamsters' attraction to female vaginal secretions. *Physiology and Behavior*, 1979, *22*, 77–84.

Price, J. L. An autoradiogaphic study of complementary laminar patterns of termination of afferent fibers to the olfactory cortex. *Journal of Comparative Neurology*, 1973, *150*, 87–108.

Raisman, G. An experimental study of the projection of the amygdala to the accessory olfactory bulb and its relationship to the concept of a dual olfactory system. *Experimental Brain Resarh*, 1972, *14*, 395–408.

Raisman, G., and Field, P. M. Sexual dimorphism in the neuropil of the preoptic area of the rat and its dependence on neonatal androgen. *Brain Research*, 1973, *54*, 1–29.

Ramsey, G. V. The sexual development of boys. *American Journal of Psychology*, 1943, *54*, 217

Rasmussen, E., Kaada, B., and Bruland, H. Effects of neocortical and limbic lesions on the sex drive in rats. *Acta Physiologica Scandinavica*, 1960, *Supplement 175*, 126–127.

Roberts, W. W. Are hypothalamic motivational mechanisms functionally and anatomically specific. *Brain, Behavior, and Evolution*, 1969, *2*, 317.

Robinson, B. W., and Mishkin, M. Ejaculation evoked by stimulation of the preoptic area in monkeys. *Physiology and Behavior*, 1966, *1*, 269–272.

Robinson, B. W., and Mishkin, M. Penile erection evoked from forebrain structures in *Macaca mulatta*. *Archives of Neurology*, 1968, *19*, 184–198.

Rodreguez-Sierra, J. F., and Terasawa, E. Lesions of the preoptic area facilitate lordosis behavior in male and female guinea pigs. *Brain Research Bulletin*, 1979, *4*, 513–517.

Roeder, F., Muller, D., and Orthner, H. Stereotaxic treatment of psychoses and neurosis. In W. Umback (Ed.), *Special Topics in Stereotaxis*. Stuttgart: Hippokrates Verlag, 1970.

Root, W. S., and Bard, P. The mediation of feline erection through sympathetic pathways with some remarks on sexual behavior after deafferentation of the genitalia. *American Journal of Physiology,* 1947, *151,* 80.

Rosner, B. S. Recovery of function and localization of function in historical perspectives. In D. G. Stein, J. J. Rosen, and N. Butters (Eds.), *Plasticity and Recovery of Function in the Central Nervous System.* New York: Academic Press, 1974.

Rowe, R. A., and Edwards, D. A. Olfactory bulb removal: Influences on the mating behavior of male mice. *Physiology and Behavior,* 1972, *8,* 37–42.

Ryan, E. L., and Frankel, A. I. Studies on the role of the medial preoptic area in sexual behavior and hormonal response to sexual behavior in the mature male laboratory rat. *Biology of Reproduction,* 1978, *19,* 971–983.

Sachs, B. D. Role of the rat's striated penile muscles in penile reflexes, copulation and the induction of pregnancy. *Journal of Reproduction and Fertility,* 1982, *66,* 433–443.

Sachs, B. D., and Barfield, R. J. Functional analysis of masculine copulatory behavior in the rat. In J. S. Rosenblatt, R. A. Hinde, E. Shaw and C. G. Beer (Eds.), *Advances in the Study of Behavior,* Vol. 7. New York: Academic Press, 1976.

Sachs, B. D., and Garinello, L. D. Interaction between penile reflexes and copulation in male rats. *Journal of Comparative and Physiological Psychology,* 1978, *92,* 759–767.

Sachs, B. D., and Garinello, L. D. Spinal pacemaker controlling sexual reflexes in male rats. *Brain Research,* 1979, *171,* 152–156.

Sainsbury, R. S., and Jason, G. W. Fimbria-fornix lesions and sexual-social behavior of the guinea pig. *Physiology and Behavior,* 1976, *17,* 963–967.

Sakuma, Y., and Pfaff, D. Electrophysiologic determination of projections from ventromedial hypothalamus to midbrain central gray: Differences between female and male rats. *Brain Research,* 1981, *225,* 184–188.

Sar, M., and Stumpf, W. E. Autoradiographic localization of radioactivity in the rat brain after the injection of 1,2-^{3}H-testosterone. *Endocrinology,* 1973, *92,* 251–256.

Sawyer, C. H., and Robinson, B. L. Separate hypothalamic areas controlling pituitary gonadotropic functions and mating behavior in female cats and rabbits. *Journal of Clinical Endocrinology and Metabolism,* 1956, *16,* 914–915.

Scalia, F., and Winans, S. S. The differential projections of the olfactory bulb and accessory olfactory bulb in mammals. *Journal of Comparative Neurology,* 1975, *161,* 31–56.

Schoenfield, T. A., and Hamilton, L. W. Theoretical review—Secondary brain changes following lesions: A new paradigm for lesion experimentation. *Physiology and Behavior,* 1977, *18,* 951–967.

Schreiner, L., and Kling, A. Behavioral changes following rhinencephalic injury in the cat. *Journal of Neurophysiology,* 1953, *16,* 643–659.

Schreiner, L., and Kling, A. Effects of castration on hypersexual behavior induced by rhinencephalic injury in the cat. *Archives of Neurological Psychiatry,* 1954, *72,* 180–186.

Schreiner, L., and Kling, A. Rhinencephalon and behavior. *American Journal of Physiology,* 1956, *184,* 486–490.

Scouten, C. W., Burrell, L., Palmer, T., and Cegavske, C. F. Lateral projections of the medial preoptic area are necessary for androgenic influence on urine marking and copulation in rats. *Physiology and Behavior,* 1980, *25,* 237–243.

Semans, J. H., and Langworthy, O. R. Observations of the neurophysiology of sexual function in the male cat. *Journal of Urology,* 1939, *40,* 836–846.

Sherrington, C. S. The spinal cord. In E. A. Sharpey-Schafer (Ed.), *Textbook of Physiology.* Edinburgh, Scotland, 1900.

Shraberg, D., and Weisberg, L. A single case study: The Kluver–Bucy syndrome in man. *Journal of Nervous and Mental Disease,* 1978, *166,* 130–134.

Singer, J. J. Hypothalamic control of male and female sexual behavior in female rats. *Journal of Comparative and Physiological Psychology,* 1968, *66,* 738–742.

Slimp, J. C., Hart, B. L., and Goy, R. W. Heterosexual, autosexual, and social behavior of adult male rhesus monkeys with medial preoptic-anterior hypothalamic lesions. *Brain Research,* 1978, *142,* 105–122.

Smith, E. R., Damassa, D. A., and Davidson, J. M. Plasma testosterone and sexual behavior following intracerebral implantation of testosterone propionate in the castrated male rat. *Hormones and Behavior,* 1977, *8,* 77–87.

Soulairac, A., and Soulairac, M. L. Effects des lésions hypothalamiques sur le comportement sexuel et le tractus génital du rat mâle. *Annales d'Endocrinologie*, 1956, *17*, 731–745.

Spaulding, W. S., and Peck, C. K. Sexual behavior of male rats following removal of the glans penis at weaning. *Developmental Psychology*, 1974, *7*, 43–46.

Spies, H. G., Norman, R. L., Clifton, D. K., Ochsner, A. J., Jensen, J. N., and Phoenix, C. H. Effects of bilateral amygdaloid lesions on gonadal and pituitary hormones in serum and on sexual behavior in female rhesus monkeys. *Physiology and Behavior*, 1976, *17*, 985–992.

Stelzner, D. J., Ershler, W. B., and Weber, E. D. Effects of spinal transection in neonatal and weanling rats: Survival of function. *Experimental Neurology*, 1975, *46*, 156–177.

Stephan, F. K., Valenstein, E. S., and Zucker, I. Copulation and eating during electrical stimulation of the rat hypothalamus. *Physiology and Behavior*, 1971, *6*, 587–594.

Stewart, J., and Atkinson, S. Effects of septal lesions on the lordotic behavior of weanling male and female rats. *Hormones and Behavior*, 1977, *9*, 99–106.

Stone, C. P. The effects of cerebral destruction on the sexual behavior of male rabbits: I. The olfactory bulbs. *American Journal of Physiology*, 1925, *71*, 430–435.

Swanson, L. W. An autoradiographic study of the efferent connections of the preoptic region in the rat. *The Journal of Comparative Neurology*, 1976, *167*, 227–256.

Swedin, G. Pre- and postganglionic denervation of the vas deferens and accessory male genital glands; Short term effects on fertility in the rat. *Andrologie*, 1971, *3*, 1–8.

Szechtman, H., Cagguila, A. R., and Wulkan, D. Preoptic knife cuts and sexual behavior in male rats. *Brain Research*, 1978, *150*, 569–591.

Talbot, H. S. A report on sexual function in paraplegics. *Journal of Urology*, 1949, *61*, 265–270.

Talbot, H. S. The sexual function in paraplegia. *Journal of Urology*, 1955, *73*, 91–100.

Terzian, H., and Dall Ore, G. Syndrome of Kluver and Bucy reproduced in man by bilateral removal of the temporal lobes. *Neurology*, 1955, *5*, 373–380.

Thor, D. H., and Flannelly, H.. Social-olfactory experience and initiation of copulation in the virgin male rat. *Physiology and Behavior*, 1977, *19*, 411–417.

Toran-Allerand, C. D. Sex steroids and the development of the newborn mouse hypothalamus and preoptic area *in vitro;* Implications for sexual differentiation. *Brain Research*, 1976, *106*, 407–412.

Tsuji, I., Nakajima, F., Morimoto, J., and Nounaka, Y. The sexual function in patients with spinal cord injury. *Urologia Internationalis*, 1961, *72*, 270–280.

Twiggs, D. D., Popolow, H. B., and Gerall, A. A. Medial preoptic lesions and male sexual behavior: Age and environmental interaction. *Science*, 1978, *200*, 1414–1415.

Van Dis, H., and Larsson, K. Spontaneous seminal discharge and preceding sexual activity. *Physiology and Behavior*, 1970, *5*, 1161–1163.

Van Dis, H., and Larsson, K. Induction of sexual arousal in the castrated male rat by intracranial stimulation. *Physiology and Behavior*, 1971, *6*, 85–86.

Vaughan, E., and Fisher, A. E. Male sexual behavior induced by intracranial electrical stimulation. *Science*, 1962, *137*, 758–760.

Walker, L. C., Gerall, A. A., and Kostrzewa, R. M. Rostral midbrain lesions and copulatory behavior in male rats. *Physiology and Behavior*, 1981, *26*, 349–353.

Wallach, S. J. R., and Hart, B. L. The role of the striated penile muscles of the male rat in seminal plug dislodgement and deposition. *Physiology and Behavior*, 1983, *31*, 815–821.

Weiskrantz, L. Behavioral changes associated with ablation of the amygdaloid complex in monkeys. *Journal of Comparative and Physiological Psychology*, 1956, *49*, 381–391.

Wenzel, B. M. The olfactory system and behavior. In: L. V. Di Cara (Ed.), *Limbic and Autonomic Nervous Systems Research.* New York: Plenum Press, 1974.

Whalen, R. E. Sexual behaviour of cats. *Behaviour*, 1963, *20*, 321–342.

Whitelaw, G. P., and Smithwick, R. H. Some secondary effects of sympathectomy: With particular reference to disturbance of sexual function. *New England Journal of Medicine*, 1951, *245*, 121–130.

Wilhelmsson, M., and Larsson, K. The development of sexual behavior in anosmic male rats reared under various social conditions. *Physiology and Behavior*, 1973, *11*, 227–232.

Winans, S. S., and Powers, J. B. Olfactory and vomeronasal deafferentation of male hamsters: Histological and behavioral analyses. *Brain Research*, 1977, *126*, 325–344.

Wysocki, C. J. Neurobehavioral evidence for the involvement of the vomeronasal system in mammalian reproduction. *Neuroscience and Biobehavioral Reviews*, 1979, *3*, 301–314.

Wysocki, C. J., Wellington, J. L., and Beauchamp, G. K. Access of urinary nonvolatiles to the mammalian vomeronasal organ. *Science*, 1980, *207*, 781–783.

BENJAMIN L. HART
AND MITZI G.
LEEDY

Wyss, J. M., and Goldstein R. Lesion artifact in brain stimulation experiments. *Physiology and Behavior,* 1976, *16,* 387–389.

Zeitlin, A. B., Cottrell, T. L., and Lloyd, F. A. Sexology of the paraplegic male. *Fertility and Sterility,* 1957, *8,* 337–344.

Zitrin, A., Jaynes, J., and Beach, F. A. Neural mediation of mating in male cats: III. Contributions of occipital, parietal and temporal cortex. *Journal of Comparative Neurology,* 1956, *105,* 111–122.

10

Neural Mechanisms of Female Reproductive Behavior

DONALD PFAFF AND DOAN MODIANOS

APPROACHES TO THE NEURAL MECHANISMS OF BEHAVIOR

The ultimate aim in the study of the nervous system is the explanation of behavior: the demonstration of how behavioral responses are produced as a function of nerve cell activity. Even for small bits of neural tissue and restricted aspects of behavior, the number of nerve cells involved is so large and their connections are so complex that large numbers of hypotheses can be imagined. As a result, in the history of behavioral studies, a great deal of "neurologizing" has occurred. Broad speculation about the overall "organization of the brain" and the manner in which it controls behavior has been entertained because, in most cases, the number of facts available to rule out hypotheses has been small. Thus, no comprehensive hypothesis really could be proved. It has seemed necessary to pick a situation where it would be fruitful to gather a great number of behavioral, neuroanatomical, and neurophysiological facts and thus to narrow down the allowable hypotheses to a small number. If enough relevant facts can be brought to bear, and hypotheses can be conclusively ruled out, then a strong inference (Platt, 1964) of the correct hypothesis will finally be possible. In turn, principles can be stated clearly and can be tested.

How can this be accomplished? Some have taken the approach of concentrating on a region of neural tissue (for instance, the septum) and doing a great deal of experimental work on the question "What does this tissue do?" Others have

---

DONALD PFAFF AND DOAN MODIANOS   Laboratory of Neurobiology and Behavior, The Rockefeller University, New York, New York 10021.

taken a single experimental technique and applied it to all the regions of neural tissue and functional problems available. In contrast, our approach is to try and see how the nervous system accomplishes a particular "job": to pick a particular pattern of adaptive behavior and learn how the nervous system elaborates and controls that behavior. From this approach, principles of the neural control of behavior may be generated and then tested with other behavior patterns.

Obviously, to learn as much as possible about a particular behavior pattern (in this case, female reproductive behavior), it is necessary to study in detail all of its sensory and motor aspects. Moreover, in studying its neural control, one may not work merely on forebrain mechanisms that seem to govern the behavior as a whole but must also study the spinal mechanisms that execute the behavior and the brainstem mechanisms that link forebrain and spinal neurons. It is likely that the reticular core of the brainstem is involved in the control of female reproductive behavior (as it is in many other behaviors), and this likelihood poses undeniably complex problems. Our strategy has been to try to approach the complete neural mechanism step by step, (1) moving from the anatomically simple toward the anatomically complex, and (2) using to advantage the hormonal control of female reproductive behavior.

HORMONES AND REPRODUCTIVE BEHAVIOR

Working with a hormone-controlled behavior pattern gives experimenters a set of powerful manipulations, based on a natural determinant of the behavior, that they can use to study the neural mechanism. Using the tools of experimental endocrinology, experimenters can remove the natural source of the hormone, replace the hormone by injection, implant it locally in brain tissue, study the distribution of radioactively labeled hormone, catalog the neurophysiological and neurochemical effects of the hormone, and so forth.

The mating behavior of rodents has obvious biological importance and is easily elicited in the laboratory in what appears to be a natural form. A history of excellent work on the reproductive behavior of rodents and its hormonal control by Beach, Young, Goy, their collaborators, and many other experimentalists has further opened the way for a detailed study of neural mechanisms.

Among rodent reproductive behavior responses, lordosis behavior by the female is essential for fertilization to occur and is under strong hormonal control by estrogens and progesterone. Lordosis appears to be stereotyped enough, its motor topography appears to be simple enough, and its stimulus–response relations appear to be reliable enough to allow detailed work on its mechanisms. For all of these reasons, the lordosis response of the female is an attractive subject for neurophysiological analysis.

THE NEW REFLEXOLOGY

Our approach, then, is to use the methods of neuroanatomy, neurophysiology, and endocrinology in a systematic study of a hormone-controlled behavior pattern,

425

NEURAL
MECHANISMS OF
FEMALE
REPRODUCTIVE
BEHAVIOR

in order to discover its neural mechanisms. In this endeavor, one can regard mating behavior responses as hormone-sensitive reflexes (Pfaff *et al.,* 1972). In this use of the term *reflex,* one does not imply a simple equivalence with the phrase "spinal reflex" or "reflex dependent solely on spinal tissue," even if the behavior pattern involved does depend on spinal tissue (as lordosis responses clearly do). Rather, we use the term *reflex* here to mean merely that a well-defined set of stimuli reliably determines a well-defined constellation of muscular responses.

An important tradition in the history of brain research has been to regard reflex action as a prototype for the study of animal behavior (Fearing, 1930). For instance, in helping to put the study of behavior on a scientific basis (and particularly, helping to link it with the physical sciences), Jacques Loeb said:

> It is probably unnecessary to emphasize the fact that it is better for the progress of science to derive the more complex phenomena from simpler components than to do the contrary. For all "explanation" consists solely in the presentation of a phenomenon as an unequivocal function of the variables by which it is determined. . . . The progress of natural science depends upon the discovery of rationalistic elements or simple natural laws. . . . Tropisms and tropism-like reactions are elements which pave the way for a rationalistic conception of the psychological reactions of animals. (Loeb, 1912, pp. 58–60)

> The understanding of complicated phenomena depends upon an analysis by which they are resolved into their simple elementary components. If we ask what the elementary components are in the physiology of the central nervous system, our attention is directed to a class of processes which are called reflexes. . . . There has been a growing tendency in physiology to make reflexes the basis of the analysis of the functions of the central nervous system. (Loeb, 1912, and Loeb, *Comparative Physiology of the Brain and Comparative Psychology,* 1899, p. 65)

Sherrington (1906) asserted that the study of reflex action would account for the integrated, coordinated flow of behavioral responses (aside from questions of conscious or volitional control):

> The study of reflexes as adapted reactions evidently, therefore, includes reactions of two ranks. With the nervous system intact, the reactions of the various parts of that system, the "simple reflexes," are ever combined into great unitary harmonies; actions which in their sequence one upon another constitute in their continuity what may be termed the behaviour of the individual as a whole. (p. 238)

William James (1890) supposed that even complex behavioral acts, at least in many of their aspects, may conform to rules based on the study of reflex action:

> In the "loop-line" along which the memories and ideas of the distant are supposed to lie, the action, so far as it is a physical process, must be interpreted after the type of the action in the lower centres. If regarded here as a reflex process it must be reflex there as well. The current in both places runs out into the muscles only after it has first run in; but whilst the path by which it runs out is determined in the lower centres by reflexions few and fixed among the cell arrangements, in the hemispheres the reflexions are many and instable. This, it will be seen, is only a difference of degree and not of kind, and does not change

the reflex type. The conception of *all* action as conforming to this type is the fundamental conception of modern nerve-physiology. (p. 23)

According to these thoughts, the differences between more complex and less complex behavioral mechanisms may be in the number, the subtlety, and the liability of the neural connections involved, rather than in the fundamental unit and mode of analysis. The physiologist Bayliss (1920) agreed:

> The difference between spinal reflexes and those in which the higher centres, and especially the cerebral cortex, take part is the regularity of the former and the ease with which the latter are modified or abolished by events in other parts of the central nervous system. (p. 508)

For some time, then, brain and behavior researchers have believed it possible that behavioral responses can be explained by the application of relatively simple principles. For instance, it is possible that most behavioral responses can be explained as functions of the interactions and sequences of reflexes. The challenge is clearly to synthesize and prove a correct set of hypotheses showing the neural mechanisms of a mammalian behavior pattern. Attempts to use the clear thinking of reflex physiology are demanded simply by the requirement for deterministic thinking, in the absence of clearly stated alternatives. Thus, using the strategies and thinking of reflexology and the methods of modern neuroanatomy, neurophysiology, and endocrinology, we have continued a systematic study of the neural mechanisms of lordosis.

ORGANIZATION OF THE CHAPTER

Two ideas guide us to appropriate points of attack on complex neural mechanisms of behavior. First, the investigator can "follow" the routes of determining influences on the behavior from their first points of contact with the nervous system toward the sites of their interactions. In the case of female reproductive behavior, we can study the influences of steroid sex hormones in brain tissue ("Mechanisms of Hormonal Control"). Likewise, with lordosis as with other behaviors, it is important to define precisely the sensory determinants of the behavior and to "follow" the sensory influence over the normal neuroanatomical routes by means of anatomic and physiological techniques ("Sensory Control of the Lordosis Reflex"). Female reproductive behavior gives exquisite examples of interactions between hormonal and sensory influences in determining the occurrence and the magnitude of the behavioral response.

The second guiding idea is to trace the neural mechanisms from the parts of relative anatomical simplicity toward the more complex neural regions. Thus, on the motor side, the investigator first defines the characteristics of the motor response and the muscles involved and then studies the relevant motor neurons. Then, the role of descending pathways, including those that originate in hypothalamus, can be studied. Increasing amounts of information on all of these topics should lead to increasingly detailed and reliable synthetic descriptions of the entire neural mechanism for lordosis behavior.

Mechanisms of Hormonal Control

427

NEURAL
MECHANISMS OF
FEMALE
REPRODUCTIVE
BEHAVIOR

Hormones Involved

Classical work showed that estrogen treatment followed by progesterone treatment could induce sexual receptivity in ovariectomized female rodents (Beach, 1948; Young, 1961). More recent work on the behavioral effects of circulating sex hormones has been reviewed (see chapters by Feder and by Whalen *et al.* in this book). Estrogenic facilitation of female reproductive behavior is found in a wide variety of vertebrate species (Kelley and Pfaff, 1977; see Chapter 5 in this book).

In the usual schedule of hormone treatment for inducing sexual receptivity in ovariectomized female rodents, "priming" treatment with estrogen is given for at least 48 hr before behavioral testing, and then progesterone is injected 2–6 hr before testing. The minimal time needed, for facilitating female reproductive behavior, between estrogen treatment and behavioral testing is at least 16 hr (Green *et al.*, 1970; McEwen *et al.*, 1975). If estrogen treatment is applied over a long enough period (at least several days), it alone can facilitate receptivity in ovariectomized female rats, without progesterone (Davidson *et al.*, 1968; Pfaff, 1970b). Relatively high doses of estrogen can partially substitute for progesterone in the usual schedule of hormone injections (Kow and Pfaff, 1975b).

The presence of the pituitary is not required for the induction of sex behavior by steroid hormones, as estrogen and progesterone injections can cause lordosis in hypophysectomized, ovariectomized female rats (Pfaff, 1970a). Likewise, pituitary hormones such as luteinizing hormone (LH) and follicle-stimulating hormone (FSH) have not been found to be effective in facilitating rodent reproductive behavior (Pfaff, 1970a; Moss and McCann, 1973). However, the decapeptide isolated from hypothalamic tissue, luteinizing-hormone-releasing factor (LRF), can facilitate reproductive behavior in estrogen-primed ovariectomized female rats (Pfaff, 1973; Moss and McCann, 1973, 1975).

Binding of Sex Hormones in the Brain

RODENTS. The consequences of estradiol action in rat brain are orderly, coordinated changes in mating behavior and pituitary gonadotropin release. These functional results of hormone action depend on (1) the concentration of estrogen by specific neuronal groups and (2) selective axonal projections of these neuronal groups.

Radioactive estrogen is concentrated highly by cells in specific limbic and hypothalamic structures in the brains of females rodents. Autoradiographic data from rats (Pfaff, 1968; Pfaff and Keiner, 1973) agree well with biochemical studies of dissected neural tissue (Eisenfeld and Axelrod, 1965, 1966, Kato and Villee, 1967a,b; Zigmond and McEwen, 1970). For a review of cell fractionation results, see McEwen *et al.* (1974) and Luine and McEwen (chapter in this book). Estrogen is accumulated specifically by cells of the medial preoptic area, the medial anterior hypothalamus, the ventromedial nucleus of the hypothalamus, the arcuate (infundibular) nucleus, and the ventral premammillary nucleus. In the limbic system, estrogen is concentrated highly by neurons of the medial and the cortical amygdala and or the lateral septum, and to a lesser extent by neurons in the ventral hippocampus. An extension of this estrogen-voncentrating system is seen posteriorly in the mesencephalon, where cells in the ventrolateral and the dorsolateral portions of the central gray accumulate radioactive estradiol. A summary picture of the estrogen-concentrating system in rat brain is given in Figure 1.

The distribution of radioactive estrogen to cells in the brain of the female hamster is very similar to that in the female rat; the only detailed differences appear to depend on functional differences between these two species (Floody and Pfaff, 1974; Krieger *et al.*, 1977b). In both rats and hamsters, peak concentrations of radioactive estrogen are seen in those nerve cell groups that have previously been implicated in the control of estrogen-dependent phenomena, in some cases behavioral changes and in other cases pituitary gonadotropic controls (for references on rats, see Pfaff and Keiner, 1973; hamsters, Krieger *et al.*, 1977b).

Neuroanatomical evidence suggests several fiber pathways that connect neurons in estrogen-concentrating regions and that may form *systems* of estrogen-sensitive structures (Figure 2) Literature references for the pathways known by 1973 are given in Pfaff and Keiner (1973). More detailed evidence on the efferent connections of preoptic and hypothalamic neurons have lent support to the idea of linked estrogen-concentrating neurons (Conrad and Pfaff, 1975, 1976a,b). Briefly, efferent connections from medial preoptic neurons, efferents from estrogen-concentrating limbic structures (for instance, the stria terminalis and the medial corticohypothalamic tract) and connections between the septum, the olfactory tubercle, and the diagonal band of Broca all serve to link estrogen-concentrating regions with other estrogen-concentrating regions. The obvious longitudinal fiber systems (periventricular and medial forebrain bundle fibers) do the same. This arrangement of anatomical connections suggests that estrogen-concentrating cells can form *chains* of two or more neurons. Thus, the possibility is raised of multiple sites of action by estrogen in particular neural circuits. One implication of this notion is that estrogen could achieve a particular behavioral effect by "cascaded" (or multiplied) actions in the appropriate neural pathways.

VERTEBRATES IN GENERAL. Autoradiographic work with a wide variety of vertebrates has shown the generality of conclusions based on studies with rodents (reviews by Morrell *et al.*, 1975b; Pfaff, 1976). Among mammals, besides the work with rodents mentioned above, autoradiographic localization of estrogen-concentrating neurons has been completed with a carnivore, the mink (Morrell *et al.*, 1977a), and a primate, the rhesus monkey (Gerlach *et al.*, 1976; Pfaff *et al.*, 1976). Among birds, autoradiographic work with tritiated testosterone has been completed for the chaffinch (Zigmond *et al.*, 1973), the zebra finch (Arnold *et al.*, 1976), and the chicken (Barfield *et al.*, 1977). The conclusions stated below encompass unpublished work with the lizard *Anolis carolinensis,* representing reptiles (Morrell *et al.*, 1977b). Work with amphibians has included studies of both tritiated estradiol and tritiated testosterone in the brains of both male and female *Xenopus laevis* (Kelley *et al.*, 1975; Morell *et al.*, 1975a), and experiments have also been

429

NEURAL
MECHANISMS OF
FEMALE
REPRODUCTIVE
BEHAVIOR

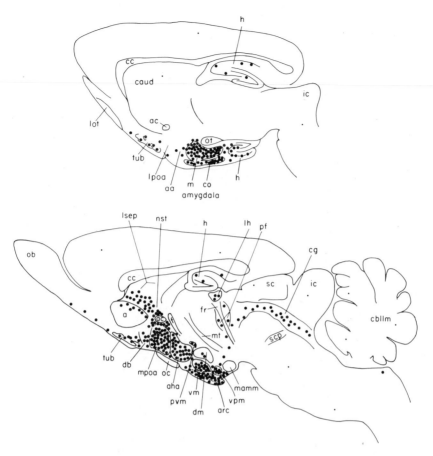

Fig. 1. Distribution of estrogen-concentrating neurons in the brain of the female rat represented schematically in two sagittal sections. Most labeled neurons could be represented in a medial plane (bottom
drawing) based primarily on Figure L740 in the atlas of König and Klippel and Figures A35 and A36
in the atlas of Zeman and Innes. Estradiol-concentrating neurons in the amygdala and the hippocampus
are represented in a more lateral plane (top drawing) based on Figure L2590 in the atlas of König and
Klippel. Locations of estradiol-concentrating neurons are represented by black dots (●). (From Pfaff
and Keiner, 1973.) Abbreviations: a, nucleus accumbens; aa, anterior amygdaloid area; ac, anterior commissure; aha, anterior hypothalamic area; arc, arcuate nucleus; caud, caudate nucleus; cbllm, cerebellum; cc, corpus callosum; cg, central gray; co, cortical nucleus of the amygdala; db, diagonal band of
Broca; dm, dorsomedial nucleus of the hypothalamus; f, fornix; fr, fasciculus retroflexus; h, hippocampus; ic, inferior colliculus; lh, lateral habenula; lot, lateral olfactory tract; lpoa, lateral preoptic area;
lsep (or ls), lateral septum; m, medial nucleus of the amygdala; mamm, mammillary bodies; mch, medial
corticohypothalamic tract; mpoa, medial preoptic area; mt, mammillothalamic tract; nst, bed nucleus of
the stria terminalis; ob, olfactory bulb; oc, optic chiasm; ot, optic tract; perivent, periventricular fibers;
pf, nucleus parafascicularis; pvm, paraventricular nucleus (magnocellular); sc, superior colliculus; scp,
superior cerebellar peduncle; sm, stria medullaris; st, stria terminalis; tub, olfactory tubercle; vm, ventromedial nucleus; vpm, ventral premammillary nucleus.

performed with *Rana* (Kelley and Pfaff, 1975). Finally, autoradiographic work
localizing estrogen and androgen-concentrating cells has been done with teleosts,
the sunfish (Morrell *et al.*, 1975b) and the paradise fish (Morrell *et al.*, 1976).

For all the vertebrates studied so far, certain conclusions seem to be univer-
sally true: (1) the binding of steroid sex hormones by neurons in the brain can

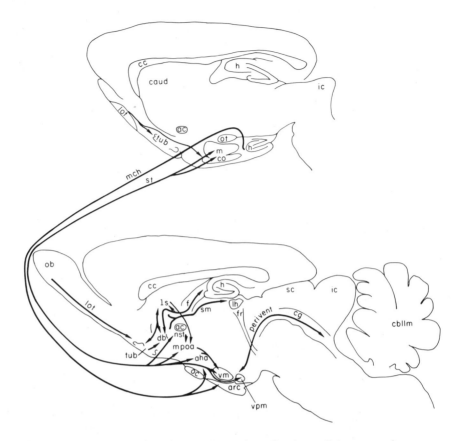

Fig. 2. Established anatomical connections among regions of peak estradiol concentration, represented schematically on the same sagittal sections as used in Figure 1. The references for each pathway drawn in this figure are given in Pfaff and Keiner (1973). The density of projections among regions containing large numbers of estrogen-concentrating neurons suggests an anatomical basis for coordinated physiological action by these neurons. Abbreviations as in Figure 1. (From Pfaff and Keiner, 1973.)

always be detected autoradiographically. (2) There are obvious similarities in the neuroanatomical pattern of sex steroid binding: there is always a strong accumulation of estrogen or androgen by cells in the medial preoptic area, in the infundibular hypothalamic nuclei, and in limbic forebrain structures such as the medial amygdala and the lateral septum (or their homologous structures). In many cases, there is also binding in a specific midbrain region below the tectum and near the mesencephalic ventricle. (3) In each species, there are good correlations between the locations of estrogen- or androgen-concentrating cells and the locations that have been implicated by other techniques in the control of neuroendocrine functions. References to the functional (nonautoradiographic) evidence for each species are given in the papers cited above. The striking anatomical correlations between the autoradiographic and the functional results suggest that estrogen and androgen binding in the brain is, indeed, one step in the physiological process by which the brain responds to these steroid hormones and controls mating behavior and pituitary function.

Certain detailed additions to or variations on the basic vertebrate neuroanatomical pattern of sex hormone binding further support the notion that hormone
binding is an important part of the hormone action in brain tissue. For instance,
detailed differences between female rats and hamsters in the binding of radioactive
estrogen in the medial preoptic area are correlated with differences in estrogen
implant results between these two species (Krieger *et al.*, 1977b). Also, the specific
binding of radioactive androgen in the motor neurons controlling the syrinx in the
zebra finch are probably related to the androgen-dependent control of song in that
bird (Arnold *et al.*, 1976). Similarly, androgen binding in a dorsal tegmental area
of the medulla and in a presumptive motor nucleus of cranial nerves IX and X in
Xenopus laevis have been related to the androgen control of courtship behavior
(Kelley *et al.*, 1975). Thus, both the nature of the vertebratewide neuroanatomical
pattern of hormone binding and the nature of the detailed additions to it support
the correlation between hormone binding and hormone action in the vertebrate
brain.

431

NEURAL
MECHANISMS OF
FEMALE
REPRODUCTIVE
BEHAVIOR

BEHAVIORAL AND PHYSIOLOGICAL EFFECTS OF ESTROGEN IN BRAIN REGIONS WHERE IT IS BOUND

Further evidence of the relationship of hormone binding to hormone action
in brain tissue comes from studies of direct estrogen effects on behavioral and
electrophysiological measures.

HORMONE IMPLANTS IN BRAIN. Implantation of estrogen in the medial preoptic
area of ovariectomized female rats leads to increased female sexual behavior (Lisk,
1962; Lisk *et al.*, 1972). This result has been confirmed (Malsbury and Pfaff, 1974,
Table 3; Barfield, 1976). Work with estrogen implants in rats later injected with
progesterone shows that basal anterior hypothalamic sites of estrogen application
can facilitate lordosis (Lisk and Barfield, 1975). Estrogen implants in the region of
the ventromedial nucleus of the hypothalamus were reported to increase female
reproductive behavior in ovariectomized rats (Dörner *et al.*, 1968), and other
results have indicated that implants here are as effective as those in the preoptic
area (Barfield, 1976). All of these effective estrogen-implant sites are within the
distribution of estrogen-concentrating neurons as defined with steroid autoradiography (Pfaff and Keiner, 1973).

In the ovariectomized female hamster, estrogen implantation in a restricted
zone of the dorsomedial anterior hypothalamus leads to the display of estrous
behavior (Ciaccio and Lisk, 1973). The efficacy of an anterior hypothalamic site,
contrasted with negative results from the preoptic area of the female hamster,
matches the autoradiographic results with tritiated estrogen in female hamsters as
opposed to rats (Krieger *et al.*, 1977a). On the basis of evidence available from
rodents and rabbits, Kelley and Pfaff (1977) suggested that hypothalamic estrogen
implants are effective in facilitating female sex behavior at hypothalamic levels at
or anterior to the locations where hypothalamic lesions lead to decreases in female
reproductive behavior.

The effectiveness of preoptic and hypothalamic estrogen implants in facilitating female sex behavior suggests a causal relationship between estrogen binding and behaviorally important estrogen action.

ELECTROPHYSIOLOGICAL EFFECTS. Several experiments have shown variations in hypothalamic unit activity during the estrous cycle of the female rat (Cross and Dyer, 1971; Dyer, 1973; Dyer *et al.*, 1972; Kawakami *et al.*, 1970; Moss and Law, 1971; Terasawa and Sawyer, 1969). These effects have been reviewed (Pfaff, 1973b) and may reflect effects of estrogen. However, because the blood levels of several hormones change during the estrous cycle, individual manipulations of estrogen alone are required for a clear analysis of that hormone's effects.

In one experiment (Bueno and Pfaff, 1976), single-unit activity was recorded with micropipettes in the medial hypothalamus and the preoptic area of urethane-anesthetized ovariectomized female rats, some treated with estrogen and others untreated. In the medial preoptic region and the bed nucleus of the stria terminalis, the estrogen-treated rats had fewer cells (compared to the untreated rats) with recordable spontaneous activity, primarily because of a loss of cells with very slow firing rates. In the basomedial hypothalamus, the estrogen-treated rats had more cells (than the untreated rats) with recordable spontaneous activity, primarily because of an increase in the number of cells with slow firing rates. The responsiveness of neurons to somatosensory stimulation was depressed by estrogen treatment in medial preoptic neurons and in the bed nucleus of the stria terminalis, whereas it tended to be elevated by estrogen treatment in the medial anterior hypothalamus and the basomedial hypothalamus.

This difference between preoptic and basomedial hypothalamic neurons in the direction of the estrogen effect was at first surprising. However, a reexamination of the literature showed that comparisons of previous studies also indicated a difference between the preoptic area and the basomedial or anterior hypothalamus in the electrophysiological effects of sytemically injected estrogen. Kawakami *et al.* (1971) showed increases in the firing rates of arcuate neurons following estrogen treatment, and Cross and Dyer (1972) found increases in the medial anterior hypothalamus. In spayed female cats, responsiveness in the ventromedial nucleus and the medial anterior hypothalamus was greater following estrogen treatment, and there was a shift in the direction of response toward excitation (Alcaraz *et al.*, 1969). In contrast, in the medial preoptic area, Whitehead and Ruf (1974) and Yagi (1970, 1973; Yagi and Sawaki, 1973) found units that decreased their resting discharge rates to very low levels for a long time after estrogen administration. Likewise, Kelly *et al.* (1976) found predominantly inhibitory effects of microelectrophoresed estradiol on medial preoptic neurons in female rats. Lincoln (1967) showed that long-term estrogen treatment of ovariectomized rats is followed by lower spontaneous activity of preoptic units, of units at the very anterior border of the medial anterior hypothalamus, and in the lateral septum. The responses of preoptic neurons to probing of the vaginal cervix were shifted significantly in the direction of inhibition following estrogen treatment (Lincoln and Cross, 1967).

Thus, estradiol has significant electrophysiological effects on nerve cells in the same brain regions where it is bound. Moreover, there are stable differences in the

433

NEURAL
MECHANISMS OF
FEMALE
REPRODUCTIVE
BEHAVIOR

nature of the effect (at least after systemic estrogen treatment) between preoptic and basomedial hypothalamic neurons. It is interesting that there are also differences in the neurochemical effects of estrogen treatment between the preoptic and the basomedial hypothalamic regions (see below, and see the Luine and McEwen chapter in this book). Most important, the preoptic-basomedial hypothalamic difference in the direction of the estrogen effect matches the differences in the effect of these two neural regions on lordosis behavior in female rats (see the section below on hypothalamic mechanisms).

The electrophysiological effects of estrogen presumably are based on hormone-caused neurochemical alterations. Therefore, it is interesting that neurochemical effects of long-term estrogen administration have been reported, and that the nature of these effects is different in the basomedial hypothalamus and in the preoptic region (see the Luine and McEwen chapter in this book). Other evidence suggests further that, following estrogen binding, a behaviorally important effect of the hormone depends on RNA synthesis. The infusion of actinomycin D into the preoptic area of estrogen-treated female rats can block the production of lordosis behavior (Quadagno et al., 1971; Hough et al., 1974; Whalen et al., 1974). Large doses of estradiol have been reported to cause obvious electron-microscopic changes in nerve cells and glial cells in the basomedial hypothalamus of female rats (Brawer and Sonnenschein, 1975).

RESULT OF DECREASED ESTROGEN BINDING

A group of compounds classified as antiestrogens has been shown to block the nuclear binding of estradiol in the uterus (Ruh and Ruh, 1974; Capony and Rochefort, 1975). Antiestrogens (for instance, CI-628) can also block the accumulation of radioactive estradiol by cell nuclei in the brain (Chazal et al., 1975; Luttge et al., 1976; Roy and Wade, 1977; Luine and McEwen, 1977).

Antiestrogens prevent the long-term uterine responses to estradiol that are essential for estrogen-stimulated uterine growth (Katzenellenbogen and Ferguson, 1975), presumably because of interference with the nuclear action of an estrogen-receptor complex (Buller and O'Malley, 1976). Comparisons between the uterine growth responses to estradiol, estriol, and long-acting estriol derivatives have shown marked differences in the effectiveness of these various compounds, and these differences have been related to the duration of binding of the various steroids (Anderson et al., 1975; Lan and Katzenellenbogen, 1976). Can a functional effect of antiestrogens be shown in studies of estrogen-stimulated mating behavior?

Systemic injections of antiestrogens have, indeed, been shown to diminish estrogen-stimulated lordosis behavior. Decreased lordosis results from the blocking action of CI-628 (Arai and Gorski, 1968; Powers, 1975; Roy and Wade, 1977) and clomiphene (Ross et al., 1973), and similar results have also been reported with the antiestrogen MER-25 (Meyerson and Lindstrom, 1968). Thus, in both peripheral and neural tissues, decreased estrogen binding is associated with decreased estrogen action. In brain tissue, this includes decreased estrogenic stimulation of lordosis. It is reasonable to conclude that estrogen binding in brain tissue is necessary for the normal estrogenic stimulation of lordosis behavior.

Summary

Estrogen is required for the normal hormonal facilitation of the lordosis reflex in female rodents. Estrogenic and androgenic steroid hormones are bound by nerve cells in specific hypothalamic and limbic structures. The presence and the neuroanatomic pattern of hormone-binding nerve cells are lawful, phylogenetically stable phenomena demonstrated in a wide variety of vertebrate species. The sites of estrogen binding in the brain are correlated with the control sites for estrogen-dependent functions. In fact, the behavioral and electrophysiological effects of estrogen are shown in brain regions where it is bound, and decreased estrogen binding results in decreased behavioral effects of the hormone. Therefore, estrogen accumulation by nerve cells, as studied autoradiographically, probably reveals a phase in the neural action of estrogen that is important in the facilitation of lordosis behavior, as well as of other estrogen-sensitive behaviors and of gonadotropin release.

Sensory Control of the Lordosis Reflex

Research on the sensory control of reproductive behavior in female rodents has been reviewed (Kow and Pfaff, 1976a). In this chapter, we consider briefly facts directly relevant to the neural mechanisms of lordosis in female rats. First, the stimuli essential for lordosis and the appearance of the estrogen effect are defined, and then, the routes followed by these stimuli are traced to successively higher levels of the central nervous system.

Definition of Stimuli That Evoke Lordosis in Female Rats

Stimuli Applied by the Male Rat. This consideration is restricted to somatosensory stimuli. With film analyses (Pfaff and Lewis, 1974), it could be seen that female rats did not do lordosis before being touched by the male, but that they promptly initiated lordosis shortly (161 msec) after contact by the male with the female's skin (Table 1). Female rats that were surgically blinded, deafened, and rendered anosmic did normal lordoses (Kow and Pfaff, 1965b). Because taste does not play a role in this behavior of the female, only somatosensory stimuli were available to support lordosis. Thus, somatosensory stimuli are necessary and sufficient for lordosis.

The locations on the female's skin contacted by the male during mating encounters were plotted by coating the entire ventral surface of the male with dyes and recording the loci of dye deposition on the female's skin after mounts by the male (Pfaff *et al.*, 1977). Heavy depositions of dye were found regularly on the flanks, the posterior rump, the tail base, and the perineum of the female. In film

435

NEURAL
MECHANISMS OF
FEMALE
REPRODUCTIVE
BEHAVIOR

TABLE 1. PERCENTAGES OF MATING ENCOUNTERS IN WHICH RUMP ELEVATION OF THE FEMALE
RAT PRECEDES, COINCIDES WITH, OR FOLLOWS BEHAVIORS OF THE MALE RAT[a]

Rump elevation of female rat begins	Relation of initial female rump elevation to male behaviors listed under it		
	Precedes	Coincides with	Follows
Male rat's paws grasp female's flanks	0	0	100
Male rat's head touches female rat's back	0	10	90
Male rat begins deepest pelvic thrust	84	8	8
Deepest point of deepest pelvic thrust	99	0	1
Male rat releases female rat	100	0	0

[a]Data from Pfaff and Lewis (1974).

analyses (Pfaff and Lewis, 1974), contacts by the male on these same regions of skin of the female rat could be seen.

The order of responses by the male and the female rat during mounts and lordoses was also informative. The females began the rump elevation of lordosis an average of 161 msec following the first contact by the male with the female's skin (Pfaff and Lewis, 1974). During that time, the male rat's paws were on the female's flanks, the male rat's head often touched the female rat's back, the male rat had stepped up with his rear feet so as to contact the posterior rump of the female, and preliminary pelvic thrusts had usually begun. However, the female had virtually always begun rump elevation or was in the lordotic posture before intravaginal stimulation by the deepest (intromittory) pelvic thrust by the male (Table 1). Thus, stimuli by the male on the flanks, the posterior rump, or (from the initial pelvic thrusts) the perineum of the female could be involved in causing the initiation of lordosis, but in the normal case, intravaginal stimulation cannot.

The onset of cutaneous stimulation from the male's mount is sudden, and once applied, these stimuli are not stationary. Some stimuli applied by the male are repetitive, such as forepaw palpation and pelvic thrusting. From film analyses, it is seen that the dominant frequencies of these repetitive stimuli are between 10 and 20 per sec (Pfaff et al., 1977).

STIMULI ESSENTIAL FOR LORDOSIS. Among the stimuli applied by the male to the female rat during mating encounters, which are essential for the initiation of lordosis? Nonsomatosensory stimuli are not essential, as lordosis occurs in blinded, deafened, anosmic female rats (Kow and Pfaff, 1976b). Among the somatosensory stimuli, intravaginal stimulation and cutaneous stimulation of the thorax are not necessary, for removal of these does not interfere with lordosis (Pfaff et al., 1972; Kow and Pfaff, 1976b).

The type of somatosensory input crucial for lordosis is cutaneous, as experimental manipulations that affect only cutaneous nerves can block lordosis (surgical cutaneous denervation, Kow and Pfaff, 1976b). Denervation of the skin on the posterior rump, the tail base, and the perineum caused a significant loss of lordosis. If to this pattern of cutaneous denervation was added bilateral denervation of the ventral flanks, lordosis was nearly abolished (Kow and Pfaff, 1976b). Lordosis decrements caused by surgical cutaneous denervation were in good agreement with the results of parallel experiments using subcutaneous injections of the local anesthetic Procaine (Pfaff *et al.*, 1972; Kow and Pfaff, 1976b).

Thus, only cutaneous stimuli from the posterior rump, the tail base, the perineum, and the flank are essential for lordosis in the female rat.

STIMULI SUFFICIENT FOR LORDOSIS. Stimuli other than somatosensory stimuli are not normally sufficient for lordosis in the female rat. The withdrawal of cutaneous input by experimental manipulations, leaving these other modalities intact, virtually abolished lordosis (Kow and Pfaff, 1976b). Also, the female rat does not do lordosis until touched by the male, in the usual case (Pfaff and Lewis, 1974).

On the skin of the hormone-primed female rat, stimulation of the flanks (bilaterally) followed by stimulation of the posterior rump, the tail base, and the perineum is sufficient to trigger strong lordosis responses (Diakow *et al.*, 1973; Pfaff *et al.*, 1977). Increased estrogen doses lead to stronger responses for a given cutaneous input from these regions and can reduce the requirement of cutaneous input for a given level of lordosis response (Diakow *et al.*, 1973; Pfaff *et al.*, 1977).

Within the skin regions where stimulation can trigger lordosis, hair deflection alone is not sufficient. Neither light stimulation of the hair by the experimenter's hand (Pfaff *et al.*, 1977) nor hair deflection caused by air puffs or a small camel-hair brush (Kow *et al.*, 1977c) caused lordosis. In the relevant regions of skin on the hormone-primed female rat, pressure on the skin is included in the minimal stimulation normally sufficient for eliciting lordosis. For instance, under good conditions, pressure on the perineal skin of between 100 and 200×10^3 dynes/cm^2 is included in the stimulation sufficient for eliciting lordosis (Kow *et al.*, 1977c; Kow and Pfaff, 1976a).

Observations of hamster mating encounters provide an interesting comparison with results from rats. The overall form of the mount by the male on the female appears to be grossly similar in hamsters and rats. In female hamsters, the skin regions whose stimulation triggers lordosis are the same as those in rats (Kow *et al.*, 1976). However, in female hamsters, circumscribed cutaneous stimulation of small spots within the generally effective skin areas can trigger lordosis, whereas female rats usually require a pattern of stimulation that affects all of these areas for lordosis normally to occur. Also, under good conditions, hair deflection on specific regions of the hindquarters, caused by light brushing (Kow *et al.*, 1976) or even by air puffs (Manogue, Kow and Pfaff, unpublished observations, 1979; Kow and Pfaff, 1976a), can cause lordosis, whereas in female rats hair deflection is not effective. Thus, in female hamsters, the same regions of skin trigger lordosis as in female rats. However, in female hamsters the "threshold" for such stimulation seems lower: stimulation of less area and of a less intense quality is sufficient in hamsters but not in rats.

CUTANEOUS RECEPTORS INVOLVED. The results reviewed above show that stimuli from the male rat primarily on the posterior rump, the tail base, and the perineum of the female, facilitated by stimuli on a vertical band of flank and lateral abdomen, are necessary and sufficient for the normal and frequent occurrence of lordosis. Because cutaneous desensitization can cause very large reductions in the

437

NEURAL
MECHANISMS OF
FEMALE
REPRODUCTIVE
BEHAVIOR

performance of lordosis, skin receptors themselves must be crucial for lordosis, and deep receptors in and between muscles are not sufficient for the normal occurrence of the lordosis reflex. The stimuli for eliciting lordosis in female rats are mechanical in nature, so chemoreceptors and thermoreceptors can be ruled out. These considerations indicate that cutaneous mechanoreceptors on the posterior rump, the tail base, and the perineum, and possibly the flanks, play a crucial role in triggering lordosis in the female rat.

All of the critical skin regions are covered with hair. However, simple hair deflection, either with air puffs, brushing, or the experimenter's fingers, cannot usually elicit lordosis in female rats. Thus, although input from hair receptors may have some role in triggering lordosis, the essential sensory information must come from pressure receptors, requiring skin contact or indentation.

Some types of cutaneous mechanoreceptors can be ruled out by considering temporal characteristics: response latency and stimulus frequency. Female rats begin the rump raise of lordosis with a mean latency of 161 msec after the male's paws grasp their flanks (Pfaff and Lewis, 1974). Receptors associated with C fibers require strong stimulation for at least 150–200 msec for response (Burgess and Perl, 1973). Because time within the 161 msec must also be allowed for impulse conduction, spinal neuronal integration, and the development of the motor response to a point where it is visible on film, it seems unlikely that mechanoreceptors associated with C fibers are fast enough to account for the initial triggering of lordosis.

All the stimuli from the male on the skin regions of the female important for lordosis appear to be moving stimuli. Film analyses of repetitive movements such as palpation by the male's forepaws (on the female's ventral flanks) and the male's pelvic thrusting (on the female's posterior rump, tail base, and perineum) reveal dominant frequencies between 10 and 20/sec (Pfaff *et al.,* 1977). With such frequencies, it seems unlikely that Pacinian corpuscles or hair receptors of the G1 type found in the cat are very important for female rat lordosis. These are very difficult to stimulate at frequencies below 40–60/sec and appear to be "tuned" to frequencies between 100 and 400/sec (Burgess and Perl, 1973).

The mechanoreceptors in the skin of the posterior rump, the tail base, and the perineum that remain in consideration for providing essential sensory input for lordosis initiation include Iggo Type II receptors and "field" receptors. Hair receptors associated with $A\alpha$ or $A\delta$ fibers could participate but are not by themselves sufficient.

INPUT–OUTPUT DEFINITION OF THE ESTROGEN EFFECT. The facilitation of lordosis by estrogen has been well documented (see the section above on the mechanisms of hormonal control). The hormone effect reveals itself in studies of the sensory determinants of lordosis. In experiments on cutaneous desensitization, higher estrogen doses or greater initial responsiveness to a given dose partially compensated for sensory loss in the production of lordosis (Kow and Pfaff, 1976b). For instance, bilateral sectioning of the pudendal nerve significantly reduced the lordosis quotient only in animals with relatively low estrogen doses. Further, surgical denervation of the posterior rump, tail-base, and perineal regions had quantitatively greater effects in animals with lower estrogen doses and lower initial "pre-operative" responsiveness. In other experiments, using manual stimulation to determine which stimuli are sufficient for lordosis, greater responsiveness to a given stimulus element was found in rats with higher estrogen doses (Diakow *et al.,* 1973; Komisaruk and Diakow, 1973; Pfaff *et al.,* 1977; Kow *et al.,* 1977c). In a variety of comparisons within these experiments, it could be seen that higher estrogen doses (or estrogen doses supplemented by progesterone) could maintain response strength with weaker stimulation. Estrogen effects of this sort could take place peripherally, for it has been demonstrated (Kow and Pfaff, 1973; Komisaruk *et al.,* 1972) that estrogen can increase the size of the pudendal-nerve receptive field and the sensitivity of perineal cutaneous receptors. It is interesting that it was the fast-adapting mechanoreceptors in the perineal skin that were the most sensitive and that were influenced by estrogen treatment (Kow and Pfaff, 1973). It is also likely, however, that central effects of estrogen play an important role in these phenomena. These interactions between sensory requirements and hormone levels are an expression of the overall hormone effects on lordosis behavior. Discovery of the mechanisms of the "trading relations" between sensory input and estrogen level in triggering lordosis may be equivalent to describing the mechanism of hormone action on this behavior.

Precise definition of the stimuli and the receptors involved in triggering lordosis allows an "input–output" definition of the estrogen effect on this response. The essential element of the lordosis response is vertebral dorsiflexion, manifest in the elevation of the rump (see the section below on the motor components of lordosis). Thus, from the information above, it is clear that the estrogen effect on lordosis can be described as an increased functional connection from pressure receptors in the skin of the rump, the tail base, and the perineum to the motor neurons for elevation of the rump.

PERIPHERAL AND CENTRAL SENSORY PATHWAYS FOR LORDOSIS

The peripheral nerves that transmit lordosis-eliciting sensory information from the skin of the female rat to the spinal cord have been partially determined. In rats, the skin on the perineal region is innervated solely by the pudendal nerves (Kow and Pfaff, 1973, 1975a). By recording from the pudendal nerve, we found sensitive, rapidly adapting receptors whose inputs are sensitive to estrogen (Kow and Pfaff, 1973). The peripheral nerves innervating other skin areas important in lordosis have not been studied systematically in the rat. For a given skin region, some nerve fibers descend directly through subcutaneous layers, whereas others travel through neighboring skin regions before traveling toward the cord (Kow and Pfaff, 1975a).

The dorsal roots that carry sensory information for lordosis-relevant skin areas were determined by recording from each of the dorsal roots T_{11} through S_3 and determining their cutaneous receptive fields (Kow and Pfaff, 1975a). Inputs from the flanks of the female rat must enter at least over dorsal roots L_1 and L_2, whereas inputs from the posterior rump, the tail base, and the perineum enter over dorsal roots L_5, L_6, and S_1.

439

NEURAL
MECHANISMS OF
FEMALE
REPRODUCTIVE
BEHAVIOR

ASCENDING PATHWAYS: LOCATION AND ROLE. Female rats with complete transections of the spinal cord at thoracic levels do not perform the lordosis reflex (Pfaff *et al.*, 1972; Kow *et al.*, 1977b; Hart, 1969). A series of partial transections of the spinal cord, aiming for individual columns, were performed in order to localize the crucial ascending and descending fiber systems. The ascending systems available are the dorsal columns, the dorsolateral columns, and the anterolateral columns. Complete transection, bilaterally, of the dorsal columns or the dorsolateral columns, and even combined bilateral transection of these two ascending systems, did not affect lordosis (Kow *et al.*, 1977b). In contrast, large transections of the anterolateral columns could seriously disrupt or abolish the lordosis reflex (Kow *et al.*, 1977b). Large transections were required. It appeared that if 20% or 25% in any portion of the anterolateral columns remained, the lordosis deficit would be modest.

The results of spinal transection studies indicate that supraspinal control of lordosis mechanisms is required for the normal elaboration of lordosis and for its hormone sensitivity. It is likely that this control takes the form of a supraspinal loop, with an ascending and a descending limb in the spinal cord. The results reviewed above show that any ascending information required must travel in the anterolateral columns. Because large anterolateral transections were required to abolish lordosis, it appears that the essential information is spread within the anterolateral columns, and that a small portion of it suffices to govern supraspinal control of the reflex.

Ascending fibers in the anterolateral columns are dispersed widely within these columns, rather than ascending in a restricted lemniscal fashion (Antonetty and Webster, 1975; Land *et al.*, 1975). These fibers terminate throughout rather large regions of the brainstem reticular formation, as well as the lateral vestibular nucleus, deep layers of the tectum (including the intercollicular region), the medial geniculate body pars medialis, and the thalamus (Mehler, 1969; Anderson and Berry, 1959; Schroeder and Jane, 1971; Goldberg and Moore, 1967; Lund and Webster, 1967). These ascending systems in the anterolateral columns are apparently phylogenetically old, having been detected in all classes of vertebrates (Herrick and Bishop, 1958; Ebbesson, 1967, 1969; Hayle, 1973). In contrast to the dorsal column system, one gets the following picture of the anterolateral column system of ascending fibers: an ancient ascending system of fibers that have rather widespread trajectories and relatively diffuse areas of termination.

Electrophysiological recordings from ascending anterolateral column axons (usually spinothalamic) have not revealed a great deal of specificity. When locations of axons in the ventrolateral white matter of the cord were mapped with respect to their cutaneous responses, there was a complete overlap among hair-activated, low-mechanical-threshold, and high-mechanical-threshold axons (Applebaum *et al.*, 1975). Individual neurons within this ascending system can respond over a wide range of mechanical stimuli from light brushing to noxious pinching (Dilly *et al.*, 1968; Willis *et al.*, 1975). In studies that included visceral as well as cutaneous stimuli, cells with ascending axons in the anterolateral columns were found to respond to both (Foreman *et al.,* 1975).

Presumably, the function of an extended ascending (sensory) system, with several internuncial neurons, is to allow for a reorganization or "reshuffling" of sensory information from a code determined entirely by receptor properties to codes appropriate for the direction of discrete behavioral responses. Recent evidence suggests that, even within modalities, different ascending systems accomplish this function in at least two different ways (Schneider, 1967, 1969; Werner, 1974). One type of sensory system, with relatively little access to the telencephalon, has relatively low discriminative capacity and mediates relatively large movements in response to stimulation. The other type of ascending system depends heavily on the telencephalon and has high discriminative capacity for stimulus shape and pattern. For somatosensory input from the rear half of the body, the anterolateral-column ascending system falls in the former category and the dorsal column system in the latter. Ascending anterolateral fibers appear to work primarily through brainstem (not telencephalic) mechanisms, appear to be widespread in their trajectories and terminations, and have relatively low specificity of response. Lordosis depends on the anterolateral columns. In fact, there are parallels between the phenomena of lordosis control and those concerned with the so-called ventral flexor reflex afferents that run in the anterolateral columns (see the section below on the reflex nature of the lordosis behavior). Lordosis-relevant fibers are apparently spread throughout the anterolateral columns, and any sizable fraction of them left intact can subserve the reflex (Kow *et al.,* 1977b). From the considerations above, it seems that ascending lordosis-relevant fibers could act to help trigger the reflex on each mount by the male or could act via a tonic mechanism. However, the facts reviewed above place a severe upper limit on the specificity of this ascending lordosis-triggering or tonic-facilitatory mechanism. That is, if the execution of the lordosis reflex depends on precise stimulus guidance ("steering") *and* facilitation of the entire stimulus–response connection through an internuncial chain ("acceleration"), the supraspinal loop acting through the anterolateral columns appears to supply more of the latter than the former.

BRAINSTEM. Projections of ascending spinal fibers that do not travel in the dorsal columns have been described for the rat (Mehler, 1969). Comparable information has been gathered for other mammalian species also (Anderson and Berry, 1959; Mehler, 1969; Schroeder and Jane, 1971). For ascending fibers from the anterolateral columns, a widespread pattern of termination in the posterior brainstem has been reported, and terminations become less dense at more anterior brainstem levels. Anterolateral fibers give a heavy projection to the lateral reticular nucleus and to the medullary reticular formation, especially the nucleus gigantocellularis. Projections to parts of the inferior olive and the facial nucleus have been reported. The lateral vestibular nucleus receives ascending spinal fibers. At pontine levels, the reticular formation gets a spinal projection, as does the nucleus subcoeruleus. At mesencephalic levels, there are projections from the anterolateral columns to cells in and around deep tectal layers, especially in the nucleus intercollicularis and the lateral portions of the central gray. A distinct projection to the medial (magnocellular) division of the medial geniculate body is seen (Lund and Webster, 1967; Goldberg and Moore, 1967).

In anesthetized female rats, it was possible with electrical stimulation to elicit movements of the tail, the tail base, the rump, and the rear legs that used the same muscle systems involved in lordosis, and some movements resembled elements of lordosis (Pfaff *et al.*, 1972). The distribution of the points in the brainstem from which these movements could be evoked closely resembled the brainstem distribution of ascending anterolateral-column fibers. This was additional preliminary evidence that the anterolateral ascending system is involved in lordosis. Within the 161-msec response latency of lordosis (Pfaff and Lewis, 1974), there is enough time for brainstem neurons to act on ascending information from the genital region. Even in the cat, which has longer conduction distances than in the rat, responses to pudendal nerve stimulation arrive at the posterior end of the brainstem within 10–15 msec (Meyer *et al.*, 1960). Among the ascending fibers not traveling in the dorsal columns, some can be eliminated from immediate consideration for lordosis control. Fibers traveling to the cerebellum need not be considered now, as female rats with massive cerebellar damage can still perform lordosis (Zemlan and Pfaff, 1975). Lesion studies have eliminated the inferior olive from immediate consideration (Modianos and Pfaff, 1976b). A role for projections to the facial nucleus can not be envisioned, as the motor systems controlled by that nucleus have no obvious role in lordosis. This process of elimination leaves spinoreticular fibers, spinovestibular fibers, and the projections to the mesencephalon (the nucleus intercollicularis, the central gray, and the magnocellular medial geniculate body) as candidates for having important roles in lordosis control.

In general, spinoreticular input to the medulla yields responses of medullary reticular neurons that have large receptive fields and little specificity for type of somatosensory stimulation. Recording from the lateral reticular nucleus, Rosén and Scheid (1973a,b,c) found that the majority of cells had wide receptive fields, responding to stimulation of all four limbs, and showed a convergence from different types of somatosensory receptors. These authors concluded that the convergence of sensory information recorded from individual medullary cells could be explained largely by convergence at segmental levels on to the neurons that form the input for the lateral reticular nucleus. Eccles *et al.* (1975a,b), recording from reticulospinal neurons, found somatosensory receptive fields that were wide, with little discrimination between ipsilateral and contralateral and between forelimb and hindlimb. Rose and Sutin (1973) recorded responses to genital stimulation in the lateral reticular nucleus and the ventrolateral reticular formation of the cat. Some of these neurons also responded to mechanical stimulation of the body surface outside the region of the genitalia. In the rat, Hornby and Rose (1976) also found responses to genital stimulation in the medullary reticular formation, but most of these responsive neurons were found in more medial tracts, in the ventral portion of the nucleus gigantocellularis. Nearly all of the genitally sensitive neurons responded to extragenital somatic or visceral stimulation.

Deep tectal layers, the intercollicular region, and the lateral portions of the central gray receive spinal fibers from the anterolateral column (Mehler, 1969;

RoBards *et al.,* 1976). In the spinotectal projection of the rat, the sacral cord projects to the most caudal deep tectum (Antonetty and Webster, 1975). The nucleus gigantocellularis of the medulla also gives ascending projections to the central gray (Basbaum *et al.,* 1976). Electrophysiological recording also shows a strong somatosensory input to deep tectal layers and to the periaqueductal region. Recording evoked potentials, Liebeskind and Mayer (1971) found good responses in the central gray to the stimulation of various body regions. There was a tendency for better responses to caudal body regions to occur at more caudal mesencephalic levels. Single-unit recording has also shown that mesencephalic neurons just deep to the tectum can be activated by somesthetic stimuli (Horn and Hill, 1966; Masland *et al.,* 1971). In fact, single-unit recording from the central gray, the intercollicular region, and the deep tectal layers of rats showed that many somatosensory responsive units are activated by tactile stimulation of the type necessary to trigger lordosis in the hormone-primed female (Malsbury *et al.,* 1972).

The likely nature of the spinotectal somesthetic system appears to be consistent with notions of the role of the anterolateral column system presented above. Receptive-field maps of bimodally responsive cells in deep layers of the superior colliculus show a close topographic correspondence between the stimulus requirements for different modalities (Wickelgren, 1971; Stein *et al.,* 1975). These neuronal response properties seem well adapted for directing immediate action (orienting or following responses) with respect to external stimuli. The responses of these mesencephalic cells deep to the tectum fall into the "nonlemniscal, low-discrimination, action-directing" category, rather than the "lemniscal, highly discriminative, pattern-recognizing" category, according to the distinction discussed above for spinal ascending systems. Thus, if information in this somesthetic system rising to deep tectal layers plays a role in lordosis control, it must be in the form of a relatively gross and immediate "trigger."

Lesion studies have not yet clearly defined the behavioral deficits following the removal of spinotectal input. Large mesencephalic lesions in cats (Sprague *et al.,* 1963) led to massive behavioral deficits. These, and also some smaller lesions that appeared to interrupt somesthetic input to deep tectal regions, often led to behavioral deficits that could be construed as abnormal responses to somesthetic input; for instance, several animals showed abnormal grooming responses toward the skin on the rear half of the body. Unpublished experiments from our own laboratory suggest that, under some conditions, large mesencephalic lesions, including destruction of the intercollicular region, can decrease the strength of the lordosis reflex in hormone-primed female rats.

If we hypothesize that the response to cutaneous input on the rear half of the body shown by neurons near the lateral borders of the central gray (Malsbury *et al.,* 1972) do have a role in triggering lordosis, then three other sets of facts fall into place. First, with autoradiography, estrogen-concentrating neurons can be found in the lateral portions of the central gray in the female rat (Pfaff and Keiner, 1973). Second, the dorsolateral and ventrolateral portions of the central gray receive interesting differential projections from estrogen-concentrating regions in the preoptic area and the anterior hypothalamus (Conrad and Pfaff, 1975,

443

NEURAL
MECHANISMS OF
FEMALE
REPRODUCTIVE
BEHAVIOR

1976a,b). Third, in the anesthetized female rat, electrical stimulation near the lateral borders of the central gray can elicit tail-base, rump, and rear-leg movements, some of which resemble components of lordosis (Pfaff *et al.,* 1972).

Over what pathways might subtectal mesencephalic neurons, responding to cutaneous stimuli, influence lordosis? Projections ascending from the central gray to the posterior hypothalamus have been reported (Hamilton and Skultety, 1970; Chi, 170; Hamilton, 1972; Ruda, 1975). However, it is not clear that the hypothalamus participates in lordosis control by virtue of its immediate response to cutaneous input (see below). Therefore, descending projections from mesencephalic regions such as the periaqueductal gray should receive special attention. Axons descending from the mesencephalic central gray to the pontine and the medullary reticular formations have been reported (Hamilton and Skultety, 1970; Ruda, 1975). Through such connections, mesencephalic neurons, responding to lordosis-relevant cutaneous input, might influence the excitability of reticulospinal neurons.

HYPOTHALAMUS. Does the hypothalamus participate in lordosis control through its responses to adequate lordosis-relevant somatosensory input? It appears not. We used micropipettes to record the single-unit activity of neurons in the hypothalamic and preoptic regions that other evidence (see below) implicated in the control of lordosis (Bueno and Pfaff, 1976). Only a small percentage of the units responded to cutaneous stimulation that, in unanesthetized, hormone-primed female rats, would elicit lordosis. Those unit responses detected were not usually very strong or prompt. (They would have to be immediate, strong responses to facilitate a reflex that begins about 160 msec after the onset of stimulation.) Finally, it might be expected that an important feature of hypothalamic participation would be the property of estrogen dependence of its neuronal activity. The single-unit recording evidence available suggests that most of the estrogen-sensitive neurons in the hypothalamus fire very slowly (for instance, $< 1/\text{sec}$). It would seem that the interspike intervals of such neurons, even during detectable responses to stimulation, would be too long to facilitate lordosis (reflex latencies of less than 200 msec) simply in response to immediate somatosensory input. Thus, there is no strong evidence for the idea that, on each mount by the male, relevant somatosensory input ascends to the hypothalamus, where neurons "make the decision" about lordosis according to the size of their responses to this input. Rather, it now seems more likely that hypothalamic neurons participate in lordosis by a tonic, hormone-dependent output, influencing the excitability of reflex loops completed at lower brainstem levels.

MOTOR COMPONENTS OF LORDOSIS

Precise behavioral descriptions of lordosis allow us to focus on certain groups of skeletal muscles, to explain how the reflex is executed. Work with these muscles themselves then shows the physical basis of the behavioral response. Knowing the identity of the muscles responsible for the lordosis reflex, we can then locate and study the motoneurons controlling those muscles (the final common path for the reflex). In turn, it is then easier to study the role of the descending system(s) that alters the excitability of the final common path (see the section below on the descending pathways). In this way, the analysis proceeds from regions of relative anatomical simplicity to regions of greater complexity.

DESCRIPTION OF THE RESPONSE

In attempting to focus on the explanation of the lordosis response itself, it is simpler to delay consideration of the prelordotic ("soliciting") behavior of the hormone-primed female rat. We have argued (Pfaff *et al.,* 1972; Pfaff and Lewis, 1974) that the "courtship" behaviors of the receptive female rat are adaptive preparations for the execution of lordosis and help to explain some of the differences between the hormone-primed and the unprimed female.

In frame-by-frame film analyses (Pfaff and Lewis, 1974; Pfaff *et al.,* 1972), it is seen that the estrous female rat usually begins the lordosis reflex from a crouching posture. Initial contact from the male, on the female's flanks and midback, are followed by a forward extension of the female's front legs and a partial extension or abduction of the rear legs. Following contact by the male's hindquarters against the posterior rump, the tail base, and the perineum of the female, before and during the initial pelvic thrusts by the male, the female lifts her perineal region by a rump elevation and begins to elevate her head. As the pelvic thrusts by the male are repeated, before penile insertion, the rump is elevated to its maximum height, as is the head; the result is the full lordosis posture of vertebral dorsiflexion. Rump elevation begins an average of 160 msec after first contact from the male and may continue for another 200 msec or more (Pfaff and Lewis, 1974).

X-ray cinematographic analyses of lordosis (Pfaff *et al.,* 1974) show the large change in the vertebral angle, with respect to horizontal, in the rump region, which comprises the lordosis reflex. As this change in angle is due to an actual elevation of the rump, the change in rump angle cannot be due simply to a lowering of the thorax. Only two forces are available to achieve this rump elevation and change in vertebral angle: leg extension and muscular action on the vertebral column itself. The X-ray analyses show that leg extension cannot by itself account for the change in rump angle. Therefore, muscles acting directly to dorsiflex the vertebral column must be responsible for the reflex.

MUSCLES

Among the epaxial deep back muscles of the female rat to be considered for the execution of lordosis (Pfaff *et al.,* 1972), the origins and insertions of the muscle lateral longissimus dorsi with respect to bones and fascia suggest that this muscle is well connected to effect rump elevation (Brink and Pfaff, 1977). Bilateral electrical stimulation of this muscle shows directly that its action results in a dorsiflexion of the rump vertebral column, which is an important component of lordosis. The insertions of the medial longissimus dorsi indicate that its action should result in an elevation of the proximal tail base. Direct bilateral electrical stimulation of the muscle confirms this prediction. The multifidus system of muscles, closely attached to the dorsal and the dorsolateral processes of the vertebrae, has shorter muscle fibers. These are responsible for local vertebral dorsiflexion (Brink and Pfaff, 1977). From these anatomical descriptions, it appears that the lateral longissimus dorsi and the multifidus system play an important role in the execution of lordosis (Figure 3).

Muscle ablation experiments were performed that directly examined the importance of individual deep back muscles for the lordosis reflex in the hormone-primed female rat (Brink *et al.*, 1977). Complete ablations of the lateral longissimus dorsi led to significant reductions in lordosis strength. Complete ablations of the lumbar multifidus system also reduced the strength of the lordosis reflex. Combining these two types of muscle ablations led to predictable additions in the weakening of the lordosis reflex. Under these testing conditions, medial longissimus dorsi ablations did not significantly affect lordosis. Thus, both from muscle anatomy and muscle ablations (Figure 4), it appears that the muscle nerve activation of

445

NEURAL
MECHANISMS OF
FEMALE
REPRODUCTIVE
BEHAVIOR

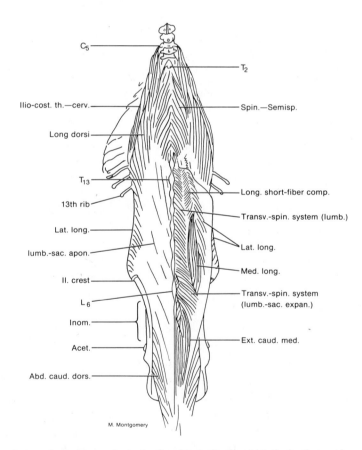

Fig. 3. Dorsal view of epaxial muscles in the female rat. On the right, the lumbosacral aponeurosis has been cut and deflected laterally to reveal underlying muscles, including fibers of the lateral longissimus (Lat. long.) taking from the medial face of the aponeurosis (apon.). (From Brink and Pfaff, 1977.)

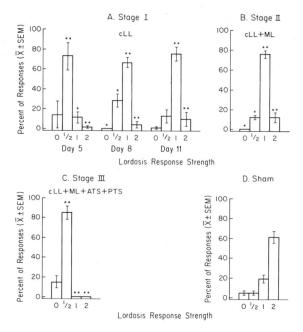

Fig. 4. Distributions of lordosis response strength in female rats following selective ablations of deep back muscles. A = following Stage I operations; B = following Stage II operations; C = following Stage III operations; D = response distribution of sham-operated animals. Values represent group mean percentages (± standard error of the mean) for each category of response (0, ½, 1, 2,). For any group, summing values across the response distribution yields 100%. 0 = no lordosis; ½ = slight lordosis; 1 = moderate lordosis; 2 = strong lordosis. cLL = complete lateral longissimus ablation; ML = medial longissimus; ATS = anterior transversospinalis; PTS = posterior transversospinalis; ML + cLL = combined longissimus ablation; ATS + PTS = combined (multifidus) transversospinalis ablation; ML + cLL + ATS + PTS = combined longissimus and (multifidus) transversospinalis ablation. * = different from sham animals, $p < .05$, two-tailed. ** = different from shams, $p < .002$, two-tailed. (From Brink *et al.*, 1977.)

the lateral longissimus dorsi and the lumbar multifidus system explains the most crucial motor component of the lordosis reflex.

MOTONEURONS

The locations of the cell bodies of motoneurons controlling the deep back muscles of the female rat have been determined by backfilling these neurons with horseradish peroxidase (HRP) from the muscle nerves (Brink *et al.*, 1977). Preliminary indications are that motoneurons for the lateral longissimus dorsi and the multifidus muscles lie on the medial, the ventral, and the ventrolateral sides of the ventral horn except in the lumbar enlargement, where they are restricted to the medial and the ventromedial sides. Studies using microstimulation in the ventral horn (Brink *et al.*, 1977) indicate that effective stimulation sites for evoking contraction in these epaxial deep-back muscles can be found in and around the medial and the ventral edges of the ventral horn.

Recordings from the motoneurons thus localized, in progress in our laboratory, should reveal some of the characteristics of the final common path for the lordosis reflex. It can already be stated, however, that the role of the descending pathways controlling lordosis must be to facilitate reflex vertebral dorsiflexion, operating eventually through these motoneurons.

447

NEURAL
MECHANISMS OF
FEMALE
REPRODUCTIVE
BEHAVIOR

Descending Pathways Controlling Lordosis

Spinal circuits presumably provide the primary reflex basis of lordosis in the intact, hormone-primed female rat. Currently, we are exploring these circuits with electrophysiological techniques. However, recent experiments show that spinal mechanisms by themselves cannot account entirely for the occurrence and the hormone dependence of lordosis. Female rats given complete spinal transections at thoracic levels cannot perform lordosis, and those responses remaining do not appear to show any estrogen dependence (Kow *et al.*, 1977b; Hart, 1969). Thus, descending control is required for the normal performance and the hormone dependence of lordosis. The net influence of the descending control is facilitatory for lordosis.

Hypothalamic Mechanisms

Because it seems unlikely that immediate hypothalamic neuronal responses to cutaneous input govern the reflex performance of lordosis, the hypothalamus should probably not be conceived of as lying on the "ascending side" of the obligatory supraspinal lordosis control loop. Rather, hormonal input into hypothalamic tissue (see the section above on hormonal control) appears to alter hypothalamic neuronal outflow, which, in turn, should influence descending systems originating at lower levels of the brainstem.

Hormone Input. In view of the strong control of lordosis behavior by estrogen (see references in section above entitled "Hormones Involved"), the question of estrogen influence in the hypothalamus is primary. The accumulation of radioactive estrogen by nerve cells in the hypothalamus has provided initial clues about the possible sites of estrogenic action in controlling female reproductive behavior (see references in the section above on the binding of sex hormones in the brain, and see Figures 1 and 2). Autoradiographic results showed that estrogen is bound by cells in the preoptic area and in the bed nucleus of the stria terminalis, by some cells in the medial anterior hypothalamus, and by cells in the basomedial hypothalamus, namely, in the ventromedial nucleus, the arcuate nucleus, and the ventral premammillary nucleus (Pfaff and Keiner, 1973). Electrophysiological results have suggested that estrogen can have physiological effects on the neural regions where it is bound. The unit activity of cells in the hypothalamus varies as a function of the estrus cycle and of estrogen treatment (see references in the section above on the behavioral and physiological effects of estrogen). An important difference in response to estrogen has been shown for cells in two different regions of the hypothalamus. Estrogen affects a population of neurons in the medial preoptic area and in the bed nucleus of the stria terminalis by depressing their resting discharge and their responses to somatosensory input; in contrast, in the basomedial hypothalamus, the resting discharge and the responsivity of a population of nerve cells are elevated (Bueno and Pfaff, 1976).

Studies using local implants of estrogen in the brain have shown that the estrogen-binding and electrophysiological phenomena summarized above may have direct importance in the estrogenic control of female reproductive behavior by the hypothalamus. Estrogen implantation in the medial preoptic area, in the base of the medial anterior hypothalamus, and in and around the ventromedial nucleus of the hypothalamus allows the performance of female reproductive behavior by ovariectomized female rats (see references in the section above on the behavioral and physiological effects of estrogen).

In a wide variety of vertebrate species, estradiol is active in restoring sexual receptivity to castrated females (reviewed by Kelley and Pfaff, 1977). Among the best studied mammalian forms, the estradiol implant sites in the hypothalamus most effective in facilitating female reproductive behavior are at or just anterior to the hypothalamic lesion sites most effective in reducing female reproductive behavior (reviewed by Kelley and Pfaff, 1977).

LESIONS OF HYPOTHALAMIC TISSUE. Lesions of the medial preoptic area in female rats are followed by no change or by a small facilitation of lordosis behavior (Law and Meagher, 1958; Singer, 1968; Powers and Valenstein, 1972). Small preoptic lesions are not likely to cause any change in lordosis in hormone-primed female rats or hamsters (Kow *et al.*, 1976; Malsbury *et al.*, 1977). Numan (1974) found small increases in the lordosis behavior of female rats following either bilateral medial preoptic lesions or bilateral parasagittal knife cuts that interrupted preoptic axons (Conrad and Pfaff, 1975, 1976a) heading toward the medial forebrain bundle. Lesion results suggesting that preoptic tissue has a small inhibitory effect on lordosis are complemented by electrical stimulation experiments. In female rats (Napoli *et al.*, 1972; Moss *et al.*, 1974) and female hamsters (Malsbury and Pfaff, 1973, 1977), electrical stimulation of the medial preoptic area suppresses lordosis.

The suprachiasmatic nuclei are necessary for the entrainment of many biological rhythms by light cycles, including circadian rhythms of activity and drinking behavior (Stephan and Zucker, 1972a,b), adrenal and pineal rhythms (Moore, 1974), and gonadal function and growth (Stetson and Watson-Whitmyre, 1976; Rusak and Morin, 1975). However, complete destruction of the suprachiasmatic nuclei in female rats (Smalstig and Clemens, 1975) or female hamsters (Malsbury *et al.*, 1977) had no effect on lordosis behavior. In cycling ewes, also, suprachiasmatic nucleus lesions had no effect on female reproductive behavior (Domanski *et al.*,1972). Suprachiasmatic nucleus neurons are affected by optic input (Koizumi *et al.*, 1975). However, suprachiasmatic nucleus cells do not bind estrogen, even though neurons nearby in the medial anterior hypothalamus do accumulate radioactive estradiol (Pfaff, 1968a; Pfaff and Keiner, 1973).

Large lesions of the medial anterior hypothalamus in female rats lead to significant decreases in lordosis, despite treatment with ovarian hormones (Greer,

1953; Averill and Purves, 1963; Law and Meagher, 1958; Singer, 1968; Herndon and Neill, 1973). Small anterior hypothalamic lesions are not usually effective in altering lordosis either in female rats (Clark, 1942) or in female hamsters (Malsbury *et al.*, 1977). Large lesions at the base of the medial anterior hypothalamus were shown to abolish estrous behavior in female guinea pigs (Brookhart *et al.*, 1940, 1941; Dey *et al.*, 1940; Dey, 1941). In a more recent study with female guinea pigs, hypothalamic lesions damaging either the anterior hypothalamic nucleus or the ventromedial nucleus caused loss of female mating behavior (Goy and Phoenix, 1963). The effects of anterior hypothalamic lesions in female guinea pigs were not due to unintended damage of the pituitary, as purposeful lesions of the pituitary did not affect mating behavior (Dey *et al.*, 1942). This inference is consistent with work on female rats showing that hypophysectomized females can respond to estrogen and progesterone injections with normal lordosis behavior (Pfaff, 1970a). In the female cat, also, anterior hypothalamic lesions caused permanent loss of female mating behavior, despite treatment with estrogen (Sawyer and Robison, 1956; Sawyer, 1960). In female cats or rabbits, tuberal lesions led to losses of mating behavior that could be restored following treatment with exogenous estrogen (Sawyer and Robison, 1956; Sawyer, 1960). Lesions deep in the anterior hypothalamic region of cycling ewes abolished female mating behavior, and the behavioral loss could not be ascribed to pituitary or ovarian changes (Clegg *et al.*, 1958).

Rodgers and Schwartz (1972) studied mating behavior in female rats with intact ovaries. Coronal transections were made at the posterior border of the optic chiasm (that is, at levels of the anterior hypothalamus just behind the suprachiasmatic nucleus, but anterior to the ventromedial nucleus of the hypothalamus). Behavioral testing was begun several days after constant vaginal estrus was confirmed. Under these conditions, the female rat showed a high, constant level of behavioral receptivity. In a similar study (Rodgers and Schwartz, 1976), in which the majority of the knife cuts were more anterior in the anterior hypothalamus, constant, high-level behavioral receptivity again was shown to accompany persistent vaginal estrus. To the extent that these cuts actually interrupted anterior hypothalamic fibers that stream posterior through the hypothalamus (Conrad and Pfaff, 1975, 1976b), it can be inferred that lordosis behavior survives a substantial reduction in anterior hypothalamic neuronal influence, under conditions of persistent estrus. Kalra and Sawyer (1970) studied cycling female rats with coronal transections under conditions that made it more difficult to achieve a positive criterion of mating behavior: with the anterior hypothalamic transections being made on the morning of proestrus, female mating behavior could be shown to be present only if male rats achieved ejaculation with those females the very night after surgery. These testing conditions served to show the contribution of anterior hypothalamic neurons. With coronal transections at very anterior levels in the hypothalamus (anterior to the suprachiasmatic nuclei), 15 of 18 females mated the night after surgery. However, with transections at posterior levels of the anterior hypothalamus (behind the suprachiasmatic nuclei), only 1 of 10 females mated. In female hamsters, also, coronal transections through the anterior hypothalamus disrupt

449

NEURAL
MECHANISMS OF
FEMALE
REPRODUCTIVE
BEHAVIOR

lordosis, even in ovariectomized females primed with estrogen and progesterone (Warner, 1975; Malsbury *et al.*, 1977).

It appears that transections or lesions that cut off the influence of a large proportion of anterior hypothalamic neurons cause a decrease in the performance of lordosis, especially under testing conditions that do not favor the appearance of lordosis. Many axons of anterior hypothalamic neurons synapse in the ventromedial nucleus of the hypothalamus, and others exit posteriorly from the hypothalamus, having swept by the ventromedial nucleus (Conrad and Pfaff, 1975, 1976b). These anatomical facts raise questions about the interpretations of lordosis changes following ventromedial nucleus lesions (see below). Because under some conditions females with sizable transections in the anterior hypothalamus can do lordosis, lesion effects on female reproductive behavior following ventromedial region lesions need not be interpreted entirely as effects due to the interruption of the fibers of passage.

Bilateral lesions in and around the ventromedial nucleus of the hypothalamus of the female rat are followed by a decreased performance of lordosis (Kennedy and Mitra, 1963; Kennedy, 1964; Dörner *et al.*, 1969, 1975; LaVaque and Rodgers, 1974, 1975; Carrer *et al.*, 1973). In female sheep (Domanski *et al.*, 1972) and female cats (Hagamen and Brooks, 1958), also, lesions in the region of the ventromedial nucleus abolished behavioral receptivity. Lesions that were effective in abolishing receptivity in female hamsters appear to have been placed at anterior-ventromedial-nucleus levels in such a way that ventromedial nucleus functions and fibers passing by would be interrupted (White, 1954). Studies with smaller lesions confined to the ventromedial nucleus itself suggest strongly that neurons in this nucleus have a primary role in facilitating the execution of lordosis. In female rats, small lesions within the ventromedial nucleus were effective in significantly reducing the performance of lordosis during estrogen treatment (Edwards and Mathews, 1977). In this study, it was not possible to define a "critical zone" within the ventromedial nucleus. Working with female hamsters, Malsbury *et al.* (1977) showed that the degree of lordosis deficit following bilateral lesions was significantly correlated with the amount of the ventromedial nucleus destroyed, especially at anterior levels of the ventromedial nucleus. Considered together, these results indicate that ventromedial nucleus neurons themselves contribute importantly to the execution of lordosis, but that to guarantee the abolition of lordosis behavior, lesions must disrupt both ventromedial nucleus function and anterior hypothalamic axons that sweep by the ventromedial nuclei.

Summary. Bilateral lesions of the ventromedial nucleus of the hypothalamus and large anterior hypothalamic lesions just anterior to the ventromedial nucleus disrupt the performance of lordosis. The action of estrogen on these cells relevant to female mating behavior as shown by demonstrations of estrogen binding and implant effects (see the sections above on hormonal control and on hormone input) leads to increased cellular activity, as shown by single-unit recording (Bueno and Pfaff, 1976). Therefore, we infer that estrogen facilitates lordosis by acting on these cells to increase their electrophysiological activity, which, in turn, increases activity in the brainstem reflex loops governing lordosis. Removal of these cells (and therefore of this estrogen influence) decreases lordosis performance. Under some conditions, preoptic lesions are followed by increased lordosis performance. Estrogen actions on these cells, shown to be relevant to lordosis by estrogen binding and implant experiments (see sections above on hormonal control and on hormone input), decrease cellular activity, as shown by single-unit recording work (Bueno and Pfaff, 1976). Therefore, estrogen, acting to decrease activity in these cells, facilitates lordosis by decreasing an influence that is inhibitory to the reflex. Apparently, there are differences between the medial preoptic area (along with the bed nucleus of the stria terminalis) and the ventromedial nucleus (along with the adjacent anterior hypothalamus) both in the physiology of estrogen action and in their participation in lordosis control. These differences tie into a pattern of anatomical and physiological differences between, on the one hand, the preoptic area and the basal forebrain and, on the other hand, the ventromedial nucleus and the posterior hypothalamus. The pattern of differences seen has generated the theory summarized below.

451

NEURAL
MECHANISMS OF
FEMALE
REPRODUCTIVE
BEHAVIOR

Theory of Hypothalamic and Preoptic Participation in Reproductive Behavior Responses

Female Reproductive Function. The net effect of medial preoptic tissue on lordosis is to decrease the performance of the reflex, whereas the net effect of hypothalamic tissue in and around the ventromedial nucleus is to facilitate lordosis (see the previous section). Studies in the control of ovulation also show differing roles for the medial preoptic area and the basomedial hypothalamus in female rats. Basomedial hypothalamic tissue is sufficient for tonic gonadotropin release and negative feedback effects of steroids in female rats, whereas the medial preoptic area is required for normal ovulation and cyclicity (Barraclough and Gorski, 1961; Halasz and Gorski, 1967; Halasz, 1969). Electrochemical stimulation of the preoptic area and of basal forebrain tissue anterior to the preoptic area (the diagonal bands of Broca and the septum) can stimulate ovulatory surges of luteinizing-hormone release in female rats (Everett, 1969; Kubo *et al.*, 1975), and the mechanism of this effect is being investigated with electrophysiological techniques (Terasawa and Sawyer, 1969; Dyer and Burnet, 1976). Recording of preoptic-area multiple-unit activity (Kawakami *et al.*, 1970) and single-unit activity (Wuttke, 1974) has shown a correlation between an increased preoptic neuronal firing rate and an increased release of luteinizing hormone. The mechanism of preoptic participation in ovulation control may also involve prolactin, as preoptic-lesioned female rats that are not cycling regularly show long periods of pseudopregnancy (Clemens *et al.*, 1976).

Regarding the control of lordosis, certain areas of the basal forebrain share with the medial preoptic area the function of suppressing female reproductive behavior. Large lesions of the lateral septum are followed by impressive increases in the lordosis performance of estrogen-primed female rats (Nance *et al.*, 1974, 1975). Conversely, electrical stimulation of the septal area suppresses lordosis (Zasorin *et al.*, 1975). Control lesions in the amygdala had neither facilitating nor suppressing effects on lordosis (Nance *et al.*, 1974), but amygdala destruction may interact with the septal effect (Nance *et al.*, 1976). Basal forebrain suppression of lordosis may be related to olfactory function, as olfactory ablations are followed by increased lordosis performance in hormone-primed female rats (Moss, 1971; Edwards and Warner, 1972; Nance *et al.*, 1976). Lordosis suppression by the basal forebrain may operate by input to the preoptic area (Yamanouchi and Arai, 1975) or by effects on a common neural substrate further downstream in lordosis circuits. In any case, medial preoptic and lateral septal neurons seem to share a role in the suppression of lordosis, and their effect is the opposite of the effect of the basomedial hypothalamus.

Estrogen from the circulation has a suppressive effect on the single-unit activity of neurons in the medial preoptic area, as shown by records of spontaneous activity and responsivity to peripheral stimuli (Bueno and Pfaff, 1976; Lincoln, 1967; Whitehead and Ruf, 1974; Yagi, 1970, 1973). In contrast, estrogen increases the single-unit activity of hypothalamic tissue posterior to the preoptic area, especially in the basomedial hypothalamus (Bueno and Pfaff, 1976; Cross and Dyer, 1972; Kawakami *et al.*, 1971). It should also be noted that the neurochemical effects of estrogen on enzyme activities are not the same in the preoptic area and in the basomedial hypothalamus (Luine *et al.*, 1974, 1975a,b). In addition, Brawer and Sonnenschein (1975) showed effects of estrogen on fine structure in arcuate neurons without such effects' showing up in the medial preoptic area, and Araki *et al.* (1975) showed different effects of ovariectomy on LRF content in basomedial as opposed to far anterior hypothalamic tissue.

Thus, with respect to lordosis control, ovulation control, and the electrophysiological and neurochemical effects of estrogen, there is a clear bifurcation of function between the medial preoptic area (along with the basal forebrain) and the basomedial hypothalamus.

Male Mating Behavior. Medial preoptic neurons stimulate masculine mating behavior and are required for its normal performance (reviewed by Malsbury and Pfaff, 1974). Electrical stimulation of the medial preoptic area of male rats facilitates various aspects of male copulatory behavior (Malsbury, 1971). Similar data have been gathered from other species. Electrical stimulation of the medial preoptic area of male opossums triggered mounting and other male mating responses (Roberts *et al.*, 1967), and similar data have been gathered in studies of the male guinea pig (Martin, 1976). At least part of the facilitatory influence of the preoptic area on male mating behavior must be due to the effects of testosterone (or testosterone metabolites) on preoptic neurons. Autoradiographic experiments show that testosterone is strongly accumulated by preoptic neurons (Pfaff, 1968a). Single-unit recording from preoptic cells has shown that systemic injections of testosterone or a direct application of testosterone to the preoptic region can alter spontaneous discharge rates and responsivity to peripheral stimuli by preoptic neurons, in anesthetized, castrated male rats (Pfaff and Pfaffmann, 1969). Indeed, the implantation of testosterone into the preoptic area of adult (Davidson, 1966; Lisk, 1967; Johnston and Davidson, 1972) and neonatal (Christensen and Gorski, 1976) rats significantly facilitates masculine reproductive behavior and can restore the male sexual behavior of castrated male rats. Preoptic facilitation is required for male mating behavior: many studies have shown that medial preoptic lesions can reduce or abolish sex behavior in male animals (see references in Malsbury and Pfaff, 1974). The importance of the medial preoptic area in masculine mating behavior is not restricted to male rodents; medial preoptic neurons have been strongly implicated in the control of male sex behavior in a wide variety of mammals, birds, amphibia, and fish (Kelley and Pfaff, 1977).

Axons of the preoptic neurons important in male sex behavior leave the preoptic area going laterally, toward the medial forebrain bundle. Parasagittal knife cuts separating the medial preoptic area from the medial forebrain bundle severely impair male sexual behavior (Paxinos and Bindra, 1972; Szechtman *et al.*, 1975). Coronal transections behind the preoptic area in the medial hypothalamus (Paxinos and Bindra, 1972) and transections that spare preoptic tissue near the base of the brain (Rodgers, 1969) do not significantly affect male mating responses, a finding showing that the critical axons run laterally, not posterior, through the medial hypothalamus, and that some critical axons run near the bottom of the brain. These results suggest that axons of the neurons important for male sexual behavior travel posterior in the medial forebrain bundle and predict the results of medial-forebrain-bundle lesions and electrical stimulation. As expected, medial-forebrain-bundle lesions disrupt or abolish masculine mating responses by male rats (Hitt *et al.*, 1970, 1973; Caggiula *et al.*, 1973, 1975; Modianos *et al.*, 1973). The same type of medial-forebrain-bundle lesions does not disrupt female reproductive behavior (Hitt *et al.*, 1970; Modianos *et al.*, 1973). In fact, under some endocrine conditions, medial-forebrain-bundle lesions can lead to an elevation of lordosis performance similar to the effect seen with preoptic area lesions (Modianos *et al.*, 1976). Masculine mating behavior is facilitated by electrical stimulation of the medial forebrain bundle (Caggiula and Hoebel, 1966; Caggiula, 1970; Eibergen and Caggiula, 1973). All of these results are consistent with the idea that medial preoptic neurons sending their axons posterior through the medial forebrain bundle are essential for the facilitation of male, but not female, reproductive behavior.

Neuroanatomical findings support the inferences from lesion and electrical stimulation studies. Medial preoptic neurons (in the preoptic regions governing male mating behavior) have been shown to send their axons through the medial portions of the medial forebrain bundle, whereas medial anterior hypothalamic and ventromedial nucleus neurons (in the regions essential for female mating behavior) do not send descending axons through the medial forebrain bundle (Conrad and Pfaff, 1975, 1976a,b). Whereas neurons from the ventromedial nucleus of the hypothalamus and the medial anterior hypothalamus project strongly to the dorsal (as well as the ventral) central gray of the mesencephalon and to the lateral midbrain reticular formation of the mesencephalon, medial preoptic neurons do not have these strong projections. Further, there are differences between preoptic neurons and medial basal hypothalamic neurons in their pattern of projection to the lateral septal nucleus. Differences in the strength and the nature of synaptic effects (Mancia, 1974), as well as the site of terminations, could underlie the differences between preoptic-basal forebrain functions and basome-

453

NEURAL
MECHANISMS OF
FEMALE
REPRODUCTIVE
BEHAVIOR

dial-posterior hypothalamic functions. However, the axonal projections demonstrated so far provide possible anatomical substrates for separate neural systems mediating male and female sexual behaviors: preoptic neurons projecting through the medial forebrain bundle, critical for male mating responses, and basomedial hypothalamic neurons projecting to the mesencephalic central gray and the lateral reticular formation, critical for female mating responses.

Autonomic Function. Hess (1954, 1957) electrically stimulated the brains of freely moving cats and observed autonomic effects. Electrical stimulation of the preoptic area, the medial thalamus, and the basal forebrain (notably, the septum) was followed by decreases in blood pressure, decreases in heart rate, decreased respiratory activity, pupillary constriction, and the initiation of micturition, defecation, and salivation. Electrical stimulation of the posterior hypothalamus and the periventricular system leading to the central gray of the mesencephalon was followed by increased blood pressure, increased heart rate, an increase in respiratory activity, and pupillary dilatation. Based on these observations, the idea was proposed that the preoptic area and the basal forebrain serve, primarily, a parasympathetic or an antisympathetic function (a "trophotropic" function). The posterior hypothalamus was conceived of as serving, primarily, a sympathetic function (an "ergotropic" function). Modern work has borne out Hess's experimental observations. Electrical stimulation of the preoptic region and the basal forebrain (especially, the septum) has consistently yielded decreases in blood pressure and heart rate (Hilton and Spyer, 1971; Zehr, 1973; Takeuchi and Manning, 1973; Ahmed, 1974; Ciriello *et al.,* 1975). Electrical stimulation of the basomedial hypothalamus (including the ventromedial and the dorsomedial nuclei) and the posterior hypothalamus reliably elicits increases in blood pressure and heart rate (Bagshaw *et al.,* 1971; Calaresu, 1974; Djojosugito *et al.,* 1970; Verrier *et al.,* 1975; Ahmed, 1974; Takeuchi and Manning, 1973).

The association of the preoptic area with parasympathetic (or antisympathetic) functions may explain two types of observations reported after preoptic lesions. First, in the preoptic lesion and transection experiments in which male mating behavior is reduced or abolished (see references above), some authors have reported that the lesioned males mounted the female, but intromissions were not achieved. In any case, where intromission was achieved, it is said that several may then be achieved by the same animal and ejaculation results. These qualitative observations are consistent with the notion that the main failure in male mating behavior after preoptic lesions is the inability to achieve an intromission, and that, in turn, could be explained by a failure of the ability to have erection. Because erection requires parasympathetic activity (or an inhibition of sympathetic function), the failure of intromission (or erection) could be described as a failure of parasympathetic function. This description is consistent with the findings in autonomic physiology reviewed above, for the preoptic area. Thus, the importance of preoptic neurons in parasympathetic (or antisympathetic) functions provides a mechanism for the participation of the preoptic region in male mating behavior: these neurons are required to allow the male to achieve an erection, and thus an intromission.

A second type of observation following preoptic lesions, usually an unintended effect, is death of the experimental animal due to pulmonary edema. Pulmonary edema is due to abnormally intense systemic vasoconstriction (Chen and Chai, 1974). It can thus be described as a failure of antisympathetic functions, namely, the expected result after preoptic lesions according to the results reviewed above. Preoptic lesions would allow abnormally high systemic vasoconstriction, which, in turn, would give abnormally high central blood pressures, leading to pulmonary edema. Thus, both the intended and the unintended aspects of the results of preoptic lesion studies on male mating behavior can be explained by the parasympathetic role of preoptic tissue.

Summary. The comparison of medial preoptic (and basal forebrain) tissue with basomedial (and posterior) hypothalamus is summarized in Table 2 and is presented schematically in Figure 5.

Briefly, preoptic neurons are required for male mating behavior but suppress female mating behavior. They have a parasympathetic (or antisympathetic) function. The types of effects of estradiol on these neurons are consistent with their role in female mating behavior. The association of preoptic tissue with parasympathetic function provides an explanation of the mechanism of preoptic control over male mating behavior and also explains deaths due to pulmonary edema following preoptic lesions. It appears that preoptic neurons exert descending influences through axons that travel in the medial part of the medial forebrain bundle.

Basomedial hypothalamic neurons are not required for male mating behavior but are required for normal female mating behavior. The effects of estrogen on these neurons are consistent with their role in governing responses in female

455

NEURAL
MECHANISMS OF
FEMALE
REPRODUCTIVE
BEHAVIOR

TABLE 2. COMPARISON OF MEDIAL PREOPTIC AREA (AND BASAL FOREBRAIN) WITH BASOMEDIAL HYPOTHALAMUS: BRIEF SUMMARY OF SOME MAJOR DIFFERENCES

	Medial preoptic area (and basal forebrain)	Basomedial hypothalamus (and posterior hypothalamus)
Reproductive function		
Male mating behavior	↑	0
Female mating behavior	↓	↑
Effect on luteinizing hormone	Ovulatory surge ("positive feedback")	Negative feedback
Effect of estradiol on unit activity	↓	↑
Autonomic function		
Blood pressure	↓	↑
Heart rate	↓	↑
Diameter of pupil	↓	↑
Micturition, defecation, and salivation	↑	
Anatomical projections		
Descending axon trajectories	Median forebrain bundle	Outside median forebrain bundle
To midbrain central gray	Weak	Strong
To lateral midbrain reticicular formation	Weak	Strong
To lateral septum	Dorsal	Midlateral

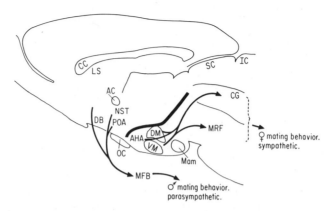

Fig. 5. Schematic representation of hypothalamus and basal forebrain to show division according to participation in reproductive behavior and autonomic function. The preoptic control of male mating behavior may be based partly in its control of parasympathetic function. No direct causal connection is implied between female mating behavior and sympathetic function. Abbreviations: AC, anterior commissure; AHA, anterior hypothalamus; CC, corpus callosum; CG, central gray; DB, diagonal bands; DM, dorsomedial nucleus; IC, inferior colliculus; LS, lateral septum; Mam, mammillary bodies; MFB, medial forebrain bundle; MRF, mesencephalic reticular formation; NST, bed nucleus of stria terminalis; OC, optic chiasm; POA, preoptic area; SC, superior colliculus; VM, ventromedial nucleus of hypothalamus.

reproductive behavior. Basomedial and posterior hypothalamic neurons serve a sympathetic autonomic function. At present, no direct causal implications can be drawn from connections between the autonomic and the behavioral roles of these neurons. Axons descending from basomedial hypothalamic neurons do not travel in the medial forebrain bundle. Those traveling directly posterior run through the medial hypothalamus. Strong projections of these neurons to the midbrain central gray and the lateral portions of the midbrain reticular formation may be important in mediating descending influences from the basomedial and the posterior hypothalamus.

NATURE AND ANATOMY OF HYPOTHALAMIC OUTFLOW. Although hypothalamic neurons can respond to cutaneous stimulation, a rather small percentage of neurons do so (Bueno and Pfaff, 1976). Those responses recorded are usually sluggish and are small in amplitude compared to peripheral cutaneous systems (Bueno and Pfaff, 1976). Thus, it is unlikely that hypothalamic neurons control lordosis by alterations in activity on a mount-by-mount basis, because their responses to cutaneous-input-mimicking mounts are not prompt or strong enough. Moreover, it appears to be neurons that fire slowly (for instance, $< 1/\text{sec}$) that show the strongest estrogen dependence in single-unit recording experiments (Bueno and Pfaff, 1976). Alterations in the activity of such neurons, whose interspike intervals would be long, could not, on a mount-by-mount basis, explain the control of the lordosis reflex, the latency of which is less than 200 msec (Pfaff and Lewis, 1974). Estrogen-sensitive neurons would thus be expected to control lordosis in a more tonic fashion. We suggest, therefore, that estrogen influences lordosis through hypothalamic mechanisms primarily by causing a tonic, hormone-dependent output, which, in turn, alters transmission in brainstem lordosis-relevant reflex loops.

A complete review of hypothalamic anatomy is outside the scope of this chapter (for review, see Conrad and Pfaff, 1977). In general, some of the older techniques for studying axonal projections had problems of sensitivity in the hypothalamus (perhaps because of the fine caliber of hypothalamic axons) and also suffered from the problem of fibers of passage (the lesioning, by accident, of fibers passing through the area whose output was to be studied). From the results of experiments using these methods, one got the impression that hypothalamic cells projected only locally; no long-axoned projections could be documented.

457

NEURAL
MECHANISMS OF
FEMALE
REPRODUCTIVE
BEHAVIOR

The newer technique of tritiated amino acid autoradiography avoids the problems of sensitivity and fibers of passage (Cowan *et al.*, 1972; Conrad and Pfaff, 1976a). Long-axoned projections of hypothalamic neurons are clearly evident, and interesting differences in the projections of different cell groups can be documented (see the previous section; also see Conrad and Pfaff, 1976b, 1977). Here, we focus on the projections of neurons whose activity appears to be essential for the normal performance of lordosis: neurons in and around the ventromedial nucleus of the hypothalamus and in the medial anterior hypothalamus (see the section above on lesions of the hypothalamus). Among the projections of these neurons, we further concentrate here on axons that run posterior, toward the mesencephalon, because the anterior projections (for instance, to limbic structures such as the amygdala and the septum) do not go to structures whose activity has been shown to be essential for lordosis.

The mesencephalic projections of neurons in the ventromedial nucleus of the hypothalamus are to the central gray (both its dorsolateral and its ventrolateral aspects) and to the lateral aspects of the mesencephalic reticular formation (including part of the cuneiform nucleus) (Conrad and Pfaff, 1975, 1976a,b; Krieger *et al.*, 1977a). Some of the axons heading toward the mesencephalon travel through the far lateral posterior thalamus and may synapse there. A site of termination on the diencephalic-mesencephalic border appears to correspond (Morest, 1964; Heath and Jones, 1971; Graybiel, 1973; Poggio and Mountcastle, 1960) to the magnocellular (medial) subsection of the medial geniculate body. All of the projections from neurons in the ventrolateral portion of the ventromedial nucleus and the adjacent basal hypothalamus are illustrated in Figure 6. The mesencephalic projections of neurons in the anterior hypothalamic nucleus are to the central gray (both its dorsolateral and ventrolateral aspects) and to the ventrolateral portion of the mesencephalic reticular formation.

The mesencephalic projections of neurons in and around the ventromedial nucleus (essential for lordosis) clearly match some of the ascending (somatosensory) projections to the mesencephalon that arise from the anterolateral columns of the spinal cord. By comparing the anatomical results above with the reports of Mehler (1969) and other workers cited in the section above on the sensory control of lordosis, it is seen that ventromedial hypothalamic efferents and anterolateral spinal somatosensory afferents converge in and around the lateral central gray and the dorsal lateral mesencephalic reticular formation, and at the medial border of the medial geniculate body. Thus, these mesencephalic regions are locations where hormone-dependent hypothalamic outputs crucial for lordosis intersect with soma-

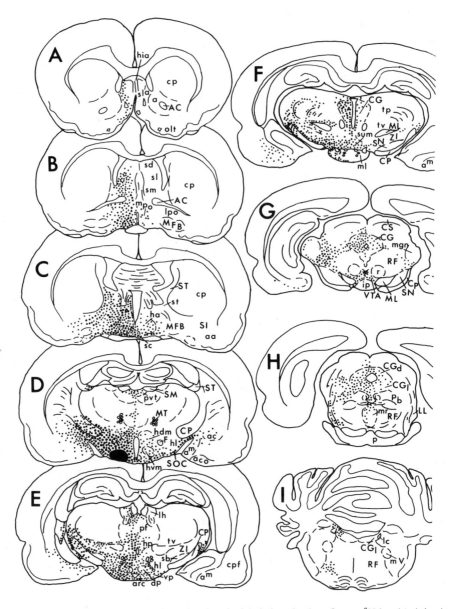

Fig. 6. Drawings from transverse sections showing the labeled projections from a ^{3}H-leucine injection in the ventromedial nucleus and the adjacent basal hypothalamus. Symbols for charting as follows: the heavier dots represent labeled fibers; the smaller dots represent the fields of individual grains. The region of labeled cell bodies is indicated in solid black. (From Conrad and Pfaff, 1976b.)

tosensory inputs also shown to be involved in triggering lordosis. The form of the somatosensory input appears to be a relatively broad zone of termination of the most primitive ascending system (see the section above on the sensory control of lordosis). The mesencephalic neuronal cell bodies in regions receiving both hypothalamic and somatosensory influences may reflect some of the interactions between sensory and hormonal influences that characterize female reproductive behavior in rodents.

Delimiting the mesencephalic sites of the ventromedial and anterior hypothalamic axonal projections also has another importance. It shows the locations of those mesencephalic neurons whose descending outputs must be considered carriers of hormone-dependent influences to lower brainstem neurons (see below).

HYPOTHALAMIC OUTPUT: PARTICIPATION IN LORDOSIS CONTROL. Lordosis has not been seen in decerebrate female rats (Kow *et al.*, 1977a) or guinea pigs (Dempsey and Rioch, 1939). In these experiments, decerebrations were accomplished by complete transection of the brainstem at mesencephalic levels. These transections cut all hypothalamic output as well as output from other diencephalic and telencephalic structures. However, none of these structures, other than hypothalamic cell groups, have been shown to be necessary for lordosis. Therefore, the absence of lordosis in decerebrate animals shows that the net influence of hypothalamic output is facilitatory for lordosis.

Among hypothalamic outputs, which pathways are important in carrying lordosis-relevant influences? The best possibility—namely, a sheet of axons from the ventromedial nucleus of the hypothalamus to the central gray and the lateral reticular formation of the midbrain—has not yet been adequately assessed. Preliminary experiments indicate that large lesions near the lateral borders of the central gray can reduce the strength and the frequency of lordosis in hormone-primed female rats (Diakow and Pfaff, unpublished observations, 1972). Further studies of mesencephalic lesions and transections are now being carried out in our lab (Manogue *et al.*, unpublished observations, 1979).

Other pathways that have been explored more thoroughly are the medial forebrain bundle and the stria medullaris. These pathways include telencephalic as well as hypothalamic output, and some of their effects could be due to fibers going in the opposite (ascending) direction. Four studies using female rats primed with suprathreshold doses of estrogen and progesterone failed to find significant reductions of female mating behavior following lesions of the medial forebrain bundle (Rodgers and Law, 1967; Singer, 1968; Hitt *et al.*, 1970; Modianos *et al.*, 1973). This was true even though the same type of lesions led to significant reductions of masculine sexual behavior (for instance, Hitt *et al.*, 1973). However, facilitation of lordosis had been reported following septal or preoptic lesions (Nance *et al.*, 1974, 1975; Powers and Valenstein, 1972), and inhibition of lordosis following septal or preoptic stimulation (Zasorin *et al.*, 1975; Napoli *et al.*, 1972; Moss *et al.*, 1974), and axons of these cell groups run in the medial forebrain bundle. Therefore, we reasoned that a medial-forebrain-bundle (MFB) lesion study using hormonal conditions designed to reveal increments in lordosis following the lesion might show a significant MFB lesion effect. In our first attempt at this type of study (Modianos *et al.*, 1976), we found that female rats relatively unresponsive in preoperative testing with standard estrogen doses showed higher lordosis quotients following lesions of the medial forebrain bundle. In subsequent experiments, with MFB lesions in female rats, we found that although the estrogen "thresholds" for evoking lordosis were unchanged after the lesion, the number of days on which lordosis could be evoked following the injection of 10 μg of estradiol benzoate and 0.5 mg of progesterone was greatly increased. Whereas in sham and unoperated rats, lor-

459

NEURAL
MECHANISMS OF
FEMALE
REPRODUCTIVE
BEHAVIOR

dosis in response to manual stimulation could be evoked for a mean of 7 days, following MFB lesions lordosis was still detectable for a mean of about 17 days. These results suggest that, like the preoptic area and the lateral septum, under some hormonal conditions, the MFB may play a role in the suppression of lordosis. The MFB lesion effect could be due to the interruption of descending septal or preoptic axons. However, because axons from the median raphe nucleus of the midbrain ascend in the medial forebrain bundle (Conrad *et al.*, 1974) and because the serotonergic raphe nuclei may be involved in the suppression of lordosis (Kow *et al.*, 1974b), it is also possible that the MFB lesion effects are due to the interruption of these ascending connections.

The stria medullaris–habenula system has also been examined for a role in female reproductive behavior. Bilateral lesions of either the stria medullaris or the habenula have been reported to produce decrements in lordosis behavior in female rats (Zouhar and DeGroot, 1962; Rodgers and Law, 1967; Modianos *et al.*, 1974, 1975). Lordosis decrements following habenular lesions apparently show up best (Modianos *et al.*, 1975) when female rats are being tested for responsiveness to the synergistic effects of estrogen and progresterone.

Brainstem Modules

Compared to the hypothalamus, mesencephalic and lower brainstem levels have not been explored as thoroughly by lesion and implant studies for their contributions to female reproductive behavior. Experiments currently being carried out in our laboratory are aimed at correcting this deficiency. However, from transections and anatomical data now available, we can deduce something of the nature of the logical brainstem "modules" that are involved in controlling lordosis.

Decerebrate female rats do not perform lordosis (Kow *et al.*, 1977a). Therefore, the net effect of hypothalamic output must be facilitatory for lordosis. The anatomical distribution of the axons of the hypothalamic cells involved in lordosis tells us the locations of those brainstem neurons that must be regarded as candidates for carrying hypothalamic hormone-dependent influences. Most descending hypothalamic axons do not extend beyond the mesencephalon (Conrad and Pfaff, 1975, 1976a,b). (An exception to this general rule is found in the fact that some neurons in the paraventricular nucleus and the dorsomedial nucleus of the hypothalamus project to the medulla and to the spinal cord—Conrad and Pfaff, 1976b; Krieger *et al.*, 1977a; Zemlan *et al.*, 1977. However, according to hypothalamic lesion studies as summarized above, these neurons are not likely to be involved in lordosis control. Also, some axons from the ventromedial nucleus of the hypothalamus descend as far as the pontine reticular formation, but this is a weak projection compared to the rich field of ventromedial axon terminations in the mesencephalon—Conrad and Pfaff, 1976b.) Therefore, there must be a "module" of mesencephalic neurons importantly involved in the control of lordosis, by carrying hypothalamic, lordosis-relevant influences. Mesencephalic neurons receiving axonal projections from the ventromedial hypothalamic nucleus and from the anterior hypothalamic region are found in the central gray, in the dorsol lateral mesencephalic reticular formation (including the cuneiform nucleus), and in the region that may correspond to the magnocellular division of the medial geniculate body (see the section on hypothalamic outflow).

461

NEURAL
MECHANISMS OF
FEMALE
REPRODUCTIVE
BEHAVIOR

Descending projections from these mesencephalic regions do not reach the spinal cord. Therefore, there must be a block of neurons in the lower brainstem that carries descending hypothalamic (by way of mesencephalic) lordosis-relevant, hormone-dependent influences. The mesencephalic neurons involved do send descending axons that terminate in the medullary reticular formation in the nucleus gigantocellularis. Anatomical work both with the mesencephalic central gray (Hamilton and Skultety, 1970; Ruda, 1975) and with the portion of the mesencephalic reticular formation named the *cuneiform nucleus* (Edwards, 1975; Edwards and De Olmos, 1976) has documented descending projections to the region of medullary reticulospinal neurons. In turn, lesion studies have shown that these reticulospinal neurons are importantly involved in the performance of lordosis (see the section below on data on lordosis). In summary, the limitations imposed by the length of the axonal projections of lordosis-relevant neurons suggest that there must be at least a mesencephalic block and a medullary block of neurons involved in descending lordosis control. It appears that the relevant hypothalamic neurons synapse on the mesencephalic neurons designated above; that these neurons, in turn, synapse in the nucleus gigantocellularis of the medulla; and that these reticulospinal neurons influence lordosis mechanisms in the spinal cord.

Because spinal rats do not normally perform lordosis (Pfaff *et al.,* 1972; Kow *et al.,* 1977b), the net influence of descending axons from the medulla on the spinal cord must be facilitatory for lordosis. This finding, together with the proof that the net influence of hypothalamic output is facilitatory for lordosis (Kow *et al.,* 1977a), places limitations on the nature of the synaptic effects in the brainstem. There must be at least two levels of synapses, one in the mesencephalon and the other in the medulla. These must preserve, from the hypothalamus to the medulla, the relationship of increased activity leading to increased performance of lordosis. Therefore, if the net synaptic effect in the mesencephalon is excitatory, that in the medulla must also be excitatory. If, on the other hand, the net synaptic effect in the mesencephalon is inhibitory, that in the medulla must also be inhibitory.

PATHWAYS FROM BRAINSTEM TO SPINAL CORD

DATA ON LORDOSIS. Because spinal female rats do not perform lordosis (Pfaff *et al.,* 1972; Kow *et al.,* 1977b), the net combined influence of the pathways leading from the brainstem to the spinal cord must be facilitatory for lordosis. We know that the cerebellum does not play a major primary role in the control of the lordosis-relevant pathways, at least as far as their participation in lordosis is concerned. Major damage to the cerebellum can be followed by strong lordosis reflexes in female rats (Zemlan and Pfaff, 1975). Destruction of the source of a major input to the cerebellum (the inferior olive) or cerebellar deep nuclei or a major cerebellar output (the superior cerebellar peduncle) had no significant effect on lordosis in hormone-primed female rats (Modianos and Pfaff, 1976b).

The descending tracts involved in the control of the lordosis reflex have been deduced by a process of elimination. Eight major descending tracts are now well recognized. The corticospinal tract cannot be involved in any important way in the facilitation of lordosis because its transection at spinal cord levels has no effect on lordosis (Kow *et al.*, 1973, 1977b) and because lesions (Beach, 1944) or spreading depression (Clemens *et al.*, 1969) of the cortical cell bodies giving rise to this tract do not reduce the performance of lordosis. Bilateral surgical transections of the entire dorsal half of the spinal cord at cervical or thoracic levels have no effects on lordosis, and transections of the medial columns can be added and still allow lordosis to be performed (Kow *et al.*, 1977b). Therefore, descending tracts running in the anterolateral columns are sufficient for lordosis. Conversely, large bilateral transections destroying all anterolateral fibers can eliminate lordosis, a finding showing that descending systems running there are necessary for the normal performance of this female reproductive behavioral reflex. Because fibers of the rubrospinal tract run in the dorsal part of the lateral columns, they cannot be involved in an important way (Nyberg-Hansen, 1966; Petras, 1967; Brown, 1974; Waldron and Gwyn, 1969). Tectospinal fibers can also be eliminated from consideration because this system runs only in the medial column and descends only to cervical levels (Nyberg-Hansen, 1966; Petras, 1967; Waldron and Gwyn, 1969). Similarly, the interstitiospinal tract is restricted to the medial columns (Nyberg-Hansen, 1966) and therefore can be transected without lordosis loss.

The tracts remaining are the vestibulospinal and the reticulospinal tracts. The medial vestibulospinal tract runs in the medial (ventral) columns and therefore is probably not primarily involved in lordosis (Nyberg-Hansen, 1964). The medial reticulospinal tract arises in the pontine reticular formation and runs in the ventral columns (Nyberg-Hansen, 1965; Petras, 1967). The tracts remaining, which are strong candidates for carrying the descending control of lordosis, are the lateral vestibulospinal and the lateral reticulospinal tracts. The lateral vestibulospinal tract arises in the lateral vestibular nucleus (Deiters's nucleus), and its fibers run in the ventral part of the anterolateral columns (Nyberg-Hansen and Mascitti, 1964; Petras, 1967). The lateral reticulospinal tract arises in the reticular formation of the medial medulla (Valverde, 1962; Fox, 1970), and its fibers run in the anterolateral columns (Nyberg-Hansen, 1965; Petras, 1967). Following this process of elimination, our subsequent efforts were directed primarily toward the vestibulospinal and the reticulospinal systems, with special emphasis on the lateral vestibulospinal tract and the lateral reticulospinal tract.

We have conducted extensive lesion studies on the contributions to lordosis of the lateral vestibular nucleus (LVN), which is the source of the lateral vestibulospinal tract (LVST), and the nucleus reticularis gigantocellularis (NGC), which is the source of the lateral reticulospinal tract (Brodal, 1969). We found that damage to LVN produced substantial decrements in lordosis in female rats (Figure 7), whereas subtotal damage to the NGC produced less pronounced but statistically significant decrements in lordosis when behavioral tests were conducted weeks after the lesion (Modianos and Pfaff, 1976a,b). Among rats with either LVN or NGC lesions, the lordosis quotient was negatively correlated with the percentage

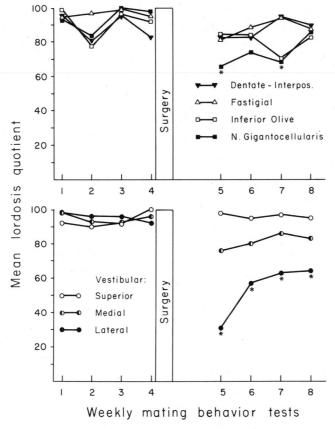

463

NEURAL
MECHANISMS OF
FEMALE
REPRODUCTIVE
BEHAVIOR

Fig. 7. Mean lordosis quotients for each lesion group in preoperative and weekly postoperative lordosis-reflex tests. Asterisk denotes the significant difference ($p < 0.025$) comparing each group's weekly postoperative lordosis quotient with that group's mean preoperative lordosis results. (From Modianos and Pfaff, 1976b.)

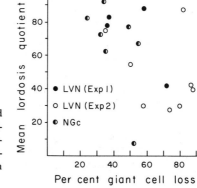

Fig. 8. Mean postoperative lordosis quotient plotted against the percentage of Deiters's cell loss in lateral-vestibular-nucleus-lesioned rats (LVN) and the percentage of giant cell loss in nucleus-gigantocellularis-lesioned rats (NGc). Each point represents data from an individual rat (From Modianos and Pfaff, 1976b.)

of giant cell loss in the LVN or the NGC (Figure 8). A further experiment (Modianos and Pfaff, 1977b) was conducted that explored the possibility that the long recovery periods in our initial experiments may have allowed for a recovery of function following LVN or NGV lesions. In this experiment, the animals were tested preoperatively, were then lightly anesthetized with halothane and were lesioned, and were postoperatively tested 4 hr after lesions. Whereas lesions of the cerebellar cortex or the spinal nucleus of the trigeminal nerve had no effect on either lordosis or general posture and locomotor ability (a finding indicating a lack of effect due to anesthetic or surgical trauma independent of lesion placement), LVN and NGC lesions had pronounced effects on lordosis. After NGC lesions, the mean lordosis quotient dropped from 100 to 21.5, and after LVN lesions, the mean lordosis quotient dropped from 100 to 48.5. The magnitude of the lordosis deficits following LVN lesions was comparable to that which we had previously found with long recovery periods. The magnitude of the lordosis deficits following NGC lesions was much greater than that which we had previously found, a finding indicating a substantial recovery of function over a period of weeks following subtotal NGC lesions. All of these results indicate that Deiters's neurons in the lateral vestibular nucleus (operating through the lateral vestibulospinal tract) and neurons in the medial medullary reticular formation (operating through the lateral reticulospinal tract) make important contributions to the descending control of lordosis.

These lesion results are complemented by electrical stimulation experiments (Modianos and Pfaff, 1975, 1977a). In this work, we found that electrical stimulation of the LVN facilitated lordosis in estrogen-primed, ovariectomized female rats. A facilitation was observed both in tests with manual somatosensory stimulation of the flank and rump-perineal skin regions (Figure 9) and in tests where somatosensory stimulation was supplied by male rats (Figure 10). Unilateral stimulation of the LVN with low current intensities (under 10 μAs) was often effective (Figure 11). In some animals, the duration of individual lordoses was greatly increased (to 10 or 15 sec), whereas in other animals the amount of somatosensory stimulation necessary for lordosis appeared to be reduced. It was obvious, however, that LVN

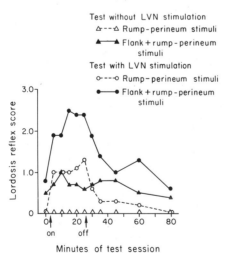

Fig. 9. Lordosis reflex scores for female rat before, during, and after electrical stimulation of the lateral vestibular nucleus (LVN) and in tests with the same time course, but without LVN stimulation. Note that two types of manual stimulation were used. (From Modianos and Pfaff, 1977a.)

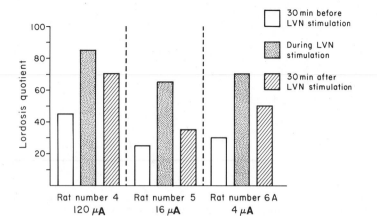

465

NEURAL
MECHANISMS OF
FEMALE
REPRODUCTIVE
BEHAVIOR

Fig. 10. Lordosis responses to mounting by male rats before, during, and after lateral-vestibular-nucleus (LVN) stimulation for female rats 4, 5, and 6A. Current values given are for the amplitude of the cathodal (leading) pulse. (From Modianos and Pfaff, 1977a.)

stimulation by itself was not a substitute for estradiol injection, for the electrical stimulation was not effective if the rat was completely unreceptive prior to testing. An interesting parallel to our lesion work emerged from this experiment: the magnitude of the stimulation effect on lordosis was inversely correlated with the distance of the electrode tip from the center of the LVN (Figure 12), whereas in our lesion experiments we found that among rats with LVN lesions the magnitude of the lordosis deficit was correlated with the percent of Deiters's cell loss in the LVN (see Figure 8).

Thus, it appears that the facilitatory effects on lordosis of pathways going from the brainstem to the spinal cord depend on the action of the lateral vestibulospinal and the lateral reticulospinal tracts.

NEUROPHYSIOLOGICAL CHARACTERISTICS OF LORDOSIS-RELEVANT DESCENDING SYSTEMS

Lateral Vestibulospinal Tract. The anatomy of the lateral vestibulospinal tract has been reviewed (Nyberg-Hansen, 1975), and additional references have been given in the previous section. The neurophysiological literature on the lateral ves-

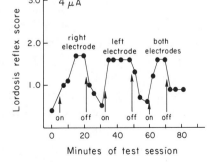

Fig. 11. Lordosis reflex scores for female rat 13 before, during, and after unilateral or bilateral LVN stimulation. All tests were conducted in the same day. Currrent used was 4 μA. (From Modianos and Pfaff, 1977a.)

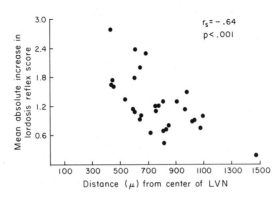

Fig. 12. Mean absolute increase in lordosis reflex score as a result of electrical stimulation of the LVN, plotted against the distance of the electrode tip from the center of the LVN. Each point represents data from an individual rat. (From Modianos and Pfaff, 1977a.)

tibulospinal tract has also been reviewed extensively (Wilson, 1975a; Pompeiano, 1975).

There is a marked tendency for lateral vestibulospinal fibers to terminate on the medial side of the ventral horn, where motoneurons for the axial muscles are mainly located (Nyberg-Hansen, 1975). In fact, the motoneurons controlling the lateral longissimus dorsi and the multifidus muscles, which are responsible for the vertebral dorsiflexion of lordosis in female rats (Brink and Pfaff, 1977; Brink et al., 1977), are located primarily on the medial and the ventromedial sides of the ventral horn in female rats (Brink et al., 1977). In a much different field of investigation, Lawrence and Kuypers (1968a,b) also found that interruption, in monkeys, of the descending pathways whose axons take a ventromedial course (including the lateral vestibulospinal tract) produced severe impairment of axial movements. Very early observations indicated that vestibular lesions were followed by a loss of postural muscle tone (Ewald, 1892, cited in Meyer, 1977). All of these facts can be summarized by saying that an important function of the lateral vestibulospinal tract is to maintain the tone of axial muscles, including the deep back muscles, which, in female rodents, are important in lordosis. In turn, this physiological characterization fits well with the facilitation of lordosis by electrical stimulation of the lateral vestibular nucleus (Modianos and Pfaff, 1977a).

Several neurophysiological characteristics of the lateral vestibulospinal tract are consonant with an important role in lordosis. The predominant facilitatory effect of the lateral vestibulospinal tract is on the motoneurons for extensor muscles (Wilson, 1975; see also Wilson and Yoshida, 1969; Wilson et al., 1970; Hongo et al., 1975; ten Bruggencate and Lundberg, 1974; Lund and Pompeiano, 1968; Grillner et al., 1970, 1971). Most important for lordosis is the fact that electrical stimulation of the lateral vestibular nucleus can produce excitatory postsynaptic potentials in the motoneurons of deep back muscles that cause dorsiflexion of the vertebral column (Wilson et al., 1970). Further, it has been suggested (ten Bruggencate and Lundberg, 1974) that the stimulation of Deiters's cells facilitates extensor motoneurons by excitation of the interneurons involved in the crossed-extensor-reflex action evoked from contralateral-flexor-reflex afferents. The involvement of flexor reflex afferents in the reflex pattern facilitated by the lateral vestibulospinal tract is another similarity to lordosis physiology, in which ascending flexor-reflex afferents in the anterolateral columns are probably involved (see the section above on sensory control of the lordosis reflex).

467

NEURAL
MECHANISMS OF
FEMALE
REPRODUCTIVE
BEHAVIOR

The organization of Deiters's nucleus itself gives additional clues to how the lateral vestibulospinal system might participate in lordosis control. Spinovestibular fibers appear to terminate preferentially in the dorsocaudal portion of the lateral vestibular nucleus (Pompeiano and Brodal, 1957). The predominant effect of cutaneous and other somatosensory input into these Deiters's neurons is excitation (Wilson *et al.*, 1966, 1967). The Deiters's neurons whose axons project as far as the lumbosacral segments of the spinal cord are located in this same portion of the lateral vestibular nucleus, that is, the dorsocaudal portion (Brodal *et al.*, 1962; Brodal, 1969; Nyberg-Hansen, 1975). Thus, in addition to the possible tonic effects of lateral vestibulospinal activity on lordosis mechanisms, another possible role in lordosis for the lateral vestibular nucleus can be construed: cutaneous input due to stimulation by the male might excite those Deiters's neurons whose axons return to lumbosacral portions of the cord, facilitating the motoneurons for the muscles involved in lordosis.

The physiological actions of lateral vestibulospinal fibers are not limited to effects on or near motoneurons. There is also evidence of effects on sensory transmission in the spinal cord and on ascending pathways. By recording potentials from dorsal roots, Cook *et al.* (1969a,b) showed that stimulation of the lateral vestibular nucleus results in negative dorsal root potentials (which correlate with primary afferent depolarization). Destruction of the vestibular nuclei abolished the effect, whereas transection of either the dorsal columns or the ventromedial columns had no effect on the dorsal root potentials. If the ventromedial transection included the medial portion of the anterolateral columns, the root potentials were abolished. Therefore, by elimination, the lateral vestibulospinal and perhaps the reticulospinal tracts mediate the effect. A vestibuloreticulospinal link might be considered, as ipsilateral hemisection of the spinal cord only partially depressed the dorsal root potentials evoked from the lateral vestibular nucleus, and the lateral vestibulospinal tract is an ipsilateral tract (Brodal, 1969). Cook *et al.* (1969a,b) further showed that the presence of the cerebellum was not required for obtaining the negative root potentials. Chan and Barnes (1972) reported that primary afferent depolarization can be obtained from stimulation points in the medial pontine and the medullary reticular formation.

In terms of ascending pathways, it is very interesting that the lateral vestibular nuclei have not been shown to influence transmission ascending through the dorsal or the dorsolateral columns, although they do influence ascending information running in the anterolateral columns (reviewed by Pompeiano, 1975). Spinoreticular neurons whose axons ascend in the anterolateral columns receive excitatory influences from the lateral vestibulospinal tract (see, for instance, Holmqvist *et al.*, 1960a,b). Thus, the pattern of lateral vestibulospinal influences on ascending pathways matches the pattern of results regarding ascending requirements for the performance of lordosis (see the section above on sensory control of the lordosis reflex): both sets of results focus on spinoreticular systems in the anterolateral columns. If lateral vestibulospinal influences in the lumbosacral cord facilitate ascending systems that, in turn, excite Deiters's neurons in the dorsocaudal lateral vestibular nucleus, it would be possible to see how rapid postural adaptations to somatosensory input are achieved.

Although the strongest effects of the lateral vestibulospinal tract are seen on the ipsilateral side of the spinal cord, qualitatively similar effects are seen on the contralateral side: predominant excitation of extensor and inhibition of flexor motoneurons (Nyberg-Hansen, 1975; Wilson, 1975b). In a study by Hongo *et al.* (1975), lateral vestibulospinal stimulation was found to evoke excitatory postsynaptic potentials in both ipsilateral and contralateral extensor motoneurons. The crossed effects were not due to activation of the contralateral vestibulospinal tract by way of crossing fibers of the ipsilateral tract, as transection of the contralateral half of the spinal cord above the level of recording did not abolish the crossed excitation effect. Whether some of the stimulation effects may be due to activation of reticulospinal cells was difficult to determine, but recording from the contralateral cord was not affected by a midsagittal transection of the brainstem extending from the lateral vestibular nucleus to the obex (Hongo *et al.,* 1975). Because of the vestibuloreticular interactions demonstrated with neuroanatomical (Ladpli and Brodal, 1968) and physiological (Peterson and Felpel, 1971) techniques, the possibility still must be considered that a vestibuloreticulospinal system could cross at a level below the obex. Nevertheless, Hongo *et al.* (1975) suggested that the contralateral interaction occurs at the spinal level and involves a crossing interneuron. Because the lordosis reflex involves bilaterally symmetrical movements, primarily a bilateral dorsiflexing action on the vertebral column, it makes sense that the lateral vestibulospinal system—shown to be involved, by lesion and stimulation experiments, in lordosis—has similar actions on both sides of the spinal cord.

Evidence has been presented (Abzug *et al.,* 1974) that individual axons of the lateral vestibulospinal tract branch to innervate more than one level of the spinal cord. For instance, in these experiments, antidromic activation of individual lateral vestibulospinal neurons could be obtained both from local branches in the gray matter of cervical or thoracic cord and from the lateral vestibulospinal tract at lumbar levels. Although the branching patterns are not homogenous and unrestricted (Wilson *et al.,* 1976), it appears that information carried by individual vestibulospinal neurons can be passed to a wide variety of spinal cord levels. On the one hand, this possibility places an upper limit on the specificity of the information likely to be carried by such neurons. On the other hand, it shows nicely how the lateral vestibulospinal system could participate in lordosis, a reflex that involves vertebral column dorsiflexion from the posterior rump and the tail base all the way forward to the neck.

Lateral (Medullary) Reticulospinal Tract. Anatomical work on the reticulospinal systems has been reviewed (Nyberg-Hansen, 1975), and references are also given in section above on lordosis data. Electrophysiological recording (Peterson *et al.,* 1975b) has shown that the reticulospinal axons running in the anterolateral col-

469

NEURAL
MECHANISMS OF
FEMALE
REPRODUCTIVE
BEHAVIOR

umns arise primarily from cells in the caudoventral part of the nucleus gigantocel-lularis in the reticular formation of the medulla. The transection and lesion exper-iments cited in the section above on lordosis data suggest that the reticulospinal axons running in more lateral positions, rather than those that run in a medial position through the medulla and the spinal cord, are most important in lordosis. Therefore, lordosis should be primarily a function of the reticulospinal neurons in the caudoventral medulla whose axons run through the anterolateral columns. Although, compared to the lateral vestibulospinal system, the medullary reticulos-pinal system is likely to be more critical for lordosis, and more likely to carry hypo-thalamic influences to the spinal cord (see the section above on lordosis data and the section that follows), the detailed physiology of the reticulospinal neuron is less well understood.

Magoun and his colleagues showed that electrical stimulation of the ventral part of the medial medullary reticular formation inhibited limb reflexes (Magoun and Rhines, 1946, 1947; Sprague *et al.,* 1948; Lindsley *et al.,* 1949). In contrast, electrical stimulation of the pontine reticular formation facilitated limb reflexes (Rhines and Magoun, 1946; Sprague *et al.,* 1948; Lindsley, 1952; Magoun, 1963). Many of the experiments reported by these investigators showed bilateral effects of reticular stimulation, with similar actions on both flexor and extensor reflexes. In later work, which included the use of threshold levels of stimulation, Sprague and Chambers (1954) reported reciprocal effects of reticular stimulation on flexor and extensor reflexes of the legs. The reticulospinal axons mediating these effects descend in the anterolateral columns (Niemer and Magoun, 1947). The locations and the size of the spinal transections needed to abolish these reticular effects appear to be similar to the transections required to abolish the lordosis reflex (Kow *et al.,* 1977b).

Reticulospinal axons are among those included in the ventromedial group of descending systems described by Kuypers (Lawrence and Kuypers, 1968a,b) as con-trolling axial musculature. Brainstem reticular neurons with different inputs and outputs are, for the most part, interspersed with each other and are not arranged according to physiological properties, in large, easily recognized nuclei (Eccles *et al.,* 1976). The synaptic arrangements underlying the reticular effects on spinal reflexes are not yet clear. Inputs into reticulospinal neurons have been character-ized by means of modern recording techniques (Peterson *et al.,* 1974, 1975a; Peterson and Abzug, 1975). Of special interest is the demonstration (Peterson *et al.,* 1974) that neurons deep to the superior colliculus, including laterally situated subtectal neurons, produce excitatory postsynaptic potentials in ipsilateral reticu-lar neurons. Cutaneous stimuli also evoked postsynaptic potentials in reticulospinal neurons (Peterson *et al.,* 1974; see also the section above on sensory control of the lordosis reflex). Reticulospinal neurons in the medulla appear to share several physiological properties with likely lordosis mechanisms: they respond to cuta-neous input, they are involved with descending excitation from subtectal regions, and they are involved in the control of axial musculature.

Like vestibulospinal neurons, individual reticulospinal neurons have been found to send branched axons to different levels of the spinal cord (Peterson *et al.,*

1975b). Individual reticular neurons that could be antidromically activated by stimulation within the gray matter of the cervical spinal cord could also be antidromically stimulated from the lumbar spinal cord. As well as sending a large contingent of axons down the ipsilateral lateral reticulospinal tract, medullary reticular neurons also send a small number of axons down the contralateral lateral reticulospinal tract (Peterson *et al.,* 1975b); such axons could participate in the bilateral actions of reticulospinal stimulation on reflexes. The longitudinal branching and bilateral crossing properties of reticulospinal axons might serve well for participation in lordosis control, as lordosis involves dorsiflexion of a large portion of the vertebral column and is bilaterally nearly symmetrical in its appearance.

Physiological experiments with monosynaptic reflexes involving hindlimb motoneuron pools have suggested that some descending effects result from the combined activity of the lateral vestibulospinal and reticulospinal tracts (Hassen and Barnes, 1975). Such is also likely to be the case in the control of lordosis.

ROLES OF THE LATERAL VESTIBULOSPINAL TRACT AND LATERAL (MEDULLARY)-RETICULOSPINAL TRACT IN LORDOSIS CONTROL. Transection, lesion, and stimulation data (see the section above on lordosis data) on lordosis and neurophysiological studies (see the previous section) indicate that the lateral vestibulospinal tract and the lateral (medullary) reticulospinal tract work together in the control of the lordosis reflex. Because no obvious anatomical connections have been discovered that link the hypothalamus to the lateral vestibular nucleus either directly or indirectly, we assume that the lateral vestibulospinal tract does not function as a carrier of descending hypothalamic influences. In contrast, reticulospinal cells in the medullary reticular formation, as well as receiving cutaneous input, also receive descending input from midbrain structures (the central gray and the dorsal lateral reticular formation), which, in turn, receive direct axonal connections from the hypothalamic regions involved in lordosis control. Therefore, we suggest that medullary reticulospinal neurons, operating through the lateral reticulospinal tracts, play an important integrative role in the control of lordosis, and, in particular, that they help to relay descending, hormone-dependent influences from the hypothalamus. Both the lateral vestibulospinal and the lateral reticulospinal tracts are known to be involved in the regulation of axial musculature and to have crossed effects similar to their ipsilateral actions. Both of these properties suit these tracts well in the regulation of lordosis, a vertebral dorsiflexion that is essentially bilaterally symmetrical. Because spinal female rats do not perform lordosis (Pfaff *et al.,* 1972; Kow *et al.,* 1977b), the net influence of these tracts must be facilitatory for lordosis. Therefore, the net action of the lateral vestibulospinal and the lateral reticulospinal tracts must be either to increase activity in the spinal cord neurons excitatory for lordosis (to enhance spinal lordotic mechanisms) or to inhibit activity in the spinal circuits inhibitory for lordosis (perhaps including the reduction of competing responses).

Regarding the lateral reticulospinal tract, when we used the horseradish peroxidase technique on female rats, we found the cell bodies of reticulospinal neurons in especially great numbers in the ventral caudal part of the medullary reticular formation (Zemlan *et al.,* 1977). Descending axons from the central gray and the dorsal lateral reticular formation of the mesencephalon also enter this region of the medullary reticular formation. At first, the emphasis for lordosis control on this part of the medullary reticular formation may seem puzzling because this region is what Magoun and his collaborators (see the previous section) designated the "inhibitory" reticular mechanism, whereas the net effect of the reticulospinal systems is facilitatory for lordosis. However, it is important to note that the studies of the Magoun laboratory were on reflexes of the limbs (see references in the previous section). Electrical stimulation of the ventromedial medullary reticular formation brings locomotion and limb movement to a halt (Magoun, 1950). This result is precisely what is needed as a part of lordosis facilitation, as the initiation of lordosis in the estrogen-primed female rat requires that she stand still on cutaneous contact from the male, just prior to active vertebral dorsiflexion (Pfaff and Lewis, 1974). Moreover, the pattern of transections of the lateral and the anterolateral columns that are needed to disrupt the descending effects of this medullary reticulospinal system (Niemer and Magoun, 1947) fit nicely with the types of transections needed to eliminate lordosis in estrogen-primed female rats (Kow et al., 1977b). Therefore, it seems likely that reticulospinal neurons in the ventral and the caudal portions of the medullary reticular formation facilitate lordosis by halting locomotor reflexes, as well as by facilitating axial reflexes, and that they do so by means of axons that are spread throughout the anterolateral columns.

Three aspects of the descending control of lordosis by the lateral vestibulospinal and the lateral reticulospinal tracts are discussed below.

Specificity. Evidence has been reviewed (see the previous section) that individual axons of the lateral vestibulospinal tract and the lateral reticulospinal tract branch to terminate at a variety of different spinal levels. Both of these tracts also have crossed effects similar to their ipsilateral actions. Both of these properties fit well with lordosis control, as the vertebral dorsiflexion of lordosis involves many different levels of the spinal cord and as lordosis is bilaterally nearly symmetrical. These properties of the lateral vestibulospinal and reticulospinal tracts also place upper limits on the specificity with which these tracts can act. For instance, it is unlikely that the descending action of these tracts have the effect specifically of facilitating lordosis without implications for any other reflex patterns. This relative lack of perfect specificity is reflected in the nature of the transection and lesion results during experiments on lordosis. With transections of spinal columns (Kow et al., 1977b), it appeared that any portion of the anterolateral columns remaining after transection, as long as that portion was larger than 25% or 30% of the cross-sectional area of the columns, would support fairly high levels of lordosis. In lesion studies of the nuclei giving rise to the lateral vestibulospinal and the lateral reticulospinal tracts (Modianos and Pfaff, 1976b), it appeared that the amount of lordosis loss following destructive lesions was a function of the number of giant cells lost, almost regardless of exactly which giant cells they were within those nuclei.

The consideration of specificity can be extended to the inputs into reticulospinal systems. The cutaneous receptive fields of reticulospinal neurons are usually very large (see the section above on sensory control of lordosis). It is likely to be through reticulospinal neurons that a variety of the facilitatory effects of stimuli on lordosis can be routed, for instance, the facilitation of lordosis in estrous female hamsters by olfactory and ultrasound stimuli from the male (Noble, 1972, 1973; Floody and Pfaff, 1977a,b). Therefore, if, as in the section above on sensory control of lordosis, we think of lordosis as requiring both a precise control of reflex form and enough amplification to generate sufficient muscular power, we can speak of both a "steering" function and an "acceleratory" function. From the data now available, it would appear that the precise "steering" control of the reflex form can best be ascribed to spinal circuits, whereas a relatively gross facilitation of these spinal circuits (an amplification or "acceleration" function) can be ascribed to the combined action of the lateral vestibulospinal and the lateral reticulospinal tracts.

Timing. The lateral vestibulospinal and reticulospinal tracts might have two modes of action in facilitating lordosis. One mode, perhaps especially important in the lateral vestibulospinal tract, could be in the form of a tonic effect. The tonic facilitation of lordosis would result in a subliminal amount of background activity in the motoneuron pools that supply the muscles important in lordosis, as well as a reduction of activity in the motoneuron pools that supply the muscles antagonistic to lordosis. Against this background, the cutaneous stimuli provided by male rats during mating encounters are able to trigger lordosis. A neurophysiological mechanism of this sort could account for the time course of our electrical stimulation results, in which lordosis facilitation due to stimulation of the lateral vestibular nucleus often took as long as 5 min to reach a peak (Modianos and Pfaff, 1977b). If the lateral reticulospinal tract also has a tonic mode of action, this might result in part from the transmission of a tonic, hormone-dependent output from the hypothalamus. An additional reason for considering tonic modes of action is that certain conspicuous prelordotic elements of female-rodent mating behavior could provide appropriate stimuli for increasing the background activity of lordosis-relevant descending systems (see the following section).

A second mode of action of the lateral vestibulospinal and the lateral reticulospinal systems could involve spinobulbospinal reflex loops. Ascending cutaneous sensory afferents are known to excite reticulospinal cells in the medial medullary reticular formation and vestibulospinal cells in the dorsocaudal part of the lateral vestibular nucleus (which projects back to the lumbosacral cord) (see the section above on the characteristics of descending systems). Thus, those brainstem neurons receiving lordosis-relevant cutaneous afferents could, in turn, facilitate lordosis by their projections back down to the spinal cord. This type of action of lordosis-relevant descending systems may best be seen as a lordosis-to-lordosis effect, that is, a triggering of the relevant spinobulbospinal reflex loops following each application of the adequate cutaneous stimuli, which evoke a new occurrence of lordosis.

Participation of Courtship Behavior in Preparing the Female Rat for Lordosis. The behavior of highly receptive female rats, well primed with estrogen and progesterone, just before lordosis in the presence of a male rat is characterized by an unusual pattern of locomotion ("hopping and darting") and by apparent "ear wiggling." The darting form of locomotion consists of sudden bursts of forward move-

473

NEURAL
MECHANISMS OF
FEMALE
REPRODUCTIVE
BEHAVIOR

ment followed by sudden stops. We have argued (Pfaff *et al.,* 1972) that the sudden
stops promote successful male–female mount–lordosis encounters (1) by increasing the chance that the male (following from the rear) will bump into the female
and (2) by putting the female in a bilaterally balanced posture prepared to support
the weight of the male. We also said (Pfaff *et al.,* 1972) that the sudden starts and
stops put the female in a state of muscular tension that, although vaguely defined,
is known to facilitate lordosis. Now we can point to physiological features of the
lateral vestibulospinal system, which are involved in lordosis, that may explain how
these "courtship behaviors"—hopping and darting and ear wiggling—facilitate
lordosis.

Hopping and darting are comprised of unusual linear accelerations along longitudinal and vertical axes. Linear acceleration is a sufficient stimulus for firing
primary afferents from the utricle of the labyrinth, and the lateral vestibular
nucleus receives utricular afferents (Brodal *et al.,* 1962). Thus, linear acceleration
during hopping and darting could stimulate the lateral vestibulospinal system and,
in a tonic manner, facilitate lordosis by increasing the background activity in this
system.

The ear-wiggling component of the courtship behavior of receptive female
rats was long ago observed to be a function of rapidly alternating head movements
(Beach, 1942). It is now evident from slow-motion movie-film analyses of mating
encounters that ear wiggling is actually a result of rapid oscillations of the head
around the longitudinal (sagittal horizontal) axis (Modianos, unpublished observations, 1976). Such stimulation strongly excites the anterior and posterior semicircular canals of the labyrinth (Mountcastle, 1974), which, in turn, send primary
afferent fibers to the lateral vestibular nucleus. Peterson (1970) reported that head
tilting results in monosynaptic and disynaptic excitatory postsynaptic potentials in
many cells in the lateral vestibular nucleus, including those that send efferents to
the spinal cord. The work of Wilson *et al.* (1970) and of Hassen and Barnes (1975)
has shown that labyrinthine stimulation can produce excitatory postsynaptic potentials in extensor motoneurons of the spinal cord. In particular, the anterior semicircular canals have a special role in exciting those Deiters's neurons whose axons
travel through the lateral vestibulospinal tract (reviewed by Wilson, 1975a). This
role is especially important because (1) such Deiters's neurons are known to have
excitatory actions on spinal motoneurons involved in vertebral dorsiflexion; (2)
their axons running through the lateral vestibulospinal tract can reach the lumbosacral spinal cord; and (3) the lateral vestibulospinal tract is known to be important in lordosis. In fact, stimulation of the anterior semicircular canals causes a
head movement upward (Suzuki and Cohen, 1964), clearly showing how this kind
of vestibular stimulation could predispose a female rat toward lordosis.

Female rat courtship behaviors such as hopping and darting and ear (head)
wiggling show a very high, and unusual, state of excitation of musculature facilitated by the lateral vestibulospinal system just prior to lordosis. In turn, the vestibular stimulation resulting from these courtship behaviors should lead to an even
higher pitch of excitation in the lateral vestibulospinal system and the axial mus-

culature stimulated by it. Thus, the nature of these courtship behaviors in the receptive female rat may give evidence of a kind of positive feedback relationship involving the lateral vestibulospinal system, the skeletal musculature, and vestibular input. The vestibulospinal and musculature excitation maintained by such courtship behavior would not cause lordosis by itself but may facilitate lordosis by maintaining higher levels of background activity in those spinal circuits important in lordosis, a given level of sensory input on stimulation by the male rat would cause a stronger lordosis reflex.

Working Models of Female Reproductive Behavior Mechanisms

Reflex Nature of Lordosis Behavior

The following sections cover three approaches to the conceptualization of lordosis. One approach is a capsule description of the reflex itself, with the intention of seeing what principles of reflex organization arise. A second approach is to seek examples of "good fit," that is, to describe the biological usefulness of lordosis itself and to give examples of adaptive relations between lordosis and other behavior patterns of the female and male. A third approach is to look for relations between lordosis behavior and previously studied reflex types.

The Reflex. Lordosis is the reflex response of the hormone-primed female rodent to bilateral cutaneous stimuli on the flanks and then on the posterior rump and the perineum. The lordosis response itself should be conceived of as a standing response combined with an elevation of the perineum. The standing component of lordosis is characterized by all four legs' being firmly placed and somewhat extended, and by the rear legs' being somewhat abducted. The perineum elevation is achieved by a vertebral dorsiflexion (Pfaff *et al.,* 1972; Pfaff and Lewis, 1974). A behavioral (input–output) definition of the estrogen effect on lordosis is that it increases the strength of functional connections between cutaneous receptors on the flanks-rump-perineum and the motoneurons controlling vertebral dorsiflexion.

Abstracting certain features of lordosis may help to induce concepts that go beyond a simple description. The standing component of lordosis must be characterized by great enough tension in the muscles of all four limbs to support the weight of the male; in this respect, it may be similar to the standing responses of females of a large number of mammalian species. Both the standing response and the perineum elevation of lordosis are characterized by *bilateral symmetry.* In fact, the hopping and darting locomotion of the receptive female rat, as well as lordosis itself, is bilaterally symmetrical, whereas several aspects of the rejection behavior of the unreceptive female rat can be described as departures from bilateral symmetry (Pfaff and Lewis, 1974). A third feature of lordosis in female rats is that the dorsiflexion that marks the response involves virtually the entire vertebral column. This total involvement may be a function of the origins and the insertions of the muscles involved, or it may be a result of the amount of cutaneous stimulation involved, or of a "radiation" of neural excitation in the spinal cord, or of some combination of these three factors.

475

NEURAL
MECHANISMS OF
FEMALE
REPRODUCTIVE
BEHAVIOR

Achievement of the full lordosis posture, taking over 300 msec as it does (Pfaff and Lewis, 1974), should not be conceived of as a unitary process. Rather, several component stimuli (on the skin of the flanks, the rump, the tail base, and the perineum) are followed by several component responses (extension of the front legs, responses of the rear legs, elevation of the rump and the tail base, and elevation of the head). Therefore, lordosis is best conceived of as a chain of component reflexes in sequence (Pfaff *et al.*, 1972, Figure 10). In fact, the female's reflex responses to stimuli early in the sequence allow the male to apply the subsequent stimuli necessary, in turn, for her subsequent responses (Pfaff *et al.*, 1972). Thus, a "cascade" effect may operate in the organization of lordosis.

CONCEPTS OF "GOOD FIT." The biological usefulness of each lordosis response component is obvious. The standing response as part of lordosis allows the male to mount and prepares the female to support the weight of the male. Elevation of the perineum (vertebral dorsiflexion) as part of lordosis exposes the perineum to pelvic thrusting by the male, allowing intromission and fertilization and leading eventually to successful reproduction. Furthermore, the fact that mating responses in the female are estrogen-dependent, and in the male androgen-dependent, allows for a hormone-controlled synchrony between the mating pair (Pfaff *et al.*, 1972, Figure 11). The effect of this coordination is to ensure that only conspecifics endocrinologically competent to reproduce will finish the mating-behavior sequence.

A "good fit" also exists between features of the pre-lordotic courtship behavior of the female rat, the requirements for lordosis, and the behavior of the male. For instance, the hopping-and-darting form of locomotion, a bilaterally symmetrical gait (as opposed to normal limb alternation), facilitates the male rat encountering the female from the rear in normal mounting position; leaves the female in a bilaterally symmetrical posture prepared to support a downward and forward force; and leaves the female in a state of muscular tension that facilitates the lordosis reflex (Pfaff *et al.*, 1972).

RELATIONS TO KNOWN REFLEX TYPES. Comparing lordosis to previously established types of reflexes is attractive because the identification of lordosis as a special case of a well-studied reflex type would be helpful in the study of lordosis circuitry and mechanisms. It would allow us to make use of a great deal of existing literature on spinal reflexes. Knowing the exact "nature" of lordosis mechanisms at the spinal level might also allow us to deduce the exact role of descending (supraspinal) influences.

Do the facts about lordosis now available force the postulation of new spinal-reflex types? So far, it appears that lordosis involves a concatenation of spinal reflex mechanisms already established in physiological studies of the spinal cord (Kow and Pfaff, 1976b). Regarding the first stimulus from the male, on the skin of the female's flanks, pressure from the male's forearms could deform muscles, triggering stretch reflexes in the deep back longissimus dorsi system that are important in lordosis. Also, this stimulus could cause enough skin deformation on the anterior thigh to initiate bilateral rear-leg extension, as can be observed in spinal animals. Finally, palpation by the male's forepaws on the lateral abdomen probably stimulates the field of the genitofemoral nerve and may help to initiate rump elevation or bilateral rear-leg extension by the mechanism of a withdrawal reflex, that is, elevation of the hindquarters away from a stimulus on the ventral side of the body.

Cutaneous stimulation on the female's tail base and posterior rump triggers lordotic movements that might be conceived as a withdrawal reflex, as elevation of that region—and, finally, slight forward movement in the dorsiflexed posture—moves the posterior rump and the tail base away from the main focus of stimulation. In addition, pressure on this region from stimulation by the male may stretch muscles or tendons of deep back muscles that help the rat to execute lordosis, thus triggering stretch reflexes in that muscle system.

A conventional reflex role in lordosis for cutaneous stimulation on the perineum is very clear (Kow and Pfaff, 1976b). Observations in decerebrate cats (Creed *et al.*, 1932) and in female rats (Pfaff *et al.*, 1977) show that light cutaneous perineal stimuli cause bilateral rear-leg extension and tail-base elevation. Such a sublordotic response fits the pattern for a withdrawal reflex because elevation of the perineum withdraws the skin surface from the direction of stimulation.

Insofar as these various conventional spinal-reflex types exist and actually operate in the execution of lordosis, when acting in parallel they may summate to add to the total lordosis-response strength.

Comparisons of lordosis to electrophysiological studies of spinal reflex mechanisms have not yielded a clear identification of lordosis with an individual reflex mechanism. For instance, from cutaneous stimuli such as those that trigger lordosis one expects ipsilateral flexion, at least in the study of limb reflexes. Yet, so-called flexor reflex afferents also lead to crossed extensor activity, and this is interesting because the lateral vestibulospinal tract has been shown to facilitate the interneuronal activity involved in the crossed extensor reflex (ten Bruggencate and Lundberg, 1974). Moreover, lordosis is triggered by cutaneous stimuli that are roughly bilaterally symmetrical. Therefore, as far as limb reflexes are concerned, on the application of cutaneous stimuli that trigger lordosis there is reason to expect excitation of motor neurons for both flexor and extensor muscles on both sides of the body. Thus, it appears that the onset of lordosis is probably accompanied by a state of high tension in several muscle systems.

A different sort of comparison can be made between lordosis behavior and "attitudinal" reflex patterns. For instance, with no direct evidence available, one might wonder if lordosis includes vestibular-dominated components that are led by head movements (head elevation). The "looking-under-the-shelf" posture is an example of an attitudinal reflex pattern that involves elevation of the hindquarters (Mountcastle, 1974). Still another possibility is that, during some portions of the lordosis reflex, some muscle systems in the female are flaccid, and thus that certain aspects of the lordosis posture might result primarily from forces generated by movements of the male rat.

Lordosis cannot be explained yet by facile identification with a single, individual, previously described reflex mechanism or reflex pattern. For instance, lordosis

477

NEURAL
MECHANISMS OF
FEMALE
REPRODUCTIVE
BEHAVIOR

shares some properties of decerebrate rigidity. In lordosis, the legs are somewhat extended, and the head and the tail base elevated. Antigravity muscles must be activated to support the weight of the male. Further, decerebrate rigidity is often ascribed to the action of the lateral vestibulospinal tract, which is supposed to predominate because inhibitory influences from higher centers are cut off. In turn, the lateral vestibulospinal tract has been shown to facilitate lordosis (see the section above on lordosis data). *Yet,* lordosis cannot be a simple example of decerebrate rigidity. First, we have been unable to obtain lordosis in animals with complete transections at various midbrain levels, in which the lower brainstem was cut off from the influences of higher centers (Kow *et al.,* 1977a). Second, decerebrate rigidity is not a real standing posture. Standing involves flexor as well as extensor muscle tension, to achieve a firm and stable limb posture (Mountcastle, 1974).

Most important, lordosis also shares properties with the ventral flexor-reflex afferent system, a system that is normally thought to oppose decerebrate rigidity. First, lordosis is triggered by cutaneous stimuli, which are normally thought of as flexor reflex afferents. Second, the ventral flexor-reflex afferent system is similar to lordosis in that the tracts ascend in the anterolateral columns, involve relatively nonspecific cutaneous responses, and do not involve the cerebellum (Lundberg and Oscarsson, 1962; Holmqvist *et al.,* 1960b). Third, lordosis reflexes and the ventral flexor-reflex afferent system may be under similar forms of descending control. Both are not active in the decerebrate preparation (Eccles and Lundberg, 1959; Job, 1953; Kow *et al.,* 1977a). Both are activated by electrical stimulation in the ventral medulla at sites that probably excite axons of both the lateral vestibulospinal and the reticulospinal tracts (Holmqvist *et al.,* 1960a; Grillner *et al.,* 1968; Pfaff *et al.,* 1972).

Therefore, it appears that lordosis reflex organization cuts across the dichotomy between decerebrate (extensor) rigidity and the ventral flexor-reflex afferent system. As regards limb muscles, lordosis must be characterized as a state in which the adequate cutaneous stimuli lead to high activity of both the extensor and the flexor systems. This activity is actually required for rigid and stable support of the weight of the male. The remaining distinguishing characteristic of lordosis is the vertebral dorsiflexion, especially the elevation of rump and perineum, that allows fertilization.

NEURAL CIRCUITRY AND DYNAMICS UNDERLYING HORMONE EFFECTS AND LORDOSIS MECHANISMS

We shall summarize information on the neural mechanisms of female-rat reproductive behavior by attempting to follow the flow of neural activity during the onset of lordosis. This attempt comprises a working model of the neural circuitry and dynamics underlying lordosis and the associated hormone effects (Figure 13).

SENSORY AND ASCENDING MECHANISMS. On contact from the male rat, cutaneous mechanoreceptors are activated first in the flanks and then in the posterior rump, the tail base, and the perineum. The resulting increased neural activity is

DONALD PFAFF AND
DOAN MODIANOS

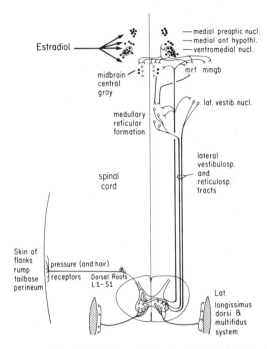

Fig. 13. Schematic representation of minimal neural circuit for lordosis behavior in the female rat: A working model. Estradiol effects are shown as mediated by estrogen-binding neurons (●) in the hypothalamus and the central gray. Ascending spinal fibers for lordosis travel in anterolateral columns of the cord. Abbreviations: mmgb, medial division of the medial geniculate body; mrf, dorsal mesencephalic reticular formation.

carried over the peripheral nerves and enters the spinal cord over dorsal roots L_1, L_2, L_5, L_6, and S_1. Responding neurons on both sides of the body synapse in the dorsal horn. Second-order somatosensory neurons in the dorsal horn send local projections to interneurons at the same and neighboring segmental levels. However, a substantial number of axons (from both dorsal horns) also cross and send ascending projections through the anterolateral columns. The local spinal mechanisms activated by cutaneous stimulation are not by themselves sufficient for the execution of lordosis. Therefore, the ascending anterolateral projections are the first part of an obligatory supraspinal loop.

The information ascending in anterolateral column axons does not give as precise and specific a record of somatosensory events as, for instance, in the dorsal columns. The ascending anterolateral system may be part of a "trigger" for lordosis, but it may also help to facilitate tonic descending activity that supports lordosis. The ascending anterolateral axons that may participate in lordosis terminate in the medulla, in the medial reticular formation, and in the caudal part of the lateral vestibular nucleus. In the midbrain, they terminate in and around the lateral borders of the central gray and in the dorsolateral reticular formation (including the region by the medial border of the medial geniculate body). It must be at these midbrain and/or medullary termination sites that the supraspinal lordosis loop is susceptible to facilitation by estrogen-dependent hypothalamic output.

HORMONE EFFECTS. If lordosis is to occur on stimulation from the male, then blood estrogen levels must have been raised for at least 20–24 hr. Estrogen is bound by neurons in the ventromedial nucleus of the hypothalamus, as well as by neurons in the arcuate nucleus and by some neurons in the anterior hypothalamic nucleus, and the action of estrogen in the basomedial hypothalamus is directly to facilitate lordosis. The electrical activity, particularly of slowly firing neurons, in the basomedial hypothalamus is elevated by estrogen. Axonal projections transmitting this tonically elevated activity, and relevant to lordosis, are to the midbrain central gray (bilaterally) and to the dorsal lateral mesencephalic reticular formation, as far lateral as the medial border of the medial geniculate body (Figure 13). Axonal projections from neurons in and around the ventromedial nucleus of the hypothalamus are to the same areas of the midbrain as receive somatosensory input through the anterolateral columns. At these points in the dorsal mesencephalon, the supraspinal lordosis loop is facilitated by the tonic, estrogen-dependent output from the basomedial hypothalamus.

In the estrogen-primed female rat, progesterone treatment facilitates lordosis. The mechanisms by which this facilitation occurs are less well specified than those for estrogen effects on lordosis. Similarities between the effects of raphe lesions, septal lesions, and serotonin depletion prompted us to formulate a hypothesis about how, in neuroanatomical terms, progesterone facilitates female reproductive behavior in rodents (Kow et al., 1974b). Briefly, this hypothesis suggests that progesterone inhibits activity in a midbrain–limbic loop which inhibits lordosis. Projections from the midbrain raphe nuclei reach the septum directly (Conrad et al., 1974). Projections from the septum indirectly reach the midbrain, including the raphe nuclei. The net effect of the descending projections from this midbrain–limbic loop, the most important of which may be serotonergic fibers from the raphe, is to inhibit responses to somatosensory input, including the inhibition of lordosis. Interruption of function in the loop facilitates responses to somatosensory input, including the facilitation of lordosis. Therefore, if progesterone decreases the activity and the responsiveness of neurons in this loop, it should facilitate lordosis (Kow et al., 1974b).

The funnel through which hormone effects on the hypothalamus and the midbrain (whether effects of estrogen effects, of progesterone, or of LRF) must pass is formed by the neurons whose cell bodies are in the lower brainstem and whose axons pass down the spinal cord.

DESCENDING AND MOTOR MECHANISMS. Among the major, well-recognized descending systems, only the lateral vestibulospinal tract and the lateral (medullary) reticulospinal tract are necessary for lordosis. Because the lateral vestibular nucleus does not receive direct inputs from relevant hypothalamic or midbrain cell groups (and the medullary reticular formation does receive such connections), it must be the medullary reticulospinal axons that carry the obligatory facilitatory influences for lordosis. Working in concert, these two systems (1) allow descending responses to adequate ascending sensory signals and/or (2) exert tonic facilitatory influences. The combined action of lateral reticulospinal and vestibulospinal axons, working through interneurons and motoneurons on both sides of the spinal cord, is to facilitate the lordosis response, that is, the standing response combined with vertebral dorsiflexion.

The vertebral dorsiflexion of lordosis depends on the lateral longissimus muscle and on the multifidus system of deep back muscles in the female rat. Therefore, hormone-dependent descending influences must combine with adequate cutaneous input to activate the motoneurons for these muscles, in order to produce the rump elevation that is unique to lordosis.

REFERENCES

Abzug, C., Maeda, M., Peterson, B. W., and Wilson, V. J. Cervical branching of lumbar vestibulospinal axons. *Jounral of Physiology* (London), 1974, *243*, 499–522.

Ahmed, S. M. Electrocardiographic changes on hypothalamic stimulation in dog. *XXVI International Congress of Physiological Sciences*, New Delhi, 1974, Abstract #597.

Alcaraz, M., Guzman-Flores, C., Salas, M., and Beyer, C. Effect of estrogen on the responsivity of hypothalamic and mesencephalic neurons in the female cat. *Brain Research*, 1969, *15*, 439–446.

Anderson, F. D., and Berry, C. M. Degeneration studies of long ascending fiber systems in the cat brain stem. *Journal of Comparative Neurology*, 1959, *111*, 195–229.

Anderson, J. N., Peck, E. J., Jr., and Clark, J. H. Estrogen-induced uterine responses and growth: relationship to receptor estrogen binding by uterine nuclei. *Endocrinology*, 1975, *96*, 160–167.

Antonetty, C. M., and Webster, K. E. The organization of the spinotectal projection. An experimental study in the rat. *Journal of Comparative Neurology*, 1975, *163*, 449–466.

Applebaum, A. E., Beall, J. E., Foreman, R. D., and Willis, W. D. Organization and receptive fields of primate spinothalamic tract neurons. *Journal of Neurophysiology*, 1975, *38*, 572–586.

Arai, Y., and Gorski, R. A. Effect of anti-estrogen on steroid induced sexual receptivity in ovariectomized rats. *Physiology and Behavior*, 1968, *3*, 351–353.

Araki, S., Ferin, M., Zimmerman, E. A., and Vande Wiele, R. L. Ovarian modulation of immunoreactive gonadotropins-releasing hormone (Gn-RH) in the rat brain: Evidence for a differential effect on the anterior and midhypothalamus. *Endocrinology*, 1975, *96*, 644–650.

Arnold, A., Nottebohm, F., and Pfaff, D. W. Hormone-concentrating cells in vocal control and other areas of the brain of the zebra finch *(Poephila guttata)*. *Journal of Comparative Neurology*, 1976, *165*, 487–512.

Averill, R. L. W., and Purves, H. D. Differential effects of permanent hypothalamic lesions on reproduction and lactation in rats. *Journal of Endocrinology*, 1963, *26*, 463–477.

Bagshaw, R. J., Iizuka, M., and Peterson, L. H. Effect of interaction of the hypothalamus and the carotid sinus mechanoreceptor system on renal hemodynamics in the anesthetized dog. *Circulation Research*, 1971, *29*, 569–585.

Barfield, R. J. Activation of estrous behavior by intracerebral implants of estradiol benzoate (EB) in ovariectomized rats.. *Federation Proceedings*, 1976, *35*, 429.

Barfield, R., Ronay, G., and Pfaff, D. W. Autoradiographic localization of androgen-concentrating cells in the chicken brain. *Neuroendocrinology*, 1977, *26*, 297–311.

Barraclough, C. A., and Gorski, R. A. Evidence that the hypothalamus is responsible for androgen-induced sterility in the female rat. *Endocrinology*, 1961, *68*, 68–76.

Basbaum, A. I., Clanton, C. H., and Fields, H. L. Ascending projections of nucleus gigantocellularis and nucleus raphe magnus in the cat: An autoradiographic study. *Anatomical Record*, 1976, *184*, 354.

Bayliss, W. M. *Principles of General Physiology* (3rd ed.). London: Putnam, 1920.

Beach, F. A. Importance of progesterone to induction of sexual receptivity in spayed female rats. *Proceedings of the Society for Experimental Biology and Medicine*, 1942, *51*, 369–371.

Beach, F. A. Effects of injury to the cerebral cortex upon sexually receptive behavior in the female rat. *Psychosomatic Medicine*, 1944, *6*, 40–55.

Beach, F. A. *Hormones and Behavior*. New York: Hoeber, 1948.

481

NEURAL
MECHANISMS OF
FEMALE
REPRODUCTIVE
BEHAVIOR

Brawer, J. R., and Sonnenschein, C. Cytopathological effects of estradiol on the arcuate nucleus of the female rat. A possible mechanism for pituitary tumorigenesis. *American Journal of Anatomy,* 1975, *144,* 57–88.

Brink, E., and Pfaff, D. W. Anatomy of epaxial deep back muscles in the rat. *Brain, Behavior and Evolution,* 1977, *17,* 1–47.

Brink, E., Modianos, D., and Pfaff, D. W. Effects of epaxial deep back muscles on lordosis behavior in the female rat. *Brain, Behavior and Evolution,* 1977, *17,* 67–88.

Brodal, A. *Neurological Anatomy.* New York: Oxford University Press, 1969.

Brodal, A., Pompeiano, O., and Walberg, F. *The Vestibular Nuclei and Their Connections, Anatomy and Functional Correlations.* Springfield, Ill.: Thomas, 1962.

Brookhart, J. M., Dey, F. L., and Ranson, S. W. Failure of ovarian hormones to cause mating reactions in spayed guinea pigs with hypothalamic lesions. *Proceedings of the Society for Experimental Biology and Medicine,* 1940, *44,* 6–64.

Brookhart, J. M., Dey, F. L., and Ranson, S. W. The abolition of mating behavior by hypothalamic lesions in guinea pigs. *Endocrinology,* 1941, *28,* 561–565.

Brown, L. T. Rubrospinal projections in the rat. *Journal of Comparative Neurology,* 1974, *154,* 169–188.

Bueno, J., and Pfaff, D. W. Single unit recording in hypothalamus and preoptic area of estrogen-treated and untreated ovariectomized female rats. *Brain Research,* 1976, *101,* 67–78.

Buller, R. E., and O'Malley, B. W. The biology and mechanism of steroid hormone interactions with the eukaryotic nucleus. *Biochemical Pharmacology,* 1976, *25,* 1–12.

Burgess, P. R., and Perl, G. R. Cutaneous mechanoreceptors and nociceptors. In A. Iggo (Ed.), *Somatosensory Systems: Handbook of Sensory Physiology,* Vol. 2. New York: Springer-Verlag, 1973.

Caggiula, A. R. Analysis of the copulation-reward properties of posterior hypothalamic stimulation in male rats. *Journal of Comparative Physiology and Psychology,* 1970, *70,* 399–412.

Caggiula, A. R., and Hoebel, B. G. "Copulation-reward site" in the posterior hypothalamus. *Science,* 1966, *153,* 1284–1285.

Caggiula, A. R., Antelman, S. M., and Zigmond, M. J. Disruption of copulation in male rats after hypothalamic lesions: a behavioral, anatomical and neurochemical analysis. *Brain Research,* 1973, *59,* 273–287.

Caggiula, A. R., Gay, V. L., Antelman, S. M., and Leggens, J. Disruption of copulation in male rats after hypothalamic lesions: A neuroendocrine analysis. *Neuroendocrinology,* 1975, *17,* 193–202.

Calaresu, F. R. Central pathways of integration of cardiovascular responses. *XXVI International Congress of Physiological Sciences,* New Delhi, 1974.

Capony, E., and Rochefort, H. In vivo effect of anti-estrogens on the localisation and replenishment of estrogen receptor. *Molecular and Cellular Biochemistry,* 1975, *3,* 233–251.

Carrer, H., Asch, G., and Aron, C. New facts concerning the role played by the ventromedial nucleus in the control of estrous cycle duration and sexual receptivity in the rat. *Neuroendocrinology,* 1973, *13,* 129–138.

Chan, S. H. H., and Barnes, C. D. A presynaptic mechanism evoked from brain stem reticular formation in the lumbar cord and its temporal significance. *Brain Research,* 1972, *45,* 101–114.

Chazal, G., Faudon, M., Gogan, E., and Rotsztejn, W. Effects of two estradiol antagonists upon the estradiol uptake in the rat brain and peripheral tissues. *Brain Research,* 1975, *89,* 245–254.

Chen, H. I., and Chai, C. Y. Pulmonary edema and hemorrhage as a consequence of systemic vasoconstriction. *American Journal of Physiology,* 1974, *227,* 144–151.

Chi, C. C. An experimental silver study of the ascending projections of the central gray substance and adjacent tegmentum in the rat with observations in the cat. *Journal of Comparative Neurology,* 1970, *139,* 259–272.

Christensen, L. W., and Gorski, R. A. Sites of neonatal gonadal steroid action in hypothalamic sexual differentiation. *V International Congress of Endocrinology,* Hamburg, 1976, Abstract 217.

Ciaccio, L. A., and Lisk, R. D. Central control of estrous behavior in the female golden hamster. *Neuroendocrinology,* 1973, *13,* 21–28.

Ciriello, J., Calaresu, F. R., and Mogenson, G. J. Neural pathways mediating cardiovascular responses elicited by stimulation of the septum in the rat. *Society for Neuroscience,* 1975, Abstract 655.

Clark, G. Sexual behavior in rats with lesions in the anterior hypothalamus. *American Journal of Physiology,* 1942, *137,* 746–749.

Clegg, M. T., Santolucito, J. A., Smith, J. D., and Ganong, W. F. The effect of hypothalamic lesions on sexual behavior and estrous cycles in the ewe. *Endocrinology,* 1958, *62,* 790–797.

Clemens, L. G., Wallen, K., and Gorski, R. A. Mating behavior: facilitation in the female rat after cortical application of potassium chloride. *Science,* 1969, *157,* 1208–1209.

Clemens, J. A., Smalstig, E. B., and Sawyer, B. D. Studies on the role of the preoptic area in the control of reproductive function in the rat. *Endocrinology,* 1976, *99,* 728–735.

Conrad, L. A., and Pfaff, D. W. Axonal projections of medial preoptic and anterior hypothalamic neurons. *Science,* 1975, *190,* 112–114.

Conrad, L. C. A., and Pfaff, D. W. Autoradiographic study of efferents from medial basal forebrain and hypothalamus in the rat: I. Medial preoptic area. *Journal of Comparative Neurology,* 1976a, *169,* 185–220.

Conrad, L. C. A., and Pfaff, D. W. Autoradiographic study of efferents from medial basal forebrain and hypothalamus in the rat: II. Medial anterior hypothalamus. *Journal of Comparative Neurology,* 1976, *169,* 221–262.

Conrad, L. C. A., and Pfaff, D. W. Hypothalamic neuroanatomy: Steroid hormone binding and axonal projections. In G. Bourne (Ed.), *International Review of Cytology.* New York: Academic Press, 1977.

Conrad, L., Leonard, C., and Pfaff, D. W. Connections of the median and dorsal raphe nuclei in the rat: an autoradiographic and degeneration study. *Journal of Comparative Neurology,* 1974, *156,* 179–206.

Cook, W. A., Jr., Cangiano, A., and Pompeiano, O. Dorsal root potentials in the lumbar cord evoked from the vestibular system. *Archives Italiennes de Biologie,* 1969a, *107,* 275–295.

Cook, W. A., Jr., Cangiano, A., and Pompeiano, O. Vestibular control of transmission in primary afferents to the lumbar spinal cord. *Archives Italiennes de Biologie,* 1969b, *107,* 296–320.

Cowan, W. M., Gottelieb, D. I., Hendrickson, A. E., Price, J. L., and Woolsey, T. A. The autoradiographic demonstration of axonal connections in the central nervous system. *Brain Research,* 1972, *37,* 21–51.

Creed, R. S., Denny-Brown, D., Eccles, J. C., Liddell, E. G. T., and Sherrington, C. S. *Reflex Activity of the Spinal Cord.* Oxford: Clarendon Press, 1932.

Cross, B. A., and Dyer, R. G. Cyclic changes in neurons of the anterior hypothalamus during the rat estrous cycle and the effect of anesthesia. In C. H. Sawyer and R. A. Gorski (Eds.), *Steroid Hormones and Brain Function, UCLA Forum in Medical Sciences, No. 15.* Los Angeles: University of California Press, 1971.

Cross, B. A., and Dyer, R. G. Ovarian modulation of unit activity in the anterior hypothalamus of the cyclic rat. *Journal of Physiology* (London), 1972, *222,* 25P.

Davidson, J. M. Activation of the male rat's sexual behavior by intracerebral implantation of androgen. *Endocrinology,* 1966, *79,* 783–794.

Davidson, J. M., Rodgers, C. H., Smith, E. R., and Bloch, G. J. Stimulation of female sex behavior in adrenalectomized rats with estrogen alone. *Endocrinology,* 1968, *82,* 193–195.

Dempsey, E. W., and Rioch, D. McK. The localization in the brain stem of the oestrous responses of the female guinea pig. *Journal of Neurophysiology,* 1939, *2,* 9–18.

Dey, F. L. Changes in ovaries and uteri in guinea pigs with hypothalamic lesions. *American Journal of Anatomy,* 1941, *69,* 61–87.

Dey, F. L., Fisher, C., Berry, C. M., and Ranson, S. W. Disturbances in reproductive functions caused by hypothalamic lesions in female guinea pigs. *American Journal of Physiology,* 1940, *129,* 39–46.

Dey, F. L., Leininger, C. R., and Ranson, S. W. The effect of hypophysial lesions on mating behavior in female guinea pigs. *Endocrinology,* 1942, *30,* 323–326.

Diakow, C., Pfaff, D. W., and Komisaruk, B. Sensory and hormonal interactions in eliciting lordosis. *Federation Proceedings,* 1973, *32,* 241 (abstr.).

Dilly, P. N., Wall, P. D., and Webster, K. E. Cells of origin of the spinothalamic tract in the cat and rat. *Experimental Neurology,* 1968, *21,* 550–562.

Djojosugito, A. M., Folkow, B., Kylstra, P. H., Lisander, B., and Tuttle, R. S. Differentiated interactions between the hypothalamic defence reaction and baroreceptor reflexes. I. Effects on heart rate and regional flow resistance. *Physiologica Scandinavica,* 1970, *78,* 376–385.

Domanski, E., Przekop, F., and Skubiszewski, B. The role of the anterior regions of the medial basal hypothalamus in the control of ovulation and sexual behavior in sheep. *Neurobiologiae Experimentalis* (Warsaw), 1972, *32,* 753–762.

Dörner, G., Döcke, F., and Moustafa, S. Differential localization of a male and a female hypothalamic mating centre. *Journal of Reproduction and Fertility,* 1968, *17,* 583–586.

Dörner, G., Döcke, F., and Hinz, G. Homo- and hypersexuality in rats with hypothalamic lesions. *Neuroendocrinology,* 1969, *4,* 20–24.

483

NEURAL
MECHANISMS OF
FEMALE
REPRODUCTIVE
BEHAVIOR

Dörner, G., Döcke, F., and Gotz, F. Male-like sexual behaviour of female rats with unilateral lesions in the hypothalamic ventromedial nuclear region. *Endokrinologie*, 1975, *65*, 133–137.

Dyer, R. G. An electrophysiological dissection of the hypothalamic regions which regulate the pre-ovulatory secretion of luteinizing hormone in the rat. *Journal of Physiology* (London), 1973, *234*, 421–442.

Dyer, R. G., and Burnet, F. Effects of ferrous ions on preoptic area neurons and luteinizing hormone secretion in the rat. *Journal of Endocrinology*, 1976, *69*, 247–254.

Dyer, R. G., Pritchett, C. J., and Cross, B. A. Unit activity in the diencephalon of female rats during the oestrous cycle. *Journal of Endocrinology*, 1972, *53*, 151–160.

Ebbesson, S. O. E. Ascending axon degeneration following hemisection of the spinal cord in the tegu lizard *(Tupinambis nigropunctatus)*. *Brain Research*, 1967, *5*, 178–206.

Ebbesson, S. O. E. Brain stem afferents from the spinal cord in a sample of reptilian and amphibian species. *Annals of the New York Academy of Science*, 1969, *167*, 80–101.

Eccles, R. M., and Lundberg, A. Supraspinal control in motoneurones mediating spinal reflexes. *Journal of Physiology* (London), 1959, *147*, 565–584.

Eccles, J. C., Nicoll, R. A., Schwarz, D. W. F., Taborikova, H., and Willey, T. J. Reticulospinal neurons with and without monosynaptic inputs from cerebellar nuclei. *Journal of Neurophysiology*, 1975a, *38*, 513–530.

Eccles, J. C., Nicoll, R. A., Taborikova, H., and Willey, T. J. Medial reticular neurons projecting rostrally. *Journal of Neurophysiology*, 1975b, *38*, 531–538.

Eccles, J. C., Nicoll, R. A., Rantucci, T., Taborikova, H., and Willey, T. J. Topographic studies on medial reticular nucleus. *Journal of Neurophysiology*, 1976, *39*, 109–118.

Edwards, D. A., and Mathews, D. The ventromedial nucleus of the hypothalamus and the hormonal arousal of sexual behaviors in the female rat. *Physiology and Behavior*, 1977, *14*, 439–453.

Edwards, D. A., and Warner, P. Olfactory bulb removal facilitates the hormonal induction of sexual receptivity in the female rat. *Hormones and Behavior*, 1972, *3*, 321–332.

Edwards, S. B. Autoradiographic studies of the projections of the midbrain reticular formation: Descending projections of nucleus cuneiformis. *Journal of Comparative Neurology*, 1975, *161*, 341–358.

Edwards, S. B., and de Olmos, J. S. Autoradiographic studies of the projections of the midbrain reticular formation: Ascending projections of nucleus cuneiformis. *Journal of Comparative Neurology*, 1976, *165*, 417–432.

Eibergen, R. D., and Caggiula, A. R. Ventral midbrain involvement in copulatory behavior of the male rat. *Physiology and Behavior*, 1973, *10*, 435–442.

Eisenfeld, A. J., and Axelrod, J. Selectivity of estrogen distribution in tissues. *Journal of Pharmacology and Experimental Therapeutics*, 1965, *150*, 469–475.

Eisenfeld, A. J., and Axelrod, J. Effect of steroid hormones, ovariectomy, estrogen pretreatment, sex and immaturity on the distribution of ^{3}H-estradiol. *Endocrinology*, 1966, *79*, 38–42.

Everett, J. W. Neuroendocrine aspects of mammalian reproduction. *Annual Review of Physiology*, 1969, *31*, 383–416.

Fearing, F. *Reflex Action*. Baltimore: Williams and Wilkins, 1930.

Floody, O., and Pfaff, D. W. Steroid hormones and aggressive behavior: Approaches to the study of hormone-sensitive brain mechanisms for behavior. *Research Publications, Association for Research in Nervous and Mental Disease*, 1974, *52*, 149–185.

Floody, O. R., and Pfaff, D. W. Communication among hamsters by high-frequency acoustic signals: I. Physical characteristics of hamster cells. *Journal of Comparative Physiology and Psychology*, 1977a, *91*, 794–806.

Floody, O. R., and Pfaff, D. W. Communication among hamsters by high-frequency acoustic signals: III. Responses evoked by natural and synthetic ultrasounds. *Journal of Comparative Physiology and Psychology*, 1977b, *91*, 820–829.

Foreman, R. D., Hancock, M. B., and Willis, W. D. Convergence of visceral and cutaneous input onto spinothalamic tract neurons in the thoracic spinal cord of the rhesus monkey. *Society for Neuroscience*, 1975, Abstract 233.

Fox, J. E. Reticulospinal neurones in the rat. *Brain Research*, 1970, *23*, 35–40.

Gerlach, J., McEwen, B., Pfaff, D., Moskovitz, S., Ferin, M., Carmel, P., and Zimmerman, E. Cells in regions of rhesus monkey brain and pituitary retain radioactive estradiol, corticosterone and cortisol differentially. *Brain Research*, 1976, *103*, 603–612.

Goldberg, J. M., and Moore, R. Y. Ascending projections of the lateral lemniscus in the cat and monkey. *Journal of Comparative Neurology*, 1967, *129*, 143–156.

Goy, R. W., and Phoenix, C. H. Hypothalamic regulation of female sexual behavior: Establishment of behavioural oestrus in spayed guinea-pigs following hypothalamic lesions. *Journal of Reproduction and Fertility*, 1963, *5*, 23–40.

Graybiel, A. M. The thalamo-cortical projection of the so-called posterior nuclear group: a study with anterograde degeneration methods in the cat. *Brain Research*, 1973, *49*, 229–244.

Green, R., Luttge, W. G., and Whalen, R. E. Induction of receptivity in ovariectomized female rats by a single intravenous injection of estradiol-17P. *Physiology and Behavior*, 1970, *5*, 137–141.

Greer, M. A. The effect of progesterone on persistent vaginal estrus produced by hypothalamic lesions in the rat. *Endocrinology*, 1953, *53*, 380–390.

Grillner, S., Hongo, T., and Lund, S. The origin of descending fibres monosynaptically activating spinoreticular neurones. *Brain Research*, 1968, *10*, 259–262.

Grillner, S., Hongo, T., and Lund, S. The vestibulospinal tract. Effects on alpha-motoneurones in the lumbosacral spinal cord in the cat. *Experimental Brain Research*, 1970, *10*, 94–120.

Grillner, S., Hongo, T., and Lund, S. Convergent effects on alpha motoneurones from the vestibulospinal tract and a pathway descending in the medial longitudinal fasciculus. *Experimental Brain Research*, 1971, *12*, 457–479.

Hagamen, W. D., and Brooks, D. C. Sexual behavior of female cats following lesions of the ventromedial nucleus of the hypothalamus. *Anatomical Record*, 1958, *130*, 414 (abst. 348).

Halasz, B. The endocrine effects of isolation of the hypothalamus from the rest of the brain. In W. F. Ganong and L. Martini (Eds.), *Frontiers in Neuroendocrinology, 1969*. New York: Oxford University Press, 1969.

Halasz, B., and Gorski, R. A. Gonadotrophic hormone secretion in female rats after partial or total interruption of neural afferents to the medial basal hypothalamus. *Endocrinology*, 1967, *80*, 608–622.

Hamilton, B. L. Projections of the subnuclear areas of the periaqueductal gray matter in the cat. *Society for Neuroscience*, 1972, Abstract 112, p. 148.

Hamilton, B. L., and Skultety, F. M. Efferent connections of the periaqueductal gray matter in the cat. *Journal of Comparative Neurology*, 1970, *139*, 105–114.

Hart, B. L. Gonadal hormones and sexual reflexes in the female rat. *Hormones and Behavior*, 1969, *1*, 65–71.

Hassen, A. H., and Barnes, C. D. Bilateral effects of vestibular nerve stimulation on activity in the lumbar spinal cord. *Brain Research*, 1975, *90*, 221–233.

Hayle, T. H. A comparative study of spinal projections to the brain (except cerebellum) in three classes of poikilothermic vertebrates. *Journal of Comparative Neurology*, 1973, 149, 463–476.

Heath, C. J., and Jones, E. G. An experimental study of ascending connections from the posterior group of thalamic nuclei in the cat. *Journal of Comparative Neurology*, 1971, *141*, 397–426.

Herndon, J. G., and Neill, D. B. Amphetamine reversal of sexual impairment following anterior hypothalamic lesions in female rats. *Pharmacology, Biochemistry and Behavior*, 1973, *1*, 285–288.

Herrick, C. J., and Bishop, G. H. A comparative survey of the spinal lemniscus systems. In H. H. Jasper, L. D. Proctor, R. S. Knighton, W. C. Noshay, and R. T. Costello (Eds.), *Reticular Formation of the Brain*. Boston: Little, Brown, 1958.

Hess, W. R. *Diencephalon: Autonomic and Extrapyramidal Functions*. New York: Grune and Stratton, 1954.

Hess, W. R. *The Functional Organization of the Diencephalon*. New York: Grune and Stratton, 1957.

Hilton, S. M., and Spyer, K. M. Participation of the anterior hypothalamus in the baroreceptor reflex. *Journal of Physiology* (London), 1971, *218*, 271–293.

Hitt, J. C., Hendricks, S. E., Ginsberg, S. I., and Lewis, J. H. Disruption of male, but not female, sexual behavior in rats by medial forebrain bundle lesions. *Journal of Comparative Physiology and Psychology*, 1970, *73*, 377–384.

Hitt, J. C., Bryon, D. M., and Modianos, D. T. Effects of rostral medial forebrain bundle and olfactory tubercle lesions upon sexual behavior of male rats. *Journal of Comparative Physiology and Psychology*, 1973, *82*, 30–36.

Holmqvist, B., Lundberg, A., and Oscarsson, O. A supraspinal control system monosynaptically connected with an ascending spinal pathway. *Archives Italiennes de Biologie*, 1960a, *98*, 402–422.

485

NEURAL
MECHANISMS OF
FEMALE
REPRODUCTIVE
BEHAVIOR

Holmqvist, B., Lundberg, A., and Oscarsson, O. Supraspinal inhibitory control of transmission to three ascending spinal pathways influenced by the flexion reflex afferents. *Archives Italiennes de Biologie,* 1960b, *98,* 60–80.

Hongo, T., Kudo, N., and Tanaka, R. The vestibulospinal tract: crossed and uncrossed effects on hindlimb motoneurones in the cat. *Experimental Brain Research,* 1975, *24,* 37–55.

Horn, G., and Hill, R. M. Responsiveness to sensory stimulation of units in the superior colliculus and subjacent tecto-tegmental regions of the rabbit. *Experimental Neurology,* 1966, *14,* 199–223.

Hornby, J. B., and Rose, J. D. Responses of caudal brain stem neurons to vaginal and somatosensory stimulation in the rat and evidence of genital-nociceptive interactions. *Experimental Neurology,* 1976, *51,* 363–376.

Hough, J. C., Jr., H., G. K-W, Cooke, P. H., and Quadagno, D. M. Actinomycin D: Reversible inhibition of lordosis behavior and correlated changes in nucleolar morphology. *Hormones and Behavior,* 1974, *5,* 367–375.

James, W. *The Principles of Psychology,* Vol. 1. New York: Holt, 1890. (Dover edition, 1950).

Job, C. Über autogene Inhibition und Reflexumkehr bei spinalisierten und decerebrierten Katzen. *Pflügers Archiv, European Journal of Physiology,* 1953, *256,* 406–418.

Johnston, P., and Davidson, J. M. Intracerebral androgens and sexual behavior in the male rat. *Hormones and Behavior,* 1972, *3,* 345–357.

Kalra, S. P., and Sawyer, C. H. Blockade of copulation-induced ovulation in the rat by anterior hypothalamic deafferentation. *Endocrinology,* 1970, *87,* 1124–1128.

Kato, J., and Villee, C. A. Preferential uptake of estradiol by the anterior hypothalamus of the rat. *Endocrinology,* 1967a, *80,* 567–575.

Kato, J., and Villee, C. A. Factors affecting uptake of estradiol-6,7-^{3}H by the hypophysis and hypothalamus. *Endocrinology,* 1967b, *80,* 1113–1138.

Katzenellenbogen, B. S., and Ferguson, E. R. Antiestrogen action in the uterus: Biological ineffectiveness of nuclear bound estradiol after antiestrogen. *Endocrinology,* 1975, *97,* 1–12.

Kawakami, M., Terasawa, E., and Ibuki, T. Changes in multiple unit activity of the brain during the estrous cycle. *Neuroendocrinology,* 1970, *6,* 30–48.

Kawakami, M., Terasawa, E., Ibuki, T., and Manaka, M. Effects of sex hormones and ovulation-blocking steroids and drugs on electrical activity of the rat brain. In C. H. Sawyer and R. A. Gorski (Eds.), *Steroid Hormones and Brain Function, UCLA Forum in Medical Sciences,* No. 15. Los Angeles: University of California Press, 1971.

Kelley, D. B., and Pfaff, D. W. Locations of steroid hormone-concentrating cells in the central nervous system of *Rana pipiens. Society for Neuroscience,* 1975, Abstract 681.

Kelley, D. B., and Pfaff, D. W. Generalizations from comparative studies on neuroanatomical and endocrine mechanisms for sex behavior. In J. Hutchison (Ed.), *Biological Determinants of Sexual Behavior.* Chichester, England: Wiley, 1977.

Kelley, D. B., Morrell, J. I., and Pfaff, D. W. Autoradiographic localization of hormone-concentrating cells in the brain of an amphibian, *Xenopus laevis:* I. Testosterone. *Journal of Comparative Neurology,* 1975, *164,* 47–62.

Kelly, M. J., Moss, R. L., and Dudley, C. A. Differential sensitivity of preoptic-septal neurons to microelectrophoresed estrogen during the estrous cycle. *Brain Research,* 1976, *114,* 152–157.

Kennedy, G. C., Hypothalamic control of the endocrine and behavioural changes associated with oestrus in the rat. *Journal of Physiology* (London), 1964, *172,* 383–392.

Kennedy, G. C., and Mitra, J. Hypothalamic control of energy balance and the reproductive cycle in the rat. *Journal of Physiology* (London), 1963, *166,* 395–407.

Koizumi, K., Nishino, H., and Colman, D. The suprachiasmatic nuclei and circadian rhythms. *Society for Neuroscience,* 1975, Abstract 692.

Komisaruk, B. R., and Diakow, C. Lordosis reflex intensity in rats in relation to the estrous cycle, ovariectomy, estrogen administration and mating behavior. *Endocrinology,* 1973, *93,* 548–557.

Komisaruk, B. R., Adler, N. T., and Hutchison, J. Genital sensory field: enlargement by estrogen treatment in female rats. *Science,* 1972, *178,* 1295–1298.

Kow, L.-M., and Pfaff, D. W. Effects of estrogen treatment on the size of receptive field and response threshold of pudendal nerve in the female rat. *Neuroendocrinology,* 1973, *13,* 299–313.

Kow, L.-M., and Pfaff, D. W. Dorsal root recording relevant for mating reflexes in female rats: Identification of receptive fields and effects of peripheral denervation. *Journal of Neurobiology,* 1975a, *6,* 23–37.

Kow, L.-M., and Pfaff, D. W. Induction of lordosis in female rats: Two modes of estrogen action and the effect of adrenalectomy. *Hormones and Behavior*, 1975b, *6*, 259–276.

Kow, L.-M., and Pfaff, D. W. Sensory control of reproductive behavior in female rodents. In B. Wenzel and H. P. Zeigler (Eds.), *Tonic Functions of Sensory Systems. Annals, N.Y. Academy of Sciences*, 1976a, *290*, 72–97.

Kow, L.-M., and Pfaff, D. W. Sensory requirements for the lordosis reflex in female rats. *Brain Research*, 1976a, *101*, 47–66.

Kow, L.-M., Montgomery, M., and Pfaff, D. W. Spinal tract transections and the lordosis reflex in female rats. *Physiologist*, 1973, *16*, 367 (abstr.).

Kow, L.-M., Malsbury, C. W., and Pfaff, D. W. Effects of medial hypothalamic lesions on the lordosis response in female hamsters. *Proceedings of the Society for Neuroscience*, 1974a, Abstract 365.

Kow, L.-M., Malsbury, C., and Pfaff, D. W. Effects of progesterone on female reproductive behavior in rats: Possible modes of action and role in behavioral sex differences. In W. Montagna and W. Sadler (Eds.), *Reproductive Behavior*. New York: Plenum Press, 1974b.

Kow, L.-M., Malsbury, C., and Pfaff, D. W. Lordosis in the male golden hamster elicited by manual stimulation: Characteristics and hormonal sensitivity. *Journal of Comparative Physiology and Psychology*, 1976, *90*, 26–40.

Kow, L.-M., Grill, H., and Pfaff, D. W. Absence of lordosis in decerebrate female rats. *Physiology and Behavior*, 1977a, *19*.

Kow, L.-M., Montgomery, M., and Pfaff, D. W. Effects of spinal cord transections on lordosis reflex in female rats. *Brain Research*, 1977b, *123*, 75–88.

Kow, L.-M., Montgomery, M., and Pfaff, D. W. Quantitative characteristics of somatosensory stimuli triggering lordosis in female rats. *Journal of Neurophysiology*, 1977c, *42*, 195–202.

Krieger, M., Conrad, L. C. A., and Pfaff, D. W. Axonal projections of neurons of ventromedial nucleus of the hypothalamus. *Anatomical Record*, 1977a, *187*, 770.

Krieger, M., Morrell, J., and Pfaff, D. W. Locations of estrogen-concentrating cells in the brain of the female hamster. *Neuroendocrinology*, 1977b, *26*.

Kubo, K., Mennin, S. P., and Gorski, R. A. Similarity of plasma LH release in androgenized and normal rats following electrochemical stimulation of the basal forebrain. *Endocrinology*, 1975, *96*, 492–500.

Ladpli, R., and Brodal, A. Experimental studies of commissural and reticular formation projections from the vestibular nuclei in the cat. *Brain Research*, 1968, *8*, 65–96.

Lan, N. C., and Katzenellenbogen, B. S. Temporal relationships between hormone receptor binding and biological responses in the uterus: Studies with short -and long-acting derivatives of estriol. *Endocrinology*, 1976, *98*, 220–227.

Land, L. J., Reese, B. A., and Whitlock, D. G. Ascending pathways in the anterolateral funiculus of the rat spinal cord. *Society for Neuroscience*, 1975, Abstract 1058, p. 685.

LaVaque, T. J., and Rodgers, C. H. Effects of ventromedial hypothalamic lesions upon mating behavior in the female rat. *Federation Proceedings*, 1974, *33*, 232.

LaVaque, T. J., and Rodgers, C. H. Recovery of mating behavior in the female rat following VMH lesions. *Physiology and Behavior*, 1975, *14*, 59–63.

Law, T., and Meagher, W. Hypothalamic lesions and sexual behavior in the female rat. *Science*, 1958, *128*, 1626–1627.

Lawrence, D. G., and Kuypers, H. G. J. M. The functional organization of the motor system in the monkey: I. The effects of bilateral pyramidal lesions. *Brain*, 1968a, *91*, 1–14.

Lawrence, D. G., and Kuypers, H. G. J. M. The functional organization of the motor system of the monkey: II. The effects of lesions of the descending brain-stem pathways. *Brain*, 1968b, *91*, 15–36.

Liebeskind, J. C., and Mayer, D. J. Somatosensory evoked responses in the mesencephalic central gray matter of the rat. *Brain Research*, 1971, *27*, 133–151.

Lincoln, D. W. Unit activity in the hypothalamus, septum and preoptic area of the rat: Characteristics of spontaneous activity and the effect of oestrogen. *Journal of Endocrinology*, 1967, *37*, 177–189.

Lincoln, D. W., and Cross, B. A. Effect of oestrogen on the responsiveness of neurones in the hypothalamus, septum and preoptic area of rats with light-induced persistent oestrus. *Journal of Endocrinology*, 1967, *37*, 191–203.

Lindsley, D. B. Brain stem influences on spinal motor activity. *Research Publications, Association for Research in Nervous and Mental Disease*, 1952, *30*, 174–195.

487

NEURAL
MECHANISMS OF
FEMALE
REPRODUCTIVE
BEHAVIOR

Lindsley, D. B., Schreiner, L. H., and Magoun, H. W. An electromyographic study of spasticity. *Journal of Neurophysiology,* 1949, *12,* 197–205.

Lisk, R. D. Diencephalic placement of estradiol and sexual receptivity in the female rat. *American Journal of Physiology,* 1962, *203,* 493–496.

Lisk, R. D. Neural localization for androgen activation of copulatory behavior in the male rat. *Endocrinology,* 1967, *80,* 754–761.

Lisk, R. D., and Barfield, M. A. Progesterone facilitation of sexual receptivity in rats with neural implantation of estrogen. *Neuroendocrinology,* 1975, *19,* 28–35.

Lisk, R. D., Ciaccio, L. A., and Reuter, L. A. Neural centers of estrogen and progesterone action in the regulation of reproduction. In J. T. Velardo and B. A. Kasprow (Eds.), *Biology of Reproduction—Basic and Clinical Studies.* 1972.

Loeb, J. *Comparative Physiology of the Brain and Comparative Psychology.* New York: Putnam, 1899.

Loeb, J. *The Mechanistic Conception of Life.* Chicago: University of Chicago Press, 1912.

Luine, V. N., and McEwen, B. S. Effects of an estrogen antagonist on enzyme activities and [^{3}H]-estradiol nuclear binding in uterus, pituitary and brain. *Endocrinology,* 1977, *100,* 386–394.

Luine, V. N., Khylchevskaya, R. I., and McEwen, B. S. Oestrogen effects on brain and pituitary enzyme activities. *Journal of Neurochemistry,* 1974, *23,* 925–934.

Luine, V. N., Khylchevskaya, R. I., and McEwen, B. S. Effect of gonadal hormones on enzyme activities in brain and pituitary of male and female rats. *Brain Research,* 1975a, *86,* 283–292.

Luine, V. N., Khylchevskaya, R. I., and McEwen, B. S. Effect of gonadal steroids on activities of monomine oxidase and choline acetylase in rat brain. *Brain Research,* 1975b, *86,* 293–306.

Lund, R. D., and Webster, K. E. Thalamic afferents from the spinal cord and trigeminal nuclei. An experimental anatomical study in the rat. *Journal of Comparative Neurology,* 1967, *130,* 313–328.

Lund, S., and Pompeiano, O. Monsynaptic excitation of alpha motoneurones from supraspinal structures in the cat. *Acta Physiologica Scandinavica,* 1968, *73,* 1–21.

Lundberg, A., and Oscarsson, O. Two ascending spinal pathways in the ventral part of the cord. *Acta Physiologica Scandinavica,* 1962, *54,* 270–286.

Luttge, W. G., Gray, H. E., and Hughes, J. R. Regional and subcellular [^{3}H] estradiol localization in selected brain regions and pituitary of female mice: effects of unlabelled estradiol and various antihormones. *Brain Research,* 1976, *104,* 273–281.

Magoun, H. W. Caudal and cephalic influences of the brain stem reticular formation. *Physiological Review,* 1950, *30,* 459–474.

Magoun, H. W. *The Waking Brain* (2nd ed.). Springfield, Ill.: Charles C Thomas, 1963.

Magoun, H. W., and Rhines, R. An inhibitory mechanism in the bulbar reticular formation. *Journal of Neurophysiology,* 1946, *9,* 165–171.

Magoun, H. W., and Rhines, R. *Spasticity: The Stretch Reflex and Extrapyramidal Systems.* Springfield, Ill.: Charles C Thomas, 1947.

Malsbury, C. W. Facilitation of male rat copulatory behavior by electrical stimulation of the medial preoptic area. *Physiology and Behavior,* 1971, *7,* 797–805.

Malsbury, C., and Pfaff, D. W. Suppression of sexual receptivity in the hormone-primed female hamster by electrical stimulation of the medial preoptic area. *Proceedings, Society for Neuroscience,* 1973, Abstract.

Malsbury, C., and Pfaff, D. W. Neural and hormonal determinants of mating behavior in adult male rats: A review. In L. DiCara (Ed.), *Limbic and Autonomic Nervous Systems Research.* New York: Plenum Press, 1974.

Malsbury, C., and Pfaff, D. W. Suppression of lordosis in female hamsters by electrical stimulation of the medial preoptic area. *Brain Research,* 1977, *181,* 267–284.

Malsbury, C., Kelley, D. B., and Pfaff, D. W. Responses of single units in the dorsal midbrain to somatonsensory stimulation in female rats. In C. Gaul (Ed.), *Progress in Endocrinology, Proc. IV International Congress Endocrinology.* Excerpta Medica International Congress Series #273. 1972.

Malsbury, C., Kow, L.-M., and Pfaff, D. W. Effects of medial hypothalamic lesions on lordosis and other behaviors in female hamsters. *Physiology and Behavior,* 1977, *19,* 223–237.

Mancia, M. Mechanisms in EEG synchronization and desynchronization. *XXVI International Congress of Physiological Sciences,* New Delhi, 1974, Abstract.

Martin, J. R. Motivated behaviors elicited from hypothalamus, midbrain, and pons of the guinea pig (*Cavia porcellus*). *Journal of Comparative Physiology and Psychology,* 1976, *90,* 1011–1034.

Masland, R. H., Chow, K. L., and Stewart, D. L. Receptive-field characteristics of superior colliculus neurons in the rabbit. *Journal of Neurophysiology*, 1971, *34*, 148–156.

McEwen, B. S., Denef, C. J., Gerlach, J. L., and Plapinger, L. Chemical studies of the brain as a steroid hormone target tissue. In F. O. Schmitt and F. G. Worden (Eds.), *The Neurosciences, Third Study Program*. Cambridge, Mass.: M.I.T. Press, 1974.

McEwen, B. S., Pfaff, D. W., Chaptal, C., and Luine, V. N. Brain cell nuclear retention of [³H] estradiol doses able to promote lordosis: Temporal and regional aspects. *Brain Research*, 1975, *86*, 155–161.

Mehler, W. R. Some neurological species differences—A posteriori. *Annals of the New York Academy of Science*, 1969, *167*, 424–468.

Meyer, D. L. Historical survey of the tonic hypothesis. In B. Wenzel and P. Ziegler (Eds.), *Tonic Functions of Sensory Systems. Annals, New York Academy of Science*, 1977.

Meyer, M., LaPlante, E. S., and Campbell, B. Ascending sensory pathways from the genitalia of the cat. *Experimental Neurology*, 1960, *2*, 186–190.

Meyerson, B. J., and Lindstrom, L. Effect of an oestrogen antagonist ethamoxytriphetol [MER-25] on oestrus behavior in rats. *Acta Endocrinologica*, 1968, *59*, 41–48.

Modianos, D., and Pfaff, D. W. Facilitation of the lordosis reflex by electrical stimulation of the lateral vestibular nucleus. *Proceedings of the Society for Neuroscience*, 1975, Abstract 710.

Modianos, D., and Pfaff, D. W. Brain stem and cerebellar lesions in female rats. I. Tests of posture and movement. *Brain Research*, 1976a, *106*, 31–46.

Modianos, D., and Pfaff, D. W. Brain stem and cerebellar lesions in female rats. II. Lordosis reflex. *Brain Research*, 1976b, *106*, 47–56.

Modianos, D., and Pfaff, D. W. Facilitation of the lordosis reflex by electrical stimulation of the lateral vestibular nucleus. *Brain Research*, 1977a, *134*, 333–339.

Modianos, D., and Pfaff, D. W. Lateral vestibular lesions, medullary reticular lesions and lordosis reflex in female rats. *Brain Research*, 1977b, *171*, 334–338.

Modianos, D., Flexman, J. E., and Hitt, J. C. Rostral medial forebrain bundle lesions produce decrements in masculine, but not feminine, sexual behavior in spayed female rats. *Behavioral Biology*, 1973, *8*, 629–636.

Modianos, D., Hitt, J. C., and Flexman, J. Habenular lesions produce decrements in feminine, but not masculine, sexual behavior in rats. *Behavioral Biology*, 1974, *10*, 75–87.

Modianos, D., Hitt, J. C., and Popolow, H. B. Habenular lesions and feminine sexual behavior of ovariectomized rats: diminished responsiveness to the synergistic effects of estrogen and progesterone. *Journal of Comparative Physiology and Psychology*, 1975, *89*, 231–237.

Modianos, D., Delia, H., and Pfaff, D. W. Lordosis in female rats following medial forebrain bundle lesions. *Behavioral Biology*, 1976, *18*, 135–141.

Moore, R. Y. Visual pathways and the central neural control of diurnal rhythms. In F. O. Schmitt and F. G. Worden (Eds.), *The Neurosciences. Third Study Program*. Cambridge: M.I.T. Press, 1974.

Morest, D. K. The neuronal architecture of the medial geniculate body of the cat. *Journal of Anatomy*, (London), 1964, *98*, 611–630.

Morrell, J. I., Kelley, D. B., and Pfaff, D. W. Autoradiographic localization of hormone-concentrating cells in the brain of an amphibian, *Xenopus laevis:* II. Estradiol. *Journal of Comparative Neurology*, 1975a, *164*, 63–78.

Morrell, J., Kelley, D., and Pfaff, D. W. Sex steroid binding in the brains of vertebrates: Studies with light microscopic autoradiography. In K. Knigge, D. Scott, H. Kobayashi and S. Ishii (Eds.), *Brain-Endocrine Interaction*, Vol. 2. Basel: Karger, 1975b.

Morrell, J. I., Davis, R. E., and Pfaff, D. W. Autoradiographic localization of sex steroid concentrating cells in the brain of the paradise fish after ³H-estradiol of ³H-testosterone administration. *Proceedings of the Fifth International Congress of Endocrinology*, Hamburg, July 1976, Abstract 278.

Morrell, J. I., Ballin, A., and Pfaff, D. W. Topography of estrogen-accumulating cells in the brain and pituitary of the female mink. *Anatomical Record*, 1977a, *189*, 609–624.

Morrell, J., Crews, D., Ballin, A., and Pfaff, D. Sex hormone concentrating neurons in the brain of a reptile, *Anolis carolinensis. Proceedings of the Society for Neuroscience*, 1977b.

Moss, R. Modification of copulatory behavior in the female rat following olfactory bulb removal. *Journal of Comparative Physiology and Psychology*, 1976, *74*, 372–382.

Moss, R. L., and Law, O. T. The estrous cycle: its influence on single unit activity in the forebrain. *Brain Research*, 1971, *30*, 435–438.

Moss, R. L., and McCann, S. M. Induction of mating behavior in rats by luteinizing hormone-releasing factor. *Science*, 1973, *181*, 177–179.

Moss, R. L., and McCann, S. M. Action of luteinizing hormone-releasing factor (LRF) in the initiation of lordosis behavior in the estrone-primed ovariectomized female rat. *Neuroendocrinology*, 1975, *17*, 309–318.

Moss, R. L., Paloutzian, R. F., and Law, O. T. Electrical stimulation of forebrain structures and its effect on copulatory as well as stimulus-bound behavior in ovariectomized hormone-primed rats. *Physiology and Behavior*, 1974, *12*, 997–1004.

Mountcastle, V. B. *Medical Physiology*, Vol. 1. (13th Ed.). St. Louis: Mosby, 1974.

Nance, D. M., Shryne, J., and Gorski, R. A. Septal lesions: effects on lordosis behavior and pattern of gonadotropin release. *Hormones and Behavior*, 1974, *5*, 73–81.

Nance, D. M., Shryne, J., and Gorski, R. A. Effects of septal lesions on behavioral sensitivity of female rats to gonadal hormones. *Hormones and Behavior*, 1975, *6*, 59–64.

Nance, D. M., McGinnis, M., and Gorski, R. A. Interaction of olfactory and amygdala destruction with septal lesions: effects on lordosis behavior. *Society for Neuroscience*, 1976, Abstract 932.

Napoli, A., Powers, J. B., and Valenstein, E. Hormonal induction of behavioral estrus modified by electrical stimulation of hypothalamus. *Physiology and Behavior*, 1972, *9*, 115–117.

Niemer, W. T., and Magoun, H. W. Reticulo-spinal tracts influencing motor activity. *Journal of Comparative Neurology*, 1947, *87*, 367–379.

Noble, R. G. Facilitation of the lordosis response of the female hamster *(Mesocricetus auratus)*. *Physiology and Behavior*, 1972, *10*, 663–666.

Noble, R. G. Sexual arousal of the female hamster. *Physiology and Behavior*, 1973, *10*, 973–975.

Numan, M. Medial preoptic area and maternal behavior in the female rat. *Journal of Comparative Physiology and Psychology*, 1974, *87*, 746–759.

Nyberg-Hansen, R. Origin and termination of fibers from the vestibular nuclei descending in the medial longitudinal fasciculus. An experimental study with silver impregnation methods in the cat. *Journal of Comparative Neurology*, 1964, *122*, 355–367.

Nyberg-Hansen, R. Sites and mode of termination of reticulo-spinal fibers in the cat. An experimental study with silver impregnation methods. *Journal of Comparative Neurology*, 1965, *124*, 71–100.

Nyberg-Hansen, R. Functional organization of descending supraspinal fibre systems to the spinal cord. Anatomical observations and physiological correlations. *Rev. Anat. Embry. Cell Biol.*, 1966, *39*, 1–48.

Nyberg-Hansen, R. Anatomical aspects of the functional organization of the vestibulospinal pathways. In R. F. Naunton (Ed.), *The Vestibular System*. New York: Academic Press, 1975.

Nyberg-Hansen, R., and Mascitti, T. A. Sites and mode of termination of fibers of the vestibulospinal tract in the cat: An experimental study with silver impregnation methods. *Journal of Comparative Neurology*, 1964, *122*, 369–387.

Paxinos, G., and Bindra, D. Hypothalamic knife cuts: effects on eating, drinking, irritability, aggression, and copulation in the male rat. *Journal of Comparative Physiology and Psychology*, 1972, *79*, 219–229.

Peterson, B. S., and Felpel, L. P. Excitation and inhibition of reticulospinal neurons by vestibular, cortical, and cutaneous stimulation. *Brain Research*, 1971, *27*, 373–376.

Peterson, B. W. Distribution of neural responses to tilting within vestibular nuclei of the cat. *Journal of Neurophysiology*, 1970, *33*, 750–767.

Peterson, B. W., and Abzug, C. Properties of projections from vestibular nuclei to medial reticular formation in the cat. *Journal of Neurophysiology*, 1975, *38*, 1421–1435.

Peterson, B. W., Anderson, M. E., and Filion, M. Responses to pontomedullary reticular neurons to cortical, tectal and cutaneous stimuli. *Experimental Brain Research*, 1974, *21*, 19–44.

Peterson, B. W., Filion, M., Felpel, L. P., and Abzug, C. Responses of medial reticular neurons to stimulation of the vestibular nerve. *Experimental Brain Research*, 1975a, *22*, 335–350.

Peterson, B. W., Maunz, R. A., Pitts, N. G., and Mackel, R. G. Patterns of projection and branching of reticulospinal neurons. *Experimental Brain Research*, 1975b, *23*, 333–351.

Petras, J. M. Cortical, tectal and tegmental fiber connections in the spinal cord of the cat. *Brain Research*, 1967, *6*, 275–324.

Pfaff, D. W. Autoradiographic localization of radioactivity in the rat brain after injection of tritiated sex hormones. *Science*, 1968a, *161*, 1355–1356.

Pfaff, D. W. Uptake of estradiol-17β-H^3 in the female rat brain: An autoradiographic study. *Endocrinology*, 1968b, *82*, 1149–1155.

Pfaff, D. W. Mating behavior of hypophysectomized rats. *Journal Comparative Physiology and Psychology*, 1970a, *72*, 45–50.

Pfaff, D. W. Nature of sex hormone effects on rat sex behavior: specificity of effects and individual patterns of response. *Journal of Comparative Physiology and Psychology*, 1970b, *73*, 349–358.

Pfaff, D. W. Interactions of steroid sex hormones with brain tissue: Studies of uptake and physiological effects. In S. Segal *et al.* (Eds.), *The Regulation of Mammalian Reproduction.* Springfield, Ill.: Thomas, 1973a.

Pfaff, D. W. Luteinizing hormone releasing factor (LRF) potentiates lordosis behavior in hypophysectomized ovariectomized female rats. *Science*, 1973b, *182*, 1148–1149.

Pfaff, D. W. The neuroanatomy of sex hormone receptors in the vertebrate brain. In T. C. A. Kumar (Ed.), *Neuroendocrine Regulation of Fertility.* Basel: Karger, 1976.

Pfaff, D. W., and Keiner, M. Atlas of estradiol-concentrating cells in the central nervous system of the female rat. *Journal of Comparative Neurology*, 1973, 151, 121–158.

Pfaff, D. W., and Lewis, C. Film analyses of lordosis in female rats. *Hormones and Behavior*, 1974, 317–335.

Pfaff, D. W., and Pfaffmann, C. Olfactory and hormonal influences on the basal forebrain of the male rat. *Brain Research*, 1969, *15*, 137–156.

Pfaff, D. W., Lewis, C., Diakow, C., and Keiner, M. Neurophysiological analysis of mating behavior responses as hormone-sensitive reflexes. In E. Stellar and J. M. Sprague (Eds.), *Progress in Physiological Psychology*, Vol. 5. New York: Academic Press, 1972.

Pfaff, D. W., Diakow, C., Zigmond, R. E., and Kow, L.-M. Neural and hormonal determinants of female mating behavior in rats. In F. O. Schmitt and F. G. Worden (Eds.), *The Neurosciences, Vol. 3.* Cambridge: M.I.T. Press, 1974.

Pfaff, D. W., Gerlach, J., McEwen, B. S., Ferin, M., Carmel, P., and Zimmerman, E. Autoradiographic localization of hormone-concentrating cells in the brain of the female rhesus monkey. *Journal of Comparative Neurology*, 1976, *170*, 279–294.

Pfaff, D. W., Montgomery, M., and Lewis, C. Somatosensory determinants of lordosis in female rats: behavioral definition of the estrogen effect. *Journal of Comparative Physiology and Psychology*, 1977, *91*, 134–145.

Platt, J. R. Strong inference. Certain systematic methods of scientific thinking may produce much more rapid progress than others. *Science*, 1964, 347–353.

Poggio, G. F., and Mountcastle, V. B. A study of the functional contributions of the lemniscal and spinothalamic systems to somatic sensibility. Central nervous mechanisms in pain. *The Johns Hopkins Hospital Bulletin*, 1960, *106*, 266–316.

Pompeiano, O. Vestibulo-spinal relationships. In R. F. Naunton (Ed.), *The Vestibular System.* New York: Academic Press, 1975.

Pompeiano, O., and Brodal, A. Spino-vestibular fibers in the cat. An experimental study. *Journal of Comparative Neurology*, 1957, *108*, 353–382.

Powers, J. B. Anti-estrogenic suppression of the lordosis response in female rats. *Hormones and Behavior*, 1975, *6*, 379–392.

Powers, B., and Valenstein, E. S. Sexual receptivity: facilitation by medial preoptic lesions in female rats. *Science*, 1972, *175*, 1003–1005.

Quadagno, D. M., Shryne, J., and Gorski, R. A. The inhibition of steroid-induced sexual behavior by intrahypothalamic actinomycin-D. *Hormones and Behavior*, 1971, *2*, 1–10.

Rhines, R., and Magoun, H. W. Brain stem facilitation of cortical motor response. *Journal of Neurophysiology*, 1946, *9*, 219–229.

RoBards, M. J., Watkins, D. W., III, and Masterton, R. B. An anatomical study of some somesthetic afferents to the intercollicular terminal zone of the midbrain of the opossum. *Journal of Comparative Neurology*, 1976, *170*, 499–524.

Roberts, W. W., Steinberg, M. L., and Means, L. W. Hypothalamic mechanisms for sexual, aggressive, and other motivational behaviors in the opossum, Didelphis virginiana. *Journal of Comparative Physiology and Psychology*, 1967, *64*, 1–15.

Rodgers, C. H. Total and partial surgical isolation of the male rat hypothalamus: Effects on reproductive behavior and physiology. *Physiology and Behavior*, 1969, *4*, 465–470.

Rodgers, C. H., and Law, O. T. The effects of habenular and medial forebrain bundle lesions on sexual behavior in female rats. *Psychonomic Science*, 1967, *8*, 1–2.

Rodgers, C. H., and Schwrtz, N. B. Diencephalic regulation of plasma LH, ovulation, and sexual behavior in the rat. *Endocrinology*, 1972, *90*; 461–465.Rodgers, C. H., and Schwartz, N. B. Differentiation between neural and hormonal control of sexual behavior and gonadotrophin secretion in the female rat. *Endocrinology*, 1976, *98*, 778–786.

491

NEURAL
MECHANISMS OF
FEMALE
REPRODUCTIVE
BEHAVIOR

Rose, J. D., and Sutin, J. Responses of single units in the medulla to genital stimulation in estrous and anestrous cats. *Brain Research*, 1973, *50*, 87–99.

Rosén, I., and Scheid, P. Patterns of afferent input to the lateral reticular nucleus of the cat. *Experimental Brain Research*, 1973a, *18*, 242–255.

Rosén, I., and Scheid, P. Responses in the spino-reticulo-cerebellar pathway to stimulation of cutaneous mechanoreceptors. *Experimental Brain Research*, 1973b, *18*, 268–278.

Rosén, I., and Scheid, P. Responses to nerve stimulation in the bilateral ventral flexor reflex tract (bVFRT) of the cat. *Experimental Brain Research*, 1973c, *18*, 256–267.

Ross, J. W., Paup, D. C., Brant-Zawadski, M., Marshall, J. R., and Gorski, R. A. Effects of cis- and trans-clomiphene in the induction of sexual behavior. *Endocrinology*, 1973, *93*, 681–685.

Roy, E. J., and Wade, G. N. Binding of ³H-estradiol by brain cell nuclei and female rat sexual behavior: Inhibition by antiestrogens. *Brain Research*, 1977, *120*.

Ruda, M. A. An autoradiographic study of the efferent connections of the midbrain central gray in the cat. *Anatomical Record*, 1975, *181*, 468.

Ruh, T. S., and Ruh, M. E. The effect of anti-estrogens on the nuclear binding of the estrogen receptor. *Steroids*, 1974, *24*, 209–224.

Rusak, B., and Morin, L. P. Testicular responses to photoperiod are blocked by lesions of the suprachiasmatic nuclei in golden hamsters. *Society for Neuroscience*, 1975, Abstract 678.

Sawyer, C. H. Reproductive behavior. In J. Field, H. W. Magoun, and V. E. Hall (Eds.), *Handbook of Physiology, Section 1: Neurophysiology, Vol 2*. Washington, D.C.: American Physiological Society, 1960.

Sawyer, C. H., and Robison, B. Separate hypothalamic areas controlling pituitary gonadotropic function and mating behavior in female cats and rabbits. *Journal of Clinical Endocrinology and Metabolism*, 1956, *16*, 914–915.

Schneider, G. Contrasting visuomotor functions of tectum and cortex in the golden hamster. *Psychologische Forschung*, 1967, *31*, 52–62.

Schneider, G. E. Two visual systems. *Science*, 1969, *163*, 895–902.

Schroeder, D. M., and Jane, J. A. Projection of dorsal column nuclei and spinal cord to brainstem and thalamus in the tree shrew, *Tupaia glis. Journal of Comparative Neurology*, 1971, *142*, 309–350.

Sherrington, C. *The Integrative Action of the Nervous System*. New Haven, Conn., Yale University Press, 1906.

Singer, J. J. Hypothalamic control of male and female sexual behavior in female rats. *Journal of Comparative Physiology and Psychology*, 1968, *66*, 738–742.

Smalstig, E. B., and Clemens, J. A. The role of the suprachiasmatic nuclei in reproductive cyclicity. *Society for Neuroscience*, 1975, Abstract 673.

Sprague, J. M., and Chambers, W. W. Control of posture by reticular formation and cerebellum in the intact, anesthetized and unanesthetized and in the decerebrated cat. *American Journal of Physiology*, 1954, *176*, 52–64.

Sprague, J. M., Schreiner, L. H., Lindsley, D. B., and Magoun, H. W. *Journal of Neurophysiology*, 1948, *11*, 501–507.

Sprague, J. M., Levitt, M., Robson, K., Liu, C. N., Stellar, E., and Chambers, W. W. A neuroanatomical and behavioral analysis of the syndromes resulting from midbrain lemniscal and reticular lesions in the cat. *Archives Italiennes de Biologie*, 1963, *101*, 225–295.

Stein, B. E., Magalhaes-Castro, B., and Kruger, L. Superior colliculus: visuotopic-somatotopic overlap. *Science*, 1975, *189*, 224–225.

Stein, B. E., Magalhaes-Castro, B., and Kruger, L. Relationship between visual and tactile representations in cat superior colliculus. *Journal of Neurophysiology*, 1976, *39*, 401–419.

Stephan, F. K., and Zuker, I. Circadian rhythms in drinking behavior and locomotor activity of rats are eliminated by hypothalamic lesions. *Proceedings of the National Academy of Sciences* USA, 1972a, *69*, 1583–1586.

Stephan, F. K., and Zuker, I. Rat drinking rhythms: central visual pathways and endocrine factors mediating responsiveness to environmental illumination. *Physiology and Behavior*, 1972b, *8*, 315–326.

Stetson, M. H., and Watson-Whitmyre, M. Nucleus suprachiasmaticus: the biological clock in the hamster? *Science*, 1976, *191*, 197–199.

Suzuki, J.-I., and Cohen, B. Head, eye, body and limb movements from semicircular canal nerves. *Experimental Neurology*, 1964, *10*, 393–405.

Szechtman, H., Caggiula, A. R., and Wulkan, D. Systematic isolation of the preoptic area with a micro-knife: a neuroanatomical and behavioral analysis of the disruption of sexual behavior in male rats. *Society for Neuroscience*, 1975, Abstract 844.

Takeuchi, T., and Manning, J. W. Hypothalamic mediation of sinus baroreceptor-evoked muscle cholinergic dilator response. *American Journal of Physiology*, 1973, *224*, 1280–1287.

ten Bruggencate, G., and Lundberg, A. Facilitatory interaction in transmission to motoneurones from vestibulospinal fibres and contralateral primary afferents. *Experimental Brain Research*, 1974, *19*, 248–270.

Terasawa, E., and Sawyer, C. H. Changes in electrical activity in the rat hypothalamus related to electrochemical stimulation of adenohypophyseal function. *Endocrinology*, 1969, *85*, 143–149.

Valverde, F. Reticular formation of the albino rat's brain stem cytoarchitecture and corticofugal connections. *Journal of Comparative Neurology*, 1962, *119*, 25–53.

Verrier, R. L., Calvert, A., and Lown, B. Effect of posterior hypothalamic stimulation on ventricular fibrillation threshold. *American Journal of Physiology*, 1975, *228*, 923–927.

Waldron, H. A., and Gwyn, D. G. Descending nerve tracts in the spinal cord of the rat. I. Fibers from the midbrain. *Journal of Comparative Neurology*, 1969, *137*, 143–154.

Warner, R. L. Long term effects of hypothalamic deafferentation in the hamster. *Anatomical Record*, 1975, *181*, 505.

Werner, G. Neural information processing with stimulus feature extractors. In F. O. Schmitt and F. G. Worden (Eds.), *The Neurosciences: Third Study Program*. Cambridge: M.I.T. Press, 1974.

Whalen, R. E., Gorzalka, B. B., DeBold, J. F., Quadagno, D. M., Ho, G. K.-W., and Hough, J. C., Jr. Studies on the effects of intracerebral antinomycin D implants on estrogen-induced receptivity in rats. *Hormones and Behavior*, 1974, *5*, 337–343.

White, G. Van S. *Certain Effects of Electrolytic Lesions in the Hypothalamus on the Mating Behavior of the Golden Hamster*. Ph.D. thesis, Louisiana State University and Agricultural and Mechanical College, 1954.

Whitehead, S. A., and Ruf, K. B. Responses of antidromically identified preoptic neurons in the rat to neurotransmitters and to estrogen. *Brain Research*, 1974, *79*, 185–198.

Wickelgren, B. G. Superior colliculus: Some receptive field properties of bimodally responsive cells. *Science*, 1971, *173*, 69–72.

Willis, W. D., Maunz, R. A., Foreman, R. D., and Coulter, J. D. Static and dynamic responses of spinothalamic tract neurons to mechanical stimuli. *Journal of Neurophysiology*, 1975, *38*, 587–600.

Wilson, V. J. The labyrinth, the brain, and posture. *American Scientist*, 1975a, *63*, 325–332.

Wilson, V. J. Physiology of the vestibular nuclei. In R. F. Naunton (Ed.), *The Vestibular System*. New York: Academic Press, 1975b.

Wilson, V. J., and Yoshida, M. Comparison of effects of stimulation of Deiters' nucleus and medial longitudinal fasciculus on neck, forelimb, and hindlimb motoneurons. *Journal of Neurophysiology*, 1969, *32*, 743–758.

Wilson, V. J., Kato, M., Thomas, R. C., and Peterson, B. W. Excitation of lateral vestibular neurons by peripheral afferent fibers. *Journal of Neurophysiology*, 1966, *29*, 508–529.

Wilson, V. J., Dato, M., Peterson, B. W., and Wylie, R. M. A single-unit analysis of the organization of Deiters' nucleus. *Journal of Neurophysiology*, 1967, *30*, 603–619.

Wilson, V. J., Yoshida, M., and Schor, R. H. Supraspinal monosynaptic excitation and inhibition of thoracic back motoneurons. *Experimental Brain Research*, 1970, *11*, 282–295.

Wilson, V. J., Uchino, Y., Susswein, A. J., and Rapoport, S. Properties of vestibular neurons projecting to the neck segments of the cat spinal cord. *Society for Neuroscience*, 1976, Abstract 1522, p. 1055.

Wuttke, W. Preoptic unit activity and gonadotropin release. *Experimental Brain Research*, 1974, *19*, 205–216.

Yagi, K. Effects of estrogen on the unit activity of the rat hypothalamus. *Journal of the Physiological Society of Japan*, 1970, *32*, 692–693.

Yagi, K. Changes in firing rates of single preoptic and hypothalamic units following an intravenous administration of estrogen in the castrated female rat. *Brain Research*, 1973, *53*, 343–352.

Yagi, K., and Sawaki, Y. Feedback of estrogen in the hypothalamic control of gonadotrophin secretion. In K. Yagi and S. Yoshida (Eds.), *Neuroendocrine Control*. Tokyo: University of Tokyo Press, 1973.

Yamanouchi, K., and Arai, Y. Female lordosis pattern in the male rat induced by estrogen and progesterone: effect of interruption of the dorsal inputs to the preoptic area and hypothalamus. *Endocrinologia Japonica*, 1975, *22*, 243–246.

Young, W. C. The hormones and mating behavior. In W. C. Young (Ed.), *Sex and Internal Secretions,* Vol. 2 (3rd ed.). Baltimore: Williams and Wilkins, 1961.

Zasorin, N., Malsbury, C., and Pfaff, D. W. Suppression of lordosis in the hormone-primed female hamster by electrical stimulation of the septal area. *Physiology and Behavior,* 1975, *14,* 595–599.

Zemlan, F., and Pfaff, D. W. Lordosis after cerebellar damage in female rats. *Hormones and Behavior,* 1975, *6,* 27–33.

Zemlan, F., Morrell, J., Kow, L.-M., and Pfaff, D. W. Anatomical features of systems descending through anterolateral columns of rat spinal cord. *Journal of anatomy* (London), *28,* 489–512.

Zigmond, R. E., and McEwen, B. S. Selective retention of estradiol by cell nuclei in specific brain regions of the ovariectomized rat. *Journal of Neurochemistry,* 1970, *17,* 889–899.

Zigmond, R. E., Nottebohm, F., and Pfaff, D. W. Androgen-concentrating cells in the midbrain of a songbird. *Science,* 1973, *179,* 1005–1007.

Zouhar, R. L., and DeGroot, J. Effects of limbic brain lesions on aspects of reproduction in female rats. *Anatomical Record,* 1962, *145,* 358.

493

NEURAL
MECHANISMS OF
FEMALE
REPRODUCTIVE
BEHAVIOR

Neuropharmacology, Neurotransmitters, and Sexual Behavior in Mammals

BENGT J. MEYERSON, CARL OLOF MALMNÄS, AND BARRY J. EVERITT

INTRODUCTION

In this chapter, we discuss the possible relationship between certain neurotransmitters, sexual behavior, and its controlling hormones. The knowledge in this field has accumulated through experiments that have utilized chemical substances that influence transmission processes in the central nervous system. To understand how drugs can be used to investigate different elements of sexual behavior, together with the possibilities and limitations of such an approach, we first survey some basic knowledge and concepts in neuropharmacology in general and behavioral pharmacology in particular. We follow this survey with a review of the experimental data within this area as they relate to the neuroendocrine basis of sexual behavior.

NEURONAL INTERACTION

Simplistically, it is possible to classify neurons in the mammalian brain as either (1) projection neurons (PNs) with long axons projecting out of a given struc-

BENGT J. MEYERSON   Department of Medical Pharmacology, University of Uppsala, Sweden.   CARL OLOF MALMNÄS   Late of the Department of Medical Pharmacology, University of Uppsala, Sweden.   BARRY J. EVERITT   Anatomy Department, University of Cambridge, Cambridge, England.

ture or (2) local circuit neurons (LCNs) connecting with other cells in the vicinity of their origin. The LCNs seem to be in higher proportions, at least in certain areas (e.g., the caudate nucleus; see Kemp and Powell, 1971; Buchwald *et al.*, 1973), and there is some evidence that the significance of LCNs increases phylogenetically (Blinkov and Glezer, 1968). (For further references see Rakic, 1975.)

The concepts that have dominated research on cerebral information-transmission have been those concerned particularly with one-way information-transmission systems, that is, a dendritic receptor surface that transfers the message to the axonal terminal by the propagation of an action (spike) potential. This model seems, therefore, to be most relevant to the PNs. However, evidence is available that suggests that dendritic networks exist, especially related to LCNs, in which information is also transferred between neurons through dendrodentritic synapses. Dendrites may be pre- and postsynaptic to each other (reciprocal synapses) or a dendrite may be postsynaptic to one dendrite and presynaptic to another (serial synapses; for references, see Schmidt *et al.*, 1976). Interneuronal information transfer may also occur by direct electrotonic coupling through gap junctions between neurons (small potential charges that synaptically influence electrical activity in other neurons; Shephard, 1974). The evidence concerning dendrodendritic and electronic interaction (which means that the message is transferred between neurons without the involvement of an axonal action potential) has mainly been coupled to the concept of local neuronal circuits. However, the possibility is not excluded that the same processes are also be important for the neuronal functioning of the more traditional "through circuits" mediated by the PNs.

CHEMICAL TRANSMISSION

PUTATIVE NEUROTRANSMITTERS. The importance, in quantitative terms, of electrotonic interaction in the total interneuronal communication process is not clear, but it is generally held today that most nerve impulses are transmitted chemically in the mammalian central nervous system (CNS). Although we do not by any means know all of the neurotransmitter substances, it is already clear that the mammalian CNS contains a much wider variety of them than does the peripheral nervous system (PNS). The criteria to be met before a chemical is identified as a neurotransmitter are also more difficult to fulfill in the CNS than in the PNS. Very few cerebral pathways are experimentally accessible enough to enable an electrophysiological and neurochemical stimulus-bound release and response relationship to be established. Nevertheless, on neurophysiological and neurochemical grounds, it is generally agreed that acetylcholine (ACh), the catecholamines dopamine (DA) and norepinephrine (NE), and the indoleamine 5-hydroxytryptamine, (serotonin, 5-HT) are transmitters in the CNS. Several other candidates exist like the amino acids γ-aminobutyric acid (GABA), glycine, glutamic acid, and aspartate. Recent data obtained by means of immunohistochemical techniques suggest that certain peptides (e.g., substance P, luteinizing hormone releasing hormone (LHRH), endorphins, and cholecystokinin) exist intraneuronally and may serve a neurotransmitter or neuromodulatory function. The observation that peptide and amine may coexist intraneuronally (Hokfelt *et al.*, 1980a,b) provides an intriguing and functionally perplexing problem. However, in this chapter, we are concerned only with the relationship between monoaminergic (5-HT, DA, and NE) and cholinergic mechanisms and sexual behavior. Because neurons containing these agents account for only 12% or so of the total number of neurons in the CNS, this seems to be, and is, a severe limitation. However, these agents are seen to be widely distributed in the mammalian brain and to be particularly concentrated in those areas (e.g., the hypothalamus and the midbrain) known to be concerned with sexual behavior.

THE SYNAPTIC TRANSMISSION PROCESS. So far as the action of psychotropic drugs is concerned (see below), the most important events controlling neuronal transmission are the biosynthesis, storage, release, binding to postsynaptic receptor sites, and elimination of the transmitter. The last-mentioned event occurs either by metabolic degradation or reuptake into the nerve terminal (Figure 1). Two findings should also be mentioned in this context. The first concerns the evidence suggesting that cyclic nucleotides, such as cAMP, are involved in at least two types of CNS transmission. The nucleotide is thought to act as a second messenger for the neurotransmitter and, by activating membrane-bound enzymatic systems, to change the postsynaptic neuronal membrane permeability and to influence enzymes involved in the biosynthesis of the neurotransmitter (see Wiegant and Gispen, 1976).

The second discovery concerns the mechanisms involved in the feedback regulation of neuronal impulse flow and transmitter turnover. Transmission itself is largely dependent on newly synthetized transmitter. Thus, the rate-limiting enzyme in the biosynthetic pathway of the transmitter plays a crucial role in the transmission process. A decreased enzymatic activity leads to a decrease in the synthesis of the transmitter and, subsequently, to impaired transmission. The rate-limiting enzyme in DA and NE synthesis is tyrosine hydroxylase, and in 5-HT synthesis, it is tryptophan hydroxylase. The activity of these enzymes varies inversely with the nerve terminal monoamine content (end-product inhibition), although the evidence for 5-HT in this context is less clear (Figure 1). Besides end-product regulation, impulse flow is probably regulated by at least two feedback mechanisms. One is a transneuronal feedback-loop system that involves the transfer of information from the postsynaptic neuron. The other one is a local feedback system involving receptors located on the presynaptic neuron sensitive to the transmitter released by the neuron itself (presynaptic receptors or autoreceptors). Stimulation of presynaptic receptors decreases or inhibits impulse flow and influences the regulation of biosynthetic enzymes (see Usdin and Bunney, 1975).

Besides the above-mentioned processes operating in monoaminergic neurons (and, to some extent, the data are also relevant to cholinergic neurons), we must recognize that one and the same postsynaptic membrane may contain receptor sites for more than one kind of transmitter substance. This is clearly so in the mammalian superior cervical ganglion, which has dopaminergic as well as cholinergic receptors, and what occurs in the PNS may also occur in the CNS, although we should not equate the two systems uncritically. To this picture we should add the

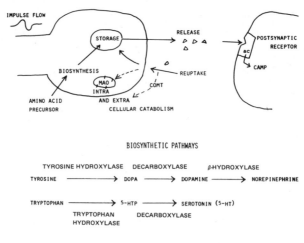

Fig. 1. The synaptic transmission process. Neurotransmitter substances mediate the information between neurons by releasing at the nerve terminal and exerting their action at pre- and postsynaptic receptor sites. Psychotropic drugs are thought to influence events like biosynthesis, storage, release, effects at receptor site, reuptake into the nerve terminal, and catabolism. The models (A,B,C) for feedback mechanisms are partly hypothetical. The feedback process means that stimulation of either pre- or postsynaptic receptors inhibits the firing rate of the presynpatic neuron. A decrease of the firing rate subsequently decreases the transmitter turnover in the presynaptic neuron. The presynaptic receptor is thought to implicate an adenylate cyclase (ac)–cAMP system influencing the enzymatic regulation of transmitter synthesis. Abbreviation: MAO, Monoamine oxidase. (See also Table 1.)

role of neuropeptides. As far as the data now suggest, neuropeptides may participate in the transmission process either intra-axonally, acting on transmitter-releasing mechanisms, or extra-axonally at pre- or postsynaptic receptor sites. The complexity of the CNS is apparent when we incorporate into the above picture of chemical transmission the fact that even very simple functions are likely to be represented by a network of pathways involving different transmitter substances that may exert excitatory or inhibitory actions. It is as well to keep this complexity in mind when interpreting the results of experiments in which drugs have been used to modulate one or several components of such a system of neuronal connections.

NEUROPSYCHOPHARMACOLOGY

Neuropsychopharmacology is the study of chemical substances (psychotropic drugs) that affect mood and behavior. It is well known that some of these drugs have selective actions on certain events within central nervous transmission processes. Such agents modulate synaptic transmission by interfering with the biosynthesis, storage, release, or elimination of the transmitter or, alternatively, by stimulating (agonists) or blocking (antagonists) postsynaptic receptor sites. As a consequence of the above effects, psychotropic drugs may also directly or indirectly alter the feedback mechanisms referred to above. Thus, there are data suggesting that higher doses of the dopamine agonist apomorphine are required to stimulate postsynaptic than to stimulate presynaptic receptors (Aghajanian and Bunney, 1974). Hypothetically, one could thus expect opposite net effects on the dopaminergic transmission of low and high doses of apomorphine, as both doses decrease firing rate, but only the higher one stimulates postsynaptically.

TABLE 1. SURVEY OF PSYCHOTROPIC DRUGS USED IN THE INVESTIGATION OF NEUROTRANSMITTERS AND SEXUAL BEHAVIOR

Drugs	

5-Hydroxytryptamine (5-HT, serotonin)
  1 5-HTP (5-hydroxytryptophan)
  2 Pargyline, nialamide, pheniprazin
  3a PCPA (*p*-chlorophenylalanine)
  3c NSD 1024 (3-hydroxy-
      benzyloxyamine)
  4 MK 486 (L-α-(3,4-dihydroxybenzyl)-α-
      hydrazine prop.acid; RO4-4602 (N-
      (DL-seryl)-N-(2,3,4-trihydroxybenzyl-
      hydrazine)
  5 Reserpine, tetrabenazine
  6 Fenfluramine
  7a LSD (lysergic acid diethylamide), DMT
      (dimethyltryptamine)
   b Metergyline, cinanserin,
      methiothepin, methysergide
  8 Imipramine and other tricyclic
      antidepressants
  9 5,6 and 5,7 Dihydroxytryptamine
	Dopamine (DA)
	  1 Dopa (dihydroxyphenylalanine)
	  2 As for 5-HT
	  3b αMT (α-methyl-p-tyrosine)
	  3c As for 5-HT
	  4 As for 5-HT
	  5 As for 5-HT
	  6 D-Amphetamine
	  7a Apomorphine, ET 495 (piribedil)
	   b Pimozide
	  8 Desmethylimipramine and other tricyclic
	      antidepressants
	  9 6-Hydroxy dopamine

Norepinephrine (NE)
  1 Dopa, DOPS (D,L threo-3,4-
      dihydroxyphenylserine)
  2 As for 5-HT
  3b As for DA
   c As for 5-HT
   d FLA-63 (bis(4-methyl-1-
      homopiperazinyl thiocarbonyl)-
      disulphide
  4 As for 5-HT
  5 As for 5-HT
  6 As for DA
  7a Clonidine (α-rec.)
   b Phenoxybenzamine; phentolamine (α-
      rec.); propranolol (β-rec.)
  8 As for DA
  9 As for DA
	Acetylcholine (ACh)
	  7a Pilocarpine, oxotremorine, arecholine
	      (muscarinic rec.), nicotine (nicotinic rec.)
	   b Atropine (muscarinic rec.)

Continued

TABLE 1. *(Continued)*

Mechanisms of action		
1. Increased substrate for synthesis Amino acid precursors to the monoamines. In contrast to the amines themselves, these precursors readily pass the blood–brain barrier.	*4. Inhibition of synthesis outside the blood–brain barrier* Decarboxylase inhibitors that do not readily pass into the brain.	*7. Receptor stimulation and blockade* a. Receptor stimulation (agonists) b. Receptor blockage (antagonists)
2. Decreased catabolism MAO (monoamine oxidase) inhibitor. A general increase of the monoamine level is achieved by decreased catabolism. A selective increase of NE, DA, or 5-HT achieved by MAO inhibitor combined with a precursor.	*5. Amine depletion* Prevents storage into storage granules. *6. Displacement of amines from storage pools* Increased activity due to release from newly synthesized storage pool. Other actions as well.	*8. Increased activity of postsynaptic receptors* Inhibits reuptake of the amines. Different compounds in this group inhibit the reuptake for the different monoamines more or less selectively.
3. Inhibition of synthesis a. Tryptophane hydroxylase b. Tyrosine hydroxylase inhibitor c. Decarboxylase inhibitor d. β-Hydroxylase inhibitor		*9. Neurotoxic agents*

The existence of substances that selectively increase or decrease activity in certain types of neurons is the basis for using psychotropic drugs as tools when investigating the roles of different types of neurons in certain elements of a behavior. The rationale behind the use of these drugs is, to a great extent, based on the assumption that they are selective in their actions on a particular category of neurons. However, this is rarely the case. The selectiveness is very much more dependent on the appropriate use of the compound, which, among other things, means a careful choice of dose- and time-dependent variables.

SELECTIVENESS OF DRUG ACTION. The majority of psychotropic drugs interact with target tissues (CNS) in a way that produces a functional response relative to the dose of the drug. The response normally increases until it reaches a maximum, after which increasing the dose often involves a loss of specificity; that is, the spectrum of actions of the drug also increases. The effects of a drug are not only dose- but also time-dependent. Thus, to achieve specificity when using a drug, it is essential to establish the right dose and also the right time interval between administration and the behavioral observation. In addition, some pharmacokinetic factors are equally important in this context. Thus, absorption of the drug from its place of administration, its uptake and distribution into different tissues, its storage within those tissues, and its subsequent metabolsim and elimination are all important factors to be taken into account. The effective dose of a drug is closely related to these pharmacokinetic factors, all of which influence the latency and the duration of its action. The ease with which a drug may cross the blood–brain barrier is also important. Monoamines like 5-HT, DA, and NE do not readily pass into the brain, whereas the monoamine-precursor amino acids do. Sometimes, the fact that a drug does not enter the brain is useful when differentiating between its central and peripheral actions. Atropine readily passes into the brain, whereas methyl atropine does not. The same rationale is used for some extracerebral decarboxylase inhibitors (see below). Furthermore, it must be remembered that many drugs can and do interact pharmacokinetically by increasing or decreasing metabolism, by competing for tissue storage sites, and so on.

Although it is known how a certain drug influences neuronal transmitter processes, it is often difficult to relate this effect directly to the observed effect on behavior. To some extent, this difficulty is overcome by measuring the effect of several compounds with a common action on the same behavioral parameter, or by using drugs with different points of action but equivalent functional effects. Besides this strategy, together with a careful consideration of the dose, time, and response relationships, the specificity of a drug's action may further be improved by using antagonists or agonists. Thus, if a drug is thought to exert its action by increasing dopaminergic activity, it should be possible to inhibit its effect with a dopamine receptor antagonist.

In addition to the general problems in behavioral pharmacology such as those mentioned above, specific problems become apparent when dealing with hormone-dependent behaviors such as sexual behavior. This point has been discussed elsewhere (Meyerson and Eliasson, 1977; Meyerson and Malmnäs, 1978) and is not covered in detail here. Let us just bring up three points: (1) Hormone-induced behavioral responses must be kept at submaximal levels if *both* stimulatory and inhibitory actions of a drug are to be observed in a situation where its effects might be unpredictable. (2) Furthermore, it cannot be emphasized enough that one is looking at the interaction between the effects achieved by the hormone(s) and drug treatment and that the effects of the latter may be different if it is given when the hormone(s) has exerted its full effect (e.g., 38 hr after estrogen) compared to what was given in connection with the hormone treatment (e.g., simultaneously with progesterone) when the hormone has not yet produced its full effect. (3) The drug may also interfere not only with the processes directly involved in the production of the behavior but indirectly by modulating the CNS (hypothalamo-)–pituitary-hormone secretory systems, thereby increasing or decreasing the endogenous secretion of hormones able to influence the behaviors under investigation.

NEUROPSYCHOPHARMACOLOGY OF SEXUAL BEHAVIOR: THE RATIONALE. As it became clear that gonadal hormones activate copulatory behavior by a local effect on the central nervous system (for references, see below), the role of transmitter functions in this connection became important. The rationale of using neuropharmacological agents in this field is (1) to determine if specific types of neurons are important in the element of sexual behavior studied, and further, (2) to investigate the notion of an intimate relationship between hormones and putative neurotransmitters in activating and maintaining the hormone-dependent behavior. Most work has been carried out in the rat, and as a working model, copulatory behavior (lordosis response in the female and mounting in the male) has mainly been utilized because these sexual responses are easy to measure and to activate to a controlled response level by exogenous hormones in the gonadectomized subject. More recently, methods that measure components of sexual behavior other than copulatory performance have been employed in order to extend our knowledge of the neurotransmitters involved in the activation of sexual responses.

THE VARIABLES STUDIED. Before we discuss in detail the effect of psychotropic drugs on sexual behavior, we might first consider the variables involved in the experimental model. As is true of all other behaviors, sexual behavior is a highly complex function of sensory inputs transmitted, modulated, and integrated to finally activate the motor patterns from which, as observers, we judge that it has a sexual context. Every such behavior is then modulated by feedback from the environment, thus establishing a continuous, dynamic process producing a sequence of behaviors that together make up a pattern specific for the species.

Sexual behavior is one of a group of behaviors that, besides sensory stimuli, is dependent on hormonal ones. The significance of the central nervous action of gonadal hormones in activating different elements of sexual behavior has been dealt with in detail elsewhere. Briefly, after gonadectomy, the capacity to display sexual behavior disappears, and the behavior can be reinduced by exogenous gonadal hormones. The hormone replacement therapy most commonly used in male mammals has been testosterone propionate, given either once daily or at longer intervals. Female copulatory behavior is activated in rodents either by estradiol benzoate alone (if given in a high enough dosage) or by estradiol benzoate followed after 48 hr by progesterone. The latter treatment schedule is the most commonly used because it probably reflects more closely the sequence of events occurring in the intact animal. When using females of other species (e.g., some carnivores or primates), this treatment schedule is inappropriate, and the therapy must therefore be modified as appropriate for the species studied.

Studies examining the biochemical basis of sexual responses have been very much restricted to coital behavior, and the species studied has mainly been the laboratory rat. To distinguish between certain elements of sexual behavior, which have been studied and probably continue to be studied, we use the following functional descriptions (Figure 2A). *Copulatory behavior,* which is commonly taken to mean the whole sequence of events that occur during sexual interaction, we define here as the characteristic coital posture (lordosis) taken by the female to permit the male to achieve intromission; the male type of copulatory behavior is the performance initiated by the mounting and includes the intromission and ejaculation patterns. The behavior repertoire displayed in connection with the coital act, *the precopulatory performance* (Figure 2B) includes invitation, rejection, and approach patterns; genital investigation; pursuit; soliciting behavior; and so on. The physiological condition that brings the subject to seek sexual contact is not reflected in any specific motor pattern such as copulatory behavior. The "urge to seek sexual contact" and the specific orientation of this contact have been measured indirectly by different *instrumental responses,* which are set up *to measure the sexual motivation*

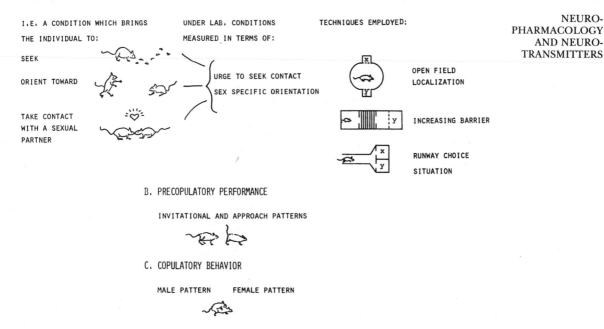

Fig. 2. Sexual responses used in the study of psychotropic drugs and sexual behavior.

of the animal (Figure 2C). Beach (1976) suggested that the terms *proceptivity, receptivity,* and *attractivity* be adopted in studies of sexual behavior. Some standardization is clearly needed. However, whatever words we use, we must be aware that we, as observers, associate a certain behavior, posture, or behavioral event with some causal factor. Our evidence, hidden behind the word *proceptivity* or *motivation* is always indirect in terms of whether the behavior recorded actually measures what we think it does. Our approach to this problem should be to be as clear as possible in our structural and functional description of the motor expression of what we measure. Further, our *associative description* of the behavior (such as sexual receptivity, proceptivity, and sexual motivation) must be based on the outcome of several different measures. A female monkey that makes a sexual invitational gesture to a male might be sexually motivated or, equally possibly, submissive; a female rat crossing an electrified grid to reach a sexually active male might do so for sexual or other reasons (e.g., agonistic, exploratory, or nonspecific locomotor activity). Not until we use a number of different test situations and responses will we be able to associate the response to a causal factor (such as sexual motivation). A detailed description of the measures of female and male sexual responses used in drug studies is given below.

Thus, the sensory stimuli, the hormonal condition, and the specific element of the sexual behavior or instrumental response that we think express what we have desired to measure are the independent and dependent variables involved. In addition, we add a chemical treatment with a selective action on neurotransmission.

METHODS

BEHAVIORAL DEFINITIONS AND MEASURES OF FEMALE SEXUAL BEHAVIOR. *Copulatory behavior* is the specific posture normally taken by the female in response to a mount. The extent to which this behavior can be elicited is measured in terms of the number of mounts followed by lordosis (expressed as a ratio L/M) or as the proportion of animals that displayed lordosis after a standardized amount of mounting stimulus. Other measures used are, for example, acceptance ratio, lordosis intensity, and duration (see Everitt *et al.*, 1975a).

Precopulatory behavior includes soliciting behavior (expressed as the presence or absence of, e.g., hopping, ear wiggling, and darting movements in the rat).

Instrumental measures of sociosexual motivation and sex-specific orientation include different techniques employed to measure the urge to seek sexual contact. (See below and Meyerson and Lindström, 1973, for details.)

Sexual attractiveness measures those qualities of an animal, both behavioral and nonbehavioral, that define its values as a sexual stimulus.

EXPERIMENTAL DESIGN AND HORMONAL REGIMENS. It is important to understand at the outset the behavioral methodologies that were used in these investigations, as well as the hormonal-pharmacological models that have been developed, before reviewing the experimental data. It is beyond the scope of this chapter to discuss different experimental designs. However, because of the complex stimulus–response situation and all the variables that could influence the results, it must be emphasized that "statistical significance" does not permit us to draw conclusions about the effects of a certain treatment unless (1) replicative tests have been conducted and (2) appropriate controls have been run in parallel.

Behavioral observations were generally made on oppositely sexed pairs of animals in the dark phase of a reversed day–night cycle. The hormonal treatment regimens were commonly used as the basis on which the effects of psychoactive drugs were investigated, and those all involved ovariectomized rats, which were (1) treated with estradiol benzoate (EB) alone, given either once 48 hr before behavioral observations or daily in a low dosage (this treatment regimen, when used carefully, induced no or very low levels of copulatory response in the female) or (2) were given combined treatment with both EB and progesterone. In the latter model, then, the facilitatory and inhibitory effects of the drugs may be observed, provided the dosage of the two hormones is adjusted to give a submaximal response level with regard to the response measured (proportion of animals that displayed lordosis, L/M ratio, etc.). In the case of the first treatment regimen, only excitatory effects can be observed.

MONOAMINES AND COPULATORY BEHAVIOR IN THE FEMALE

It is well to restate here that a fundamental principle for using drugs as tools in the investigation of behavior is that meaningful conclusions can be drawn only when dose–response and time–response relationships have been established and the specificity of the drug action has been established by assessing the effects of analogous drug treatments.

Various approaches can be employed to investigate the implication of monoaminergic neurotransmission in copulatory behavior. Let us first consider the effect of a general increase and decrease in the central nervous content of monoamines achieved by, respectively, decreased intra-axonal catabolism (monoamine oxidase inhibition) or monoamine depletion (mainly by preventing the storage of amine in storage granules).

In the early days (1964) of psychopharmacological experiments on sexual behavior, it was demonstrated that raising the intracerebral levels of monoamines by treatment with different monoamine oxidase (MAO) inhibitors inhibited the EB + progesterone-induced lordosis response in ovariectomized rats (Meyerson, 1964a,b). This inhibition was shown by means of three structurally different monoamine oxidase inhibitors and was demonstrated in different species (rat, mouse, rabbit, and hamster) (see Meyerson *et al.*, 1973). The increased lordosis-inhibitory effect at different times after the treatment with the MAO inhibitor correlated with the increase of cerebral monoamine levels, especially the increase of 5-HT (Meyerson, 1964a). More recently, intrahypothalamic implants of the MAO inhibitor pargyline led to a reduction in estrogen + progesterone-activated lordosis behavior (Luine and Fischette, 1982). The most effective sites were dorsal to or within the lateral ventromedial nucleus. Work with selectively acting MAO inhibitors suggests that inhibition of both the A and the B forms of MAO is necessary to reduce lordosis behavior (Luine and Paden, 1982).

Conversely, amine depletors like reserpine and tetrabenazine were seen to increase the copulatory response of female rats given submaximal doses of estrogen and progesterone, as well as of animals that were treated with estrogen alone (Meyerson, 1964c). An analysis of the intracerebral levels of the amines showed that the increase in copulatory response was correlated with an increased depletion of NE, DA, and 5-HT (Meyerson, 1964a). Later investigations have provided evidence that the main effect of reserpine on the lordosis response is exerted indirectly by an endogenous release of progestogenes effective in facilitating the behavior (Paris *et al.*, 1971; Ahlenius *et al.*, 1972b). This topic will be further discussed below.

Thus, we can see from these experiments that 5-HT and catecholamines may be implicated in the control of hormone-activated copulatory response in the female rat. We now consider each of these transmitter substances separately and deal with the evidence relating to each that has accrued through the use of a variety of compounds such as precursor amino acids and receptor agonists and antagonists.

5-HYDROXYTRYPTAMINE. Treatment with the 5-HT precursor 5-HTP after pretreatment with MAO inhibitors, the dosage of the latter drug adjusted so as not to give an effect on its own, inhibited the lordosis response (Meyerson, 1964b, 1975). The precursor of catecholamines, l-dopa, and the precursor of NE, DOPS (dihydrophenylserine), did not have this effect in analogous experiments (Meyerson,

1964a,b). A transient reduction of the response was, however, seen, and later experiments with *l*-dopa treatment in combination with an extracerebral decarboxylase inhibitor inhibited the lordosis response (see below). Selective blockade of 5-HT reuptake by drugs like Lilly 110140, ORG 6582, and GEA 654 (Everitt and Fuxe, 1977b; Meyerson, 1966; Smith and Meyerson, 1984a,b) causes an inhibition of copulatory behavior in estrogen- and progesterone-treated females. Similar results follow the stimulation of 5-HT release with fenfluramine (Everitt *et al.*, 1974; Michanek and Meyerson, 1975, 1977a,b). Blockade of axonal reuptake mechanisms and displacement of endogenous 5-HT provide quite specific ways to increase the 5-HT postsynaptic activity. 5-HT agonists such as LSD, dimethyltryptamine (DMT), and 5-methoxy-DMT all decrease copulatory behavior induced by estrogen and progesterone if given in doses that cause a stimulation of postsynaptic receptors (Eliasson and Meyerson, 1973; Meyerson *et al.*, 1974; Everitt *et al.*, 1974, 1975a; Everitt and Fuxe, 1977b; Eliasson and Meyerson, 1976, 1977; Eliasson, 1976). It is important to remember here the complicated dose–response relationship of these compounds, which were mentioned in the introduction. Low doses of LSD, DMT, and 5-methoxy-DMT preferentially stimulate presynaptic 5-HT receptors (autoreceptors) and hence have the effect of turning off 5-HT neurons. These drugs were found to cause a significant increase in copulatory behavior, the changes in behavior appearing within 5 min after intraperitoneal drug injection (Everitt and Fuxe, 1977b). The "turning off" of 5-HT neurons brings us to the effects of depletors of 5-HT. The compound *p*-chlorophenylalanine (PCPA), which blocks the biosynthesis of 5-HT by an inhibitory effect on the rate-limiting enzyme of 5-HT biosynthesis, tryptophan hydroxylase, was seen to cause an increase in copulatory response not only in the rat but in a variety of species, including monkeys, an effect later seen to be reversed by 5-HTP (Hoyland *et al.*, 1970; Meyerson and Lewander, 1970; Hansult *et al.*, 1972; Ahlenius *et al.*, 1972a; Zemlan *et al.*, 1973; Everitt *et al.*, 1975a; Gradwell *et al.*, 1975).

Another group of compounds that should be relevant to study in this connection are the 5-HT antagonists. Compounds like methysergide, methiothepin, cinanserin, and metergoline reportedly have a 5-HT antagonistic effect, although data to indicate the specificity of these drugs are rather poor. Metergoline seems to have the best specificity. This agent increases copulatory activity in estrogen-treated female rats (Fuxe *et al.*, 1976b). Ward *et al.* (1975) found that the intracranial application of methysergide or cinanserin facilitated lordosis in estrogen-treated female rats, and Sietnicks (1984) demonstrated the specific 5HT$_2$ antagonist altanserin to prevent the LSD-induced inhibition of lordosis.

Taken together, these data, which have accumulated through the use of a wide variety of drugs with various points of action, demonstrate that modulating 5-HT transmission in the CNS affects sexual behavior. A hypothesis has been proposed about the possible action of progesterone on 5-HT mechanisms involved in the lordosis response (see Sietnieks and Meyerson, 1982 and 1983). As mentioned above, LSD and 5-HTP inhibit the estrogen + progesterone-activated lordosis response. This effect was much less evident when the lordosis response was activated by estrogen alone and was found to be enhanced by raising the progesterone dose (Sietnieks and Meyerson, 1980). The effect was specific to progesterone, as no analogous effect was achieved by altered estrogen treatment. Adding to the specificity of these findings is the fact that, in an exploratory test situation, the effects of 5-HTP or LSD were not altered by different progesterone doses. In addition, the stimulatory effect of low doses of LSD, achieved by the proposed presynaptic effect, was enhanced by progesterone (Sietnieks and Meyerson, 1983).

These data, taken together, led to the proposal that progesterone facilitates serotonergic mechanisms. Presynaptic 5-HT stimulation results in a subsequent decrease in postsynaptic activity and thereby in the removal of a tonic 5-HT inhibition of the lordosis response. Under nondrug conditions, with hormones only, progesterone stimulates copulatory behavior by acting on presynaptic 5-HT mechanisms. (For details see Meyerson *et al.*, 1979; Sietnieks and Meyerson, 1980, 1982a,b; see also the section in this chapter on neuroanatomy and monoamines) (Figure 3).

DOPAMINE. Experiments analogous to those described above concerning the implications of 5-HT have been conducted to investigate the possible implications of catecholamines in female copulatory behavior. The early MAO inhibitor and precursor experiments indicated a very slight and nonsignificant reduction of the hormone-induced lordosis response in the female rat, but a significant effect in the female hamster (Meyerson, 1970). Studies in which extracerebral decarboxylase inhibitors have been given as well resulted in a clear-cut inhibition of dopa treatment also in the rat (Meyerson and Malmnäs, 1976). Dopamine receptor stimulators like ET495 (Everitt and Fluxe, 1977b; Everitt *et al.*, 1974) and apomorphine (Eliasson and Meyerson, 1973, 1976; Meyerson *et al.*, 1974) inhibit the estrogen–progesterone-induced lordosis response. So, too, does the catecholamine releaser

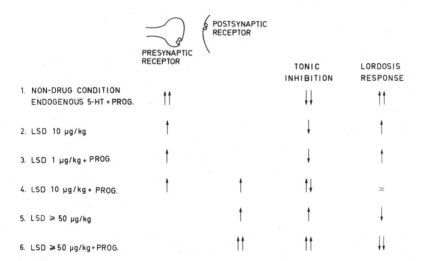

Fig. 3. The serotonin–progesterone model. Progesterone is proposed to facilitate the lordosis response by increasing presynaptic 5-HT activity and thereby turning off a postsynaptic 5-HT tonic inhibition. The figure, based on pharmacological data (see text), depicts how the 5-HT agonist LSD, depending on the dose, inhibits or facilitates the lordosis response by a pre- or postsynaptic action. (From Sietnieks and Meyerson, 1982b.)

amphetamine (Meyerson, 1964c, 1968a; Michanek and Meyerson, 1975, 1977a,b; Everitt *et al.,* 1975a). That these effects are activated by dopamine stimulation is suggested by the fact that the effect of dopamine receptor agonists as well as that of *d*-amphetamine is prevented by the dopamine antagonist pimozide (Meyerson *et al.,* 1974; Michanek and Meyerson, 1977a; Eliasson and Meyerson, 1976). Pimozide and the dopamine antagonist spiroperidol have also been shown to cause a marked increase in the lordosis/mount ratios of estrogen-treated female rats (Everitt *et al.,* 1974).

The fact that PCPA is known to cause a large decrease in catecholamine levels during the first 12 hr after treatment, a period during which the animals show increases in sexual behavior, prompted Ahlenius *et al.* (1972a) to look at the effects of α-methyl-p-tyrosine, an inhibitor of catecholamine synthesis, on sexual behavior. This drug also caused a very clear increase in the lordotic behavior of estrogen-treated female rats, thus adding weight to the suggestion that catecholamines may be important in the regulation of sexual behavior. However, the direct stimulatory effect of α-methyl-p-tyrosine (as well as that of PCPA) has been questioned, as no effect was seen in adrenalectomized rats, a finding suggesting that the stimulatory effect seen after these drugs might be due to endogenous progesterone release from the adrenal cortex (Eriksson and Södersten, 1973; see below).

It has been shown that, as for the 5-HT receptor agonists, dopamine agonists also exert an effect at presynaptic receptors at low doses, thus inhibiting impulse flow in DA neurons (Aghajanian and Bunney, 1974). ET495 and apomorphine in low dosages induce copulatory behavior in estrogen-treated female rats that show no responses after estrogen treatment alone (Everitt and Fuxe, 1977a). The compound sulpiride has been suggested as being a blocker of dopamine receptors preferentially at presynaptic sites (Tagliamonte *et al.,* 1975), and this compound was found to block the lordosis-stimulatory but not the lordosis-inhibitory effect of apomorphine (Everitt and Fuxe, 1977a).

There is no evidence that progesterone alters dopamine activity. Progesterone did not enhance the dopa inhibition of the lordosis response (Sietnieks and Meyerson, 1982a). Apomorphine, however, was slightly more effective in inhibiting the lordosis response after estrogen + progesterone than after estrogen alone (Michanek and Meyerson, 1982), but no progesterone-dose-dependent effect was seen.

What is important to realize here is that drugs that decrease DA transmission only enhance the immobile, lordosis response. They simultaneously inhibit active, proceptive responses. The nature of the response, then, is critical when discussing the actions of dopaminergic drugs, "inhibitory" or "excitatory," on sexual behavior in the female. This topic is discussed further in the section on neuroanatomy and the monoamines.

NOREPINEPHRINE. Of the amines, norepinephrine has proved to be the most difficult to study selectively, and therefore, many conflicting data have appeared concerning its role in sexual behavior. The fact that amphetamine given to estrogen-treated female rats treated with pimozide cuased an even larger increase in lordosis/mount ratios than after pimozide alone suggested to us (Everitt *et al.,* 1974) that norepinephrine may have an excitatory role. The basis of the suggestion is that amphetamine causes the release of both DA and NE, but with DA receptors blocked (by pimozide), only NE can act at its receptor sites.

Consistent with this view was the finding that phenoxybenzamine (an α-noradrenaline receptor blocker) inhibits lordosis induced by estrogen and progesterone, as does FLA63, which inhibits NE synthesis (by inhibiting dopamine betahydroxylase enzyme; see Figure 1) (Everitt, 1977). The latter result was challenged, however, in a study using different doses (Ahlenius *et al.*, 1975). Using clonidine (a presynaptic NE receptor stimulator), Crowley *et al.* (1976) came to a similar conclusion regarding NE in guinea pigs. But this drug inhibits copulatory behavior in estrogen–progesterone-treated female rats (Michanek, 1979). Hansen *et al.* (1980), using electrolytic and 6-hydroxydopamine lesions, provided good evidence that ascending noradrenergic neurons distributed to subcortical areas of the brain are involved in the response to tactile information generated during sexual interaction (see the section on neuroanatomy and the monoamines).

The problem may be resolved by a much more careful analysis of the involvement of alpha and beta receptor stimulation in this context. Michanek (1979), testing different α- and β-adrenergic-receptor-blocking drugs discussed the possibility that α_1 receptors are implicated in a facilitatory system. However, interactions between 5-HT and NA may be important. Whereas alpha blockers cause a decrease in copulatory responses in females treated with estrogen and progesterone, beta blockers, in some reports, do the opposite; that is, they increase sexual receptivity in estrogen-treated females when given either systemically (Fuxe *et al.*, 1976a) or by intracerebral administration (Ward *et al.*, 1975). Of importance in this context is also the idea that beta receptors are stimulated by epinephrine (E). E neurons (i.e., a phenylethanolamine-N-methyl-transferase-positive neuron, mapped by means of immunofluorescence techniques) are present in the brainstem and project *inter alia* into the hypothalamus (Hökfelt *et al.*, 1974). Histochemical studies have demonstrated NA nerve terminals in the pontine raphe complex (Chu and Bloom, 1974). The activity of 5-HT neurons in raphe nuclei has been shown to be regulated by NA neurons (Morgane *et al.*, 1974) with NA-inhibiting 5-HT activity (Key and Krzywosinski, 1977). A decreased NA influence (by presynaptic α-adrenergic stimulation) on raphe nuclei would activate 5-HT neurons and thereby explain inhibition of the lordosis response (Michanek, 1979). Much more work is needed to define even the questions that must be asked here, such as differentiation between alpha and beta NE receptor mechanisms in terms of their role in sexual behavior, as well as the potential role of E neurons.

ACETYLCHOLINE AND COPULATORY BEHAVIOR

There have been comparatively few studies of cholinergic mechanisms in the sexual behavior of female mammals. The effects of systemically administered muscarinic compounds on copulatory activity have been investigated by Lindström (1970, 1971, 1972; Lindström and Meyerson, 1967), who showed that pilocarpine, oxotremorine, and arecoline (which stimulate cholinergic muscarinic receptors) all inhibited estrogen- and progesterone-induced lordotic behavior in female rats. The effect appears after short time intervals (within 30 min) and is itself blocked by atropine, but not by methyl atropine, a finding suggesting a central site of action of the cholinergic drugs. The behavioral effects of muscarinic stimulation seem to be mediated by a monoaminergic mechanism, as the inhibition of MAO prolongs the action of pilocarpine, whereas amine depletion (with reserpine or tetrabenazine) prevents its inhibiting actions on lordosis behavior. A further examination of this aminergic involvement using α-methyl-p-tyrosine (to inhibit catecholamine synthesis) and PCPA (to inhibit 5-HT synthesis) demonstrated that 5-HT is a more important transmitter than the catecholamines, as blockade of the inhibitory effects of pilocarpine was obtained only after decreasing 5-HT synthesis.

These experiments also emphasize the importance of considering time–response relationships, as both pilocarpine and oxotremorine can induce lordosis responses in estrogen-treated females (i.e., they replace progesterone) if observations are made 3 hr after injection of the drug. This effect is probably due to the release of adrenocortical steroids that effectively substitute for progesterone (Lindstrom, 1973).

Clemens *et al.* (1980) have demonstrated that the microinfusion of cholinergic agonists such as carbachol into the preoptic and the ventromedial hypothalamic area facilitates estrogen-induced lordosis behavior in female rats (Clemens *et al.*, 1980; Dohanich and Clemens, 1981). Infusions into the mesencephalic reticular formation or the cortex were ineffective. The priming dose of EB necessary to achieve a response was very low (0.13 μg for 2 days). Carbachol was preferentially effective in areas known to concentrate estrogen. The authors proposed a relationship between cholinergic stimulation and the ability to concentrate estrogen. However, it may be that the stimulating effects of cholinergic agonists are related to an estrogen mechanism, whereas the inhibitory effects reported from systemic administration are related to a progesterone–5-HT interaction.

The treatment of ovariectomized female rats receiving estrogen alone with nicotine also causes a significant increase in copulatory activity within 10 min in the absence of progesterone, a finding suggesting that a "nicotinic" cholinergic receptor may be involved in the activation of lordotic behavior. The effects of nicotine were blocked by treatment with the ganglion-blocking agent mecamylamine (Fuxe *et al.*, 1977). Nicotine, 0.5 mg/kg subcutaneously 5 hr after progesterone treatment, induced a significant increase (lordosis percentage: experimentals = 23%, controls = 0%, $n = 48$, $p < 0.001$) also in ovariectomized female rats treated with a low dose of estradiol benzoate (2.5 μg/kg) and progesterone (0.4 mg/rat), the hormones given 48 hr apart. (Lindström and Meyerson, unpublished). This effect was present 30 and 90 min after the nicotine injection and had declined at 3 hr. A similar effect was seen when nicotine was given to subjects treated with EB alone (10 μg/kg) (lordosis percentage at 30 and 60 min, experimentals = 38%, controls = 0%, $n = 24$, $p < 0.05$). In some doses, nicotine is seen significantly to reduce 5-HT turnover, a finding suggesting that the effects of nicotine receptor stimulation may be mediated by 5-HT neurons. This possibility must be explored more carefully because an adrenal involvement must be considered. However, the time interval between drug treatment and response is short, and the response declines after 2 hr, a finding that does not fit into the picture of a progesterone effect (see also the conclusions to this chapter).

Thus, both muscarinic and nicotinic cholinergic mechanisms have been implicated in the control of lordosis. Muscarinic mechanisms seem to be inhibitory, whereas nicotinic mechanisms are facilitatory in this context. Both effects could operate indirectly via a serotoninergic mechanism.

MONOAMINES AND NONCOPULATORY SEXUAL RESPONSES

PRECOPULATORY PERFORMANCE. There are no systemic studies of precopulatory performance and the effect of psychotropic drugs in the female. There are scattered reports indicating that stimulation of copulatory behavior is not increased by soliciting behaviors (see Zemlan *et al.*, 1973; Everitt *et al.*, 1975a).

MONOAMINES AND ATTRACTIVENESS. *Sexual attractiveness* means, as mentioned above, those qualities of an animal, both behavioral and nonbehavioral, that define its value as a sexual stimulus. The influence of drugs on attractiveness is an interesting field, but very little research has been reported. In an investigation of the effect of monoaminergic drugs on the attractiveness of the female partner, the approach of an intact, sexually vigorous male was observed in an open-field arena (Meyerson, 1978). An intact male normally spends significantly more time in the vicinity of an estrogen + progesterone-treated, ovariectomized female than in the vicinity of a sexually active male (Hetta and Meyerson, 1975, 1978; Eliasson and Meyerson, 1981). When two females, both with the same hormonal treatment, were used as incentive animals but one was treated with PCPA, the male spent significantly more time with the non-PCPA-treated female. These data are preliminary, and we cannot yet exclude nonspecific effects and the possibility that other dose regimens may give other results. Nevertheless, the fact that a drug influences attractiveness should be considered in studies on drug action and sexual responses.

MONOAMINES AND INSTRUMENTAL MEASURES OF SOCIOSEXUAL MOTIVATION AND SEX-SPECIFIC ORIENTATION. To study the effect of hormones and drugs on heterosexual approach, different techniques were used to measure how the female rat would seek contact with a sexually active male (Figure 2; for details, see Meyerson and Lindström, 1973). The methods differed with respect to the behavior that the subject had to perform to reach contact with the incentive object. So far, psychotropic drugs have been tested only in female rats. We first give a brief summary of the relationship between ovarian hormones and the responses measured.

In ovariectomized subjects, estrogen induced an obvious increase in the amount of aversive stimulus (crossing an electrified grid) that the subject was willing to tolerate in order to reach contact with a sexually active male. The female also spent more time sitting close to the male in an open-field arena, and in a runway-choice situation, there was increase in the number of trials in which the male was chosen in preference to an estrous female. In intact, 5-day cyclic females, there was a cyclic variation in these measures, with a clear peak in responding when the female was in a proestrous condition (Meyerson and Lindström, 1973; Eliasson and Meyerson, 1975). Decreased biosynthesis of serotonin by means of PCPA treatment increased the amount of aversive stimulus the female was willing to take to reach a sexually active male (Meyerson *et al.*, 1974). The specificity of this effect was suggested by the fact that it was obtained only in the estrogen-treated female. The same PCPA treatment decreased the estrogen-induced increase in time that the female spent in the vicinity of the male in the open field and the preference of the male versus an estrous female measured in a runway-choice situation (Meyerson *et al.*, 1974; Meyerson, 1975). α-Methyl-p-tyrosine decreased the grid-crossing response but did not affect the runway-choice response. The serotonin agonist LSD and the dopamine agonist apomorphine both inhibited the grid-crossing response (Meyerson *et al.*, 1974). The role of serotonin and dopamine in the regulation of heterosexual approach in the female rat must be more completely elucidated by further experiments before any conclusion can be drawn about the implication of monoamines in sociosexual motivation and heterosexual approach.

Neuropsychopharmacology of Sexual Behavior in the Male

Methods

Behavioral Definition and Measures of Male Sexual Behavior. Studies of the effects of drugs on male sexual behavior have largely been restricted to the laboratory rat. Irrespective of subsequent mounting or nonmounting, male rats ordinarily approach the female immediately after her introduction into the observation cage and start to investigate her anogenital region and to pursue her when she darts away. In sexually inexperienced or castrated males, anogenital sniffing and pursuit may be the only components of heterosexual interaction to be displayed. However, in sexually experienced, intact males, initiation of copulation is often a matter of a few seconds. Copulatory behavior in the male rat consists of repetitive mounting of the female partner, with or without penile intromission into the female's vagina, terminated by ejaculation. On mounting, which elicits lordosis in the female, the male executes rapid pelvic thrustings and clasps the flanks of the female between his forelegs. If intromission is achieved, a vigorous backward lunge is displayed by the male during dismount. After a certain number of mounts with intromission (5–15), the ejaculatory reflex is triggered, characterized by a prolonged mount with intense clasping of the female, followed by a slow dismount and subsequent genital licking. Once copulation is initiated, ejaculation is very likely to occur within the next 3–10 min, followed by a postejaculatory interval of around 5 min.

Copulatory behavior—the display of mounting, with or without intromission, terminated by ejaculation:

1. The percentage of males that display the different components of coital behavior: mount percentage, intromission percentage, and ejaculation percentage.

2. The latency from the introduction of the female to first mount and intromission and from the first intromission to ejaculation (mount latency, intromission latency, and ejaculation latency).

3. The time from ejaculation to the next intromission (postejaculatory interval).

4. Number of mounts and intromissions per minute.

5. Number of mounts and intromissions preceding ejaculation (mount frequency and intromission frequency).

6. The proportion of mounts with intromission in the total number of mounts (intromission ratio).

Precopulatory behavior—the display of pursuit and anogenital sniffing.

EXPERIMENTAL DESIGN, SPECIFICITY REQUIREMENTS, AND LABORATORY CONDITIONS. Whereas the basic pattern of sexual behavior is very similar in individuals of the same species, the temporal pattern of sexual behavior differs greatly in individuals but is fairly stable from one test to another in each individual subject. Therefore, each subject could very well serve as its own control. However, because some environmental factors may vary from one test to the next, in this type of experimental design blank-treated controls must always be tested in parallel with drug-treated subjects as a check on the stability of the baseline response.

Although most investigators in the field have used intact male rats, there are certain methodological advantages in using testosterone-treated, castrated males in neuropharmacological studies of male sexual behavior. Thus, with testosterone-treated castrates, one gets around the problem of a possible drug effect on gonadal hormone production. A dose of testosterone can be chosen that induces a response that permits both facilitatory and inhibitory effects of the drug on the behavior to be detected. (For an applied study, see Malmnäs, 1973.) Male sexual behavior involves an approach of the male toward the partner, and for this reason, the choice of drug dosage in neuropharmacological studies of this behavior should be based on pilot experiments in which particular attention is paid to the drug's influence on motor capability and reaction to environmental stimuli. A simple and sensitive screening test is to score whether the male approaches an estrous female (Malmnäs, 1973). If this approach is lacking because of motor incapability or competing behavioral responses like compulsive stereotypies, copulation is not be expected, and no meaningful conclusion can be drawn about the drug's specific influence on male sexual behavior. These methodological considerations have been further elaborated by Meyerson and Höglund (1981) in a multivariate analysis of exploratory and sociosexual behavior.

The methods used in various laboratories differ with respect to (1) testing procedure (light conditions, adaptation time, test duration, etc.); (2) the hormonal state of the experimental subject (intact or castrated with or without testosterone); and (3) the stimulus object (sex, hormone, or drug treatment). Ordinarily, sexual behavior tests are carried out in dimmed light during the dark phase of the diurnal light–dark cycle, and the male is allowed around 5 min to adapt in the observation cage before a sexually receptive female (treated with extradiol benzoate followed by progesterone) is introduced. In some studies, male-to-male mounting behavior in groups of saline- or drug-treated males have been investigated. In these studies, no determination has been made of whether the drug-induced behavioral effect is due primarily to an action on the mounting subject or to an action on the mounted object.

MONOAMINES AND COPULATORY BEHAVIOR IN THE MALE

SEROTONIN. The effects of MAO inhibitors on copulatory behavior in male rats have been studied by Soulairac (1963), Butcher *et al.*, (1969), Leavitt (1969), Malmnäs and Meyerson (1970, 1971), Tagliamonte *et al.*, (1971), Dewsbury *et al.*, (1972), and Malmnäs (1973). A consistent finding in these studies is that MAO inhibitors inhibit different aspects of copulatory behavior. By what mechanism is this inhibition achieved? The following evidence suggests that the inhibition is due to increased central nervous serotonergic activity: when monoamine precursors were given after a small dose of the MAO inhibitor pargyline, inhibition of copulatory behavior was induced by the serotonin precursor 5-HTP, but not by the catecholamine precursor *l*-dopa (Malmnäs and Meyerson, 1970). The inhibitory effects of 5-HTP on copulatory behavior (decreased mount and intromission percentages and increased mount and intromission latencies) persisted when an extracerebral decarboxylase inhibitor (MK486) was added to the pargyline pretreatment, a finding demonstrating that the inhibitory action of the serotonin precursor was exerted within the CNS (Malmnäs and Meyerson, 1972; Malmnäs, 1973). The significance of serotonin in the inhibitory effect of MAO inhibitors on male-rat copulatory behavior is further demonstrated by the fact that this inhibitory effect is prevented by pretreatment with the serotonin synthesis inhibitor PCPA (Malmnäs and Meyerson, 1971; Tagliamonte *et al.*, 1971). Bignami (1966) demonstrated that minute amounts of the serotonin-receptor-stimulating agent LSD decreased intromission latencies and postejaculatory intervals in male rats, whereas the opposite effects were obtained with higher doses. Inhibitory effects of LSD were also found by Malmnäs (1973) on mount and intromission percentages as well as on the number of mounts per minute. Further evidence for an inhibitory role of cerebral 5-HT in male rat sexual behavior is the fact that the 5-HT reuptake inhibitor zimelidine enhanced the 5-HTP-induced prolongation of ejaculation latency. This effect was counteracted by the 5-HT receptor-blocking agent melergoline (Ahlenius *et al.*, 1980a).

Several components of copulatory behavior are facilitated by the serotonin-synthesis-inhibitor PCPA. This drug was first reported to increase male-to-male mounting behavior in groups of intact male rats, all receiving the same treatment (Sheard, 1969; Shillito, 1969, 1970; Tagliamonte *et al.*, 1969). Furthermore, the display of heterosexual copulatory behavior in castrated and castrated + adrenalectomized male rats was facilitated by PCPA (increased mount, intromission, and ejaculation percentages; decreased mount latency and increased number of mounts per minute; increased intromission ratio) (Malmnäs and Meyerson, 1971; Malmnäs, 1973). In intact male rats, PCPA has been shown to increase the number of mounts and intromissions per minute and to decrease ejaculation latency (Ahlenius *et al.*, 1971; Salis and Dewsbury, 1971). In agreement with the forementioned studies, Whalen and Luttge (1970) found no effect of PCPA on mount and intromission frequency or on the number of ejaculations preceding satiation, parameters different from those in which facilitatory effects have been reported. Interestingly, Gawienowski and Hodgen (1971) reported that PCPA, in combination with pargyline, no longer facilitated male-to-male mounting behavior in rats after hypophysectomy. No doubt, the possible contribution of hypophyseal factors to the facilitatory effects of PCPA on mounting behavior deserves further study. Increased male-to-male mounting behavior after PCPA has been reported in the cat (Hoyland *et al.*, 1970; Ferguson *et al.*, 1970) and the rabbit (Perez-Cruet *et al.*, 1971). However, no facilitation of copulatory behavior was found after PCPA in sexually vigorous cats (Zitrin *et al.*, 1970) or in *Macaca speciosa* (Redmond *et al.*, 1971).

Compared to PCPA the neurotoxic drug 5,6-dihydroxytryptamine (5,6-DHT), which causes damage to serotonergic and noradrenergic neurons (see below), has been little used in studies of male sexual behavior. In intact male rats, this drug has been shown to facilitate male-to-male mounting behavior after injection into the lateral ventricles (Da Prada *et al.*, 1972) and, after intracerebral injection (Gessa and Tagliamonte, 1975), to increase the number of sexually inexperienced male rats that achieved ejaculation with estrous females. Carlsson and coworkers reported on a new group of synthetic agents, aminotetralines, which produce a reduction in ejaculation latencies in the male rat (Ahlenius *et al.*, 1981). An increase in the initiation of mounting was also seen. These compounds are labeled as 5-HT receptor agonists, presumably acting at 5-HT presynaptic receptors. However, the mechanism of action has not been fully elucidated.

DOPAMINE. Reserpine and tetrabenazine, which prevent granular monoamine storage, decreased intromission frequency in sexually vigorous intact male rats (Soulairac; 1963; Dewsbury and Davis, 1970; Dewsbury, 1972). However, these drugs decreased the mount and intromission percentages, increased mount latency (reserpine), and decreased the number of mounts per minute (reserpine) in testosterone-treated, castrated male rats (Malmnäs, 1973). The inhibitory effects of amine depletors found in castrated subjects are probably due to decreased dopaminergic activity. This conclusion is partly based on the fact that similar inhibitory effects were obtained after α-methyl-p-tyrosine, which inhibits the synthesis of both dopamine and norepinephrine, but not after FLA-63, which inhibits the synthesis of norepinephrine only (Malmnäs, 1973).

The neuroleptics chlorpromazine and haloperidol, which block dopamine and norepinephrine receptors, inhibited the display of copulatory behavior in male rats (Malmnäs, 1973). Chlorpromazine inhibited copulatory behavior both after a dose that increased (1 mg/kg) and after a dose that decreased (2.5 mg/kg) motor activity (Malmnäs, 1973). Phenoxybenzamine, which blocks norepinephrine but not dopamine receptors, did not affect the copulatory behavior in doses with which some motor capability was still maintained (Malmnäs, 1973). In contrast, pimozide, which blocks dopamine but not norepinephrine receptors, decreased the mount and intromission percentages as well as the number of mounts per minute and prolonged mount latency; these effects were obtained by a dose (0.25 mg/kg) that did not affect motor activity (Malmnäs, 1973).

Small doses of the catecholamine precursor *l*-dopa, given after pargyline and MK486, increased mount, intromission, and ejaculation percentages; decreased mount and intromission latencies; and increased intromission ratios (Malmnäs and Meyerson, 1972; Malmnäs, 1973). In contrast, no facilitation was obtained by any dose of the selective norepinephrine precursor DOPS (Malmnäs and Meyerson, 1972; Malmnäs, 1973). The facilitatory effects of *l*-dopa and the lack of facilitatory effects of DOPS suggest that dopamine is the catecholamine of major importance in mediating the facilitatory effects of low doses of *l*-dopa. This suggestion is strongly supported by the fact that most of the facilitatory effects of *l*-dopa on copulatory behavior are blocked by the selective dopamine-receptor-antagonist pimozide, given in a dose that by itself does not interfere with the behavior (Malmnäs, 1976). When *l*-dopa was given after combined pargyline + MK486 pretreatment, facilitatory effects were obtained after quite low doses (≤ 2.5 mg/kg), whereas higher doses of *l*-dopa inhibited copulatory behavior (Malmnäs, 1976). The fact that *l*-dopa inhibits copulatory behavior at certain dose levels might explain the inconsistent results obtained with the drug when it is given after pretreatment with an extracerebral decarboxylase inhibitor (Ro4-4602) but without any MAO inhibitor. Thus, facilitatory effects of *l*-dopa were obtained by Tagliamonte *et al.*, (1974) on copulatory behavior and by Da Prada *et al.* (1973) on male-to-male mounting behavior, whereas slight inhibitory effects on copulatory behavior of *l*-dopa were found by Hyyppä *et al.* (1971) and Gray *et al.* (1974) (decreased number of intromissions per minute and increased ejaculation latency). Although it appears that *l*-dopa facilitation of copulatory behavior in the male rat occurs by increasing the central nervous dopaminergic-receptor activity (Malmnäs and Meyerson, 1972; Malmnäs, 1973, 1976), the mechanism by which *l*-dopa inhibits copulatory behavior remains to be elucidated.

The dopamine-receptor-stimulating agent apomorphine increased mount and intromission percentages and decreased mount latency (Malmnäs and Meyerson, 1972; Malmnäs, 1973; Tagliamonte *et al.*, 1974). The number of mounts per minute was increased by apomorphine in doses of 0.030 mg/kg and was decreased by 0.30 mg/kg (Malmnäs, 1973). A decrease in intromission frequency after apomorphine (0.3 mg/kg) was reported by Butcher *et al.* (1969), but an interpretation of this finding is not possible because no data were presented on the number of mounts and intromissions per minute.

A decrease in the number of intromissions preceding ejaculation and a shortening of the ejaculation latency are seen after lisuride treatment (Ahlenius *et al.*, 1980b; Ahlenius *et al.*, 1982; Da Prada *et al.*, 1977). Lisuride is a dopamine agonist but influences 5-HT mechanisms as well (probably serotin 1 receptor agonist; Peroutka *et al.*, 1981).

NOREPINEPHRINE. The norepinephrine precursor DOPS, given after combined pargyline and MK486 pretreatment, increased mount latency and decreased the number of mounts per minute (Malmnäs, 1973). When, in addition to the pargyline and MK486 pretreatment, dopamine receptors were protected by pimozide, the only significant effects of *l*-dopa on copulatory behavior were a decreased mount latency and a decreased number of mounts per minute (Malmnäs, 1976). As in the treatment with *l*-dopa after pimozide, the mount latency and the number of mounts per minute were decreased by the norepinephrine (and epinephrine?) agonist clonidine (Malmnäs, 1973).

Decreased norepinephrine synthesis by means of FLA-63 increased the number of mounts per minute (Malmnäs, 1973). However, no significant effects on copulatory behavior were obtained by the adrenergic α-receptor antagonist phenoxybenzamine (Malmnäs, 1973). In contrast, copulatory behavior was inhibited by the adrenergic β-receptor antagonist propranolol (Malmnäs, 1973). Of the two optical isomers of propranolol, 1-propranolol is the only one that blocks β-receptors (Barrett and Cullum, 1968) and the only one that inhibits male-rat copulatory behavior (Malmnäs, 1975).

ACETYLCHOLINE AND COPULATORY BEHAVIOR IN THE MALE

Data on the effects of cholinergic drugs on male sexual behavior are sparse. An increase in the ejaculation latency and the postejaculatory interval was obtained by the cholinesterase inhibitor physostigmine in toxic doses (Soulairac, 1963); whether this drug had any effect on sexual behavior in nontoxic doses was not reported. Nicotine, which stimulates one of the two main types of cholinergic receptors, was demonstrated by Bignami (1966) to increase intromission latencies and the postejaculatory interval. In contrast, Soulairac and Soulairac (1972) found a reduction of the ejaculation latency and the postejaculatory interval after nicotine. A complete disruption of male sexual behavior was induced by a fairly high dose of atropine (Soulairac, 1963). In moderate doses, atropine and other antimuscarinic compounds increased intromission latency and the postejaculatory interval (Bignami, 1966).

MONOAMINES AND NONCOPULATORY SEXUAL RESPONSES

The effect of monoamingergic drugs on precopulatory behavior in male rats has been studied in castrated animals given small doses of testosterone, insufficient to activate mounting (Malmnäs, 1973). Of the drugs that affect dopaminergic neurotransmission, apomorphine increased, and pimozide decreased, pursuit and anogenital sniffing; the apomorphine-induced facilitation of precopulatory behavior was antagonized by pimozide. No significant effect on precopulatory behavior was found after treatment with drugs that affect norepinephrinergic neurotransmission (clonidine, DOPS, phenoxybenzamine, and FLA-63). Of drugs affecting serotonergic neurotransmission, PCPA increased (Malmnäs, unpublished) and 5-HTP and LSD decreased, pursuit and anogential sniffing.

Increased dopaminergic activity influences female and male copulatory behavior differently (Table 2). Female behavior is inhibited and male copulatory behavior is stimulated. Increased serotonergic activity, however, increases both female and male copulatory responses. Are the monoaminergic systems bound to the behavior rather than to the genetic sex?

Reserpine, tetrabenazine, PCPA, and α-methyl-p-tyrosine facilitate the occurrence of lordosis responses in male rats castrated as adults and given estradiol benzoate (Meyerson, 1968b; Larsson and Södersten, 1971; Meyerson and Malmnäs, 1977). This effect was also achieved by intrahypothalamic placement of methysergide and cinanserin in EB-treated male rats (Crowley *et al.*, 1975). Female copulatory behavior shown by EB-treated castrated males was inhibited by increased dopaminergic activity achieved with apomorphine (0.15–0.30 mg/kg) (Meyerson and Malmnäs, 1976).

Emery and Sach (1975) reported facilitation of ejaculatory patterns in EB-implanted female rats by PCPA treatment. In the study of Meyerson and Malmnäs (1977), mounting was induced by testosterone treatment, and in the same animal, 1 day later, progesterone was given as well, to activate lordotic behavior. Tested with receptive females, the experimental subjects showed mounting (pelvic thrusting and clasps), and after progesterone treatment and tested with males, the females responded to mounting with lordosis responses. Mounting behavior in the female was increased by the dopamine agonist apomorphine (0.1 mg/kg), while the testosterone- and progesterone-induced female type of copulatory behavior was inhibited by the same drug (0.10–0.30 mg/kg). By testing both behaviors in the same sex, it is possible to exclude sex-linked metabolic differences, and the dose effective to induce and inhibit the two types of copulatory behavior could be compared (see Meyerson, 1978; Everitt, 1978; Meyerson and Malmnäs, 1978). Thus apomorphine facilitated mounting and inhibited lordotic behavior in genetic males as well as in genetic female rats. This finding is consistent with the hypothesis that the monoaminergic system is bound to the behavior (lordosis and initiation of mounting) rather than to the genetic sex.

NEUROPEPTIDES, MONOAMINES, AND SEXUAL RESPONSES

The discovery of endogenous neuropeptides, which in pharmacological experiments produced behavioral effects, led to the assumption that neuropeptides may participate in central nervous transmitter processes. The interaction between monoamines and neuropeptides is a new field that no doubt will become a focus of interest in the future. The implication of neuropeptides in different aspects of the biochemical basis of behavior is beyond the limits of the present chapter. However, we will here mention a few interesting reports within the scope of neuropharmacology, neurotransmitters and sexual behavior.

TABLE 2. SCHEMATIC REPRESENTATION OF THE INFLUENCE OF MONOAMINES ON DIFFERENT COMPONENTS OF SEXUAL BEHAVIOR IN MALE AND FEMALE RATS

Behavioral component	Amine postsynaptic receptor activity						References[a]
	Dopamine		Norepinephrine		Serotonin		
	Increased	Decreased	Increased	Decreased	Increased	Decreased	
Male							
Precopulatory behavior							
Pursuit and anogenital sniffing	Increased	Decreased	No effect	No effect	Decreased	Increased	A
Copulatory behavior							
Mounting							
1. Initiation	Increased	Decreased	No effect	No effect	Decreased	Increased	A,B,C,D,E,F
a. Mount percentage	Decreased	Increased	Decreased	No effect (?)	Increased	Decreased	A,D
b. Mount latency							
2. Maintenance							
3. No. of mounts/minute	Increased	Decreased	Decreased	Increased	Decreased	Increased	A,B,D,G,H
Intromission							
1. Intromission percentage	Increased	Decreased	No effect (?)	No effect (?)	Decreased	Increased	A,D,E,F
2. Intromission latency	Decreased	Increased	No effect (?)	No effect (?)	Increased	Decreased	A,D
3. Intromission ratio	Increased	No effect (?)	No effect (?)	No effect (?)	No effect (?)	Increased	A,D
Ejaculation							
1. Ejaculation latency	Decreased				Increased (?)	Decreased	A,D,F,G,H,I
Postejaculatory interval	Decreased (?)				Increased (?)	Decreased	F,G,I
Female							
Copulatory behavior							
Lordosis response							
Activated by EB + progesterone (submaximal response level)	Decreased		Decreased (?) α-Rec.		Decreased		J,K,L,O,P,R
After treatment with EB alone (only increase could be seen)		Increased		Increased (?) β-Rec.		Increased	M,N
Heterotype copulatory behavior							
Mount percentage (female)	Increased						
Lordosis percentage (male)	Decreased					Increased	Q

[a]A. Malmnäs (1973); B. Malmnäs and Meyerson (1971); C. Malmnäs and Meyerson (1972); D. Malmnäs (1976); E. Tagliamonte et al. (1971); F. Tagliamonte et al. (1974); G. Salis and Dewsbury (1971); H. Ahlenius et al. (1971); I. Bignami (1966); J. Meyerson (1974); K. Eliasson and Meyerson (1973, 1976); L. Everitt et al. (1974, 1975a); M. Ward et al. (1975); N. Zemlan et al. (1973); O. Meyerson and Lewander (1970); P. Ahlenius et al. (1972b); Q. Meyerson and Malmnäs (1977); R. Sietnieks and Meyerson (1980, 1982a,b). For further references, see text.

LHRH elicits lordosis behavior in ovariectomized estrogen-primed rats (Moss and McCann, 1973; Pfaff, 1973). Beyer *et al.* (1981) reported also that dibutyryl-cAMP can elicit lordosis in ovariectomized, estrogen-primed rats. The possibility that LHRH acts by means of raising cAMP levels in neurons related to the expression of lordosis was investigated by Beyer *et al.* (1982). The combined action of phosphodiestrase inhibitors and LHRH revealed a synergistic effect. These findings are obviously interesting, although more has to be done to relate the LHRH action to a cAMP mechanism relevant to the expression of the female's copulatory response and also to associate this cAMP mechanism to a transmitter system.

β-Endorphin was found to impair male copulatory performance in the male rat (Meyerson and Terenius, 1977; Meyerson, 1978, 1981). The effects seem different from those seen after morphine. The relationship between the β-endorphin effects and monoaminergic mechanisms has been reported only in brief (Meyerson, 1980, 1982). The possibility of antagonizing the effects of β-endorphin by a dopamine agonist (apomorphine) or 5-HT depletion (PCPA) was tested. The unexpected finding was that the β-endorphin inhibition of copulatory behavior was enhanced by pretreatment with PCPA. The effects of exploratory behavior caused by β-endorphin were not influenced by PCPA.

NEUROANATOMICAL APPROACHES TO THE STUDY OF MONOAMINES IN SEXUAL BEHAVIOR

The often profound effects of drugs reviewed in the previous sections suggest that monoaminergic (MA) neurons in the brain are concerned with the expression of sexual behavior. Clearly, the psychopharmacological approach gives no clue to the neural loci where MA neurons are important in this context. Other techniques must be employed to this end, and there is a wealth of data on the distribution of MA neurons in the brain that make these techniques feasible. It should be emphasized here, however, that cerebral MA neurons are involved in the regulation of many behaviors (e.g., eating, drinking, and learning), and that the special consideration in neuroanatommical studies of sexual behavior is that hormones play a critical role in its display. Much is known about the sites of action of these hormones from behavioral, biochemical, and anatomical studies. All point to the medial preoptic area (in males) and the ventromedial-anterior hypothalamic region (in females) as the sites where estradiol and testosterone (or its aromatized metabolites) act and induce sexual behavior (see McEwen *et al.*, 1979, for review and Figure 4 for an overview of monoaminergic pathways).

How do monoamine neurons fit into this picture, and what is the relationship, if any, between steroid hormone-dependent structures and these MA neurons? Although some answer to the former question is possible, little in the way of an answer can be given to the latter, as will be seen.

The methods for investigating MA neurons directly in the brain depend on the availability of the selective neurotoxins 6-hydroxydopamine (6-OHDA) and 5,7- or 5,6-dihydroxytryptamine (5,7-DHT or 5,6-DHT). These compounds are taken up into CA or 5-HT neurons, respectively, and therein exert their toxic effects, so destroying the neurons (see Jonsson *et al.*, 1975). 6-OHDA may be made selective for DA neurons by pretreating animals with an NA-uptake inhibitor to prevent entry of the neurotoxin into NA neurons. The same pretreatment also increases the selectivity of 5,7-DHT and 5,6-DHT for 5-HT neurons. However, the absence of a reasonably specific inhibitor of DA reuptake means that it is very difficult to destroy only DA neurons when a mixed population of DA and NA neurons exists. Increasing intraneuronal levels of the amines with pargyline increase the potency of the neurotoxins through a not entirely clear mechanism.

The ultimate aid to selectivity when injecting these compounds into the brain is precise information on the distribution of the MA neurons. This has been systematically built up by neuroanatomical studies using, first, the Falck-Hillarp technique to visualize intraneuronal amines (Fuxe, 1965; Dahlstrom and Fuxe, 1964; Ungerstedt, 1971) and, more recently, immunofluorescent and immunoperoxidase

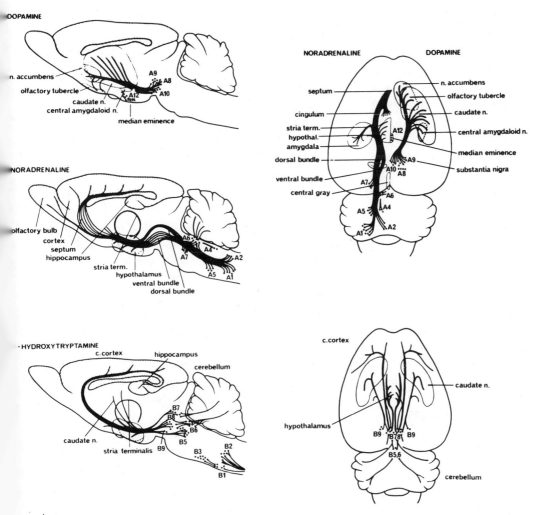

Fig. 4. Monoaminergic pathways. (From Ungerstedt 1971.)

techniques that have a very high degree of resolution (Hökfelt *et al.*, 1980a,b). The projections of NA, 5-HT, and DA neurons are summarized very briefly below, but greater detail may be found in reviews by Moore and Bloom (1978, 1979) and by Steinbusch (1981).

NE cell bodies (groups A1 to A7; Dahlstrom and Fuxe, 1965) are located exclusively in the reticular formation of the medulla and the pons. The locus coeruleus (A6) innervate hippocampus, cingulate, prepyriform, entorhinal, and periamygdaloid cortex, as well as neocortex, projecting via the dorsal noradrenergic bundle (DNAB), which joins the medial forebrain bundle (mfb) at the level of the mammillary nuclei to run toward the septum, where it turns dorsally and caudally in the cingulum. Subcortical NE terminals in, for example, the hypothalamus, the septal nuclei, and areas of the limbic forebrain arise largely from the lateral tegmental cell groups A1 and A2, but also from A5 and A7. Their axons run in the ventral noradrenergic bundle (VNAB), which is not as discrete as was once thought and also joins the mfb in the midbrain to run rostrally in the lateral hypothalamus. There is a substantial NA innervation of the spinal cord from both the locus coeruleus and the lateral tegmental cell groups.

The DA cell groups are located primarily in the midbrain. Cell groups A8, A9 (the substantia nigra and the pars compacta), and A10 form a continuous band of DA neurons in the ventral midbrain, extending up into the raphe, and project, via the mfb in the lateral hypothalamus, to innervate the neostriatum (mainly A8/A9), the nucleus accumbens, the entorhinal and medial frontal cortex olfactory tubercle, and the septal nuclei (mainly A10). Although DA neurons are present periventricularly throughout the hypothalamus, cell group A12 is worth a special mention because it lies within the arcuate nucleus of the hypothalamus, its neurons projecting to the region of the primary capillary plexus in the median eminence. This groups of neurons (the tuberoinfundibular dopamine neurons, or TIDA) seems to play an important role in the regulation of anterior pituitary function.

The 5-HT cell groups B1–B9 lie within the raphe nuclei of the brain stem. Ascending 5-HT axons that run in the mfb originate primarily in the dorsal and median raphe nuclei (B7 and B8) in the midbrain, but there are also contributions from cell groups B9 (in the midbrain) and B5 and B6 (in the pons). They innervate large areas of the brain, including the neocortex, the limbic cortex, the septal nuclei, the hypothalamus (particularly the suprachiasmatic nucleus), the amygdala, and the hippocampus. There appear to be, in the rostral midbrain before the axons have joined the mfb, two partially separate 5-HT bundles—one medial (which runs largely to subcortical structures) and one lateral (which runs to the cortex and the striatum). As is true of NA neurons, the caudally placed groups of 5-HT neurons also project throughout the spinal cord. Indeed, 5-HT and NA projections are much more similar to each other than to the DA projections, which are more restricted in their innervation of the neuraxis. It has been suggested that these NA and 5-HT systems are two chemically defined subsystems of the brainstem reticular formation. This is a most important point when considering functional interpretations of lesion studies because, as with functions of the reticular formation in general, it must be appreciated that such systems have a rather pervasive role in the neural regulation of behavior rather than a specific one (such as might be the case for the highly topically organized sensory systems). It will also be apparent, from the above account, that all ascending components of the MA projections run through the lateral hypothalamus in the mfb. Thus, many early experiments in which the mfb was lesioned were, inadvertently, lesions of MA neurons, among other things. The effects of such lesions on sexual behavior have been reviewed elsewhere (Everitt, 1978), and all that needs to be emphasized here is that, for obvious reasons, the effects of such lesions are very difficult to interpret in aminergic terms.

EFFECTS OF LESIONS TO 5-HT NEURONS USING 5,7-DIHYDROXYTRYPTAMINE

Although it has been reported that electrolytic lesions of the dorsal and the median raphe have little or no effect on sexual behavior, lesions to the ascending axons of these neurons in the midbrain achieved with 5,7-DHT do. Thus, in ovariectomized female rats treated daily with a low dose of estradiol, a long-lasting increase in the display of lordosis and proceptive responses occurred when these rats were compared with their sham-operated controls (Everitt *et al.*, 1975b). The lesions caused a substantial loss of 5-HT in the hypothalamus and the septal nuclei, but not in the cerebral cortex, and it is not possible to define the neuroanatomical locus of this effect more discretely. Although there was an unsuccessful attempt to replicate this finding (Södersten *et al.*, 1978), the reason is not entirely clear (perhaps it was a different regional specificity of the lesion); however, the result was confirmed in the Cambridge laboratory (Everitt, unpublished observations). Similar, long-term potentiation of sexual behavior in the male rat, treated with a low dose of or no testosterone, has also been reported (Larsson *et al.*, 1978; Södersten *et al.*, 1978).

The interpretation of these data remains contentious. Clearly, 5-HT depletion achieved in this way produced results analogous to those described above for the female, in that progesterone was unnecessary for the display of proceptivity and receptivity in 5-HT-depleted, estradiol-treated females. But can it really be stated that 5-HT depletion has "substituted" for progesterone? Another view is that the female has been made *more sensitive to estradiol,* as it is well known that long-term treatment with higher doses of the hormone also induces high levels of proceptivity and receptivity without the addition of progesterone (see Johnson and Everitt, 1980). This interpretation may also be applied to the results in the male without proposing yet another, idiosyncratic mechanism.

Interestingly, such an interpretation may suggest a line of investigation that might throw light on the function of 5-HT neurons in the brain—and this, ultimately, is the purpose of all such functional studies. Thus, another maneuver that increases sensitivity to estradiol in behavioral terms is ablation of the suprachiasmatic nuclei (SCN) (Hansen *et al*, 1978). The effect is an interesting one in that the circadian rhythm of sensitivity to estradiol is abolished so that female rats bearing these lesions show as high levels of receptivity in the light as in the dark period of the day–night cycle. Now, a most prominent input to the SCN is serotonergic, arising from the midbrain raphe and running in the medial ascending 5-HT bundle there (i.e, the axons lesioned in the above experiments). Although a direct relationship between the two effects has not been established, the enormous literature on 5-HT, the SCN, and the rhythmic phenomena in neuroendocrinology suggests that research effort here may be rewarding and may also shed light on a fundamental function of the 5-HT-containing neural system.

EFFECTS OF LESIONS TO NA NEURONS USING 6-HYDROXYDOPAMINE

Lesion of the dorsomedial tegmentum in male rats markedly shortens their PEI (Clark *et al.*, 1975). At the time, this effect was, quite justifiably, potentially attributed to destruction of the DNAB, which runs through the midbrain at the site of the lesion, as there was considerable loss of telencephalic NA. However, in a subsequent study (Clark, 1980), the DNAB was lesioned specifically by injection of 6-OHDA. Telencephalic NA levels were significantly reduced, but there was no alteration in PEI. Of course, the lesson of interpreting the effects of electrolytic lesions has been learned before (e.g., from the lateral hypothalamic syndrome) but is well illustrated here. The interpretation of the results of both experiments has, however, been taken one stage further. Clearly, either a nonnoradrenergic projection running in the dorsomedial midbrain, or a group of neurons there, must have been the critical structures damaged in the original experiment. Judicious use of the cell-body neurotoxin ibotenic acid in the midbrain, in parallel with electrolytic and noradrenergic lesions (Hansen *et al.*, 1982,a,b,), has allowed the latter possibility to be ruled out. The original, intriguing finding of reduced PEI following midbrain lesions, then, appears to be the result of damage to a nonnoradrenergic projection running in the vicinity of the DNAB. Few studies of lesions to the DNAB have been performed in female rats (see, e.g., Alton *et al.*, 1979), and their interpretation is too speculative to include here.

It might be reasonably argued that NA projections via the VNAB, which originates in lateral tegmental cell groups of the medulla, have a disposition in the brain appropriate to a neuroendocrine function. Many of the cell bodies take up and bind estradiol (Heritage *et al.*, 1977), and the projections are particularly channeled to the hypothalamus (see Sawchenko and Swanson, 1981). Although the effects of VNAB lesions on reproductive cycles and gonadotropin secretion are slight, the neuroendocrine consequences of these lesions are considerable when examined in an appropriate context. Ovariectomized, estradiol- and progesterone-treated rats bearing VNAB lesions display normal or even exaggerated levels of proceptive behaviors; that is, their sexual motivation appears to be unimpaired, but they show a much reduced or even absent capacity to display lordosis (Hansen *et al.*, 1980, 1981). A similar picture emerges in the intact female, which shows normal, 4-day reproductive cycles according to vaginal cytology and proceptivity, but very low levels of receptivity during estrus following the VNAB lesion (Hansen *et al.*, 1981). This unique dissociation of proceptive and receptive behaviors appears to be related both to the critical stimuli eliciting the lordosis posture (which are tactile) and to the immobile nature of the response. Thus, the stimuli resulting from a mounting attempt by the male are known to be carried in the anterolateral columns (Pfaff, 1980), which have a heavy termination in the ventrolateral medulla around cell group A1 (Zemlan *et al.*, 1978). In addition, visceral afferents running in the vagus terminate heavily in cell group A2 (Sawchenko and Swanson, 1981). It seems reasonable to suggest, therefore, that the reason that VNAB lesions disrupt lordosis is that they interfere in some way with the receipt of, or the response to, somatosensory information carried in the ascending somatosensory pathways.

The nature of the influence of the NA pathways in this situation has yet to be specified, except that such a diffusely projecting system is unlikely to carry highly spatially and temporally coded sensory information. That is the function of the lemniscal and other spinothalamic pathways, which are organized somatotopically to perform such a role. Analogy with suggested functions of the locus coeruleus–DNAB NA system would imply some form of "enabling" role, that is, that the system is in some way concerned with responsiveness to sensory events, perhaps invoking some form of arousal mechanism. This implication awaits further sutdy. However, that cues generated during mounting or coitus are less effective in eliciting appropriate resonses is further emphasized in endocrine studies. Thus, cervical stimulation during proestrous fails to induce pseudopregnancy in female rats bearing VNAB lesions (Hansen *et al.*, 1981). Similarly, the olfactory block to pregnancy in mice, which also requires a coital event to induce the olfactory specificity that characterizes the phenomenon, is rendered nonspecific by noradrenergic denervations of the main and the accessory olfactory bulbs (Keverne and de la Riva, 1982). Much more work is required before the neural processes that NA systems are concerned with become clear.

EFFECTS OF LESIONS TO DA NEURONS USING 6-HYDROXYDOPAMINE

The now well-known reinterpretation of the lateral hypothalamic syndrome of adipsia and aphagia following the introduction of 6-OHDA as a dopaminergic neurotoxin need not be reviewed here (but see Robbins and Everitt, 1982). Suffice it to say that large DA depletions in the striatal complex, including the ventral striatum (nucleus accumbens) results in severe akinesia so that the organism is aphagic, adipsic, somnolent, and sometimes cataleptic and, not surprisingly, displays no sexual behavior. The interpretation of this syndrome and, therefore, of DA-dependent functions of the striatum has occupied the literature and has led to the concept of "sensorimotor integration" or, more recently, "activation" as an explanation of the pervasive role of DA in the control of behavior (Teitelbaum *et al.*, 1975; Robbins and Everitt, 1982).

Because it is the main theme of this chapter, let us focus briefly on the effects of these nigrostriatal-mesolimbic DA neurons on sexual behavior, as to do so emphasizes an important point in such studies: the nature of the behavioral response.

In the male rat, for example, sexual behavior is expressed in a most active way: a rapid sequence of investigation, mounting, thrusting, and eventually, ejaculation. Lesions of DA neurons, which render the male akinetic, of course abolish this behavior sequence—just as lateral hypothalamic lesions do (Caggiula *et al.,* 1973). There is little that one can do with this preparation to investigate the neural mechanisms further.

In the female, such lesions have an identical effect as far as active, proceptive responses are concerned (Robbins and Everitt, 1982). However, the same lesion actually *enhances* the display of lordosis, the immobile—even cataleptic—posture (Alton *et al.,* 1979; Robbins and Everitt, 1982). The same pattern of response change follows treatment with neuroleptic drugs (Everitt *et al.,* 1974, 1975a; Caggiula *et al.,* 1979). Thus, it would appear inadvisable now to speak of DA as an "inhibitory" system in the regulation of female sexual behavior and as an "excitatory" one in the male. Such an artificial distinction ignores the important and determining nature of the response, active or passive. Conversely, notions of "activation" that do take into account the nature of the response are more readily compatible with a fundamental and pervasive function of the DA system. For further discussion on this subject, see Robbins and Everitt (1982).

In the context of dopamine, let us return briefly to the relationship between hormone-dependent and monoaminergic mechanisms in the regulation of sexual behavior. Clearly, DA neurons in the midbrain are not prime sites for the action of sex steroids, whereas the preoptic area and the ventromedial areas of the hypothalamus are. Is there any relationship between the two? Anatomically, there is (see Hansen *et al.,* 1982) because, via the mfb, these structures project both to the A9–A10 region and to the peripeduncular nucleus in the midbrain. Interestingly, lesions of the medial preoptic area achieved electrolytically or with ibotenic acid, which so disrupt copulatory performance in the male rat, are effectively reversed behaviorally by treatment with the DA agonist lisuride (Hansen *et al.,* 1982a). Although this finding may be interpreted in several ways, one possible and attractive hypothesis is that the motivational specificity of the hypothalamic, sex-hormone-dependent mechanism gains access to the more general activational–motor-output DA system of the midbrain (the mesostriatal system) via these descending projections. Ingenious experiments are required to study this problem because the bilaterally DA-depleted animal is a very difficult one to observe and manipulate behaviorally.

SUMMARY AND CONCLUSIONS

From the above data, it can be seen that neuropharmacological compounds that influence monoaminergic and cholinergic mechanisms induce alterations in female and male copulatory behavior. Increasing dopaminergic and serotonergic activity impairs the display of receptive behaviors, although the opposite is also true. Proceptivity is clearly diminished by decreasing DA activity but is enhanced by decreasing 5-HT activity. These effects of changing DA transmission are related to the activation–immobility-inducing actions of drugs that, respectively, increase and decrease DA-mediated responses. Data concerning cholinergic and noradrenergic manipulations have also been reviewed. The actions of many of the drugs that affect sexual behavior seem to be explainable by direct effects on the CNS. Occasionally, controversy has arisen in the case of some drugs that facilitate female copulatory behavior (e.g., reserpine, PCPA, and α-MPT), which are thought to act indirectly by a release of progestogenic steroids from the adrenal cortex. This action has been quite persuasively shown for reserpine (Paris *et al.*, 1971). Treatments such as ether anesthesia, sodium hexobarbital, diphenylhydantoin, strychnine, and picrotoxin are also effective in inducing lordosis responses in ovariectomized rats after treatment with EB alone. This effect is not seen after adrenalectomy (Carrer and Meyerson, 1976), a finding suggesting similar mechanisms to that for reserpine. However, PCPA and α-MPT, methysergide, and cinanserin have been reported to induce changes in copulatory activity after adrenalectomy (Everitt *et al.*, 1975a; Ward *et al.*, 1975), although contradictory data were reported by Eriksson and Södersten (1973). But adrenalectomy is itself an unsatisfactory procedure for establishing or refuting the hypothesis that a drug acts via the adrenal, as adrenalectomy itself modifes cerebral monoamine metabolism (see Javoy *et al.*, 1968).

The time course of action of a drug might give some information about whether its action is direct or is mediated by progestogen release. Although the response after progesterone given intravenously does not appear until after 30–90 min, there are certain progestins with a latency of around 10–30 min (Meyerson, 1972). We do not know what steroids from adrenocortical sources are effective, but a latency of 10 min cannot be excluded. Furthermore, certain CNS stimulants, like strychnine, decreased the time between progestogen injection and the appearance of lordosis in adrenalectomized females (Carrer and Meyerson, 1976). In other words, the drug treatment may influence the time that it takes a progestin to produce its effect. However, for all the progestins tested so far, it is apparent that, once the behavior is activated, it will have a duration of several hours. Drugs that facilitate the lordosis response for an interval shorter than 1–2 hr are unlikely, therefore, to influence the response through progestin release. Unfortunately, such data are not available from studies using short-acting drugs facilitating the behavior. Thus, adrenal involvement can never be disregarded as being involved in a drug effect, but if the effect is dose-dependent and has a latency from injection to the occurrence of the behavior of 5–10 min and a duration of less than an hour, a direct effect is suggested. In addition, although adrenocortical secretion may be one factor that facilitates estrogen-induced female copulatory behavior, this should not exclude the involvement of other mechanisms as well.

The picture of the monoaminergic influence on sexual behavior in the male is much more complete than the one in the female. For male sexual behavior, there is no doubt that a clear-cut facilitation of all the main components of precopulatory and copulatory behavior can be achieved by increased dopaminergic and decreased serotonergic receptor activity. These effects are mediated by neither the gonads nor the adrenals. Inhibitory effects on male sexual behavior are obtained by decreased dopaminergic and increased serotonergic receptor activity. For details of the monoaminergic influence on male sexual behavior, see Table 2. Further studies are required to reveal the exact role of norepinephrine and epinephrine.

It appears, however, that β receptors are of greater significance in male sexual behavior than are α receptors.

If we consider heterotypical behavior (i.e., the male types of behavior displayed by a female and vice versa), it appears that mounting behavior is facilitated and lordotic behavior is inhibited by increased dopaminergic activity in genetic males as well as in genetic female rats. This finding is consistent with the hypothesis that the monoaminergic system is bound to the behavior rather than to the genetic sex.

Precopulatory behavior in the female has been too little studied to allow any conclusions about the neurotransmitters involved. In the male, nonmounting subjects had higher female-oriented activity after dopamine-stimulating agents, and the same behavior was decreased by 5-HT stimulation, a finding indicating that the monoaminergic control of the male's interaction with the female is not restricted to the copulatory act and includes elements preceding the copulatory act as well.

How is the urge to seek sexual contact influenced by psychotropic drugs? As yet, too few studies have been conducted to permit any conclusions, and the role of different neurotransmitters in the regulation of "sexual motivation" has to be elucidated by further experiments. However, an analysis of the data obtained in the different methods employed to measure sociosexual motivation and sex-specific orientation demonstrates that psychotropic drugs may have different actions on different appetitive elements of the behavior. For instance, PCPA increased the eagerness of a female to seek contact with a male, but the hormone-induced preference for a male in a choice situation was no longer present. In addition, the attractiveness of the female rat to a sexually active male was decreased, whereas PCPA seemed to facilitate the copulatory posture of the female and the mounting behavior of the male rat. This finding could serve as a good example of how one and the same drug may modulate different elements of sexual responses.

Given that it is possible to replace or modify the actions of hormone-induced behavior with a drug, we are still left with the serious dilemma of establishing whether the hormone and drug affect the behavior by common or completely different mechanisms. Very little direct information is available to solve this problem. It has been approached in two different ways, first, by searching the CNS for likely sites where hormones may affect aminergic neurons, and second, by attempting to demonstrate that hormones affect certain aspects of amine biosynthesis or metabolism or aminergic neuronal function. So far, however, the effects of neurotoxic compounds that destroy aminergic neurons have not been possible to relate directly to the hormone-dependent elements of sexual behavior. Hormones have been shown to affect monoaminergic neuronal functions. The changes in monoamines that hitherto have been demonstrated might be specifically related to the hormone activation of sexual behavior. However, although there are many neuroendocrine functions in which monoamines are implicated, it is, for the time being, not feasible to directly relate the hormone-induced changes in cerebral monoamines to the production of, for example, copulatory behavior.

It is with great regret that we record here that soon after this chapter had been completed, Carl Olof Malmnäs was killed in a car accident. Calle impressed everyone who came in contact with him. He was a dedicated scientist, full of enthusiasm, and with an intuitive feel for this unusually complicated area of research, monoamines and behavior. Scientists in this area must feel the great loss of this incisive investigator at the start of his research career. We also sadly acknowledge the loss of a dear personal friend.

B.J.M.

B.J.E.

References

Aghajanian, G. K., and Bunney, B. S. Pre- and post-synaptic feedback mechanisms in central dopaminergic neurons. In P. Seeman, and G. M. Brown (Eds.), *Frontiers of Neurology and Neuroscience Research*. Toronto: University of Toronto Press, 1974.

Ahlenius, S., Eriksson, H., Larsson, K., Modigh, K., and Södersten, P. Mating behavior in the male rat treated with p-chlorophenylalanine methyl ester alone and in combination with pargyline. *Psychopharmacologia* (Berlin), 1971, *20*, 383–388.

Ahlenius, S., Engel, J., Eriksson, J., Modigh, K., and Södersten, P. Importance of central catecholamines in the mediation of lordosis behavior in ovariectomized rats treated with estrogen and inhibitors of monoamine synthesis. *Journal of Neural Transmission*, 1972a, *33*, 247–256.

Ahlenius, S., Engel, J., Eriksson, H., and Södersten, P., Effects of tetrabenazine on lordosis behaviour and on brain monoamines in the female rat. *Journal of Neural Transmission*, 1972b, *33*, 155–162.

Ahlenius, S., Engel, J., Eriksson, J., Modigh, K., and Södersten, P. Involvement of monoamines in the mediation of lordosis behavior. In M. Sandler and G. L. Gessa (Eds.), *Sexual Behaviour, Pharmacology and Biochemistry*. New York: Raven Press, 1975.

Ahlenius, S., Engel, J., Larsson, K., and Svensson, L. Effects of pergolicle and bromocriphine on male rat sexual behavior. *Neural Transmission*, 1982, *54*, 165–170.

Ahlenius, S., Larsson, K., Svensson, L. Further evidence for an inhibitory role of central 5-HT in male rat sexual behavior. *Psychopharmacology* (1980a), *68*, 217–220.

Ahlenius, S., Larsson, K., and Svensson, L. Stimulating effects of Lisuride on masculine sexual behavior or rats. *European Journal of Pharmacology*, 1980b, *64*, 47–51.

Ahlenius, S., Larsson, K., Svensson, L., Hjort, S., Carlsson, A., Lindberg, P. Wikström, H., Sanchez, D., Arvidsson, L.-E., Hacksell, U., and Nilsson, J. L. G. Effects of a new type of 5-HT receptor agonist on male rat sexual behavior. *Pharmacology Biochemistry and Behavior*, 1981, *15*, 785–792.

Alton, E. F. W., Wright, Ch. W., and Everitt, B. J. Catecholamine mechanisms in the hormonal regulation of sexual behaviour. In E. Usdin (Ed.), *Catecholamines: Basic and Clinical Frontiers*. New York: Pergamon, 1979.

Barfield, R. J., Wilson, C., and McDonald, P. G. Sexual behaviour: Extreme reduction of postejaculatory refractory period by midbrain lesions in male rats. *Science*, 1975, *189*, 147–149.

Barrett, A. M., and Cullum, V. A. The biological properties of the optical isomers of propranolol and their effects on cardiac arrhytmias. *British Journal of Pharmacology*, 1968, *34*, 43–45.

Beach, F. A. Sexual attractivity, proceptivity and receptivity in female mammals. *Hormones and Behavior*, 1976, *7*, 105–138.

Beyer, C., Canchola, E., and Larsson, K. Facilitation of lordosis behavior in the ovariectomized estrogen primed rat by dibutyryl cAMP. *Physiology and Behavior*, 1981, *26*, 249–251.

Beyer, C. Gomora, P., Canchola, E., and Sandoval, Y. Pharmacological evidence that LH-RH action on lordosis behavior is mediated through a rise in cAMP. *Hormones and Behavior*, 1982, *16*, 107–112.

Bignami, G. Pharmacological influence on mating behavior in the male rat. *Psychopharmacologia* (Berlin), 1966, *10,* 44–58.

Blinkov, S. M., and Glezer, I. I. *The Human Brain in Figures and Tables: A Quantitative Handbook.* New York: Basic Books, Plenum Press, 1968.

Buchwald, N. A., Price, D. D., Vernon, L., and Hull, C. D. Caudate intracellular response to thalamic and cortical inputs. *Experimental Neurology,* 1973, *38,* 311–323.

Butcher, L. L., Butcher, S. G., and Larsson, K. Effects of apomorphine, (+)-amphetamine and nialamide on tetrabenazine-induced suppression of sexual behavior in the male rat. *European Journal of Pharmacology,* 1969, *7,* 283–288.

Caggiula, A. R., Antelman, S. M., and Zigmond, M. J. Disruption of copulation in male rats after hypothalamic lesions: A behavioural, anatomical and neurochemical analysis. *Brain Research,* 1973, *59,* 273–287.

Caggiula, A. R., Gay, V. L., Antelman, S. M., and Leggens, J. Disruption of copulation in male rats after hypothalamic lesions: A neuroendocrine analysis. *Neuroendocrinology,* 1975, *17,* 193–202.

Caggiula, A. R., Herndon, J. G., Scanlon, R., Greenstone, D., Bradshaw, W., and Sharp, D. Dissociation of active from immobility components of sexual behaviour in female rats by central 6-hydroxydopamine: Implications for CA involvement in sexual behaviour and sensorimotor responsiveness. *Brain Research,* 1979, *172,* 505–520.

Carrer, H., and Meyerson, B. J. Effects of CNS-depressants and stimulants on lordosis response in the female rat. *Pharmacology, Biochemistry and Behavior,* 1976, *4,* 497–505.

Chu, N.-S., and Bloom, F. E. The catecholamine-containing neurons in the cat dorso-lateral pontine tegmentum: distribution of the cell bodies and some axonal projections. *Brain Research,* 1974, *66,* 1–21.

Clark, T. K. Male rat sexual behaviour compared after 6-OHDA and electrolytic lesions in the dorsal NA bundle region of the midbrain. *Brain Research,* 1980, *202,* 429–443.

Clark, T. K., Caggiula, A. R., McConnell, R. A., and Antelman, S. M. Sexual inhibition is reduced by rostral midbrain lesions in the male rat. *Science,* 1975, *190,* 169–171.

Clemens, L. G., Humphrys, R. R., and Dohanich, G. P. Cholinergic brain mechanisms and the hormone regulation of female sexual behavior in the rat. *Pharmacology Biochemistry and Behavior,* 1980, *13,* 81–88.

Crowley, W. R., Ward, I. L., and Margules, D. L. Female lordotic behavior mediated by monoamines in male rats. *Journal of Comparative and Physiological Psychology,* 1975, *88,* 62–68.

Crowley, W. R., Feder, H. H., and Morin, L. P. The role of monoamines in sexual behavior of the female guinea pig. *Pharmacology, Biochemistry and Behavior,* 1976, *4,* 67–71.

Dahlström, A., and Fuxe, K. Evidence for the existence of monoamine neurons in the central nervous system: IV. Distribution of monoamine nerve terminals in the central nervous system. *Acta Physiologica Scandinavica,* 1965, *64,* 39–85.

Da Prada, M., Carruba, M., O'Brien, R. A., Saner, A., and Pletscher, A. Effect of 5,6-dihydroxytryptamine on sexual behavior of male rats. *European Journal of Pharmacology,* 1972, *19,* 288–290.

Da Prada, M., Carruba, M., Saner, A., O'Brien, R. A., and Pletscher, A. Action of L-dopa on sexual behavior of male rats. *Brain Research,* 1973, *55,* 383–389.

Da Prada, M., Bonetti, E. P., and Keller, H. H. Induction of mounting behavior in female and male rats by lisuride. *Neuroscience Letters,* 1977, *6,* 349–353.

Dewsbury, D. A. Effects of tetrabenazide on the copulatory behavior of male rats. *European Journal of Pharmacology,* 1972, *17,* 221–226.

Dewsbury, D. A., and Davis, H. N. Effects of reserpine on the copulatory behavior of male rats. *Physiology and Behavior,* 1970, *5,* 1331–1333.

Dewsbury, D. A., Davis, H. N., and Jansen, P. E. Effects of monoamine oxidase inhibitors on the copulatory behavior of male rats. *Psychopharmacologia* (Berlin), 1972, *24,* 209–217.

Dohanich, G. P., and Clemens, L. G. Brain areas implicated in cholinergic regulation of sexual behavior. *Hormones and Behavior,* 1981, *15,* 157–167.

Eliasson, M. Action of repeated injection of LSD and apomorphine on the copulatory response of female rats. *Pharmacology, Biochemistry and Behavior,* 1976, *5,* 621–625.

Eliasson, M., and Meyerson, B. J. Influence of LSD and apomorphine on hormone-activated copulatory behavior in the female rat. *Acta Physiologica Scandinavica,* 1973, *Suppl. 396,* Abstr. 129.

Eliasson, M., and Meyerson, B. J. Sexual preference in female rats during estrous cycle, pregnancy and lactation. *Physiology and Behavior,* 1975, *14,* 705–710.

Eliasson, M., and Meyerson, B. J. Comparison of the action of lysergic acid diethylamide and apomorphine on the copulatory response in the female rat. *Psychopharmacology* (Kbh), 1976, *49*, 301–306.

Eliasson, M., and Meyerson, B. J. The effects of lysergic acid diethylamide on copulatory behavior in the female rat. *Neuropharmacology,* 1977, *16*, 37–44.

Eliasson, M., and Meyerson, B. J. Development of socio-sexual approach behavior in male laboratory rats. *Journal of Comparative and Physiological Psychology,* 1981, *95*, 160–165.

Emery, D. E., and Sachs, B. D. Ejaculatory pattern in female rats without androgen treatment. *Science,* 1975, *190*, 484–485.

Eriksson, H., and Södersten, P. Failure to facilitate lordosis behavior in adrenal-ectomized and gonadectomized estrogen-primed rats with monoamine-synthesis inhibitors. *Hormones and Behavior,* 1973, *4*, 89–98.

Everitt, B. J. Adrenergic mechanisms in the control of sexual behaviour: The relationship to depression and its treatment. In A. Jukes (Ed.), *Depression—The Biochemical and Physiological Role of Ludionil.* New York: Ciba Publ., 1977.

Everitt, B. J. A neuroanatomical approach to the study of monoamines and sexual behavior. In J. Hutchison (Ed.), *Biological Determinants of Sexual Behaviour.* New York: Wiley, 1978.

Everitt, B. J. Cerebral monoamines and sexual behavior. In J. Money and H. Musagh (Eds.), *Textbook of Sexology.* New York: Elsevier, 1981.

Everitt, B. J., and Fuxe, K. Dopamine and sexual behavior of female rats: Effects of dopamine receptor agonists and sulpiride. *Neuroscience Letters,* 1977a, *4*, 209–213.

Everitt, B. J., and Fuxe, K. Serotonin and the sexual behavior of female rats: Effects of hallucinogenic indolealkylamines and phenylethylamines. *Neuroscience Letters,* 1977b, *4*, 213–220.

Everitt, B. J., Fuxe, K., and Hökfelt, T. Inhibitory role of dopamine and 5-hydroxytryptamine in the sexual behavior of female rats. *European Journal of Pharmacology,* 1974, *29*, 187–191.

Everitt, B. J., Fuxe, K., Hökfelt, T., and Jonsson, G. Role of monoamines in the control by hormones of sexual receptivity in the female rat. *Journal of Comparative and Physiological Psychology,* 1975a, *89*, 556–572.

Everitt, B. J., Fuxe, K., and Jonsson, G. The effects of 5,7-dihydroxytryptamine lesions of ascending 5-hydroxytryptamine pathways on the sexual and aggressive behavior of female rats. *Journal of Pharmacology* (Paris), 1975b, *6*, 25–32.

Ferguson, J., Henriksen, S., Cohen, H., and Mitchell, G. Hypersexuality and behavioral changes in cats caused by administration of p-chlorophenylalanine. *Science,* 1970, *168*, 499–501.

Fuxe, K. Evidence for the existence of monoamine neurons in the central nervous system: IV. Distribution of monoamine nerve terminals in the central rat nervous system. *Acta Physiologica Scandinavica,* 1965, *64*, 37–85.

Fuxe, K., Bolme, P., and Everitt, B. J. The effect of beta-adrenergic blocking agents on arterial blood pressure and central monoamine turnover: Possibility of agonist activity at central adrenaline receptors. In D. Ganten, R. Dietz, B. Löth, and F. Gross (Eds.), *Beta Adrenergic Blockers and Hypertension.* Stuttgart: G. Thieme, 1976a.

Fuxe, K., Everitt, B. J., Agnati, L., Fredholm, B., and Jansson, B. On the biochemistry and pharmacology of hallucinogens. In D. Kermal, G. Bartholini, and P. Richter, (Eds.), *Schizophrenia Today.* New York: Pergamon Press, 1976b.

Fuxe, K., Everitt, B. J., and Hökfelt, T. Enhancement of sexual behavior in the female rat by nicotine. *Pharmacology, Biochemistry and Behavior,* 1977, *7*, 147–151.

Gawienowski, A. M., and Hodgen, G. D. Homosexual activity in male rats after p-chlorophenylalanine-effects of hypophysectomy and testosterone. *Physiology and Behavior,* 1971, *7*, 551–555.

Gessa, G. L., and Tagliamonte, A. Role of brain serotonin and dopamine in male sexual behavior. In M. Sandler and G. L. Gessa (Eds.), *Sexual Behavior, Pharmacology and Biochemistry.* New York: Raven Press, 1975.

Gradwell, P. B., Everitt, B. J., and Herbert, J. 5-Hydroxytryptamine in the central nervous system and sexual receptivity of female rhesus monkeys. *Brain Research,* 1975, *88*, 281–293.

Gray, G. D., Davis, H. N., and Dewsbury, D. A. Effects of *l*-DOPA on the heterosexual copulatory behavior of male rats. *European Journal of Pharmacology,* 1974, *27*, 367–370.

Green, A. G., and Graham-Smith, D. G., (−)-Propranolol inhibits the behavioural responses of rats to increased 5-hydroxytryptamine in the central nervous system. *Nature,* 1976, *262*, 594–596.

Hansen, S., Södersten, P., and Srebro, B. A daily rhythm in the behavioural sensitivity of the female rat to oestradiol. *Journal of Endocrinology,* 1978, *77*, 381–388.

Hansen, S., Stanfield, E. J., and Everitt, B. J. The role of ventral bundle noradrenergic neurones in sensory components of sexual behaviour and coitus-induced pseudopregnancy. *Nature,* 1980, *286,* 152–154.

Hansen, S., Stanfield, E. J., and Everitt, B. J. The effects of lesion of lateral tegmental noradrenergic neurons on components of sexual behaviour and pseudopregnancy in female rats. *Neuroscience,* 1981, *6,* 1105–1107.

Hansen, S., Köhler, C., Goldstein, M., and Steinbusch, H. U. M. Effects of ibotenic acid-induced neuronal degeneration in the medial preoptic area and the lateral hypothalamic area on sexual behaviour in the male rat. *Brain Research,* 1982a, *239,* 213–232.

Hansen, S., Köhler, C., and Ross, S. B. On the role of the dorsal mesencephalic tegmentum in the control of masculine sexual behaviour in the rat: Effects of electrolytic lesions, ibotenic acid and DSP4. *Brain Research,* 1982b, *240,* 311–320.

Hansult, C. D., Uphouse, L. L., Schlesinger, K., and Wilson, J. R. Induction of estrus in mice: Hypophyseal-adrenal effects. *Hormones and Behavior,* 1972, *3,* 113–121.

Heritage, A. S., Grant, L. D., and Stumpf, W. E. ^{3}H-estradiol in catecholamine neurons of the rat brain stem: combined localizations by autoradiography and formalin-induced fluorescence. *Journal of Comparative Neurology,* 1977, *176,* 607–630.

Hetta, J., and Meyerson, B. J. Neonatal castration and sex dependent orientation in the male rat. *Experimental Brain Research,* 1975, *Suppl. to Vol. 23,* 89.

Hetta, J., and Meyerson, B. J. Sexual motivation in the male rat. A methodological study of sex-specific orientation, and the effects of gonadal hormones. *Acta Physiologica Scandinavica,* 1978, *Suppl. 453,* 1–68.

Hökfelt, T., Johansson, O., Ljungdahl, Å., Lundberg, J. M., and Schultzberg, M. Peptidergic neurons. *Nature* (London), 1980a, *284,* 515–522.

Hökfelt, T., Skirboll, L., Rehfeld, J. F., Goldstein, M., Markey, K., and Dann, O. A subpopulation of mesencephalic dopamine neurons projecting to limbic area contains a cholecystokininlike peptide: Evidence from immunohistochemistry combined with retrograde tracing. *Neuroscience,* 1980b, *5,* 2093–2124.

Hoyland, V. J., Shillito, E. E., and Vogt, M. Tje effect of parachlorophenylalanine on the behavior of cats. *British Journal of Pharmacology,* 1970, *40,* 659–667.

Hyyppä, M., Lehtinen, P., and Rinne, U. K. Effect of L-DOPA on hypothalamic, pineal and striatal monoamine and on the sexual behaviour of the rat. *Brain Research,* 1971, *30,* 265–272.

Javoy, F., Glowinski, J., and Kordon, C. Effects of adrenalectomy on the turnover of norepinephrine in the rat brain. *European Journal of Pharmacology,* 1968, *4,* 103–104.

Johnson, M. H., and Everitt, B. J. *Essential Reproduction.* Oxford, England: Blackwell, 1980.

Jonsson, G., Malmfors, T., and Sachs, C. (Eds.) *Chemical Tools in Catecholamine research: I. 6-Hydroxydopamine as a Denervation Tool in Catecholamine Research.* Amsterdam: Elsevier, 1975.

Kemp, J. M., and Powell, T. P. S. The structure of the caudate nucleus of the cat: Light and electron microscopy. *Philosophical Transactions of the Royal Society of London* B, 1971, *262,* 383–401.

Keverne, E. B., and de la Riva, C. Pheromones in mice: Reciprocal interaction between the nose and brain. *Nature,* 1982, *296,* 148–150.

Key, B. J., and Krzywosinski, L. Electrocortical changes induced by the perfusion of noradrenaline, acetylcholine and their antagonists directly into the dorsal raphé nucleus of the cat. *British Journal of Pharmacology,* 1977, *61,* 297–306.

Larsson, K., and Södersten, P. Lordosis behavior in male rats treated with estrogen in combination with tetrabenazine and malamide. *Psychopharmacologia,* 1971, *21,* 13.

Larsson, K., Fuxe, K., Everitt, B. J., Holmgren, M., and Södersten, P. Sexual behavior of male rats after intracerebral injection of 5,7-dihydroxytryptamine. *Brain Research* 1978, *141,* 292–303.

Leavitt, F. I. Drug-induced modifications in sexual behavior and open field lodomotion of male rats. *Physiology and Behavior,* 1969, *4,* 677–683.

Lindström, L. H. The effect of pilocarpine in combination with monoamine oxidase inhibitors, imipramine or desmethylimipramine on oestrous behaviour in female rats. *Psychopharmacologia* (Berlin), 1970, *17,* 160–168.

Lindström, L. H. The effect of pilocarpine and oxotremorine on oestrous behaviour in female rats after treatment with monoamine depletors or monoamine synthesis inhibitors. *European Journal of Pharmacology,* 1971, *15,* 60–65.

Lindström, L. H. The effect of pilocarpine and oxotremorine on hormone-activated copulatory behavior in the ovariectomized hamster. *Naunyn-Schmiedeberg's Archives of Pharmacology*, 1972, *275*, 233–241.

Lindström, L. H. Further studies on cholinergic mechanisms and hormone-activated copulatory behaviour in the female rat. *Journal of Endocrinology*, 1973, *56*, 275–283.

Lindström, L. H., and Meyerson, B. J. The effect of pilocarpine, oxotremorine and arecoline in combination with methyl-atropine of atropine on hormone activated oestrous behaviour in ovariectomized rats. *Psychopharmacologia*, 1967, *11*, 405–413.

Luine, V. N., and Fischette, C. T. Inhibition of lordosis behavior by intrahypothalamic implants of pargyline. *Neuroendocrinology*, 1982, *34*, 237–244.

Luine, V. N., and Paden, C. M. Effects of monoamine oxidase inhibition on female sexual behavior, serotonin levels and type A and B monoamine oxidase activity. *Neuroendocrinology*, 1982, *34*, 245–251.

Malmnäs, C. O. Monoaminergic influence on testosterone-activated copulatory behavior in the castrated male rat. *Acta Physiologica Scandinavica*, 1973, *Suppl. 395*, 1–128.

Malmnäs, C. O. Effects of D- and L-propranolol on sexual behavior in male rats. *Experimental Brain Research*, 1975, *Suppl. to Vol. 23*, 133.

Malmnäs, C. O. The significance of dopamine, versus other catecholamines, for L-DOPA induced facilitation of sexual behavior in the castrated male rat. *Pharmacology Biochemistry and Behavior*, 1976, *4*, 521–526.

Malmnäs, C. O., and Meyerson, B. J. Monoamines and testosterone activated copulatory behaviour in the castrated male rat. *Acta Pharmacologica et Toxicologica*, 1970, *28*, Suppl. 1.

Malmnäs, C. O., and Meyerson, B. J. p-Chlorophenylalanine and copulatory behaviour in the male rat. *Nature*, 1971, *232*, 398–400.

Malmnäs, C. O., and Meyerson, B. J. Monoamines and copulatory activation in the castrated male rat. *Acta Pharmacologica et Toxicologica*, 1972, *31*, Suppl. 1.

McEwen, B. S., Davis, P. G., Parsons, B., and Pfaff, D. W. The brain as a target for steroid hormone action. *Annual Review of Neuroscience*, 1979, *2*, 65–112.

Meyerson, B. J. Central nervous monoamines and hormoned induced estrus behaviour in the spayed rat. *Acta Physiologica Scandinavica* 1964a, *63*, Suppl. 241, 3–32.

Meyerson, B. J. The effect of neuropharmacological agents on hormone-activated estrus behaviour in ovariectomized rats. *Archives Internationales de Pharmacodynamie et de Thérapie*, 1964b, *150*, 4–33.

Meyerson, B. J. Estrus behaviour in spayed rats after estrogen or progesterone treatment in combination with reserpine or tetrabenazine. *Psychopharmacologia*, 1964c, *6*, 210–218.

Meyerson, B. J. The effect of imipramine and related antidepressive drugs on estrus behaviour in ovariectomized rats activated by progesterone, reserpine or tetrabenazine in combination with estrogen. *Acta Physiologica Scandinavica*, 1966, *67*, 411–422.

Meyerson, B. J. Amphetamine and 5-hydroxytryptamine inhibition of copulatory behaviour in the female rat. *Annales Medicinae Experimentalis et Biologiae Fenniae*, 1968a, *46*, 394–398.

Meyerson, B. J. Female copulatory behaviour in male and androgenized female rats after oestrogen/amine depletor treatment. *Nature* 1968b, *217*, 683–684.

Meyerson, B. J. Monoamines and hormone activated oestrous behaviour in the ovariectomized hamster. *Psychopharmacologia*, 1970, *18*, 50–57.

Meyerson, B. J. Latency between intravenous injection of progestines and appearance of estrous behavior in the estrogen treated ovariectomized rat. *Hormones and Behaviour*, 1972, *3*, 1–9.

Meyerson, B. J. Drugs and sexual motivation in the female rat. In M. Sandler and G. L. Gessa (Eds.), *Sexual Behavior, Pharmacology and Biochemistry* New York: Raven Press, 1975.

Meyerson, B. J. Psychotropic drugs and sexual behavior. Proceedings of the 10th CINP Congress, Quebec. In P. Veniker, C. Radouco-Thomas, and A. Villeneuve (Eds.), *Neuro-Psychopharmacology*. New York: Pergamon Press, 1978.

Meyerson, B. J. Hypothalamic hormones and behavior. *Medical Biology*, 1978, *57*, 69–83.

Meyerson, B. J. The relationship between β-endorphin induced effects and monominergic mechanisms on socio-sexual behaviors in the male rat. *Neuroscience Letters*, 1980, *Suppl. 5*, 356.

Meyerson, B. J. Comparison of the effects of β-endorphin and morphine on exploratory and socio-sexual behaviour in the male rat. *European Journal of Pharmacology*, 1981, *69*, 453–463.

Meyerson, B. J. Endorphin-monoamine interaction and steroid dependent behaviour. In Proc. E.S.C.P.B. 4th Conference on Hormones and Behaviour in Higher Vertebrates, 1982.

Meyerson, B. J., and Eliasson, M. Pharmacological and hormonal control of reproductive behavior. In L. L. Iversen, S. D. Iversen, and S. H. Snyder (Eds.), *Handbook of Psychopharmacology*. New York: Plenum Press, 1977.

Meyerson, B. J., and Höglund, A. U. Exploratory and socio-sexual behaviour in the male laboratory rat: A methodological approach for the investigation of drug action. *Acta Pharmacologica et Toxicologica*, 1981, *48*, 168–180.

Meyerson, B. J., and Lewander, T. Serotonin synthesis inhibition and estrous behaviour in female rats. *Life Sciences*, 1970, *9*, 661–671.

Meyerson, B. J., and Lindström, L. H. Sexual motivation in the female rat. A methodological study applied to the investigation of the effect of estradiol benzoate. *Acta Physiologica Scandinavica*, 1973, *Suppl. 389*, 1–80.

Meyerson, B. J., and Malmnäs, C. O. Monoaminergic implication in different elements of sexual behaviour. *Brain Research*, 1977, *127*, 377–378.

Meyerson, B. J., and Malmnäs, C. O. Brain monoamines and sexual behaviour. In J. Hutchison (Ed.), *Biological Determinants of Sexual Behaviour*. London: Wiley, 1978.

Meyerson, B. J., and Terenius, L. β-Endorphin and male sexual behavior. *European Journal of Pharmacology*, 1977, *42*, 191.

Meyerson, B. J., Eliasson, M., Lindström, L., Michanek, A., and Söderlund, A. C. Monoamines and female sexual behaviour. In T. A. Ban, J. R. Boissier, G. J. Gessa, H. Heimann, L. Hollister, H. E. Lehmann, I. Munkvad, H. Steinberg, F. Sulser, A. Sundwall, and O. Vinař (Eds.), *Psychopharmacology, Sexual Disorders and Drug Abuse*. Amsterdam: North Holland Publication Company, 1973.

Meyerson, B. J., Carrer, H., and Eliasson, M. 5-Hydroxytryptamine and sexual behavior in the female rat. In E. Costa, M. Gessa, and M. Sandler (Eds.), *Advances in Biochemistry and Psychopharmacology*, Vol. 2. New York: Raven Press, 1974.

Meyerson, B. J., Palis, A., and Sietnieks, A. Hormone-monoamine interactions and sexual behavior. In C. Beyer (Ed.), *Endocrine Control of Sexual Behavior*. New York: Raven Press, 1979.

Michanek, A. Potentiation of D- and L-amphetamine effects on copulatory behavior in female rats by treatment with α-adrenoreceptor blocking drugs. *Archives internationales de Pharmacodynamie et de Thérapie*, 1979, *239*, 241–256.

Michanek, A., and Meyerson, B. J. Copulatory behavior in the female rat after amphetamine and amphetamine derivatives. In M. Sandler and G. L. Gessa (Eds.), *Sexual Behavior, Pharmacology and Biochemistry*. New York: Raven Press, 1975.

Michanek, A., and Meyerson, B. J. A comparative study of different amphetamines on copulatory behavior and stereotype activity in the female rat. *Psychopharmacologia*, 1977a, *53*, 175–183.

Michanek, A., and Meyerson, B. J. The effects of different amphetamines on copulatory behaviour and stereotype activity in the female rat after treatment with monoamine depletors and synthesis inhibitors. *Archives Internationales de Pharmacodynamie et de Thérapie*, 1977b, *229*, 301–312.

Michanek, A., and Meyerson, B. J. Influence of estrogen and progesterone on behavioral effects of apomorphine and amphetamine. *Pharmacology Biochemistry and Behavior*, 1982, *16*, 875–879.

Moore, R. Y., and Bloom, F. E. Central catecholamine neuron systems: Anatomy and physiology of dopamine systems. *Annual Review of Neuroscience*, 1978, *1*, 129–169.

Moore, R. Y., and Bloom, F. E. Central catecholamine neuron systems: Anatomy and physiology of the norepinephrine and epinephrine systems. *Annual Review of Neuroscience*, 1979, *2*, 113–168.

Morgane, P. J., Stern, W. C., and Berman, E. Inhibition of unit activity in the anterior raphé by stimulation of the locus coeruleus. *Anatomical Record*, 1974, *178*, 421.

Moss, R. L., and McCann, S. M. Induction of mating behavior in rats by luteinizing-hormone-releasing factor. *Science*, 1973, *181*, 177–179.

Paris, C. A., Resko, J. A., and Goy, R. W. A possible mechanism for the induction of lordosis by reserpine in spayed rats. *Biology of Reproduction*, 1971, *4*, 23–30.

Perez-Cruet, J., Taligamonte, A., and Gessa, G. L. Differential effects of p-chlorophenylalanine (PCPA) on sexual behavior and on sleep patterns of male rabbits. *Rivista Farmacologia e Terapeutica*, 1971, *11*, 27–34.

Peroutka, S. J., Lebovitz, R. N., and Snyder, S. H. Two distinct central serotonin receptors with different physiological functions. *Science*, 1981, *212*, 827–829.

Pfaff, D. W. Luteinizing-hormone-releasing factor potentiates lordosis behavior in hypophysectomized ovariectomized female rats. *Science*, 1973, *182*, 1148–1149.

Pfaff, D. W. *Estrogens and Brain Function*. New York: Springer, 1980.

Rakic, P. Role of cell interactions in development of dendritic patterns. In G. Kreutzberg (Ed.), *Advances in Neurology*, Vol. 12. *Physiology and Pathology of Dendrites*. New York: Raven Press, 1975.

Redmond, E. E., Moas, J. W., Kling, A., and Graham, C. W. Social behavior of monkeys selectively depleted of monoamines. *Science*, 1971, *174*, 428–430.

Robbins, T. W., and Everitt, B. J. Functional studies of the central catecholamines. *International Review of Neurobiology*, 1982, *23*, 303–365.

Salis, P. J., and Dewsbury, D. A. p-Chlorophenylalanine facilitates copulatory behaviour in male rats. *Nature*, 1971, *232*, 400–401.

Sawchenko, P. E., and Swanson, L. W. Central noradrenergic pathways for the integration of hypothalamic neuroendocrine and autonomic responses. *Science*, 1981, *214*, 685–687.

Schmidt, F. O., Dev, P., and Smith, B. H. Electronic processing of information by brain cells. *Sciences*, 1976, *193*, 114–120.

Sheard, M. The effect of p-chlorophenylalanine on behaviour in rats: Relation to brain serotonin and 5-hydroxyindole acetic acid. *Brain Research*, 1969, *15*, 524–528.

Sheard, M. Brain serotonin depletion by p-chlorophenylalanine or lesions of raphe neurons in rats. *Physiology and Behavior*, 1973, *10*, 809–811.

Shepard, G. W. *The Synaptic Organization of the Brain*. New York: Oxford Press, 1974.

Shillito, E. E. The effect of p-chlorophenylalanine on social interactions of male rats. *British Journal of Pharmacology*, 1969, *36*, 193–194P.

Shillito, E. E. The effect of parachlorophenylalanine on social interactions of male rats. *British Journal of Pharmacology*, 1970, *38*, 305–315.

Sietnieks, A. Involvement of 5-HT$_2$ receptors in the LSD-induced suppression of lordosic behavior in the female rat. *Journal of Neural Transmission*, in press.

Sietnieks, A., and Meyerson, B. J. Enhancement by progesterone by progesterone of lysergic acid diethylamide inhibition of copulatory response in the female rat. *European Journal of Pharmacology*, 1980, *63*, 57–64.

Sietnieks, A., and Meyerson, B. J. Enhancement by progesterone of 5-hydroxytryptophan inhibition of the copulatory response in the female rat. *Neuroendocrinology*, 1982, *35*, 321–326.

Sietnieks, A., and Meyerson, B. J. Progesterone enhancement of lysergic acid diethylamide and levo-5-hydroxytryptophan stimulation of the copulatory response in the female rat. *Neuroendocrinology*, 1983, *36*, 462–467.

Smith, L. A., and Meyerson, B. J. Long-term effects of two antidepressants on copulatory and exploratory behaviour in the female rat. *Acta Pharmaceutica Et Toxicologica*, 1984a, *55*, 188–193.

Smith, L. A., and Meyerson, B. J. Influence of long-term zimeldine treatment on LSD-induced behavioural effects. *Acta Pharmaceutica Et Toxicologica*, 1984b, *55*, 194–198.

Södersten, P., Berge, O. G., and Hole, K. Effects of p-chloroamphetamine and 5,7-dihydroxytryptamine on the sexual behaviour of gonadectomized male and female rats. *Pharmacology, Biochemistry and Behavior*, 1978, *9*, 499–508.

Soulairac, M. L. Étude expérimentale des régulations hormone-nerveuses du comportement sexuel du rat mâle. *Annales d'Endocrinologie* (Paris), 1963, *166*, 798–802.

Soulairac, M. L., and Soulairac, A. Action de la nicotine sur le comportement sexuel du rat mâle. *Comptes Rendus Des Séances De La Société De Biologie et de Ses Filiales (Paris)*, 1972, *166*, 798–802.

Steinbusch, H. W. M. Distribution of serotonin-immunoreactivity in the central nervous system of the rat: Cell bodies and terminals. *Neuroscience*, 1981, *6*, 557–618.

Tagliamonte, A., Tagliamonte, P., Gessa, G. L., and Brodie, B. B. Compulsive sexual activity induced by p-chlorophenylalanine in normal and pinelaectomized male rats. *Science*, 1969, *166*, 1433–1435.

Tagliamonte, A., Tagliamonte, P., and Gessa, G. L. Reversal of pargyline-induced inhibition of sexual behaviour in male rats by p-chlorophenylalanine. *Nature* (London), 1971, *230*, 244–245.

Tagliamonte, A., De Montis, G., Olianos, M., Vargin, L., Corsini, G. V., and Gessa, G. L. Selective increase of brain dopamine synthesis by sulpiride. *Journal of Neurochemistry*, 1975, *24*, 707–710.

Tagliamonte, A., Frata, W., del Fiacco, M., Gessa, G. L. Possible stimulatory role of brain dopamine in the copulatory behavior of male rats. *Pharmacology, Biochemistry and Behavior*, 1977, *2*, 257–260.

Teitelbaum, P., and Wolgin, D. L. Neurotransmitters and the regulation of food intake. *Progress in Brain Research*, 1975, *42*, 235–249.

Ungerstedt, U. Stereotaxic napping of the monoamine pathways in the rat brain. *Acta Physiologica Scandinavica*, 1971, *367*, 1–48.

Usdin, E., and Bunney, W. E. *Pre and Postsynaptic Receptors*. New York: Dekker, 1975.

Ward, L. L., Crowley, W. R., Zemlan, F. P., and Margules, D. L. Monoaminergic mediation of female sexual behavior. *Journal of Comparative Physiology and Psychology,* 1975, *88,* 53–61.

Whalen, R. E., and Luttge, W. G. p-Chlorophenylalanine methyl ester: An aphrodiasiac? *Science,* 1970, *169,* 1000–1001.

Wiegant, V. M., and Gispen, W. H. Cyclic nucleotides and nerve function. In W. H. Gispen (Ed.), *Molecular and Functional Neurobiology.* Amsterdam: Elsevier, 1976.

Zemlan, F. P., Ward, I. L. L., Crowley, W. R., and Margules, D. L. Activation of lordotic responding in female rats by suppression of serotonergic activity. *Science,* 1973, *179,* 1010–1011.

Zemlan, F. P., Leonard, C. M., Kow, L-M., and Pfaff, D. W. Ascending tracts of the lateral columns of rat spinal cord: A study using the silver impregnation and horseradish peroxidase techniques. *Experimental Neurology,* 1978, *62,* 298–334.

Zitrin, A, Beach, F. A., Barchas, J. D., and Dement, W. C. Sexual behavior of male cats after administration of parachlorophenylalanine. *Science,* 1970, *170,* 1970.

12

Brain Mechanisms and Parental Behavior

MICHAEL NUMAN

INTRODUCTION

The present review concerns itself with a discussion of the brain mechanisms underlying parental behavior in mammals. Until recently, little research has been directed at attempting to elucidate the neural mechanisms underlying such behavior. Additionally, most of the research has investigated the neural mechanisms underlying maternal behavior in the rat and the mouse, and the majority of these studies deal with the rat. It is hoped that the present review will be useful in stimulating research that will eventually result in a truly comparative understanding of the brain mechanisms underlying mammalian parental behavior.

Because most of the studies on the neural basis of parental behavior in mammals have been restricted to maternal behavior in the laboratory rat, a brief description of the normal course of such behavior is presented now. Information on various aspects of parental behavior in other species is included in other sections of this review when it is considered essential to a proper understanding of the experimental results. The reader is referred to Rosenblatt and Lehrman (1963) for a more extensive description of maternal behavior in the rat and to the volume edited by Rheingold (1963) for descriptions of maternal behavior in several other mammalian species.

The duration of pregnancy in the rat is 22–23 days. Near the time of parturition, the female begins to build a maternal nest, in which she subsequently gives

MICHAEL NUMAN   Department of Psychology, Boston College, Chestnut Hill, Massachussetts 02167.

birth (paper strips are provided as nesting material in several laboratories). This nest is more elaborate than the sleeping nests typically built by nonpregnant females (Kinder, 1927). Nursing begins almost immediately after parturition. During nursing, the mother crouches over the young in order to expose her mammary region, in this way allowing the pups to attach to her nipples and suckle. During nursing, the female appears relatively motionlesss, but close observation indicates that postural adjustments are often made in response to stimulation from the nuzzling pups. Retrieving can also be observed beginning at parturition. If a pup becomes displaced from the nest, for example, the mother, on encountering it, picks it up in her mouth and carries it back to the nest, where it is deposited. Retrieving can also be observed when a mother changes the site of her nesting area. Retrieving can be elicited in the laboratory by removing the young from the nest and placing them in a quadrant of the observation cage opposite the nest. Lastly, beginning at parturition, the female can also be seen to lick the pups, particularly their anogenital region; this licking stimulates elimination.

Once initiated, maternal behavior is maintained for a period of about 4 weeks, until the young are weaned. The level of maternal responsivity, however, does not remain constant; instead, it decreases as the young grow older. Both retrieving and nest-building behavior begin to decline during the second postpartum week (Moltz and Robbins, 1965; Reisbick et al., 1975; Rosenblatt and Lehrman, 1963). According to Moltz and Robbins (1965), nursing does not decline during the first 20 or 21 days postpartum, although beginning at about Day 14 postpartum the pups rather than the mother initiate nursing bouts (Rosenblatt and Lehrman, 1963; cf. Grota and Ader, 1974). Nursing usually ceases by the end of the fourth postpartum week.

The various components of maternal behavior in the rat (retrieving, nursing, nest building, and licking) should not be viewed as being determined by a single underlying physiological mechanism. Behavioral observations have indicated that the various components of maternal behavior can occur independently of one another (Slotnick, 1967b), and Slotnick (1969) showed that different forebrain lesions can selectively interfere with different aspects of maternal behavior. Therefore, in studies dealing with the neural basis of maternal behavior, all of its components should be measured. Such measurement helps to determine the extent to which the various components share a common neural basis.

At this point, it is worthwhile to briefly discuss some pertinent findings concerning the hormonal and the nonhormonal determinants of maternal behavior in the rat. This discussion will serve as useful background material and will underscore the complexities involved in interpreting studies dealing with the neural basis of maternal behavior in the rat. For more extensive information on these topics, the reader is referred to the chapter by Rosenblatt, Mayer, and Siegel in this book and to the following reviews: Moltz (1971), Rosenblatt (1970), and Slotnick (1975).

The female rat shows immediate maternal attention toward her young at the time of parturition, and this immediate display of maternal responsivity is also exhibited toward foster young offered at this time (Moltz et al., 1966; Rosenblatt and Siegel, 1975). Research accumulated over the past several years has shown that this immediate onset of maternal behavior at parturition is hormonally mediated (Moltz *et al.,* 1970; Rosenblatt and Siegel, 1975; Siegel and Rosenblatt, 1975b; Terkel and Rosenblatt, 1972; Zarrow *et al.,* 1971a). The point to stress is that the exact nature of the hormones involved has not been conclusively determined. Three hormones—estrogen, progesterone, and prolactin—have received the most attention. Moltz *et al.* (1970) noted that, during the latter part of pregnancy, blood progesterone levels decline, whereas estrogen and prolactin concentrations increase, and they provided evidence that these particular hormonal changes were important to the immediate onset of maternal behavior at parturition. More recent findings have indicated a major role for the near-term increase in estrogen levels in the hormonal activation of maternal behavior (Rosenblatt and Siegel, 1975; Siegel and Rosenblatt, 1975a,b).

Unlike the onset of maternal behavior at parturition, the maintenance of maternal behavior during the postpartum period is believed to have a nonhormonal basis (Numan *et al.,* 1972; Rosenblatt, 1970, 1975). The hypothesis has been advanced that the hormonal changes that occur around the time of parturition affect neural mechanisms, with the result that the parturient female is immediately responsive to pup-related stimuli. Maternal behavior is then maintained and subsequently declines during the postpartum period in response to sensory stimuli from the pups that act directly on the neural mechanisms underlying maternal behavior. Hormonal mediation is not regarded as being involved in this postpartum phase of maternal behavior.

In contrast to the parturient and lactating female rat, virgin female rats are not immediately maternally responsive to foster young. This finding is what one should expect if the hormonal changes associated with the termination of pregnancy are essential for the immediate onset of maternal behavior at parturition. In fact, research on the hormonal basis of the onset of maternal behavior in the rat took as its starting point this difference between the virgin and the parturient female (Stone, 1925; Wiesner and Sheard, 1933). Under certain experimental conditions, however, the virgin female can be induced to show maternal behavior. If such females are continuously exposed to (cohabit with) foster young, after a latency of approximately 5–7 days they will begin to show a pattern of maternal behavior toward these young that is quite similar to the one that occurs in the postpartum female (Fleming and Rosenblatt, 1974a; Reisbick *et al.,* 1975; Rosenblatt, 1967). This pup-induced or pup-stimulated maternal behavior in virgin females does not seem to be dependent on pup-induced hormone release, as it occurs in either ovariectomized or hypophysectomized virgins (Rosenblatt, 1967; also see Terkel and Rosenblatt, 1971). Therefore, we can conceive of sensory stimuli from the pups as affecting the neural mechanisms underlying maternal behavior, with the result that such behavior is induced in the absence of hormonal activation. This process takes some 5–7 days, and therefore, the hormonal changes of pregnancy can be viewed as affecting neural mechanisms and thus resulting in the parturient female's immediate responsiveness to her own or foster pups. Indeed, hormonal treatments have been developed that simulate the endocrine changes associated with the termination of pregnancy, and when such treatments are administered to virgin females, they result in what can be called short-latency maternal behavior. Hormonally treated virgins can be induced to show maternal behavior after 24–48 hr of pup exposure rather than after the 5–7 days of pup exposure that is necessary in the absence of hormonal treatment (Moltz *et al.,* 1970; Siegel and Rosenblatt, 1975a; Zarrow *et al.,* 1971a).

In summary, then, the immediate onset of maternal behavior at parturition is hormonally mediated. Maternal behavior is then maintained during the postpartum period in the absence of hormonal mediation, being regulated instead by sensory stimuli from the pups. Maternal behavior can be induced in virgin females through pup stimulation, and this onset of maternal behavior is not dependent on hormones for its activation. With respect to the neural basis of maternal behavior in the rat, certain points can now be briefly discussed.

First, the question can be asked whether equivalent neural mechanisms are involved in all these aspects of maternal behavior. Only a few studies have provided evidence on this point, and they are reviewed below.

The second question to arise is how one should interpret the disruptions of maternal behavior resulting from brain lesions that have been inflicted prior to parturition. Such a disruption could result from a direct interference with the neural mechanisms underlying the onset of maternal behavior at parturition. Alternatively, such a disruption could be the indirect result of an interference with neuroendocrine mechanisms. Estrogen, progesterone, and prolactin are the hormones that have received the most attention as mediators of the immediate onset of maternal behavior at parturition. The secretion of estrogen and progesterone from the ovaries is dependent on gonadotropin secretion from the anterior pituitary. Prolactin is also secreted by the anterior pituitary. Hypothalamic regulation and limbic system modulation of anterior pituitary function is now well established (Raisman and Rield, 1971). Therefore, a lesion in the central nervous system could possibly interfere with anterior pituitary function. If such a lesion were performed before parturition, a disruption of maternal behavior at parturition might not be the result of directly interfering with the neural mechanisms underlying the behavior; instead, it could be secondary to a neuroendocrine disruption. This is a serious problem that is complicated by the fact that the exact nature of the hormones involved in the onset of maternal behavior at parturition has not been conclusively determined. One way to attempt to answer this question is to provide the lesioned females with healthy foster pups (supplied by donor females) each day for at least the first 7 postpartum days. In this way, one could determine whether such females could be induced to show maternal behavior as a result of pup stimulation, which is not mediated by hormones. If maternal behavior were exhibited after such treatment, the case would be strengthened that the disruption was the result of an interference with neuroendocrine mechanisms. However, an alternative possibility would still exist: such a lesion may have directly interfered with the neural mechanisms underlying the hormonal onset of maternal behavior (perhaps by destroying a site where the hormone acts), while sparing those essential for pup-induced maternal behavior.

This question is raised here because many of the studies to be reviewed involved brain lesions that were performed before parturition. Many of these studies were undertaken well before hormonal treatments were developed that were capable of reliably inducing short-latency maternal behavior in virgin females. It might be useful to use such hormone-treated virgins in future studies in order to answer some of the questions and the related problems that are discussed below.

SENSORY MECHANISMS

Traditionally, maternal behavior in mammals has been viewed as being under multisensory control(Lehrman, 1961). Simply put, this view states that no one sensory modality is essential for the recognition of young and the performance of maternal activities by the mother. Research on rodents has indicated that olfactory, visual, auditory, and other sensory modalities are all used by mothers in the efficient performance of their maternal activities (Allin and Banks, 1972; Beach and Jaynes, 1956a,b; Noirot, 1972a; Smotherman *et al.*, 1974). The multisensory view would hold that the elimination of any one of these sensory modalities would not eliminate maternal behavior.

THE DEVELOPMENT OF THE CONCEPT OF MULTISENSORY CONTROL

The first systematic investigation of the sensory cues utilized by maternal females was that of Beach and Jaynes (1956b), who investigated the sensory cues involved in the elicitation of the retrieving response in postpartum, lactating female rats. Beach and Jaynes provided evidence that visual, chemical, tactile, and thermal cues are involved in the response of the mother to her young. In addition, the surgical elimination of either vision, olfaction, or tactile sensitivity of the snout and the lip region leaves maternal retrieving intact. Elimination of any two, or all three, of these sensory systems, although causing deficits in maternal retrieving, does not completely abolish the behavior (see Figure 1).

As Beach and Jaynes's studies were performed on lactating female rats, these females had prior experience with young. Furthermore, although the authors were aware of the possible importance of audition, this sensory modality was not directly investigated. Recent findings have suggested that ultrasonic vocalizations emitted by young can influence the maternal behavior of rodents (Allin and Banks, 1972; Noirot, 1972a; Smotherman *et al.*, 1974). Herrenkohl and Rosenberg (1972) investigated the possibility that the maternal behavior of inexperienced female rats is under unisensory control. They also investigated the contribution of audition. Different groups of primiparous rats were either deafened by destruction of the basilar membrane, rendered anosmic by olfactory bulbectomy, or blinded by orbital enucleation during pregnancy. The subsequent maternal behavior of these females was studied over a 3-day period. In addition to observing retrieving behavior, the other components of maternal behavior were also studied. Although slight changes in maternal responsivity were observed in some of the groups, the major finding

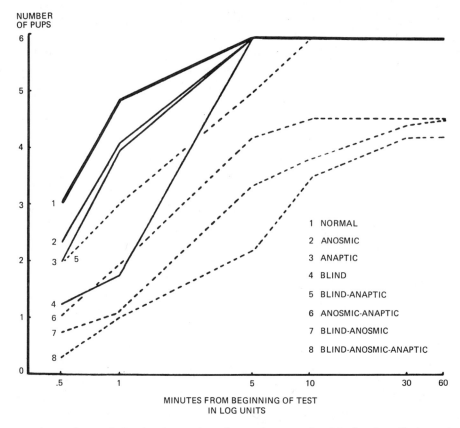

Fig. 1. Composite graph showing the number of normal pups retrieved by females suffering various types of peripheral desensitization. (From Beach and Jaynes, 1956b. Copyright, 1956, by E. J. Brill-Leiden. Reprinted by permission.)

was that maternal behavior remained intact in most of the rats. These investigators concluded that their findings supported the concept of a multisensory control of maternal behavior, in that no one sensory modality was found to be essential for either the onset or the maintenance of maternal behavior in rats.

OLFACTORY INPUT AND THE ROLE OF THE OLFACTORY BULB

The studies reviewed above provide support for the concept that maternal behavior is under multisensory control. They were performed on the rat and indicated that the maternal behavior of females that have undergone pregnancy does not seem to be dependent on any single sensory input. More recent studies have concentrated on the role of olfaction and the olfactory bulbs in the control of maternal behavior in the rat and other species. These studies indicate that we may have to modify the statement that the maternal behavior of mammals is under multisensory control. They also provide some interesting possibilities concerning the neural mechanisms underlying maternal behavior.

THE RAT. The studies cited above indicate that olfactory bulbectomy, when performed either during pregnancy or during the postpartum period, does not interfere with the initiation or the maintenance of maternal behavior in the rat. Additional studies support these findings (Fleming and Rosenblatt, 1974b; Pollak and Sachs, 1975; Schlein *et al.*, 1972). However, as Benuck and Rowe (1975) noted, most of these studies have observed the maternal behavior of the bulbectomized females for only a few days following parturition. Therefore, Benuck and Rowe bulbectomized primiparous female rats during pregnancy and examined the effects of such treatment on the subsequent maternal behavior of these females, which were allowed to rear their pups from birth to weaning. These investigators observed larger effects of olfactory bulbectomy on maternal behavior than those previously reported. The major deficits observed were in parturitional behaviors (cleaning fetal membranes and eating the placentas) and in retrieving. Nursing behavior and nest building were not disrupted, and bulbectomized females were capable of rearing their litters to weaning, although their pups weighed less at weaning than those of control females. Therefore, although deficits were observed, they were relatively mild.

Traditionally, olfactory bulbectomy has been the method of producing anosmia in mammals to allow an investigation of the role of olfaction in the control of behavior. This procedure, of course, involves damage to the central nervous system, and recent findings have indicated that many of the effects of bulbectomy on behavior are not the result of the production of anosmia; rather they are the result of an interference with nonolfactory functions of the bulb (for reviews, see Cain, 1974b; Edwards, 1974). In order to dissociate these two functions of the bulb, investigators have turned to the use of methods that produce anosmia peripherally rather than centrally, such as the intranasal application of zinc sulfate, which destroys the nasal epithelium and results in a temporary anosmia (Alberts and Galef, 1971; for a critical review of the various methods of producing olfactory deficits, see Alberts, 1974). Benuck and Rowe (1975) studied the effects of the intranasal application of zinc sulfate administered to late pregnant primiparous female rats on their subsequent maternal behavior toward their litters. Behavioral tests of anosmia indicated that the zinc-sulfate-treated females met Benuck and Rowe's criterion for anosmia during the period of behavioral testing. Most of the deficits observed after olfactory bulbectomy were not observed in zinc-sulfate-treated females, although such females did display slight deficits in retrieving behavior. These results indicate that the peripheral elimination of olfaction does not eliminate maternal behavior, although small deficits in retrieving are produced, presumably because of the decreased ability of females to locate pups displaced from the nest (cf. Smotherman *et al.*, 1974). Slightly larger deficits in maternal behavior are produced as the result of bulbectomy, but maternal behavior is not eliminated. These larger deficits are presumably due to a disruption of some non-olfactory function of the olfactory bulbs.

Up to this point, we have been concerned with the sensory cues from the young that elicit the various components of maternal behavior in lactating female rats and that are necessary for the normal initiation of maternal behavior at parturition and its maintenance during the postpartum period. Sensory stimuli from the pups, however, have other influences on maternal behavior in the rat and other rodents (for reviews, see Richards, 1967; Noirot, 1972b), and an understanding of some of these influences may provide us with some insight into the physiological and neural mechanisms underlying maternal behavior. In particular, the induction of maternal behavior in virgin female rats that results from pup stimulation indicates that sensory stimuli from young pups have motivational influences on virgin females, in some way changing them from maternally unresponsive to maternally responsive females. Several recent studies have explored the sensory mechanisms involved in this process.

Schlein et al. (1972) have reported that bilateral olfactory bulbectomy results in virgin female rats that consistently cannibalize young pups to which they are exposed. Importantly, it was found that previous maternal experience with pups prevents this bulbectomy-induced cannibalism. Virgin female rats that were first induced to show maternal behavior through pup stimulation and were then bulbectomized did not cannibalize young following the bulbectomy; instead, they acted maternally toward these young. This experience effect is presumably independent of hormonal influences, as pup-induced maternal behavior is believed to have a nonhormonal basis.

The fact that inexperienced primiparous females that are bulbectomized during pregnancy do not cannibalize their own young on giving birth (reviewed above) and do not cannibalize foster young presented to them around the time of parturition (Schlein et al., 1972) suggests that the hormonal events of pregnancy, which are involved in the initiation of maternal behavior at parturition, prime the female in such a manner as to preclude cannibalism and to favor maternal behavior.

Recent findings by Fleming and Rosenblatt (1974b,c) have considerably increased our understanding of the bulbectomy-induced cannibalism that occurs in inexperienced virgin female rats. This work suggests that the cannibalism observed in inexperienced, cycling virgin females after olfactory bulbectomy is not the result of anosmia but results from a disruption of nonolfactory influences of the bulb. Indeed, these findings make the significant contribution that anosmia, and therefore a presumed inability to smell the odors of young pups, leads to a short-latency onset of maternal behavior in virgin females exposed to young rather than the typical 5- to 7-day latency normally observed in such females.

In their first study, Fleming and Rosenblatt (1974b) found that bilateral olfactory bulbectomy resulted in a twofold effect when performed on inexperienced, cycling virgin female rats prior to pup exposure. Unlike in the findings of Schlein et al. (1972), not all virgins cannibalized the young: this response occurred in approximately 50% of the females. The remaining bulbectomized females showed a remarkable reduction in latencies to onset of maternal behavior following pup exposure; most of these females retrieved and adopted a nursing posture over the foster young within the first 24 hr of pup exposure. The difference between the two studies may be related to the fact that Schlein et al. (1972) presented foster young to the bulbectomized virgins 24 hr postoperatively, whereas Fleming and Rosenblatt (1974b) allowed their females a 1- to 3-week postoperative recovery period. Interestingly, and possibly related to the previous remark, bulbectomies performed in two stages, with 1 week intervening between each stage, resulted in females that exhibited short-latency maternal behavior in the absence of cannibalism.

In their second study, Fleming and Rosenblatt (1974c) reported that cycling virgin females pretreated intranasally with zinc sulfate showed a short latency to the onset of maternal behavior following pup exposure. None of the virgins cannibalized the young, and most exhibited maternal behavior during the first 24 hr of pup exposure. Similar results were reported for females that received lesions of the lateral olfactory tract, the major efferent pathway, but not the only efferent pathway, which connects the olfactory bulbs with the rest of the brain (see Devor, 1976). Behavioral tests of olfactory sensitivity indicated that females receiving either zinc sulfate treatment or olfactory bulbectomy met Fleming and Rosenblatt's criterion for anosmia. The effect of lateral-olfactory-tract lesions on olfactory sensitivity was not examined, although others have reported that such lesions result in anosmia (Long and Tapp, 1970).

The findings that single-stage olfactory bulbectomy results in short-latency maternal behavior in approximately 50% of the treated virgins and that peripherally induced anosmia, lateral olfactory tract lesions, and two-stage bulbectomies result in short-latency maternal behavior in nearly all of the treated virgins support the contention of Fleming and Rosenblatt that anosmia facilitates the onset of maternal behavior in virgins. The high incidence of cannibalism observed in virgins receiving single-stage bulbectomies is probably the result of disrupting a nonolfactory influence of the bulbs that does not seem to be mediated by the lateral olfactory tracts. Fleming and Rosenblatt argued that pup odors delay the onset of maternal behavior in virgins. It remains to be determined, however, whether the observed reduced latency to onset of maternal behavior is specifically related to the inability of the virgins to smell pup odors. Furthermore, how pup exposure can override this presumed inhibitory effect of pup odors remains to be elucidated. If pup odors do indeed inhibit the onset of maternal behavior in virgins, the possibility arises that one of the ways in which the hormonal events of pregnancy may facilitate the normal onset of maternal behavior at parturition is by altering the female's olfactory sensitivity. A recent report by Pietras and Moulton (1974) shows that hormones can influence odor detection in rats and, significantly, that odor detection performance is markedly depressed during pseudopregnancy. However, it remains to be shown whether the hormonal events that underlie the onset of maternal behavior near the time of parturition actually depress olfactory sensitivity. Because olfactory cues from pups have been shown to play a positive role in the maternal behavior of rats (Beach and Jaynes, 1956b; Benuck and Rowe, 1975; Smotherman *et al.*, 1974), an alternative possibility is that the hormonal events of pregnancy modify the female's preference for pup odors (cf. Carr *et al.*, 1965), rather than depressing her sensitivity to such odors. These two possibilities are not necessarily mutually exclusive (see Pietras and Moulton, 1974).

If the physiological events of pregnancy do modify the female's perference for pup odors, this is probably not their only influence, as such events presumably also inhibit single-stage bulbectomy-induced cannibalism. The present author suggests that the hormonal events involved in the initiation of maternal behavior in the rat probably have multiple effects on the neural systems underlying maternal behavior, biasing the parturient female in the direction of a high level of immediate maternal responsivity.

Although the work of Fleming and Rosenblatt argues in favor of the hypothesis that anosmia results in short-latency maternal behavior in virgins, a few words of caution are in order. A recent report by Mayer and Rosenblatt (1975) has indicated that intranasal zinc sulfate can result in short-latency maternal behavior in virgin females without producing anosmia. Although researchers have been interested in the nonolfactory effects of bulbectomy, more attention should be given to the nonolfactory effects of zinc sulfate treatment. Some of the effects of zinc sulfate on behavior may be due to nonsensory systemic effects (Sieck and Baumbach, 1974). Ingestion of zinc sulfate after intranasal application is unavoidable, and therefore, systemic controls should be employed. Furthermore, intranasal application of zinc sulfate may have effects on the animal's endocrine system, and therefore, it is possible that the short-latency maternal behavior observed in such females is hormonally induced. Shelesynak and Rosen (1938) have shown that the intranasal application of both irritants and anesthetics can influence the estrous cycle in rats (see also Rosen and Shelesnyak, 1937; Rosen et al., 1940). With these constraints in mind, the most parsimonious explanation is that zinc sulfate treatment results in short-latency maternal behavior by producing anosmia, or at least hyposmia (cf. Mayer and Rosenblatt, 1975), because other methods that produce anosmia (lateral olfactory tract lesions and two-stage bulbectomies) also result in short-latency maternal behavior in virgins.

A question that arises at this point is to what extent disruption of sensory input from the vomeronasal organ, rather than from the primary olfactory epithelium, is involved in the facilitation of maternal behavior in virgin rats. Much attention has recently been focused on both the anatomy and the function of the vomeronasal system (Broadwell, 1975; Estes, 1972; Powers and Winans, 1975; Raisman, 1972; Scalia and Winans, 1975; Winans and Scalia, 1970). The vomeronasal organ is situated in the nasal cavity, and the vomeronasal nerves terminate in the accessory olfactory bulb. Although the vomeronasal–accessory-olfactory-bulb system has been suggested as being involved in the mediation of chemoreceptive and behavioral functions (Estes, 1972), only recently has evidence been provided for a functional role of this system in the sexual behavior of male hamsters (Powers and Winans, 1975). The point to note here is that both lateral-olfactory-tract lesions and two-stage bulbectomies, procedures that result in short-latency maternal behavior in virgins, would disrupt centripetal input from the vomeronasal–accessory-olfactory-bulb system (Broadwell, 1975; Scalia and Winans, 1975; Winans and Scalia, 1970). It is still controversial, however, to what extent the intranasal application of zinc sulfate destroys the sensory epithelium of the vomeronasal organ (Powers and Winans, 1975).

We should now turn our attention to the possible mechanisms underlying single-stage-bulbectomy-induced cannibalism in virgin female rats. The fact that neither lateral-olfactory-tract lesions nor two-stage bulbectomies, procedures that presumably result in anosmia and the elimination of both primary olfactory and vomeronasal input, result in a high incidence of cannibalism but result in short-latency maternal behavior supports the contention of Fleming and Rosenblatt (1974b,c) that the high incidence of cannibalism that occurs after a single-stage bulbectomy is the result of a disruption of nonsensory influences of the olfactory bulb. This contention is further supported by the zinc sulfate findings. The high incidence of cannibalism observed in virgin females after single-stage bulbectomies is most likely a reflection of a nonspecific increase in irritability that seems to result from a disruption of normal brain functioning rather than from anosmia (Alberts and Friedman, 1972; Cain, 1974a,b; Cain and Paxinos, 1974; Fleming and Rosenblatt, 1974b; Spector and Hull, 1972). Recently, Cain (1974a) has demonstrated that this increase in irritability is most likely the result of damage to the olfactory peduncle and the anterior olfactory nucleus rather than of damage to the olfactory bulb. He concluded that irritability in rats is therefore not due to the loss of olfaction but is the result of damage to specific brain areas caudal to the bulb.

The manner in which (1) the physiological events of pregnancy, (2) previous maternal experience, and (3) two-stage bulbectomies can overcome this increase in irritability remains to be determined.

THE MOUSE. The mouse presents a picture considerably different from that of the rat, and the evidence available suggests that the maternal behavior of the mouse is much more dependent on the olfactory sense than is that of the rat. Before we examine this literature, we should first note that the mechanisms underlying maternal behavior in the mouse are different from those of the rat. Inexperienced virgin female mice show pup retrieval, pup licking, and nursing behavior within minutes of their first exposure to foster young (Noirot, 1972b). They contrast with the virgin rat, which, under normal circumstances, requires 5–7 days of constant exposure to pups before behaving maternally. This nearly immediate maternal response of virgin mice does not seem to be dependent on hormonal stimulation (Leblond, 1940). These results would seem to suggest that, unlike in the rat, the neural mechanisms underlying the normal initiation of maternal behavior at parturition in the mouse are not dependent on facilitation by the hormonal events of pregnancy; instead they seem to be directly elicitable by pup-related stimuli (cf. Gandelman *et al.*, 1970; Rosenson, 1975; Voci and Carlson, 1973). One aspect of maternal behavior in the mouse, however, is dependent on the hormonal events of pregnancy. It has been shown that the maternal nest-building that occurs during pregnancy, and therefore prior to pup stimulation, in the mouse is dependent on progesterone, or both progesterone and estrogen secretion from the ovary (Koller, 1952, 1955; Lisk, 1971; Lisk *et al.*, 1969). For the most part, however, maternal behavior in the mouse seems to be dependent on stimuli derived from the pups for both its initiation and its maintenance, and high levels of maternal nest-building can also be elicited in mice by pup stimulation alone in the absence of hormonal mediation (Gandelman, 1973; Koller, 1952, 1955). We therefore turn our attention to those stimulative properties of the young that seem essential for maternal responsivity in the mouse.

Gandelman *et al.*, (1971a) reported that bilateral olfactory bulbectomy effectively eliminates the maternal behavior of multiparous female mice. On giving birth, such females eat the placentas and clean their young, but retrieving, nest building, and nursing do not occur. Most of the females eventually cannibalize their young, and those not exhibiting cannibalism ignore their young. Additionally, removal of the olfactory bulbs in virgin female mice similarly eliminates the maternal behavior typically observed in these females. These findings have been confirmed and extended to include primiparous female mice (Gandelman *et al.*, 1972; Zarrow *et al.*, 1971b). Zarrow *et al.*, (1971b) also reported that bilateral removal of the olfactory bulbs (1) eliminates the normal increase in maternal nest-building that occurs during pregnancy in the mouse; (2) depresses the increase in nest building induced in virgin mice by progesterone administration; and (3) eliminates the nest building induced in virgin mice by exposing them to low ambient temperatures (cf. Edwards and Roberts, 1972; cf. Kinder, 1927; cf. Koller, 1955). The third finding suggests that the deficits in nest building observed after bulbectomy in mice may not be specific to maternal behavior.

In the above studies on primiparous and multiparous mice, the bulbectomies were performed either before or during pregnancy. In another study, Gandelman *et al.* (1971b) performed bulbectomies on primiparous lactating female mice. Their major finding was that bulbectomy does not necessarily result in the elimination of maternal behavior and the occurrence of cannibalism. Whereas most females bulbectomized on Day 3 of lactation cannibalized their young, none of the females bulbectomized on Day 14 of lactation did so; instead they successfully reared their litters to weaning. Evidence was provided that indicated that this difference was a function not of maternal experience, but of the changing stimulative properties of the young. A clear interpretation of these findings, however, is limited by the fact that the maternal behavior of the females bulbectomized on Day 14 of lactation was not described in detail.

These experiments imply that olfaction is necessary for the initiation and the maintenance of maternal behavior during the early postpartum period in the mouse. Gandelman *et al.* (1971b) proposed that maternal behavior displayed toward young pups may be under unisensory control, whereas that displayed toward older pups may be under multisensory control. Although Gandelman *et al.* (1971b) suggested that the mouse may not be able to identify or recognize young pups when olfactory cues are absent, Gandelman *et al.* (1972) did raise the alternative possibility that the olfactory bulb may serve a nonsensory function, the disruption of which leads to an elimination of maternal behavior under certain conditions. Vandenbergh (1973) explored this question by examining the effects of central (bulbectomy) and peripheral (zinc-sulfate-treatment) anosmia on the maternal behavior of female mice. Both central and peripheral anosmia disrupted maternal nest-building during pregnancy and maternal care during the early postpartum period, as indicated by an increased incidence of cannibalism or maternal neglect, in primiparous female mice, but the severity of the effect was greater among centrally anosmic females. The important point is that zinc sulfate was administered to the females at least 30 days prior to pregnancy. Because behavioral tests for anosmia were not conducted during the last week of pregnancy and during lactation, a question can be raised about whether the disruption of olfactory sensitivity was equivalent in the bulbectomized and zinc-sulfate-treated females. More recently, Seegal and Denenberg (1974) found that, when intranasal zinc sulfate is administered to primiparous female mice near the end of pregnancy, 13 of the 15 treated females cannibalized their pups during the early postpartum period. Observations on nest building were not reported. Interestingly, only 2 of 15 similarly treated multiparous females cannibalized their young. The behavior of the remaining multiparous females toward their litters was not reported, although the implication was that maternal behavior remained unaffected. These findings should be contrasted with those of Gandelman *et al.* (1971a), who found that bulbectomy reliably results in cannibalism of young by multiparous females.

These findings are difficult to interpret. They do indicate that olfactory sensitivity is important in the normal expression of maternal behavior in inexperienced primiparous mice, as peripheral anosmia disrupts the maternal behavior of such females. The fact that Vandenbergh (1973) observed more severe effects in bulbectomized females and that the previous experience of pregnancy, parturition, and rearing a litter can overcome zinc-sulfate-induced cannibalism (Seegal and Denenberg, 1974) but not bulbectomy-induced cannibalism (Gandelman *et al.*, 1971a) suggests the involvement of a disruption of nonsensory functions as a result of bulbectomy. It is possible, however, that bulbectomy results in anosmia whereas zinc sulfate results in hyposmia, and that the effects of reduced olfactory sensitivity as a result of zinc sulfate can be overcome by experiential effects. Additionally, the involvement of sensory input from the vomeronasal–accessory-olfactory-bulb system remains to be determined.

It is clear that more research has to be done to elucidate the nature of the deficits and the mechanisms involved in the disruptions of maternal behavior that occur in both bulbectomized and zinc-sulfate-treated mice. The weight of the evidence does indicate that the inexperienced primiparous female mouse is highly dependent on olfactory sensitivity for the expression of maternal behavior during the early postpartum period.

ADDITIONAL SPECIES. Not much research has been undertaken on the sensory basis of parental behavior in other mammalian species. Leonard (1972) reported on the effects of neonatal (Day 10) olfactory-bulb ablations on the maternal behavior of female hamsters in adulthood. After parturition, only 10 of 15 neonatally bulbectomized females gathered their pups into the nest and nursed them. The remaining females left the young scattered and eventually cannibalized them. All sham-operated females (12 of 12) and 5 of 6 females receiving unilateral bulbectomies reared their litters to weaning. Because neonatal neural damage may be followed by some recovery of function, the effects of removal of the olfactory bulbs in adulthood should be investigated.

The role of olfaction in the maternal behavior of goats and sheep has also been investigated. Observational and experimental studies have indicated that olfactory, gustatory, auditory, and visual input may all play some role in the maternal behavior of ungulates (Altmann, 1963; Lindsay and Fletcher, 1968). Baldwin and Shillito

(1974) performed olfactory bulbectomies during pregnancy on primiparous and multiparous Soay ewes. Their results indicated that the olfactory bulbs are not essential for maternal behavior to be displayed by ewes, although anosmia presumably disrupted the ability of the ewe to distinguish her lamb from alien lambs (cf. Beach and Jaynes, 1956b). Similar findings have been reported for goats (Klopfer and Gamble, 1966; Klopfer and Klopfer, 1968).

The Role of Afferent Feedback from the Mammary Gland and Nipples

In 1961, Lehrman reviewed the literature dealing with the relationship between the condition of the mammary gland and nursing behavior. It was suggested that, in certain instances, a relationship exists between mammary engorgement and the motivation to nurse the young, although in other instances a clear dissociation could be observed between nursing behavior and mammary distension. Earlier, in 1933, Wiesner and Sheard raised a related question: Whether the afferent feedback from the nipples that accompanies suckling stimulation plays a role in the regulation of maternal behavior in the rat. It is the purpose of this section to briefly review the role of peripheral feedback from the mammary region in the regulation of maternal behavior.

In 1933, Wiesner and Sheard found that removal of the nipples (thelectomy) prior to parturition did not interfere with the immediate onset of maternal behavior at parturition in female rats. In a more definitive study, Moltz et al., (1967) performed total mammectomies on immature female rats, a surgical procedure that resulted in the removal of all mammary tissue in addition to the removal of each nipple. These females were mated in adulthood, and their subsequent maternal behavior was compared to that of sham-operated controls. A fostering procedure was established so that these females could be observed from parturition through Day 21 postpartum. The nursing behavior of the mammectomized females did not differ from that of the controls. In addition, such females also showed normal retrieving and nest-building activities. It can be concluded that neither mammary distension nor suckling stimulation is essential for the initiation and maintenance of nursing behavior, as well as other maternal activities, in primiparous postpartum female rats. This conclusion is substantiated by findings that have shown that (1) disruption of the hormonal conditions necessary for the maintenance of lactation in postpartum rats does not disrupt maternal behavior (Obias, 1957; Numan et al., 1972); (2) hormonal treatments that induce maternal behavior in virgin rats do not induce lactation (Moltz et al., 1970; Siegel and Rosenblatt, 1975a; Zarrow et al., 1971a); and (3) virgin rats induced to show maternal behavior through pup stimulation exhibit a pattern of maternal behavior remarkably similar to that observed in postpartum females, although such virgins are not capable of lactating (Fleming and Rosenblatt, 1974a; Reisbick et al., 1975). These studies, of course, do not rule out a role for somatosensory stimulation of the ventral body surface in the regulation of nursing behavior.

The rabbit is interesting in that postpartum females of this species normally nurse their young for only a single period of approximately 3-min duration each day (Cross, 1952; Zarrow *et al.*, 1965). Several investigators have studied the relationship between the condition of the mammary gland and nursing behavior in postpartum female rabbits. Cross (1952) observed that lactating postpartum rabbits could be induced to nurse their litters more than once in a single day if milk removal was prevented by sealing the teats with collodion. He suggested that mammary distension plays a role in the regulation of nursing behavior in the postpartum doe. Findlay (1969) confirmed this observation and also presented evidence that shows that mammary engorgement is not essential for the normal occurrence of daily nursing bouts in the rabbit, and Findlay and Roth (1970) provided evidence indicating that mammary distension is neither necessary nor sufficient for the daily display of nursing behavior in the rabbit. In one experiment, rabbits were found to display normal nursing behavior after their mammary glands had been emptied during a prior period of anesthesia. Lincoln (1974) added the observation that the duration of each daily nursing bout is not altered by preventing the normal loss of milk from the mammary glands during the bout. These findings indicate that, although afferent feedback from the mammary region may play a role in the regulation of nursing behavior in the rabbit, it is not essential for the normal occurrence of such behavior (also see Findlay and Tallal, 1971).

ADDITIONAL AFFERENT INPUT

Rats show an increase in self-licking of the mammary and vaginal region during pregnancy (Roth and Rosenblatt, 1967), but it has been shown that these events, and their consequent afferent feedback, are not essential for the normal onset of maternal behavior at parturition (Christophersen and Wagman, 1965; Kirby and Horvath, 1968). The experience of parturition is also not essential for the normal occurrence of maternal behavior in the rat. Female rats receiving cesarean sections near term and subsequently presented with foster young show normal maternal behavior (Moltz *et al.*, 1966; Wiesner and Sheard, 1933). Possibly related to these findings are those of Bacq (1931), which indicated that abdominal sympathectomies performed prior to mating did not interfere with the subsequent maternal behavior of female rats. Such an operation would disrupt some of the afferents from the vagina and the uterus. Similar observations have been reported for the rabbit (Labate, 1940), and sympathetic innervation is probably also not necessary to the maternal behavior of cats and dogs (Cannon and Bright, 1931; Cannon *et al.*, 1929; Simeone and Ross, 1938). The fact that virgin mice show nearly immediate maternal attention to foster young also indicates that the afferent neural feedback that is a consequence of pregnancy and parturition does not play a major role in the maternal behavior of this species (see Gandelman and Svare, 1974).

These findings, obtained mainly in the rat, fit well with those findings indicating that short-latency maternal behavior can be induced in nonpregnant female rats by hormonal treatment. Evidence is presented below that shows that the critical hormonal events necessary for the immediate onset of maternal behavior in the rat have a central site of action.

COMMENTS AND CONCLUSIONS

The position taken here with respect to the sensory basis of maternal behavior was to examine the question of whether any one sensory modality is essential for the display of maternal behavior. The research reviewed for the rat indicates that maternal behavior during the normal maternal behavior cycle has a multisensory control system. The observations from work with virgin rats also raise the interesting possibility that some alteration in olfactory preference may be involved in the normal onset of maternal behavior at parturition. The research on the mouse seems to indicate that one cannot extend a multisensory control system to all mammals. Olfactory sensitivity seems to be essential for the display of maternal behavior by primiparous female mice during the early postpartum period. Interestingly enough, not much research has examined the question of whether other sensory modalities are just as critical for the normal display of maternal behavior in the mouse. Gandelman *et al.* (1971a) cited an unpublished finding that enucleation of the eyes does not eliminate maternal behavior in the mouse. Experimental observations have shown that auditory stimulation can influence the maternal behavior of mice (Noirot, 1972a), and Noirot (1972a) reported that congenitally deaf mice show severe deficits in maternal behavior.

The olfactory system has probably received the most attention because electrophysiological and neuroanatomical findings have demonstrated that olfactory input can influence those areas of the brain believed to be involved in the control of species-typical behavior (cf. Heimer, 1972; Pfaff and Pfaffman, 1969; Scott and Leonard, 1971). The present author thinks that the vomeronasal-organ–accessory-olfactory-bulb system should receive more attention, as neuroanatomical findings have shown that input from this system can reach the medial preoptic region (Scalia and Winans, 1975). This region is critically involved in the maternal behavior of rats (see the section below on the medial preoptic area). Finally, one sensory modality that has, surprisingly, received very little attention with respect to its role in maternal behavior is taste. Female mammals of many species are known to lick their young at parturition and during the postpartum period, and gustatory input may provide important information to the central mechanisms underlying maternal behavior (see Norgren, 1976).

Although the data were organized in this section in order to cover the multisensory control question, the present author agrees with others (e.g., Smotherman *et al.,* 1974) that more research should be directed at relating specific stimulus properties of the young to specific aspects of maternal responsiveness. The reader is referred to a recent short review on this subject that illustrates some of the questions involved: Herrenkohl and Sachs (1972).

EXPERIMENTAL FINDINGS

Beach (1937) explored the role of neocortical mechanisms in the maternal behavior of the female rat. Neocortical lesions of various sizes were produced in adult virgin female rats. Such females were subsequently mated, and their maternal behavior was studied during the first 4 postpartum days. Besides testing for retrieval and nest building, Beach also examined each female's response to air and heat blasts directed at the nest and the litter. It was found that lesions involving less than 20% of the total neocortical surface produced only slight deficits in maternal behavior. Larger lesions produced more severe deficits, and when more than 40% of the cortex was destroyed, maternal behavior was almost completely abolished (see Figure 2). Examination of the localization and the extent of cortical damage suggested that the degree of disruption of maternal behavior was not significantly related to the site of the lesion but was directly related to the amount of neocortex destroyed. In a second study, Beach (1938) reported that females receiving neocortical ablations prepuberally also show deficits in maternal behavior when tested in adulthood, although the deficits are not as severe.

The finding that small neocortical ablations have only minor effects on the maternal behavior of rats has been confirmed (Slotnick, 1967a; Stamm, 1955; Stone, 1938) and extended to include mice (Carlson and Thomas, 1968; Slotnick and Nigrosh, 1975). Stone (1938), Davis (1939), and Stamm (1955) have confirmed Beach's finding that large neocortical ablations, when performed in adulthood,

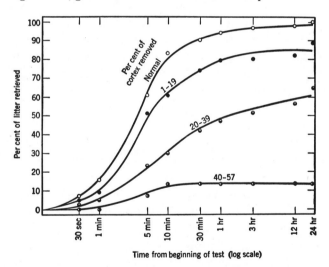

Fig. 2. Records of retrieving behavior in normal and cortically operated female rats. The numbers on the curves show the percentage of cortex removed. The litter is scattered by the experimenter, and the female retrieves the pups to the nest, one at a time. The speed and the completeness of retrieving behavior decrease with increasing size of the cortical lesion. (From Beach, 1951 [redrawn from Beach, 1937]. Copyright, 1951, by John Wiley and Sons, Inc. Reprinted by permission.)

result in an almost complete abolition of maternal responsivity in female rats. In all these studies, the lesions were performed prior to mating. Stone (1938) raised the significant question of whether the disruption of maternal behavior following large neocortical lesions was the indirect result of a lesion-produced endocrine dysfunction. The fact that the females in all these studies mated and became pregnant, carried to term, and had normal parturitions suggests that hormonal mechanisms were not seriously interfered with.

If the neocortex is directly involved in the neural mediation of maternal behavior, the manner in which it is involved is not quite clear. It is possible that the deficits in maternal behavior derive from the fact that as the amount of neocortex destroyed increases, more cortical projection sites of the various sensory modalities are included within the lesioned area. In fact, Beach and Jaynes (1956b) implied that a relationship exists between their findings concerning the multisensory control of maternal behavior in the rat and the effects of neocortical lesions on maternal behavior. Alternatively, it is possible that the neocortex serves some nonsensory functions with respect to maternal behavior (see Beach, 1937). Beach (1937) did not regard the disruption of maternal behavior observed after neocortical lesions as being the result of a motor dysfunction.

Interestingly, Colombo et al. (1973–1974) reported that cortical spreading depression produced by application of potassium chloride to the neocortex (a procedure that results in a temporary cortical deactivation) also results in a decreased multiunit activity recorded from the medial preoptic area (MPOA) of female rats. Because MPOA lesions abolish maternal behavior in the rat (see below), there may be some relationships here. In any event, cortical spreading depression would seem to be a useful technique for substantiating the role of cortical mechanisms in the control of maternal behavior (cf. Larsson, 1962b). There is only one study in the literature that has examined the effects of cortical spreading depression on maternal behavior, and the species investigated was the guinea pig (Stern and Siepierski, 1971). The results indicated that temporary cortical deactivation did not modify the nursing behavior of primiparous guinea pigs during the postpartum period. In comparison to rat pups, guinea piglets are quite mature at birth and are ambulatory. It is the young that initiate most of the nursing bouts, and the mother simply permits or prevents suckling. Stern and Siepierski (1971) suggested that perhaps cortical deactivation does not disrupt maternal nursing in the guinea pig because the female's role is a passive one. This passivity, of course, should be contrasted with the active role of the maternal female rat during the early postpartum period, a role that includes not only initiating nursing bouts but also maintaining a nest and transporting pups if the necessity arises.

Recent research on rhesus monkeys has provided some evidence on neocortical mechanisms and maternal behavior in subhuman primates. These studies indicate that circumscribed areas of the neocortex may play a role in the maternal behavior of rhesus monkeys. Ablations of either the prefrontal cortex or the anterior temporal cortex at approximately 2 months postpartum resulted in severe and long-term deficits in maternal behavior (Franzen and Myers, 1973; also see Bucher et al., 1970; Myers et al., 1973). Some of the deficits included (1) loss of protective retrieval of the infants in threatening situations; (2) maternal withdrawal from the infant; (3) increases in infant punishment by the mother; and (4) decreases in the amount of active infant cuddling. In some cases, the presence of the infant was tolerated and it was allowed to nurse at its own initiative, the mother assuming a completely passive role. Lesions of the cingulate cortex or the visual association cortex did not affect mother–infant relationships. The prefrontal ablations included all the cortical tissue lying anterior to the frontal eye fields and involved the orbitofrontal cortex in its entirety. The anterior temporal ablations included all the neocortex of the anterior one third of the temporal lobe. The damage was restricted to the temporal cortex and did not extend into underlying structures such as the hippocampus and the amygdala.

The effects of these lesions were not restricted to maternal behavior deficits. Severe deficits in most aspects of social behavior were observed in both males and females. The most apparent change in the behavior of these animals was a reduction in those activities that serve to maintain social bonds within rhesus monkey groups (also see Deets *et al.*, 1970). Myers *et al.* (1973) pointed out that the prefrontal and the anterior temporal cortex are anatomically interconnected and suggested that they are also functionally related, both contributing to the regulation of social behavior in subhuman primates.

COMMENTS

Kimble *et al.* (1967) questioned the finding that the neocortex is involved in the neural mediation of maternal behavior in rats. They found that relatively large lesions of the dorsal neocortex left maternal behavior unaffected. It should be pointed out that the behavioral tests employed by Kimble *et al.* were not as demanding as those used by Beach (1937). Furthermore, their neocortical lesions were not as large as those that Beach (1937) and Stone (1938) found to result in the greatest impairments of maternal responsivity. For that matter, Kimble *et al.* (1967) did not present any reconstructions of their neocortical ablations, and therefore, comparisons between the studies are difficult to make.

It should be indicated, however, that a large proportion of Beach's (1937) subjects had damage to the dorsal hippocampus, although this damage was described as being superficial and unilateral. The subjects in Stone's study (1938) that showed the greatest deficits in maternal behavior also sustained dorsal hippocampal damage and/or damage to the cingulate cortex (as most of Stone's largest lesions extended to the midline cortex). The fact that both the hippocampus and the cingulate cortex have been implicated in the maternal behavior of rats (see the next section) may indicate that at least some of the deficits in maternal behavior observed after cortical lesions were the result of damage to these structures.

It is suggested that a reinvestigation of the role of the neocortex in the maternal behavior of rats be undertaken to resolve some of these problems as well as possibly to localize more circumscribed areas underlying such behavior (cf. Beach, 1940; Larsson, 1962a, 1964).

LIMBIC MECHANISMS

CINGULATE CORTEX

Several studies have explored the role of the cingulate cortex in the maternal behavior of both rats and mice. What seems clear is that, although this cortical limbic structure may be involved in the neural machinery underlying maternal behavior in the rat, it probably does not play a critical role in the neural mediation of maternal behavior in the mouse.

Three studies reported that rats with lesions of the cingulate cortex show deficits in postpartum maternal behavior (Slotnick, 1967a; Stamm, 1955; Wilsoncroft, 1963). The study by Slotnick is described here in detail because it provides the clearest indication of the nature of the deficits involved. Female rats with previous maternal experience received either full cingulate cortex lesions, partial cingulate lesions, small dorsolateral neocortical lesions, or sham operations and were mated, and their maternal behavior was studied over the first 5 postpartum days. The maternal behavior of females receiving small neocortical ablations was comparable to that of controls. Females receiving partial cingulate lesions, which destroyed either the anterior or the posterior cingulate cortex, showed slight deficits in maternal behavior. The most severe deficits were exhibited by those females that received lesions of the entire cingulate cortex, and it is the behavior of these females that is described here.

The major components of maternal behavior were present near the time of parturition in these females, and therefore, the initiation of maternal behavior was not disrupted. The major deficit exhibited by cingulate-lesioned females can be described as a disruption of the sequential organization of the various maternal responses. During retrieving tests, pups were repeatedly brought into and carried out of the nest area or dropped randomly about the cage. The nursing position was adopted over one or two pups, in or out of the nest site, whereas nearby pups were ignored. Although such females attempted to utilize nest material to build nests, the nest-building responses were also disorganized, so that when nests were built, they were of poor quality. Slotnick (1967a) suggested that maternal motivation was not disrupted in these females, as they attempted to perform the various maternal activities. What seemed to be disrupted was the integration of the various maternal responses into an effective sequence leading to efficient pup care.

Importantly, the females that received full cingulate lesions were capable of keeping their pups alive over the 5-day test period, although their pups did not gain weight normally. Also, Slotnick (1967a) found that, by the third postpartum day, the maternal behavior of the cingulate-lesioned rats had improved to nearly normal levels, most females retrieving their pups to the nest site with little disorganization and adopting a nursing posture over them. This improvement was also observed by Stamm (1955). That at least some of the deficits in maternal behavior are short-lasting deserves further attention. Slotnick presented some evidence that adaptation to the novelty of the test situation may have been involved in the observed improvement.

In contrast to these findings on the rat, it has been reported that mice with cingulate cortical lesions show few or no deficits in maternal behavior (Carlson and Thomas, 1968; Slotnick and Nigrosh, 1975). Therefore, a clear species difference appears to exist between the mouse and the rat with regard to a role for the cingulate cortex in the control of maternal behavior.

An interesting question that arises here is whether the deficits observed in the maternal behavior of rats are entirely attributable to destruction of the cingulate cortex. The cingulate cortex lies along the interhemispheric medial cerebral cortex and has sometimes been referred to as the dorsal limbic cortex. The cingulate cortex is a cortical projection site of the anterior thalamic nuclei, and Slotnick did observe retrograde degeneration in these nuclei after cingulate lesions. However, more recent anatomical findings have indicated that the thalamocortical projections of the mediodorsal nucleus and the anteromedial nucleus overlap each other extensively in the anterior medial cortex of the rat (Beckstead, 1976; Domesick, 1972; Leonard, 1969). Rose and Woolsey (1948) suggested that the region of the cortex that receives projections from the mediodorsal thalamic nucleus in subprimates should be considered homologous to the prefrontal cortex of primates. If we accept this anatomical proposal, perhaps some of the deficits observed in the maternal behavior of rats following medial cortical ablations are the result of damage to the prefrontal cortex rather than to the cingulate cortex. This possibility is interesting because deficits—of a qualitatively different nature, however—are observed in monkeys following prefrontal ablations (see above). At the present time, this suggestion is difficult to confirm. Slotnick (1967a) did not observe any differences in the maternal behavior of rats with anterior cingulate or posterior cingulate lesions, and the deficits observed in these partially lesioned groups were slight in comparison to those observed in animals receiving full cingulate lesions. In contrast, Wilsoncroft (1963) observed that rats receiving anterior cingulate lesions showed disorganized retrieving, whereas those receiving posterior lesions did not.

Leonard (1969) also found that the mediodorsal thalamic nucleus, in addition in projecting to the anterior medial cortex, projects to the cortex that forms the dorsal lip of the rhinal sulcus (the sulcal cortex). This area, then, should also be considered homologous to the primate prefrontal cortex. In examining the Lashley diagrams of both Beach (1937) and Stone (1938), it can be seen that this area was probably not damaged bilaterally in any of their rats receiving neocortical ablations. In light of the monkey findings, it might be interesting to explore the role of this cortical area in the maternal behavior of rodents.

HIPPOCAMPUS

Kimble *et al.* (1967) examined the role of the hippocampus in the maternal behavior of female rats. Virgin female rats received either large bilateral dorsal hippocampal lesions, which also destroyed the fimbria, or dorsal neocortical ablations. An additional group served as unoperated controls. These females were subsequently mated, and their maternal behavior was observed during the first 7 postpartum days. Only females receiving dorsal hippocampal lesions exhibited deficits in maternal behavior. This impairment included fewer pups surviving to weaning, increased maternal cannibalism, less frequent nursing, and inferior nest-building. The clearest impairment was that hippocampal-lesioned females spent significantly less time nursing. This reduction in nursing, in turn, was the probable cause of their pups' not gaining weight normally during the first 6 postpartum days. Perhaps weak or dying pups were cannibalized. Whether the various maternal responses were disorganized and the result was inefficient nest-building, retrieving, and nursing is not clear from the presentation of Kimble *et al.* They did note that, on several occasions, nursing behavior was abnormal, females assuming a nursing posture several inches away from any of the pups. These females did display some normal nursing, however, and were only slightly inferior to control females in their retrieving. They were able to keep 41% of their pups alive to the time of weaning, and by Day 14 postpartum there were no differences between groups in the mean weight of their pups.

One of the major efferent pathways from the hippocampus, which also carries afferent input into the hippocampus, is the fimbria (Raisman *et al.*, 1965, 1966). Brown-Grant and Raisman (1972) provided evidence that destruction of the fimbria probably was not involved in the deficits in maternal behavior observed by Kimble *et al.* The fimbria was destroyed near the anterior pole of the hippocampus in female rats. Such females were then mated, and although their subsequent maternal behavior was not directly observed, their pups were weighed on Days 4, 10, 15, and 21 of the postpartum period. The weights of the pups raised by lesioned and sham-operated mothers did not differ significantly on any of these days, and the number of pups weaned by each of these groups was also comparable.

This finding fits well with that of Numan (1974), who investigated the role of the medial corticohypothalamic tract (MCHT) in the maternal behavior of rats. The MCHT arises from the subiculum (part of the hippocampal formation), passes through the fimbria, and then leaves the main part of the postcommissural fornix to terminate in the hypothalamus (Nauta, 1956; Raisman *et al.*, 1966). Destruction of the MCHT in postpartum lactating female rats did not affect the maternal behavior of such females (Numan, 1974; see the section below on maintenance of maternal behavior).

In addition to the fimbria, another important fiber pathway that interconnects the hippocampus with other brain regions is the dorsal fornix (Raisman *et al.*, 1966). Perhaps it is this pathway that is important in the normal expression of maternal behavior in the rat. Interestingly, the dorsal fornix sends projections to the mammillary bodies of the hypothalamus, and Slotnick (1969, 1975) reported that lesions of the mammillary bodies result in maternal behavior deficits in rats quite similar to those observed by Kimble *et al.* after dorsal hippocampal lesions.

Most animals with mammillary body lesions showed no deficits in retrieving, but nursing and nest building were interfered with.

Amygdala

Only a few studies have been concerned with the role of the amygdala in maternal behavior, and they indicate that this limbic structure probably does not play a critical role in the maternal behavior of either rats or mice. Slotnick and Nigrosh (1975) reported that lesions of the amygdala, performed prior to impregnation, have little or no effect on the subsequent maternal behavior of primiparous postpartum mice. The lesions destroyed the lateral amygdaloid nuclei and in most cases extended into the surrounding pyriform cortex. In abstract form, Slotnick (1969) also reported that similar lesions have no effect on the maternal behavior of primiparous rats.

The major efferent pathway that connects the amygdala with other brain regions is the stria terminalis (DeOlmos and Ingram, 1972). Numan (1974) bilaterally destroyed the stria terminalis in its horizontal course over the internal capsule in postpartum lactating female rats. Such lesions had no effect on the maintenance of nest building, retrieving, and nursing during the postpartum period (see the section below on maintenance of maternal behavior). These findings indicate that the total amygdaloid projection traveling through the stria terminalis is not essential for the normal maintenance of maternal behavior in the rat during the postpartum period (see Brown-Grant and Raisman, 1972, whose findings suggest that the stria terminalis is also not essential for the normal onset of maternal behavior at parturition in rats).

More research has to be done to completely rule out a role for the amygdala in the maternal behavior of mice and rats. In the mouse, the medial amygdaloid nuclei should be examined, particularly because these nuclei receive a projection from the accessory olfactory bulb (Scalia and Winans, 1975). In the rat, lesioning the entire stria terminalis, which has no effect on maternal behavior, may be different from selectively eliminating amygdaloid input into one of the several components of the stria terminalis (see DeOlmos and Ingram, 1972). Finally, the experiments with rats were not designed to detect a possible inhibitory influence of the stria terminalis on maternal behavior.

Septum

Probably the most severe effects of limbic lesions on maternal behavior have resulted from lesions of the septal area; these effects have been demonstrated in the rabbit (Cruz and Beyer, 1972), the mouse (Carlson and Thomas, 1968; Slotnick and Nigrosh, 1975); and the rat (Fleischer, 1972; Slotnick, 1969).

Cruz and Beyer (1972) reported that lesions of the medial septum, which also damaged the fornix, almost completely abolished maternal behavior in the rabbit. The lesions were produced in virgin rabbits prior to mating, and although mating, ovulation, pregnancy, and parturition appeared to be normal, maternal behavior was disrupted. The rabbit typically does not show retrieving behavior (Ross *et al.*, 1959). Rather, near the time of parturition, the female builds a maternal nest, gives birth in this nest, and then nurses its litter for approximately 5 min each day during the postpartum period. There is good evidence that maternal nest-building is regulated by the hormonal changes that occur near the time of parturition (Zarrow *et al.*, 1963). Rabbits receiving medial septal lesions did not build maternal nests and failed to voluntarily nurse their litters, although nursing under restraint indicated that lactation was probably not interfered with. A high incidence of maternal cannibalism was also noted. These disturbances in maternal behavior were not associated with noticeable motor or emotional alterations.

Septal lesions in rabbits, therefore, result in a female that shows a lack of interest in her young. This result contrasts with the effects of such lesions in mice and rats. In rodents, septal lesions result in a disorganized pattern of maternal responses rather than in elimination of maternal behavior. Carlson and Thomas (1968) and Slotnick and Nigrosh (1975) reported on the effects of septal lesions in mice. In both studies, the lesions were performed prior to pregnancy, and histological analysis subsequently revealed that the lesions destroyed most of the medial and the lateral septal nuclei and also damaged the fornix. The retrieving behavior of such mice during the early postpartum period received the most attention in both studies, and retrieving was severely disorganized in a manner quite similar to that previously reported by Slotnick (1967a) for rats with cingulate lesions. Nest building was also disrupted in these females. Slotnick and Nigrosh (1975) noted that only 59% of the pups born to mice with septal lesions survived over the 6-day postpartum observation period. This finding is difficult to interpret, however, because observations on nursing behavior were completely lacking or deficient in both studies. Carlson and Thomas (1968) concluded that the maternal behavior deficit in mice with septal lesions resulted from a disturbance of the sequential organization of the various responses involved in pup care and not from impaired motivation. In both studies, females with septal lesions were found to be hyperemotional, but not irritable, during the period of behavioral testing. It is not clear to what extent this hyperemotionality contributed to the observed deficits in maternal behavior.

Slotnick (1969) and Fleischer (1972) reported that septal lesions also result in severe deficits in the maternal behavior of rats. The maternal behavior of primiparous rats that had received septal lesions prior to pregnancy was studied over the first 3 or 4 postpartum days. Parturition appeared normal in these females, but their pups were not completely cleaned at this time. Nest building was disrupted, retrieving was disorganized, and nursing was not observed. All of the pups of these females died within 48 hr of parturition and were replaced by healthy foster young. During retrieving tests, the septal mothers carried pups repeatedly and dropped them randomly about the cage. No improvement in maternal behavior was observed over successive retrieving tests. As Fleischer (1972) noted, the behavior of these rats indicated that they were interested in the pups and that retrieving was initiated, but such females showed little evidence of a progression to other maternal activities. Tests for hyperemotionality indicated that these females were hyperemotional and irritable during the immediate postoperative period, but their emotionality ratings returned to normal levels by the time of the maternal behavior tests. Histological analysis indicated that the septal lesions destroyed most of the medial and the lateral septal nuclei and also damaged the fornix.

In a subsequent experiment, Fleischer (1972) reported on the effects of septal lesions on pup-induced maternal behavior in virgin female rats. The latency to onset of retrieving behavior in these females did not differ from that of control females, averaging 4–6 days in length. However, nest building was disrupted, and the same disorganized pattern of retrieving was observed. Interestingly, some of these females were observed adopting nursing postures over their litters.

The studies reviewed here show that septal lesions disrupt maternal behavior in several species, although the nature of the deficits in the rabbit are different from those in rodents. In all of these studies, the septal lesions damaged parts of both the pre- and postcommissural fornix. Some of the deficits in maternal behavior observed after such lesions may therefore have resulted from damage to the afferent and efferent connections of the hippocampal formation. In most of these studies, the septal lesions were large. Research should now be directed at determining whether a more exact localization of function exists.

CONCLUSIONS

To summarize the results of the studies in this section, lesions of the medial cortex (which includes the cingulate cortex) result in deficits in the maternal behavior of rats but do not seriously alter the maternal behavior of mice. The deficits in rats, although initially severe, appear to be transient, and such rats are capable of keeping most of their pups alive. Dorsal hippocampal lesions primarily disrupt the nursing behavior of rats, whereas amygdala and stria terminalis lesions have not been found effective in disrupting the maternal behavior of rats and mice. Septal lesions have severe effects on maternal behavior. Rabbits seem to completely lose interest in their pups. Mice show disorganized retrieving, but they can keep some of their pups alive, a finding suggesting that some nursing occurs. Rats with septal lesions show such a disorganized pattern of retrieving that they do not progress to other maternal activities, and they cannot keep their pups alive during the postpartum period.

Nearly all of these limbic lesions were inflicted prior to parturition. Thus, a question is raised about whether the resultant deficits in maternal behavior were secondary to an interference with neuroendocrine mechanisms. Several lines of evidence indicate that limbic lesions interfere with maternal behavior by directly affecting the neural mechanisms underlying behavior, rather than by disrupting neuroendocrine mechanisms:

1. The hormonal changes of pregnancy seem essential for the immediate onset of maternal behavior at parturition in the rat; their absence or modification delays the onset of maternal behavior. Rats with limbic lesions exhibited maternal behavior at parturition, they were motivated and interested in their pups. What seemed disrupted was the organization of maternal response patterns into an appropriate

sequence. In other words, maternal behavior seemed initiated at parturition, but was not organized properly.

2. Septal lesions result in similar deficits in the maternal behavior of both mice and rats. Mice are considered less dependent upon hormones for the activation of their maternal behavior.

3. Septal and cingulate lesions result in deficits in the maternal behavior stimulated in nonpregnant rats through pup exposure and these deficits are similar to those observed in postpartum females (Fleischer, 1972; Slotnick, 1967a). Pup-induced maternal behavior is believed to have a nonhormonal basis.

4. The limbic lesions inflicted in the above studies did not interfere with the hormonal mechanisms necessary for the initiation, maintenance, and termination of pregnancy.

In all of the studies cited above, except for the study on septal lesions in rabbits, maternal motivation or the willingness to engage in maternal activities appeared to be intact at the time of parturition. The question we raise here, then, is what behavioral function the limbic system serves with respect to maternal behavior. The first point to note is that the cingulate cortex, the septal area, the hippocampus, and the mammillary bodies are all anatomically related and can reciprocally influence each other. Although these are anatomically complex structures that do not serve unitary functions, a large body of evidence does suggest that they are all involved in the regulation of complex behavioral response systems. Lubar and R. Numan (1973) proposed that the medial septum plays a role in "the organism's ability to modulate its response output based on the consequences of action" (p. 18). They also suggested that the cingulate cortex is involved in the sequencing of behavioral responses. Ranck (1973) reviewed evidence indicating that the hippocampus and its connections are involved in the appropriate sequencing of behavior patterns and in shifts from one behavior to another. Kimble (1968) suggested that the hippocampus and anatomically related structures enable an organism "to decouple its attention from one stimulus and shift its attention to new or more consequential environmental events" (p. 293). Vanderwolf (1971) argued that the septal-hippocampal system is concerned with organizing motor activity effectively. Commenting on the disorganization of maternal behavior that results from limbic lesions, Vanderwolf speculated that such lesions may prevent neural systems underlying maternal motivation from gaining access to those neural systems involved in regulating the efficient organization of response patterns.

What this analysis suggests is that the effects of limbic system lesions are not specific to maternal behavior, but that normal limbic function is essential for efficient and appropriately organized maternal response patterns. Although this hypothesis seems to describe some of the deficits in maternal behavior observed after limbic lesions, a unitary explanation of such deficits should not be emphasized too strongly. It is not clear from Kimble et al.'s report whether hippocampal lesions disorganize the maternal behavior of rats. Cingulate lesions temporarily disorganize the maternal behavior of rats, but not mice. The clearest and most severe disorganization of maternal behavior occurs after septal lesions in mice and rats, but in rabbits, septal lesions seem to eliminate rather than to disorganize maternal behavior. Carlson and Vallante (1974) reported that septal lesions result in an increased reactivity to olfactory stimuli in mice, and they suggested that perhaps septal lesions disturb the maternal behavior of mice and rats by altering the animal's reactivity to olfactory stimuli. Finally, it is also possible that some of the effects of septal lesions on maternal behavior are the result of an increased behavioral sensitivity to estrogen (Nance *et al.,* 1975). More research is certainly necessary to elucidate the mechanisms involved in the disturbances of maternal behavior following limbic lesions.

HYPOTHALAMIC MECHANISMS

THE LATERAL HYPOTHALAMUS

In a series of studies, Avar and Monos (1966, 1967, 1969b)) provided evidence that lesions of the midlateral parafornical hypothalamus eliminate maternal behavior in female rats. Primiparous female rats received lesions of the lateral hypothalamus (LH) during late pregnancy. The effective lesions were described as destroying an area just lateral and ventral to the fornix in the tuberoinfundibular region and included damage to a substantial part of the medial forebrain bundle. The internal capsule was not involved in the lesion.

As is well known, lesions of the LH result in disturbances of food and water intake (Epstein, 1971), and the LH lesions that disrupted maternal behavior in the above studies were found to result in a temporary but marked hypophagia and hypodipsia. By the time of parturition, food and water intake and body weight had returned to preoperative levels. That this disruption of food and water intake did not directly contribute to the disturbances in maternal behavior was suggested by the finding that maternal behavior was not disrupted in a group of intact pregnant rats given the same amount of food and water that had been ingested by the LH-lesioned rats (Avar and Monos, 1969b).

The duration of pregnancy and the course of parturition were normal in the LH-lesioned mothers. In addition, histological analysis of the mammary glands during the early postpartum period indicated normal gland development in these females. Such females also ate the placentas of the young and cleaned them of fetal membranes. All other aspects of maternal behavior were absent. Prepartum and postpartum nest-building did not occur. The young were not retrieved and gathered together in a single area of the observation cage but were ignored, with the result that they became scattered about the cage. Nursing was not observed to occur, and by 48 hr postpartum, most of the young of the females receiving LH lesions were dead. Although control females showed normal maternal behavior toward healthy foster young presented within 24 hr of parturition, LH-lesioned females did not; instead, they were indifferent to the healthy young, which eventually died because of maternal neglect. Additional findings indicated that the LH lesions that interfered with maternal behavior did not affect general locomotor activity or body temperature, thus providing some evidence for the specificity of the behavior effect. More research should be done along these lines (see below).

In a subsequent experiment, Avar *et al.* (1973) studied the effects of smaller parafornical LH lesions on maternal behavior and endocrine function in rats. The extent of the lesion was described as being only half that of the lesions produced in the previous studies. Maternal behavior was observed for only the first 24 hr following parturition, and it was observed that "only one third of the operated animals displayed retrieving and nest building activities and nursing was also of a reduced degree" (p. 47). Evidence was also presented indicating a reduced release of prolactin and an increased release of ACTH from the anterior pituitaries of the lesioned rats.

Avar and Monos (1969a) contrasted the behavioral changes produced in late-pregnant rats following far-lateral hypothalamic lesions with those changes observed after midlateral parafornical hypothalamic lesions. The far-lateral lesions were described as extending in the lateral hypothalamic area to the internal capsule at the level of the ventromedial nucleus of the hypothalamus. Such lesions resulted in a persistant aphagia and hypodipsia, and no recovery occurred. Body weight decreased continuously, general locomotor activity was severely depressed, and body temperature fell. Prepartum nest-building did not occur in these females. Most of these females were near death at the time of parturition and were laparatomized at this time. This procedure precluded observations of postpartum maternal behavior. Most of the offspring of the lesioned females had died *in utero*. The differential effects of midlateral and far-lateral hypothalamic lesions on food and water intake have been known for some time (Morgane, 1961) and are commented on below.

Although these studies suffer from a lack of detail both in histological analysis of lesion localization and in observations of maternal behavior, they have made an important contribution to our understanding of the neural basis of maternal behavior. The question that arises here is whether destruction of the midlateral hypothalamus disrupts maternal behavior directly or disrupts it indirectly through a resultant hormonal deficiency. Hypothalamic regulation of pituitary gland function is well established, and recent studies have indicated that neural pathways located in the lateral hypothalamus may play a role in this regulation (Carrer and Taleisnik, 1972; Phelps *et al.*, 1976). Avar and Monos (1967) are of the opinion that the elimination of maternal behavior after midlateral LH lesions is the result of a direct effect on neural structures essential for the "organization of maternal behavior." Evidence was presented, however, that suggested that such lesions may result in alterations of endocrine function (Avar *et al.*, 1973). Future research should be directed at answering this question. There are several ways in which this problem can be attacked (see the introduction to this chapter and the next section below).

With this limitation on the interpretation of the above results, the present author also suggests that the lateral hypothalamus and its associated medial forebrain bundle are directly involved in the neural mediation of maternal behavior in the rat. The reason for such a statement is that the LH and the fiber pathways which pass through it have been shown to be involved in the regulation of a variety of motivated behaviors. In particular, lesions similar to those of Avar and Monos have been found to disrupt male sexual behavior in the rat, and this disruption has been shown to be independent of hormonal dysfunction (Caggiula *et al.*, 1975; Hitt *et al.*, 1970).

THE MEDIAL PREOPTIC AREA AND MATERNAL BEHAVIOR IN THE RAT

The medial preoptic area lies in the medial, rostral hypothalamus. Some have included this structure as part of the telencephalon, others as part of the diencephalon. The latter is the position taken here. The MPOA is critically involved in the mediation of reproductive physiology and behavior in rats. The MPOA and its posterior connections with the medial basal hypothalamus are necessary for the occurrence of ovulation (Barraclough *et al.*, 1964; Koves and Halász, 1970; Tejasen and Everett, 1967). Hormones, particularly estrogen, are selectively taken up in this part of the brain (Pfaff and Keiner, 1973). Intracerebral implants of small amounts of estrogen in this area are capable of inducing sexual behavior in ovariectomized female rats (Lisk, 1962). Lesions of the MPOA abolish male sexual behavior (Heimer and Larsson, 1966–1967), and androgen implants in this region restore sexual behavior in castrated male rats (Davidson, 1966; Lisk, 1967).

THE MAINTENANCE OF MATERNAL BEHAVIOR. Because of the central importance of the MPOA in the reproductive function in the rat, Numan (1974) investigated the effects of lesions restricted to this region on maternal behavior. The lesions were produced during the postpartum period, a time during which maternal responsivity is considered free of hormonal control. Numan also investigated the effects of stria terminalis lesions (ST) and destruction of the medial corticohypothalamic tract (MCHT) on postpartum maternal behavior. These two fiber pathways are the major limbic efferents that pass through the MPOA, the ST arising from the amygdala, and the MCHT arising from the hippocampal formation (see the section above on limbic mechanisms). Both fiber pathways were destroyed at a point outside the MPOA. To sever the MCHT, a special brain knife assembly similar to that described by Sclafani and Grossman (1969) was employed. The knife-cut technique is useful for cutting fiber pathways without producing major cellular damage.

Virgin female rats were mated, and following parturition, their maternal behavior was studied on each of the first 12 postpartum days. Daily observations of nursing, retrieving, and nest building were taken, in that order. The surgical procedures were carried out after the completion of all behavioral observations on Day 5 postpartum. Females received either MPOA lesions, sham MPOA lesions, ST lesions, or knife cuts of the MCHT. On Day 6 postpartum, the females were reunited with their pups, and observations of maternal behavior continued through Day 12 postpartum. A fostering procedure was adopted during the postoperative period so that whenever a litter grew weak because of inadequate maternal care, it was replaced by a healthy litter of six pups.

Figure 3A shows the outlines and the extent of the individual MPOA lesions. Figure 3B depicts an approximation of the common area destroyed by most of the MPOA lesions. The MPOA lesions were generally small, involving partial destruction of this region. The behavioral results of this experiment are shown in Table 1. Preoperatively, all females showed good maternal behavior, and there were no differences between the groups. Postoperatively, all females were in good physical condition. The females that received MPOA lesions groomed themselves well and lost only an average of 6.5g in body weight as determined on Day 6 postpartum. By Day 12 postpartum, their body weight had returned to preoperative levels. Females receiving MPOA lesions, however, showed a complete lack of maternal responsivity. Although such females approached, licked, and sniffed the young, the three major components of maternal care were almost completely eliminated. None of these females built nests, and retrieving was never observed either during the daily 15-min retrieving tests or during the daily 30-min nursing observations.

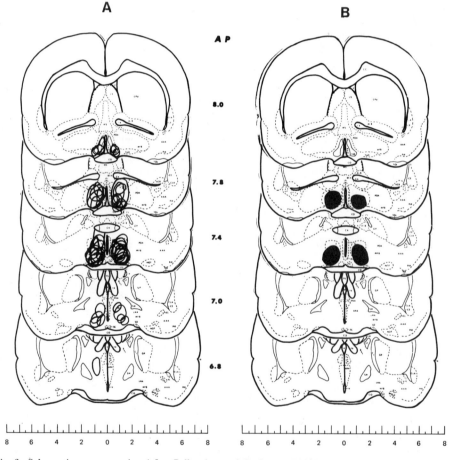

Fig. 3. Schematic representation (after Pellegrino and Cushman, 1967) of MPOA lesions: (A) outlines and extent of individual lesions and (B) an approximation of the common area destroyed by most of the lesions. Abbreviation: AP, anterior-posterior plane of the coronal sections. (From Numan, 1974. Copyright, 1974, by the American Psychological Association. Reprinted by permission.)

TABLE 1. AVERAGE PREOPERATIVE AND POSTOPERATIVE MATERNAL BEHAVIOR SCORES FOR
FEMALE RATS WITH MEDIAL-PREOPTIC (MPOA), SHAM-MPOA, AND STRIA TERMINALIS (ST)
LESIONS AND WITH KNIFE CUTS[a]

	Measure					
	Nursing (in sec)		% Retrieving		% Nest building	
Group	Preoperative	Postoperative	Preoperative	Postoperative	Preoperative	Postoperative
MPOA	1,551	4	100	0	96	0
Sham						
MPOA	1,578	1,323	100	89	94	75.7
ST	1,554	1,290	98	89	94	80
MCHT	1,445	1,188	98	93	88	63

[a]Knife cuts were made in the medial corticohypothalamic tract (MCHT). From Numan (1974). Copyright (1974) by the American Psychological Association. Reprinted by permission.

Of the 10 females receiving MPOA lesions, 9 were never observed to nurse their young. Only 1 female was observed to adopt a nursing posture over her litter, and this occurred for only 5 min on the seventh postoperative day. The litters of females receiving MPOA lesions lost weight daily and were periodically replaced by fresh pups. Because of the lack of maternal attention, these pups were usually found scattered about the observation cage floor each day.

The females in each of the remaining three groups continued to show good maternal behavior postoperatively and did not differ from one another on any of the measures of maternal behavior. Their litters gained weight daily and remained in good physical condition.

This experiment shows that an acute lesion of the MPOA abolishes the maintenance of maternal behavior in postpartum lactating rats. Because the lesions were produced during the maintenance phase of maternal behavior, which is not dependent on hormones for its mediation, the observed disruption was a direct effect of the MPOA lesion, rather than an indirect effect resulting from an interference with pituitary gland function. Additionally, the absence of maternal behavior in females with MPOA lesions cannot be attributed to damage to the ST or the MCHT. In fact, Brown-Grant and Raisman (1972) presented evidence that, even when both of these fiber pathways are destroyed in the same rat, maternal behavior appears to be normal. Therefore, the two major limbic efferents that pass through the MPOA are not involved in the disruption of maternal behavior resulting from MPOA lesions, a finding suggesting that it is the MPOA neurons and their connections, rather than fibers of passage, that are essential for maternal behavior.

A lesion of the MPOA, of course, destroys a large part of the afferent and efferent connections of this region that enter or leave the MPOA from a dorsal, anterior, posterior, or lateral direction. A few experiments have been reported that are pertinent to the problem of localizing the critical connections of the MPOA essential for the maintenance of maternal behavior. Yokoyama et al. (1967) studied the effects of various knife cuts on lactation in rats. The knife cuts were performed during the postpartum period with a brain knife similar to that designed by Halász and Pupp (1965). Although some of the cuts interfered with lactation, maternal behavior was reported to be normal in all cases. Importantly, coronal cuts posterior to the medial-preoptic–anterior-hypothalamic area (MPO-AHA) did not disrupt maternal behavior, a finding indicating that the posterior connections of the MPO-AHA with the medial basal hypothalamus (MBH) are not essential for the maintenance of maternal behavior. An experiment by Herrenkohl and Rosenberg (1974) provided the additional finding that the connections of the entire MBH with other parts of the brain are not critical for either the onset of maternal behavior at parturition or its maintenance during the postpartum period. These investigators completely isolated the MBH of pregnant primiparous female rats using a brain knife similar to that designed by Halász and Pupp (1965). Although histological analysis of the hypothalamic islands produced by these cuts was not presented in detail, the descriptions did indicate that a large portion of the MBH was isolated from the rest of the brain. Although such cuts interfered with lactation, maternal behavior was normal. These results suggest that the connections of the MBH with the rest of the brain, and with the MPOA in particular, are not essential for the display of maternal behavior.

Because the posterior connections of the MPOA with the MBH do not seem critical for maternal behavior, the possibility arises that the lateral connections of the MPOA with the lateral preoptic area and the lateral hypothalamus (LPOA-LH) are critical for the expression of maternal behavior. The medial forebrain bundle (MFB) runs through the LPOA and the LH, and there is extensive dendritic and axonal overlap between MPOA and LPOA neurons (Millhouse, 1969). As the lateral hypothalamus and the MFB had been implicated in the control of maternal behavior (Avar and Monos, 1969b), Numan (1974) investigated the possibility that the neural connections between the MPO-AHA and the LPOA-LH are essential for maternal behavior. The design of this experiment was similar to his preceding one, except that only two groups of postpartum lactating female rats were studied. One group received parasagittal knife cuts that severed the mediolateral connections of the MPO-AHA on Day 5 postpartum, and the other group received sham knife cuts. Histological analysis revealed that the cuts severed most of the mediolateral connections throughout most of the anterior-posterior extent of the preoptic and the anterior hypothalamic nuclei (see Figure 4). Table 2 shows the behavioral findings. Preoperatively, the maternal behavior of the two groups was comparable. Postoperatively, the maternal behavior of the females receiving the knife cuts was severely disrupted in a manner similar to that described for females receiving MPOA lesions. The control females continued to show good maternal behavior postoperatively.

These findings indicate that the MPOA and its lateral connections, possibly with the MFB, are essential for the normal display of maternal behavior in postpartum lactating female rats. Numan (1973, 1974) carried out a third experiment in order to gain some information on the specificity of the behavioral effect produced by MPOA lesions or knife cuts that sever the mediolateral connections of the MPO-AHA. To this end, the effects of MPOA lesions and the disruption of the mediolateral connections of the preoptic-AHA on maternal behavior, sexual

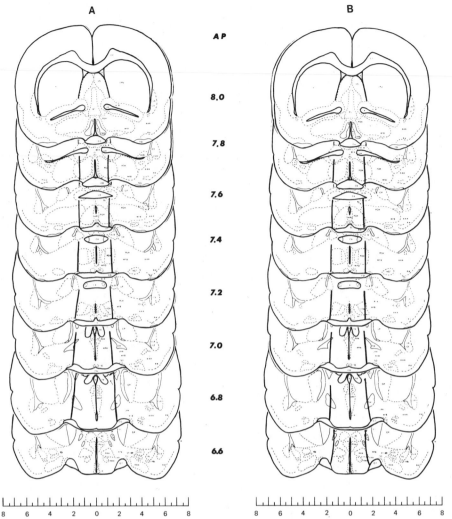

Fig. 4. Schematic representation (after Pellegrino and Cushman, 1967) of the extent of two typical parasagittal knife cuts. (From Numan, 1974. Copyright, 1974, by the American Psychological Association. Reprinted by permission.)

TABLE 2. AVERAGE PREOPERATIVE AND POSTOPERATIVE MATERNAL BEHAVIOR SCORES FOR FEMALE RATS WITH PARASAGITTAL KNIFE CUTS (M-L) AND SHAM OPERATIONS[a]

| | Measure | | | | | |
| | Nursing (in sec) | | % Retrieving | | % Nest building | |
Group	Preoperative	Postoperative	Preoperative	Postoperative	Preoperative	Postoperative
M-L	1,543	47.9	100	0	88	0
Sham	1,590	1,351	100	97.1	90	64

[a]Knife cut was mediolateral. From Numan (1974). Copyright by the American Psychological Association. Reprinted by permission.

behavior, and general locomotor activity were examined in the same animal. The general design of the experiment was similar to that reported for the previous two experiments, except that general locomotor activity was measured on Day 5 postpartum preoperatively and Day 7 postpartum postoperatively by measuring each animal's movement in a stabilimeter cage for 30 min on each of the test days. In addition, each female was injected with estrogen and progesterone and was tested for sexual receptivity on Day 16 postpartum. Although the lesions and the knife cuts disrupted maternal behavior, confirming the previous results, they did not interfere with sexual behavior. It was suggested that independent neural mechanisms for the control of maternal behavior and of female sexual behavior exist within the hypothalamus of female rats. This experiment also indicated that general locomotor impairment could not be advanced as the cause of the disruption of maternal behavior resulting from MPOA lesions or from severing the mediolateral connections of the MPO-AHA region.

THE HORMONAL ONSET OF MATERNAL BEHAVIOR. The mechanisms underlying maternal behavior in the rat can be separated into those controlling its onset and maintenance. As reviewed in the introduction to this chapter, the hormonal changes occurring at about the time of parturition are necessary for the onset of maternal behavior at parturition, but the postpartum maintenance of maternal behavior has been shown to have a nonhormonal basis. In the experiments of Numan (1973, 1974), the MPOA lesions and knife cuts were done during the postpartum period and consequently during the maintenance phase of maternal behavior. A question arises, therefore, about whether the MPOA is also essential for the onset of maternal behavior. In this section, evidence is reviewed indicating that the MPOA is involved in the hormonally mediated onset of maternal behavior in the rat. In particular, it is shown that estrogen acts on the MPOA to facilitate the onset of maternal behavior in the rat.

Rosenblatt and Siegel (1975) showed that terminating pregnancy by hysterectomy facilitates the onset of maternal behavior in the rat when the hysterectomies are performed during the second half of pregnancy. In comparison to non-hysterectomized pregnant females, such females show a reduced latency to onset of maternal behavior following exposure to foster pups. If the ovaries are removed at the time of the hysterectomy, however, this facilitation of maternal behavior fails to appear. It was suggested that hysterectomy alters the pattern of ovarian hormone secretion, and that this hormonal change facilitates the onset of maternal behavior. In a subsequent study, Siegel and Rosenblatt (1975b) found that pregnant rats that had been hysterectomized and ovareictomized on Day 16 of pregnancy and injected subcutaneously with estradiol benzoate (EB) showed a shorter latency to the onset of maternal behavior when exposed to foster pups than control females injected with oil. These studies indicated the importance of estrogen in the facilitation of maternal behavior that results from pregnancy termination. Therefore, Numan et al. (1977) investigated whether estrogen acts on the MPOA in facilitating the onset of maternal behavior in 16-day pregnant, hysterectomized, and ovariectomized female rats.

Virgin female rats were mated, and on Day 16 of pregnancy, all females were hysterectomized and ovariectomized. The following five groups were then formed: The experimental group reveived unilateral implants of undiluted crystalline estradiol benzoate aimed at the MPOA on Day 16 of pregnancy. This procedure was accomplished by inserting a hormone-loaded inner cannula into an outer guide cannula that had previously been implanted in the MPOA on Day 2 of pregnancy. Control females were implanted with cholesterol in the MPOA, or with EB in the ventromedial hypothalamus (VMH) or the mammillary bodies (MB), or subcutaneously (SC). Following the implantation of EB or cholesterol on Day 16 of pregnancy, daily vaginal smears began and continued for the next 7 days. This procedure was done so that we could get an estimate of the extent of leakage of the implanted estrogen from the hypothalamus into the systemic circulation. Vaginal smears were rated on a 4-point scale, with a rating of 4 indicating full vaginal cornification. Two days after the implantation of EB or cholesterol (Day 0 of testing), the cannulas containing these substances were removed, each female was presented with nesting material and a group of four test pups, and daily observations of maternal behavior commenced. A fresh group of foster pups was presented daily, and this procedure continued until maternal behavior was displayed or until 5 days had elapsed. A female was regarded as behaving maternally if she built a nest, retrieved, and assumed a nursing posture over the pups on 2 consecutive days. The first day on which these behaviors were observed was used to compute the latency of maternal behavior.

This experiment showed that, of the areas tested, the MPOA was the most effective site at which estrogen acted to facilitate the onset of maternal behavior (see Table 3 and Figure 5). Females receiving estrogen implants in the MPOA had latencies to onset of maternal behavior that were significantly shorter than those of the females in each of the remaining four groups. These latter groups did not differ significantly among themselves. Of the 12 females in the EB-MPOA group, 10 responded maternally on their first day of exposure to pups (Day 0). The fact

TABLE 3. MEDIAN LATENCY TO ONSET OF MATERNAL BEHAVIOR IN 16-DAY-PREGNANT, HYSTERECTOMIZED, AND OVARIECTOMIZED RATS[a]

Group	n	Mdn latency (in days)
EB-MPOA	12	0^b
Chol-MPOA	11	3
EB-VMH	11	1
EB-MB	12	1.5
EB-SC	8	2

[a]Latency was measured from the time pups were introduced at 48 hr after surgery. Abbreviations for identification of groups: EB = estradiol benzoate; MPOA = medial preoptic area; Chol = cholesterol; VMH = ventromedial hypothalamus; MB = mammillary bodies; SC = subcutaneous. From Numan *et al.* (1977). Copyright (1977) by the American Psychological Association. Reprinted by permission.
[b]Significantly different from each of the other four groups.

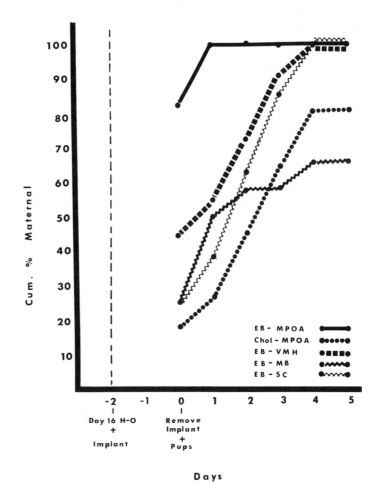

Fig. 5. Cumulative (Cum.) percentage of 16-day-pregnant, hysterectomized (H), and ovariectomized (O) female rats showing maternal behavior over the 5-day test period. Abbreviations for identification of groups: EB, estradiol benzoate; MPOA, medial preoptic area; Chol, cholesterol; VMH, ventromedial hypothalamus; MB, mammillary bodies; SC, subcutaneously implanted control. (From Numan *et al.*, 1977. Copyright, 1977, by the American Psychological Association. Reprinted by permission.)

that females receiving cholesterol implants in the MPOA displayed longer latencies to onset of maternal behavior than females that received EB implants in this region suggests that the short-latency maternal behavior observed in the EB-MPOA group cannot be explained on the basis of a nonspecific irritation of the MPOA resulting from the implantation procedure. The evidence presented also indicated that the short-latency maternal behavior observed in females receiving EB implants in the MPOA cannot be explained on the basis of leakage of the implanted estrogen into the systemic circulation. In particular, the vaginal smear ratings indicated that the EB implants resulted in very little estrogenic stimulation of the vagina. More important, there were no significant differences in the overall vaginal-smear ratings among the five groups receiving either EB or cholesterol implants. This evidence is further substantiated by the latencies to onset of maternal behavior exhibited by females that received subcutaneous implants of estrogen.

Figure 6 shows sagittal sections of the rat diencephalon indicating the location of the hormone implants. Solid circles indicate latencies of Day 1 or less, and open circles denote latencies greater than 1 day. It can be seen that each of the 12 EB-MPOA females responded maternally by Day 1 of testing (the second day of pup exposure), whereas only 3 of the 11 Chol-MPOA females, 6 of 11 EB-VMH females, 6 of 12 EB-MB females, and 3 of 8 EB-SC females did so. The EB-SC females are represented by the implant locations situated outside the brain.

Rosenblatt and Siegel (1975) showed that hysterectomy during the second half of pregnancy in the rat facilitates the onset of maternal behavior and related the hormonal changes that occur as a result of hysterectomy to the hormonal changes that normally occur in the days prior to parturition. Their analysis indicates that hysterectomy-induced maternal behavior and normally occurring maternal behavior may have a common hormonal basis. More specifically, Siegel and Rosenblatt (1975b) presented evidence that estrogen may be the hormone responsible both for hysterectomy-induced maternal behavior and for the normal onset of maternal behavior at parturition. The results of Numan et al. (1977) support the contention that the presence of estrogen is essential for hysterectomy-induced short-latency maternal behavior, and they also provide evidence that the MPOA is the site at which estrogen acts in order to facilitate the onset of maternal behavior in such females. This evidence suggests that the action of estrogen on the MPOA may be one of the events responsible for the normal onset of maternal behavior at parturition. The fact that the MPOA contains estrogen-concentrating neurons (Pfaff and

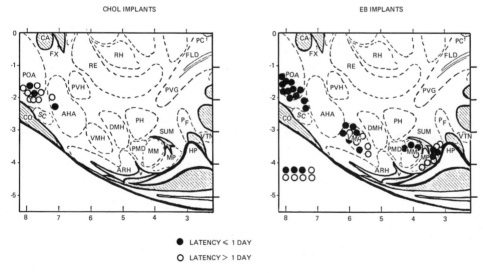

Fig. 6. Sagittal sections of the rat diencephalon (after De Groot, 1960) showing location of hormone implants in 16-day-pregnant, hysterectomized, ovariectomized female rats. Abbreviations: Chol, cholesterol; EB, estradiol benzoate. (From Numan et al., 1977. Copyright, 1977, by the American Psychological Association. Reprinted by permission.)

Keiner, 1973) favors such a position. In addition to estrogen, other hormonal events may be involved in the immediate onset of maternal behavior at parturition in the rat (see the introduction). Whether such events influence the function of the MPOA in producing their effects remains to be determined. In any event, it can be concluded from the results just reviewed that the MPOA is not only involved in the maintenance of maternal behavior (Numan, 1974) but is also involved in the hormonally mediated onset of maternal behavior in the rat.

THE ONSET OF MATERNAL BEHAVIOR INDUCED BY PUP STIMULATION. In another experiment, Numan *et al.* (1977) investigated whether lesions of the MPOA disrupt the onset of maternal behavior that is induced by pup stimulation in virgin females. It should be recalled that this onset of maternal behavior in virgin females is not dependent on hormones for its activation (Rosenblatt, 1967; Terkel and Rosenblatt, 1971). Virgin female rats were ovariectomized and then received either lesions of the MPOA or control lesions in which the lesioning electrodes were lowered to a point just above the MPOA but no current was passed through the electrodes. Two weeks following the brain operation and ovariectomy, each female was supplied with nesting material and a group of test pups. Maternal behavior observations then commenced and continued for the next 14 days, with a fresh group of test pups being supplied daily.

Histological analysis indicated that the MPOA lesions were generally large, and in all animals they damaged parts of both the MPOA and the AHA. All animals were in good physical condition at the time of behavioral testing, the females in both groups showing comparable body-weight gains during the postoperative period. Lesions of the MPOA, however, severely disrupted the onset of maternal behavior induced in virgin females by pup stimulation (see Table 4). An animal was regarded as showing complete maternal behavior if it built a nest, retrieved, and nursed its pups on 2 consecutive days. None of the 12 females that received MPOA lesions met this criterion for complete maternal behavior. In contrast, 10 of the 12 control females did so. It should be noted that the median latencies to onset of retrieving, nursing, and complete maternal behavior in the control females coincided well with the typical 5- to 7-day average latency reported by others for the onset of maternal behavior induced in virgin females by pup stimulation. In con-

TABLE 4. NUMBER OF FEMALES SHOWING COMPLETE MATERNAL BEHAVIOR AND MEDIAN LATENCIES IN DAYS TO ONSET OF VARIOUS COMPONENTS OF MATERNAL BEHAVIOR IN VIRGIN RATS[a]

Group	n	No. of females showing complete maternal behavior	*Mdn* latency to onset of behavior		
			Retrieving	Nursing	Complete maternal
MPOA lesion	12	0	15	15	15
Control lesion	12	10	5	5	5.5

[a]Abbreviation for identification of groups: MPOA = medial preoptic area. Latency was measured from time pups were introduced. Modified from Numan *et al.* (1977). Copyright (1977) by the American Psychological Association. Reprinted by permission.

trast, the median latencies for females receiving MPOA lesions were 15 days. Females were assigned latencies of 15 days if the particular component of maternal behavior was not observed during the 14-day test period. Thus, most of the females receiving MPOA lesions showed a complete lack of maternal responsivity.

A major problem in lesion experiments concerns the question of whether the lesioned area is directly involved in the neural mediation of the behavior under investigation. A lesion destroys a geographical area that may include neural systems involved in more than one function, and therefore, it is possible that the disruption of maternal behavior observed in the present experiment after MPOA lesions was the indirect result of some other effect of the lesions. For example, lesions of the preoptic–anterior-hypothalamic region have been found to impair heat loss mechanisms (Hammel, 1968), and subjecting lactating rats to high ambient temperatures (33°C) has been reported to disrupt maternal behavior (Hearnshaw and Wodzicka-Tomaszewska, 1973). It is therefore possible that the effect of MPOA lesions on maternal behavior was the result of disturbances in the ability of the rat to thermoregulate. This criticism, of course, is applicable not only to the present experiment but also to those of Numan (1974).

Although this general problem clearly deserves further attention, the following points should be considered in its evaluation. The fact that females receiving MPOA lesions were in good physical condition, as indicated by body weight gains, provides some support for the specificity of the lesion effect. Findings previously reviewed have indicated that the disruption of maternal behavior after MPOA lesions in postpartum female rats is the result of interfering with the mediolateral connections of the preoptic region, and that this disruption of maternal behavior cannot be explained on the basis of an impairment of general locomotor activity. Also, MPOA lesions that disrupt maternal behavior do not affect the sexual behavior of such females. Finally, and importantly, the fact that lesions of the MPOA disrupt maternal behavior whereas estrogen stimulation of this region activates maternal behavior strengthens the position that the MPOA is directly involved in the mediation of the behavior.

DISCUSSION. The studies reviewed above indicate that the MPOA is critically involved in all aspects of maternal behavior in the female rat. To briefly summarize the findings, the MPOA and its lateral connections have been shown to be essential for the maintenance of maternal behavior in postpartum lactating rats. Estrogen acts on the MPOA to facilitate the onset of maternal behavior in the rat. The MPOA is also involved in the onset of maternal behavior induced in virgin females by pup stimulation. Therefore, the MPOA is involved in both the onset and the maintenance of maternal behavior in the rat.

Maternal behavior in the rat is interesting in that it is a behavior that is activated by hormones, particularly estrogen, at about the time of parturition and is thereafter maintained in the absence of hormonal mediation during the postpartum period, such maintenance being based solely on the stimulation that the female receives from pups. Furthermore, the onset of maternal behavior induced in virgin females by pup stimulation, like the maintenance of maternal behavior during the postpartum period, occurs in the absence of hormonal activation. The MPOA has now been shown to be involved in all these functions: it is not only the site that is responsive to estrogen activation, it is also essential for the activation and maintenance of maternal behavior by sensory stimuli from pups. Therefore, although the MPOA is a neural substrate on which hormones act to facilitate maternal behavior, it is also involved in those aspects of the neural mediation of maternal behavior that are not themselves directly dependent on hormonal activation.

The MPOA has been shown to be one of the sites at which hormones act to facilitate the onset of maternal behavior in the rat. Some suggestive evidence exists that the MPO-AHA may be a neural substrate on which hormones act to facilitate components of maternal behavior in other species as well. The reader is referred to the paper by Voci and Carlson (1973), which suggests that prolactin implants in the anterior-medial hypothalamus may facilitate certain aspects of maternal responsivity in mice, and to papers by Wallace et al. (1973) and Owen et al. (1974), which indicate that estrogen implants in the MPO-AHA facilitate maternal scent-marking in gerbils.

Numan (1974) suggested that the MPOA exerts its influence on maternal behavior through its lateral connections with the MFB. This suggestion is supported by the findings that both MFB lesions (Avar and Monos, 1969b) and MPOA lesions (Numan, 1974; Numan et al., 1977; also see Ho et al., 1974) disrupt maternal behavior. Furthermore, coronal cuts posterior to the MPO-AHA do not disrupt maternal behavior (Yokoyama et al., 1967), but cuts lateral to this continuum do (Numan, 1974). Such lateral cuts would sever the connections between the MPOA and the MFB at a rostral level (Millhouse, 1969), while causing little cellular damage to surrounding structures. All of these points, of course, are in need of further confirmation. It is possible that the cuts that severed the mediolateral connections of the MPO-AHA actually severed fibers of passage that enter or leave the MPOA from a lateral direction. One would like to see experiments performed to determine whether, in fact, the MPOA and the MFB represent a continuous functional system with respect to maternal behavior (cf. Hendricks and Scheetz, 1973). Also of relevance would be experiments designed to determine whether the anterior and dorsal connections of the MPOA play a role in the regulation of maternal behavior (see Velasco et al., 1974). Finally, because the knife cut experiments were performed on postpartum female rats, one would like to see experiments that examine the effects of such cuts on other aspects of maternal behavior.

The MPOA may exert its influence on maternal behavior through its lateral connections with the ascending components of the MFB, thereby affecting limbic mechanisms, or through its lateral connections with the descending components of the MFB, in this way affecting lower brainstem mechanisms. Alternatively, incoming information to the MPOA from the ascending or descending components of the MFB may be essential for the normal expression of maternal behavior. These two alternatives are not mutually exclusive.

Neuroanatomical findings on the efferent connections of the MPO-AHA have been reported by Conrad and Pfaff (1975) and deserve some attention here. Evidence has existed that the medial preoptic region has short local projections to other brain regions, including lateral connections with the MFB at the level of the LPOA (Millhouse, 1969). Conrad and Pfaff (1975), however, demonstrated the presence of long axonal projections from the MPOA-AHA of the rat to other brain regions through the use of tritiated amino acid autoradiography. Conrad and Pfaff found that the efferent projections of the MPOA reach both the limbic forebrain and the midbrain reticular formation. Ascending projections leave the MPOA, presumably from a lateral direction, and then ascend to terminate in the nuclei of the diagonal band as well as other septal nuclei. Importantly, axons from the MPOA were also found to travel within the MFB. Fibers were observed to travel postero-laterally from the MPOA and to collect just lateral to the fornix within the MFB on their descent to the lower brainstem.

These findings, therefore, suggest that the MPOA has direct projections to both the limbic system and the brainstem. Furthermore, it seems probable that severing the mediolateral connections of the preoptic region would sever these efferent connections of the MPOA. Significantly, the parafornical MFB lesions of Avar and Monos (1969b) probably also destroyed the descending projections of the MPOA to the midbrain reticular formation. As the septal-hippocampal system has been implicated in maternal behavior, direct MPOA projections to this region gain added significance. Following Vanderwolf (1971), perhaps this may be the route by which the neural systems underlying maternal motivation gain access to those neural systems involved in regulating the efficient organization of response patterns. The fact that the descending projections of the MPOA, which travel in the parafornical MFB, eventually reach the midbrain reticular formation suggests that the MPOA may be capable of influencing the ascending aminergic neural systems of the brainstem (Ungerstedt, 1971b). The role of these systems in maternal behavior is commented on below.

In a autoradiographic study of the efferent connections of the preoptic region in the rat, Swanson (1976) did not observe projections of the MPOA to the septum and the brainstem. Many of the long axonal projections that Conrad and Pfaff (1975) reported as arising from the MPOA Swanson reported as arising from the LPOA. Future studies should clarify these differences. In any event, as the MPOA and LPOA are anatomically related through mediolateral connections (Millhouse, 1969), these differences probably do not put heavy constraints on functional neuroanatomical interpretations.

The final question that remains, of course, is what behavioral function the MPO-AHA serves with respect to maternal behavior in the rat. Evidence has been presented that this region is directly involved in the mediation of maternal behavior and that the disruption of such behavior after lesions and knife cuts was not the indirect result of some other effect of these lesions. The MPOA has been found to be involved in all phases and aspects of maternal behavior, and the mechanisms by which this region mediates these functions deserves further attention (see the section below on a neurobehavioral model).

578

MICHAEL NUMAN

COMPARISONS BETWEEN THE NEUROHORMONAL MECHANISMS UNDERLYING FEMALE
MATERNAL BEHAVIOR WITH THOSE UNDERLYING MALE SEXUAL BEHAVIOR AND FEMALE
SEXUAL BEHAVIOR IN THE RAT

The previous section stressed the importance of the MPOA in maternal behavior in the rat. From a comparative perspective, however, it should be indicated that the MPO-AHA has been found to occupy a central position in the neural mechanisms underlying reproductive behavior in several species of vertebrates represented throughout several classes. The present section restricts itself to a comparison of the similarities and differences in the hypothalamic mechanisms underlying female maternal behavior, male sexual behavior, and female sexual behavior in the rat. It is hoped that such an analysis will shed some light on MPOA function with respect to maternal behavior.

The evidence suggests that the hypothalamic mechanisms underlying maternal behavior in the female and sexual behavior in the male are similar, both differing from those mechanisms underlying sexual behavior in the female rat. Lesions of the MPOA disrupt both female maternal behavior (Numan, 1974; Numan et al., 1977) and male sexual behavior (Giantonio et al., 1970; Heimer and Larsson, 1966–1967), but they leave female sexual behavior unaffected (Law and Meagher, 1958; Numan, 1974; Singer, 1968). Indeed, MPOA lesions may even enhance female sexual behavior (Powers and Valenstein, 1972). Lesions of the MFB disrupt both maternal behavior (Avar and Monos, 1969b) and male sexual behavior (Caggiula et al., 1973, 1975; Hitt et al., 1970), leaving female sexual behavior unaffected (Hitt et al., 1970; Rodgers and Law, 1967). Parasagittal knife cuts severing the mediolateral connections of the preoptic–anterior-hypothalamic region disrupt both maternal behavior (Numan, 1974) and male sexual behavior (Paxinos and Bindra, 1972, 1973; Szechtman, 1975), but leave female sexual behavior unaffected. Estrogen acts on the MPOA to facilitate the onset of maternal behavior in female rats. Estrogen implants in the MPOA have also been found to be effective in activating female sexual behavior in ovariectomized rats (Chambers and Howe, 1968; Lisk, 1962; Lisk and M. Barfield, 1975), although other studies have questioned this finding (Dörner et al., 1968; R. Barfield et al., 1975). Lesions of the MPOA, however, disrupt female maternal behavior but may actually facilitate female sexual behavior. This possibility suggests that the MPOA is excitatory of female maternal behavior while perhaps being inhibitory of female sexual behavior. It is possible, therefore, that two different populations of MPOA neurons, with different responses to estrogen, are involved in the control of the two behaviors. Perhaps the population of MPOA neurons involved in the control of maternal behavior is activated by estrogen, whereas the population of MPOA neurons involved in the regulation of female sexual behavior is inhibited by estrogen (cf. Powers and Valenstein, 1972). In support of such a proposition, Yagi (1973) and Yagi and Sawaki (1973) presented electrophysiological evidence of two populations of estrogen-responsive neurons within the MPOA, one group being excited and the other inhibited by estrogen. Whitehead and Ruf (1974) electrophysiologically identified, by antidromic stimulation techniques, neurons in the MPOA with axons terminating in the ventromedial-arcuate region of the hypothalamus, and they found that the firing rates of the majority of such caudally projecting MPOA neurons were inhibited by systemic estrogen treatment. Perhaps the posterior-medial connections of the MPOA serve inhibitory functions with respect to female sexual behavior. In support of this hypothesis, Rodgers and Schwartz (1976) reported that coronal cuts posterior to the MPO-AHA facilitate female sexual behavior. It would be interesting to determine whether the laterally projecting neurons of the MPOA are excited by estrogen, as cutting these connections disrupts female maternal behavior without affecting female sexual behavior. Importantly, Christensen and Clemens (1974) reported that estrogen implants in the MPOA can activate male sexual behavior in the castrated male rat. It should be recalled that both lesions of the MPOA and knife cuts that sever the mediolateral connections of the preoptic-AH region disrupt male sexual behavior.

These findings, taken together, indicate that the neurohormonal mechanisms underlying female maternal behavior are more similar to those underlying male sexual behavior than to those underlying female sexual behavior. Estrogen seems capable of facilitating both female maternal behavior and male sexual behavior by *activating* neurons in the MPOA, and it is possible that such neurons leave the MPOA from a lateral direction. In comparison, estrogen may *inhibit* posteromedially projecting MPOA neurons in facilitating female sexual behavior. In conclusion, one can suggest that the mediolateral connections of the MPOA are excitatory of both female maternal behavior and male sexual behavior, and that these neurons are activated by hormones. The posteromedially projecting neurons may serve inhibitory functions with respect to female sexual behavior and may be inhibited by estrogen.

With respect to these similarities and differences of the three behaviors, it is interesting to note that both male sexual behavior and female maternal behavior are complex behaviors oriented toward a stimulus object, that object being manipulated in some way. Maternal behavior in the female and sexual behavior in the male consist of behavior patterns that are more appetitive and goal-directed than the reflexlike nature of the motor responses involved in lordosis, which is the most frequent response measured in studies of female sexual behavior. In 1951, Beach also commented on these similarities and differences, noting that large neocortical ablations disrupt both female maternal behavior and male sexual behavior, while leaving female sexual behavior relatively unaffected. He noted that the maternal reactions of female rats and the sexual responses of male rats are comparable in several ways, and the first point he noted was that maternal females and sexually active males are required to take the initiative in responding to another animal, this not being the case for female sexual behavior.

Although the hypothalamic and neocortical mechanisms underlying male sexual behavior and female maternal behavior are similar in many respects, this similarity should not be taken to mean that the neural basis of maternal behavior is equivalent to that of male sexual behavior. For example, lesions of the stria terminalis or the medial amygdala have disruptive effects on male sexual behavior (Giantonio *et al.,* 1970; Harris and Sachs, 1975), while having no effect on female maternal behavior (Numan, 1974). Future research should be directed at unraveling the implications of the findings that indicate that the hypothalamic mechanisms underlying female maternal behavior and male sexual behavior are similar. The possibility should certainly be considered that these mechanisms serve a function that is common to both behaviors.

NEUROCHEMICAL CORRELATES OF MATERNAL BEHAVIOR

The present section reviews the research that has concerned itself with the neurochemical basis of maternal behavior. Not much research of this kind has been done, and all such studies have concentrated on the possible role of the monoamines, which include dopamine, norepinephrine (the catecholamines), and serotonin, in the neurochemical control of maternal behavior in the rat.

Recent findings have greatly enhanced our knowledge of the neuroanatomy of the central monoaminergic systems in the rat, and it will be useful to briefly review these findings. The reader is referred to the following sources for more information on this topic: Beckstead (1976); Jacobowitz and Palkovits (1974); Lindvall and Björklund (1974); Lindvall et al. (1974); Palkovits and Jacobowitz (1974); Swanson and Hartman (1975); and Ungerstedt (1971b).

The majority of the cell bodies of monoaminergic neurons are found in the lower brainstem. These neurons have both ascending and descending projections, but most of the research on the role of these systems in behavior has stressed the ascending systems. A large part of the ascending monoaminergic systems has been found to project from the brainstem through the lateral hypothalamus to widespread areas of the diencephalon and the telencephalon, including the hypothalamus, the septum, the hippocampus, the striatum, and the cerebral cortex. Two major ascending noradrenergic (NE) bundles have been described. The dorsal NE bundle arises from cell bodies of the locus coeruleus in the pons; the hippocampus, the thalamus, and the neocortex are among the areas that it innervates. The ventral NE bundle arises from other cell bodies in the pons and the medulla and innervates a large part of the preoptic-hypothalamic region and the basal telencephalic structures. Two major ascending dopaminergic (DA) systems have been described: the nigrostriatal pathway and the mesolimbic pathway. The nigrostriatal pathway arises from the substantia nigra in the midbrain and terminates mainly in the caudate-putamen. The mesolimbic DA system arises from cell bodies near the interpeduncular nucleus of the midbrain and among the areas that it innervates are the nucleus accumbens and the olfactory tubercle. Other areas receiving ascending DA input include the cingulate and the prefrontal cortex. Finally, ascending serotonergic neurons arise from the raphe nuclei in the brainstem and terminate in the preoptic-hypothalamic region, the striatum, the septum, the neocortex, and the thalamus.

Two points are worth noting at this time. First, monoaminergic ascending input is widespread, and therefore, it is not surprising that most of the areas implicated in maternal behavior in the rat receive such input. This fact, in turn, suggests that alterations in ascending monoaminergic activity may influence maternal behavior. The second point is that a large part of the ascending monoaminergic systems travels within the lateral hypothalamus (LH) and the MFB. Ascending NE and serotonergic systems travel within the hypothalamic MFB, and at the level of the midhypothalamus, the ascending DA systems travel within the far-lateral hypothalamus. Therefore, a lesion of the lateral hypothalamus would remove much of the monoaminergic influence on forebrain structures. Previously, the studies of Avar and Monos were reviewed that indicated that LH lesions performed during pregnancy disrupt the subsequent maternal behavior of female rats. Interference with monoaminergic neural mechanisms may have played a role in the observed deficits, and attention will now be turned to those studies that may illuminate this possibility.

One approach to the study of the role of particular neurochemical systems in maternal behavior would be to measure the concentration and the metabolism of particular neurochemicals in the brain in an attempt to determine whether any significant changes occur that are correlated with a particular phase of maternal behavior. This approach has been taken in a study by Moltz *et al.* (1975) (cf. Fuxe *et al.*, 1969; Greengrass and Tonge, 1972, 1974; Voogt and Carr, 1974). Moltz *et al.* measured the concentration of NE and one of its metabolites, 3-methoxy-4-hydroxyphenylglycol (MHPG), in rat hypothalamus on Days 5, 15, and 21 of pregnancy and on Days 1 and 5 postpartum. The mean concentrations of both NE and MHPG did not change significantly during pregnancy, but on Day 1 postpartum, NE showed a significant decrease in concentration, whereas MHPG showed a corresponding increase. This increase in the metabolism of NE on Day 1 postpartum was interpreted as reflecting an increase in the activity of noradrenergic neurons within the hypothalamus. Moltz *et al.* suggested that an increase in noradrenergic activity within the hypothalamus (which included the preoptic region) around the time of parturition may underlie the onset of maternal behavior at parturition, although they were aware of the possibility that the observed changes in NE metabolism may have reflected other functional processes. In particular, these changes may form the basis of neuroendocrine, rather than neurobehavioral, events. Recent studies have shown that catecholaminergic neural mechanisms participate in the hypothalamic regulation of anterior pituitary function (for reviews, see Kalra and McCann, 1973; McCann *et al.*, 1972). Ascending NE neurons innervate those areas of the hypothalamus known to be involved in the regulation of anterior pituitary function, the median eminence being included in these areas of innervation (Björklund *et al.*, 1973). The rat has a postpartum estrus and ovulates within 24 hr of parturition (Johnson *et al.*, 1974). Perhaps the changes in NE metabolism measured by Moltz *et al.* underlie these events rather than the onset of maternal behavior. Several investigators have suggested the NE plays a role in the preovulatory release of gonadotropins (Kalra and McCann, 1974; Terasawa *et al.*, 1975).

In abstract form, Rosenberg (1976) and Steele (1976) reported on research that attempted to extend the findings of Moltz *et al.* (1975). Steele reported that lesions of the lateral hypothalamus, performed on Day 8 of pregnancy, did not affect pregnancy or parturition, but that maternal behavior was disrupted in some of the rats. The animals were classified as either showing or not showing maternal behavior, and it was found that the absence of maternal behavior was correlated with a significant decrease of hypothalamic NE levels. The results were interpreted as indicating that the ascending ventral NE bundle, which innervates the hypothalamus, makes a significant contribution to those mechanisms underlying the onset of maternal behavior at parturition.

Rosenberg (1976) examined the effects of intraventricular infusions of 6-hydroxydopamine (6-OHDA) on both maternal behavior and hypothalamic NE levels. When used properly, 6-OHDA has been reported to be a specific catecholaminergic neurotoxin that selectively destroys both dopaminergic and noradrenergic neurons (Sachs and Jonsson, 1975). The intraventricular infusions of 6-OHDA were performed either 2 days prior to parturition or 4 days after parturition. Prepartum treatment disrupted the onset of maternal behavior at parturition and partially depleted hypothalamic NE levels. Equivalent depletion of hypothalamic NE levels in postpartum females had no significant effect on the maintenance of maternal behavior. Rosenberg suggested that hypothalamic NE is involved in the initiation of maternal behavior, but not in its maintenance. It is possible, however, that a more complete depletion of hypothalamic NE would have also disrupted the maintenance of maternal behavior.

Interestingly, the maternal behavior of those females administered 6-OHDA prior to parturition was observed to improve during the postpartum period. Such females were provided with healthy foster pups each day, and although the disruption of their maternal behavior was initially severe, by Day 7 postpartum 70% of these females were caring for their young. This finding suggests that pup-induced maternal behavior was not interfered with in these females. As both pup-stimulated maternal behavior and the maintenance of maternal behavior have a nonhormonal basis, whereas the onset of maternal behavior at parturition has a hormonal basis, these findings raise the possibility that the disruption of maternal behavior observed by Rosenberg was the result of an interference with neuroendocrine mechanisms. It is also possible, and should not be excluded, that the 6-OHDA treatment selectively disrupted those central mechanisms essential for the hormonal onset of maternal behavior but left those mechanisms underlying pup-induction and maintenance of maternal behavior unaffected.

The findings of Rosenberg should be compared with those of Sorenson and Gordon (1975). In this study, the effects of intraventricularly injected 6-OHDA on maternal retrieving and on whole-brain levels of both NE and DA were examined. It was found that 6-OHDA that was administered prior to pregnancy did not disrupt the onset of maternal behavior at parturition. Indeed, evidence was provided that maternal retrieving was facilitated. Compared to control females, 6-OHDA-treated females had shorter latencies to the retrieval of displaced pups back to their nests. It should be pointed out that maternal behavior was not studied in detail by these investigators; only one retrieving test was given, and this was undertaken during the first postpartum day. The results do indicate, though, that maternal behavior was not disrupted and that retrieving may have been facilitated. The 6-OHDA treatment partially depleted whole-brain levels of both DA and NE.

The major procedural difference between these two studies was the time at which the 6-OHDA was administered. Rosenberg administered 6-OHDA 2 days prior to parturition, whereas Sorenson and Gordon (1975) administered the neurotoxin prior to pregnancy. Perhaps the contrasting behavioral results derived from the possibility that, in the rats administered 6-OHDA prior to pregnancy, enough time had elapsed to allow for the development of a supersensitivity in the catecholaminergic neural systems not destroyed by 6-OHDA. This supersensitivity could result in a recovery of function, and possibly even in a facilitation of the neural mechanisms underlying maternal behavior. If this is the case, then the findings of Sorenson and Gordon also open up the possibility that the disruptive effects of both LH lesions (Steele, 1976) and 6-OHDA treatment (Rosenberg, 1976) on maternal behavior in the rat are not the result of interfering with hypothalamic NE systems but are the result of disrupting ascending DA systems.

The findings reviewed up to this point suggest that the ascending monoaminergic systems may be involved in the expression of maternal behavior in the rat, and the role of the ventral NE bundle that innervates the hypothalamus has been stressed. Except for the study by Sorenson and Gordon (1975), assays for DA were not reported after either LH lesions or after 6-OHDA treatment, although both of these procedures may have affected DA systems. There are a few studies that are relevant to the question of the involvement of DA systems in maternal behavior. The first concerns itself with the effects of a mild tail pinch on the onset of maternal behavior induced in virgin female rats by pup stimulation. Szechtman and Siegel (1976) found that the repeated application of a tail pinch reduced the latency to onset of maternal behavior in virgin females exposed to foster young. It should be noted that tail-pinch-induced behavior is not specific to maternal behavior. A variety of behaviors have been facilitated through tail pinch or tail shock, including eating, gnawing, licking, copulatory behavior, and aggression (Antelman and Szechtman, 1975; Antelman et al., 1975; Caggiula and Eibergen, 1969; also see Caggiula et al., 1974). Antelman et al. (1975) suggested that the tail pinch can induce a variety of behaviors, and that the particular behavior that is potentiated is dependent on the particular stimuli available. Most important, Antelman et al. (1975) and Antelman and Szechtman (1975) provided good evidence that tail-pinch-induced behavior is dependent on the nigrostriatal DA system. They hypothesized that this system is involved in the processing of sensorimotor information and in regulating the organism's responsiveness to a wide variety of environmental stimuli. The peripheral tail-pinch is conceived of as activating this system, with the result that the organism is more responsive to external stimuli.

Although these findings raise the possibility that the nigrostriatal DA system may be involved in maternal behavior, other interpretations of the results of Szechtman and Siegel are possible. Virgin female rats receiving the tail pinch were observed to immediately lick and manipulate pups. The tail-pinch may therefore have indirectly influenced maternal behavior. The increased pup contact and licking of pups may have potentiated the onset of maternal behavior. Also, it is possible that the tail-pinch induced hormonal changes that facilitated the onset of maternal behavior.

If the ascending DA nigrostriatal system plays a role in maternal behavior, one might expect lesions of the caudate putamen to have some effect on maternal behavior. Kirkby (1967) reported that lesions of the dorsal aspect of the head of the caudate nucleus, which were inflicted prior to pregnancy, had no effect on the maternal behavior of rats during the postpartum period, although such lesions did increase general activity levels. In interpreting this finding, it should be pointed out that destruction of the DA input to the caudate may not be equivalent to direct destruction of the caudate, which not only interferes with input but also interferes with output. Several investigators have suggested that the ascending CA systems inhibit neural systems that have an inhibitory effect on behavior (Roberts, 1974; Stricker and Zigmond, 1976). With respect to the nigrostriatal system, this disinhibitory process would mean that caudate-putamen lesions and lesions of the nigrostriatal pathway do not produce the same results. Indeed, caudate lesions have been found to result in hyperactivity (Whittier and Orr, 1962), whereas 6-OHDA destruction of the nigrostriatal pathway has been found to result in hypokinesia and decreased activity (Ungerstedt, 1971a). Therefore, the fact that caudate nucleus lesions did not affect maternal behavior does not eliminate the possibility that the nigrostriatal system plays a role in its control. It should also be mentioned that the caudate nucleus is not homogeneous with regard to function. Different striatal regions have been found to be involved in different behavioral functions (Neill and Linn, 1975; Neill et al., 1974, 1975). Therefore, the finding that lesions of the dorsal aspect of the head of the caudate putamen had no effect on the maternal behavior of rats does not preclude the possibility that lesions of other parts of this region would be effective.

Only one study exists that is pertinent to the possible involvement of the mesolimbic DA system in maternal behavior. One of the structures innervated by the ascending mesolimbic DA system is the nucleus accumbens, and Smith and Holland (1975) reported on the effects of lesions of the nucleus accumbens on the maternal behavior of rats. Lesions of the nucleus accumbens performed prior to mating were found to disrupt both maternal behavior and lactation following a normal parturition. Unfortunately, detailed descriptions of the deficits in maternal behavior were not provided; thus, the results are difficult to evaluate. A high rate of maternal cannibalism was observed, and those females that did not cannibalize their litters showed inferior maternal behavior in comparison to controls. Smith and Holland suggested that these maternal behavior deficits were not secondary to hormonal imbalance but were the result of a lesion produced hyperemotionality and hyperactivity. In support of this suggestion, it was found that lesioning the posterior hypothalamus in females that had previously received lesions of the nucleus accumbens decreased emotionality and reversed the effects of nucleus accumbens lesions on maternal behavior, but lactation was still impaired. Lesions of the posterior hypothalamus alone affected neither maternal behavior nor lactation. These findings are interesting and deserve further research attention.

Only two reports are available concerning the possible role of serotonin in the maternal behavior of the rat. Rowland (1976) presented evidence that partial interference with serotonergic neural mechanisms does not disrupt the maintenance of maternal behavior in postpartum lactating rats. *P*-chloramphetamine, a drug that selectively destroys serotonergic neurons (Harvey *et al.*, 1975) was administered intraventricularly to lactating rats. This treatment resulted in a 50% depletion of brain serotonin and disrupted lactation but had no effect on maternal behavior (also see Gallo *et al.*, 1975; Kordon *et al.*, 1973–1974). These findings should be contrasted with those of Moore and Hampton (1974), who studied the effects of systemically administered para-chlorophenylaline (PCPA), a serotonin-synthesis inhibitor, on pregnancy and parturition. PCPA treatment resulted in a 95% depletion of brain serotonin levels. Although primarily interested in the role of 5-HT in pregnancy maintenance and parturition, these investigators did report that all females treated with PCPA cannibalized their own young and foster young at parturition. The mechanisms responsible for this disruption of maternal behavior are not clear.

To conclude this section, it should be indicated that the ascending catecholaminergic systems have been shown to subserve many of those functions formerly attributed to the classical ascending reticular-activating system. The dorsal NE bundle is believed to be involved in electrocortical arousal, and the ascending DA systems play a role in behavioral arousal (Bolme *et al.*, 1972; Jones *et al.*, 1973; Lidbrink, 1974). Such systems, of course, are necessary for the effective display of a variety of motivated behaviors, and therefore, it is not surprising that evidence exists indicating the involvement of such systems in maternal behavior in the rat. Branchey and Branchey (1970) studied the electrocortical sleep and wakefulness patterns in female rats during pregnancy. Beginning approximately 15 hr prior to the onset of parturition, an increase in electrocortical wakefulness and a decrease in electrocortical sleep were observed. Behavioral observations were not undertaken. Although the possibility exists that such changes in sleep–wakefulness patterns were the result of an increase in afferent feedback from the uterus as parturition approached, Branchey and Branchey provided suggestive evidence that the observed modifications were the result of the dramatic hormonal changes taking place at or near term. Perhaps these changes reflect processes essential for the hormonal activation of the onset of maternal behavior (cf. Moltz *et al.*, 1970).

Zoloth (1974) compared the response of sleeping postpartum female rats and virgin female rats to pup vocalizations. The maternal females were more easily aroused, both electrocortically and behaviorally, from sleep than virgin females. This finding indicates that an increased responsiveness and/or sensitivity to pup-related stimuli is associated with the maternal condition. This finding can be related to those of Allin and Banks (1972), who played back recorded pup calls to adult virgin females and lactating female rats. Only the lactating rats were found to leave their nests and to engage in searching behavior in response to pup calls.

The evidence reviewed in this section suggests that the ascending monoaminergic neural systems may be involved in maternal behavior, although the particular systems involved have not been clearly determined. The results are certainly inconclusive and open to multiple interpretations. Perhaps the best evidence available suggesting the involvement of these systems in maternal behavior is that lesions of the LH disrupt maternal behavior. Future research should be devoted to a thorough analysis of the involvement of these systems in maternal behavior.

A Neurobehavioral Model

Research on the neural basis of parental behavior is clearly not as advanced as is our understanding of the neural basis of other behavioral patterns. It is therefore difficult to integrate the findings reviewed above in order to reach some definitive conclusions. In what follows, an attempt is made to present a limited neurobehavioral model of the mechanisms underlying maternal behavior in the rat. The model is clearly speculative and is presented in the hope that it will stimulate research that will result in a fuller understanding of the neural mechanisms underlying this biologically significant behavioral complex. The model centers on the following findings: (1) lesions of the MPOA disrupt maternal behavior; (2) estrogen acts on the MPOA to facilitate the onset of maternal behavior; (3) severing the mediolateral connections of the preoptic region disrupts maternal behavior; (4) midlateral LH lesions disrupt maternal behavior; and (5) the ascending catecholaminergic systems may be involved in maternal behavior.

The first proposition to be offered is that the *efferent* projections of the MPOA, which leave this region laterally, are essential for maternal behavior in the rat. That the mediolateral connections of the preoptic region are being stressed is based on the fact that cutting these connections disrupts maternal behavior in postpartum lactating rats. The suggestion that these essential mediolateral connections are efferent projections from the MPOA neurons is based on both neuroanatomical and neurobehavioral findings. Conrad and Pfaff (1975) found that the efferent projections of the MPOA, which presumably leave this region laterally, reach both the limbic forebrain and the midbrain reticular formation. Millhouse (1969) demonstrated that efferent projections of the MPOA leave this region laterally to synapse on LPOA neurons. Swanson (1976) described efferents from the LPOA that travel within the MFB and project to the limbic system and the lower brainstem (which included projections to the substantia nigra and the locus coeruleus). Therefore, the efferent projections of the MPOA, which leave this region laterally, are capable of influencing the activity of those structures suggested as being involved in the neural control of maternal behavior.

Lesions of the lateral hypothalamus have been found to disrupt maternal behavior in the rat. The second proposition to be offered is that this disruption is the result of an interference with the functioning of those neural structures that subserve nonspecific motivational functions and/or with the utilization of these nonspecific motivational systems by more specific motivational systems. Lesions of the lateral hypothalamus disrupt a variety of motivated behaviors, including feeding and drinking behavior (Epstein, 1971); male sexual behavior (Caggiula *et al.*, 1975; Hitt *et al.*, 1970); and maternal behavior. Marshall *et al.* (1971) and Marshall and Teitelbaum (1974) reported that LH lesions in rats resulted in severe impairments in orienting to sensory stimuli. The deficits observed can be described as a disruption of sensorimotor integration that results in an inability of the organism to voluntarily initiate and maintain behavior in response to sensory stimulation. Subsequent findings have suggested that a large part of this deficit may be the result of damage to the ascending nigrostriatal DA pathway (Marshall *et al.*, 1974). In line with this hypothesis, Stricker and Zigmond (1976) suggested that damage to the ascending catecholaminergic systems that results from LH lesions should result in behavioral impairments that "include any voluntary activity that demands orientation to sensory stimuli and sustained alertness" (p. 143).

Electrical stimulation of the lateral hypothalamus and the MFB has been found to be capable of eliciting or potentiating a wide variety of species-typical responses in rats (Valenstein *et al.*, 1970); included among these are object carrying (Phillips *et al.*, 1969), male sexual behavior (Caggiula, 1970), feeding, drinking, and gnawing (Phillips and Fibiger, 1976). Evidence has also implicated the ascending catecholaminergic systems in some of these stimulation-induced behaviors (Phillips, 1975; Phillips and Fibiger, 1976). A wide variety of behavior patterns can also be potentiated by a tail pinch, including maternal behavior, and the nigrostriatal DA system has been implicated in the mediation of tail-pinch-induced behavior. Antelman *et al.* (1975) suggested that both LH stimulation and tail-pinch-induced behavior may share a neural basis in which the nigrostriatal pathway is activated with the result that the organism's sensitivity to external stimuli is increased. Finally, the ascending catecholaminergic systems, which would be disrupted by LH lesions, have been found to be involved in behavioral arousal, in electrocortical arousal, and in the regulation of extrapyramidal motor activity (Marshall *et al.*, 1974). All of these functions would seem to contribute to a variety of behavioral patterns.

It is suggested, then, that LH lesions disrupt maternal behavior by interfering with the utilization of these nonspecific motivational mechanisms by those neural systems more specifically involved in the mediation of maternal behavior. Lesions of the lateral hypothalamus could impair these systems directly; the result would be deficits in maternal behavior. An alternative possibility is that certain LH lesions may impair descending influences that activate the nonspecific ascending catecholaminergic motivational systems at their point of origin in the brainstem, with the result that the ascending systems would remain intact but could not be utilized by those neural systems more specifically related to maternal behavior. This latter possibility may be the most accurate with respect to maternal behavior. Lesions of the far-lateral hypothalamus have been found to have the most severe effects on motivated behaviors, resulting, for example, in a severe aphagia and adipsia and in sensory neglect, and these disturbances have been attributed, at least in part, to nigrostriatal DA system damage (see review by Stricker and Zigmond, 1976). The lesions of Avar and Monos (1969b), which disrupted maternal behavior, were located in the midlateral parafornical hypothalamus and did not result in severe aphagia and adipsia (cf. Avar and Monos, 1969a). Although such lesions may have disrupted maternal behavior by damaging the ascending NE systems, the alternative possibility is offered that such lesions disrupted descending influences from the preoptic region that pass through the parafornical lateral hypothalamus before terminating in the lower brainstem (Conrad and Pfaff, 1975; Swanson, 1976). This interference would prevent the utilization of the nonspecific motivational systems by the more specific neural systems underlying maternal behavior.

In a similar manner, it can be suggested that severing the mediolateral connections of the preoptic region prevents the MPOA from influencing the nonspecific systems ascending from the brainstem. It is proposed here that the preoptic region belongs to a neural system specifically related to the mediation of maternal behavior, but that, for maternal behavior to occur, this region must utilize the nonspecific systems. Mediolateral cuts would prevent this utilization without interfering with the ability of other specific motivational systems to utilize nonspecific systems. Indeed, severing these connections does not result in aphagia or adipsia, and female sexual behavior and general activity levels are not affected (Numan, 1973, 1974). Szechtman (1975) showed that similar cuts do not result in sensorimotor impairments in male rats.

The interesting question that arises, however, is if, as suggested above, the mechanisms underlying male sexual behavior and female maternal behavior are similar and share a common function, how specific is the preoptic region and its connections to maternal behavior. Perhaps limbic and cortical mechanisms, as well as environmental conditions, contribute to the specificity of maternal behavior mechanisms.

This speculative and limited model can now be summarized as follows: Hormones, particularly estrogen, act on the MPOA to alter its activity near the time of parturition. Efferent lateral projections descend to the brainstem to activate the ascending nonspecific catecholaminergic systems so that they can be utilized by those neural systems underlying maternal behavior. Perhaps this activation is necessary for the immediate onset of maternal behavior at parturition. The descending projections from the preoptic region may also influence the brainstem motor mechanisms essential for the display of certain aspects of maternal behavior. The ascending projections from the MPOA alter the activity of limbic forebrain mechanisms and contribute to the efficient organization of maternal response patterns. During the nonhormonal maintenance of maternal behavior during the postpartum period, these same mechanisms may be affected by sensory stimuli from the pups that ultimately influence the MPOA. Perhaps this afferent input into the MPOA is also dependent on the mediolateral connections of the preoptic region. Finally, in the absence of hormonal stimulation, prolonged pup stimulation can influence these same mechanisms.

Acknowledgments

The author would like to thank Marilyn Grant of the Boston College reference librarian staff for bibliographic assistance.

APPENDIX

589

BRAIN
MECHANISMS AND
PARENTAL
BEHAVIOR

The preceding review was completed in August 1976. A delay between the submission of my manuscript and its publication required that I present an update. This appendix, which was prepared in May 1982, is meant to provide information on research into the neural basis of maternal behavior that appeared after August 1976. The material is organized into three sections. The first section selectively discusses some recent research developments. The second section lists pertinent publications that appeared after the completion of the original review, whereas a third section lists additional readings. The discussion in the first section refers to publications listed in the second section and to articles cited in the original review.

DISCUSSION OF RECENT RESEARCH DEVELOPMENTS

HORMONAL MECHANISMS

The hormonal changes that accompany the termination of pregnancy are important in promoting the immediate onset of maternal behavior at parturition in rats. The near-term decline in progesterone secretion and the rise in estrogen secretion have been shown to be important hormonal events stimulating maternal behavior (Numan *et al.*, 1977; Numan, 1978). Two recent studies (Pedersen and Prange, 1979; Pedersen *et al.*, 1982) have also suggested that oxytocin may act with estrogen in promoting the onset of maternal behavior in rats. The finding that intraventricular, but not systemic, injections of oxytocin were found to facilitate maternal behavior suggests that the central release of oxytocin by the hypothalamic neurons that produce this neuropeptide may be important in the onset of maternal behavior. Oxytocin-containing neurons within the hypothalamus have long been known to project to the posterior pituitary gland, and as a neurohormone, oxytocin is known to be involved in parturition and milk ejection. Recent findings have also suggested that oxytocin may serve as a neurotransmitter in neurons that originate in the hypothalamus and project to the limbic system and to the brainstem (Armstrong *et al.*, 1980; Buijs *et al.*, 1978; Swanson, 1977). With respect to a possible relationship between oxytocin's facilitatory effect on maternal behavior and the importance of the MPO-AHA in maternal behavior (Numan, 1974; Numan *et al.*, 1977), it should be noted that immunohistochemical studies have indicated that oxytocin-containing neurons are located in the MPO-AHA (Rhodes *et al.*, 1981).

Virgin female rats are not immediately maternally responsive to young, whereas parturient females are. Treating virgin rats with hormones brings their maternal responsiveness closer to that of the parturient female. A hypothesis that has been advanced recently is that the endocrine changes associated with parturition facilitate maternal behavior, in part, by reducing the female's fear of novel stimuli, which may include the odor of pups (Fleming and Luebke, 1981; Rosenblatt *et al.*, 1979). Concerning oxytocin, in a passive-avoidance paradigm it has been shown that intracranial oxytocin treatment is capable of producing an effect that can be interpreted as being based on fear reduction (Kovacs *et al.*, 1979).

SENSORY MECHANISMS

In the original review, evidence was presented that olfactory input inhibits maternal behavior in virgin rats, this input being responsible, in part, for the fact that such females are not immediately maternally responsive to young but, instead, require 5–7 days of exposure to pups before showing maternal behavior. The major finding that supported this view was that olfactory deafferentation facilitates pup-stimulated maternal behavior in cycling virgin rats (Fleming and Rosenblatt, 1974b,c). A question that arose at that time was the extent to which disruption of olfactory input from the vomeronasal organ, rather than from the primary olfactory epithelium, was involved in this facilitation of maternal behavior. Fleming *et al.* (1979) reported that vomeronasal deafferentation alone is capable of facilitating maternal behavior in virgin rats when compared to sham-operated controls. Selective damage to the main olfactory bulb also facilitated maternal behavior. Rats sustaining combined damage to both the vomeronasal and the main olfactory system showed the greatest facilitation of maternal behavior. These results were interpreted as showing that input from both the vomeronasal organ and the main olfactory system exerts an inhibitory effect on rat maternal behavior. Marques (1979) also reported that vomeronasal deafferentation facilitates the maternal behavior of virgin female hamsters.

The role of the vomeronasal-organ–accessory-olfactory-bulb system in maternal behavior gains added significance because input from this system can reach the MPOA. The accessory olfactory bulb projects to the corticomedial amygdala, which, in turn, can influence the MPOA via the stria terminalis (Scalia and Winans, 1975). It should also be noted that the MPOA projects to the amygdala, one component of this projection passing through the stria terminalis (Conrad and Pfaff, 1976).

If vomeronasal input inhibits maternal behavior by activating the amygdala, then one might expect that damage to the amygdala would facilitate maternal behavior. Fleming *et al.* (1980) reported that corticomedial amygdala lesions and lesions of the stria terminalis facilitate pup-stimulated maternal behavior in cycling virgin female rats.

How can we interrelate these findings with the important role of the MPOA in the control of maternal behavior? Two possibilities immediately present themselves: (1) Perhaps the MPOA inhibits the amygdala, this being the mechanism by which the MPOA regulates the onset and the maintenance of maternal behavior. (2) Perhaps the amygdala can inhibit the MPOA, and this is the mechanism by which vomeronasal input exerts an inhibitory effect on maternal behavior. Perhaps estrogen action on the MPOA (Numan *et al.*, 1977) stimulates maternal behavior, in part, by reducing this inhibitory effect. Concerning these two possibilities, the fact that amygdala lesions no longer facilitate the maternal behavior of virgin rats that also have MPOA lesions (Fleming and Miceli, 1981) suggests that possibility Number 2 is more likely than possibility Number 1.

THERMOREGULATORY PROCESSES AND MATERNAL BEHAVIOR

A series of studies by Leon *et al.* (1978; Woodside and Leon, 1980; Woodside *et al.*, 1980) indicated that an important relationship exists between thermoregulation and maternal behavior. Their basic finding is that lactating rats terminate nursing bouts as a result of the elevated maternal body temperatures that occur at the end of each bout.

These findings are important in several respects. In the first place, they indicate that it is not completely accurate to state that the maintenance of maternal behavior during the postpartum period in the rat is free from hormonal control (Numan *et al.*, 1972; Rosenblatt, 1970, 1975). Adrenal steroids have been shown to influence maternal body temperature and thus, in turn, the duration of individual nursing bouts (Woodside and Leon, 1980). Therefore, although hormones do not appear to be required for the occurrence of maternal behavior during the postpartum period, their presence at this time certainly modulates maternal activities.

The findings concerning a relationship between maternal behavior and thermoregulation gain added importance with respect to the neural and hormonal basis of maternal behavior when the following facts are considered:

1. The MPOA is critically involved in thermoregulation (Hammel, 1968).
2. The absence of maternal behavior in female rats with knife cuts severing the lateral connections of the MPO-AHA is associated with a transient hyperthermia (Numan and Callahan, 1980). The maternal behavior deficits, however, outlast the hyperthermia.
3. Estrogen, a hormone that facilitates the onset of maternal behavior in the rat (Siegel and Rosenblatt, 1975b), is also capable of lowering the body temperature of rats (Wilkinson et al., 1980).
4. Progesterone, a hormone that inhibits the onset of maternal behavior in the rat (Numan, 1978), is also capable of raising the body temperature of rats (Woodside et al., 1981).

Clearly, more work needs to be done to unravel the possible significance of these relationships. At the present time, I would like to present the following question: Since MPOA damage promotes hyperthermia by impairing autonomic temperature regulation, leaving behavioral temperature regulation intact (see Szymusiak and Satinoff, 1982, for a recent discussion), is it possible that some of the maternal behavior deficits (nursing deficits, in particular) observed in female rats after MPOA damage represent a form of behavioral thermoregulation (since Leon et al., 1978, found that engaging in nursing raises the mother's body temperature)? The reader is referred to the paper by Numan and Callahan (1980) for a preliminary analysis of this issue.

FURTHER STUDIES ON THE ROLE OF THE MPOA IN THE CONTROL OF MATERNAL BEHAVIOR

My original studies on the role of the MPOA and its lateral connections in the control of maternal behavior suggested that the lateral connections of the MPOA-AHA were essential to all components of maternal behavior in the rat (Numan, 1974). Three recent studies have found, however, that lesions of the MPOA or cutting the lateral connections of the MPOA disrupts retrieving and nest building more than nursing (Jacobson et al., 1980; Numan and Callahan, 1980; Terkel et al., 1979). Terkel et al. (1979) suggested, therefore, that the lateral connections of the MPOA are more important in the active components (retrieving and nest building) than in the passive components (nursing behavior) of maternal behavior. It should be emphasized, however, that although cutting the lateral connections of the MPOA eliminates retrieval, but not nursing, in all rats, nursing behavior is, in fact, decreased in all rats (Numan and Callahan, 1980).

Lesions of the MPOA or cutting its lateral connections disrupts maternal behavior. Although this finding suggests the particular importance of the lateral connections of the MPOA in maternal behavior, this suggestion would be more convincing if it were shown that the dorsal, posterior, and anterior connections of the MPOA are not essential for maternal behavior. Numan and Callahan (1980) explored this issue. They found that severing the lateral connections of the MPOA severely disrupted maternal behavior, whereas severing the dorsal or the posterior connections of the MPOA produced either minor deficits or no deficits. Severing the anterior connections of the MPOA did produce large deficits in maternal behavior, but this result was associated with hypoactivity and loss of body weight, suggesting that the maternal behavior deficits in this group were the result of a general physical debilitation. These results, therefore, emphasize the unique importance of the lateral connections of the MPOA in maternal behavior.

Terkel et al. (1979) and Jacobson et al. (1980) presented evidence that the lateral connections of the dorsocaudal MPOA are the ones critical for maternal behavior. Knife cuts that included these lateral connections interfered with maternal behavior in postpartum lactating rats, whereas severing the lateral connections of either the anterior MPOA or the ventrocaudal MPOA was without effect. Lesions restricted to the AHA also did not disrupt maternal behavior.

Concerning species other than the rat, Marques *et al.* (1979) and Miceli and Malsbury (1982) found that cutting the lateral connections of the MPO-AHA also disrupts maternal behavior in the hamster. This disruptive effect, however, appears to be qualitatively distinct from that in the rat. In the hamster, such cuts appear to promote cannibalism. Cannibalism is not observed after similar cuts in the rat.

Neurochemical Correlates of Maternal Behavior

Given that the lateral connections of the MPOA are important in maternal behavior, the following question can be asked: With what neural structure(s) does this region interact in order to influence maternal behavior? In the original review (see the last section), it was suggested that the efferent lateral projections of the MPOA may influence maternal behavior through projections of brainstem catecholaminergic systems. Neuroanatomical evidence has indicated that the descending projections of the preoptic region reach the locus coeruleus (the origin of the ascending dorsal NE system), the substantia nigra (the origin of the ascending DA nigrostriatal system), and the ventral tegmental area (the origin of the ascending mesolimbic DA system) (Conrad and Pfaff, 1976; Phillipson, 1979; Swanson, 1976). Recent findings, by implicating some of these brainstem systems in the control of maternal behavior, are consistent with the possibility that preoptic projections to these regions are important in maternal behavior. Steele *et al.* (1979) reported that electrolytic lesions of the dorsal NE bundle disrupt maternal behavior, and that this disruption is correlated with a decrease in hippocampal NE levels. Hypothalamic NE levels were not affected. The interpretation of these results is complicated by the fact that electrolytic lesions do not selectively destroy NE neurons, so that the maternal behavior deficits may have been caused by an effect of the lesion other than damage to NE systems in this region.

Gafori and LeMoal (1979) reported that radiofrequency lesions of the ventral tegmental area, performed prior to mating, completely eliminated the subsequent maternal behavior of female rats. Here, too, further studies are needed. Injections of 6-OHDA into the ventral tegmental area could be used to determine whether the lesion effect is the result of damaging DA systems in this region. It is possible, for example, that the observed maternal behavior deficits were the result of damage to MPOA efferents that were simply passing through this area on their way to other brainstem regions (see Conrad and Pfaff, 1976).

Publications Relevant to the Neural Basis of Maternal Behavior That Appeared between August 1976 and May 1982

Armstrong, W. E., Warach, S., Hatton, G. I., and McNeill, T. H. Subnuclei in the rat hypothalamic paraventricular nucleus: A cytoarchitectural, horseradish peroxidase and immunocytochemical analysis. *Neuroscience*, 1980, *5*, 1931–1958.

Buijs, R. M., Swaab, D. F., Dogterom, J., and van Leeuwen, F. W. Intra- and extrahypothalamic vasopressin and oxytocin pathways in the rat. *Cell and Tissue Research*, 1978, *186*, 423–433.

Conrad, L. C. A., and Pfaff, D. W. Efferents from medial basal forebrain and hypothalamus in the rat: I. An autoradiographic study of the medial preoptic area. *Journal of Comparative Neurology*, 1976, *169*, 185–220.

Fleming, A. S., and Luebke, C. Timidity prevents the virgin female rat from being a good mother: Emotionality differences between nulliparous and parturient females. *Physiology and Behavior*, 1981, *27*, 863–868.

Fleming, A. S., and Miceli, M. *Preoptic Lesions Prevent the Facilitation of Maternal Behavior Produced by Amygdaloid Lesions.* Paper presented at Conference on Reproductive Behavior, Vanderbilt University, Nashville, 1981.

Fleming, A. S., Vaccarino, F., Tambosso, L., and Chee, P. Vomeronasal and olfactory system modulation of maternal behavior in the rat. *Science,* 1979, *203,* 372–374.

Fleming, A. S., Vaccarino, F., and Luebke, C. Amygdaloid inhibition of maternal behavior in the nulliparous female rat. *Physiology and Behavior,* 1980, *25,* 731–743.

Gaffori, O., and Le Moal, M. Disruption of maternal behavior and appearance of cannibalism after ventral mesencephalic tegmentum lesions. *Physiology and Behavior,* 1979, *23,* 317–323.

Jacobson, C. D., Terkel, J., Gorski, R. A., and Sawyer, C. H. Effects of small medial preoptic area lesions on maternal behavior: Retrieving and nest building in the rat. *Brain Research,* 1980, *194,* 471–478.

Kovacs, G. L., Bohus, B., Versteeg, D. H. G., De Kloet, E. R., and De Wied, D. Effect of oxytocin and vasopressin on memory consolidation: Sites of action and catecholaminergic correlates after local microinjection into limbic-midbrain structures. *Brain Research,* 1979, *175,* 303–314.

Leon, M., Croskerry, P. G., and Smith, G. K. Thermal control of mother-young contact in rats. *Physiology and Behavior,* 1978, *21,* 793–811.

Marques, D. M. Roles of the main olfactory and vomeronasal systems in the response of the female hamster to young. *Behavioral and Neural Biology,* 1979, *26,* 311–329.

Marques, D. M., Malsbury, C. W., and Daood, J. Hypothalamic knife cuts dissociate maternal behaviors, sexual receptivity, and estrous cyclicity in female hamsters. *Physiology and Behavior,* 1979, *23,* 347–355.

Miceli, M. O., and Malsbury, C. W. Sagittal knife cuts in the near and far lateral preoptic area-hypothalamus disrupt maternal behavior in female hamsters. *Physiology and Behavior,* 1982, *28,* 857–867.

Numan, M. Progesterone inhibition of maternal behavior in the rat. *Hormones and Behavior,* 1978, *11,* 209–231.

Numan, M. and Callahan, E. C. The connections of the medial preoptic region and maternal behavior in the rat. *Physiology and Behavior,* 1980, *25,* 653–665.

Numan, M., Rosenblatt, J. S., and Komisaruk, B. R. Medial preoptic area and onset of maternal behavior in the rat. *Journal of Comparative and Physiological Psychology,* 1977, *91,* 146–164.

Pedersen, C. A., and Prange, A. J., Jr. Induction of maternal behavior in virgin rats after intracerebroventricular adminstration of oxytocin. *Proceedings of the National Academy of Science,* 1979, *76,* 6661–6665.

Pedersen, C. A., Ascher, J. A., Monroe, Y. L., and Prange, A. J., Jr. Oxytocin induces maternal behavior in virgin female rats. *Science,* 1982, *216,* 648–649.

Phillipson, O. T. Afferent projections to the ventral tegmental area of Tsai and interfascicular nucleus: A horseradish peroxidase study in the rat. *Journal of Comparative Neurology,* 1979, *187,* 117–144.

Rhodes, C. H., Morrell, J. I., and Pfaff, D. W. Immunohistochemical analysis of magnocellular elements in rat hypothalamus: Distribution and numbers of cells containing neurophysin, oxytocin, and vasopressin. *Journal of Comparative Neurology,* 1981, *198,* 45–64.

Rosenblatt, J. S., Siegel, H. I., and Mayer, A. D. Progress in the study of maternal behavior in the rat: Hormonal, nonhormonal, sensory, and developmental aspects. In J. S. Rosenblatt, R. A. Hinde, E. Shaw, and C. Beer (Eds.), *Advances in the Study of Behavior,* Vol. 10. New York: Academic Press, 1979.

Steele, M., Rowland, D., and Moltz, H. Initiation of maternal behavior in the rat: Possible involvement of limbic norepinephrine. *Pharmacology, Biochemistry and Behavior,* 1979, *11,* 123–130.

Swanson, L. W. Immunohistochemical evidence for a neurophysin-containing autonomic pathway arising in the paraventricular nucleus of the hypothalamus. *Brain Research,* 1977, *128,* 346–353.

Szymusiak, R., and Satinoff, E. Acute thermoregulatory effects of unilateral electrolytic lesions of the medial and lateral preoptic area in rats. *Physiology and Behavior,* 1982, *28,* 161–170.

Terkel, J., Bridges, R. S., and Sawyer, C. H. Effects of transecting lateral neural connections of the medial preoptic area on maternal behavior in the rat: Nest building, pup retrieval and prolactin secretion. *Brain Research,* 1979, *169,* 369–380.

Wilkinson, C. W., Carlisle, H. J., and Reynolds, R. W. Estrogenic effects on behavioral thermoregulation and body temperature of rats. *Physiology and Behavior,* 1980, *24,* 337–340.

Woodside, B., and Leon, M. Thermoendocrine influences on maternal nesting behavior in rats. *Journal of Comparative and Physiological Psychology,* 1980, *94,* 41–60.

Woodside, B., Pelchat, R., and Leon, M. Acute elevation of the heat load of mother rats curtails maternal nest bouts. *Journal of Comparative and Physiological Psychology*, 1980, *94*, 61–68.

Woodside, B., Leon, M., Attard, M., Feder, H. H., Siegel, H. I., and Fischette, C. Prolactin-steroid influences on the thermal basis for mother-young contact in Norway rats. *Journal of Comparative and Physiological Psychology*, 1981, *95*, 771–780.

Selected Readings

Barofsky, A.-L., and Harney, J. W. Impairments in lactation in the rat following destruction of the median raphe nucleus. *Neuroendocrinology*, 1978, *26*, 333–351.

Copenhaver, J. H., Schalock, R. L., and Carver, M. J., Para-Chloro-D, L-phenylalanine induced filicidal behavior in the female rat. *Pharmacology, Biochemistry and Behavior*, 1978, *8*, 263–270.

Fleischer, S., and Slotnick, B. M. Disruption of maternal behavior in rats with lesions of the septal area. *Physiology and Behavior*, 1978, *21*, 189–200.

Fleischer, S., Kordower, J. H., Kaplan, B., Dicker, R., Smerling, R., and Ilgner, J. Olfactory bulbectomy and gender differences in maternal behaviors of rats. *Physiology and Behavior*, 1981, *26*, 957–959.

Friedman, M. I., Bruno, J. P., and Alberts, J. R. Physiological and behavioral consequences in rats of water recycling during lactation. *Journal of Comparative and Physiological Psychology*, 1981, *95*, 26–35.

Kenyon, P., Cronin, P., and Keeble, S. Disruption of maternal retrieving by perioral anesthesia. *Physiology and Behavior*, 1981, *27*, 313–321.

Mayer, A. D., and Rosenblatt, J. S. Effects of intranasal zinc sulfate on open field and maternal behavior in female rats. *Physiology and Behavior*, 1977, *18*, 101–109.

Mayer, A. D., and Rosenblatt, J. S. Hormonal influences during the ontogeny of maternal behavior in female rats. *Journal of Comparative and Physiological Psychology*, 1979, *93*, 879–898.

Murphy, M. R., MacLean, P. D., and Hamilton, S. C. Species-typical behavior of hamsters deprived from birth of the neocortex. *Science*, 1981, *213*, 459–461.

Noonan, M., Kristal, M. B. Effects of medial preoptic lesions on placentophagia and on the onset of maternal behavior in the rat. *Physiology and Behavior*, 1979, *22*, 1197–1202.

Piccirillo, M., Alpert, J. E., Cohen, D. J., and Shaywitz, B. A. Effects of 6-hydroxydopamine and amphetamine on rat mothering behavior and offspring development. *Pharmacology, Biochemistry and Behavior*, 1980, *13*, 391–395.

Prilusky, J. Induction of maternal behavior in the virgin rat by lactating-rat brain extracts. *Physiology and Behavior*, 1981, *26*, 149–152.

Rosenberg, P., Leidahl, L., Halaris, A., and Moltz, H. Changes in the metabolism of hypothalamic norepinephrine associated with the onset of maternal behavior in the nulliparous rat. *Pharmacology, Biochemistry and Behavior*, 1976, *4*, 647–649.

Rosenberg, P., Halaris, A., and Moltz, H. Effects of central norepinephrine depletion on the initiation and maintenance of maternal behavior in the rat. *Pharmacology, Biochemistry and Behavior*, 1977, *6*, 21–24.

Rowland, D., Steele, M., and Moltz, H. Serotonergic mediation of the suckling-induced release of prolactin in the lactating rat. *Neuroendocrinology*, 1978, *26*, 8–14.

Schwartz, E., and Rowe, F. A. Olfactory bulbectomy: Influences on maternal behavior in primiparous and multiparous rats. *Physiology and Behavior*, 1976, *17*, 879–883.

Slotnick, B. M., and Berman, E. J. Transection of the lateral olfactory tract does not produce anosmia. *Brain Research Bulletin*, 1980, 5, 141–145.

Slotnick, B. M., and Gutman, L. A. Evaluation of intranasal zinc sulfate treatment of olfactory discrimination in rats. *Journal of Comparative and Physiological Psychology*, 1977, *91*, 942–950.

Smotherman, W. P., Hennessy, J. W., and Levine, S. Medial preoptic area knife cuts in the lactating female rat: Effects on maternal behavior and pituitary-adrenal activity. *Physiological Psychology*, 1977, *5*, 243–246.

Smotherman, W. P., Bell, R. W., Hershberger, W. A., and Coover, G. D. Orientation to rat pup cues: Effects of maternal experiential history. *Animal Behavior*, 1978, *26*, 265–273.

Szechtman, H., Siegel, H. I., Rosenblatt, J. S., and Komisaruk, B. R. Tail-pinch facilitates onset of maternal behavior in rats. *Physiology and Behavior*, 1977, *19*, 807–809.

Terlecki, L. J., and Sainsbury, R. S. Effects of fimbria lesions on maternal behavior in the rat. *Physiology and Behavior,* 1978, *21,* 89–97.

REFERENCES

Alberts, J. R. Producing and interpreting experimental olfactory deficits. *Physiology and Behavior,* 1974, *12,* 657–670.

Alberts, J. R., and Friedman, M. I. Olfactory bulb removal but not anosmia increases emotionality and mouse-killing. *Nature,* 1972, *238,* 454–455.

Alberts, J. R., and Galef, B. G. Acute anosmia in the rat: A behavioral test of peripherally induced anosmia. *Physiology and Behavior,* 1971, *6,* 619–621.

Allin, J. T., and Banks, E. M. Functional aspects of ultrasound production by infant albino rats *(Rattus norvegicus). Animal Behavior,* 1972, *20,* 175–185.

Altmann, M. Naturalistic studies of maternal care in moose and elk. In Harriet L. Rheingold (Ed.), *Maternal Behavior in Mammals.* New York: Wiley, 1963.

Antelman, S. M., and Szechtman, H. Tail pinch induces eating in sated rats which appears to depend on nigrostriatal dopamine. *Science,* 1975, *189,* 731–733.

Antelman, S. M., Szechtman, H., Chin, P., and Fisher, A. E. Tail pinch induced eating, gnawing and licking behavior in rats: Dependence on the nigrostriatal dopamine system. *Brain Research,* 1975, *99,* 319–337.

Avar, Z., and Monos, E. Effect of lateral hypothalamic lesion on pregnant rats and foetal mortality. *Acta Medica Academiae Scientiarum Hungaricae,* 1966, *22,* 259–264.

Avar, Z., and Monos, E. Effect of lateral hypothalamic lesion on maternal behavior and foetal vitality in the rat. *Acta Medica Academiae Scientiarum Hungaricae,* 1967, *23,* 255–261.

Avar, Z., and Monos, E. Behavioural changes in pregnant rats following far-lateral hypothalamic lesions. *Acta Physiologica Academiae Scientiarum Hungaricae,* 1969a, *35,* 295–303.

Avar, Z., and Monos, E. Biological role of lateral hypothalamic structures participating in the control of maternal behaviour in the rat. *Acta Physiologica Academiae Scientiarum Hungaricae,* 1969b, *35,* 285–294.

Avar, Z., Monos, E., Kurcz, M., Nagy, I., and Bukulya, B. Mechanism of perinatal reproductive disorders induced by parafornical hypothalamic lesions in the rat. *Acta Physiologica Academiae Scientarium Hungaricae,* 1973, *44,* 45–49.

Bacq, Z. M. The effect of sympathectomy on sexual functions, lactation, and the maternal behavior of the albino rat. *American Journal of Physiology,* 1931, *99,* 444–453.

Baldwin, B. A., and Shillito, E. The effects of ablation of the olfactory bulbs on parturition and maternal behaviour in soay sheep. *Animal Behaviour,* 1974, *22,* 220–223.

Barfield, R. J., Chen, J. J., and McDonald, P. G. *Activation of Estrous Behavior by Intracerebral Implants of Estradiol Benzoate in Ovariectomized Rats.* Paper presented at Eastern Conference on Reproductive Behavior, Nags Head, N.C., 1975.

Barraclough, C. A., Yrarrazaval, S., and Hatton, R. A possible hypothalamic site of action of progesterone in the facilitation of ovulation in the rat. *Endocrinology,* 1964, *75,* 838–845.

Beach, F. A. The neural basis of innate behavior: I. Effects of cortical lesions upon the maternal behavior pattern in the rat. *Journal of Comparative Psychology,* 1937, *24,* 393–436.

Beach, F. A. The neural basis of innate behavior: II. Relative effects of partial decortication in adulthood and infancy upon the maternal behavior of the primiparous rat. *Journal of Genetic Psychology,* 1938, *53,* 109–148.

Beach, F. A. Effects of cortical lesions upon copulatory behavior in male rats. *Journal of Comparative Psychology,* 1940, *29,* 193–244.

Beach, F. A. Instinctive behavior: Reproductive activities. In S. S. Stevens (Ed.), *Handbook of Experimental Psychology.* New York: Wiley, 1951.

Beach, F. A., and Jaynes, J. Studies of maternal retrieving in rats: I. Recognition of young. *Journal of Mammalogy,* 1956a, *37,* 177–180.

Beach, F. A., and Jaynes, J. Studies of maternal retrieving in rats: III. Sensory cues involved in the lactating female's response to her young. *Behavior,* 1956b, *10,* 104–125.

Beckstead, R. M. Convergent thalamic and mesencephalic projections to the anterior medial cortex in the rat. *Journal of Comparative Neurology,* 1976, *166,* 403–416.

Benuck, I., and Rowe, F. A. Centrally and peripherally induced anosmia: Influences on maternal behavior in lactating female rats. *Physiology and Behavior,* 1975, *14,* 439–447.

Björklund, A., Moore, R. Y., Nobin, A., and Stenevi, U. The organization of tubero-hypophyseal and reticulo-infundibular catecholamine neuron systems in the rat brain. *Brain Research,* 1973, *51,* 171–191.

Bolme, P., Fuxe, K., and Lidbrink, P. On the function of central catecholamine neurons: Their role in cardiovascular and arousal mechanisms. *Research Communications in Chemical Pathology and Pharmacology,* 1972, *4,* 657–697.

Branchey, M., and Branchey, L. Sleep and wakefulness in female rats during pregnancy. *Physiology and Behavior,* 1970, *5,* 365–368.

Broadwell, R. D. Olfactory relationships of the telencephalon and diencephalon in the rabbit: I. An autoradiographic study of the efferent connections of the main and accessory olfactory bulbs. *Journal of Comparative Neurology,* 1975, *163,* 329–346.

Brown-Grant, K., and Raisman, G. Reproductive function in the rat following selective destruction of afferent fibres to the hypothalamus from the limbic system. *Brain Research,* 1972, *46,* 23–42.

Bucher, K., Myers, R. E., and Southwick, C. Anterior temporal cortex and maternal behavior in monkey. *Neurology,* 1970, *20,* 415.

Caggiula, A. R. Analysis of the copulation-reward properties of posterior hypothalamic stimulation in male rats. *Journal of Comparative and Physiological Psychology,* 1970, *70,* 399–412.

Caggiula, A. R., and Eibergen, R. Copulation of virgin male rats evoked by painful peripheral stimulation. *Journal of Comparative and Physiological Psychology,* 1969, *69,* 414–419.

Caggiula, A. R., Antelman, S. M., and Zigmond, M. J. Disruption of copulation in male rats after hypothalamic lesions: A behavioral, anatomical and neurochemical analysis. *Brain Research,* 1973, *59,* 273–287.

Caggiula, A. R., Antelman, S. M., and Zigmond, M. J. Ineffectiveness of sexually arousing stimulation after hypothalamic lesions in the rat. *Physiology and Behavior,* 1974, *12,* 313–316.

Caggiula, A. R., Gay, V. L., Antelman, S. M., and Leggens, J. Disruption of copulation in male rats after hypothalamic lesions: A neuroendocrine analysis. *Neuroendocrinology,* 1975, *17,* 193–202.

Cain, D. P. Olfactory bulbectomy: Neural structures involved in irritability and aggression in the male rat. *Journal of Comparative and Physiological Psychology,* 1974a, *86,* 213–220.

Cain, D. P. The role of the olfactory bulb in limbic mechanisms. *Psychological Bulletin,* 1974b, *81,* 654–671.

Cain, D. P., and Paxinos, G. Olfactory bulbectomy and mucosal damage: Effects on copulation, irritability, and interspecific aggression in male rats. *Journal of Comparative and Physiological Psychology,* 1974, *86,* 202–212.

Cannon, W. B., and Bright, E. M. A belated effect of sympathectomy on lactation. *American Journal of Physiology,* 1931, *97,* 319–321.

Cannon, W. B., Newton, H. F., Bright, E. M., Menkin, V., and Moore, R. A. Some aspects of the physiology of animals surviving complete exclusion of sympathetic nerve impulses. *American Journal of Physiology,* 1929, *89,* 84–107.

Carlson, N. R., and Thomas, G. J. Maternal behavior of mice with limbic lesions. *Journal of Comparative and Physiological Psychology,* 1968, *66,* 731–737.

Carlson, N. R., and Vallante, M. A. Enhanced cue function of olfactory stimulation in mice with septal lesions. *Journal of Comparative and Physiological Psychology,* 1974, *87,* 237–248.

Carr, W. J., Loeb, L. S., and Dissinger, M. L. Responses of rats to sex odors. *Journal of Comparative and Physiological Psychology,* 1965, *59,* 370–377.

Carrer, H. F., and Taleisnik, S. Neural pathways associated with the mesencephalic inhibitory influence on gonadotrophic secretion. *Brain Research,* 1972, *38,* 299–313.

Chambers, W. F., and Howe, G. A study of estrogen-sensitive hypothalamic centers using a technique for rapid application and removal of estradiol. *Proceedings of the Society for Experimental Biology and Medicine,* 1968, *128,* 292–294.

Christensen, L. W., & Clemens, L. G. Intrahypothalamic implants of testosterone or estradiol and resumption of masculine sexual behavior in long-term castrated male rats. *Endocrinology,* 1974, *95,* 983–990.

Christophersen, E. R., and Wagman, W. Maternal behavior in the albino rat as a function of self-licking deprivation. *Journal of Comparative and Physiological Psychology,* 1965, *60,* 142–144.

Colombo, J. A., Whitmoyer, D. I., Ellendorff, F., and Sawyer, C. H. Effects of cortical spreading depression on multiunit activity in the preoptic area and hypothalamus of the female rat. *Neuroendocrinology,* 1973–1974, *13,* 189–200.

Conrad, L. C. A., and Pfaff, D. W. Axonal projections of medial preoptic and anterior hypothalamic neurons. *Science,* 1975, *190,* 1112–1114.

Cross, B. A. Nursing behavior and the milk ejection reflex in rabbits. *Journal of Endocrinology,* 1953, *8,* xiii–xiv.

Cruz, M. L., and Beyer, C. Effect of septal lesions on maternal behavior and lactation in the rabbit. *Physiology and Behavior,* 1972, *9,* 361–365.

Davidson, J. M. Activation of the male rat's sexual behavior by intracerebral implantation of androgens. *Endocrinology,* 1966, *79,* 783–794.

Davis, C. D. The effect of ablations of the neocortex on mating, maternal behavior and the production of pseudopregnancy in the female rat and on copulatory activity in the male. *American Journal of Physiology,* 1939, *127,* 374–380.

Deets, A. C., Harlow, H. F., Singh, S. D., and Blomquist, A. J. Effects of bilateral lesions of the frontal granular cortex on the social behavior of rhesus monkeys. *Journal of Comparative and Physiological Psychology,* 1970, *72,* 452–461.

De Groot, J. The rat hypothalamus in stereotaxic coordinates. *Journal of Comparative Neurology,* 1960, *113,* 389–400.

DeOlmos, J. S., and Ingram, W. R. The projection field of the stria terminalis in rat brain: An experimental study. *Journal of Comparative Neurology,* 1972, *146,* 303–334.

Devor, M. Fiber trajectories of olfactory bulb efferents in the hamster. *Journal of Comparative Neurology,* 1976, *166,* 31–48.

Domesick, V. B. Thalamic relationships of the medial cortex in the rat. *Brain, Behavior and Evolution,* 1972, *6,* 457–483.

Dörner, G., Döcke, F., and Moustafa, S. Differential localization of a male and a female hypothalamic mating centre. *Journal of Reproduction and Fertility,* 1968, *17,* 583–586.

Edwards, D. A. Non-sensory involvement of the olfactory bulbs in the mediation of social behaviors. *Behavioral Biology,* 1974, *11,* 287–302.

Edwards, D. A., and Roberts, R. L. Olfactory bulb removal produces a selective deficit in behavioral thermoregulation. *Physiology and Behavior,* 1972, *9,* 747–752.

Epstein, A. N. The lateral hypothalamic syndrome: Its implications for the physiological psychology of hunger and thirst. In E. Stellar and J. M. Sprague (Eds.), *Progress in Physiological Psychology,* Vol 4. New York: Academic Press, 1971.

Estes, R. D. The role of the vomeronasal organ in mammalian reproduction. *Mammalia,* 1972, *36,* 315–341.

Findlay, A. L. R. Nursing behavior and the condition of the mammary gland in the rabbit. *Journal of Comparative and Physiological Psychology,* 1969, *69,* 115–118.

Findlay, A. L. R., and Roth, L. L. Longterm dissociation of nursing behavior and the condition of the mammary gland in the rabbit. *Journal of Comparative and Physiological Psychology,* 1970, *72,* 341–344.

Findlay, A. L. R., and Tallal, P. A. Effect of reduced suckling stimulation on the duration of nursing in the rabbit. *Journal of Comparative and Physiological Psychology,* 1971, *76,* 236–241.

Fleischer, S. F. *Deficits in Maternal Behavior of Rats with Lesions of the Septal Area.* Unpublished doctoral dissertation, Columbia University, 1972.

Fleming, A. S., and Rosenblatt, J. S. Maternal behavior in the virgin and lactating rat. *Journal of Comparative and Physiological Psychology,* 1974a, *86,* 957–972.

Fleming, A. S., and Rosenblatt, J. S. Olfactory regulation of maternal behavior in rats: I. Effects of olfactory bulb removal in experienced and inexperienced lactating and cycling females. *Journal of Comparative and Physiological Psychology,* 1974, *86b,* 221–232.

Fleming, A. S., and Rosenblatt, J. S. Olfactory regulation of maternal behavior in rats: II. Effects of peripherally induced anosmia and lesions of the lateral olfactory tract in pup-induced virgins. *Journal of Comparative and Physiological Psychology,* 1974c, *86,* 233–246.

Franzen, E. A., and Myers, R. E. Neural control of social behavior: Prefrontal and anterior temporal cortex. *Neuropsychologia,* 1973, *11,* 141–157.

Fuxe, K., Hökfelt, T., and Nilsson, O. Factors involved in the control of the activity of the tubero-infundibular dopamine neurons during pregnancy and lactation. *Neuroendocrinology,* 1969, *5,* 257–270.

Gallo, R. V., Rabii, J., and Mobert, G. P. Effect of methysergide, a blocker of serotonin receptors, on plasma prolactin levels in lactating and ovariectomized rats. *Endocrinology,* 1975, *95,* 1096–1105.

Gandelman, R. Induction of maternal nest building in virgin female mice by the presentation of young. *Hormones and Behavior,* 1973, *4,* 191–198.

Gandelman, R., and Svare, B. Mice: Pregnancy termination, lactation, and aggression. *Hormones and Behavior,* 1974, *5,* 397–405.

Gandelman, R., Zarrow, M. X., and Denenberg, V. H. Maternal behavior: Differences between mother and virgin mice as a function of testing procedure. *Developmental Psychobiology,* 1970, *3,* 207–214.

Gandelman, R., Zarrow, M. X., Denenberg, V. H., and Myers, M. Olfactory bulb removal eliminates maternal behavior in the mouse. *Science,* 1971a, *171,* 210–211.

Gandelman, R., Zarrow, M. X., and Denenberg, V. H. Stimulus control of cannibalism and maternal behavior in anosmic mice. *Physiology and Behavior,* 1971b, *7,* 583–586.

Gandelman, R., Zarrow, M. X., and Denenberg, V. H. Reproductive and maternal performance in the mouse following removal of the olfactory bulbs. *Journal of Reproduction and Fertility,* 1972, *28,* 453–456.

Giantonio, G. W., Lund, N. L., and Gerall, A. A. Effect of diencephalic and rhinencephalic lesions on the male rat's sexual behavior. *Journal of Comparative and Physiological Psychology,* 1970, *73,* 38–46.

Greengrass, P. M., and Tonge, S. R. Brain monoamine metabolism in the mouse drain during the immediate postpartum period. *British Journal of Pharmacology,* 1972, *46,* 533p–534p.

Greengrass, P. M., and Tonge, S. R. Further studies on monoamine metabolism in three regions of mouse brain during pregnancy: Monoamine metabolite concentrations and the effects of injected hormones. *Archives Internationales de Pharmacodynamie et de Therapie,* 1974, *212,* 48–59.

Grota, L. J., and Ader, R. Behavior of lactating rats in a dual-chambered maternity cage. *Hormones and Behavior,* 1974, *5,* 275–282.

Halász, B., and Pupp, L. Hormone secretion of the anterior pituitary gland after physical interruption of all nervous pathways to the hypophysiotropic area. *Endocrinology,* 1965, *77,* 553–562.

Hammel, H. T. Regulation of internal body temperature. *Annual Review of Physiology,* 1968, *30,* 641–710.

Harris, V. S., and Sachs, B. D. Copulatory behavior in male rats following amygdaloid lesions. *Brain Research,* 1975, *86,* 514–518.

Harvey, J. A., McMaster, S. E., and Yunger, L. M. p-Chloroamphetamine: Selective neurotoxic action in the brain. *Science,* 1975, *187,* 841–843.

Hearnshaw, H., and Wodzicka-Tomaszewska, M. Effect of high ambient temperatures in early and late lactation on litter growth and survival in rats. *Australian Journal of Biological Science,* 1973, *26,* 1171–1178.

Heimer, L. The olfactory connections of the diencephalon in the rat. *Brain, Behavior and Evolution,* 1972, *6,* 484–523.

Heimer, L., and Larsson, K. Impairment of mating behavior in male rats following lesions in the preoptic-anterior hypothalamic continuum. *Brain Research,* 1966–1967, *3,* 248–263.

Hendricks, S. E., and Scheetz, H. A. Interaction of hypothalamic structures in the mediation of male sexual behavior. *Physiology and Behavior,* 1973, *10,* 711–716.

Herrenkohl, R. L., and Rosenberg, P. A. Exteroceptive stimulation of maternal behavior in the naive rat. *Physiology and Behavior,* 1972, *8,* 595–598.

Herrenkohl, L. R., & Rosenberg, P. A. Effects of hypothalamic deafferentation late in gestation on lactation and nursing behavior in the rat. *Hormones and Behavior,* 1974, *5,* 33–41.

Herrenkohl, L. R., & Sachs, B. D. Sensory regulation of maternal behavior in mammals. *Physiology and Behavior,* 1972, *9,* 689–692.

Hitt, J. C., Hendricks, S. E., Ginsberg, S. I., and Lewis, J. H. Disruption of male, but not female, sexual behavior in rats by medial forebrain bundle lesions. *Journal of Comparative and Physiological Psychology,* 1970, *73,* 377–384.

Ho, G. K., Quadagno, D., and Moltz, H. Intracranial cycloheximide: Effects on maternal behavior in the postpartum rat. *Pharmacology, Biochemistry and Behavior,* 1974, *2,* 455–458.

Jacobowitz, D. M., and Palkovits, M. Topographic atlas of catecholamine and acetylcholinesterase-containing neurons in the rat brain: I. Forebrain (telencephalon, diencephalon). *Journal of Comparative Neurology,* 1974, 13–28.

Johnson, N. P., Zarrow, M. X., and Denenberg, V. H. Postceasarean ovulation in the rat. *Journal of Reproduction and Fertility*, 1974, *40*, 1–5.

Jones, B. E., Bobillier, P., Pin, C., and Jouvet, M. The effect of lesions of catecholamine-containing neurons upon monoamine content of the brain and EEG and behavioral waking in the cat. *Brain Research*, 1973, *58*, 157–177.

Kalra, S. P., and McCann, S. M. Involvement of catecholamines in feedback mechanisms. *Progress in Brain Research*, 1973, *39*, 185–198.

Kalra, S. P., and McCann, S. M. Effects of drugs modifying catecholamine synthesis on plasma LH and ovulation in the rat. *Neuroendocrinology*, 1974, *15*, 79–91.

Kimble, D. P. Hippocampus and internal inhibition. *Psychological Bulletin*, 1968, *70*, 285–295.

Kimble, D. P., Rodgers, L., and Hendrickson, C. W. Hippocampal lesions disrupt maternal, not sexual, behavior in the albino rat. *Journal of Comparative and Physiological Psychology*, 1967, *63*, 401–407.

Kinder, E. F. A study of the nest-building activity of the albino rat. *Journal of Experimental Zoology*, 1927, *47*, 117–161.

Kirby, H. W., and Horvath, T. Self-licking deprivation and maternal behaviour in the primiparous rat. *Canadian Journal of Psychology*, 1968, *22*, 369–376.

Kirkby, R. J. Caudate nucleus lesions and maternal behavior in the rat. *Psychonomic Science*, 1967, *9*, 601–602.

Klopfer, P. H., and Gamble, J. Maternal "imprinting" in goats: the role of chemical senses. *Zeitschrift für Tierpsychologie*, 1966, *23*, 588–592.

Klopfer, P. H., and Klopfer, M. S. Maternal "imprinting" in goats: Fostering of alien young. *Zeitschrift für Tierpsychologie*, 1968, *25*, 862–866.

Koller, G. Der Nestbau der weissen Maus und seine hormonale Auslösung. *Verhandlungen der Deutschen Zoologischen Gesellschaft, Freiburg*, 1952, 160–168.

Koller, G. Hormonale und psychische steuerung beim nestbau weissen mäuse. *Verhandlungen der Deutschen Zoologischen Gesellschaft*, 1955, 123–132.

Kordon, C., Blake, C. A., Terkel, J., and Sawyer, C. H. Participation of serotonin-containing neurons in the suckling-induced rise in plasma prolactin levels in lactating rats. *Neuroendocrinology*, 1973–1974, *13*, 213–223.

Koves, K., and Halász, B. Location of the neural structures triggering ovulation in the rat. *Neuroendocrinology*, 1970, *6*, 180–193.

Labate, J. S. Influence of uterine and ovarian nerves on lactation. *Endocrinology*, 1940, *27*, 342–344.

Larsson, K. Mating behavior in male rats after cerebral cortex ablation: I. Effects of lesions in the dorsolateral and the median cortex. *Journal of Experimental Zoology*, 1962a, *151*, 167–176.

Larsson, K. Spreading cortical depression and the mating behavior in male and female rats. *Zeitschrift für Tierpsychologie*, 1962b, *19*, 321–331.

Larsson, K. Mating behavior in male rats after cerebral cortex ablation: II. Effects of lesions in the frontal lobes compared to lesions in the posterior half of the hemispheres. *Journal of Experimental Zoology*, 1964, *155*, 203–214.

Law, T., and Meagher, W. Hypothalamic lesions and sexual behavior in the female rat. *Science*, 1958, *128*, 1626–1627.

Leblond, C. P. Nervous and hormonal factors in the maternal behavior of the mouse. *Journal of Genetic Psychology*, 1940, *57*, 327–344.

Lehrman, D. S. Hormonal regulation of parental behavior in birds and infrahuman mammals. In W. C. Young (Ed.), *Sex and Internal Secretions*, Vol. 2. Baltimore: Williams and Wilkins, 1961.

Leonard, C. M. The prefrontal cortex in the rat: I. Cortical projections of the mediodorsal nucleus. II. Efferent connections. *Brain Research*, 1969, *12*, 321–343.

Leonard, C. M. Effects of neonatal (day 10) olfactory bulb lesions on social behavior of female golden hamsters *(Mesocricetus auratus)*. *Journal of Comparative and Physiological Psychology*, 1972, *80*, 208–215.

Lidbrink, P. The effect of lesions of ascending noradrenaline pathways on sleep and waking in the rat. *Brain Research*, 1974, *74*, 19–40.

Lincoln, D. W. Suckling: A time constant in the nursing behavior of the rabbit. *Physiology and Behavior*, 1974, *13*, 711–714.

Lindsay, D. R., and Fletcher, I. C. Sensory involvement in the recognition of lambs by their dams. *Animal Behaviour*, 1968, *16*, 415–417.

Lindvall, O., and Björklund, A. The organization of the ascending catecholamine neuron systems in the rat brain. *Acta Physiologica Scandinavica*, 1974, *Suppl. 412*, 1–48.

Lindvall, O., Björklund, A., Nobin, A., and Stenevi, U. The adrenergic innervation of the rat thalamus as revealed by the glyoxylic acid fluorescence method. *Journal of Comparative Neurology*, 1974, *154*, 317–348.

Lisk, R. D. Diencephalic placement of estradiol and sexual receptivity in the female rat. *American Journal of Physiology*, 1962, *203*, 493–496.

Lisk, R. D. Neural localization for androgen activation of copulatory behavior in the male rat. *Endocrinology*, 1967, *80*, 754–761.

Lisk, R. D. Oestrogen and progesterone synergism and elicitation of maternal nest-building in the mouse *(Mus musculus)*. *Animal Behaviour*, 1971, *19*, 606–610.

Lisk, R. D., and Barfield, M. A. Progesterone facilitation of sexual receptivity in rats with neural implantation of estrogen. *Neuroendocrinology*, 1975, *19*, 28–35.

Lisk, R. D., Pretlow, R. A., and Friedman, S. M. Hormonal stimulation necessary for elicitation of maternal nest-building in the mouse *(Mus musculus)*. *Animal Behaviour*, 1969, *17*, 730–737.

Long, C. J., and Tapp, J. T. Significance of olfactory tracts in mediating response to odors in the rat. *Journal of Comparative and Physiological Psychology*, 1970, *72*, 435–443.

Lubar, J. F., and Numan, R. Behavioral and physiological studies of septal function and related medial cortical structures. *Behavioral Biology*, 1973, *8*, 1–25.

Marshall, J. F., and Teitelbaum, P. Further analysis of sensory inattention following lateral hypothalamic damage in rats. *Journal of Comparative and Physiological Psychology*, 1974, *86*, 375–395.

Marshall, J. F., Turner, B. H., and Teitelbaum, P. Sensory neglect produced by lateral hypothalamic damage. *Science*, 1971, *174*, 523–525.

Marshall, J. F., Richardson, J. S., and Teitelbaum, P. Nigrostriatal bundle damage and the lateral hypothalamic syndrome. *Journal of Comparative and Physiological Psychology*, 1974, *87*, 808–830.

Mayer, A. D., and Rosenblatt, J. S. Olfactory basis for the delayed onset of maternal behavior in virgin female rats: Experimental effects. *Journal of Comparative and Physiological Psychology*, 1975, *89*, 701–710.

McCann, S. M., Kalra, P. S., Donoso, A. O., Bishop, W., Schneider, H. P. G., Fawcett, C. P., and Krulich, L. The role of monoamines in the control of gonadotropin and prolactin secretion. In K. Knigge, D. Scott, and A. Weindl (Eds.), *Brain-Encdorine Interaction. Median Eminence: Structure and Function.* Basel: Karger, 1972.

Millhouse, O. E. A golgi study of the descending medial forebrain bundle. *Brain Research*, 1969, *15*, 341–363.

Moltz, H. The ontogeny of maternal behavior in some selected mammalian species. In H. Moltz (Ed.), *The Ontogeny of Vertebrate Behavior.* New York: Academic Press, 1971.

Moltz, H., and Robbins, D. Maternal behavior of primiparous and multiparous rats. *Journal of Comparative and Physiological Psychology*, 1965, *60*, 417–421.

Moltz, H., Robbins, D., and Parks, M. Caesarean delivery and the maternal behavior of primiparous and multiparous rats. *Journal of Comparative and Physiological Psychology*, 1966, *61*, 455–460.

Moltz, H., Geller, D., and Levin, R. Maternal behavior in the totally mammectomized rat. *Journal of Comparative and Physiological Psychology*, 1967, *64*, 225–229.

Moltz, H., Lubin, M., Leon, M., and Numan, M. Hormonal induction of maternal behavior in the ovariectomized nulliparous rat. *Physiology and Behavior*, 1970, *5*, 1373–1377.

Moltz, H., Rowland, D., Steele, M., and Halaris, A. Hypothalamic norepinephrine: Concentration and metabolism during pregnancy and lactation in the rat. *Neuroendocrinology*, 1975, *19*, 252–258.

Moore, W. T., and Hampton, J. K. Effects of parachlorophenylalanine on pregnancy in the rat. *Biology of Reproduction*, 1974, *11*, 280–287.

Morgane, P. J. Medial forebrain bundle and "feeding centers" of the hypothalamus. *Journal of Comparative Neurology*, 1961, *117*, 1–25.

Myers, R. E., Swett, C., and Miller, M. Loss of social group affinity following prefrontal lesions in free-ranging macaques. *Brain Research*, 1973, *64*, 257–269.

Nance, D. M., Shryne, J., and Gorski, R. A. Effects of septal lesions on behavioral sensitivity of female rats to gonadal hormones. *Hormones and Behavior*, 1975, *6*, 59–64.

Nauta, W. J. H. An experimental study of the fornix in the rat. *Journal of Comparative Neurology*, 1956, *104*, 247–273.

Neill, D. B., and Linn, C. L. Deficits in consummatory responses to regulatory challenges following basal ganglia lesions in rats. *Physiology and Behavior*, 1975, *14*, 617–624.

Neill, D. B., Ross, J. F., and Grossman, S. P. Comparisons of the effects of frontal, striatal, and septal lesions in paradigms thought to measure incentive motivation or behavioral inhibition. *Physiology and Behavior,* 1974, *13,* 297–305.

Neill, D. B., Parker, S. D., and Gold, M. S. Striatal dopaminergic modulation of lateral hypothalamic self-stimulation. *Pharmacology, Biochemistry and Behavior,* 1975, *3,* 485–491.

Noirot, E. Ultrasounds and maternal behavior in small rodents. *Developmental Psychobiology,* 1972a, *5,* 371–387.

Noirot, E. The onset of maternal behavior in rats, hamsters, and mice. A selective review. In D. S. Lehrman, R. A. Hinde, and E. Shaw (Eds.), *Advances in the Study of Behavior,* Vol. 4. New York: Academic Press, 1972b.

Norgren, R. Taste pathways to hypothalamus and amygdala. *Journal of Comparative Neurology,* 1976, *166,* 17–30.

Numan, M. *The Role of the Medial Preoptic Area in the Regulation of Maternal Behavior.* Unpublished doctoral dissertation, University of Chicago, 1973.

Numan, M. Medial preoptic area and maternal behavior in the female rat. *Journal of Comparative and Physiological Psychology,* 1974, *87,* 746–759.

Numan, M., Leon, M., and Moltz, H. Interference with prolactin release and the maternal behavior of female rats. *Hormones and Behavior,* 1972, *3,* 29–38.

Numan, M., Rosenblatt, J. S., and Komisaruk, B. R. The medial preoptic area and the onset of maternal behavior in the rat. *Journal of Comparative and Physiological Psychology,* 1977, *91,* 146–164.

Obias, M. D. Maternal behavior of hypophysectomized gravid albino rats and the development and performance of their young. *Journal of Comparative and Physiological Psychology,* 1957, *50,* 120–124.

Owen, K., Wallace, P., and Thiessen, D. Effects of intracerebral implants of steroid hormones on scent marking in the ovariectomized female gerbil *(Meriones unguiculatus). Physiology and Behavior,* 1974, *12,* 755–760.

Palkovits, M., and Jacobowitz. D. M. Topographic atlas of catecholamine and acetylcholinesterase-containing neurons in the rat brain: II. Hindbrain (mesencephalon, rhombencephalon). *Journal of Comparative Neurology,* 1974, *157,* 29–42.

Paxinos, G., and Bindra, D. Hypothalamic knife cuts: Effects on eating, drinking, irritability, aggression, and copulation in the male rat. *Journal of Comparative and Physiological Psychology,* 1972, *79,* 219–229.

Paxinos, G., and Bindra, D. Hypothalamic and midbrain pathways involved in eating, drinking, irritability, aggression, and copulation in the male rat. *Journal of Comparative and Physiological Psychology,* 1973, *82,* 1–14.

Pellegrino, L. J., and Cushman, A. J. *A stereotaxic atlas of the rat brain.* New York: Appleton-Century-Crofts, 1967.

Pfaff, D., and Keiner, M. Atlas of estradiol-concentrating cells in the central nervous system of the female rat. *Journal of Comparative Neurology,* 1973, *151,* 121–158.

Pfaff, D. W., and Pfaffman, C. Olfactory and hormonal influences on the basal forebrain of the male rat. *Brain Research,* 1969, *15,* 137–156.

Phelps, C. P., Kreig, R. J., and Sawyer, C. H. Spontaneous and electrochemically stimulated changes in plasma LH in the female rat following hypothalamic deafferentation. *Brain Research,* 1976, *101,* 239–249.

Phillips, A. G. Object-carrying by rats: Disruption by ventral mesencephalic lesions. *Canadian Journal of Psychology,* 1975, *29,* 250–262.

Phillips, A. G., and Fibiger, H. C. Long-term deficits in stimulation-induced behaviors and self-stimulation after 6-hyroxydopamine administration in rats. *Behavioral Biology,* 1976, *16,* 127–143.

Phillips, A. G., Cox, V. C., Kakolewski, J. W., and Valenstein, E. S. Object-carrying by rats: An approach to the behavior produced by brain stimulation. *Science,* 1969, *166,* 903–905.

Pietras, R. J., and Moulton, D. G. Hormonal influences on odor detection in rats: Changes associated with the estrous cycle, pseudopregnancy, ovariectomy, and administration of testosterone propionate. *Physiology and Behavior,* 1974, *12,* 475–491.

Pollak, E. I., and Sachs, B. D. Male copulatory behavior and female maternal behavior in neonatally bulbectomized rats. *Physiology and Behavior,* 1975, *14,* 337–343.

Powers, J. B., and Valenstein, E. S. Sexual receptivity: Facilitation by medial preoptic lesions in female rats. *Science,* 1972, *175,* 1003–1005.

Powers, J. B., and Winans, S. S. Vomeronasal organ: Critical role in mediating sexual behavior of the male hamster. *Science*, 1975, *187*, 961–963.

Raisman, G. An experimental study of the projection of the amygdala to the accessory olfactory bulb and its relationship to the concept of a dual olfactory system. *Experimental Brain Research*, 1972, *14*, 395–408.

Raisman, G., and Field, P. M. Anatomical considerations relevant to the interpretation of neuroendocrine experiments. In L. Martini and W. F. Ganong (Eds.), *Frontiers in Neuroendocrinology, 1971*. New York: Oxford University Press, 1971.

Raisman, G., Cowan, W. M., and Powell, T. P. S. The extrinsic afferent, commissural and association fibres of the hippocampus. *Brain*, 1965, *88*, 963–996.

Raisman, G., Cowan, W. M., and Powell, T. P. S. The experimental analysis of the efferent projection of the hippocampus. *Brain*, 1966, *89*, 83–108.

Ranck, J. B. Studies on single neurons in dorsal hippocampal formation and septum in unrestrained rats: I. Behavioral correlates and firing repertoires. *Experimental Neurology*, 1973, *41*, 461–531.

Reisbick, S., Rosenblatt, J. S., and Mayer, A. D. Decline of maternal behavior in the virgin and lactating rat. *Journal of Comparative and Physiological Psychology*, 1975, *89*, 722–732.

Rheingold, H. L. (Ed.). *Maternal Behavior in Mammals*. New York: Wiley, 1963.

Richards, M. P. M. Maternal behavior in rodents and lagomorphs: A review. In A. McLaren (Ed.), *Advances in Reproductive Physiology*, Vol. 2. New York: Academic Press, 1967.

Roberts, E. A model of the vertebrate nervous system based largely on disinhibition: A key role of the GABA system. *Advances in Behavioral Biology*, 1974, *10*, 419–449.

Rodgers, C. H., and Law, O. T. The effects of habenula and medial forebrain bundle lesions on sexual behavior in female rats. Psychonomic Science, 1967, *8*, 1–2.

Rodgers, C. H., and Schwartz, N. B. Differentiation between neural and hormonal control of sexual behavior and gonadotropin secretion in the female rat. *Endocrinology*, 1976, *98*, 778–786.

Rose, J. E., and Woolsey, C. N. The orbito frontal cortex and its connections with the mediodorsal nucleus in rabbit, sheep and cat. *Research Publications of the Association for Nervous and Mental Diseases*, 1948, *27*, 210–232.

Rosen, S., and Shelesnyak, M. C. Induction of pseudopregnancy in rat by silver nitrate on nasal mucosa. *Proceedings of the Society for Experimental Biology and Medicine*, 1937, *36*, 832–834.

Rosen, S., Shelesnyak, M. C., and Zacharias, L. R. Naso-genital relationship: II. Pseudopregnancy following extirpation of the sphenopalatine ganglion in the rat. *Endocrinology*, 1940, *27*, 463–468.

Rosenberg, P. *Intraventricular Infusion of 6-Hydroxydopamine and the Onset of Maternal Behavior in the Rat*. Paper presented at Eastern Conference on Reproductive Behavior, Saratoga Springs, N.Y., 1976.

Rosenblatt, J. S. Nonhormonal basis of maternal behavior in the rat. *Science*, 1967, *156*, 1512–1513.

Rosenblatt, J. S. Views on the onset and maintenance of maternal behavior in the rat. In L. R. Aronson, E. Tobach, D. S. Lehrman, and J. S. Rosenblatt (Eds.), *Development and Evolution of Behavior*. San Francisco: Freeman, 1970.

Rosenblatt, J. S. Prepartum and postpartum regulation of maternal behavior in the rat. *Parent-Infant Interaction Ciba Foundation Symposium 33*. Amsterdam: Elsevier, 1975.

Rosenblatt, J. S., and Lehrman, D. S. Maternal behavior in the laboratory rat. In H. L. Rheingold (Ed.), *Maternal Behavior in Mammals*. New York: Wiley, 1963.

Rosenblatt, J. S., and Siegel, H. I. Hysterectomy-induced maternal behavior during pregnancy in the rat. *Journal of Comparative Physiological Psychology*, 1975, *89*, 685–700.

Rosenson, L. M. Responses of virgin mice, and of maternal females, to normal and caesarian-section delivered pups. *Journal of Comparative and Physiological Psychology*, 1975, *88*, 670–677.

Ross, S., Denenberg, V. H., Frommer, C. P., and Sawin, P. B. Genetic, physiological and behavioral background of reproduction in the rabbit: V. Nonretrieving of neonates. *Journal of Mammology*, 1959, *40*, 91–96.

Roth, L. L., and Rosenblatt, J. S. Changes in self-licking during pregnancy in the rat. *Journal of Comparative and Physiological Psychology*, 1967, *63*, 397–400.

Rowland, D. *Effects of Intraventricular p-Chloroamphetamine on the Suckling-Induced Maintenance of Lactation in the Rat*. Paper presented at Eastern Conference on Reproductive Behavior, Saratoga Springs, N.Y., 1976.

Sachs, C., and Jonsson, G. Mechanisms of action of 6-hydroxydopamine. *Biochemical Pharmacology*, 1975, *24*, 1–8.

Scalia, F., and Winans, S. S. The differential projections of the olfactory bulb and accessory olfactory bulb in mammals. *Journal of Comparative Neurology*, 1975, *161*, 31–55.

Schlein, P. A., Zarrow, M. X., Cohen, H. A., Denenberg, V. H., and Johnson, N. P. The differential effect of anosmia on maternal behavior in the virgin and primiparous rat. *Journal of Reproduction and Fertility,* 1972, *30,* 139–142.

Sclafani, A., and Grossman, S. P. Hyperphagia produced by knife cuts between the medial and lateral hypothalamus in the rat. *Physiology and Behavior,* 1969, *4,* 533–538.

Scott, J. W., and Leonard, C. M. The olfactory connections of the lateral hypothalamus in the rat, mouse, and hamster. *Journal of Comparative Neurology,* 1971, *141,* 331–344.

Seegal, R. F., and Denemberg, V. H. Maternal experience prevents pup-killing in mice induced by peripheral anosmia. *Physiology and Behavior,* 1974, *13,* 339–341.

Shelesnyak, M. C., and Rosen, S. Naso-genital relationship: Induction of pseudopregnancy in the rat by nasal treatment. *Endocrinology,* 1938, *23,* 58–63.

Sieck, M. H., and Baumbach, H. D. Differential effects of peripheral and central anosmia producing techniques on spontaneous behavior patterns. *Physiology and Behavior,* 1974, *13,* 407–425.

Siegel, H. I., and Rosenblatt, J. S. Estrogen-induced maternal behavior in hysterectomized-ovariectomized virgin rats. *Physiology and Behavior,* 1975a, *14,* 465–471.

Siegel, H. I., and Rosenblatt, J. S. Hormonal basis of hysterectomy induced maternal behavior during pregnancy in the rat. *Hormones and Behavior,* 1975b, *6,* 211–222.

Simeone, F. A., and Ross, J. F. The effect of sympathectomy on gestation and lactation in the cat. *American Journal of Physiology,* 1938, *122,* 659–667.

Singer, J. J. Hypothalamic control of male and female sexual behavior in female rats. *Journal of Comparative and Physiological Psychology,* 1968, *66,* 738–742.

Slotnick, B. M. Disturbances of maternal behavior in the rat following lesions of the cingulate cortex. *Behaviour* 1967a, *29,* 204–236.

Slotnick, B. M. Intercorrelations of maternal activities in the rat. *Animal Behaviour,* 1967b, *15,* 267–269.

Slotnick, B. M. Maternal behavior deficits following forebrain lesions in the rat. *American Zoologist,* 1969, *9,* 1068.

Slotnick, B. M. Neural and hormonal basis of maternal behavior in the rat. In B. E. Eleftherious and R. L. Sprott (Eds.), *Hormonal Correlates of Behavior, Vol. 2: An Organismic View.* New York: Plenum Press, 1975.

Slotnick, B. M., and Nigrosh, B. J. Maternal behavior of mice with cingulate cortical, amygdala, or septal lesions. *Journal of Comparative and Physiological Psychology,* 1975, *88,* 118–127.

Smith, M. O., and Holland, R. C. Effects of lesions of the nucleus accumbens on lactation and postpartum behavior. *Physiological Psychology,* 1975, *3,* 331–336.

Smotherman, W. P., Bell, R. W., Starzec, J., Elias, J., and Zachman, T. A. Maternal responses to infant vocalizations and olfactory cues in rats and mice. *Behavioral Biology,* 1974, *12,* 55–66.

Sorenson, C. A., and Gordon, M. Effects of 6-hydroxydopamine on shock-elicited aggression, emotionality and maternal behavior in female rats. *Pharmacology, Biochemistry and Behavior,* 1975, *3,* 331–335.

Spector, S. A., and Hull, E. M. Anosmia and mouse killing by rats: A nonolfactory role for the olfactory bulbs. *Journal of Comparative and Physiological Psychology,* 1972, *80,* 354–456.

Stamm, J. S. The function of the median cerebral cortex in maternal behavior in rats. *Journal of Comparative and Physiological Psychology,* 1955, *48,* 347–356.

Steele, M. *Onset of Maternal Behavior in the Primiparous Rat: Neuroanatomical and Neurochemical Analysis.* Paper presented at Eastern Conference on Reproductive Behavior, Saratoga Springs, N.Y. 1976.

Stern, J. J., and Siepierski, L. Spreading cortical depression and the maternal behavior of guinea pigs. *Psychonomic Science,* 1971, *25,* 301–302.

Stone, C. P. Preliminary note on maternal behavior of rats living in parabiosis. *Endocrinology,* 1925, *9,* 505–512.

Stone, C. P. Effects of cortical destruction on reproductive behavior and maze learning in albino rats. *Journal of Comparative Psychology,* 1938, *26,* 217–236.

Stricker, E. M., and Zigmond, M. J. Recovery of function after damage to central catecholamine-containing neurons: A neurochemical model for the lateral hypothalamic syndrome. In J. M. Sprague and A. N. Epstein (Eds.), *Progress in Psychobiology and Physiological Psychology,* Vol. 6. New York: Academic Press, 1976.

Swanson, L. W. An autoradiographic study of the efferent connections of the preoptic region in the rat. *Journal of Comparative Neurology,* 1976, *167,* 227–256.

Swanson, L. W., and Hartman, B. K. The central adrenergic system. An immunofluorescence study of the location of cell bodies and their efferent connections in the rat utilizing dopamine-B-hydroxylase as a marker. *Journal of Comparative Neurology*, 1975, *163*, 467–506.

Szechtman, H. *Systematic, Surgical Isolation of the Preoptic Area with a Microknife: A Neuroanatomical and Behavioral Analysis of the Disruption of Sexual Behaviour in Male Rats.* Unpublished doctoral dissertation, University of Pittsburgh, 1975.

Szechtman, H., and Siegel, H. I. *Stimulation-Bound Maternal Behavior in the Rat.* Paper presented at Eastern Conference on Reproductive Behavior, Saratoga Springs, N.Y., 1976.

Tejasen, T., and Everett, J. W. Surgical analysis of the preoptic-tuberal pathway controlling ovulatory release of gonadotrophins in the rat. *Endocrinology*, 1967, *81*, 1387–1396.

Terasawa, E., Bridson, W. E., Davenport, J. W., and Goy, R. W. Role of brain monoamines in release of gonadotropin before proestrus in the cyclic rat. *Neuroendocrinology*, 1975, *18*, 345–358.

Terkel, J., and Rosenblatt, J. S. Aspects of non-hormonal maternal behavior in the rat. *Hormones and Behavior*, 1971, *2*, 161–171.

Terkel, J., and Rosenblatt, J. S. Humoral factors underlying maternal behavior at parturition: Cross transfusion between freely moving rats. *Journal of Comparative and Physiological Psychology*, 1972, *80*, 365–371.

Ungerstedt, U. Aphagia and adipsia after 6-hydroxydopamine induced degeneration of the nigro-striatal dopamine system. *Acts Physiologica Scandinavica*, 1971a, *Suppl. 367*, 97–122.

Ungerstedt, U. Stereotaxic mapping of the monoamine pathways in the rat brain. *Acta Physiologica Scandinavica*, 1971b, *Suppl. 367*, 1–48.

Valenstein, E. S., Cox, V. C., and Kakolewski, J. W. Reexamination of the role of the hypothalamus in motivation. *Psychological Review*, 1970, *77*, 16–31.

Vandenbergh, J. G. Effects of central and peripheral anosmia on reproduction of female mice. *Physiology and Behavior*, 1973, *10*, 257–261.

Vanderwolf, C. H. Limbic-diencephalic mechanisms of voluntary movement. *Psychological Review*, 1971, *78*, 83–113.

Velasco, M. E., Castro-Vazquez, A., and Rothchild, I. Effects of hypothalamic deafferentation on criteria of prolactin secretion during pregnancy and lactation in the rat. *Journal of Reproduction and Fertility*, 1974, *41*, 385–395.

Voci, V. E., and Carlson, N. R. Enhancement of maternal behavior and nest building following systemic and diencephalic administration of prolactin and progesterone in the mouse. *Journal of Comparative and Physiological Psychology*, 1973, *83*, 388–393.

Voogt, J. L., and Carr, L. A. Plasma prolactin levels and hypothalamic catecholamine synthesis during suckling. *Neuroendocrinology*, 1974, *16*, 108–118.

Wallace, P., Owen, K., and Thiessen, D. D. The control and function of maternal scent marking in the mongolian gerbil. *Physiology and Behavior*, 1973, *10*, 463–466.

Whitehead, S. A., and Ruf, K. B. Responses of antidromically identified preoptic neurons in the rat to neurotransmitters and to estrogen. *Brain Research*, 1974, *79*, 185–198.

Whittier, J. R., and Orr, H. Hyperkinesis and other physiological effects of caudate deficit in the adult albino rat. *Neurology*, 1962, *12*, 529–534.

Wiesner, B. P., and Sheard, N. M. *Maternal Behaviour in the Rat.* Edinburgh: Oliver and Boyd, 1933.

Wilsoncroft, W. E. Effects of median cortex lesions on the maternal behavior of the rat. *Psychological Reports*, 1963, *13*, 835–838.

Winans, S. S., and Scalia, F. Amygdaloid nucleus: New afferent input from the vomeronasal organ. *Science*, 1970, *170*, 330–332.

Yagi, K. Changes in firing rates of single preoptic and hypothalamic units following an intravenous administration of estrogen in the castrated female rat. *Brain Research*, 1973, *53*, 343–352.

Yagi, K., and Sawaki, Y. Feedback of estrogen in the hypothalamic control of gonadotrophin secretion. In K. Yagi and S. Yoshida (Eds.), *Neuroendocrine Control.* Tokyo: University of Tokyo Press, 1973.

Yokoyama, A., Halász, B., and Sawyer, C. H. Effect of hypothalamic deafferentation on lactation in rats. *Proceedings of the Society for Experimental Biology and Medicine*, 1967, *125*, 623–626.

Zarrow, M. X., Farooq, A., Denenberg, V. H., Sawin, P. B., and Ross, S. Maternal Behavior in the rabbit: Endocrine control of maternal nest-building. *Journal of Reproduction and Fertility*, 1963, *6*, 375.

Zarrow, M. X., Denenberg, V. H., and Anderson, C. O. Rabbit: Frequency of suckling in the pup. *Science*, 1965, *150*, 1835–1836.

Zarrow, M. X., Gandelman, R., and Denenberg, V. H. Prolactin: Is it an essential hormone for maternal behavior in the mammal? *Hormones and Behavior*, 1971a, *2*, 343–354.

Zarrow, M. X., Gandelman, R., and Denenberg, V. H. Lack of nest building and maternal behavior in the mouse following olfactory bulb removal. *Hormones and Behavior,* 1971b, *2,* 227–238.

Zoloth, S. R. *Arousal in the Female Rat: The Effect of Stimulus Relevance and Motivational State.* Paper presented at Eastern Conference on Reproductive Behavior, Atlanta, Ga., 1974.

PART IV
Biochemical Models of Hormonal Action

13

The Role of Metabolism in Hormonal Control of Sexual Behavior

RICHARD E. WHALEN, PAULINE YAHR, AND
WILLIAM G. LUTTGE

THE PROHORMONE CONCEPT

In 1970, Baulieu published a paper entitled "The Action of Hormone Metabolites: A New Concept in Endocrinology." The new concept advanced by Baulieu was that the secretions of endocrine glands may exert their biological actions only after intracellular metabolism to active agents. Specifically, Baulieu suggested that testosterone (T) maintains growth and secretion of target tissues (e.g., the prostate) by local metabolism to dihydrotestosterone (DHT) and 3α, 5α-androstanediol (abbreviated *Adiol;* the chemical names, trivial names, and abbreviations of most of the steroids and drugs referred to in this chapter are given in Table 1).

The concept that glandular secretions act as "prohormones" has received increasing attention in recent years. In this chapter, we review information on the effects of gonadal steroids and their metabolites on the homotypical sexual behaviors of mammals. We also assess how the prohormone concept applies to neurobehavioral systems.

RICHARD E. WHALEN Department of Psychology, University of California, Riverside, California 92521. PAULINE YAHR Department of Psychobiology, University of California, Irvine, California 92717. WILLIAM G. LUTTGE Department of Neuroscience and Center for Neurobiological Sciences, University of Florida, College of Medicine, Gainesville, Florida 32610. Unpublished research cited here was supported by grants HD-00893 to Richard E. Whalen and MH-26481 to Pauline Yahr.

TABLE 1. STEROID AND DRUG NOMENCLATURE

Abbreviation	Trivial name	Systematic name
Androgens		
T	Testosterone	17β-Hydroxy-4-androsten-3-one
A	Androstenedione	4-Androstene-3, 17-dione
DHT	Dihydrotestosterone	17β-Hydroxy-5α-androstan-3-one
Adiol	3α, 5α-Androstanediol	5α-Androstane-3α, 17β-diol
3β-Adiol	3β, 5α-Androstanediol	5α-Androstane-3β, 17β-diol
	3α, 5β-Androstanediol	5β-Androstane-3α, 17β-diol
	4-Androstenediol	4-Androstene-3β,17β-diol
	5-Androstenediol	5-Androstene-3β, 17β-diol
	Androstanedione	5α-Androstane-3, 17-dione
	Androsterone	3α-Hydroxy-5α-androstan-17-one
DHEA	Dehydroepiandrosterone	3β-Hydroxy-5-androsten-17-one
epiT	Epitestosterone	17α-Hydroxy-4-androsten-3-one
11β-OHA	11β-Hydroxyandrostenedione	11β-Hydroxy-4-androstene-3, 17-dione
19-OHT	19-Hydroxytestosterone	17β, 19-Dihydroxy-4-androsten-3-one
19-OHA	19-Hydroxyandrostenedione	19-Hydroxy-4-androstene-3, 17-dione
19-OHDHT	19-Hydroxydihydrotestosterone	17β, 19-Dihydroxy-5α-androstan-3-one
	19-Oxoandrostenedione	4-Androstene-3, 17, 19-trione
	2β-Hydroxy-19-oxoandrostenedione	2β-Hydroxy-4-androstene-3, 17, 19-trione
19-norT	19-Nortestosterone	19-Nor-17β-hydroxy-4-androsten-3-one
	17α-Ethynyl-19-nortestosterone	17α-Ethynyl-19-nor-17β-hydroxy-4-androsten-3-one
4-OHA	4-Hydroxyandrostenedione	4-Hydroxy-4-androstene-3,17-dione
6α-fluoroT	6α-Fluorotestosterone	6α-Fluoro-17β-hydroxy-4-androsten-3-one
	6α-Bromoandrostenedione	6α-Bromo-4-androstene-3,17-dione
ATD		1, 4, 6-Androstatriene-3, 17-dione
	Fluoxymesterone	9α-Fluoro-11β, 17β-dihydroxy-17α-methyl-4-androsten-3-one
ADT		4-Androstene-3, 6, 17-trione
Estrogens		
E_2	Estradiol	1, 3, 5(10)-Estratriene-3, 17β-diol
E_1	Estrone	3-Hydroxy-1, 3, 5(10)-estratrien-17-one
E_3	Estriol	1, 3, 5(10)-Estratriene-3, 16α, 17β-triol
2-OHE$_2$	2-Hydroxyestradiol	1, 3, 5(10)-Estratriene-2, 3,17β-triol
2-OHE$_1$	2-Hydroxyestrone	2, 3-Dihydroxy-1, 3, 5(10)-estratrien-17-one
	6α-Fluoroestradiol	6α-Fluoro-1, 3, 5(10)-estratriene-3, 17β-diol
	Ethynyl estradiol	17α-Ethynyl-1, 3, 5(10)-estratriene-3,17β-diol
	Diethylstilbestrol	3, 4- Bis(4-hydroxyphenyl)-3-hexane
Progestins		
P	Progesterone	4-Pregnene-3, 20-dione
DHP	Dihydroprogesterone	5α-Pregnane-3, 20-dione
5β-DHP	5β-Dihydroprogesterone	5α-Pregnane-3, 20-dione
	5α-Pregnan-3α-ol-20-one	3α-Hydroxy-5α-pregnan-20-one
	5α-Pregnan-20α-ol-3-one	20α-Hydroxy-5α-pregnane-3-one
20α-OHP	20α-Hydroxyprogesterone	20α-Hydroxy-4-pregnen-3, 20-dione
		5α-Pregnane-3α, 20α-diol
		3α-Hydroxy-5α-pregnan-20-one
		3β-Hydroxy-5α-pregnan-20-one
	17α-Hydroxyprogesterone	17α-Hydroxy-4-pregnen-3, 20-dione
	Pregnenolone	3β-Hydroxy-5-pregnen-20-one
Other Agents		
	Corticosterone	11β, 21-Dihydroxy-4-pregnene-3, 20-dione
	Desoxycorticosterone	21-Hydroxy-4-pregnene-3, 20-dione
	Prednisolone	11β, 17α, 21-Trihydroxy-1, 4-pregnadiene
	RU-2858	11β-Methoxy-17α-ethynyl-1,3,5(10)-estratriene-3,17β-diol

Continued

Abbreviation	Trivial name	Systematic name
	CI-628	α-(4-Pyrrolidionoethyoxy) phenyl-4-methoxy-α-nitrostilbene
	MER-25	1-(*p*-2-Diethylaminoethoxyphenyl)-1-phenyl-2-p-methoxyphenyl ethanol
	Cis-clomiphene (enclomiphene)	1-(*p*)(β-diethylaminoethoxy)phenyl)-1, 2-diphenyl-2-chloroethylene
	Tamoxifen (ICI-46474)	*trans*-1-(p-β-dimethylaminoethoxyphenyl)-12-diphenylbut-1-ene
	Metapirone	2-Methyl-1, 2-bis-(3-pyridyl)-1-propanone
	Aminoglutethimide	β-(4-Aminophenyl)-3-ethyl-piperidine-2, 6-dione
	Cyproterone	1, 2α-Methylene-6-chloro-17α-hydroxy-4, 6-pregnadiene-3, 20-dione
	Flutamide	α-α-α-Trifluoro-2-methyl-4-nitro-m-propionotoluidide

Two lines of evidence lead to the prohormone concept. The first concerns the biological activity of various hormones and their metabolites. As long ago as 1938, Bottomley and Folley demonstrated that DHT is more potent than T in maintaining the weight of the prostate and seminal vesicles. More recently, Baulieu *et al.* (1968) demonstrated that DHT very effectively induces hyperplasia, (i.e., cell division) in prostate tissue explants, whereas Adiol, normally a cytoplasmic metabolite of DHT, selectively induces cellular hypertrophy and secretion. Such studies suggest that the metabolism of T to DHT and Adiol may be necessary to express the many effects of T.

The second line of evidence leading to the prohormone concept concerns the cellular metabolism and the selective retention of hormones. Anderson and Liao (1968) and Bruchovsky and Wilson (1968) independently reported that the primary steroid in prostatic nuclei after administration of radiolabeled T is DHT, not T. Bruchovsky and Wilson also found high levels of Adiol in prostatic cytoplasm. Both sets of workers speculated that metabolic conversion of T to DHT is important for T's action and that DHT might indeed be the active form of T.

In 1970, McDonald and colleagues extended the prohormone concept to the hormonal control of behavior. They compared the effectiveness of testosterone propionate (TP) and dihydrotestosterone propionate (DHTP) in restoring mating behavior in castrated male rats. DHTP stimulated peripheral target tissues (although less effectively than TP) but had no effect on behavior. McDonald *et al.* noted that DHT, unlike T, cannot be metabolized (aromatized) to an estrogen, and they suggested that "aromatization of testosterone may be an important step in the activation of sexual behaviour in both the male and the female" (p. 964). Thus, they hypothesized that T is a prohormone that is active behaviorally only after metabolic conversion to estradiol (E_2).

Shortly after the "aromatization hypothesis" was formulated, Naftolin and colleagues published a series of papers (reviewed in Naftolin and Ryan, 1975) demonstrating that central neural tissues of various mammalian species can aromatize T and androstenedione (A) to E_2 and estrone (E_1). These studies added impetus to the aromatization hypothesis that Naftolin and colleagues feel may account for most, if not all, of the CNS actions of androgens (Naftolin *et al.*, 1975).

One difficulty in the aromatization hypothesis of male sexual behavior is that the full mating pattern, including ejaculation, is not readily induced by estrogen alone (Davidson, 1969). This realization, combined with the observation that neural tissue can metabolize T to both E_2 and DHT, led Baum and Vreeburg (1973) to hypothesize that both of these metabolites are important in the hormonal mediation of mating behavior in males. They found that doses of estradiol benzoate (EB) and DHTP that are ineffective alone stimulate the full mating response, including ejaculation, when combined. These findings support what might be termed the *aromatization-reduction hypothesis* of androgen action.

Although the prohormone concept has been directed primarily toward the primary androgens, it need not be so limited. One can also ask whether the primary secretions of the ovaries, the estrogens and progestins, are prohormones. As this review indicates, it is unlikely that estrogens act as prohormones to stimulate female sexual behavior. E_2, at least, is metabolized very little in the brain (Feder *et al.*, 1974b; Kato and Villee, 1967; Luttge and Whalen, 1970b), although the formation of catechol estrogens in central neural tissues (Fishman and Norton, 1975) may be physiologically important.

Metabolism is more likely to be involved in the actions of progestins. Females of some species readily metabolize progesterone (P) to 5α-reduced products, for example, dihydroprogesterone (DHP) and 5α-pregnan-3α-ol-20-one (Karavolis and Herf, 1971). As Karavolis *et al.* (1976) suggested, the reduction of P to DHP may parallel the reduction of T to DHT and may be important in the central neural action of P. The metabolites 20α-hydroxyprogesterone (20α-OHP) and 5α-pregnan-20α-ol-3-one can also be detected in brain tissue after the administration of P (Mickan, 1972; Nowak *et al.*, 1976). Moreover, several metabolites of P stimulate female mating responses (Czaja *et al.*, 1974; Kubli-Garfias and Whalen, 1977; Whalen and Gorzalka, 1972). These studies do not imply, however, that metabolism is necessary for P's action.

The prohormone concept has also been used to analyze hormonal effects on sexual differentiation. In 1959, Phoenix *et al.* showed that administering TP to maturing female mammals during sensitive periods of development causes permanent changes in their later behavioral responses to gonadal hormones. These changes can be characterized as enhanced responsiveness to androgens (masculinization) and diminished responsiveness to estrogens and progestins (defeminization). Similarly, castrating male mammals early in the sensitive period permits them to become responsive to estrogens and progestins, as normal females are (Feder and Whalen, 1965). These masculinization and defeminization processes seem to be distinct, as responsiveness to ovarian and testicular hormones can be independently affected by hormonal treatment during development (Whalen, 1974).

Because the testes secrete T during differentiation (Resko *et al.,* 1968), and because one can alter sexual differentiation by administering T, it has been assumed that differentiation is controlled directly by T. The prohormone concept demands a reexamination of that view. Although the 5α-reduction of T may not play a role in sexual differentiation (Whalen and Rezek, 1974), aromatization of T to E_2 could be critical for some of T's differentiating effects (Doughty *et al.,* 1975; McEwen *et al.,* 1977a; Whalen and Nadler, 1963). The data relevant to this view are also examined here.

Throughout this chapter, we assume that the sexual differentiation of the nervous system in developing animals and the activation of sexual behavior in mature animals occur when target neurons are stimulated by particular steroids or combinations of steroids. These steroids may be secretory products of the gonads (or adrenals); circulating metabolites produced in the blood, the liver, or other organs; or metabolites produced within the CNS target tissues. According to the prohormone concept, target tissue cells do not respond directly to the parent hormone secreted by an endocrine gland. Rather, they respond to one or more of its metabolites. Therefore, to test the prohormone concept, one must study both hormone metabolism and the interaction of metabolites with target tissues. The questions that can be asked include: (1) Can target tissue cells metabolize the hormone? (2) Do target tissue cells accumulate the metabolite *in vivo*? (3) Is the metabolite more physiologically active than the parent hormone? (4) Do the physiological effects of related hormones correspond to the effects of their metabolites? (5) Do drugs that block the action of a metabolite block the physiological effects of its parent hormone? (6) Do drugs that block the metabolism of a hormone inhibit its physiological effects without inhibiting the activity of its metabolite?

In our first section, we review evidence on these questions for the hypothesis that T activates male sexual behavior via metabolism to estrogens and/or other androgens. More specifically, we consider the possibilities that T acts via conversion to A, to DHT, to various other androgens, to E_2, or to both E_2 and DHT. In later sections, we consider the hypotheses that E_2 and/or P affects female sexual behavior via metabolism to other estrogens or progestins. In particular, we discuss the possibility that E_2 acts via metabolism to E_1 or estriol (E_3) and the possibility that P must be metabolized to DHP, 20α-OHP, or some other progestin to affect sexual behavior. Finally, we review evidence concerning the role of metabolism in testis- or androgen-induced sexual differentiation.

For most of these hypotheses, only a few of the six questions listed above have been tested. Nonetheless, in some cases, the available data seem sufficient for us to reject the hypothesis. For example, if target neurons convert very little T to a particular metabolite, and if that metabolite is no more potent behaviorally than T, then it seems unlikely that the metabolite mediates T's behavioral effects. For other hypotheses, data pertinent to each question have been gathered. Yet, even when the experimental answer to each question seems consistent with a particular hypothesis, some room for skepticism remains. Unfortunately, the "tools" available for testing some of the questions may not be adequate. It is difficult, for example, to interpret the behavioral effects of drugs used to inhibit either the metabolism of T or the actions of T's metabolites. Typically, drugs are referred to as *antimetabolites* or as *metabolic inhibitors* because of their effects on one or more peripheral target tissues (Figure 1). Because different tissues may respond to the same hormone by different mechanisms, a drug that disrupts hormone action in one tissue may not do so in another. Thus, if a metabolic inhibitor, for example, does not disrupt T-induced sexual behavior in males, it could be because the drug does not actually inhibit the metabolism of T in the brain. Or dose–response relationships may be very different for drug action in the brain versus that in other tissues. Alternatively, if the drug does attenuate T-induced sexual behavior, it may do so via nonspecific effects on the brain rather than by blocking T metabolism, which it may also do. The metabolic inhibitors offer a distinct advantage in this respect because one can test their effects on mating behavior induced by the metabolite itself.

Even the various steroids used to evaluate the prohormone concept may be imperfect tools, particularly if tissue differences are overlooked. Although a given steroid may not be metabolized by one tissue, it may be metabolized by another. There is also the inherent problem of confounding variables when comparing the behavioral effects of steroids that differ in their ability to be metabolized. The same or correlated structural differences that determine metabolism may directly, and independently, determine the differences in behavioral activity.

MALE SEXUAL BEHAVIOR

CIRCULATING ANDROGENS AND ESTROGENS IN MALES

Before considering specific hypotheses of how T stimulates male sexual behavior, we must briefly consider what steroid hormones arrive at neural target cells.

Testosterone (T)
(17β-hydroxy-4-androsten-3-one)

Dihydrotestosterone (DHT)	19-Hydroxytestosterone (19-OHT)	Androstenedione (A)
(17β-Hydroxy-5α-androstan-3-one)	(17β, 19-Dihydroxy-4-androsten-3-one)	(4-Androstene-3, 17-dione)

3α, 5α-Androstanediol (Adiol)	19-Oxotestosterone	19-Hydroxyandrostenedione (19-OHA)
(5α-Androstane-3α, 17β-diol)	(17β-Hydroxy-4-androstene-3, 19-dione)	(19-Hydroxy-4-androstene-3, 17-dione)

2β-Hydroxy-19-oxotestosterone
(2β, 17β-Dihydroxy-4-androstene-3, 19-dione

19-oxoandrostenedione
(4-Androstene-3, 17, 19-trione)

Estradiol (E₂)
(1, 3, 5(10)-Estratriene-3, 17(β-diol)

2β-Hydroxy-19-oxoandrostenedione
(2β-Hydroxy-4-androstene-3, 17, 19-trione)

2-Hydroxyestradiol (2-OHE₂)
(1, 3, 5(10)-Estratriene-2, 3, 17β-triol)

Estrone (E₁)
(3-Hydroxy-1, 3, 5(10)-estratrien-17-one)

2-Hydroxytestrone (2-OHE₁)
(2, 3-Dihydroxy-1, 3, 5(10)-estratrien-17-one)

Fig. 1. Metabolic pathways of testosterone.

These are determined primarily by the secretory activity of the testes and the adrenals but can also be affected by the peripheral metabolism of circulating steroids.

T is the predominant androgen in the testes and the blood of adult male rodents (Coyotupa *et al.*, 1973; Hashimoto and Suzuki, 1966; Resko, 1970b; Resko *et al.*, 1968); rabbits (Ewing *et al.*, 1975; Schanbacher and Ewing, 1975); dogs (Ibayashi *et al.*, 1965; Mason and Samuels, 1961; Tremblay *et al.*, 1972); domestic ruminants (Lindner, 1961a,c; Lindner and Mann, 1960); rhesus monkeys (Resko, 1967); and humans (Baulieu and Robel, 1970). Usually, A is also present in the testes and/or the blood of adult males, although in dogs (Ibayashi *et al.*, 1965) and boars (Baulieu *et al.*, 1967), dehydroepiandrosterone (DHEA) is more prevalent than A.

The T/A ratio, as well as the absolute level of T in the blood, increases considerably at puberty. In fact, in some species, notably guinea pigs (Resko, 1970b), cattle (Lindner, 1961a), and rhesus monkeys (Resko, 1967), the spermatic vein or peripheral blood of prepuberal males contains more A than T. This may or may not be true of rats (Hashimoto and Suzuki, 1966; Resko *et al.*, 1968). In boars and rams (Lindner, 1961c), T is dominant both before and after puberty, but the T/A ratio changes.

The blood of neonatal male rats contains more T than A (Resko *et al.*, 1968), but this is not true of newborn guinea pigs or rhesus monkeys (Resko, 1970a,b). Androgen titers seem to be elevated after birth in male rats and rhesus monkeys relative to the rest of the prepuberal period (Dohler and Wuttke, 1975; Resko, 1970a; Resko *et al.*, 1968), but this is apparently not the case in guinea pigs or mice (Resko, 1970a; Selmanoff *et al.*, 1977b).

Androgens can be metabolized in blood. When radiolabeled A is incubated with blood from rabbits, dogs, ruminants, or humans (Blaquier *et al.*, 1967; Lindner, 1961b; van der Molen and Groen, 1968), 1%–19% is converted to T, and as much as 20%–36% may be converted to epitestosterone (epiT). Blood enzymes from some of these species also oxidize T to A, but this conversion occurs at a low rate (1%–3%) unless glucose and cofactors are provided. Blood from human males produces T from 4-androstenediol (6%–16%) but produces very little T from either 5-androstenediol or DHEA. T can also be produced from precursors by other tissues, especially the liver, but little, if any, of this T ever reaches the blood (Baulieu and Robel, 1970).

Significant amounts of DHT, Adiol, and 3β, 5α-androstanediol (3β-Adiol) are secreted by the testes of male rabbits (Ewing *et al.*, 1975), and each of these androgens can be detected in peripheral plasma of this species (Schanbacher and Ewing, 1975). Some researchers have detected DHT in the blood of rats, dogs, and men (Folman *et al.*, 1972; Ito and Horton, 1970; Lewis *et al.*, 1976; Podesta and Rivarola, 1974; Tremblay *et al.*, 1972), but others have reported that the circulating levels of this androgen are negligible or nondetectable in rats, mice, men, bulls, boars, rams, and shrews (Badr and Bartke, 1974; Coyotupa *et al.*, 1973; Falvo and Nalbandov, 1974; Frankel *et al.*, 1975). In the blood of prepubertal male rats, Adiol is reportedly 5.75 to 8 times as prevalent as the combined levels of T and DHT (Moger, 1977).

The only mammals known in which the testes of normal adult males secrete significant amounts of estrogen are pigs and stallions (Eik-Nes, 1970). According to Eik-Nes, estrogen secretion by the testes of other mammalian species probably has little physiological consequence, although androgen secretion by the testes can contribute significantly to blood titers of estrogen in males as a result of extragonadal metabolism. However, under pathological conditions, the secretions of the testes may change. For example, the labial or inguinal testes of humans with the testicular feminization syndrome (genotypic male, phenotypic female) secrete relatively large amounts of estrogen (French *et al.*, 1965) compared to the testes of normal men. Because the aromatization of androgens to estrogens may be involved in sexual differentiation of behavior (see below), it is interesting to note that serum levels of E_2 are elevated in male rats during the frst 2 days after birth (Dohler and Wuttke, 1975).

HYPOTHESIS: T MEDIATES MALE SEXUAL BEHAVIOR VIA METABOLISM TO A

A is less potent than T in promoting growth of the chick comb, the ventral prostate, and the seminal vesicles, but recent experiments have challenged the generalization that A is a weak androgen behaviorally. In fact, T and A seem to be equally potent in stimulating sexual behavior in male rats. Sexually experienced, castrated male rats recover full sexual activity when injected with increasing doses of A (0–800 μg/day), and the pattern of recovery parallels that shown by males given T (Luttge and Whalen, 1970a). Large doses of A and T (800–1000 μg/day) are also equipotent in maintaining mating behavior after castration (Whalen and Luttge, 1971c) and in initiating behavior in sexually inexperienced castrates (Paup *et al.*, 1975). Androstenedione enol propionate (150 μg/day) maintains normal postejaculatory intervals (PEI; Parrott, 1975a). One research group, however, obtained conflicting results in studying A. Beyer and coworkers initially found that A and T (1 mg/day) induced similar levels of sexual behavior in naive, castrated male rats (Beyer *et al.*, 1973) but later reported that A (0.3–3 mg/day) stimulates relatively little sexual behavior in such males (Morali *et al.*, 1974). Taken as a whole, the data suggest that circulating A could influence sexual behavior in male rats.

A formed from T within the brain could also affect male sexual behavior. ^{3}H-A can be detected in the brains of male rats injected systemically with ^{3}H-T (Rezek and Whalen, 1975; Stern and Eisenfeld, 1971; Whalen and Rezek, 1972), and *in vitro* studies indicate that this conversion can occur within the brain itself (Denef *et al.*, 1973; Jaffe, 1969; Loras *et al.*, 1974; Massa *et al.*, 1972; Rommerts and van der Molen, 1971). The enzyme involved is a soluable protein found uniformly, in relatively small amounts, throughout the brain.

Despite A's behavioral potency, it is unlikely that the metabolism of T to A mediates T's effects on sexual behavior in male rats. Because much less A than T appears in brain cells exposed to T, either *in vivo* or *in vitro*, A could mediate T's behavioral effects only if it were considerably more potent than T. The behavorial data do not support this possibility.

Oxidation of T to A by brain tissue has been demonstrated *in vitro* for male guinea pigs (Sholl *et al.*, 1975); rabbits (Reddy *et al.*, 1973, 1974b); dogs (Perez-Palacios *et al.*, 1970); and human fetuses (Jaffe, 1969); and *in vivo* in rhesus monkeys (Sholiton *et al.*, 1974). ^{3}H-A has also been detected in the brains of castrated male mice (Applegren, 1970) and guinea pigs (Sholl *et al.*, 1975) injected systemically with ^{3}H-T. Unfortunately, the only one of these species in which the behavioral effects of A have been studied is mice. In mice, A (100-500 μg/day) is relatively ineffective in restoring male sexual behavior after castration (Luttge and Hall, 1973).

In contrast, A is at least as potent as T, and possibly more so, in stimulating copulatory behavior in castrated male hamsters with previous sexual experience. Castrated male hamsters recover sexual activity at the same rate whether they are injected with A or with T (100 μg/day; Tiefer and Johnson, 1973). Mating behavior also declines at a similar rate when male hamsters are given decreasing doses of either T or A (1000–0 μg/day) after castration (Whalen and DeBold, 1974). Similarly, A (500 μg/day) maintains mating behavior in castrates previously treated with T (Christensen *et al.*, 1973). Whether A can initiate mating behavior in sexually naive, castrated male hamsters is unknown. It is clear, though, that A is not a weak androgen behaviorally in either hamsters or rats. A may even be the active metabolite mediating T's effects on mating behavior in male hamsters. It is disappointing that data on neural metabolism of T to A are not available for this species.

Hypothesis: T Mediates Male Sexual Behavior via Metabolism to DHT

Both the behavioral effects of DHT and the regional distribution of 5α-reductase activity within the brain have been studied in male rats. The 5α-reduction of T to DHT occurs in hypothalamic and/or whole brain tissue from male rats *in vitro* (Denef *et al.*, 1973; Genot *et al.*, 1975; Jaffe, 1969; Kniewald *et al.*, 1970; Loras *et al.*, 1974; Massa *et al.*, 1972, 1975; Rommerts and van der Molen, 1971; Sholiton *et al.*, 1966; Shore and Snipes, 1971; Verhoeven *et al.*, 1974). DHT is present in the hypothalamus of male rats endogenously (Robel *et al.*, 1973), and radiolabeled DHT can be detected in the hypothalamus after systemic injections of radiolabeled T (Lieberburg and McEwen, 1977; Rezek and Whalen, 1975; Sholiton *et al.*, 1972; Stern and Eisenfeld, 1971; Whalen and Rezek, 1972).

If DHT mediates sexual behavior in male rats, then the regional distribution of DHT within the brain should reflect this mediation. Because T elicits mating behavior most effectively when implanted in the anterior hypothalamic–preoptic-area region (Davidson, 1966), the cells in this region should accumulate DHT. Indeed, the hypothalamus does accumulate more total radioactivity than the cortex when ^{3}H-DHT is injected systemically into castrated male rats (Lieberburg *et al.*, 1977a; Lieberburg and McEwen, 1977; Perez-Palacios *et al.*, 1973). Similarly, analyses of whole tissue homogenates indicate that the hypothalamus and/or the limbic structures accumulate more ^{3}H-DHT than the cortex after systemic injections of ^{3}H-T (Liederburg and McEwen, 1977; Rezek and Whalen, 1975). The hypothalamic-cortical differences are small, though, and cannot always be detected (Stern and Eisenfeld, 1971). Somewhat larger differences were detected by Robel *et al.* (1973) using a gas-chromatographic electron-capture technique to assay endogenous DHT.

5α-Reductase activity is not confined to the hypothalamus. It occurs throughout the brain in male rats. In fact, both *in vivo* and *in vitro,* the enzyme activity seems higher in the midbrain, the cerebellum, and/or the brainstem than in the hypothalamus (Denef *et al.,* 1973; Rommerts and van der Molen, 1971; Verhoeven *et al.,* 1974). The hypothalamus, in turn, displays more 5α-reductase activity than the cortex and/or the amygdala when tissue slices or minces or cell nuclear preparations are incubated with ^{3}H-T *in vitro* (Denef *et al.,* 1973; Jaffe, 1969; Rommerts and van der Molen, 1971; Selmanoff *et al.,* 1977a; Shore and Snipes, 1971; Verhoeven *et al.,* 1974). Yet, when ^{3}H-T or ^{3}H-DHT is given to male rats *in vivo,* the nuclei of hypothalamic cells accumulate more ^{3}H-DHT than do nuclei of cells in either the cortex, the amygdala, or the midbrain (Lieberburg *et al.,* 1977a; Lieberburg and McEwen, 1977).

In assessing the possible role of DHT in male sexual behavior, it is particularly interesting to compare metabolism of T to that of DHT in the preoptic area versus the hypothalamus. The preoptic area is the brain region most often implicated in the control of sexual behavior in male rats, whereas the hypothalamus is involved in both behavior and negative feedback regulation of gonadotropin secretion (Davidson and Block, 1969). Studies comparing 5α-reductase activity or the accumulation of DHT in these two brain areas of male rats have provided conflicting data. Several studies report that preoptic area cells and their nuclei accumulate less DHT *in vivo* (Lieberburg *et al.,* 1977a; Lieberburg and McEwen, 1977; Rezek and Whalen, 1975) and possess less 5α-reductase activity *in vitro* (Denef *et al.,* 1973) than hypothalamic cells do. Verhoeven *et al.* (1974) reported similar differences using isolated cell nuclei; however, their data are difficult to evaluate because their preoptic area samples included septal tissue and their hypothalamic samples included thalamic tissues. In contrast, Selmanoff *et al.* (1977a) found 10 times as much 5α-reductase activity in the lateral preoptic nucleus, the medial-preoptic–anterior-hypothalamic nuclei, and the lateral hypothalamic nucleus as in the medial basal hypothalamus. They also reported that the three former areas did not differ from each other in 5α-reductase activity and that the regional distribution of enzyme activity was the same whether the substrate was ^{3}H-T or ^{3}H-A. T was reduced more efficiently than A, however.

Discrepancies between studies regarding 5α-reductase activity in preoptic area cells may reflect procedural differences. The amount of DHT isolated from brain tissue probably depends on a variety of parameters, including the length of *in vivo* or *in vitro* incubation, the endocrine condition of the animal when the hormone is administered or the tissue isolated (e.g., the presence or absence of the gonads and/or the adrenals), the time since castration, the dose of T used, and the particular neural regions and/or subcellular fractions isolated. A study by Sholiton *et al.* (1972) highlights the problems involved. After administering ^{3}H-T to male rats that were either castrated or castrated plus totally eviscerated, they studied the ratio of ^{3}H-DHT to ^{3}H-T in the brain. The blood ratio of DHT to T was the same in both groups (0.52 versus 0.50); however, the neural DHT/T ratio was higher after evisceration (1.70 versus 0.46). Clearly, the neural differences were not due solely to changes in the amount of ^{3}H-DHT produced peripherally.

Subcellular analyses indicate that 5α-reductase activity is associated with the microsomal and mitochondrial fractions of brain tissue from male rats (Rommerts and van der Molen, 1971; Sholiton *et al.*, 1966; Verhoeven *et al.*, 1974). This enzyme may be the same one that reduces P to DHP (see later sections of this chapter). However, after directly comparing 5α-reductions of T and P in female rats, Cheng and Karavolis (1975) suggested that either two enzymes exist or else P is the preferred substrate for the single 5α-reductase. Whether T or P is involved, 5α-reduction seems to be an irreversible process. For example, 30 min after injecting male rats with ^{3}H-DHT, Whalen and Rezek (1972) found no ^{3}H-T in the brain. The DHT was metabolized to Adiol and androstanedione.

Based on the data presented above, it is plausible that DHT acts on the brain to influence sexual behavior in male rats. However, behavioral studies suggest that this metabolite alone cannot account for T's effects. McDonald *et al.* (1970) were the first to reach this conclusion. They reported that sexually naive, castrated male rats rarely mounted when injected with DHTP (125 μg/day) for 8 days, whereas males injected with TP (100 μg/day) intromitted and ejaculated. Since that study, many laboratories have verified the observation that DHT(P) is less effective than T(P) in stimulating sexual behavior in male rats of most strains.[1]

Several studies indicate that neither DHT (400–1000 μg/day) nor DHTP (50–150 μg/day) maintains intromissions or ejaculations after castration or withdrawal of TP treatment in sexually experienced male rats, although DHTP may retard the decline in mating behavior (Johnston and Davidson, 1972; Parrott, 1974, 1976; Whalen and Luttge, 1971d; Yahr and Gerling, 1978). Parrott (1975a,b) reported that DHTP (150 μg/day) can maintain ejaculations in 70%–80% of males for 5 weeks after castration, but the PEI doubles in this time. The gradual decline in sexual behavior in DHT-treated castrates occurs whether the hormone is contained in silastic capsules (20 mm, 1.57 mm id, 3.18 mm od) or is dissolved (decreasing doses of 250–100 μg/day) in safflower oil, propylene glycol, or a mixture of Tween 80 (polyoxyethylene sorbitan mono-oleate), ethanol, and saline (Yahr and Jackson, unpublished observations, 1976). The behavioral effects of DHP in female rats and of DHT in male mice are profoundly affected by the vehicle in which the hormone is dissolved (see below), but this apparently is not the case for DHT in male rats. DHT is somewhat less effective in propylene glycol than in other vehicles, but none of the vehicles tested make DHT more behaviorally effective than it is in oil.

DHT (50 μg/day) is less effective than TP (200 μg/day) in maintaining penile reflexes (Hart, 1973), but this difference may be due to differences in the dose or form (propionate versus free alcohol) of the hormone used. DHTP (200 μg/day) and TP (200 μg/day) are equally potent in reinstating penile reflexes in castrated male rats (Hart, 1978).

---

[1]Strain differences in response to DHT occur in rats. K. L. Olsen (1979) has recently found that male rats of the King-Holtzman strain all ejaculate when given 1.0 mg of DHT daily.

Surprisingly, DHT(P) seems more effective in reinstating and/or initiating sexual behavior in male rats than in maintaining it (Yahr, 1979). More than a dozen studies have been published, including the one by McDonald *et al.* (1970), in which DHT(P) was administered to male rats beginning several weeks after castration (Baum and Vreeburg, 1976; Beyer *et al.*, 1973; Feder, 1971; Feder *et al.*, 1974a; Johnston and Davidson, 1972; Johnston *et al.*, 1975; Larsson *et al.*, 1973a,b, 1976; Luttge *et al.*, 1975; McDonald *et al.*, 1970; Paup *et al.*, 1975; Södersten, 1975). These studies agree that DHT(P) is much less potent than T(P) in inducing or restoring sexual behavior in castrated males. Yet, only three studies reported complete failure of DHT(P)-treated males to ejaculate (Beyer *et al.*, 1973; Luttge *et al.*, 1975; McDonald *et al.*, 1970). Each of these three studies used sexually naive male rats, a procedure that could have reduced the probability of ejaculation. Two also used short test periods. Both McDonald *et al.* and Luttge *et al.* termined sexual behavior tests after 15 min, although Luttge *et al.* extended the tests by 5 min if the male had intromitted five or more times. These brief tests could have decreased the probability of ejaculation, particularly if the intromission latencies were long. Beyer *et al.* also allowed only 15 min for intromission but tested for an additional 30 min if any intromissions occurred. In the remaining studies, 10%–55% of the DHT(P)-treated castrates ejaculated. This level of sexual behavior is low compared to the levels induced by the same doses of T(P), but it seems high for an androgen that is often described as behaviorally ineffective in rats.

An experiment by Whalen and Luttge (1971d) is particularly pertinent for comparing the effects of DHT on the maintenance versus the reinstatement of male sexual behavior. Beginning at castration, male rats were injected with DHT (400 μg) daily until they stopped mounting. Then, the dose of DHT was increased (800 μg/day); within 4 weeks, three of the five surviving males ejaculated. Unfortunately, one cannot determine whether the dosage increase was necessary to elicit the change in behavior. Prolonged exposure to DHT may make male rats more sensitive to it. This study does show, however, that DHT can effectively stimulate the complete mating pattern of male rats, including ejaculation. Paup *et al.* (1975) replicated this observation. In their study, five of nine castrated males ejaculated after treatment with DHT (1 mg/day). Despite the similarities in the data of these two studies, they are often cited as if they support different conclusions about the behavioral effects of DHT. The Whalen and Luttge (1971d) study is usually cited to support the statement that DHT does not stimulate mating behavior in male rats, whereas the study by Paup *et al.* (1975) is interpreted as showing that DHT can promote mating behavior in this species.

Because T(P) is more potent in maintenance than in reinstatement paradigms, it seems paradoxical that the opposite could be true for DHT(P). DHT(P) may only appear more potent during reinstatment because such paradigms usually employ larger doses and/or longer treatment periods. It is also difficult to determine if the sexual behavior observed in some studies was actually induced by DHT(P) therapy. Many of the studies (e.g., Baum and Vreeburg, 1976; Feder *et al.*, 1974a; Larsson *et al.*, 1973a,b, 1976; Södersten, 1975) did not test vehicle-treated controls or did not present control data. In Feder's (1971) study, the control data suggest that

DHT(P) treatment was not the critical variable, but in the remaining studies, increased sexual behavior apparently was due to DHT(P) stimulation. It may also be pertinent that the two studies (Paup *et al.*, 1975; Whalen and Luttge, 1971c) reporting the highest levels of sexual behavior used DHT rather than DHTP. Although esterifying T to TP increases its behavioral potency, this may not be true of DHT. For stimulating growth of the seminal vesicles, DHT is more potent than T and is sometimes more potent than TP, but DHTP is less potent than TP (Feder, 1971; Larsson *et al.*, 1973b; McDonald *et al.*, 1970; Whalen and Luttge, 1971d; Yahr and Gerling, 1978).

Exposing gonadally intact prepubertal male rats to DHT delays the onset of mounting behavior, apparently as a result of decreased gonadotropin, and hence T secretion (Södersten *et al.*, 1977). If both hormones are given exogenously, preexposure to DHT hastens the onset of mating behavior in sexually naive, castrated male rats given T or TP. Mounting, intromission, and ejaculatory behaviors begin sooner, whether DHT treatment continues (Larsson *et al.*, 1976; Södersten, 1975) or stops (Larsson *et al.*, 1975) when T(P) therapy begins. Similarly, data presented by Baum and Vreeburg (1973) suggest that concurrent treatment with DHTP may enhance the behavioral effectiveness of small TP doses. In contrast, pretreatment with DHTP (125 μg/day) does not facilitate the recovery of sexual behavior in sexually experienced male rats that are then switched to TP (125–250 μg/day) treatment (Feder, 1971). Also, Feder *et al.* (1974a) observed few ejaculations in sexually experienced male rats given TP plus DHTP (150 μg + 125 μg/day), despite preexposure to DHTP for 2 weeks. It may be important, though, that, in both of the latter two studies, the copulatory tests were only 10 min long. Any facilitation of T(P)'s effects by DHT(P) could reflect actions of DHT on the penis, even though DHTP does not precisely mimic TP's effects on penile spines (Phoenix *et al.*, 1976). Two studies suggest, however, that increased development or sensitivity of the penis may not account for DHT(P)'s effects. For example, Davidson and Trupin (1975) reported that systemically administered DHT does not potentiate mating behavior induced by TP implants into the preoptic area. Also, the data of Lodder and Baum (1977) suggest that DHTP may facilitate TP-induced mounting behavior even after pudendectomy.

As DHT is so much less potent than T behaviorally, it alone could not be the active form of T in male rats. Most attempts to disrupt sexual behavior in male rats by means of antiandrogens support this conclusion. Although both cyproterone acetate and flutamide block DHT's action in the seminal vesicles and the prostate (see reviews by Liao, 1976; Mainwaring, 1975), neither impairs spontaneous or T(P)-induced mounting behavior in male rats, according to published reports (Beach and Westbrook, 1968b; Bloch and Davidson, 1971; Södersten *et al.*, 1975; Whalen and Edwards, 1969). Flutamide does inhibit ejaculatory behavior, but this inhibition has been attributed to decreased sensory input from the penis (Södersten *et al.*, 1975). More recently, Clemens and Gladue (personal communication, 1978) observed that flutamide also interferes with T's ability to reinstate mounting behavior in male rats that stop mounting after castration. This finding suggests that flutamide may counteract central as well as peripheral effects of T.

The hypothesis that DHT is the active metabolite mediating T's effects on sexual behavior has been tested in other mammalian species. Brain tissues from male guinea pigs (Sholl *et al.*, 1975), dogs (Perez-Palacios *et al.*, 1970), and human fetuses (Jaffe, 1969; Mickan, 1972; Schindler, 1976) reduce T to DHT *in vitro*. Hypothalamic cells of male guinea pigs (Sholl *et al.*, 1975), rhesus monkeys (Sholiton *et al.*, 1974), and mice (Applegren, 1970) also accumulate DHT after *in vitro* administration of radiolabeled T. Thus, the potential for DHT to mediate T's behavioral effects exists in each of these species.

The effects of DHT(P) on male sexual behavior have been studied in hamsters, gerbils, mice, guinea pigs, rabbits, and rhesus monkeys. In hamsters, large doses of DHT (1 mg/day) fully maintain mating behavior (Whalen and DeBold, 1974). When smaller doses (500 µg/day or less) are used, sexual behavior decreases after castration, despite DHT treatment, but DHT may retard the decline (Christensen *et al.*, 1973; Clemens, 1974; Whalen and DeBold, 1974). DHTP (300 µg/day) also reinstates intromissive behavior after castration and occasionally stimulates ejaculation (Noble and Alsum, 1975). Thus, DHT(P) may be more effective in hamsters than in rats, but differences in the dose of DHT(P) used relative to body weight make this comparison difficult.

Sexually experienced male gerbils implanted with silastic capsules (2 mm, 1.47 mm id, 1.96 mm od) of DHT at castration stop mounting within 4 weeks, as do controls given no hormone (Yahr and Stephens, unpublished observations, 1978). However, in a reinstatement paradigm, a larger dose (5 mm) of DHT can stimulate ejaculation within 4 weeks in some males.

Some sexually inexperienced, castrated male rabbits begin copulating after DHT (2.5 mg/day) treatment, particularly if they are castrated postpuberally (Beyer *et al.*, 1975). Although T-treated males were not tested in this study, the percentage of males responding to DHT and the frequency of responses observed seem smaller than would be induced by T. DHTP (1–9 mg/day) is clearly less behaviorally effective than TP in prepuberally castrated male rabbits (Beyer and Rivaud, 1973). Silastic implants (20 mm, 3.35 mm id, 4.65 mm od) of DHT are also less potent than T implants in maintaining sexual behavior in castrated male rabbits (Foote *et al.*, 1977). In this study, sexually naive male rabbits implanted with T or DHT at castration had comparable mount latencies for 5 weeks, but then the mount latencies of the DHT-treated males increased. The decline in mating behavior apparently did not reflect a decreased release of DHT over time. According to Foote *et al.*, the release rate decreased relatively little over even longer time periods. DHT was released more slowly than T (0.14 versus 0.18 mg/day), however. Perhaps this difference had cumulative effects on behavior across weeks.

Cyproterone acetate decreases ejaculation frequency in intact male rabbits (Ağmo, 1975) whether they are sexually experienced or naive. The decrease could be related to DHT's ability to stimulate low levels of sexual behavior in this species. Alternatively, it could reflect cyproterone acetate's ability to depress gonadotropin secretion or aromatase activity (Johnson and Naqvi, 1969; Schwarzel *et al.*, 1973).

Among male mice, the ability to respond behaviorally to DHT varies across strains. Increasing doses of DHT (100–500 µg/day) initiate intromissive behavior in 50% of sexually naive, castrated Swiss-Webster males. The same doses of T elicit behavior more often (80% of males) and more rapidly, but once the behavior has developed, intromission duration and frequency are equivalent with either androgen in this strain (Luttge and Hall, 1973; Luttge *et al.*, 1974a). In contrast, sexually naive CD-1 castrates rarely display mounting behavior with this treatment. DHT (200 μg/day) can nonetheless maintain intromissive behavior after castration in about 70% of sexually experienced CD-1 males, although the behavior disappears within 2 weeks when the DHT dose is halved (Wallis and Luttge, 1975). This strain difference in responsiveness to DHT may reflect a more general difference in sensitivity to androgens between CD-1 and Swiss-Webster mice. Sexually inexperienced CD-1 castrates are also less responsive to T and A (100–500 μg/day) than Swiss-Webster males are (Luttge and Hall, 1973). In Swiss-Webster mice, responsiveness to DHT depends on the vehicle in which the hormone is administered. DHT (200–400 μg/day), but not T, elicits much less mating behavior when given in propylene glycol instead of oil (Luttge *et al.*, 1974a). The ability of DHP to facilitate lordosis in female rats also declines when it is given in propylene glycol instead of oil (see below).

DHTP (1 mg/kg/day) is quite potent in reinstating sexual behavior in castrated male rhesus monkeys (Cochran and Perachio, 1977; Phoenix, 1974a,b). This dose restores ejaculation frequency to the levels seen in intact males, although the rate of recovery is about half that seen after TP treatment.

Although the behavioral effects of DHT and T have not been systematically compared in human males, men apparently do not require the metabolism of T to DHT in order to display heterosexual copulatory interests and behavior. Men with a genetic loss of 5α-reductase activity display both these traits (Imperato-McGinley *et al.*, 1974; Peterson *et al.*, 1977).

Guinea pigs are the only species known in which DHTP seems fully as potent as TP in stimulating male sexual behavior. In prepuberally castrated males, both DHTP and TP (50 μg/100g/day) elicit ejaculation in all subjects (Alsum and Goy, 1974). The developmental patterns of intromissive and ejaculatory behaviors are the same with both hormones, although TP stimulates a more rapid onset of mounting and a higher mounting rate, particularly for mounts without intromission. As Alsum and Goy pointed out, the mounting rates of DHTP-treated castrates are similar to the mounting rates of intact males (Phoenix, 1961). Thus, the higher mounting rates with TP treatment could reflect a pharmacological effect of this hormone. Given that DHTP is so potent in male guinea pigs, it is surprising that cyproterone acetate (3–15 mg/day) does not inhibit male sexual behavior in this species (Zucker, 1966a).

In summary, the behavioral effects of DHT vary as a function of species, strain, vehicle, dose, duration of hormone treatment, form of the hormone (DHT or DHTP), length of test, time since castration, age at castration, and/or previous sexual experience. With the possible exceptions of guinea pigs and rhesus monkeys, it is unlikely that DHT mediates the effects of endogenous T on male sexual behavior. DHT may interact with other metabolites of T in the brain. This possiblity is considered in more detail below.

Beyer *et al.* (1973) tested various androgens, in addition to T, DHT, and A, for their ability to initiate sexual behavior in sexually inexperienced, castrated male rats. These included androstanedione, androsterone, 5-androstenediol, Adiol, $3\alpha,5\beta$-androstanediol, DHEA, and 11β-hydroxyandrostenedione (11β-OHA). The first four of these can be detected in the brain tissue of male rats as metabolites of T or DHT either *in vivo* or *in vitro* (Denef *et al.*, 1973; Genot *et al.*, 1975; Jaffe, 1969; Loras *et al.*, 1974; Massa *et al.*, 1972, 1975; Perez *et al.*, 1975; Rezek and Whalen, 1975; Sholiton and Werk, 1969; Sholiton *et al.*, 1970, 1972; Stern and Eisenfeld, 1971; van Doorn *et al.*, 1975; Whalen and Rezek, 1972).

In the Beyer *et al.* (1973) study, only T, A, and 5-androstenediol (1 mg/day) stimulated ejaculation. 5-Androstenediol was as effective as A on all reported measures of sexual behavior, and both were as effective as T on all measures except mount latency. Using the same dose, Morali *et al.* (1974) found that fewer males mounted, intromitted, or ejaculated when injected with either A or 5-androstenediol than when injected with T. The males that displayed these behaviors obtained similar frequency and latency scores, however, regardless of the hormone they received. At higher doses (3 mg/day), 5-androstenediol and T were both more effective than A in this study. Except for males given androsterone, some males in each of the androgen treatment groups displayed mounting behavior (Beyer *et al.*, 1973). Some males given DHEA or 11β-OHA also intromitted. Unfortunately, it is difficult to determine if these copulatory behaviors resulted from hormonal stimulation, as data on oil-treated controls were not provided.

The propionate esters of androsterone, Adiol, and 3β-Adiol do not maintain sexual behavior in sexually experienced male rats after castration as assessed by increases in the PEI or by the frequency of ejaculations (Parrott, 1975a). Androstanedione enol propionate did maintain ejaculatory behavior in this study, but this result was attributed to contamination (50%) with A.

Androstanedione (0–500 μg/day) does not initiate mounting behavior in either CD-1 or Swiss-Webster mice that are castrated and lack sexual experience (Luttge and Hall, 1973). Some sexually naive, castrated Swiss-Webster males do mount and intromit, however, when injected with Adiol or 3β-Adiol (Luttge *et al.*, 1974a). These two androgens are about equally effective at each of two dose levels (200 or 400 μg/day). Both are considerably less potent than T but are more potent than DHT when administered in propylene glycol.

Androsterone (500 μg/day) maintains mounting and intromissions in sexually experienced, castrated male hamsters previously treated with T (500 μg/day), but it seems less effective than either T or A (Christensen *et al.*, 1973).

As was noted for DHT, these various androgenic metabolites may intereact with other metabolites of T in the brain to influence male sexual behavior. However, considering the behavioral dose–response relationships for these androgens and their low levels within the brain, they probably do not mediate the behavioral effects of T.

Three other androgens, 19-hydroxytestosterone (19-OHT), 19-hydroxyandrostenedione (19-OHA), and 19-hydroxydihydrotestosterone (19-OHDHT), have been tested for their ability to activate male sexual behavior in rhesus monkeys and/or rats. Because 19-OHT and 19-OHA are the first metabolites formed when T or A is aromatized, we will discuss their behavioral effects and those of 19-OHDHT in the next section. The effects of several other androgens (e.g., fluoxymesterone) are also discussed below.

Hypothesis: T Mediates Male Sexual Behavior via Metabolism to Estrogen

More than 40 years ago, Ball (1937, 1939) and Beach (1942) observed that systemic treatment with estrogens stimulated male copulatory behavior in castrated male rats. For many years, these data were interpreted as reflecting a lack of strict hormonal specificity in the control of male sexual behavior. More recently, these data have been interpreted as meaning that the central neural effects of T are mediated by aromatization. This concept was first postulated by McDonald *et al.* (1970) to account for the failure of DHTP to initiate sexual behavior in castrated male rats. They noted that DHT, unlike T, cannot be aromatized in a human placental assay system and speculated that this metabolic difference between T and DHT accounted for the difference in their behavioral effects. This aromatization hypothesis stimulated much research, and most of the available data are consistent with it for rats.

Numerous studies have demonstrated that brain tissues of male mammals can aromatize androgens to estrogens. The conversion of A to E_1 occurs *in vitro* in hypothalamic tissue from adult male rats (Kobayashi and Reed, 1977; Naftolin *et al.*, 1972; Reddy *et al.*, 1974a; Selmanoff *et al.*, 1977a); adult male rabbits (Reddy *et al.*, 1973); immature male rhesus monkeys (Flores *et al.*, 1973a); and human male fetuses (Naftolin *et al.*, 1971a,b). The aromatization of T to E_2 occurs *in vitro* in hypothalamic tissue of male rabbits (Reddy *et al.*, 1973); rhesus monkeys (Flores *et al.*, 1973a); and rats (Selmanoff *et al.*, 1977a). The hypothalami of male rats and human fetuses also form E_2 from A *in vitro* (Naftolin *et al.*, 1971a,b; Selmanoff *et al.*, 1977a). ^{3}H-estrogens can be detected in whole tissue homogenates and cell nuclei of adult (Lieberburg and McEwen, 1975a; Rezek, 1977) and neonatal (Lieberburg and McEwen, 1975a) male rats after the *in vivo* administration of ^{3}H-T. Aromatization to both E_1 and E_2 also occurs when isolated male monkey brains are perfused with either A or T (Flores *et al.*, 1973a,b; Naftolin *et al.*, 1975). The aromatization of T apparently does *not* occur in hypothalamic tissue of male guinea pigs, either *in vivo* or *in vitro* (Sholl *et al.*, 1975).

In males, the preoptic area and the anterior hypothalamus have the highest levels of aromatase activity. After bisecting hypothalamic samples at the midmedian eminence, Naftolin and coworkers (1975) found little (human fetuses) or no (rats, rabbits) aromatase activity in the posterior segments. By assaying the amount of ^{3}H-water released when $[1\beta\text{-}^3\text{H}]$ A is converted to E_1, Kobayashi and Reed (1977) found the highest levels of aromatase activity in the medial preoptic area nucleus, the periventricular preoptic area nucleus, and the medial amygdaloid nucleus of male rats. Enzyme activity in the medial-preoptic-area nucleus exceeds activity in the lateral-preoptic-area nucleus by 11- to 20-fold. Assaying the conversion of A to E_1 by male rat brain tissue *in vitro*, Selmanoff *et al.* (1977a) found 6–30 times as much aromatase activity in the medial-preoptic–anterior-hypothalamic region as in the medial basal hypothalamus, the lateral preoptic nucleus, the lateral hypothalamic nucleus, or the amygdala. The conversion rate was 0.001%–0.03%. When T was the substrate, the lateral hypothalamic and lateral preoptic nuclei did not display aromatase activity, but the other regions did.

With *in vitro* techniques, no aromatase activity has been detected outside the limbic system in adult males, but only rabbits have been studied (Reddy *et al.*, 1973). Nonetheless, this observation agrees with Lieberburg and McEwen's findings (1975a) that hypothalamic, but not cortical, cell nuclei of adult male rats accumulate ^{3}H-E_2 after the systemic administration of ^{3}H-T *in vivo*. In contrast, cortical nuclei of neonatal male rats do accumulate ^{3}H-E_2 after ^{3}H-T treatment *in vivo* (Lieberburg and McEwen, 1975b). Similarly cortical tissue from human fetuses aromatizes A to E_1 *in vitro* (Naftolin *et al.*, 1975).

Thus, except in guinea pigs, estrogens could mediate T's effects on male sexual behavior in many species. This hypothesis has been tested in many behavioral studies. Some have assessed the ability of estrogens to stimulate male sexual behavior. Others have compared the behavioral effects of androgens that vary in aromatizability. Still others have assessed the ability of antiestrogens and/or aromatase inhibitors to block T(P)-induced mating behavior in males. All of these approaches have been used in male rats.

As noted above, the ability of estrogens to stimulate copulatory behavior in male rats has been known for decades. More recently, others have verified these observations. In castrated males with little or no prior sexual experience, EB (10 μg/day) readily elicits mounts and intromissions (Pfaff, 1970; Pfaff and Zigmond, 1971). Sexual behavior, particularly intromissive and ejaculatory behavior, declines after castration despite EB (50–70 μg/day) treatment (Davidson, 1969; Johnson and Tiefer, 1974), but larger EB doses (100–200 μg/day) can reinstate sexual behavior, including ejaculation, particularly if the treatment is prolonged (Davidson, 1969; Södersten, 1973). Estradiol dipropionate (150 μg/day, 5 days/week) appears to be less effective than TP (75 μg/day, 5 days/week) in maintaining ejaculations after castration, but both hormones maintain short PEIs in males that do ejaculate (Parrott, 1975a). In sexually experienced, intact males, EB can inhibit (10–20 μg/day) or maintain (5–200 μg/kg/day) sexual behavior despite the fact that large EB doses (50–200 μg/kg/day) cause testicular atrophy and hence presumably decrease T secretion (Davidson, 1969; Södersten, 1973).

Prepuberally castrated male rats also mount and intromit, but rarely ejaculate, when injected with EB (50 μg/day) in adulthood (Södersten, 1973). Södersten suggested that their failure to display ejaculatory behavior could be due to poor development of penile spines; however, data presented by Baum (1972, 1973) conflict with this interpretation. Baum found that treating intact males with EB (5 μg/100 g/day) starting at 11 days of age hastens the onset of intromissive and ejaculatory behaviors yet retards penile development. In EB-treated males, the frenulum remains attached to the glans penis, preventing full erections and making insertion of the penis into the vagina problematic. Nevertheless, EB-treated males are as precocious behaviorally as TP-treated (1 mg/100 g/day) males in which penile development is enhanced. Both larger (100 μg/100 g/day) and smaller (50 ng/100 g/day) EB doses are less effective in inducing precocious mating. The effective EB dose is as potent in males castrated at 10 days of age as it is in intact males, a finding indicating that androgens from the atrophied testes do not contribute to the precocious behavioral development. Males castrated at 10 days of age and given EB (5 μg/100 g/day) for 20 days are also more behaviorally sensitive to later TP (50 μg/day) treatment than are males given TP (5 μg/100 g/day) after castration.

Two research groups studied how pretreatment with EB (1 μg/day) affects the ability of TP to initiate mating behavior in sexually naive, castrated male rats. Södersten (1975) found that 30 days of EB treatment facilitated TP's (200 μg/day) effects. In contrast, Larsson *et al.* (1975) found that 21 days of pretreatment did not alter TP's (1 mg/day) effects. The discrepancies between these two studies could reflect differences in either the TP doses used or the length of EB pretreatment. They could also be due to the fact that Södersten continued the EB injections when TP treatment began, whereas Larsson *et al.* did not.

Although EB's ability to elicit the full copulatory pattern in male rats lends support to the aromatization hypothesis, the need for large doses suggests that aromatization alone cannot account for all of T's actions. In fact, it is not even clear that the stimulatory effects of large EB doses are mediated directly by EB. Sexually experienced male rats that are adrenalectomized as well as castrated do not ejaculate when given EB (100–150 μg/day; Gorzalka *et al.*, 1975). Hence, adrenal steroids, possibly androgens, may mediate EB-induced ejaculatory behavior. However, T's effects on mounting may reflect aromatization.

A small controversy exists about eliciting sexual behavior in castrated male rats by implanting estrogen in the brain. Davidson (unpublished observations, cited in Davidson and Trupin, 1975) found that "large" implants of EB would not restore intromissions or ejaculations in castrates. Then, Christensen and Clemens (1974) reinstated both behavior patterns quite effectively by applying E_2 (10 μg) to each side of the preoptic area every 3 days for 2 weeks. Repeated applications of E_2 may have made the critical difference. Bilateral injections of E_2 (2.5 μg/day/side) for 2 weeks also induced 1 of 10 males to ejaculate, although untreated controls were not tested (Christensen and Clemens, 1975). It seems unlikely that adrenal activation could have mediated these effects, but as the males tested were not adrenalectomized, this possibility must remain open.

In sexually experienced CD-1 mice, EB (1 μg/day) maintains sexual behavior, including ejaculation, after castration, but the behavior declines when the hormone dose is reduced (0.5 μg/day). The adrenal apparently does not mediate EB's effects on mating behavior in this strain, as castrates continue to respond to EB therapy after adrenalectomy (Wallis and Luttge, 1975). Increasing doses of EB (0.5–100 μg/day) also initiate mounting behavior in sexually naive, castrated Swiss-Webster mice (Edwards and Burge, 1971).

EB (6 μg/day) rarely stimulates intromissions in sexually experienced male hamsters castrated in adulthood (Noble and Alsum, 1975; Tiefer, 1970). It can induce mounting behavior, whether males are castrated before or after puberty (Clemens, 1974; Noble and Alsum, 1975), but this stimulatory effect is not always observed (Tiefer, 1970).

Sexual behavior, particularly ejaculation, declines more quickly in castrated male gerbils implanted with silastic capsules of E_2 (2 mm, 1.47 mm id, 1.96 mm od), but E_2 retards the decline relative to controls (Yahr and Stephens, unpublished, 1978). Preliminary observations suggest, however, that larger capsules (5 mm) of E_2 can occasionally reinstate sexual behavior in this species.

Daily injections of EB (0.33 mg) reinstated both mounting and ejaculation in 20% of sexually experienced male rabbits after castration (Aǧmo and Södersten, 1975). Although control animals were not tested in this study, the authors pointed out that this level of sexual activity was not much higher than that usually seen without hormone treatment (20% mounting, 10% ejaculating). Larger doses of EB (1 mg/day) increased these percentages to 75% and 64%, respectively, but the same dose of testosterone benzoate was more potent. Similarly, silastic implants (5 mm, 3.35 mm id, 4.65 mm od) of E_1 do not maintain mounting behavior or ejaculation in castrated male rabbits as effectively as T implants (20 mm) do, but E_1 may retard the decline in sexual behavior when compared to that of controls (Foote *et al.*, 1977). Beyer *et al.* (1975) stated that EB (5 μg/day) does not affect sexual behavior in prepuberally castrated male rabbits, but no data were provided.

EB (2 μg/100 g/day) is no more effective than oil in initiating male sexual behavior in prepuberally castrated guinea pigs (Alsum and Goy, 1974). Similarly, Antliff and Young (1956) found that large doses of "α-estradiol benzoate" (100 I.U./100 gm/day) neither maintained nor reinstated male sexual behavior in this species, but the same doses of E_1 were partially effective. As Alsum and Goy (1974) suggested, it is unlikely that aromatization mediates T's behavioral effects in guinea pigs. As discussed earlier, male guinea pigs do not aromatize T within the brain; moreover, they are very sensitive behaviorally to DHT. Reduction of T to DHT could adequately account for T's behavioral actions in this species.

In castrated male rhesus monkeys, EB (5 μg/day) does not activate male sexual behavior (Phoenix, unpublished observations, cited in Phoenix, 1976). EB implants (100 mg) do activate sexual behavior in castrated red deer stags (Fletcher and Short, 1974).

The data on neural aromatization of androgens by males and on estrogen's ability to stimulate male sexual behavior suggest that the aromatization hypothesis is more plausible for some species than for others. But before we evaluate the aromatization hypothesis further, it will be helpful to (re)consider the behavioral effects of classes or categories of related androgens that have been studied behaviorally.

The most commonly used classification, at least in behavioral endocrinology, is the distinction between "aromatizable" and "nonaromatizable" androgens. Androgens are classified in this way based on the ability of enzymes from the female reproductive tract to convert them to estrogens. The usual assay system utilizes human placental microsomes. In this system, aromatization proceeds by a series of three hydroxylations (A → 19-OHA → 19-oxoandrostenedione → 2β-hydroxy, 19-oxoandrostenedione) apparently catalyzed by one enzyme. Then, the product collapses nonenzymatically into an estrogen (Goto and Fishman, 1977; Kelly *et al.*, 1977). Surprisingly, few research papers on the behavioral effects of aromatizable versus nonaromatizable androgens provide any references for this classification.

To our knowledge, no studies have examined the ability or the inability of neural enzymes to aromatize androgens that the human placenta cannot. Thus, most conclusions about the behavioral effects of nonaromatizable androgens are based on the tacit assumption that the aromatase system in the brain and that in the placenta are the same. Some authors have faced this problem more directly and cautiously, though they have reached the same conclusions. Selmanoff *et al.* (1977a) stated that, "While the low levels of activity in rat brain preclude extensive characterization of the aromatase enzyme, there is no reason to believe that it differs markedly from the human enzyme" (p. 842). Similarly, Goto and Fishman (1977) stated:

> Available evidence suggests that ovarian estrogen biosynthesis proceeds by the same pathway, and it is likely that peripheral aromatizations including those in brain . . . share the same mechanism. The biosynthetic scheme that we describe need not be the only aromatization route. It is, however, the only one for which experimental support is provided, and there is at present no evidence that alternative aromatization sequences do indeed exist. (p. 81)

Hopefully, the validity of such assumptions and conclusions will be tested soon.

Placental aromatase activity differs across species. In a comparative analysis, Ainsworth and Ryan (1966) found that placental preparations from ewes, cows, sows, and mares converted A and/or DHEA to estrogens, whereas placental preparations from rabbits and guinea pigs did not. Grossman and Bloch (1973) replicated the observation that placental tissues from guinea pigs do not aromatize androgens. More recently, however, an aromatase system has been detected in the microsomes of guinea pig placentae (Adessi *et al.*, 1978). As noted earlier, hypothalamic tissue from male guinea pigs apparently does not possess this enzyme system (Sholl *et al.*, 1975). Thus, the existence of an aromatization system in the placenta does not imply that the system also exists in the brain. Conversely, the data on rabbits suggest that, in some species, neural tissues can aromatize androgens even when the placenta cannot. As discussed earlier, aromatization has been detected in hypothalamic tissue of male rabbits, but not in placental preparations from female rabbits (Ainsworth and Ryan, 1966; Reddy *et al.*, 1973).

The ability to detect aromatase activity also varies with the method of tissue preparation. For example, Ainsworth and Ryan detected aromatase activity in microsomal preparations from bovine placentae, but not in cell homogenate preparations of the same tissue. They attributed this difference to a possible inhibitory system that is removed when the microsomal fraction is prepared. The possibility that different processes of aromatization occur is also suggested by studies comparing the aromatization of C-19 steroids with the aromatization of 19-nortestosterone (19-norT) (Thompson and Siiteri, 1973).

Within the classification system of aromatizable versus nonaromatizable androgens, one can distinguish ring A or B unsaturated androgens from saturated androgens. (For this purpose, we will treat 19-nor compounds as androgens). The saturated androgens (e.g., androsterone, androstanedione, DHT, 19-OHDHT, and the androstanediols) are all presumably nonaromatizable in placental tissue. This fact has actually been demonstrated for DHT (Ryan, 1960) and androstanedione (Gual *et al.*, 1962).

The unsaturated androgens include T and A and their respective intermediates, 19-OHA and 19-OHT, in the aromatization pathway (Goto and Fishman, 1977; Siiteri and Thompson, 1975; Thompson and Siiteri, 1973). They also include various naturally occurring and synthetic hormones—for example, 5-androstenediol, DHEA, 19-norT, 6α-fluorotestosterone (6α-fluoroT), fluoxymesterone, and 11β-OHA—that have been tested behaviorally. These vary in aromatizability (Gual *et al.*, 1962; Ryan, 1959). DHEA is aromatized 66%–100% as efficiently as T or A, whereas 19-norT, like 19-norA, is aromatized 16%–20% as efficiently as T or A by human placental microsomes. 11β-OHA and 6α-fluoroT are not aromatized at all in the same assay system. The aromatizability of fluoxymesterone has not been determined directly. Based on the failure of 11β-OHA to form estrogen, Gual *et al.* concluded that axial substitutents at carbon 11 interfere with aromatization. It is apparently on the weight of this conclusion that other researchers (e.g., Beyer *et al.*, 1973); Aǧmo, 1977) concluded that fluoxymesterone cannot be aromatized. Nonetheless, Ryan (1960) found that 11β-OHT produces small amounts of estrogen. To our knowledge, the aromatizability of 5-androstenediol has not been tested directly.

As suggested earlier, the prohormone concept can be tested by comparing the effects of hormones that differ in their ability to form the presumably active metabolite. The aromatization hypothesis has accordingly been tested by studying the effects of aromatizable versus nonaromatizable androgens on male sexual behavior. The results of most of these studies (e.g., those comparing T and A with DHT) have already been discussed.

In addition, the presumably nonaromatizable androgen fluoxymesterone has been tested for its effects on mating behavior in rats and rabbits. Fluoxymesterone (500 μg/day) does not reinstate or maintain (400 μg/day) sexual behavior in castrated, sexually experienced male rats (Beach and Westbrook, 1968a; Johnson and Tiefer, 1974), but when castrates receiving TP are switched to fluoxymesterone (doses not specified), the decline in sexual activity is attenuated (Clemens, 1974). The latter effect is apparently due to fluoxymesterone's androgenic action on the penis, as sexual activity declines at the same rate in fluoxymesterone- and oil-treated castrates if the dorsal nerve to the penis is cut. Large doses of fluoxymesterone (4–12 mg/day) can induce mounting behavior in rabbits (Aǧmo, 1977).

The behavioral effects of 19-hydroxylated androgens have been studied in rats and rhesus monkeys. As noted above, 19-OHT is an intermediate product in the aromatization of T to E_2, whereas 19-OHDHT presumably cannot be converted to estrogen in the brain. Indeed, the diacetate ester of 19-OHDHT (19-OHDHTA) does not increase uterine weight in immature female mice, whereas 19-OHT is more potent than T in this assay system (Johnston *et al.*, 1975). At doses of 1 μg/kg/day, neither 19-OHT nor 19-OHDHTA stimulates mating behavior in sexually experienced, castrated male rhesus monkeys (Phoenix, 1976, 1978).

The effects of 19-hydroxylated androgens on male rats are more confusing. In 1974, Parrott reported that sexually-experienced male rats maintained on 19-hydroxytestosterone propionate (19-OHTP; 150 μg/day, 5 days/week) after castration have PEIs equal to those of males maintained on TP (75 μg/day, 5 days/week), although 19-OHTP (1.8 mg/2 days) does not maintain peripheral reproductive tissues. In both behavioral experiments of this report, ejaculatory behavior decreased in 19-OHTP-treated castrates, but not in castrates given TP. In one experiment, the mount and intromission frequencies increased with 19-OHTP, but in the other, they did not. Considering the data as a whole, Parrott (1974) concluded that "it would seem safest at present to assume that 19-hydroxytestosterone propionate has little effect on copulatory performance" (p. 113). Later, when comparing the abilities of T and 19-OHT (1 mg/day, 5 days/week for 4 weeks) to maintain mating behavior in sexually experienced castrates, Parrott (1976) found that the two androgens were equally effective in nearly all aspects of sexual behavior. Parrott did not speculate on the basis for these discrepant findings. In fact, he concluded (Parrott, 1976) that "these results are in agreement with previous findings (Parrott, 1974)" (p. 213).

The behavioral effects of 19-OHT in rats were also studied by Johnston *et al.* (1975). They compared 19-OHT, DHT, and TP (500 μg/kg/day) for their abilities to reinstate sexual behavior in castrated (C) or castrated-adrenalectomized (C/A) males. They reported that "Only testosterone and 19-OHT, but not 5α-DHT, established near-normal copulatory patterns by day 15 in C- and C/A-prepared rats" (p. 233). Actually, the "copulatory profiles" of their castrated rats receiving 19-OHT and DHT look fairly similar; however, the castrated-adrenalectomized males were clearly more responsive to 19-OHT and TP than to DHT. These authors also studied the effects of various doses of 19-OHT (500, 300, 100, 30, and 10 μg/kg/day subcutaneously and 5 and 50 mg/kg/day orally) on the sexual behavior of castrated-adrenalectomized males. The copulatory profile of males given 500 μg 19-OHT/kg appeared similar to that of males given the comparison dose of TP (500 μ/kg). It appears that some males receiving 19-OHT by injection were ejaculating after 2 weeks, regardless of the dose administered, although Johnston *et al.* stated that "treatment with 100 μg/kg or greater elicited normal copulatory patterns in C/A rats. At lower doses, the number of mounts and intromission increased, but not ejaculations" (p. 234). Given orally, 50 but not 5 mg 19-OHT/kg reinstated mating behavior. Reasons that it is difficult to evaluate the behavioral effects of 19-OHT on the basis of these data are that the authors did not compare similar forms of androgens (e.g., 19-OHT and T or 19-OHTP and TP), did not study the dose–response relations for both androgens being compared, did not explain the copulatory profiles used to present the data, and did not provide statistical support for their conclusions.

In sexually experienced male rats, the enol propionate ester of 19-OHA (15 μg/day, 50 days/week) seems to be less effective than the ester of A for maintaining ejaculations after castration, but males that do ejaculate have essentially normal PEIs (Parrott, 1975a). 19-OHDHT (1 mg/day, 5 days/week) does not maintain mounting or intromissions (Parrott, 1976). In contrast, 19-OHDHTA (500 μg/kg/day) reinstates ejaculations in castrated, adrenalectomized males; its 5β-isomer does not (Johnston *et al.*, 1975). It is not known whether castrated-adrenalectomized male rats are more responsive than castrated males to 19-OHDHT (A), as they seem to be to 19-OHT.

As we discussed earlier, there are problems in using diverse androgens to assess a prohormone concept such as the aromatization hypothesis because the same or correlated structural changes that affect aromatizability may directly and independently alter the behavioral activity of the hormone. Yahr and Gerling (1978) addressed this problem by studying the ability of low doses (12.5 μg/100 g/day) of 19-nortestosterone propionate (19-norTP) and 6α-fluorotestosterone propionate (6α-fluoroTP) to maintain sexual behavior in male rats after castration. Both these steroids have an A-ring structure similar to that of TP, and both were more potent than the aromatization hypothesis would predict. Although 19-norT is aromatized only 20% as efficiently as T by placental microsomes and 6α-fluoroT is not aromatized at all, 19-norTP maintained ejaculation in 50% of castrated males and 6α-fluoroTP was as effective as TP. 6α-FluoroTP maintains nearly all parameters of sexual behavior as well as TP even at half this dose (6.25 μg/100 g/day; Nordeen and Yahr, 1981). 6α-FluoroTP and TP (125–50 μg/day) were also equally potent in reinstating ejaculatory behavior in castrates. These data suggest that structural characteristics other than those affecting aromatizability determine an androgen's ability to stimulate sexual behavior in male rats.

Parrott's (1975b) report that larger doses of 19-norTP (150 μg/day) maintain ejaculation frequency and PEI length as effectively as TP (75 μg/day) is consistent with this suggestion, although Parrott interpreted his data as supporting the hypothesis that 19-norTP is aromatized in the brain. Interestingly, Parrott also found that DHTP (75 μg/day) antagonizes the behavioral effects of 19-norTP when the two are given together. If 19-norTP acts via aromatization, DHTP may curtail its effects by blocking aromatization, because DHT blocks conversion of A to E_1 in placental microsomes (Siiteri and Thompson, 1975; Schwarzel *et al.*, 1973).

In sexually experienced male gerbils, silastic implants (2 mm, 1.47 mm id, 1.96 mm od) of 19-norT are at least as effective as T implants in maintaining and reinstating ejaculatory behavior after castration (Yahr and Stephens, unpublished observations, 1978).

Of course, the behavioral effects of 6α-fluoroTP and 19-norT(P) may not provide a true test of the aromatization hypothesis because the brain and the placenta may metabolize androgens differently. For example, enzymes in the brain or elsewhere may cleave the fluoro group from 6α-fluoroTP, converting it to TP. Such cleavage apparently does not occur in human placental microsomes because no phenolic steroids (estrogens) are detectable when the microsomes are incubated with 6α-fluoroT (Gual *et al.*, 1962). In fact, 6α-fluoroT may inactivate the aromatase enzyme. The related halogenated steroid, 6α-bromoandrostenedione, inactivates the placental aromatase, apparently by binding to the active enzyme site (Bellino *et al.*, 1976). It seems unlikely, however, that the fluoro group would be cleaved because a fluoro group would be the most difficult halogen to remove from a carbon bond. Thus, the fluoro group may even protect 6α-fluoroT from liver metabolism.

Another possibility is that 6α-fluoroT is aromatized in the brain, perhaps to 6α-fluoroestradiol. Pretreatment with 6α-fluoroTP does reduce cell nuclear accumulation of ^{3}H-E$_2$ in the hypothalamus-preoptic area (Nordeen and Yahr, 1981), but does not affect ^{3}H-E$_2$ uptake in the pituitary (Nordeen and Yahr, unpublished observations, 1982). Similarly, 19-norT may be aromatized more efficiently in the brain than in the placenta. The latter seems unlikely, though, because the related steroid, 17α-ethynyl-19-nortestosterone, is aromatized to ethynyl estradiol at a low rate (0.008%) by hypothalamic and limbic tissues from human fetuses (Naftolin *et al.*, 1975). Nonetheless, the aromatization of presumably nonaromatizable androgens cannot be ruled out because the bacterial flora of the human gut may aromatize DHT (Hill *et al.*, 1971; see also the discussion of R. J. Sherin's question, pp. 315–316, following Naftolin *et al.*, 1975).

Research on the neural metabolism of nonaromatizable androgens could provide convincing support for the aromatization hypothesis or could extend its generality. For example, strain differences in the responsiveness of mice to DHT may reflect strain differences in aromatization. Even mating in male rats treated with DHT could result from the induction of an enzyme for aromatizing DHT. On the other hand, such research could generate data that conflict with the aromatization hypothesis. For example, rats may not aromatize 6α-fluoroT. Alternatively, they may aromatize fluoxymesterone. If so, their relative insensitivity to this androgen requires a different explanation. Yet another possibility is that some androgens (e.g., 6α-fluoroT) can interact directly with estrogen receptors in the brain without being aromatized. This interaction has already been demonstrated for 3β-Adiol (Vreeburg *et al.*, 1975).

Tests of the aromatization hypothesis in rats with the use of antiestrogens and aromatase inhibitors have provided conflicting data. In an initial study, four male rats treated with TP (300 μg/day) showed no decrease in sexual activity when later treated concurrently with TP plus the antiestrogen CI-628 (1 mg/day; Whalen *et al.*, 1972). Later, however, Luttge (1975) noted that daily pretreatment with CI-628 (4 mg/day) reduced ejaculation frequency in sexually naive castrates treated with T (800 μg/day). He went on to show that pre- plus concurrent treatment with CI-628 (2.5 mg twice daily) reduces mount, intromission, and ejaculatory behaviors in castrated males receiving T (500 μg/day). However, the same CI-628 treatment (2.5 mg twice daily) does not retard reinstatement of mating in sexually experienced male rats given TP (125 μg/day), nor does it accentuate the decline in behavior when the TP (50 μg/day) dose is reduced (Yahr and Gerling, 1978). The difference in the effects of CI-628 across studies may reflect differences in the form of the hormone used and/or in the sexual experience of the males studied. They may also reflect differences in adrenal secretion, as CI-628 can cause weight loss and adrenal hypertrophy (Luttge, 1975; Yahr and Gerling, 1978). If the hypertrophied adrenals secrete androgens, this secretion could account for the observation that CI-628 occasionally potentiates T-induced changes in peripheral structures (Luttge, 1975). Another problem in interpreting these data is that it is not known whether CI-628 inhibits cells nuclear uptake of estrogen in the brains of adult male rats treated with either ^{3}H-E$_2$ or ^{3}H-T. However, pretreatment with CI-628 does inhibit the cell nuclear accumulation of radioactivity in the brains of adult females given ^{3}H-E$_2$ or ^{3}H-EB (Chazal *et al.*, 1975; Landau, 1977; Luttge *et al.*, 1976; Roy and Wade, 1977) and in the brains of neonatal female rats given ^{3}H-T (Lieberburg *et al.*, 1977b).

Three other antiestrogens, MER-25, cis-clomiphene, and tamoxifen, have been studied in similar paradigms. In each case, sexually naive, castrated male rats were given T(P) daily to induce sexual activity but were pretreated each day with antiestrogen. In sufficiently large doses, tamoxifen (400 μg/day, but not 4–130 μg/day, orally) inhibited the onset of sexual behavior in males given 1 mg T daily, but MER-25 (2–28 mg/day) and cis-clomiphene (250–1000 μg/day) did not (Beyer *et al.*, 1976). Similarly, in other laboratories, MER-25 (10–20 mg/day) pretreatment did not block the behavioral effects of T (800 μg/day; Luttge, 1975) or TP (200 μg; Baum and Vreeburg, 1976). The failure of MER-25 and cis-clomiphene to block T(P)'s behavioral effects may result from these drugs' interacting directly with cytoplasmic receptors for E$_2$ in the hypothalamus of male rats (Vreeburg *et al.*, 1975), thereby stimulating the behavior themselves. Alternatively, these antiestrogens may simply be less behaviorally potent than CI-628, as they are in tissues of the female reproductive tract.

To date, five aromatase inhibitors—4-hydroxyandrostenedione (4-OHA); androstanedione; 1,4,6-androstatriene-3,17-dione (ATD); metapirone; and aminoglutethimide—have been tested for their ability to disrupt T(P)-induced sexual behavior in male rats. Each of these drugs inhibits conversion of A to E$_1$ in placental microsomes, and 4-OHA inhibits ovarian E$_2$ synthesis in female rats *in vivo* (Brodie *et al.*, 1977; Giles and Griffiths, 1964; Schwarzel *et al.*, 1973; Thompson and Siiteri, 1973). In two reports on the behavioral effects of these drugs, mating behavior was induced in sexually inexperienced, castrated male rats by a single injection of TP (6 mg). Aromatase inhibitors were given 1–3 hr before the TP injection and again every 12 hr for 8 days in an attempt to block the behavioral effects of the androgen (Beyer *et al.*, 1976; Morali *et al.*, 1977). In this paradigm, aminoglutethimide (5–15 mg/12 hr), 4-OHA (4.5 mg/12 hr), and ATD (9 mg/12 hr) severely inhibit or eliminate TP's effects on mounts, intromissions, and ejaculations, but metapirone (10–22 mg/12 hr) and androstanedione (2.5–5 mg/12 hr) do not. ATD (60 μg/day) also blocks reinstatement of mounting behavior in sexually experienced castrates when it is infused directly into the preoptic area 20 min before the infusion of T (10 μg/day; Christensen and Clemens, 1975). Infusing metapirone (300 μg twice daily) into the preoptic area also inhibits mounting behavior induced by systemically administered T (150 μg/day; Clemens, 1974).

These data seem to support the aromatization hypothesis in male rats. Unfortunately, only one aromatase inhibitor has been tested for its ability to decrease the amount of estrogen in the preoptic area of adult male rats given T(P). ATD decreases the neural aromatization of T, as assessed by a decreased translocation of estrogen receptors into cell nuclei (Krey *et al.*, 1979). The effects of other aromatase inhibitors on neural metabolism are unknown. Even in the case of ATD, it is not clear whether the dose–response curves for inhibition of aromatization and inhibition of T(P)-induced male sexual behavior correspond. Without more information on the neural actions of the aromatase inhibitors, the behavioral data are difficult to interpret. For example, if systemically administered androstanedione and metapirone block aromatization of T in the brain, then their failure to block TP-induced male sexual behavior is unexplained. Alternatively, behaviorally effective aromatase inhibitors may not block neural aromatization in the doses used. Studies of how aromatase inhibitors affect androgen metabolism in the brain could provide stronger support for the aromatization hypothesis, if inhibition of male sexual behavior and inhibition of neural aromatization are highly correlated. Such research has been done in neonates. Pretreatment with ATD blocks the defeminizing effects of the testes or of exogenous T on the behavioral development of male and female rats, respectively (McEwen *et al.*, 1977a). It also reduces cell nuclear accumulation of ^{3}H-E_2 in the brains of neonatal female rats injected with ^{3}H-T (Lieberburg *et al.*, 1977b).

The possibility that aromatase inhibitors block T(P)'s behavioral effects via nonspecific actions has been tested by studying how these drugs affect male sexual behavior induced by estrogens. Infusions of ATD (600 μg/day) into the brain do not inhibit mounting behavior induced by infusions of E_2 (5 μg/day; Christensen and Clemens, 1975). Similarly, castrated males given E_2 (50 μg/day) systemically and metapirone (300 μg/day) centrally mount as often as castrates receiving T (150 μg/day) systemically and sucrose, the metapirone vehicle, centrally (Clemens, 1974). Morali *et al.* (1977) also showed that inhibition of TP-induced sexual behavior by ATD, 4-OHA, or aminoglutethimide can be prevented or reduced by giving EB (1–3 μg) concurrently with the aromatase inhibitors.

HYPOTHESIS: T MEDIATES MALE SEXUAL BEHAVIOR VIA METABOLISM TO ESTROGEN PLUS DHT

Baum and Vreeburg (1973) and Larsson *et al.* (1973a) independently demonstrated that a high percentage of sexually inexperienced, castrated male rats ejaculate if treated with EB plus DHT(P). When castrates pretreated with DHT (1 mg/day) for 10 days are given EB (0.5–50 μg/day) plus DHT, 70%–90% of the males ejaculate within 4 weeks (Larsson *et al.*, 1973a). This equals the performance of males given T (1 mg/day). Nearly 50% of males given 0.05 μg EB plus DHT also ejaculate. Similarly, Baum and Vreeburg (1973) found that all males given EB (2 μg/day) plus DHTP (200 μg/day) ejaculated at least once in 4 weeks of testing, and the behavior of these males (intromission and ejaculation latencies and PEI) resembled that of males given TP (200 μg/day). Because only 37% of males given TP (4 μg/day) plus DHTP ejaculate, the behavior of males given EB plus DHTP is not due solely to increased levels of total steroids. These data suggest that two metabolites of T, E_2 and DHT, synergize to mediate T's effects on sexual behavior in male rats.

Pre- and concurrent exposure to DHT (500 μg/day) also enhances sexual activity in sexually naive, castrated males given E_1 (1–5 μg/day) or E_3 (1–5 μg/day), although DHT plus E_2 (1–5 μg/day) is the most effective combination (Larsson *et al.*, 1976). Relatively little sexual behavior occurs when DHT is given with larger doses of E_3 (25 μg/day). Davidson and Trupin (1975) also noted high levels of sexual activity in prepuberally castrated male rats given EB (0.5 μg/day) injections and implanted with silastic capsules (30 mm, 1.6 mm id) of DHT (pre- and concurrent treatment). In contrast, Feder *et al.* (1974a) reported that relatively few (14%–57%) sexually experienced castrates ejaculated when given EB (10–167 μg/day) plus pre- and concurrent treatment with DHTP (125 μg/day). They did not observe any ejaculations when they used a smaller dose of EB (0.5 μg/day). Because none of the males in this study received EB alone, it is not clear that the hormones synergized at any dose. The total amount of steroids administered could account for the ejaculation frequencies in this study, as 60% of the males given TP (150 μg/day) plus DHTP instead of EB plus DHTP ejaculated.

Alternatively, DHT(P) may synergize most effectively with low doses of estrogen (e.g., < 10 μg EB/day). By this hypothesis, the sexual behavior of males given DHTP plus TP reflects the actions of DHTP plus low levels of E_2 formed by the aromatization of T. It is possible, however, that very little aromatization occurs when T(P) and DHT(P) are administered together. In the human placenta, DHT is a potent aromatase inhibitor (Siiteri and Thompson, 1975; Schwarzel *et al.*, 1973). Whether DHT inhibits the aromatization of T in the brain is unknown. Commenting on this problem, Selmanoff *et al.* (1977a) stated:

> Since ring-A reduced androgens have been shown to inhibit aromatization, areas active in 5-α-reduction might be expected to show little aromatase activity. However, the MPN-AHN [medial-preoptic–anterior-hypothalamic nuclei] exhibited substantial levels of both activities. This may suggest heterogeneous neuronal subpopulations in the MPN-AHN, involved, perhaps specifically, in aromatizaiton *or* 5α-reduction. (p. 845; comment in brackets added to quote)

One hypothesis about the synergistic activity of estrogens and DHT is that estrogens activate central neural processes underlying sexual behavior, whereas DHT influences penile sensitivity. The observation that EB (50 μg/day) and fluoxymesterone (400 μg/day) synergistically maintain ejaculations after castration in sexually experienced males (Johnson and Tiefer, 1974) is consistent with this suggestion. Yet, even after transection of the pudendal nerve, sexually experienced castrates display more mounting behavior when given EB (0.5–5 μg/day) plus DHTP (200 μg/day) than when given EB alone (Lodder and Baum, 1977). Because pudendal nerve transection prevents sensory input from the genitalia, these data suggest that DHT interacts with estrogen in the brain to stimulate sexual behavior. Unfortunately, a control group given DHTP alone was not included in this study, so the extent of the hormonal interaction could not be fully assessed. However, males given TP (5 μg/day) plus DHTP mounted less often than males given EB (5 μg/day) plus DHTP.

Pre- and concurrent treatment with DHT in silastic capsules enhances the sexual activity of prepuberally castrated male rats given EB implants in the anterior-hypothalamus–preoptic area (Davidson and Trupin, 1975). Similarly, Parrott and McDonald (1975) found that castrated males with hypothalamic implants of 19-OHTP, an estrogen precursor, showed a distinct increase in sexual behavior when they received supplemental injections of DHT subcutaneously. The ability of DHTP (200 μg/day) to initiate ejaculation in sexually naive castrates can also be enhanced by administering either MER-25 (10 mg/day) or 3β-Adiol (1 mg/day), both of which can bind to estradiol receptors in the brain (Baum and Vreeburg, 1976; Vreeburg et al., 1975). Each of these observations is consistent with either hypothesis, peripheral or central, about where DHT acts to synergize with estrogen. However, other observations by Davidson and Trupin (1975) are not easily compatible with either hypothesis. They found that central EB implants continued to stimulate high levels of sexual behavior for a month after the silastic capsules of DHT were removed. Similarly, subcutaneous injections of EB (0.5 μg/day) continued to stimulate ejaculations for a month after concurrent DHTP treatment had stopped.

Cyproterone acetate blocks the synergistic effects of EB plus DHT. Luttge et al. (1975) found that concurrent treatment with cyproterone acetate (100 mg/day) prevented the development of ejaculatory behavior and reduced mount and intromission frequency in sexually inexperienced, castrated male rats given EB (1 μg/day) plus DHT (1 mg/day). The drug did not affect the number of penile spines supported by the steroids. However, cyproterone acetate is an antiestrogen, as well as an antiandrogen, and this fact may account for its inhibitory effects (Luttge et al., 1975). In contrast, the aromatase inhibitor aminoglutethimide (15 mg/day) does not inhibit the ability of EB (2.5 μg/day) plus DHT (1 mg/day) to stimulate ejaculations in sexually inexperienced castrates (Morali et al., 1977). This finding indicates again that the ability of this drug to inhibit TP-induced sexual behavior is not due to nonspecific effects on the brain.

The hypothesis that T(P) activates sexual behavior via metabolism to E_2 plus DHT cannot easily account for the behavioral effects of 6α-fluoroTP and 19-norTP. These steroids are only half as potent as TP in promoting growth of the seminal vesicles, a fact suggesting either that they are not reduced very efficiently or that their 5α-reduced metabolites are not very potent (Yahr and Gerling, 1978). Thus, their ability to maintain and/or reinstate sexual behavior in male rats seems even less compatible with the aromatization-reduction hypothesis than it is with an hypothesis based on aromatization alone.

Another problem in the aromatization-reduction hypothesis was noted by Gorzalka (1977). He found that EB plus DHT does not stimulate copulatory behavior in sexually experienced male rats if they are both castrated and adrenalectomized. Moreover, he reported that EB plus DHT synergistically causes hypertrophy of the adrenal cortex. This finding suggests that the synergistic behavioral effects of these hormones may be mediated by adrenal steroids. Adrenal activation may not account, however, for the behavioral effects of 6α-fluoroTP, as it does not affect adrenal size (Yahr and Gerling, 1978). Clearly, more research is needed before either the aromatization or the aromatization-reduction hypothesis of androgen control of male sexual behavior can be accepted for rats.

The effects of estrogens plus DHT on sexual behavior in male mammals have also been tested in mice and rabbits. In sexually experienced male mice, EB (1 μg/day) plus DHT (200 μg/day) maintains sexual behavior after castration and continues to do so in most males even when the hormones doses are reduced (0.5 μg EB + 100 μg DHT/day; Wallis and Luttge, 1975). Yet again, the adrenals may be involved. In this study, the males that were still displaying intromissions were adrenalectomized and retested. After adrenalectomy, sexual activity of males receiving EB plus DHT dropped below the levels shown by adrenalectomized males that were gonadally intact or that were receiving EB. However, the absence of sham-operated control groups makes it difficult to interpret these data.

In sexually inexperienced male rabbits, therapy with EB (5 μg/day) plus DHT (2.5 mg/day) stimulates intromissions (in rabbits, ejaculation usually occurs on the first intromission) in 85%–100% of males, whether they are castrated pre- or post-puberally (Beyer *et al.*, 1975). The synergistic effects are more pronounced in prepuberally castrated males, as 62% of postpuberally castrated males ejaculate when given this dose of DHT alone. In sexually experienced, castrated male rabbits, treatment with EB (0.33 mg/day) plus dihydrotestosterone benzoate (1 mg/day) is no more effective than treatment with dihydrotestosterone benzoate alone (Ağmo and Södersten, 1975). Similarly, DHT and E_1 have very small, if any, synergistic effects on the onset or maintenance of sexual behavior in sexually inexperienced male rabbits when both hormones are administered in silastic capsules (2 cm DHT, 0.5 cm E_1, 3.35 mm id, 4.65 mm od) at castration (Foote *et al.*, 1977). It is more likely that their effects are simply additive.

In boars, E_1, E_2, and diethylstilbestrol all seem to enhance the ability of T to induce male sexual behavior (Joshi and Raeside, 1973).

Female Sexual Behavior

As with male behavior, one can ask whether the hormones secreted by the ovaries stimulate female sexual behavior directly or via metabolic conversion to an active form. We begin this analysis by examining the hormones secreted by the ovaries.

Secretion Products of the Ovaries: Estrogens

It is generally accepted that E_2 and E_1 are the principal estrogens secreted by the mammalian ovary; however, the evidence for this conclusion is not clear. Most early contemporary studies assayed total estrogens (e.g., Hori *et al.*, 1968; Yoshinaga *et al.*, 1969) and did not distinguish between E_2 and E_1, even though the source (ovarian versus adrenal), the levels, and the pattern of secretion of these estrogens differ (Atkinson *et al.*, 1975; Baranczuk and Greenwald, 1973; Chamley *et al.*, 1973; Labhsetwar *et al.*, 1973; Shaikh, 1971; Shaikh and Harper, 1972;

Shaikh and Shaikh, 1975). Moreover, studies measuring E_2 (Brown-Grant *et al.*, 1970; Butcher *et al.*, 1974; Kalra and Kalra, 1974; Preslock *et al.*, 1973) do not report the proportion of total estrogens accounted for by this steroid. E_3 is also present in the plasma of some primate species, such as chimpanzees (Reyes *et al.*, 1975) and humans (Tulchensky and Korenman, 1970), though it is not considered an ovarian secretion in most species.

The assumption that E_2 and E_1 are the major (or sole) functional estrogens must be carefully scrutinized. Catechol estrogens, such as 2-hydroxyestradiol (2-OHE_2) and 2-hydroxyestrone (2-OHE_1), are present in plasma (Paul and Axelrod, 1977) and may have potent physiological effects (Parvizi and Ellendorff, 1975). Whether these estrogens are secreted by the ovaries or the adrenals or are metabolites of E_1 and E_2 is not known. In either case, they may activate estrogen-sensitive systems.

Studies of estrogenic steroids in neural target tissues suggest that both E_2 and E_1 could be behaviorally active. E_1 can be detected in the brains of female rats administered ^3H-E_2, although more than 90% of the radioactivity chromatographs as E_2 (Kato and Villee, 1967, Luttge and Whalen, 1970b). Only a small fraction of the radioactivity is found in the polar zones, which would include E_3. E_1 levels are somewhat higher in the posterior than in the anterior hypothalamus. Cell nuclei from brain samples containing the septum, the preoptic region, the hypothalamus, the olfactory bulbs, and the amygdala contain predominantly E_2 (86%) after ^3H-E_2 injection, but E_1 (4%) and E_3 (2%) are also detectable (Zigmond and McEwen, 1970). Hippocampal and cortical nuclei contain less E_2 (75%), but more E_3 (8%) and E_1 (4%), than nuclei from basal brain samples. The retention of E_2 metabolites by the brain is the same in males and females (Luttge and Whalen, 1970b).

This pattern of estrogen metabolism seems to hold across species. For example, in female guinea pigs, E_2 is the primary steroid (76%) found in the hypothalamus after treatment with ^3H-E_2 or ^3H-EB (Eaton *et al.*, 1975; Wade and Feder, 1972c). E_1 accounts for most of the remaining radioactivity. In female mice, 90% of the radioactivity in the hypothalamus after an injection of ^3H-E_2 chromatographs as E_2 (Luttge *et al.*, 1976). Feder *et al.* (1974b) specifically compared the metabolism of E_2 in female rats, guinea pigs, and hamsters. E_2 is the predominant steroid in hypothalamic tissue in each of these species 30 min to 1 hr after injection of ^3H-E_2. The species do differ, however. The hamster hypothalamus accumulates the least and the rat hypothalamus the most E_2 24 hr after injection. In each of the three species, most of the remaining radioactivity in the hypothalamus chromatographs as E_1. The cortex and midbrain tissues of guinea pigs accumulate less unmetabolized E_2 than do the same tissues of rats and hamsters. Nearly 50% of the radioactivity in these tissues in guinea pigs is E_1. In all three species, E_2 and E_1 account for 76%–88% of the total radioactivity in the three brain tissues 8 hr after an E_2 injection. The remaining radioactivity was not identified but could, of course, represent steroids with important biological activities. E_3, for example, is reported to be a significant metabolite of E_2 in rat brain (Presl *et al.*, 1973).

These studies clearly indicate that brain tissues could accumulate E_1 and E_3 as well as E_2 as a result of an ovarian secretion of E_2. Similarly, E_1 secreted by the ovaries could be a source of E_2 and other steroids within the brain. For example, 2 hr after the administration of ^3H-E_1 to female rats, Luttge and Whalen (1972) found the following ratios of E_1 to E_2 in the brain: anterior hypothalamus, 57 to 29; posterior hypothalamus, 68 to 17; anterior mesencephalon, 62 to 12; and cortex 53 to 12. These values represent the ether-soluble radioactivity retained by these tissues. It should be noted, however, that although over 90% of the radioactivity was either soluble in the hypothalamus and the cortex, 27% of the mesencephalic radioactivity was not. The non-ether-soluble radioactivity may be conjugated estrogens, which may be active behaviorally (Beyer *et al.*, 1971).

Radiolabeled E_3 is apparently not metabolized to E_2 or E_1 but forms compounds that are not extractable from brain tissue by ethyl acetate. Using this solvent, Landau and Feder (1977) found that 93% of the radioactivity extracted from the hypothalamus, the cortex, the amygdala, and the septum after injection of E_3 chromatographs as E_3. The E_1 and E_2 zones contain less than 5% of the soluble radioactivity. However, the ethyl-acetate-soluble radioactivity rarely exceeds 1% of the total radioactivity, a finding suggesting that E_3 is readily conjugated.

To summarize, it appears that E_2 and E_1 are the major estrogenic secretions of the ovaries, that each produces significant amounts of the other in neural tissue, and that either may lead to neural accumulation of both free and complexed (conjugated?) E_3. As mentioned earlier, catechol estrogens are also found in blood and brain (Paul and Axelrod, 1977). Moreover, neural tissue can metabolize E_2 and E_1 to 2-OHE_2 and 2-OHE_1, respectively (Fishman *et al.*, 1976; Fishman and Norton, 1975). These catechol estrogens have a relatively high affinity for cytoplasmic estrogen receptors in the hypothalamus (Davies *et al.*, 1975). Because the amounts of catechol estrogens found in rat hypothalamus endogenously (15.5 ng/g) and in human hypothalamus after metabolic conversion *in vitro* 5%–6% of the substrate) are surprisingly high, these steroids may mediate estrogen action on behavior.

Little research has been done on the behavioral effects of estrogens other than E_2 and its ester, EB. Both of these agents induce receptive behavior in all the mammalian species studied, except humans (Money, 1961) and Asian musk shrews (Dryden and Anderson, 1977). Women and female musk shrews show no changes in the frequency of copulation after ovariectomy. In rats, the most frequently studied species, the threshold dose of EB for inducing sexual receptivity varies from 0.5 to 2 μg/kg (Powers and Valenstein, 1972). Females of other species are much less sensitive to E_2 or EB (Feder *et al.*, 1974b).

The first study to systematically compare the behavioral effects of various estrogens (Beyer *et al.*, 1971) reported that low doses (1 μg/day/10 days) of E_2, E_1, or estrone-3-sulfate (E_1-sulfate), but not E_3, induced sexual receptivity in female rats. At this dose, E_2 was significantly more effective than E_1. At a higher dose (4 μg/day), all four estrogens induced lordosis behavior. E_3 was less effective than E_2 or E_1, but the latter two were equipotent. E_1-sulfate was more active than free E_1 at the lower dose and was less active at the higher dose. In fact, the higher dose of E_1-sulfate was less potent than the lower dose.

In female guinea pigs, E_1, E_2, and E_3 can induce lordosis behavior (Feder and Silver, 1974). When the free alcohol form of these steroids is administered, the effective dose is approximately the same for all three estrogens; however, EB is appreciably more potent than estrone benzoate. Feder and Silver suggested that retention of the free and esterified estrogens may differ, and that this difference could account for their different potencies. This explanation is quite reasonable because both E_2 and E_1 are retained longer in neural tissues following administration of esterified estrogens than following administration of free estrogens (Eaton *et al.*, 1975; Landau, 1977; Luttge and Whalen, 1972). Feder and Silver also suggested that E_1, E_2, and E_3 are each behaviorally active in their own right. They noted that E_1 is readily converted to E_2, which could account for its effects. However, this argument cannot apply to E_3 because E_3 apparently is not metabolized to E_2 (Landau and Feder, 1977).

The available data suggest a qualitative and perhaps a quantitative relationship between neural accumulation and retention of naturally occurring estrogens and their behavioral effects. The one exception seems to be the catechol estrogens. As noted earlier, 2-OHE_2 and 2-OHE_1 are endogenous estrogens and may be 10 times more concentrated than their parent estrogens in hypothalamic tissue (Paul and Axelrod, 1977). However, 2-OHE_2 does not readily stimulate lordotic behavior in female rats, whether it is administered subcutaneously, intravenously, or intracerebrally (Luttge and Jasper, 1977). In female guinea pigs, 2-OHE_2 facilitates receptivity only when combined with E_2 (Marrone *et al.*, 1977).

Secretion Products of the Ovaries: Progestins

Little is known about the progestins secreted by the ovaries of mammals. P is the major circulating progestin in most species studied (van Tienhoven, 1968), although female rats, rabbits, and baboons have relatively high levels of $20\alpha\text{-OHP}$ (Eto *et al.*, 1962; Goncharov *et al.*, 1976; Hilliard *et al.*, 1961; Telegdy and Endroczi, 1963).

The levels of plasma P fluctuate systematically during the estrous or menstrual cycles of rats (Butcher *et al.*, 1974; Feder *et al.*, 1968b; Kalra and Kalra, 1974; Smith *et al.*, 1975); mice (Michael, 1976); guinea pigs (Feder *et al.*, 1968a); hamsters (Labhsetwar *et al.*, 1973; Shaikh, 1972); dogs (Smith and McDonald, 1974); sheep (Chamley *et al.*, 1973); pigs (Masuda *et al.*, 1967); Asian and African elephants (Plotka *et al.*, 1975); rhesus monkeys (Neill *et al.*, 1967); marmosets (Preslock *et al.*, 1973); crab-eating macaques (Stabenfeld and Hendricks, 1973); baboons (Goncharov *et al.*, 1976); chimpanzees (Reyes *et al.*, 1975); and humans (Abraham *et al.*, 1972). Cyclic fluctuations of progestins other than P and $20\alpha\text{-}$OHP have not been examined to any great extent.

A study of Ichikawa *et al.* (1974) is an exception. These investigators tracked fluctuations of P and five of its major metabolites in ovarian venous plasma of rats across the estrous cycle. They, as others, found a high level of P during late proestrus and a secondary peak during diestrus Day 1 of the 4-day cycle (Figure 2). DHP showed a proestrous, but not a diestrous, peak. $5\alpha\text{-Pregnan-}3\alpha\text{-ol-}20\text{-one}$ showed a pattern similar to that of DHP. The next metabolite in this series, $5\alpha\text{-pregnan-}3\alpha,20\alpha\text{-diol}$, showed a low and relatively broad peak that extended from late proes-

Progesterone (P)
(4-pregnane-3, 20-dione)

20α-Hydroxyprogesterone (20α-OHP)　　Dihydroprogesterone (DHP)　　　　5β-Dihydroprogesterone (5β-DHP)
(20α-Dihydroprogesterone)　　　　　　 (5α-Pregnane-3, 20-dione)　　　　　(5β-pregnane-3, 20-dione)
(20α-Hydroxy-4-pregnene-3,20-dione)

　　　　　　　　　　　　　　　　　　　3α-Hydroxy-5α-pregnan-20-one　　　3α-Hydroxy-5β-pregnan-20-one
20α-Hydroxy-5α-pregnan-3-one

　　　　　　　　　　　　　　　　　　　5α-Pregnane-3α, 20α-diol　　　　　　Pregnanediol
　　　　　　　　　　　　　　　　　　　　　　　　　　　　　　　　　　　　　(5β-Pregnane-3α, 20α-diol)

Fig. 2. Metabolic pathways of progesterone.

trus into diestrus Day 1. The most abundant progestin, 20α-OHP, showed multiple peaks throughout the cycle, reaching approximately 380 μg/100 ml during mid-proestrus, 480 μg/100 ml during late estrus, and over 540 μg/100 ml at middiestrus Day 2. The troughs between these peaks were relatively stable at approximately 280 μg/100 ml. The metabolite of 20α-OHP, 5α-pregnan-20α-ol-3-one, showed a broad, but modest, peak (20 μg/100 ml) from late proestrus to middiestrus Day 1. The important feature of these findings is that significant amounts of DHP, 5α-pregnan-3α-ol-20-one and 20α-OHP enter the circulation at the same time as the proestrous surge of P. This fact suggests that 5α- and 20α-reduced progestins could have direct effects on the central nervous system quite independent of any intracellular metabolism within the brain.

PROGESTIN METABOLISM

METABOLISM OF P TO DHP. A major metabolic pathway for P is initiated by a 5α-reductase enzyme that converts P to DHP. DHP production, like DHT production, may be an important metabolic event mediating physiological processes. For example, DHP is 10–20 times as potent as P in stimulating the uterine implantation of blastocysts (M. Sanyal, as cited by Villee, 1971). DHP is also more effective than P in inhibiting the release of luteinizing hormone (LH) in response to luteinizing-hormone-releasing hormone (LRH) *in vitro* (Schally *et al.*, 1973). Thus, the 5α-reduction of P to DHP may be integral to P's action.

Brain tissues of female rats and guinea pigs selectively accumulate radioactivity following the administration of radiolabeled P (Laumas, 1967; Laumas and Farooq, 1966; Seiki *et al.*, 1968, 1969; Wade and Feder, 1972b,c). The radioactivity is localized more in mesencephalic than in diencephalic tissue (Luttge *et al.*, 1974; Whalen and Luttge, 1971a; Wade and Feder, 1972b) and is higher after adrenalectomy in rats (Whalen and Luttge, 1971c).

Neural tissues metabolize P to DHP and other progestins (Snipes and Shore, 1972). When hypothalamic tissue from female rats is incubated with radiolabeled P *in vitro*, 33%–50% of the recovered radioactivity chromatographs as DHP (Karavolis and Herf, 1971; Rommerts and van der Mollen, 1971; Tabei *et al.*, 1974). Traces of 3α- and 3β-hydroxy-5α-pregnan-20-one are also found. DHP is also the major neural metabolite of P formed *in vitro* by mesencephalic tissue, which reduces P more actively than either the hypothalamus or the cortex. Hypothalamic tissue also produces 20α-OHP from P *in vitro* (Tabei *et al.*, 1974). Other studies confirm these findings (Cheng and Karavolis, 1973; Hanukoglu *et al.*, 1977; Karavolis *et al.*, 1976).

Species differ in P metabolism. Female guinea pigs, for example, do not readily metabolize P *in vivo*. Wade and Feder (1972b) noted that, 4 hr after the injection of radiolabeled P, 60%–70% of the radioactivity in the brain tissue of guinea pigs was identifiable as P, whereas only 20%–30% of the neural radioactivity remained as unmetabolized P in rats. In mice, the metabolic pattern is similar to that found in rats (Luttge *et al.*, 1974b). Human brain tissue produces little if any 5β-dihydroprogesterone (5β-DHP) (Mickan, 1972), whereas 5β-DHP is readily detectable in dogs (Kawahara *et al.*, 1975) and baboons (Tabei and Heinrichs, 1974). In rhesus monkeys and baboons, 20α-hydroxysteroid dehydrogenaose activity exceeds 5α-reductase activity in the cortex, but the opposite pattern occurs in rats (Tabei and Heinrichs, 1974). These data suggest that DHP or other metabolites could mediate some of P's effects on female sexual behavior.

Behavioral data are consistent with this suggestion. DHP facilitates lordosis in estrogen-primed female rats (Henrik and Gerall, 1976; Meyerson, 1972; Whalen and Gorzalka, 1972), although the effectiveness of DHP relative to P largely depends on the carrier vehicle and the route of hormone administration. DHP administered subcutaneously in oil is about 60% as potent as P (Whalen and Gorzalka, 1972). Given subcutaneously in Tween-80, DHP and P are equipotent (Gorzalka and Whalen, 1977). DHP is almost completely ineffective given intravenously in propylene glycol (Kubli-Garfias and Whalen, 1977). These different treatment conditions do not alter the potency of P.

DHP also stimulates receptive behavior in female guinea pigs but is less potent than P (Czaja *et al.*, 1974; Wade and Feder, 1972a). In this species, the threshold dose of DHP (subcutaneous in oil) for eliciting lordosis is 627 μg, whereas the threshold dose for P is 32.5 μg. Similar findings have been obtained with hamsters (Johnson *et al.*, 1976).

In mice, the problem of progestin action is more complex. Following ovariectomy and estrogen treatment, female mice, unlike rats, hamsters, or guinea pigs, are initially unresponsive to P. Repeated injections of estrogen and P are needed to induce lordotic behavior. For example, Gorzalka and Whalen (1974) treated estrogen-primed Swiss-Webster and CD-1 female mice with EB and P weekly for 5 weeks before maximal receptivity was induced. EB plus P treatment continued for 7 weeks in this study. When the females were treated with EB and DHP during the eighth week, lordosis was not induced. As weekly treatment with EB plus DHP continued, females of the CD-1 strain became progressively more receptive, but Swiss-Webster females did not. The differential responsiveness of these strains to DHP has been confirmed by Luttge and Hall (1976).

Subsequent research showed that CD-1 mice responded to either P or DHP and that priming with either steroid did not affect the animals' responsiveness to the other (Gorzalka, 1974). Hybrid offspring of Swiss-Webster and CD-1 mice behave as their CD-1 parents do; that is, they respond to both P and DHP (Gorzalka and Whalen, 1976). Swiss-Webster females still fail to respond to DHP when the steroid is given in Tween-80 instead of oil (Carrozzo and Whalen, unpublished,

1978). Thus, in mice, the behavioral effects of DHP depend largely on the genotype of the animal. Nonetheless, some strains do respond to DHP.

METABOLITES OF DHP. The next step in the metabolism of P yields 5α-pregnan-3α-ol-20-one. This steroid is about 60% as potent as DHP in inducing lordosis in female rats when given subcutaneously in oil (Whalen and Gorzalka, 1972) and is half as potent as P when given intravenously in propylene glycol (Kubli-Garfias and Whalen, 1977). Similarly, this progestin is less potent than DHP in guinea pigs (Czaja *et al.*, 1974) and is behaviorally inactive in hamsters (Johnson *et al.*, 1976).

Another metabolite in this series, 5α-pregnane-3α-20α-diol, is no more effective than the vehicle in rats (Kubli-Garfias and Whalen, 1977).

METABOLITES OF 20α-OHP. The first step in the second major metabolic pathway of P is the formation of 20α-OHP. Most studies report that 20α-OHP is not behaviorally active in female rats (Henrik and Gerall, 1976; Meyerson, 1972; Whalen and Gorzalka, 1972; Zucker, 1967). Langford and Hilliard (1967) found that 20α-OHP occasionally induces lordosis, but the effect is transient. In contrast, Kubli-Garfias and Whalen (1977) found that intravenously administered 20α-OHP is even more effective than P, particularly shortly (5 min) after treatment; under these conditions, receptivity remains high for 2 hr. The next metabolite in this series, 5α-pregnan-20α-ol-3-one, also induces lordosis when given intravenously and is approximately 60% as potent as 20α-OHP.

OTHER STEROIDS. Steroids other than P, including corticosterone (Gorzalka and Whalen, 1977; Wade and Feder, 1972a), 17α-hydroxyprogesterone (Wade and Feder, 1972a; Zucker and Goy, 1967), and desoxycorticosterone (Bosley and Leavitt, 1972; Byrnes and Shipley, 1955; Gorzalka and Whalen, 1977), can facilitate receptivity in estrogen-primed female rodents; however, because these steroids are not natural metabolites of P and are not unusually effective in facilitating lordosis, they are not considered further here.

INHIBITORY EFFECTS OF PROGESTINS

In 1966, Goy *et al.* reported that female guinea pigs treated with estrogen and P during the luteal phase of the estrous cycle do not become sexually receptive. They suggested that endogenous P inhibits the response to estrogen and that, during normal estrous cycles, P may exert biphasic effects, first facilitating estrogen's action, then inhibiting it. Since that time, various studies have shown that P can both facilitate and inhibit the behavioral effects of estrogen in guinea pigs (Zucker, 1966b); rats (Nadler, 1970); mice (Edwards, 1970); and hamsters (Ciaccio and Lisk, 1967; DeBold *et al.*, 1976).

The two independent effects of P on behavior may be mediated by P and one of its metabolites or by two different metabolites of the parent compound, but this hypothesis has not received much attention. Zucker and Goy (1967) did not detect any inhibitory effects of 17α-hydroxyprogesterone, pregnenolone, or 20α-DHP in female guinea pigs. More recently, Czaja *et al.* (1974) found that DHP is inhibitory in this species, but it is not as potent as P. In the same study 5α-pregnan-3α-ol-20-

one and 5β-DHP did not inhibit hormone-induced lordosis. Unfortunately, little else has been done on this problem.

SEXUAL DIFFERENTIATION

In many mammalian species, male–female differences in genital morphology and sexual behavior are brought about by hormones during sensitive periods of fetal and/or neonatal development. This phenomenon has been termed the *organizing* (Young *et al.*, 1964), *developmental* (Beach, 1971), or *differentiating* (Whalen, 1974) *effect* of hormones. With respect to behavioral control systems, there appear to be two independent dimensions of sexual differentiation: masculinization and defeminization (Whalen, 1974). An animal is masculinized when it develops the potential to show male-type mounting responses and is feminized when it develops the potential to show female-type receptive responses. Each of these dimensions of sexuality is influenced by the hormonal environment of the developing organism.

Male rats, for example, appear to be masculinized by prenatal and defeminized by postnatal testicular secretions. Male rats castrated at birth readily display lordosis when given estrogen in adulthood, whereas males castrated 10 days after birth rarely do so. Thus, male rats are defeminized by their testes postnatally. The idea that masculinization occurs prenatally in rats is supported by observations that normal adult females show mounting behavior when injected with T or TP. Moreover, their behavioral sensitivity to TP depends on their proximity to male siblings during gestation (Clemens, 1974).

Hamsters, in contrast, undergo behavioral masculinization postnatally. Female hamsters rarely show mounting in adulthood when given T, unless they are treated with androgens at birth or shortly thereafter. Thus, although they are not normally masculinized, they can be masculinized by hormone treatment neonatally. The precise timing of the hormonal events underlying sexual differentiation and their later behavioral effects have been studied extensively in the past 20 years. Because these studies have been thoroughly reviewed (Beach, 1971; Goldman, 1978; Goy and Goldfoot, 1973; Whalen, 1974), we do not detail them here. Instead, we focus on the nature of the differentiating substance and on the possible role of steroid metabolism in the differentiation process.

Because early studies identified the testes rather than the ovaries as the source of differentiation, investigators concentrated on the role of androgens in sexual differentiation. Most published studies have reported on the effects of T or TP. All of these studies indicate that, if a sufficient amount of T is present at the appropriate time and for an appropriate duration, it induces sexual differentiation in those species in which differentiation is normally induced by the testes. Nonetheless, the conclusion that T is the active agent is, of course, unwarranted.

As early as 1943, Wilson demonstrated that estrogen as well as androgen can defeminize female rats. Wilson treated female rats with estradiol dipropionate for 4 weeks starting at birth or 5, 10, 15, 20, 30, or 40 days after birth. If treatment started before the fifteenth postnatal day, the females did not show spontaneous mating responses as adults and would not display lordosis even when given estrogen and P. In the same year, Koster (1943) reported that female rats treated with estrogen in infancy would, as adults, display intense masculine-type sexual responses. Neither of these studies, however, suggested that an estrogen could normally be the active differentiating agent. It was not until the work of Anderson and Liao (1968) and Bruchovsky and Wilson (1968) that steroid metabolism became an important issue. Their work suggested that the 5α-reduction of T to DHT might be critical.

THE ROLE OF 5α-REDUCTION

DEFEMINIZATION. Although DHT plays a major role in differentiation of the external genitalia, it plays little, if any, role in the differentiation of behavior. The hypothalamus of 5- to 7-day-old female rats accumulates DHT after injection of radiolabeled T or A (Weisz and Gibbs, 1974a,b), but DHT does not inhibit the development of lordosis in rats (Arai, 1972; Booth, 1977; Luttge and Whalen, 1970a; McDonald and Doughty, 1972, 1974; Whalen and Luttge, 1971b) even when the steroid is given continuously during the first 10 postnatal days via a silastic capsule (Whalen and Rezek, 1974). DHT is also much less effective than T in hamsters. Some studies (Coniglio *et al.*, 1973a; Paup *et al.*, 1972) report that DHT has no effect on behavioral development in this species, whereas others (Gerall *et al.*, 1975; Payne, 1976) find that it does reduce lordotic behavior somewhat. It seems unlikely that DHT mediates T's defeminizing actions in either rats or hamsters. Whether DHT contributes to sexual differentiation in humans is not known, but hypothalamic tissue from both male and female human fetuses can convert T to DHT *in vitro* (Schindler, 1975, 1976).

MASCULINIZATION. DHT-induced enhancement of masculine responses has received little attention, possibly because masculinization occurs primarily before birth in rats. The hypothalamus of 7-day-old male rats does contain DHT receptors (Kato, 1976), but their role in behavioral development is not known. In hamsters, masculinization occurs postnatally and can be manipulated rather easily, yet mounting behavior is not enhanced in female hamsters given DHT in infancy (Coniglio *et al.*, 1973a; Payne, 1976). Thus, it appears that T does not affect masculinization via metabolism to DHT.

THE ROLE OF OTHER ANDROGENS

Although DHT apparently does not make a major contribution to the differentiation of behavior, other androgens might. A, for example, is produced both by the testes and as a metabolite of T. A defeminizes rats (Stern, 1969; Whalen and Rezek, 1974), mice (Edwards, 1971), and hamsters (Tiefer and Johnson, 1975) but is a relatively weak androgen in this system (Goldfoot *et al.*, 1969; Luttge and Whalen, 1969). A can also enhance masculine behavior in both rats (Popolow and

Ward, 1978) and hamsters (Tiefer and Johnson, 1975). Whether A acts directly in these systems or via metabolism to T is unknown.

Another androgen that has received some attention is androsterone. Androsterone neither suppresses lordosis nor enhances mounting behavior in hamsters (Coniglio *et al.,* 1973a), but androsterone propionate does inhibit lordotic behavior in rats (Gerall *et al.,* 1975; McDonald and Doughty, 1974). Differences in dose, in the form of the hormone used, or in the species could account for this discrepancy.

McDonald and Doughty (1974) found no inhibition of lordosis in female rats treated daily with 5α-androstanedione enol propionate, $3\alpha,5\alpha$-androstanediol dipropionate, $3\beta,5\alpha$-androstanediol dipropionate, or 19-OHTP for the first 5 days after birth. Androsterone, however, did defeminize the females. The differing effects of androsterone and 19-OHTP are particularly interesting. Androsterone is presumably not aromatizable, whereas 19-OHTP represents the first metabolic step toward the formation of E_2 from T. Thus, the developmental effects of these hormones are not consistent with an aromatization hypothesis of androgen-induced sexual differentiation. Another presumably nonaromatizable androgen, 6α-fluoroTP, defeminizes sexual development in rats but is less potent than TP. The relative potencies of TP and 6α-fluoroTP correlate with their abilities to inhibit estrogen uptake by cell nuclei in the hypothalamus–preoptic area (Nordeen and Yahr, 1981).

THE ROLE OF AROMATIZATION

With the exception of the McDonald and Doughty (1974) report, most published studies indicate that sexual differentiation is induced by aromatizable, but not by nonaromatizable, androgens (see Plapinger and McEwen, 1978, for a review). Indeed, estrogens themselves are quite potent in this regard.

DEFEMINIZATION. As mentioned earlier, Wilson (1943) found that neonatal stimulation with estrogen defeminizes female rats. Whalen and Nadler (1963) replicated that finding, and several investigators have since reported similar data on both rats (see Plapinger and McEwen, 1978, for a review of those studies) and hamsters (Coniglio *et al.,* 1973a,b: Whalen and Etgen, 1978).

Because both natural and synthetic estrogens cause defeminization, it is possible that aromatizable androgens defeminize *because* they are metabolized to estrogen. Consistent with this hypothesis are observations that the estrogen antagonist CI-628 and various aromatase inhibitors, including ATD and 4-androstene-3,6,17-trione (ADT), can prevent testis-induced or androgen-induced defeminization (Booth, 1977; McEwen *et al.,* 1977a; Vreeburg *et al.,* 1977). The hypothalamus of 5- to 7-day-old female rats also accumulates E_2 and E_1 after injections of T or A (McEwen *et al.,* 1977b; Weisz and Gibbs, 1974a,b). Hypothalamic tissue from rat fetuses or neonates of both sexes can also convert A to E_1 *in vitro* (Reddy *et al.,* 1974a). This is also true of human fetuses (Schindler, 1975).

If androgens defeminize behavior via aromatization, estrogens should defeminize at low doses because, in rats at least, the yield of estrogen from androgen in the brain is less than 0.5% (Reddy *et al.,* 1974a). However, according to Brown-

Grant (1973), EB is only three times as potent as TP in inducing defeminization in rats. This discrepancy may be accounted for by the presence in rats of a plasma E_2-binding protein, α-fetoneonatal protein, which protects the animal from the actions of estrogens but not of androgens. The observation that prednisolone, which inhibits the synthesis of α-fetoneonatal protein (Belanger *et al.*, 1975), can enhance the defeminizing potency of EB (Whalen and Olsen, 1978) supports this suggestion. Thus, the evidence favors the aromatization hypothesis of testes- or androgen-induced defeminization in rats.

MASCULINIZATION. Koster's (1943) findings cited above suggest that early exposure to estrogen could facilitate the later display of masculine behavior. Further studies of rats indicate that neonatal EB treatment suppresses rather than enhances masculine behavior in males (Levine and Mullins, 1964; Whalen, 1964). Some researchers have found that EB facilitates the development of masculine behavior in females (Hendricks and Gerall, 1970; Levine and Mullins, 1964), but others have not (Whalen and Edwards, 1967). This discrepancy suggests that the facilitating actions of estrogen critically depend on the dose of steroid given (Mullins and Levine, 1968).

The masculinizing effects of estrogens can be seen more clearly in hamsters than in rats because untreated female hamsters rarely mount. Mounting is facilitated in female hamsters by neonatal exposure to diethylstilbestrol, E_2, or EB (Coniglio *et al.*, 1973a; Paup *et al.*, 1972). Moreover, Whalen and Etgen (1978) recently reported that as little as 500 ng of EB induces this change. They found that the synthetic estrogen RU-2858, which does not bind to α-fetoneonatal protein, was even more potent in this regard. Only 0.05 ng of this drug was needed to increase mounting behavior. Thus, rather small doses of estrogen can masculinize hamsters. This finding is consistent with the aromatization hypothesis.

CONCLUSIONS

The present review suggests the conclusion that the metabolism of the androgenic, estrogenic, and progestational secretions of the gonads *may*, but *need not*, be a critical step for the action of these steroids in the regulation of sexual behavior. All of the major secretions of the gonads undergo metabolic changes in the blood, the liver, and/or the brain; however, it is not clear that the products of metabolism, rather than the precursors, are the active agents in the hormonal action. On the other hand, there is no convincing evidence that gonadal secretions are behaviorally active without metabolism. The evidence is clear that the metabolites of gonadal hormones can stimulate changes in behavior. In most cases, however, they are less efficacious than the parent steroid.

Even if one accepts the hypothesis that metabolism is obligatory for hormone action, one must accept the conclusion that no single metabolic event (or pattern of metabolic events) is likely to explain hormone action in general. Species specificity and even strain specificity become readily apparent with even a cursory examination of the literature. In spite of this truism, it has become popular to state, for example, that testicular androgens must be aromatized to become behaviorally active. Our review indicates that this conclusion is unwarranted, even for rats, the species on which the "universal" aromatization hypothesis is based. We feel, therefore, that investigators must become highly sensitive to genotype and genotype–steroid interactions in the development of hypotheses about the role of metabolism in hormone action.

We must also be sensitive to the limitations inherent in the tools used to obtain data on the validity of particular metabolic hypotheses. For example, the aromatization hypothesis was based on the finding that DHT failed to stimulate mounting behavior in rats. DHT rapidly became the "test steroid" for the assessment of aromatization, even though the metabolism of DHT by the brain remained uninvestigated, as did the interaction of DHT with the CNS cytoplasmic-receptor and nuclear-steroid-binding systems that are thought to mediate hormonal events.

Assuming that DHT undergoes no neural aromatization or interaction with brain estrogen receptor systems, it is indeed a valuable tool, but, unfortunately, it is inadequate by itself to determine whether aromatization is critical for T's action. We now know that DHT is behaviorally active in a variety of species, including rats. Yet, this does not mean that aromatization plays no role in T's action.

An adequate assessment of any metabolic hypothesis must employ a variety of approaches. For example, support for the role (not necessarily exclusive) of aromatization in androgen action has been obtained by means of agents such as ADT and ATD, which block aromatization. However, we still know little about other effects that these agents may have on the neural systems that control sexual behavior. The effect of the aromatase inhibitors is supportive, but not conclusive. We should be reluctant to readily conclude that ADT blocks T-induced mounting behavior *because* it blocks aromatization, just as we should have been reluctant to conclude that aromatization is critical for T's action because rats (of a given strain), with a particular sexual history (inexperienced), castrated for a particular time (6 weeks), tested for a limited time (15 min), and given DHT (a single dose) failed to mate (McDonald *et al.*, 1970).

We must also be sensitive to the pharmacology of hormone action. Steroids, both gonadal hormones and their metabolites, may be differentially efficacious for a variety of reasons others than differences in their intrinsic activity. Dose, solvent, route of administration, plasma binding, metabolic clearance rate, and so forth can have major effects on the action of a steroid. These variables should not be overlooked, particularly when we find that particular steroids, such as DHT and 20α-OHP, lack behavioral activity.

There is no need to belabor the point that both our technology and our experimental design must be adequate to the task that we face in determining the critical metabolic events underlying the hormonal control of behavior. This review attests to the fact that valid, and hopefully general, conclusions will not come easily. Nonetheless, a surprising amount of progress has been made since 1970.

Abraham, G. E., Odell, W. D., Swerdloff, R. S., and Hopper, K. Simultaneous radioimmunoassay of plasma FSH, LH, progesterone, 17-hydroxyprogesterone and estradiol-17β during the menstrual cycle. *Journal of Clinical Endocrinology and Metabolism*, 1972, *34*, 312–318.

Adessi, G. L., Goutte-Coussieu, C., Tran Quang Nhuan, Eichenberger, D., and Jayle, M. F. Conversion, in vitro, of (7n-^{3}H) testosterone to estrone and estradiol-17β and their 3-sulfate conjugate by the guinea pig placenta. *Steroids*, 1978, *32*, 295–306.

Ağmo, A. Cyproterone acetate diminishes sexual activity in male rabbits. *Journal of Reproduction and Fertility*, 1975, *44*, 69–75.

Ağmo, A. The comparative actions of fluoxymesterone and testosterone on sexual behavior and accessory sexual glands in castrated rabbits. *Hormones and Behavior*, 1977, *9*, 112–119.

Ağmo, A., and Södersten, P. Sexual behavior in castrated rabbits treated with testosterone, oestradiol, dihydrotestosterone or oestradiol in combination with dihydrotestosterone. *Journal of Endocrinology*, 1975, *67*, 327–332.

Ainsworth, L., and Ryan, K. J. Steroid hormone transformations by endocrine organs from pregnant mammals: I. Estrogen biosynthesis by mammalian placental preparations *in vitro. Endocrinology*, 1966, *79*, 875–883.

Alsum, P., and Goy, R. W. Actions of esters of testosterone, dihydrotestosterone or estradiol on sexual behavior in castrated male guinea pigs. *Hormones and Behavior*, 1974, *5*, 207–217.

Anderson, K, M., and Liao, S. Selective retention of dihydrotestosterone by prostatic nuclei. *Nature*, 1968, *219*, 277–279.

Antliff, H. R., and Young, W. C. Behavioral and tissue responses of male guinea pigs to estrogens and the problem of hormone specificity. *Endocrinology*, 1956, *59*, 74–82.

Applegren, L. E. Chromatographic studies and scintillation counting of selected tissues of mice injected with labeled testosterone. *Biology of Reproduction*, 1970, *3*, 128–133.

Arai, Y. Effect of 5α-dihydrotestosterone on differentiation of masculine pattern of the brain in the rat. *Endocrinologia Japonica*, 1972, *19*, 389–393.

Atkinson, L. E., Hotchkiss, J., Fritz, G. R., Surve, A. H., Neill, J. D., and Knobil, E. Circulating levels of steroids and chorionic gonadotropin during pregnancy in the rhesus monkey with special attention to the rescue of the corpus luteum in early pregnancy. *Biology of Reproduction*, 1975, *12*, 335–345.

Badr, F. M., and Bartke, A. Effect of ethyl alcohol on plasma testosterone level in mice. *Steroids*, 1974, *23*, 921–927.

Ball, J. Sex activity of castrated male rats increased by estrin administration. *Journal of Comparative Psychology*, 1937, *24*, 135–144.

Ball, J. Male and female mating behavior in prepubertally castrated male rats receiving estrogens. *Journal of Comparative Psychology*, 1939, *28*, 273–283.

Baranczuk, R., and Greenwald, G. S. peripheral levels of estrogen in the acyclic hamster. *Endocrinology*, 1973, *92*, 805–812.

Baulieu, E. E. The action of hormone metabolites: A new concept in endocrinology. *Annals of Clincial Research*, 1970, *2*, 246–250.

Baulieu, E. E., and Robel, P. Catabolism of testosterone and androstenedione. In K. B. Eik-Nes (Ed.), *The Androgens of the Testis.* New York: Marcel Dekker, 1970.

Baulieu, E. E., Fabre-Jung, I., and Huis in't Veld, L. G. Dehydroepiandrosterone sulfate: A secretory product of the boar testis. *Endocrinology*, 1967, *81*, 34–38.

Baulieu, E. E., Lasnitzki, I., and Robel, P. Metabolism of testosterone and actions of metabolites on prostate glands grown in organ culture. *Nature*, 1968, *219*, 1155–1156.

Baum, M. J. Precocious mating in male rats following treatment with androgen or estrogen. *Journal of Comparative and Physiological Psychology*, 1972, *78*, 356–367.

Baum, M. J. Hormonal stimulation of precocious mating in male rats without anticedent effects on sexual clasping or ambulation. *Physiology and Behavior*, 1973, *10*, 137–140.

Baum, M. J., and Vreeburg, J. T. M. Copulation in castrated male rats following combined treatment with estradiol and dihydrotestosterone. *Science*, 1973, *182*, 283–285.

Baum, M. J., and Vreeburg, J. T. M. Differential effects of the anti-estrogen MER-25 and of three 5α-reduced androgens on mounting and lordosis behavior in the rat. *Hormones and Behavior*, 1976, *7*, 87–104.

Beach, F. A. Copulatory behavior in prepubertally castrated male rats and its modification by estrogen administration. *Endocrinology*, 1942, *31*, 679–683.

Beach, F. A. Hormonal factors controlling the differentiation, development and display of copulatory behavior in the ramstergig and related species. In E. Tobach, L. R. Aronson, and E. Shaw (Eds.), *The Biopsychology of Development*. New York: Academic Press, 1971.

Beach, F. A., and Westbrook, W. H. Dissociation of androgenic effects on sexual morphology and behavior in male rats. *Endocrinology*, 1968a, *83*, 395–398.

Beach, F. A., and Westbrook, W. H. Morphological and behavioral effects of an "antiandrogen" in male rats. *Journal of Endocrinology*, 1968b, *42*, 379–382.

Belanger, L., Hamel, D., Lachance, L., Dufour, D., Tremblay, M., and Gagnon, P. M. Hormonal regulation of α_1 foetoprotein. *Nature*, 1975, *256*, 657–659.

Bellino, F. L., Gilani, S. S. H., Eng, S. S., Osawa, Y., and Duax, W. L. Active-site-directed inactivation of aromatase from human placental microsomes by brominated androgen derivatives. *Biochemistry*, 1976, *15*, 4730–4736.

Beyer, C., and Rivaud, N. Differential effect of testosterone and dihydrotestosterone on the sexual behavior of prepuberally castrated male rabbits, *Hormones and Behavior*, 1973, *4*, 175–180.

Beyer, C., Morali, G., and Vargas, R. Effect of diverse estrogens on estrous behavior and genital tract development in ovariectomized rats. *Hormones and Behavior*, 1971, *2*, 273–277.

Beyer, C., Larsson, K., Perez-Palacios, G., and Morali, G. Androgen structure and male sexual behavior in the castrated rat. *Hormones and Behavior*, 1973, *4*, 99–108.

Beyer, C., de la Torre, L., Larsson, K., and Perez-Palacios, G. Synergistic actions of estrogen and androgen on the sexual behavior of the castrated male rabbit. *Hormones and Behavior*, 1975, *6*, 301–306.

Beyer, C., Morali, G., Naftolin, F., Larsson, K., and Perez-Palacios, G. Effect of some antiestrogens and aromatase inhibitors on androgen induced sexual behavior in castrated male rats. *Hormones and Behavior*, 1976, *7*, 353–363.

Blaquier, J., Forchielli, E., and Dorfman, R. I. *In vitro* metabolism of androgens in whole human blood. *Acta Endocrinologica*, 1967, *55*, 697–704.

Bloch, G. J., and Davidson, J. M. Behavioral and somatic responses to the antiandrogen cyproterone. *Hormones and Behavior*, 1971, *2*, 11–25.

Booth, J. E. Effects of an aromatization inhibitor, androst-4-ene-3,6,17-trione on sexual differentiation induced by testosterone in the neonatally castrated rat. *Journal of Endocrinology*, 1977, *72*, 53P–54P.

Bosley, C. G., and Leavitt, W. W. Specificity of progesterone action during the preovulatory period in the cyclic hamster. *Federation Proceedings*, 1972, *31*, 257.

Bottomley, A. C., and Folley, S. J. The effect of high doses of androgenic substances on the weights of the testes, accessory reproductive organs and endocrine glands of young male guinea pigs. *Journal of Physiology*, 1938, *94*, 26–39.

Brodie, A. M. H., Schwarzel, W. C., Shaikh, A. A., and Brodie, A. H. The effect of an aromatase inhibitor, 4-hydroxy-4-androstene-3, 17-dione, on estrogen-dependent processes in reproduction and breast cancer. *Endocrinology*, 1977, *100*, 1684–1695.

Brown-Grant, K. Recent studies on the sexual differentiation of the brain. In K. S. Comline, K. W. Cross, G. S. Dawes, and P. W. Nathanielsz (Eds.), *Foetal and Neonatal Physiology*. Cambridge, England: Cambridge University Press, 1973.

Brown-Grant, K., Exley, D., and Naftolin, F. Plasma peripheral oestradiol and luteinizing hormone concentrations during the oestrus cycle of the rat. *Journal of Endocrinology*, 1970, *48*, 295–296.

Bruchovsky, N., and Wilson, J. D. The conversation of testosterone to 5α-androstan-17β-ol-3-one by rat prostate *in vivo* and *in vitro*. *Journal of Biological Chemistry*, 1968, *243*, 2012–2021.

Butcher, R. L., Collins, W. E., and Fugo, N. W. Plasma concentration of LH, FSH, prolactin, progesterone and estradiol-17β throughout the 4-day estrous cycle of the rat. *Endocrinology*, 1974, *94*, 1704–1708.

Byrnes, W. W., and Shipley, E. G. Guinea pig copulatory reflex in response to adrenal steroids and similar compounds. *Endocrinology*, 1955, *57*, 5–9.

Chamley, W. A., Buckmaster, J. M., Cerini, M. E., Cumming, I. A., Goding, J. R., Obst, J. M., Williams, A., and Winfield, C. Changes in the level of progesterone, corticoids, estrone, estradiol-17β, luteinizing hormone and prolactin in the peripheral plasma of the ewe during the late pregnancy and parturition. *Biology of Reproduction*, 1973, *9*, 30–35.

Chazal, G., Faudon, M., Gogan, F., and Rotsztejn, W. Effects of two estradiol antagonists upon the estradiol uptake in the rat brain and peripheral tissues. *Brain Research*, 1975, *89*, 245–254.

Cheng, Y.-J., and Karavolis, H. J. Conversion of progesterone to 5α-pregnane-3,20-dione and 3α-hydroxy-5α-pregnan-20-one by rat medial basal hypothalami and the effects of estradiol and stage of estrous cycle on the conversion. *Endocrinology,* 1973, *93,* 1157–1162.

Cheng, Y.-J., and Karavolis, H. J. Subcellular distribution and properties of progesterone (Δ⁴-steroid) 5α-reductase in rat medial basal hypothalamus. *Journal of Biological Chemistry,* 1975, *250,* 7997–8003.

Christensen, L. W., and Clemens, L. G. Intrahypothalamic implants of testosterone or estradiol and resumption of masculine sexual behavior in long-term castrated male rats. *Endocrinology,* 1974, *95,* 984–990.

Christensen, L. W., and Clemens, L. G. Blockade of testosterone-induced mounting behavior in the male rat with intracranial application of the aromatization inhibitor, androst-1,4,6-triene,-3,17-dione. *Endocrinology,* 1975, *97,* 1545–1551.

Christensen, L. W., Coniglio, L. P., Paup, D. C., and Clemens, L. G. Sexual behavior of male golden hamsters receiving diverse androgen treatments. *Hormones and Behavior,* 1973, *4,* 223–229.

Ciaccio, L. A., and Lisk, R. D. Facilitation and inhibition of estrous behavior in the spayed golden hamster. *American Zoologist,* 1967, *7,* 712.

Clemens, L. G. Neurohormonal control of male sexual behavior. In W. Montagna and W. A. Sadler (Ed.), *Reproductive Behavior.* New York: Plenum Press, 1974.

Cochran, C. A., and Perachio, A. A. Dihydrotestosterone propionate effects on dominance and sexual behaviors in gonadectomized male and female rhesus monkeys. *Hormones and Behavior,* 1977, *8,* 175–187.

Coniglio, L. P., Paup, D. G., and Clemens, L. G. Hormonal factors controlling the development of sexual behavior in the male golden hamster. *Physiology and Behavior,* 1973a, *10,* 1087–1094.

Coniglio, L. P., Paup, D. G., and Clemens, L. G. Hormonal specificity in the suppression of sexual receptivity of the female golden hamster. *Journal of Endocrinology,* 1973b, *57,* 55–61.

Coyotupa, J., Farlow, A. F., and Kovacic, N. Serum testosterone and dihydrotestosterone levels following orchiectomy in the adult rat. *Endocrinology,* 1973, *92,* 1579–1581.

Czaja, J. A., Goldfoot, D. A., and Karavolis, H. Comparative facilitation and inhibition of lordosis in the guinea pig with progesterone, 5α-pregnane-3,20-dione or 3α-hydroxy-5α-pregnan-20-one. *Hormones and Behavior,* 1974, *5,* 261–274.

Davidson, J. M. Activation of the male rat's sexual behavior by intracerebral implantation of androgen. *Endocrinology,* 1966, *79,* 783–794.

Davidson, J. M. Effects of estrogen on the sexual behavior of male rats. *Endocrinology,* 1969, *84,* 1365–1372.

Davidson, J. M., and Bloch, G. J. Neuroendocrine aspects of male reproduction. *Biology of Reproduction,* 1969, *1,* 67–92.

Davidson, J. M., and Trupin, S. Neural mediation of steroid-induced sexual behavior in rats. In M. Sandler and G. L. Gessa (Eds.), *Sexual Behavior: Paramacology and Biochemistry.* New York: Raven Press, 1975.

Davies, I. J., Naftolin, F., Ryan, K. J., Fishman, J., and Siu, J. The affinity of catechol estrogens for estrogen receptors in the pituitary and anterior hypothalamus of the rat. *Endocrinology,* 1975, *97,* 554–557.

DeBold, J. F., Martin, J. V., and Whalen, R. E. The excitation and inhibition of sexual receptivity in female hamsters by progesterone: Time and dose relationships, neural localization and mechanisms of action. *Endocrinology,* 1976, *99,* 1519–1527.

Danef, C., Magnus, C., and McEwen, B. S. Sex differences and hormonal control of testosterone metabolism in rat pituitary and brain. *Journal of Endocrinology,* 1973, *59,* 605–621.

Dohler, K. S., and Wuttke, W. Changes with age in levels of serum gonadotropins, prolactin, and gonadal steroids in prepubertal male and female rats. *Endocrinology,* 1975, *97,* 898–907.

Doughty, C., Booth, J. E., McDonald, P. G., and Parrott, R. F. Effects of oestradiol-17β, oestradiol benzoate and the synthetic oestrogen RU-2858 on sexual differentiation in the neonatal female rat. *Journal of Endocrinology,* 1975, *67,* 419–424.

Dryden, G. L., and Anderson, J. M. Ovarian hormone: Lack of effect on reproductive structures of female Asian musk shrews. *Science,* 1977, *197,* 782–784.

Eaton, G. G., Goy, R. W., and Resko, J. A. Brain uptake and metabolism of estradiol benzoate and estrous behavior in ovariectomized guinea pigs. *Hormones and Behavior,* 1975, *6,* 81–97.

Edwards, D. A. Induction of estrus in female mice: Estrogen-progesterone interactions. *Hormones and Behavior,* 1970, *1,* 299–304.

Edwards, D. A. Neonatal administration of androstenedione, testosterone or testosterone propionate: Effects on ovulation, sexual receptivity and aggressive behavior in female mice. *Physiology and Behavior,* 1971, *6,* 223–228.

Edwards, D. A., and Burge, K. G. Estrogenic arousal of aggressive behavior and masculine sexual behavior in male and female mice. *Hormones and Behavior,* 1971, *2,* 239–245.

Eik-Nes, K. B. Synthesis and secretion of androstenedione and testosterone. In K. B. Eik-Nes (Ed.), *The Androgens of the Testis.* New York: Marcel Dekker, 1970.

Eto, T., Masuda, H., Shzuki, Y., and Hosi, T. Progesterone and pregn-4-ene-20α-ol-3-one in rat ovarian venous blood at different stages in reproductive cycle. *Japanese Journal of Reproduction,* 1962, *8,* 34–40.

Ewing, L., Brown, B., Irby, D. C., and Jardine, I. Testosterone and 5α-reduced androgen secretion by rabbit testes-epididymides perfused *in vitro. Endocrinology,* 1975, *96,* 610–617.

Falvo, R. E., and Nalbandov, A. V. Radioimmunoassay of peripheral plasma testosterone in males from eight species using a specific antibody without chromatography. *Endocrinology,* 1974, *95,* 1466–1472.

Feder, H. H. The comparative actions of testosterone propionate and 5α-androstan-17β-ol-3-one propionate on the reproductive behavior, physiology and morphology of male rats. *Journal of Endocrinology,* 1971, *51,* 241–252.

Feder, H. H., and Silver, R. Activation of lordosis in ovariectomized guinea pigs by free and esterified forms of estrone, estradiol-17β and estriol. *Physiology and Behavior,* 1974, *13,* 251–255.

Feder, H. H., and Whalen, R. E. Feminine behavior in neonatally castrated and estrogen-treated male rats. *Science,* 1965, *147,* 306–307.

Feder, H. H., Resko, J. A., and Goy, R. W. Progesterone concentrations in the arterial plasma of guinea pigs during the oestrous cycle. *Journal of Endocrinology,* 1968a, *40,* 505–513.

Feder, H. H., Resko, J. A., and Goy, R. W. Progesterone levels in the arterial plasma of preovulatory and ovariectomized rats. *Journal of Endocrinology,* 1968b, *41,* 563–569.

Feder, H. H., Naftolin, F., and Ryan, K. J. Male and female neural responses in male rats given estradiol benzoate and 5α-androstan-17β-ol-3-one propionate. *Endocrinology,* 1974a, *94,* 136–141.

Feder, H. H., Siegel, H., and Wade, G. N. Uptake of [6, 7-³H] estradiol-17β in ovariectomized rats, guinea pigs and hamsters: Correlation with species differences in behavioral responsiveness to estradiol. *Brain Research,* 1974b, *71,* 93–103.

Fishman, J., and Norton, B. Catechol estrogen formation in the central nervous system of the rat. *Endocrinology,* 1975, *96,* 1054–1059.

Fishman, J., Naftolin, F., Davies, I. J., Ryan, K. J., and Petro, Z. Catechol estrogen formation by the human fetal brain and pituitary. *Journal of Clinical Endocrinology and Metabolism,* 1976, *42,* 177–180.

Fletcher, T. J., and Short, R. V. Restoration of libido in castrated red deer stags *(Cervus elaphus)* with oestradiol-17β. *Nature,* 1974, *248,* 616–618.

Flores, F., Naftolin, F., and Ryan, K. J. Aromatization of androstenedione and testosterone by rhesus monkey hypothalamus and limbic system. *Neuroendocrinology,* 1973a, *11,* 177–182.

Flores, F., Naftolin, F., Ryan, K. J., and White, R. J. Estrogen formation by the isolated perfused rhesus monkey brain. *Science,* 1973b, *180,* 1074–1075.

Folman, Y., Haltmeyer, G. C., and Eik-Nes, K. B. Production and secretion of 5α-dihydrotestosterone by the dog testis. *American Journal of Physiology,* 1972, *222,* 653–656.

Foote, R. H., Draddy, P. J., Breite, M., and Oltenacu, E. A. B. Action of androgen and estrone implants on sexual behavior and reproductive organs of castrated male rabbits. *Hormones and Behavior,* 1977, *9,* 57–68.

Frankel, A. I., Mock, E. J., Wright, W. W., and Kamel, F. Characterization and physiological validation of a radioimmunoassay for plasma testosterone in the male rat. *Steroids,* 1975, *25,* 73–98.

French, F. S., Baggett, B., Van Wyk, J. J., Talbert, L. M., Hubbard, W. R., Johnston, F. R., Weaver, R. P., Forchielli, E., Rao, G. S., and Sarda, I. R. Testicular feminization: Clinical, morphological and biochemical studies. *Journal of Clinical Endocrinology and Metabolism,* 1965, *25,* 661–677.

Genot, A., Loras, B., Monbon, M., and Bertrand, J. *In vitro* metabolism of testosterone in the rat brain during sexual maturation: III. Studies of the formation of main androstane-diols and androstene-diols. *Journal of Steroid Biochemistry,* 1975, *6,* 1247–1252.

Gerall, A. A., McMurray, M. M., and Farrell, A. Suppression of the development of female hamster behaviour by implants of testosterone and non-aromatizable androgens administered neonatally. *Journal of Endocrinology.* 1975, *67,* 439–445.

Giles, C., and Griffiths, K. Inhibition of the aromatizing activity of human placenta by SU-4885 (Meta-pirone). *Journal of Endocrinology*, 1964, *28*, 343–344.

Goldfoot, D. A., Feder, H. H., and Goy, R. W. Development of bisexuality in the male rat treated neonatally with androstenedione. *Journal of Comparative and Physiological Psychology*, 1969, *67*, 41–45.

Goldman, B. D. Developmental influences of hormones on neuroendocrine mechanisms of sexual behaviour: Comparisons with other sexually dimorphic behaviour. In J. B. Hutchison (Ed.), *Biological Determinants of Sexual Behaviour*. Chichester, England: Wiley, 1978.

Goncharov, N., Aso, T., Cekan, Z., Pachalisi, N., and Diczfalusy, E. Hormonal changes during the menstrual cycle of the baboon *(Papio hamadryas)*. *Acta Endocrinologica*, 1976, *82*, 396–412.

Gorzalka, B. B. *Neural Mechanisms of Progesterone Action*. Unpublished doctoral dissertation, University of California, Irvine, 1974.

Gorzalka, B. B. *Copulatory Behavior: Interactions between Adrenal and Exogenous Gonadal Steroids*. Paper presented at Third Annual Meeting. International Academy of Sex Research, Bloomington, Indiana, 1977.

Gorzalka, B. B., and Whalen, R. E. Genetic regulation of hormone action: Selective effects of progesterone and dihydroprogesterone (5α-pregnane-3, 20-dione) on sexual receptivity in mice. *Steroids*, 1974, *23*, 499–505.

Gorzalka, B. B., and Whalen, R. E. Effects of genotype on differential behavioral responsiveness to progesterone and 5α-dihydroprogesterone in mice. *Behavioral Genetics*, 1976, *6*, 7–15.

Gorzalka, B. B., and Whalen, R. E. The effects of progestins, mineralocorticoids, glucocorticoids and steroid solubility on the induction of sexual receptivity in rats. *Hormones and Behavior*, 1977, *8*, 94–99.

Gorzalka, B. B., Rezek, D. L., and Whalen, R. E. Adrenal mediation of estrogen-induced ejaculatory behavior in the male rat. *Physiology and Behavior*, 1975, *14*, 373–376.

Goto, J., and Fishman, J. Participation of a nonenzymatic transformation in the biosynthesis of estrogens from androgens. *Science*, 1977, *195*, 80–81.

Goy, R. W., and Goldfoot, D. A. Hormonal influences on sexually dimorphic behavior. In R. O. Greep and E. B. Astwood (Eds.), *Handbook of Physiology*, Section, 7, Vol. 2. Washington, D.C.: American Physiological Society, 1973.

Goy, R. W., Phoenix, C. H., and Young, W. C. Inhibitory action of the corpus luteum on the hormonal induction of estrous behavior in the guinea pig. *General and Comparative Endocrinology*, 1966, *6*, 267–275.

Grossman, S. B., and Bloch, E. Comparative placental steroid biosynthesis: II. C_{19} steroid metabolism by guinea-pig placentas and fetal adrenals *in vitro*. *Steroids*, 1973, *21*, 813–832.

Gual, C., Morato, T., Hayano, M., Gut, M., and Dorfman, R. I. Biosynthesis of estrogens. *Endocrinology*, 1962 *71*, 920–925.

Hanukoglu, I., Karavolis, H. J., and Goy, R. W. Progesterone metabolism in the pineal, brain stem, thalamus and corpus collosum of the female rat. *Brain Research*, 1977, *125*, 313–324.

Hashimoto, I., and Suzuki, Y. Androgens in testicular venous blood in the rat, with special reference to puberal change in the secretory pattern. *Endocrinologica Japonica*. 1966. *13*, 326–337.

Hendricks, S. E., and Gerall, A. A. Effects of neonatally administered estrogen on development of male and female rats. *Endocrinology*, 1970, *87*, 435–439.

Henrik, E., and Gerall, A. A. Facilitation of receptivity in estrogen-primed rats during successive mating tests with progestins and methysergide. *Journal of Comparative and Physiological Psychology*, 1976, *90*, 590–600.

Hill, M. J., Goddard, P., and Williams, R. E. O. Gut bacteria and aetiology of cancer of the breast. *Lancet*, 1971, *2*, 472–473.

Hilliard, J., Endroczi, E., and Sawyer, C. H. Stimulation of progestin release from rabbit ovary *in vivo*, 20α-OH-preg-4-en-3-one released into ovarian vein. *Proceedings of the Society for Experimental Biology and Medicine*. 1961, *108*, 154.

Hori, T., Ide, M., and Miyake, T. Ovarian estrogen secretion during the estrous cycle and under the influence of exogenous gonadotropins in rats. *Endocrinologica Japonica*, 1968, *15*, 215–222.

Ibayashi, H., Nakamura, M., Uchikawa, T., Murakawa, S., Yoshida, S., Nakao, K., and Okinaka, S. C19 steroids in canine spermatic venous blood following gonadotropin administration. *Endocrinology*, 1965, *76*, 347–352.

Ichikawa, S., Sawada, T., Nakamura, Y., and Morioka, H. Ovarian secretion of pregnane compounds during the estrous cycle and pregnancy in rats. *Endocrinology*, 1974, *94*, 1615–1620.

Imperato-McGinley, J., Guerrero, L., Gautier, T., and Peterson, R. E. Steroid 5α-reductase deficiency in man: An inherited form of pseudohermaphroditism. *Science,* 1974, *186,* 1213–1215.

Ito, T., and Horton, R. Dihydrotestosterone in human peripheral plasma. *Journal of Clinical Endocrinology and Metabolism,* 1970, *31,* 362–368.

Jaffe, R. B. Testosterone metabolism in target tissues: Hypothalamic and pituitary tissues of the adult rat and human fetus, and the immature rat ephiphysis. *Steroids,* 1969, *14,* 483–498.

Johnson, D. C., and Naqvi, R. H. Effect of cyproterone acetate on LH in immature rats. *Endocrinology,* 1969, *84,* 421–425.

Johnson, W. A., and Tiefer, L. Mating in castrated male rats during combined treatment with estradiol benzoate and fluoxymesterone. *Endocrinology,* 1974, *95,* 912–915.

Johnson, W. A., Billiar, R., and Little, B. Progesterone and 5α-reduced metabolites: Facilitation of lordosis behavior and brain uptake in female hamsters. *Behavioral Biology,* 1976, *18,* 489–497.

Johnston, J. O., Greenwell, J. F., Benson, H. D., Kandel, A., and Petrow, W. Behavioral effects of 19-hydroxytestosterone. In M. Sandler and G. L. Gessa (Eds.), *Sexual Behavior: Pharmacology and Biochemistry.* New York: Raven Press, 1975.

Johnston, P., and Davidson, J. M. Intracerebral androgens and sexual behavior in the male rat. *Hormones and Behavior,* 1972, *3,* 345–357.

Joshi, H. S., and Raeside, J. I. Synergistic effect of testosterone and oestrogens on accessory sex glands and sexual behaviour of the boar. *Journal of Reproduction and Fertility,* 1973, *33,* 411–423.

Kalra, S. P., and Kalra, P. S. Temporal interrelationships among circulating levels of estradiol, progesterone, and LH during the rat estrous cycle: Effect of exogenous progesterone. *Endocrinology,* 1974, *95,* 1711–1718.

Karavolis, H. J., and Herf, S. M. Conversion of progesterone by rat medial basal hypothalamic tissue to 5α-pregnane-3, 20-dione. *Endocrinology,* 1971, *89,* 940–942.

Karavolis, H. J., Hodges, D., and O'Brien, D. Uptake of [³H] progesterone and [³H] 5α-dihydroprogesterone by rat tissues *in vivo* and analysis of accumulated radioactivity: Accumulation of 5α-dihydroprogesterone by pituitary and hypothalamic tissues. *Endocrinology,* 1976, *98,* 164–175.

Kato, J. Ontogeny of 5α-dihydrotestosterone receptors in the hypothalamus of the rat. *Annales de Biologie Animale Biochimie Biophysique,* 1976, *16,* 467–469.

Kato, J., and Villee, C. A. Preferential uptake of estradiol by the anterior hypothalamus of the rat. *Endocrinology,* 1967, *80,* 567–575.

Kawahara, F. S., Berman, M. L., and Green, O. C. Conversion of progesterone 1-2-³H to 5α-pregnane-3, 20-dione by brain tissue. *Steroids,* 1975, *25,* 459–463.

Kelly, W. G., Judd, D., and Stolee, A. Aromatization of Δ⁴-androstene-3, 17-dione, 19-hydroxy-Δ⁴-androstene-3, 17-dione, and 19-oxo-Δ⁴-androstene-3, 17-dione at a common catalytic site in human placental microsomes. *Biochemistry,* 1977, *16,* 140–145.

Kniewald, Z., Massa, R., and Martini, L. Conversion of testosterone into 5α-androstan-17β-ol-3-one at the anterior pituitary and hypothalamic level. *Hormonal Steroids, Excerpta Medica International Congress Series,* 1970, *No. 219,* 784–791.

Kobayaski, R. M., and Reed, K. C. Conversion of androgens to estrogens (aromatization) in discrete regions of the rat brain: Sexual differences and effects of castration. *Society for Neuroscience Abstracts,* 1977, *3,* 348.

Koster, R. Hormone factors in male behavior of the female rat. *Endocrinology,* 1943, *33,* 337–348.

Krey, L. C., Lieberburg, I., Roy, E., and McEwen, B. S. Oestradiol plus receptor complexes in the brain and anterior pituitary gland: Quantitation and neuroendocrine significance. *Journal of Steroid Biochemistry,* 1979, *11,* 279–284.

Kubli-Garfias, C, and Whalen, R. E. Induction of lordosis behavior in female rats by intravenous administration of progestins. *Hormones and Behavior,* 1977, *9,* 380–386.

Labhsetwar, A. P., Joshi, H. S., and Watson, D. Temporal relationship between estradiol, estrone and progesterone secretion in ovarian venous blood of cyclic hamsters. *Biology of Reproduction,* 1973, *8,* 321–326.

Landau, I. T. Relationships between the effects of the anti-estrogen, CI-628, on sexual behavior, uterine growth, and cell nuclear estrogen retention after estradiol-17β-benzoate administration in the ovariectomized rat. *Brain Research,* 1977, *133,* 119–138.

Landau, I. T., and Feder, H. H. Whole cell and nuclear uptake of [³H] estriol in neural and peripheral tissues of the ovariectomized guinea pig. *Brain Research,* 1977, *121,* 190–195.

Langford, J., and Hilliard, J. Effect of 20α-hydroxypregn-4-en-3-one on mating behavior in spayed female rats. *Endocrinology,* 1967, *80,* 381–383.

Larsson, K., Södersten, P., and Beyer, C. Induction of male sexual behaviour by oestradiol benzoate in combination with dihydrotestosterone. *Journal of Endocrinology*, 1973a, *57*, 563–564.

Larsson, K., Södersten, P., and Beyer, C. Sexual behavior in male rats treated with estrogen in combination with dihydrotestosterone. *Hormone and Behavior*, 1973b, *4*, 289–299.

Larsson, K., Perez-Palacios, G., Morali, G., and Beyer, C. Effects of dihydrotestosterone and estradiol benzoate pretreatment upon testosterone-induced sexual behavior in the castrated male. rat. *Hormones and Behavior*, 1975, *6*, 1–8.

Larsson, K., Södersten, P., Beyer, C., Morali, G., and Perez-Palacios, G. Effects of estrone, estradiol and estriol combined with dihydrotestosterone on mounting and lordosis behavior in castrated male rats. *Hormones and Behavior*, 1976, *7*, 379–390.

Laumas, K. R. The distribution and localization in tissue of tritium labelled steroid sex hormones. *Proceedings of the Third Asia and Oceania Congress on Endocrinology*, 1967, 124–135.

Laumas, K. R., and Farooq, A. The uptake *in vivo* of 1,2-^{3}H-progesterone by the brain and genital tract of the rat. *Journal of Endocrinology*, 1966, *36*, 95–96.

Levine, S., and Mullins, R. F., Jr. Estrogen administered neonatally affects adult sexual behavior in male and female rats. *Science*. 1964, *144*, 185–187.

Lewis, J. G., Ghanadin, R., and Chisholm, G. D. Serum 5α-dihydrotestosterone and testosterone changes with age in man. *Acta Endocrinologica*, 1976, *82*, 444–448.

Liao, S. Receptors and the mechanism of action of androgens. In J. Pasqualini (Ed.), *Receptors and Mechanism of Action of Steroid Hormones*, Part I. New York: Marcel Dekker, 1976.

Lieberburg, I., and McEwen, B. S. Estradiol-17β: A metabolite of testosterone recovered in cell nuclei from limbic areas of adult male rat brains. *Brain Research*, 1975a, *91*, 171–174.

Lieberburg, I., and McEwen, B. S. Estradiol-17β: A metabolite of testosterone recovered in cell nuclei from limbic areas of neonatal rat brains. *Brain Research*, 1975b, *85*, 165–170.

Lieberburg, I., and McEwen, B. S. Brain cell nuclear retention of testosterone metabolites, 5α-dihydrotestosterone and estradiol-17β in adult rats. *Endocrinology*, 1977, *100*, 588–597.

Lieberburg, I., MacLusky, N. J., and McEwen, B. S. 5α-dihydrotestosterone (DHT) receptors in rat brain and pituitary cell nuclei. *Endocrinology*, 1977a, *100*, 598–607.

Lieberburg, I., Wallach, G., and McEwen, B. S. The effects of an inhibitor of aromatization (1,4,6-androstatriene-3,17-dione) and an anti-estrogen (CI-628) on *in vivo* formed testosterone metabolites recovered from neonatal rat brain tissues and purified cell nuclei. Implications for sexual differentiation of the rat brain. *Brain Research*, 1977b, *128*, 176–181.

Lindner, H. R. Androgens and related compounds in the spermatic vein blood of domestic animals: I. Neutral steroids secreted by the bull testis. *Journal of Endocrinology*, 1961a, *23*, 139–159.

Lindner, H. R. Androgens and related compounds in the spermatic vein blood of domestic animals: II. Species-linked differences in the metabolism of androstenedione in the blood. *Journal of Endocrinology*, 1961b, *23*, 161–166.

Lindner, H. R. Androgens and related compounds in the spermatic vein blood of domestic animals: IV. Testicular androgens in the ram, board and stallion. *Journal of Endocrinology*, 1961c, *23*, 171–178.

Lindner, H. R., and Mann, T. Relationship between the content of androgenic steroids in the testes and the secretory activity of the seminal vesicles in the bull. *Journal of Endocrinology*, 1960. *21*, 341–360.

Lodder, J., and Baum, M. J. Facilitation of mounting behavior by dihydrotestosterone propionate in castrated estriol benzoate-treated male rats following pudendectomy. *Behavioral Biology*, 1977, *20*, 141–148.

Loras, B., Genot, A., Monbon, M., Buecher, F., Reboud, J. P., and Bertrand, J. Binding and metabolism of testosterone in the rat brain during sexual maturation: II. Testosterone metabolism. *Journal of Steroid Biochemistry*, 1974, *5*, 425–431.

Luttge, W. G. Effects of anti-estrogens on testosterone stimulated male sexual behavior and peripheral target tissue in the castrated male rat. *Physiology and Behavior*, 1975, *14*, 839–840.

Luttge, W. G., and Hall, N. R. Differential effectiveness of testosterone and its metabolites in the induction of male sexual behavior in two strains of albino mice. *Hormones and Behavior*, 1973, *4*, 31–43.

Luttge, W. G., and Hall, N. R. Interactions of progesterone and dihydroprogesterone and dihydrotestosterone on estrogen activated sexual receptivity in female mice. *Hormones and Behavior*, 1976, *7*, 253–257.

Luttge, W. G., and Jasper, T. W. Studies on the possible role of 2-OH-estradiol in the control of sexual behavior in female rats. *Life Sciences*, 1977, *20*, 419–426.

Luttge, W. G., and Whalen, R. E. Partial defeminization by administration of androstenedione to neonatal female rats. *Life Sciences*, 1969, *8*, 1003–1008.

Luttge, W. G., and Whalen, R. E. Dihydrotestosterone, androstenedione, testosterone: Comparative effectiveness in masculinizing and defeminizing reproductive systems in male and female rats. *Hormones and Behavior*, 1970a, *1*, 265–281.

Luttge, W. G., and Whalen, R. E. Regional localization of estrogenic metabolites in the brains of male and female rats. *Steroids*, 1970b, *15*, 605–612.

Luttge, W. G., and Whalen, R. E. The accumulation, retention and interaction of oestradiol and oestrone in central neural and peripheral tissues of gonadectomized female rats. *Journal of Endocrinology*, 1972, *52*, 379–395.

Luttge, W. G., Hall, N. R., and Wallis, C. J. Studies on the neuroendocrine, somatic and behavioral effectiveness of testosterone and its 5α-reduced metabolites in Swiss-Webster mice. *Physiology and Behavior*, 1974a, *13*, 553–561.

Luttge, W. G., Wallis, C. J., and Hall, N. R. Effects of pre- and post-treatment with unlabelled steroids on the *in vivo* uptake of [³H] progestins in selected brain regions, uterus and plasma of the female mouse. *Brain Research*, 1974b, *71*, 105–115.

Luttge, W. G., Hall, N. R., Wallis, C. J., and Campbell, J. C. Stimulation of male and female sexual behavior in gonadectomized rats with estrogen and androgen therapy and its inhibition with concurrent anti-hormone therapy. *Physiology and Behavior*, 1975, *14*, 65–73.

Luttge, W. G., Gray, H. L., and Hughes, J. R. Regional and subcellular [³H] estradiol localization in selected brain regions and pituitary of female mice: Effects of unlabeled estradiol and various antihormones. *Brain Research*, 1976, *104*, 273–281.

Mainwaring, W. I. P. A review of the formation and binding of 5α-dihydrotestosterone in the prostate of the rat and other species. *Journal of Reproduction and Fertility*, 1975, *44*, 377–393.

Marrone, B. L., Rodriguez-Sierra, J. F., and Feder, H. H. Role of catechol estrogens in activation of lordosis in female rats and guinea pigs. *Pharmacology and Biochemistry of Behavior*, 1977, *7*, 13–17.

Mason, N. R., and Samuels, L. T. Incorporation of acetate-1-C¹⁴ into testosterone and 3β-hydroxysterols by the canine testis. *Endocrinology*, 1961, *68*, 899–907.

Massa, R., Stupnicka, E., Kniewald, Z., and Martini, L. The transformation of testosterone into dihydrotestosterone by the brain and the anterior pituitary. *Journal of Steroid Biochemistry*, 1972, *3*, 385–399.

Massa, R., Justo, S., and Martini, L. Conversion of testosterone into 5α-reduced metabolites in the anterior pituitary and in the brain of maturing rats. *Journal of Steroid Biochemistry*, 1975, *6*, 567–571.

Masuda, H., Anderson, L. L., Henricks, D. M., and Melampy, R. M. Progesterone in ovarian venous plasma and corpora lutea of the pig. *Endocrinology*, 1967, *80*, 240–246.

McDonald, P. G., and Doughty, C. Comparison of the effect of neonatal administration of testosterone and dihydrotestosterone in the female rat. *Journal of Reproduction and Fertility*, 1972, *30*, 55–62.

McDonald, P. G., and Doughty, C. Effect of neonatal administration of different androgens in the female rat: Correlation between aromatization and the induction of sterilization. *Journal of Endocrinology*, 1974, *61*, 95–103.

McDonald, P., Beyer, C., Newton, F., Brien, B., Baker, R., Tan, H. S., Sampson, C., Kitching, P., Greenhill, R., and Pritchard, D. Failure of 5α-dihydrotestosterone to initiate sexual behavior in the castrated male rat. *Nature* 1970, *227*, 964–965.

McEwen, B. S., Lieberburg, I., Chaptal, C., and Krey, L. C. Aromatization: Important for sexual differentiation of the neonatal rat brain. *Hormones and Behavior*, 1977a, *9*, 249–263.

McEwen, B. S., Lieberburg, I., MacLusky, N., and Plapinger, L. Do estrogen receptors play a role in the sexual differentiation of the rat brain? *Journal of Steroid Biochemistry*, 1977b, *8*, 593–598.

Meyerson, B. J. Latency between intravenous injection of progestins and the appearance of estrous behavior in estrogen-treated ovariectomized rats. *Hormones and Behavior*, 1972, *3*, 1–9.

Michael, S. D. Plasma prolactin and progesterone during the estrous cycle in the mouse. *Proceedings of the Society for Experimental Biology and Medicine*, 1976, *153*, 254–257.

Mickan, H. Metabolism of 4-¹⁴C-progesterone and 4-¹⁴C-testosterone in brain of the previable human fetus. *Steroids*, 1972, *19*, 659–668.

Moger, W. H. Serum 5α-androstane-3α, 17β-diol, androsterone, and testosterone concentrations in the male rat: Influence of age and gonadotropin stimulation. *Endocrinology*, 1977, *100*, 1027–1032.

Money, J. Sex hormones and other variables in human eroticism. In W. C. Young (Ed.), *Sex and Internal Secretions*. Baltimore: Williams and Wilkins, 1961.

Morali, G., Larsson, K., Perez-Palacios, G., and Beyer, C. Testosterone, androstenedione, and andros-tenediol: Effects on the initiation of mating behavior of inexperienced castrated male rats. *Hormones and Behavior,* 1974, *5*, 103–110.

Morali, G., Larsson, K., and Beyer, C. Inhibition of testosterone induced sexual behavior in the cas-trated male rat by aromatase inhibitors. *Hormones and Behavior,* 1977, *9*, 203–213.

Mullins, R. F., Jr., and Levine, S. Hormonal determinants during infancy of adult sexual behavior in the female rat. *Physiology and Behavior,* 1968, *3*, 333–338.

Nadler, R. D. A biphasic influence of progesterone on sexual receptivity of spayed female rats. *Physi-ology and Behavior,* 1970, *5*, 95–97.

Naftolin, R., and Ryan, K. J. The metabolism of androgens in central neuroendocrine tissue. *Journal of Steroid Biochemistry,* 1975, *6*, 993–997.

Naftolin, F., Ryan, K. J., and Petro, Z. Aromatization of androstenedione by the diencephalon. *Journal of Clinical Endocrinology and Metabolism,* 1971a, *33*, 368–370.

Naftolin, F., Ryan, K. J., and Petro, Z. Aromatization of androstenedione by limbic system from human foetuses. *Journal of Endocrinology,* 1971b, *51*, 795–796.

Naftolin, F., Ryan, K. J., and Petro, Z. Aromatization of androstenedione by the anterior hypothalamus of adult male and female rats. *Endocrinology,* 1972, *90*, 295–298.

Naftolin, F., Ryan, K. J., Davies, I. J., Reddy, V. V., Flores, F., Petro, Z., Kuhn, M., White, R. J., Tak-aoka, Y., and Wolin, L. The formation of estrogens by central neuroendocrine tissues. *Recent Prog-ress in Hormone Research,* 1975, *31*, 295–319.

Neill, J. D., Johansson, E. D. B., and Knobil, E. Levels of progesterone in peripheral plasma during the menstrual cycle of the rhesus monkey. *Endocrinology,* 1967, *81*, 1161–1164.

Noble, R. G., and Alsum, P. B. Hormone-dependent sex dimorphisms in the golden hamster *(Mesocri-cetus auratus). Physiology and Behavior,* 1975, *14*, 567–574.

Nordeen, E. J., and Yahr, P. Activation and differentiation of sexual behavior and translocation of hypo-thalamic estrogen receptors in rats by 6α-fluorotestosterone. *Hormones and Behavior,* 1981, *15*, 123–140.

Olsen, K. L. Induction of male mating behavior in androgen-insensitive (tfm) and normal (King-Holtz-man) male rats: Effect of testosterone propionate, estradiol benzoate, and dihydrotestosterone. *Hormones and Behavior,* 1979, *13*, 66–84.

Parrott, R. F. Effects of 17β-hydroxy-4-androsten-19-ol-3-one (19-hydroxytestosterone) and 5α-andros-tan-17β-ol-3-one (dihydrotestosterone) on aspects of sexual behavior in castrated male rats. *Journal of Endocrinology,* 1974, *61*, 105–115.

Parrott, R. F. Aromatizable and 5α-reduced androgens: Differentiation between central and peripheral effects on male rat sexual behavior. *Hormones and Behavior,* 1975a, *6*, 99–108.

Parrott, R. F. Stimulation of sexual behaviour in male and female rats with synthetic androgen, 17β-hydroxyoestra-4-en-3-one (19-nortestosterone). *Journal of Endocrinology,* 1975b, *65*, 285–286.

Parrott, R. F. Homotypical sexual behavior in gonadectomized female and male rats treated with 5α-19-hydroxytestosterone: Comparison with related androgens. *Hormones and Behavior,* 1976, *7*, 207–215.

Parrott, R. F., and McDonald, P. G. Sexual behaviour of male rats, implanted in the brain with 19-hydroxytestosterone. *Journal of Endocrinology,* 1975, *64*, 37P–39P.

Parvizi, N., and Ellendorff, I. 2-Hydroxy-oestradiol-17β as a possible link in steroid brain interaction. *Nature,* 1975, *256*, 59–60.

Paul, S. M., and Axelrod, J. Catechol estrogens: Presence in brain and endocrine tissues. *Science,* 1977, *197*, 657–659.

Paup, D. G., Coniglio, L. P., and Clemens, L. G. Masculinization of the female golden hamster by neo-natal treatment with androgen or estrogen. *Hormones and Behavior,* 1972, *3*, 121–131.

Paup, D. C., Mennin, S. P., and Gorski, R. A. Androgen- and estrogen-induced copulatory behavior and inhibition of luteinizing hormone (LH) secretion in the male rat. *Hormones and Behavior,* 1975, *6*, 35–46.

Payne, A. P. A comparison of the effects of neonatally administered testosterone, testosterone propio-nate and dihydrotesterone on aggression and sexual behavior in the female golden hamster. *Journal of Endocrinology,* 1976, *69*, 23–31.

Perez, A. E., Ortiz, A., Cabeza, M., Beyer, C., and Perez-Palacios, G. *In vitro* metabolism of ^{3}H-andro-stenedione by the male rat pituitary, hypothalamus, and hippocampus. *Steroids,* 1975, *25*, 53–62.

Perez-Palacios, G., Casteneda, E., Gomez-Perez, A. E., and Gual, C. *In vitro* metabolism of androgens in dog hypothalamus, pituitary, and limbic system. *Biology of Reproduction,* 1970, *3,* 205–213.

Perez-Palacios, G., Perez, A. E., Cruz, M. L., and Beyer, C. Comparative uptake of [³H] androgens by the brain and the pituitary of castrated male rats. *Biology of Reproduction,* 1973, *8,* 395–399.

Peterson, R. E., Imperato-McGinley, J., Gautier, T., and Sturla, E. Male pseudohermaphroditism due to steroid 5α-reductase deficiency. *American Journal of Medicine,* 1977, *62,* 170–191.

Pfaff, D. W. Nature of sex hormone effects on rat sex behavior: Specificity of effects and individual patterns of response. *Journal of Comparative and Physiological Psychology,* 1970, *73,* 349–358.

Pfaff, D. W., and Zigmond, R. E. Neonatal androgen effects on sexual and nonsexual behavior of adult rats tested under various hormone regimes. *Neuroendocrinology,* 1971, *7,* 129–145.

Phoenix, C. H. Hypothalamic regulation of sexual behavior in male guinea pigs. *Journal of Comparative Physiology and Psychology,* 1961, *54,* 72–77.

Phoenix, C. H. Effects of dihydrotestosterone on sexual behavior of castrated male rhesus monkeys. *Physiology and Behavior,* 1974a *12,* 1045–1055.

Phoenix, C. H. The role of androgens in the sexual behavior of adult male rhesus monkeys. In W. Montagna and W. A. Sadler (Eds.), *Reproductive Behavior.* New York: Plenum Press, 1974b.

Phoenix, C. H. Sexual behavior of castrated male rhesus monkeys treated with 19-hydroxytestosterone. *Physiology and Behavior,* 1976, *16,* 305–310.

Phoenix, C. H. Steroids and sexual behavior in castrated male rhesus monkeys. *Hormones and Behavior,* 1978, *10,* 1–9.

Phoenix, C. H., Goy, R. W., Gerall, A. A., and Young, W. C. Organizing action of prenatally administered testosterone propionate on the tissues mediating mating behavior in the female guinea pig. *Endocrinology,* 1959, *65,* 369–382.

Phoenix, C. H., Copenhaver, K. H., and Brenner, R. M. Scanning electron microscopy of penile papillae in intact and castrated rats. *Hormones and Behavior,* 1976, *7,* 217–227.

Plapinger, L., and McEwen, B. S. Gonadal steroid-brain interactions in sexual differentiation. In J. B. Hutchison (Ed.), *Biological Determinants of Sexual Behaviour.* Chichester, England: Wiley, 1978.

Plotka, E. D., Seal, V. S., Schobert, E. E., and Schomoller, G. C. Serum progesterone and estrogens in elephants. *Endocrinology,* 1975, *97,* 485–487.

Podesta, E. J., and Rivarola, M. A. Concentrations of androgens in whole testis, seminiferous tabules and interstitial tissue of rats at different stages of development. *Endocrinology,* 1974, *95,* 455–461.

Popolow, H. B., and Ward, I. L. Effects of perinatal androstenedione on sexual differentiation in female rats. *Journal of Comparative and Physiological Psychology,* 1978, *92,* 13–21.

Powers, J. B., and Valenstein, E. S. Individual differences in sexual responsiveness to estrogen and progesterone in ovariectomized rats. *Physiology and Behavior,* 1972, *8,* 673–676.

Presl, J., Herzmann, J., Rohling, G., and Hornsky, J. Regional distribution of estrogenic metabolites in the female rat hypothalamus. *Endocrinologia Experimentalis,* 1973, *7,* 119–123.

Preslock, J. P., Hampton, S. H., and Hampton, J. K., Jr. Cyclic variations of serum progestins and immunoreactive estrogens in marmosets. *Endocrinology,* 1973, *92,* 1096–1101.

Reddy, V. V. R., Naftolin, F., and Ryan, K. J. Aromatization in the central nervous system of rabbits: Effects of castration and hormone treatment. *Endocrinology,* 1973, *92,* 589–594.

Reddy, V. V. R., Naftolin, F., and Ryan, K. J. Conversion of androstenedione to estrone by neural tissues from fetal and neonatal rats. *Endocrinology,* 1974a, *94,* 117–121.

Reddy, V. V., Naftolin, F., and Ryan, K. J. Steroid 17β-oxidoreductase activity in the rabbit central nervous system and adenohypophysis. *Journal of Endocrinology,* 1974b, *62,* 401–402.

Resko, J. A. Plasma androgen levels of the rhesus monkey: Effects of age and season. *Endocrinology,* 1967, *81,* 1203–1212.

Resko, J. A. Androgen secretion by the fetal and neonatal rhesus monkey. *Endocrinology,* 1970a, *87,* 680–687.

Resko, J. A. Androgens in systemic plasma of male guinea pigs during development and after castration in adulthood. *Endocrinology,* 1970b, *86,* 1444–1447.

Resko, J. A., Feder, H. H., and Goy, R. W. Androgen concentrations in plasma and testis of developing rats. *Journal of Endocrinology,* 1968, *40,* 485–491.

Reyes, F. I., Winter, J. S. D., Faiman, C., and Hobson, W. C. Serial serum levels of gonadotropins, prolactin and sex steroids in the nonpregnant and pregnant chimpanzee. *Endocrinology,* 1975, *96,* 1447–1455.

Rezek, D. L. Nuclear localization of testosterone, dihydrotestosterone and estradiol-17β in basal rat brain. *Psychoneuroendocrinology*, 1977, *2*, 173–178.

Rezek, D. L., and Whalen, R. E. Localization of intravenously administered [³H] testosterone and its metabolites in the brain of the male rat: The absence of a major effect related to the time of day of the injection. *Journal of Steroid Biochemistry*, 1975, *6*, 1193–1199.

Robel, P., Corpechot, C., and Baulieu, E. E. Testosterone and androstanolone in rat plasma and tissues. *FEBS Letters*, 1973, *33*, 218–220.

Rommerts, F. F. G., and van der Molen, H. J. Occurrence and localization of 5α-steroid reductase, 3α- and 17β-hydroxysteroid dehydrogenase in hypothalamus and other brain tissues of the male rat. *Biochimica et Biophysica Acta*, 1971, *248*, 489–502.

Roy, E. J., and Wade, G. N. Binding of [³H] estradiol by brain cell nuclei and female rat sexual behavior: Inhibition by antiestrogens. *Brain Research*, 1977, *126*, 73–87.

Ryan, K. J. Biological aromatization of steroids. *Journal of Biological Chemistry*, 1959, *234*, 268–272.

Ryan, K. J. Estrogen formation by the human placenta: Studies on the mechanisms of steroid aromatization by mammalian tissue. *Acta Endocrinologica*, 1960, *35*, (Suppl. 51), 697–698.

Schally, A. V., Redding, T. W., and Arimura, A. Effect of sex steroids on pituitary responses to LH- and FSH-releasing hormone *in vitro*. *Endocrinology*, 1973, *93*, 893–902.

Schanbacher, B. D., and Ewing, L. L. Simultaneous determination of testosterone, 5α-androstan-17β-ol-3-one, 5α-androstane-3α, 17β-diol and 5α-androstane-3β, 17β-diol in plasma of adult male rabbits by radioimmunoassay. *Endocrinology*, 1975, *97*, 787–792.

Schindler, A. E. Metabolism of androstenedione and testosterone in human fetal brain. *Progress in Brain Research*, 1975, *42*, 330.

Schindler, A. E. Steroid metabolism in foetal tissues: IV. Conversion of testosterone to 5α-dihydrotestosterone in human foetal brain. *Journal of Steroid Biochemistry*. 1976, *7*, 97–100.

Schwarzel, W. C., Kruggel, W. G., and Brodie, H. J. Studies on the mechanisms of estrogen biosynthesis: VIII. The development of inhibitors of the enzyme system in the human placenta. *Endocrinology*, 1973, *92*, 866–880.

Seiki, K., Higashida, M., Imanishi, Y., Miyamoto, M., Kitagawa, T., and Kotani, M. Radioactivity in the rat hypothalamus and pituitary after injection of labelled progesterone. *Journal of Endocrinology*, 1968, *41*, 109–110.

Seiki, K., Miyamoto, M., Yamashita, A., and Kotani, M. Further studies on the uptake of labelled progesterone by hypothalamus and pituitary of rats. *Journal of Endocrinology*, 1969, *43*, 129–130.

Selmanoff, M. K., Brodkin, L. D., Weiner, R. I., and Siiteri P. K. Aromatization and 5α-reduction of androgens in discrete hypothalamic and limbic regions of the male and female rat. *Endocrinology*, 1977a, *101*, 841–848.

Selmanoff, M. K., Goldman, B. S., and Ginsburg, B. E. Developmental changes in serum luteinizing hormone, follicle stimulating hormone and androgen levels in males of two inbred mouse strains. *Endocrinology*, 1977b, *100*, 122–127.

Shaikh, A. A. Estrone and estradiol levels in the ovarian venous blood from rats during the estrous cycle and pregnancy. *Biology of Reproduction*, 1971, *5*, 297–307.

Shaikh, A. A. Estrone, estradiol, progesterone and 17α-hydroxyprogesterone in the ovarian venous plasma during the estrous cycle of the hamster. *Endocrinology*, 1972, *91*, 1136–1140.

Shaikh, A. A., and Harper, M. J. K. Ovarian steroid secretion in estrus, mated and HCG-treated rabbits, determined by concurrent cannulation of both ovarian veins. *Biology of Reproduction*, 1972, *7*, 387–397.

Shaikh, A. A., and Shaikh, S. Adrenal and ovarian steroid secretion in the rat estrous cycle temporally related to gonadotropins and steroid levels found in peripheral plasma. *Endocrinology*, 1975, *96*, 37–44.

Sholiton, L. J., and Werk, E. E. The less-polar metabolites produced by incubation of testosterone-4-¹⁴C with rat and bovine brain. *Acta Endocrinologica*, 1969, *61*, 641–648.

Sholiton, L. J., Marnell, R. T., and Werk, E. E. Metabolism of testosterone-4-C¹⁴ by rat brain homogenates and subcellular fractions. *Steroids*, 1966, *8*, 265.

Sholiton, L. J., Hall, I. L., and Werk, E. E. The iso-polar metabolites produced by incubation of [4-¹⁴C] testosterone with rat and bovine brain. *Acta Endocrinologica*, 1970, *63*, 512–518.

Sholiton, L. J., Jones, E. E., and Werk, E. E. The uptake and metabolism of (1,2-³H)-testosterone by the brain of functionally hepatectomized and totally eviscerated male rats. *Steroids*, 1972, *20*, 399–415.

Sholiton, L. J., Taylor, B. B., and Lewis, H. P. The uptake and metabolism of labelled testosterone by the brain and pituitary of the male rhesus monkey *(Macaca mulatta)*. *Steroids*, 1974, *24*, 537–547.

Sholl, S. A., Robinson, J. A., and Goy, R. W. Neural uptake and metabolism of testosterone and dihydrotestosterone in the guinea pig. *Steroids*, 1975, *25*, 203–215.

Shore, L. S., and Snipes, C. A. Metabolism of testosterone *in vitro* by hypothalamus and other areas of rat brain. *Federation Proceedings*, 1971, *30*, 363.

Siiteri, P. K., and Thompson, E. A. Studies of human placental aromatase. *Journal of Steroid Biochemistry*, 1975, *6*, 317–322.

Smith, M. S., and McDonald, L. B. Serum levels of luteinizing hormone and progesterone during the estrous cycle, pseudopregnancy and pregnancy in the dog. *Endocrinology*, 1974, *94*, 404–412.

Smith, M. S., Freeman, M. E., and Neill, J. D. The control of progesterone secretion during the estrous cycle and early pseudopregnancy in the rat: Prolactin, gonadotropin and steroid levels association with rescue of the corpus-luteum of pseudopregnancy, *Endocrinology*, 1975, *96*, 219–226.

Snipes, C. A., and Shore, L. S. Metabolism of progesterone *in vitro* by neural and uterine tissues. *Federation Proceedings*, 1972, *146*, 31.

Södersten, P. Estrogen-activated sexual behavior in male rats. *Hormones and Behavior*, 1973, *4*, 247–256.

Södersten, P. Mounting behavior and lordosis behavior in castrated male rats treated with testosterone propionate, or with estradiol benzoate or dihydrotestosterone in combination with testosterone propionate. *Hormones and Behavior*, 1975, *6*, 109–126.

Södersten, P., Gray, C., Damassa, D. A., Smith, E. R., and Davidson, J. M. Effects of a non-steroidal antiandrogen on sexual behavior and pituitary-gonadal function in the male rat. *Endocrinology*, 1975, *97*, 1468–1475.

Södersten, P., Damassa, D. A., and Smith, E. R. Sexual behavior in developing male rats. *Hormones and Behavior*, 1977, *8*, 320–341.

Stabenfeldt, G. H., and Hendricks, A. G. Progesterone studies in the *Macaca fascicularis*. *Endocrinology*, 1973, *92*, 1296–1300.

Stern, J. J. Neonatal castration, androstenedione and mating behavior of the male rat. *Journal of Comparative and Physiological Psychology*, 1969, *69*, 608–612.

Stern, J. M., and Eisenfeld, A. J. Distribution and metabolism of ^{3}H-testosterone in castrated male rats: Effects of cyproterone, progesterone and unlabelled testosterone. *Endocrinology*, 1971, *88*, 1117–1125.

Tabei, T., and Henrichs, W. L. Metabolism of progesterone by the brain and pituitary gland of subhuman primates. *Neuroendocrinology*, 1974, *15*, 281–289.

Tabei, T., Haga. H., Heinrichs, W. L., and Herrmann, W. L. Metabolism of progesterone by rat brain pituitary gland and other tissues. *Steroids*, 1974, *23*, 651–666.

Telegdy, G., and Endroczi, E. The ovarian secretion of progesterone and 20α-hydroxypregn-4-en-3-one in rats during the estrous cycle. *Steroids*, 1963, *2*, 119–123.

Thompson, E. A., and Siiteri, P. K. Studies on the aromatization of C-19 androgens. *Annals of the New York Academy of Science*, 1973, *212*, 378–388.

Tiefer, L. Gonadal hormones and mating behavior in the adult golden hamster. *Hormones and Behavior*, 1970, *1*, 189–202.

Tiefer, L., and Johnson, W. A. Restorative effect of various androgens on copulatory behavior of the male golden hamster. *Hormones and Behavior*, 1973, *4*, 359–364.

Tiefer, L., and Johnson, W. A. Neonatal androstenedione and adult sexual behavior in golden hamsters. *Journal of Comparative and Physiological Psychology*, 1975, *88*, 239–247.

Tremblay, R. R., Forest, M. G., Shalf, J., Martel, J. G., Kowarski, A., and Migeon, C. J. Studies on the dynamics of plasma androgens and on the origin of dihydrotestosterone in dogs. *Endocrinology*, 1972, *91*, 556–561.

Tulchensky, D., and Korenman, S. H. A radio-ligand assay for plasma estrone: Normal values and variations during the menstrual cycle. *Journal of Endocrinology and Metabolism*, 1970,, *31*, 76–80.

van der Molen, H. J., and Groen, D. Interconversion of progesterone and 20α-dihydroprogesterone and of androstenedione and testosterone *in vitro* by blood erythrocytes. *Acta Endocrinologica*, 1968, *58*, 419–444.

van Doorn, E. J., Burns, B., Wood, D., Bird, C. E., and Clark, A. F. *In vivo* metabolism of ^{3}H-dihydrotestosterone and ^{3}H-androstanediol in adult male rats. *Journal of Steroid Biochemistry*, 1975, *6*, 1549–1554.

van Tienhoven, A. *Reproductive Physiology of Vertebrates*. Philadelphia: W. B. Saunders, 1968.

Verhoeven, G., Lamberigts, G., and de Moor, F. Nucleus-associated steroid 5α-reductase activity and androgen responsiveness: A study in various organs and brain regions of rats. *Journal of Steroid Biochemistry*, 1974, *5*, 93–100.

Villee, C. A. Effects of sex hormones on the genetic mechanism. In K. W. McKerns (Ed.), *The Sex Steroids*. New York: Appleton-Century-Crofts, 1971.

Vreeburg, J. T. M., Schretlen, P. J. M., and Baum, M. J. Specific, high-affinity binding of 17β-estradiol in cytosols from several brain regions and pituitary of intact and castrated adult male rats. *Endocrinology*, 1975, *97*, 969–977.

Vreeburg, J. T. M., van der Vaart, P. D. M., and van der Schoot, P. Prevention of central defeminization but not masculinization in male rats by inhibition neonatally of oestrogen biosynthesis. *Journal of Endocrinology*, 1977, *74*, 375–382.

Wade, G. N., and Feder, H. H. Effects of several pregnane and pregnene steroids on estrous behavior in ovariectomized guinea pigs. *Physiology and Behavior*, 1972a, *9*, 773–775.

Wade, G. N., and Feder, H. H. [1,2-³H] Progesterone uptake by guinea pig brain and uterus: Differential localization, time course of uptake and metabolism and effects of age, sex, estrogen-priming and competing steroids. *Brain Research*, 1972b, *45*, 525–543.

Wade, G. N., and Feder, H. H. Uptake of [1,2-³H] 20α-hydroxypregn-4-ene-3-one, [1,2-³H] corticosterone and [6,7-³H] estradiol-17β by guinea pig brain and uterus: Comparison with uptake of [1,2-³H] progesterone. *Brain Research*, 1972c, *45*, 545–554.

Wallis, C. J., and Luttge, W. G. Maintenance of male sexual behavior by combined treatment with oestrogen and dihydrotestosterone in CD-1 mice. *Journal of Endocrinology*, 1975, *66*, 257–262.

Weisz, J., and Gibbs, C. Conversion of testosterone and androstenedione to estrogens *in vitro* by the brain of female rats. *Endocrinology*, 1974a, *94*, 616–620.

Weisz, J., and Gibbs, C. Metabolites of testosterone in the brain of the newborn female rat after an injection of tritiated testosterone. *Neuroendocrinology*, 1974b, *14*, 72–86.

Whalen, R. E. Hormone induced changes in the organization of sexual behavior in the male rat. *Journal of Comparative and Physiological Psychology*, 1964, *57*, 175–182.

Whalen, R. E. Sexual differentiation: Models, methods and mechanisms. In R. C. Friedman, R. M. Richart, and R. L. Vande Wiele (Eds.), *Sex Differences in Behavior*. New York: Wiley, 1974.

Whalen, R. E., and DeBold, J. F. Comparative effectiveness of testosterone, androstenedione and dihydrotestosterone in maintaining mating behavior in the castrated male hamster. *Endocrinology*, 1974, *95*, 1674–1679.

Whalen, R. E., and Edwards, D. A. Hormonal determinants of the development of masculine and feminine behavior in male and female rats. *Anatomical Record*, 1967, *157*, 173–180.

Whalen, R. E., and Edwards, D. A. Effects of the antiandrogen cyproterone acetate on mating behavior and seminal vesicle tissue in male rats. *Endocrinology*, 1969, *84*, 155–156.

Whalen, R. E., and Etgen, A. M. Masculation and defeminization induced in female hamsters by neonatal treatment with estradiol benzoate and RU-2858. *Hormones and Behavior*, 1978, *10*, 170–177.

Whalen, R. E., and Gorzalka, B. B. The effects of progesterone and its metabolites on the induction of sexual receptivity in rats. *Hormones and Behavior*, 1972, *3*, 221–226.

Whalen, R. E., and Luttge, W. C. Differential localization of progesterone uptake in brain: Role of sex, estrogen pretreatment and adrenalectomy. *Brain Research*, 1971a, *33*, 147–155.

Whalen, R. E., and Luttge, W. G. Perinatal administration of dihydrotestosterone to female rats and the development of reproductive function. *Endocrinology*, 1971b, *89*, 1320–1322.

Whalen, R. E., and Luttge, W. G. Role of the adrenal in the preferential accumulation of progestin by mesencephalic structures. *Steroids*, 1971c, *18*, 141–146.

Whalen, R. E., and Luttge, W. G. Testosterone, androstenedione and dihydrotestosterone: Effects on mating behavior of male rats. *Hormones and Behavior*, 1971d, *2*, 117–125.

Whalen, R. E., and Nadler, R. D. Suppression of the development of female mating behavior by estrogen administered in infancy. *Science*, 1963, *141*, 273–274.

Whalen, R. E., and Olsen, K. L. Prednisolone modifies estrogen-induced sexual differentiation. *Behavioral Biology*, 1978, *24*, 549–553.

Whalen, R. E., and Rezek, D. L. Localization of androgenic metabolites in the brain of rats administered testosterone or dihydrotestosterone. *Steroids*, 1972, *20*, 717–724.

Whalen, R. E., and Rezek, D. L. Inhibition of lordosis in female rats by subcutaneous implants of testosterone, androstenedione and dihydrotestosterone in infancy. *Hormones and Behavior*, 1974, *5*, 125–128.

Whalen, R. E., Battie, C., and Luttge, W. G. Anti-estrogen inhibition of androgen induced sexual receptivity in rats. *Behavioral Biology,* 1972, *7,* 311–321.

Wilson, J. G. Reproductive capacity of adult female rats treated prepuberally with estrogenic hormone. *Anatomical Record,* 1943, *86,* 341–359.

Yahr, P. Data and hypotheses in tales of dihydrotestosterone. *Hormones and Behavior,* 1979, *13,* 92–96.

Yahr, P., and Gerling, S. A. Aromatization and androgen stimulation of sexual behavior in male and female rats. *Hormones and Behavior,* 1978, *10,* 128–142..

Yoshinaga, K., Hawkins, R. A., and Stocker, J. F. Estrogen secretion by the rat ovary *in vivo* during the estrous cycle and pregnancy. *Endocrinology,* 1969, *85,* 103–112.

Young, W. C., Goy, R. W., and Phoenix, C. H. Hormones and sexual behavior. *Science,* 1964, *143,* 212–218.

Zigmond, R. E., and McEwen, B. S. Selective retention of oestradiol by cell nuclei in specific brain regions by the ovariectomized rat. *Journal of Neurochemistry,* 1970, *17,* 889–899.

Zucker, I. Effects of an antiandrogen on the mating behavior of male guinea pigs and rats. *Journal of Endocrinology,* 1966a, *35,* 209–210.

Zucker, I. Facilitatory and inhibitory effects of progesterone on sexual responses of spayed guinea pigs. *Journal of Comparative and Physiological Psychology,* 1966b, *62,* 376–381.

Zucker, I. Actions of progesterone in the control of sexual receptivity of the spayed female rat. *Journal of Comparative and Physiological Psychology,* 1967, *63,* 313–316.

Zucker, I., and Goy, R. W. Sexual receptivity in the guinea pig: Inhibitory and facilitatory actions of progesterone and related compounds. *Journal of Comparative and Physiological Psychology,* 1967, *64,* 378–383.

14

Steroid Hormone Receptors in Brain and Pituitary

Topography and Possible Functions

VICTORIA N. LUINE AND BRUCE S. MCEWEN

INTRODUCTION

That steroid hormones influence brain function and behavior has been an experimentally documented observation since the work of Berthold (1849) on the testicular control of mating and aggressive behavior in roosters. The role of gonadal steroids in the control of mating behavior is perhaps best described as a permissive action in which physical stimuli from the sexual partner actually trigger the behavior. However, the primary permissive role of gonadal steroids such as estradiol and testosterone does not appear to be that of a costimulus in that the hormone does not necessarily have to be present at the time that the behavior is elicited. This circumstance can now be better understood in the light of new information concerning a cellular mechanism of steroid hormone action that appears to be universal for all steroid target tissues, neural and nonneural. It is the purpose of this chapter to describe this cellular mechanism of steroid hormone action and to review evidence of its operation in the central nervous system and the pituitary gland in relation to the specific behavioral and neuroendocrine processes that are regulated by steroid hormones.

VICTORIA N. LUINE AND BRUCE S. MCEWEN   The Rockefeller University, New York, New York 10021. The work performed in this laboratory was supported by research grant NS07080 from the National Institutes of Health, and by Rockefeller Foundation grant RF70095 for research in reproductive biology.

VICTORIA N. LUINE
AND BRUCE S.
McEWEN

VISUALIZATION OF GENOMIC EFFECTS OF A STEROID HORMONE

The giant polytene salivary-gland chromosomes of *Diptera* respond to a steroid hormone, ecdysone, with a specific pattern of puffs of a few among many chromosomal bands (Clever and Karlson, 1960). This particular event recapitulates the puffing of salivary gland chromosomes that takes place toward the end of the larval phase of development, a period during which ecdysone is secreted (Berendes, 1967). The puffing is believed to involve the unwinding of some of the identical units of DNA, many copies of which are lined up together to form the band. The puffs characteristically show an enhanced incorporation of labeled uridine into RNA and an accumulation of acidic proteins. Interestingly, there are, at least in one species, *Drosophila heidei,* puffs that regress as a result of ecdysone treatment, besides those puffs that form (Berendes, 1967). Thus, the action of ecdysone on these particularly visible chromosomes is to alter an entire *pattern* of chromosomal activity.

CELLULAR UPTAKE AND RETENTION OF RADIOACTIVE STEROID HORMONES

Among vertebrates, tissues such as the reproductive tract, the liver, the kidney, and the thymus each respond to particular steroid hormones in characteristic ways. For example, estradiol causes the uterus to increase in size through imbibation of water, followed by cell proliferation and by cell growth and maturation (Hamilton, 1968). Glucocorticoids stimulate lymphocytes of the thymus gland and other lymphatic tissues to "self-destruct" (Makman *et al.,* 1967) while inducing the formation of certain enzymes in the liver (Feigelson and Feigelson, 1965). These specific responses appear to involve hormone action at the genomic level, and their complexity and tissue specificity suggest a pattern of tissue-specific chromosomal activity under hormonal control.

The brain is also a target for steroid hormone action but is especially interesting and possibly unique from several standpoints. First, this single organ appears to be responsive to five major groups of steroid hormones: estrogens, androgens, progestins, glucocorticoids, and mineralocorticoids. Second, the target area for each group of hormones, where it is known, appears to be a discrete area (or several discrete areas) of the brain.

A common feature of brain and nonneural tissue as target tissues for steroid hormones is that they contain binding sites for particular hormones that may well function as receptors in mediating the hormone effect on the cell. These putative receptor sites have a high affinity and a low capacity for the hormone and were discovered only after the successful synthesis in the early 1960s of radioactive steroids labeled with high specific radioactivity tritium. The high specific activity and the low decay energy of the tritium isotope permitted the demonstration of these limited number of binding sites not only by biochemical techniques but also by autoradiography.

CELL NUCLEAR RETENTION OF RADIOACTIVE STEROID HORMONES

667

HORMONE
RECEPTORS IN
BRAIN AND
PITUITARY

As far as can be determined, steroids cross the cell membrane in a passive manner and thus, traveling in the blood, have access to virtually all cells. In certain tissues, the hormone is accumulated and retained. For example, the work of Jensen and Jacobson (1962) established for the uterus the accumulation and the retention of ^{3}H estradiol compared to another muscle, the diaphragm. Thus, retention was shown to be due to the presence of stereospecific binding proteins in two cellular compartments: the cell nuclear fraction (Noteboom and Gorski, 1965), which in the uterus is heavily contaminated with myofibrils; and the cytosol or supernatant from a high-speed centrifugation of uterine homogenate (Talwar *et al.*, 1964; Toft and Gorski, 1966). The subsequent autoradiographic study of uteri labeled with ^{3}H estradiol confirmed the cell nuclear localization of much of the accumulated isotope (Stumpf and Roth, 1966; Stumpf, 1968c). At 2°C, uterine fragments or cells exposed to ^{3}H estradiol show little cell nuclear accumulation and substantial labeling of the cytoplasmic portion of the cell. On warning of the tissue, there is a migration of the label into the cell nucleus, which led Jensen *et al.* (1968) and Shyamala and Gorski (1969) to propose a "two-step" transfer mechanism: (1) passive entry into the cell and binding to a cytoplasmic receptor and (2) temperature-dependent transfer into the cell nucleus. It should be noted that, although the weight of opinion favors a cytoplasmic origin for cytosol hormone-binding protein, there are numerous examples of proteins that leak out of cell nuclei or other cell organelles after cell rupture. Except for the autoradiographic observations alluded to above concerning cells exposed to ^{3}H estradiol at 2°C, there is very little direct evidence on this important point.

THE SEARCH FOR CELL NUCLEAR ACCEPTOR SITES

It is clear in any event that there are two cellular compartments and a transfer of hormone between them. What is the driving force and mechanism of this transfer? It has been known for a number of years that uterine cytosol estradiol-receptor complexes undergo a transformation in size and shape from 4 to 5S on warming (see Jensen *et al.*, 1968; Gorski *et al.*, 1978; Jensen and deSombre, 1973). The transformation is catalyzed by the presence of DNA (Yamamato and Alberts, 1972) and appears to involve the addition of a $\approx$ 50,000 Dalton subunit "X" to the $\approx$ 60–70,000 Dalton estrogen-binding subunit (Notides and Nielson, 1974; Yamamato, 1974). Subunit "X" apparently does not remain associated with the estrogen-binding subunit during purification procedures (see Bresciani *et al.*, 1973). Puca and coworkers have also described the formation of this estrogen-binding subunit (a monomer of 60–70,000 Daltons) from its dimer through the action of Ca^{++}-dependent "receptor transforming factor" (Bresciani *et al.*, 1973).

Is there an intranuclear "acceptor" site for the cytosol hormone-receptor complex? Initial experiments with progesterone-receptor complexes indicated a saturation of binding to oviduct chromatin as a function of an increasing concentration of the complex and a limited degree of tissue specificity (O'Malley *et al.*,

1972). The chromatin-binding sites called *acceptors* appear to be a class of acidic cell nuclear proteins (O'Malley *et al.*, 1972). Studies on intact uterine cells, however, revealed that, as a function of increasing saturation of the total cell "receptor" from 5% to 95%, a constant proportion, around 85%, of labeled hormone was found in the cell nuclear compartment (Williams and Gorski, 1972). This finding does not appear to be consistent with the "acceptor" concept, unless there are very few acceptor sites relative to the total amound of "receptor" in the nucleus. Further contradiction is found in two other experiments dealing with cell-free cytosol transferred to the nucleus (see also Williams, 1975): first, the loading of "acceptor" nuclei with unlabeled hormone-receptor complex before isolation failed to reduce the binding of labeled hormone-receptor complex as would be expected from the notion of a limited-capacity acceptor site (Higgins *et al.*, 1973b; Shepherd *et al.*, 1974); second, running *in vitro* cytosol-nuclear transfer experiments in the presence of constant cytosol protein produced a straight-line binding curve, indicating that the appearance of saturation might have been an artifact of variable protein concentration (Chamness *et al.*, 1974). In fairness to the original observations, done with progesterone and chick oviduct, it should be pointed out that these contradictory experiments were done with other steroid hormones and other target tissues. More recent evidence indicates that the saturable binding of cytosol progesterone receptors to nuclear acceptor sites in the oviduct occurs even in the presence of constant cytosol protein concentrations (Buller *et al.*, 1975a,b). Moreover, the notion of a nuclear "acceptor" also persists in the form of possible receptor interactions with DNA, as will now be discussed.

Because in many cases deoxyribonuclease releases bound steroid hormones from isolated target-cell nuclei, DNA appears to be somehow involved in the cell nuclear retention process (Musliner and Chader, 1971; Shyamala-Harris, 1971; King and Gordon, 1972; Higgins *et al.*, 1973a). Estrogen, progesterone, and glucocorticoid receptors, in crude cytosols as well as in those partially purified, bind to purified DNA in sucrose gradients or on DNA–cellulose columns (Clemens and Kleinsmith, 1972; Toft, 1972; Yamamoto and Alberts, 1972; Schrader *et al.*, 1972; King and Gordon, 1972; Mainwaring and Irving, 1973; Yamamoto, 1974; Yamamoto *et al.*, 1974). The interaction appears to have relatively low specificity for type of DNA (Yamamoto and Alberts, 1974), although another report indicates some selectivity of the 4S uterine "receptor" for eukaryote as opposed to bacterial DNA (Clemens and Kleinsmith, 1972). According to Yamamoto (1974), it is the factor X (see earlier discussion) that confers DNA-binding affinity on the estrogen receptor when it attaches to the 4S subunit, and factor X appears to be present in nontarget as well as target tissues. In this connection, Yamamoto *et al.* (1974) reported that certain glucocorticoid-insensitive variants of lymphoma cells contain glucocorticoid "receptors" that bind to DNA either less well or better than parental "receptors." It is not clear yet whether another "factor X" is involved in this interaction. Nor is it yet clear how the kind of interaction of the "receptor" with DNA as seen *in vitro* will lead to selective alterations of gene expression. Possibly, the specificity of the effect is conferred by the acidic and basic chromosomal proteins that apparently mask large portions of the genome. Or as suggested by Yamamoto and Alberts (1974), it may be that the nonspecific interactions of the receptor with DNA actually mask, by dilution, a smaller number of highly specific interactions.

EFFECTS OF STEROID HORMONES ON GENE TRANSCRIPTION

In spite of such ambiguities, the assay of hormone-dependent alterations in RNA metabolism reveal that, by whatever mechanism, the earliest events in estrogen action on uterus and oviduct involve the increase of mRNA production (see O'Malley and Means, 1974). In uterus, the activation of nucleoplasmic RNA polymerase II within 15 min of estrogen administration appears to be responsible for early-induced protein synthesis and for subsequent increases in polymerase I and II activity (Glasser *et al.*, 1972). In oviduct, the mRNA for ovalbumin is known to increase within 60 min of estrogen treatment from almost undetectable levels at time zero (see O'Malley and Means, 1974). The increase in mRNA synthesis appears to result from an increase in chromatin initiation sites, rather than from an increased rate of transcription or release of RNA product.

Glucocorticoid administration has been shown to increase in the liver by almost twofold the amount of mRNA for the enzyme trytophan oxygenase, the concentration of which is regulated by this class of hormones (Schutz *et al.*, 1973). Less direct evidence, from inhibitor studies, points to rapid mRNA production as a very early event in thymus lymphocytes treated with glucocorticoids, leading to decreased glucose transport in these cells (Young *et al.*, 1974).

Because in no case are purified receptors for one hormone available from different target tissues of one species, it is impossible to say if, for example, pituitary, brain, and uterine estrogen receptors are identical or different proteins. It is nevertheless important to understand this question, which can also be posed as the distinction between "smart" and "dumb" receptors (G. Tompkins, personal communication, 1974). The "smart" receptor would recognize its specific target sites on chromatin and thus be the primary determinant of which genes were to be activated. A tissue-specific estrogen receptor would, quite likely, also be a "smart" receptor, and there might exist within a given target cell multiple "isomers" of such receptors, the number depending on the number of genes to be activated. The "dumb" receptor would perhaps recognize only a single site (e.g., a regulatory site on an RNA polymerase molecule or a common regulatory DNA codon). It would not have to be tissue-specific. The specificity each tissue's response would be built into the genome during differentiation in much the same way that a particular cell is programmed to produce (or to be unable to produce) a cell-specific proteinlike hemoglobin.

CRITERIA FOR "RECEPTORS"

In the preceding remarks and in the rest of this chapter, steroid-binding protein are often referred to as *receptors* or *putative receptors*. It is essential to keep in mind the nature of the evidence supporting the use of these terms to refer to a binding protein (for fuller discussion, see McEwen *et al.*, 1974). First, there is the evidence that hormone-sensitive tissues contain binding proteins for the effective hormone, whereas nonsensitive tissues lack such proteins (e.g., uterus vs. diaphragm, as first shown by Jensen and Jacobson, 1962). In the brain, a corollary is that brain regions responsive to hormone implants have been shown to possesss a hormone-binding mechanism, whereas nonresponsive regions do not. Second, there are exceptions to the above that prove the rule, that is, mutant forms of normal hormone-responsive cells or tissues that lack normal hormone responsiveness and at the same time are deficient in "receptors" (for examples, see McEwen *et al.*, 1974). Third, active agonists or antagonists for particular steroid hormone effects bind to the putative receptors, whereas inactive steroids do not.

PROPERTIES AND TOPOGRAPHY OF PUTATIVE STEROID RECEPTORS IN BRAIN

ESTRADIOL

The first neural steroid hormone-binding sites to be recognized were those for estradiol (for reviews, see McEwen *et al.*, 1972; Kato, 1973; Zigmond, 1975). The initial studies of the *in vivo* accumulation of ^{3}H estradiol by whole itssue from the blood revealed extremely high accumulations in pituitary as well as uterus, as well as lower but substantial accumulations in the hypothalamic region of the brain. Within brain, the ^{3}H estradiol accumulation is highest in the hypothalamus and the preoptic region. A substantial proportion of tissue uptake into these brain regions and into pituitary and uterus can be blocked by concurrent administration with the ^{3}H estradiol of 100- or 1000-fold excesses of unlabeled 17β-estradiol, but not by similar excesses of unlabeled testosterone or 17α-estradiol (an inactive estrogen). Such competition at the tissue level establishes a substantial portion of the hormone uptake as a limited-capacity phenomenon with specificity for active estrogens.

Autoradiography has provided more detailed information about the distribution within the brain of binding sites for ^{3}H estradiol (Glascock and Michael, 1962; Michael, 1964, 1965a,b; Attramadal, 1964; Pfaff, 1968c; Anderson and Greenwald, 1969; Stumpf, 1968b, 1970; Warembourg, 1970a,b; Tuohimaa, 1971; Pfaff and Keiner, 1973). Neurons, rather than glial cells, appear to contain the highest concentrations of these putative receptor sites, and these neurons are concentrated within the hypophysiotropic area (the medial preoptic area and the anterior and medial-basal hypothalamus) and the cortical and medial nuclei of the amygdala. To a lesser extent, labeled cells are found in the midbrain central gray (Pfaff and Keiner, 1973), in the spinal cord (Keefer *et al.*, 1973), and in the ventral hippocampus (Pfaff and Keiner, 1973) of the female rat. Not all neurons within these areas bind estradiol, but many of the cells that do bind the hormone have an intensity of labeling comparable to that found in cells of the uterus and the pituitary. Autoradiographic studies have also established that ^{3}H estradiol-binding sites in the preoptic area and the tuberal hypothalamus have a phylogenetically stable distribution from fish to primates (Morrell *et al.*, 1975).

Cell fractionation studies of the pituitary and the hypothalamus, the preoptic area, and the amygdala demonstrated the existence of soluble (cytosol) binding sites that resemble those found in the uterus on the basis of sedimentation rate in sucrose-density gradients (approximately 8S at low ionic strength) and specificity of binding toward active estrogens such as 17β-estradiol and diethylstilbestrol (Kahwanango *et al.*, 1970; Eisenfeld, 1970; Mowles *et al.*, 1971; Korach and Muldoon, 1973, 1974; Maurer, 1974; Ginsburg *et al.*, 1974b; Plapinger and McEwen, 1973; Vertes and King, 1971; Notides, 1970; Kato *et al.*, 1970a; McGuire *et al.*, 1973). Quantitative estimates of the binding parameters indicate that the affinity for estradiol is high (Kd $\cong 1 \times 10^{-10}$M) and uniform among estrogen-sensitive tissues (see Table 1), and that the maximum concentration of binding sites is as much as 30 times higher in uterus and pituitary than in whole hypothalamus (Table 1). In spite of similarities in sedimentation behavior, affinity constant, and hormonal specificity, no one knows for sure if the estrogen-binding proteins are identical in these various estrogen target tissues.

If we take into consideration the DNA content of these tissues, the capacity of uterus and pituitary cytosol to bind ^{3}H estradiol corresponds to 12–19,000 molecules per cell (Notides, 1970; Leavitt *et al.*, 1973), whereas the capacity of cytosol from the entire hypothalamus is around 2–3,000 molecules per cell. These estimates are, of course, averages that do not take into account the proportion of cells in each tissue that bind the hormone.

The estradiol that attaches to cytosol estrogen-binding sites is transferred into the cell nuclei, and substantial amounts of ^{3}H estradiol can be recovered in isolated nuclei from these target tissues (Chader and Vilee, 1970; Zigmond and McEwen, 1970; Kato *et al.*, 1970b; Mowles *et al.*, 1971; Vertes and King, 1971; Friend and Leavitt, 1972; Payne *et al.*, 1973; McEwen *et al.*, 1975b). The relative magnitude of cell nuclear binding closely parallels the relative magnitude of cytosol "receptor" concentration in uterus, pituitary, and various brain regions. There is good agreement within the brain between autoradiographic demonstration of ^{3}H estradiol binding and cell nuclear isolation experiments. As shown in Figure 1 (McEwen *et al.*, 1975b), a dissection scheme was designed to remove discrete subregions of

TABLE 1. ESTRADIOL BINDING IN CYTOSOL OF FEMALE RAT TISSUES[a]

Source of cytosol	Kd (M $\times 10^{10}$)	Binding capacity (sites/ mg tissue $\times 10^{-8}$)
Uterus	1.76 ± 0.21	61 ± 9
Pituitary	0.69 ± 0.05	65 ± 6
Hypothalamus	1.48 ± 0.27	1.98 ± 0.12
Amygdala	1.08 ± 0.16	0.83 ± 0.04
Cerebral cortex	1.66 ± 0.65	0.18 ± 0.03

[a]Data from Ginsburg *et al.* (1974a). Determination by filtration on Sephadex LH-20. Results are from Scatchard plots of concentration dependence of ^{3}H-estradiol-17β binding.

VICTORIA N. LUINE
AND BRUCE S.
McEWEN

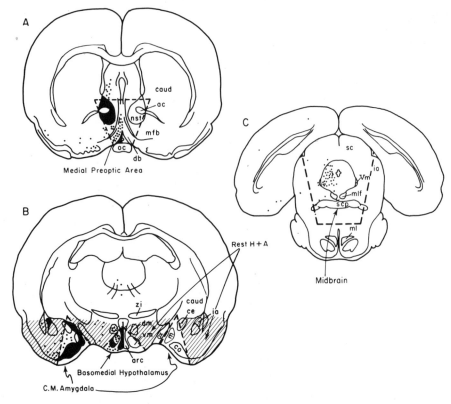

Fig. 1. Coronal planes through rat brain. The rostral-caudal positions of Planes A and B are illustrated in Luine *et al.* (1974). Details of the midbrain dissection (Plane C) may be found in McEwen and Pfaff (1970). Broken lines delineate the areas removed as the medial preoptic area, the basomedial hypothalamus, the corticomedial amygdala, and the midbrain. In B, the areas adjacent to basomedial hypothalamus and the corticomedial amygdala are called "Rest H plus A." The dots and the blackened areas (drawn on the left side only) indicate the location of the estradiol-concentrating cells found by autoradiography, and all three planes are adapted from Pfaff and Keiner (1973). Planes A and B are further modified from Luine *et al.* (1974). Abbreviations: ac, anterior commissure; arc, arcuate nucleus; caud, caudate nucleus; co, cortical nucleus of amygdala; db, diagonal bands; dm, dorsomedial nucleus of hypothalamus; mfb, medial forebrain bundle; ml, medial lemniscus; mlf, medial longitudinal fasciculus; nst, bed nucleus of stria terminalis; oc, optic chiasm; scp, superior cerebellar peduncle; vm, ventromedial nucleus of hypothalamus; zi, zona incerta. (From McEwen *et al.*, 1975b, by permission.)

hypothalamus, preoptic area, amygdala, and midbrain corresponding to the areas having the highest densities of labeling in the autoradiograms (Pfaff and Keiner, 1973). Ovariectomized female rats were injected intraperitonially with 10 μg of ^{3}H estradiol or ^{3}H diethylstilbestrol and sacrificed 1, 2, or 4 hr later. Cell nuclei were isolated from the above-mentioned subregions and from additional brain regions such as hippocampus, cerebral cortex, and the "rest" of the hypothalamus and amygdala. (See Figure 1 for details concerning the "rest" sample.) As shown in Figure 2, besides the pituitary, the highest concentrations of radioactivity were found in the basomedial hypothalamus (BMH), the medial preoptic area (MPO), and the corticomedial amygdala (CMA) samples. The concentrations in the "rest" sample were considerably lower than the BMH and the CMA but higher than the

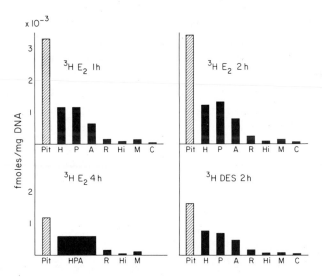

Fig. 2. Concentration in cell nuclei of radioactivity injected as [³H]estradiol-17β (E₂), or [³H]diethylstilbestrol (DES), expressed as fmoles/mg DNA. The time interval from labeling to sacrifice is indicated in the figure. Abbreviations: Pit, pituitary; H, basomedial hypothalamus; P, medial preoptic area; A, corticomedial amygdala; R, rest of H plus A; Hi, hippocampus; M, midbrain; C, cortex. Each panel of the figure presents data from one experiment using pooled tissue from three identically treated, ovariectomized female rats, each given 10 μg of [³H]estradiol. (From McEwen *et al.*, 1975b, by permission.)

other brain regions sampled. Among the other brain regions, the midbrain was highest, in agreement with the autoradiographic results of Pfaff and Keiner (1973). This result can be seen in Table 2, where nuclear binding is expressed as the molecules bound per cell. It should also be noted from Table 2 that the figure for the pituitary is within the range of estimates quoted above from data for cytosol binding. (In the uterus, it is known that 85% of the total cell receptor capacity ends up

TABLE 2. ESTIMATED CELL NUCLEAR BINDING CAPACITY FOR ESTRADIOL BINDING IN VIVO[a]

Tissue	Cell nuclear binding (molecules/cell)
Pituitary	12,500
Basomedial hypothalamus	4,400
Medial preoptic area	4,700
Corticomedial amygdala	2,850
Remainder of hypothalamus and amygdala	810
Midbrain	440
Hippocampus	200
Cerebral cortex	120

[a]Data from McEwen *et al.* (1975a). Cell nuclear binding was measured 2 hr after an intraperitoneal injection of 10μg ³H-estradiol-17β into ovariectomized, adult female rats. The brain dissection is depicted in Figure 1. Note that the binding values are the average value for all cells in a particular tissue.

in the cell nuclei; Williams and Gorski, 1972.) The estimates of molecules per cell Table 2 for the basomedial hypothalamus are higher than the estimate quoted above for the whole hypothalamus, and this difference is undoubtedly a reflection of the more restricted and specific sampling procedure. Warembourg (1970a) presented illuminating comparative data from autoradiography concerning the average number of silver grains per labeled cell in brain regions of the ovariectomized mouse after labeling with ^{3}H estradiol. In the bed nucleus of the stria terminalis, in the medial preoptic area, and in the arcuate nucleus, there were 90–120 grains per labeled cell. In the amygdaloid complex and the nucleus triangularis septi, there were 35–50 grains per labeled cell. It must be pointed out that differences in cell size may account for some of these differences in apparent "receptor" content.

Cell-nuclear-bound ^{3}H estradiol can be extracted as a complex with protein by KC1 in concentrations of 0.3 M or greater. These complexes have sedimentation coefficients in linear 5%–20% sucrose-density gradients run in the presence of 0.3 M salt, which are reported as 6–7S for the pituitary (Kato *et al.*, 1970b; Mowles *et al.*, 1971) and 5–7S for the hypothalamus (Vertes and King, 1971; Mowles *et al.*, 1971). As reported for calf uterus (Bresciani *et al.*, 1973), the 8.6 cytosol estradiol-binding protein is believed to be a tetrameric form of a subunit of molecular weight around 61,000 Daltons. Sedimentation coefficients of around 4S are reported for cytosol receptor treated with 0.2 M or greater KC1, indicating dissociation to the monomeric form. The cell nuclear form, which is often described as 5S, is believed to be a stabilized form of the 60–70,000 Dalton monomer (about 4S), to which a factor X of about 50,000 Daltons has attached (see earlier discussion). An implication of this sequence of events, which has not been finally proved, is that the nuclear "receptor" is derived from a "receptor" protein residing in the cytoplasm of unstimulated cells.

An interesting and novel assay procedure has been developed for bound estradiol in uterus, pituitary, and brain tissue by Clark and associates (Anderson *et al.*, 1972, 1973; Clark *et al.*, 1972). It utilizes the ability of the estradiol-receptor complex in a crude nuclear fraction of the tissue to exchange bound estrogen with ^{3}H estradiol during a 30 min incubation at 37°C *in vitro* without a significant loss of receptor capacity. In studies with uterus from intact adult rats, the levels of endogenous receptor-estrogen complex varied cyclically: metestrus, 220; diestrus, 750; proestrus, 120; and estrus, 310 fmoles/mg DNA. Fluctuations of ^{3}H estradiol uptake during the estrous cycle that are inversely related to these endogenous levels have been reported for hypothalamus tissue by McGuire and Lisk (1968) and by Kato (1970).

TESTOSTERONE

The story of testosterone "receptors" in neural and pituitary tissue must be considered in the light of two primary enzymatic conversions to which this hormone is subjected in these tissues: 5α-reduction and aromatization (Figure 3).

METABOLISM: 5α-REDUCTION. The 5α-reduction of testosterone is catalyzed by Δ4-3-ketosteroid-5α reductase, EC1.3.1.99 (Figure 3). The product, 17β-hydroxy-5α-androstan-3-one (5α-dihydrotestosterone, DHT), is more active than testosterone (T) in tests of the growth of chick comb, rat ventral prostate, seminal vesicle, and levator ani muscle (Liao and Fang, 1969). Further reduction products of DHT, 3β and 3α, 17β-dihydroxy-5α-androstane, appear to be less potent than testosterone in the same tests (Liao and Fang, 1969). DHT is highly effective in suppressing gonadotropin release, but not necessarily more so than T (Beyer *et al.*, 1971, 1972; Feder, 1971; Davidson, 1972; Swerdloff *et al.*, 1972; Paup *et al.*, 1975). However, DHT is less effective than T in restoring male sexual behavior in the castrated male rat (Beyer *et al.*, 1970a; McDonald *et al.*, 1970; Whalen and Luttge, 1971b; Johnston and Davidson, 1972; Paup *et al.*, 1975). The relative differences in potency between DHT and T with respect to gonadotropin secretion and sexual behavior in the male rat raise the possibility that different receptors may be involved. This possibility will be discussed again below in connection with the aromatization of testosterone. It should be noted that the relative ineffectiveness of DHT in restoring male sexual behavior is not observed in all mammals, as DHT is highly effective in stimulating male sexual behavior in castrated male guinea pigs and rhesus monkeys (Phoenix, 1974; Alsum and Goy, 1974).

If we can confine our discussion to the rat, it is instructive to compare the relative amounts of T and DHT in various target tissues. Table 3 summarizes the concentrations of T and DHT recovered and measured by gas–liquid chromatography from a variety of tissues of intact rats, and from plasma (Robel *et al.*, 1973), and the proportion of ³HT and ³HDHT recovered from such tissues of castrated rats following injections of ³HT (Stern and Eisenfeld, 1971; Jouan *et al.*, 1973; Monbon *et al.*, 1973). It can be seen that only in seminal vesicles and prostate is the concentration of DHT (likewise the percentage of ³HDHT) higher than that of T. Significant amounts of DHT are found in pituitary and hypothalamus and kidney but in lesser amounts than T. DHT is difficult to detect in levator ani muscle and thigh muscle, even though these tissues show an anabolic response to androgens, and the former tissue is more responsive to DHT than to T (Liao and Fang,

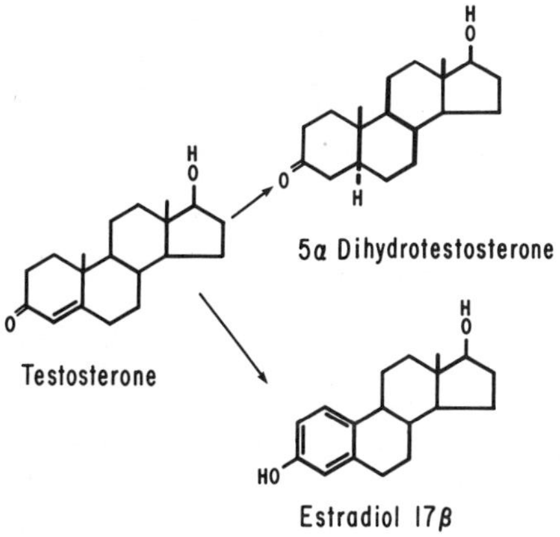

Fig. 3. Two enzymatic conversions of testosterone occurring in brain tissue.

TABLE 3. TESTOSTERONE AND 5α-DIHYDROTESTOSTERONE (DHT) IN
RAT PLASMA AND TISSUES

Sample	Testosterone[a]	DHT[a]	Testosterone[b]	DHT[b]
Plasma	2.5	<0.2	31	6
Seminal vesicles	2.2	3.0	23	72
Ventral prostate	2.0	2.8	—	—
Levator ani muscle	7.9	<0.2	—	—
Thigh muscles	2.9	<0.4	—	—
Hypothalamus	13.7	2.0	41	23
Pituitary	60.5	≤6.0	29	61
Parietal cortex	1.3	<0.4	39[c]	15[c]
Kidney	12.5	3.0	—	—
Small intestine	<0.2	<0.2	—	—

[a]Ng/gram tissue or ml plasma. Data from Robel *et al.* (1973).
[b]Percentage of total recovered radioactivity 1 hr after i.v. ^{3}H-testosterone.
Data from Stern and Eisenfeld (1971).
[c]Entire cerebrum.

1969). DHT, together with 5α-androstanediol and androstenedione, have been
detected as T metabolites in pituitary and brain of male rhesus monkies following
ventricular perfusion by ^{3}HT (Sholiton *et al.*, 1974).

It appears likely that DHT found in pituitary and brain tissue may have originated from 5α-reduction within that tissue. As shown in Figure 4, there is within
the brain and in the pituitary tissue a distinctive pattern of *in vitro* conversion of
^{3}H testosterone to ^{3}HDHT (black bars), 3α-17β-dihydroxy-5α-androstane DIOL
(gray bars), and androstenedione (white bars) (cf. Denef *et al.*, 1973, for data and

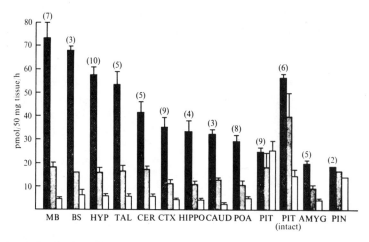

Fig. 4. Regional distribution of testosterone metabolism in the brain of the adult male rat. [1,2-
^{3}H]Testosterone was incubated at a concentration of 1.2×10^{-6} mol/l with brain slices. Preparations
with intact pituitary are also given. Values are means + S.E.M.; the number of measurements made for
each mean value is given in parentheses. Abbreviations: MB, midbrain; BS, brainstem; HYP, hypothalamus; TAL, thalamus; CER, cerebellum; CTX, parietal cortex; HIPPO, hippocampus; CAUD, caudate
nucleus; POA, preoptic area; PIT, anterior pituitary; AMYG, amygdala and overlying cortex; PIN, intact
pineal gland. Black bars, 5α-dihydrotestosterone; white bars, androstenedione; gray bars, 3α-androsta-
nediol. (From Denef *et al.*, 1973, by permission.)

references). Conversion to DHT is higher in slices of midbrain, brainstem, hypothalamus, and thalamus and also in intact pituitary than in cortex, hippocampus, amygdala, pineal gland, or cerebellum. DIOL formation is high in pituitary. DIOL and androstenedione formation is lower and less differentiated among brain regions than the formation of DHT. Sex differences in DHT and DIOL formation are seen in intact pituitaries (but not in brain, except for a small difference in hypothalamus), males having higher conversion rates than females (Denef *et al.*, 1973). DHT and DIOL formation in pituitary increases several fold after gonadectomy in both sexes, and the sex difference disappears (Stahl *et al.*, 1971; Massa *et al.*, 1972; Thien *et al.*, 1972, 1974; Denef *et al.*, 1973; Thieulant *et al.*, 1974b; Kniewald and Milkovic, 1973), but thyroidectomy and adrenalectomy are without effect on DHT formation (Denef *et al.*, 1973). Replacement therapy with testosterone propionate prevents the postcastration rise in pituitary DHT formation in male and female rats (Denef *et al.*, 1973; Thien *et al.*, 1974; Kniewald and Milkovic, 1973), but estradiol benzoate (0.1 μg/100 g body weight/day) is less effective in males than testosterone propionate (Denef *et al.*, 1973; Thien *et al.*, 1974) and less effective in males than in females (Denef *et el.*, 1973). In contrast to these results, high doses of estradiol benzoate (2 mg/100 g body weight/day) were found ineffective in attenuating the postcastration rise of DHT formation in female rats (Kniewald and Milkovic, 1973).

Correlations of these results with changes in gonadotropin secretion following castration and correlations between DHT formation and gonadotropin levels in postnatal development (Denef *et al.*, 1974) have led several laboratories to propose a relationship between DHT formation and follicle-stimulating-hormone (FSH) secretion (Kniewald *et al.*, 1971; Denef *et al.*, 1973, 1974). However, one should be cautioned against equating DHT formation with the DHT responsiveness of a particular tissue (Verhoeven *et al.*, 1974), as is illustrated by the low level of DHT in the levator ani muscle (Table 3) and the high responsiveness of this muscle to DHT (Liao and Fang, 1969). The discrepancies may be resolved by information regarding the presence or absence of DHT "receptors" in various tissues (see discussion below).

METABOLISM: AROMATIZATION. The biological conversion of androgens to estrogens has been recognized for many years not only as the pathway for estrogen formation in the gonad and the adrenal but also as an active process in the placenta (cf. Gual *et al.*, 1962, and Ryan, 1962, for references). According to this work the pathway for aromatization of Ring A appears to involve the hydroxylation of C-19 as a prelude to the removal of this carbon (cf. Schwarzel *et al.*, 1973). In 1971, Naftolin and co-workers reported conversion of ^{14}C androstenedione to estrone and estradiol by homogenates of the diencephalon (and not by the cerebral cortex) from 10- and 22-week human male fetuses. Subsequent work reported similar results from the adult as well as the fetal limbic (hypothalamus, amygdala, and hippocampus) tissue from rat, rabbit, and macaque (Naftolin *et al.*, 1971, 1972; Ryan *et al.*, 1972; Reddy *et al.*, 1973, 1974; Flores *et al.*, 1973). In these studies, the pituitary was consistently without significant conversion. Estrogen formation was also observed in two isolated perfused rhesus monkey brains from infused ^{3}H androstenedione (Flores *et al.*, 1973).

Whereas the major precursor and product in the hands of Naftolin and Ryan have been androstenedione (A) and estrone (E_1), respectively, Weisz and Gibbs (1974) found that by using tissue fragments instead of homogenates conversions of testosterone-7α-^{3}H to ^{3}H estradiol (E_2) could be observed in 7-day-old female rat hypothalamus and amygdala and in 120-day-old female rat hypothalamus. Moreover, conversion rates of 0.1%–0.4% (for T to E_2) and 0.2%–0.8% (for A to E_1) were observed during a 1-hr incubation of tissue fragments (Weisz and Gibbs, 1974), although Reddy et al. (1974) had reported lower conversion rates (for A to E_1) in homogenates. Both groups reported lower conversion of both T and A to estrogen in cerebral cortex than in hypothalamus or limbic structures.

The significance of the aromatization of testosterone and androstenedione may well be related to the reported facilitation effects of these androgens on female sexual receptivity and of estrogens on male sexual behavior. First, it has been clear since 1970 that, when given systemically, DHT is ineffective compared to testosterone in restoring sexual behavior in castrated male rats (McDonald et al., 1970; Feder, 1971; Whalen and Luttge, 1971b; Beyer et al., 1973; Beyer and Rivaud, 1973) and hamsters (Christensen et al., 1973) and marginally effective in doing so when implanted intracranially (Johnston and Davidson, 1972). However, besides testosterone, estradiol is also able to restore at least part of the male sexual behavior lost following castration (Ball, 1937; Beach, 1942a; Davidson, 1969; Pfaff, 1970; Södersten, 1973; Christensen and Clemens, 1974), and 19-hydroxytestosterone, a presumed intermediate in aromatization, is also effective in doing so even though it was without a maintenance effect on penile spines (Parrott, 1974). A particularly striking effect of microgram amounts of estradiol in mimicking a effect of milligram amounts of testosterone was reported by Fletcher and Short (1974) to be the restoration of full sexual behavior and the introduction of antler changes in the castrated red deer stag *(Cervus elaphus)*. Another impressive example of estrogenic effects if the demonstration that implants of estradiol in the preoptic area of castrated male rats are more effective than testosterone in restoring male copulatory activity (Christensen and Clemens, 1974).

Further support for the effectiveness of estrogen in restoring components of male sexual behavior comes from reports that the combined treatment of castrated male rats with estradiol benzoate and dihydrotestosterone propionate restores male copulatory behavior more effectively than either hormone alone (Baum and Vreeburg, 1973; Larsson et al., 1973a,b; Feder et al., 1974; Noble, 1974). Curiously, however, antiestrogens that block estrogen effects on female sexual behavior (Shirley et al., 1968; Meyerson and Lindstrom, 1968; Arai and Gorski, 1968; Komisaruk and Beyer, 1972; Whalen and Gorzalka, 1973; Södersten and Larsson, 1974; Södersten, 1974) are ineffective in blocking the testosterone or estradiol induction of male behavior (Beyer and Vidal, 1971; Whalen et al., 1972; Södersten, 1974).

With respect to female mating behavior, testosterone has a significant effect in restoring lordosis responding in spayed female rats, cats, and rabbits (Beach, 1942b; McDonald et al., 1970a; Beyer et al., 1970; Whalen and Hardy, 1970; Pfaff, 1970; Beyer and Komisaruk, 1971; Södersten, 1972; McDonald and Meyerson,

1973). DHT is without effect, but 19-hydroxyandrostenedione (an estrogen precursor) is effective (Beyer *et al.*, 1970b; Beyer and Komisaruk, 1971; McDonald and Meyerson, 1973). Further support for the conversion of androgen to estrogen in mediating female sexual receptivity is the observation that two estrogen antagonists, MER-25 and CI-628, attenuate testosterone as well as estrogen induction of lordosis responding (Whalen *et al.*, 1972; McDonald and Meyerson, 1973). Finally, both testosterone propionate and estradiol benzoate facilitate gonadotropin secretion under conditions in which DHT-propionate is without effect (Brown-Grant, 1974).

It is important to repeat the cautionary remarks made above in connection with the discussion of DHT, namely, that, in species such as guinea pig (Alsum and Goy, 1974) and rhesus monkey (Phoenix, 1974), DHT is a behaviorally active androgen, and that, in the former case at least, estradiol is without a restorative effect on male sexual behavior. Moreover, in the rat, DHT has proved to have some, albeit small, effect on lordosis behavior (Beyer *et al.*, 1971) and male copulatory behavior (Johnston and Davidson, 1972) after intracranial implantation or systemic administration (Paup *et al.*, 1975). These results are consistent with the conclusions of Baum *et al.* (1974), based on the effects of genital anesthetization, that DHT does not act solely on genital sensory receptors when acting synergistically with estradiol; rather, it probably also acts centrally. A study by Parrot (1975) further defines the nature of the central action of aromatizable androgens by showing that these steroids, and not 5α-reduced androgens, prevented significant increases in the postejaculatory refractory-period durations that followed castration. The data of Paup *et al.* (1975) are consistent with these observations.

It would thus appear that mammalian species differ in the degree to which neurons influenced by estradiol, by DHT, and even by testosterone itself may be linked together into networks that regulate sexual behavior. In the remainder of this section, we consider evidence for the existence of neural androgen "receptor" sites.

ESTROGEN RECEPTORS IN THE MALE BRAIN. The male rat pituitary and brain contain binding sites for ^{3}H estradiol with a regional distribution similar to that seen in the female brain (Eisenfeld and Axelrod, 1966; Pfaff, 1968b; Anderson and Greenwald, 1969; McEwen and Pfaff, 1970; Maurer and Woolley, 1974), and it is apparent from the evidence cited above that these sites may function in males in relation to the activation of male copulatory behavior.

TOPOGRAPHY OF RADIOACTIVITY INJECTED AS ^{3}H TESTOSTERONE. The distribution of cells labeled with an injection of ^{3}H testosterone in castrated male rats is similar but not identical to the distribution of cells labeled with ^{3}H estradiol (Pfaff, 1968a,b; Tuohimaa, 1971; Sar and Stumpf, 1972, 1973a; see also discussion by Zigmond, 1975), although the intensity of cellular labeling is generally lower for ^{3}HT than for ^{3}HE$_2$. The highest concentrations of labeled cells are seen in the hypophysiotrophic area, in the amygdala and the pituitary, and also in the lateral septum and hippocampus. ^{3}HT labels more cells than ^{3}HE$_2$ in lateral septum, hippocampus, dentate gyrus, subiculum, and ventromedial nucleus, whereas ^{3}HE$_2$ labels more cells than ^{3}HT in arcuate nucleus (Stumpf, 1970; Sar and Stumpf,

1973a). In the medial preoptic area, a majority of the cells are labeled by both ^{3}HT and ^{3}HE (Sar and Stumpf, 1973a). This finding clearly indicates an overlap of the cells binding the two labels, but it may well be explained by the known conversion of ^{3}HT to ^{3}HE$_2$ (see above). Between 35% and 50% of the cell-nuclear-bound radioactivity in adult male rat limbic structures has, in fact, been identified as ^{3}HE$_2$ following a ^{3}HT injection (Lieberburg and McEwen, 1975). Thus, the conversion of testosterone either to estradiol or to 5α-reduced metabolites makes difficult the interpretation of any labeling experiment with ^{3}HT. The pituitary is somewhat less complicated because no ^{3}HE$_2$ has been identified in cell nuclear extracts following ^{3}HT (Lieberburg and McEwen, 1975). Autoradiographic studies reveal that $\simeq$ 15% of rat pituitary cells, mostly basophils, are labeled after ^{3}HT, whereas 60%–80% of the cells are labeled after ^{3}HE$_2$ (Stumpf, 1968a; Sar and Stumpf, 1973b,e).

In birds, the cellular accumulation of radioactivity injected as ^{3}H testosterone has been reported in periventricular areas of the hypophysiotrophic area (Zigmond *et al.*, 1972; Meyer, 1973). In addition, the accumulation of radioactivity after ^{3}HT administration has been demonstrated in the midbrain of the chaffinch, primarily in the nucleus intercollicularis, an area from which vocalizations can be stimulated in birds (Zigmond *et al.*, 1973). Most of the radioactivity retained in the cell nuclei in this last study appears to be testosterone or its 5α- or 5β-reduced metabolites.

ANDROGEN UPTAKE AND BINDING IN THE MALE BRAIN. The tissue uptake of radioactivity injected as ^{3}H testosterone into castrated guinea pigs and rats is highest in the prostate and the seminal vesicles. Concentrations of radioactivity in pituitary and neural tissues, although lower than those in accessory sex glands, are nevertheless equal to or higher than those in serum or plasma. Pituitary generally is highest, followed by hypothalamus and cerebral cortex, although the concentrations of radioactivity among CNS structures are similar, and the regional differences that have been reported by one group have not always been found by another (Resko *et al.*, 1967; Roy and Laumus, 1969; McEwen *et al.*, 1970a,b; Stern and Eisenfeld, 1971; Phuong and Sauer, 1971; Perez-Palacios *et al.*, 1973; Sar and Stumpf, 1973c; Dixit and Niemi, 1974). Perez-Palacios *et al.*, (1973) compared the CNS tissue uptake of ^{3}H-labeled testosterone, 5α-DHT, and androstenedione and found that 5α-DHT was accumulated significantly more than the other two steroids relative to cerebral cortex by pituitary and by hippocampus and midbrain tegmentum. In this connection, it will be recalled that 5α-DHT is detected as a metabolite of T in pituitary and in brain regions (see Table 3 and discussion above). Estradiol has also been detected as a T metabolite in brain tissue (see discussion of aromatization above).

The uptake of radioactivity injected as ^{3}HT is significantly reduced by unlabeled testosterone in pituitary and in septum and less strongly inhibited in amygdala, preoptic area, hypothalamus, and olfactory bulb (McEwen *et al.*, 1970b; Stern and Eisenfeld, 1971). The antiandrogenic steroid cyproterone reduces the tissue uptake of radioactivity injected as ^{3}HT in pituitary, septum, preoptic area, amygdala, and hypothalamus (McEwen *et al.*, 1970b; Stern and Eisenfeld, 1971; Sar and Stumpf, 1973c). Progesterone also reduces the uptake of ^{3}HT radioactivity in pituitary, preoptic area, and central hypothalamus (Stern and Eisenfeld, 1971; Sar and

Stumpf, 1973c), and this effect may be due to the inhibition of Δ 4, 3-ketosteroid 5α-reductase by progesterone, a preferred substrate, and the resulting reduction of the formation of ^{3}H-5α-DHT (Stern and Eisenfeld, 1971; Massa and Martini, 1971–1972.

It should be noted that cyproterone is an inhibitor of the aromatizing enzyme complex of placenta (Schwarzel *et al.*, 1973), and thus, its effects on ^{3}HT uptake noted above might be due in part to its ability to block ^{3}HE$_2$ formation. In this connection, it was noted by McEwen *et al.* (1970b) that unlabeled E$_2$ competed as well as, or better than, unlabeled T for the tissue uptake of ^{3}HT radioactivity by preoptic area, septum, and olfactory bulb. In contrast, T competed better than E for the uptake of ^{3}HT radioactivity by pituitary (McEwen *et al.*, 1970b). These observations are consistent with the detection of ^{3}HE$_2$ as a metabolite of ^{3}HT in cell nuclei from limbic rat-brain structures and failure to detect ^{3}HE$_2$ in pituitaries of the same animals (see above).

Besides the effects noted above on the uptake of ^{3}HT radioactivity, several laboratories have reported that neonatal castration reduces the uptake of ^{3}HT radioactivity, compared to that in adult castrate males, in pituitary and in all brain structures (McEwen *et al.*, 1970b), whereas androgen treatment of neonatal castrates selectively increases the uptake of ^{3}HT radioactivity in anterior hypothalamus, pituitary, seminal vesicles, and ventral prostate (Dixit and Niemi, 1974). Another treatment that reduces brainwide the uptake of ^{3}HT radioactivity is hypophysectomy (McEwen *et al.*, 1970a).

The binding of ^{3}H testosterone or ^{3}H5αDHT to soluble macromolecules from the brain regions and the pituitary of castrated male rats has been described in a number of laboratories (Samperez *et al.*, 1960a,b; Jouan *et al.*, 1971a,b, 1973; Thieulant *et al.*, 1974a; Monbon *et al.*, 1973, 1974; Loras *et al.*, 1974; Kato and Onouchi, 1973; Naess and Attramadal, 1974; Naess *et al.*, 1975). According to one laboratory, the soluble form of the "receptor" from hypothalamus has a sedimentation rate constant of 8.6S and a dissociation constant of 7×10^{-10}M and appears to bind 5αDHT preferentially (Kato and Onouchi, 1973). Another group reported that such molecules sediment at 6–7S and bind T and cyproterone as well as DHT but do not bind estradiol or cortisol (Naess *et al.*, 1975). A third report indicates that cytosol DHT-binding sites are found in the following descending order of abundance in castrate male rats: ventral prostate, 100; pituitary, 32; hypothalamus, 14.5; amygdala, 7.4; and cortex, 8.8 (Ginsburg *et al.*, 1974a). Dissociation constants in that study were on the order of $1 - 2 \times 10^{-9}$ M.

The regional distribution of these androgen-binding molecules agrees generally with the reported distribution by autoradiography of cells accumulating radioactivity injected as ^{3}HT (see above). Besides cytosol binding, there also appears to be evidence for the cell nuclear retention of radioactivity injected as ^{3}HT (Jouan *et al.*, 1973; Monbon *et al.*, 1973; Zigmond *et al.*, 1972). In all of the above-mentioned studies, bound radioactivity in either cell nuclear or cytosol fraction contains some 5αDHT, but T is the major form of the recovered radioactivity. As noted above, ^{3}HE$_2$ was identified in purified cell nuclei after the administration of ^{3}H-7-testosterone (Lieberburg and McEwen, 1975). In that study, isotopic hydrogen in the 7 position was not lost during aromatization, whereas in studies cited above, where 1, 2 labeled ^{3}HT was used (Naess *et al.*, 1975), as much as 75% of the ^{3}H was lost during aromatization (Kelly, 1974).

PROGESTINS

Progesterone (P), like testosterone, is likely to undergo one of a number of metabolic transformations in the body. Its conversion to 5αdihydroprogesterone (5αDHP) and to 20αdihydroprogesterone has been demonstrated to occur *in vitro* in pituitary and brain tissue (Massa and Martini, 1971–1972; Karavolas and Herf, 1971; Robinson and Karavolas, 1973; Cheng and Karavolas, 1973; Tabei *et al.*, 1974. The further transformation of 5αdihydroprogesterone to 3αhydroxypregnan-20-one and of 20αhydroxyprogesterone to 20αhydroxy-5αpregnane-3-one and 5αpregnane-3α,20α-diol has also been demonstrated (Robinson and Karavolas, 1973; Cheng and Karavolas, 1973; Nowak and Karavolas, 1974). Although differences in hypothalamic 5α-reduction *in vitro* occur as a function of the stage of the estrous cycle of the donor animal, estradiol priming of ovariectomized females did not significantly alter the *in vitro* hypothalamic 5α-reduction of progesterone (Cheng and Karavolas, 1973). 5αDHP and its 3αOH derivative have been reported to be active, but less so than P itself, in inducing lordosis behavior in estrogen-primed rodents (Meyerson, 1967; Whalen and Gorzalka, 1972; Wade and Feder, 1972a; Czaja *et al.*, 1974). 5αDHP, and not 5βDHP, appears to mimic the effect of progesterone in inhibiting ovulation under rigidly defined experimental conditions in immature rats treated with PMSG (Sanyal and Todd, 1972). 20α-Dihydroxyprogesterone is effective in maintaining luteinizing hormone (LH) release following mating in the rabbit (Hilliard *et al.*, 1976) but is without effect on lordosis behavior in the estrogen-primed rat (Whalen and Gorzalka, 1972).

Progesterone is accumulated by the brain from the blood rapidly and in large amounts (Laumas and Farooq, 1966; Seiki *et al.*, 1968, 1969; Raisinghani *et al.*, 1968; Whalen and Luttge, 1971a; Wade and Feder, 1972b,c; Luttge *et al.*, 1973, 1974; Wade *et al.*, 1973; Whalen and Gorzalka, 1974). There appear to be very little difference in the accumulation of this hormone between pituitary and various brain structures, except for a higher accumulation in midbrain and brainstem relative to forebrain structures (Seiki *et al.*, 1968, 1969; Raisinghani *et al.*, 1968; Whalen and Luttge, 1971a; Luttge *et al.*, 1973, 1974; Whalen and Gorzalka, 1974; Wade and Feder, 1972b,c; Wade *et al.*, 1973. The estrogen priming of ovariectomized or ovariectomized-adrenalectomized animals is without effect on pituitary or neural ^{3}H progesterone uptake (Whalen and Luttge, 1971a; Whalen and Gorzalka, 1974; Luttge *et al.*, 1974; Wade and Feder, 1972b), whereas adrenalectomy has been reported to increase both the plasma and the brain levels of radioactivity remaining after ^{3}H progesterone injection (Whalen and Luttge, 1971a; Whalen and Gorzalka, 1974) or to have no consistent effect on retention (Luttge *et al.*, 1974).

As judged by competition experiments with unlabeled progesterone and other steroids, there appears to be no evidence of limited-capacity binding sites at the tissue level (Luttge *et al.*, 1974; Wade and Feder, 1972b). Instead, the neural affinity for steroids such as progesterone, 20α-hydroxyprogesterone, and corticosterone in nonadrenalectomized animals seems to be inversely related to the polarity of these steroids and seems to be positively related to the reported effectiveness of these steroids in inducing estrous behavior in estrogen-primed, gonadectomized rodents (Wade and Feder, 1972c). In this connection, the guinea pig brain has a higher "affinity" for progesterone than the hamster brain or the rat brain, and both guinea pig brain and hamster brain retain progesterone radioactivity for longer times after injection than rat brain. These species differences may be related to the differences among these three species in sensitivity to the facilitatory effects of progesterone on estrous behavior (Wade *et al.*, 1973). It is finally of interest to note that the midbrain, a site of high accumulation of ^{3}HP radioactivity, is the site of inhibition, by implanted progesterone of lordosis behavior in guinea pigs (Morin and Feder, 1974b), whereas the hypothalamus is the site of progesterone facilitation of lordosis in guinea pigs and rats (Morin and Feder, 1974a; Powers, 1972).

Attempts to saturate and thereby reveal limited-capacity binding sites for ^{3}H progesterone by means of unlabeled progesterone have proved uniformly unsuccessful and have tended to suggest that specific progesterone-binding sites either may not exist at all or may be so few in number as to escape detection. There are, however, several pieces of evidence that tend to support the idea of limited-capacity progesterone-binding sites in brain and pituitary. First, Sar and Stumpf (1973d) reported autoradiographic localization of the binding of radioactivity injected as ^{3}H progesterone in neurons of the basomedial hypothalamus of the spayed, estrogen-primed guinea pig. Unlabeled progesterone abolishes this localization, whereas unlabeled cortisol is without effect. According to these authors, estrogen priming is important if one is to see ^{3}H progesterone accumulation. Similar results have been reported for the cytosol binding of ^{3}H progesterone in the pituitary and the median eminence region of spayed, estrogen-primed rats: a progesterone-saturable binding component was detected that could not be saturated by unlabeled corticosterone (Seiki and Hattori, 1973).

Second, adrenalectomy may, under some circumstances, produce an increase in the tissue uptake of ^{3}H progesterone in brain (see above). Not only does adrenalectomy remove one source of endogenous progesterone, it also unmasks glucocorticoid-binding sites in the brain to which progesterone is able to bind (Grosser *et al.*, 1971, 1973; McEwen *et al.*, 1976). One peculiarity about the interaction of progesterone with glucocorticoid receptors is that the progesterone is not extensively translocated to the cell nuclei, as is the case for corticosteroids (McEwen and Wallach, 1973; McEwen *et al.*, 1976). Progesterone can, in fact, block the nuclear translocation of ^{3}H corticosterone, presumably by occupying cytosol receptor sites (McEwen and Wallach, 1973). Progesterone is therefore an antiglucocorticoid and has been shown to block a number of glucocorticoid effects on thymus, chick neural retina, and hepatoma tissue culture cells (see McEwen, 1974). Thus, progesterone may exert some neural effects via an antiglucorticoid action. A major argument against this hypothesis is that glucocorticoids are normally present in large excess over circulating progesterone, thus reducing the efficacy of progesterone as a competitive inhibitor.

The failure to adrenalectomize the guinea pigs used to study the cytosol binding sites for ^{3}H progesterone in the brain and the pituitary may have contributed to the failure of one report to detect any saturable binding (Atger *et al.*, 1974). In this connection, it should be noted that the adrenal appears to contribute substantially to the circulating progesterone, as well as to serve as the source of glucocorticoids (Fajer *et al.*, 1971; Piva *et al.*, 1973; Mann and Barraclough, 1972; Shaikh and Shaikh, 1975).

It should be noted that ^{3}H progesterone is not retained by highly purified pituitary or brain cell nuclei after the infusion of low doses into estrogen-primed, ovariectomized-adrenalectomized female rats under conditions in which comparable doses of ^{3}H corticosterone and ^{3}H dexamethasone are retained strongly by isolated nuclei (McEwen *et al.*, 1976). Published reports of cell nuclear retention of ^{3}H progesterone or its metabolites are based on radioactivity in crude nuclear pellets (Robinson and Karavolas, 1973; Cheng and Karavolas, 1973) and do not necessarily indicate the presence of cell nuclear receptors. In view of the lack of conclusive evidence for such receptors and the indication of short-latency effects of progesterone on brain EEG-activity thresholds (Kawakamini and Sawyer, 1959; Ramirez *et al.*, 1967), it is conceivable that progesterone may act at the synaptic level to alter neural activity. It should be noted, however, that cycloheximide, a protein-synthesis inhibitor, has been shown to prevent progesterone-dependent postestrus refractoriness in guinea pigs (Wallen *et al.*, 1972), and actinomycin D implants in the arcuate nucleus and the ventromedial hypothalamus (but not in the pituitary of the amygdala) block progesterone-induced LH release in testosterone-pretreated ovariectomized rats (Jackson, 1975). These observations tend to reinforce the notion that cell nuclear receptors for progesterone may actually exist.

GLUCOCORTICOIDS

Attempts to find glucocorticoid receptors in the brain were initiated in 1967 in this laboratory, with the expectation that such sites would be found predominantly in the hypothalamus, where the feedback regulation of ACTH secretion was presumed to occur (see Mangili *et al.*, 1966). The first surprise was that, in adrenalectomized rats, ^{3}H corticosterone (the predominant glucocorticoid in this species) is concentrated predominantly in hippocampus and septum, and that the concentrations of radioactivity in hypothalamus is among the lowest of those in all brain regions (see McEwen *et al.*, 1972; Knizely, 1972; Lemaire *et al.*, 1974). The superiority of hippocampus and septum over other brain regions in concentrating ^{3}H corticosterone was shown to be due to the presence of limited-capacity binding sites residing in cell nuclei and in the cytosol fractions of the tissue (see McEwen *et al.*, 1972; Grosser *et al.*, 1971, 1973). Cell-fractionation and autoradiographic studies revealed the presence of neuronal ^{3}H corticosterone-binding sites in amygdala, anterior and lateral septum, indusium griseum, recommisural hippocampus, subiculum, and anterior olfactory nucleus, as well as in postcommisural hippocampus (Stumpf, 1972; Gerlach and McEwen, 1972; Warembourg, 1975; Rhees *et al.*, 1975; McEwen *et al.*, 1975a).

685

HORMONE
RECEPTORS IN
BRAIN AND
PITUITARY

A second unexpected finding in studies of *in vivo* ^{3}H dexamethasone uptake in brains of adrenalectomized rats was that this synthetic glucocorticoid does not accumulate in hippocampus and that levels of binding in this structure are similar to those observed in other brain regions, including hypothalamus (deKloet *et al.*, 1975). The structure that accumulates the largest amounts of ^{3}H dexamethasone bound to cell nuclei is the anterior pituitary gland (deKloet *et al.*, 1975). This observation is consistent with the widely accepted notion that dexamethasone produces its primary ACTH-release-inhibiting effect at the level of the pituitary. The more than fivefold greater uptake of ^{3}H dexamethasone than of ^{3}H corticosterone by pituitary cell nuclei is consistent with the greater potency of dexamethasone, compared to corticosterone, in level-sensitive, delayed feedback on ACTH release (see deKloet *et al.*, 1975; deKloet and McEwen, 1976). As to the physiological significance of brain receptors for glucocorticoids, the subject is reviewed elsewhere, and the reader is referred to a number of review articles on the subject (McEwen and Pfaff, 1973; McEwen, 1974; McEwen *et al.*, 1975b; deKloet and McEwen, 1976). The fact that ^{3}H corticosterone receptors have been found in the hippocampus and the septum of the Peking duck (Rhees *et al.*, 1972), the guinea pig (Warembourg, 1974), and the rhesus monkey (Gerlach *et al.*, 1976), besides the rat, argues for an ancestral origin and a phylogenetically stable role for neural glucocorticoid receptors among vertebrates.

MINERALOCORTICOIDS

Proteins capable of binding ^{3}H deoxycorticosterone (DOC) have been described in brain cytosols from adrenalectomized rats (Lassman and Mulrow, 1974), but these proteins also bind glucocorticoids and progesterone, and ^{3}H DOC is not extensively accumulated by brain cell nuclei *in vivo* (McEwen *et al.*, 1976). Levels of the cytosol and cell nuclear binding of ^{3}H aldosterone, lower than in kidney but comparable to that in liver, have been detected in brains of adrenalectomized rats, and cell nuclear "^{3}H aldosterone binding proteins" have been extracted from purified brain-cell nuclei (Swanek *et al.*, 1969). Brain-cell nuclear radioactivity after ^{3}H aldosterone is comparable to that seen after ^{3}H deoxycorticosterone administration (McEwen *et al.*, 1976).

With respect to the possible neural significance of mineralocorticoid binding sites, it is known that salt-taste-preference thresholds are decreased and salt appetite is increased in rats by deoxycorticosterone acetate (DOCA) (Rice and Richter, 1943; Herxheimer and Woodbury, 1960). DOCA has the effect in adrenalectomized rats of returning salt intake to normal (Richter, 1941). With respect to the neural sites of this action, Wolf (1967) abolished DOC-dependent increases in salt appetite with lesions in the lateral hypothalamus. It may therefore be significant that Lassman and Mulrow (1974) found a deficiency in the hypothalamus of Long-Evans rats (compared to Sprague-Dawleys) of ^{3}H-DOC-binding to cytosols, as the

Long-Evans rat is resistant to the induction of increased salt appetite by DOC (Hall *et al.*, 1972).

Neurochemical Consequences of Steroid Interactions with Putative Neural Receptors as Illustrated for Estradiol

Temporal Relationships of Receptor Occupation to Physiological Effects

The behavioral or neuroendocrine consequences of steroid hormone action may occur some hours after the interaction of the hormone with the putative neural or pituitary "receptor" sites. This delay is most clearly illustrated by the examples of estradiol induction of the LH surge and facilitation of the lordosis response in female rats, which occur around 30 hr and 20–24 hr, respectively, after a single estradiol injection. In the latter case, as little as 2–10 μg of estradiol 17β is able to promote lordosis, and even as much as 100 μg does not shorten the minimum lag period below 16–20 hr (Green *et al.*, 1970; Whalen and Gorzalka, 1973). We capitalized on the low dose of estradiol sufficient to elicit increased lordosis responding by measuring the time course of retention of radioactivity from a 10-μg dose of $^3HE_2$ by cell nuclear sites in pituitary and pooled hypothalamus, preoptic area, and amygdala. To our surprise, cell-nuclear-estradiol radioactivity, which is maximal at 1–2 hr, declines by 4 hr and is largely gone from "receptor" sites by 12 hr after the injection (Figure 5; McEwen *et al.*, 1975a). Thus, 16–18 hr elapse from full "receptor" occupation to the onset of lordosis responding, and at least 9 hr elapse during which cell nuclear sites are virtually unoccupied. Such a lag period prior to the occurrence of the specific hormone-mediated functional response must involve specific changes in cellular metabolism in these target areas.

In agreement with this interpretation are the observations of several laboratories that actinomycin D (an inhibitor of DNA-dependent RNA synthesis) and cycloheximide (a protein-synthesis inhibitor) reversibly inhibit the estrogen facilitation of lordosis responding when given intracranially in the anterior hypothalamus–preoptic area 6 hr before or up to 12 hr after estradiol benzoate (Terkel *et al.*, Whalen *et al.*, 1974; Quadagno and Ho, 1975). Given these implications that changes in RNA and protein synthesis underlie estrogen effects on behavioral estrus, let us examine the evidence for estrogen-induced changes in these parameters of neuronal function.

Biochemical studies of hormone action in the CNS have been limited in number and scope by the nature of the brain itself. Unlike the uterus, which differentiates and grows some twofold within 24 hr after estrogen administration (Baquer and McLean, 1972), the adult brain neither grows nor differentiates (except perhaps microscopically) as a response to estrogen. The application of biochemical techniques is further strained because, as indicated earlier, the interaction of hormones is confined to limited areas of the brain, and interactions do not occur with all cells in these particular regions. Despite these inherent limitations of nervous tissue as an object of investigation, a number of biochemical studies have been

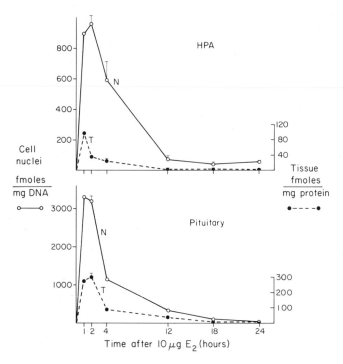

Fig. 5. Time course of retention of radioactivity injected as [³H]estradiol-17β by cell nuclei (N; data expressed as fmoles/mg DNA) and aliquots of homogenized tissue from which nuclei were isolated (T; data expressed as fmoles/mg protein). Abbreviation: HPA, pooled medial preoptic area, basomedial hypothalamus, and corticomedial amygdala. Each experiment analyzed HPA tissue from one ovariectomized female rat given 10 μg [³H]estradiol. Number of experiments at indicated times: 1 hr, 1; 2 hr, 4; 4 hr, 3; 12 hr, 3; 18 hr, 2; 24 hr, 3. Vertical bars, where appropriate, indicate a standard error of mean. Pituitary: each experiment analyzed pituitary tissue pooled from three identically treated ovariectomized female rats, each given 10 μg [³H]estradiol. One experiment is presented at time intervals 1 hr, 4 hr, 12 hr, 18 hr, and 24 hr. Data for 2 hr are from two experiments ± S.E.M. (From McEwen *et al.*, 1975b, by permission.) Abbreviations: MDH, malate dehydrogenase; ICOH, isocitrate dehydrogenase; MAO, monoamine theoxidase.

carried out with estradiol. This hormone is particularly advantageous because there are distinct physiological endpoints—namely, the lordosis response and the release of pituitary hormones—and there is a concordance of information about the brain sites where these effects may be produced and the sites where putative estrogen receptors are found.

ALTERATIONS OF RNA FORMATION

Association of estrogen with the cell nucleus in neural and hypophyseal tissue suggests that biochemical changes might be initiated by way of alterations in genomic activity. Indeed, investigation of the rate of synthesis and levels of RNA as a function of estrogen in neural and hypophyseal areas indicates that estrogen may have an activating effect. Eleftheriou and Church (1967) found higher RNA levels at estrus and proestrus than at diestrus in the hypothalamus, the amygdala, and the cortex. However, Litteria and Schapiro (1972) found no changes in RNA or

DNA content during the estrous cycle in the hypothalamus, the cerebellum, the cerebral cortex, the hippocampus, the caudate nucleus, or the olfactory bulbs.

Hypophyseal RNA levels were reported by Convey and Reece (1969) and by Robinson and Leavitt (1971) to be the lowest during diestrus. The highest concentrations were found at proestrus by Robinson and Leavitt (1971), and Convey and Reece (1969) reported the estrus values to be the highest. The changes in RNA concentrations were ascribed to the action of estrogen, as ovariectomized animals showed the lowest concentrations of RNA and replacement therapy with estrogen-restored normal RNA levels (Robinson and Leavitt, 1971).

One study in brain utilizing the *in vivo* administration of ^{14}C-adenine to cycling females showed an opposite effect, namely, that the density of autoradiographic grains (an index of RNA incorporation into RNA) in supraoptic, paraventricular, and infundibular neurons was greater when labeling occurred during diestrus than when it occurred during estrus (Belajev *et al.*, 1967). However, Salaman (1970) measured the rate of RNA synthesis with an *in vitro* system containing ^{3}H-uridine and anterior hypothalamic tissue or anterior pituitaries from female rats in different stages of the estrus cycle. Both the pituitary and the hypothalamic rates of ^{3}H-uridine incorporation varied in the same manner throughout the cycle, showing the lowest incorporation rates in diestrus. Thus, the results of these biochemical studies, although somehwat contradictory, suggest that increases in nucleic acid metabolism may occur during the lag period between the occupation of nuclear sites by estrogen and the initiation of the functional response.

The involvement of RNA synthesis in the mediation of the lordosis response has been shown in a number of experiments utilizing the inhibitor of DNA-directed RNA synthesis, actinomycin D. Terkel *et al.* (1973) investigated in detail the parameters of the actinomycin D inhibition of the lordosis response. Ovariectomized females were given 3 μg of estradiol benzoate, followed by 0.5 mg of progesterone 48 hr later. At 4–6 hr after progesterone, lordosis quotients were obtained in the females. These authors found that the infusion of actinomycin D into the preoptic area (POA) or the third ventricle could depress lordosis responding when given simultaneously with estrogen or 6 hr later. The inhibition was reversible because repriming with hormones 10–14 days later again increased lordosis responding. No changes in lordosis responding were found when actinomycin D was given into the POA 12, 18, or 26 hr after estrogen or 12 hr before estrogen. Whalen *et al.* (1974) confirmed that infusions of actinomycin D into the POA would inhibit the estrogen facilitation of lordosis, and they also reported that inhibition was evident when actinomycin D was infused up to 12 hr after subcutaneous administration of estradiol benzoate. An attempt was made to examine the neuronal specificity of the effect by infusing actinomycin D into the caudate nucleus and into the third ventricle near the POA. Caudate nucleus infusions were ineffective and third-ventricular infusions were effective. The marked but reversible suppression of lordosis behavior in these studies is consistent with the actinomycin-D-inhibiting estrogen facilitation of lordosis behavior by means of a biochemical rather than a cytotoxic action.

In support of these conclusions, Ho *et al.* (1973–1974) studied lordosis responding and also examined ultrastructural features of cells in the POA after actinomycin D treatment. The preoptic cells did not show nonspecific cytotoxic effects but only the structural changes typical of actinomycin D treatment, and these were reversible in parallel with the estrogen sensitivity of the behavior. A further, reversible effect of actinomycin D on behavior is the inhibition of the increased wheel-running activity usually seen when estrogen is given to ovariectomized females (Stern and Jankowiak, 1972).

Mechanisms involved in the synthesis and the release of gonadotropins have also been explored in several studies with a biochemical viewpoint: The chronic release of small amounts of estrogen from the ovary exerts a tonic negative effect on the release of LH from the pituitary. However, when a single large surge of estrogen is secreted (or administered), the system is affected positively and responds with an outpouring of LH leading to ovulation. Jackson (1972, 1973) utilized actinomycin D to explore the mechanism of the positive effect of estrogen on LH release. Ovariectomized females were treated with small amounts of estradiol benzoate for 5 days. On Day 6, 50 μg of estradiol benzoate was given at varying times in a single injection resulting in the release of LH. Actinomycin D, given either 4 hr prior to or up to 13 hr after estradiol benzoate, blocked LH release. Actinomycin D given 20 hr after estradiol benzoate did not block LH release. These and other results indicate that RNA synthesis occurring up to 13 hr after EB is needed to induce LH release. Actinomycin D was not able to block the pituitary response to LHRF (Jackson, 1973), and this result suggests but does not prove that neural processes subserving LHRF synthesis and/or release were altered by this inhibitor. However, considering the systemic route of Actinomycin D administration, it is uncertain whether substantial inhibition of RNA synthesis actually occurred in the brain.

CHANGES IN PROTEIN FORMATION

As changes in RNA synthesis have been demonstrated in a number of experimental situations, comparable changes should occur in the rate of protein synthesis in neural and pituitary target cells as a consequnce of estrogen action. A number of groups have explored protein synthesis utilizing autoradiographic techniques. In general, the results from these experiments show that estrogen, whether given exogenously or secreted during the estrous cycle, results in an enhanced rate of incorporation of amino acids into protein. Seiki *et al.* (1972) and Seiki and Hattori (1972), using ^{3}H leucine, compared females ovariectomized for 30 days with ovariectomized females given estrogen 50 hr and 3 ½ hr before incorporation was measured. Yaginuma *et al.* (1969) compared 1-month ovariectomized females with proestrus females that had high blood titers of estrogen. Litteria (1973) compared proestrus females with diestrus females. All these investigators found no differences among the groups in the incorporation of ^{3}H-labeled amino acids into the neurons of the cerebral cortex (Table 4). In addition, Litteria (1973) examined the anterior hypothalamic nucleus, the dorsomedial nucleus, and the lateral preoptic

TABLE 4. CHANGES IN AUTORADIOGRAPHIC LABELING OVER CELL BODIES IN THE FEMALE RAT BRAIN AFTER INJECTION OF ^{3}H-LYSINE OR LEUCINE

Brain nuclei	Treatment			
	Proestrus vs. diestrus[1]	Proestrus vs. OVX[2]	OVX + EB vs. OVX (50 hr [3])	OVX + EB vs. OVX (3.5 hr)[4]
Paraventricular*	↑	↑	↑	↑
Arcuate*	↑	↓	↑	↑
Medial preoptic area*	↑		↑	—
Periventricular*	↑		—	—
Ventromedial*	—	↓	↑	↑
Supraoptic	↑	↑	↑	↑
Cortical cells	—	—	—	

*Areas that show cell nuclear ^{3}H-estradiol accumulation.
↑ = increase; ↓ = decrease; — = no change; OVX = ovariectomy; EB = estradiol benzoate.
[1]Litteria (1973); [2]Yaginuma et al. (1969); [3]Seiki et al. (1972); [4]Seiki and Hattori (1972).

area and found unchanged incorporation in these regions, which do not show cell nuclear binding of estrogen.

Of the regions that do show cell nuclear estrogen binding, two areas, the paraventricular nucleus of the hypothalamus and the medial preoptic area, show a consistent increase in the incorporation of amino acid into protein after the presence of estrogen (Seiki et al., 1972; Yaginuma et al., 1969; Litteria, 1973). In addition, MacKinnon (1970), using ^{35}S-methionine, observed that labeled cells were present in a narrow band in the periventricular region of the medial preoptic area during diestrus, but at proestrus, labeled cells covered the entire central region.

Effects on amino acid incorporation were found by all groups in the arcuate region, an estrogen-binding area. However, Litteria (1973), Seiki and Hattori (1972), and Seiki et al. (1972) found that estrogen increased incorporation, whereas Yaginuma et al. (1969) found decreased incorporation. Inconsistent changes were reported in the ventromedial and the periventricular nuclei.

One region where estrogen binding is not pronounced but nonetheless shows an increased incorporation of amino acid by autoradiography after estrogen is the supraoptic nucleus (Seiki and Hattori, 1972; Seiki et al., 1972; Yaginuma et al., 1969; Litteria, 1973.) This nucleus appears to be a locus of oxytocin- and vasopressin-producing neurons.

Quantitative measurement of the incorporation of labeled amino acids into protein in the anterior pituitary, the median eminence, the preoptic area, the amygdala, the putamen, and the thalamus were made by TerHaar and MacKinnon (1973). They found that the rates of incorporation were significantly highest at proestrus in the anterior pituitary, the median eminence, the amygdala, and the preoptic area, whereas in the putamen and the thalamus, no significant changes were seen during the estrus cycle. If exogenous estrogen was given, then all brain regions showed an increased incorporation rate and a reappearance of a circadian variation in ^{35}S methionine incorporation (TerHaar and MacKinnon, 1973; Burnet and MacKinnon, 1975). However, the anterior pituitary and the median eminence showed the greatest increases. Similar results were obtained in ovariectomized guinea pigs by Wade and Feder (1974). When animals that had been ovariectomized for 10–014 days were given estradiol benzoate, and amino acid incorporation was measured 24 hr later, an approximately 50% increase in amino acid incorporation occurred in the hypothalamus, the midbrain, and the cerebral cortex. Whether such widespread changes in incorporation are the result of changing amino-acid-pool sizes is not known.

Another group investigated the effect of pharmacological doses of estrogen, given for 4 months, on the rate of protein synthesis and found results opposite to those of the previous studies (Litteria and Timiras, 1970). Utilizing both *in vitro* quantitative measurements and autoradiography, they reported that ^{3}H-lysine incorporation into protein was decreased in specific nuclei of the hypothalamus but not in the cerebellum or the motor cortex. Because large estradiol doses were given for an extended period, these results may reflect a toxic pharmacological effect of estrogen.

It would seem, therefore, that estrogen-dependent increases of amino acid incorporation into proteins are restricted, for the most part, to those brain regions that contain putative estradiol receptor sites when physiological fluctuations in estrogen levels are studied, but that, if estradiol is given exogenously, other brain regions respond with an increased or decreased incorporation rate. It is, of course, unknown whether these more widespread changes in incorporation represent changes in the pool size of amino acid rather than altered protein synthesis.

Agents known to inhibit protein synthesis, as is the case with drugs that inhibit RNA synthesis, interfere with estrogen facilitation of the lordosis response (Meyerson, 1973; Quadagno and Ho, 1975). Quadagno and Ho showed that cycloheximide, when infused bilaterally into the preoptic area or into the third ventricle of ovariectomized rats simultaneously with estrogen injections, results in a dramatic lowering of the lordosis quotient when compared with saline infusions into these areas. The neural speicificity of the effect is evident from the observation that cycloheximide infusions into the caudate nucleus, the lateral ventricle, or the unilateral preoptic area, or a subcutaneous injection of cycloheximide does not result in significant reductions in lordosis responding as compared to that in the saline-infused group. Action on protein synthesis, not general destruction of tissue, seems to account for the effect because repriming of the drug-treated animals 7 days later resulted in normal lordosis quotients. Further experiments indicated, as in actinomycin D studies, that, for inhibition of the lordosis response, cycloheximide is effective only during a circumscribed period beginning from 6 hr before estrogen injection and extending up to 12 hr after estrogen (Quadagno and Ho, 1975).

ESTROGEN EFFECTS ON PITUITARY SENSITIVITY TO LHRH

Recent characterization of the polypeptide sequence of luteinizing-hormone-releasing hormone (LHRH) and the availability of synthetic LHRH have allowed more sophisticated studies of its action on the release of LH and FSH and have further indicated that gonadal steroids may modulate the ability of the pituitary to respond to LHRH. If synthetic LHRH or a purified stalk-median-eminence fraction of LHRH is injected into females at different stages of the estrous cycle or into an incubation of pituitaries from female rats, the pituitary responds with an increased output of LH into the serum or into the incubation medium. Several investigators have shown that the release of LH is augmented at proestrus and estrus and that the greatest release of LHRH-induced LH occurs on the afternoon of proestrus (Gordon and Reichlin, 1974; Cooper *et al.*, 1974).

Because the greatest release occurs at proestrus, it seemed possible that the rise in plasma estrogen titers occurring before and during proestrus might be involved in the sensitization of the pituitary to LHRH. To investigate this possibility, Vilchez-Martinez *et al.* (1974b) gave sesame oil or estradiol benzoate dissolved in sesame oil at 9 A.M. each day of the estrous cycle, and 24 hr later, they measured the release of LH after LHRH. Estradiol administration did not alter the basal levels of LH but augmented the response to LHRH on all days when compared to the oil-injected group. As was seen in cycling studies, a greater LHRH sensitivity still occurred on proestrous and estrous days.

Libertun *et al.* (1974) measured LHRH-stimulated LH release *in vivo* and *in vitro* in proestrus females, and in 3–6 weeks ovariectomized females given vehicle, estradiol, progesterone, or estradiol plus progesterone 72 hr before LHRH. Pretreatment with estradiol or estradiol and progesterone significantly increased pituitary sensitivity to LHRH.

The results of administering LHRH at shorter intervals after estradiol indicate the time dependence of these effects. If proestrus rats are given LHRH 4–14 hr after estradiol, release of LH occurs, but the amount released is smaller than in similarly oil-injected females (Vilchez-Martinez *et al.*, 1974b). In a diestrous rat, no LHRH-stimulated LH release occurs 2–6 hr after estrogen treatment, but at 14 hr, an augmented response, compared to that of oil-injected females, is seen (Vilchez-Martinez *et al.*, 1974a). Van Dieten *et al.* (1974) also found (in the diestrous rat) no LHRH- stimulated LH release 3 hr after estradiol, but at 27 hr after EB, an extremely large LH release occurred in response to LHRH.

The results on the effects of progesterone in altering LHRH-stimulated LH release are conflicting. Libertun *et al.* (1974) found that progesterone alone given to ovariectomized rats (72 hr previously) did not modify the response to LHRH, wheras Nakano *et al.* (1974) found that in rats pretreated 48 hr earlier with sesame oil, estradiol, or progesterone, the basal LH levels were lowest in hormone-treated rats. At 30 min after LHRH administration, the estradiol-treated group showed the greatest release of LH, whereas the oil and progesterone groups showed a similar release of LH. At 120 min after LHRH, the estradiol and oil group had lower LH levels than at 30 min, but the progesterone group had an even lower LH level.

At this time, it seems reasonable to conclude that estrogen (after at least 14 hr) can increase the sensitivity of the pituitary to administered LHRH. The involvement of progesterone in the response is not clear. Because pentobarbital-blocked proestrous animals do not show as great an increase in LHRH-stimulated LH release as normal proestrous animals, other hypothalamic or hypophyseal hormones may also be involved in the heightened sensitivity (Gordon and Reichlin, 1974).

Spona (1973a,b, 1974) investigated a binding protein in the pituitary plasma membrane as a possible site for the action of hypothalamic releasing factors. The binding of (^{125}I)-LHRH is proportional to the pituitary-membrane protein concentration and is reduced by unlabeled LHRF, but not by other hypothalamic releasing factors (Spona, 1973a). A very rapid association of LHRH occurs with the protein that is similar to the rapid release of LH seen after LHRH administration. A relationship between the binding protein and the gonadal hormones is suggested because intact female rats show two distinct binding sites for (^{125})-LHRH, one with a high affinity and a low capacity and a second site with lower affinity and a 6 times higher capacity (Spona, 1973b). At 30 days after ovariectomy, only one binding site is evident that resembles the higher-capacity, lower-affinity site (Spona, 1973b). If estradiol or estradiol and progesterone are given to ovariectomized rats for 6 days, the second binding site appears to have been regenerated (Spona, 1974). Although further study of the properties of the pituitary membrane LHRH-binding protein is required, the levels of this "receptor" protein may provide the mechanism by which changes in sensitivity to LHRH are regulated by gonadal hormones. This system also illustrates that the levels of proteins, other than enzymes, may be dependent on circulating estrogen.

HYPOTHALAMIC PEPTIDASES

The enzymes that are implicated in the synthesis and the degradation of hypothalamic releasing factors respond to changes in the level of circulating estrogen. The activity of cystine-aminopeptidase (CAPA) in the hypothalamus changes during the estrous cycle, reaching a peak at metestrus (Heil *et al.,* 1971). Increased CAPA activity is seen in ovariectomized females given ethinylestradiol or estradiol. The increased activity is first measurable 10 hr after estrogen administration and reaches a maximum in 15–18 hr. Increases of 60%–80% occur in the hypothalamus, but the greatest changes are seen in the amygdala (180%–200%). No other brain regions respond to estrogen, and testosterone is not effective. These results were confirmed by Bickel *et al.* (1972), who also showed that estrogen or testosterone administration to males increases CAPA. The activity of arylamidases also increases in the hypothalamus 16 hr after estrogen administration (Kuhl *et al.,* 1974).

Griffiths and colleagues (1974) investigated the activity of a soluble peptidase from hypothalamus that they showed to inactivate synthetic LHRH. In earlier studies (Griffiths and Hooper, 1973), they showed that the activity of this enzyme decreases on the afternoon of proestrus, a finding that would make it a candidate for regulating the realease of LH by allowing LHRH to remain active. However, they also reported in other studies (Griffiths and Hooper, 1974; Griffiths *et al.,* 1975) that ovariectomy decreased peptidase, and estrogen replacement therapy increased the peptidase activity, a finding that appears to be at variance with the decreased enzyme activity observed on the afternoon of proestrus, when estrogen titers are elevated. One reason for the apparent discrepancy may be that the two studies were performed with different assays for the inactivation of LHRH. It must also be borne in mind that estradiol is capable of exerting both positive and negative effects on LH release, which may differ from each other in their time and dose dependency. Thus, differential time- and dose-dependent effects of estradiol on hypothalamic peptidase activity are not out of the question.

OXIDATIVE METABOLISM

Estrogen has been shown to modulate oxygen consumption in specific brain areas. This is an important consideration in neural function because the brain must maintain an enormous rate of oxygen consumption in order to provide the energy required for sustaining membrane electrical activity. Moguilevsky and Malinow (1964) measured, with manometric techniques, *in vitro* oxygen uptake in the cortex, the hypothalamus, and the pituitary from rats at various stages of the estrous cycle. No changes in oxygen uptake were found in the cortex, but in the hypothalamus, the rates at proestrus and estrus were higher than at metestrus and diestrus. The pituitary showed the lowest rates at diestrus. Schiaffini *et al.* (1969), using the same method, confirmed the lack of cyclicity on the cortex and also reported that the amygdala was similar to the pituitary and the hypothalamus, showing higher rates of oxygen uptake at estrus than at diestrus.

Because oxygen is utilized in the brain almost entirely for the oxidation of carbohydrate, many studies have considered, as a function of estrogen treatment, the activity of enzymes that are present in pathways that metabolize glucose. The glycolytic enzyme, lactate dehydrogenzse (LDH), was measured during the estrus cycle in the anterior, the middle, and the posterior hypothalamus. A 30% increase in activity was found at proestrus in the anterior hypothalamus (Morishita *et al.*, 1972). Estrogen administration to ovariectomized female rats did not result in increased LDH activity in the ventromedial hypothalamus or the medial preoptic area. However, activities of the enzyme were increased 150% in the pituitary (Luine *et al.*, 1974, 1975b).

The activity of glucose-6-phosphate dehydrogenase (G6PDH), the regulatory enzyme of the pentose phosphate pathway, changes in the hypothalamic area during the estrous cycle (Hunter and Hagy, 1969). Luine *et al.* (1974) found that this change could be attributed to a direct action of estrogen, as increased activity was found after estrogen administration to ovariectomized-adrenalectomized females and to hypophysectomized females, and the magnitude of the increased activity was dependent on the administered dose of estrogen. The pituitary also responds to estrogen with increases in G6PDH activity as well as with increases in the second enzyme of the pentose phosphate pathway, 6-phosphogluconate dehydrogenase. Two enzymes that generate reduced triphosphopyridine nucleotide, isocitrate dehydrogenase (6PGDH) and malate dehydrogenase (MDH), are also increased in the basomedial hypothalamus and the corticomedial amygdala by the direct action of estrogen (Luine *et al.*, 1974, 1975a).

CATECHOLAMINES AND SEROTONIN

695

HORMONE
RECEPTORS IN
BRAIN AND
PITUITARY

FUNCTIONAL INVOLVEMENT. The hypothalamus and related limbic brain regions receive a major portion of catecholaminergic and serotonergic innervation of the brain (Wurtman, 1971). Substantial amounts of hypothalamic dopamine arise from rostral, dorsal, and caudal cell groups (Björklund and Nobin, 1973) and from the tuberoinfundibular dopaminergic system that innervates the external layer of the median eminence (Fuxe and Hökfelt, 1969). An increasing amount of experimental work is devoted to the study of the functional involvement of monoaminergic innervation in the control of neuroendocrine secretion and sexual behavior.

Studies of the monoaminergic control of gonadotropin secretion have centered primarily on the possible involvement of dopamine and noradrenaline (McCann and Moss, 1975). In spite of its localization in the median eminence adjacent to portal blood vessels, dopamine (DA) has not been clearly implicated in the control of gonadotropin secretion, and it remains a possibility that the stimulatory effects of DA obtained *in vitro* and *in vivo* systems are due to reuptake and conversion to noradrenaline (see McCann and Moss, 1975).

Noradrenaline appears to be more convincingly implicated in the control of gonadotropin secretion. For example, the suppression of catecholamine synthesis by α methyl$p$ tyrosine (an inhibitor of tyrosine hydroxylase) and by diethyldithiocarbamate (a dopamine β hydroxylase inhibitor) blocks estrogen-facilitated gonadotropin release, and this blockade by either inhibitor can be relieved by treatment with dihydroxyphenylserine (DOPS), which is directly decarboxylated to form NA (see McCann and Moss, 1975). The noradrenergic innervation involved in gonadotropin release appears to occur between the preoptic area and the median eminence, as catecholamine-synthesis inhibition interferes with the effects of electrical stimulation in the preoptic area and not in the arcuate median eminence region (see McCann and Moss, 1975).

Catecholamines and serotonin are also implicated in the regulation of female and male sexual behavior (Gessa and Tagliamonte, 1974a,b; Meyerson *et al.*, 1974; Zemlan *et al.*, 1973; Malmnäs, 1974; Ward *et al.*, 1975; Everitt *et al.*, 1975; Gradwell *et al.*, 1975; see also Crowley and Zemlan, 1981). This work with rats and (in one case) rhesus monkeys utilizes ovariectomized (and in some cases also adrenalectomized) animals primed with estrogen. For rats, estrogen is supplemented with progesterone, thus producing maximum lordosis responding, in cases where the inhibitory effects of neurotransmitter-related drugs are being studied. Where facilitatory drug effects are under investigation, estrogen priming alone is used so as to bring about suboptimal lordosis in the absence of drug treatment.

Adrenergic involvement appears to be negligible in the control of male sexual behavior (Malmnäs, 1974). In female behavior, lesioning of the ascending NA pathways *inhibits,* and treatment with NA agonists in the presence of DA receptor blockers *facilitates* lordosis responding, a finding suggesting adrenergic *facilitation* of female sexual behavior (Crowley and Zemlan, 1981). However, β adrenergic antagonists facilitate lordosis responding (while α-blockers have no effect), a finding suggesting also an adrenergic inhibitory influence (Ward *et al.*, 1975). Possibly, there are two *opposing* behavioral systems dependent on adrenergic transmissions (Crowley and Zemlan, 1981).

Studies of dopaminergic involvement in sexual behavior indicate that this system *facilitates* male behavior and *inhibits* female behavior (Gessa and Tagliamonte, 1974b; Malmnäs, 1974; Everitt *et al.*, 1975; Meyerson *et al.*, 1974). This finding is interesting in view of the opposite effects of preoptic area lesions on male and female sexual behavior, inhibiting the former and facilitating or having no effect on the latter (Singer, 1968; Powers and Valenstein, 1972). There is no evidence implicating DA transmission in the preoptic area in sexual behavior, but this would appear to be an important brain region in which to study this transmitter system in male and female rats.

Serotonin has been implicated as an inhibitory system with respect to *both* male and female sexual behavior (Meyerson *et al.*, 1974; Zemlan *et al.*, 1973; Ward *et al.*, 1975; Everitt *et al.*, 1975; Malmnäs, 1974; Gessa and Tagliamonte, 1974a). Several studies have demonstrated facilitatory effects of reduced serotonergic transmission in adrenalectomized-ovariectomized (and estrogen-primed) animals (Zemlan *et al.*, 1973; Gradwell *et al.*, 1975). In the case of rats, behavioral effects in an adrenalectomized subject is an indication that adrenal progesterone is not an essential component of the drug effects (Zemlan *et al.*, 1973). In studies on rhesus monkeys, there appears to be an involvement of adrenal androgens, as adrenalectomy increases female refusals and decreases female presentations to sexually active males (in spite of continued estrogen priming), and androstenedione and testosterone both reverse the effects of adrenalectomy (Everitt *et al.*, 1972; Everitt and Herbert, 1975). These androgen effects are interesting in view of previous reports by Michael (1972) that, whereas female attractiveness to male rhesus monkeys is facilitated by estrogen (and inhibited by progesterone), female responsiveness to the presence of males is under the control of androgens.

NEUROCHEMICAL EFFECTS OF GONADAL STEROIDS ON MONOAMINE TURNOVER. The administration of pCPA, a serotonin synthesis blocker, to adrenalectomized-ovariectomized, estrogen-primed rhesus monkeys is reported to facilitate female responding to males and, in this respect, to substitue for the presence of adrenal androgens (Gradwell *et al.*, 1975). Both estrogen and testosterone are reported to *decrease* the turnover of cerebral serotonin, an effect consistent with the notion that serotonergic transmission exerts an inhibitory influence on sexual behavior (Gradwell *et al.*, 1975). In apparent contradiction to this report for rhesus monkeys is the experiment of Bapna *et al.* (1971), who showed that serotonin turnover in anterior and posterior hypothalamus did not differ between ovariectomized rats given estradiol 56 hr before and progesterone 6 hr before testing and ovariectomized rats treated with the injection vehicle alone. However, the time dependence of estradiol and progesterone effects on serotonin turnover have not been systematically investigated in either species, nor have the regional changes in serotonin metabolism been thoroughly investigated, and there may, in fact, be no real contradiction in these two sets of observations.

Studies of noradrenaline turnover have indicated increased turnover associated with states of high gonadotropin release, for example, during proestrus and as a result of gonadectomy (see McCann and Moss, 1975). Estrogen plus progesterone replacement therapy, as for example in the studies of Bapna *et al.* (1971) cited above, is reported to decrease noradrenaline turnover, especially in the preoptic area–anterior hypothalamic region of the brain. In contrast, the turnover of tuberoinfundibular dopamine appears to be decreased during proestrus, and it has been proposed that this decrease is related to increased prolactin secretion at this time, as DA has an inhibitory effect on prolactin release (see McCann and Moss, 1975). Low doses of estrogen for 3 days have been reported to increase the turnover of tuberoinfundibular dopamine (Fuxe *et al.,* 1969). This observation is troublesome in view of the decreased DA turnover attributed to proestrus, when estrogen levels are highest, but the time dependence of the estrogen effects has not been systematically investigated, and it is conceivable that estradiol may even decrease DA turnover at short time intervals.

It is beyond the intended scope of this review to further summarize the evidence pertaining to the effects of endogenous or exogenous gonadal steroids on the *in vivo* turnover of catecholamines and serotonin (see Crowley and Zemlan, 1981). Aside from the inherent difficulties of assessing turnover, there are the problems alluded to above of the time and dose dependencies of hormonal effects and the necessity of measuring the turnover in individual brain regions. The time dependency of estrogen effects was illustrated in an earlier section for the sensitivity of the pituitary to LHRH. The elucidation of such time and dose dependencies for monoamine metabolism may provide the key to understanding the complex events underlying the positive and negative feedback of gonadal steroids, especially of estradiol and progesterone, and is clearly an important avenue for future research. In what follows below, we consider some evidence for estradiol effects on the activities of enzymes involved in catecholamine metabolism, which effects may underlie and ultimately explain changes in the amine turnover resulting from hormone treatment.

EFFECTS OF GONADAL STEROIDS ON ENZYMES OF MONOAMINE METABOLISM. Beattie *et al.* (1972) examined the activity of tyrosine hydroxylase (TH), the rate-limiting enzyme for norepinephrine synthesis, in the hypothalamus of intact and ovariectomized rats. Ovariectomy led to a two- to threefold increase in TH between 4 and 60 days after ovariectomy, a finding that agress with the turnover studies (noted above) showing an increased metabolism of norepinephrine after ovariectomy. Similar results were obtained by Kizer *et al.* (1974), who, using a new method for sampling discreet neuronal groupings, were able to measure TH in different nuclei in the hypothalamus. In castrated males, they found that TH activity was increased about 60% in the median eminence. The activity of dopamine β hydroxylase was unaltered in any nuclear grouping by gonadectomy.

The activity of monoamine oxidase (MAO), an enzyme present in the mitochondria of neural tissue and responsible for the inactivation of catecholamines and serotonin by the formation of deaminated metabolites, has been reported by a number of groups to change during the estrous cycle. Kobayashi *et al.* (1964), using tyramine as a substrate, first reported changes in activity during the estrous cycle. They found the highest activity at proestrus in the posterior hypothalamus (from the center of the infundibulum to the mammillary bodies). The anterior hypothalamus (the optic chiasm to the center of the infundibulum) and the frontal cortex showed no changes. Zolovick *et al.* (1966) measured MAO in whole hypothalamus, whole amygdala, and cerebral cortex and reported that the values were higher in all areas at proestrus and estrus than at diestrus. In contrast to these studies, Kamberi and Kobayashi (1970) found a biphasic change in MAO activity in the median eminence and the anterior hypothalamus (including the optic chiasm to the median eminence). The highest activities were found early in proestrus and were decreased later in proestrus 3, only to rise again at estrus. In the lateral and posterior hypothalamus, a monophasic change in activity occurred, with the highest values early in proestrus, and in the amygdala, the highest values occurred in the middle of proestrus. In the cerebral cortex, the pituitary, and the uterus, no changes in MAO activity occurred during the estrous cycle.

Holzbauer and Youdim (1973), using kynuramine as a substrate, measured MAO in the hypothalamus, the caudate nucleus, the septum, and the peripheral organs and found low values at estrus and metestrus and high values at diestrus and proestrus. In order to determine the hormones that might be responsible for the changes, they administered either estradiol or progesterone to intact females for 8 days and measured MAO activities in the ovaries, the uterus, and the adrenal glands. Estrogen administration led to decreased MAO activity in the ovaries, the uterus, and the adrenal glands. Estrogen administration led to decreased MAO activity in these organs, whereas progesterone led to increased activity.

Although it is clear that MAO activity changes in brain regions as a function of the estrous cycle, the results are conflicting and have not clearly indicated the hormones or the cellular mechanisms that may be responsible for the CNS changes. A number of factors have contributed to these discrepancies. First, the use of different substrates for the reaction may have led to different results because it has been shows that MAO is comprised of a number of enzymes that show different specificities for various physiologically occurring substrates as well as for nonphysiological monoamines (Yang and Neff, 1973; White and Wu, 1975). There are also differences among laboratories in anatomical landmarks for the sampling of brain regions. Finally, the levels of a number of gonadal hormones, as well as of gonadotropins, change during the estrous cycle. In an effort to control some of these factors, Luine *et al.* (1975a) utilized ovariectomized females and the physiologically occurring monoamine serotonin for study. When estradiol benzoate was given to ovariectomized females for 3–7 days, 30%–50% reductions in MAO activity occurred in the basomedial hypothalamus (the rostral border of optic chiasm to the mammillary bodies and the area medial to the third ventricle) and the corticomedial amygdala (same cross-section as hypothalamus, area medial to optic tract), whereas no changes were found in the hippocampus, the cerebral cortex, or the preoptic area. The decreased activity did not seem to be a result of a direct interaction of estradiol with the enzyme protein because the addition of estradiol to the homogenate and incubation mixture did not result in decreased MAO activity. Because estradiol is known to affect the circulating levels of a number of pituitary hormones, estradiol benzoate was given to females that had been hypophysectomized for at least 1 week. The activity of MAO was decreased in the corticomedial amygdala; however, activity did not change in the basomedial hypothalamus, a finding suggesting that the changes in the hypothalamus were due to pituitary factors. However, as MAO is undoubtedly present in nerve endings in the median eminence, the possibility exists that hypophysectomy may have damaged these nerve endings and may thereby have altered the function of the cell mechanisms responsible for generating the response to estradiol. The results obtained thus far would tend to favor an estrogen action that reduces either the biosynthetic rate of the enzyme or of an essential catalytic component or that lowers the conversion rate of an inactive enzyme to an active form.

OTHER NEUROCHEMICAL EFFECTS OF GONADAL STEROIDS RELATED TO MONOAMINE METABOLISM. Another parameter of monoamine metabolism in neural tissue is the reuptake or accumulation of these amines. For example, in the work of Endersby and Wilson (1974), the accumulation of ^{3}H noradrenaline was increased in hypothalamic slices taken from ovariectomized rats that had received 3 days of estrogen treatment. The accumulation of ^{3}H dopamine was decreased in ovariectomized rats that were given estrogen and progesterone, although the hormonal manipulation did not alter the accumulation of ^{3}H serotonin. These results cannot be interpreted fully until the synthesis and release rates for this system are known. This approach, nonetheless, may provide a simple system for studying hormonal effects on an important parameter of monoamine metabolism.

The second messenger for the action of certain hormones, adenosine 3^15^1-monophosphate (cyclic AMP), has not been shown to be the mediator of estrogen-induced changes in either peripheral or CNS target tissues. However, two studies in brain suggest that a relationship may exist between estrogen treatment, monoamines, and cyclic AMP. *In vitro* incubation of hypothalamus from 21-day-old female rats with norepinephrine, epinephrine, or dopamine leads to 50%–75% increases in cyclic AMP levels within 5 min. Likewise, incubation with estrogen also leads to increased cyclic AMP levels, but with a delay of 1 hr. Increases in cyclic AMP levels brought about by estrogen can be blocked by preincubation with α or β adrenergic receptor blockers (Gunaga and Menon, 1973). Intraperitoneal injection of estradiol benzoate into 21-day-old rats also leads to 25%–50% increases in hypothalamic cyclic AMP levels, which can be blocked by prior treatment with the antiestrogen clomiphene (Gunaga *et al.*, 1974). This latter observation is somewhat suggestive of a role for the estrogen receptors described in earlier sections of this chapter, but the connection between these receptors and the relatively rapid change in cAMP levels, as well as the involvement of adrenergic receptors, is far from apparent.

Besides containing estrogen receptor sites, the rat hypothalamus is a primary site of the conversion of estradiol and estrone to catecholestrogens (Fishman and Norton, 1975). Catecholestrogens are preferential substrates for cerebral as well as hepatic catechol-0-methyltransferases and thus are able to act as inhibitors of the 0-methylation of catecholamines (Breuer and Koster, 1974). The physiological significance of this inhibition in the metabolism of catecholamines is obscure at present. From a neuroendocrine point of view, it has been reported that 2-hydroxyestradiol and 2-hydroxyestrone elevate serum LH levels in immature male rats within 6 hr of administration (Naftolin *et al.*, 1975). It remains to be seen whether this effect occurs in females as well as in males and whether the effect is an estrogenic effect or is related to interference with catecholamine metabolism. It is, of course, conceivable that catecholestrogens may be involved in the elevations of cAMP cited above.

ACETYLCHOLINE

FUNCTIONAL INVOLVEMENT. Cholinergic involvement in female sexual behavior in rats and hamsters was first indicated by the work of Lindström and Meyerson (1967) and of Lindström (1970, 1971, 1972, 1973). The effects of systemically administered muscarinic cholinergic agonists to estrogen–progesterone-primed ovariectomized (OVX) female rats revealed inhibitory effects that could be blocked by the centrally acting antagonist, atropine (Lindström and Meyerson, 1967). Increased central monoaminergic activity brought about by MAO inhibitors or by reuptake blockers augmented these inhibitory effects (Lindström, 1970). The monoaminergic involvement was further traced to serotonin (Lindström, 1971). Studies on the estrogen–progesterone-primed OVX hamster indicated that a similar inhibitory effect of cholinergic agonists did not involve monoaminergic neurons (Lindström, 1972). Furthermore, when cholinergic agonists were applied to estrogen-primed OVX hamsters and rats not treated with progesterone, the drugs *facilitated* lordosis responding and thus substituted for the progesterone treatment (Lindström, 1972, 1973). These facilitatory effects in the rat were prevented by adrenalectomy or hypophysectomy, a finding thus indicating a possible role of pituitary-adrenal secretions (Lindström, 1973).

In order to counteract possible adrenal influences, Humphreys and Clemens (1975) used dexamethasone treatment to suppress ACTH secretion and then applied cholinergic agonists intracranially to OVX, estrogen-primed rats. They obtained evidence implicating facilitatory cholinergic action on lordosis behavior in the mesencephalic reticular formation, which is both nicotinic and muscarinic. They also found facilitatory muscarinic (and not nicotinic cholinergic effects in the preoptic region.

NEUROCHEMICAL STUDIES. Experiments in this laboratory (Luine *et al.*, 1975a) have indicated that treatment of OVX or hypophysectomized rats with 5–30 μg/day of EB for 3–7 days increases the activity of choline acetyltransferase (CAT) in two estrogen-receptor-containing brain regions (the corticomedial amygdala and the medial preoptic area) and not in another area (the basomedial hypothalamus). (For the dissection of these brain regions, the reader should consult Figure 1.) Estrogen treatment failed to increase hippocampal CAT activity at a lower does (5 μg/day $\times$ 7 days) but did increase activity significantly at the higher dose (30 μg/day $\times$ 7 days). This effect may be a reflection of estrogen sensitivity among neurons of the septum, which are believed to be the source of most hippocampal CAT (see

Luine *et al.,* 1975a). The estrogen effects on CAT activity could be blocked by concurrent treatment with an antiestrogen, MER-25. The percentage increase of CAT activity is on the order of only 25%–30% but might be increased further by using the microdissection techniques described by Brownstein *et al.* (1975).

The physiological significance of these changes in CAT activity and the estrogen sensitivity of other components of the cholinergic system of the brain remain to be elucidated. In this connection, an unusually rapid (4-hr), brainwide, estrogen-dependent increase of acetylcholinesterase activity was reported by Moudgil and Kanungo (1973). This particular estrogen effect appeared to decline as the age of the experimental animal increased, and it remains to be established if this effect involves estrogen receptors of the type described in this chapter.

ENZYMES AS NEUROCHEMICAL MARKERS FOR ESTROGEN RECEPTOR FUNCTION

Enzymes and other proteins are the final products of gene expression, and the measurement of specific proteins is one of the most sensitive end points for the assessment of hormone effect on genomic activity. For example, changes in the level of a specific protein may be detected under conditions in which changes in overall protein synthesis, measured by the incorporation of radioactive amino acids, are masked by high levels of ongoing protein synthesis unrelated to the hormonal stimulus.

Brain and pituitary enzymes, activities of which are altered by estrogen treatment, are therefore useful markers for the exploration of estrogen receptor function. The earlier discussion of such estrogen-sensitive enzymes focused on their possible functional roles in the cell. In this section, we consider them simply as markers for estrogen action and summarize the evidence indicating that they reflect the operation of the estrogen receptors.

In the first place, the specific brain regions where estradiol alters enzyme activities are those that contain the putative estrogen-receptor sites. An enzyme that is changed by estradiol in a receptor-containing region does not necessarily change in another receptor-containing region. However, no estrogen effects attributable to those regions have been found in brain areas, such as the cerebral cortex, where estrogen receptors are absent or barely detectable. Table 5 compares the activities of three selected enzymes—MAO, glucose-6-phosphate dehydrogenase (G6PDH), and isocitrate dehydrogenase (ICDH)—in a number of brain regions and in the pituitaries of vehicle-treated and estrogen-treated ovariectomized rats. Also shown in Table 5 are the percentage changes in enzyme activities, which indicate that the estrogen effects are generally larger in pituitary than in hypothalamus, preoptic area, or amygdala. The larger estrogen effects in pituitary are consistent with the greater concentration of estrogen receptor sites in this tissue, compared to their concentration in brain regions.

Second, the enzyme changes appear to be a direct "inductive" effect of estradiol, as changes in enzyme activities can still be elicited by estradiol in adrenalec-

TABLE 5. ACTIVITY OF ENZYMES IN BRAIN REGIONS FROM OVARIECTOMIZED RATS (OVX) AND
OVARIECTOMIZED RATS TREATED WITH ESTRADIOL BENZOATE (EB)[a]

Brain region	Monoamine oxidase		Glucose-6-phosphate dehydrogenase		Isocitrate dehydrogenase	
	OVX	+EB	OVX	+EB	OVX	+EB
Cortex	135 ± 11	142 ± 9	135 ± 9	135 ± 9	355 ± 19	343 ± 19
Preoptic area	97.3 ± 10	92.3 ± 3	118 ± 12	122 ± 8	331 ± 34	324 ± 34
Hypothalamus	104 ± 11	$73.9 \pm 6*$	119 ± 9	$149 \pm 7*$	320 ± 24	$454 \pm 16**$
Amygdala	125 ± 14	$62.6 \pm 5**$	105 ± 8	111 ± 10	290 ± 14	$382 \pm 28*$
Hippocampus	164 ± 15	156 ± 7	118 ± 9	112 ± 6	263 ± 11	244 ± 12
Pituitary	—	—	382 ± 29	$799 \pm 36***$	345 ± 23	387 ± 31

[a]Data from Luine *et al.* (1974, 1975a,b). EB was given daily for 7 days at 30 µg/220 g body weight.
Values are the average ±S.E.M. for determinations in 6–9 animals. Differences between OVX and
+EB were tested by Stud.'s *t*-test where:
*$p < .05$.
**$p < .01$.
***$p < .001$.

tomized-ovariectomized and in hypophysectomized female rats (Luine *et al.*, 1974,
1975a). Moreover, these changes could not be reproduced by food deprivation, a
finding indicating that the estrogen was not acting indirectly via its ability to
depress food intake and to decrease body weight (Luine *et al.*, 1974, 1975a).

Third, the magnitude of the changes in enzyme activities were in most cases
directly related to the dose of administered estradiol benzoate. Selected examples
are shown in Figure 6 for malate dehydrogenase (MDH) and isocitrate dehydrogenase (ICDH) in hypothalamus and monoamine oxidase (MAO) in amygdala. It
should be noted that the percentage change in activity is plotted in Figure 6 against
the total administered dose of estradiol benzoate. The decreasing slope as a function of increasing dose indicates a marked tendency for the estrogen effect to reach
a plateau at higher doses, and this result is reminiscent of the saturation of estrogen receptors with increasing doses of estradiol 17β.

Fourth, the effectiveness of estradiol 17β and diethylstilbestrol and the ineffectiveness of estradiol 17α and testosterone in altering pituitary glucose-6-phos-
phate dehydrogenase (G6PDH) activity (Figure 7) are reminiscent of the binding
specificity of the estrogen receptor sites. In contrast to pituitary, cerebral-cortex
G6PDH activity is unaltered by any of the four steroids. In a fashion parallel to
that in the pituitary, brain enzyme activities altered in ovariectomized rats by estradiol benzoate treatment are not similarly altered in ovariectomized females by testosterone propionate treatment (Luine *et al.*, 1974, 1975b).

Fifth, the nonsteroidal "antiestrogens" have been useful tools for examining
the relationship between estrogen binding and estrogen-mediated functional
changes in target tissues. As discussed earlier, MER-25 is known to inhibit ovulation and lordosis responding in intact females (Shirley *et al.*, 1968) and in estradiol–progesterone-primed, ovariectomized female rats (Meyerson and Lindström,
1968). MER-25 does not by itself, except in very high doses, increase uterine
weight (Lerner, 1964) or alter lordosis responding when given with progesterone

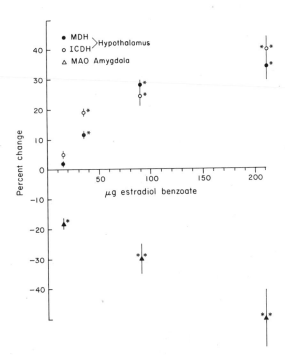

Fig. 6. Percentage change in enzyme activities in subregions of hypothalamus and amygdala (see Figure 1) as a function of the total cumulative dose of estradiol benzoate given to ovariectomized female rats. Statistical significance of estrogen effects was tested by Stud.'s t-test where $*p < .05$; $**p < .01$. (Data from Luine *et al.*, 1974, 1975b, by permission.)

(Meyerson and Lindström, 1968). Antiestrogens like MER-25, CI-628, and nafox-idine are known to interact with putative estrogen receptors in uterus, pituitary, and brain, and to exert their effects by blocking the access of estradiol to these sites. It appears that some of these antiestrogens may actually reside for long periods of time in the cell nuclear compartment and may prevent the target cell

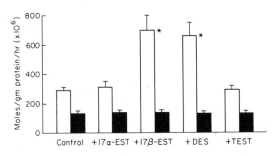

Fig. 7. Activity of glucose-6-phosphate dehydrogenase in the cortex and the pituitary of ovariectomized rats given various hormones. Values are the mean $\pm$ S.E.M. for determinations of four animals in ovariectomized rats (control) and ovariectomized rats given estradiol-17α (17α-EST), estradiol-17β benzoate (17β-EST), diethylstilbestrol (DES), and testosterone propionate (TEST). Open bars indicate activity in the pituitary and closed bars activity in the cortex. Rats received 30 μg EB and equimolar amounts of the other hormones daily for 7 days. The differences between control and hormone-treated rats were tested by Stud.'s t-test, where $*p < .01$. (Reprinted from Luine *et al.*, 1974, with permission of Pergamon Press, copyright 1974.)

from repleting its supply of cytoplasmic receptor sites (Rochefort and Capony, 1973; Katzenellenbogen and Katzenellenbogen, 1973; Clark *et al.*, 1973, 1974; Ruh and Ruh, 1974; Chazal *et al.*, 1975). Whatever the cellular mechanism of action, the treatment of ovariectomized rats with MER-25 in combination with estradiol benzoate attenuates or blocks the effect of the latter treatment on G6PDH and 6PGDH in the pituitary, MAO in the amygdala and the hypothalamus, and choline acetyltransferase in the preoptic area and the amygdala (Luine *et al.*, 1975a,b).

Finally, in one case where the minimum lag time was determined for detecting increases in an enzyme activity, this was on the order of 10–15 hr after estrogen administration (Heil *et al.*, 1971). Moreover, the direct addition of estradiol to the enzyme incubation mixture does not significantly alter enzyme activites (see Luine *et al.*, 1974, 1975a).

Whereas the changes in enzyme activities described above can be ascribed to a direct priming or an inductive action of estrogen, and whereas the characteristics of the changes are consonant with a wide range of experimental data on the properties of the estrogen receptor system, it cannot be stated at this time whether changed enzyme activities are actually a reflection of altered levels of enzyme proteins arising from altered rates of enzyme biosynthesis. Proof of such a relationship ultimately requires the use of specific antibodies to quantitatively determine enzyme levels.

Summary and Prospectus: A Model for Steroid Hormone Action on Neuronal Functioning

We have reviewed the evidence for the highly localized neuronal binding of radioactive steroid hormones in brain tissue and for the cellular binding of hormones in the pituitary. The final destination for these steroids in the cell appears to be the cell nucleus, and a model that appears to be widely applicable to steroid hormone action in invertebrates as well as among vertebrates predicts that the primary action of a cell nuclear hormone is the alteration of genomic function leading to specific cellular changes in RNA and protein formation that mediate the cellular responses to the hormone. Evidence for the role of this "inductive" mechanism in the neuroendocrine and behavioral effects of steroid hormones, which so far are documented for estrogens for the most part, may be classified into three categories:

1. *Temporal.* A time lag of 20–30 hr occurs between estrogen administration and the appearance of behavioral estrus or an ovulatory LH surge. In correlative studies with the rat, estradiol seems to have largely disappeared from the putative pituitary and neural receptors by the time the physiological effects are manifested. Thus, the estrogen effect is not that of a "stimulus" (i.e., required at the time that the response occurs) but that of a "permissive agent," increasing the probability that appropriate stimuli, occurring after an induction period, will elicit the response. It is important to emphasize the essential role of the appropriate stimuli

(e.g., the male palpating the female's flanks, eliciting lordosis; the day–night light cycle, leading to ovulation), for the hormone does not by itself induce the physiological response.

2. *Pharmacological.* Inhibitors of RNA and protein synthesis can reversibly block the estrogen facilitation of lordosis, locomotor activity, and the ovulatory LH surge. Likewise, so-called antiestrogens, which have the ability to occupy and render inactive the estrogen receptor mechanism, also appear to block the estrogen facilitation of lordosis and the ovulatory LH surge. These effects are consistent with the cellular mechanism of steroid hormone action, which is the theme of this chapter, and are indeed predicted by it.

3. *Neurochemical.* Correlative studies of estrogen effects on pituitary and brain RNA and protein metabolism provide some evidence for altered metabolic states resulting from enhanced estrogen secretion or from estrogen administration. Measurements of brain and pituitary enzyme activities as a function of estrogen administration also point to a variety of "inductive" effects on cellular metabolism. Where these effects have been examined in some detail, they appear to be the direct result of estrogen action at the receptor level, although the definitive proof of this relationship is lacking. Thus, an "inductive" neurochemical effect of estrogen may underlie a "permissive" action at the physiological level. The resolution of any paradox implicit in the juxtaposition of the terms *inductive* and *permissive* may rest on the fact that none of the neurochemical changes are increases from undetectable levels of a gene product. Rather, they appear to be increases (or decreases) in the level or the activity of a constituent that is already present in a substantial amount in the absence of the hormonal stimulus. Modulation in the amount of these constituents might be thought of as a means of "tuning" neuronal systems or of increasing (or decreasing) the functional efficiency of specific neural circuits. For a discussion of the neural circuits that are involved in the mediation of the estrogen-dependent lordosis response, the reader is referred to the work of Pfaff (Pfaff *et al.,* 1973; McEwen and Pfaff, 1973).

A variety of possible neurochemical mechanisms for estrogen effects on reproductive function have been described in this chapter. They may be categorized and generalized into a working hypothesis or model of steroid hormone action, as shown in Figure 8. First, there are alterations in the oxidative metabolism of neural tissue and of the pituitary that may provide increased amounts of energy for neuronal function, as well as providing reducing equivalents for the reductive biosynthesis of lipids and pentose sugars for RNA synthesis. Second, there may be hormone-induced alterations in biosynthetic and degradative enzymes for neurotransmitters and releasing hormones. In such cases, it might be expected that part of the "inductive" lag in the manifestation of hormonal effects would be occupied by the time required for the axonal or dendritic transport of the newly synthesized enzymes. Third, there may also be hormonally induced alterations in the amount of postsynaptic "receptors" for neurotransmitters. At the present time, the only evidence pointing in this direction deals with the sensitivity of the pituitary of LHRH (see above), but future work will undoubtedly yield interesting results as techniques for quantitatively measuring these receptors become available.

VICTORIA N. LUINE
AND BRUCE S.
McEWEN

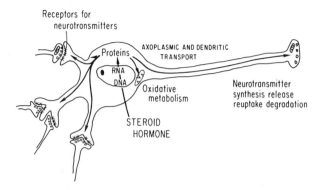

Fig. 8. Speculative model of the neurochemical consequences of estrogen action on neurons.

Beyond this already complicated and incomplete subject lies the even more difficult job of integrating individual neurochemical changes into a meaningful sequence of neurochemical and neurophysiological events by which we can describe and understand the hormonal facilitation of neuroendocrine and behavioral changes.

Acknowledgment

We would like to thank Ms. Freddi Berg for editorial assistance in the preparation of this manuscript.

References

Alsum, P., and Goy, R. W. Actions of esters of testosterone, dihydrotestosterone, or estradiol on sexual behavior in castrated male guinea pigs. *Hormones and Behavior,* 1974, *5,* 207–218.

Anderson, C. H., and Greenwald, S. S. Autoradiographic analysis of estradiol uptake in the brain and pituitary of the female rat. *Endocrinology,* 1969, *85,* 1160–1165.

Anderson, J. N., Peck, E. J., Jr., and Clark, J. H. Oestrogen and nuclear binding sites: Determination of specific sties by [³H] oestradiol exchange. *Biochemistry Journal,* 1972, *126,* 561–567.

Anderson, J. N., Peck, E. J., Jr., and Clark, J. H. Nuclear receptor estrogen complex: Accumulation, retention and localization in the hypothalamus and pituitary. *Endocrinology,* 1973, *93,* 711–717.

Arai, Y., and Gorski, R. A. Effect of anti-estrogen on steroid induced sexual receptivity in ovariectomized rats. *Physiology and Behavior,* 1968, *3,* 351–353.

Atger, M., Baulieu, E., and Milgrom, E. An investigation of progesterone receptors in guinea pig vagina, uterine cervix, mammary glands, pituitary and hypothalamus. *Endocrinology* 1974, *94,* 161–167.

Attramadal, A. Distribution and site of action of oestradiol in the brain and pituitary gland of rat following intramuscular administration. *Proceedings of the Second International Congress on Endocrinology, Excerpta Medica International Congress,* Series No. 83. Amsterdam; Excerpta Medica, 1964.

Ball, J. Sex activity of castrated male rats increased by estrin adminstration. *Journal of Comparative Psychology,* 1937, *24,* 135–144.

Bapna, J., Neff, N. H., Costa, E. A method for studying norepinephrine and serotonin metabolism in small regions of rat brain: Effect of ovariectomy on amine metabolism in anterior and posterior hypothalamus. *Endocrinology,* 1971, *89,* 1345–1349.

Baquer, N. Z., and McLean, P. The effect of oestradiol on the profile of constant and specific proportion groups of enzymes in rat uterus. *Biochemical and Biophysical Research Communications,* 1972, *48,* 729–734.

Baum, M. J., and Vreeburg, J. T. M. Copulation in castrated male rats following combined treatment with estradiol and dihydrotestosterone. *Science,* 1973, *182,* 283–285.

Baum, M. J., Södersten, P., and Vreeburg, J. T. M. Mounting and receptive behavior in the ovariectomized female rat: Influence of estradiol, dihydrotestosterone and genital anesthetization. *Hormones and Behavior,* 1974, *5,* 175–190.

Beach, F. A. Male and female mating behavior in prepuberally castrated female rats treated with androgen. *Endocrinology,* 1942b, *31,* 673–678.

Beach, F. A. Copulatory behavior in prepuberally castrated male rats and its modification by estrogen administration. *Endocrinology,* 1942a, *31,* 679–683.

Beattie, C. W., Rogers, C. H., and Soyka, L. F. Influence of ovariectomy and ovarian steroids on hypothalamic tyrosine hydroxylase activity in the rat. *Endocrinology,* 1972, *91,* 276–279.

Belajev, D. K., Korotchkin, L. I., Bajev, A. A., Golubitsa, A. N., Korotchkina, L. S., Maksimovsky, L. F., Davidovskaja, A. E. Activity of the genetic apparatus of hypothalamic nerve cells at various stages of the oestrous cycle in the albino rat. *Nature,* 1967, *214,* 201–202.

Berendes, H. D. The hormone ecdysone as effector of specific changes in the pattern of gene activities of *Drosophila hydei. Chromosoma* (Berlin, 1967, *22,* 274–293.

Berthold, A. A. Transplantation der Hoden. *Archiv für Anatomie, Physiologie, Wissenschaftliche Medicin,* 1849, *16,* 42–46.

Beyer, C., and Komisaruk, B. Effects of diverse androgens on estrous behavior, lordosis reflex, and genital tract morphology in the rat. *Hormones and Behavior,* 1971, *2,* 217–225.

Beyer, C., and Rivaud, N. Differential effect of testosterone and dihydrotestosterone on the behavior of prepuberally castrated male rabbits. *Hormones and Behavior,* 1973, *4,* 175–180.

Beyer, C., and Vidal, N. Inhibitory action of MER-25 on androgen-induced oestrous behaviour in the ovariectomized rabbit. *Journal of Endocrinology,* 1971, *51,* 401–402.

Beyer, C., McDonald, P., and Vidal, N. Failure of 5-α-dihydrotestosterone to elicit estrous behavior in the ovariectomized rabbit. *Endocrinology,* 1970a, *86,* 939–941.

Beyer, C., Vidal, N., and Mijares, A. Probable role of aromatization in the induction of estrous behavior by androgens in the ovariectomized rabbit. *Endocrinology,* 1970b, *87,* 1386–1389.

Beyer, C., Morali, G., and Cruz, M. L. Effect of 5-alpha-dihydrotestosterone on gonadotropin secretion and estrous behavior in the female Wistar rat. *Endocrinology,* 1971, *89,* 1158–1161.

Beyer, C., Jaffe, R. B., and Gay, V. L. Testosterone metabolism in target tissues: Effects of testosterone and dihydrotestosterone injection and hypothalamic implantation on serum LH in ovariectomized rats. *Endocrinology,* 1972, *91,* 1372–1375.

Beyer, C., Larsson, K., Perez-Palacios, G., and Morali, G. Androgen structure and male sexual behavior in the castrated rat. *Hormones and Behavior,* 1973, *4,* 99–108.

Bickel, M., Kuhl, H., Tan, J. S. E., and Taubert, H.-D. Evidence of a sex-specific effect of testosterone and progesterone upon L-cystine-aminopeptidase activity in the hypothalamus and paleopallium of the rat. *Neuroendocrinology,* 1972, *9,* 321–331.

Björklund, A., and Nobin, A. Fluorescence, histochemical and microspectrofluorometric mapping of dopamine and noradrenaline cell groups in the rat diencephalon. *Brain Research,* 1973, *51,* 193–205.

Bresciani, F., Nola, E., Sica, V., and Puca, G. Early stages in estrogen control of gene expression and its derangement in cancer. *Federation Proceedings,* 1973, *32,* 2126–2132.

Breuer, H., and Koster, G. Interactions between oestrogens and neurotransmitters at the hypophyseal-hypothalamic level, *Journal of Steroid Biochemistry,* 1974, *5,* 961–967.

Brown-Grant, K. Steroid hormone administration and gonadotrophin secretion in the gonadectomized rat. *Journal of Endocrinology,* 1974, *62,* 319–332.

Brownstein, M., Kobayashi, R., Palkovits, M., and Saavedra, J. M. Choline acetyltransferase levels in diencephalic nuclei of the rat. *Journal of Neurochemistry,* 1975, *24,* 35–38.

Buller, R. E., Toft, D. O., Schrader, W. T., and O'Malley, B. W. Progesterone-binding components of chick oviduct: VIII. Receptor activation and hormone-dependent binding to purified nuclei. *Journal of Biological Chemistry,* 1975a, *250,* 801–808.

Buller, R. E., Toft, D. O., Schrader, W. T., and O'Malley, B. W. Progesterone-binding components of the chick oviduct: IX. The kinetics of nuclear binding. *Journal of Biological Chemistry,* 1975b, *250,* 809–818.

Burnet, F. R., and MacKinnon, P. C. B. Restoration by oestradiol benzoate of a neural and hormonal rhythm in the ovariectomized rat. *Journal of Endocrinology,* 1975, *64,* 27–35.

Chader, G. J., and Villee, C. A. Uptake of oestradiol by the rabbit hypothalamus. *Biochemistry Journal,* 1970, *118,* 93–97.

Chamness, G. C., Jennings, A. W., and McGuire, W. L. Estrogen receptor binding to isolated nuclei, a nonsaturable process. *Biochemistry,* 1974, *13,* 327–331.

Chazal, G., Faudon, M., Gogan, F., and Rotsztejn, W. Effects of two estradiol antagonists upon the estradiol uptake in the rat brain and peripheral tissues. *Brain Research,* 1975, *89,* 245–254.

Cheng, Yi-J., Karavolas, H. J. Conversion of progesterone to 5α-pregnane-3,20-dione and 3α-hydroxy-5α-pregnan-30-one by rat medial basal hypothalami and the effects of estradiol and stage estrous cycle on the conversion. *Endocrinology,* 1973, *93,* 1157–1162.

Christensen, L. W., and Clemens, L. G. Intrahypothalamic implants of testosterone or estradiol and resumption of masculine sexual behavior in long-term castrated male rats. *Endocrinology,* 1974, *95,* 984–990.

Christensen, L. W., Coniglio, L. P., Paup, D. C., and Clemens, L. G. Sexual behavior of male golden hamsters receiving diverse androgen treatments. *Hormones and Behavior,* 1973, *4,* 223–230.

Clark, J. H., Anderson, J., Peck, E. J., Jr. Receptor-estrogen complex in the nuclear fraction of rat uterine cells during the estrous cycle. *Science,* 1972, *176,* 528–530.

Clark, J. H., Anderson, J., and Peck, E. J., Jr. Estrogen receptor-anti-estrogen complex: Atypical binding by uterine nuclei and effects on uterine growth. *Steroids,* 1973, *22,* 707–718.

Clark, J. H., Peck, E. J., and Anderson, J. N. Oestrogen receptors and antagonism of steroid hormone action. *Nature,* 1974, *251,* 446–448.

Clemens, L. E., and Kleinsmith, L. J. Specific binding of the oestradiol-receptor complex to DNA. *Nature New Biology,* 1972, *237,* 204–206.

Clever, U., and Karlson, P. Induktion von puff-veränderungen in den Speicheldrusenchromosomen von *Chironomus Tentans* durch Ecdyson. *Experimental Cell Research,* 1960, *20,* 623–626.

Convey, E. M., and Reece, R. P. Influence of the estrous cycle on the nucleic acid content of the rat anterior pituitary. *Proceedings of the Society for Experimental Biology and Medicine,* 1969, *132,* 878–880.

Cooper, K. J., Fawcett, C. P., and McCann, S. M. Variations in pituitary responsiveness to a luteinizing hormone/follicle stimulating hormone releasing factor (LH-RF/FSH-RF) preparation during the rat estrous cycle. *Endocrinology,* 1974, *95,* 1293–1299.

Crowley, W. R., and Zemlan, F. F. The neurochemical control of mating behavior. In N. T. Adler (Ed.), *Neuroendocrinology of Reproduction.* New York: Plenum Press, 1981.

Czaja, J. A., Goldfoot, D. A., and Karavolas, H. J. Comparative facilitation and inhibition of lordosis in the guinea pig with progesterone, 5α-pregnane-3,20-dione, or 3α-hydroxy-5α-pregnan-20-one. *Hormones and Behavior,* 1974, *5,* 261–274.

Davidson, J. M. Effects of estrogen on the sexual behavior of male rats. *Endocrinology,* 1969, *84,* 1365–1372.

Davidson, J. M. Hormones and reproductive behavior. In H. Balin and S. Glasser, (Eds.), *Reproductive Biology.* Amsterdam: Excerpta Medica, 1972.

deKloet, R., and McEwen, B. S. Glucocorticoid interactions with brain and pituitary. In W. H. Gispen, (Ed.), *Chemical Neurobiology* Amsterdam: Elsevier, 1976.

deKloet, R., Wallach, G., and McEwen, B. S. Differences in corticosterone and dexamethasone binding to rat brain and pituitary. *Endocrinology,* 1975, *96,* 598–609.

Denef C., Magnus, C., and McEwen, B. S. Sex differences and hormonal control of testosterone metabolism in rat pituitary and brain. *Journal of Endocrinology,* 1973, *59,* 605–621.

Denef, C., Magnus, C., and McEwen, B. S. Sex-dependent changes in pituitary 5α-dihydrotestosterone and 3α-androstanediol formation during postnatal development and puberty in the rat. *Endocrinology,* 1974, *94,* 1265–1274.

Dixit, V. P., and Niemi, M. Uptake of exogenous ³H-1,2-testosterone in the hypothalamus, endocrine glands and sex-accessory organs in neonatally castrated and androgenized male rats. *Endocrinologica Experimentalis,* 1974, *8,* 39–43.

Eisenfeld, A. J. ³H-Estradiol: *In vitro* binding to macromolecules from the rat hypothalamus, anterior pituitary and uterus. *Endocrinology,* 1970, *86,* 1313–1318.

Eisenfeld, A. J., and Axelrod, J. Effect of steroid hormones, ovariectomy, estrogen pretreatment, sex and immaturity on the distribution of ³H-estradiol. *Endocrinology,* 1966, *79,* 38–42.

Eleftheriou, B. E., and Church, R. L., Concentrations of RNA in the brain during oestrus in the deer-mouse. *Nature,* 1967, *215,* 1195–1196.

Endersby, C. A., and Wilson, C. A. The effect of ovarian steroids on the accumulation of ^{3}H-labeled monoamines by hypothalamic tissue *in vitro*. *Brain Research*, 1974, *73*, 321–331.

Everitt, B. J., and Herbert, J. The effects of implanting testosterone propionate into the central nervous system on the sexual behavior of adrenalectomized female rhesus monkeys. *Brain Research*, 1975, *86*, 109–120.

Everitt, B. J., Herbert, J., and Harnes, J. D. Sexual receptivity of bilaterally adrenalectomized female rhesus monkeys. *Physiology and Behavior*, 1972, *8*, 409–415.

Everitt, B. J., Fuxe, K., Hökfelt, T., and Jonsson, G. Pharmacological and biochemical studies on the role of monoamines in the control by hormones of sexual receptivity in the female rat. *Journal of Comparative Physiology and Psychology*, 1975, 556–572.

Fajer, A. B., Holzbauer, M., and Newport, H. M. The contribution of the adrenal gland to the total amount of progesterone produced in the female rat. *Journal of Physiology*, 1971, *214*, 115–126.

Feder, H. H. The comparative actions of testosterone propionate and 5α-androstan-17β-ol-3-one propionate on the reproductive behavior physiology and morphology of male rats. *Journal of Endocrinology*, 1971, *51*, 241–252.

Feder, H. H., Naftolin, F., and Ryan, K. J. Male and female sexual responses in male rats given estradiol benzoate and 5α-androstan-17β-ol-3-one propionate. *Endocrinology*, 1974, *94*, 136–141.

Feigelson, M., and Feigelson, P. Metabolic effects of glucocorticoids as related to enzyme induction. In G. Weber, (Ed.), *Advances in Enzyme Regulation*, Vol. 3. New York: Pergamon Press, 1965.

Fishman, J., and Norton, B. Catechol estrogen formation in the central nervous system of the rat. *Endocrinology*, 1975, *96*, 1054–1059.

Fletcher, T. J., and Short, R. V. Restoration of libido in castrated red deer stag (Cervus elaphus) with oestradiol-17β. *Nature*, 1974, *248*, 616–618.

Flores, F., Naftolin, F., Ryan, K. J., and White, R. J. Estrogen formation by the isolated perfused rhesus monkey brain. *Science*, 1973, *180*, 1074–1075.

Friend, J. P., and Leavitt, W. W. Interaction of ^{3}H-oestradiol with anterior pituitary nuclei in a cell-free system. *Acta Endocrinologica*, 1972, *69*, 230–240.

Fuxe, K., and Hökfelt, T. Catecholamines in the hypothalamus and the pituitary gland. In L. Martini and W. F. Ganong (Eds.), *Frontiers in Neuroendocrinology*. New York: Oxford Press, 1969.

Fuxe, K., Hökfelt, T., and Nilsson, O. Castration, sex hormones and tubero-infundibular dopamine neurons. *Neuroendocrinology*, 1969, *5*, 107–120.

Gerlach, J. L., and McEwen, B. S. Rat brain binds adrenal steroid hormone: Radioautography of hippocampus with corticosterone. *Science*, 1972, *175*, 1133–1136.

Gerlach, J. L., McEwen, B. S., Pfaff, D. W., Moskowitz, S., Ferin, M., Carmel, P. W., and Zimmerman, E. A. Cells in regions in monkey brain and pituitary retain radioactive estradiol, corticosterone and cortisol differentially. *Brain Research*, 1976, *103*, 603–612.

Gessa, G. L., and Tagliamonte, A. Possible role of brain serotonin and dopamine in controlling male sexual behavior. In E. Costa, G. L. Gessa, and M. Sandler, (Eds.), *Advances in Biochemical Psychopharmacology. Vol. 11: Serotonin—New Vistas—Biochemistry and Behavioral and Clinical Studies*. New York: Raven Press, 1974a.

Gessa, G. L., and Tagliamonte, A. Role of brain monoamines in male sexual behavior. *Life Sciences*, 1974b, *14*, 425–436.

Ginsburg, M., Greenstein, B. D., MacLusky, N. J., Morris, I. D., and Thomas, P. J. Dihydrotestosterone binding in brain and pituitary cytosol of rats. *Journal of Endocrinology*, 1974a, *61*, 24.

Ginsburg, M., Greenstein, B. D., MacLusky, N. J., Morris, I. D., and Thomas, P. J. An improved method for the study of high affinity steroid binding: Oestradiol binding in brain and pituitary. *Steroids*, 1974b, *23*, 773–792.

Glascock, R. F., and Michael, R. P. Localization of oestrogen in a neurological system in the brain of the female cat. *Journal of Physiology*, 1962, *163P*, 38–39.

Glasser, S. R., Chytil, F. C., and Spelsberg, T. C. Early effects of oestradiol-17β on the chromatin and activity of the deoxyribonucleic acid-dependent ribonucleic acid polymerases (I and II) of the rat uterus. *Biochemistry Journal*, 1972, *130*, 947–957.

Gordon, J. H., and Reichlin, S. Changes in pituitary responsiveness to luteinizing hormone-releasing factor during the rat estrous cycle. *Endocrinology*, 1974, *94*, 974–977.

Gorski, J., Toft, D., Shyamala, G., Smith, D., and Notides, A. Hormone receptors: Studies on the interaction of estrogen with the uterus. *Recent Progress in Hormone Research*, 1968, *24*, 45–72.

Gradwell, P. B., Everitt, B. J., and Herbert, J. 5-Hydroxytryptamine in the central nervous system and sexual receptivity of female rhesus monkeys. *Brain Research*, 1975, *88*, 281–293.

Green, R., Luttge, W. G., and Whalen, R. E. Induction of receptivity in ovariectomized rats by a single intravenous injection of estradiol-17 beta. *Physiology and Behavior*, 1970, *5*, 137–141.

Griffiths, E. C., and Hooper, K. C. Changes in hypothalamic peptidase activity during the oestrous cycle in the adult female rat. *Acta Endocrinologica*, 1973, *74*, 41–48.

Griffiths, E. C., and Hooper, K. C. Peptidase activity in different areas of the rat hypothalamus. *Acta Endocrinologica*, 1974, *77*, 10–18.

Griffiths, E. C., Hooper, K. C., Jeffcoate, S. L., and Holland, D. T. The presence of peptidases in the rat hypothalamus inactivating luteinizing hormone-releasing hormone (LH-RH). *Acta Endocrinologica*, 1974, *77*, 435–442.

Griffiths, E. C., Hooper, K. C., Jeffcoate, S. L., and Holland, D. T. The effects of gonadectomy and gonadal steroids on the activity of hypothalamic peptidases inactivating luteinizing hormone-releasing hormone (LH-RH). *Brain Research*, 1975, *88*, 384–388.

Grosser, B. I., Stevens, W., Bruenger, F. W., and Reed, D. J. Corticosterone binding in rat brain cytosol. *Journal of Neurochemistry*, 1971, *18*, 1725–1732.

Grosser, B. I., Stevens, W., and Reed, D. J. Properties of corticosterone binding macromolecules from rat brain cytosol. *Brain Research*, 1973, *57*, 387–395.

Gual, C., Morato, T., Hayano, M., Gut, M., and Dorfman, R. I. Biosynthesis of estrogens. *Endocrinology*, 1962, *71*, 920–925.

Gunaga, K. P., and Menon, K. M. J. Effect of catecholamines and ovarian hormones on cyclic AMP accumulation in rat hypothalamus. *Biochemical and Biophysical Research Communications*, 1973, *54*, 440–448.

Gunaga, K. P., Kawano, A., and Menon, K. M. J. *In vivo* effect of estradiol benzoate on the accumulation of adenosine $3'5'$-cyclic monophosphate in the rat hypothalamus. *Neuroendocrinology*, 1974, *16*, 273–281.

Hall, C. E., Ayachi, S., and Hull, O. Salt appetite and hypertensive response to salt and to desoxycorticosterone in Sprague-Dawley and Long-Evans rats. *Texas Reports on Biology and Medicine*, 1972, *30*, 143–153.

Hamilton, T. H. Control by estrogen of genetic transcription and translation. *Science*, 1968, *161*, 649–661.

Heil, H., Meltzer, V., Kuhl, H., Abraham, R., and Taubert, H. D. Stimulation of L-cystine-aminopeptidase activity by hormonal steroids and steroid analogs in the hypothalamus and other tissues of the female rat. *Fertility and Sterility*, 1971, *22*, 181–187.

Herxheimer, A., and Woodbury, D. M. The effect of deoxycorticosterone on salt and sucrose taste preference thresholds and drinking behavior in rats. *Journal of Physiology*, 1960, *151*, 253–260.

Higgins, S. J., Rousseau, G. G., Baxter, J. D., and Tomkins, G. M. Nature of nuclear acceptor sites for glucocorticoid and estrogen receptor complexes. *Journal of Biological Chemistry*, 1973a, *248*, 5873–5879.

Higgins, S. J., Rousseau, G. G., Baxter, J. D., and Tomkins, G. M. Nuclear binding of steroid receptors: comparison in intact cells and cell-free systems. *Proceedings of the National Academy of Science* (USA), 1973b, *70*, 3415–3418.

Hilliard, J., Penardi, R., and Sawyer, C. H. A functional role for 20α-hydroxypregn-4-en-3-one in the rabbit. *Endocrinology*, 1967, *80*, 901–909.

Ho, G. K., Quadagno, D. M., Cooke, P. H., and Gorski, R. A. Intracranial implants of actinomycin-D: Effects on sexual behavior and nucleolar ultrastructure in the rat. *Neuroendocrinology*, 1973–1974, *13*, 47–55.

Holzbauer, M., and Youdim, M. B. H. The oestrous cycle and monoamine oxidase activity. *British Journal of Pharmacology*, 1973, *48*, 600–608.

Humphreys, R. R., and Clemens, L. G. *Lordosis in the Female Rat Resultant from Direct Chemical Stimulation of Central Cholinergic Mechanisms*. Abstract, Winter Conference for Brain Research, Steamboat Springs, Colorado, January 1975.

Hunter, F., and Hagy, G. W. Interactions of tissue G6PDH activity with age and sexual development in the rat. *Endokrinologie*, 1969, *54*, 85–97.

Jackson, G. L. Effect of actinomycin D on estrogen-induced release of luteinizing hormone in ovariectomized rats. *Endocrinology*, 1972, *91*, 1284–1287.

Jackson, G. L. Time interval between injection of estradiol benzoate and LH release in the rat and effect of actinomycin D or cycloheximide. *Endocrinology*, 1973, *93*, 887–891.

Jackson, G. L. Blockage of progesterone-induced release of LH by intrabrain implants of actinomycin D. *Neuroendocrinology*, 1975, *17*, 236–244.

Jensen, E. V., and deSombre, E. R. Estrogen-receptor interaction. *Science*, 1973, *182*, 126–134.

Jensen, E. V., and Jacobson, H. I. Basic guides to the mechanism of estrogen action. *Recent Progress in Hormone Research*, 1962, *18*, 387–408.

Jensen, E. V., Suzuki, T., Kawashima, T., Stumpf, W. E., Jungblut, P. W., and deSombre, E. R. A two-step mechanism for the interaction of estradiol with rat uterus. *Proceedings of the National Academy of Sciences* (USA), 1968, *59*, 632–638.

Johnston, P., and Davidson, J. M. Intracerebral androgens and sexual behavior in the male rat. *Hormones and Behavior*, 1972, *3*, 345–357.

Jouan, P., Samperez, S., Thieulant, M. L., and Mercier, L. Étude du récepteur cytoplasmique de la (1,2 ^{3}H) testostérone dans l'hypophyse antérieure et l'hypothalamus du rat. *Journal of Steroid Biochemistry*, 1971a, *2*, 223–236.

Jouan, P., Samperez, S., Thieulant, M. L., and Mercier, L. Neurobiologie moléculaire: Étude de la composition des stéroides liés au "récepteurs" cytoplasmiques et de l'hypophyse antérieure et de l'hypothalamus après incubation en présence dê testostérone-^{3}H. *Comptes Rendus Hebdomadaires des Séances de l'Académie des Sciences* (Paris), 1971b, *272*, 2368–2371.

Jouan, P., Samperez, S., and Thieulant, M. L. Testosterone "receptors" in purified nuclei of rat anterior hypophysis. *Journal of Steroid Biochemistry*, 1973, *4*, 65–74.

Kahwanago, I., Heinrichs, W. L., and Herrmann, W. L. Oestradiol "receptors" in the hypothalamus and anterior pituitary gland: Inhibition of oestradiol binding by SH group blocking agents and clomiphene citrate. *Endocrinology*, 1970, *86*, 1319–1326.

Kamberi, A. I., and Kobayashi, Y. Monoamine oxidase activity in the hypothalamus and various other brain areas and in some endocrine glands of the rat during the estrous cycle. *Journal of Neurochemistry*, 1970, *17*, 216–268.

Karavolas, H. J., and Herf, S. M. Conversion of progesterone by rat medial basal hypothalamic tissue to 5α-pregnane-3,20-dione. *Endocrinology*, 1971, *89*, 940–942.

Kato, J. *In vitro* uptake of tritiated oestradiol by the rat anterior hypothalamus during the oestrous cycle. *Acta Endocrinologica* (Copenh.) 1970, *63*, 577–584.

Kato, J. Localization of oestradiol receptors in the rat hypothalamus. *Acta Endocrinologica*. 1973, *72*, 663–670.

Kato, J., and Onouchi, T. 5α-Dihydrotestosterone receptor in the rat hypothalamus. *Journal of Biochemistry Endocrinologia Japonica*, 1973, *20*, 429–432.

Kato, J., Atsumi, Y., and Inaba, M. A soluble receptor for estradiol in rat anterior hypophysis. *Journal of Biochemistry* (Tokyo), 1970a, *68*, 759–761.

Kato, J., Atsumi, Y., and Muramatsu, M. Nuclear estradiol receptor in rat anterior hypophysis. *Journal of Biochemistry* (Tokyo), 1970b, *67*, 871–872.

Katzenellenbogen, B. S., and Katzenellenbogen, J. A. Antiestrogens: Studies using an *in vitro* estrogen-responsive uterine system. *Biochemical and Biophysical Research Communications*, 1973, *50*, 1152–1159.

Kawakami, M., and Sawyer, C. H. Neuroendocrine correlates of changes in brain activity thresholds by sex steroids and pituitary hormones. *Endocrinology*, 1959, *65*, 652–668.

Keefer, D. A., Stumpf, W. E., and Sar, M. Estrogen-topographical localization of estrogen-concentrating cells in the rat spinal cord following ^{3}H-estradiol administration. *Proceedings of the Society for Experimental Biology and Medicine*, 1973, *143*, 414–417.

Kelly, W. G. On the mechanism of conversion of circulating Δ4-androstene-3,17-dione to estrone: Stereochemistry of the elimination of hydrogen from ring A. *Endocrinology*, 1974, *95*, 308–310.

King, R. J. B., and Gordon, J. Involvement of DNA in the acceptor mechanism for uterine oestradiol receptor. *Nature New Biology*, 1972, *240*, 185–187.

Kizer, J. S., Palkovits, M., Zivin, J., Brownstein, M., Saavedra, J. M., and Kopin, I. J. The effect of endocrinological manipulations on tyrosine hydroxylase and dopamine βhydroxylase activities in individual hypothalamic nuclei of the adult male rat. *Endocrinology*, 1974, *95*, 799–812.

Kniewald, Z., and Milkovic, S. Testosterone; a regulator of 5α-reductase activity in the pituitary of male and female rats. *Endocrinology*, 1973, *92*, 1772–1775.

Kniewald, Z., Massa, R., and Martini, L. Conversion of testosterone into 5α-androstan-17β-ol-3-one at the anterior pituitary and hypothalamic level. In V. H. T. James and L. Martini (Eds.), *Hormonal Steroids*. Amsterdam: Excerpta Medica, 1971.

Knizley, H., Jr. The hippocampus and septal area as primary target sites for corticosterone. *Journal of Neurochemistry*, 1972, *19*, 2737–2745.

Kobayashi, T., Kobayashi, T., Kato, J., and Minaguchi, H. Fluctuations in monoamine oxidase activity in the hypothalamus of rat during the estrous cycle and after castration. *Endocrinologia Japonica*, 1964, *11*, 283–290.

Komisaruk, B. R., and Beyer, C. Differential antagonism, by MER-25, of behavioral and morphological effects of estradiol benzoate in rats. *Hormones and Behavior*, 1972, *3*, 63–70.

Korach, K. S., and Muldoon, T. G. Comparison of specific 17β-estradiol-receptor interactions in the anterior pituitary of male and female rats. *Endocrinology*, 1973, *92*, 322–326.

Korach, K. S., and Muldoon, T. G. Studies on the nature of the hypothalamic estradiol-concentrating mechanism in the male and female rat. *Endocrinology*, 1974, *94*, 785–793.

Kuhl, H., Rosniatowski, C., Oen, S., and Taubert, H. Sex steroids stimulate the activity of hypothalamic arylamidases in the rat. *Acta Endocrinologica*, 1974, *76*, 1–14.

Larsson, K., Södersten, P., and Beyer, C. Induction of male sexual behaviour by oestradiol benzoate in combination with dihydrotestosterone. *Journal of Endocrinology*, 1973a, *57*, 563–564.

Larsson, K., Södersten, P., and Beyer, C. Sexual behavior in male rats treated with estrogen in combination with dihydrotestosterone. *Hormones and Behavior*, 1973b, *4*, 289–300.

Lassman, M. N., and Mulrow, P. J. Deficiency of deoxycorticosterone-binding protein in the hypothalamus of rats resistant to deoxycorticosterone-induced hypertension. *Endocrinology*, 1974, *94*, 1541–1546.

Laumas, K. R., and Farooq, A. The uptake *in vivo* of [1,2-³H] progesterone by the brain and genital tract of the rat. *Journal of Endocrinology*, 1966, *36*, 95–96.

Leavitt, W. W., Kimmel, G. L., & Friend, J. P. Steroid hormone uptake by anterior pituitary cell suspensions. *Endocrinology*, 1973, *92*, 94–102.

Lemaire, I., Dupont, A., Bastarache, E., Labrie, F., and Fortier, C. Some characteristics of corticosterone uptake by the dorsal hippocampus and the adenohypophysis in the rat. *Canadian Journal of Physiology and Pharmacology*, 1974, *52*, 451–457.

Lerner, L. J. Hormone antagonists: Inhibitors of specific activities of estrogen and androgen. *Recent Progress in Hormone Research*, 1964, *20*, 435–490.

Liao, S., and Fang, S. Receptor proteins for androgens and the mode of action of androgens on gene transcription in ventral prostate. *Vitamins and Hormones*, 1969, *27*, 17–90.

Libertun, C., Cooper, K. J., Fawcett, C. P., and McCann, S. M. Effects of ovariectomy and steroid treatment on hypophyseal sensitivity to purified LH-releasing factor (LRF). *Endocrinology*, 1974, *94*, 518–525.

Lieberburg, I., and McEwen, B. S. Estradiol-17β: A metabolite of testosterone recovered in cell nuclei from limbic areas of adult male rat brains. *Brain Research*, 1975, *91*, 171–174.

Lindström, L. H. The effect of pilocarpine in combination with monoamine oxidase inhibitors, imipramine and desmethyl imipramine on oestrous behavior in female rats. *Psychopharmacologia* (Berlin), 1970, *17*, 160–168.

Lindström, L. H. The effect of pilocarpine and oxotremorine on oestrous behavior in female rats after treatment with monoamine depletors or monoamine synthesis inhibitors. *European Journal of Pharmacology*, 1971, *15*, 60–65.

Lindström, L. H. The effect of pilocarpine and oxotremorine on hormone-activated copulatory behavior in the ovariectomized hamster. *Naunyn-Schmiederberg's Archives of Pharmacology*, 1972, *275*, 233–241.

Lindström, L. H. Further studies on cholinergic mechanisms and hormone-activated copulatory behavior in the female rat. *Journal of Endocrinology*, 1973, *56*, 275–283.

Lindström, L. H., and Meyerson, B. J. The effect of pilocarpine, oxotremorine and arecoline in combination with methyl-atropine or atropine on hormone activated oestrous behavior in ovariectomized rats. *Psychopharmacologia* (Berlin), 1967, *11*, 405–413.

Litteria, M. *In vivo* alterations in the incorporation of [³H]lysine into the medial preoptic nucleus and specific nuclei during the estrous cycle of the rat. *Research*, 1973, *55*, 234–237.

Litteria, M., and Schapiro, S. Brain nucleic acid content during the estrous cycle in the rat. *Experientia*, 1972, *28*, 898.

Litteria, M., and Timiras, P. S. *In vivo* inhibition of protein synthesis in specific hypothalamic nuclei by 17β-estradiol. *Proceedings of the Society for Experimental Biology and Medicine*, 1970, *134*, 256–261.

Loras, B., Genott, A., Monbon, M., Beucher, F., Reboud, J. P., and Bertrand, J. Binding and metabolism of testosterone in the rat brain during sexual maturation: II. Testosterone metabolism. *Journal of Steroid Biochemistry*, 1974, *5*, 425–431.

Luine, V. N., Khylchevskaya, R. I., and McEwen, B. S. Oestrogen effects on brain and pituitary enzyme activities. *Journal of Neurochemistry*, 1974, *23*, 925–934.

Luine, V. N., Khylchevskaya, R. I., and McEwen, B. S. Effect of gonadal steroids on activities of monoamine oxidase and choline acetylase in the rat brain. *Brain Research*, 1975a, *86*, 293–306.

Luine, V. N., Khylchevskaya, R. I., and McEwen, B. S. Effect of gonadal hormones on enzyme activities in brain and pituitary of male and female rats. *Brain Research*, 1975b, *86*, 283–292.

Luttge, W. G., Chronister, R. B., and Hall, N. R. Accumulation of ^{3}H-progestins in limbic, diencephalic, and mesencephalic regions of mouse brain. *Life Sciences*, 1973, *12*, 419–424.

Luttge, W. G., Wallis, C. J., and Hall, N. R. Effects of pre- and post-treatment with unlabeled steroids on the *in vivo* uptake of [^{3}H] progestins in selected brain regions, uterus and plasma of the female mouse. *Brain Research*, 1974, *21*, 105–115.

MacKinnon, P. C. B. Some observations of protein synthesis in the medial preoptic area of mice before and after puberty and of female rats at different phases of the oestrous cycle. *Journal of Endocrinology*, 1970, *48*, xliv.

Mainwaring, W. I. P., and Irving, R. The use of deoxyribonucleic acid-cellulose chromatography and isoelectric focusing for the characterization and partial purification of steroid-receptor complexes. *Biochemistry Journal*, 1973, *134*, 113–127.

Makman, M. H., Nakagawa, S., and White, A. Studies of the mode of action of adrenal steroids on lymphocytes. *Recent Progress in Hormone Research*, 1967, *23*, 195–219.

Malmnäs, C. O. Opposite effects of serotonin on copulatory activation in castrated male rats. In: E. Costa, G. L. Gessa, and M. Sandler (Eds.), *Advances in Biochemical Psychopharmacology, Vol. II: Serotonin—New Vistas—Biochemistry and Behavioral and Clinical Studies.* New York: Raven, Press, 1974.

Mangili, G., Motta, M., and Martini, L. Control of adrenocorticotrophic hormone secretion. In: L. Martini and W. F. Ganong (Eds.), *Neuroendocrinology* Vol. 1. New York: Academic Press, 1966.

Massa, R., and Martini, L. Interference with the 5α-reductase system: A new approach for developing antiandrogens. *Gynecologic Investigation*, 1971–1972, *2*, 253–270.

Massa, R., Stupnicka, E., Kniewald, Z., and Martini, L. The transformation of testosterone into dihydrotestosterone by the brain and the anterior pituitary. *Journal of Steroid Biochemistry*, 1972, *3*, 385–399.

Maurer, R. A. [^{3}H]Estradiol binding macromolecules in the hypothalamus and anterior pituitary of normal female, androgenized female and male rats. *Brain Research*, 1974, *67*, 175–177.

Maurer, R. A., and Woolley, D. E. Demonstration of nuclear ^{3}H-estradiol binding in hypothalamus and amygdala of female, androgenized female, and male rats. *Neuroendocrinology*, 1974, *16*, 137–147.

McCann, S. M., and Moss, R. L. Putative neurotransmitters involved in discharging gonadotrophin-releasing neurohormones and the action of LH-releasing hormone on the CNS. *Life Sciences*, 1975, *16*, 833–852.

McDonald, P. G., and Meyerson, B. J. The effect of oestradiol, testosterone and dihydrotestosterone on sexual motivation in the ovariectomized female rat. *Physiology and Behavior*, 1973, *11*, 515–520.

McDonald, P. G., Beyer, C., Newton, F., Brien, B., Baker, R., Tan, H. S., Sampson, C., Kitching, P., Greenhill, R., and Pritchard, D. Failure of 5α-dihydrotestosterone to initiate sexual behavior in the castrated male rat. *Nature*, 1970, *227*, 964–965.

McEwen, B. S. Adrenal steroid binding to presumptive receptors in the limbic brain of the rat. In: P. Dell (Ed.), *Neuroendocrinologie de l'Axe Corticotrope*, INSERM Colloque, Vol. 22. Paris: INSERM, 1974.

McEwen, B. S., and Pfaff, D. W. Factors influencing sex hormone uptake by rat brain regions: I. Effects of neonatal treatment, hypophysectomy, and competing steroid on estradiol uptake. *Brain Research*, 1970, *21*, 1–16.

McEwen, B. S., and Pfaff, D. W. Chemical and physiological approaches to neuroendocrine mechanisms: Attempts at integration. In: W. F. Ganong and L. Martini, (Eds.), *Frontiers in Neuroendocrinology*. New York: Oxford University Press, 1973.

McEwen, B. S., and Wallach, G. Corticosterone binding to hippocampus: nuclear and cytosol binding *in vitro*. *Brain Research*, 1973, *57*, 373–386.

McEwen, B. S., Pfaff, D. W., and Zigmond, R. E. Factors influencing sex hormone uptake by rat brain regions: II. Effects of neonatal treatment and hypophysectomy on testosterone uptake. *Brain Research*, 1970a, *21*, 17–28.

McEwen, B. S., Pfaff, D. W., and Zigmond, R. E. Factors influencing sex hormone uptake by rat brain regions: III. Effects of competing steroids on testosterone uptake. *Brain Research*, 1970b, *21*, 29–38.

McEwen, B. S., Zigmond, R. E., and Gerlach, J. L. Sites of steroid binding and action in the brain. In: G. H. Bourne (Ed.), *Structure and Function of Nervous Tissue*, Vol. 5. New York: Academic Press, 1972.

McEwen, B. S., Denef, C. J., Gerlach, J. L., and Plapinger, L. Chemical studies of the brain as a steroid hormone target tissue. In: F. O. Schmitt and F. G. Worden (Eds.), *The Neurosciences: Third Study Program* Boston: M.I.T. Press, 1974.

McEwen, B. S., Gerlach, J. L., and Micco, D. J., Jr. Putative glucocorticoid receptors in hippocampus and other regions of the rat brain. In: R. Isaacson and K. Pribram (Eds.), *The Hippocampus: A Comprehensive Treatise*. New York: Plenum Press, 1975a.

McEwen, B. S., Pfaff, D. W., Chaptal, C., and Luine, V. N. Brain cell nuclear retention of ^{3}H estradiol doses able to promote lordosis: Temporal and regional aspects. *Brain Research*, 1975b, *86*, 155–161.

McEwen, B. S., deKloet, E. R., and Wallach, G. Interactions *in vivo* and *in vitro* of corticoids and progesterone with cell nuclei and soluble macromolecules from rat brain regions and pituitary. *Brain Research*, 1976, *105*, 129–136.

McGuire, J. L., and Lisk, R. D. Estrogen receptors in the intact rat. *Proceedings of the National Academy of Sciences* (USA), 1968, *61*, 497–503.

McGuire, W. L., DeLaGarza, M., and Chamness, G. C. Estrogen receptor in a prolactin-secreting pituitary tumor (MtTW5). *Endocrinology*, 1973, *93*, 810–813.

Meyer, C. C., Testosterone concentration in the male chick brain: An autoradiographic survey. 1973, *Science, 180*, 1381–1382.

Meyerson, B. J. Relationship between the anaesthetic and gestagenic action and estrous behavior-inducing activity of different progestins. *Endocrinology*, 1967, *81*, 369–374.

Meyerson, B. J. Mechanism of action of sex steroids on behavior: Inhibition of estrogen-activated behavior by MER-25, colchicine and cycloheximide. *Progress in Brain Research*, 1973, *39*, 135–147.

Meyerson, B. J., and Lindström, L. Effects of an oestrogen antagonist ethanmoxytriphentol (MER-25) on oestrous behaviour in rats. *Acta Endocrinologica*, 1968, *59*, 41–48.

Meyerson, B. J., Carrer, H., and Eliasson, M. 5-Hydroxytryptamine and sexual behavior in the female rat. In E. Costa, G. L. Gessa, and M. Sandler (Eds.), *Advances in Biochemical Psychopharmacology. Vol. II: Serotonin—New Vistas—Biochemistry and Behavioral and Clinical Studies*. New York: Raven Press, 1974.

Michael, R. P. Action of hormones in the cat brain. In R. A. Gorski and R. E. Whalen (Eds.), *Brain and Behavior*, Vol. 3. Los Angeles: University of California Press, 1964.

Michael, R. P. Oestrogens in the central nervous system. *British Medical Bulletin*, 1965a, *21*, 87–90.

Michael, R. P. The selective accumulation of estrogens in the neural and genital tissues of the cat. In L. Martini and A. Pecile (Eds.), *Hormonal Steroids*, Vol. 2. New York: Academic Press, 1965b.

Michael R. P. Determinants of primate reproductive behavior. *Acta Endocrinologica*, 1972, *(Suppl.), 66*, 322–361.

Moguilevsky, J. A., and Malinow, M. R. Endogenous oxygen uptake of the hypothalamus in female rats. *American Journal of Physiology*, 1964, *206*, 855–857.

Monbon, M., Loras, B., Reboud, J. P., and Bertrand, J. Uptake, binding and metabolism of testosterone in rat brain tissues. *Brain Research*, 1973, *53*, 139–150.

Monbon, M., Loras, B., Reboud, J. P., and Bertrand, J. Binding and metabolism of testosterone in the rat brain during sexual maturation: I. Macromolecular binding of androgens. *Journal of Steroid Biochemistry*, 1974, *5*, 417–423.

Morin, L. P., and Feder, H. H. Hypothalamic progesterone implants and facilitation of lordosis behavior in estrogen-primed ovariectomized guinea pigs. *Brain Research*, 1974a, *70*, 81–93.

Morin, L. P., and Feder, H. H. Inhibition of lordosis behavior in ovariectomized guinea pigs by mesencephalic implants of progesterone. *Brain Research*, 1974b, *70*, 71–80.

Morishita, H., Nagamachi, N., Kawamoto, M., Yoshida, J., Ozasa, T., and Adachi, H. Cyclic change in hypothalamic lactic dehydrogenase in the adult female rat. *Acta Endocrinologica*, 1972, *71*, 225–232.

Morrell, J. I., Kelley, D. B., and Pfaff, D. W. Sex steroid binding in the brains of vertebrates: studies with light microscopic autoradiography. In K. Knigge, D. S. Scott, and K. Kobayashi (Eds.), *The Ventricular System in Neuroendocrine Mechanisms, Proceedings of Second Brain-Endocrine Interaction Symposium*. Basel: S. Karger, 1975.

Moudgil, V. K., and Kanungo, M. L. Induction of acetylcholinesterase by 17β-estradiol in the brain of rats of various ages. *Biochemical and Biophysical Research Communications*, 1973, *52*, 725–730.

Mowles, T. F., Ashkanazy, B., Mix, E., Jr., and Sheppard, H. Hypothalamic and hypophyseal estradiol-binding complexes. *Endocrinology,* 1971, *89,* 484–491.

Musliner, T. A., and Chader, G. J. A role for DNA in the formation of nuclear-receptor complex in a cell-free system. *Biochemical and Biophysical Research Communications,* 1971, *45,* 998–1003.

Naess, O., and Attramadal, A. Uptake and binding of androgens by the anterior pituitary gland, hypothalamus, preoptic area and brain cortex of rats. *Acta Endocrinologica,* 1974, *76,* 417–430.

Naess, O., Attramadal, A., and Aakvaag, A. Androgen binding proteins in the anterior pituitary, hypothalamus, preoptic area and brain cortex of the rat. *Endocrinology,* 1975, *96,* 1–9.

Naftolin, F., Ryan, K. J., and Petro, Z. Aromatization of androstenedione by the diencephalon. *Journal of Clinical Endocrinology,* 1971, *368–370.*

Naftolin, F., Ryan, K. J., and Petro, Z., Aromatization of androstenedione by the anterior hypothalamus of adult male and female rat. *Endocrinology,* 1972, *90,* 295–298.

Naftolin, F., Morishita, H., Davies, I. J., Todd, R., Ryan, K. J., and Fishman, J. 2-Hydroxyestrone induced rise in serum luteinizing hormone in the immature male rat. *Biochemical and Biophysical Research Communications,* 1975, *64,* 905–910.

Nakano, R., Kayashima, F., Kotsuji, F., and Tojo, S. Effect of gonadal steroids on the pituitary gonadotropin response to luteinizing hormone releasing factor (LRF) in the rat. *Endokrinologie,* 1974, *63,* 147–154.

Noble, R. Estrogen plus androgen induced mounting in adult female hamsters. *Hormones and Behavior,* 1974, *5,* 227–234.

Noteboom, W. D., and Gorski, J. Stereospecific binding of estrogens in the rat uterus. *Archives of Biochemistry and Biophysics,* 1965, *111,* 559–568.

Notides, A. C. Binding affinity and specificity of the estrogen receptor of the rat uterus and anterior pituitary. *Endocrinology,* 1970, *87,* 987–992.

Notides, A. C., and Nielsen, S. The molecular mechanism of the *in vitro* 4S to 5S transformation of the uterine estrogen receptor. *Journal of Biology and Chemistry,* 1974, *249,* 1866–1873.

Nowak, F. V., and Karavolas, H. J. Conversion of 20α-hydroxypregn-4-en-3-one to 20α-hydroxy-5α-pregnan-3-one and 5α-pregnane-3α,20α-diol by rat medial basal hypothalamus. *Endocrinology,* 1974, *94,* 994–997.

O'Malley, B. W., and Means, A. R. Female steroid hormones and target cell nuclei. *Science,* 1974, *183,* 610–620.

O'Malley, B. W., Spelsburg, T. C., Schrader, W. T., Chytil, F., and Steggles, A. W. Mechanisms of interaction of a hormone-receptor complex with the genome of a eukaryotic target cell. 1972, *Nature,* *235,* 141–144.

Parrott, R. F. Effects of 17β-hydroxy-4-androsten-19-ol-3-one (19-hydroxytestosterone) and 5α-androstan-17β-ol-3-one (dihydrotestosterone) on aspects of sexual behavior in castrated male rats. *Journal of Endocrinology,* 1974, *61,* 105–115.

Parrott, R. F. Aromatizable and 5α-reduced androgens: Differentiation between central and peripheral effects of male rat sexual behavior. *Hormones and Behavior,* 1975, *6,* 99–108.

Paup, D. C., Mennin, S. P., and Gorski, R. A. Androgen- and estrogen-induced copulatory behavior and inhibition of luteinizing hormone (LH) secretion in the male rat. *Hormones and Behavior,* 1975, *6,* 35–46.

Payne, A. H., Lawrence, C. C., Foster, D. L., and Jaffe, R. B. Intranuclear binding of 17β-estradiol and estrone in female ovine pituitaries following incubation with estrone sulfate. *Journal of Biology and Chemistry,* 1973, *248,* 1598–1602.

Perez-Palacios, B., Perez, A. E., Cruz, M. L. and Beyer, C. Comparative uptake of [³H] androgens by the brain and the pituitary of castrated male rats. *Biology of Reproduction,* 1973, *8,* 395–399.

Pfaff, D. W. Autoradiographic localization of radioactivity in rat brain after injection of tritiated sex hormones. *Science,* 1968a, *161,* 1355–1356.

Pfaff, D. W. Autoradiographic localization of testosterone-³H in the female rat brain and estradiol-³H in the male rat brain. *Experientia,* 1968b, *24,* 958–959.

Pfaff, D. W. Uptake of estradiol-17β-³H in the female rat brain: An autoradiographic study. *Endocrinology,* 1968c, *82,* 1149–1155.

Pfaff, D. Nature of sex hormone effects on rat sex behavior. *Journal of Comparative Physiology and Psychology,* 1970, *73,* 349–358.

Pfaff, D. W., and Keiner, M. Atlas of estradiol-concentrating cells in the central nervous system of the female rat. *Journal of Comparative Neurology,* 1973, *151,* 121–158.

Pfaff, D., Lewis, C., Diakow, C., and Keiner, M. Neurophysiological analysis of mating behavior responses as hormone-sensitive reflexes. In E. Stellar and J. M. Sprague, (Eds.), *Progress in Physiological Psychology*, Vol. 5. New York: Academic Press, 1973.

Phoenix, C. H. Effects of dihydrotestosterone on sexual behavior of castrated male rhesus monkeys. *Physiology and Behavior*, 1974, *12*, 1045–1055.

Phuong, N. T., and Sauer, G. Die specifische Aufnahme von markiertem Testeron im Hypothalamus. *Acta Biologica et Medica Germanica*, 1971, *26*, 1247–1249.

Piva, F., Gagliano, P., Motta, M., and Martini, L. Adrenal progesterone: Factors controlling its secretion. *Endocrinology*, 1973, *93*, 1178–1184.

Plapinger, L., and McEwen, B. S. Ontogeny of estradiol-binding sites in rat brain: I. Appearance of presumptive adult receptors in cytosol and nuclei. *Endocrinology*, 1973, *93*, 1119–1128.

Powers, B., and Valenstein, E. L. Sexual receptivity: Facilitation by medial preoptic lesions in female rats. *Science*, 1972, *175*, 1003–1005.

Powers, J. B. Facilitation of lordosis in ovariectomized rats by intracerebral progesterone implants. *Brain Research*, 1972, *48*, 311–325.

Quadagno, D. M., and Ho, G. K. W. The reversible inhibition of steroid-induced sexual behavior by intracranial cycloheximide. *Hormones and Behavior*, 1975, *6*, 19–26.

Raisinghani, K. H., Dorfman, R. I., Forchielli, E., Gyermek, L., and Genther, G. Uptake of intravenously administered progesterone, pregnanedione and pregnanolone by the rat brain. *Acta Endocrinologica*, 1968, *57*, 395–404.

Ramirez, V. D., Komisaruk, B. R., Whitmoyer, D. I., and Sawyer, C. H. Effects of hormones and vaginal stimulation on the EEG and hypothalamic units in rats. *American Journal of Physiology*, 1967, *212*, 1376–1384.

Reddy, V., Naftolin, F., and Ryan, K. J. Aromatization in the central nervous system of rabbits: Effects of castration and hormone treatment. *Endocrinology*, 1973, *92*, 589–594.

Reddy, V. V. R., Naftolin, F., and Ryan, K. J. Conversion of androstenedione to estrone by neural tissues from fetal and neonatal rats. *Endocrinology*, 1974, *94*, 117–121.

Resko, J. A., Goy, R. W., and Phoenix, C. H. Uptake and distribution of exogenous testosterone-1,2-[3H] in neural and genital tissues of the castrate guinea pig. *Endocrinology*, 1967, *80*, 490–498.

Rhees, R. W., Abel, J. H., Jr., and Haack, D. W. Uptake of tritiated steroids in the brain of the duck *(Anas platyrhynchos)*, an autoradiographic study. *General and Comparative Endocrinology*, 1972, *18*, 292–300.

Rhees, R. W., Grosser, B. I., and Stevens, W. Effect of steroid competition and time on the uptake of [3H]corticosterone in the rat brain: An autoradiographic study. *Brain Research*, 1975, *83*, 293–300.

Rice, K. K., and Richter, C. P. Increased sodium chloride and water intake of normal rats treated with deoxycorticosterone acetate. *Endocrinology*, 1943, *33*, 106–115.

Richter, C. P. Sodium chloride and dextrose appetite of untreated and treated adrenalectomized rats. *Endocrinology*, 1941, *29*, 115–129.

Robel, P., Corprechot, C., and Baulieu, E. E. Testosterone and androstanolone in rat plasma and tissues. *Federation of European Biochemical Societies Letters*, 1973, *33*, 219–220.

Robinson, J. A., and Karavolas, H. J. Conversion of progesterone by rat anterior pituitary tissue to 5α-pregnane-3,20-dione and 3α-hydroxy-5α-pregnan-20-one. *Endocrinology*, 1973, *93*, 430–434.

Robinson, J. A., and Leavitt, W. W. Estrogen related changes in anterior pituitary RNA levels. *P.S.E.B.M.*, 1971, *139*, 471–475.

Rochefort, H., and Capony, F. Étude comparée du comportement d'un anti-oestrogène de l'oestradiol dans les cellules uterine. *Comptes Rendus Hebdomadaires des Séances de l'Académie des Sciences* (Paris), 1973, *276*, 2321–2324.

Roy, S. K., Jr., and Laumas, K. R. 1,2-[3H]-Testosterone: Distribution and uptake in neural and genital tissues of intact male, castrate male and female rats. *Acta Endocrinologica*, 1969, *61*, 629–640.

Ruh, T. S., and Ruh, M. F. The effect of antiestrogens on the nuclear binding of the estrogen receptor. *Steroids*, 1974, *24*, 209–224.

Ryan, K. J. Hormones of the placenta. *American Journal of Obstetrics and Gynecology*, 1962, *84*, 1695–1713.

Ryan, K. J., Naftolin, F., Reddy, V., Flores, F., and Petro, Z. Estrogen formation in the brain. *American Journal of Obstetrics and Gynecology*, 1972, *114*, 454–460.

Salaman, D. F. RNA synthesis in the rat anterior hypothalamus and pituitary: Relation to neonatal androgen and the oestrous cycle. *Journal of Endocrinology*, 1970, *48*, 125–137.

Samperez, S., Thieulant, M. L., and Jouan, P. Mise en évidence d'une association macromoléculaire de la testostérone 1-2-³H dans l'hypophyse antérieure et l'hypothalamus du rat normal et castré. *Comptes Rendus Hebdomadaires des Séances de l'Académie des Sciences* (Paris), 1969a, *628*, 2965–2968.

Samperez, S., Thieulant, M. L., Poupon, R., Duval, J., and Jouan, P. Étude de la pénétration de la testostérone-1,2-³H dans l'hypophyse antérieure, l'hypothalamus, et le cortex cérébral du rat castré et normal. *Bulletin de la Société de Chimie Biologique*, 1969b, *51*, 117–131.

Sanyal, M. K., and Todd, R. B. 5α-Dihydrotestosterone influence on ovulation of prepuberal rats. *Proceedings of the Society for Experimental Biology and Medicine*, 1972, *141*, 622–624.

Sar, M., and Stumpf, W. E. Cellular localization of androgen in the brain and pituitary after the injection of tritiated testosterone. *Experientia*, 1972, *28*, 1364–1366.

Sar, M., and Stumpf, W. E. Autoradiographic localization of radioactivity in the rat brain after the injection of 1,2-³H-testosterone. *Endocrinology*, 1973a, *92*, 251–256.

Sar, M., and Stumpf, W. E. Cellular and subcellular localization of radioactivity in the rat pituitary after injection of 1,2-³H-testosterone using dry-autoradiography. *Endocrinology*, 1973b, *92*, 631–635.

Sar, M., and Stumpf, W. E. Effects of progesterone or cyproterone acetate on androgen uptake in the brain, pituitary and peripheral tissues. *Proceedings of the Society of Experimental Biology and Medicine*, 1973c, *144*, 26–29.

Sar, M., and Stumpf, W. E. Neurons of the hypothalamus concentrate ³H progesterone or its metabolite. *Science*, 1973d, *182*, 1266–1268.

Sar, M., and Stumpf, W. E. Pituitary gonadotrophs: nuclear concentration of radioactivity after injection of [³H]testosterone. *Science*, 1973a, *179*, 389–391.

Schiaffini, O., Marin, B., and Gallego, A. Oxidative activity of limbic structure during the sexual cycle in the rat. *Experentia*, 1969, *25*, 1255–1256.

Schrader, W. T., Toft, D. O., and O'Malley, B. W. Progesterone-binding protein of chick oviduct: VI. Interactions of purified progesterone-receptor components with nuclear constituents. *Journal of Biological Chemistry*, 1972, *247*, 2401–2407.

Schutz, G., Beato, M., and Feigelson, P. Messenger RNA for hepatic tryptophan oxygenase: its partial purification, its translation in a heterologous cell-free system, and its control by glucocorticoid hormones. *Proceedings of the National Academy of Sciences* (USA), 1973, *70*, 1218–1221.

Schwarzel, W. C., Kruggel, W. G., and Brodie, H. J. Studies on the mechanism of estrogen biosynthesis: VII. The development of inhibitors of the enzyme system in human placenta. *Endocrinology*, 1973, *92*, 866–880.

Seiki, K., and Hattori, M. An autoradiographic study on the incorporation of ³H-leucine in the brain of ovariectomized rats after administration of sex steroids. *Endocrinologia Japonica*, 1972, *19*, 269–276.

Seiki, K., and Hattori, M. *In vivo* uptake of progesterone by the hypothalamus and pituitary of the female ovariectomized rat and its relationship to cytoplasmic progesterone-binding protein. *Endocrinologia Japonica*, 1973, *20*, 111–119.

Seiki, K., Higashida, M., Imanishi, Y., Miyamoto, M., Kitagawa, T., and Kotani, M. Radioactivity in the rat hypothalamus and pituitary after injection of labelled progesterone. *Journal of Endocrinology*, 1968, *41*, 109–110.

Seiki, K., Miyamoto, M., Yamashita, A., and Kotani, M. Further studies on the uptake of labelled progesterone by the hypothalamus and pituitary of rats. *Journal of Endocrinology*, 1969, *43*, 129–130.

Seiki, K., Hattori, M., Okada, T., and Machida, S. Effect of ovarian hormones on the uptake of ³H-leucine by the castrated rat brain: An autoradiographic study. *Endocrinologia Japonica*, 1972, *19*, 375–382.

Shaikh, A. A., and Shaikh, S. A. Adrenal and ovarian steroid secretion in the rat estrous cycle temporally related to gonadotropins and steroid levels found in peripheral plasma. *Endocrinology*, 1975, *96*, 37–44.

Shepherd, R. E., Huff, K., and McGuire, W. L. Non-interaction between *in vivo* and cell free nuclear binding of estrogen receptor. *Endocrine Reserach Communications*, 1974, *1*, 73–85.

Shirley, B., Wolinsky, J., and Schwartz, N. B. Effects of a single injection of an estrogen antagonist on the estrous cycle of the rat. *Endocrinology*, 1968, *82*, 959–968.

Sholiton, L. J., Taylor, B. B., and Lewis, H. P. The uptake and metabolism of labelled testosterone by the brain and pituitary of the male rhesus monkey *(Macaca mulatta)*. *Steroids*, 1974, *24*, 537–547.

Shyamala-Harris, G. Nature of oestrogen specific binding sites in the nuclei of mouse uteri. *Nature New Biology*, 1971, *213*, 246–248.

Shyamala, G., and Gorski, J. Estrogen receptors in the rat uterus. *Journal of Biological Chemistry*, 1969, *244*, 1097–1103.

Singer, J. Hypothalamic control of male and female sexual behavior in female rats. *Journal of Comparative Physiology and Psychology*, 1968, *66*, 738–742.

Södersten, P. Mounting behavior in the female rat during the estrous cycle, after ovariectomy, and after estrogen or testosterone administration. *Hormones and Behavior*, 1972, *3*, 307–320.

Södersten, P. Estrogen-activated sexual behavior in male rats. *Hormones and Behavior*, 1973, *4*, 247–256.

Södersten, P. Effects of an estrogen antagonist, MER-25, on mounting behavior and lordosis behavior in the female rat. *Hormones and Behavior*, 1974, *5*, 111–121.

Södersten, P., and Larsson, K. Lordosis behavior in castrated male rats treated with estradiol benzoate or testosterone propionate in combination with an estrogen antagonist, MER-25, and intact male rats. *Hormones and Behavior*, 1974, *5*, 13–18.

Spona, J. LH-RH interaction with the pituitary plasma membrane. *Federation of European Biochemical Societies Letters*, 1973a, *34*, 24–26.

Spona, J. LH-RH stimulated gonadotropin release mediated by two distinct pituitary receptors. 1973b, *35*, 59–62.

Spona, J. LH-RH interaction with the pituitary plasma membrane is affected by sex steroids. 1974, *39*, 221–225.

Stahl, F., Poppe, I., and Dorner, G. Unwandlugen von Testosterone zu Dihydrotesteron und Androsteron in Hypophyse, Hypothalamus und Kortex mannlicher und weiblicher Ratten. *Acta Biologica et Medica Germanica*, 1971, *26*, 855–858.

Stern, J., and Eisenfeld, A. Distribution and metabolism of ^{3}H-testosterone in castrated male rats: Effects of cyproterone, progesterone and unlabelled testosterone. *Endocrinology*, 1971, *88*, 1117–1125.

Stern, J., and Janowiak, R. Effects of actinomycin-D implanted in the anterior hypothalamic-preoptic region of the diencephalon on spontaneous activity in ovariectomized rats. *Journal of Endocrinology*, 1972, *55*, 465–466.

Stumpf, W. E. Cellular and subcellular ^{3}H-estradiol localization in the pituitary by autoradiographs. *Zeitschrift für Zellforschung*, 1968a, *92*, 23–33.

Stumpf, W. E. Estradiol-concentrating neurons: Topography in the hypothalamus by dry mount autoradiography. *Science*, 1968b, *162*, 1001–1003.

Stumpf, W. E. Subcellular distribution of ^{3}H-estradiol in rat uterus by quantitative autoradiography—A comparison between ^{3}H-estradiol and ^{3}H-norethynodrel. *Endocrinology*, 1968c, *83*, 777–782.

Stumpf, W. E. Estrogen-neurons and estrogen-neuron systems in the periventricular brain. *American Journal of Anatomy*, 1970, *129*, 207–217.

Stumpf, W. E. Estrogen, androgen, and glucocorticosteroid concentrating neurons in the amygdala, studied by autoradiography. In B. E. Eleftheriou (Ed.), *Advances in Behavioral Biology, Vol. 2: The Neurobiology of the Amygdala*. New York: Plenum Press, 1972.

Stumpf, W. E., and Roth, L. J. High resolution autoradiography with dry-mounted, freeze-dried, frozen sections: Comparative study of six methods using two diffusible compounds, ^{3}H-estradiol and ^{3}H-mesobilirubinogen. *Journal of Histochemistry and Cytochemistry*, 1966, *14*, 274–287.

Swaneck, G. E., Highland, E., and Edelman, I. S. Stereospecific nuclear and cytosol aldosterone-binding proteins of various tissues. *Nephron*, 1969, *6*, 297–316.

Swerdloff, R. S., Walsh, P. C., and Odell, W. D. Control of LH and FSH secretion in the male: Evidence that aromatization of androgens to estradiol is not required for inhibition of gonadotropin secretion. *Steroids*, 1972, *20*, 13–22.

Tabei, T., Haga, H., Heinrichs, W. L., and Herrmann, W. L. Metabolism of progesterone by rat brain, pituitary gland and other tissues. *Steroids*, 1974, *23*, 651–666.

Talwar, G. P., Segal, S. J., Evans, A., and Davidson, O. W. The binding of estradiol in the uterus: A mechanism for depression of RNA synthesis. *Proceedings of the National Academy of Sciences* (Washington), 1964, *52*, 1059.

TerHaar, M. B., and McKinnon, P. C. B. Changes in serum gonadotropin levels and in protein levels and *in vivo* incorporation of [^{35}S] methionine into protein of discrete brain areas and pituitary of the rat during the oestrous cycle. *Journal of Endocrinology*, 1973, *58*, 563–579.

Terkel, A. S., Shryne, J., and Gorski, R. A. Inhibition of estrogen facilitation of sexual behavior by the intracerebral infusion of actinomycin-D. *Hormones and Behavior*, 1973, *4*, 377–386.

Thien, N. C., Thieulant, M. L., Samperez, S., and Jouan, P. Augmentation de l'activité de la 5α-stéroide réductase et de la 3α-hydroxystéroide deshydrogénase dans le cytoplasme de l'hypophyse antérieure du rat castré. *Comptes Rendus Hebdomadaires des Séances de l'Académie des Sciences* (Paris), 1972, *275*, 1927–1930.

Thien, N. C., Duval, J., Samperez, S., and Jouan, P. Testosterone 5α-reductase of microsomes from rat anterior hypophysis: Properties, increase by castration, and hormonal control. *Biochimie*, 1974, *56*, 899–906.

Thieulant, M. L., Mercier, M. L., Samperez, S., and Jouan, P. Mise en évidence d'un récepteur spécifique de la testostérone dans le cytoplasme de l'hypophyse antérieure du rat male prépubère. *Comptes Rendus Hebdomadaires des Séances de l'Académie des Sciences* (Paris), 1974a, *278*, 2569–2572.

Thieulant, M. L., Pelle, G., Samperez, S., and Jouan, P. Augmentation de l'activité de la 5α-stéroide réductase des noyaux cellulaires purifiés d'hypophyses de rats males castrés. *Comptes Rendus Hebdomadaires des Séances de l'Académie des Sciences* (Paris), 1974b, *278*, 1281–1284.

Toft, D. O. The interaction of uterine estrogen receptors with DNA. *Journal of Steroid Biochemistry*, 1972, *3*, 515–522.

Toft, D., and Gorski, J. A receptor molecule for estrogens: Isolation from the rat uterus and preliminary characterization. *Proceedings of the National Academy of Sciences* (USA), 1966, *55*, 1574–1581.

Tuohimaa, P. The radioautographic localization of exogenous tritiated dihydrotestosterone, testosterone, and oestradiol in the target organs of female and male rats. In P. O. Hubinot, F. Leroy, and P. Galand (Eds.), *Basic Actions of Sex Steroids on Target Organs*. Basel: Karger, 1971.

Van Dieten, J. A. M. J., Steijsiger, J., Dullaart, J., and van Rees, G. P. The effect of estradiol benzoate on the pituitary responsiveness to LH-RH in male and female rats. *Neuroendocrinology*, 1974, *15*, 182–188.

Verhoeven, G., Lambergits, G., and deMoor, P. Nucleus-associated steroid 5α-reductase activity and androgen responsiveness: A study in various organs and brain regions of rats. *Journal of Steroid Biochemistry*, 1974, *5*, 93–100.

Vertes, M., and King, R. J. B. The mechanism of oestradiol binding in the rat hypothalamus: Effect of androgenization. *Journal of Endocrinology*, 1971, *51*, 271–282.

Vilchez-Martinez, J. A., Arimura, A., Debeljuk, L., and Schally, A. V. Biphasic effect of estradiol benzoate on the pituitary responsiveness to LH-RH. *Endocrinology*, 1974a, *94*, 1300–1303.

Vilchez-Martinez, J. A., Arimura, A., and Schally, A. V. Influence of estradiol benzoate on pituitary responsiveness to LH-RH at different stages of the estrous cycle in rats (38206). *Proceedings of the Society for Experimental Biology and Medicine*, 1974b, *146*, 859–862.

Wade, G. N., and Feder, H. H. Effects of several pregnane and pregnene steroids on estrous behaviour in ovariectomized estrogen-primed guinea pigs. *Physiology and Behavior*, 1972a, *9*, 773–775.

Wade, G. N., and Feder, H. H. [1,2-³H]Progesterone uptake by guinea-pig brain and uterus: Differential localization, time-course of uptake and metabolism and effects of age, sex, estrogen-priming and competing steroids. *Brain Research*, 1972b, *45*, 525–543.

Wade, G. N., and Feder, H. H. Uptake of [1,2-³H]20α-hydroxypregn-4-en-3-one, [1,2-³H] corticosterone, and [6,7-³H]estradiol-17β by guinea pig brain and uterus: Comparison with uptake of [1,2-³H] progesterone. *Brain Research*, 1972c, *45*, 545–554.

Wade, G. N., and Feder, H. H. Stimulation of [³H] leucine incorporation into protein by estradiol-17β or progesterone in brain tissues of ovariectomized guinea pigs. *Brain Research*, 1974, *73*, 545–549.

Wade, G. N., Harding, C. F., and Feder, H. H. Neural uptake of [1,2-³H] progesterone in ovariectomized rats, guinea pigs and hamsters: Correlation with species differences in behavioural responsiveness. *Brain Research*, 1973, *61*, 357–367.

Wallen, K., Goldfoot, D. A., Joslyn, W. D., and Paris, C. A. Modification of behavioral estrus in the guinea pig following intracranial cycloheximide. *Physiology and Behavior*, 1972, *8*, 221–223.

Ward, I. L., Crowley, W. R., Zemlan, F. P. and Margules, D. L. Monoaminergic mediation of female sexual behavior. *Journal of Comparative Physiology and Psychology*, 1975, *88*, 53–61.

Warembourg, M. Fixation de l'oestradiol ³H au niveau des noyaux amygdaliens, septaux et du système hypothalamo-hypophysaire chez la souris fémelle. *Comptes Rendus Hebdomadaires des Séances de l'Académie des Sciences* (Paris), 1970a, *270*, 152–154.

Warembourg, M. Fixation de l'oestradiol ³H dans le telencephale et le diencephale chez la souris femelle. *Comptes Rendus des Séances de la Société de Biologie et de Ses Filiales* (Paris), 1970b, *164*, 126–129.

Warembourg, M. Etude radioautographique des retroactions centrales des corticostéroides ^{3}H chez le rat et le cobaye. In Paul Dell (Ed.), *Neuroendocrinologie de l'axe corticotrope: Brain-Adrenal Interactions.* Paris: INSERM, 1974.

Warembourg, M. Radioautographic study of the rat brain after injection of [1,2-^{3}H]corticosterone. *Brain Research,* 1975, *89,* 61–70.

Weisz, J., and Gibbs, C. Conversion of testosterone and androstenedione to estrogen *in vitro* by the brain of female rats. *Endocrinology,* 1974, *94,* 616–620.

Whalen, R. E., and Gorzalka, B. B. The effects of progesterone and its metabolites on the induction of sexual receptivity in rats. *Hormones and Behavior,* 1972, *3,* 221–226.

Whalen, R. E., and Gorzalka, B. B. Effects of an estrogen antagonist on behavior and on estrogen retention in neural and peripheral target tissues. *Physiology and Behavior,* 1973, *10,* 35–40.

Whalen, R. E., and Gorzalka, B. B. Estrogen-progesterone interactions in uterus and brain of intact and adrenalectomized immature and adult rats. *Endocrinology,* 1974, *94,* 214–223.

Whalen, R. E., and Hardy, D. F. Induction of receptivity in female rats and cats with estrogen and testosterone. *Physiology and Behavior,* 1970, *5,* 529–533.

Whalen, R. E., and Luttge, W. G. Differential localization of progesterone uptake in brain, role of sex, estrogen pretreatment and adrenalectomy. *Brain Research,* 1971a, *33,* 147–155.

Whalen, R. E., and Luttge, W. G. Testosterone, androstenedione and dihydrotestosterone: Effects on mating behavior of male rats. *Hormones and Behavior,* 1971b, *2,* 117–125.

Whalen, R. E., Battie, C., and Luttge, W. G. Anti-estrogen inhibition of androgen induced sexual receptivity in rats. *Behavioural Biology,* 1972, *7,* 311–320.

Whalen, R. E., Gorzalka, B. B., DeBold, J. F., Quadagno, D. M., Kan-Wha Ho, G., and Hough, J. C., Jr. Studies on the effects of intracerebral actinomycin-D implants on estrogen-induced receptivity in rats. *Hormones and Behavior,* 1974, *5,* 337–343.

White, H. L., and Wu, J. C. Multiple binding sites of human brain monoamine oxidase as indicated by substrate competition. *Journal of Neurochemistry,* 1975, *25,* 21–26.

Williams, D. L. The estrogen receptor: A minireview. *Life Sciences,* 1975, *15,* 583–597.

Williams, D., and Gorski, J. Kinetic and equilibrium analysis of estradiol in uterus: A model of binding-site distribution in uterine cells. *Proceedings of the National Academy of Sciences* (USA), 1972, *69,* 3454–3468.

Wolf, G. Hypothalamic regulation of sodium intake: Relation to preoptic and tegmental function. *American Journal of Physiology,* 1967, *213,* 1433–1438.

Wurtman, R. J. (Ed.). *Brain Monoamines and Endocrine Function,* Neurosciences Research Program Bulletin 9, No. 2, 1971.

Yaginuma, T., Watanabe, T., Kigawa, T., Nakai, T., Kobayashi, T, and Kobayashi, T. Uptake of ^{3}H-leucine in the brain of the female rat and its change after castration. *Endocrinologia Japonica,* 1969, *16,* 591–598.

Yamamoto, K. R. Characterization of the 4S and 5S forms of the estradiol receptor protein and their interaction with deoxyribonucleic acid. *Journal of Biological Chemistry,* 1974, *249,* 7068–7075.

Yamamoto, K. R., and Alberts, B. M. *In vitro* conversion of estradiol-receptor protein to its nuclear form: Dependence on hormone and DNA. *Proceedings of the National Academy of Sciences* (USA), 1972, *69,* 2105–2109.

Yamamoto, K. R., and Alberts, B. On the specificity of the binding of the estradiol receptor protein to deoxyribonucleic acid. *Journal of Biological Chemistry,* 1974, *249,* 7076–7086.

Yamamoto, K. R., Stampfer, M. R., and Tomkins, G. M. Receptors from glucocorticoid-sensitive lymphoma cells and two classes of insensitive clones: Physical and DNA-binding properties. *Proceedings of the National Academy of Sciences* (USA), 1974, *71,* 3901–3905.

Yang, H. Y. T., and Neff, N. H. β-Phenylethylamine: A specific substrate for type B monoamine oxidase of brain. *Journal of Pharmacology and Experimental Therapeutics,* 1973, *187,* 365–371.

Young, D. A., Barnard, T., Mendelsohn, S., and Giddings, S. An early cordycepin-sensitive event in the action of glucocorticoid hormones on rat thymus cells *in vitro:* Evidence that synthesis of new mRNA initiates the earliest metabolic effects of steroid hormones. *Endocrine Research Communications,* 1974, *1,* 63–72.

Zemlan, F. P., Ward, I. L., Crowley, W. R., and Margules, D. L. Activation of lordotic responding in female rats by suppression of serotonergic activity. *Science,* 1973, *179,* 1010–1011.

Zigmond, R. E. Binding and metabolism of steroid hormones in the central nervous system. In L. L. Iversen, S. D. Iversen, and S. H. Snyder (Eds.), *Handbook of Psycho-pharmacology.* New York: Plenum Press, 1975.

Zigmond, R. E., and McEwen, B. S. Selective retention of oestradiol by cell nuclei in specific brain regions of the ovariectomized rat. *Journal of Neurochemistry,* 1970, *17,* 889–899.

Zigmond, R. E., Stern, J., and McEwen, B. S. Retention of radioactivity in cell nuclei in the hypothalamus of the ring dove after injections of ^{3}H-testosterone. *General of Endocrinology,* 1972, *18,* 450–453.

Zigmond, R. E., Nottebohm, F., and Pfaff, D. W. Androgen concentrating cells in the midbrain of a songbird. *Science,* 1973, *179,* 1005–1007.

Zolovick, A. J., Pearse, R., Boehlke, K. W., and Eleftheriou, B. E. Monoamine oxidase activity in various parts of the rat brain during the estrous cycle. *Science,* 1966, *154,* 649.

PART V

Social and Experiential
Influences on Reproduction

15

Neuroendocrine Consequences of Sexual Behavior

T. O. ALLEN AND N. T. ADLER

INTRODUCTION

HORMONE–BEHAVIOR INTERACTIONS

Reproduction is the product of behavioral and physiological integration. Classically, investigations of reproduction treated behavior as resulting from internal (physiological) and external (environmental) factors. More recently, behavior has been recognized as a causal agent in the control of reproductive physiology as well.

The dependence of copulatory behavior on the hormonal condition of an organism has long been recognized. From the classical work of Young (1961) and of Beach (McGill *et al.*, 1978), as well as of their coworkers, it is known that the ovarian hormones estrogen and progesterone stimulate female mating behavior in many mammals and that, analogously, androgen stimulates copulatory behavior in males. Although the theme has been elaborated extensively in recent years, the basic truth still obtains: reproductive behavior is a product of an animal's endogenous biological machinery.

Yet, as the authors of popular psychological material are fond of saying, there is more to the story. Hormones do not directly release reproductive behavior patterns: gonadal hormones, at least in the adult, act in a "permissive" way (Ingle, 1954) to allow the full expression of copulatory events when the external physical

---

T. O. ALLEN AND N. T. ADLER   Department of Psychology, University of Pennsylvania, Philadelphia, Pennsylvania 19104.

and social environments are appropriate (Komisaruk and Diakow, 1973). During copulation in rats, for example, the soliciting behavior of the hormonally primed female provokes mounting by the male (McClintock and Adler, 1978). The mounting, in turn, elicits a postural reflex (the lordosis response) from the female that allows the male to insert his penis into the vaginal orifice, and that thus permits the progression of mating (Pfaff *et al.,* 1973). These few examples, and the very large literature from which they were drawn, demonstrate that behavior is the joint product of endogenous physiological workings and exogenous (behavioral) stimulation.

Reproductive physiology is also jointly controlled by endogenous and exogenous (behavioral) events. Behavior may be necessary for ovulation itself (as in rabbits) or for the facilitation of sperm transport and the initiation of the hormonal state necessary for pregnancy (as in rats; Adler, 1974).

In this chapter, we discuss the consequences of sexual behavior patterns in reproductive physiology. These behavior–endocrine relationships have, through evolutionary time, become fixed parts of the species' reproductive repertoire. Consequently, the study of the reproductive consequences of behavioral stimulation is part of the larger enquiry into the mechanisms subserving biological adaptation.

The study of the behavioral control of endocrine function draws us into the realm of neuroendocrinology because behavior effects physiological and behavioral changes via the recipient animal's nervous system. The peripheral nervous system receives the stimuli of a copulatory encounter, and the central nervous system transduces these stimuli into hormonal results. For example, the pelvic nerves that innervate the cervix mediate the copulation-induced surges of prolactin (Smith *et al.,* 1976).

In the following discussion, we try to highlight both the functional and the mechanistic aspects of the behavioral control of neuroendocrine processes.

PATTERNS OF NEUROENDOCRINE CONTROL: OVULATION AND LUTEAL FUNCTION

The physiological processes of ovulation, fertilization, and pregnancy do not proceed automatically in all species. Behavior, and the stimulation derived from behavior, often plays a critical role in their induction. For pregnancy to be established after a successful copulation, two conditions must prevail. Sperm and egg must meet, and this union must occur in an environment suitable for implantation and embryonic development.

Of the three focal physiological consequences of reproductive behavior (induction of ovulation, facilitation of sperm transport, and initiation of the progestational state), more research has been addressed to behavioral influences on ovulation and luteal formation. Conaway (1971) and Dewsbury (1975) classified species into categories depending on whether ovulation and the luteal phase are spontaneous or are induced by mating. For example, humans have both a spontaneous ovulation and a spontaneous luteal phase, whereas the laboratory rat has a spontaneous ovulation and an induced luteal phase.

Although corpus luteum formation (and the consequent secretion of progestational hormones) is dependent on a previous ovulation, the two ovarian events, release of the ovum and luteal maintenance, can sometimes be separated (normally in some species and experimentally in others). For example, in rats, ovulation is dependent on a surge of luteinizing hormone (Everett, 1964), whereas luteal formation is dependent on surges of prolactin (Smith *et al.*, 1975). In addition, the appropriate release of these two peptide hormones is controlled by different areas of the brain (see Figures 2 and 7). Finally, when these two neuroendocrine reflexes are induced by behavior, they may be further distinguishable by their different stimulus requirements. Consequently, we will discuss the release of the ovum and the formation of the corpus luteum as separate categories—although normally, in the lifetime of an animal, the two are obviously linked.

BEHAVIORAL EFFECTS ON OVULATION

As first postulated by Dempsey (1937), ovulation (whether spontaneous or induced by mating) is dependent on a surge of luteinizing hormone (LH). The apparently qualitative differences between spontaneous and induced ovulators may, however, more truly reflect quantitative differences in neuroendocrine thresholds, rather than radically different patterns of physiological organization (for reviews, see Conaway, 1971; Everett, 1964; Jöchle, 1975). In the following paragraphs, we briefly review the influence of behavioral events on ovulation in both reflexively and spontaneously ovulating species.

In order to analyze the mechanics by which the neuroendocrine reflex is triggered, it is necessary to "dissect" the behavior pattern to see what components produce the physiologically relevant stimulation. In the following sections, we discuss the role of individual sensory systems in the induction of ovulation. We concentrate on the chemo- and the somatosensory systems, both of which can be subdivided into at least two subsystems. The somatosensory system, for example, may have separate neural pathways that are differentially stimulated by mounts and intromissions and by ejaculation.

REFLEXIVE OVULATION

The females of some species, like rabbits, cats, and voles, normally require coital stimulation before they ovulate. We summarize here studies on two reflexively ovulating species, voles *(Microtus)* and rabbits *(Oryctolagus)*. These animals were selected for discussion because extensive work has been done on either the peripheral stimulus requirements or the central mediation of reflexive ovulation.

PERIPHERAL STIMULUS REQUIREMENTS: VOLES. Members of the genus *Microtus* include a wide variety of species that differ in social structure and habitat: *M. ochrogaster*, the prairie vole (Gray *et al.*, 1974b; Richmond and Conaway, 1969); *M. pennsylvanicus*, the meadow vole (Cluclow and Mallory, 1970); *M. californicus*, the California vole (Greenwald, 1956); *M. montanus*, the montane vole (Cross, 1972;

Davis *et al.,* 1974); and *M. agrestis,* the field vole (Austin, 1957; Breed, 1967)—
these are all reflexive ovulators. Both LH and progesterone rise significantly 1 hr
postcoitum (Charlton *et al.,* 1975a; Gray *et al.,* 1976; Milligan and Mackinnon,
1976). Identifiable corpora lutea first appear 8–10 hr after mating (Breed and
Clarke, 1970; Richmond and Conaway, 1969). In general, the species that have
been closely observed show a copulatory pattern characterized by intravaginal
thrusting, no lock, and multiple ejaculations (*M. agrestis,* Milligan, 1975a; *M. californicus,* Kenney *et al.,* 1979; *M. montanus,* Dewsbury, 1973; *M. ochrogaster,* Gray
and Dewsbury, 1973; *M. pennsylvanicus,* Gray and Dewsbury, 1975).

Dewsbury and his colleagues have studied the relationship between the
female's reproductive pattern and the male's copulatory behavior. Males of *M. ochrogaster* usually attain only two ejaculations before satiety (Gray and Dewsbury,
1973); females of this species require only a single ejaculatory series for ovulation
and implantation (Gray *et al.,* 1974b). *M. montanus* males, in contrast, display five
ejaculations before satiety (Dewsbury, 1973); and the females require more than
two ejaculatory series for ovulation (Davis *et al.,* 1974). These adaptive correlations
illustrate a general evolutionary process by which copulatory behavior, and the
physiological processes that it stimulates, has become coadapted.

Chemosensory Control of Estrus. In our attempt to "dissect" the copulatory
behavior pattern into its components relevant to ovulation, we first analyzed the
influence of a male's odor on the female's neuroendocrine status.

When housed in isolation or in all-female groups, the females in some species
of *Microtus* do not spontaneously display either behavioral or vaginal estrus (Cluclow and Mallory, 1970). The presence of an intact male, even when no physical
contact is allowed, can induce estrus in these females (Gray *et al.,* 1974a; Hasler
and Conaway, 1973; Richmond and Conaway, 1969). This effect may be similar to
the well-known "Whitten effect" displayed by mice (Whitten, 1956, 1957, 1958,
1959). In *M. ochrogaster,* male-induced estrus is probably mediated by olfaction, as
females whose main and accessory olfactory bulbs had been completely removed
failed to display estrus after exposure to males (Horton and Shepherd, 1979; Richmond and Stehn, 1976). Because airborne olfactory cues from males were not sufficient to activate female reproduction in *M. ochrogaster,* the accessory olfactory
system may be the critical component (Carter *et al.,* 1980).

Chemosensory Control of Ovulation. Once estrus has been induced, ovulation
can occur. Ovulation has also been reported following the mere presence of a
male. Gray *et al.* (1974) induced ovulation in 17% of *M. montanus* females by exposing them to the presence of a male. When caged next to a male or when allowed
direct access to bedding soiled by males, 7%–13% of *M. agrestis* females ovulated
(Milligan, 1975c). In *M. agrestis,* ovulation occurred in 22%–83% of group-housed
females exposed to a male behind a wire barrier. In this study, the average number
of corpora lutea formed after this exposure was equivalent to that found after full
copulation (Milligan, 1974).

In another group of experiments, however, ovulation did not occur following
the mere presence of males, that is, when copulation was prevented (*M. montanus,*
Cross, 1972; *M. ochrogaster,* Gray *et al.,* 1974b; Richmond and Conaway, 1969).

Because there were differences in experimental designs, in the reproductive history of the animals, as well as in the particular species of *Microtus* examined, any comparisons between studies yielding positive and negative results must be made cautiously. It is possible that when negative results were obtained, the conditions may not have been optimal for inducing ovulation. Milligan (1974) stated that 4 hr of exposure to a male is needed to induce ovulation. Hence, the single hour of exposure to a male allowed by Cross (1972) may not have been sufficiently long. In the studies by Gray *et al.* (1974a,b) and by Richmond and Conaway (1969), the experimenters determined ovulation rates by examining the ovaries for the presence of corpora lutea 8 days after copulation. By this time, corpora lutea may have formed but would already have regressed if the female had not also become progestational (Milligan, 1975b).

Although males may sometimes induce ovulation, the precise nature of the chemosensory channel may be complex. Milligan (1974) found, for example, that the addition of a second wire barrier reduced the percentage of females that ovulated in response to the presence of a male from 83% to 14%. The second barrier (25 cm from the first) prevented direct contact with feces or with soiled bedding. Nonvolatile olfactory or gustatory cues may therefore be important.

As we have already mentioned, it is difficult to draw general conclusions concerning stimulus control when considering different species—even when these species are contained within the same genus. It seems clear that olfaction has a role in the play of reproductive events, but it may only provide a necessary baseline of stimulation.

Puberty in female rodents, measured by uterine weight, day of vaginal opening, and so on, also reflects the capacity of the hypothalamus–pituitary–ovary axis for ovulation. The influence of male odors on female ovulatory competence is seen in two other phenomena first described in mice, the male-induced acceleration of puberty (Drickamer and Murphy, 1978; Lombardi and Vandenbergh, 1977; Lombardi *et al.*, 1976; Vandenbergh, 1967, 1969; Vandenbergh *et al.*, 1972) and the retardation of sexual maturity in all-female groups (Christian *et al.*, 1965; Drickamer, 1974, 1977; Drickamer *et al.*, 1978; McIntosh and Drickamer, 1977; Vandenbergh *et al.*, 1972). These phenomena are mediated through the accessory olfactory system (Kameko *et al,.* 1980). Female *Microtus* housed with males display an acceleration of sexual maturity (Hasler and Nalbandov, 1974). Conversely, grouping *Microtus* females together from weaning to puberty reduces their reproductive potential: previously grouped females produced 44% fewer corpora lutea following copulation than did previously isolated females (Pasley and MacKinney, 1973).

The "Bruce effect," first described in mice, is a blockade of pregnancy on exposure to the odors of a strange male (Bruce, 1959, 1960; Bruce and Parrott, 1960). When studying this phenomenon, investigators usually focus on the progestational consequences of the male's odor, rather than on the correlated induction of a fresh ovulation. This effect, mediated by reductions in prolactin secretion (Dominic, 1966), presumably through an increased release of gonadotropin in the female (Chapman *et al.*, 1970; Hoppe and Whitten, 1972), is dependent on an intact accessory olfactory system (Bellringer *et al.,* 1980). *Microtus* females are susceptible to pregnancy blockade when exposed to strange males (*M. agrestic,* Cluclow and Clarke, 1968; *M. pennsylvanicus,* Cluclow and Langford, 1971). The Bruce effect in voles differs somewhat from that in mice however: *Microtus* females respond to the presence of a strange male even in the later stages of pregnancy (Kenney *et al.,* 1977; Stehn and Richmond, 1975).

Somatosensory Control: Mating. For optimal rates of ovulation, additional somatosensory input from mounts, intromissions, and ejaculation may be necessary. The high rates of ovulation (up to 83%) in group-housed female voles exposed to male odors may have been due to additional tactile contact between the females (Milligan, 1974), as singly housed females exposed to male odorants show ovulation rates of only 7%–13% (Milligan, 1975c).

As is the case with olfaction, the specific level of tactile stimulation required for ovulation varies from case to case. Fifty percent of parous *M. agrestis* ovulated following only 3 min of mounts, even without intromission by a male (Milligan, 1975b). However, in another study, increases in circulating LH could not be demonstrated following mounts alone in receptive *M. agrestis* females (Charlton *et al.,* 1975a). In another species of vole *(M. ochrogaster),* 8–33 mounts by a male were not sufficient to induce ovulation even though the females were parous (Gray *et al.,* 1974b). More intense stimulation (as from intromissions) was necessary.

The addition of cervical stimulation supplied by even a single intromission induced ovulation in 92% of female *M. agrestis,* and a complete ejaculatory series produced an ovulation rate of 100% (Milligan, 1975b). Based on work with *M. montanus,* Dewsbury and coworkers concluded that the greater the amount of stimulation received by *Microtus* females, the greater the number of females ovulating (Davis *et al.,* 1974).

Ejaculation increases the female's physiological response to mating. Both the percentage of females ovulating and the number of corpora lutea formed are greater in female *M. ochrogaster* receiving ejaculation than in females receiving only multiple intromissions (100% vs. 70%) (Gray *et al.,* 1974b; Kenney *et al.,* 1978).

Somatosensory Control: Artificial Stimulation. Although the vaginocervical stimulation supplied by the male's intromissions and ejaculations elicited ovulation in the majority of females, vaginocervical stimulation applied by the experimenter did not always mimic the effects of mating. Mechanical stimulation of the vaginocervical area in and of itself was not sufficient to induce either ovulation (Breed, 1967; Charlton *et al.,* 1975a; Greenwald, 1956) or LH release (Charlton *et al.,* 1975a; Milligan and Mackinnon, 1976). Additional stimulation from a male was needed (Kenney *et al.,* 1977; Milligan, 1975b).

Mechanical stimulation provided by the experimenter may fail to provide an essential feature of the male's natural copulatory stimulation. That is, the experimental stimulation does not mimic what Diamond (1970) described as the "species-specific vaginal code." This term highlights a major biological principle: what may seem at first sight to be minor variations in the physical nature of the stimulus are, in fact, important biological parameters leading to profound effects on the organism. Before one can induce ovulation in the laboratory, it may be necessary to decode the species-specific requirements.

Kenny *et al.* (1978) found that, when the total amount of copulatory stimulation was held constant (but the timing was varied by the use of interspecific matings), more females ovulated following mating with a conspecific male (80%, vs. 40% or 50%). Of course, this is only one interpretation of the data. It is also possible that incompatible olfactory cues from heterospecific males or the assessment of ovulation by 8-day-old corpora lutea contributed to this finding.

PERIPHERAL STIMULUS REQUIREMENTS: RABBITS. Rabbits are reflexive ovulators, with ovulation occurring about 10 hr after coitus (Fee and Parkes, 1929; Heape, 1905). LH is released following mating (Brown-Grant *et al.*, 1968; Dufy-Barbe *et al.*, 1973; Hilliard *et al.*, 1964; Scaramuzzi *et al.*, 1972).

Sexual behavior in the rabbit is multimodal: It starts with courtship, which includes chasing, urination, and tail flagging (during which the male elevates his hindquarters in order to display the white underside of his tail) Denenberg *et al.*, 1969). During copulation, the male mounts and intromits quickly after 8–12 exploratory pelvic movements (Hagen, 1974). Intromissions are characterized by a cessation of pelvic thrusting and last less than 1 sec (Contreras and Beyer, 1979). After intromission, the male falls off the female and utters a cry, hops about, stamps, and then remounts. Receptivity in the female, which is necessary for intromission, consists of raising her tail and hindquarters (Hammond, 1925). According to the schema of Dewsbury (1972), the copulatory behavior pattern is one of no lock, no intravaginal thrusting, and multiple ejaculation.

Brooks (1937) found that the removal of the senses of sight, smell, or hearing—alone or in combination—did not affect the ability of the rabbit to ovulate in response to coitus. He also found that the removal of both sympathetic chains was compatible with ovulation (Brooks, 1935) but that sectioning of the spinal cord at the upper lumbar or the thoracic level *did* block ovulation (Brooks, 1935, 1937; Marshall and Verney, 1936). Furthermore, electrical stimulation of the cord at lumbar levels elicited ovulation (Marshall and Verney, 1936). So stimulation of peripheral nerves that enter the spinal cord below the thoracic level seems to be *necessary* for ovulation.

Despite the importance of caudal somatosensory input, vaginal stimulation itself seems not to be necessary, as does ovulated when mounted by other females or when the vulva was rubbed by the experimenter (Hammond, 1925; Sanford, 1979). Elimination of sensory input from this area by anesthesia, by denervation, or by extirpation of the genital tract did not block ovulation in response to mating (Brooks, 1937; Fee and Parkes, 1930; Friedman, 1929).

Also, vaginal stimulation was sometimes not *sufficient* for ovulation in estrous females (Carlyle and Williams, 1961; Hammond and Asdell, 1926; Heape, 1905; Sawyer and Markee, 1959). Even when females received some exogenous estrogen, mechanical stimulation of the vagina elicited ovulation in less than half of them (Sawyer, 1949; Sawyer and Markee, 1959). Electrical stimulation of the vulva was, however, sufficient for ovulation—if the females were in "full estrus" (Carlyle and Williams, 1961).

What is the "normal" stimulus for copulation-induced ovulation in rabbits? It is not a direct humoral signal from the semen to the ovary: artificial insemination of female rabbits failed to elicit ovulation (Hammond, 1925; Heape, 1905). A neural signal is involved because general anesthesia prevented ovulation (Carlyle and Williams, 1961; Sawyer and Markee, 1959), whereas electrical stimulation of the head or the spinal cord did elicit ovulation (Harris, 1936; Marshall and Verney, 1936).

This sensory input reaches the CNS quickly; an administration of Nembutal within 12 sec of copulation failed to block ovulation in four of six rabbits. This signal must remain undisrupted for just under 1 min in order to be effective; ovulation is prevented by the infusion of adrenergic or cholingeric blocking agents within 1 min following copulation (Sawyer et al., 1947, 1949a, 1950). The neural signal is translated into a humoral event because transplanted ovaries with completely disrupted nervous connections still ovulated in response to coitus (Friedman, 1929).

The importance of the adenohypophysis was established early. Fee and Parkes (1929) demonstrated that removal of the pituitary within 1 hr of coitus blocked ovulation. Further research established that ovulation occurred in response to direct electrical stimulation of the pituitary (Harris, 1937; Markee et al., 1946). Yet, the signal for ovulation at the hypophyseal level did not seem to be a neural one, for sectioning of the sympathetic (Brooks, 1935; Hinsey and Markee, 1933) or the parasympathetic (Vogt, 1942) innervation of the pituitary failed to block coitus-induced ovulation. In addition, electrical stimulation of the pituitary afferents did not elicit ovulation (Haterius, 1934; Markee et al., 1946). Because the integrity of the pituitary stalk was known to be essential for ovulation (Brooks, 1938), a humoral signal (now known to be luteinizing-hormone-releasing factor—LHRH) must be responsible for activating the anterior pituitary (Markee et al., 1946; Sawyer et al., 1947). Hence, the "translation" from a neural to an endocrine mode in the coital triggering of ovulation was established in this species.

CNS MEDIATION. In recent times, the role of specific CNS sites controlling ovulation has been investigated. Hypothalamic control of the adenohypophysis was first demonstrated by Harris and coworkers, who discovered that electrical stimulation of the basal hypothalamus produced ovulation (Harris, 1937, 1948). In fact, ovulation could be induced with lower levels of stimulation in this area than when the stimulus was applied to the adenohypophysis itself (Harris, 1948; Markee et al., 1946). Conversely, lesions of the tuberal area block coitus-induced ovulation (Sawyer, 1956, 1959; Sawyer and Robison, 1956) or LH release (Charlton et al., 1975a) (see Figure 1).

Only some of the nuclear areas in the medial basal hypothalamus are involved in ovulation. Electrical stimulation of the posterior median eminence, the basal tuberal region, the ventromedial hypothalamus, the arcuate area, and the dorsomedial or posterior hypothalamus elicited both immediate release of LH and subsequent ovulation. Stimulation of the lateral or dorsal hypothalamus or of the mammillary region failed to provoke either LH release or ovulation (Hayward et al., 1964; Kawakami et al., 1967; Sawyer et al., 1963).

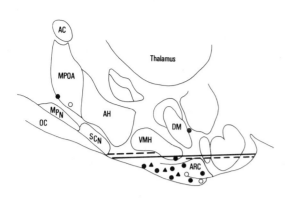

Fig. 1. Sagittal brain section depicting areas in the hypothalamus involved in *reflexive* ovulation in the *rabbit*: (——) knife cuts that blocked reflexive ovulation; (– – –) knife cuts that failed to block reflexive ovulation; (▲) lesions that blocked reflexive ovulation; (●) sites at which electrical stimulation invoked ovulation; (○) sites at which electrical stimulation failed to provoke ovulation. Abbr.: AH, anterior hypothalamus; ARC, arcuate nucleus; DM, dorsomedial hypothalamic nucleus; OC, optic chiasm; MPN, medial preoptic nucleus; MPOA, median preoptic area; SCN, suprachiasm nucleas; VMH, ventromedial hypothalamus.

Copulation-derived stimulation may converge on the medial basal hypothalamus. Voloschin and Gallardo (1976) found that cutting around the mediobasal hypothalamus to form a complete "hypothalamic island" blocked ovulation in rabbits, but that hemicircular cuts in either the rostral or the caudal or lateral arcs did not prevent ovulation. Electrophysiological recordings from the arcuate nucleus substantiate the involvement of this area in the induction of reflexive ovulation. In response to ovulation-inducing vaginal stimulation, the multiunit activity of the arcuate nucleus, the premammillary, and the posteriolateral hypothalamus increases whereas that of the medial preoptic area (MPOA) and the ventromedial hypothalamus (VMH) decreases (Vincent *et al.*, 1970).

Brain areas outside the basal hypothalamus may also influence ovulation. For example, stimulation of the basal, anterior portions of the MPOA leads to ovulation (Haterius and Derbyshire, 1937). Within 1 hr following coitus, neurons in the medial preoptic nucleus of the rabbit display ultrastructural changes, suggestive of synthetic activity, that may be responsible for the replenishment of LHRH (Clattenburg *et al.*, 1971). However, a later report found that stimulation of this area did not produce ovulatory amounts of LH (Kanematsu *et al.*, 1974).

Stimulation of the amygdala is sufficient to elicit an immediate rise in LH and subsequent ovulation (Hayward *et al.*, 1964; Kanematsu *et al.*, 1974; Kawakami *et al.*, 1966, 1967). The amygdala may, however, not be a necessary component of the ovulatory circuit because lesioning of this area could not block copulation-induced ovulation (Sawyer, 1956, 1959). Finally, stimulation of the hippocampus elicited ovulation (Kawakami *et al.*, 1966, 1967).

Only a few studies have directly investigated the brain areas responsible for ovulation in *Microtus*. Breed and Charlton (1968, 1971) found that electrical stimulation either of the anterior hypothalamus or of the median eminence elicited ovulation, whereas stimulation of the posterior hypothalamus did not. Electrical stimulation of the anterior hypothalamus (including the POA) or of the VMH–arcuate area caused a significant release of LH (Charlton *et al.*, 1975b).

T. O. ALLEN AND
N. T. ADLER

In this chapter, as well as in the majority of the research literature on which it is based, spontaneous and induced ovulation are treated as separate categories of reproductive organization. It is certainly true that spontaneous ovulation is controlled by an endogenous neural triggering mechanism—and need not be reflexively evoked by a behavioral trigger (Sawyer et al., 1949b). It is also true, however, that there are great similarities between these two forms of ovulation. As was pointed out in the previous section, qualitative differences in the patterns of female reproduction may result from quantitative differences in neuroendocrine thresholds. Although spontaneous ovulation is not triggered by the external environment, exogenous stimuli (e.g., light cycles) do regulate the timing of the ovarian cycle (Elliot and Goldman, 1981).

Female hamsters, for example, display estrous cycles in constant light (Alleva et al., 1971). Under these conditions, though, the ovarian cycle (like the circadian activity cycle on which it is based) is approximately, but not precisely, timed to 24-hr units. Because such females are not synchronized with the sidereal day, they go out of phase with the earth's daily geophysical cycle—an interesting experimental preparation but a disastrous condition if it occurred in nature. The process by which the animal's biological rhythms are normally "entrained" to light cycles involves a phase shifting of the biological oscillator by the external (photic) cycle. (The details of this fascinating process are not reviewed here but are discussed in Elliott and Goldman, 1981).

Other effects of the light cycle are displayed by female rats exposed to constant light. In this preparation, ovulation is blocked (Dempsey and Searles, 1943). In both the rat and the hamster, it seems that although ovulation is normally not "triggered" by exogenous stimulation, spontaneous ovulation (and the ovarian cycle of which it is a part) does depend on a permissive, cyclic photic environment.

The parallel between induced and spontaneous ovulators becomes even stronger when one recognizes that coitus can affect the rate of ovulation in some species of spontaneous ovulators. Even though the rat normally ovulates spontaneously, coital activity may increase the number of eggs released by a cycling female (Rodgers, 1971; Rodgers and Schwartz, 1972); the amount of LHRH released (Takahashi et al., 1975); and the amount of LH released (Blake and Sawyer, 1972; Herlyn et al., 1965; Moss and Cooper, 1973; Moss et al., 1973, 1977; Spies and Niswender, 1971; Taleisnik et al., 1966; Wuttke and Meites, 1972).

By increasing the number of eggs released by a spontaneously ovulating organism, copulatory activity may facilitate reproduction at times in the life cycle when LH is depressed (Rodgers and Schwartz, 1972, 1973). The phasic release of LH was attenuated even in middle-aged female rats (8–14 months old) that still showed regular estrus cycles (Cooper et al., 1980; Gray et al., 1980). In old females, the number of ova shed was directly related to the amount of copulatory stimulation received (Davis et al., 1977).

Perhaps the most striking similarity between the two reproductive patterns is that, under some circumstances, a spontaneous ovulator can actually become an induced ovulator. The LH surge necessary for spontaneous ovulation is blocked following treatment with exogenous estrogen (Aron *et al.*, 1966, 1968); with anesthesia (Everett, 1952, 1967; Harrington *et al.*, 1966; Kalra and Sawyer, 1970; Zarrow and Clark, 1968); or with continuous illumination (Brown-Grant *et al.*, 1973; Davidson *et al.*, 1973; Dempsey and Searles, 1943). Under these experimental conditions coitus induces ovulation—in a demonstration that females of spontaneously ovulating species like the rat possess the neuroanatomical and neuroendocrine machinery for induced ovulation.

The evolutionary implications of these data are that phylogenetic shifts between spontaneous and induced ovulation can be effected by quantitative shifts in neuroendocrine thresholds. Conaway (1971) and Zarrow and Clark (1968) proposed that induced ovulation is the more basic pattern and that spontaneous ovulation is a specialization. They noted that induced ovulation is the more widespread phenomenon and that it is common in the "primitive" eutherian order Insectivora. Because the pattern of induced ovulation is often accompanied by a synchronization of the estrous period among females, Conaway suggested that spontaneous ovulation may have evolved as an adaptation that spreads the period of estrus over a longer time. In populations with a limited number of breeding males, this alteration in the timing of estrus would increase the likelihood of the fertilization of all the females.

Shifts between global patterns of reproductive physiology (like induced and spontaneous ovulation) can occur over relatively brief spans of evolutionary time because of the hierarchical organization of physiological systems (Gallistel, 1980; Simon, 1973). In the case of mammalian reproduction, the ovary is at the lowest level of the hierarchy and is controlled by (and communicates with) the pituitary. The pituitary (the "master gland" of classical endocrinology) is, in turn, controlled by the CNS. Exogenous influences—as from the light cycle or from behaviorally derived stimulation—have access to the ovary via the central neuroendocrine control axis. The shift from induced to spontaneous ovulation follows the emancipation of the LH surge from environmental triggering and its dependence on endogenous humoral events.

Peripheral Stimulus Requirements. The female rat, perhaps in response to ultrasonic vocalizations of the male (McIntosh *et al.*, 1978; Barfield *et al.*, 1979), initiates the copulatory encounter by soliciting the male's pursuit (McClintock and Adler, 1978).

In the rat, mounts may sometimes induce ovulation, even without accompanying intromissions. Moss (1974) showed that mounts alone produced a significant elevation of LH release in normally cycling proestrus females. Following the blockade of ovulation with exogenous hormones, 47% of the female rats ovulated in response to mounts by a male (Aron *et al.*, 1968). With the inhibition of ovulation by constant light, 54%–67% ovulated (Brown-Grant *et al.*, 1973; Johns *et al.*, 1980; Davidson *et al.*, 1973). Nembutal blockade of ovulation was overcome in most cases by exposure to pudendectomized males that could not achieve intromission (Van Riemsdijk and Zeilmaker, 1979).

How is the behavioral stimulation of a mount transduced by the female's nervous system to culminate in ovulation? When mounted by a male, the female receives both olfactory and tactile stimulation. In analyzing the stimulus components, it is clearly necessary to dissect the role of these two modalities, as these two cues act synergistically to induce ovulation, as modeled by precocious puberty (Bronson and Maruniak, 1975).

Chemosensory Control. The olfactory input may involve both the vomeronasal and the main olfactory systems. Removal of the olfactory bulbs (and the accessory olfactory bulbs) significantly reduced the number of ova shed by mated females following Nembutal blockade of ovulation (Curry, 1974). Up to 50% of the female rats in light-induced, persistent estrus ovulated when allowed contact with bedding soiled by males; the response was abolished when the vomeronasal organ was occluded (Johns *et al.*, 1978). Of chlorpromazine-blocked females, 14% ovulated after one night of cohabitation with a noncopulating male, whereas only 4% of CPZ-blocked females that were isolated overnight did so (Harrington *et al.*, 1966).

Male odors can elicit surges of LH in mice (Bronson and Desjardin, 1969; Chapman *et al.*, 1970). The main and accessory olfactory bulbs project to numerous libic-system structures, including the medial and cortical amygdaloid nuclei (Winans and Scalia, 1970).

Somatosensory Control. In the female rat, mounts without intromission stimulate the female rat's perineum, flanks, and rump (Adler *et al.*, 1977; Pfaff *et al.*, 1977). In addition, the male's penis is thrust several times around the vaginal orifice before finally achieving insertion (Adler and Bermant, 1966; Bermant, 1965). The somatosensory information derived from these mounts-without-intromission is relayed to the spinal cord via peripheral nerves, including the pudendal, the genitofemoral, and the posterior cutaneous (of the thigh) nerves (Komisaruk *et al.*, 1972; Kow and Pfaff, 1976; Reiner *et al.*, 1981). The pelvic nerve that innervates the vaginocervical area—but not the perineum (Komisaruk *et al.*, 1972)—can be dispensed with: females whose pelvic nerves have been cut continue to show a significant elevation in LH release following coitus (Spies and Niswender, 1971). Deep cervical stimulation, without the perineal stimulation that would normally accompany the male's mounting, does not induce LH release in rats (Beach *et al.*, 1975; Blake and Sawyer, 1972; Moss, 1974; Taleisnik *et al.*, 1966; Tyler and Gorski, 1980). However, the additional cervical stimulation from intromissions can potentiate the ovulatory response that would follow mounts. When intromissions were allowed, the secretion of LH (Moss, 1974) and the percentage of females ovulating (Brown-Grant *et al.*, 1973) increased to higher levels than were seen with mounts alone. Moreover, in the chlorpromazine-blocked rat, intact pelvic nerves were necessary in order to release amounts of LH sufficient for ovulation (Harrington *et al.*, 1967; Zarrow and Clark, 1968).

The supplemental role of intromission-derived stimulation in inducing ovulation may be analogous to the effects of intromissions on the lordosis behavior of the female: intromissions potentiate lordosis responding (Diakow, 1975; Kuehn and Beach, 1963), although cervical stimulation alone is not sufficient to induce lordosis (Komisaruk and Diakow, 1973).

mounting male enters the spinal cord and probably travels in the anterolateral columns: transection of these columns severely reduces the lordosis response (Kow *et al.*, 1977). Midbrain areas are involved as relay stations. The unit activity of cells in the midbrain reticular formation and in the nucleus reticularis gigantocellularis changes in response to genital stimulation (Hornby and Rose, 1976; Kawakami and Kubo, 1971).

Translation from a neural to an endocrine signal takes place in the hypothalamus. Hypothalamic control of ovulation is not a unitary process. It has long been known that the neural system controlling LH release has both tonic and phasic components. Tonic release of low levels of LH from the pituitary is necessary to maintain all major aspects of ovarian function except ovulation itself, which requires a phasic or episodic release of LH.

Barraclough and Gorski (1961) localized the tonic controls in the arcuate-ventromedial region and the phasic controls in the suprachiasmatic-preoptic area. Halasz *et al.* (1962) identified a "hypophysiotrophic area," extending from the posterior border of the anterior hypothalamus through ventral portions of the ventromedial nucleus to the arcuate region. This area sustains the normal function of grafted tissue from the anterior pituitary.

Two basic approaches have been devised for the study of central nervous system controls on ovulation. First, one can remove the influence of a structure by a lesion or a deafferentation. Alternatively, one can either directly stimulate the activity of an area or record its activity during an ovulation-provoking stimulation.

Electrical stimulation of the medial basal hypothalamus (MBH) induces ovulation (Critchlow, 1958; Everett *et al.*, 1976; Kalra and Sawyer, 1970; Kawakami *et al.*, 1970).

Direct stimulation of the MPOA induces the release of LHRH (Barr and Barraclough, 1978; Chiappa *et al.*, 1977; Eskay *et al.*, 1977; Fink and Jamieson, 1976; Kalra *et al.*, 1973). In animals whose spontaneous ovulation has been blocked by anesthesia, POA stimulation induces LH release and/or ovulation (Barr and Barraclough, 1978; Cramer and Barraclough, 1971; Everett, 1964, 1965; Everett *et al.*, 1973, 1976; Everett and Radford, 1961; Kalra *et al.*, 1971, 1973; Kawakami *et al.*, 1970, 1971; Quinn, 1969; Terasawa and Sawyer, 1969a,b; Turgeon and Barraclough, 1973).

Moreover, when the POA and the MBH were surgically separated medially and laterally, electrical stimulation of the POA had no effect (Everett and Quinn, 1966; Phelps *et al.*, 1976; Phelps and Sawyer, 1977; Tejasen and Everett, 1967).

Manual stimulation of the vagina and the cervix of rats produces changes in the electrophysiological responses of the MPOA (Barraclough, 1960; Barraclough and Cross, 1963; Blake and Sawyer, 1972; Kawakami and Kubo, 1971; Wuttke, 1974) and of the MBH (Blake and Sawyer, 1972; Haller and Barraclough, 1970; Kawakami and Kubo, 1971; Lincoln, 1969; Margherita *et al.*, 1965; Ramirez *et al.*, 1967; Terasawa and Sawyer, 1969a).

The relationship between the MPOA and the MBH in controlling ovulation has been explored by using stimulation and recording techniques simultaneously.

All the rats that ovulated in response to an electrochemical stimulation of the POA showed an immediate elevation of multi-unit activity (MUA) in the arcuate-median eminence (ARC-ME) region, whereas none of the rats that failed to ovulate showed a rise (Terasawa and Sawyer, 1969a). With an adequate level of stimulation, 85%–90% of the cells in the VMH responded when the MPOA was stimulated (Haller and Barraclough, 1970). Dyer and Cross (1972) identified cells projecting from the POA to the MBH by antidromic means. Tritiated amino acid autoradiography has been used to find anatomical projections from the POA to the MBH (Conrad and Pfaff, 1975, 1976; Swanson, 1976).

Electrophysiological recordings suggest that the MPOA is critical in spontaneous ovulation. During a female rate's normal estrous cycle, there is a "critical period" when pentobarbital can block the ovulation; in the rat, this period occurs on the afternoon of proestrus (Everett and Sawyer, 1950; Everett et al., 1949). During this critical period, both the multiple-unit and the single-unit activity of the MPOA increased on the afternoon of proestrus (Cross, 1973; Dyer et al., 1972; Kawakami et al., 1970; Wuttke, 1974).

Results from lesion studies have often corroborated the results from work with stimulation and recording. Total destruction of the MBH was not necessary to block the phasic secretion of LH: simply isolating the MBH from its rostral input (medially and laterally) was sufficient to block both spontaneous and reflexive ovulation (Blake et al., 1972; Halasz and Pupp, 1965; Halasz and Gorski, 1967; Kalra and Sawyer, 1970; Kimura and Kawakami, 1978; Koves and Halasz, 1970; Phelps et al., 1976; Phelps and Sawyer, 1977; Rodgers and Schwartz, 1972; Taleisnik et al., 1979; Tejasen and Everett, 1967). However, spontaneous ovulation was not affected when the deafferentation was made rostral to the MPOA (Koves and Halasz, 1970). Thus, for both spontaneous ovulation and reflexive ovulation in the rat, the neural trigger for the ovulatory surge of LH is routed through the POA.

Research on ovulatory mechanisms has centered on the suprachiasmatic nucleus (SCN). This area lies immediately adjacent to the preoptic and is perhaps responsible for the blocking effects of preoptic lesions on *spontaneous* ovulation (Clemens et al., 1976; Kimura and Kawakami, 1978; Taleisnik and McCann, 1961). Brown-Grant and Raisman (1977), using a lateral approach to avoid damaging the POA, found that damage outside the SCN was neither necessary nor sufficient to block spontaneous ovulations.

It should be pointed out though that not all manifestations of the ovulatory mechanism are blocked by SCN lesions. *Reflexive ovulation* still follows coitus in female rats with lesions of the SCN (Brown-Grant and Raisman, 1977; Taleisnik et al., 1979). In addition, although SCN lesions attenuate the steroid-induced release of LH (Gray et al., 1978; Wiegand et al., 1978), damage to more anterior suprachiasmatic areas is necessary to abolish this response (Wiegand et al., 1980).

The olfactory stimulation received during mating that induces ovulation in light-blocked females (Johns et al., 1978) may reach the POA through extrahypothalamic neural circuits. The main olfactory bulbs project to the anterior cortical nucleus of the amygdala and the anterior hippocampus, whereas the accessory olfactory bulbs project both to the medial and the posterior cortical nuclei of the amygdala and to the bed nucleus of the stria terminalis (Scalia and Winans, 1975).

The amygdala, a heterogeneous structure, has been shown to influence ovulation in opposing ways. The *corticomedial nuclei,* which receive olfactory information, facilitate ovulation. Acute lesions of the corticomedial complex or sectioning of the stria terminalis blocks spontaneous ovulation in rats (Kimura and Kawakami, 1978; Velasco and Taleisnik, 1971). Stimulation of corticomedial amygdaloid nuclei facilitates ovulation (Bunn and Everett, 1957; Kawakami *et al.,* 1973) and the release of LH (Velasco and Taleisnik, 1969b). The rise in LH release, however, is not dependable (Ellendorff *et al.,* 1973).

On the other hand, stimulation of the *basolateral amygdala* inhibits both ovulation and LH release (Beltramino and Taleisnik, 1978; Kawakami and Kimura, 1975; Velasco and Taleisnik, 1969b).

The hippocampus also has a negative influence on ovulation. Electrical stimulation of hippocampus blocks ovulation and LH release, including that usually initiated by electrical stimulation of the MPOA or of the amygdala (Gallo *et al.,* 1971; Kawakami *et al.,* 1973; Rabii *et al.,* 1977; Velasco and Taleisnik, 1969a,b). Kawakami and Kimura (1976) hold that the role of the MPOA is that of an integration and timing center, and that the LH surge leading to spontaneous ovulation originates in the medial amygdala and the bed nucleus of the stria terminalis. Other findings support this notion. During the estrous cycle, the MUA of the corticomedial amygdaloid nuclei rises and precedes the rise in activity seen in the arcuate nucleus (Kawakami *et al.,* 1970). Direct electrical stimulation of the medial amygdala induces a rise in the MUA of the POA (Kawakami *et al.,* 1973). Single units in the POA that project to the ARC-ME respond to stimulation of the amygdala and of the hippocampus in an opposite manner (Carrer *et al.,* 1978).

Areas in the brainstem are possible afferent "way stations" involved in reflexive ovulation. Stimulation of the dorsal tegmentum facilitated ovulation in females exposed to constant light (Carrer and Taleisnik, 1970).

Figure 2 is a sagittal view of the rat brain and summarizes the neural control of spontaneous ovulation.

BEHAVIORAL INDUCTION OF PROGESTATIONAL HORMONE SECRETION

Once a female mammal has successfully ovulated, one of two neuroendocrine sequences can follow: she can proceed through the luteal phase of the estrous cycle and return to the follicular phase of another; or her eggs are fertilized and she enters the prolonged luteal phase characteristic of pregnancy. In the females of some species (e.g., in some primates), the spontaneous luteal phase of each ovarian cycle is long; and there is enough progesterone to permit the uterine implantation of a fertilized egg. If no egg implants, the uterus, which has been developing under the influence of progesterone, sheds its inner lining (the endometrium), and another ovarian cycle begins.

Unlike the ovarian cycle of most primates, however, the estrous cycle of some species (the rat, the hamster, and several species of mice) does not spontaneously

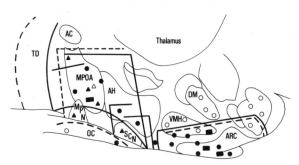

Fig. 2. Sagittal brain section depicting areas of the hypothalamus involved in *spontaneous* ovulation in the *rat:* (——) knife cuts that blocked spontaneous ovulation; (– – –) knife cuts that failed to block spontaneous ovulation; (▲) lesions that blocked spontaneous ovulation; (△) lesions that failed to block spontaneous ovulation; (●) sites at which electrical (or electrochemical) stimulation invoked ovulation; (○) sites at which stimulation failed to provoke ovulation; (■) sites at which brain activity changed during the critical period for spontaneous ovulation (or after ovulation-provoking stimulation). Abbr.: AC, anterior commissure; AH, anterior hypothalamus; ARC, arcuate nucleus; DM, dorsomedial hypothalamic nucleus; OC, optic chiasm; MPN, medial preoptic nucleus; MPOA, median preoptic area; SCN, suprachiasmatic nucleus; TD, tractus diagonalus of Broca; VMH, ventromedial hypothalamus.

include a luteal phase of any functional consequence; in these species, if the female is to become pregnant, some event must trigger the progestational state. Often, it is some aspect of the male's copulation that triggers the progestational state underlying pregnancy (Adler, 1974; Dewsbury, 1975; Diamond, 1970; Diamond and Yanagimachi, 1968; McGill, 1970).

The progestational state is similar to ovulation in that both are physiological responses that can either occur spontaneously or be induced by mating, depending on the natural history of the species and/or the experimental manipulation.

In this section, we focus on how the male rat's behavior facilitates the induction of pregnancy in the female, and we describe some of the mechanisms that produce these adaptive behavioral patterns.

For well over half a century, it has been known that mating induces the progestational state underlying pregnancy in the female rat (Long and Evans, 1922). The total pattern of stimulation received by the female during coitus, however, is quite complex; it consists of tactile, visual, olfactory, and auditory cues, and consequently, the behavioral induction of progesterone secretion could depend on one or on many of these stimuli (Diakow, 1974).

As in our treatment of ovulation, one of the purposes of the present section is to examine the specific characteristics of the mating situation that are necessary and sufficient for the induction of the progestational state. We also treat the behavioral adaptations that facilitate the induction of the progestational state, as well as the central nervous system mechanisms that generate this state.

COMPONENTS OF MATING

Because one of the most consistent features of the rat's copulatory pattern is the series of multiple intromissions preceding each ejaculation (Adler, 1978; Dewsbury, 1975), the question arises: What function do these multiple intromissions serve in successful reproduction?

In several experiments, the incidence of pregnancy was determined for females that received a normal complement of intromissions preceding the male's ejaculation (high-intromission group); these data were compared to the incidence of inducing pregnancy in a group of females that were permitted only a reduced number of preejaculatory intromissions (low-intromission group) (Adler, 1969; Wilson *et al.*, 1965). About 20 days later, the females in both groups were sacrificed, and their uteri were examined for the presence of viable fetuses. In one study (Wilson *et al.*, 1965), approximately 90% of the females in the high-intromission group were pregnant, whereas only 20% of the females in the low-intromission group were pregnant. Thus, multiple intromissions appear to be necessary for the induction of pregnancy in rats.

Because the female rat does not spontaneously produce a progestational state as part of her estrous cycle although she does ovulate spontaneously, it seemed more likely that the multiple copulatory intromissions facilitate pregnancy by invoking progesterone secretion rather than by affecting ovulation. We hypothesized that the stimulation derived from multiple intromissions triggers a neuroendocrine reflex that results in the secretion of progesterone. This interpretation was supported by progesterone assays (Adler *et al.*, 1970): within 24 hr after mating, the females in the high-intromission group had higher, sustained elevations of progesterone in their peripheral blood than did the females in the low-intromission group. The majority of the females in the high-intromission group also displayed the nocturnal surges of prolactin characteristic of pregnancy, whereas those receiving low intromission did not (Terkel and Sawyer, 1978).

Later in this section, we examine the neural and chemical hormonal links in the neuroendocrine reflex leading to pregnancy induction, and we discuss how the copulatory stimulation is transduced through these various stages. The major point for now is that stimulation from the male rat's copulatory intromissions initiates a neuroendocrine reflex in the female rat; the result of this reflex is repeated surges of prolactin, which, in turn, induce an elevated concentration of progesterone in the female.

Although there was a difference in the number of intromissions received by the females in the high- and the low-intromission groups, all of them received an ejaculation. Thus, although multiple intromissions may have been *necessary* for the induction of the progestational state, they may not have been sufficient. Therefore, we performed a series of experiments to evaluate the role of the ejaculation, the vaginal plug, and the mounts. When the vaginal orifice was sutured closed, the females receiving mounts alone (over 40) did not become progestational (Adler, 1969). Therefore, vaginocervical stimulation is necessary. In another study (Adler, 1974), males were treated with Ismelin (guanethidine sulfate). This treatment permitted them to copulate but prevented the formation of a vaginal plug, the enzymatic coagulate of seminal-vesicle and prostate fluid in the semen. Females that copulated with treated males received multiple copulatory intromissions prior to ejaculation but did not receive the vaginal plug. Despite the absence of the vaginal plug, these females entered the progestational state.

In another study, we found that with increasing numbers of intromissions (without ejaculation), the proportion of females becoming progestational increased (Adler, 1969) (Figure 3). With 4 or fewer intromissions, fewer than 10% of the females were progestational; with 13–16 intromissions, approximately 85% of the fermales became progestational. The occurrence of multiple copulatory intromissions is necessary and sufficient to trigger the hormonal state of pregnancy, even though the deposition of a vaginal plug adds an increment to the stimulation (Brown-Grant, 1977; Chester and Zucker, 1970).

For the sake of analytic simplicity, the induction of the progestational state is often treated as an isolated system, in which other organismic variables are held constant. In this research, the subjects were young, virgin female rats. The triggering of this neuroendocrine reflex is, however, normally integrated with other behavioral and physiological systems. For example, female rats come into heat within a few hours after giving birth to a litter. Because it has long been assumed that females, in the wild, are virtually always pregnant during their reproductive seasons (Sadleir, 1969), the stimulus requirements for pregnancy induction in postparturient female rats may differ from those for virgin females; a postpartum female seems to require more intromissions than does a virgin female in order to become pregnant (Davis and Connor, 1980).

The induction of pregnancy in an immediately postpartum female is also complicated by the demands of motherhood. The time spent with her newly born litter is time during which she is unavailable for copulation. Conversely, time spent copulating may increase the probability of her producing a new litter but takes time away from the altricial pups that have just been born. To examine how this conflict is handled, Gilbert *et al.* (1980) recorded sexual and maternal behavior during a postpartum estrus. Sexual behavior during the postpartum mating was similar to that during the prepartum mating (except that the latter was a bit more compressed in time); the amount of time spent with the young was significantly less than when no mating occurred. However, the most striking feature of the female's behavior was an alternation. She did not interrupt mating (ejaculatory series) and returned to the nest only after ejaculation (during the PEI). In the five females studied, 216 of the 220 intromissions occurred in uninterrupted ejaculatory series,

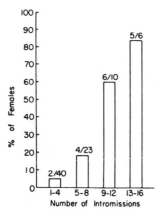

Fig. 3. Number of female rats showing cessation of behavioral receptivity and therefore, by inference, secretions of progesterone after different numbers of intromissions. (From Adler, 1969. Copyright 1969 by the American Psychological Association. Reprinted by permission of the publisher and the author.)

with bouts of maternal behavior restricted to the post-ejaculatory interval. These data are illustrated in Figure 4.

Although there are many different species that have induced ovulatory and/or progestational changes, the necessary and sufficient behavioral events that trigger these physiological events vary. In some, a single ejaculation (with its antecedent mounts and/or intromissions) is sufficient (e.g., rats, Adler, 1969; mice, Land and McGill, 1967; McGill and Coughlin, 1970; cactus mice, Dewsbury and Estep, 1975). In others, multiple ejaculatory series are required (peromyscus, Dewsbury, 1979; Dewsbury and Lanier, 1976). The ecological causes of each pattern (and the specific evolutionary path) remain to be worked out (Dewsbury, 1975).

Along with comparative studies and work on the relationship between behavioral pregnancy induction, there has been an increasing amount of attention devoted to "tracing" the effects of the behavioral stimulus through successive levels of the female's neuroendocrine system to determine the mechanism by which copulatory intromissions trigger the secretion of the gestational hormones.

RECEPTORS AND AFFERENT PATHWAYS

From the studies that we discussed in the preceding paragraphs, it seems that stimulation of the vaginocervical area is the definitive component in the induction of pregnancy in the female rat.

Manual tapping or electrical stimulation of the cervix provided adequate stimuli for inducing the progestational state (Carlson and DeFeo, 1965; Friedgood and Bevin, 1938; Greep and Hisaw, 1938; Haterius, 1933; Kollar, 1953; Long and Evans, 1922; Meyer *et al.*, 1929; Shelesnyak, 1931; Sloanaker, 1929). Furthermore, desensitization of the genital area with the local anesthetic lidocaine just prior to copulation prevented the occurrence of pregnancy (Adler, 1974; Long and Evans, 1922).

From the anatomical and electrophysiological study of the peripheral genital nervous system, we confirmed earlier findings that it is the pelvic nerve that inner-

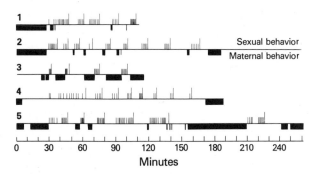

Fig. 4. The temporal patterning of maternal and sexual behavior during the mating interval of five female Norway rats. Each solid block below the baseline represents a single nest bout. Each vertical line above the baseline represents an intromission; the taller vertical lines are ejaculations. Each mating interval began at the 30-min mark on the time scale and extended to the end of the last ejaculatory series. (From Gilbert, Pelchat, and Adler, 1980. Copyright 1980 by the Association for the Study of Animal Behavior. Reprinted by permission of the publisher and the author.)

vates the vaginocervical area (Komisaruk *et al.*, 1972; Reiner *et al.*, 1981). When the pelvic nerves had been cut, female rats did not become progestational following mechanical stimulation of the cervix (Kollar, 1953; Carlson and DeFeo, 1965; Spies *et al.*, 1971; Spies and Niswender, 1971). The data reviewed in this section support the notion that the pelvic nerve is the afferent channel for the induction of progesterone secretion.

PITUITARY INVOLVEMENT

Vaginocervical stimulation in the female rat is not relayed by direct neural connection to the ovary, as the progestational state can be elicited after transplanting the ovary (Long and Evans, 1922). The central mediation of this effect was demonstrated in 1936 by Harris, who showed that direct electrical stimulation through the head was sufficient to induce the progestational reaction. Another interesting early experiment that pointed to a central mediation of the progestational response was performed by Meyer *et al.* (1929), who found that general anesthesia reduced the response to glass rod stimulation of the cervix.

A variety of studies have documented the nature of the next stage in the neuroendocrine reflex: pituitary secretion. Vaginocervical stimulation leads to an immediate release of pituitary prolactin when measured by bioassay (Herlyn *et al.*, 1965). This pituitary activation results in mammary development within a few days after the stimulation: Dilley and Adler (1968) showed that female rats receiving copulatory stimulation had lobuloalveolar development 3–5 days after they received a large number of copulatory intromissions. With the aid of more refined assay procedures, it has been possible to measure directly the pituitary output following various kinds of exteroceptive stimulation. Radioimmunoassay reveals changes in FSH (follicle stimulating hormone), LH (leutinizing hormone), PRL (prolactin), and MSH (melanocyte stimulating hormone) within the first few minutes after stimulation of the vaginocervical area (Alonso and Deis, 1973–1974; Amenomori *et al.*, 1970; Blake and Sawyer, 1972; Davidson *et al.*, 1973; Linkie and Niswender, 1972; Moss, 1974; Moss and Cooper, 1973; Moss *et al.*, 1973, 1977; Rodgers, 1971; Spies and Niswender, 1972; Taleisnik *et al.*, 1966; Taleisnik and Tomatis, 1970).

Precisely which of these many pituitary hormonal responses mediate(s) the male's copulatory stimulation of progesterone secretion? Both prolactin and LH play a role in regulating progesterone secretion. Although LH is important in the maintenance of progesterone secretion in the later stages of pregnancy (Madhwa Raj and Moudgal, 1970; Morishige and Rothchild, 1974), the initial luteotropic stimulus seems to be prolactin (McLean and Nikitovitch-Winer, 1973; Smith *et al.*, 1975).

THE PROLACTIN SURGES. The idea that prolactin was an essential ingredient in the induction of pregnancy in the female rat is relatively old. Because prolactin injections maintained corpora lutea in hypophysectomized rats (Astwood, 1941; Evans *et al.*, 1941), and because cervical stimulation acutely depleted prolactin stores in the pituitary (Herlyn *et al.*, 1965), it was generally accepted that the initial stages of the progestational response resulted from a continuously elevated level of prolactin secretion.

The data cited in the previous paragraph strongly implicated prolactin in the induction of pregnancy. When the radioimmunoassay technique was first applied to blood samples collected once daily during the progestational state, there seemed to be no major increase in prolactin, compared to the levels of this hormone seen during the estrous cycle.

The paradoxically low levels of pituitary luteotrophic hormone during the initial stages of the progestational response became understandable when it was shown that, following more frequent blood sampling, there is indeed an elevation of circulating prolactin (Butcher *et al.,* 1972; Freeman and Neill, 1972; Freeman *et al.,* 1974). This increase is, however, not a tonic elevation; rather, it takes the form of two daily surges (Figure 5). The so-called diurnal surge begins while the lights are on, reaches maximal levels as the lights go off, and returns to basal levels by midnight. The other daily prolactin peak, the nocturnal surge, begins while the lights are off, reaches its maximum as the lights come on, and decreases to baseline by the middle of the day. These surges are labeled *diurnal* and *nocturnal* on the basis of when they begin, rather than when they peak.

In the previous paragraphs, we have seen that vaginocervical stimulation is transduced by the nervous system into a pituitary response (prolactin surges), which, in turn, stimulates the ovary. Much of the physiological analysis that follows is devoted to unraveling the way in which the graded stimulation provided by a series of copulatory intromissions is converted into the state of pregnancy. At what stage of the neuroendocrine axis—CNS, pituitary, or ovary,—is the graded stimulus of mating converted to the digital state of pregnancy? In more colloquial terms, at what stage of the neuroendocrine reflex is it possible for a female rat to be a little bit pregnant?

As we mentioned in the previous section, Terkel and Sawyer (1978) found that there is a relationship between the number of intromissions a female rat receives

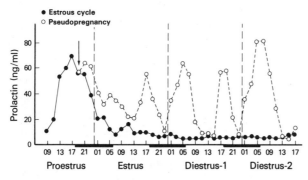

Fig. 5. Serum prolactin levels during the estrous cycle and early pseudopregnancy. The numbers along the bottom abscissa represent the time of day in terms of a 24-hr clock. The dark bars represent the dark cycle (1800–0600 hr), and the dashed line bisecting each bar represents midnight. Cervical stimulation was performed at 1900 hr of proestrus, and each point represents the mean of 5–6 decapitated animals. (From Smith, Freeman, and Neill, 1975. Copyright 1975 by The Endocrine Society. Reprinted by permission of the publisher and the author.)

and the *probability* that she will display the prolactin surges characteristic of pregnancy. That is, the majority of the females in the high-intromission group displayed the surges, whereas only 6% of the females in the low-intromission group did so. There was, however, no correlation between the magnitude of the surges and the number of intromissions. With reference to the pituitary, at least, the female rat cannot be a little bit pregnant—the surges are all-or-none.

A second basic question concerning the relationship between mating and pregnancy is where and how the neural stimulation is processed before pituitary prolactin surges are initiated. It is possible that the information derived from the cervical stimulation was retained and expressed at regular intervals over a number of days. Alternatively, it is possible that the prolactin secretion was only initiated by cervical stimulation and that feedback from the resultant secretion of ovarian steroids maintained prolactin surges during the ensuing progestational state.

To determine which of these two hypotheses is correct, Neill and his coworkers removed the ovaries of female rats 18 hr after cervical stimulation and measured their plasma prolactin levels 1–2, 5–6, and 9–10 days later. Restricting our analysis to the progestationally relevant nocturnal surge (the diurnal surge is not needed for pregnancy), it seems that these surges were still present 6 days after ovariectomy, although they did disappear by Day 10 (Freeman *et al.*, 1974). In fact, the surges can be initiated even without an ovary present (Smith and Neill, 1976b). Thus, it seems that the secretion of nocturnal surges of prolactin can continue for at least 6 days without the support of ovarian steroids, although the ovary is necessary for the prolonged expression of these surges.

In the next section, we are concerned with the presumed neural mechanism responsible for the rhythmic secretion of pituitary luteotrophic hormone. Where and how is the initial cervically derived stimulation stored so that the pituitary "clock" is activated?

THE PROLACTIN CLOCK. A rhythmic, bicircadian surge of prolactin suggests that a biological oscillator or clock is at work. With such clocks, rhythmicity *per se* is not actively triggered by environmental cues but is generated by a "self-sustaining biological oscillator" (Aschoff, 1960).

Further evidence of the copulatory triggering of a "prolactin clock" was provided by Smith and Neill (1976a) when they showed that, although the time of mating could be experimentally varied, the surges of prolactin occurred at the same time of day, regardless of when the stimulation occurred. In one experiment, even though cervical stimulation was delivered at different times of day (1900 hr, 2400 hr, or 0400 hr), the prolactin surge always occurred between 0300 and 0700 hr. For these three times of stimulation in this study, the latencies from cervical stimulation to the respective prolactin surge were 8 hr, 5 hr, and 3 hr.

The environmental rhythm entrains the biological rhythm by continually modifying the phase of the biological oscillator. In the absence of an external cycle (generally light), the biological rhythm "free-runs" and gradually goes out of phase with the solar day. True circadian rhythms persist in constant photic conditions (either constant light [LL] and/or constant dark [DD]). (See the volume on biological rhythms in this series.)

Three studies examined the pattern of plasma prolactin secretion under constant photic conditions (Bethea and Neill, 1979; Pieper and Gala, 1979; Yogev and Terkel, 1980). These investigators compared the prolactin surge in female rats housed under light–dark cycles with those in females housed under constant conditions of light or dark. They found that animals housed in DD or blinded animals continued to show prolactin surges following cervical stimulation. These surges were not diminished in size and were still circadian in timing. In two of these studies, the prolactin surges were not synchronized between animals; that is, the rhythm was free-running (Bethea and Neill, 1979; Pieper and Gala, 1979). In the third study, the surges wre synchronized; however, the nonphotic timing cues may have entrained prolactin secretion (Yogev and Terkel, 1980). Constant light (LL) reduced the magnitude of the prolactin surges and altered their periodicity (Bethea and Neill, 1979; Pieper and Gala, 1979; Yogev and Terkel, 1980). However, constant light not only removed periodic photic cues but may additionally have suppressed the expression of periodic hormonal events, for example, ovulation in rats (Dempsey and Searles, 1943).

The oscillatory basis of the prolactin surges is also demonstrated by the fact that the surges can be advanced by 6 hr when the light cycle is phase-shifted by this amount (Freeman *et al.,* 1974; Pieper and Gala, 1979; Yogev and Terkel, 1980; Bethea and Neill, 1979).

The results of all of these experiments imply that cervical stimulation activates an endogenous oscillator (perhaps in the central nervous system). At least initially, this oscillator requires neither the support of ovarian steroids nor the temporal information from environmental light cycles. Of course, like other biological rhythms, the environmental LD cycle does have an influence on the endogenous oscillator. The light cycle entrains the approximately 24-hr endogenous oscillator so that prolactin secretions occur at a precise 24-hr periodicity.

STORAGE OF THE COPULATORY STIMULUS

The experiment by Smith and Neill showed that a "dose" of cervical stimulation, which was effective in triggering the progestational state, was held for subsequent pituitary "processing" until the correct phase of the circadian cycle was reached. There are other illustrations of storage in the pregnancy systems, examples from both shorter and longer time spans.

SHORT-TERM STORAGE. Because a number of discrete intromissions, spaced over several minutes, are required to trigger the progestational state, there must be a storage mechanism that integrates the stimulation from each brief (250 msec) penile intromission. To determine the limits of this storage, we performed an experiment in which the rate of stimulation was manipulated (Edmonds *et al.,* 1972). For different groups of female rats, the number of intromissions was varied (2, 5, or 10). Within each of these groups, the females were placed in subgroups that were defined by the intervals between successive intromissions. The interintromission-interval values ranged from the control rate of ad libitum copulation (approximately 40 sec) up to 1 hr. The results are summarized in Figure 6.

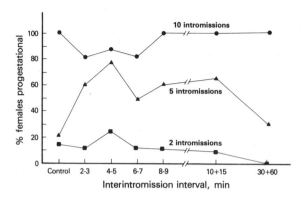

Fig. 6. Effects of different interintromission intervals on the induction of progestational hormone secretion in female rats. (From Edmonds, Zoloth, and Adler, 1972, in Adler, 1974. Copyright 1974 by Plenum Press. Reprinted by permission of the publisher and the author.)

With 10 intromissions, 100% of the females became progestational, even if the intromissions were distributed at the rate of 1 every half hour. Even when females received fewer than 10 intromissions, one could impose interintromission intervals that were longer than normal without reducing the effectiveness of the copulatory stimulation. When female rats received a total of 5 intromissions at the rate of 1 every 4 or 5 min, the stimulation was more effective in inducing the progestational state than when females received copulatory stimulation at the "control" rate.

In fact, the true "ad libitum" rate of copulation is longer than 1 intromission every 40 sec, for when the female is allowed to control the pacing of the copulation, the interval between intromissions is longer (Bermant, 1961; Gilman *et al.*, 1979; McClintock and Adler, 1978; Pierce and Nuttal, 1961). With a limited number of intromissions (5), more females became progestational when they were allowed to pace the copulation (Gilman *et al.*, 1979).

The ability of copulatory intromissions to stimulate the progestational state—even when these intromissions are widely spaced—points to an exquisitely adapted form of neuroendocrine integration by which species-specific stimuli (the multiple intromissions) are stored by an adaptively specialized neural mechanism. For example, 1 intromission delivered every half hour provided a density of stimulation of only 1 part per 7,200 (that is 1 250-msec intromission per half hour).

Long-Term Storage. In another variant of this storage experiment (Edmonds *et al.*, 1972), we compared the rate of inducing progestational responses in female rats that received 8 consecutive intromissions (with about 1 min between successive intromissions) with females that received 2 batches of 4 intromissions each, the 2 bursts of intromissions being separated by intervals that could be as long as 6 hr. With even 4 hr between the 2 blocks of 4 intromissions, female rats showed no diminution of the progestational response.

The phenomenon of delayed pseudopregnancy (Greep and Hisaw, 1938) is another example of the storage of copulatory stimulation. In this paradigm, a female rat receives vaginocervical stimulation hours or days before ovulation is to occur. Under these conditions, the progestational state occurs but is delayed until after the succeeding estrus. Following Everett's (1967) postulation that the central neuroendocrine system retains a trace of the stimulus until the ovary contains corpora lutea that are competent to respond to the stimulation, Neill and his coworkers (Freeman *et al.*, 1974) proposed that the prolactin surges could bridge the gap between cervical stimulation and the delayed ovarian response. However, Beach *et al.* (1978) found that prolactin levels surged during the delay interval only when the pseudopregnancy was initiated by direct electrical stimulation of the dorsomedial-ventromedial hypothalamus. It is not *necessary* for prolactin surges to occur during the delay interval because, under normal conditions, even the surges of prolactin are postponed until after the next estrus (Beach *et al.*, 1975; deGreef and Zeilmaker, 1976). The prolactin surges therefore represent the "hands of the clock" rather than the clock itself.

These data are consistent with a dual mechanism for processing cervical stimulation. First, the stimulation is received and stored in the CNS; then, it activates a clock that generates the twice-daily prolactin surges characteristic of pregnancy. Developmentally, these two aspects of the neural processing are separable. The ability to retain the information from cervical stimulation develops by Day 23 in the rat, but the ability to express the information as prolactin surges does not develop until Day 25 (Smith and Ramaley, 1978).

CNS MEDIATION

STIMULATION STUDIES. Although the progestational state is normally induced following ovulation, these two neuroendocrine events have at least partially separable neural mechanisms. The progestational state (and/or prolactin surges) can be elicited without a concomitant ovulation if one stimulates portions of the medial hypothalamus, including the anterior hypothalamus–paraventricular nuclei, the dorsomedial-ventromedial nuclei, and the premammillary complex (Beach *et al.*, 1978; Everett, 1968; Everett and Quinn, 1966; Freeman and Banks, 1980; Quinn and Everett, 1967; Figure 7). On the other hand, neither the progestational state nor increaesd prolactin release followed stimulation of the POA (Clemens *et al.*, 1971; Everett, 1962; Everett and Quinn, 1966; Kalra *et al.*, 1971)—even though ovulation does follow this kind of stimulation (see the previous section on ovulation).

LESION AND DISCONNECTION STUDIES. Bilateral electrolytic lesions of the MPOA induce repeated bouts of pseudopregnancy (Brown-Grant *et al.*, 1977; Clemens and Bennett, 1977; Clemens *et al.*, 1976). In contrast to the stimulation experiments in which MPOA stimulation was ineffective, the MPOA lesions induced the progestational state, characterized by daily, nocturnal surges of prolactin (Freeman and Banks, 1980). Ventral noradrenergic-bundle lesions prevent pseudopregnancy in response to vaginocervical stimulation (Hansen *et al.*, 1980). These lesion studies imply that the MPOA is an inhibitory area, whereas the ventral noradrenegic bundle is a facilatatory structure.

Although lesioning a CNS structure can show that it is involved (as a facilitator or an inhibitor) in a specific neuroendocrine function, the relationship between

T. O. ALLEN AND
N. T. ADLER

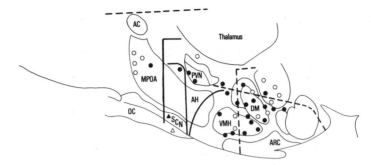

Fig. 7. Sagittal brain section depicting areas in the hypothalamus involved in induction of the progestational state: (——) knife cuts that block pseudopregnancy (PSP) or prolactin (PRL) surges in response to vaginal-cervical stimulation; (– – –) knife cuts that failed to block PSP or PRL surges in response to vaginal-cervical stimulation; (▲) lesions blocking PRL surges in reaction to vaginal-cervical stimulation; (△) lesions that failed to block PRL surges in reaction to vaginal-cervical stimulation; (●) sites at which electrical (or electrochemical) stimulation invoked PSP or PRL surges; (○) sites at which stimulation failed to invoke PSP or PRL surges.

the lesioned structure and other CNS sites is determined by other techniques, for example, surgically disconnecting areas of the brain that are normally connected. Severing the posterior and the lateral connections to the MBH did not interfere with the induction of pseudopregnancy (Arai, 1969). However, when the anterior connections to the hypothalamus had been severed in animals that were primed with LH, the progestational response could not be elicited by cervical stimulation (Carrer and Taleisnik, 1970). Retrochiasmatic cuts separating the anterior hypothalamus (AH) and the POA prevented both the nocturnal and the diurnal surge of prolactin that normally follows cervical stimulation (Freeman *et al.*, 1974).

This anterior deafferentation also disrupted the ongoing progestational state that had been initiated prior to making the lesions (Carrer and Taleisnik, 1970; Velasco *et al.*, 1974), and with both surges of prolactin were abolished (Kato *et al.*, 1978).

More medical deafferentation, separating the AH from the ME, did not disrupt a previously initiated pregnancy (Velasco *et al.*, 1974); in this case, the lesion abolished the diurnal prolactin surge while only reducing the magnitude of the nocturnal surge (Kato *et al.*, 1978), giving further evidence that the nocturnal surge is the one necessary for pregnancy. These data also demonstrate that the diurnal and the nocturnal peaks of prolactin are generated in different areas of the hypothalamus.

At this point, it may be possible to put these various pieces of anatomical information together into a working hypothesis of the central nervous system's role in triggering the progestational state. Although cervical stimulation normally induces twice-daily surges of prolactin, the SCN lesion blocked the response (Bethea and Neill, 1980; Yogev and Terkel, 1980). As was discussed previously, the SCN is a major circadian pacemaker in the rat. Bethea and Neill postulated that (1) the SCN is a circadian pacemaker for prolactin surges in the female rat; (2) the MPOA acts as a tonic inhibitory area that blocks the SCN's oscillatory input of prolactin secretion; and (3) the MPOA's inhibition is turned off when the female rat receives vaginocervical stimulation, thereby allowing the SCN to drive the prolactin surges.

When it was first discovered that retrochiasmatic cuts abolished the prolactin surges, it was thought that the MPOA itself generated the prolactin secretory rhythm. It seems more likely now that the retrochiasmatic cuts removed the influence of *both* the SCN (stimulatory) and the MPOA (inhibitory) on the prolactin-regulating neural elements. Other data also fit with this model. Lesions of the MPOA would produce the progestational state because the SCN's facilitory output is no longer opposed. Stimulation of the MPOA reduces prolactin output because it increases the inhibition by the MPOA.

RECORDING STUDIES. Although lesion and disconnection studies can demonstrate what the system is capable of doing, to understand what normally occurs when a female goes into the progestational state, it is also necessary to record from (assay) the CNS.

Electrophysiological recording has identified brain areas responsive to cervical stimulation. The areas that receive copulatory information in female rats include the anterior, the lateral, the ventromedial, and the preoptic areas of the hypothalamus; the medial and basolateral complexes of the amygdala; the nucleus reticularis gigantocellularis of the medulla; the dorsal midbrain; the dorsal hippocampus; the lateral septum; and the reticular formation (Barraclough, 1960; Barraclough and Cross, 1963; Blake and Sawyer, 1972; Cross and Silver, 1965; Hornby and Rose, 1976; Kawakami and Kubo, 1971; Lincoln, 1969; Malsbury *et al.*, 1972; Margherita *et l.*, 1965; Ramirez *et al.*, 1967). These areas that are responsive to vaginocervical stimulation may be involved in the storage of the stimulation.

We are using a new technique, the (^{14}C) deoxyglucose (2DG) method of neural mapping, to investigate brain areas responsive to vaginocervical stimulation (Allen *et al.*, 1981). A major advantage of the 2DG method over other techniques is the ability to assay concurrently a virtually unlimited number of brain areas in an individual animal—without invading or destroying the integrity of the brain. This advantage becomes especially critical in discriminating among neural systems that overlap in part of their extent. The mediobasal-hypothalamus–median-eminence complex, for example, is a final common path for two behaviorally initiated neuroendocrine reflexes: induced ovulation and the progestational state.

This method is based on the fact that nerve cells preferentially use glucose for energy, and on the assumption that the functional activity and energy requirements of neural tissue are coupled. (^{14}C) Deoxyglucose is taken up by active cells and metabolized to deoxyglucose-6-phosphate. It is not metabolized further and is essentially trapped in the cell. The rate of glucose utilization, and hence the activity of that structure, can be estimated by autoradiographic measurement of the (^{14}C) deoxyglucose-6-phosphate concentration (for a description of the procedure, see Sokoloff *et al.*, 1977). Areas of high metabolic activity are directly visualized as areas of high optical density. Histological staining of the tissue after it has been autoradiographed permits a precise identification of anatomical structures.

In this study, female rats were ovariectomized and injected with exogenous estrogen and progesterone in order to equalize the physiological status of females in each of the groups. Only females displaying an excellent lordotic posture in response to manual palpation of the flanks were used.

The experiment was conducted on the day of behavioral estrus one week after a jugular catheter had been inserted. All females were awake and gently restrained with cloth wrappings. Four females received intermittent vaginocervical stimulation: a smooth-tipped metal rod (2 mm in diameter) attached to a vibrator was repeatedly inserted into the vagina until it just contacted the cervix. Four females heard the sound of the vibrator but did not receive vaginocervical stimulation (yoked control). Vaginocervical stimulation started 5 min before a pulse of 2DG (15–20 microcuries/100 g) was infused through the catheter, and it was followed with a flush of saline. The stimulation continued for the next 45 min.

At the end of the experimental period, the animals were killed by an overdose of anesthetic. The brains were removed, frozen, cut in a cryostat, dried quickly, and autoradiographed on x-ray film along with a set of six methylmethacrylate standards of known concentration ^{14}C. The optical densities of the autoradiographs were measured with a microdensitometer for 44 brain structures, and automatically converted to ^{14}C concentration by comparison to the six calibrated standards. Following autoradiography, the sections were stained to facilitate anatomical identification.

In order to compare the ^{14}C concentration in a gray matter structure between animals in these groups, we needed to control for individual variation in the brain's metabolism of glucose. Accordingly, the average concentration of ^{14}C in fiber tracts of the brain was used as an internal baseline.

Of the 44 CNS structures, 8 increased their metabolic activity by 20% or more in response to vaginocervical stimulation (Table 1). The MPOA showed the greatest increase (37% over control values). This increase was statistically significant ($p <$

TABLE 1. RELATIVE CONCENTRATION OF ^{14}C IN SELECTED BRAIN AREAS[a]

Structure	Relative concentration ^{14}C		% Increase over control values
	Vaginal stimulation	No stimulation	
Medial preoptic area	1.89 ± 0.18	1.38 ± 0.21	37.0**
Reticular formation	2.05 ± 0.26	1.67 ± 0.19	22.8*
N. stria terminalis	1.98 ± 0.16	1.62 ± 0.15	22.2*
Locus coeruleus	1.89 ± 0.51	1.55 ± 0.29	21.9
Lateral preoptic area	2.17 ± 0.24	1.79 ± 0.29	21.2
Anterior hypothalamus	1.60 ± 0.08	1.32 ± 0.18	21.2
Dorsal raphe	2.24 ± 0.17	1.85 ± 0.21	21.1*
Habenula	2.89 ± 0.23	2.40 ± 0.37	20.4

[a]These data are derived from optical density measurements of (^{14}C) deoxyglucose autoradiographs. Each value is the ratio of the ^{14}C concentration in each gray-matter structure to the ^{14}C concentration in the average white matter for that animal. For each structure, the increase of relative ^{14}C concentration in the stimulated females is expressed as a percentage of control.
*$p < .029$, by Mann Whitney U test (one-tailed).
**$p < .014$.

.014). On the autoradiograms, this area is visible as a bilateral wedge of darkening, lateral to the third ventricle and extending from the anterior commissure to the optic chiasm. It is bordered laterally by darker areas in the lateral preoptic (Figure 8A). Autoradiographs from control females did not show this wedge-shaped area of increased activity (Figure 8B). In addition to heightened activity in the MPOA, the stimulated females also showed statistically significant increases in activity in the mesencephalic reticular formation, the nucleus of the stria terminalis, and the dorsal raphe. Although the locus coeruleus, the lateral preoptic area, the anterior hypothalamus, and the habenula showed an elevated uptake of 2DG, these increases were not statistically significant.

The other 36 nuclei exhibited less than 18% difference in relative ^{14}C concentration between stimulated and unstimulated animals. None of these comparisons

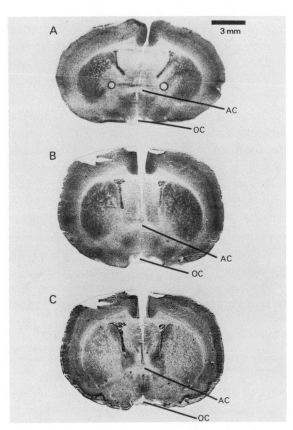

Fig. 8. (A) (^{14}C) 2-deoxyglucose autoradiograph from a female rat that received vaginocervical stimulation. The preoptic area is visible as the bilateral wedge of darkening extending from the anterior commissure (AC) to the optic chiasm (OC). The average relative ^{14}C concentration in the medial preoptic area in this animal was 2.15. (B) Autoradiograph from a female that did not receive the stimulation. The medial preoptic area is pale. The average relative ^{14}C concentration in the medial preoptic area in this animal was 1.70. (C) Cresyl-violet stained section. This is the same section from which the autoradiograph in (B) was taken. The stained sections were used to confirm the anatomical identification in the autoradiographs; note the medial preoptic area here. (From Allen, Adler, Greenberg, and Reivich, 1981. Copyright 1981 by the American Association for the Advancement of Science. Reprinted by permission of the publisher and the author.)

was statistically significant, except for the 13% increase over control values in the globus pallidus ($p < .014$).

These findings provide a map of brain areas that are activated by vaginocervical stimulation. At least some of these areas may be involved in the induction of the progestational state. The CNS areas that showed increased metabolism following vaginocervical stimulation have been implicated in the control of reproduction by experiments that used recording, lesion, and/or stimulation techniques (see the sections above on CNS mediation). The 2DG method of autoradiography can therefore validate the results of other experimental methods. However, several brain structures implicated in reproductive physiology by other techniques did not increase their metabolic activity in this study. At least in part, the discrepancy may have been due to 2DG autoradiography's assessment of activity in aggregates of neurons. This method may not detect increases in a diffusely organized nucleus, such as the nucleus reticularis gigantocellularis, in which only 4% of the genitally sensitive neurons respond specifically to vaginocervical stimulation (Hornby and Rose, 1976).

At this point, we can only speculate about the specific role of the medial preoptic area in the induction of the progestational state. One function of the CNs in the induction of the progestational state is that of a storage system. This system integrates the brief stimulation from each intromission even when the intromissions are widely spaced (Edmonds *et al.,* 1972), and it delays the expression of the prolactin surges until the appropriate time (Smith and Neill, 1976a). The medial preoptic area fulfills a similar timing function in another reproductive event, ovulation. The neural trigger for spontaneous ovulation arises here and precedes the ovulatory surge of LH by a number of hours (Everett and Sawyer, 1950). The MPOA has been characterized as a receiving area for exteroceptive stimulation related to ovulation (Blake and Sawyer, 1972). In this schema, the medial preoptic area evaluates the amount of copulatory stimulation, then at a later time (Smith and Neill, 1976a) is inhibited to allow the progestationally relevant nocturnal surge of prolactin (Bethea and Neill, 1980; Freeman and Banks, 1980).

SUMMARY

In this chapter, we have examined some of the machinery of the mammalian body, reviewing studies on the neuroendocrine organization of reproduction in that class. As important as each individual neuroanatomical and neurophysiological link may be, the overall integration—into efficient patterns of reproduction—is the "driving force" of evolutionary adaptation.

Although endogenous feedback between brain, pituitary, and gonad provides the basic framework for reproduction, we have concentrated on the endocrine events that are controlled by exogenous, behavioral stimulation. Our thesis is that such behavioral influences are not merely added onto an autonomous neuroendocrine apparatus but provide the fine-tuning by which members of a particular species are adapted to their local environments. The physiological events of ovulation and pregnancy must be coordinated with reproductive behavior patterns for the sexually reproducing animal to survive.

REFERENCES

Adler, N. T. Effects of the male's copulatory behavior in successful pregnancy of the female rat. *Journal of Comparative and Physiological Psychology,* 1969, *69,* 613–622.

Adler, N. T. The behavioral control of reproductive physiology. In W. Montagna and W. A. Sadler (Eds.), *Reproductive Behavior.* New York: Plenum Press, 1974.

Adler, N. T. Social and environmental control of reproductive processes in animals. In T. McGill, D. Dewsbury, and B. Sachs (Eds.), *Sex and Behavior: Status and Prospectus.* New York: Plenum Press, 1978.

Adler, N. T., and Bermant, G. Sexual behavior of male rats: Effects of reduced sensory feedback. *Journal of Comparative and Physiological Psychology,* 1966, *61,* 240–243.

Adler, N. T., Resko, J. A., and Goy, R. W. The effect of copulatory behavior on hormonal change in the female rat prior to implantation. *Physiology and Behavior,* 1970, *5,* 1003–1007.

Adler, N. T., Davis, P. G., and Komisaruk, B. K. Variation in the size and sensitivity of a genital sensory field in relation to the estrous cycle in rats. *Hormones and Behavior,* 1977, *9,* 334–344.

Allen, T. O., Adler, N. T., Greenberg, J. H., and Reivich, M. Vaginocervical stimulation selectively increases metabolic activity in the rat brain. *Science,* 1981, *211,* 1070–1072.

Alleva, J. J., Waleski, M. V., and Alleva,F. R. A biological clock controlling the estrous cycle of the hamster. *Endocrinology,* 1971, *88,* 1368–1379.

Alonso, N., and Deis, R. P. Hypophysial and serum concentration of prolactin in the rat after vaginal stimulation. *Neuroendocrinology,* 1973–1974, *13,* 63–68.

Amenomori, Y., Chen, C. L., and Meites, J. Serum prolactin levels in rats during different reproductive state. *Endocrinology,* 1970, *86,* 506–510.

Arai, Y. Effect of hypothalamic deafferentation on induction of pseudopregnancy by vaginal-cervical stimulation in the rat. *Journal of Reproduction and Fertility,* 1969, *19,* 573–575.

Aron, C., Asch, G., and Roos, J. Triggering of ovulation by coitus in the rat. *International Review of Cytology,* 1966, *20,* 139–172.

Aron, C., Roos, J., and Asch, G. New facts concerning the afferent stimuli that trigger ovulation by coitus in the rat. *Neuroendocrinology,* 1968, *3,* 47–54.

Aschoff, J. Exogenous and endogenous components in circadian rhythms. *Cold Spring Harbor Symposia on Quantitative Biology,* 1960, *25,* 11–28.

Astwood, E. B. The regulation of corpus luteum function by hypophysial luteotrophin. *Endocrinology,* 1941, *28,* 309–320.

Austin, C. R. Oestrus and ovulation in the field vole *(Microtus agrestis). Journal of Endocrinology,* 1957, *15,* iv.

Barfield, B. J., Auer, P., Geyer, L. A., and McIntosh, T. K. Ultrasonic vocalizations in rat sexual behavior. *American Zoologist,* 1979, *19,* 469–480.

Barr, G. D., and Barraclough, C. A. Temporal changes in medial basal hypothalamic LH-RH correlated with plasma LH during the rat estrous cycle and following electrochemical stimulation of the medial preoptic area in pentobarbital-treated proestrous rats. *Brain Research,* 1978, *148,* 413–423.

Barraclough, C. A. Hypothalamic activity associated with stimulation of vaginal cervix in rats. *Anatomical Record,* 1960, *136,* 159.

Barraclough, C. A., and Cross, B. A. Unit activity in the hypothalamus of the cyclic female rat: Effect of genital stimuli and progesterone. *Endocrinology,* 1963, *26,* 339–359.

Barraclough, C. A., and Gorski, R. A. Evidence that the hypothalamus is responsible for androgen-induced sterility in the female rat. *Endocrinology,* 1961, *68,* 68–79.

Beach, J. E., Tyrey, L., and Everett, J. W. Serum prolactin and LH in early phases of delayed us direct pseudopregnancy in the rat. *Endocrinology,* 1975, *96,* 1241–1246.

Beach, J. E., Tyrey, L., and Everett, J. W. Prolactin secretion preceding delayed pseudopregnancy in rats after electrical stimulation of the hypothalamus. *Endocrinology,* 1978, *103,* 2247–2251.

Bellringer, J., Pratt, H., and Keverne, E. Involvement of the vomero nasal organ and prolactin in pheromonal induction of delayed implantation in mice. *Journal of Reproduction and Fertility,* 1980, *59,* 223–228.

Beltramino, C., and Taleisnik, S. Facilitatory and inhibitory effects of electrical stimulation of the amygdala on the release of luteinizing hormone. *Brain Research*, 1978, *144*, 95–107.

Bermant, G. Response latencies of female rats during sexual intercourse. *Science*, 1961, *133*, 1771.

Bermant, G. Sexual behavior of male rats: Photographic analysis of the intromission response. *Psychonomic Science*, 1965, *2*, 65–66.

Bethea, C. L., and Neill, J. D. Prolactin secretion after cervical stimulation of rats maintained in constant dark or constant light. *Endocrinology*, 1979, *104*, 870–876.

Bethea, C. L., and Neill, J. D. Lesions of the suprachiasmatic nuclei abolish the cervically stimulated prolactin surges in the rat. *Endocrinology*, 1980, *107*, 1–5.

Blake, C. A., and C. H. Sawyer. Effects of vaginal stimulation on hypothalamic multiple-unit activity and pituitary LH release in the rat. *Neuroendocrinology*, 1972, *10*, 358–370.

Blake, C. A., Weiner, R. I., and Gorski, R. A. Secretion of pituitary luteinizing hormone and follicle stimulating hormone in female rats made persistently estrous or diestrus by hypothalamic deafferentation. *Endocrinology*, 1972, *90*, 855–861.

Breed, W. G. Ovulation in the genus *Microtus*. *Nature*, 1967, *214*, 826.

Breed, W. G., and Charlton, H. M. The control of ovulation in the vole *(Microtus agrestis)*. *Journal of Psychology*, 1968, *198*, 2.

Breed, W. G., and Charlton, H. M. Hypothalamo-hypophysial control of ovulation in the vole, *Microtus agrestis*. *Journal of Reproduction and Fertility*, 1971, *25*, 225–229.

Breed, W. G., and Clarke, J. R. Ovulation and associated histological changes in the ovary following coitus in the vole *(Microtus agrestis)*. *Journal of Reproduction and Fertility*, 1970, *22*, 173–175.

Bronson, F., and Desjardin, C. The relationship of gonadotrophins in ovariectomized mice after exposure to a male. *Journal of Endocrinology*, 1969, *44*, 293–297.

Bronson, F. H., and Maruniak, J. A. Male-induced puberty in female mice: Evidence for a synergistic action of social cues. *Biology of Reproduction*, 1975, *13*, 94–98.

Brooks, C. Studies on the neural basis of ovulation in the rabbit. *American Journal of Physiology*, 1935, *113*, 18–19.

Brooks, C. M. The role of the cerebral cortex and of various sense organs in the excitation and execution of mating activity in the rabbit. *American Journal of Physiology*, 1937, *120*, 544–553.

Brooks, C. M. A study of the mechanism whereby coitus excites the ovulation-producing activity of the rabbit's pituitary. *American Journal of Physiology*, 1938, *121*, 157–177.

Brown-Grant, K. The induction of pseudopregnancy and pregnancy by mating in albino rats exposed to constant light. *Hormones and Behavior*, 1977, *8*, 62–76.

Brown-Grant, K., and Raisman, G. Abnormalities in reproductive function associated with the destruction of the SCN in female rats. *Proceedings of Royal Society London*, 1977, *198*, 279–296.

Brown-Grant, K., El Kabir, D. J., and Sink, G. The effect of mating on pituitary luteinizing hormone and thyrotrophic hormone content in the female rabbit. *Journal of Endocrinology*, 1968, *41*, 91–94.

Brown-Grant, K., Davidson, J., and Grieg, F. Induced ovulation in albino rats exposed to constant light. *Journal of Endocrinology*, 1973, *32*, 119–128.

Brown-Grant, K., Murray, M. A., Raisman, G., and Sood, M. C. Reproductive function in male and female rats following extra- and intra-hypothalamic lesions. *Proceedings of Royal Society London*, 1977, *198*, 267–278.

Bruce, H. An exteroceptive block to pregnancy in the mouse. *Nature*, 1959, *184*, 105.

Bruce, H. A block to pregnancy in the mouse caused by proximity to strange males. *Journal of Reproduction and Fertility*, 1960, *1*, 96–103.

Bruce, H., and Parrott, D. The role of olfactory sense in pregnancy-block by strange males. *Science*, 1960, *131*, 1526.

Bunn, J. P., and Everett, J. W. Ovulation in persistent estrus rats after electrical stimulation of the brain. *Proceedings of Society for Experimental Biology and Medicine*, 1957, *96*, 369–371.

Butcher, R. L., Fugo, H. W., and Collins, W. E. Semicircadian rhythm in plasma levels of prolactin during early gestation in the rat. *Endocrinology*, 1972, *90*, 1125–1127.

Carlson, R. R., and DeFeo, V. J. Role of the pelvic nerve vs. the abdominal sympathetic nerves in the reproductive function of the female rat. *Endocrinology*, 1965, *77*, 1014–1022.

Carlyle, A., and Williams, T. D. Artificially induced ovulation in the rabbit. *Journal of Physiology*, 1961, *157*, 42.

Carrer, H. F., and Taleisnik, S. Induction and maintenance of pseudopregnancy after interruption of preoptic hypothalamic connections. *Endocrinology*, 1970, *86*, 231–236.

Carrer, H. F., Whitmoyer, D. I., and Sawyer, C. H. Effects of hippocampal and amygdaloid stimulation on the firing of preoptic neurons in the proestrus female rat. *Brain Research*, 1978, *142*, 363–367.

Carter, C. S. Getz, L., Gavish, L., McDermott, J., and Arnold, P. Male-related phermones and the activation of female reproduction in the prairie vole *(M. ochrogaster)*. *Biology and Reproduction*, 1980, *23*, 1038–1045.

Chapman, V. M., Desjardin, C., and Whitten, W. K. Pregnancy block in mice: changes in pituitary LH and LTH and plasma progestin levels. *Journal of Reproduction and Fertility*, 1970, *21*, 333–337.

Charlton, H. M., Naftolin, F., Sood, M. C., and Worth, R. W. The effect of mating upon LH release in male and female voles of the species, *Microtus agrestis. Journal of Reproduction and Fertility*, 1975a, *42*, 167–170.

Charlton, H. M., Naftolin, F., Sood, M. C., and Worth, R. W. Electrical stimulation of the hypothalamus and LH release in the vole, *Microtus agrestis. Journal of Reproduction and Fertility*, 1975b, *42*, 171–174.

Chester, R. V., and Zucker, I. Influence of male copulatory behavior on sperm transport, pregnancy, and pseudopregnancy in female rats. *Physiology and Behavior*, 1970, *5*, 35–43.

Chiappa, S. A., Fink, G., and Sherwood, N. M. Immunoreactive luteinizing hormone releasing factor (LRF) in pituitary stalk plasma from female rats: Effects of stimulating diencephalon, hippocampus and amygdalas. *Journal of Physiology*, 1977, *267*, 625–640.

Christian, J., Lloyd J., and Davis, D. The role of endocrines in self regulation of mammalian populations. *Recent Progress Hormone Research*, 1965, *21*, 501–578.

Clattenburg, R. E., Singh, R. P., and Montemurro, D. G. Ultrastructural changes in the preoptic nucleus of the rabbit following coitus. *Neuroendocrinology*, 1971, *8*, 289–306.

Clemens, J. A., and Bennett, D. R. Do aging changes in the POA contribute to the loss of cyclic endocrine functions? *Journal of Gerontology*, 1977, *32*, 19–24.

Clemens, J. A., Shaar, C. J., Kleber, J. W., and Tandy, W. A. Areas of the brain stimulatory to LH and FSH secretion. *Endocrinology*, 1971, *88*, 180–184.

Clemens, J. A., Smalstig, E. B., and Sawyer, B. D. Studies on the role of the preoptic area in the control of reproductive function in the rat. *Endocrinology*, 1976, *99*, 728–735.

Cluclow, F., and Clarke, J. Pregnancy-block in *Microtus agrestis*, an induced ovulator. *Nature*, 1968, *219*, 511.

Cluclow, F., and Langford, P. Pregnancy-block in the meadow vole, *Microtus pennsylvanicus. Journal of Reproduction and Fertility*, 1971, *24*, 275–277.

Cluclow, F. V., and Mallory, F. F. Oestrus and induced ovulation in the meadow vole, *Microtus pennsylvanicus. Journal of Reproduction and Fertility*, 1970, *23*, 341–343.

Conaway, C. H. Ecological adaptation and mammalian reproduction. *Biology of Reproduction*, 1971, *4*, 239–247.

Conrad, L. C., and Pfaff, D. W. Axonal projections of medial preoptic and anterior hypothalamic neurons. *Science*, 1975, *190*, 1112–1114.

Conrad, L. C., and Pfaff, D. W. Efferents from medial basal forebrain and hypothalamus in the rat: I. An autoradiographic study of the medial preoptic area. *Journal of Comparative Neurology*, 1976, *169*, 185–220.

Contreras, J. L., and Beyer, C. A polygraphic analysis of mounting and ejaculation in the New Zealand white rabbit. *Physiology and Behavior*, 1979, *23*, 939–943.

Cooper, R. L., Conn, P. M., and Walker, R. F. Characterization of the LH surge in middle-aged female rats. *Biology and Reproduction*, 1980, *23*, 611–615.

Cramer, O. M., and Barraclough, C. A. Effect of electrical stimulation of the preoptic area on plasma LH concentrations in proestrous rats. *Endocrinology*, 1971, *88*, 1175–1183.

Critchlow, V. Blockade of ovulation in the rat by mesencephalic lesions. *Endocrinology*, 1958, *63*, 596–610.

Cross, B. A. Towards a neurophysiological basis for ovulation. *Journal of Reproduction and Fertility*, 1973 (supplement), *20*, 97–117.

Cross, B. A., and Silver, I. A. Effect of luteal hormone on the behavior of hypothalamic neurons in pseudopregnant rats. *Journal of Endocrinology*, 1965, *31*, 251–263.

Cross, P. C. Observations on the induction of ovulation in *Microtus montanus. Journal of Mammalogy*, 1972, *53*, 210–212.

Curry, J. J. Alterations in incidence of mating and copulation-induced ovulation after olfactory bulb ablation in female rats. *Journal of Endocrinology*, 1974, *62*, 245–250.

Davidson, J. M., Smith, E. R., and Bowers, C. Y. Effects of mating on gonadotropin release in female rat. *Endocrinology,* 1973, *93,* 1185–1192.

Davis, H. N., and Conner, J. R. Male modulation of female reproductive physiology in Norway rats: Effects of mating during post partum estrus. *Behavior and Neural Biology,* 1980, *29,* 128–131.

Davis, H. N., Gray, G. D., Zerylnick, M., and Dewsbury, D. A. Ovulation and implantation in montane voles *(Microtus montanus)* as a function of varying amounts of copulatory stimulation. *Hormones and Behavior,* 1974, *5,* 383–388.

Davis, H. N., Gray, G. D., and Dewsbury, D. A. Maternal age and male behavior in relation to successful reproduction by female rats *(Rattus norvegicus). Journal of Comparative and Physiological Psychology,* 1977, *91,* 281–289.

deGreef, W. J., and Zeilmaker, G. H. Prolactin and delayed pseudopregnancy in the rat. *Endocrinology,* 1976, *98,* 305–310.

Dempsey, E. W. Follicular growth rate and ovulation after various experimental procedures in the guinea pig. *American Journal of Physiology,* 1937, *120,* 126–132.

Dempsey, E. W., and Searles, H. F. Environmental modification of certain endocrine phenomena. *Endocrinology,* 1943, *32,* 119–128.

Denenberg, V. H., Zarrow, M. X., and Ross, S. The behavior of rabbits. In E. S. E. Hafez (Ed.), *The Behavior of Domestic Animals.* Baltimore: Williams and Williams, 1969.

Dewsbury, D. A. Copulatory behavior of montane voles *(Microtus montanus). Behavior,* 1973, *44,* 186–202.

Dewsbury, D. A. Diversity and adaptation in rodent copulatory behavior. *Science,* 1975, *190,* 949–954.

Dewsbury, D. A. Copulatory behavior of deer mice *(Peromyscus maniculatus):* III. Effects on pregnancy initiation. *Journal of Comparative Physiological Psychology,* 1979, *93,* 178–188.

Dewsbury, D. A., and Estep, D. Q. Pregnancy in cactus mice: effects of prolonged copulation. *Science,* 1975, *187,* 552–553.

Dewsbury, D. A., and Lanier, D. L. Effects of variations in copulatory behavior on pregnancy in two species of *Peromyscus. Physiology and Behavior,* 1976, *17,* 921–924.

Diakow, C. Male-female interactions and the organization of mammalian mating patterns. In *Advances in the Study of Behavior,* Vol. 5. New York: Academic Press, 1974.

Diakow, C. Motion picture analysis of rat mating behavior. *Journal of Comparative and Physiological Psychology,* 1975, *88,* 704–712.

Diamond, M. Intromission pattern and species vaginal code in relation to induction of pseudopregnancy. *Science,* 1970, *169,* 995–997.

Diamond, M., and Yanagimachi, R. Induction of pseudopregnacy in the golden hamster. *Journal of Reproduction and Fertility,* 1968, *17,* 165–168.

Dilley, W. G., and Adler, N. T. Post-copulatory mammary gland secretion in rats. *Proceeding of the Society for Experimental Biology and Medicine,* 1968, *129,* 964–966.

Dominic, C. Observations of the reproductive pheromones of mice. II. Neuroendocrine mechanisms involved in the olfactory block to pregnancy. *Journal of Reproduction and Fertility,* 1966, *11,* 415–421.

Drickamer, L. Sexual motivation of female house mice: Social inhibition. *Developmental Psychobiology,* 1974, *7,* 257–265.

Drickamer, L. Delay of sexual maturation in female house mice by exposure to grouped females or urine from grouped females. *Journal of Reproduction and Fertility,* 1977, *51,* 77–81.

Drickamer, L., and Murphy, R. Female mouse maturation effects of excreted and bladder urine from juvenile and adult males. *Developmental Psychobiology,* 1978, *11,* 63–72.

Drickamer, L., McIntosh, T., and Rose, E. Effects of ovariectomy on the presence of a maturation delaying phermone in the urine of female mice. *Hormones and Behavior,* 1978, *11,* 135–138.

Dufy-Barbe, L., Franchimont, P., and Faure, J. M. A. Time course of LH and FSH release after mating in the female rabbit. *Endocrinology,* 1973, *92,* 1318–1321.

Dyer, R. G., and Cross, B. A. Antidromic identification of units in the preoptic and anterior hypothalamic areas projecting directly to the ventromedial and arcuate nuclei. *Brain Research,* 1972, *43,* 254–258.

Dyer, R. G., Pritchett, C., and Cross, B. A. Unit activity in the diencephalon of female rats during the estrus cycle. *Journal of Endocrinology,* 1972, *53,* 151–160.

Edmonds, S., Zoloth, S. R., and Adler, N. T. Storage of copulatory stimulation in the female rat. *Physiology and Behavior,* 1972, *8,* 161–164.

Ellendorff, F., Colombo, J. A., Blake, C. A., Whitmoyer, D. I., and Sawyer, C. H. Effects of electrical stimulation of the amygdala on gonadotropin release and ovulation in the rat. *Proceedings of Society for Experimental Biology and Medicine*, 1973, *142*, 417–420.

Elliot, J., and Goldman, B. Seasonal reproduction: Photoperiodism and biological clocks. In N. Adler (Ed.), *Neuroendocrinology of Reproduction*. New York: Plenum Press, 1981.

Eskay, R. L., Mical, R. S., and Porter, J. C. Relationship between luteinizing hormone releasing hormone concentration in hypophysial portal blood and luteinizing hormone release in intact, castrated, and electrochemically-stimulated rats. *Endocrinology*, 1977, *100*, 263–270.

Evans, H. M., Simpson, M. E., Lyons, W. R., and Turpeinen, K. Anterior pituitary hormones which favor the production of traumatic uterine placentomata. *Endocrinology*, 1941, *28*, 933–945.

Everett, J. W. Presumptive hypothalamic control of spontaneous ovulation. *Ciba Foundation Colloquium of Endocrinology*, 1952, *4*, 167–177.

Everett, J. W. Absence of pseudopregnancy in rats after ovulation-inducing stimulation of the preoptic brain. *Physiologist*, 1962, *5*, 137.

Everett, J. W. Central neural control of reproductive functions of the adenohypophysis. *Physiological Review*, 1964, *44*, 373–431.

Everett, J. W. Ovulation in rats from preoptic stimulation through platinum electrodes: Importance of duration and spread of stimulus. *Endocrinology*, 1965, *76*, 1195–1201.

Everett, J. W. Provoked ovulation or long-delayed pseudopregnancy from coital stimuli in barbiturate-blocked rats. *Endocrinology*, 1967, *80*, 145–154.

Everett, J. W. Delayed pseudopregnancy in the rat, a tool for the study of central neural mechanisms in reproduction. In M. Diamond (Ed.), *Perspectives in Reproduction and Sexual Behavior*. Bloomington and London: Indiana University Press, 1968.

Everett, J. W., and Radford, H. M. Irritative deposits from stainless steel electrodes in the preoptic rat brain causing release of pituitary gonadotropin. *Proceedings of the Society for Experimental Biology and Medicine*, 1961, *108*, 604–609.

Everett, J. W., and Sawyer, C. H. A 24-hour periodicity in the LH-release apparatus of female rats, disclosed by barbiturate sedation. *Endocrinology*, 1950, *47*, 198–218.

Everett, J. W., Sawyer, C. H., and Markee, J. E. A neurogenic timing factor in control of the ovulatory discharge of luteinizing hormone in the cyclic rat. *Endocrinolgoy*, 1949, *44*, 234–250.

Everett, J. W., and Quinn, D. L. Differential hypothalamic mechanisms inciting ovulation and pseudopregnancy in the rat. *Endocrinology*, 1966, *78*, 141–150.

Everett, J. W., Krey, L. C., and Tyrey, L. The quantitative relationshp between electrochemical preoptic stimulation and LH release in proestrous versus late diestrous rats. *Endocriology*, 1973, *93*, 947–953.

Everett, J. W., Quinn, D. L., and Tyrey, L. Comparative effectiveness of preoptic and tuberal stimulation for luteinizing hormone release and ovulation in two strains of rats. *Endocrinology*, 1976, *98*, 1302–1308.

Fee, A. R., and Parkes, A. S. Studies on ovulation. I. The relation of the anterior pituitary body to ovulation in the rabbit. *Journal of Physiology*, 1929, *67*, 383–389.

Fee, A. R., and Parkes, A. S. Studies on ovulation III. Effect of vaginal anesthesia on ovulation in the rabbit. *Journal of Physiology*, 1930, *70*, 385–388.

Fink, G., and Jamieson, M. G. Immunoreactive luteinizing hormone releasing factor in rat pituitary stalk blood: effects of electrical stimulation of the medial preoptic area. *Journal of Endocrinology*, 1976, *68*, 71–87.

Freeman, M. E., and Banks, J. A. Hypothalamic sites which control the surges of prolactin secretion induced by cervical stimulation. *Endocrinology*, 1980, *106*, 668–673.

Freeman, M. E., and Neill, J. D. The pattern of prolactin secretion during pseudopregnancy in the rat: A daily nocturnal surge. *Endocrinology*, 1972, *90*, 1292–1294.

Freeman, M. E., Smith, M. S., Nazian, S. J., and Neill, J. D. Ovarian and hypothalamic control of the daily surges of prolactin secretion during pseudopregnancy in the rat. *Endocrinology*, 1974, *94*, 875–882.

Friedgood, H. B., and Bevin, S. Relationship of cervical sympathetics to pseudopregnancy in the rat. *American Journal of Physiology*, 1938, *123*, 71.

Friedman, M. H. The mechanism of ovulation in the rabbit: I. The demonstration of a humoral mechanism. *American Journal of Physiology*, 1929, *89*, 438–442.

Gallistel, C. R. *The Organization of Action: A New Synthesis*. New Jersey: Erlbaum, 1980.

Gallo, R. V., Johnson, J. H., Goldman, B. D., Whitmoyer, D. I., and Sawyer, C. H. Effects of electrochemical stimulation of the ventral hippocampus on hypothalamic electrical activity and pituitary gonadotrophin secretion in female rats. *Endocrinology*, 1971, *89*, 704–713.

Gilbert, A. W., Pelchat, R. J., and Adler, N. T. Postpartum copulatory and maternal behavior in Norway rats under seminatural conditions. *Animal Behaviour*, 1980, *28*, 989–995.

Gilman, D. P., Mercer, L. F., and Hitt, J. C. Influence of female copulatory behavior on the induction of pseudopregnancy in the female rat. *Physiology of Behavior*, 1979, *22*, 675–678.

Gray, G. D., and Dewsbury, D. A. A quantitative description of copulatory behavior in prairie voles *(Microtus ochrogaster)*. *Brain, Behavior and Evolution*, 1973, *8*, 437–452.

Gray, G. D., and Dewsbury, D. A. A quantitative description of the copulatory behavior of meadow voles *(Microtus pennsylvanicus)*. *Animal Behavior*, 1975, *23*, 261–267.

Gray, G. D., Davis, H. N., Zerylnick, M., and Dewsbury, D. A. Oestrus and induced ovulation in montane voles. *Journal of Reproduction and Fertility*, 1974a, *38*, 193–196.

Gray, G. D., Zerylnick, M., Davis, H. N., and Dewsbury, D. A. Effects of variations in male copulatory behavior on ovulation and implantation in prairie voles *(Microtus ochrogaster)*. *Hormones and Behavior*, 1974b, *5*, 389–396.

Gray, G. D., Davis, H. N., Kenney, A. M., and Dewsbury, D. A. Effect of mating on plasma levels of LH and progesterone in montane voles *(Microtus montanus)*. *Journal of Reproduction and Fertility*, 1976, *47*, 89–91.

Gray, G. D., Sodersten, P., Tallentine, D., and Davidson, J. M. Effects of lesions in various structures of the SCN-PO region on LH regulation and sexual behavior in female rats. *Neuroendocrinology*, 1978, *25*, 174–191.

Gray, G. D., Tennent, B., Smith, E. R., and Davidson, J. M. Luteinizing hormone regulation and sexual behavior in middle-aged female rats. *Endocrinology*, 1980, *107*, 187–194.

Greenwald, G. S. The reproductive cycle of the field mouse *(Microtus californicus)*. *Journal of Mammology*, 1956, *37*, 213–222.

Greep, R. O., and Hisaw, R. L. Pseudopregnancies from electrical stimulation of the cervix in the diestrum. *Proceedings of the Society for Experimental Biology and Medicine*, 1938, *39*, 359–360.

Hagen, K. W. Colony husbandry. In S. H. Weisbroth, R. E. Flatt, and A. L. Kraus (Eds.), *The Biology of the Laboratory Rabbit*. New York: Academic Press, 1974.

Halasz, B., and Gorski, R. Gonadotrophic hormone secretion in female rats after partial or total interruption of neural afferents to the medial basal hypothalamus. *Endocrinology*, 1967, *80*, 608–622.

Halasz, B., and Pupp, L. Hormone secretion of the anterior pituitary gland after physical interruption of all nervous pathways to the hypophysiotrophic area. *Endocrinology*, 1965, *77*, 553–561.

Halasz, B., Pupp, L., and Uhlarik, S. Hypophysiotrophic area in the hypothalamus. *Journal of Endocrinology*, 1962, *25*, 147–153.

Haller, E. W., and Barraclough, C. A. Alterations in unit activity of hypothalamic ventromedial nuclei by stimuli which affect gonadotrophic hormone secretion. *Experimental Neurology*, 1970, *29*, 111–120.

Hammond, J. *Reproduction in the Rabbit*. Edinburgh: Oliver and Boyd, 1925.

Hammond, J., and Asdell, S. A. The vitality of the spermatoza in the male and female reproductive tracts. *British Journal of Experimental Biology*, 1926, *4*, 155–185.

Hansen, S., Stanfield, E. J., and Everitt, B. J. The role of the ventral bundle noradrenergic neurons in sensory components of sexual behavior and coitus-induced pseudopregnancy. *Nature*, 1980, *286*, 151–154.

Harrington, F. E., Eggert, R. G., Wilbur, R. D., and Linkenheimer, W. H. Effect of coitus on chloropromazine inhibition of ovulation in the rat. *Endocrinology*, 1966, *79*, 1130–1134.

Harrington, F. E., Eggert, R. G., and Wilbur, R. D. Induction of ovulation in chloropromazine-blocked rats. *Endocrinology*, 1967, *81*, 877–881.

Harris, G. W. The induction of pseudopregnancy in the rat by electrical stimulation through the head. *Journal of Physiology* (London), 1936, *88*, 361–367.

Harris, G. W. The induction of ovulation in the rabbit by electrical stimulation of the hypothalamus-hypophysial mechanism. *Proceedings of the Royal Society of London*, 1937, *B122*, 374–394.

Harris, G. W. Electrical stimulation of the hypothalamus and the mechanism of neural control of the adenohypophysis. *Journal of Physiology*, 1948, *107*, 418–429.

Hasler, M. J., and Conaway, C. H. The effects of males on the reproductive state of female *Microtus ochrogaster*. *Biology of Reproduction*, 1973, *9*, 426–436.

Hasler, M., and Nalbandov, A. The effect of weanling and adult males on sexual maturation in female voles *(Microtus ochrogaster)*. *General and Comparative Endocrinology,* 1974, *23,* 237–238.

Haterius, H. O. Partial sympathectomy and induction of pseudopregnancy. *American Journal of Physiology,* 1933, *103,* 97–103.

Haterius, H. O. The genitopituitary pathway. Non effect of stimulation of superior cervical sympathetic ganglia. *Proceedings of the Society for Experimental Biology and Medicine,* 1934, *31,* 1112–1113.

Haterius, H. O., and Debyshire, A. J. Ovulation in the rabbit following upon stimulation of the hypothalamus. *American Journal of Physiology,* 1937, *119,* 329–330.

Hayward, J. N., Hilliard, J., and Sawyer, C. H. Time of release of pituitary gonadotropin induced by electrical stimulation of the rabbit brain. *Endocrinology,* 1964, *74,* 108–113.

Heape, W. Ovulation and degeneration of ova in the rabbit. *Proceedings of the Royal Society London* (Series B), 1905, *76,* 260–268.

Herlyn, U., Geller, H. F., v. Berswordt-Wallrabe, I., and v. Berswordt-Wallrabe, R. Pituitary lactogenic hormone release during onset of pseudopregnancy in intact rats. *Acta Endocrinologica,* 1965, *48,* 220–224.

Hilliard, J., Hayward, J. N., and Sawyer, C. H. Postcoital patterns of secretion of pituitary gonadotropin and ovarian progestin in the rabbit. *Endocrinology,* 1964, *75,* 957–963.

Hinsey, J. C., and Markee, J. E. Pregnancy following bilateral section of the cervical sympathetic trunk in the rabbit. *Proceedings of the Society for Expermental Biology and Medicine,* 1933, *31,* 270–271.

Hoppe, P., and Whitten, W. Pregnancy block: imitation by administered gonadotropin. *Biology and Reproduction,* 1972, *7,* 254–259.

Hornby, J. B., and Rose, J. D. Responses of caudal brain stem neurons to vaginal and somatosensory stimulation in the rat and evidence of genital-nociceptive interactions. *Experimental Neurology,* 1976, *51,* 363–376.

Horton, L. W., and Shepherd, B. A. Effects of olfactory bulb ablation on estrus-induction and frequency of pregnancy. *Physiology and Behavior,* 1979, *22,* 847–850.

Ingle, D. J. Permissibility of hormone action: A review. *Acta Endocrinologica,* 1954, *17,* 172–186.

Jöchle, W. Current research in coitus-induced ovulation: A review. *Journal of Reproduction and Fertility,* 1975, *Suppl. 22,* 165–207.

Johns, M. A., Feder, H. H., Komisaruk, B. R., and Mayer, A. D. Urine-Induced reflex ovulation in an ovulatory rats may be a vomeronasal effect. *Nature,* 1978, *272,* 446–448.

Johns, M. A., Feder, H. H., and Komisaruk, B. R. Reflex ovulation in light-induced persistent estrus rats: Role of sensory stimuli and the adrenals. *Hormones and Behavior,* 1980, *14,* 7–19.

Kalra, P. S., and Sawyer, C. H. Blockade of copulation-induced ovulation in the rat by anterior hypothalamic deafferentation. *Endocrinology,* 1970, *87,* 1124–1128.

Kalra, S. P., Ajika, K., Krulich, L., Fawcett, C. P., Quijada, M., and McCann, S. M. Effects of hypothalamic and preoptic electrochemical stimulation on gonadotropin and prolactin release in proestrous rats. *Endocrinology,* 1971, *88,* 1150–1158.

Kalra, S. P., Krulich, L., and McCann, S. M. Changes in gonadotropin-releasing factor content in the rat hypothalamus following electrochemical stimulation of anterior hypothalamic area and during the estrous cycle. *Neuroendocrinology,* 1973, *12,* 321–333.

Kameko, W., Debski, E., Wilson, M., and Whitten, W. Puberty acceleration in mice. 2. Evidence that the vomeronasal organ is a receptor for the primer pheromone in male mouse urine. *Biology and Reproduction,* 1980, *22,* 873–878.

Kanematsu, S., Scaramuzzi, R. J., Hilliard, J., and Sawyer, C. H. Patterns of ovulation inducing LH release following coitus, electrical stimulation and exogeneous LH-RH in the rabbit. *Endocrinology,* 1974, *95,* 247–252.

Kato, H., Velasco, M. E., and Rothchild, I. Effects of medial hypothalamic deafferentation on prolactin secretion in pseudopregnant rats. *Acta Endocrinology,* 1978, *89,* 425–431.

Kawakami, M., and Kimura, F. Inhibition of ovulation in the rat by electrical stimulation of the lateral amygdala. *Endocrinology,* 1975, *22,* 61–65.

Kawakami, M., and Kimura, F. Limbic-preoptic responses to estrogens and catecholamines in relation to cyclic LH secretion. In F. Naftolin, K. J. Ryan, and J. Davies (Eds.), *Subcellular Mechanisms in Reproductive Neuroendocrinology.* Amsterdam: Elsevier, 1976.

Kawakami, M., and Kubo, K. Neuro-correlate of limbic-hypothalmo-pituitary-gonadal axis in the rat: change in limbic-hypothalmic unit activity induced by vaginal and electrical stimulation. *Neuroendocrinology,* 1971, *7,* 65–89.

Kawakami, M., Seto, K., and Yoshida, K. Influence of the limbic system on ovulation and on progesterone and oestrogen formation in rabbits ovary. *Japanese Journal of Physiology*, 1966, *16*, 254–273.

Kawakami, M., Seto, K., Terasawa, E., and Yoshida, K. Mechanisms in the limbic system controlling reproductive functions of the ovary with special reference to the positive feedback of progestin to the hippocampus. In W. R. Adey and T. Tokizane (Eds.), *Progress in Brain Research*, Vol. 27. Amsterdam: Elsevier, 1967.

Kawakami, K., Terasawa, E., and Ibuki, T. Changes in multiple unit activity of the brain during the estrous cycle. *Neuroendocrinology*, 1970, *6*, 30–48.

Kawakami, M., Terasawa, E., Seto, K., and Wakabayashi, K. Effect of electrical stimulation of the medial preoptic area on hypothalamic multiple unit activity in relation to LH release. *Endocrinologica Japonica*, 1971, *18*, 13–21.

Kawakami, M., Terasawa, E., Kimura, F., and Wakabayashi, K. Modulation effect of limbic structures on gonadotropin release. *Neuroendocrinology*, 1973, *12*, 1–16.

Kenney, A., Evans, R., and Dewsbury, D. Postimplantation pregnancy disruption in *Microtus ochrogaster, M. pennsylvanicus* and *Peromyscus maniculatus. Journal of Reproduction and Fertility*, 1977, *49*, 365–367.

Kenney, A. M., Hartung, T. G., Davis, H. W., Gray, G. D., Zerylnick, M., and Dewsbury, D. A. Male copulatory behavior and the induction of ovulation in female voles: A quest for species specificity. *Hormones and Behavior*, 1978, *11*, 123–130.

Kenney, A. M., Hartung, T. G., and Dewsbury, D. A. Copulatory behavior and the initiation of pregnancy in California voles *(M. californicus). Brain Behavior and Evolution*, 1979, *16*, 176–191.

Kimura, F., and Kawakami, M. Reanalysis of the preoptic afferents and efferents involved in the surge of LH, FSH and prolactin release in the proestrous rat. *Neuroendocrinology*, 1978, *27*, 74–85.

Kollar, E. J. Reproduction in the female rat after pelvic nerve neurectomy. *Anatomical Record*, 1953, *115*, 641–658.

Komisaruk, B. R., and Diakow, C. Lordosis reflex intensity in rats in relation to the estrous cycle, ovariectomy, estrogen administration and mating behavior. *Endocrinology*, 1973, *93*, 548–557.

Komisaruk, B. R., Adler, N. T., and Hutchinson, J. B. Genital sensory field: Enlargement by estrogen treatment in female rats. *Science*, 1972, *178*, 1295–1298.

Koves, K., and Halasz, B. Location of the neural structures triggering ovulation in the rat. *Neuroendocrinology*, 1970, *6*, 180–193.

Kow, L. M., and Pfaff, D. W. Sensory requirements for the lordosis reflex in female rats. *Brain Research*, 1976, *101*, 47–66.

Kow, L. M., Montgomery, M. O., and Pfaff, D. W. Effects of spinal cord transactions on lordosis reflex in female rats. *Brain Research*, 1977, *123*, 75–88.

Kuehn, R. E., and Beach, F. A. Quantitative measurement of sexual receptivity in female rats. *Behavior*, 1963, *21*, 282–299.

Land, R. B., and McGill, T. E. The effect of the mating pattern of the mouse on the formation of corpora lutea. *Journal of Reproduction and Fertility*, 1967, *13*, 121–125.

Lincoln, D. W. Response of hypothalamic units to stimulation of the vaginal cervix: Specific vs. nonspecific effects. *Journal of Endocrinology*, 1969, *43*, 683–684.

Linkie, D. M., and Niswender, G. D. Serum levels of prolactin, luteinizing hormone and follicle stimulating hormone during pregnancy in the rat. *Endocrinology*, 1972, *90*, 632–637.

Lombardi, J., and Vandenbergh, J. Pheromonally-induced sexual maturation in females: Regulation by the social environment of the male. *Science*, 1977, *196*, 545–546.

Lombardi, J., Vandenbergh, J., and Whitsett, M. Androgen control of the sexual maturation pheromone in house mouse urine. *Biology and Reproduction*, 1976, *15*, 179–186.

Long, J. A., and Evans, H. Mc. The oestrous cycle in the rat and its associated phenomena. *Memoirs of the University of California*, 1922, *6*, 1–148.

Madhwa Raj, H. G., and Moudgal, N. R. Hormonal control of gestation in the intact rate. *Endocrinology*, 1970, *86*, 874–889.

Malsbury, C. W., Kelley, D. B., and Pfaff, D. W. Responses of single units in the dorsal midbrain to somatosensory stimulation in female rats. *Proceedings of the 4th International Congress of Endocrinology*, 1972, 205–209.

Margherita, G., Albritton, D. MacInnes, R. Hayward, J., and Gorski, R. A. Electroencephalographic changes in ventromedial hypothalmus and amygdala induced by vaginal and other periperal stimuli. *Experimental Neurology*, 1965, *13*, 96–108.

Markee, J. E., Sawyer, C. H., and Hollinshead, W. H. Activation of the anterior hypophysis by electrical stimulation in the rabbit. *Endocrinology,* 1946, *38,* 345–357.

Marshall, F. H. A., and Verney, E. B. The occurrence of ovulation and pseudopregnancy in the rabbit as a result of CNS stimulation. *Journal of Physiology,* 1936, *86,* 327–336.

McClintock, M. K., and Adler, N. T. The role of the female during copulation in the wild and domestic Norway rat. *Behaviour,* 1978, *67,* 67–96.

McGill, T. E. Induction of luteal activity in female house mice. *Hormones and Behavior,* 1970, *1,* 211–222.

McGill, T. E., and Coughlin, R. C. Ejaculatory reflex and luteal activity induction in mus musculus. *Journal of Reproduction and Fertility,* 1970, *21,* 215–220.

McGill, T. E., Dewsbury, D., and Sachs, B. (Eds.). *Sex and Behavior: Status and Prospectus.* New York: Plenum Press, 1978.

McIntosh, T., and Drickamer, L. Excreted urine, bladder urine and the delay of sexual maturation in female house mice. *Animal Behaviour,* 1977, *25,* 999–1004.

McIntosh, T. K., Barfield, R. J., and Geyer, L. A. Ultrasonic vocalizations facilitate sexual behavior of female rats. *Nature,* 1978, *272,* 163–164.

McLean, B. K., and Nikitovitch-Winer, M. B. Corpus luteum function in the rat: A critical period for luteal activation and the control of luteal maintenance. *Endocrinology,* 1973, *93,* 316–322.

Meyer, R. K., Leonard, S. L., and Hisaw, F. L. Effect of anesthesia on artificial production of pseudopregnancy in the rat. *Proceedings of the Society for Experimental Biology and Medicine,* 1929, *27,* 340–342.

Milligan, S. R. Social environment and ovulation in the vole, *Microtus agrestis. Journal of Reproduction and Fertility,* 1974, *41,* 35–47.

Milligan, S. R. The copulatory behavior of *Microtus agrestis. Journal of Mammalogy,* 1975a, *56,* 220–224.

Milligan, S. R. Mating, ovulation and corpus luteum function in the vole, *Microtus agrestis. Journal of Reproduction and Fertility,* 1975b, *42,* 35–44.

Milligan, S. R. Further observations on the influence of the social environment on ovulation in the vole, *Microtus agrestis. Journal of Reproduction and Fertility,* 1975c, *44,* 543–544.

Milligan, S. R., and MacKinnon, P. C. B. Correlation of plasma LH and prolactin levels with the fate of the corpus luteum in the vole, *Microtus agrestis. Journal of Reproduction and Fertility,* 1976, *47,* 111–113.

Morishige, W. K., and Rothchild, I. A paradoxical inhibiting effect of ether on prolactin release in the rat: Comparison with effect of ether on LH and FSH. *Neuroendocrinology,* 1974, *16,* 93–107.

Moss, R. L. Relationships between the central regulation of gonadotropins and mating behavior in female rats. In W. Montagna and W. Sadler (Eds.), *Reproductive Behavior.* New York: Plenum Press, 1974.

Moss, R. L., and Cooper, K. J. Temporal relationship of spontaneous and coitus-induced release of luteinizing hormone in the normal cyclic rat. *Endocrinology,* 1973, *92,* 1748–1753.

Moss, R. L., Cooper, K. J., and Danhof, I. E. Coitus-induced release of LH in normal cyclic female rats. *Federal Proceedings,* 1973, *32,* 239.

Moss, R. L., Dudley, C. A., and Schwartz, N. B. Coitus-induced release of luteinizing hormone in the proestrus rat: Fantasy or fact? *Endocrinology,* 1977, *100,* 394–397.

Pasley, J., and MacKinney, T. Grouping and ovulation in *Microtus pennsylvanicus. Journal of Reproduction and Fertility,* 1973, *34,* 527–530.

Pfaff, D., Lewis, C., Diakow, C., and Keiner, M. Neurophysiological analyses of mating behavior responses as hormone sensitive reflexes. In E. Stellar and J. Sprague (Eds.), *Progress in Physiological Psychology,* Vol. 5, New York: Academic Press, 1973.

Pfaff, D., Montgomery, M., and Lewis, C. Somatosensory determinants of lordosis in female rats: Behavioral definition of the estrogen effect. *Journal of Comparative and Physiological Psychology,* 1977, *91,* 134–145.

Phelps, C. P., and Sawyer, C. H. Electrochemically stimulated release of luteinizing hormone and ovulation after surgical interruption of lateral hypothalamic connections in the rat. *Brain Research,* 1977, *131,* 335–344.

Phelps, C. P., Krieg, R. J., and Sawyer, C. H. Spontaneous and electrochemically stimulated changes in plasma LH in the female rat following hypothalamic deafferentation. *Brain Research,* 1976, *101,* 239–249.

Pieper, D. R., and Gala, R. R. The effect of light on the prolactin surges of pseudopregnant and ovariectomized estrogenized rats. *Biology of Reproduction,* 1979, *20,* 727–732.

Pierce, J. T., and Nutall, R. L. Self-paced sexual behavior in the female rat. *Journal of Comparative Physiological Psychology*, 1961, *54*, 310–313.

Quinn, D. L. Neural activation of gonadotrophic hormone release by electrical stimulation in the hypothalamus of the guinea pig and the rat. *Neuroendocrinology*, 1969, *4*, 254–263.

Quinn, D. L., and Everett, J. W. Delayed pseudopregnancy induced by selective hypothalamic stimulation. *Endocrinology*, 1967, *80*, 155–162.

Rabii, J., Koranyi, L., Carrillo, A. J., and Sawyer, C. H. Inhibition of brain stem neuronal activity by electrochemical stimulation of the dorsal hippocampus in the freely moving proestrous rat. *Experimental Neurology*, 1977, *56*, 421–430.

Ramirez, V. D., Komisaruk, B. R., Whitmoyer, D. I., and Sawyer, C. H. Effects of hormones and vaginal stimulation on the EEG and hypothalamic units in rats. *American Journal of Physiology*, 1967, *212*, 1376–1384.

Reiner, P., Woolsey, J., Adler, N., and Morrison, A. Appendix: a gross anatomical study of the peripheral nerves associated with reproductive function in the female albino rat. In N. T. Adler (Ed.), *Neuroendocrinology of Reproduction*. New York: Plenum Press, 1981.

Richmond, M., and Conaway, C. H. Induced ovulation and oestrus in *Microtus ochrogaster*. *Journal of Reproduction and Fertility*, 1969, *6*, 357–376.

Richmond, M., and Stehn, R. Olfaction and reproductive behavior in microtine rodents: Mammalian olfaction. In R. L. Doty (Ed.), *Reproductive Processes and Behavior*. New York: Academic Press, 1976.

Rodgers, C. H. Influence of copulation on ovulation in the cycling rat. *Endocrinology*, 1971, *88*, 433–436.

Rodgers, C. H., and Schwartz, N. B. Diencephalic regulation of plasma LH, ovulation, and sexual behavior in the rat. *Endocrinology*, 1972, *90*, 461–465.

Rodgers, H., and Schwartz, N. B. Serum LH and FSH levels in mated and unmated proestrous female rats. *Endocrinology*, 1973, *92*, 1475.

Sadleir, R. M. *The Ecology of Reproduction in Wild and Domestic Animals*. London: Methuen, 1969.

Sanford, J. C. *The Domestic Rabbit*. New York: Wiley, 1979.

Sawyer, C. H. Reflex induction of ovulation in the estrogen-treated rabbit by artificial vaginal stimulation. *Anatomical Record*, 1949, *103*, 502.

Sawyer, C. H. Effects of central nervous system lesions on ovulation in the rabbit. *Anatomical Record*, 1956, *124*, 358.

Sawyer, C. H. Effects of brain lesions on estrous behavior and reflexogenous ovulation in the rabbit. *Journal of Experimental Zoology*, 1959, *142*, 227–246.

Sawyer, C. H., and Markee, J. E. Estrogen facilitation of release of pituitary ovulating hormone in the rabbit in response to vaginal stimulation. *Endocrinology*, 1959, *65*, 614–621.

Sawyer, C. H., and Robison, B. A. Separate hypothalamic areas controlling pituitary gonadotrophic function and mating behavior in female cats and rabbits. *Journal of Clinical Endocrinology and Metabolism*, 1956, *16*, 914–915.

Sawyer, C. H., Markee, J. E., and Hollinshead, W. H. Inhibition of ovulation in the rabbit by the adrenergic-blocking agent dibenamine. *Endocrinology*, 1947, *41*, 395–402.

Sawyer, C. H., Markee, J. E., and Townsend, B. F. Cholinergic and adrenergic components in the neurohumoral control of the release of LH in the rabbit. *Endocrinology*, 1949a, *44*, 18–37.

Sawyer, C. H., Everett, J. W., and Markee, J. E. A neural factor in the mechanism by which estrogen induces the release of luteinizing hormone in the rat. *Endocrinology*, 1949b, *44*, 218–233.

Sawyer, C. H., Markee, J. E., and Everett, J. W. Further experiments on blocking pituitary activation in the rabbit and the rat. *Journal of Experimental Zoology*, 1950, *113*, 659–682.

Sawyer, C. H., Haun, C. K., Hilliard, J., Radford, H. M., and Kanematsu, S. Further evidence for the identity of hypothalamic areas controlling ovulation and lactation in the rabbit. *Endocrinology*, 1963, *73*, 338–344.

Scalia, F., and Winans, S. The differential projections of the olfactory bulbs and accessory olfactory bulbs in mammals. *Journal of Comparative Neurology*, 1975, *161*, 31–55.

Scaramuzzi, R. J., Blake, C. A., Papkoff, H., Hilliard, J., and Sawyer, C. H. Radioimmunoassay of rabbit luteinizing hormone: Serum levels during various reproductive states. *Endocrinology*, 1972, *90*, 1285–1291.

Shelesnyak, M. C. The induction of pseudopregnancy in the rat by means of electrical stimulation. *Anatomical Research*, 1931, *49*, 179–183.

Simon, H. A. The organization of complex systems. In H. H. Pattee (Ed.), *Hierarchy Theory*. New York: Braziller, 1973.

Slonaker, J. R. Pseudopregnancy in the albino rat. *American Journal of Physiology*, 1929, *89*, 406–416.

Smith, M. S., and Neill, J. D. A critical period for cervically-stimulated prolactin release. *Endocrinology*, 1976a, *98*, 324–328.

Smith, M. S., and Neill, J. D. Termination at midpregnancy of the two daily surges of plasma prolactin initiated by mating in the rat. *Endocrinology*, 1976b, *98*, 696–701.

Smith, M. S., and Ramaley, J. A. Development of ability to initiate and maintain prolactin surges induced by uterine cervical stimulation in immature rats. *Endocrinology*, 1978, *102*, 351–357.

Smith, M. S., Freeman, M. E., and Neill, J. D. The control of progesterone secretion during the estrous cycle and early pseudopregnancy in the rat: Prolactin, gonadotropin, and steroid levels associated with rescue of the corpus luteum of pseudopregnancy. *Endocrinology*, 1975, *96*, 219–226.

Smith, M. S., McLean, B. K., and Neill, J. D. Prolactin: The initial luteotrophic stimulus of pseudo-pregnancy in the rat. *Endocrinology*, 1976, *98*, 1370–1377.

Sokoloff, L., Reivich, M., Kennedy, C., Des Rosiers, M. H., Patlak, C. S., Pettigrew, K. D., Sakurada, O., and Shinohara, M. The [14C] deoxyglucose method for the measurement of local cerebral glucose utilization: Theory, procedure, and normal values in the conscious and anesthetized albino rat. *Journal of Neurochemistry*, 1977, *28*, 897–916.

Spies, H. G., and Niswender, G. D. Levels of prolactin, LH and FSH in the serum of intact and pelvic-neurectomized rats. *Endocrinology*, 1971, *88*, 937–943.

Spies, H. G., Forbes, Y. M., and Clegg, M. T. The influence of coitus, suckling, and prolactin injections on pregnancy in pelvic neurectomized rats. *Proceedings of the Society for Experimental Biology and Medicine*, 1971, *138*, 470–474.

Stehn, R., and Richmond, M. Male-induced pregnancy termination in the prairie vole, *Microtus ochrogaster*. *Science*, 1975, *187*, 1211–1213.

Swanson, L. W. An autoradiographic study of the efferent connections of the preoptic region in the rat. *Journal of Comparative Neurology*, 1976, *167*, 227–256.

Takahashi, M., Ford, J., Yoshinaga, K., and Greep, R. O. Effects of cervical stimulation and anti-LH releasing hormone serum on LH releasing hormone content in the hypothalamus. *Endocrinology*, 1975, *96*, 453–457.

Taleisnik, S., and McCann, S. M. Effects of hypothalamic lesions on the secretion and storage of hypophysial luteinizing hormone. *Endocrinology*, 1961, *68*, 263-272.

Taleisnik, S., and Tomatis, M. Mechanisms that determine the changes in pituitary MSH activity during pseudopregnancy induced by vaginal stimulation in the rat. *Neuroendocrinology*, 1970, *6*, 368–377.

Taleisnik, S., Caligaris, L., and Astrada, J. J. Effect of copulation on the release of pituitary gonadotrophins in male and female rat. *Neuroendocrinology*, 1966, *79*, 49–54.

Taleisnik, S., Sherwood, M. R. C., and Raisman, G. Dissociation of spontaneous and mating induced ovulation by frontal hypothalamic deafferentations in the rat. *Brain Research*, 1979, *169*, 155–162.

Tejasen, T., and Everett, J. W. Surgical analysis of the preoptico-tuberal pathway controlling ovulatory release of gonadotropins in the rat. *Endocrinology*, 1967, *81*, 1387–1396.

Terasawa, E., and Sawyer, C. H. Changes in electrical activity in the rat hypothalamus related to electrochemical stimulation of adenohypophyseal function. *Edncorinology*, 1969a, *85*, 143–149.

Terasawa, E., and Sawyer, C. H. Electrical and electrochemical stimulation of the hypothalamo-adenohypophysial system with stainless steel electrodes. *Endocrinology*, 1969b, *84*, 918–925.

Terkel, J., and Sawyer, C. H. Male copulatory behavior triggers nightly prolactin surges resulting in successful pregnancy in rats. *Hormones and Behavior*, 1978, *11*, 304–309.

Turgeon, J., and Barraclough, C. A. Temporal patterns of LH release following graded preoptic electrochemical stimulation in proestrous rats. *Endocrinology*, 1973, *92*, 755–761.

Tyler, J. L., and Gorski, R. A. Temporal limits for copulation-induced ovulation in the pentobarbital-blocked proestrous rat. *Endocrinology*, 1980, *106*, 1815–1819.

Vandenbergh, J. Effect of the presence of a male on the sexual maturation of female mice. *Endocrinology*, 1967, *81*, 345–349.

Vandenbergh, J. Male odor accelerates female sexual maturation in mich. *Endocrinology*, 1969, *84*, 658–660.

Vandenbergh, J., Drickamer, L., and Colby, D. Social and dietary factors in the sexual maturation of female mice. *Journal of Reproduction and Fertility*, 1972, *28*, 397–405.

Van Riemsdijk, A., and Zeilmaker, G. H. Ovulation blockade and receptivity in the cylcic rat. *Acta Endocrinologica*, 1979, *90*, 711–717.

Velasco, M. E., and Taleisnik, S. Effect of hippocampal stimulation on the release of gonadotropin. *Endocrinology*, 1969a, *85*, 1154–1159.

Velasco, M. E., and Taleisnik, S. Release of gonadotropins induced by amygdaloid stimulation in the rat. *Endocrinology*, 1969b, *84*, 132–139.

Velasco, M. E., and Taleisnik, S. Effects of interruption of amygdaloid and hippocampal afferents to the medial hypothalamus on gonadotropin release. *Journal of Endocrinology*, 1971, *51*, 41–55.

Velasco, M. E., Castro-Vazquez, A., and Rothchild, I. Effects of hypothalamic deafferentation on criteria of prolactin secretion during pregnancy and lactation in the rat. *Journal of Reproduction and Fertility*, 1974, *41*, 385–395.

Vincent, J. D., Dufy, B., and Faure, J. M. A. Effects of vaginal stimulation on hypothalamic single units in unanesthetized rabbits. *Experentia*, 1970, *26*, 1266–1267.

Vogt, M. Ovulation in the rabbit after destruction of the greater superficial petrosal nerves. *Journal of Physiology*, 1942, *100*, 410–416.

Voloschin, L. M., and Gallardo, E. A. Effect of surgical disconnection of the medial basal hypothalamus on postcoital reflex ovulation in the rabbit. *Endocrinology*, 1976, *99*, 959–962.

Whitten, W. Modification of the estrus cycle of the mouse by external stimuli associated with the male. *Journal of Endocrinology*, 1956, *13*, 399–404.

Whitten, W. Effect of exteroceptive factors on estrus cycle of mice. *Nature*, 1957, *180*, 1436.

Whitten, W. Modification of the estrus cycle of the mouse by external stimuli associated with the male: Changes in estrus cycle determined by vaginal smears. *Journal of Endocrinology*, 1958, *17*, 307–313.

Whitten, W. Occurrence of anestrus in mice caged in groups. *Journal of Endocrinology*, 1959, *18*, 102–107.

Wiegand, S. J., Terasawa, E., and Bridson, W. E. Persistent estrus and blockade of progesterone-induced LH release following lesions which do not damage the suprachiasmatic nucleus. *Endocrinology*, 1978, *102*, 1645–1648.

Wiegand, S. J., Terasawa, E., Bridson, W., and Goy, R. Effects of discrete lesions of preoptic and suprachiasmatic structures in the female rat. *Neuroendocrinology*, 1980, *31*, 147–157.

Wilson, J. R., Adler, N., and Leboeuf, B. The effects of intromission frequency on successful pregnancy in the female rat. *Proceedings of the National Academy of Science*, 1965, *53*, 1392–1395.

Winans, S., and Scalia, F. Amygdaloid nucleus: New afferent input from the vomeronasal organ. *Science*, 1970, *170*, 330–332.

Wuttke, W. Preoptic unit activity and gonadotropin release. *Experimental Brain Research*, 1974, *19*, 205–216.

Wuttke, W., and Meites, J. Induction of pseudopregnancy in the rat with no rise in serum prolactin. *Endocrinology*, 1972, *90*, 438–443.

Yogev, L., and Terkel, J. Effects of photoperiod, absence of photicues, and suprachiasmatic nucleus lesions on nocturnal prolactin surges in pregnant and pseudopregnant rats. *Neuroendocrinology*, 1980, *31*, 26–33.

Young, W. C. The hormones and mating behavior. In W. C. Young (Ed.), *Sex and Internal Secretions*, Vol. 2. Baltimore: Williams and Wilkins, 1961.

Zarrow, M. X., and Clark, J. H. Ovulation following vaginal stimulation in a spontaneous ovulator and its implications. *Journal of Endocrinology*, 1968, *40*, 343–352.

16

On Measuring Behavioral Sex Differences in Social Contexts

DAVID A. GOLDFOOT AND DEBORAH A. NEFF

INTRODUCTION

This chapter examines specific methodological limitations associated with the study of behavioral sex differences in social contexts, identifies potential social or environmental sources of influence that might affect the development of sex differences, and gives suggestions for evaluating biological and/or social mechanisms involved in the production of sex differences in social groups. In addition, the initial section of the chapter addresses criticism advanced against current studies of sex differences. A rather extensive and growing set of papers has raised serious objections not only to aspects of methodology, but also to the more basic issues of sexual bias, to improper generalization of results, and essentially to much of the conduct and interpretation of the research in which sex differences are measured or explored.

It is beyond the scope of this chapter to evaluate or respond in detail to the criticism, but because some of the objections speak directly to issues considered in this report, two complaints relevant to the major theme of this chapter are examined.

DAVID A. GOLDFOOT AND DEBORAH A. NEFF   Wisconsin Regional Primate Research Center, 1223 Capitol Court, Madison, Wisconsin 53706.   This chapter, publication 24-018 of the Wisconsin Regional Primate Research Center, was supported in part by grants from NIH (RR-00167) and NIMH (MH-21312).

DAVID A.
GOLDFOOT AND
DEBORAH A. NEFF

Many researchers have not been aware that studies of sex differences, including those dealing with animals, have been the source of much heated criticism. Consider the following summary assessment of the entire field of sex difference research:

> Much of the research on sex differences is replete with methodological and sexist errors. Conclusions often attribute differences to sex when no effort has been made to control other relevant factors. Notions that fit popular bias are accepted without question and often without statistical test. Results pass through an interpretive process that ends with platitudes that could have been pronounced before the research began . . . these scientific inaccuracies are not surprising since the study of sex differences is an intellectual garbage heap with many a banana peel for the unwary to slip on. (Sherman, 1976, p. 119)

It is possible to identify two related themes that characterize much of the objection to studies of sex differences. The first theme deals with the issue of dangers of generalization: that is, because each species has a unique evolutionary history, behavioral findings from studies of one species should not be generalized to any other species, and certainly not to human beings (Kass-Simon, 1976). This argument encompasses the notion that each species has had its own set of survival pressures, which have resulted in its particular morphology and behavior. Beach (1971) invented the mythical "ramstergig" not only to make a similar point, but to balance the problem with directives suggesting appropriate and necessary uses of comparative data. The second theme, spelled out by Bleier (1976, 1979), Sherman (1976), Hubbard and Lowe (1979), and several other writers (e.g., see Tobach and Rosoff, 1978) concerns the lack of emphasis on the potential social mechanisms that might contribute to sex differences. These authors have stated that behavior should never be considered in abstraction from its social context. They contend that animal behavior studies often give the impression that a sex difference is due to "nature" and not "nurture," and that "aggressiveness," "leadership," and other valued "strong" traits are represented as masculine and are said to be fixed exclusively by genes and biology. Because virtually all behavioral scientists abhor the old saw of the nature–nurture controversy and subscribe instead to the epigenetic viewpoint that behavior represents the joint actions of biology interacting with environment, this rather forceful admonishment comes as a surprise and might be too easily dismissed as being off the mark. What we believe is being addressed, however, goes beyond this issue and is a very important complaint, namely, that species with far more limited social inputs than humans are being studied as models for human behavioral mechanisms. When endocrine or other physiological substrates are found to be major behavioral control devices for these species, we (scientists) are too hasty in assuming or implying the prepotence of the same mechanisms in people. We are not convinced that the scientific literature frequently makes the bridges implied here, but there are some notable examples (see later sections), and it is certainly appropriate to consider these criticisms in considerably more detail.

Kass-Simon (1976) expressed the problem of inappropriate generalization as follows:

> to put it in terms of women and their social and biological roles: is it possible to argue from examples in the animal kingdom about what is and what is not appropriate to the nature of something called the "female animal"? (p. 76)

Like other students of ethology, Kass-Simon argued that it is only from the broad comparative approach that sex differences begin to make sense. Patterns of dimorphic behavior for the two sexes must be understood in terms of the particular problems faced by the species under consideration and in terms of the adaptive significance for each sex of the behavioral difference.

A related issue spelled out by Leibowitz (1979), and more recently by Weisstein (1982), deals with interpretations of the behavioral sex differences found in primate species. It is the case that primate studies are often framed in terms of monkey and human genetic similarity, so that findings of sex differences in given attributes may carry more suggested implications for our own species than do findings from rodents or carnivores. Usually, monkey–human comparisons are not explicitly expressed in articles, but most of us are aware that such bridges are tacitly drawn by a significant percentage of those who read the research reports, and by some of us who write them. However, with more than 80 genera of primates to consider, and with species ranging from the mouse lemur to the gorilla, there are no easily formed generalizations regarding dimorphic behavior among the nonhuman primates, let alone generalizations that can be extended from simians to humans (Goldfoot, 1977; Mitchell, 1979). We *should* take greater care in making sure that our findings are not prematurely generalized, and that knowledgeable use is made of the limits and potential of the comparative method. To judge "female" behavioral tendencies or "male" behavioral tendencies from the results of studies of one or two conveniently available species makes little biological sense in view of comparative data, which demonstrate an impressive between-species variance.

On the other hand, the assertion that comparative animal research, and in particular, primate research, has no bearing whatsoever on elucidating behavioral principles applicable to our own species seems equally unwarranted. In spite of considerable genetic and behavioral variance among primates, many monkey and ape species share with humans not only physiological, endocrine, and anatomical similarity, but also several general behavioral characteristics, which include wide-ranging sensitivities to social conditions (Goldfoot *et al.,* 1984; Epple *et al.,* 1982; Keverne, 1982; Bernstein *et al.,* 1979; Dixson and Herbert, 1977; Goldfoot, 1977; Missakian, 1972; de Waal, 1982; Stephenson, 1973; Carpenter, 1942). Such data are invaluable in generating hypotheses or model systems concerning general principles of behavior and biology. Empirical tests of relationships found in the animal data can be conducted in order to evaluate the model and to guide further research on new species of interest, and on *Homo sapiens* as well. This is the power of com-

parative analyses, and if it is used appropriately, there is much that we can learn
about ourselves, male and female.

THE "ASSUMED BIOLOGICAL MEDIATION" PROBLEM

When a sex difference is described in an animal study, it is often assumed,
without appropriate assessment, to exist because of biological (neural?) differences. The assumption is, of course, premature. Moreover, even when the sex difference in the animal study is proved to be related to a biological mechanism, a
similar prepotent biological mechanism is then often assumed to account for an
analogous behavioral difference in humans. This error of analogy would be similar
to finding specific mechanisms for the actions of sexual releaser pheromones in
insects, for example, and then assuming that the mechanism governing sexually
arousing odors in primates or humans is the same. Odor could play a role in the
sexual activities of both insects and primates, but the way in which the olfactant
comes to have its effects may be dependent on totally different processes. As a
second example, the fact that wings in birds and insects serve analogous functions
(i.e., flight) is not sufficient for imputing a similarity in mechanism. In the flight
case, the two structures are not homologous, nor are the ways in which the structures are used for flight exactly similar. Bleier (1976) outlined her objections to
this form of reasoning as follows:

> There are, of course, measureable biological and measurable psychological and
> behavioral differences (viewed statistically) between men and women. The serious logical fallacy often made, however, is to assume that there is a necessary
> *causal* relationship between these differences; for example between androgen
> and "aggressiveness." For this assumption there is no convincing evidence whatsoever in humans, and to use animal data as evidence that particular psychological or behavioral sex differences in humans are caused by particular *biological*
> sex differences is to ignore a fact that in other contexts is recognized to be of
> supreme importance: that humans are *qualitatively* different from all other animals. They are different precisely because of their brain and their culture which
> is the unique product of that brain. (p. 68)

We disagree with the contention that humans are so qualitatively different
from all other animals as to render comparative information invalid, but we do
agree that there are inappropriate uses of analogy. There is also a failure to adequately consider nonbiological factors in the mediation of sex differences in animal
studies. This failure has left some with the impression that animals do not have
social experiences that contribute in major ways to behavioral sex differences. The
lack of a data base regarding social influences on sex differences in animals cripples
a broadly based comparative approach that *might* highlight similarities between
humans and animals. In the same way, there are few data regarding biological influences on sex differences in humans. Thus, such differences in humans are currently
assumed to have arisen from social and cultural influences, with a far more obscure
role being played by biology (Money and Ehrhardt, 1972). Money (1981) has discussed this situation extensively and has argued for more complete theoretical and
experimental syntheses of biological and social determinants.

It should be acknowledged that most animal research has not analyzed the influence of social context on the production of sex differences. In fact, it is quite difficult to find animal studies that identify the social or environmental variables that might differentially influence the behavior of males and females. Most investigators of animal behavior are aware of the current imbalance of information concerning nonbiological variables and their effects on target behavior, however, and would object to the use of findings from biologically oriented studies to make direct statements regarding the behavior of human beings. Once again, we seem to be grappling with the issue of finding an appropriate level of analysis for these biological findings from animal research.

BEHAVIORAL ENDOCRINOLOGY AND SEX DIFFERENCES RESEARCH

The discipline of behavioral endocrinology in many ways stands squarely in the middle of the issues discussed up to this point. The phenomenon of sexually dimorphic behavior is a major area of research for this field, data from animals rather than people comprise the majority of information generated within this field, and in the animal studies, biological rather than social mechanisms of sex differences have been more frequently pursued.

The keystone of classic behavioral endocrinology is the *organization–activation* hypothesis, in which it is held that target structures, including certain neural components, develop embryologically or neonatally, along different lines as a function of the presence or absence of certain steroids (see review by Goy and Goldfoot, 1973). This early "organizational" hormonal history induces structural and/or functional neural changes that are believed to be permanent, and that result in differential behavioral reactivity to later hormones circulating in adulthood (the activational phase of the hypothesis). Some behavioral sex differences (e.g., the rough-and-tumble play of rhesus monkey infants) are not dependent on the activation phase, whereas other types of dimorphic behavior need both early hormonal differences and concurrent hormonal activity in later life to be expressed (e.g., lordosis in most rodents).

General principles of endocrine involvement, during either larval, fetal, or neonatal life stages in the establishment of behavioral predispositions, have been demonstrated in virtually every mammal experimentally investigated, as well as in an increasing sample of birds, reptiles, and fish.

The data base is now so extensive that it is difficult to conceive that one could disagree with the generalization that hormones presented early in the development of an organism can influence the probability of certain behavioral characteristics later in life. Major questions of critical importance remain to be answered regarding this generalization, however, including the degree to which other, nonhormonal processes interact with this system. First, there is the question of the species and behavioral specificity of hormonal effects: we know that, depending on the species, the precise nature of the active hormone may differ, the exact behavioral patterns influenced by hormones may differ, and the degree to which the behavior predispositions can be modified by subsequent social experience may differ. Second, there is the question of the strength, the duration, and the permanence of the hormonal organizing influences. That is, are the early endocrine influences permanent, as initially suggested (Young, 1961); or can social context and/or social experience shift hormonally mediated predispositions significantly? Third, and of major importance to extensions of animal studies to human concerns, what is the relationship between behavior influenced by early hormones and such concepts as gender role, gender identity, or sexual preference?

The major sources of controversy potentially faced by the behavioral endocrinologist are exactly those outlined earlier: generalization difficulties from one species to another and the assertion of biological mediation in the absence of appropriate manipulation of nonbiological factors. Recently published objections to some medical interventions reflect the painful fact that neither of these issues seems close to resolution (see the Diamond vs. Money controversy *re* the origins of gender identity in Diamond, 1982; see German Society for Sex Research vs. Dorner *re* the origins of homosexuality in Sigursch *et al.*, 1982).

Perhaps the *jargon* of behavioral endocrinology itself has partially obscured some of these issues, as the literature is admittedly filled with emotionally charged terms like *masculinize* and *defeminize*. These terms can—and apparently do—give the impression to some that what is being studied is a particular behavioral response of the "female animal" or the "male animal." Behavioral endocrinologists using the concept of *masculinize,* for example, mean something quite different from "to cause stereotyped male human attributes." They mean, in the most restricted sense, that procedures of the experiment caused the behavior displayed by a genetic female to resemble the behavior displayed by the conspecific male. If the male of a species shows less aggressive behavior than the female of the species, then a hormonal administration to females during early development that reduces subsequent aggression would be an example of masculinization. Behavioral endocrinologists, often concentrating their efforts on biological mechanisms, do need to take greater care to guard against the possibility that their work can be misinterpreted. Readers of the work, in turn, should take care not to anthropomorphize, even when some of the terms used have rather anthropomorphic associations.

It is also possible that the goal of many behavioral endocrinology studies is sometimes misunderstood. Many hormone–behavior experiments are better construed as *behavioral bioassays* of endocrine function rather than as studies in which the primary purpose is to explore behavioral sex differences. Often, the strategy has been to find a dimorphic response that can be shown to be modifiable by endocrine manipulations, so that additional target organs influenced by the endocrine system (e.g., the brain) can be studied, or so that more information concerning the endocrine system can be elucidated. It is rarely the intent of the investigator to assert that the endocrine functions identified in this manner are the *sole* determinants of the dimorphic response under study. In many cases, behavioral responses or experimental strategies chosen for this approach are purposely artificial (e.g., eliciting lordosis by "fingering").

Most investigators of animal behavior are aware of the current imbalance of information concerning nonbiological variables and their effects on target behavior and would object to the use of findings from biologically oriented studies to make direct statements regarding the behavior of human beings. However, greater clarity could be obtained by carefully discriminating between "behavioral bioassay" and "behavioral description." In addition, there is a need for studies that investigate the multiply determined nature of behavioral sex differences, by expanding experimental designs to vary the biological and the social factors simultaneously.

Measuring Behavioral Sex Differences

Definition of Behavioral Sex Differences

The identification of responses that are displayed differentially (by statistical comparison) by males and females is a major component of many studies assessing the influences of hormonal, somatic, social, or environmental variables on behavior. An accepted procedure is to observe both sexes under identical conditions for a variety of behavioral measures, and to classify as sexually dimorphic those responses that differ significantly between the sexes in frequency, in duration, in intensity, or on other meaningful dimensions. The phrase "sexually dimorphic behavior" is therefore a statistical and relational concept that implies a differential, but not necessarily an exclusive, display by one sex as compared to the other (Goy and Goldfoot, 1973). The concept is based on grouped data and has the same limitations with regard to the prediction of an individual's behavior as any other concept derived from a comparison of group means.

To measure behavior, some type of physical and social environment must be supplied. There is no specific set of guidelines that can be offered for the most reasonable set of social variables to consider regarding studies of dimorphic behavior. Two points can be made, however. First, it seems clear that the identification of a behavioral dimorphism cannot be legitimately separated from the environment used to study it. Second, both earlier social experiences and the current testing environments must be regarded as possible mediating factors, just as both earlier and current hormonal states must be considered. Emphasizing the social factors, for example, one could ask the following types of questions: Is a hormonally mediated dimorphism seen under several different rearing (and/or testing) conditions, or is it restricted to special environmental or social circumstances? If several sex differences are found under one set of testing (rearing) conditions, will they all be present under a different testing (rearing) regimen, or is there evidence of a separate social mediation for different classes of behavior? Such questions raise the issue of how past and current social conditions channel, modify, or in other ways influence the expression of sexually dimorphic behavior.

Unfortunately, although the term *dimorphic behavior* is used throughout the literature, there are a number of different definitions (explicit and implicit) of this single term. These differences have undoubtedly led to confusion and need correction. The usage problem is one of not differentiating between measurements of the mean frequency of display between males and females and measurements of the proportion of males and females displaying the behavior. It is possible to find no difference with regard to the proportion of males and females displaying a given behavior, and yet to find a significant difference in the mean frequency of display of that behavior (e.g., presenting behavior in rhesus monkeys). Alternatively, there are behavioral classes in which both a significantly higher proportion of one sex than the other displays a response and frequency differences exist between the responders of each sex (e.g., the rough play behavior of rhesus monkeys). Finally, there may be cases in which significantly different proportions of individuals display the response, but responders of both sexes have similar frequencies of display (e.g., substance abuse in humans). Each of these would be examples of sexually dimorphic behavior, even though the social and/or biological causes of such distinctly different statistical representations of the concept could be expected to have separate origins.

In many studies dealing with biological determinants of sex differences, subjects of both sexes are experimentally manipulated so that their concurrent hormonal levels are equated prior to behavioral assessments. For example, both adult male and adult female guinea pigs may be gonadectomized and then given the same exogenous treatment of hormones prior to a behavioral evaluation. Such studies ask whether a sex difference is present even when the circulating levels of hormones are roughly equivalent. Although this seems to be a logical starting point for many studies, it is, in fact, one step beyond the initial determination of the existence of a sex difference under "intact" conditions. Equating for concurrent hormonal conditions is one appropriate way of beginning to explore the mechanisms responsible for the sex difference, but it automatically implies that the difference is due to endocrinology.

DIMORPHIC BEHAVIOR ASSESSED INDIVIDUALLY

The least complicated way to investigate dimorphic behavior is to study individuals of each sex, one at a time, in nonsocial environments, such as in activity wheels, open-field arenas, or other standardized physical environments. In humans, standardized tests of vocabulary, mathematics, dexterity, pupil dilation, and so on are analogous forms of individual assessments in which each sex is presented with the identical set of physical stimuli. In both human and nonhuman studies, when subjects of both sexes are individually tested under identical concurrent situations, the measurements of differences are straightforward, at least from a purely statistical viewpoint. Uncovering the developmental mechanisms of sex differences is, of course, not at all a trivial matter because they may be related to a host of factors, including prenatal or concurrent hormonal differences, differential rearing histories, differing social expectations, and so on. However, at least the fact that the test reveals a statistically significant difference between the sexes can be agreed on as a starting point.

Although sex differences measured under nonsocial conditions are easy to quantify and have some experimental advantages, it is on social interactions that most of the interest in dimorphic behavior has been focused. Yet, whenever the behavior of an individual is observed within a social context, the accurate identification of sex differences becomes considerably more difficult, and problems of procedure and analysis that are not encountered in individual testing must be recognized.

In dyadic interaction tests, two individuals are placed together, and their social interactions are recorded, usually by designating one of the animals as the "experimental" and the other as the "stimulus" subject. An immediate problem, however, is that the behavioral or nonbehavioral (e.g., sex, size, color, or odor) characteristics of one individual can influence the behavior of the other individual from the outset. Thus, a sequence of mutually influential behavioral events is initiated in which the stimulus characteristics for both subjects are constantly changing.

A partial realization that dyadic situations need "special handling" in sex difference studies occurred several years ago, and it was suggested that a reasonable way to proceed was to test each sex with exactly the same kind of partner to keep the stimulus conditions constant. For example, if an experimental female is to be tested with a male partner, then an experimental male must also be tested with a male partner (Goy and Goldfoot, 1973). Although controlling for nonbehavioral stimulus characteristics presented to experimental subjects, this strategy unfortunately cannot control for the fact that the male *partners* are confronted with quite different stimulus configurations (one male has a male and the other a female with which to interact). Accordingly, it must be expected that the partners will act differently from one another (one might aggress and the other attempt to mount). The experimental animals are being assessed, therefore, in situations that are not at all similar, and the stimulus–response chain gets more complicated and divergent from there on out. Thus, the basic scientific method of holding all but the experimental variables constant is compromised in such designs: How does one maintain stimulus equivalence for both sexes when each sex induces differing patterns of response in partners of each sex? Paradigms of this type must be considered dynamic, interactive, and not entirely within the control of the experimenter (Larssen, 1973). For an elegant demonstration of these factors in action in human behavior studies, see Ickes (1981).

There does not seem to be a completely satisfactory solution to this problem. Some investigators have abandoned the strategy altogether by relying on observations of the reflexive response patterns incorporated in natural behavioral displays (e.g., the elicitation of lordosis by manual stimulation). An important dimension of sexual dimorphism—stimulus selectivity and processing—is short-circuited by such nonsocial, reflexological tests. In studies of aggression, some investigators have attempted to use a "neutral" stimulus equivalent for both sexes. Either a "trained fighter" or a noncombative animal (e.g., a bulbectomized mouse) has been used to avoid the problem. (See Leshner, 1978). These studies are severely limited by their lack of generalizability to other conditions, but they do accomplish a reasonable degree of stimulus equivalence.

Even with its shortcomings, perhaps the best strategy is to test animals of both sexes with both male and female partners, thus generating four comparison groups: males × males; females × females; males × females; and females × males. Common attributes for each sex could then be identified in terms of their qualities (1) as initiators of behavior; (2) as recipients of behavior; and (3) in terms of the specific patterns of response of each dyadic type rather than in terms of sex differences *per se*. This approach, in fact, represents a strategy of regarding the unit of study as the *pair* rather than as the *individual* (see Patterson and Moore, 1979, for a complete description of this data-collection methodology). Even if these suggestions are implemented, however, readers can recognize that the dynamic characteristics of such tests cannot be factored out or counterbalanced. The animals are going to behave differently depending on the animal with which they are paired. Tests of individual characteristics linked to sex can be suggestive, but not entirely definitive.

Certain behavioral patterns would appear to be immune from such considerations, but it is not ever appropriate to assume that the factors referred to here have no effect, without directly testing for such influences. For example, given the published data, it is hard to imagine a social condition that would facilitate lordosis in adult male guinea pigs. Nonetheless, Thornton (1979) found in Strain 2 guinea pigs that isolation greatly enhances lordosis in males but has no such effect on females. Moreover, prenatal testosterone eliminates lordosis in group-housed females, as would be expected, but *facilitates* the response in isolated females. Because males show facilitated lordosis following isolation, Thornton concluded that she had obtained another "masculinization" effect of prenatal androgen. Used in this way, the term conveys the meaning that a change in the female's behavior resembling a characteristic of the conspecific male has occurred. Here, then, *masculinization* translates to a facilitation of lordosis under specific environmental conditions! The point to be made for the current discussion is that, without experiments modifying the social conditions of rearing and testing, this entirely unexpected set of findings would never have been observed.

DIMORPHIC BEHAVIOR ASSESSED IN GROUPS

Observing animals in social contexts in which more than two individuals are present is often a desirable step when one is attempting to see the animal in more naturalistic contexts, or when one expects that a social group is a necessary environment to elicit for certain types of behavior. Studies by McClintock (1981) and by McClintock *et al.* (1982) specifically illustrate how greatly such experimental manipulations of social variables add to our knowledge of the sexual behavior of species previously studied in laboratory pair test paradigms. However, several additional levels of complexity are confronted when more than two animals are observed in social contexts. Obviously, the dyadic encounters and the dynamic qualities inherent in them increase geometrically as a function of the number of animals in the group. In addition, other phenomena also come into play that further limit the interpretation of sex differences.

Sex Ratio Effects. The sex ratio of the group may influence the overall output of behavior of each sex in complex ways. For example, in five-member groups of rhesus monkeys, rough-and-tumble play among females is very low in all-female groups; moderate among females in one-male, four-female groups; and lower again in two-male, three-female groups (Goldfoot, Neff, and Goy, manuscript in preparation). It seems that, when no male is present, this particular form of play is not often displayed by females, but when a single male is a group member, females both receive and initiate more rough-and-tumble play, mostly with the male. When *two* males are in the group, however, the males engage in this type of play mostly with each other, and the females, although still showing marginally higher play than when in all-female groups, display the activity less frequently. As it turns out, a sex difference can be found in each of these cases, with males showing higher levels than females, but the degree of variance between the sexes in each situation can be traced in part to the overall social conditions. Other behavior patterns have been found to be even more susceptible to the influence of sex ratio, for example, mounting and presenting behavior (see Goldfoot *et al.*, 1984), so that it would seem prudent to test specifically for such possibilities regardless of the outcome of the distribution of display of the behavior between sexes in one particular social context.

Statistical approaches to the problem, such as dividing each animal's scores by the number of partners of each sex available, may be partial remedies for roughly equating groups of differing sex ratios, but such data-handling tactics obscure the fact that sex ratios may influence the results of sex differences studies.

Indirect Effects of Other Animals. There are now a number of examples of the fact that animals themselves "assess" social situations before engaging in particular courses of action. The most dramatic demonstration comes from a study by Perachio (1978), in which a male rhesus monkey implanted with a stimulatory electrode in his hypothalamus, mounted a female partner each time an electrical impulse was delivered by the experimenter. In the presence of a more dominant male, however, the implanted male did not mount the female, even though repeated electrical signals were given. This remarkable inhibition occurred even though no direct aggression between the two males was observed. With regard to sex difference studies, two examples make the point very clearly that impressions of sex differences based on a limited number of social conditions can be quite misleading.

Gibber (1981) was interested in comparing the parental behavior of male and female rhesus monkeys. She devised a laboratory test in which an animal in an experimental cage could see and hear a newborn rhesus infant that was "stranded" in the adjacent cage. A passageway led from the experimental cage to the infant's cage. Gibber observed that both adult males and adult females oriented, approached, and picked up the stranded infant with equivalent latencies and frequencies. She then repeated the tests but, this time, placed an adult male and female together in the experimental cage, again with a stranded infant adjacent to them. Under these conditions, the females displayed nearly 100% of all orientation, approach, and holding or carrying behavior directed toward the infant. In fact, many of the males that showed caregiving responses when tested alone kept as far away from the infant as the caging situation permitted. Thus, the amount of parental behavior displayed by males under these two experimental conditions was very much a function of the specific social conditions of testing that were sampled. Depending on those conditions, it could be concluded that (1) male and female rhesus monkeys do not differ in regard to their tendencies to pick up abandoned infants or, alternatively, that (2) males are very unlikely to come to the aid of a stranded infant. This example illustrates the concept of situation-specific dimorphic behavior, because obviously the caregiving response measured in this study is dimorphic in one condition, but not in another. As is true in newer ways of viewing personality constructs (see Mischel and Peake, 1982), the behavior under consideration is *situation*-specific rather than *trait*-specific.

French and Snowdon (1981) provided a second example of a situation-specific dimorphism in which the behavior of a focal animal changed as a function of the stimulus qualities and the behavior of a third animal. Studying pair-bonded cotton-topped tamarins, these investigators found significantly higher displays of aggression by males than by females toward conspecific territorial intruders. However, this sex difference was maintained only when the intruder was male. When a female intruder was used, the behavioral dimorphism was no longer apparent. In such conditions, the male was less likely to attack, whereas his mate was more likely to do so. The tendency to attack in this species is therefore a function of the sex of the actor, the sex of the intruder, and probably, the presence of a pair-bonded mate. This possibility is supported by the results of Evans (1983), who showed that a male pair-bonded common marmoset shows agression toward male or female intruders if his mate is nearby. If his mate is some distance away, however, he is likely to display aggression to male intruders, but sexual behavior toward female intruders. Here, the probability of the display of two alternative response classes becomes a function of social mediation involving third parties. Similar illustrations of situation-specific behavior have been found with regard to maternal versus sexual behavior in postpartum rats (Gilbert *et al.,* 1983), and to different sex-typical patterns of copulation in Norway rats (McClintock *et al.,* 1982). Also, Lamb (1979) provided specific examples and additional suggestions for research designs that focus on the effects of third parties on interactive dyadic relationships.

ASSOCIATIVE SEX DIFFERENCES. It is sometimes the case that a sex difference arises from factors correlated with, but not dependent on, the genetic sex of an individual. This phenomenon has received very little attention from researchers studying animals, and it needs much additional effort, in part because of its centrality to objections voiced by the critics of sexual dimorphism research, and also because it may very well represent an important source of variability in many kinds of behavioral research efforts. Examples of associative (or noncausal) sex differences include size or strength dimorphisms, correlated with, but not dependent on, genetic sex. It is likely, for example, that a controlling variable in some studies is that large animals elicit fear and associated withdrawal responses from smaller animals. If largeness and male sex are correlated, then it could be that, without manipulating the dimension of size, one could prematurely conclude that the observed behavioral sex difference in fear elicitation is due to "maleness" rather than "largeness." In many social situations, social rank, which may or may not be dependent on physical strength, can also act in a similar way to generate associative sex-difference conclusions. For example, the occupation of a high or a low position of social rank, regardless of sex, may have specific influence on the response probability of Behavior X. In many situations, social rank and sex are correlated, so that a behavioral sex difference may be found even though the relative expression of the behavior may have to do with the animal's social rank as well as, or instead of, his or her genetic sex.

In a real sense, both of these hypothetical conditions (social rank or physical size rather than sex being the controlling variable) are examples of legitimate behavioral sex differences. The sex differences are related to the specific situation in which they are observed and can be understood in terms of social or physical stimulus mediation. However, without a detailed identification of the social and/ or physical stimulus variables that support such a behavioral dimorphism, both predictive and explanatory validity will be considerably lessened, and a more thorough understanding of the situation will not be obtained. These kinds of behavioral sex differences certainly have limited value to the neurobiologist looking for sex differences mediated primarily by structures within the central nervous system. With this point in mind, we have used the term *associative* to refer to such situations.

Surprisingly, the mounting behavior of preadolescent rhesus monkeys is an example of behavior that partially falls within the category of an "associative" sex difference. Although prenatal hormones can be shown to have direct effects on the display of this behavior, social rank is also a mediating variable for both sexes. When monkeys are reared in mixed-sex groups and separated from mothers at 3 months of age, (1) males are far more likely than females to be dominant in social groups; (2) dominant males mount more than subordinate males; and (3) females are seen not even to display the double-foot clasp mount (Goy and Goldfoot, 1974; Goldfoot, 1977). When reared in mixed-sex groups under less restricted social conditions (i.e., left with their mothers for the first year in communal pens containing other peers), (1) about 25% of the females show double-foot clasp mounts, and (2) there is an increased probability that the females will attain high levels of dominance in their social groups. Further analyses reveal that those females that display foot-clasp mounting are most often high-ranking within their groups. Finally, when female infants are reared with their mothers and all female peers for the first year of life, 67% of these females display foot-clasp mounting, and again, there is a relationship between a position of high dominance and a display of the behavior (Goldfoot *et al.*, 1984). It is therefore possible to manipulate the percentage of females displaying malelike foot-clasp mounting from 0% in mixed-sex, peer-reared groups to 67% in isosexual mother–peer groups. One identified factor mediating the differential display of the response is social rank.

DAVID A.
GOLDFOOT AND
DEBORAH A. NEFF

It must be borne in mind that these are *social* and not *hormonal* manipulations. Both the removal of males from a mixed-sex group and rearing females in the complete absence of males result in an increased likelihood that a significant proportion of the females will show this "male" behavior. Nonetheless, these results do not contradict earlier findings that prenatal hormones augment this same behavior. On the contrary, it appears that social and hormonal conditions work either synergistically or additively to produce the behavioral result. If we had not proceeded with experiments that investigated the social end of the putative synergism, however, we might still be entirely misjudging the behavioral potential of females for the display of this behavioral response. It is anticlimactic to point out that analogous social variables may exist for virtually all the behavioral measures we take.

CONCLUSION

This chapter has emphasized that social and experiential variables are factors that can mediate behavioral sex differences in animals, and that the factors can be operative in the majority of studies in which biological regulating mechanisms for such differences are sought. Far from being easily controlled by simple experimental manipulations, the social variables are such that they limit the number and kinds of questions that can be legitimately pursued. That is not to say that such questions are not answerable, but that special strategies are necessary for a full evaluation of the *interactive* nature of biological and social determinants. Identifying the interplay of social and biological variables is quite doable science, however, and is one of the most fascinating of activities.

Rather than responding to social factors as a type of "noise," which should be rigidly "factored out" or otherwise controlled for, it may be more prudent in the long run to include them within research design protocols from the outset, even if one is interested in biology more than in social psychology. There can be little doubt that social factors are important mediators of sexually dimorphic behavior, and that they can influence the interpretation of any biological result. It seems clear, therefore, that for a full evaluation of an animal's potential to display a given behavior, one must observe him or her in many different social environments of rearing and/or testing. As we have seen, finding that an animal never mounts in Condition A does not rule out the possibility that he or she will mount in Condition B. Using strategies of "comparative socialization," one could characterize more accurately the degree to which social factors participate in the regulation of a given sex-typed behavior and, at the same time, could have a better idea of the extent to which biological variables within a study have first-line influences.

Preliminary results that are available suggest that manipulations of the social environment are distinctly advantageous to biological theory-builders. It is now becoming clearer that different behaviors believed to be under the same prenatal endocrine "organization" influences do not necessarily respond in parallel to given changes in social conditions. In rhesus monkeys, prenatal testosterone augments mounting and rough-and-tumble play in genetic females (Goy, 1981). Isosexual rearing of females also augments mounting but has no facilitatory effect on rough play (Goldfoot *et al.,* 1984; Goldfoot and Wallen, 1978). Because a set of dimorphic behaviors can constitute a behavioral definition of gender role, the results would imply that the separate behavioral components of male and female gender-role behavior can be separately and independently sensitive to changes in particular social conditions. Even in animal studies, then, it does not seem possible or desirable to define a sex difference without regard to the particular details of the social condition in which it was measured.

The issue has never been to separate nature from nurture, but to understand the intricate way in which animals, including human beings, are influenced by both biology and sociality.

Acknowledgments

The authors thank Jane Piliavin, Charles Snowdon, Janice Thornton, and Frans de Waal for reading and commenting on drafts of the manuscript. Special thanks are extended to R. W. Goy for years of discussion and exchange of opinion.

References

Beach, F. A. Hormonal factors controlling the differentiation, development and display of copulatory behavior in the ramstergig and related species. In L. Aronson and E. Tobach (Eds.), *Biopsychology of Development.* New York: Academic Press, 1971.

Bernstein, I. S., Gordon, T. P., and Peterson, M. Role behavior of an agonadal alpha male rhesus monkey in a heterosexual group. *Folia Primatologica,* 1979, *32,* 263–267.

Bleier, R. H. Brain, body, and behavior. In J. I. Roberts (Ed.), *Beyond Intellectual Sexism: A New Woman, A New Reality.* New York: David McKay, 1976.

Bleier, R. Social and political bias in science: An examination of animal studies and their generalizations to human behavior and evolution. In R. E. Hubbard and M. Lowe (Eds.), *Genes and Gender, Vol. 2: Pitfalls in research on Sex and Gender.* New York: Gordian Press, 1979.

Carpenter, C. R. Sexual behavior of free-ranging rhesus monkeys *(Macaca mulatta):* II. Periodicity of oestrus, homosexual, autoerotic and nonconformist behavior. *Journal of Comparative Psychology,* 1942, *33,* 143–162.

de Waal, F. *Chimpanzee Politics.* London: Jonathan Cape, 1982.

Diamond, M. Sexual identity, monozygotic twins reared in discordant sex roles and the BBC follow-up. *Archives of Sexual Behavior,* 1982, *11,* 181–186.

Dixson, A. F., and Herbert, J. Testosterone, aggressive behavior and dominance rank in captive adult male talapoin monkeys *(Miophithecus talapoin). Physiology and Behavior,* 1977, *18,* 539–543.

Epple, G., Alveario, M. C., and Katz, Y. The role of chemical communication in aggressive behavior and its gonadal control in the tamarin *(Saguinus fuscicollis).* In C. T. Snowdon, C. H. Brown, and M. R. Petersen (Eds.), *Primate Communication.* London: Cambridge University Press, 1982.

Evans, S. The pair-bond of the common marmoset *Callithrix jacchus jacchus:* An experimental investigation. *Animal Behaviour,* 1983, *31,* 651–658.

French, J., and Snowdon, C. Sexual dimorphism in responses to unfamiliar intruders in the tamarin *Saguinus oedipus. Animal Behaviour,* 1981, *29,* 822–829.

Gibber, J. R. *Infant-Directed Behaviors in Male and Female Rhesus Monkeys.* Unpublished doctoral dissertation, University of Wisconsin–Madison, Department of Psychology, 1981.

Gilbert, A. N., Burgoon, D. A., Sullivan, K. A., and Adler, N. T. Mother-weanling interactions in Norway rats in the presence of a successive litter produced by postpartum mating. *Physiology and Behavior,* 1983, *30,* 267–271.

Goldfoot, D. A. Sociosexual behavior of nonhuman primates during development and maturity: Social and hormonal relationships. In A. M. Schrier (Ed.), *Behavioral Primatology: Advances in Research and Theory*, Vol. 1. Hillsdale, N. J.: Lawrence Erlbaum, 1977.

Goldfoot, D. A., and Wallen, K. Development of gender role behaviors in heterosexual and isosexual groups of infant rhesus monkeys. In D. J. Chivers and J. Herbert (Eds.), *Recent Advances in Primatology, Vol. 1: Behavior. Proceedings of the Sixth Congress of the International Primatological Society.* New York: Academic Press, 1978.

Goldfoot, D. A., Goy, R. W., Neff, D. A., Wallen, K., and McBrair, M. C. Social influences upon the display of sexually dimorphic behavior in rhesus monkeys: Isosexual rearing. *Archives of Sexual Behavior*, 1984, *13*, 395–412.

Goy, R. W., and Goldfoot, D. A. Hormonal influences on sexually dimorphic behavior. In R. O. Greep and E. B. Astwood (Eds.), *Handbook of Physiology: Endocrinology*, Vol. 2, Part 1. Baltimore: Williams and Wilkins, 1973.

Goy, R. W., and Goldfoot, D. A. Experiential and hormonal factors influencing development of sexual behavior in the male rhesus monkey. In F. O. Schmidt and F. G. Worden (Eds.), *The Neurosciences: Third Study Program.* Cambridge: M.I.T. Press, 1974.

Hubbard, R., and Lowe, M. (Eds.). *Genes and Genders, Vol. 2: Pitfalls in Research on Sex and Gender.* New York: Gordian Press, 1979.

Ickes, W. Sex role influences in dyadic interactions: A theoretical model. In C. Mayo and N. Henley (Eds.), *Gender and Nonverbal Behavior.* New York: Springer-Verlag, 1981.

Kass-Simon, G. Female strategies: Animal adaptations and adaptive significance. In J. I. Roberts (Ed.), *Beyond Intellectual Sexism: A New Woman, A New Reality.* New York: David McKay, 1976.

Keverne, E. B. Olfaction and the reproductive behavior of nonhuman primates. In C. T. Snowdon, C. H. Brown, and M. R. Petersen (Eds.), *Primate Communication.* London: Cambridge University Press, 1982.

Lamb, M. E. The effects of the social context on dyadic social interaction. In M. E. Lamb, S. J. Suomi, and G. R. Stephenson (Eds.), *Social Interaction Analysis: Methodological Issues.* Madison: University of Wisconsin Press, 1979.

Larsson, K. Sexual behavior: The result of an interaction. In J. Zubin and J. Money (Eds.), *Contemporary Sexual Behavior: Critical Issues in the 1970s.* Baltimore: Johns Hopkins University Press, 1973.

Leibowitz, L. "Universals" and male dominance among primates: A critical examination. In R. Hubbard and M. Lowe (Eds.), *Genes and Gender, Vol. 2: Pitfalls in Research on Sex and Gender.* New York: Gordian Press, 1979.

Leshner, A. E. *An Introduction to Behavioral Endocrinology.* New York: Oxford University Press, 1978.

McClintock, M. K. Simplicity from complexity: A naturalistic approach to behavior and neuroendocrine function. In I. Silverman (Ed.), *Laboratory and Life: New Directions for Methology of Social and Behavioral Research*, No. 8. San Francisco: Jossey-Bass, 1981.

McClintock, M., Anisko, J. J., and Adler, N. T. Group mating among Norway rats: II. The social dynamics of copulation: Competition, cooperation and mate choice. *Animal Behavior*, 1982, *30*, 410–425.

Mischel, W., and Peake, P. K. Beyond déja vu in the search for cross-situational consistency. *Psychological Review*, 1982, *89*, 730–755.

Missakian, E. A. Genealogical and cross-genealogical dominance relations in a group of free-ranging rhesus monkeys *(Macaca mulatta)* on Cayo Santiago. *Primates*, 1972. *13*, 169–180.

Mitchell, G. *Behavioral Sex Differences in Nonhuman Primates.* New York: Van Nostrand Reinhold, 1979.

Money, J. The development of sexuality and eroticism in humankind. *Quarterly Review of Biology*, 1981, *56*, 379–404.

Money, J., and Ehrhardt, A. A. *Man and Woman, Boy and Girl.* Baltimore: Johns Hopkins University Press, 1972.

Patterson, G. R., and Moore, D. Interactive patterns as units of behavior. In M. E. Lamb, S. J. Suomi, and G. R. Stephenson (Eds.), *Social Interaction Analysis: Methodological Issues.* Madison: University of Wisconsin Press, 1979.

Perachio, A. A. Hypothalamic regulation of behavioral and hormonal aspects of aggression and sexual performance. In D. J. Chivers and J. Herbert (Eds.), *Recent Advances in Primatology, Vol. 1: Behavior. Proceedings of the Sixth Congress of the International Primatological Society.* New York: Academic Press, 1978.

Sherman, J. A. Some psychological "facts" about women: Will the real Ms. please stand up? In J. I. Roberts (Ed.), *Beyond Intellectual Sexism: A New Woman, A New Reality.* New York: David McKay, 1976.

Sigursch, V., Schorsch, E., Dannecker, M., and Schmidt, G. Official statement by the German Society for Sex Research (Deutsche Gesellschaft für Sexualforschung e. V.) on the research of Prof. Dr. Gunter Dorner on the subject of homosexuality. *Archives of Sexual Behavior,* 1982, *11,* 445–449.

Thornton, J. E. *Some Factors Affecting the Display of Bisexual Behavior by Males from Two Inbred Strains of Guinea Pig.* Unpublished master's thesis, University of Wisconsin–Madison, Department of Psychology, 1979.

Tobach, E., and Rosoff, B. (Eds.). *Genes and Gender.* New York: Gordian Press, 1978.

Weisstein, N. Tired of arguing about biological inferiority? *Ms.,* November 1982, *85,* 41–46.

Young, W. C. The hormones and mating behavior. In W. C. Young (Ed.), *Sex and Internal Secretions,* Vol. 2. Baltimore: Williams and Wilkins, 1961.

Index